Ecology and Management
of Large Mammals
in North America

Stephen Demarais
Mississippi State University

Paul R. Krausman
The University of Arizona

Prentice Hall
Upper Saddle River, NJ 07458

Library of Congress Cataloging-in-Publication Data
Ecology and management of large mammals in North America / [editors]
   Stephen Demarais, Paul R. Krausman.
        p.   cm.
   Includes bibliographical references.
   ISBN 0-13-717422-5
      1. Mammals—Ecology—North America.   2. Wildlife management—North
America.   I. Demarais, Stephen.   II. Krausman, Paul R., 1946–   .
   QL739.8.E36   2000
   599. 17'097—dc21
                                                             99-40881
                                                             CIP

Publisher: Charles E. Stewart, Jr.
Associate Editor: Kate Linsner
Managing Editor: Mary Carnis
Production Liaison: Eileen O'Sullivan
Production/Editorial Supervision: Lori Harvey, Carlisle Publishers Services
Director of Production and Manufacturing: Bruce Johnson
Production Manager: Marc Bove
Marketing Manager: Ben Leonard
Cover Designer: Liz Nemeth
Cover Photo: Bob Zaiglin
Formatting/Page Make-up: Carlisle Communications, Ltd.
Interior Designer: Carlisle Communications, Ltd.
Printer/Binder: Courier Westford

© 2000 by Prentice-Hall, Inc.
Upper Saddle River, New Jersey 07458

Printed in the United States of America
10 9 8 7 6 5 4 3 2 1

ISBN 0-13-717422-5

Prentice-Hall International (UK) Limited, *London*
Prentice-Hall of Australia Pty. Limited, *Sydney*
Prentice-Hall Canada Inc., *Toronto*
Prentice-Hall Hispanoamericana, S.A., *Mexico*
Prentice-Hall of India Private Limited, *New Delhi*
Prentice-Hall of Japan, Inc., *Tokyo*
Pearson Education Asia Pte. Ltd., *Singapore*
Editora Prentice-Hall do Brasil, Ltda., *Rio de Janeiro*

This book is dedicated to our children:
Jennifer, Christopher, and Brian Demarais
Curtis Krausman and Julie Wegner

Contents

Contents

Contents

Preface

Managing big game species has been an important activity of agency biologists since the early restocking efforts of the 1940s. Teaching eager student-professionals about ecology and management of big game has been an important role of the academic community since the 1960s, and has been an important part of our professional lives since the early 1980s. Plenty has changed on the biological and political scenes in the short time since we began our university careers. Volumes of data and vast experiences have accumulated on the big game species covered in this text. Moreover, the field of wildlife ecology and management has been molded into the amorphous mass commonly called *ecosystem management.* Big game species are referred to as large mammals, habitats are referred to as landscapes, and those that hunt and manage the hunted spend additional time justifying their recreational pursuits and professions.

Like many university faculty, we used the Wildlife Management Institute's *Big Game of North America* as our textbook of choice for our big game management classes until it went out of print. We were frustrated by the lack of an appropriate replacement textbook. Richard McCabe with the Wildlife Management Institute suggested we pursue publication of a replacement text with a commercial publisher.

Many agreed that a completely new and updated text would be advantageous to everyone interested in large mammals. And who isn't? Large, wild mammals excite anyone fortunate enough to come in contact with them, whether the contact is for hunting, observational recreation, photography, serendipitous sightings, nature study, scientific study, or any other avenue that brings humans and large mammals together. All associations do not yield the same results. A rancher who has lost livestock to predators will respond differently than an outdoors person who hears the clash of rams' horns, but the reaction will still be one of excitement and the more these interest groups know about big game, the more valuable their encounters will be.

The concepts of "ecosystem management," "biodiversity," and "conservation biology" are critically important to effective wildlife management in today's complex world. However, we believe there is still much value in the study of individual species and their habitats. And although large mammals have been studied more than other species, the exploration into their life history is not yet complete nor does it include an analysis of the consequences that emerge when humans are added to the equation. Natural areas are rapidly being overrun by urbanization—a relatively recent addition to the human–animal–habitat equation. We believe that students, practicing biologists, and interested laymen will benefit by having a single source for contemporary thought about large mammal ecology and management. The individual species chapters are by no means meant to be complete treatises on their respective species. Individuals in need of more extensive material should consult the growing list of single-species texts produced by the Wildlife Management Institute. Our compilation from the continent's experts on single species and topics provides a valuable textbook and reference for students and a record of contemporary practices for wildlife historians.

Thus, we prepared this book for students, professionals, and laypeople with more than just an idle curiosity about large mammals. We solicited a select group of authors who were actively involved with the species or topics, were familiar with the literature, and were willing to share their time and expertise. During the process of editing this volume, we have become increasingly awed and honored at having the opportunity to participate with these outstanding authors in the preparation of a truly outstanding summary of large mammal ecology and management in North America.

Wildlife biologists working with big game are eager to share their knowledge. This is exemplified by the many organizations made up of laypeople and professionals alike that contribute to the conservation of large mammals in North America. The Desert Bighorn Council, Rocky Mountain Elk Foundation, Mule Deer Foundation, Foundation for North American Wild Sheep, Western States and Provinces Elk and Mule Deer Workshop, Southeast and Northeast Deer Study Groups, Boone and Crockett Club, Quality Deer Management Association, North American Pronghorn Foundation, and Interstate Antelope Conference are just a few of the organizations dedicated to conservation and research on regional and national scales. There are proportionally larger numbers of organizations at the state and local levels dedicated to a better understanding of the interactions among large mammals, humans, and their habitats. These and numerous similar organizations are testimony to the interest humans have in large, wild animals.

With the exponentially increasing knowledge base on the ecology and management of large mammals in North America, it will not be long before this volume will require revision and improvement. Until then we hope it serves as a useful text and reference book, a source of additional references, and an accurate record of how large mammals successfully interact with humans and the habitats we have altered.

Stephen Demarais, Associate Professor　　　　　　　Paul R. Krausman, Professor
Wildlife Management　　　　　　　Wildlife and Fisheries Ecology
Mississippi State University　　　　　　　The University of Arizona

Acknowledgments

Numerous people contributed to this publication, in addition to the editors and chapter authors. The editors express their warmest appreciation to Natalie Hayes and Bronson Strickland for editorial assistance, and to Robert E. Zaiglin for photographic contributions. The editors express our appreciation to our administrators' support of our activities: at Mississippi State University - H. R. Robinette, Head, Department of Wildlife and Fisheries and J. E. Gunter, Dean/Director and W. S. Thompson, Interim Dean/Director, College of Forest Resources/Forest and Wildlife Research Center; at The University of Arizona - W. W. Shaw, Program Leader, Wildlife and Fisheries, C. P. P. Reid, Director, School of Renewable Natural Resources, and C. C. Kaltenbach, Director, Arizona Agricultural Experiment Station. We appreciate the professionalism of Kate Linsner and Lori Harvey. The chapter authors provide the following partial list of acknowledgments to individuals and agencies: J. W. Bickham, Y. Lou, C. W. Walker, J. Coady, J. Dau, P. Reynolds, R. Seavoy, H. Thing, B. Fournier, A. Gunn, J. Huot, J. Nagy, and M. Forchhhammer for making available unpublished data and general status reports; J. deVos, T. K. Fuller, R. Hayes, L. D. Mech, R. Nowak, K. P. Taylor, H. A. Whitlaw, D. Houston, E. D. Ables, C. Vogel, R. T. Bowyer, F. S. Danks, L. Kuck, J. A. Bissonette, W. W. Shaw, R. C. Etchberger, and W. B. Ballard for valuable editorial assistance and critical evaluation of earlier chapter drafts; U.S. Department of Agriculture, Forest Service, North Central Forest Experiment Station, whose research funding supported development of some chapter portions.

AUTHOR AFFILIATIONS

Editors

Demarais, Stephen, Associate Professor of Wildlife Management, Department of Wildlife and Fisheries, Mississippi State University, Mississippi State, MS 39762 sdemarais@cfr.msstate.edu.

Krausman, Paul R., Professor of Wildlife and Fisheries Science, School of Renewable Natural Resources, The University of Arizona, Tucson, AZ 85721 krausman@ag.arizona.edu

Chapter Authors

Ballard, Warren B., Associate Professor of Wildlife Ecology, Department of Range, Wildlife and Fisheries Management, Texas Tech University, Box 42125, Lubbock, TX 79409 c7bwb@ttacs.ttu.edu

Bergerud, Arthur T., 1233 Isabell Road, Salt Spring Island, British Columbia, Canada V8K1T5

Bowyer, R. Terry, Professor of Wildlife Ecology, Institute of Arctic Biology, Department of Biology and Wildlife, University of Alaska, Fairbanks, AK 99775-7000 ffrtb@uaf.edu

Carpenter, Len H., Field Representative, Wildlife Management Institute, 4015 Cheney Drive, Fort Collins, CO 80526 lenc@verinet.com

Conner, Mark C., Manager of Chesapeake Farms, DuPont Agricultural Enterprise, Chesapeake Farms, Chestertown, MD 21620 connermc@a1.csag1.umc.dupont.com

Cook, John G., Research Wildlife Biologist, National Council of the Paper Industry for Air and Stream Improvement, Forestry and Range Sciences Laboratory, La Grande, OR 97850 cookjg@eou.edu

Czech, Brian, Conservation Biologist, U.S. Fish and Wildlife Service, Division of Refuges, 4401 N. Fairfax Drive, MS 670, Arlington, VA 22203 Brian_Czech@fws.gov

Franzmann, Albert W., Retired Director of the Alaska Department of Fish and Game, Moose Research Center, Consultant with the International Wildlife Veterinary Service, P.O. Box 666 Soldotna, AK 99669

Fryxell, John M., Associate Professor, Department of Zoology, Department of Zoology, University of Guelph, Ontario, Canada, N1G 2W1 jfryxell@uoguelph.ca

Geist, Valerius, Professor Emeritus of Environmental Science, Faculty of Environmental Design, University of Calgary, Calgary, Alberta, Canada, T2N 1N4 geistvr@cedar.alberni.net

Gibson, Philip S., Leader, Kansas Cooperative Fish and Wildlife Research Unit, U.S. Geological Survey, 205 Leasure Hall, Kansas State University, Manhattan, KS 66506 gipson@ksuvm.ksu.edu

Guynn, David C., Jr., Professor, Department of Forest Resources, Clemson University, Clemson, SC 29634-1003 dguynn@clemson.edu

Guynn, Dwight E., Senior Project Leader, U.S. Fish and Wildlife Service Management Assistance Team, Fort Collins, CO 80525 dwight@mat.fws.gov

Heffelfinger, James R., Regional Game Specialist, Arizona Game and Fish Department, 555 N. Greasewood Road, Tucson, AZ 85745 cervidnut@aol.com

Hellgren, Eric C., Assistant Professor of Zoology, Department of Zoology, Oklahoma State University, Stillwater, OK 74078 ehellgr@okway.okstate.edu

Hirth, David H., Associate Professor, Wildlife and Fisheries Biology Program, University of Vermont, Burlington, VT 05405 dhirth@nature.snr.uvm.edu

Hoffman, Robert S., Senior Scientist, Division of Mammals, National Museum of Natural History, Smithsonian Institution, MRC 108, Washington, DC 20560 rhoffman@sivm.si.edu

Honeycutt, Rodney L., Faculty of Genetics and Department of Wildlife and Fisheries Sciences, Texas A&M University, College Station, TX 77843-2258 rhoneycutt@ tamu.edu

Jacobson, Harry A., Professor Emeritus, Department of Wildlife and Fisheries, Mississippi State University. Current address: 10661 Davis Lane, Athens, TX 75751

Kellert, Stephen R., Professor of Social Ecology, School of Forestry and Environmental Studies, Yale University, New Haven, CT 06511 stephen.kellert@yale.edu

Kie, John G., Research Wildlife Biologist and Team Leader. Forestry and Range Sciences Lab. 1401 Gekeler Lane, La Grande, OR 97850 kien@eou.edu

Klein, David R., Professor Emeritus, Institute of Arctic Biology and Department of Biology and Wildlife, University of Alaska–Fairbanks, Fairbanks, AK 99709

Lancia, Richard A., University Alumni Distinguished Professor, Forestry Department, North Carolina State University, Raleigh, NC 27695 lancia@unity.ncsu.edu

Leslie, David M., Jr., Unit Leader and Adjunct Professor, Oklahoma Cooperative Fish and Wildlife Research Unit, United States Geological Survey, Biological Division, Department of Zoology, Oklahoma State University, 404 Life Sciences West, Stillwater, OK 74078-3051 r8cuok@usgs.gov

Lochmiller, Robert L., Regent's Professor of Zoology, Department of Zoology, Oklahoma State University, Stillwater, OK 74078 rllzool@okway.okstate.edu

Logan, Kenneth A., Research Scientist, Hornocker Wildlife Institute, P.O. Box 3246, University of Idaho, Moscow, ID 83843-1908 hwi@uidaho.edu

Mackie, Richard J., Professor Emeritus, Fish and Wildlife Program, Department of Biology, Montana State University, Bozeman, MT 59717 rjmackie@montana.campuscwix.net

Maegher, Mary, Research Biologist (retired), 634 Cinnabar Road, Gardiner, MT 59030

Messier, François, Professor of Wildlife Ecology, Department of Biology, University of Saskatchewan, 112 Science Place, Saskatoon, Canada SK S7N 5E2, francois_messier@usask.ca

Miller, Karl V., Associate Professor, Wildlife Ecology and Management, Warnell School of Forest Resources, University of Georgia, Athens, GA 30602 kmiller@smokey.forestry.uga.edu

Mungall, Elizabeth Cary, Exotic Wildlife Association Wildlife Research Consultant, and Adjunct Professor, Department of Biology, Texas Woman's University, Denton, TX 76204 Chris_Mungall@interoz.com

O'Gara, Bart W., Wildlife Professor Emeritus and Research Biologist (retired), U.S. Fish and Wildlife Service, Montana Cooperative Wildlife Research Unit, University of Montana, Missoula, MT 59812

Pasitschniak-Arts, Maria, Research Associate, Department of Biology, University of Saskatchewan, 112 Science Place, Saskatoon, Canada SK S7N 5E2 arts@sask.usask.ca

Patterson, Michael E., Assistant Professor, School of Forestry, University of Montana, Missoula, MT 59812 mike@forestry.umt.edu

Peek, James M., College of Forestry, Wildlife, and Range Sciences, University of Idaho, Moscow, ID 83844-1136

Pelton, Michael R., Professor of Wildlife Science, Department of Forestry, Wildlife, and Fisheries, University of Tennessee, Knoxville, TN 37901 mpelton@utk.edu

Rachlow, Janet L., Postdoctoral Fellow, Institute of Zoology, Zoological Society of London, Regent's Park, NW1 4RY, London, UK jrachlow@hotmail.com

Rosenberry, Christopher S., Research Assistant, Department of Agriculture and Natural Resources, Delaware State University, Dover, DE 19901 crosenberry@dnrec.de.us

Shackleton, David M., Department of Animal Science, Faculty of Animal Sciences, Suite 208-2357 Mail Mall, University of British Columbia, Vancouver, British Columbia, Canada V6T 1Z4

Shaw, James H. Professor and Head, Department of Zoology, Oklahoma State University, Stillwater, OK 74078 shawjh@okway.okstate.edu

Sinclair, A. R. E., Director, Centre for Biodiversity Research, 6270 University Boulevard, University of British Columbia, Vancouver, British Columbia, Canada V6T 1Z4 sinclair@zoology.ubc.ca

Smith, Carter P., South Texas Field Representative, The Nature Conservancy, P.O. Box 2873, Harlinger, TX 78551-2873 carter_smith@TNC.org

Spalinger, Donald E., Alaska Department of Fish and Game, Division of Wildlife Conservation, 333 Raspberry Road, Anchorage, AK 99518 Don_Spalinger@ fishgame.state.ak.us

Sweanor, Linda L., Research Scientist, Hornocker Wildlife Institute, P.O. Box 3246, University of Idaho, Moscow, ID 83843-1908 hwi@uidaho.edu

Valdez, Raul, Professor of Wildlife Science, Department of Fishery and Wildlife Sciences, New Mexico State University, Las Cruces, NM 88003 rvaldez@leopold.nmsu.edu

Wentworth, James M., Zone Wildlife Biologist, U.S. Forest Service, P.O. Box 9, Blairsville, GA 30514 jwentworth r8,chattoconee,brasstown@fs.fed.es

White, Gary C. Department of Fishery and Wildlife Biology, Colorado State University, Fort Collins, CO 80523 gwhite@cnr.colostate.edu

White, Ronald J. Director, Division of Agricultural Programs and Resources, New Mexico Department of Agriculture - New Mexico State University, MSC APR Box 30005, Las Cruces, NM 88003 rwhite@nmda-bubba.nmsu.edu

Wisdom, Michael J., Research Wildlife Biologist, USDA Forest Service Pacific Northwest Research Station, Forestry and Range Sciences Laboratory, La Grande, OR 97850 wisdomm@eou.edu

Yoakum, Jim D., Wildlife Consultant, Western Wildlife, P. O. Box 369, Verdi, NV 89439-0369

Foreword

In 1978, the Wildlife Management Institute (and Stackpole Books) released *Big Game of North America: Ecology and Management,* compiled and edited by John L. Schmidt and Douglas L. Gilbert. For nearly 15 years and through three reprintings, that reference volume served as the definitive text on the continent's large terrestrial mammals. Despite continuing sales of the title, the institute declined to permit a fourth reprinting, because too much of the information contained therein was dated or misleading.

Despite a clear professional need to update the 1978 book, the Wildlife Management Institute was too committed to other volumes in preparation to contemplate the additional investment of time and resources in revising that particular work. However, even before *Big Game of North America: Ecology and Management* was fully out of circulation, Steve Demarais and Paul R. Krausman approached us about taking on the daunting task of coordinating a new edition. They expressed their belief that, with the Schmidt/Gilbert book out of print, too big a void was being created in the professional literature on such an important topic. We concurred, but could not divert attention from its other publication projects in progress. We suggested that Steve and Paul produce an updated and expanded big game book of their own. And based on a review of the subsequent proposal's preliminary outline and excellent line-up of authors, the Wildlife Management Institute strongly encouraged and endorsed the undertaking. The Wildlife Management Institute was confident that the art and science of big game management would be properly served by these editors and their contributors. This work, *Ecology and Management of Large Mammals in North America,* is a superb achievement, and more than evidences that the institute's confidence was well placed.

* * * *

In the foreword of the 1978 book, then-president of the Wildlife Management Institute Daniel A. Poole wrote: "As new understanding [of the forces and events that affect big game populations] is gained, it will be blended into management strategies and field programs. In some cases, this may involve only relatively simple changes. More extensive management adjustments may be required in others. But in all cases, the objective of wildlife management and all that it entails is to replace unplanned events and uncertain results with planned actions and known results." I doubt that Dan's insight could have been stated any more eloquently. And I know that readers of this volume will find similiar eloquence in the new understanding of big game ecology and management represented in the pages of each chapter.

Richard E. McCabe
Secretary
Wildlife Management Institute

CHAPTER 1

Taxonomy and the Conservation of Biodiversity

Valerius Geist, Bart O'Gara, and Robert S. Hoffmann

INTRODUCTION

Taxonomy is concerned with the ordering and naming of living organisms. It attempts to illustrate the evolution of life (phylogeny) by developing a hierarchical system of relationships among named organisms. As such, taxonomy is basic to biological sciences. It is composed of two complementary elements, taxonomy proper and systematics. The former is the theory and practice of describing and classifying organisms and bestowing internationally accepted names. The latter is an interdisciplinary, comparative synthesis of organismic diversity to determine evolutionary relationships among organisms. Systematics is thus the scientific component within taxonomy (1). We have little to say here about the taxonomic component, which is regulated by the International Commission on Zoological Nomenclature. Instead, we focus on systematics, in particular on the species and subspecies levels, because it is here that the application of taxonomy is crucial, both scientifically and legally, to the conservation of wildlife and biodiversity.

The "father of taxonomy," Swedish botanist Carl von Linné, better known under his latinized name as Linnaeus, developed the system of binomial nomenclature used today. All taxonomically valid names date from the 10th edition of his book *Systema naturae,* published in 1758. He designated each species with a generic and a specific name; a further level of refinement, the subspecies, was introduced later. For example, the correct taxonomic name for the elk or wapiti of North America is *Cervus elaphus canadensis* Erxleben, 1777. Its genus is *Cervus,* named by Linnaeus in 1758; the species is *elaphus,* also named by Linnaeus in 1758; the subspecies is *canadensis,* so labeled by Erxleben in the year 1777. The label thus illustrates the elk as subspecies of red deer (*Cervus elaphus* Linnaeus, 1758).

The Linnaean classification hierarchy has above the genus the following principal levels, in ascending order: tribe, family, order, class, phylum, and kingdom. Due to the enormous number of organisms under classification, these levels have been subdivided into more manageable portions by adding super-, sub-, and infra- to the basic levels (1). Also, the ending of each label identifies the level of classification. Thus the ending *oidea* identifies the superfamily (i.e., Cervoidea = deer-like ruminants); *idae* identifies family (Cervidae = all deer); *inae* identifies subfamily (Cervinae = Old World deer); *ini* identifies Tribe (Cervini = the red deer lineage and its branches), and *ina* identifies the subtribe (Cervina = the red deer lineage without its side branches).

Taxonomy has value as a tool in conservation and in wildlife management because, in reflecting phylogeny, it may give guidance for the conservation of genetic diversity. This is crucial, because conservation should focus on the preservation of genes, or hereditary potential, rather than on the diversity of their expressions (phenotypes, ecotypes). The aim is to conserve adaptive potential and evolutionary flexibility for the future. This increases the chances for survival of organisms when they have to deal with inevitable environmental change. Genetic diversity theoretically permits a maximum of adjustment to environmental change, which is the key to long-term survival. Taxonomy that accurately labels genetically different populations allows managers to pick appropriate stock for reintroductions, to safeguard unique relict populations, to prevent inadvertent mongrelizing, and to conserve widespread and superficially diverse populations of the same genetic stock with economy of resources.

A word of caution: There are ecologically unique populations within subspecies whose preservation ensures the continuity of unique ecological and behavioral traditions that reintroduced populations are unlikely to master. Conservation depends on much more than genetics. This last point is not trivial.

Prior to the enactment of national laws and international treaties on conservation, taxonomic nomenclature was of concern primarily to systematists. In practice, taxonomic debate pertaining to large mammals did not have much practical importance. However, taxonomy moved beyond science once taxonomic names were "enshrined" in legislation such as the Convention on International Trade in Endangered Species (CITES) or the United States Endangered Species Act, and became actionable in courts of law (2, 3, 4, 5). Taxonomic practices and conclusions are now subject to more than scientific peer review; they are now examined and judged by the courts. Taxonomic questions need to be translated into lay language, because its theories and practices are being scrutinized in detail and subject to hostile cross-examination in public. Therefore, an expert witness preparing to testify has to reexamine diligently the scientific basis by which taxa are defined and explain this in plain language to judges and lawyers. Expert witnesses may have to answer questions posed by peers opposed to their view, but asked through lawyers. Such questions are likely to be posed with great skill and require considered answers in kind. This can entrap the unwary. Assumptions previously hidden become unearthed, as are documents published obscurely, and errors published by the expert witnesses are exposed to public view. In short, taxonomic conclusions may now be scrutinized publicly and in a manner more adversarial and harsh than the anonymous peer review commonly used in science. Through it all, the expert witness must never forget to whom his or her testimony is addressed: to the patiently listening judge and not the questioning lawyer.

It must be understood that conclusions reached by the courts are likely to set precedence and may be binding on future cases, potentially limiting the input of new scientific information. Consequently, it is crucial that the science presented in court be sound and that assumptions, weaknesses, and alternatives be spelled out clearly.

Despite the adversarial nature of court procedures, the science needs to be presented in a visibly disinterested, nonpartisan manner. That is, the expert witness must strive to be a friend of the court, be that formally recognized or not. In the British (Canadian) tradition expert witnesses are treated as friends of the court, a privilege not automatically extended in U.S. courts. Because this difference affects how an expert witness may answer, it is imperative that the manner of giving testimony be discussed in some detail with the pertinent lawyer well before the trial.

Thus, taxonomy is no longer an obscure science. It is imperative to scrutinize taxonomy to ensure that the authority of science is not invoked in vain, which, unfortunately, has happened in the recent past (3, 4). When taxonomy is questioned in court and presented in the public limelight, it must not embarrass science nor invalidate good conservation legislation or management practices. Unfortunately, the current taxonomy of large mammals would not fulfill this requirement. Clearly, poor science must not be enshrined by the courts in legal decisions binding on management for conservation. There is, consequently, more at stake than merely revitalizing an old, fundamental, but currently neglected branch of biological science. The downsizing of museums and curatorial responsibilities for collections has already begun to haunt conservation (5).

What are the problems with the taxonomy of large mammal species in North America? The major problem is that most species have been classified on the basis of flawed methodologies combined with superficial reviews of materials essential to the classification of organisms. Consequently, current classifications do not reflect the true genetic diversity of large mammals, and the use of current nomenclature may endanger wildlife conservation rather than promote it. The flaws necessitate a taxonomic reassessment of all North American species of large mammals. In practical terms it means that biologists who rely on "accepted" taxonomies may mishandle their mandate and face public challenges. This is not a theoretical possibility, but a reality that we have personally experienced in formal public hearings and in law enforcement cases. However, before addressing the particular, we need to look at some general problems.

DESCENT VERSUS ADAPTATION

Taxonomy suffers from a profound, apparently irreconcilable contradiction, namely, to classify by genetic descent or by adaptation. Taxonomy strives to reflect evolution and the diversity of life, and descent and adaptation inform us about evolution. With the advance of molecular biology, this contradiction has been brought into sharp focus. Claims have been made, for instance, that genetically the order Cetacea, the whales, is so closely related to Artiodactyla that they are a mere branch within the Suiformes, placed between pigs and hippos (6, 7). Basing taxonomy on genetics, the order Cetacea vanishes, whales become the marine equivalent of pigs, and pigs become the terrestrial equivalent of whales. This emphasizes the adaptability of the order Artiodactyla, which evolved extremes such as tree-adapted, cat-like climbing forms (i.e., the Agriochoeridae in the early Tertiary), tropical to Arctic herbivorous cursors, subterranean diggers such as warthogs, freshwater-adapted hippos, and marine-adapted whales.

Conversely, the orders Artiodactyla and Cetacea are conventionally segregated by adaptations. The unique foot structure of the Artiodactyla, an ancient adaptation linking members of this order, is missing in modern Cetacea. However, take classification by adaptation to its extreme and genetic adaptation grades into environmental adjustment, and into epistasis or nonhereditary genetic variation, with no ready means of distinction at hand. Consequently, genetically meaningless ecotypic and epistatic variations have often been employed in formal taxonomy. The same basic genotype is classified differently, depending on where and how it developed into a phenotype. No longer is such a classification a reflection of evolution, contrary to the basic tenet of taxonomy. The ideal solution would be to base classification at the species level and below on genetically fixed adaptations. This is, however, a very difficult task and one that will require much more research and revision. Phenotypic characteristics are only rarely the result of environmentally insensitive genetic penetrance, so that all adaptations have elements of adjustment (8, 9). Although molecular biology can specify descent with greater precision, it cannot shed light on why evolution took place (10). Conversely, adaptation provides an answer concerning why a phenotype evolved, but says nothing about when it happened or who might be the relatives.

SPECIES DEFINITIONS

Linnaeus in 1758 distinguished between species by their external appearance. He even named the male and female mallards two different species, *Anas boschas* and *A. platyrhynchos*, respectively. In other contexts, what Linnaeus labeled as species we would label today as subspecies. His approach was typological, ignoring within-species variation. It is not surprising, therefore, that the typological species concept led in the past two centuries to many names being applied by eager naturalists to minor variations of the same species or subspecies. Modern taxonomy, consequently, has taken a more rigorous approach to the species definition, defining such as populations that are actually or potentially reproductively isolated (11). This is the biological species concept. At first glance this appears straightforward and satisfactory. However, in practice, this definition has been applied in dubious, troubling ways and has suffered from inadequacies in systematics.

For instance, some taxonomists declared as a "biological species" two different forms if these hybridize in captivity. A hybrid, even one marginally viable despite human care, was considered as proof that its parents belonged to the same species. Some allowance was made to the contrary if the hybrids proved to be sterile (12). The "zoo criterion" is most dubious because related forms that hybridize in zoos may fail to do so in nature, despite the opportunity to do so. That is, there are reproductive barriers under natural conditions between species that break down in captivity, and a species designation based on mere hybridization in captivity is based on flawed biology. Nevertheless, as can be seen later in the discussions pertaining to species and subspecies, the zoo criterion is still invoked.

Moreover, free-living forms that normally do not hybridize may do so if they are moved to another location, particularly if they are no longer exposed to effective predators. Two examples may suffice. Bezoar goats *(Capra aegagrus)* and Caucasian tur *(Capra caucasica)* live sympatrically without hybridization and are, consequently, good species. However, when bezoar goats were introduced to a region where previously only turs were found, these two good species hybridized (13). Another example is the hybridization of red and sika deer *(Cervus nippon)* in several European localities (14, 15). Such hybridization, whether in captivity or in nature where both species are sympatric, such as in Manchuria, is exceptional (13). This suggests hybrid disadvantage. The parent species differs significantly in antipredator strategies, so that hybrids are likely to be incompetent in their antipredator behavior as are, for instance, hybrids of mule deer and white-tailed deer (16, 17, 18, 19). The "biological species" can thus be determined only by studying free-living populations—provided these have the natural complement of predators and competitors, and are not feral populations formed from individuals that escaped from captivity or were deliberately introduced. These conditions are currently very difficult, if not impossible, to meet for many large mammals.

What happens when using the "zoo criterion" can be illustrated with wild sheep. Because different forms of wild sheep readily hybridize in captivity, Old World sheep, no matter how divergent morphologically, behaviorally, or ecologically, could be placed into the same "species," *Ovis ammon* Linnaeus, 1758 (20, 21). This is an extreme, for in nature different forms of wild sheep are geographically or ecologically isolated. They do not meet, and if they did are likely to ignore one another. In most cases, reliable information about hybridization in nature has not been obtained. With the possible exception of one hybrid zone in Iran, we do not know if hybrids of different subspecies are viable in nature (22). There is currently no feasible way to determine what could comprise "biological species" among wild sheep. We do know, however, that the different forms of wild sheep have retained their identity. They could not have done this if hybridization between races were common and successful. Nevertheless, some species or subspecies have apparently arisen from earlier crosses as is recorded in their genetic structure. One example is the crossing of $2n=58$ urials *(O. vignei)* and $2n=54$ mouflons *(O. musimon)* to form a hybrid population with $2n = 55/56$ (22). Another example is the mule deer, which may have originated in past crossings of male black-tailed and female white-tailed deer, as mule deer have mtDNA very close to that of white-tailed deer (23, 23a, 24, 25, 25a).

There are practical implications to any decision to apply the "zoo criterion"; calls may be heard today within international conservation organizations to ignore all subspecies and conserve only at the "species level." This would mean that conserving one race of wild sheep is all that is required to safeguard the species. Here the uncritical application of captivity observations to a complex "species problem" could even lead to the absurdity of scientists supporting, or at least ignoring, the extinction of all the diverse wild sheep, except for *one* population. The same would apply to most species of large mammals, if survival of hybrids in zoos is the criterion for inclusion in a species. It is thus not trivial where and how observations were gathered to determine what is and what is not a "biological species."

The many difficulties with the "biological species" concept led some systematists to classify caprid species not by reproductive isolation, but by the "evolutionary species" concept (3, 24, 25, 26, 27, 28, 29, 30). Ironically, even where only one species, *Ovis ammon*, was proposed for all Old World sheep, these were still divided into taxonomically unsanctioned, but very obvious natural groupings (13, 20). Wild sheep also separate into two natural groups represented by the subgenera *Ovis* (Old World sheep) and *Pachyceros* (Beringian and New World sheep). However, Heptner and others designated these levels not as subgenera, but as species (13). The Old World sheep, in turn, segregate into three "real groups," the mouflons *(musimon)*, the urials *(vignei)*, and the argalis *(ammon)*. The New World sheep also segregate into three "real groups," the Siberian snow sheep *(nivicola)*, thinhorn sheep *(dalli)*, and bighorn sheep *(canadensis)*. Modern students of mountain sheep have labeled these groups with Latin species names. These species, in turn, segregate into distinct geographic populations that have received subspecies designations. The species here is equated to an *adaptive radiation*, and the taxonomic classification thus follows natural lines and closely reflects phylogeny.

Note that the current species criterion used for sheep is not the same as that currently used for red deer. Here consensus continues to favor a taxonomy based on the biological species concept, uniting all red deer under *Cervus elaphus*. European authors do not describe the basis for this determination (13, 31, 32). Some authors accept North American wapiti and European red deer as one species based on the fact that they hybridized in captivity and that both forms, introduced to New Zealand, formed hybrid populations (12, 33, 34). One must add here: *in the absence of predators.* This is an unnatural situation and, given the maladaptations of red deer × wapiti hybrids, it is doubtful that they would survive predation by large predators (35, 36). Moreover, in Asia, some subspecies of red deer live in proximity and do not hybridize or, at most, form narrow hybrid zones, such as advanced wapiti *(canadensis)* in the Tien Shan Mountains and the primitive Lop Nor stag *(yarkandensis)* in the valleys below, or advanced wapiti meeting primitive Manchurian wapiti *(xanthopygos)* in southern Siberia (13). This implies that these morphologically, behaviorally, and ecologically distinct races form maladapted hybrids. However, East and West European red deer form a broad hybrid zone in eastern Europe, implying reproductive compatibility in the presence of natural predators (37).

Were red deer to be classified by the "evolutionary species" concept, the criteria used by most students of wild sheep, then three adaptive radiations would have been designated as species: the western European elaphines *(Cervus elaphus* Linnaeus 1758), the primitive central Asian red deer *(Cervus affinis* Hodgson, 1841), and the Siberian/American wapiti/maral forms *(Cervus canadensis* Erxleben, 1777). Each of these radiations forms a cluster of distinct subspecies.

For conservation purposes, the classification of species as adaptive radiations, the "evolutionary species" concept, is superior to any guesswork about "biological species" based on conjectures about what might, or might not, happen to hybrids in nature should the different forms ever meet. In short, while the biological species concept is theoretically elegant, it is in practice so difficult to apply to large mammals as to be virtually useless. Classification and nomenclature based on the evolutionary species concept is more consistent with phylogeny and thus scientifically more significant than an unnatural misapplication of the biological species concept. An expert witness should be aware of the above when explaining "the species concept" to a judge or a public hearing panel.

Serious obstacles to classification exist at the subspecies level. Classical taxonomy describes species and subspecies from a few specimens kept in museums. The specimen to which the scientific name is applied, together with a description that distinguishes the specimen from others, is labeled the type specimen. In view of the great variability even within a population, the type specimen can only serve as a voucher for the scientific name, and cannot do justice to the natural within- or between-population variation. Some of the variability, however, can be captured by measurements and the application of statistics; in particular, dimensions of mass and space. Some important variables, such as differences in color, shape, and behavior, which are readily apparent to the eye are difficult to capture quantitatively. Nevertheless, it has become normal practice to use metric measurements to characterize populations, a logical outgrowth of attempts to capture variability objectively. One can thus compare populations morphometrically and very precisely depict differences in measurable factors.

This apparently objective approach to describing within- and between-population variability cannot be used alone to draw taxonomic conclusions and may become an example of the misapplication of quantitative methods, a case of misplaced and misleading concreteness. This problem in much of taxonomy can be traced to a deficiency, namely, a general inability to distinguish between (genetic) adaptation and (phenotypic) adjustment. Morphometric comparisons may confuse not only adaptation and adjustment, but also homology and analogy, genotype and phenotype, ecotype and subspecies, or the effects of nature versus nurture. Comparative morphometrics thus confounds genetic, epistatic, environmental, and statistical variations.

Comparative morphometrics, as exemplified by statistical comparisons of skull measurements from specimens collected in the field, cannot in principle determine genetic differences. Only experimental morphometrics, performed on specimens that were raised under conditions in which environmental factors affecting growth were controlled, could do so, in part (3, 4, 5). Even where relevant environmental factors can be controlled, the effect of epistasis or nonhereditary genetic effects needs to be determined. Hybrid vigor, for instance, is a well-known epistatic effect. In short, comparative measurements between populations can only be used to determine differences between populations. However, such quantitative comparisons tell nothing about the origin of the differences. The differences could be due to hereditary and nonhereditary genetic effects, or due to environmental effects such as the quality, amount, seasonal patterning, or toxicity of nutrition, or due to the kind and intensity of physical work the organism performed, or due to powerful climatic stimuli such as heat, cold, and water shortages. All of these factors shape the individual organism simultaneously and, depending on the region and population history, develop widely different phenotypes from genetically similar animals. Conversely, the same environment generates similar phenotypes from different genotypes (35, 38). Although largely neglected in mainstream biology, significant work on this problem has been done in a variety of biological disciplines (39). How hereditary and environmental factors shape individuals also has had application in wildlife management. For instance, populations in superior habitat, in particular colonizing populations that exploit virgin ranges, generate very large-bodied, vivacious, luxuriant individuals (*dispersal phenotype individuals*). These are quite different from the meager, listless efficiency type of individuals (*maintenance phenotype individuals*) forced to live on overpopulated ranges (40, 41). Although differing in body mass by as much as fivefold, and very different in behavior, the two normally still can be recognized as belonging to the same race by the identity of their hair coat and markings (3, 37).

How bodies change in size and proportions in response to nutrition and heredity has been studied in detail in the agricultural discipline of animal science and there codified into the *centripetal theory of growth* (42). This discipline focused on the body and less on the skull that is of so much interest to taxonomists. Moreover, with changes in adult body size go predictable changes in skull and body proportions. To these changes the term *allometry* has been applied, although neoteny or paedomorphism would be more precise. The term *allometry* specifies no direction in which change takes place, as do neoteny and paedomorphism. In paedomorphism, with shrinking body size the adult body increasingly resembles a juvenile. The first use of comparative morphometrics on red deer by Ingebrigtsen demonstrated an increase in paedomorphic features with declining body size (43). Thus stags from small-bodied island races were more juvenile-like, not due to hereditary factors, but because they were poorly nourished. It is ironic that Ingebrigtsen used this quantitative method not to argue that interpopulation differences were of taxonomic significance, but that they were not (43). He reduced the multitude of so-called red deer subspecies by pointing out that paedomorphic dwarfs and hypermorphic giants were a consequence of nurture (environment) not nature (genes). Wallace demonstrated that the facial portion of the skull in domestic sheep is quite sensitive to nutrition compared to the cerebral portion (42). Excellent work has been done in Central Europe on nutrition in cervids and its application to quality deer management and more recently also in North America (35, 44, 45, 46, 47, 48, 49, 50, 51, 52, 53, 54).

One recent branch of taxonomy, phenetics, completely ignored the origin of within- and between-population variations and, for each population, took all possible measurements irrespective of origin or weight. Using sophisticated computer programs, a multitude of variables could be compared. Those practicing phenetics insisted that this approach was more objective and rigorous. It is not. The computed differences have no objective meaning, because they are a mixture of different, noncommensurable variances. The computed decision trees are thus void of meaning. Fortunately, phenetics need not detain us further because apparently there are no phenetic analyses species covered in this volume.

CRITERIA BY WHICH TO DIFFERENTIATE SUBSPECIES

If comparative morphometrics is not useful as a guide to genetic differences between populations, what objective taxonomic criteria are there to segregate subspecies? One criterion, the color and pattern of hair, has received much attention in taxonomy. If description is careful to differentiate between the completely grown nuptial coat and the changes in appearance due to the wear on hairs, shedding, the regrowth of pelage, and the effects of nutrition and bright sunlight on the appearance of the pelage, then differences in hair coats can be used taxonomically. Nuptial coats change little with environment. Raised in captivity or the wild, fed well or poorly, the nuptial coat remains fundamentally recognizable. We now know, however, that bison vary their nuptial coat like deer vary antlers (4). Consequently, caution is advisable. Nevertheless, the hair coat remains one of the better criteria for identifying and classifying individuals correctly. Unfortunately, individuals may be similar in nuptial coat, but differ greatly in genetics, as is the case for white-tailed deer (see later discussion). What has been said of the hair coat applies in principle to secondary sexual organs such as horn-like structures.

In conjunction with the hair coat, the number and structure of chromosomes can be a valuable tool in taxonomy. While urials and mouflons are very similar in external appearance, they differ in chromosome numbers, mouflons with 2n=54 and urials 2n=58. There are forms with 2n=55 and 56, which have been interpreted as hybrids (22, 26, 27). Morphologically Severtsov's sheep appeared to be close to urials, but its chromosome numbers placed it among the argalis (55).

Molecular biology promises the best classification by descent, bypassing as it does at the DNA level the noise from adaptation and phenotypic adjustment. It also promises precise identification down to the individual level (DNA fingerprints) (6). Still, as a developing field it is not without problems (56; see pronghorn discussion later in chapter). Unfortunately, in the zoo, field, market, or at the customs counter, where recognition by nongenetic features is important, molecular methods sometimes are impractical. Also, higher ruminants are closely related genetically; that is, all species share some alleles at some loci and biochemical studies may call for large sample sizes to derive statistically significant results and avoid typology (57). Moreover, any method that identifies gene products, as opposed to genes, can only capture those genes that have been activated by environmental circumstances. Consequently, such a method has potentially the same problems as comparative morphometrics, discussed earlier. Also, some forms can be quite distinct morphologically, but have nearly identical DNA sequences, as happens for the mitochondrial DNA of mule deer and white-tailed deer, or polar bears and brown bears (58, 59). The analysis of DNA variation is thus quite complex and is rapidly evolving into its own science. We can expect significant changes in our current taxonomy as a consequence.

The subspecies thus poses grave difficulties that are not resolved by simplistic lumping or splitting, but by thoughtful application of taxonomic criteria based on inherited differences (60). In practice, however, one runs into additional difficulties. For instance, in an investigation into taxonomy for law enforcement purposes the following was found:

1. Type specimen had been altered or fabricated from several specimens.
2. Skull collections lay in disarray with loose labels, and horns, skulls, and jaws heaped into one box.
3. Seasonal, sex, and age-related differences in the nuptial coat had been mistakenly used to create different subspecies or species.
4. Comparative morphometrics had been misused, thereby confounding genetic, epistatic, environmental, and statistical variations.
5. A typological approach had been used in classification.
6. Artistic renditions of subspecies were inaccurate and thus no guide to differences between forms.
7. Inconsistent criteria for classification were inconsistently applied.
8. Inappropriate criteria of classification were used.
9. We found a seemingly authoritative discussion of an important paper, although the author could not possibly have read that paper.
10. We found mislabeled forms in zoological gardens (3, 5).

Similarly, Ernst Eick in meticulous research on sika deer discovered additional sources of misinformation:

11. Type specimens had been bought in markets without any information about whether they came from the wild or from captivity, where they came from, if hybridized or not.
12. Type specimens were taken from zoological gardens, without knowledge of origin or breeding history.
13. New subspecies were described way out of the historical range of the species, and thus are likely to be feral populations formed from escaped semidomestic forms (61).

Clearly, there is cause to be alert when examining taxonomic publications and material.

COMMENTARY ON THE TAXONOMY OF BIG GAME SPECIES

Family Cervidae Gray, 1821

Cervidae, the deer family, are antlered ruminants, except for one antlerless species, the water deer *(Hydropotes inermis* Swinhoe, 1870) from China and Korea. The family can be divided into two subfamilies, based on the manner in which the second and fifth metapodia are retained (Table 1–1). The Old World deer, subfamily Plesiometecarpalinae, retain the upper ends of these metapodia, the only group of ruminants to do so. The New World deer retain the lower end of the second and fifth metapodia, the common conditon in cursorial ruminants. The family Cervidae contains about 16 genera and 44 recent species (62).

TABLE 1–1
Taxonomic Relationships of Large Mammals Covered in This Volume

Cervidae Gray, 1821	
Plesiometecarpalinae Brooke, 1878	
Cervus elaphus canadensis Erxleben, 1777	Wapiti or American elk
Telemetecarpalinae Brooke, 1878	
Odocoileus virginianus Boddaert, 1785	White-tailed deer
Odocoileus hemionus Rafinesque, 1817	Black-tailed and mule deer
Rangifer tarandus Linnaeus, 1758	Caribou
Alces alces Linnaeus, 1758	Moose
Bovidae Gray, 1821	
Bovinae Gill, 1872	
Bison bison Linnaeus, 1758	American bison
Caprinae Gill, 1872	
Ovibos moschatus De Blainville, 1816	Muskoxen
Oreamnos americanus Rafinesque, 1817	North American mountain goat
Ovis [Pachyceros] dalli Nelson, 1884	Thinhorn sheep
O. d. dalli	Dall's sheep
O. d. stonei	Stone's sheep
Ovis [Pachyceros] canadensis Shaw, 1804	Bighorn sheep
O. c. canadensis	Rocky Mountain bighorn
O. c. nelsoni	Desert bighorn
Antilocapridae Gray, 1866	
Antilocapra americana Ord, 1818	Pronghorn
Tayassuidae Palmer, 1897	
Tayassu [Pecari] tajacu Fischer, 1814	Collared peccary
Ursidae Gray, 1825	
Ursus maritimus Phipps, 1774	Polar bear
Ursus arctos Linnaeus, 1758	Brown bear
Ursus americanus Pallas, 1780	American black bear
Canidae Gray, 1821	
Canis lupus Linnaeus, 1758	Gray wolf or timber wolf
Felidae Gray, 1821	
Felis [Puma] concolor Jardine, 1834	Mountain lion
Panthera [Jaguarius] onca Fitzinger, 1869	Jaguar

The Old World Deer, Subfamily Plesiometecarpalinae Brooke, 1878

Wapiti or American Elk (Cervus elaphus canadensis Erxleben, 1777).

There is merit in dividing the red deer into three evolutionary species, each with its cluster of distinct subspecies. Emerging evidence indicates that Eurasian subspecies of red deer may act ecologically like good species; reproductive problems arise when hybridizing geographically distant subspecies; and the success of such hybrids in nature is limited. However, we have chosen to follow current consensus and regard all red deer as one spec*ies, Cervus elaphus* Linnaeus, 1758. We have taken this position pending a thorough reevaluation of this taxon.

The wapiti of North America are thus a subspecies of red deer whose evolutionary radiations center on Eurasia. The most primitive cluster of subspecies is the central Asiatic red deer, including the mountain-adapted Kashmir stag *(hanglu)*, MacNeill's deer *(macneilli)*, shou *(affinis)*, the riparian-adapted Buchara *(bactrianus)*, and Lop Nor stags *(yarkandensis)*. They have five-pronged antlers and are adapted to life in thickets. They are closest in body shape and security adaptations (saltors) to the ancestral sika deer. The second radiation are the European red deer, which have five-pronged antlers with complex distal antler portions or *crowns*. They have long tails and bicolored rump patches with a lower white and an upper yellow portion that extend well above the root of the tail. In body shape they are intermediate (cursorial/saltatorial) between the primitive (saltatorial) Asiatic forms and the advanced (cursorial) wapiti. The latter form the third radiation, which contains three subspecies *(canadensis, xanthopygos, alashanicus)*. They have six-pronged antlers, very short tails, uniformly colored yellow rump patches, and some elongation of the neck hair in females. The most highly evolved of these subspecies, the wapiti *(canadensis)*, is found in central Asia and in North America. It is the largest, the most cold-adapted, the most cursorial of all living Old World deer, the least sexually dimorphic of the red deer, and the most adapted to life in open plains (37).

Splitting of the advanced wapiti of Asia and North America into further subspecies cannot be upheld with the data at hand (37, 63). All advanced wapiti should therefore be labeled *Cervus elaphus canadensis* Erxleben, 1777, as was proposed by Flerov (32). The rationale is that all populations of this form on both continents carry the same basic coat pattern, dentition, and antler form and have the same rutting calls, compared to the primitive wapiti adjacent to these deer in Asia, the Manchurian or Izubr stag *(C. e. xanthopygos)* and the Ala Shan stag *(C. e. alashanicus)*. Both have six-pronged antler structures, but differ in their nuptial coats and rutting calls. A limited number of published and unpublished genetic probes indicate that all subspecies of the wapiti radiation are closely related, but fairly distant from European red deer (63).

The differences that exist among geographically distant populations of North American wapiti, such as differences in body size and antler characteristics, are in the first instance ecotypic. Genetically, they are all very close (63). This may be due to a number of factors, including a limited entry of wapiti into North America in late Pleistocene times, the mixing of local remnant populations with introduced stock from Yellowstone Park and Jackson Hole, or small populations in pre-Colombian times (37).

Elk became common only post glacially in North America as they colonized a land depleted by extinction of its native megafauna. There is little evidence that wapiti were part of the native Rancholabrean fauna; rather they are recent East Siberian immigrants along with grizzly be*ar (Ursus arctos)*, gray wolf *(Canis lupus)*, wolverine *(Gulo luscus)*, moose *(Alces alces)*, bison *(Bison occidentalis)*, and humans (37). In their southern distribution, some of these cold-adapted Siberians show deficits in adaptation (37, 64). Elk suffered a near demise late in the 19th century. The largest remnant population came under U.S. cavalry protection in Yellowstone National Park. From here elk were widely reintroduced throughout the United States and Canada (65). Therefore, what we find today may reflect genetically this recent pattern of reintroductions, rather than an old pattern of local adaptations.

Moreover, the pattern of wapiti distribution in North America prior to the great wildlife eradications before 1900 may in itself have been an historic artifact. Evidence is mounting that in pre-Colombian North America a dense human population, largely of agricultural cultures, kept large mammals low in numbers and spotty in distributions. With the post-Colombian introduction of European diseases and warfare, the native human population dropped rapidly, freeing "nature" from their controlling grip (66). Consequently, "wilderness" rebounded and wildlife populations expanded and spread rapidly, a condition that lasted about 250 years until the heavy hand of "white man" put an end to it (67). There is good evidence that bison expanded their range rapidly in North America beginning about 1600 (68). It is likely that elk did the same. The widespread elk populations of the late 18th century were probably a post-Colombian artifact of colonization. The low genetic diversity of wapiti could thus also be a function of barely escaping extinction following the expansion of human populations post-glacially on this continent.

The differences between populations appear to be ecotypic, with a few, but questionable, exceptions for the California wapiti. The common names (e.g., Roosevelt elk, Rocky Mountain elk) retain validity for popular use because elk do vary regionally in a recognizable manner. West Coast elk usually are large or very large, as are the elk in northern prairie regions. Some differences in antler structure are deserved that apparently reflect the toxicity of forage or patterns of seasonal food availability. Some experts can differentiate California or tule elk skulls from those originating elsewhere. Different valleys within the Rockies reflect differences in the skull morphometry of females, with fewer differences between males (68a). However, difference in size notwithstanding, all American elk carry the same six-pronged antlers, the same pelage markings and hair coat patterns, and bugle in the same fashion, which other subspecies of Asiatic wapiti do not.

California elk (*C. e. nannodes*) are taxonomically troublesome because they suffered a severe population bottleneck at the turn of the century and may thus reflect phenodeviance (biologically meaningless variation due to inbreeding akin to the "Habsburg jaw" among European royalty) (33, 69). Secondly, at small body size, we suspect individuals to suffer from resource shortages, and their shape and structure should show paedomorphic features. This, however, is not a taxonomic distinction (43).

There is currently no evidence for more than one subspecies of North American elk, *Cervus elaphus canadensis*, with California elk as a possible courtesy subspecies, pending more definitive genetic and ontogenetic research.

New World Deer, Subfamily Telemetecarpalinae Brooke, 1878

***White-Tailed Deer* (Odocoileus virginianus *Boddaert, 1785*).** The genus *Odocoileus* defines the North American Blancan, the long cooling period in the late Pliocene epoch before the major glaciations of the Pleistocene (70). The earliest of these deer were, according to the late Björn Kurtén, white-tailed deer, which also was the case in South America (71). The whitetail is likely the oldest deer species in existence. Its geographic range is large, covering some 79 degrees of latitude within North and South America. Its geographic variation, however, is small, making its taxonomy complex. Some 38 subspecies have been listed, 30 in North and Central America and 8 in South America (72, 73, 74). However, the major differences between purported subspecies is body size and size-related proportions. There is little variation in nuptial coats from North to South America. Some such differences in coat color and patterns have been studied in western whitetails in North America by Cowan and less precisely by Kellogg for other forms (72, 75). A taxonomic difference is the size of the metatarsal glands; they are large in northern, but small or absent in southern forms. Plains-adapted whitetails may have enhanced antler development and enlarged tails (37). In tropical latitudes, lowland forms may have near-permanent summer coats, while those at high elevations may have near-permanent winter coats, doubtless a genetic distinction

(73, 74). South American savannah whitetails may have relatively larger teeth and many have upper canines. While the phenotypic differences of white-tailed deer are not great, the genotypic differences are profound; South American whitetails are more different genetically from North American whitetails than the latter are from black-tailed deer *(O. hemionus)* (76).

Moreover, there is great phenotypic plasticity so that deer feeding on forages from fertilized farm fields are huge, compared to those eking out a living in other habitats. Molecular genetics indicates great regional differences in genetic composition, sometimes over very short distances and quite unrelated to currently accepted subspecies (77). This is likely due to two factors: a genetic effect due to rapid recolonization of North America following its near demise early in the century, and, due to its popularity as a game animal, genetic mixing following widespread introductions of non-native stock (78, 79, 80). Fortunately the latter effect appears to be minimal (81, 82). Its taxonomy, based largely on geographic differences in size, is not tenable and a revision based on an examination of genetic features of taxonomic significance is needed (72). Cowan's scrutiny of regional differences in western white-tailed deer may be a model to begin with (75). Regional studies of mtDNA may shed some light on recent relationships, but how that will translate into a taxonomy of subspecies remains a discussion for the future.

Black-Tailed and Mule Deer *(Odocoileus hemionus Rafinesque, 1817)*.

The modern taxonomy of black-tailed deer goes back to Cowan (75, 83). It remains largely valid, because Cowan's judgments were based primarily on notable regional differences in secondary sexual, gland, and pelage characteristics, and on qualitative skull characteristics; morphometrics played a descriptive role. Subsequently, Wallmo questioned if the subspecies *O. h. inyoensis* was based on an intergradation between the subspecies *O. h. californicus* and *O. h. hemionus* (84). Molecular genetics will tell, eventually. The desert mule deer subspecies *(O. h. crooki* Mearns, 1897) is based on a hybrid specimen, a cross between mule deer and white-tailed deer. J. R. Heffelfinger (Arizona Game and Fish Department, unpublished data) documents the history of this invalid taxon in detail, which was suspected to be a hybrid soon after publication. It should be replaced by *O. h. eremicus* Mearns, 1897. The Mexican Tiburon Island mule deer *(O. h. sheldoni)* may be no more than a regular *O. h. eremicus* living under marginal environmental conditions and thus of small body size. Unrecognized taxonomically is the great genetic division between mule deer and black-tailed deer proper. Mule deer contain the mtDNA of white-tailed deer; the Columbian and Sitka black-tailed deer do not (23, 24, 25, 58, 85). This indicates that mule deer arose out of crosses of black-tailed bucks and white-tailed mothers, while the closeness of the mtDNA of mule and white-tailed deer indicates a recent origin of mule deer. The most likely time for a successful speciation event is following the post-glacial megafaunal extinctions 12,500 to 7,000 years ago (19, 37).

Caribou *(Rangifer tarandus Linnaeus, 1758)*.

The taxonomy currently accepted is that of Banfield; it reflects the great difficulties encountered in classifying a highly variable species (86). Banfield provides a detailed review of the taxonomic history of caribou, showing that early naturalists took differences between caribou as a given without considering if these characteristics were acceptable taxonomically. Banfield introduced skull morphometrics as his primary criterion for segregating subspecies, with some emphasis on antler morphology and geography. There is no attempt to deal with variation in nuptial coats. Oddly, the coat characteristics of a woodland male are, mistakenly, given as a norm for Newfoundland, Ungava, Osborn, and other large-bodied caribou. Antlers are merely segregated into woodland and barren-ground types. In practice, this means that caribou are lumped or segregated basically by skull size, unless geographic position imparts distinction. Consequently, some rather dissimilar caribou are united into subspecies. For instance, the American woodland caribou *(R. t. caribou* Gmelin, 1788), according to Banfield, includes the large, dark, small-maned woodland forms from the mainland and the colorful Newfound-land caribou *(terranovae)*, the distinct Ungava caribou *(sylvestris)*, the big Osborn's caribou *(osborni)* from northwestern British Columbia, intermediate between wood-land and barrenground forms *(groenlandicus)*, and the sessile, and therefore large-bodied, barrenground caribou in Alaska, and the Yukon and North West Territories with their characteristic coat patterns and antler shapes. All of these forms are quite distinct in color, hair patterns, and antler shapes, and can be recognized as distinct by the unaided eye. They are, however, of much the same size. Conversely, the large, ses-sile barrenground caribou from the Alaska peninsula is segregated as a separate sub-species from the smaller, but migratory, barrenground form.

Banfield's search for objectivity and statistical measurements to capture variation is laudable, and his avoidance of dealing with hair coat differences is understandable. Swift and substantial changes occur as caribou change from the dark, uniform summer coat to the contrasting, complex nuptial coat, which changes in the male after the rut, due to loss of display hair, into the winter coat, which in turn is followed by an erosion of the hair surfaces due to severe wear into the light late-winter/spring coat. Superim-posed on this are differences of sex and age, plus great individual variation.

An attempt to segregate North American caribou by nuptial coat, body, and antler form independent of body size unites the southern woodland form and the western mountain caribou as *R. t. caribou*. It segregates the Newfoundland caribou as a conver-gent woodland form *(R. t. terranovae)*. The Labrador/Ungava caribou *(R. t. sylvestris)* retains its distinction as a convergent barrenground form. The Osborn caribou retain their distinctness as *R. t. osborni*. Sessile and migratory barrenground caribou are united into the same subspecies so that *R. t. granti* is collapsed into *R. t. groenlandicus*. The rest of Banfield's proposed subspecies *(dawsoni, pareyi, eogroenlandicus)* remain unal-tered (37).

In practice, this means that "woodland caribou" are not the same across North America, and that Newfoundland caribou are not a correct source of individuals for reintroductions on the mainland. It points to the plight of the woodland caribou on the mainland because this classification sets it apart from the Newfoundland caribou, which, thanks to decades of good management, is abundant. Nor may populations of sessile large-bodied barrenground caribou be mistaken for the rare and endangered woodland form.

Moose (Alces alces Linnaeus 1758). American moose *(Alces alces americana* Clinton, 1822) are an extension of East Siberian moose into North Amer-ica (32). One can distinguish two groups of moose by coat color pattern and antler form, the European/West Siberian moose and the East Siberian/North American va-riety. There is a broad hybrid zone centered about Lake Baikal and northern Outer Mongolia. This distinction is taxonomically obscured by a plethora of purported sub-species. Moose differ in size and allometry with latitude in Europe, Asia, and in North America, with populations of small, southern moose invariably assigned to dif-ferent subspecies than large-bodied northern forms. Thus, the now extinct Caucasian mountain moose *(A. a. caucasica)* is the small, southern European moose. In Asia it is *A. a. cameloides* from Manchuria, and in North America it is *A. a. shirasi*. The large northern counterparts are *A. a. alces* in Europe, *A. a. burtulini* in Eastern Siberia, and *A. a. gigas* in North America. Genetically, North American moose, de-spite considerable phenotypic variation, are very close. The genetic distance between European and American moose, however, is considerable and there are chromosome differences (87).

American moose differ from European ones by having an ornate, contrasting hair coat in which a light dorsal saddle is delineated from a dark underpart. The legs are lighter than the body, but unlike European moose are a medium-brownish gray rather than a dirty white. Males have a dark rostrum bordered on the front by reddish-blond hair, whereas females have the reddish-blond extending over the entire face and rostrum. Consequently, male and female American moose in winter coats, but not European moose, can be identified at a glance. The antlers of European moose are three pronged; those of the American moose are four pronged. The relative antler mass of European moose is about half that of American moose, which also have longer antler beams, a primitive feature. However, the narrower premaxillary, the slightly longer, higher rostrum, and smaller teeth suggest an advanced adaptation to feeding on aquatic vegetation and foliage in American moose (37).

There is considerable regional differentiation of moose, a sign of ability to adjust. Although body and antler size vary latitudinally, coat characteristics stay fairly constant. Peterson labeled American moose in accordance with a vision of Pleistocene glaciations and survival (88). However, his model of glaciations and differentiation has not been verified by the paleontological/archaeological record. Rather, modern moose are recent immigrants into North America, the oldest finds in Alaska dating only some 9,500 years (89). Although there is a significant change in body size with latitude, moose as large as Alaska moose may appear also for environmental reasons within their southern distribution. Because moose have evolved to exploit the ecological successions caused by forest fires and other destructive agents, the appearance of excellent moose habitat may soon be followed by a local population of very large moose. We are aware of such a case in Ontario that came up in court testimony, but was based on an unpublished report.

Family Bovidae Gray, 1821

This family comprises the hollow-horned ruminants (Table 1–1). The horns are never branched and the sheaths are not replaced annually as are the horn sheaths of pronghorns or the antlers of most forms of deer. This is the largest extant ruminant family with about 45 genera and 137 modern species (62).

Subfamily Bovinae Gill, 1872

The American Bison (*Bison bison *Linnaeus, 1758). The conventional view is that bison are divided into two subspecies, the plains bison (*Bison bison bison* Linnaeus, 1758), and the wood bison (*B. b. athabascae* Rhoads, 1897) (90). An unconventional view divides the species into three subspecies: southern plains bison (*B. b. bison*), northern plains bison (*B. b. montanae* Krumbiegel, 1980), and the wood bison. Neither position is tenable in fact or theory (4, 91, 92, 93). An excellent review of the history of naming bison is given by Roe (68, pp. 26–68). Reduced to its essence there are four results.

1. Rhoads, who had not seen a wood bison, used the secondhand description of a single (male) specimen by Professor D. J. Macoun to justify the subspecific distinction (*B. b. athabascae*) (94). The description by Macoun, "Size large, colors dark, horns slender, much longer and incurved and hair more dense and silky than in *B. bison*," is not diagnostic, but could apply to any number of plains bison. There is no basis here for any taxonomic distinction.

2. The pelage differences described by Geist and Karsten between "plains" bison and "wood" bison held captive in 1975 in Elk Island National Park turned out to be ecotypic captivity effects (95). They were confined entirely to bison from that park and disappeared in transplanted "wood" bison. Also, moose and elk in this park suffered antler distortions parallel to the pelage distortions in bison (4). There are no differences in the nuptial pelage of plains and wood bison. However, the nuptial coat varies with sex, age, and nutrition. This insight was new.

3. The distinctions between plains and wood bison according to Van Zyll de Jong were the (invalid) pelage characteristics referred to earlier, and morphometric differences, an invalid taxonomic methodology (4, 5, 90). Although skull shape and size varied between populations, the tooth row lengths did not. Because tooth row varies least with environment, this suggests that there was no genetic

difference in size between populations. Wood and plains bison raised in feed trials grew to the same size (96). Consequently, no valid taxonomic distinctions between plains and wood bison have been reported.

4. Krumbiegel and Krumbiegel and Sehm based splitting of plains bison into two subspecies on differences in pelage characteristics (91, 92). Zoo experiences in Germany and historic sketches suggested different pelage patterns in southern and northern plains bison. Krumbiegel declared the image of a captive bison male held in Hamburg Zoo in 1923 as the type of the southern plains bison (91). The photo, taken during early summer (hair shedding), is of an *old* male. Krumbiegel and others were not aware that the nuptial coat of the bison changes with age and with environment and season analogous to antlers in deer (4). The bison in question shows the old age coat. Attempts to find characteristics in southern bison pelage comparable to Krumbiegel's type specimen failed (91). However, similar coat characteristic were seen in old male plains buffalo.

Thus, subspecific distinctions in bison are unwarranted. However, there are bison populations that are more valuable than others for conservation. Genetically, the most important population are the bison of Wood Buffalo National Park. They are native there and have been subject to incessant wolf predation. They are, unfortunately, carriers of bovine tuberculosis and brucellosis. Next in importance are the two populations of Nyarling River bison that originated from the northern portion of the same park: the free-living herd in the Makenzie Bison Sanctuary and a captive herd in Elk Island National Park. Both are mere genetic samples of the park bison. They are both inbred, but they are free living, preyed on by wolves, and are free of tuberculosis and brucellosis. A third population now expanding under wilderness conditions with a full complement of predators is the Pink Mountain herd in British Columbia, a feral herd formed from escaped "plains" bison that originated in Elk Island National Park. These may soon link with bison released as "wood" bison along the Liard River in northwestern Canada. Bison herds under long-term human care and free of predation are the least valuable gene pools for conservation, in particular those that carry genes from domestic cattle.

Subfamily Caprinae Gill, 1872

The Muskoxen (Ovibos moschatus *De Blainville, 1816).* The muskoxen is currently segregated into two tenuous subspecies, which differ somewhat in color patterns (97). Muskoxen decrease in size from south to north, as do other large mammals that live north of 60°N latitude (98). Consequently, the northern form, the Greenland, white-faced or high Arctic muskoxen (*O. m. wardi*), is smaller than the southern form, the dark barrenground muskoxen (*O. m. moschatus*).

The North American Mountain Goat (Oreamnos americanus *Rafinesque, 1817).* The mountain goat, like the American bison, is a species with some regional variations in body size, but apparently without taxonomically valid differences. An examination of skull variation by Cowan and McCrory did not uphold the earlier splitting into four subspecies (99). We consider it a species without subspecies.

North American Mountain Sheep, Thinhorn Sheep (Ovis [Pachyceros] dalli *Nelson, 1884), and Bighorn Sheep (Ovis [Pachyceros] canadensis *Shaw, 1804).* Two species of mountain sheep in North America, the thinhorn and bighorn sheep, plus the snow sheep of Siberia (*Ovis [Pachyceros] nivicola* Eschscholtz, 1829) comprise the subgenus *Pachyceros*. This is a distinct branch of the genus *Ovis*, characterized by 54–52 chromosomes (26, 27). Each species name defines a radiation of subspecies, the criterion for labeling the species as natural units in traditional caprid taxonomy. Biologically, these sheep are very similar when found in comparable environments. Thinhorn sheep, in their behavior, differ from bighorns by being less neotenous (that is, more adult-like), emphasizing ceremony over overt aggression and sex (25).

The most important taxonomic revision of American sheep is that by Cowan (24). However, much has been learned of mountain sheep since then and revisions are required. The thinhorn sheep segregate into two good subspecies, the dark Stone's sheep *(O. d. stoni)* and the white Dall's sheep *(O. d. dalli)*. Stone's sheep have a color pattern; Dall's sheep do not. There is a zone of integration between the two forms. Stone's sheep become increasingly lighter from south to north, while a belt of Dall's sheep, surrounding the northern wedge of Stone's sheep, have black tails and black hairs on the saddle, about the eye, and on the bridge of the nose. Thus, the last of the integration is seen in the western Yukon, in Central Alaska, and in the east on the Nahanni River in the extreme southwestern corner of the North West Territory. As Stone's sheep become lighter the pigment concentrates in the saddle and flank stripes, giving rise to a form once described as a separate species, *Ovis fannin,* by Hornaday in 1901. Today we consider these sheep a color variant of Stone's sheep. The distinction between the two subspecies would be the presence of a recognizable rump patch on Stone's sheep; black tails and a sprinkling of dark hair on the body are thus no criterion for exclusion from the Dall's sheep subspecies. *Ovis d. kenaiensis* is relegated to the Dall's sheep proper because its distinguishing features appear to be phenotypic adjustments rather than genetic adaptations.

Among bighorn sheep, the Rocky Mountain form *(O. c. canadensis)* is distinct from desert bighorns *(O. c. nelsoni)*, morphologically and in their mitochondrial and nuclear DNA (100, 101). There is good subspecific distinction at this level. Desert bighorns have relatively and absolutely large horns in females, a continuous stripe of dark fur bisecting the rump patch dorsally, larger ears, longer tooth rows, and a shorter pelage. Body size and proportions vary between regions as does horn morphology, but have no taxonomic significance. Subspecies based on morphometric differences in sizes and proportions within the desert bighorns do not match meta-populations as established by differences in mitochondrial and nuclear DNA (101). The desert bighorns thus encompass a real, but taxonomically troublesome grouping of southern bighorns (102). Following Geist these should be older phylogenetically than Rocky Mountain bighorns (25, 103). California bighorns *(O. c. californiana)* segregate weakly from Rocky Mountain bighorns and are close genetically where they are close geographically.

Family Antilocapridae Gray, 1866

This is a unique North American ruminant family with hollow horn sheaths that branch, are shed annually like deer antlers, and regrow from skin covering the horn cores. There is only one extant species (Table 1–1).

The Pronghorn (Antilocapra americana Ord, 1818). The relationship of pronghorn to other ruminants, seen by some taxonomists as the possible link between bovids and cervids, has caused disagreement for years. Biochemical studies have done little to clear up the confusion. Cross-reactivity of blood antigens of five generally recognized families of artiodactyls showed distinct differences between all other families, but not between antilocaprids and bovids (104). Studies of pancreatic ribonucleases indicated a close relationship among antilocaprids, bovids, and giraffids, leaving the cervids a separate group. Sequences of the mitochondrial cytochrome *b* gene placed antilocaprids, cervids, and giraffids together, separate from bovids (105). Sequences of mitochondrial 12s and 16s genes placed antilocaprids and bovids together and cervids and giraffids together. Other mitochondrial ribosomal results indicated that the families of higher ruminants each descended from a somewhat different group (106). Analysis of kappa-casein DNA sequences suggested that antilocaprids were a sister group to the bovid/cervid/ giraffid group (57). Regardless of these conflicting results, these recent studies support the traditional position of antilocaprids in their own family, as reported here.

The ancestor of pronghorn apparently reached the New World long before bovids, which were not apparent in North America until the Pliocene. A diversity of hornless pecoran taxa is found in the early Miocene of Eurasia. However, without horn cores, similar-sized antilocaprids and bovids cannot be distinguished. In contrast, both cervoids and giraffoids have a number of unique identifying features (107).

Frick proposed that the Merycodontinae, a North American fossil family dating from about 18 to 10 million years before present, was an ancestral subfamily of Antilocapridae (108). However, no intermediate fossils linking the two groups are known. No proof is available that Merycodontinae were or were not ancestral pronghorn.

Five subspecies of pronghorn have been named and generally recognized, *Antilocapra americana americana* Ord, 1818; *A. a. mexicana* Merriam, 1901; *A. a. sonoriensis* Goldman, 1945; *A. a. peninsularis* Nelson, 1912; and *A. a. oregona* Bailey, 1932. All were named on the basis of minor differences in color, size, or skull proportions. This confirms Bailey's notation, that when naming *A. a. oregona*: "The animals show only slight and gradual variations over their entire range, and no sharp lines of differences between described forms can be found" (109, p. 45). Lee did find minor genetic differences between *A. a. mexicana* and *A. a. americana*, but also an apparently natural intergrade zone between the two groups in southern New Mexico and western Texas (110). Such integration would preclude considering *A. a. mexicana* a valid subspecies by the criteria of many taxonomists (111). Pronghorn have been extirpated from the area where the type specimen of *A. a. sonoriensis* was collected (112). Whether animals along the Arizona–Mexico border, now called *A. a. sonoriensis*, are similar is unknown. Further study of possible subspecies is complicated by extirpation of some populations and mixing of presumed subspecies through translocations.

Pronghorn move long distances between seasonal ranges when human impediments do not prevent it. When those impediments did not exist and some 40 times as many pronghorn occupied North America, gene flow between animals throughout their range seems likely. Thus a mosaic of slightly differing pronghorn populations would be expected, rather than distinct subspecies.

Family Tayassuidae Palmer, 1897

The pig-like peccaries are an ancient North American family of ungulates that spread to South America just prior to the major glaciations of the Pleistocene (Table 1–1). They differ from pigs, among others, in having their metatarsals reduced to two functioning digits and fused at their distal ends; a complex (two-chambered) stomach, but lacking a gall bladder; a dentition reduced to 38 teeth, but the cheek teeth are larger and more complex favoring herbivory; a specialized rump gland; and a social system based on kin-defense with its mandatory throttling of reproduction. Peccaries, compared to pigs, have reduced birth numbers, neonatal mass, and number of teats. Late Pleistocene peccaries in North America were large and fleet footed, but failed to survive the general demise of the megafauna post-glacially. The more primitive collared peccary *(Tayassu tajacu)* moved subsequently north into the ecological vacuum from Central America. Two genera and three species are found today; only one species occurs in North America (113).

Collared Peccary (*Tayassu [Pecari] tajacu *Fischer, 1814). This species has apparently extended its range in historic times in the southern United States. Ten purported subspecies have been described on the basis of size and shades of color for North America, of which two (*sonoriensis, angulatus*) extend into the United States (113, 114). While there is geographic variation, minor size differences can be phenotypic adjustments, rather than genetic adaptations. Thus the subspecific designations are suspect.

Family Ursidae Gray, 1825

The bear family is closely related to the dog family (Table 1–1). They differ morphologically in that bears originally used trees as escape terrain, ascending such with speed and skill when confronted by danger. Dogs, on the other hand, chose to escape by speedy running and bounding. A very successful family, bears radiated from the tropics to colonize with distinct species in temperate forests, plains, alpine, tundra areas, and the north polar oceans. There are about six genera and nine living species, three of which are found in North America (62).

North American bears include *polar bear (Ursus maritimus* Phipps, 1774*), brown bear (Ursus arctos* Linnaeus, 1758), and *American black bear (Ursus americanus* Pallas, 1780). Members of the bear family, Ursidae, inhabit Eurasia and North and South America. Ursidae apparently evolved from early canids some 20 to 25 million years before present (115*). Ursus minimus* was a small, primitive bear, possibly the earliest of the Ursidae, found in many European localities (116). A similar bear *(U. abstruses)* spread to North America by the early Blancan 3.5 million years ago (70). It may have given rise to both, Asiatic *(U. tibetanus)* and American black bears. Timing of divergence of American black bears as estimated from fossils is close to the 4.4 million years derived through two-dimensional electrophoresis and the 3.8 million years estimated from mitochondrial DNA (59, 117). These bears rely on trees for the security of their young, whereas advanced bears, which may occupy treeless landscapes, such as brown or grizzly bears *(U. arctos)* and polar bears, rely on an aggressive defense of the cubs by the female (Stephen Herrero, University of Calgary, personal communication). The greatest threat to the cubs are large males.

Brown bear fossils upward of 0.5 million years old are found in China, where their fossil record is continuous (116). They entered Europe about 0.25 million years ago, and Alaska about 100,000 years ago, but they did not move south until about 13,000 years ago. There may have been two independent immigrations, narrow-skulled bears from southern Siberia through central Alaska to the rest of North America, and broad-skulled bears from Kamchatka to the Alaska peninsula and adjoining islands. Brown bears vary greatly in size and color, while their skull proportions vary with age and individually (118). This may have led Merriam to conclude that 86 species of brown bear occupied North America (119, 120). Rausch studied a large series of skulls and demonstrated that geographic variation was clinal, decreasing in size north, east, and south from a maximum on the Alaska peninsula and Kodiak and Afognak islands (121). He differentiated American *Ursus arctos* to two subspecies, the large, broad-skulled coastal bears of Alaska and British Columbia as *U. a. middendoerffi* Merriam, 1896) and the smaller, narrow-skulled bears in the rest of North America as *U. a. horribilis* Ord, 1815. Hall, in a more recent study, generally recognized the coastal bears as distinct, but split these into five subspecies, and the interior bears into three (122).

The polar bear apparently is a recent offshoot of brown bears. Evidence of recent divergence includes the rarity of fossil remains, although subfossils are common, and that these species produce fertile offspring in captivity (115). Mitochondrial DNA divergence and two-dimensional electrophoresis also suggest recent divergence that may date back to the huge Penultimate or Riss glaciation, beginning about 225,000 years ago (59, 117). By the next interglacial (125,000 years ago) the polar bear had emerged in Eurasia (123). A female polar bear radiocollared in the Canadian Arctic wandered over some 77,700 square kilometers; the researchers professed not to know where the males wintered (123). Subspeciation seems unlikely in animals that travel so extensively while subsisting on the same prey, caught in the same habitat.

Family Canidae Gray, 1821

Members of the dog family are fairly primitive, but very successful carnivores that evolved numerous species filling a great diversity of niches from the tropics to the Arctic. There are about 14 genera and 34 living species, with some eight species in North America (62). This volume limits presentation to the gray wolf (Table 1–1).

Gray Wolf or Timber Wolf (Canis lupus Linnaeus, 1758). The gray wolf in North America is the Siberian "snow dog," large-pawed compared to domestic dogs of comparable size, adapted to the soft substrates of northern latitudes as is the

"snow deer" of the North, the caribou or reindeer. It is likely that, just like the grizzly bear, the gray wolf is a post-Pleistocene colonizer of the southern latitudes of North America. Ecologically, it probably replaced the dire-wolf (*Canis diurus*) of the late Pleistocene Rancholabrean fauna.

Gray wolves differ in size, color, and skull proportions within their vast geographic range, but aside from color, none of these differences are striking (124). Body size increases from south to about 60 to 65°N, and decreases rapidly with higher latitudes (98). Based on morphological studies, 24 subspecies were recognized in North America some 50 years ago. Sokolov and Rossolimo recognized nine additional subspecies in Eurasia, but reduced the number in the New World to seven (125, 126). Using multivariate analysis, molecular genetics, and large samples, Brewster and Fritts suggested that five North American subspecies were more reasonable (124). Nowak also recognized five subspecies, namely, *Canis lupus arctos*, a large-toothed Arctic wolf; *C. l. occidentalis*, a large wolf of Alaska and western Canada; *C. l. nubilus*, a moderate-sized wolf originally found from Oregon to Newfoundland, and from Hudson Bay to Texas; *C. l. bailey*, an unusually small wolf from the southwest; and *C. l. lycaon*, a small subspecies now restricted to southeastern Canada (126). Nowak noted that some Eurasian wolves resemble their North American counterparts (apparently meaning wolves in similar habitats) more than they resemble other Eurasian wolves.

Mech summed up the wolf subspecies issue: "In reality one race (subspecies) of wolf is pretty much the same as any other. The behavior and natural history are similar among the various races and between North American and Eurasian wolves. In fact, any real differences seem to be more related to the precise living conditions such as food type, climate and geographic area. Physically, too the races are similar. Only a real expert measuring many skulls can distinguish among most races. Subspecific names, like Mackenzie Valley wolf, northern Rocky Mountain wolf, Great Plains wolf, or eastern timber wolf, are more descriptors of where a given wolf comes from than of any real differences among the animals" (127, p. 18). One may note that while comparative morphometrics describes very accurately regional differences in wolves, such differences can be due to many factors and do not reflect genetic differences alone.

Few barriers to movements of wolves in North America existed between the last Ice Age and their extirpation over much of North America. Wolves are notorious travelers. One male wolf was known to move 886 kilometers and a female wolf moved 840 kilometers (128, 129). Also, North American natives moved wolf-dog hybrids or semi-domesticated wolves considerable distances and some would breed with resident wolves. Thus, movements among presumed subspecies seem likely. The characteristics of wolves that formerly occupied many areas are poorly known or unknown. Whether characteristics of some presently presumed subspecies gradually grade into another or change abruptly can only be guessed. This, coupled with the similarity of Eurasian and North American wolves in similar habitats, makes the concept of a mosaic of ecotypes seem more logical than that of distinct subspecies.

Family Felidae Gray, 1821

Felidae lost many specialized genera and species with the global demise of the megafauna post-glacially, especially the large-bodied cat species (Table 1–1). Cats are a very specialized group of carnivores. They are a predator "model" that a number of unrelated carnivorous families adopted by convergent evolution. There are 18 genera and about 36 recent species (62).

Mountain Lion* (Felis [Puma] concolor *Jardine, 1834). Historically, cougars, mountain lions, panthers, or pumas—whatever one chooses to call them—were distributed throughout the Americas, from the southern Yukon Territory in Canada to southern Argentina and Chile. They have been extirpated from most of eastern North America, but persist in reduced numbers in most of South America. As in any broadly distributed species, regional variations in appearance occur. However, individual variation is also considerable. Early taxonomists named 26 subspecies scattered across the Americas with 12 recognized north of the Mexican border with the United States (130). How such subspeciation took place in a species that seems at home in deserts, swamps, jungles, mountains, deciduous forests, grasslands, and practically every habitat in the Americas from northern Canada south is hard to understand. A taxonomic review of this species is needed.

Jaguar (*Panthera [Jaguarius] onca *Fitzinger, 1869). The jaguar, unlike the mountain lion, belongs to the "big cats," which form morphologically and behaviorally a distinct group of rather successful cats. The puma is a "small cat," grown big. The "big cats," according to Ingrid Weigel, differ by having an "incompletely ossified hyoid apparatus, with an elastic cartilaginous band replacing the bony intermediate protuberances found in the hyoid apparatus of small cats" (131). This modification allows big cats to roar, but limits purring to exhaling. Big cats, when feeding, use incisors, canines, and molars to *rip* meat from a carcass, whereas small cats use their molars to shear meat. During resting, big cats tend to extend their front legs and their tails in a characteristic resting posture.

The jaguar and the puma are both big winners in the post-glacial megafaunal extinctions that swept North and South America, probably because of their ubiquitous food habits and little dependence on the megaherbivores that became extinct. Jaguars in particular take everything from small rodents, to large ungulates, to large and small reptiles, to fishes. Many prey species of the jaguar and puma probably prospered and spread after the competing megaherbivores became extinct, as exemplified in particular by deer *(Odocoileus)* and peccaries. Jaguar are excellent swimmers and aquatic hunters. More stocky in body build and larger than leopards, jaguars are more terrestrial in habit. They were found in historic times from Arizona all the way to the tip of southern South America. Their distribution coincides, roughly, with that of peccary, whereas fleet-footed prey such as deer are, apparently, left to the puma. While the arboreal puma coincides geographically with gray wolves, the more terrestrial jaguar does not. Ecologically they are tied to standing water, but may otherwise exist in fairly dry terrestrial habitats. They occur in spotted and in melanistic phases. The spots of the jaguar are larger than those of leopards.

Jaguars are originally Eurasian in origin. They have been residents of North America since the beginning of the major glaciations of the Pleistocene, almost 2 million years ago. Here jaguars became part of an unusually diversified, specialized predator fauna. They were earlier much more northern in distribution; their fossils have been found in the northern states. However, this distribution was already reduced southward during the last, the Wisconsinian glaciation, and became reduced even further southward following the last glaciation. Moreover, over the course of the Pleistocene, jaguars shrank in body size, indicating displacement to small prey by the larger and specialized predators that focused on big prey (70). Today there are noticeable regional variations in body size, with the smallest jaguars in Central America and the largest in tropical South America. A number of subspecies have been recognized traditionally, with the northernmost being the Arizona jaguar *(P. o. arizonensis* Goldman, 1932). Several subspecies were described for Mexico (131). While regional differences in size are real, it remains to be seen what genetic differences there are justifying the segregation into subspecies, as opposed to mere phenotypic (environmental) or minor genetic population differences.

SUMMARY

With the implementation of national and international conservation legislation that incorporates formal taxonomy, making such nomenclature actionable in courts of law, this old and essential biological discipline has assumed great significance for the conservation of wildlife and biodiversity. It has forced reassessment in the face of legal challenges and policy initiatives. We have examined the taxonomy of big game species and found much of it wanting, among others, due to inappropriate methodologies, unresolved differences over taxonomic categories and approaches, and by curatorial difficulties brought on by neglect of collections. In addition, molecular tracing of phylogeny, which bypasses both adaptations and adjustments by phenotypes, has cast additional doubt on accepted taxonomies. Biologists can expect to be questioned on matters of taxonomy in courts of law or before formal environmental assessment panels and be subject in public to hostile and expert cross-examination. Some familiarity with appropriate conduct as a formal expert witness is thus essential.

We focus on the species and subspecies levels because these are not only of paramount importance in the conservation of biodiversity, but are also the source of greatest controversy and legal challenges. Based on our experience we propose the following alternatives to conventional taxonomies. The splitting of most North American large mammals into a multitude of subspecies is unwarranted, because such taxonomies fail to distinguish between genetic adaptation and phenotypic (environmental) adjustment. In American bison, mountain goat, polar bears, black bears, and cougars there is no evidence for subspecies. While we follow current convention (biological species) and accept the elk as a subspecies of the red deer, we point out that if red deer were classified by the same criteria as wild sheep (evolutionary species), there would be three species of red deer. We suggest one subspecies of elk, with the California elk a possible second subspecies, pending reexamination. We find evidence for only one subspecies of moose. Classifying caribou by nuptial pelage reduces the number of subspecies to eight. The classical taxonomy of the mule deer we affirm with minor changes, but that of the white-tailed deer requires original rethinking. In the taxonomy of mountain sheep and pronghorns, there are legitimate doubts about the desert subspecies. We retain two subspecies of muskoxen and brown bears. A serious, well-funded effort is required to reexamine the taxonomy of North American large mammals, segregating adaptation from adjustment, and unifying criteria for taxonomic judgment.

LITERATURE CITED

1. SIMPSON, G. G. 1961. Principles of animal taxonomy. Columbia University Press, New York.

2. O'BRIEN, S. J., and E. MAYR. 1991. Bureaucratic mischief: Recognizing endangered species and subspecies. *Science* 231(4998):1187–1188.

3. GEIST, V. 1991. On the taxonomy of giant sheep (*Ovis ammon* Linnaeus, 1766). *Canadian Journal Zoology* 69:706–723.

4. GEIST, V. 1991. Phantom subspecies: The wood bison *Bison bison "athabascae"* Rhoads 1897, is not a valid taxon, but an ecotype. *Arctic* 44:283–300.

5. GEIST, V. 1992. Endangered species and the law. *Nature (London)* 357:247–276.

6. GATESY, J. 1997. More DNA support for a Cetacea/Hippopotamidae Clade: The blood-clotting protein gene Y-fibrinogen. *Molecular Biol. Evolution* 14:357–543.

7. GATESY, J., C. HAYASHI, M. CRONIN, and P. ARCTANDER. 1996. Evidence from milk casein genes that Cetaceans are close relatives of Hippopotamid artiodactyls. *Molecular Biol. Evolution* 13:954–963.

8. GOLDSCHMIDT, R. 1940. *The Material Basis for Evolution.* Yale University Press, New Haven, CT.

9. WADDINGTON, C. H. 1957. *The Strategy of the Gene.* Allan and Unwin, London.

10. GATESY, J., D. YELON, R. DESALLE, and E. S. VRBA. 1992. Phylogeny of Bovidae (Artiodactyla, Mammalia), based on mitochondrial ribosomal DNA sequences. *Molecular Biol. Evolution* 9(3):433–446.

11. MAYR, E. 1966. *Animal Species and Evolution.* Belknap and Harvard University Press, Cambridge, MA.

12. BRYANT, L. D., and C. MASER. 1982. Classification and distribution. Pages 1–59 *in* J. W. Thomas and D. E. Toweill, eds., *Elk of North America.* Stackpole Books, Harrisburg, PA.

13. HEPTNER, W. G., A. A. NASIMOVITCH, and A. A. BANNIKOV. 1961. *Mammals of the Soviet Union,* Vol. 1 (German translation). Gustav Fischer Verlag, Jena, Germany.

14. HARRINGTON, R. 1973. Hybridisation among deer and its implication for conservation. *Inst. Forestry* 30:64–78.

15. BARTOS, L. 1992. Sika/red deer hybridization—recognition and present status. Pages 191–195 *in* N. Maruyama, B. Bobek, Y. Ono, W. Regelin, L. Bartos, and P. R. Ratcliffe, eds., *Present Trends and Perspectives for the 21st Century.* Paper presented at INTERCOL '90, Yokahama, August 24, 1990. Japan Wildlife Center, Tokyo, Japan.

16. LINGLE, S., 1989. Limb coordination and body configuration in the fast gaits of white-tailed deer, mule deer and their hybrids: Adaptive significance and management implications. Masters degree project. Faculty of Environmental Design, University of Calgary, Alberta, Canada.

17. LINGLE, S. 1992. Escape gaits of white-tailed deer, mule deer and their hybrids: Gaits observed and patterns of limb coordination. *Behaviour* 122:153–181.

18. LINGLE, S. 1993. Escape gaits of white-tailed deer, mule deer and their hybrids: Body configuration, biomechanics, and function. *Canadian Journal of Zoology* 71:708–724.

19. GEIST, and M. FRANCIS. 1990. Mule deer country. NorthWord Press, Minocqua, Wisconsin. USA.

20. HALTENORTH, T. 1963. Klassification der Säugetiere: Artiodactyla. *Handbuch der Zoologie* 1(18):1–167.

21. PFEFFER, P. 1967. Le mouflon de Corse (*Ovis ammon musimon* Schreber, 1782); position systematique, ecologie et ethologie comparees. *Mammalia* 31 (supplement).

22. VALDEZ, R., C. F. NADLER, and T. D. BUNCH. 1978. Evolution of wild sheep in Iran. *Evolution* 32:56–72.

23. COWAN, I. MCTAGGART. 1940. Distribution and variation in the native sheep of North America. *The American Midland Naturalist* 24:505–580.

23a. CRONIN, M. A. 1986. Genetic relationship between white-tailed deer, mule deer and other large mammals inferred from mitochondrial DNA analysis. Dissertation. Montana State University, Bozeman.

24. CRONIN, M. A. 1989. Molecular evolutionary genetics and phylogeny of cervids. Ph.D. Dissertation, Yale University, New Haven, CT.

25. CRONIN, M. A., E. R. VYSE, and D. G. CAMERON, 1988. Genetic relationship between mule deer and white-tailed deer in Montana. *Journal of Wildlife Management* 52:320–328.

25a. GEIST, V. 1971. *Mountain Sheep.* University of Chicago Press, Chicago, IL.

26. NADLER, C. F., K. V. KOROBITSINA, R. S. HOFFMANN, and N. N. VORONTSOV. 1973. Cytogenetic differentiation, geographic distribution, and domestication in Palearctic sheep (*Ovis*). *Zeitschrift für Säugetierkunde* 38(2):109–125.

27. NADLER, C. F., R. S. HOFFMANN, and A. WOOLF. 1973. G-band patterns as chromosome markers, and the interpretation of chromosomal evolution in wild sheep (*Ovis*). *Experientia* 29:117–119.

28. SCHALLER, G. 1977. *Mountain Monarchs.* University of Chicago Press, Chicago, IL.

29. VALDEZ, R. 1982. *The Wild Sheep of the World.* Wild Sheep and Goat International, Box 244, Mesilla, NM.

30. SHACKLETON, D. M., ed., and the IUCN/SSC CAPRINAE SPECIALIST GROUP. 1997. Wild sheep and goats and their relatives. Status plan and conservation action plan for the Caprinae. IUCN Gland, Switzerland and Cambridge, UK.

31. ELLERMAN, J. R., and T. C. S. MORRISON-SCOTT. 1951. *Checklist of Palaearctic and Indian Mammals 1758–1946.* British Museum of Natural History, London.

32. FLEROV, K. K. 1952. *Musk Deer and Deer. Fauna of USSR. Mammals,* Vol. 2. Academy of Sciences USSR. Moscow (English translation U.S. Department of Commerce).

33. MCCULLOUGH, D. R. 1969. *The Tule Elk. Its History, Behavior and Ecology.* University of California Press, Berkeley, CA.

34. CAUGHLEY, G. 1971. An investigation of hybridization between free-ranging wapiti and red deer in New Zealand. *New Zealand Journal of Science* 14:993–1008.

35. BENINDE, J. 1937. *Zur Naturgeschichte des Rothirsches.* Monographie der Wildsäugetiere, Vol. 4. P. Schöps, Leipzig, Germany.

36. SCHOENWALD-COX, C. H., J. BAYLESS, and J. SCHOENWALD. 1985. Cranial morphometry of Pacific coast elk (*Cervus elaphus*). *Journal of Mammalogy* 66:63–74.

37. GEIST, V. 1998. *Deer of the World: Their Evolution, Behaviour and Ecology.* Stackpole Books, Mechanicsburg, PA.

38. JAMES, F. C. 1983. Environmental component of morphological differentiation in birds. *Science* 211:184–186.

39. BRUTON, N. M., ed. 1989. *Alternative Styles of Animals.* Kluwer Academic Publisher, Dordrecht, Netherlands.

40. GEIST, V. 1978. *Life Strategies, Human Evolution, Environmental Design.* Springer-Verlag, New York.

41. GEIST, V. 1989. Environmentally guided phenotype plasticity in mammals and some of its consequences to theoretical and applied biology. Pages 153–176 *in* N. M. Bruton, ed., *Alternative Styles of Animals.* Kluwer Academic Publisher, Dordrecht, Netherlands.

42. WALLACE, L. R. 1948. The growth of lambs before and after birth in relation to level of nutrition. Part III. *Journal of Agricultural Science* 38:367–401.

43. INGEBRIGTSEN, O. 1923. *Das Norwegische Rotwild.* Bergens Museum Aarbok. Naturvidensk, raekher Nr. 7. Bergen, Norway.

44. VOGT, F. 1936. *Neue Wege der Hege.* Neumann-Neudamm, Germany.

45. VOGT, F. 1948. *Das Rotwild.* Österreichischer Jagd- und Fischereiverlag, Vienna, Austria.

46. VOGT, F., and F. SCHMID, with an appendix by H. KOHLER. 1951. *Das Rehwild.* Österreichischer Jagd und Fischerei Verlag, Vienna, Austria.

47. FREVERT, W. 1957. *Rominten.* Bayerischer Landwirtschafts Verlag, Munich, Germany (1977 edition).

48. REINDERS, E. 1960. Das Rotwild in Krongut Het Loo. International Union of Game Biologists Congress, *Arnhem/Oosterbeak* 4:216–218.

49. VON BAYERN, A., and J. VON BAYERN. 1977. Über Rehe in einen Steirischen Gebirgsrevier, 2nd ed. BLV Verlagsgesellschaft, Munich, Germany.

50. ELLENBERG, H. 1978. Zur Populationsökologie des Rehes (*Capreolus capreolus* L. Cervidae) in Mitteleuropa. *Spixiana Journal of Zoology* (Supplement 2). Staatssammlung Munchen, Munich, Germany.

51. SCHAEFER, E. 1982. *Hegen und Ansprechen von Rehwild.* Bayerischer Landwirtschafts Verlag, Munich, Germany.

52. WOOD, A. J., I. McT. COWAN, and H. C. NORDAN. 1962. Periodicity of growth in ungulates as shown by deer of the genus *Odocoileus. Canadian Journal of Zoology* 40:593–603.

53. MECH, L. D., M. E. NELSON, and R. E. McROBERTS. 1991. Effect of maternal and grandmaternal nutrition on deer mass and vulnerability to wolf predation. *Journal of Mammalogy* 72(1): 146–151.

54. WINNER, C. 1996. The grandmother effect. *Wyoming Wildlife* 60:10–15.

55. LYAPUNOVA, E. A., T. B. BUNCH, N. N. VORONSOV, and R. S. HOFFMANN. 1997. Chromosome sets and the taxonomy of Severtsov wild sheep. *Russian Journal of Zoology* 1:387–396.

56. PAGE, R. D. M., and M. A. CHARLESTON. 1997. From gene to organismal phylogeny: Reconciling trees and the gene tree/species problem. *Molecular Phylogenetics and Evolution* 7:231–240.

57. CRONIN, M. A., R. STUART, B. J. PIERSON, and J. C. PATTON. 1996. Kappa-casein gene phylogeny of higher ruminants (Pecora, Artiodactyla). *Molecular Phylogenetics and Evolution* 6: 295–311.

58. CRONIN, M. A. 1991. Mitochondrial and nuclear genetic relationships of deer (*Odocoileus* spp.) in western North America. *Canadian Journal of Zoology* 69:1270–1279.

59. SHIELDS, G. F., and T. D. KOCHER. 1991. Phylogenetic relationship of North American ursids based on analysis of mitochondrial DNA. *Evolution* 45:218–221.

60. RYDER, O. A. 1986. Species conservation and systematics: The dilemma of subspecies. *Trends in Ecology and Evolution* 1:9–10.

61. EICK, E. 1992. Zur Frage der Unterarten. Unpublished manuscript. Internationale Gesellschaft Sikawild, D 4773 Moehnesee, Germany.

62. WILSON, D. E., and D. M. REEDER. 1993. *Mammal Species of the World.* Smithsonian Institution Press, Washington, DC.

63. O'GARA, B. In press. Taxonomy. *In* D. Toweill and J. W. Thomas, eds., *Elk of North America,* 2nd ed., The Wildlife Management Institute, Washington, DC.

64. GEIST, V. 1982. Adaptive behavioral strategies. Pages 219–278 *in* J. W. Thomas and D. Toweill, eds., *Elk of North America.* Stackpole Books, Harrisburg, PA.

65. ROBBINS, R. L., D. E. REDFEARN, and C. P. STONE. 1982. Refuges and elk management. Pages 479–508 *in* J. W. Thomas and D. Toweill, eds., *Elk of North America.* Stackpole Books, Harrisburg, PA.

66. KAY, C. E. 1994. Aboriginal overkill: The role of native Americans in structuring western ecosystems. *Human Nature* 5:359–398.

67. GEIST, V. 1996. *Bison Nation.* Voyageur Press, Stillwater, MN.

68. HUTTON, D. A. 1972. Variation in the skulls and antlers of wapiti (*Cervus elaphus nelsoni* Bailey). M.Sc. Dissertation. University of Calgary, Alberta, Canada.

68a. ROE, F. G. 1970. *The North American Buffalo,* 2nd ed. University of Toronto Press, Ontario, Canada.

69. VAN DYNE, T. S. 1902. The elk of the Pacific coast. Pages 167–191 *in* T. Roosevelt, ed., *The Deer Family.* Macmillan, New York.

70. KURTÉN, B., and E. ANDERSON. 1980. *Pleistocene Mammals of North America.* Columbia University Press, New York.

71. CHURCHER, C. S. 1962. *Odocoileus salinae* and *Mazama* sp. from the Talar tar seeps, Peru. Contribution No. 57. Royal Ontario Museum, Toronto, Ontario, Canada.

72. KELLOGG, R. G. 1956. What and where are the whitetails? Pages 31–35 *in* W. P. Taylor, ed., *The Deer of North America.* Stackpole Books, Harrisburg, PA.

73. BAKER, R. H. 1984. Origin, classification and distribution. Pages 1–18 *in* L. K. Halls, ed., *White-Tailed Deer.* Stackpole Books, Harrisburg, PA.

74. BROKX, P. A. 1984. South America. Pages 525–546 *in* L. K. Halls, ed., *White-Tailed Deer.* Stackpole Books, Harrisburg, PA.

75. COWAN, I. McTAGGART. 1936. Distribution and variation in deer (genus *Odocoileus*) of the Pacific coastal region of North America. *California Fish and Game* 22:155–246.

76. BACCUS, R., N. RYMAN, M. H. SMITH, C. REUTERWALL, and D. CAMRON. 1983. Genetic variability and differentiation of large grazing mammals. *Journal of Mammalogy* 64:109–120.

77. SMITH, M. H., R. BACCUS, H. O. HILLSTEAD, and M. N. MANLOVE. 1985. Population genetics. Pages 119–128 *in* L. Halls, ed. *White-Tailed Deer.* Stackpole Books, Harrisburg, PA.

78. McCABE, R. E., and McCABE, T. R. 1985. Of slings and arrows: An historical retrospection. Pages 19–72 *in* L. K. Halls, ed., *White-Tailed Deer, Ecology and Management.* Stackpole Books, Harrisburg, PA.

79. HOSLEY, N. W. 1956. Management of the white-tailed deer in its environment. Pages 197–206 *in* W. P. Taylor, ed., *The Deer of North America.* Stackpole Books, Harrisburg, PA.

80. ETLING, K. 1985. Can science produce a race of super bucks? *Outdoor Life,* January, p. 21.

81. ELLSWORTH, D. L., R. L. HONEYCUTT, N. J. SILVY, M. H. SMITH, J. W. BICKHAM, and W. D. KLIMSTRA. 1994. Historical biogeography and contemporary patterns of mitochondrial DNA variation in white-tailed deer from the southeastern United States. *Evolution* 48:122–126.

82. ELLSWORTH, D. L., R. L. HONEYCUTT, N. J. SILVY, M. H. SMITH, J. W. BICKHAM, and W. D. KLIMSTRA. 1994. White-tailed deer restoration to the southeastern United States: Evaluating genetic variation. *Journal of Wildlife Management* 58(4):686–697.

83. COWAN, I. McTAGGART. 1956. What and where are the mule and black-tailed deer? Pages 334–359 *in* W. P. Taylor, ed., *The Deer of North America.* Stackpole books, Harrisburg, PA.

84. WALLMO, O. C. 1981. Mule and black-tailed deer distribution and habits. Pages 1–25 *in* O. C. Wallmo, ed., *Mule and Black-Tailed Deer of North America.* Stackpole Books, Harrisburg, PA.

85. CRONIN, M. A. 1992. Intraspecific mitochondrial DNA variation in North American cervids. *Journal of Mammalogy* 73:70–82.

86. BANFIELD, A. W. F. 1961. A revision of the reindeer and caribou genus *Rangifer.* Bulletin No. 117. National Museum of Canada. Ottawa, Ontario, Canada.

87. GRIPENBERG, U. 1984. Characterization of the karyotype of reindeer (*Rangifer tarandus*); the distribution of heterochromatin in the reindeer and Scandinavian moose (*Alces alces*). *European Congress on Cytogenetics of Domestic Animals* 6:68–79.

88. PETERSON, R. L. 1955. *North American Moose.* University of Toronto Press, Ontario, Canada.

89. GUTHRIE, R. D. 1990. New dates on Alaska Quaternary moose, *Cervalces–Alces* archeological, evolutionary and ecological implications. *Current Research in the Pleistocene* 7:111–112.

90. VAN ZYLL DE JONG, C. G. 1986. A systematic study of recent bison with particular consideration of the wood bison. Publications in Natural Sciences No. 6. National Museum of Canada, Ottawa, Canada.

91. KRUMBIEGEL, I. 1980. Die unterartliche Trennung des Bisons, *Bison bison* (Linne, 1788), und seine Rückzüchtung. *Säugetierkundliche Mitteilungen* 28:148–160

92. KRUMBIEGEL, I., and G. G. SEHM. 1989. The geographic variability of the plains bison. A reconstruction using the earliest European illustrations of both subspecies. *Archives of Natural History* 16:169–190.

93. BORK, A. M., C. M. STROBECK, C. M. YEH, R. J. HUDSON, and R. K. ON. 1991. Genetic relationships of wood and plains bison on restriction fragment length polymorphisms. *Canadian Journal of Zoology* 69:43–48.

94. RHOADS, S. N. 1897. Notes on living and extinct species of American Bovidae. *Proceedings of the Academy of Natural Sciences of Philadelphia* 49:483–502.

95. GEIST, V. 1977. The wood bison (*Bison bison athabascae* Rhoads) in relation to hypotheses on the origin of the American bison (*Bison bison* Linnaeus). *Zeitschrift für Säugetierkunde* 42:119–127

96. RENECKER, L. A., BLYTH, C. B., and C. C. GATES. 1989. Game production in western Canada. Pages 248–267 *in Wildlife Production Systems: Economic Utilization of Wild Ungulates*. Cambridge University Press, Cambridge, UK.

97. TENER, J. S. 1965. *Muskoxen.* Canadian Wildlife Service Monograph Series No. 2. Queens Printers, Ottawa, Ontario, Canada.

98. GEIST, V. 1987. Bergmann's rule is invalid. *Canadian Journal of Zoology* 65:1035–1038.

99. COWAN, I. McTAGGART, and W. McCRORY. 1970. Variation in the mountain goat, *Oreamnos americanus* (Blainville). *Journal of Mammalogy* 51:60–73.

100. RAMEY, R. R. II 1995. Mitochondrial DNA variation, population structure, and evolution of mountain sheep in the south-western United States and Mexico. *Molecular Ecology* 4:429–439.

101. BOYCE, W., P. W. HENDICK, N. E. MUGGLI-COCKETT, STEVEN KALINOWSKI, M. C. T. PENEDO, and R. R. RAMEY II. 1996. Genetic variation of major histocompatible complex and microsatellite loci: A comparison in bighorn sheep. *Genetics* 145:421–433.

102. WEHAUSEN, J. D., and R. R. RAMEY. 1993. A morphometric reevaluation of the Peninsular bighorn subspecies. *Desert Bighorn Council Transactions* 37:1–10.

103. GEIST, V. 1985. On Pleistocene bighorn sheep: Some problems of adaptation, and its relevance to today's American megafauna. *Wildlife Society Bulletin* 13:351–359.

104. CURTAIN, C. C. , and H. H. FUDENBERG. 1973. Evolution of the immunoglobulin antigens in the ruminantia. *Biochemical Genetics* 8(3):301–308.

105. IRWIN, D. M., T. D. KOCHER, and A. C. WILSON. 1991. Evolution of cytochrome *b* genome of mammals. *Journal of Molecular Evolution* 32:128–144.

106. ALLARD, M. W., M. M. MIYAMOTO, L. JARECKI, F. KRAUS, and M. R. TENNANT. 1992. DNA systematics and evolution of the artiodactyl family Bovidae. *Proceedings National Academy of Science* 89:3972–3976.

107. JANIS, C. M., and K. M. SCOTT. 1987. The interrelationship of higher ruminant families with special emphasis on members of the Cervoidea. *American Museum Novitates* 3:1–85.

108. FRICK, C. 1937. Horned ruminants of North America. *Bulletin of the American Museum of Natural History* 69:1–699.

109. BAILEY, V. 1932. Oregon antelope. *Proceedings of the Biological Society* 45:45–46.

110. LEE, T. H. 1992. Mitochondrial DNA and allozyme analysis of pronghorn populations in North America. Ph.D. dissertation. Texas A&M University, College Station.

111. WHITAKER, J. O., JR. 1970. The biological subspecies: An adjunct of the biological species. *Biologist* 52:12–15.

112. COCKRUM, E. L. 1981. Taxonomy of the Sonoran pronghorn. *In* J. S. Phelps, ed. *The Sonoran Pronghorn.* Species Report 10. Arizona Game and Fish Department, Phoenix, AZ.

113. SOWLES, L. K. 1984. *The Peccaries.* University of Arizona Press, Tucson.

114. SOWLES, L. K. 1978. Collared peccary. Pages 191–205 *in* J. L. Schmidt and D. L. Gilbert, eds., *Big Game of North America,* Stackpole Books, Harrisburg, PA.

115. McLENNAN, B., and D. C. REINER. 1994. A review of bear evolution. Bears—their biology and management. *International Conference on Bear Research and Management* 91(1):95–96.

116. KURTÉN, B. 1968. *Pleistocene Mammals of Europe.* Aldine Publishing Company, Chicago, IL.

117. GOLDMAN, D., P. R. GIRI, and S. J. O'BRIEN. 1989. Molecular genetic-distances estimates among the ursidae as indicated by one- and two-dimensional protein electrophoresis. *Evolution* 43:282–295.

118. SHERWOOD, H. W. 1981. Morphological variation of grizzly bear skulls from Yellowstone National Park. M.Sc. dissertation. University of Montana, Missoula.

119. MERRIAM, C. H. 1914. Description of thirty apparently new grizzly and brown bears from North America. *Proceedings of the Biological Society,* 27:173–196.

120. MERRIAM, C. H. 1918. Review of the grizzly and big brown bears of North America. *North American Fauna* 41. U.S. Government Printing Office, Washington, DC.

121. RAUSCH, R. L. 1963. Geographic variation in size in North American brown bears *Ursus arctos* L., as indicated by condylobasal length. *Canadian Journal of Zoology* 41(1):33–45.

122. HALL, E. R. 1984. Geographic variation among brown and grizzly bears (*Ursus arctos*) in North America. Special Publication 13–1–16. Museum of Natural History, University of Kansas.

123. ELIOT, J. L. 1998. Polar bears: Stalkers of the Arctic. *National Geographic* 193:57–71.

124. BREWSTER, W. G., and S. H. FRITTS. 1995. Taxonomy and genetics of the gray wolf in western North America: A review. Pages 353–373 *in* L. N. Carbyn, S. H. Fritts, and D. R. Seip, eds., *Ecology and Conservation of Wolves in Changing World.* Occasional Publication No. 35. Canadian Circumpolar Institute, University of Alberta, Edmonton, Alberta.

125. SOKOLOV, V. E., and O. L. ROSSOLIMO. 1985. Taxonomy and validity. Pages 21–50 *in* D. I. Bibikov, ed., *The Wolf. History, Systematics, Morphology, and Ecology.* USSR Academy of Science Nauka, Moscow, Russia.

126. NOWAK, R. M. 1995. Another look at wolf taxonomy. Pages 375–397 *in* L. N. Carbyn, S. H. Fritts, and D. R. Seip, eds., *Ecology and Conservation of Wolves in a Changing World.* Occasional Publication No. 35. Canadian Circumpolar Institute, Edmonton, Alberta.

127. MECH, L. D. 1991. Returning the wolf to Yellowstone. Pages 309–322 *in* R. B. Keiter and M. S. Boyce, eds., *The Greater Yellowstone Ecosystem: Redefining America's Wilderness Heritage.* Yale University Press, New Haven, CT.

128. FRITTS, S. H. 1983. Record dispersal by a wolf from Minnesota. *Journal of Mammalogy* 64:166–167.

129. REAM, R. R., M. W. FAIRCHILD, D. K. BOYD, and D. H. PLETCHER. 1991. Population dynamics and home range changes in a colonizing wolf population. Pages 346–366 *in* R. B. Keiter and M. S. Boyce, eds., *The Greater Yellowstone Ecosystem: Redefining America's Wilderness Heritage.* Yale University Press, New Haven, CT.

130. HANSEN, K. 1992. *Cougar: The American Lion.* Northland Publishing, Flagstaff, AZ.

131. WEIGEL, I. 1975. Big felids and cheetah. Pages 333–372 *in* B. Grzimek, ed., *Grzimek's Animal Encyclopedia,* Vol. 12, *Mannals.* Van Nostrand and Reinhold, New York.

CHAPTER 2

Hybridization in Large Mammals

James R. Heffelfinger

INTRODUCTION

Like the half man and half bull "Minotaur" of Greek mythology, hybrids have always fascinated humans as popular subjects of myth and legend. The obsession with creatures that are half one thing and half another extends to our enjoyment of wildlife. Early naturalists often described new animals as a combination of parts from animals already known to science. For example, the mule deer was described by John J. Audubon in 1846 as having fur like an elk but hooves like a whitetail.

Hybrids have been the subject of fact and fancy since Aristotle wrote about the most famous of all hybrids, the mule. The term *hybridization* has been used in various ways to describe everything from matings between animals in different genera to the normal pairing of two individuals in the same population (1, 2). Hybridization is most commonly used to describe the successful reproduction of two individuals belonging to different species (3). Intraspecific matings (between two different races or subspecies) is simply a mixing of the natural variation within that species.

SPECIES CONCEPTS

Any discussion of hybridization between species should be based on a thorough understanding of the concept of species (4). Several species concepts have been proposed; however, natural hybridization greatly complicates the search for a universal definition of species.

The most widely used concept of species is *biological species*. Mayr proposed; this concept based in part on the earlier work of Dobzhansky that delineates individual species by stressing reproductive isolation (5, 6). With this concept, species are ". . . groups of interbreeding natural populations that are reproductively isolated from other such groups" (5, p. 12). Under the strictest interpretation, if two previously recognized species produce offspring in captivity, they are to be recognized as the same species. This obviously creates a taxonomic problem if hybridization is documented between two very different species such as the mountain lion and ocelot. Most taxonomists, however, do not consider unnatural hybridization in captivity to be a breach of species distinctions.

Although apparently rare in mammals, relatively high levels of natural hybridization have been identified recently in birds (1). In addition, reproductive compatibility is not always correlated with how closely related two species are (7). These concerns spawned a new species concept based on similar physical characteristics and common evolutionary descent. Under the *phylogenetic species concept,* species are "the smallest diagnosable cluster of individual organisms within which there is a parental pattern of ancestry and descent" (8). This deemphasizes reproductive isolation between species but the ambiguity of exactly what constitutes a "diagnosable cluster" is controversial (7, 9, 10). Moreover, the taxonomist must subjectively select one or more diagnostic characters to identify the species while ignoring many others that vary geographically and may be more important from an evolutionary perspective. Defining species as diagnosable clusters may lead to the same nebulous descriptions that have plagued the subspecies concept (11, 12).

Avise and Ball synthesized the strong points of the biological and phylogenetic species concepts and developed the concept of *genealogical concordance* (9). This uses the biological species concept as a philosophical framework but recognizes phylogenetically distinct lineages (i.e., diagnosable clusters) as subspecies only if they cluster together on the basis of several concordant differences.

Other species concepts (i.e., evolutionary, recognition, cohesion) have been proposed that are mainly variations based on reproductive isolation or evolutionary relatedness; all are ultimately related to limitations of gene flow among species (10). None of the species concepts provides a completely workable solution to the problem of delineating species and all must be relaxed to allow for natural hybridization (4).

Reproductive Isolating Mechanisms

The discreteness of individual species is dependant on the maintenance of one or more of what Dobzhansky termed *isolating mechanisms* (6). Species are prevented from interbreeding by a variety of means. They may not occupy the same area at the same time (i.e., temporal/spatial barrier), or they may possess behavioral differences, usually in courtship behavior, so that one species does not "recognize" the visual, auditory, or olfactory mating cues of the other (i.e., behavioral barrier). In some cases reproductive organs or body size are too different to allow successful fertilization (i.e., mechanical barrier).

If mating does occur, sperm may die quickly or fail to penetrate the egg. Also, if a hybrid embryo begins to form it may be unable to survive the uterine environment. If the embryo does survive gestation, it may experience a reduced rate of survival (i.e., hybrid inviability). Any of these factors would contribute to a lack of successful reproduction or a greatly reduced contribution to future generations.

In cases of successful reproduction, the offspring may have a high rate of survival but be partially or completely sterile. The sterility of the mule is well known, and many mistakenly assume that all hybrids are sterile. Hybrid sterility in mammals is more common in males and increases with the unrelatedness of the parental species (13). These isolating mechanisms are generally sufficient to keep most North American large mammal species from hybridizing. Natural hybridization among mammals is uncommon in nature but has historically occurred in captivity in surprising combinations.

HYBRIDIZATION IN LARGE MAMMALS

Most large mammal hybridization has been documented in captivity. Zoos and private breeders sometimes developed hybrids in strange combinations to boost visitation, or they may have lacked the proper facilities to keep all related species separate. This is of little consequence to the management of these species but sometimes aids in our understanding of relationships between groups of animal species.

Bear

In 1859, a male black bear and female European brown bear held at the London Zoological Gardens produced a litter of three hybrid cubs (14). Hybrids of brown bears and polar bears have also been produced in captivity (15, 16). First-generation (F_1) hybrids assumed the shape of polar bears with very light brown fur. One female brown $\times$ polar bear hybrid born in a Polish zoo was successfully bred back to her polar bear father, resulting in a male cub with nearly white fur. Another hybrid female was bred to a brown bear, resulting in two dark-colored male cubs.

Wolf

Wolves and domestic dogs interbreed in captivity. Extensive hybridization of wolves and domestic dogs was reported among the Native American tribes at the time of European contact, especially those in the arctic where wolf-like traits were useful in sled dogs (17, 18). Only in Italy is hybridization thought to present a conservation problem for a wild wolf population (19, 20). Wolf $\times$ dog hybrids have become popular, albeit controversial, in the pet trade. Although estimates vary, there are more than 300,000 wolf $\times$ dog hybrids being kept as pets in the United States (21).

Wolves also interbreed with coyotes in captivity and in the wild in the Midwest and eastern Canada but apparently not in the western United States and western Canada (22, 23, 24). This has resulted in a blurring of species distinctions in the southeastern and northeastern United States.

After the gray wolf was extirpated from the northeast, a new canid began arriving in the 1930s. These new canids were as much as 50% larger than western coyotes, which elicited speculation about hybridization with domestic dogs or wolves (25). Through an intensive morphological and genetic analysis, it was determined that wolves hybridized with the expanding coyote population in the northeast (23, 26). This genetic introgression seemed to occur only one way, because there was no evidence of wolf genes being passed into the coyote population. Hybrid individuals and their descendants were apparently being assimilated into the wolf population but not backcrossed into the coyote gene pool (23, 27).

The origins of the red wolf have been debated for decades (26, 28, 29, 30, 31). An extensive analysis of red wolf skulls by Nowak showed that red wolves were intermediate between coyotes and gray wolves up to the 1930s, when the rare red wolves began to interbreed with increasing populations of coyotes (26). The red wolf became smaller and more coyote-like because of extensive hybridization.

Analysis of mitochondrial DNA confirmed the occurrence of extensive hybridization, when researchers found that red wolves possessed only coyote DNA or wolf DNA; no red wolves had unique DNA segments to separate them as a distinct species (32). Recent mitochondrial DNA and nuclear DNA (microsatellite) analyses of red wolf samples collected prior to 1940 (when extensive hybridization is thought to have begun) again failed to reveal any unique genetic markers that would identify the red wolf as an independent species of canid (33). In light of these findings, several authorities theorize that the red wolf arose out of an extensive hybrid zone between gray wolves and coyotes in the southeast, rather than having evolved as an independent species of canid with a Pleistocene origin (26, 27, 28, 31, 32, 33, 34).

Mountain Lion

A male mountain lion and female ocelot held in a private facility produced four litters between 1990 and 1992 (35). None of the young survived more than 12 days, due ultimately to a lack of parental care.

Jaguar

Popular Mexican legend maintains that jaguars and mountain lions hybridize to produce the infamous "Onza" of western Mexico. Investigations of this mythical cat have failed to uncover any physical evidence of its existence (36).

Jaguars and leopards have been hybridized at the Salzburg Zoo in Austria and elsewhere in private collections (C. Walzer, Salzburg Zoo, personal communication). Two jaguar × leopard hybrids escaped in 1987 from a private breeder in Florida (B. Cook, Florida Game and Fresh Water Fish Commission, personal communication).

Bison

North American bison have been deliberately hybridized with European bison to bolster the latter's declining numbers (3). North American bison have also been hybridized with wild cattle (e.g., yaks) and various domestic cattle breeds (3). Offspring of domestic bulls with female bison exhibit good quality meat, a dorsal hump, some curly hair, thick hide, and the disease resistance of bison. The first-generation hybrid males are sterile, but female hybrids are fertile to some extent. This hybridization, along with genetic analyses, has led some to classify bison and domestic cattle in the same genus (37, 38).

Bighorn Sheep

Hybrids resulting from wild bighorn sheep rams breeding pastured domestic ewes have been reported in Arizona, Colorado, Nevada, Wyoming, and Utah (39, 40, 41). These hybrids differ from domestic sheep in that they have bald faces, shorter tails, longer legs, and their coat is a mixture of wool and straight hair. Hybrid offspring have a low survival rate, but those that do survive to puberty are fertile to some degree. In captivity, bighorn sheep have also hybridized with mouflon sheep from the Mediterranean (3).

North American Elk

The relationship between elk of North America and European red deer has been debated for more than a century (42). Elk were considered a different species than red deer, but current authorities now recognize them merely as subspecies belonging to one circumpolar species (43). Recent developments in genetic analysis techniques may provide additional information with which to examine the classification of this and other species.

The European form of this species has been farmed and researched for a longer period of time than the North American elk. As a result, much more information regarding hybridization with other species is available from Europe. In the deer farming industry, North American elk and red deer are frequently hybridized because the F_1 hybrids show faster growth rates in body and antlers due to hybrid vigor (44).

Hybridization between red deer and sika deer has occurred extensively in parts of Europe, endangering the genetic integrity of the native red deer and confusing the already weak subspecies distinctions of sika deer (45). Artificial insemination of red deer and sambar deer produced hybrid offspring in less than 1% of inseminations (46). After natural mating was unsuccessful, fertile male and female hybrids were produced from Père David's deer and red deer by artificial insemination and by raising the two species together in captivity from birth (47, 48).

Mule Deer and White-tailed Deer

White-tailed deer × mule deer hybrids were reported from captive facilities as early as 1865 (3). Occurrences were later reported from zoos in North Dakota and Ohio; and in captive facilities in Arizona, Alberta, Colorado, Illinois, and Texas

(3, 42, 49, 50, 51, 52, 53, 54). Researchers in Tennessee also successfully produced whitetailed deer × blacktailed deer hybrids in captivity (55). Female F$_1$ hybrids are fertile, in contrast to F$_1$ males, which are usually sterile. Survival of these hybrids is poor, even in pampered captive facilities.

White-tailed deer × mule deer hybrids have also been confirmed in the wild from Alberta, Arizona (Arizona Game and Fish Department files), British Columbia, Kansas, Montana, Nebraska, Texas, Washington, and Wyoming (51, 56, 57, 58, 59, 60, 61, 62, 63, 64). Interestingly, the desert mule deer "type" specimen that served as the representative of this subspecies for more than 100 years is a white-tailed deer × mule deer hybrid (65) (J. R. Heffelfinger, unpublished data, Arizona Game and Fish Department, 1999).

Every year numerous reports are received of "hybrid" deer from areas where both species are found, and yet very few are verified by those who know what to look for. The rarity of true hybrids is illustrated by the relative scarcity of confirmed hybrids among the thousands of deer that are seen annually throughout the area of range overlap. A few morphologic and genetic characteristics can be used to determine if a deer is a hybrid (Figure 2–1).

The most diagnostic physical characteristic to determine if a deer is a hybrid is the size and location of the metatarsal gland, which is located on the outside of the lower portion of the rear legs (Figure 2–2). The metatarsal glands on mule deer sit high on the lower leg, are 75 to 150 mm (3 to 6 inches) long, and surrounded by brown fur only. The whitetail's metatarsals are at or below the midpoint of the lower leg, less than 38 mm (1.5 inches), and always surrounded by white hairs. A whitetailed deer × mule deer hybrid has metatarsal glands that split the difference, usually measuring 50 to 100 mm (2 to 4 inches) and sometimes encircled with white hair. Curiously, black-tailed deer of the Pacific Northwest also have intermediate metatarsal glands but this is not due to hybridization.

Figure 2–1 White-tailed deer × mule deer hybrids have been produced in captivity numerous times, like this buck in Arizona. (Photograph by G. I. Day, Arizona Game and Fish Department.)

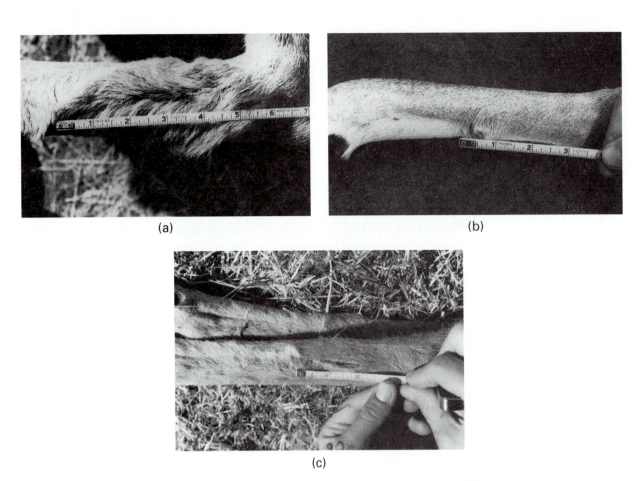

(a)

(b)

(c)

Figure 2–2   (a) Mule deer metatarsal glands measure 75 to 150 mm (3 to 6 inches), are high on the lower leg, and are covered with brown hair. (b) White-tailed deer metatarsal glands measure less than 38 mm (1.5 inches), are low on the lower leg, and are surrounded by white hairs. (c) F_1 hybrid metatarsal glands are intermediate in size (50 to 100 mm or 2 to 4 inches) and location, sometimes being surrounded by white hair. [Photographs (a) and (b) by J. R. Heffelfinger; photograph (c) by G. I. Day, Arizona Game and Fish Department.]

The preorbital gland of mule deer is deep (about 20 mm or 0.75 inch) and larger than the shallow depression found in white-tailed deer, providing yet another species-specific characteristic. Hybrids possess preorbital glands that are intermediate in size (J. R. Heffelfinger, unpublished data, Arizona Game and Fish Department, 1999).

Mule deer tails are entirely white with a distinctive black tip. They are shorter than a whitetail's and appear more rope-like. Whitetails have a flattened, fan-shaped tail that is white on the underside and brown on the dorsal (back) surface. Hybrids produced in captivity normally possess a tail that appears more whitetail-like, but often has a very dark dorsal surface (Figure 2–3).

As their name implies, mule deer ears are much longer; about half the length of their head, whereas the whitetail's are only about one-third the length of the head. Hybrids may have ears that are as long as their mule deer parent.

Hybrids appear to inherit predator avoidance strategies from both parents (52). This creates a problem because whitetail and mule deer have very different techniques for escaping predators. The whitetail's key to escaping is speed; they try to put as much distance between themselves and the predator as possible, as fast as possible. Mule deer

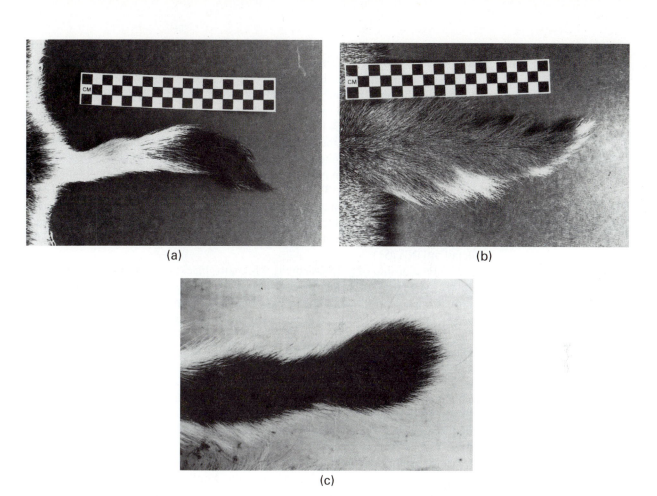

(a)

(b)

(c)

Figure 2–3   (a) Mule deer tails are rope-like and covered with white hair except for the conspicuous black tip. (b) White-tailed deer tails are flattened and V-shaped with a white underside and brown dorsal surface. (c) F_1 hybrids usually possess a tail that is more whitetail-like, but is typically much darker than a pure whitetail. [Photographs (a) and (b) by J. R. Heffelfinger; photograph (c) by G. I. Day, Arizona Game and Fish Department.]

have developed a pogostick-like bounding called *stotting* that allows them to quickly negotiate the rugged terrain they occupy.

Lingle used captive animals to show that stotting is so specialized that even a $\frac{1}{8}$ whitetail $\times$ $\frac{7}{8}$ mule deer cannot stott properly (52). The hybrid's escape behavior was a chaotic mixture of whitetail and mule deer responses, which results in a lower survival rate and consequently an evolutionary selection against hybridization.

Two-year-old mule deer are most frequently mistaken for hybrids. This is because of their smaller antler development and because dichotomous branching (forked tines) usually does not occur until the buck is at least 3 years old. Young mule deer sometimes give the appearance of very large white-tailed deer. Also, some whitetails, such as those in the Carmen Mountains of northern Mexico, exhibit a high degree of forked antlers like mule deer (66).

Recent advances in DNA analysis have allowed researchers to look at more definitive things than ears and antlers. The production of proteins and enzymes in the body is regulated by genes, so by analyzing differences between these proteins and enzymes, researchers can identify different species from only a sample of tissue. Serum albumin and erythrocyte acid phosphatase have proven particularly useful. Analyzing these compounds with a process called electrophoresis produces horizontal bands on a gel surface, which appear in a different location for whitetails and mule deer. When a hybrid is tested, both the whitetail and the mule deer bands are present.

In west Texas, managers have reported an increasing trend in the number of suspected hybrids. Stubblefield and others analyzed whitetails and mule deer in a five-county area using serum albumin and found that 5.6% of the deer they tested were hybrids. Individual ranches ranged from 0% to 24% in showing evidence of hybridization (59).

Carr et al. analyzed mitochondrial DNA of whitetails and mule deer on a ranch in west Texas (58). Animals inherit mtDNA from only their mother. This is a useful tool to examine hybridization. A hybrid that has a whitetail mother and mule deer father will have only whitetail mtDNA. (The nuclear DNA will still be from both mother and father.) If the hybrid was a female, she would continue to pass whitetail mtDNA through her daughters and their daughters even if bred by mule deer bucks. After a few generations, the results of these matings would look like mule deer but would have pure whitetail mtDNA (51).

The relationship of mtDNA among whitetails, blacktails, and mule deer remains enigmatic. Mule deer on a ranch in west Texas had mtDNA that was indistinguishable from the whitetails in this area and it was concluded that past hybridization was common (58). In addition, a different type of analysis showed that the mtDNA of mule deer on this ranch was more closely related to *whitetails* from South Carolina than to blacktails (the same species) from northern California! Protein and mtDNA analyses of 70 deer in west Texas indicated extensive past hybridization in both directions, although none of the deer analyzed were F_1 hybrids (60). Derr used protein electrophoresis of serum albumin and erythrocyte acid phosphatase to test deer from 31 locations in the southwestern United States and found very little evidence of recent hybridization (54). In addition, mtDNA variation from 33 locations in the Southwest showed that mule deer had mtDNA that was indistinguishable from that of whitetails, indicating extensive historical hybridization.

Cronin also used protein electrophoresis and mtDNA analysis to determine the extent of hybridization in 667 white-tailed and mule deer in northwestern North America and found only low levels of hybridization (57). Reproductive barriers appear to be sufficient to maintain the genetic integrity of these two species in that region.

MANAGEMENT IMPLICATIONS

Under natural conditions, hybridization is not a serious management problem in wild North American large mammal species. The introduction of domesticated hoof-stock, however, has provided the opportunity for hybridization with some wild ungulates. The widespread farming of captive red deer and the hybridization of red deer × elk throughout native elk habitat represent the greatest potential for disaster. Captive animals commonly escape confinement and the introgression of red deer genes into our North American elk herds will have unknown but surely negative consequences owing to the different behavior and appearance of these two forms.

Whitetail × mule deer hybridization is more common than previously recognized; however, it is rare and is not an important consideration in the management of either species (54, 60). Confusion arises because of the large variation in characteristics in each species. Some whitetails have features that look like those of mule deer and some mule deer may have very whitetail-like characteristics (e.g., no antler forks, black line on the back of the tail). Deer hybrids cannot be identified with certainty at a distance; the low number of interspecies matings and the low survival of hybrid offspring greatly reduce the chance of encountering one in the wild. In the future, increasing habitat disruption and fragmentation by human development may result in changes that could increase the occurrence of hybridization between these two deer species.

SUMMARY

The concept of species not only serves to categorize various forms of wildlife, but also aids in the study of the process of evolution. As species evolve differently along their own evolutionary pathways, they diverge in phenotype and genotype. Reproductive isolating mechanisms develop, leading to a decreased ability to exchange genes and ultimately to reproductive incompatibility. Species that do not hybridize under any circumstances are termed "good species;" however, all species concepts must allow for the occasional case of hybridization.

Hybridization in large mammals in North America is not currently a serious management issue; however, the widespread use of red deer and red deer × elk hybrids in captive deer farms may allow for genetic introgression of red deer genes into native elk populations. Differences in body size, disease resistance, and behavior have the potential to wreak havoc with some North American elk herds. It is unclear if the enigmatic red wolf is an ancient Pleistocene survivor or a more recent coyote × wolf hybrid. Regardless of its origin, its future as a separate species will depend on protecting it from further hybridization with coyotes. Recent genetic analyses have shown that hybridization of white-tailed and mule deer is occurring in some isolated areas. It is unclear what effect these pockets of hybridization are having on either species in areas of sympatry. Human disturbance and further fragmentation of the habitat may exacerbate this breakdown of isolating mechanisms. More work is needed on the ecological relationships of these two species.

LITERATURE CITED

1. GRANT, P. R., and B. R. GRANT. 1992. Hybridization of bird species. *Science* 256:193–197.
2. HARRISON, R. G. 1993. Hybrids and hybrid zones: Historical perspective. Pages 3–12 *in* R. G. Harrison, ed., *Hybrid Zones and the Evolutionary Process.* Oxford University Press, New York.
3. GRAY, A. P. 1972. *Mammalian Hybrids: A Check-List with Bibliography.* Commonwealth Agricultural Bureaux, Farnham Royal, England.
4. ARNOLD, M. L. 1997. *Natural Hybridization and Evolution.* Oxford University Press, New York.
5. MAYR, E. 1970. *Populations, Species, and Evolution: An Abridgment of Animal Species and Evolution.* Belknap Press, Cambridge, MA.
6. DOBZHANSKY, T. 1951. *Genetics and the Origin of Species.* Columbia University Press, New York.
7. BAUM, D. 1992. Phylogenetic species concepts. *Trends in Ecology and Evolution* 7:1–2.
8. CRACRAFT, J. 1983. Species concepts and speciation analysis. Pages 159–187 *in* R. F. Johnston, ed., *Current Ornithology.* Plenum Press, New York.
9. AVISE, J. C., and R. M. BALL, JR. 1990. Principles of genealogical concordance in species concepts and biological taxonomy. *Oxford Surveys in Evolutionary Biology* 7:45–67.
10. AVISE, J. C. 1994. *Molecular Markers, Natural History and Evolution.* Chapman and Hall, New York.
11. WILSON, E. O., and W. L. BROWN, JR. 1953. The subspecies concept and its taxonomic application. *Systematic Zoology* 2:97–111.
12. GILLHAM, N. W. 1956. Geographic variation and the subspecies concept in butterflies. *Systematic Zoology* 5:110–120.
13. HALDANE, J. B. S. 1922. Sex ratio and unisexual sterility in hybrid animals. *Journal of Genetics* 12:101–109.
14. BARTLETT, A. D. 1860. Notes on some young hybrid bears bred in the gardens of the Zoological Society. *Proceedings of the Zoological Society of London* 28:130–131.
15. COOK, R. 1950. "Gene"—the hybrid bear. *Journal of Heredity* 41:30–34.
16. KOWALSKA, Z. 1967. A note on bear hybrids at Lodz Zoo. *International Zoo Yearbook* 9:89.
17. YOUNG, S. P., and E. A. GOLDMAN. 1944. *The Wolves of North America.* Dover Publications, New York.

18. MECH, L. D. 1970. *The Wolf: The Ecology and Behavior of an Endangered Species.* The Natural History Press, Garden City, New York.

19. BOITANI, L. 1982. Wolf management in intensively used areas of Italy. Pages 158–171 *in* F. H. Harrington and P. C. Paquet, eds., *Wolves of the World: Perspectives of Behavior, Ecology, and Conservation.* Noyes Publications, Park Ridge, NJ.

20. BOITANI, L. 1992. Wolf research and conservation in Italy. *Biological Conservation* 61:125–132.

21. POLSKY, R. H. 1995. Wolf hybrids: Are they suitable as pets? *Veterinary Medicine* 90:1122–1124.

22. KOLENOSKY, G. B. 1971. Hybridization between wolf and coyote. *Journal of Mammalogy* 52:446–449.

23. LEHMAN, N., A. EISENHAWER, K. HANSEN, L. DAVID MECH, R. O. PETERSON, P. J. P. GOFAN, and R. K. WAYNE. 1991. Introgression of coyote mitochondrial DNA into sympatric North American gray wolf populations. *Evolution* 45:104–119.

24. PILGRIM, K. L., D. K. BOYD, and S. H. FORBES. 1998. Testing for wolf-coyote hybridization in the Rocky Mountains using mitochondrial DNA. *Journal of Wildlife Management* 62:683–689.

25. SILVER, H. S., and W. T. SILVER. 1969. Growth and behavior of the coyote-like canid of northern New England with observations on canid hybrids. Wildlife Monograph 17.

26. NOWAK, R. M. 1979. North American quaternary *Canis.* Monograph Number 6. Museum of Natural History, University of Kansas, Lawrence.

27. ROY, M. S., E. GEFFEN, D. SMITH, E. A. OSTRANDER, and R. K. WAYNE. 1994. Patterns of differentiation and hybridization in North American wolflike canids, revealed by analysis of microsatellite loci. *Molecular Biological Evolution* 11:553–570.

28. NOWAK, R. M. 1992. The red wolf is not a hybrid. *Conservation Biology* 6:593–595.

29. GITTLEMAN, J. L., and S. L. PIMM. 1991. Crying wolf in North America. *Nature* 351:524–525.

30. DOWLING, T. E., W. L. MINCKLEY, M. E. DOUGLAS, P. C. MARSH, and B. D. DEMARAIS. 1992. Response to Wayne, Nowak, and Phillips and Henry: Use of molecular characters in conservation biology. *Conservation Biology* 6:600–603.

31. BROWNLOW, C. A. 1996. Molecular taxonomy and the conservation of the red wolf and other endangered carnivores. *Conservation Biology* 10:390–396.

32. WAYNE, R. K., and S. M. JENKS. 1991. Mitochondrial DNA analysis implying extensive hybridization of the endangered red wolf (*Canis rufus*). *Nature* 351:565–568.

33. ROY, M. S., E. GEFFEN, D. SMITH, and R. K. WAYNE. 1996. Molecular genetics of pre-1940 red wolves. *Conservation Biology* 10:1413–1424.

34. WAYNE, R. K., and J. L. GITTLEMAN. 1995. The problematic red wolf. *Scientific American* July: 36–39.

35. DUBOST, G., and J. ROYÈRE, 1993. Hybridization between ocelot (*Felis pardalis*) and puma (*Felis concolor*). *Zoo Biology* 12:277–283.

36. CARMONY, N. B. 1995. *Onza!—The Hunt for the Legendary Cat.* High Lonesome Books, Silver City, NM.

37. STORMONT, C., J. W. MILLER, and Y. SUZUKI. 1961. Blood groups and the taxonomic status of the American Buffalo and domestic cattle. *Evolution* 15:196–208.

38. BACCUS, R., N. RYMAN, M. H. SMITH, C. REUTERWALL, and D. CAMERON. 1983. Genetic variability and differentiation of large grazing mammals. *Journal of Mammalogy* 64:109–120.

39. YOUNG, S. P., and R. H. MANVILLE. 1960. Records of bighorn hybrids. *Journal of Mammalogy* 41:523–525.

40. PILLMORE, R. E., and R. D. TEAGUE. 1955. Bighorn cross. *Colorado Conservation* 4:22–26.

41. PULLING, A. V. S. 1945. Hybridization of bighorn and domestic sheep. *Journal of Wildlife Management* 9:82–83.

42. CATON, J. D. 1877. *The Antelope and Deer of America.* Hurd and Houghton Publisher, New York.

43. BRYANT, L. D., and C. MASER. 1982. Classification and distribution. Pages 1–59 *in* J. W. Thomas and D. E. Toweill, eds., *Elk of North America.* Stackpole Books, Harrisburg, PA.

44. PEARSE, A. J. 1993. The recent status of deer farming in New Zealand. Pages 401–412 *in* N. Ohtaishi and H. I. Sheng, eds., *Deer of China.* Elsevier Science Publishers, New York.

45. BARTOS, L., J. HYANEK, and J. ZIROVNICKY. 1981. Hybridization between red and sika deer. *Zoologischer Anzeiger Jena* 207:260–270.

46. MUIR, P. D., G. SEMIADI, G. W. ASHER, T. E. BROAD, M. L. TATE, and T. N. BARRY. 1997. Sambar deer (*Cervus unicolor*) × red deer (*C. elaphus*) interspecies hybrids. *Journal of Heredity* 88:366–372.

47. TATE, M. L., G. J. GOOSEN, H. PATENE, A. J. PEARSE, K. M. MCEWAN, and P. F. FENNESSY. 1997. Genetic analysis of Père David's × red deer interspecies hybrids. *Journal of Heredity* 88:361–365.

48. KRZYWINSKI, A. 1993. Hybridization of milu stags with red deer hinds using the imprinting phenomenon. Pages 242–246 *in* N. Ohtaishi and H. I. Sheng, eds., *Deer of China.* Elsevier Science Publishers, New York.

49. NICHOLS, A. A. 1938. Experimental feeding of deer. *Agriculture Bulletin, University of Arizona Experiment Station* 75:1–39.

50. DAY, G. I. 1980. Characteristics and measurements of captive hybrid deer in Arizona. *Southwestern Naturalist* 25:434–438.

51. WISHART, W. D. 1980. Hybrids of white-tailed and mule deer in Alberta. *Journal of Mammalogy* 61:716–720.

52. LINGLE, S. 1992. Escape gaits of white-tailed deer, mule deer and their hybrids: Gaits observed and patterns of limb coordination. *Behaviour* 122:154–181.

53. SPRAKER, T. R., M. W. MILLER, E. S. WILLIAMS, D. M. GETZY, W. J. ADRIAN, G. G. SCHOONVELD, R. A. SPOWART, K. I. O'ROURKE, J. M. MILLER, and P. A. MERZ. 1997. Spongiform encephalopathy in free-ranging mule deer (*Odocoileus hemionus*), white-tailed deer (*Odocoileus virginianus*) and Rocky Mountain elk (*Cervus elaphus nelsoni*) in northcentral Colorado. *Journal of Wildlife Diseases* 33:1–6.

54. DERR, J. N. 1990. Genetic interactions between two species of North American deer, *Odocoileus virginianus* and *Odocoileus hemionus.* Ph.D. dissertation. Texas A&M University, College Station.

55. WHITEHEAD, C. J., JR. 1972. A preliminary report on white-tailed and black-tailed deer crossbreeding studies in Tennessee. *Proceedings of the Annual Conference of the Southeastern Association of Game and Fish Commissions* 25:65–69.

56. COWAN, I. M. 1962. Hybridization between the black-tail deer and the white-tail deer. *Journal of Mammalogy* 43:539–541.

57. CRONIN, M. A. 1991. Mitochondrial and nuclear genetic relationships between deer (*Odocoileus* spp.) in Western North America. *Canadian Journal of Zoology* 69:1270–1279.

58. CARR, S. M., S. W. BALLINGER, J. N. DERR, L. H. BLANKENSHIP, and J. W. BICKHAM. 1986. Mitochondrial DNA analysis of hybridization between sympatric white-tailed deer and mule deer in West Texas. *Proceedings of the National Academy of Sciences* 83:9576–9580.

59. STUBBLEFIELD, S. S., R. J. WARREN, and B. R. MURPHY. 1986. Hybridization of free-ranging whitetail and mule deer in Texas. *Journal of Wildlife Management* 50:688–690.

60. BALLINGER, S. W., L. H. BLANKENSHIP, J. W. BICKHAM, and S. M. CARR. 1992. Allozyme and mitochondrial DNA analysis of a hybrid zone between white-tailed deer and mule deer (*Odocoileus*) in West Texas. *Biochemical Genetics* 30:1–11.

61. GAVIN, T. A., and B. MAY. 1988. Taxonomic status and genetic purity of Columbian white-tailed deer. *Journal of Wildlife Management* 52:1–10.

62. KAY, C. E., and E. BOE. 1992. Hybrids of white-tailed and mule deer in western Wyoming. *Great Basin Naturalist* 52:290–292.

63. KNIPE, T. 1977. The Arizona whitetail deer. Special Report Number 6. Arizona Game and Fish Department, Phoenix, AZ.

64. KRÄMER, A. 1973. Interspecific behavior and dispersion of two sympatric deer species. *Journal of Wildlife Management* 37:288–300.

65. MEARNS, E. A. 1897. Preliminary diagnosis of new mammals of the genera Mephitis, Dorcelaphus, and Dicotyles from the Mexican border of the United States. *Proceedings of the United States National Museum Bulletin* 20:467–471.

66. LEOPOLD, A. S. 1954. Dichotomous forking in the antlers of white-tailed deer. *Journal of Mammalogy* 35:599–600.

Human Values Toward Large Mammals

Stephen R. Kellert and Carter P. Smith

INTRODUCTION

Large mammals represent far more than biological realities for people. As a large mammal ourselves, we evolved in close association with many of these creatures, some constituting critical food sources and other material advantages; some functioning as objects of considerable fear and anxiety; some our fiercest competitors; some the subject of our most cherished symbols, myths, and stories. We continue today, and probably always will, to attach great meaning and importance to these creatures. Sometimes these values reflect the particular biological character of these animals. At other times, their importance resides more in the peculiar human capacity to construct a world of meaning stretching far beyond the physical constraints of empirical reality.

Values of large mammals are the focus of this chapter. Two considerations make this subject more than a matter of curiosity or historical interest. First, values of large mammals are usually quite stable and deeply held features of human personality and society. Consequently, they greatly influence attitudes and behaviors affecting the conservation of these creatures and their habitats. Second, although we believe these values originate in human biology, we recognize that their content and importance to individuals and groups often vary considerably depending on experience, learning, and culture. Moreover, individual and group value differences are often the basis for considerable debate and sometimes conflict about the most appropriate management of these creatures. Thus, understanding how we value large mammals can tell us much about our biological and social characteristics, and hopefully help us to manage, protect, and conserve these species and their habitats.

As suggested, values toward large mammals reflect the meanings, preferences, and presumptions of worth and importance we attach to the natural world. These views combine what we know, feel, and believe about other species and nature, more generally. Managing large mammals and their habitats inevitably involves choosing among values and interests, and these choices affect diverse interests and needs. Information on human values can be relevant to a wide range of conservation objectives including allocation decisions, policy choices, damage assessments, mitigation strategies, conflict resolutions, and educational programs.

In this chapter, we first consider nine basic values of large mammals. We then examine their occurrence and variation in North American history and contemporary society. To illustrate these differences, we describe three current controversies involving deer, bison, and wolves. Finally, we conclude with a consideration of the practical utility of this information to managing and conserving large mammals.

BASIC VALUES OF LARGE MAMMALS

Various schemes have been developed to classify people's values of nature and wildlife (1). In this chapter, we rely on a typology developed by Kellert, brief definitions of which are provided in Table 3–1 (2). We view these values as biological in origin, reflecting adaptive benefits developed during the long course of human evolution, a perspective explained elsewhere (3, 4). Yet, we also regard these values as "weak" biological tendencies, the occurrence of which among individuals and groups greatly reflects people's variable experience, learning, and culture. We first describe each of these values of large mammals as they normatively occur in North American society. Subsequent sections offer historical and cultural perspectives, and variations among groups distinguished by experience (e.g., hunters and nonhunters), demography (e.g., age, gender, education), and species (e.g., deer, bison, wolves).

TABLE 3–1

A Typology of Values of Large Mammals in North America[a]

Label	Description
Naturalistic	Focus on direct experience and contact with large mammals.
Scientific	Focus on knowledge and study of large mammals.
Aesthetic	Focus on physical attraction and appeal of large mammals.
Utilitarian	Focus on material and practical benefits of large mammals.
Humanistic	Focus on emotional affection and attachment to large mammals.
Dominionistic	Focus on mastery and control of large mammals.
Moralistic	Focus on moral and spiritual importance of large mammals.
Negativistic	Focus on fear and aversion of large mammals.
Symbolic	Focus on metaphorical and figurative significance of large mammals.

[a]*Source:* From Ref. 2. Order of presentation follows occurrence in text. Used with permission of Island Press.

Naturalistic

The naturalistic value focuses on the personal pleasure and satisfaction derived from direct experience and contact with large mammals in their natural habitats. Natural diversity functions as an unrivaled context for engaging the human spirit of curiosity, imagination, and discovery. People take pleasure in encountering and immersing themselves in nature, and the pursuit and visual experience of large mammals in their natural habitats can be particularly satisfying. Humans derive a host of physical and intellectual benefits from exploring nature's rich tapestry of shapes and forms, with the more conspicuously and emotionally charged large mammals reigning above all others.

The naturalistic experience of large mammals in North American society often occurs through a variety of outdoor recreational activities, such as hunting and wildlife observation. Recreational hunting is pursued by some 10% to 15% of the American population, with naturalistic motivations constituting an important element of the experience (2). A much larger proportion of Americans reports strong interest in nonconsumptive outdoor recreation often associated with the chance to see large mammals in their natural habitats (5). The presence of large mammals can add critical elements of stimulation and excitement to the outdoors experience, even in the absence of actually observing these creatures. Heightened awareness of large mammals can prompt a feeling of adventure and an accompanying interest in exploration and discovery. As Ortega y Gasset remarked with respect to hunting, but could be related to the nonconsumptive enjoyment of seeing large mammals in their natural habitats: "When one is hunting, the air has another, more exquisite feel as it glides over the skin and enters the lungs; the rocks acquire a more expressive physiognomy, and the vegetation becomes loaded with meaning. All this is due to the fact that the hunter, while he advances or waits crouching, feels tied through the earth to the animal he pursues" (6, p. 119).

In engaging this inclination for deep involvement in nature, people foster creativity and intellectual development. The more they directly experience the intricacies of natural diversity, the more people typically engage and stimulate their capacity for wonder and discovery. Advantages also accrue in any experience that fosters the capacity for reacting quickly, resolving new and challenging situations, and overcoming difficulties. Demonstrating skill and accomplishment whether in harvesting or observing large mammals can foster these tendencies and, in the process, nurture self-confidence and self-esteem.

Scientific

The scientific value of large mammals emphasizes the empirical study and understanding of these animals. This focus frequently occurs in the absence of formal professional training. Whether among laypersons or specialists, a scientific perspective can foster intellectual growth and development. Moreover, accumulating knowledge and understanding of any element of the natural world can yield, over time and simply by chance, practical advantages, and also advance an attitude of respect dissuading people from heedlessly destroying and depleting nature.

Large mammals have been particularly instrumental in fostering a scientific perspective of nature. Few animals have stimulated as much empirical interest and objective inquiry among professionals and laypeople as deer, bear, wolves, and other large mammals. The study of these creatures has greatly expanded human knowledge of natural history, anatomy, population biology, and animal behavior. This scrutiny has also advanced the modern understanding of ecology, particularly the dynamics of prey–predator relations and ecosystem structure and function.

A scientific perspective encourages precise observation and empirical study, necessitating care, patience, and impartial inquiry. The study of natural phenomena can, thus, foster critical thinking, problem solving, and analytical skills and abilities. These intellectual skills can develop in other contexts. Yet, observing and studying nature, especially its more salient creatures, provides a stimulating and relatively accessible means for building, refining, and honing intellectual capacity.

Astute observation of even a fraction of life's diversity can encourage the recognition of how creatures and natural processes can benefit people. This knowledge often inspires a desire to protect and conserve nature. The more we learn about other life forms, the more we tend to respect and work for their maintenance and preservation. Reflecting the importance of large mammals in this regard, creatures like deer, elk, moose, wolf, and bear have been among the most significant recipients of conservation and restoration efforts.

Aesthetic

Few experiences in life exert as much consistent impact on people as the physical attraction of nature, and large mammals have been among the most aesthetically favored of creatures. Most preference surveys routinely rank large mammals among the most favored of North American species (2). Reflecting this preference, deer, bear, wolf, elk, moose, bison, and other large mammals are among the most prominently featured animals in pictures, posters, toys, paintings, sculptures, film, and other visual media.

In their idealized form or state, these animals convey an impression of particular refinement, harmony, and balance. Nature can suggest a model of perfection and form.

People discern a particular unity and symmetry in a flowering rose, the grandeur of a snow-capped mountain, and the appearance of a large mammal in its ideal or archetypal state (e.g., an elk bugling at the height of its breeding prowess, a bear standing erect, a wolf running down its prey). Each image intimates perfection in a world where frailty and shortcoming often seem normative.

These images can also inspire and instruct. Humans incorporate expressions of nature's symmetry into their lives, often through imagery and metaphor, but also through modeling and mimicry. By discerning beauty, harmony, and balance in charismatic creatures like large mammals, humans advance their understanding of how certain configurations of line, texture, space, light, contrast, movement, color, and form can be employed to produce analogous results in the human experience.

Human aesthetic perceptions of nature may have developed over evolutionary time because they enhanced the likelihood of achieving safety, sustenance, and security (7). For example, people aesthetically favor landscapes that include water, that enhance sight and mobility, or foster the prospect for seeing danger or locating shelter. Aesthetic preferences for large herbivores may reflect a deep recognition of the historic role of these creatures in meeting human needs for sustenance and security. Large predators like wolves and bears have constituted major threats to human survival, and their pelts have been highly prized and coveted. Despite the historic danger posed by these species, their hold on the human imagination has been strong and continues to evoke pronounced aesthetic responses in most people. In a modern world of increasing insulation from nature, the emotional significance of these creatures has often been transformed, especially among urban dwellers and higher socioeconomic groups, into a highly positive aesthetic view.

Utilitarian

The utilitarian value emphasizes the practical and material importance of nature and, in this respect, large mammals have long provided people with a wide variety of tangible benefits. Food is certainly among the most important of these benefits and, even today, a majority of North American hunters pursue large mammals primarily for their meat value although few actually rely on this bounty for survival (2). Reflecting this economic importance, the harvest of deer alone is reported to yield some 150,000 tons annually worth approximately one-half billion dollars (8). Beyond North America, particularly in many developing nations, wild animals continue to provide a substantial proportion—and in some cases a majority—of the food consumed (9). Although large mammals constitute only a small faction of the total, it is nonetheless instructive to note that the estimated total world economic value of wild game in 1996 was $200 billion (10).

Wildlife also provides various medical, industrial, clothing, and decorative benefits. An estimated one-quarter to one-half of all modern pharmaceuticals, for example, are reported to originate in a wild plant or animal (9, 11). Large mammals account for only a tiny fraction of these medicines, yet these species are still used in various pharmaceutical and other products. Large mammals, for example, are exploited for their skins, pelts, horns, and other bodily parts for use in many medicines, clothing, and decorative products. The horns and hides of elk and other deer species are frequently employed in apparel and decoration; the pelts of wolves, bears, and other large furbearers for clothing; the bodily parts of species like bear and deer for medicinal purposes. The internal organs and secretions of bear have, for example, been used in Chinese medicine for more than 5,000 years, and the medicinal value of bear bile has been confirmed by Western science and is the basis for a synthetic drug (12, 13). Unfortunately, continuing demand for wild bear bile has resulted in a worldwide decline of many bear species, the victims of extensive poaching to meet the demands of this trade (12).

Humans also obtain a number of utilitarian benefits from the healthy functioning of ecological systems, including pollination that assists agriculture; decomposition, which removes humans wastes and controls pollution; and nutrient cycling, which is essential for soil formation and fertility. Most of these benefits can be traced to the activities of an enormous number of organisms, particularly invertebrates. Large mammals also provide these benefits; for instance, large predators can keep herbivore populations in check to help minimize agricultural damage.

Humanistic

The humanistic value underscores the emotional affinity people have for nature, particularly other creatures. The natural world offers an important source for bonding and companionship, and large mammals have frequently filled this role in human culture and imagination. The orbit of human fellowship is extended to include nonhuman life, with other creatures becoming the subjects of deep affection and even sentiments of love and kinship. This emotional connection can be tied to human maturational development, fostering feelings of belonging and security, sociability and affiliation, and sometimes promoting physical healing and mental restoration. For many North Americans, large herbivores such as deer, elk, and bison are often the subject of strong affection and emotional interest. Even predators like wolves and bears can elicit strong positive sentiments of emotional attachment and concern, as reflected in the frequent use of bear images in children's toys and stories, or the affinity of wolves with "man's best friend," the dog.

Identifying with the presumed emotional and mental experience of individual animals represents a critical basis for a humanistic perspective of nature. This emotional identification figures in many people's perceptions of large mammals. Occasionally, difficulties and distortions arise from strong humanistic associations, such as when extreme affection for individual large mammals results in undue and dysfunctional anthropomorphism (e.g., the so-called "Bambi" syndrome). Still, close relationships with other creatures can nurture a number of adaptive benefits including emotional maturation and well-being. Moreover, any of the nine values of large mammals can potentially be manifest in distorted and maladaptive ways (e.g., unsustainable material exploitation of large mammals).

Adaptive benefits of a humanistic value of large mammals stem from the pronounced capacity of these creatures to elicit affection and form intimate ties of attachment and bonding with people. These familiar relationships often counter feelings of separation and isolation. Relationships with other creatures can sometimes provide an important means of expressing and receiving affection. This degree of connection is often associated with domesticated pets like cats and dogs. Still, many large mammals at least symbolically and sometimes in reality continue to be used in this way.

The human animal has been extraordinarily successful despite lacking the particular speed, strength, stamina, and stealth characteristics of many other species. We do possess in abundance, however, an enhanced capacity for cooperation and sociability. Bonding and affiliation have been especially critical in the development of these aptitudes, and this highly developed social capacity has been facilitated by cultivating relationships with others, even nonhumans. Family and friends represent the most important source for developing social ties. Yet, caring for another creature can sometimes be a highly salient and effective means for expressing as well as giving and receiving affection. These humanistic benefits accrue under normal circumstances, but become especially pronounced during moments of crisis and uncertainty. The vitality of nature, particularly the emotional joy derived from caring for and expressing affection for other creatures, can be restorative and healing. Among wild animals, large mammals often serve as the focus for this degree of affection, concern, and relationship.

Dominionistic

The dominionistic value emphasizes the human inclination to subdue and master nature. Although we covet kinship and affection in our lives, people also seek opportunities for outcompeting, outwitting, and overcoming challenge and adversity. The natural world, particularly large mammals, has often served as an unrivaled context for developing this more competitive and adversarial aspect of human nature. People no longer rely on besting prey, eluding menacing predators, or surviving in the wild, but the physical and mental strengths derived from challenging nature still remain an important pathway for developing fitness and adaptive capacity.

From a dominionistic perspective, nature is principally valued as an arena of contest and control. People achieve physical strength, mental fortitude, and associated feelings of self-reliance and self-confidence by nurturing their capacities for ingenuity, perseverance, and prowess through challenge and adversity, particularly against formidable opponents like large mammals. By demonstrating an ability to function effectively under difficult and trying circumstances, people can emerge more sure and more certain of themselves.

The pursuit and conquest of large mammals can foster these dominionistic objectives. Historically, hunting has provided this opportunity for contest and challenge in the wild, although today this experience is increasingly achieved through such nonconsumptive pursuits as wildlife photography, hiking, and backcountry camping. Seeking and stalking a large mammal can develop strength, prowess, cleverness, and technique. Together, the body and mind generate an enhanced physical and mental capacity. Through competition and challenge with large mammals, people can become better able to face adversity, take risks, and cope with uncertain and unfamiliar circumstances.

Moralistic

The moralistic value emphasizes a sense of ethical and moral responsibility for conserving, protecting, and properly treating nature and animals. Human society is sustained by its willingness to conserve natural process and diversity and to treat other creatures with kindness and respect. Social orders sustain their natural resource base when motivated by shared ethical standards. Too often, we rely on formal regulations and laws as the only means for protecting and conserving nature, underestimating the human tendency to act prudently when inspired by shared moral belief.

A moral concern and respect for nature has been especially facilitated in North America by the movement to conserve, protect, and restore many large mammal species. For example, a code of sportsmanship and ethical restraint in taking wild animals developed around a concern for large mammals, and a related ethical commitment arose to protect and restore the health of associated habitats and ecosystems. This moral code encouraged many hunters to act with restraint and respect despite immediate practical advantages often stemming from behaving otherwise.

Shared moral belief in an underlying order can also provide an ethical and spiritual foundation that gives definition and shape to human existence. This view of an underlying unity can be rationalized by observing numerous affinities existing among life on earth. Despite an estimated 10 to 100 million species, and numerous other examples of natural difference and diversity, people can also observe among most creatures shared molecular and genetic structures, similar cellular characteristics, analogous circulatory and reproductive features, and parallel bodily parts. A remarkable web of relationship connects a fish in the ocean, a bird in the air, a beetle in the forest, an antelope on the plains, and a human in the modern metropolis. A picture of unity and order emerges, transcending the individuality and aloneness of the single person, culture, species, and moment in time. As individuals and groups, we reap personal faith and confidence from perceiving this moral order and meaning.

This sense of moral and spiritual connection has been facilitated by a strong affinity for large mammals. The totems and religious symbols of ancient and modern North Americans often feature bears, wolves, deer, and other large mammals. Commenting on the particular role of bears, Shepard and Sanders note: "Bear myths and rituals have centered on themes of renewal whether . . . the reincarnation of the soul, the symbolic replenishment of human food, the passage of initiation, or the renewal of clan power in its heraldic image and brave deeds" (14, p. 67). Although this degree of expression is rarely encountered in modern society, we still note many associations of bears, wolves, and other charismatic large mammals with spiritual and moral values.

Negativistic

Nature also serves as a powerful source of human fears and anxieties. Many large mammals can function in this way, particularly species like wolves and bears that provoke highly aversive reactions in most people even when slightly provoked and under widely varying circumstances. Although fear and avoidance of wildlife can sometimes assume irrational proportions, in most situations these aversive reactions represent functional and adaptive aspects of human behavior, including the promotion of safety, survival, and even awe and respect for nature.

The avoidance of injury, harm, and death is a basic characteristic of all organisms. People developed a disposition to respond adversely to certain species and landscapes because they posed, over the course of human evolution, common threats to survival (15, 16). Human well-being has depended and continues to depend on skills and emotions derived from a healthy distancing from potentially injurious aspects of nature. Lacking an appreciation of risk and threat, people often behave naïvely and ignore their continuing vulnerability in an uncertain, unpredictable, and powerful natural world.

Moreover, aversion and distancing oneself from nature does not always foster disdain, dislike, or destructive tendencies. Some of nature's most feared elements, especially large mammals, also provoke awe and wonder. Respect for nature arises as much from recognizing its power to defeat us as from appreciating its nurturing and comforting qualities. Large mammals stripped of their power and fearful elements often become superficial objects of amusement and condescension. Utterly subdued and mastered, large mammals typically inspire little appreciation and esteem. Fear and awe can be as essential as deep affection in cultivating an ethic for nature and wildlife.

Certain large mammals have been especially prominent in our fears and anxieties of nature. Wolves and bears often provoke deep-seated anxieties. These fears also represent, however, an essential aspect of their power and magnificence. Visitors to Yellowstone and Glacier National Parks have a far greater chance of being injured or killed by a car than by a wolf or bear. Yet, stories of wolf and bear attacks are common among visitors, and the presence of bear and wolves is an important source of pride and satisfaction among hikers and campers who experience these wilderness areas.

Symbolic

The symbolic value reflects nature's importance as a source for human communication and expression. People employ the natural world, particularly its charismatic large mammals, as a means for developing and expediting the exchange of information and understanding. This is accomplished through language, story, myth, analogy, and abstraction. Nature as symbol has been especially important in language acquisition, psychosocial development, and everyday communication and thought.

Nature as a symbol in language development is especially important for young children (17). Acquiring human language relies heavily on the capacity for sorting objects into progressively more refined categories and classifications. For young children, this capacity is facilitated by rendering clear distinctions, separating and distinguishing objects from one another. The natural world, especially its large and emotionally salient creatures like large mammals, often provides this opportunity for classification and categorization. We encounter in most children's books on language and counting pictures of animals, frequently large mammals, which assist young children in learning the ordering, sorting, and, most of all, naming skills so integral to language development.

Symbolizing and fantasizing about nature also help the young adolescent to confront and resolve basic dilemmas of identity and selfhood. We encounter many large mammals in the myths, fairy tales, and fantasies of most if not all cultures. These narratives give children assistance in resolving difficult aspects of development such as authority and independence, order and chaos, and good and bad in a tolerable, yet disguised manner. These images assist children in confronting difficult issues obliquely, often employing anthropomorphism to render more acceptable the challenging and enigmatic matters of conflict, need, and desire (18).

Symbolizing nature also occurs in more mundane communication and thought. People frequently employ the imagery of nature, particularly of large mammals, in the language of the street, in the metaphors of the marketplace, in oratory and debate. These symbols often render normal communication and discourse more vivid and persuasive. This imagery can sometimes be trite, but occasionally moving and eloquent. Its pervasiveness in all cultures suggests a universal and indispensable role. Natural diversity provides a kind of substrate for symbolic creation, not unlike nature's role as a biochemical template for laboratory discovery. In each case, the natural world provides clay from which we mold and fabricate solutions to life's varied challenges.

Large mammals have been especially prominent in our symbolic uses of nature. Few creatures figure as often in our myths, stories, legends, images, and everyday discourse. We encounter images of wolves, bears, deer, moose, buffalo, and other large mammals in children's books, historical tales, native myths and legends, and modern advertising and marketing.

VARIATIONS IN VALUES OF LARGE MAMMALS

As indicated, we believe values of large mammals are deeply rooted in human biology, having emerged during the long course of our species evolution because of their adaptive and beneficial significance. Nonetheless, we regard these values as "weak" biological tendencies greatly influenced by human experience, learning, and culture. This "biocultural" framework regards values of large mammals as the consequence of human biological heritage and the free will and choice of a species peculiarly able to choose and shape its destiny (19). In the next sections, we examine this variation in North American values of large mammals. We first review historical and cultural differences in North America, then variations among groups in contemporary society, and finally contrasting views of deer, bison, and wolves as reflected in current controversies involving these animals.

HISTORICAL AND CULTURAL VARIATIONS

Public values of wildlife profoundly changed as a consequence of European settlement of North America. Briefly reviewing this historical context helps us appreciate the evolutionary significance of these values in American consciousness and assists in our understanding of the cultural, social, political, and economic dimensions of current policies and approaches to managing large mammals (20, 21).

Indigenous Values

The presence of large mammals undoubtedly constituted a major reason for the arrival of the first inhabitants to North America, who crossed the Bering Strait from Asia to Alaska (22). Bountiful populations of large mammals occurred around the strait and beyond in the "New World." Early indigenous peoples were big game hunters who relied largely on the abundance of species such as wooly mammoth, bison, wild sheep, moose, muskox, elk, and caribou for survival (22). Pursuing these creatures likely established the basis for a strong utilitarian value of wildlife lasting well into the 20th century.

More is known about the wildlife perceptions of subsequent Native American cultures, who harbored a complexity of intensely held values toward large mammals. These values were strongly manifest in tribal custom, religion, and folklore, reflecting a powerful mix of utilitarian, symbolic, and moralistic perspectives. These values continue to figure prominently among modern Native Americans as suggested by various customs involving large mammals.

Early nomadic tribes of the plains, prairies, and forests of North America especially depended on bison, pronghorn antelope, white-tailed deer, and caribou for sustenance, clothing, and shelter. These animals nourished an utilitarian emphasis among peoples who relied on the meat for food, the hides for clothing and teepee coverings, the bones for knife handles and other weapons, and the sinew for sewing thread and bowstrings (23, 24). Although this reliance constituted a necessity, it also served as the basis for a powerful expression of moral and spiritual belief, as illustrated by the Blackfeet who viewed bison as created by the "Old-Man World Maker" (24). The hunting of large mammals was often performed according to strict ceremonial rites reflecting a kinship and reverence Native Americans held for their presumed brethren in the animal kingdom (25, 26).

Although early Native Americans depended on the consumptive benefits derived from hunting large mammals, this emphasis also included a profound appreciation of the animals' ecological and moral worth. Sioux elders, for example, expressed this orientation in their belief that ". . . all living things are tied together with a common navel chord" (25, p. 4). A sense of kinship with large mammals was pervasive, as suggested by the Cree who believed animals killed in the hunt were immortal, having voluntarily sacrificed themselves to provide sustenance to their tribal relatives (26). Among native tribes of the Northwest coast, animal totems were used to depict family crests, frequently adorned with symbols of large mammals such as wolves and bears, which revealed the tribe's descent or clanship with the animal world (25). Native American folklore has been and continues to be rich in anthropomorphic imagery, with many stories involving large mammals speaking and interacting with humans and possessing the capacity to transform their appearance from one life form to another (25).

Charismatic large mammals represented powerful spiritual symbols for most Native Americans and played highly visible roles in tribal religious traditions. These beliefs expressed a strongly animist perspective that all sentient and even inanimate objects possessed spiritual lives (27). Large mammals were frequently viewed as intermediaries between natives and the gods; spiritual contact with higher beings often depended on visions involving animals and other natural forms (28). For example, among the plains Indians, the wolf was revered as a guardian spirit who presided over tribal hunting and war parties (26).

The grizzly bear was among the most revered and venerated of all large mammals (29). Among the Inuits, the bear was an omnipotent spiritual symbol and guardian spirit (26, 30). Tribal rites involving the passage of youth into adulthood, the initiation of new inductees into secret societies, and the sacred induction into shamanism were patterned after the bears' hibernation, thought to symbolize life and death (26). Although reverence for bears reflected genuine fear of the animal, respect also derived from a perceived closeness of bears and humans. This resemblance was reflected in the recognition that bears like humans consumed similar foods, possessed great dexterity, had the ability to stand upright on two legs, and resembled a person when skinned (20).

Most tribes refused to hunt bears because of the animal's prominence in mythology, religion, and folklore. Southwestern tribes such as the Pueblo and Navajo feared that killing a bear would invite reprisal from the animal's spirit and eventually make the offending hunter insane or infirm. The Blackfeet enforced strong taboos against killing bears, as did the Cheyenne who considered the slaying of bears tantamount to cannibalism (26). Tribes that did kill bears often performed ceremonies of deference and atonement in an attempt to make peace with the animal's spirit, the bear having presumably willingly sacrificed itself to the human hunter (26).

European Settlement Values

The first European settlers to North America confronted an unprecedented abundance of wildlife that represented to them both a luxury and a challenge. The colonial period has been called an "era of abundance" and a "biological paradise" (22, 31, 32). Extraordinary numbers of large mammals were viewed by the early settlers as symbolic of the New World's God-given fertility (31). Still, the early settlers viewed large mammals mainly through a utilitarian perspective. Meat from white-tailed deer was often necessary for survival, sometimes constituting the difference between a settlement's success and failure (33). The hides, pelts, and bones of large mammals were also extensively used in fashioning clothing, shelter, and implements.

Hunting for subsistence during colonial times also provided the historical basis for an egalitarian sporting tradition that still prevails in North American culture. Prior to European settlement in North America, hunting was largely the exclusive privilege of a ruling class and landed gentry (22). Hunting as a necessity in colonial North American evolved away from this tradition of private privilege to become a "free-born" right of citizenship accessible to all settlers regardless of socioeconomic background. This early democratization resulted in hunting becoming an important activity and recreational pastime throughout North America (22, 34).

The seemingly inexhaustible populations of large mammals fostered a climate of "prodigal disregard," as creatures like the white-tailed deer became largely a commodity for trade and barter and often exploited in an unsustainable manner. For example, in 1748, South Carolina traders alone shipped 160,000 deerskins to England (35). This period of excessive exploitation produced the first movement to protect large mammals and, more generally, wildlife. Colonial officials interceded, motivated by the desire to avoid the exhaustion of a valuable resource. By the time of the American Revolution, closed seasons on deer had been implemented in every colony but Georgia (32).

Although deer and other large ungulates were regarded as an asset, attitudes toward large predators were far less positive or charitable. Reverend Cotton Mather's dictum that ". . . what is not useful is vicious . . ." reflected a clear hierarchy of values prevailing among settlers towards ungulates, predators, and other wildlife (32). Wolves especially bore the brunt of the colonists' hostility toward large predators.

Divergent attitudes toward large mammals were also rooted in Old World superstitions, folklore, and Judeo-Christian teachings about dominion, mastery, and subjugation of nature (20, 21). Nash details the prevailing European cultural bias against wilderness as a bastion of evil and diabolical forces (36). Colonial Puritan leaders often depicted wildness as the antithesis of good and moral virtue. Wilderness was considered an impediment to progress necessitating its transformation into a tamer, more genteel setting suitable for human habitation (36).

These highly negativistic and dominionistic views were often extended to large mammals, particularly predators like wolves, mountain lions, and bears (20, 37, 38). Although bear and mountain lion were often persecuted and harvested opportunistically in association with the fur trade, these species generally did not garner as intense an animosity among settlers as did wolves (20). Puritan Roger Williams, founder of the Rhode Island colony, reflected the prevailing view of the wolf when he described this animal as ". . . a fierce bloodsucking persecutor . . ." (38, p. 7). This hostility reflected the settlers' antipathy toward wolf depredation of livestock, but it also expressed a popular perception of wolves as symbolic of the human need to tame the wilderness and root out evil (20). The first predator control initiative implemented by colonial lawmakers occurred in Massachusetts and granted a bounty of one penny for every dead wolf (32). Perceived as worth more dead than alive, the wolf was eventually extirpated from all but the most remote regions of New England by the end of the 18th century (38).

Western Frontier Values

The Louisiana Purchase of 1803 stimulated a dramatic migration of settlers from the East Coast. Accompanying this westward movement were a host of attitudes and values toward large mammals characteristic of settlers in earlier times. Yet, this period also witnessed the first major broadening of wildlife values beyond the mainly utilitarian, dominionistic, and negativistic ones. Indeed, the roots of more appreciative sentiments toward large mammals emerged during this time, focusing on scientific, aesthetic, naturalistic, and moralistic attributes of these species (31, 39).

The presence or absence of large mammals was often a critical influence in the pace and distribution of early westward expansion. Western pioneers initially placed primary emphasis on the material benefits of species such as bison, deer, elk, and pronghorn antelope. Explorers, travelers, soldiers, gold prospectors, and laborers frequently depended on these species for their meat and hides (31, 33, 39). Texas pioneer settler Noah Smithwick expressed the prevailing wisdom when he remarked, "Game, which was the principal source of food supply, was so abundant that we never thought of taking anything but salt when we went out on duty" (31, p. 29). Similarly, Sherwood suggested in his chronicles of big game in Alaska: "References to the complex and abstract psychological, sociological, anthropological, biological, theological, and historical motives for hunting is intellectually entertaining, but Alaskans hunted for food" (33, p. 101).

Reliance on large mammals to satisfy essential material needs eventually evolved into commercial markets motivated by profit and often greed and avarice. This view of large mammals as mainly a commodity originated with the fur trade. As commercial trapping companies like Hudson's Bay decimated beaver populations, they switched their emphasis to large mammals like bison.

Bison symbolized the biological wealth of the frontier and the opportunity for exploiting its economic riches (31). The bison was further perceived as an obstacle to the expanding livestock industry and an impediment to other forms of frontier economic development. The emergence of tanning technologies allowed bison hides to be used for industrial leathers back east. The completion of the Union Pacific and Northern Pacific rail lines in the 1880s also provided reliable east–west transportation and contributed to the bison's demise (32, 40). Indeed, from 1868 to 1881, an estimated 31 million bison were harvested primarily for the hides, garnering approximately $3.50 per hide. By 1890, this prominent wildlife symbol of the western frontier had been largely extirpated (32).

Other large mammals were similarly exploited, although few suffered as severely as the bison. In the 1840s, German immigrants to Texas sold bear meat for $2 per pound and venison for $1.50 per pound (31). California and Alaska miners during the 1849 and 1890s gold rushes often relied on professional hunters to supply camps with wildlife because of the paucity and expense of domestic meat (39). In the southeastern Alaska gold rushes of the 1890s, deer carcasses sold by professional meat hunters for $1 per pound "were piled up on the wharves like cordwood" (33, p. 27). Not surprisingly, these excesses led to precipitous declines of deer, antelope, sheep, bear, and other large mammals (32, 33).

As the remnant herds of bison were consigned to private preserves largely outside the plains, the rangelands became opened for livestock grazing. This precipitated another confrontation between large mammals and livestock, this time with the wolf becoming the enemy. In the absence of bison, wolves largely preyed on domestic cows and sheep, quickly becoming a focus of antagonistic feelings. An entire profession developed around the destruction of wolves during the period 1860–1885. Professional "wolfers" extensively persecuted this predator employing intensive trapping and indiscriminate poisoning (38). Even as far west as Oregon, the wolf's relationship to livestock figured prominently in the territory's early political discourse, as settlers preparing for statehood gathered to discuss ways to resolve their "wolf problem" (41, 42).

The widespread development of market hunting during the 19th century eventually led to a schism in the hunting community between commercial market hunters and sport hunters (22). The latter reacted strongly to the excesses of commercial hunting and the dominance of economic values in the treatment of large mammals. Sport hunters stressed the noncommodity benefits of hunting, including promotion of familial and social bonds and learning about the natural world (31). Henry William Herbert, an immigrant English aristocrat writing under the pen name "Frank Forester," condemned the wanton slaughter of large mammals for commercial profit, issuing a call among sportsmen to develop a new hunting ethic. His writings on behalf of conservation particularly inspired notable hunter-conservationists such as Theodore Roosevelt, George Grinnell, and C. Hart Merriam (22). Other values of large mammals beyond the utilitarian, dominionistic, and negativistic, thus started to develop during this era of western frontier expansion. Intellectual curiosity and affection for large mammals and their habits reflected an emerging scientific, naturalistic, and even moralistic perspective.

Early naturalists like Thomas Say and Titian Peale played a role in this change, as they accompanied government surveyors and explorers documenting the abundance, distribution, and biology of the region's wildlife, particularly large mammals (32). Naturalists like Alexander Wilson, John James Audubon, Thomas Nuttall, and Spencer Baird also wrote for the general public about the unusual and mysterious life histories of mountain lion, jaguar, wolf, bear, and other large mammals (32). Dr. Elliott Coues reflected this emphasis in his 1867 book, *Quadrupeds of Arizona,* when he remarked: "The traveler meets at each successive day's journey, new and strange objects, which must interest him, if only through the wonder and astonishment they execute" (39, p. 165). Several scholarly treatises addressed the biology and ecology of large mammals including the pioneering writings of John Bachman in his 1848 book, *Viviparous Quadrupeds of North America,* and John Goodman's *American Natural History.*

The Romantic Movement of the early 19th century also fostered scientific and humanistic values among writers, artists, and philosophers who celebrated the intrinsic worth, beauty, and goodness of wilderness and wildlife. Writers like Thoreau, Emerson, and Muir helped reverse the legacy of Puritan antipathy toward wilderness as a reservoir of evil and savage beasts (36). This celebration of wildness as a source of spiritual, moral, and intellectual wisdom largely occurred among an urban intellectual class who rarely confronted the hardships of frontier life. This focus accentuated differences in wildlife values among urban and rural residents (36). The Romantic Movement, nonetheless, helped initiate more appreciative sentiments towards wildlife and large mammals, especially carnivores, among the public, and this greatly influenced the development of modern wildlife conservation and management.

Early Twentieth Century Values

Historian Fredrick Turner pronounced the end of the American frontier in 1893, writing about the frontier's influence on American society and culture (36). The attitudes of the American public toward wildlife and wilderness shifted dramatically during the early 1900s with changing immigrant patterns and an exodus of rural dwellers to the city. For the first time in North American history, a majority were liberated from a largely utilitarian dependence on wildlife and came to appreciate nature from a more remote and distant setting (36).

This "kinder and gentler" attitude, however, hardly prevailed toward large carnivores. A relentless crusade instead commenced among western livestock producers to extirpate large predators from the rangelands, assisted by the U.S. Bureau of Biological Survey (43). Government trappers and stockmen from Texas to Alaska employed bounties, traps, and poisons to eliminate large carnivores, particularly wolves (38). Wolves who managed to evade this persecution assumed legendary names like "Three-toes," "Old Lefty," "Big Foot," and "Rags the Digger," and were memorialized in western campfire lore, earning the grudging respect of those who attempted to kill them (38).

This antipredator sentiment was rooted in a value system that placed greatest importance on the economic needs of livestock producers. Even conservationists like Teddy Roosevelt declared the gray wolf "a veritable scourge to the stockmen" (38, p. 264).

Stockmen such as Joseph Neal of Meeker, Colorado, remarked: "The history of the gray wolf in the west is a chronicle of the struggle between the wolf and the livestock industry for supremacy with the success or failure of the livestock business depending upon its outcome" (38, p. 263). Lopez noted a near "pathological hatred" of the wolf among western livestock producers indicative of a ". . . deficient understanding of human dependence on nature for ecological, emotional, and intellectual sustenance . . ." (20, p. 980, 44).

Other large predators were similarly vilified for their alleged complicity in obstructing economic development and posing a threat to human safety and property. These characterizations often associated large carnivores with pestilence, death, violence, and destruction. Governor DeWitt Clinton of New York, for example, made reference to the grizzly bear in 1914 as "the ferocious tyrant of the American woods" (33, p. 28). Even as recently as 1950, an Alaska legislator and fisherman blamed bears for the failure of the Kodiak Island cannery, remarking: "These bears can't vote and they don't pay taxes. I see no reason why we shouldn't get rid of them whenever they endanger our lives or our livelihood" (33, p. 28). The mountain lion also faired poorly, even among those with strong conservationist leanings like Theodore Roosevelt who declared the cougar the ". . . lord of stealthy murder, facing his doom with a heart both craven and cruel . . ." (45, p. 55).

Reflecting a deep public divide of values toward large ungulates and large carnivores, early conservationist sentiment primarily focused on assisting members of the deer family. Most sportsmen viewed big bull moose, heavily antlered white-tailed buck, and majestic mountain sheep as compelling symbols of wilderness worthy of protection. Politically prominent sportsmen like George Grinnell, Henry Lodge, and Theodore Roosevelt established the Boone and Crockett Club in 1888 to foster the virtues of big game trophy hunting and wildlife conservation. This affluent group, often inspired by the earlier writings of "Frank Forester," influenced many early conservation measures, heralding an era of 20th century wildlife protection. This group also promoted in the hunting community a "primitivist philosophy" of "moral rectitude" emphasizing reverence and respect for large mammals (36).

This period further witnessed the early development of the wildlife management profession. Aldo Leopold especially helped create a scientific basis for managing wildlife as a renewable resource that could produce sustained yields for harvesting purposes (46). Although Leopold initially championed a largely utilitarian justification for managing large mammals and controlling predators, particularly to enhance sport hunter interests, he later broadened his value perspective. In his seminal book, *Sand County Almanac* (47), Leopold eloquently argued for wildlife conservation based on an animal's scientific, naturalistic, and moralistic worth (20, 30). He helped to establish a growing recognition that people were an integral component of a broader, intertwined biotic community, and he powerfully advocated a view of humans as stewards rather than conquerors of the land. These perspectives encouraged an appreciation of the ecological role of wildlife and a growing consciousness of the multiple values of large mammals and nature, more generally.

CONSUMPTIVE AND NONCONSUMPTIVE VALUES AND USES OF LARGE MAMMALS IN CONTEMPORARY SOCIETY

Recreation and Leisure Activities

Outdoor recreation provides important benefits to people in modern society by affording enthusiasts the chance to participate in activities that enrich their lives through direct and indirect interactions with wild animals. These pursuits provide a suite of environmental, social, educational, spiritual, health, and cultural benefits (48). Feelings of kinship and closeness to large mammals in the wild can elevate the quality of the experience far beyond simple participation (48, 49).

One measure of the importance of wildlife recreation is found in public participation statistics. Recreational hunting of large mammals, for instance, is still a significant pursuit of many Americans and a substantial contributor to the national economy. In 1996, for example, approximately 11.3 million Americans over the age of 16 spent 154 million days hunting large mammals, particularly deer (5). Trip and equipment-related expenditures involving large mammals totaled $9.7 billion (5). Although wildlife professionals express concern about declining hunter numbers, the total hunting population has remained relatively stable since 1991, although as a proportion of the population the numbers of hunters have declined (5, 50, 51).

The motivation for hunting large mammals is complex, involving many social, psychological, and cultural factors (52, 53, 54). Although "the kill" is a necessary part of the hunting experience, many hunters emphasize nonharvest and nonconsumptive motivations as their primary reason for hunting. These motives include spending time with friends and family, reinforcing social ties, being close to nature, getting outdoors, providing rites of passage for children into adulthood, and others (8, 34, 54, 55, 56, 57). In the chapter's next section, we provide a more thorough discussion of the motivations of hunters.

Although the right to hunt, especially for meat, continues to enjoy extensive public support, the number of persons opposed to hunting is considerable and a growing concern among hunters and wildlife managers (55, 58). Hunting opponents often cite the philosophical perspectives of Albert Schweitzer, Joseph Krutch, Cleveland Amory, and Peter Singer (59, 60, 61, 62). An association among guns, hunting, and violence prompts many to object to hunting (22, 63). The emerging field of environmental philosophy has also stimulated discussion about the ethical, moral, religious, and ideological justifications for hunting. The antihunting movement has emphasized animal welfare and rights issues, particularly concern about inflicting pain on sentient nonhuman animals (64). In the chapter's next section, we address the extent of antihunting sentiment and the primary values associated with its occurrence. For now, we wish to emphasize that in an increasingly urban society where many people lack an appreciation of hunting as a recreational pastime or wildlife management tool, managers of large mammals should be prepared to address the ethical concerns of antihunters and other members of the public.

The importance of nonconsumptive recreation involving large mammals is also reflected in participation rates. These activities can be as casual as incidentally observing deer from a kitchen window or engaging in organized trips for the express purpose of seeing, hearing, or photographing large mammals in their natural settings. Although wildlife observation declined 17% from 1991 to 1996, 63 million Americans over the age of 16 still reported participating in this activity in 1996 (5).

Despite problems associated with using recreational expenditures as an indicator of the economic or market value of wildlife, wildlife watching participants nonetheless reported spending $28.9 billion in 1996 (5, 65). The economic importance of tourism involving large mammals is illustrated by the multimillion dollar industry centered around polar bear watching in Churchill, Manitoba. Other examples of interest in observing and experiencing large mammals include organized "wolf howls," grizzly bear viewing in Alaska, or watching elk at the National Elk Refuge in Jackson Hole, Wyoming (66, 67, 68). Moreover, Flewelling and Johnson report that wolf and bear sightings substantially improve backcountry visitor satisfaction at Denali National Park (69). Similar studies in Texas, Colorado, and North Carolina indicate that sightings of white-tailed deer are a very important element in visitor enjoyment (68, 70, 71).

The mere existence of large mammals can be highly significant for wildlife enthusiasts and park visitors apart from actually seeing the animal. The knowledge that large ungulates and predators are present in an area often enhances the recreational experience, creating the sense of a truly wilderness experience (20, 33). Aesthetic, naturalistic, scientific, and dominionistic values associated with enjoying scenery, escaping civilization, coping in the wild, and learning about wildlife can be critical to an outdoor enthusiast's experience of large mammals (72, 73). These recreational activities often produce considerable educational, spiritual, and cultural benefits, including the opportunity for experiencing a sense of renewal through atavistic ties to the land (4).

Indirect and Vicarious Uses

Large mammals also occupy a prominent role in North American culture as reflected in literature, art, and other media. Some studies suggest the mass media provide people in modern society with their primary opportunity for experiencing nature, thus greatly shaping, altering, and reinforcing public perceptions and values of wildlife (2, 74, 75). The popularity of wildlife and nature-related television programming is indicative of public appreciation and fascination for the natural world (2, 76). Media influence on public values of large mammals can be positive, increasing appreciation, admiration, and respect. This may be illustrated by the media's treatment of wolves, which has helped reduce traditional antipathies and fears of this animal (20, 43). On the other hand, the media can sometimes create confusing, contradictory, and distorted impressions of large mammals and other wildlife (20).

Children often encounter large mammals in myths and children's stories, and the symbolic portrayal of these animals has been instrumental in enhancing support for broad social values, customs, and mores (4, 74). Well-known children's stories like Winnie the Pooh, Goldilocks, Bambi, and Little Red Riding Hood, for example, have depicted bears, wolves, and deer in highly anthropomorphic ways intended to promote various morals and codes of conduct (42).

Children's contact with large mammals also occurs today through cartoon imagery in television and film (77, 78). Individuals develop many of their attitudes and values toward wildlife during their formative childhood years (76). The depiction of large mammals in cartoon imagery can potentially promote false understandings and distorted values as well as reinforce negative stereotypes (20, 34, 76, 79). Yet, Pomerantz's study of children exposed to wildlife through the National Wildlife Federation's youth-oriented *Ranger Rick* magazine did not substantiate the dangers of anthropomorphic imagery (80). Still, the pervasiveness of the visual media in modern society has important impacts on prevailing views of large mammals. Wildlife managers must monitor this exposure, seeking to eliminate distorted depictions and impressions (79).

Large mammals also appear in adult literature, language, legend, and art. For example, many references to wolves occur in the classical writings of Greece and Rome including Aristotle, Homer, Plutarch, popular folklore like Aesop's fables, and the legend of Romulus and Remus, all still in wide circulation today (42). Books, magazines, and periodicals devoted to nature-based tourism, hunting, and natural history underscore the prominence of large mammals in popular literature. These stories often emphasize "trophy" or "peak" experiences involving encounters with large mammals in the wild, experiences that often can be as rewarding for the participants as actually possessing a physical trophy of the animal (33).

Artistic renderings of large mammals additionally reflect the public's aesthetic, naturalistic, and humanistic values. These depictions tend to emphasize the majestic, wild, and spiritual attractions of these creatures. The widespread popularity of wildlife art and prints depicting large mammals illustrates how important the vicarious experience of nature can be for wildlife agencies attempting to cultivate public support for their programs (34).

VALUES OF LARGE MAMMALS AMONG CONTEMPORARY NORTH AMERICANS

Although we previously described values of large mammals as rooted in human biology, we also indicated considerable variation in the content and intensity of these values as a consequence of experience, learning, and culture. Insufficient space prevents a lengthy discussion of these differences, and thus we offer only brief illustrations. Specifically, we explore variations in values of large mammals among hunters, anti-hunters, and societal groups distinguished by age, gender, urban–rural residence, and socioeconomic status. We conclude by describing three current controversies involving deer, bison, and wolves to illustrate how these value differences can affect the conservation and management of large mammals.

Values of Large Mammals Among Hunters and Antihunters

Duda reviewed the literature on hunter participation in the United States, and notes that the "multiple satisfactions" approach pioneered by Hendee is the most widely cited method used for distinguishing hunter values, and the need for managers to recognize this variation in managing consumptive wildlife uses (8, 52). These hunter satisfactions can be roughly subsumed within the broad value categories used in this chapter. Specifically, three major types of hunters can be distinguished, primarily motivated by naturalistic, utilitarian, and dominionistic values (2).

Naturalistically oriented hunters stress the desire for deep participatory involvement in natural settings, including a chance to learn more about biological systems. The hunt offers an intimate experience with nature, with the role of predator viewed as providing the chance for participating in the complexity of ecological systems. The kill, while necessary, is not the main focus for the naturalistic hunter, but rather the excuse for deep immersion in natural settings, where large hunted mammals are often the subject of considerable appreciation, affection, and respect.

By contrast, the utilitarian hunter is motivated by the practical and material benefits of the activity. Large mammals are especially valued for their meat or, in some cases, pelt, hide, or medicinal value. This utilitarian emphasis may seem anachronistic today with few North Americans relying on wild meat as a major food source. Still, a surprisingly large number of rural residents continue to exploit large mammals as a valuable dietary supplement, especially during winter. From a utilitarian perspective, harvesting large mammals is analogous to exploiting farm animals or picking a crop. The kill is the overriding consideration, and an unsuccessful hunt often constitutes a meaningless event. While not indifferent to the animal or its surroundings, the utilitarian hunter typically subordinates these matters to the creature's practical utility and material significance. Large mammals are viewed mainly as a renewable resource managed for annual yields. Hunting is typically valued as part of a lifestyle that stresses extracting part of one's living from the land and its resources.

The dominionistic hunter mainly values the activity for its competitive and sporting attractions. The hunt offers the opportunity for demonstrating skill and mastering a challenging opponent. Important aspects of the experience include success and achievement. Taking and displaying game often attest to the presumed skill, courage, and fortitude of the hunter. Strength and prowess are important attributes of the experience. The dominionistic hunter seeks the excitement of competition and the opportunity to be tested and proven worthy. A stress on masculinity underscores qualities of hardiness, toughness, and endurance.

As indicated, hunting large mammals has become the focus of considerable debate in contemporary North America (2). This may be surprising given the presumed importance of the hunter-gatherer way of life in human evolution and the significance of hunting in North America history and settlement. Many studies report that a substantial fraction of the American public today opposes the hunting of large mammals, especially bear and wolf, and less so, various large ungulates (2, 8). Relatively few Americans, perhaps 15%, support trophy hunting, although approximately 85% approve of hunting for meat (2). Hunting large mammals for recreation and meat is supported by an estimated 60% of the American population.

Two value dimensions, the moralistic and humanistic, represent important motivational bases for antihunting sentiment. The humanistic perspective tends to be more prevalent, emphasizing strong affection for individual animals projected to include large mammals. The humanistic antihunter identifies with the presumed experience of the hunted creature, viewing the pain and suffering presumably experienced by the hunted animal as analogous to what humans would feel in a similar situation. This emphasis on the individual animal typically renders irrelevant justifications for hunting based on issues such as prey population levels and ecological health and stability.

The moralistic value provides a more philosophical basis for antihunting sentiment. This perspective emphasizes ethical objections to killing large mammals in the absence of necessity. Sport and recreational hunting is viewed as immoral, diminishing the hunter who presumably inflicts gratuitous harm and suffering on other creatures (61). Recreational hunters are considered morally culpable because of their willingness to deliberately kill for sport. Krutch emphasizes this view when he suggests: "Killing for sport is the perfect type of that pure evil for which metaphysicians have sometimes sought. . . . Most wicked deeds are done because the doer proposes some good to himself. The liar lies to gain some end; the swindler and burglar want things which, if honestly got, might be good in themselves. . . . The killer for sport, however, has no such comprehensible motive. He prefers death to life, darkness to light" (60, p. 9). The moralistic antihunter views killing for sport as the essential ethical difference between the hunter and meat-eating nonhunter. Large mammals are considered as possessing an inherent right to live, and empathy and compassion cited as reasons for extending moral rights to animals. Moralistic antihunting is often an element of a more general pacifism, with hunting viewed as illustrative of antisocial violent conduct (62).

Demographic Variations

These variations are illustrated by briefly examining value differences among gender, geographic, and socioeconomic groups. Gender differences in perspectives of large mammals generally reflect variations in wildlife values among males and females. Research has largely found stronger humanistic and moralistic wildlife values among women than men in North American society (2). Females generally express greater affection and emotional attachment to large mammals, and a greater inclination to object to hunting and trapping. By contrast, males often reveal stronger utilitarian and dominionistic values, and related support for the practical exploitation and mastery of large mammals. Males constitute a much larger fraction of hunters (approximately 90%) and members of hunting and fishing organizations (2, 8). By contrast, females comprise a much larger percentage of antihunters (approximately 70%) and members of groups opposed to hunting and trapping (approximately 80% of animal welfare and animal rights organizations).

Occupational and residential findings generally reveal stronger utilitarian and dominionistic values of large mammals among farmers, loggers, miners, commercial fishermen, and residents of open country (2). By contrast, persons owning little or no land, professionally employed, and residents of large cities frequently express stronger moralistic, humanistic, and scientific values of large mammals (2). These differences tend to be more pronounced when comparing people on the basis of where they were raised and their parents' occupation than where they currently reside or their present occupation.

Socioeconomic assessments typically involve education and income comparisons. Interestingly, few major value differences are found among people distinguished by income in views of large mammals (2). More affluent persons usually express stronger naturalistic values of large mammals, although utilitarian and dominionistic differences are rarely significant among income groups. Few substantial variations occur among income groups in opposition and support for hunting large mammals. Far more pronounced value differences occur among North Americans distinguished by education (2). College-educated persons generally express stronger naturalistic, humanistic, moralistic, and scientific values of large mammals, while those with less than a high school education tend to be more utilitarian, dominionistic, and negativistic. The college-educated population reports greater nonconsumptive outdoor recreational interest in large mammals, and also reads books and watches films and television programs about these animals.

Species Variations

Contemporary North American values of large mammals are illustrated by briefly describing three controversies: urban deer control, bison management in Wyoming, and wolf restoration in various areas of North America.

The Urban Ungulate: Deer–Human Interactions. Resource managers confront new challenges in North America's increasingly urban environment, illustrated by the problem of managing expanding deer herds in areas of high human population density. Deer have become an expanding part of the North American suburban landscape, readily adapting to edge environments on the urban–rural interface. Traditional biological management concepts such as carrying capacities, sex ratios, and harvest schedules are often marginally relevant in these urban settings, with managers forced to focus as much on managing humans as deer. An especially critical challenge for wildlife managers is being aware of urban wildlife values and learning to communicate with nontraditional wildlife groups.

Growing numbers of white-tailed deer along the urban fringe continue to outpace biological projections and pose challenges for wildlife agencies. For example, in heavily populated Princeton Township in New Jersey, Keiser and Applegate estimate the deer population at more than 50 per square mile (81). Suburban deer populations have been enhanced by hunting restrictions instituted by local municipalities and private landowners concerned about gun safety and animal rights issues, lack of predators, and the ability of deer to adapt to neighborhood habitats (82, 83, 84, 85).

Many urban and suburban residents value white-tailed deer for their aesthetic and humanistic qualities (86). Attractive in appearance, graceful in movement, and nonthreatening in demeanor, deer are often welcome additions to residents who enjoy viewing them in backyards, adjacent fields, and woodlots. Deliberate efforts are sometimes made by homeowners to attract deer by offering supplemental food and water (87). The presence of deer for many suburban residents often reflects their reason for moving "to the country," symbolizing a rural quality of life that attracted many from the city. Residential developers sometimes appeal to these sentiments by using housing development names such as "Deer Creek" and "Buck Hollow."

Public perception of deer as a valuable aesthetic, humanistic, and ecological resource can change dramatically, however, when deer populations become excessive and property damage and human health are affected. Negativistic and utilitarian values of deer typically develop with concerns over human health, safety, and property. Suburban antagonism toward deer has been prompted by anxiety about Lyme and other disease transmission, damage to ornamental plants, and motor vehicle accidents (86, 88, 89). Despite these antagonisms, the urban public often objects to lethal deer control, including sport hunting.

A nonhunted deer herd is not the same, however, as a nonmanaged one, and innovative control strategies can be employed (90). Few easy solutions, however, exist for reconciling conflicting values with the need for regulating deer overpopulation in urban areas. Managers must be aware and appreciative of the perceptions of nontraditional constituencies, and be prepared to initiate extensive public education and outreach programs (85). Citizen task forces consisting of relevant stakeholders can be used as occurred, for example, in New York State, as a way of involving communities in deer management decision making (91).

Brucellosis, Bison, and Bovines in Jackson Hole, Wyoming. Free-ranging bison were first introduced to the Jackson Hole, Wyoming, area as a means of restoring a large mammal with considerable historical, cultural, and ecological appeal. The Jackson Hole herd has become today a part of a larger controversy involving the interaction of people, bison, elk, and livestock in the Greater Yellowstone ecosystem. Since its introduction in 1969, bison have expanded from 16 to approximately 200 animals (92). This increase has prompted concerns among wildlife managers about the herd's abundance, its winter migration from Grand Teton National Park to the National Elk Refuge, competition with elk for supplemental feed during winter, and private property damage (92).

Central to the controversy is the allegation by agricultural officials that bison are a reservoir for brucellosis, a bacterial disease potentially causing infected cows to abort prematurely (93, 94). This assumed threat to the region's livestock precipitated the formation of an interagency team to develop a biologically, socially, and politically acceptable plan for regulating bison abundance, distribution, and interactions with livestock (92). A number of proposals have resulted in considerable controversy among varying stakeholders, each complaining their concerns and values have been insufficiently considered.

Agricultural officials believe brucellosis-carrying bison threaten the economic livelihood and vitality of Wyoming's beef industry. The livestock community is particularly concerned that bison will jeopardize Wyoming's "brucellosis-free" designation, a development that could seriously harm the state's ability to market beef outside the state (95). As rancher Glenn Taylor remarked: "We have to sell livestock. We can't do anything with cattle we can't sell" (96).

Opponents of bison control, including animal rights advocates and Native American groups, have expressed indignation at a Wyoming Game and Fish proposal to allow limited recreational hunting to regulate bison numbers below 200 animals (96, 97). Arguing for the rights of bison, animal welfare groups have derided the sport hunting proposal as morally repugnant and inhumane (98). Tribal groups have also opposed the hunt for different reasons, arguing the proposed means of reducing bison will result in wanton slaughter and denouncing a recreational hunt as an affront to the bison's religious and cultural significance (99).

The conservation community has taken a position in between. They have rejected as unsubstantiated the economic arguments offered by government agricultural officials, given that no documented cases of brucellosis transmission from bison to cattle have occurred in the wild, and ranchers have peacefully coexisted with infected bison and elk for years (100). Conservationists have also stressed the bison's historic role as a large herbivore in the Greater Yellowstone ecosystem and as a symbol of the Western wilderness experience (101). They argue a major priority should be restoring a complete and healthy ecosystem to the area, replete with a viable, self-sustaining bison herd.

Wildlife tourism interests have promoted the bison's economic importance. Bison represent a significant attraction to park visitors, who frequently cause "bison jams" when the animals are seen along the highway. Concessionaires at Grand Teton National Park have encouraged this interest with billboards indicating "where the buffalo roam" to assist wildlife watchers in seeing one of the world's largest land mammals (92).

This controversy reflects a major clash of values. A challenge for government officials is to achieve a balance of diverse national and local interests that equitably considers the full range of cultural, scientific, economic, and ecological perspectives (102, 103). In this way, managers can hopefully avoid the criticism of one long-term Jackson Hole resident who observed that "what's being proposed is designed to support a special interest and not a common cause" (96).

Wolf Restoration and Management. Attitudes toward wolves in North America are, as we have seen, complex and dynamic (20, 32, 44). Well into the 20th century, the wolf was perceived as a threat to livestock and humans, and as an impediment to civilization. A major shift in public values of wolves occurred mainly after World War II, with many coming to view this animal as especially worthy, noble, and ecologically significant. Considerable variation, however, remains in public perception, an ambivalence reflected in controversies regarding efforts at wolf reintroduction and restoration.

National studies suggest a roughly even split among Americans who view this animal with fear and antagonism versus considerable affection and admiration (2, 20). On the other hand, more negativistic, dominionistic, and utilitarian perceptions generally occur among rural, elderly, and especially resource-dependent groups including farmers and livestock producers. More naturalistic, humanistic, moralistic, and scientific sentiments prevail among younger, better educated, and urban residents. Hunters and trappers often express ambivalent views, citing considerable ecological respect and admiration for wolves, but also support for this animal's harvest and control.

These value differences have been especially apparent in wolf restoration and management efforts, the best known being the controversy over wolf reintroduction to Yellowstone National Park. Various studies have revealed largely negative views of wolves among resource-dependent and rural populations in proximity to the park (20, 104). By contrast, Yellowstone National Park visitors, and the American public more generally (especially younger, higher socioeconomic, and urban residents), have expressed highly positive views of wolves and strong support for Yellowstone wolf reintroduction (20).

Anecdotal expressions of these value differences are reflected in the following comments regarding proposed wolf reintroduction. Opponents have remarked: "The wolf is like a cockroach and will creep outside of Yellowstone and devour wildlife. . . . It's like inviting the AIDS virus." "Only a brain dead son-of-a-bitch would favor reintroduction of wolves." "Wolves don't feed and water the livestock and they don't help raise food for people to eat so what good are they?" By contrast, reintroduction advocates have suggested: "Only a fool would not agree to the placement [in the park] of this beautiful and essential animal." "Restoring the wolf to Yellowstone would be like restoring the American flag to Iwo Jima." "Wolves do not kill people. Fatty beef does." (20, p. 979; 105, p. 3)

VALUES OF LARGE MAMMALS IN WILDLIFE MANAGEMENT AND POLICY

Policy involving large mammals should be based on a thorough consideration of all relevant biophysical, socioeconomic, institutional-regulatory, and value dimensions, understood in the context of competing stakeholder interests, and a dynamic process over time (2, 106). Biophysical considerations underscore the importance of considering all important biotic and abiotic factors in an ecological context; socioeconomic factors; the importance of understanding diverse power, property, and community interests and relations; institutional-regulatory dimensions; and the relevance of various legal and organizational forces in policy formulation and implementation. The consideration of human values stresses the additional importance of incorporating an adequate understanding of the diverse meanings people attach to and the benefits they derive from large terrestrial mammals. Wildlife managers lacking this information are at a serious disadvantage in promulgating equitable, efficient, and effective policies, and in educating various user groups and in enlisting public support for their policies.

Striving for Management Effectiveness, Efficiency, and Equity

Wildlife management involving large mammals is typically fraught with scientific uncertainty and competing human preferences and values. This policy-making environment renders it difficult for managers to develop biologically sound and socially acceptable policies. Managers must be prepared to address not only questions of a biological nature, but also human dimensions such as prevailing public values, attitudes, and the current sociopolitical climate. This delicate balancing necessitates an interdisciplinary approach, with social research being an indispensable tool for understanding the cultural, sociological, and attitudinal factors affecting policy and decision making.

Resource agencies have the challenging task of managing wildlife for diverse public benefits. Managers must confront the need to make decisions involving trade-offs, whether balancing conflicting values among livestock producers and wolf reintroduction advocates, or resolving issues of control over public wildlife resources on private lands. Managers of large mammals need to understand and consider various value perspectives among relevant stakeholders in an objective and dispassionate manner when formulating policy decisions. This necessitates an explicit recognition of all values of large mammals, not just those predicated on economic or political considerations. Conducting issue-related contextual analysis of stakeholder perspectives and values represents an essential step in defining and resolving management decisions (107, 108).

Improving the dialogue between wildlife management agencies and affected stakeholder groups is also critical to achieving equitable and effective policies. The public should be engaged in a meaningful discourse with managers from the beginning of the decision-making process. Public participation can be facilitated through surveys, public meetings, and other methods intended to ensure a fair and balanced consideration of all stakeholder values (109). Managers must also strive to evaluate and revise programs to address changing needs and public preferences. Management flexibility and commitment to analyzing programs can help agencies be responsive to shifting values and perspectives of diverse constituents (110).

Managing for Biological and Sociological Diversity

The goal of managing game as opposed to nongame species has become increasingly irrelevant, allied less with biological reality than with an anachronistic idea of tangible benefits and economic returns. Species once considered useless are more and more recognized as possessing a diversity of present and potential future values. Wildlife managers must focus on all species, whether large mammals or not, and shift from a single species focus to a greater emphasis on species interactions and biological communities. This does not require the abandonment of a wildlife management tradition that recognizes public concern for particular large mammals or the importance of these species to certain user groups. It does suggest, however, that these management considerations should not occur at the expense of other wildlife or species viewed as direct or indirect competitors of the target animal.

Managing for sociological diversity should also be a goal of wildlife management. As discussed, managers need to consider all wildlife values among diverse users and stakeholders. Ignoring competing perspectives encourages conflict, particularly when managers fail to recognize the diverse motivations that prompt people to behave in different ways. When professions or agencies ally themselves too closely with the values of particular user groups, the impression emerges of their being captured by special interests and isolated from important constituents. A major challenge of contemporary management is recognizing all wildlife values among all relevant stakeholders, while not compromising the necessity of basing policies on the best scientific information.

SUMMARY

This chapter described past and present values of large mammals in North America. We treated large mammal values as an integration of feelings, thoughts, and beliefs, reflecting both meanings people attach to and the benefits they derive from these animals.

We viewed basic human values of large mammals as a reflection of human biology, but influenced greatly by people's experience, learning, and culture.

We initially described nine basic values of large mammals including the aesthetic, dominionistic, humanistic, moralistic, naturalistic, negativistic, scientific, symbolic, and utilitarian. We then reviewed historic North American variations in perceptions, uses, and values of large mammals among indigenous, European settler, Western frontier, and early 20th century Americans. We next examined contemporary differences in uses and values of large mammals among hunters, antihunters, wildlife observers, and Americans distinguished by gender, education, income, and population of residence. We offered three case studies to illustrate variations and controversies in values of large mammals including urban deer conflicts, bison livestock conflicts in Wyoming, and wolf restoration. We concluded with a brief discussion of the relevance of large mammal values to wildlife policy and management.

LITERATURE CITED

1. DECKER, D. J., and G. GOFF, eds. 1987. *Valuing Wildlife: Economic and Social Perspectives.* Westview Press, Boulder, CO.

2. KELLERT, S. R. 1996. *The Value of Life: Biological Diversity and Human Society.* Island Press, Washington, DC.

3. KELLERT, S. R., and E. O. WILSON, eds. 1993. *The Biophilia Hypothesis.* Island Press, Washington, DC.

4. KELLERT, S. R. 1997. *Kinship to Mastery: Biophilia in Human Evolution and Development.* Island Press, Washington, DC.

5. U.S. FISH AND WILDLIFE SERVICE. 1996. National survey of fishing, hunting, and wildlife associated recreation. Department of Interior, Washington, DC.

6. ORTEGA Y GASSET, J. 1986. *Meditations on Hunting.* Macmillan, New York.

7. ORIANS, G., and J. HEERWAGEN. 1992. Evolved responses to landscapes. Pages 133–152 *in* J. Barlow et al., eds., *The Adapted Mind: Evolutionary Psychology and the Generation of Culture.* Oxford University Press, New York.

8. DUDA, M. D. 1993. Phase 1: Literature review. Factors related to hunting and fishing participation in the United States. Federal Aid in Sport Fish and Wildlife Restoration Project No. 14-48-0009-92-1252. U.S. Fish and Wildlife Service, Harrisonburg, VA.

9. GROOMBRIDGE, G. 1992. *Global Biodiversity: Status of the Earth's Living Resources.* Chapman and Hall, London.

10. PIMENTEL, D, C. WILSON, C. McCULLUM, R. HUANG, P. DWEN, J. FLACK, Q. TRAN, T. SALTMAN, and B. CLIFF. 1997. Environmental and economic benefits of biodiversity. *BioScience* 47:747–757.

11. MYERS, N. 1983. *A Wealth of Wild Species: Storehouse for Human Welfare.* Westview Press, Boulder, CO.

12. MILLS, J., and C. SERVHEEN. 1991. The Asian trade in bears and bear parts. World Wildlife Fund/Traffic USA, Washington, DC.

13. COOPER, A. 1996. Finding our bearings in the trade of American black bear (*Ursus americanus*) parts: Are we on a course for disaster? *Human Dimensions of Wildlife* 4:69–80.

14. SHEPARD, P., and B. SANDERS. 1985. *The Sacred Paw: The Bear in Nature, Myth, and Literature.* Viking Penguin, New York.

15. ULRICH, R. 1993. Biophilia, biophobia, and natural landscapes. Pages 73–137 *in* S. Kellert and E. Wilson, eds., *The Biophilia Hypothesis.* Island Press, Washington, DC.

16. OHMAN, A. 1986. Face the beast and fear the face: Animal and social fears of prototypes for evolutionary analyses of emotion. *Psychophysiology* 23:123–145.

17. SHEPARD, P. 1996. *The Others: How Animals Made Us Human.* Island Press, Washington, DC.

18. LAWRENCE, E. 1993. The sacred bee, the filthy pig, and the bat out of hell: Animal symbolism as cognitive biophilia. Pages 301–344 *in* S. Kellert and E. Wilson, eds. *The Biophilia Hypothesis.* Island Press, Washington, DC.

19. LUMSDEN, C., and E. O. WILSON. 1981. *Genes, Mind, and Culture.* Harvard University Press, Cambridge, MA.

20. KELLERT, S. R., M. BLACK, C. R. RUSH, and A. J. BATH. 1996. Human culture and large carnivore conservation in North America. *Conservation Biology* 10:977–990.

21. CURLEE, A. P., T. W. CLARK, D. CASEY, and R. P. READING. 1994. Large carnivore conservation: Back to the future. *Endangered Species Update* 11(3–4):1–4.

22. GRAY, G. C. 1993. *Wildlife and People: The Human Dimensions of Wildlife Ecology.* University of Illinois Press, Urbana, IL.

23. SPENCE, L. 1914. *The Myths of the North American Indians.* George C. Harrap and Company, London.

24. SCHULTZ, J. W. 1962. *Blackfeet and Buffalo: Memories of Life Among the Indians.* University of Oklahoma Press, Norman.

25. ERDOES, R., and A. ORTIZ, eds. 1984. *American Indian Myths and Legends.* Pantheon Books, New York.

26. ROCKWELL, D. 1991. *Giving Voice to Bear: North American Indian Rituals, Myths, and Images of the Bear.* Roberts Rinehart Publishers, Niwot, CO.

27. HUGHES, J. D. 1983. *American Indian Ecology.* Texas Western Press, El Paso, TX.

28. BROWN, J. E. 1982. *The Spiritual Legacy of the American Indian.* Crossroad Publishing Company, New York.

29. CLARK, T. W., and D. CASEY, eds. 1992. *Tales of the Grizzly: Thirty-Nine Stories of Grizzly Bear Encounters in the Wilderness.* Homestead Publishing, Moose, WY.

30. HUMMEL, M. and S. PETTIGREW. 1991. *Wild Hunters: Predators in Peril.* Robert Rinehart Publishers, Niwot, CO.

31. DOUGHTY, R. W. 1983. *Wildlife and Man in Texas: Environmental Change and Conservation.* Texas A&M University Press, College Station.

32. MATTHIESSEN, P. 1987. *Wildlife in America.* Viking Penguin, New York.

33. SHERWOOD, M. 1981. *Big Game in Alaska: A History of Wildlife and People.* Yale University Press, New Haven, CT.

34. LANGENAU, E. E., S. R. KELLERT, and J. E. APPLEGATE. 1984. Values in management. Pages 699–720 *in* L. K. Halls, ed., *White-Tailed Deer: Ecology and Management.* Stackpole Books, Harrisburg, PA.

35. BORLAND, H. G. 1975. *The History of Wildlife in America.* National Wildlife Federation, Washington, DC.

36. NASH, R. 1967. *Wilderness and the American Mind.* Yale University Press, New Haven, CT.

37. PRIEST, J. 1837. *Stories of Early Settlers in the Wilderness.* H.S. Stone and Company, New York.

38. YOUNG, S. P., and E. A. GOLDMAN. 1944. *The Wolves of North America.* The American Wildlife Institute, Washington, DC.

39. DAVIS, G. P. 1982. Man and wildlife in Arizona: The American exploration period: 1824–1865. N. B. Carmony and D. E. Brown, eds. Arizona Game and Fish Department, Scottsdale. See also Coues, E. 1867. The quadrupeds of Arizona. *American Naturalist* 1: (August, September, October, December): 281–292, 351–363, 393–400, 531–541.

40. ARTHUR, D., and J. L. HOLECHEK. 1982. The North American bison. *Rangelands* 4:123–125.

41. HASTINGS, L. W. 1932. *The Emigrants' Guide to Oregon and California.* Princeton, NJ.

42. YOUNG, S. P. 1946. *The Wolf in North American History.* The Caxton Printers, Caldwell, ID.

43. DUNLAP, R. T. 1988. *Saving America's Wildlife.* Princeton University Press, Princeton, NJ.

44. LOPEZ, B. 1978. *Of Wolves and Men.* Charles Scribner's and Sons, New York.

45. HORNOCKER, M. G. 1992. Learning to live with mountain lions. *National Geographic* 182:52–56.

46. LEOPOLD, A. 1933. *Game Management.* The University of Wisconsin Press, Madison, WI.

47. LEOPOLD, A. 1949. *A Sand County Almanac.* Charles W. Schwartz, Oxford University Press, New York.

48. SHAW, W. W. 1987. The recreational benefits of wildlife to people. Pages 208–213 *in* D. J. Decker and G. R. Goff, eds., *Valuing Wildlife: Economic and Social Perspectives.* Westview Press, Boulder, CO.

49. KELLERT, S. R. 1987. The contributions of wildlife to human quality of life. Pages 222–229 *in* D. J. Decker and G. R. Goff, eds. *Valuing Wildlife: Economic and Social Perspectives.* Westview Press, Boulder, CO.

50. APPLEGATE, J. E. 1984. Attitudes towards deer hunting in New Jersey: 1972–1982. *Wildlife Society Bulletin* 12:19–22.

51. HEBERLEIN, T. A. 1992. Hunters predicted to be extinct by year 2050. *Milwaukee Sentinel,* January 2, p. 7A.

52. HENDEE, J. C. 1974. A multiple-satisfaction approach to game management. *Wildlife Society Bulletin* 2:104–113.

53. KELLERT, S. R. 1978. Attitudes and characteristics of hunters and anti-hunters and related policy suggestions. Speech delivered at Hunter Safety Conference, Charleston, SC, January 24, 1978.

54. DECKER, D. J., R. W. PROVENCHER, and T. L. BROWN. 1984. Antecedents to hunting participation: An exploratory study of the social-psychological determinants of initiation, continuation, and desertion in hunting. Outdoor Recreation Research Unit Publication 84–6. Department of Natural Resources, New York State College Agriculture and Life Sciences, Cornell University, Ithaca, NY.

55. KELLERT, S. R. 1980. Public attitudes toward critical wildlife and natural habitat issues. Phase I of U.S. Fish and Wildlife Service Study, Government Printing Office #024-010-00-623-4, Washington, DC.

56. APPLEGATE, J. E., and R. A. OTTO. 1982. Characteristics of first year hunters in New Jersey. Publication R-12381-(1)-82. New Jersey Agricultural Experiment Station.

57. DECKER, D. J., T. L. BROWN, and J. W. ENCK. 1992. Factors affecting the recruitment and retention of hunters: Insights from New York. Pages 670–677 in S. Csanyi and Ernhaft, eds., *Proceedings Twentieth Congress of the International Union of Game Biologists.* University of Agricultural Science, Godollo, Hungary.

58. *USA TODAY.* 1992. Hunting: a closer look. July 16, p. 12C.

59. REGAN, T., and P. SINGER, eds. 1976. *Animal Rights and Human Obligations.* Prentice Hall, Englewood Cliffs, NJ.

60. KRUTCH, J. W. 1957. The sportsman or the predator? A damnable pleasure. *Saturday Review,* August 19, p. 7.

61. AMORY, C. 1974. *Mankind? Our Incredible War on Wildlife.* Harper and Row, New York.

62. SINGER, P. 1975. *Animal Liberation: A New Ethics for Our Treatment of Animals.* New York Review, New York.

63. KLEIN, D. R. 1973. The ethics of hunting and the anti-hunter movement. *Transactions of the North American Wildlife and Natural Resources Conference* 38:256–266.

64. CALLICOTT, J. B. 1987. The philosophical value of wildlife. Pages 214–221 in D. J. Decker and G. R. Goff, eds., *Valuing Wildlife: Economic and Social Perspectives.* Westview Press, Boulder, CO.

65. BISHOP, R. C. 1987. Economic values defined. Pages 24–33 in D. J. Decker and G. R. Goff, eds., *Valuing Wildlife: Economic and Social Perspectives.* Westview Press, Boulder, CO.

66. BROWN, P. J., G. H. HAUS, and B. L. DRIVER. 1980. Value of wildlife to wilderness users. *Proceedings of the Second Conference on Scientific Research in the National Parks* 6:168–179.

67. MECH, L. D. 1981. *The Wolf: The Ecology and Behavior of an Endangered Species.* Doubleday, New York.

68. STANDAGE ACCUREACH, INC. 1990. Watchable wildlife recreation in Colorado. Report prepared for Colorado Division of Wildlife, Standage Accureach, Denver, CO.

69. FLEWELLING, B., and D. JOHNSON. 1981. Wildlife experience and trip satisfaction in the backcountry. A preliminary summary of Denali backcountry survey data. University of Washington Cooperative Park Studies Unit, Seattle.

70. ADAMS, C. E., and J. K. THOMAS. 1989. The public and Texas wildlife. Texas Agricultural Experiment Station, Texas A&M University, College Station.

71. HASTINGS, B. C. 1986. Wildlife-related perceptions of visitors in Cades Cove, Great Smoky Mountains National Park. Dissertation, University of Tennessee, Knoxville.

72. LITTLEJOHN, M. 1991. Visitor services project in Glacier National Park. Visitor Services Project Report 35. Cooperative Park Studies Unit, University of Idaho, Moscow.

73. DUDA, M. D., and K. C. YOUNG. 1994. Americans and wildlife diversity: Public opinion, attitudes, interest, and participation in wildlife viewing and wildlife diversity programs. Federal Aid in Sport Fish and Wildlife Restoration Grant Report. U.S. Fish and Wildlife Service, Harrisonburg, VA.

74. MORE, T. A. 1977. An analysis of wildlife in children's stories. in Children, nature, and the urban environment: Proceedings of a symposium-fair. Pages 89–94 in U.S. Forest Service General Technical Report. No. N-E-30.

75. KELLERT, S. R., and M. D. WESTERVELT. 1983. Phase V. Children's attitudes, knowledge, and behaviors towards animals. U.S. Fish and Wildlife Service Report, Washington, DC.

76. HAIR, J. D., and G. A. POMERANTZ. 1987. The educational value of wildlife. Pages 197–207 *in* D. J. Decker and G. R. Goff, eds. *Valuing Wildlife: Economic and Social Perspectives.* Westview Press, Boulder, CO.

77. POMERANTZ, G. A. 1977. Young people's attitudes towards wildlife. Report Number 2781. Michigan Department of Natural Resources.

78. GILBERT, F. F. 1982. Public attitudes towards urban wildlife: A pilot study in Guelph, Ontario. *Wildlife Society Bulletin* 10:245–253.

79. WONG-LEONARD, C. J., and R. B. PEYTON. 1992. Effects of wildlife cartoons on children's perceptions of wildlife and their use of conservation education material. *Transactions of the North American Wildlife and Natural Resources Conference* 57:197–206.

80. POMERANTZ, G. A. 1985. The influence of *Ranger Rick* magazine on children's perceptions of natural resource issues. Dissertation. North Carolina State University, Raleigh.

81. KEISER, J. E., and J. E. APPLEGATE. 1986. Princeton township: the history of a no-discharge ordinance's effect on deer and people. *Transactions Northeast Section.* The Wildlife Society, Hartford, CT.

82. KELLERT, S. R. 1993. Public views of deer management. *In* R. Donald, ed., *Deer Management in an Urbanizing Region: Problems and Alternatives to Traditional Management.* Humane Society of the United States, Washington, DC.

83. CURTIS, P. D., R. J. STOUT, B. A. KNUTH, L. A. MYERS, and T. M. ROCKWELL. 1993. Selecting deer management options in a suburban environment: A case study from New York.

84. PARKHURST, J. A., and R. W. O'CONNOR. 1992. The Quabbin Reservation white-tailed deer impact management plan: A case history. *Proceedings of the Eastern Wildlife Damage Control Conference* 5:173–181.

85. LITWIN, T., T. GAVIN, and M. CAPKANIS. 1987. White-tailed deer in a suburban environment: Reconciling wildlife management and human perceptions. Pages 366–370 *in* D. Decker and G. Goff, eds., *Valuing Wildlife: Economic and Social Perspectives.* Westview Press, Boulder, CO.

86. DECKER, D., and T. GAVIN. 1987. Public attitudes toward a suburban deer herd. *Wildlife Society Bulletin* 15:173–180.

87. BELLANTINI, E. S., P. R. KRAUSMAN, and W. W. SHAW. 1993. Desert mule deer use of an urban environment. *Transactions 58th North American Wildlife and Natural Resources Conference* 58:92–101.

88. CONNELLY, N., D. DECKER, and S. WEAR. 1987. Public tolerance of deer in a suburban environment. *Proceedings of the Eastern Wildlife Damage Control Conference* 3:207–218.

89. SAYRE, R., and D. DECKER. 1990. Deer damage to the ornamental horticultural industry in suburban New York: Extent, nature and economic impact. Human Dimensions Research Unit Series 90-1. New York State College of Agriculture and Life Sciences, Cornell University, Ithaca, NY.

90. KANIA, G. and M. CONOVER. 1991. How government agencies should respond to local governments that pass antihunting legislation—A response. *Wildlife Society Bulletin* 19:224–225.

91. STOUT, R., J. DECKER, and B. KNUTH. 1992. Agency and stakeholder evaluations of citizen participation in deer management decisions: Implications for damage control. *Proceedings of the Eastern Wildlife Control Conference* 5:142.

92. GRAND TETON NATIONAL PARK, NATIONAL ELK REFUGE, WYOMING GAME AND FISH DEPARTMENT, NATIONAL WILDLIFE HEALTH CENTER, and BRIDGER TETON NATIONAL FOREST. 1994. The Jackson bison herd: Long term management plan and environmental assessment (draft). Grand Teton National Park, Moose, WY.

93. THORNE, E. T., M. MEAGHER, and R. HILLMAN. 1991. Brucellosis in free ranging bison: Three perspectives. Pages 275–287 *in* R. B. Keiter and M. S. Boyce, eds., *The Greater Yellowstone Ecosystem: Redefining America's Wilderness Heritage.* Yale University Press, New Haven, CT.

94. MEYER, M., and M. MEAGHER. 1995. Brucellosis in free-ranging bison (*Bison bison*) in Yellowstone, Grand Teton, and Wood Buffalo National Parks: A review. *Journal of Wildlife Diseases* 31:579–598

95. THUERBER, A. M. 1995. Cattlemen don't mind bison but fear ruin. *Jackson Hole News,* May 31, p. A18.

96. SHELTON, C. 1995. Politics guides bison slaughter. *Jackson Hole Guide,* May 31.

97. STELLER. B. 1995. Bison slaughter unnecessary and violates public trust. *Jackson Hole Guide,* June 7.

98. THUERBER, A. M. 1994. Bison plain is criticized. *Jackson Hole News,* December 7, p. A1.

99. *JACKSON HOLE NEWS.* 1995. Shoshone will fast for tribes' right to bison. May 31.

100. KEITER, R. B., and P. FROELICHER. 1993. Bison, brucellosis and law in the Greater Yellowstone Ecosystem. *Land and Water Law Review* 28:1–75.

101. SCHULLERY, P. 1986. Drawing the lines in Yellowstone: The American bison as symbol and scourge. *Orion Nature Quarterly* 5:33–45.

102. CLARK, T. W., and S. C. MINTA. 1994. *Greater Yellowstone's Future: Prospects for Ecosystem Science, Management and Policy.* Homestead Publishing, Moose, WY.

103. REITER, R. 1997. Greater Yellowstone's bison: Unraveling of an American wildlife conservation achievement. *Journal of Wildlife Management* 61:1–11.

104. BATH, A. 1991. Public attitudes in Wyoming, Montana, and Idaho toward wolf restoration in Yellowstone national park. *Transactions of the North American Wildlife and Natural Resources Conference* 56:91–95.

105. DEVLIN, S. 1994. Wolf provokes inadvertent howlers. *High Country News,* September 19, p. 13.

106. KELLERT, S. R., and T. CLARK. 1991. The theory and application of a wildlife policy framework. Pages 36–75 *in* W. Mangun and S. Nagel, eds., *Public Policy and Wildlife Conservation.* Greenwood Press, New York.

107. LASSWELL, H. D. 1971. *A Pre-View of the Policy Sciences.* American Elsevier, New York.

108. CLARK, T. W. 1992. Practicing natural resource management with a policy orientation. *Environmental Management* 16:423–433.

109. CLARK, T. W., and R. WESTRUM. 1989. High-performance teams in wildlife conservation: A species reintroduction and recovery example. *Environmental Management* 13:663–670.

110. CLARK, T. W., and J. R. CRAGUN. 1991. Organization and management of endangered species programs. *Endangered Species Update* 8:1–4.

CHAPTER 4

Population Parameters and Their Estimation

Richard A. Lancia, Christopher S. Rosenberry, and Mark C. Conner

INTRODUCTION

Estimating population size and demographic rates for large mammals can be expensive and time consuming, so how is this effort justified? The obvious answer is that the effectiveness of management cannot be assessed without reliable data on population parameters. Although some managers might claim that seat-of-the-pants management is justifiable because accurate population data are too expensive and time consuming to acquire, the public, including hunters and antihunters, will no longer accept this line of reasoning. There is simply too much at stake. Managers must be able to defend management plans and activities with corroborating evidence of population responses.

We further contend that the process of population management should simultaneously seek the dual goals of achieving a desired population response and learning more about how populations function. This dual strategy requires implementing management with underlying explanations for population responses explicitly stated as testable hypotheses. In other words, management should be implemented in a context where predicted population responses can be rigorously evaluated, and new understanding can be gained, so management will improve with successive iterations. This adaptive management paradigm should guide management in the decades to come (1).

This chapter presents a basic overview of population parameters relevant to management of large mammal populations and briefly introduces procedures that can be used to estimate population size and fundamental demographic rates. Literature on this topic is voluminous. For entrée into the literature, the reader should consult more thorough synopses such as Seber, Pollock and others, White and Garrott, Johnson, and Lancia and others (2, 3, 4, 5, 6).

PARAMETERS OF INTEREST

Population Size and Structure

Abundance and Density. Abundance or population size is the number of individual animals in an area of interest defined by the manager or other interested person. Whereas boundaries of a management unit are usually sharp, population boundaries are "fuzzy" because animals move in and out around the periphery (Figure 4–1).

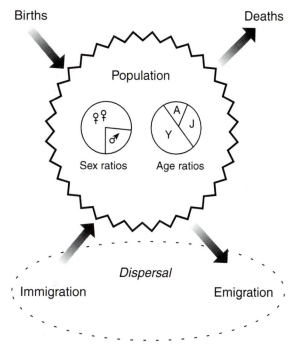

Figure 4–1 Sex and age ratios comprise a population with "fuzzy" boundaries. Births, deaths, immigration, and emigration result in changes in population size.

The degree to which this movement zone affects the identity of the population is a function of spatial and temporal scales, and mobility of animals. The smaller the area is and the more mobile the animals are, the fuzzier the boundary.

Defining a population and, hence, population size, should but does not necessarily have to encompass behavioral, ecological, and genetic dimensions. Obviously, an understanding of how a population functions in these contexts is crucial to the success of population management plans.

Population density is the number of individual animals per unit area. Estimating density requires spatial measurements such as distance or area, or use of radiotelemetry to document movements in addition to estimates of abundance derived from counts or captures (7). This makes density more time consuming and difficult to estimate than abundance. In most management applications estimates of abundance will suffice.

Sex and Age Ratios. Sex ratio is simply the proportion of each sex in a population. It can be expressed in a variety of ways, the most common being the number of males:females (or males:100 females) or the percentage of a population that is males or females. Analogously, age structure is the proportion of the population in each age class. Age classes can be as general as immature and adult or as specific as proportions among many year classes. Finally, sex and age ratios can be combined in sex-specific age ratios. The degree to which sex and age structure is discriminated depends on the biology of the species and management needs. These ratios are often used in management to assess the effects of hunting regulations and to provide a measure of young produced; however, surveys that yield accurate estimates can be difficult to accomplish, expensive, and time consuming (8, 9, 10).

Demographic Rates

Birth rates are usually defined for management purposes as births per individual (sometimes births per female) during a year at a given point in the annual cycle. For birth pulse populations, in which births occur during one well-defined and relatively short period, this point is typically immediately before onset of the birth pulse. For consistency, death rate would be measured at the same point in time. As an aside, when modeling populations this point is termed the post-harvest or prerecruitment population. This is the population on which sustained yield models are based (11, 12).

Birth Rates.   Potential and realized birth rates are a function of the species of interest and its environment. Both of these rates have management applications. Potential reproduction reflects the maximum ability of a population to reproduce and varies directly with environmental quality such as soil fertility, availability of forage or prey, and with sex and age structure of a population. Potential reproduction can influence large-scale management decisions because concentrating scarce management dollars in areas with the highest potential productivity is likely to achieve the best success.

Realized birth rate is the actual number of births or young at a given point in the annual cycle. The rate is obviously higher the earlier it is measured in the reproductive cycle because mortality can occur at any time. For example, conception rates (i.e., number of fetuses:female) would be higher than fertility rates (i.e., live births:female). The chosen point is related to management needs. If realized births seem too low, then management might assess birth rates as early as possible, and at several times in the reproductive cycle, to identify potential problems.

Death Rates.   Mortality rate is the number of deaths over a time period (defined per year above) relative to the number of individuals at the beginning of that time period. Sometimes mortality rate is over shorter periods such as seasons (e.g., hunting mortality), months, or days depending on management needs. Survival rate is the complement of mortality (i.e., $1.0 -$ mortality rate).

Dispersal, Emigration, and Immigration.   Dispersal is the movement of animals from their birth (natal) ranges to locations where they reproduce or would reproduce had they survived. For a population, dispersal represents the only means of exchange of individuals with surrounding areas. Within the general context of dispersal, emigration and immigration can be defined at the scale of the population or individual. At the population scale they are respectively defined as permanent movement out of or into a population of interest. These movements are frequently ignored in management because they are difficult to measure, yet they can have significant influences on a population, particularly when average dispersal distance is large relative to the area inhabited by the population of interest. According to Johnson, "A biologist carrying out a population analysis typically ignores dispersal, assumes it to be nonexistent, or blithely hopes that immigration and emigration cancel one another" (5, p. 441). Unless a management area is very large relative to average dispersal distance, ignoring dispersal could yield poor results in an adaptive management paradigm (13).

Whether an individual has emigrated is a function of dispersal distance and the size of the area occupied by the population. Only individuals around the borders of the population emigrate with respect to the population; the rest move around within the population. This movement and temporary wanderings create the "fuzzy" population boundary noted earlier.

At the scale of the individual, definitions of emigration and immigration are based solely on their relationship to known points. Emigration is an animal's permanent movement *from a known origin.* Immigration, on the other hand, is an animal's permanent movement *to a known destination.* Whether movement of an animal is termed emigration or immigration depends on whether the origin or destination is known. If the natal range is known, dispersal distance can be measured as the distance between the geometric center of the natal range and the point where the animal dies or reproduces (13).

Movements into and out of a population affect the composition of animals remaining after dispersal is completed. Rosenberry showed about 50% of the post-dispersal population of yearling male white-tailed deer on a 1,300-hectare farm in Maryland was immigrants

(13). He pointed out that assessing management efficacy for these males, based on characteristics of individuals in the post-dispersal herd, could be very misleading because half the deer in the assessment population were not subject to management on the farm.

Demographic Synthesis—Rate of Population Growth

Realized.   Populations change in size as a result of four fundamental demographic processes: births, deaths, immigration, and emigration (5). The change in population (lambda, λ) between time intervals (years in this example) is the ratio of the size of the population next year ($t + 1$) to the size this year (t), or

$$\lambda = N_{t+1}/N_t \ .$$

Lambda is called the finite rate of population increase. For convenience lambda is often expressed as an exponent, or

$$\lambda = e^r \ ,$$

where r is the exponential rate of increase and e is the base of natural logarithms. Both measure the same growth expressed in two different ways. This realized rate of increase reflects the net effect of each of the four demographic rates operating simultaneously. Realized growth rate, which would vary between time intervals, reflects population responses to management.

Intrinsic.   To predict population responses to management, an underlying model of population growth is needed. No population can continue to grow at the same rate forever. Beginning from a small founder population in favorable habitat, it is likely that per capita growth will slow as the population gets larger, and there are fewer resources per individual. Everything else remaining the same, growth could eventually level off when the population reaches a steady-state size where growth is zero. If a population follows this pattern of growth (or this is an underlying pattern that might not be explicitly expressed in observed growth), it can exhibit "S-shaped" or sigmoid growth in population size over time. One pattern of sigmoid growth is logistic growth.

If we continue the logic that as the population gets larger and larger, per capita growth continuously declines, then the highest per capita growth would occur when population size is smallest (Figure 4–2). This maximum per capita rate is termed the maximum intrinsic rate of increase (r_m).

The maximum intrinsic rate for a population reflects the inherent proclivity for reproduction of a species and the quality of the environment in which the population resides. The former is relatively fixed for each species, whereas the latter is of more interest to management because it is a measure of the ability of a given environment to

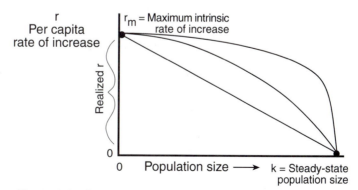

Figure 4–2   As population size increases, per capita rate of increase decreases from a maximum r_m to zero when the population size reaches a steady state.

produce large mammals. It is also a measure of the maximum potentially sustainable per capita harvest rate. Thus, populations with higher maximum rates of increase can support higher sustainable harvest rates. In theory, a population can support a sustainable harvest and remain constant in size when per capita growth equals per capita harvest.

ESTIMATES OF POPULATION SIZE

This section presents a basic conceptual framework for population size estimates appropriate for large mammals. Categorizing methods based on similarities is stressed (Figure 4–3), rather than simply giving a list of techniques without any underlying organization. Most of this section is derived from an earlier work by Lancia and others (6).

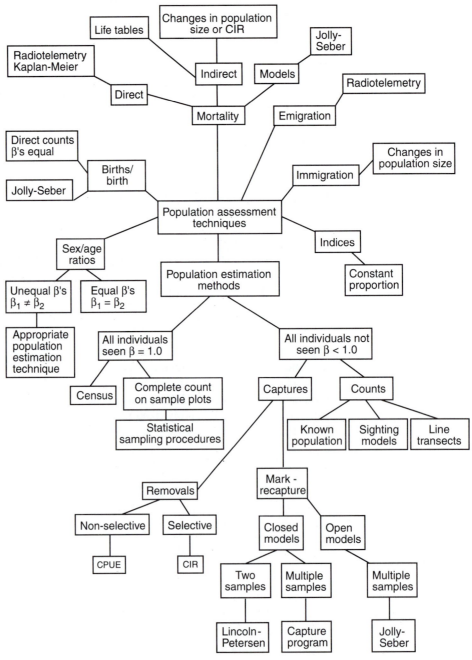

Figure 4–3 Diagramatic relationships among population assessment techniques.

Closed and Open Populations

A basic assumption that underlies population estimation procedures is whether a population is closed or open during the period when a population estimation technique is applied. When closed, a population does not change in size or composition while the estimation procedure is being applied. Closure has two components: demographic and geographic. Demographic closure means there are no births or deaths; geographic closure means there are no movements into or out of the population. Hence, only population size or densities are estimated when closed models are used.

Closed populations are covered in this section. Open models are discussed in the following section on estimates of demographic parameters because open models also estimate number of births and survival rates. The distinction between open and closed simplifies modeling, but in application the biological situation in which the procedures are applied must dictate which approach to use. Both open and closed models can be used together as in Pollock's robust design; however, this design requires more effort than may be available in most applications to big game management (14).

Open models permit increases or decreases in population size between sampling periods. However, in many cases additional information about the phenology of reproductive behavior must be known to separate births from immigration. Separating deaths from emigration is more difficult because in both cases animals usually disappear, never to be found again. Knowledge of emigration behavior can help distinguish between them.

Observability and Sampling

Two basic problems, observability and sampling, confront a manager who is trying to estimate population size. Observability problems arise from not being able to see or capture all animals within an area of interest. Sampling problems arise from not being able to survey an entire area of interest at one time, so representative samples must be taken. Population estimation procedures attempt to deal with limitations associated with one or both of these problems. This chapter principally addresses observability concerns; for statistical sampling theory consult a basic text or statistician (15).

When surveying animals, not all individuals in the population are seen or captured. In other words, the probability (β) of sighting or capturing an animal is less than 1. We can write the relationship between a count (C), which is the number of animals seen or captured, and the true population size (N) as:

$$C = \beta N .$$

Estimating the size of the population is simply a matter of dividing the count by an estimated observability, such that,

$$N = C/\beta .$$

Most population estimation procedures are clever attempts to estimate observability (β).

In many applications an entire area of interest cannot be surveyed because time and money are limited. Therefore, some subareas (samples), thought to represent the entire area, are randomly selected and surveyed. Intuitively, if 10% of an area, which correctly represents the entire area, is surveyed and all animals are seen or captured ($\beta = 1.0$), then the population in the entire area is 10 times larger than the population in the sample area. If C_s represents the count of animals on the sample area, then

$$N = C_s/\propto ,$$

where $\propto$ is the proportion of the entire area that is sampled.

These two concerns, observability and sampling, can be combined into one fundamental relationship

$$N = C_s/\propto \beta .$$

This basic relationship underlies most techniques to estimate population size.

Throughout this chapter we will be concerned primarily with population size rather than population density because harvests and harvesting models usually work with totals. If the area is known, which is a significant assumption, then converting between the two is easy.

Abundance and Density

All Individuals Observed.

Census. If all the animals in an area of interest can be counted, then both $\beta = 1$ and $\propto = 1$; and $N = C$, or the count is the population. A census is a complete count, but rarely is this possible. Some methods such as drive counts or aerial surveys by people, cameras, or thermal scanners purport to be a census, but rarely are all individuals counted. For example, McCullough used drive counts to census fenced-in white-tailed deer on the George Reserve, Michigan, and found errors were as much as 20% to 30% (11). Wolfe and Kimball compared a total count of bison from a roundup to an aerial "census" (16). They found aerial counts of adults were about 94% accurate, but counts of calves and adult males were low. However unlikely it is that all individuals are counted, the "census" might be accurate enough to assume a complete count for management purposes. In contrast, Jacobson and others used automatic cameras at bait sites to photograph essentially all marked and branch-antlered deer on a farm in Mississippi (17).

Thermal infrared sensors are another method for counting animals. Despite sensor improvements that permit distinguishing age and sex classes of animals, thermal sensors do not ensure a complete count in the survey area, although the technique continues to hold promise (18, 19, 20, 21, 22, 23). Therefore, sensor counts are likely to provide incomplete counts on sample areas and should be treated as discussed in the following sections. Problems associated with infrared sensing include detecting animals under foliage canopy cover, species identification, interpretation of sensor data, adherence to a defined sampling (flight) plan, and cost effectiveness.

If a census is possible on sample plots, then estimating the population is only a sampling problem involving representative samples and sampling error. Basic statistical sampling theory can be applied (15). Kufeld and others used total counts on sample areas to estimate the wintering population of mule deer on the Uncompahgre Plateau of Colorado (24). They "censused" plots by observing deer with a helicopter and thought the population estimate for the entire area was within 20% of the true population with 90% confidence, which was accurate enough for management purposes.

All Individuals Not Observed.

Counts. An incomplete count occurs when all individuals present in an area of interest cannot be seen and therefore are not counted. In this case $\beta < 1.0$, and some additional field and mathematical procedures are necessary to estimate what proportion of the true population is being counted. Several general methods are applicable to management of large mammals.

Double sampling is a statistical technique that requires a known population which is marked by one method and then sampled with a counting technique applicable to a particular management situation. If possible, a census could be the known population or it could be a known number of marked individuals. In essence, the observers use an appropriate observational technique to count the population, and then compare it to the known population to estimate observability (β) for their counting technique. In experiments of this sort, from about 30% to nearly 100% can be counted depending on conditions and habitats (10, 25, 26, 27, 28, 29). Sightability models, described later, use a similar approach in that a known population is used to estimate sightability.

An example of a known population that is used to estimate observability would be a radiomarked population for which all marked individuals were triangulated in the area where counting was taking place (4, 7). If the objective is to estimate this population from aerial surveys or roadside observations, then let

N = the total population size on the survey area,

n_1 = the number of radiomarked animals known to be in the survey area,

n_2 = the number of marked and unmarked animals counted during the aerial or roadside survey, and

m = the number of marked and unmarked animals seen during the aerial or roadside survey.

Intuitively, observability (β) would be the ratio of the number of marked animals seen (m) to the number of marked animals (n_1) known to be in the area and hence potentially observable, or

$$\beta = m/n_1 \ .$$

Following the general model presented earlier, the population estimate would simply be the number of animals counted divided by their observability, or

$$N = n_2/\beta \ .$$

In this case an estimate of observability is based on whether animals known to be present are actually observed. An implicit, simplifying assumption is that all the animals are equally sightable, which in practical application is rarely true.

Substituting $\beta = m/n_1$ in the above equations yields the Lincoln–Petersen estimator ($N = n_1 n_2/m$) developed from mark–recapture methods, which are discussed later in this chapter (2, 3). The marked animals are the known number of radiomarked individuals, and the "recaptures" are actually resightings and not actual recaptures.

Another approach is to develop a statistical model to predict whether an animal would be observed based on factors thought to affect observability (30, 31, 32). For example, being in a group of animals could increase the probability that animals were observed during a survey, or dense vegetation could decrease observability.

Again, a known population is required to develop these sightability models. For example, Conner used two observers to simultaneously observe deer feeding in fields (31). One observer was in a blind and the other drove a vehicle in a simulated roadside survey. Each observer counted the size and number of groups seen in fields. Deer counted by the observer in the blind were considered the known population. These data yielded the probability of observing deer as a function of group size. To illustrate, Conner estimated the probability of observing a single deer to be about 0.77 (31). In application, the number of groups would be divided by their respective observabilities to estimate the number of groups of each size actually present in the surveyed fields. The total for all fields would be estimated from the sample proportion α. Using the same rationale and employing logistic regression, Samuel and others developed sightability models for elk groups in Idaho, Anderson and Lindzey for moose in Wyoming, and Anderson and others for summer elk surveys (32, 33, 34). Steinhorst and Samuel and Cogan and Dieffenbach presented methods for sightability adjustments in aerial surveys, and Eberhardt and others compared sighting models with Petersen estimates (35, 36, 37).

Jacobson and others employed automatic cameras to estimate sightability (17). They used cameras to identify individual branch-antlered white-tailed deer and to record fawns, females, and spikes. Nearly 100% of the branch-antlered males known to be in the area were photographed. Therefore, Jacobson and others used this essentially known class of animals to determine total numbers of males, females, and fawns based on the ratio of the known class to all other classes (17). Although the initial expense of the cameras was high, the cost amortized over 5 years was less than $1.30/hectare/year.

The line transect method, particularly aerial transects, is another counting method applicable to big game population estimation. In this method lines of known length are set out within an area, according to appropriate statistical sampling theory, and an observer counts all animals seen while traversing these lines. If the transects have a fixed width within which all the animals are counted (i.e., a strip transect or strip census) then it is simply a complete count on a sample area.

If the counts are incomplete, then β must be estimated. Either measurements of perpendicular sighting distances, or angles and sighting distances to the observed animals, are required to estimate β. Intuitively, the probability of sighting animals should decline as sighting distance from the transect line increases. Thus, a sighting function that relates sighting probability to distance from the transect line must be developed. Several fundamental assumptions are required: (1) all animals on the line are detected (i.e., β = 1.0 for animals directly on the transect line), (2) animals do not move before they are detected and are not counted twice, (3) distances (and/or angles) are measured accurately, and (4) sightings are independent events. The mathematics for developing sighting functions are beyond the scope of this chapter, but suffice it to say that density is estimated directly, rather than population size, so that transects need not have a predetermined width or area (38, p. 14).

In application to large mammals, behavior and sightability of species and the habitats they inhabit must be taken into consideration. Frequently animals occur in groups, and density of groups can be estimated (39). Then group sizes must also be recorded or estimated to determine population size or density. Guidelines for line transect sampling can be found in Anderson and others and Burnham and others (38, 40).

Aerial surveys are common practice in assessment of large-mammal populations, particularly in the West and in open areas with snow cover where animals tend to be easier to observe from the air than animals in dense evergreen forests. For example, counting moose from aircraft on their winter ranges is the most common method of estimating moose populations in North America (10, p. 35). Observabilities have been determined by numerous investigators and range from about 0.97 to 0.4 (10, Table 1). If properly conducted by standardizing conditions and stratifying sampling, aerial moose surveys can be highly precise (i.e., repeatable with narrow confidence limits), but accuracy depends on how well observability is estimated.

Removals. For large mammals, removal methods are appealing because animals removed by the harvest, in conjunction with other data about the population, are part of the estimation procedure. This often makes removal methods less expensive to implement. Usually harvests alone are only a crude reflection of population size and trends because without additional information the proportion removed is not reliably known and can vary substantially.

Removal methods are categorized by whether the removals are selective— (i.e., change-in-ratio techniques [CIR]) or nonselective (i.e., catch-per-unit-effort [CPUE]). Removals are selective when types of animals, such as sexes, age classes, or sex-age classes, are removed disproportionate to their representation in the population. Removals must be selective for CIR to work.

Change-in-ratio equations are essentially algebraic solutions to simultaneous equations relating proportions of classes of animals before and after a removal, in this case hunting, to population sizes before and after harvest. This single removal, or single-stage CIR, can be expanded to multiple stages but these are variations on a theme and are beyond the scope of this chapter (41, 42).

To apply CIR, classes of animals that are relevant to management, such as antlered and antlerless deer, or males and females, or adults and juveniles, must be readily identifiable by observations in the field. Proportions of these classes in the whole population are based on field observations, such as 0.5 (50%) and 0.2 (20%) antlered deer in the population before and after removals, respectively. In addition, removals of both classes from the population must also be known or estimated accurately. If these conditions are met, then population estimates are simply solutions

Estimates of Population Size

of algebraic equations that relate proportions and removals to population size before the removals. For CIR to yield reasonably accurate results, changes in the proportions resulting from the removal must be large (43). Relative to other estimation methods, Rosenberry and Woolf found that CIR yielded the least satisfactory results when evaluated against a "reference" population, which they assumed was a known population (44).

Recall our earlier discussion that all population estimation techniques are imaginative attempts to estimate observability. In the simplest CIR this problem is circumvented by assuming observabilities of the two classes; in the above example antlered and antlerless deer are equal. Observability, which appears in both the numerator and denominator, cancels out of the equation to estimate population size. If only one type is removed, such as antlered deer, observability of the two types need not be equal; but in this case only the number of antlered animals can be estimated. Total population cannot be estimated because the relationship between the observabilities of the two classes is not known. More elaborate multiple-stage and multiple-class CIRs have been developed (42).

Change-per-unit-effort has been applied routinely in fisheries for decades, but not generally for large mammals (45). The CPUE approach is based on the premise that as more and more animals are removed from a population, CPUE should decline and approach zero when nearly all animals are removed. In essence CPUE methods estimate the total removal, which is equivalent to the initial population size, when the expected catch is zero.

Three types of data, observed over a number of time intervals such as days during a short hunting season (so as not to violate the closure assumption), are required to use CPUE: (1) units of effort (e.g., hunter-days or observation-hours), (2) number caught or "catch," and (3) number removed. The terms *caught* and *catch* are misnomers and need clarification. *Catch* relates to the type of effort expended and not necessarily to the animals removed. For example, if effort is deer observed per hour, then catch is the number of deer seen; if effort is deer killed per day, then catch is the number harvested. Bishir and Lancia, in an overview of CPUE, present a model that uses both sight and kill data together as catch (46). Lancia and others give examples of field applications with white-tailed deer and wild hogs (47). Roseberry and Woolf found CPUE closely monitored trends with a minimum of data (44).

Captures. The basic rationale behind mark–recapture is to determine observability of marked animals and apply it to the entire population. In its simplest form all animals are assumed to be equally observable, and there is only one marking occasion and one recapture occasion (i.e., Lincoln–Petersen). In the traditional application, animals would be captured by some appropriate trapping technique, marked, and released to be possibly captured again. For large mammals this is usually not feasible because capturing enough animals to yield a reasonable estimate is often too expensive and time consuming.

As an alternative, animals can be "captured" if they have body characteristics that allow individuals to be recognized by sight and recaptured by being reobserved. For example, Rosenberry "marked" individual white-tailed males in autumn by carefully observing antler configuration and drawing reference sketches in a field book (13). After an initial marking period, observations of marked and unmarked deer were used to estimate the population of antlered males. Automatic cameras, which are triggered when an animal approaches, are another means for marking and resighting animals (17). Finally, NOREMARK is a computer program that computes estimates of population size with a known number of marked animals and one or more resighting occasions (48). This model and others developed by researchers at Colorado State University are at the web site www.cwr.colostate.edu/~gwhite.

Many mark–recapture models that allow unequal catchabilities are available in the CAPTURE computer program (Colorado Cooperative Fish and Wildlife Research

Unit, Ft. Collins). Models most applicable to large mammals allow capture probabilities to vary over time and among individuals, or a combination of the two. To implement these more complex models, a capture history of each marked animal must be developed. Therefore, each animal must be marked with a unique identifier, and recaptures must occur over repeated recapture occasions.

Capture probabilities can vary significantly among individuals as was shown by Lancia and others for radiomarked white-tailed deer (49). This heterogeneity often results in overestimates of capture probabilities because individuals with high observabilities are overrepresented in observations. Overestimates of observability lead to underestimates of population size (49).

Constant Proportion Population Index

By definition a population index is inexpensive and simple to apply and is thought to be related to population size. An index is used to make comparisons between the same population at different times or between different populations at the same time. To make these comparisons valid, the relationship between the index and population must be the same for the populations being compared. Indices are used because they may be the only feasible option in a world of limited budgets and time. However, if the index misrepresents the population, then it is not cost effective. Misleading information is worthless, even if inexpensive.

For example, harvests are often used as an index of population size, following the rationale that if harvests are continuously going up, then the population must be increasing also. For this to be true, the same proportion of the population must be harvested each year, or

$$H_1 = \beta_1 N_1 \qquad \text{and} \qquad H_2 = \beta_2 N_2 \; ,$$

where the subscripts indicate the years or areas being compared, H is the harvest, β is the proportion harvested, and N is population size. To compare the population at time 2 to the population at time 1 to see if it is increasing or decreasing, then

$$\frac{H_2 = \beta_2 N_2}{H_1 = \beta_1 N_1} \; .$$

Only when $\beta_1 = \beta_2$ is the ratio of the harvests the same as the ratio of the populations. Thus, the only index that has validity in this application is a constant proportion index.

In nearly every case, the proportion harvested (β) is not constant but varies with a plethora of conditions including the number and selectivity of hunters, amount of time spent hunting, weather, and behavior of the animals. To assess populations with harvest data, additional information is needed to estimate β.

Recall that most population estimation procedures involve estimates of observability β and counts or captures of a population. In the preceding example we can think of the harvest (H) as a count (C). For many of these procedures, statistical tests of the assumption that betas are equal are available or can be developed through consultation with a statistician (50, 51, 52, 53). If betas are not different (i.e., the statistical null hypothesis is not rejected), then counts or captures are an index of abundance and can be used to make comparisons. Harvested animals could be used as an accurate and inexpensive index, if betas could be estimated and tested for equality.

An obvious question arises: If population estimation procedures are used to estimate β, and they also yield an estimate of population size (N), why not compare estimates of population size instead of counts or captures (C) as an index? The answer is that C has less variance than the associated population estimates (53; 6, p. 221). The smaller variance of C than N means statistical tests that use C will have more power (i.e., have a better chance of finding differences if they are there) than will tests with N. The advantage of using counts or captures as an index, when associated with population estimation procedures that also estimate β, are (1) the assumption that betas are equal can be tested and (2) if the test fails, the estimates of N can be used instead, but they would have less statistical power. One caution is that a poorly designed test or insufficient data could fail to detect a difference in betas when they were actually different.

Another commonly used index is the track count method, which is simply a different kind of count that is used to compare populations of white-tailed deer in the southeastern United States (54). Fritzen and others suggested modifications to account for violations of the assumptions of the method (55). However, it still remains to be proven that comparisons over time or between areas reflect true differences in populations and not just differences in the indices (i.e., different β's).

Courtois and Crete used multiple regression to relate hunting statistics collected at harvest check stations to population estimates from aerial surveys (56). In Quebec, moose populations are estimated in management zones at 5-year intervals because it is too expensive to do surveys more frequently. These 5-year estimates were used as the dependent variable. Various harvest statistics such as total harvest, productivity indices (e.g., calves : female), and hunting characteristics (e.g., length of the hunting season) were the independent variables. Regressions to predict winter density ($R^2 = 0.74$) and harvest rate ($R^2 = 0.68$) were reasonably good, so the regressions were used to track moose populations between 5-year surveys.

ESTIMATES OF POPULATION STRUCTURE AND DEMOGRAPHIC RATES

Sex and Age Ratios

Estimates of sex and age ratios comprising a large mammal population are made from counts of the population or harvest (9, 44). In both cases the underlying assumptions are that observability of sex and age classes (β) is <1.0, and observabilities of the two classes being compared in the ratio are equal. (If $\beta = 1$, the ratios would be based on a complete count and would be correct.) If observabilities are unequal, then β will not cancel out of the ratio, and differences in observability will obscure the true sex or age ratio (57). If sex and/or age ratios are compared among years, betas would have to be equal within each year, but could be different among years. Of course, in this situation, only the ratios would be correct; comparisons of individual sex and/or age groups that made up the ratios among years would be inappropriate because betas would be unequal.

Observability in counts of the population or harvest requires some additional clarification. Counts of the population would be field observations of animals whose sex and/or age can be recognized accurately from a distance. In practice, identification varies among observers and other factors such as group size and observational technique (58). During observations, sex and/or age classes in the ratios would have to be equally detectable. For pronghorns, Woolley and Lindzey recommend the same observer and same method for comparisons between years, and Bowden and others present observational sampling plans for mule deer sex and age ratios (8, 58). In counts of the harvest, selectivity by hunters is an additional factor that affects determination of sex and/or age ratios. Unless animals are harvested in proportion to their true representation in the population (equal β), ratios based on the harvest would not be accurate. Rosenberry and Woolf observed that adult male:female ratios were highest in the harvest and lowest in roadside observations, and they discuss whether fawn:female ratios in the harvest accurately reflect recruitment (44, p. 14).

In most applications, betas are unknown and most likely unequal, so estimates from counts, for example, are biased (9). To circumvent this problem, sex and/or age classes should be considered as subpopulations and estimated separately with an appropriate population estimation procedure. Multiple population estimates, one for each sex and/or age class, must be made rather than one for the entire population. Then comparisons (ratios) between classes could be tested for equal betas, or if that test fails, population estimates could be used to calculate ratios. Obviously this requires significantly more work. Another approach is to correct for visibility bias (different betas) by developing models that provide correction factors (57).

For large game mammals, population CPUEs can be a feasible approach for estimating abundance of sex and/or age classes separately. Lancia and others used hunter kills and observations to estimate a population of antlered white-tailed deer (47). Estimates were biased low, but did accurately reflect major trends over a 10-year period. The antlerless population was not estimated because few antlerless deer were removed. Based on computer simulation, removals greater than 70% are needed for accurate CPUE estimates (J. W. Bishir, North Carolina State University, personal communication), which in this application was true because males were heavily harvested.

Births and Birth Rates

Both potential and realized birth rates are a concern in large mammal management. Birth rates are expressed as births per female (or individual) per year or other appropriate time interval, whereas births are the total number of young born or recruited into the population. Births are the product of birthrate times number of females (or individuals) in the population. Potential rates can be determined by examination of reproductive tracts of females. Realized births are the number of young at a particular point in the annual cycle. Often this point occurs immediately following the peak of births in a birth pulse population. Realized births can be thought of as net recruitment into the preharvest population. Realized births and birth rates in large mammals can be determined by direct counts or by open mark–recapture methods.

Direct Counts. Counts of young per female (e.g., fawns per female) are a common measure of birthrate. The same shortcomings apply here as previously discussed for sex and/or age ratios because birthrate is an age ratio. Observabilities of fawns and females, including females with and without fawns, must be equal (i.e., equal betas). If comparisons of ratios are made between years to detect changes in reproduction, then betas must be equal within each year, but could change among years. If fawn : female ratios are used to estimate total births, then the total number of females must also be known.

Open Mark-Recapture Models. Open population models allow for changes in population size. Jolly–Seber is the basic open population model. JOLLY is a computer program that calculates Jolly–Seber estimates (2, 3, 59, 60). J. D. Nichols wrote an easily understood overview in Lancia and others (6, pp. 244–246).

The Jolly–Seber model requires individually marking (or recognizing) animals that are recaptured (or reobserved) on multiple occasions. Population size, survival rate, and births are estimated. The Jolly–Seber model assumes that observability is equal for all individuals and survival is the same for all individuals for an interval between two occasions, but it can change between intervals. Furthermore, the biology and behavioral ecology of the species and population must be known to separate immigration from births and emigration from deaths. Because of assumptions underlying the method and the effort to mark and recapture individuals, Jolly–Seber has limited application for big game management.

Births, as defined in the Jolly–Seber model, are the number of animals recruited into the population between one occasion and the next, and are calculated as the difference between the population estimate at one occasion and the number expected to survive to the next occasion. Births are estimated for all sample occasions except the first and last two.

Mortality

Mortality rate (or survival rate $= 1.0 -$ mortality rate) is the proportion of animals dying within a specified time period (i.e., the number that died during a period divided by the number alive at the beginning of that period). An obvious problem in assessing mortality rates is determining when an animal dies. Unless they are harvested, the fate of most animals is not known. Even if dead animals are found, determining exactly when they died can be difficult or impossible. Furthermore, emigration is difficult to separate from mortality.

Methods for estimating mortality rates can be categorized in four basic ways: (1) direct observations, which involve individually marking animals; (2) indirect observations of changes in population size or changes in ratios of sex and/or age classes; (3) open population models such as Jolly–Seber; and (4) life tables.

Two direct methods are most applicable to large mammals: (1) marking and recapturing animals when they die such as during a hunting season and (2) following the fate of radiomarked individuals. The choice between these two depends to some degree on when mortality occurs. For big game mammals, determining mortality during the hunting season is straightforward if all harvested animals are processed through a check station and unretrieved kills and other sources of mortality are minimal. A direct method, such as recovering marked animals in the harvest, is readily applied. The mortality rate would be the ratio of the number of marked animals that were harvested to the total number of marked animals before the hunting season. More complicated Jolly–Seber methods or banding models (MARK program, G. C. White, Colorado State University) also might be appropriate. However, the preceding methods can require unreasonable assumptions (e.g., harvest is the only source of mortality) or need multiple marking and capture occasions, which makes them less applicable for big game management.

The most practical direct method for determining mortality rates for large mammals occurring at other times of the year, often termed *natural mortality,* is to follow the fate of radiomarked animals. In theory, the advantages of telemetry are that animals can be relocated at will and transmitters with mortality modes signal when an animal has stopped moving and presumably has died. Thus, time of death can be determined accurately. In practice, capturing and radiomarking an adequate sample quickly can be difficult and expensive, radios are prone to fail, and animals move out of receiver range. In the latter case the fate of the animals is unknown, and no information on mortality is gained.

Survival rate of radiomarked animals can be estimated with the staggered entry, Kaplan–Meier estimator, which allows newly radiomarked animals to be added as the study progresses (61). In this approach survival is calculated sequentially at each time interval, typically days or weeks, based on whether individual animals die. Survival over subsequent periods is the product of previous interval survivals. This yields a stairstep pattern of survival over time (Figure 4–4). If no animals die in several time intervals, survival is a flat line; but if a number die such as during the hunting season, survival drops dramatically. This graphical display of survival through time shows how mortality patterns vary. Samuel and Fuller provide a clear overview of methods to estimate survival with radiotelemetry (62, pp. 403–408).

Indirect methods for determining large mammal mortality rates are likely to be less expensive to implement, and probably less accurate than direct methods. If the only change in population size between two periods is caused by deaths, or other changes such as additions of young or dispersal are known, then the difference between the populations is mortality. This simple approach might be appropriate to assess mortality during a harvest.

Change-in-ratio methods, previously presented in association with estimating population size, can also be used to estimate rates (ratios) such as mortality or harvest

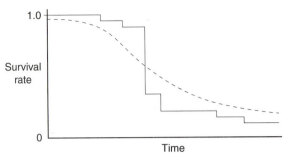

Figure 4–4 Kaplan-Meier survival rate (—) and underlying survival curve (---).

rates. When CIR is used to estimate population size, changes in proportions (ratios) of classes before and after a known number of removals are required to execute the method. However, when CIR is used to estimate mortality or harvest rates, proportions before and after removal are needed as before, but in place of removals, only the proportion of classes in the harvest is needed. In essence, proportions (rates) can be used to estimate proportions (rates), but a number is needed to "scale" the ratios so that population size can also be estimated (2). Change-in-ratio has potential application for big game mammals if ratios can be assessed accurately from observations of the population and harvest, and if ratios change markedly as a result of harvest. For example, Nelson showed how to estimate harvest rate using CIR when total harvest and population size were unknown and applied it to white-tailed deer management (63).

In the Jolly–Seber model, survival rate is the proportion of individuals still alive at the end of a period (occasion) that were alive at the beginning of that period. It is calculated as the ratio of marked animals at one occasion to those same marked animals at the previous occasion. Survival is estimated for all but the last two sample occasions.

Life tables portray the progressive depletion over time of a cohort of individuals (i.e., all born at about the same time) due to mortality (64, Chapter 8). The table most commonly consists of several columns depicting for each age class the number alive, survival probability, mortality probability, survival rate, and mortality rate. The columns are totally dependent, so each reflects the same information in different ways. Any one column can be used to construct the others.

Life tables could easily be constructed for a cohort if the number born were known, and if their survival could be traced until the last member of the cohort died. If this cohort was representative of all other cohorts in the population, then age-specific survival for the population could be approximated. In practice, knowing all births and following the fate of a cohort through time is not practical for wild, large mammals that are long lived. A biologist could spend half a career or more before all the data were in!

To circumvent this problem, a biologist would like to obtain a sample of the age structure of the population (S_x), which is the number of individuals alive in each age class (f_x) relative to the number born (f_0) at a given point in time (i.e., $S_x = f_x/f_0$), and use this information to construct a life table. In essence, age structure is used to estimate age-specific survival. In theory, a snapshot of the age structure could be derived from a harvest or in some other way. Unfortunately, the following assumptions that underlay this approach are rarely met in field applications.

Four fundamental assumptions are made when constructing a life table from age distribution data. First, animals must be aged accurately to year classes or some other appropriate age interval. For many species this can be done for younger ages, but not all age classes. Second, the sample of the age distribution must accurately represent the population. For example, when following a cohort through time it would be impossible for 3-year-olds to outnumber 2-year-olds, but this can easily occur in a sample for a variety of reasons that are not easily corrected. Third, patterns of age-specific survival and births must remain constant for at least a generation or two so that the age struc-

Estimates of Population Structure and Demographic Rates

ture becomes stable (i.e., S_x remains the same every year). Fourth, something must be known or assumed about the number born because age structure and survival are expressed relative to the number of births (recall that $S_x = f_x/f_0$). In practice, knowing births is difficult. This assumption can be circumvented by assuming that the growth rate of a population is zero (a stationary age distribution) or is known (a stable age distribution with known rate of increase). In summary, these assumptions appear too restrictive to make life tables of much practical use to managers of large mammals. Recently, new models that relax these assumptions offer some promise (65).

Dispersal, Emigration, and Immigration

To determine emigration or immigration rates, animals must be individually marked, or be identifiable from individual characteristics, and then either followed from a known origin (usually their natal range) as they disperse or relocated at a known destination after dispersal is complete. Difficulties in monitoring emigration over long distances and measuring immigration of animals from unknown locations combine to make reliable estimates of these movements difficult, especially if dispersal occurs over a long period of time.

Emigration can be assessed by marking animals with individually identifiable tags or with radiotransmitters. Emigration distance and location are determined when tags are returned after animals are found dead or are harvested. Tagging usually underestimates emigration because tags are more likely to be returned from nearby locales than from farther away, and nonhunting mortality is not likely to be discovered. Radiotelemetry is preferred because (1) death of an animal is not required to determine emigration, (2) radiomarked animals can be continuously monitored, if desired, to follow emigration routes, and (3) exact departure dates and emigration locations can be determined.

Estimates of emigration probabilities for radiomarked animals are made with the Kaplan–Meier method described earlier to estimate survival. In this approach emigration is analogous to death because an animal is no longer in the population (i.e., the probability of emigration equals one minus the probability of survival). The probability of an animal emigrating in a time period is a binomial of the conditional probability for that time period, or

$$P_{\text{emigration}} = n_{\text{emigrants}} / n_{\text{at risk}} ,$$

where $n_{\text{emigrants}}$ is the number of animals emigrating during the time period and $n_{\text{at risk}}$ is the number at risk in that time period. An animal can be censored if its transmitter fails (61).

The difficulty in estimating the number of immigrants into a particular area is marking enough individuals from outlying environs to get a reasonable estimate. Rosenberry circumvented this difficulty by taking advantage of known dispersal behavior of white-tailed deer, by using an innovative mark-resight design, and by simultaneously studying emigration of telemetered animals and immigration of "unmarked" deer (13). For the former he determined that all fall dispersal occurred within a 7-week period and only yearling males dispersed. Thus, before and after dispersal the population was closed, so differences in populations before and after would yield the number of immigrants. For the latter he used antler configuration to identify individual yearling males.

To estimate the number of immigrants, periods before and after dispersal were each subdivided into separate mark and resight segments. Pre-dispersal and post-dispersal populations were estimated with the Lincoln–Petersen method. Immigration was estimated as:

$$I = N_{\text{pre}} - \phi N_{\text{post}} ,$$

where I is the number of immigrants, N_{pre} is the estimated pre-dispersal population, ϕ is the probability of a yearling male not dispersing and surviving the dispersal period, and N_{post} is the post-dispersal population estimate. Survival ϕ was estimated with the Kaplan–Meier procedure (61). Rosenberry found that roughly equal numbers of immigrant and philopatric yearling males comprised the post-dispersal population he studied (13).

Rate of Increase

Realized rate of increase (lambda, λ) is estimated by comparing population size from one year (or some other time interval) to the next as:

$$\lambda = N_{t+1} / N_t \qquad \text{and} \qquad \lambda = e^r ,$$

where λ is the realized rate of increase, r is the exponential rate of increase, N is population size, and the subscripts represent years. Growth could be estimated each year, without assuming an underlying pattern (i.e., no underlying model of growth), and it would simply represent relative change from one year to the next over successive years. If some underlying pattern of growth is assumed, then an appropriate mathematical representation (model) of this pattern must be selected.

If estimates of growth rate are intended to be used to develop a harvest strategy, then an appropriate underlying model of population growth is required. In this chapter we briefly present a sigmoid growth model for simplicity; other models could be used.

Recall from earlier in the chapter when we postulated that per capita growth would be highest when a population was very small (Figure 4–2). As population size increases, per capita growth would decrease as resources available per individual decreased. This pattern of declining per capita growth as population size increases is the underlying basis for a sustained yield harvesting strategy (see Chapters 5 and 7). If this model is selected, then the maximum intrinsic rate of increase r_m, which would occur when the population was very small, would be estimated from population growth data.

Estimates of r_m are made by fitting an appropriate growth model to observations of population growth through time beginning from a founder population. For example, Lancia and others estimated r_m for white-tailed deer using data from McCullough for the George Reserve herd in Michigan (11, 12, 66, 67). They fit a sigmoid growth model (logistic) to population estimates from 1928, when a small founder population was introduced to the George Reserve, to 1934 and estimated r_m to about 0.9. This rate corresponds closely to the expected realized rate in the first year after introduction, which McCullough estimated to be $r = 0.847$ (67). Thus, r_m represents the maximum growth rate for deer in this environment, and can be substituted into a sigmoid growth model to predict growth rates at various population sizes and sustainable harvest levels.

SUMMARY

In this chapter we provided an overview of population parameters we believe are needed to make informed decisions in management of large mammal populations, particularly those that are hunted. We selected population size, sex and age ratios, the four fundamental demographic rates (i.e., births, deaths, immigration, and emigration) and the rate of population growth as the principal demographic parameters of interest to population managers. Of these, the rate of population growth is particularly important because it reflects a synthesis of the fundamental demographic rates. Knowledge of the other parameters helps explain changes in population growth and size. Changes in population size are the currency by which management decisions are evaluated.

A significant portion of the chapter is devoted to methods for estimating population size. Only basic estimation procedures most applicable to large mammal management are given. The chapter is organized by grouping approaches according to assumptions about observability and sampling concerns. When all individuals can be observed ($\beta = 1.0$), then sampling concerns dominate. When all individuals cannot be observed ($\beta < 1.0$), then observability concerns are also important. If all individuals can be observed, a census can be done. If all individuals cannot be observed, then methods that involve counts (e.g., double sampling, sightability models, line transects), removals (e.g., change-in-ratio, catch-per-unit-effort), or captures (e.g., mark–recapture) are needed. Sometimes, population indices can be used to compare populations in place of population estimates.

Methods to estimate sex and age ratios, the four fundamental demographic rates, and the rate of increase often involve estimates of the size of certain classes of animals within a population. These classes can then be related with ratios. Sometimes, open population models (e.g., Jolly–Seber) can be used to estimate births and deaths, and radiotelemetry can be used to estimate mortality and emigration rates. Rate in increase, when considered in the context of an appropriate pattern of underlying population growth, can be used to predict population responses associated with harvesting.

Estimating population parameters provides a challenge for managers of large mammal populations. Animals tend to be elusive, which makes accurate estimates of population size and demographic characteristics difficult to achieve. Limited budgets, personnel, and time compound the problem. Nevertheless, responsible stewardship of our wildlife resources demands that management decisions be made with accurate information and not speculation. In the short run the public, hunters and nonhunters alike, will not tolerate decisions without corroborating data. In the long run, the preservation and conservation of our rich biological heritage depends on making sound decisions.

When faced with a need to estimate population parameters, managers should carefully review methods that are currently available. A method should be selected that (1) has assumptions that can be met, so estimates are more likely to be accurate; (2) has a small variance, so estimates will be more precise; (3) provides adequate information to make an informed decision; and (4) is cost effective. Finally, managers should implement management within an adaptive resource management paradigm to accomplish the dual goals of achieving management objectives and understanding population demography.

LITERATURE CITED

1. LANCIA, R. A., C. E. BRAUN, M. W. COLLOPY, R. D. DUESER, J. G. KIE, C. J. MARTINKA, J. D. NICHOLS, T. D. NUDDS, W. R. PORATH, and N. G. TILGHMAN. 1996. ARM! For the future. *Wildlife Society Bulletin* 24:436–442.

2. SEBER, G. A. F. 1982. *The Estimation of Animal Abundance and Related Parameters,* 2nd ed. Macmillan Publishing Co., New York.

3. POLLOCK, K. H., J. D. NICHOLS, C. BROWNIE, and J. E. HINES. 1990. Statistical inference for capture–recapture experiments. *Wildlife Monographs* 107:1–97.

4. WHITE, G. C., and R. A. GARROTT. 1990. *Analysis of Wildlife Radio-Tracking Data.* Academic Press, San Diego, CA.

5. JOHNSON, D. H. 1994. Population analysis. Pages 419–444 *in* T. A. Bookhout, ed., *Research and Management Techniques for Wildlife and Habitats,* 5th ed. The Wildlife Society, Bethesda, MD.

6. LANCIA, R. A., J. D. NICHOLS, and K. H. POLLOCK. 1994. Estimating the number of animals in wildlife populations. Pages 215–253 *in* T. A. Bookhout, ed., *Research and Management Techniques for Wildlife and Habitats,* 5th ed. The Wildlife Society, Bethesda, MD.

7. MILLER, S. D., G. C. WHITE, R. A. SELLERS, H. V. REYNOLDS, S. W. SCHOEN, K. TITUS, V. G. BARNES, JR., R. B. SMITH, R. R. NELSON, W. B. BALLARD, and C. C. SCHWARTZ. 1997. Brown and black bear density estimation in Alaska using radio telemetry and replicated mark–resight techniques. *Wildlife Monographs* 133:1–55.

8. BOWDEN, D. C., A. E. ANDERSON, and D. E. MEDIN. 1984. Sampling plans for mule deer sex and age ratios. *Journal of Wildlife Management* 48:500–509.

9. MCCULLOUGH, D. R. 1993. Variation in black-tailed deer herd composition counts. *Journal of Wildlife Management* 57:890–897.

10. TIMMERMANN, H. R. 1993. Use of aerial surveys for estimating and monitoring moose populations—A review. *Alces* 29:35–46.

11. MCCULLOUGH, D. R. 1979. *The George Reserve Deer Herd.* University of Michigan Press, Ann Arbor.

12. LANCIA, R. A., K. H. POLLOCK, J. W. BISHIR, and M. C. CONNER. 1988. A white-tailed deer harvesting strategy. *Journal of Wildlife Management* 52:589–595.

13. ROSENBERRY, C. S. 1997. Dispersal ecology and behavior of yearling male white-tailed deer. Dissertation. North Carolina State University, Raleigh.

14. POLLOCK, K. H. 1982. A capture–recapture design robust to unequal probability of capture. *Journal of Wildlife Management* 46:752–757.

15. COCHRAN, W. G. 1977. *Sampling Techniques,* 3rd ed. John Wiley & Sons, New York.

16. WOLFE, M. L., and J. F. KIMBALL. 1989. Comparison of bison population estimates with a total count. *Journal of Wildlife Management* 53:593–596.

17. JACOBSON, H. A., J. C. KROLL, R. W. BROWNING, B. H. KOERTH, and M. H. CONWAY. 1997. Infrared-triggered cameras for censuring white-tailed deer. *Wildlife Society Bulletin* 25:547–556.

18. WIGGERS, E. P., and S. F. BECKERMAN. 1993. Use of thermal infrared sensing to survey white-tailed deer populations. *Wildlife Society Bulletin* 21:263–268.

19. CROON, G. W., D. R. MCCULLOUGH, C. E. OLSEN, JR., and L. M. QUEAL. 1968. Infrared scanning techniques for big game censuring. *Journal of Wildlife Management* 32:751–759.

20. GRAVES, H. B., E. D. BELLIS, and W. M. KNUTH. 1972. Censuring white-tailed deer by airborne thermal infrared imagery. *Journal of Wildlife Management* 36:875–884.

21. GARNER, D. L., H. B. UNDERWOOD, and W. F. PORTER. 1995. Use of modern infrared thermography for wildlife population surveys. *Environmental Management* 19:233–238.

22. NAUGLE, D. E., J. A. JENKINS, and B. J. KERNOHAN. 1996. Use of thermal infrared sensing to estimate density of white-tailed deer. *Wildlife Society Bulletin* 24:37–43.

23. HAVENS, K. J., and E. J. SHARP. 1998. Using thermal imagery in the aerial survey of animals. *Wildlife Society Bulletin* 26:17–23.

24. KUFELD, R. C., J. H. OLTERMAN, and D. C. BOWDEN. 1980. A helicopter quadrat census for mule deer on Uncompahgre Plateau, Colorado. *Journal of Wildlife Management* 44:632–639.

25. LUDWIG, J. 1981. Proportion of deer seen in aerial counts. *Minnesota Wildlife Research Quarterly* 41:11–19.

26. BARTMANN, R. M., L. H. CARPENTER, R. A. GARROTT, and D. C. BOWDEN. 1986. Accuracy of helicopter counts of mule deer in pinyon-juniper woodland. *Wildlife Society Bulletin* 14:356–363.

27. BEASOM, S. L., F. G. LEON III, and D. R. SYNATZSKE. 1986. Accuracy and precision of counting white-tailed deer with helicopters at different sampling intensities. *Wildlife Society Bulletin* 14:364–368.

28. STOLL, R. J., JR., W. MCCLAIN, J. C. CLEM, and T. PLAGEMAN. 1991. Accuracy of helicopter counts of white-tailed deer in western Ohio farmland. *Wildlife Society Bulletin* 19:309–314.

29. BERINGER, J., L. P. HANSEN, and O. SEXTON. 1998. Detection rates of white-tailed deer with a helicopter over snow. *Wildlife Society Bulletin* 26:24–28.

30. CAUGHLEY, G., R. SINCLAIR, and D. SCOTT-KEMMIS. 1976. Experiments in aerial survey. *Journal of Wildlife Management* 40:290–300.

31. CONNER, M. C. 1986. Refinement of the change-in ratio technique for estimating abundance of white-tailed deer. Dissertation. North Carolina State University, Raleigh.

32. SAMUEL, M. D., E. O. GARTON, M. W. SCHLEGEL, and R. G. CARSON. 1987. Visibility bias during aerial surveys of elk in north central Idaho. *Journal of Wildlife Management* 51:622–630.

33. ANDERSON, C. R., and F. G. LINDZEY. 1996. A sightability model for moose developed from helicopter surveys. *Wildlife Society Bulletin* 24:247–259.

34. ANDERSON, C. R., D. S. MOODY, B. L. SMITH, F. G. LINDZEY, and R. P. LANKA. 1998. Development and evaluation of sightability models for summer elk surveys. *Journal of Wildlife Management* 62:1055–1066.

35. STEINHORST, R. K., and M. D. SAMUEL. 1989. Sightability adjustment methods for aerial surveys of wildlife populations. *Biometrics* 45:415–425.

36. COGAN, R. D., and D. R. DIEFFENBACH. 1998. Effect of undercounting and model selection on a sightability-adjustment estimator for elk. *Journal of Wildlife Management* 62:269–279.

37. EBERHARDT, L. L., R. A. GARROTT, P. J. WHITE, and P. J. GOGAN. 1998. Alternative approaches to aerial censuring of elk. *Journal of Wildlife Management* 62:1046–1055.

38. BURNHAM, K. P., D. R. ANDERSON, and J. L. LAAKE. 1980. Estimation of density from line transect sampling of biological populations. *Wildlife Monographs* 72:1–202.

39. QUINN, T. J., II. 1981. The effect of group size on line transect estimators of abundance. Pages 502–508 *in* C. J. Ralph and J. M. Scott, eds., *Estimating Numbers of Terrestrial Birds. Studies in Avian Biology* 6. Cooper Ornithological Society, Sacramento, CA.

40. ANDERSON, D. R., J. L. LAAKE, B. R. CRAIN, and K. P. BURNHAM. 1979. Guidelines for line transect sampling of biological populations. *Journal of Wildlife Management* 43:70–78.

41. POLLOCK, K. H., R. A. LANCIA, M. C. CONNER, and B. L. WOOD. 1985. A new change-in-ratio procedure robust to unequal catchability of types of animal. *Biometrics* 41:653–662.

42. UDEVITZ, M. S. 1989. Change-in-ratio methods for estimating the size of closed populations. Dissertation. North Carolina State University, Raleigh.

43. CONNER, M. C., R. A. LANCIA, and K. H. POLLOCK. 1986. Precision of the change-in-ratio technique for deer population management. *Journal of Wildlife Management* 50:125–129.

44. ROSENBERRY, J. L., and A. WOOLF. 1991. A comparative evaluation of techniques for analyzing white-tailed deer harvest data. *Wildlife Monographs* 117:1–59.

45. RICKER, W. R. 1958. *Handbook of Computations for Biological Statistics of Fish Populations.* Bulletin 119. Fisheries Research Board of Canada.

46. BISHIR, J., and R. A. LANCIA. 1996. On catch-effort methods of estimating animal abundance. *Biometrics* 52:1457–1466.

47. LANCIA, R. A., J. W. BISHIR, M. C. CONNER, and C. S. ROSENBERRY. 1996. Use of catch-effort to estimate population size. *Wildlife Society Bulletin* 24:731–737.

48. WHITE, G. C. 1996. NOREMARK: Population estimation from mark–resighting surveys. *Wildlife Society Bulletin* 24:50-52.§

49. LANCIA, R. A., M. C. CONNER, and B. D. WALLINGFORD. 1995. Heterogeneity in observability of white-tailed deer on Remington Farms. *Proceedings of the Annual Conference of the Southeastern Association of Fish and Wildlife Agencies* 49:423–431.

50. JOLLY, G. M. 1982. Mark–recapture models with parameters constant in time. *Biometrics* 38:302–321.

51. SKALSKI, J. R., D. S. ROBSON, and M. A. SIMMONS. 1983. Comparative census procedures using single mark–recapture methods. *Ecology* 64:752–760.

52. SKALSKI, J. R., M. A. SIMMONS, and D. S. ROBSON. 1984. Use of removal sampling in comparative censuses. *Ecology* 65:1006–1015.

53. SKALSKI, J. R., and D. S. ROBSON. 1992. *Techniques for Wildlife Investigations Design and Analysis of Capture Data. Academic Press,* San Diego, CA.

54. TYSON, E. L. 1952. Estimating deer populations from tracks. *Proceedings of the Annual Conference of the Southeastern Association of Game and Fish Commission* 6:507–517.

55. FRITZEN, D. E., R. F. LABISKY, D. E. EASTON, and J. C. KILGO. 1995. Nocturnal movements of white-tailed deer: Implications for refinement of track-count surveys. *Wildlife Society Bulletin* 23:187–193.

56. COURTOIS, R., and M. CRETE. 1993. Predicting moose population parameters from hunting statistics. *Alces* 29:75–90.

57. SAMUEL, M. D., R. K. STEINHORST, E. O. GARTON, and J. W. UNSWORTH. 1992. Estimation of wildlife population ratios incorporating survey design and visibility bias. *Journal of Wildlife Management* 56:718–725.

58. WOOLLEY, T. P., and F. G. LINDZEY. 1997. Relative precision and sources of bias in pronghorn sex and age composition surveys. *Journal of Wildlife Management* 61:57–63.

59. JOLLY, G. M. 1965. Explicit estimates from capture–recapture data with both death and immigration-stochastic models. *Biometrika* 52:225–247.

60. SEBER, G. A. F. 1965. A note on the multiple-recapture census. *Biometrika* 52:249–259.

61. POLLOCK, K. H., S. R. WINTERSTEIN, C. M. BUNCK, and P. D. CURTIS. 1989. Survival analyses in telemetry studies: The staggered entry design. *Journal of Wildlife Management* 53:7–15.

62. SAMUEL, M. D., and M. R. FULLER. 1994. Wildlife radio telemetry. Pages 370–418 *in* T. A. Bookhout, ed., *Research and Management Techniques for Wildlife and Habitats,* 5th ed. The Wildlife Society, Bethesda, MD.

63. NELSON, G. P. 1997. A flexible model and proportional harvest strategy for density dependent ungulate populations. Dissertation. North Carolina State University, Raleigh.

64. CAUGHLEY, G. 1977. *Analysis of Vertebrate Populations.* John Wiley & Sons, New York.

65. UDEVITZ, M. S., and B. E. BALLACHEY. 1998. Estimating survival rates with age-structure data. *Journal of Wildlife Management* 62:779–792.

66. MCCULLOUGH, D. R. 1982. Population growth rate of the George Reserve deer herd. *Journal of Wildlife Management* 46:1079–1083.

67. MCCULLOUGH, D. R. 1983. Rate of increase of white-tailed deer on the George Reserve: A response. *Journal of Wildlife Management* 47:1248–1250.

Modeling Population Dynamics

Gary C. White

INTRODUCTION

Population modeling is a tool used by wildlife managers. At its simplest, population modeling is a bookkeeping system to keep track of the four components of population change: births, deaths, immigration, and emigration. In mathematical symbolism, a population model can be expressed as

$$N_{t+1} = N_t + B_t - D_t + I_t - E_t \ .$$

Population size (N) at time $t + 1$ is equal to the population size at time t plus births (B) minus deaths (D) plus immigrants (I) minus emigrants (E). The simplicity of the relationship belies its usefulness. If you were managing a business, you would be interested in knowing how much money was being paid out of the business relative to how much money was coming in. The net difference represents your profit. The same is true for managing a wildlife population. By knowing the net change in the size of the wildlife population from one year to the next, you as a wildlife manager know whether the population is profitable (i.e., gaining in size) or headed toward bankruptcy (i.e., becoming smaller).

Two examples will demonstrate typical uses of population models in big game management. In November 1995, voters in Colorado passed an amendment to the state's constitution that outlawed the hunting of black bears during spring. Wildlife managers were asked if the bear population would increase if there were no spring harvest. A population model can be used to answer this question. The second example of the use of a model is to determine which of two management schemes (e.g., no antlerless harvest versus significant antlerless harvest) will result in the largest buck harvest in a mule deer population. Both of these examples require specific information to answer the questions posed. The models in both cases are important for conceptualizing the problem and for defining exactly what information is needed to answer the question.

The main use of the model in the previous examples is to predict the result of some management action, such as stopping spring bear hunting or implementing an antlerless harvest. Another use of population models is to conceptualize the dynamics of a population in rigorous mathematical notation. Such a conceptual model allows biologists to think more clearly about the dynamics of a population. A third use of models is to test hypotheses about population dynamics from observed data. For example, biologists might entertain competing models, one based on density independence, the other on density dependence. Which of these models best explains our observations of a mule deer population?

Computational tools are needed to build population models on a computer. Spreadsheets are a natural vehicle for building population models. They were developed for modeling the functions of a business (i.e., keeping track of income and outflow) and hence are particularly suited for modeling population dynamics. Modern spreadsheets include a natural language to specify a population model, graphics to display the results, random number functions to model stochasticity, and functions useful for optimizing some attribute of the model's output (e.g., annual harvest) or for estimating parameter values from data. Complex models can be built and then made available via the WWW (see http://www.cnr.colostate.edu/~bruce/software.htm for these spreadsheet models of linked-sex harvest strategies) (1).

A major advantage of using spreadsheets to develop the model is that you fully understand the mechanics of the model developed. You know what relations are assumed in the model because you put them in the model. As a user of another's model, spreadsheet models allow you to dissect the model to see what relations were programmed. Most of the earlier models of population dynamics were built with predefined models, such as ONEPOP and later POP-II (2, 3). I describe these models as black box models, because as a model developer and a model user, you are not quite sure what is in the model. You cannot peer into the black box of a model to see the inner workings. Even if you can study the computer code, the coding is arcane enough to make interpretation challenging. Further, the limitations of the original model framework limit what you could do with your model. Thus, spreadsheets provide openness and flexibility not provided by black box models.

The goal of this chapter is to show how to develop and use population models with spreadsheet software such as Quattro$^{©}$, Lotus$^{©}$, or Excel$^{©}$. At an elementary level, density-independent versus density-dependent population growth models will be compared. Still at an elementary level, I will describe models that incorporate density dependence into an age-structured model. At a more advanced level, sources of stochasticity in population dynamics will be acknowledged, and I will develop stochastic models to mimic this observed variation in the data. Finally, spreadsheets will be used to fit models and estimate parameters from observed data on populations. To fully understand the last two topics, deeper training in mathematics and statistics will be required. After all, population modeling is a quantitative subject.

DENSITY-INDEPENDENT POPULATION GROWTH

Density-independent population growth is based on the concept that the population grows at the same rate, no matter how large or small it has become. With the big game populations considered here, population growth generally is not continuous as it is in bacteria or protozoans. Rather, a birth pulse takes place at a defined time of year. Thus, I will build models based on difference equations, as opposed to differential equations that assume continuous population growth (see Box 5–1). To define density-independent population growth with a difference equation, assume that the population always increases by $1 + R$ times each year. That is,

$$N_{t+1} = N_t + N_t \times R = N_t \times (1 + R) \ ,$$

so that

$$N_t = N_0 (1 + R)^t \ .$$

Finite rate of growth (R) is composed of all the processes that change the population size (i.e., R = finite birth rate − finite death rate + finite immigration rate − finite emigration rate).

Box 5–1. The differential equation for density-independent population growth is given by

$$\frac{dN}{dt} = rN \; .$$

With integration, $N_t = N_0 \, e^{rt}$ where $e^r = \lambda = N_t + {}_1/N_t$. Which do you use: differential or difference equation? Biological reasons suggest using difference equations: North American big game populations have discrete birth pulses, not continuous births as is assumed by differential equations. Mathematical reasons also suggest using difference equations: they are easier to construct and solve in a computer spreadsheet.

The population model given can be thought of as a unisex, uni-age population (i.e., all the animals in the population are the same age and sex). The annual rate of change of the population is λ, with

$$N_{t+1}/N_t = 1 + R = \lambda \; .$$

For increasing populations, $\lambda > 1$, whereas for decreasing populations, $\lambda < 1$. The concept of the annual rate of change of a population is larger than just density-independent population growth and is generally used to describe the observed annual changes in population size.

The key point about density-independent population growth is that R and, hence, λ do not change with the size of the population. For density-independent population growth, the values of R and λ are fixed, so that the population grows the same relative amount each year. This is hardly a realistic model for biological populations, except maybe for the observed human population expansion! The consequences of density-independent, or exponential population growth, is that the population expands exponentially (Figure 5–1).

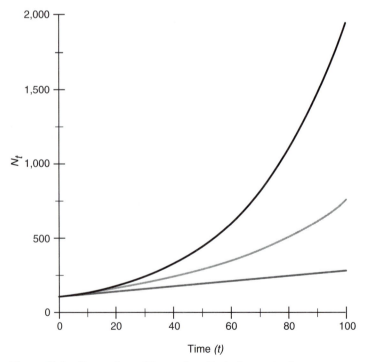

Figure 5–1 Examples of three density-independent populations with $N_0 = 2$, and different values of $R = 0.01$, 0.02, and 0.03, resulting in three different rates of growth.

TABLE 5–1 Spreadsheet Showing How the Equations are Iterative on the Previous Value

Row/Column	A (= time)	B (= N)
1	0	100 (= N_0)
2	+A1+1	+B1*(1+R)
3	+A2+1	+B2*(1+R)
4	+A3+1	+B3*(1+R)
5	+A4+1	+B4*(1+R)
6	+A5+1	+B5*(1+R)
7	+A6+1	+B6*(1+R)

Table 5–1 shows how spreadsheet equations are iterative on the previous value, that is, iterative equations are created by copying a formula. Time is in column A, and population size is in column B. Row 1 initializes time to 0 and the population to 100. Row 2 represents the population at time 1, row 3 at time 2, etc.

Each entry in column A is 1 added to the previous value of t. Each entry in column B is just the previous row's value multiplied by $(1 + R)$, or λ. If you are unfamiliar with spreadsheets, you are advised to purchase one of the beginner's guides to the particular flavor of spreadsheet you will be using. Most are very good at explaining the basics of entering formulas and copying the formulas to new cells to create an iterative model. Particularly important is the concept of relative versus absolute addressing of cell values.

Problem 5–1. Develop a spreadsheet model of density-independent population growth. Use an initial population of $N_0 = 2$, so that your results can be compared to the graph in Figure 5–1. To implement the model in a spreadsheet, define a row in the upper portion of the sheet as the parameter R, followed by the values of R that you want to model. The next row provides the headings of your model. The first column should be time. Initialize each column to the initial population size, that is, 2. Then, enter the formulas to compute the population size based on the value of R for the column in the $t = 1$ row. Each formula will refer to the value of R in the first row with absolute addressing and the cell immediately above it with relative addressing: +B3* (1+B$1). Once you have the formula correct, and are generating the correct value, you can copy it down each of the three population columns. The result should look something like this:

	A	B	C	D
1	R	0.01	0.02	0.03
2	Time	Population 1	Population 2	Population 3
3	0	2	2	2
4	1	2.02	2.04	2.06
5	2	2.0402	2.0808	2.1218
6	3	2.060602	2.122416	2.185454
7	4	2.081208	2.164864	2.251018
8	5	2.102020	2.208162	2.318548

With the graphics capabilities of the spreadsheet package you are using, develop the following plots.

A. Graph population size versus time as is done in Figure 5–1.

B. Graph change in population size versus population size, where change in population size is N_{t+1}/Nt. What sort of simple graph results?

Solutions to this problem, and other problems in this chapter, are available as Quattro Pro spreadsheets at http://www.cnr.colostate.edu/~gwhite/bgmodel.

To achieve density-dependent population growth, we must let R be a function of population size, N_t [and hence time, $R(t)$]. The simplest possibility is a linear relationship,

$$R(N_t) = R(t) = R_0\left(1 - \frac{N_t}{K}\right),$$

which is generally known as *logistic* population growth because of the linear relationship between growth rate and population size (Figure 5–2). Note that the population growth rate equals zero when N_t equals carrying capacity (K), the threshold at which population growth goes to zero. The population growth rate is negative for population sizes above K.

With this function, $R(t) = f(N_t, K)$. Then, substituting the new $R(t)$ function into the equation for population growth, $N_{t+1} = N_t[1 + R(t)]$, the population growth curve in Figure 5–3 results.

The difference equation for the logistic equation is then

$$N_{t+1} = N_t\left[1 + R_0\left(1 - \frac{N_t}{K}\right)\right],$$

where the parameter R_0 is the maximum rate of population growth for an infinitesimally small population size and K is the size of the population that results in zero population growth. See Box 5–2 for a discussion of the differential form of the logistic equation. The plot of N_t versus t is shown in Figure 5–4, where the population stabilizes in time at a level of K.

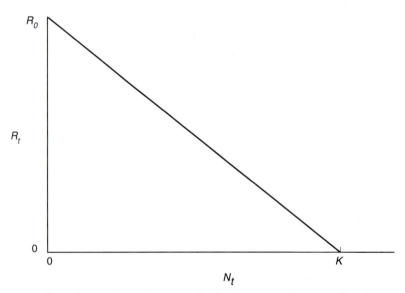

Figure 5–2 The relation between the finite population growth rate and population size that results in logistic growth. At population size K, the population growth rate is zero. At zero population size, the population growth rate is R_0.

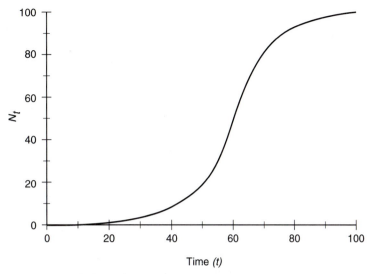

Figure 5–3 Graph of population size versus time for the logistic population growth model. The logistic model is a special case of density-dependent population growth where the finite population growth rate is a linear function of population size.

Box 5–2. The differential equation version of the logistic growth curve is in all basic ecology textbooks:

$$\frac{dN}{dt} = rN\left(1 - \frac{N}{K}\right).$$

The solution of this differential equation is

$$N = \frac{K}{1 + \left(\dfrac{K}{N_0} - 1\right)e^{-rt}}.$$

North American large mammal populations do not have continuous population growth, but rather discrete birth pulses, so the differential equation form of the logistic equation will not be considered further in this chapter.

The per capita finite rate of increase for the logistic equation is:

$$\frac{N_{t+1} - N_t}{N_t} = R_0 - \frac{R_0 N_t}{K}.$$

The concept of maximum sustained yield (MSY) can be derived from the logistic function. The population size that results in the maximum increase in the population is the population size that generates the MSY. Sustainable yield is defined as the harvest that can be taken each year (i.e., a sustained harvest). This harvest is maximized when the increase in the population is maximized (i.e., the maximum value of $N_{t+1} - N_t$).

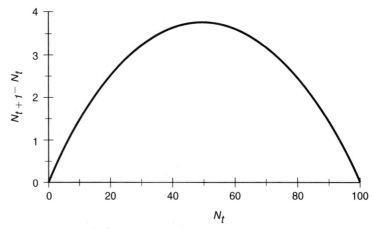

Figure 5–4 Relation of population size to annual yield for the logistic population growth model.

This maximum is achieved for $N_t = K/2$ (Figure 5–4). However, this relation only holds for the logistic model. Other models, with different relations between the finite population rate of growth and population size, result in different population sizes that produce the MSY.

To solve for the population size that maximizes yield, you must use calculus to differentiate the equation for yield with respect to N_t:

$$N_{t+1} - N_t = N_t \left[1 + R_0 \left(1 - \frac{N_t}{K} \right) \right] - N_t \; ,$$

which produces this solution:

$$N_t = K/2 \; .$$

Substituting the value $K/2$ into the equation for yield produces the result that the annual yield is

$$\frac{KR_0}{4} \; .$$

Once more, these results only apply to the logistic population growth model. Other models result in different population levels that produce MSY (see Box 5–3).

Problem 5–2. Develop a spreadsheet model of the density-dependent population growth as a difference equation model. You can assume that the rate of change of the finite growth rate is linear in terms of population size, that is, logistic population growth. Reasonable parameter values are $K = 300$ and $R_0 = 0.12$.

A. Graph population size versus time.
B. Graph change in population size versus population size.
C. Graph per capita change in population size versus population size.

Box 5–3. The mathematical expression of density dependence can take many forms. W. E. Ricker invented a discrete population model for fishery stocks (4, also see 5, p. 282):

$$N_{t+1} + N_t \exp\left[R_0 \left(1 - \frac{N_t}{K} \right) \right] .$$

Note that the density dependence in this model becomes stronger at higher densities, due to the exponential function relating density to carrying capacity.

Another example is provided by assuming a linear relationship between N_t / N_{t+1} and N_t [instead of between $(N_{t+1} - N_t) / N_t$ and N_t as in the logistic equation], so that the following model can be defined:

$$\frac{N_t}{N_{t+1}} = \beta + \frac{1-\beta}{K} N_t ,$$

with intercept β and slope $(1 - \beta)/K$. The resulting population growth model is

$$N_{t+1} = \frac{KN_t}{K\beta + (1-\beta)N_t} .$$

By taking the limit of the per capita rate of population growth as N_t approaches zero, we find that R_0 can be specified as a function of the parameter β as

$$R_0 = \frac{1-\beta}{\beta} ,$$

giving the following parameterization of the model:

$$N_{t+1} = \frac{KN_t(R_0 + 1)}{(K + N_t R_0)} .$$

This model is generalized by Hassell, Hassell and others, and May (6, 7, 8) as

$$N_{t+1} = \frac{\lambda N_t}{(1 + aN_t)^b} .$$

These three models predict MSY at values other than $K/2$. As an exercise, you might graph these functions with the logistic model in a spreadsheet to compare the differences.

AGE-STRUCTURED POPULATION MODELS

Age-structured models add complexity to a population model, but make the model more realistic in that essential features of the population growth process are captured by the model. I will continue to use difference equations to define the population model because discrete age classes require difference equations for simple solutions. Let $N_{i,t}$ be the number of individuals of age class i at time t. I define time t as the start of the biological year, or the instance in time immediately following the birth pulse or reproduction. The finite rate of change parameter (R) of the density-independent model is now broken into component parts: survival (1 − mortality), reproduction, immigration, and emigration. As an example, consider a mule deer population with four age classes: fawns (F), yearlings (Y), 2-year-old adults ($A2$), and adults older than 2 years ($A>2$). Only females are considered in the following example. I need to define the following parameters to develop the model. Estimates of these parameters to use in the model would be obtained with the methods described in Chapter 4.

1. Fawn survival rate (S_F) is 0.35 (i.e., only 35% of the fawns that are born live to their first birthday).
2. Yearling survival rate (S_Y) is 0.8 [i.e., 80% of the yearlings (fawns that celebrate their first birthday) live to celebrate their second birthday].
3. Adult survival rate (S_A) is 0.85 [i.e., 85% of the adults (both 2 years old and older) survive to celebrate their next birthday].
4. Each adult > 2 years old gives birth to 1.6 fawns (birth rate of adults > 2 years old, $B_{A>2}$), of which half are female and half are male (sex ratio defined as the proportion of births that are female, $SR = 0.5$).
5. Each adult exactly 2 years old has 0.8 fawns (birth rate of yearlings, B_{A2}), of which half are female and half are male.
6. Fawns on their first birthday (i.e., having just become yearlings) are assumed to not reproduce.
7. Immigration and emigration are assumed to be zero for all age classes.
8. Start with 10 adult females > 2 years old, so that $N_{F,0} = 0$ (number of fawns at time 0), $N_{Y,0} = 0$ (number of yearlings at time 0), $N_{A2,0} = 0$, and $N_{A>2,0} = 10$.

With these specifications, the number of adults that survive to time 1 is $N_{A>2,1} = N_{A>2,0} \times S_A$, giving $N_{A>2,1} = 10 \times 0.85$. The number of fawns born at time 1 is $N_{F,1} = N_{A>2,0} \times S_A \times B_A \times SR$, giving $N_{F,1} = 10 \times 0.85 \times 1.6 \times 0.5 = 6.8$. Because we assumed only 10 adults > 2 years old at time 0, our population at time 1 consists of only fawns and adults > 2.

At time 2, our population consists of three different age classes. The number of yearlings will be $N_{Y,2} = N_{F,1} \times S_F$ the number of adults > 2 years old will be $N_{A>2,2} = N_{A>2,1} \times S_A$, and the number of fawns born at time 2 is $N_{F,2} = N_{A>2,1} \times S_A \times B_{A>2} \times SR$. The resulting population sizes are $N_{F,2} = 5.78$, $N_{Y,2} = 2.38$, $N_{A2,2} = 0$, and $N_{A>2,2} = 7.225$.

By time 3, our population consists of all four age classes. The number of yearlings will be $N_{Y,3} = N_{F,2} \times S_F$, the number of 2-year-old adults will be $N_{A2,3} = N_{Y,2} \times S_Y$, the number of adults > 2 years old will be $N_{A>2,2} = N_{A>2,1} \times S_A$, and the number of fawns born at time 3 is $N_{F,3} = N_{A>2,2} \times S_A \times B_{A>2} \times SR + N_{Y,2} \times S_Y \times B_{A2} \times SR$. The resulting population sizes are $N_{F,3} = 5.5746$, $N_{Y,3} = 2.023$, $N_{A2,3} = 1.904$, and $N_{A>2,3} = 6.14125$.

From this beginning, I can write the general set of iterative difference equations that defines the population model:

$$N_{A>2,t+1} = N_{A>2,t} S_A + N_{A2,t} S_A,$$

$$N_{A2,t+1} = N_{Y,t} S_Y,$$

$$N_{Y,t+1} = N_{F,t} S_F, \text{ and}$$

$$N_{F,t+1} = (N_{A>2,t} + N_{A2,t}) S_A B_{A>2} SR + N_{Y,t} S_Y B_{A2} SR, \text{ or equivalently}$$

$$N_{F,t+1} = N_{A>2,t+1} B_{A>2} SR + N_{A2,t+1} B_{A2} SR.$$

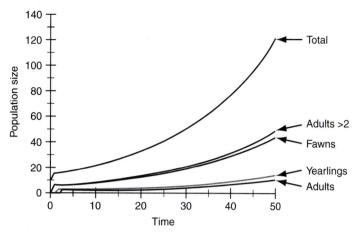

Figure 5–5 Population growth in an age-structured model of mule deer.

Because of the different reproductive rate of adults exactly 2 years old, and those older, I have to keep track of the number of animals in both of these age classes, even though they have the same survival rate. Yearlings and fawns have a different survival rate than the two adult age classes, so they also have to be followed separately.

After you have solved Problem 5–3, you will get a graph like that shown in Figure 5–5. After the initial fluctuations because only adults > 2 years old were in the population, we see a curve identical to density-independent population growth (Figure 5–1). This result should not surprise you—nowhere in the model did we include a density-dependent relation. As a result, the population stabilizes and grows at a constant rate (i.e., $\lambda = 1.044492$). To obtain this value, you compute the ratio of $N_{Total, t+1}/N_{Total, t}$ for each time period, until the effects of the initial population size have worn off and the growth rate stabilizes.

Problem 5–3. Develop a spreadsheet model of the example mule deer population.

  A. Graph population size for each age class, plus the total population, versus time.
  B. Graph change in population size versus population size—what kind of population growth is being modeled?
  C. Estimate λ.
  D. Estimate the stable age distribution (i.e., the proportion of the population that will be fawns, yearlings, adults 2 years old, and adults > 2 years old).

Another feature of this model is that the ratios of the various age classes stabilize. That is, the ratios of the numbers of animals in each age class are constant, regardless of the population size. This property of age-structured models was first formalized by Leslie in his development of the use of matrices in population modeling (see Box 5–4) (9). Caughley cautioned against the use of age ratios to infer population change for this reason (10).

Box 5–4. The Leslie matrix (also known as a projection matrix or transition matrix), is often used to model mathematically age-structured populations (9, 11). The Leslie matrix for the mule deer example presented in the text would be

$$\mathbf{L} = \begin{bmatrix} 0 & 0 & S_A B_{A2} & S_A B_{A>2} \\ S_F & 0 & 0 & 0 \\ 0 & S_Y & 0 & 0 \\ 0 & 0 & S_A & S_A \end{bmatrix},$$

where the boldface $\mathbf{L}$ indicates a matrix. Then, the vector of population sizes at time $t + 1$ is defined recursively as

$$\mathbf{N}_{t+1} = \mathbf{L} \times \mathbf{N}_t \; .$$

Other references pertinent to matrix projections are Lefkovitch, Usher, Caswell, and Manly (12, 13, 14, 15, 16).

The omission of density dependence in our model is a serious violation of biological logic—nobody expects a deer population to grow indefinitely. Thus, we need to consider how to incorporate density dependence into our model. As discussed in Chapter 8, various mechanisms regulate large mammal populations. One of the most important is increased mortality of the young in response to increased population density. For example, Bartmann and others demonstrated that overwinter fawn survival of mule deer in northwest Colorado declined with increasing density in a pasture experiment (17). The logistic regression function they fit to their observed survival was

$$S_F = \frac{\exp(1.1906 - 0.0195 \times \text{December density})}{1 + \exp(1.1906 - 0.0195 \times \text{December density})} \; ,$$

which results in a decline in fawn survival with increasing December density of deer. Implementing this function into our age-structured model results in the population growth shown in Figure 5–6. Now, the population stops growing when it reaches its carrying capacity, K, because λ approaches 1 and population growth ceases.

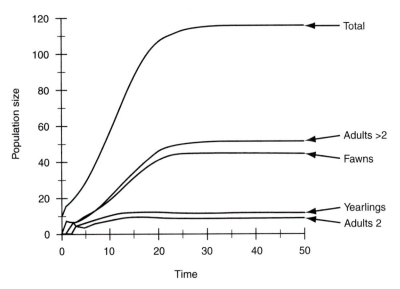

Figure 5–6 Population growth as a result of incorporating density dependence in fawn survival in the age-structured model of a mule deer population.

Age-Structured Population Models

Problem 5–4. Insert the density-dependent function of Bartmann and others for S_F in your spreadsheet model created in Problem 5–3 (17).

    A. How does λ change with time? What is the final value of λ?

    B. Does your graph of population size resemble Figure 5–6?

    C. Start with your spreadsheet from Problem 5–3 and incorporate density dependence into B_{A2} and $B_{A>2}$. Reasonable functions might be $B_{A>2} = 2.3 - 0.01\,N_{\text{Total}}$ and $B_{A2} = 1.5 - 0.01\,N_{\text{Total}}$. Do the resulting curves of population growth resemble Figure 5–6 also? Why is K not the same for this function as for density dependence in S_F?

Several important points are made with this example. First, density dependence was only incorporated into one parameter of the model (i.e., S_F). Even though all the other parameters are still density independent, the population size converges to a constant value, K. Thus, our model assumes that the birth rate in the population is constant across density, even though biologically this might not be the most reasonable assumption.

Second, incorporation of density dependence into any single parameter will result in the population converging to a constant size if the population would go extinct by making this parameter zero. That is, the effect of density must be great enough to force the value of λ to equal one. If making the parameter zero is not enough to make $\lambda = 1$, the effect of density via this parameter will not result in the population approaching K. An example of a parameter that would not result in a constant population if it were made a function of density is the birth rate of 2-year-old females in our age-structured model. Changing B_{A2} to 0 results in $\lambda = 1.0296$, which is not adequate to limit population growth. Thus, making B_{A2} a function of population density would not, by itself, be enough to regulate population growth. Other parameters in the model also would have to be made a function of density to achieve regulation. However, biologically, B_{A2} is very likely a function of density.

Third, the relationship between S_F and density was not linear, in contrast to the logistic population growth model discussed earlier. Evidence has accumulated that the effects of density in ungulates are strongest near K and are not constant. That is, the relationship between survival and/or birth rates and density is not linear (18).

Finally, a caution about developing age-structured models. Modelers have to be extremely careful to define the time of year when they census the model population. In the example presented here, the census occurred just after the birth pulse. Fawns in this model were just born. An alternative model would be to census the population just prior to the birth pulse. Then, the fawn age class would consist of animals almost 1 year old. The structure of the model would look quite different, but the results, such as the estimate of λ, would be identical (assuming you have structured the model correctly). Noon and Sauer discuss how to structure models with different census times (19). Constantly be aware of how you have defined the population measured by N, and structure your equations appropriately.

A "shortcut" often used to build age-structured models is to ignore a portion of the life cycle. A common example of this trick used in big game models is to measure recruitment to the population as the number of young alive at the start of the winter, when age ratio surveys are performed (see Chapter 4). For example, the Colorado Division of Wildlife estimates December fawn : female ratios for mule deer via helicopter surveys. The values of B_{A2} and $B_{A>2}$ cannot be distinguished, because the age of adults cannot be distinguished from the air. In fact, yearlings cannot be reliably distinguished from adults, so the estimate of fawns : females reflects the number of fawns produced for the three age classes N_Y, N_{A2}, and $N_{A>2}$, even though N_Y animals do not produce fawns. The reason for making this biologically incorrect assumption is that estimation of the true recruitment for each of these age classes would require monitoring the fate of marked, known-aged animals, and is not economically feasible. Thus, by taking this shortcut, reasonable population models can be developed from field data.

Another shortcut used in the model presented here is that no senility is assumed (i.e., adults 3 years old have the same survival rate as adults > 20 years old). This assumption is made because of the difficulty in obtaining year-specific and age-specific survival rates. The assumption is not really all that unrealistic, in that few animals live to the age where senility becomes a factor, even at high survival rates such as 0.85 used in our mule deer model. Senility is not difficult to model; it just requires different survival rates as a function of age. The problem is obtaining the data to support senility effects, particularly given that survival likely varies by year. Incorporating age-specific and year-specific effects into a model requires a large amount of data to support the complexity. Similarly, age-specific reproductive rates can be extended to more than four age classes by increasing the number of age classes in the model. The limitation is data to support the assumptions.

Another extension to the basic model presented here is to include multiple areas, so that immigration and emigration can take place between areas. You can think of emigration as the death of an animal on the first area, and its immigration to a second area as a birth, albeit older than age 0. Models that incorporate multiple areas with immigration and emigration between areas approach the general concept of a metapopulation (20).

STOCHASTIC POPULATION MODELS

Real populations never exhibit deterministic population growth (i.e., population dynamics as modeled by a deterministic equation that has a constant rate of growth such as that presented earlier). Rather, population levels fluctuate from year to year and from site to site. Being able to predict the amount of random variation that we would expect to see in a population is useful for making predictions about the future size of a population. Specifically, predictions about the persistence of a population are part of a specialized modeling field called *population viability analysis* (21).

Variation in population sizes can be classified into two general mechanisms. The true population may vary through time and space, even though a deterministic model predicts a constant population. Variation in the true population size is termed *process variation,* because of stochasticity in the population growth process. Several mechanisms can cause process variation.

Demographic variation is the randomness that exists as to whether an individual reproduces or survives. For example, the outcome of whether an individual with a specified survival probability survives is based on a Bernoulli distribution. I like to call this source of variation "penny flipping variation" because the variation about the expected number of survivors parallels the variation about the observed number of heads from flipping coins. To illustrate demographic variation, suppose the probability of survival of each individual in a population is 0.8. Then on average, 80% of the population will survive. However, random variation precludes exactly 80% surviving each time this survival rate is applied. From purely bad luck on the part of the population, a much lower proportion of the population may survive for a series of years, resulting in extinction. Because such bad luck is most likely to happen in small populations, this source of variation is particularly important for small populations, and hence the name demographic variation. The impact is small for large populations. As the population size becomes large, the relative variation decreases to zero. That is, the variance of N_{t+1}/N_t goes to zero as N_t goes to infinity. Thus, demographic variation is generally not an issue for persistence or modeling of larger populations.

To illustrate further how demographic variation operates, consider a small population with $N = 100$ and a second population with $N = 10,000$. Assume both populations have identical survival rates of 0.8. With a binomial model of the process, the probability that only 75% or less of the small population survives is 0.1314 for the small population, but $3.194E - 34$ (i.e., a very small probability) for the larger population. Thus, the likelihood that up to 25% of the small population is lost in one year is much higher than for the large population.

Demographic variation can be incorporated into spreadsheet models by using the binomial distribution. To use an example from the age-structured model of mule deer described earlier, instead of multiplying $N_{Y,t} \times S_Y$ to get $N_{A2,\,t+1}$, we assume that on average S_Y animals live, but for any specific year, some random value is observed. The CRITBINOM function of the spreadsheet software is used, that is, $N_{A2,\,t+1} = $ @CRITBINOM($N_{Y,t}$, S_Y, @RAND). The @RAND function provides a random variable distributed uniformly between 0 and 1, with the result that the observed value of $N_{A2,\,t+1}$ will vary around a mean of $N_{Y,t} S_Y$ with variance $N_{Y,t} S_Y (1 - S_Y)$. Thus, demographic variation could be included in all the survival rates of the age-structured model. Demographic variation should also be incorporated into the reproduction process. Instead of applying the birth rate of B_{A2} to $N_{A2,\,t+1}$ directly, use the binomial distribution: CRITBINOM($N_{A2,\,t+1}$, B_{A2}, @RAND). Reproduction for the adults > 2 age class is more problematic, but a reasonable model would be to assume all adults > 2 years old have 1 fawn, and some random proportion have 2 fawns. The following statement would achieve this result:

$$N_{A>2,\,t+1} + \text{@CRITBINOM}(N_{A>2,\,t+1}, B_{A>2} - 1, \text{@RAND}).$$

Knight and Eberhardt developed a stochastic model of grizzly bear populations in Yellowstone National Park (22). Their model only incorporated demographic variation (see Box 5–5).

Box 5–5: Stochastic Models. Knight and Eberhardt (22) developed a stochastic model of the grizzly bear population in Yellowstone National Park. Their model assumed only demographic variation (i.e., no other sources of stochasticity were included). The model included age structure, with age-specific rates of survival and reproduction. The purpose of the model was to assess the probability of persistence of the population for 30 years, and estimate likely future populations. The critical assumption made with their approach is that the population parameters used in the model are not changing with time or density. I consider these assumptions acceptable for the intended purpose of the model.

With Tom Beck and Bruce Gill, Colorado Division of Wildlife, I extended the Knight and Eberhardt approach to examine the question of how much the Colorado black bear population would expand with the closure of spring bear hunting in the state. A range of hunting scenarios was explored with the model, including no harvest, two spring seasons with 30% and 35% of the harvest consisting of females, and three fall seasons with 35%, 40%, and 45% of the harvest consisting of females. I incorporated temporal variation to account for changes in survival and reproduction in response to failures in mast and berry crops from late spring frosts, plus temporal variation in annual harvest rates because of hunting conditions. One year in 10 was assumed to have optimal hunting conditions. At the time I built this model, spreadsheet technology had not yet reached the level where stochastic models could be easily implemented. Thus I used SAS to implement the model. The SAS code is available at http://www.cnr.colostate.edu/~gwhite/bgmodel, along with a description of the model (23).

Although our black bear harvest model was interesting and fun to build and play with, I found the results of little value in answering the questions that motivated the model's construction. The nagging question that the model could not answer for lack of data was the role of density dependence in a burgeoning bear population. This question became increasingly apparent as we examined the various scenarios predicted by the model, and thus focused our thinking on a critical issue where more research and data are required. Because our model was looking at increases in the population, density dependence was an important issue. In contrast, Knight and Eberhardt were more concerned with declines in an already small population, where the role of density dependence is likely negligible. Thus, in one case, a stochastic model was useful in answering the motivating question, in another, useful in focusing our attention on a deeper question.

Another form of stochasticity to incorporate into a model is *temporal variation* (i.e., making the parameters random variables that assume new values each year). Such variation would be exemplified by weather in real populations. Some years, winters are mild and survival and reproduction are high. Other years, winters are harsh and survival and reproduction are poor. Another form of environmental stochasticity is *spatial variation* (i.e., variation of population parameters across the landscape). Factors causing geographic variation include geologic differences that affect soil type, and thus habitat, and weather patterns (e.g., differences in rainfall across the landscape). If the immigration and emigration rates are high across the landscape, so that subpopulations are depleted because of local conditions, high spatial variation can lead to higher persistence. This is because the probability of all the subpopulations of a population being affected simultaneously by some catastrophe is low when high spatial variation exists. In contrast, with low spatial variation, the likelihood of a bad year affecting the entire population is high. Thus, in contrast to temporal variation, where increased variation leads to lowered persistence, increased spatial variation leads to increased persistence given that immigration and emigration are effectively mixing the subpopulations. If immigration and emigration are negligible, then spatial variation divides the population into smaller subpopulations, which are more likely to suffer extinction from the effect of demographic variation on small populations.

The combination of temporal and spatial variation is termed *environmental variation*. Both dictate the animal's environment, one in time, one in space.

Environmental stochasticity can be incorporated into a model by making the parameters random variables. As an example, consider making the reproductive rate of adults > 2 years old a random variable with a normal distribution with mean $B_{A>2}$ and standard deviation $SD(B_{A>2})$. The NORMINV function of spreadsheet software can produce random normal deviates with the statement: @NORMINV(@RAND, $B_{A>2}$, SD[$B_{A>2}$]). Thus, year-specific reproductive rates would be generated randomly for each year, and then applied to the population. The result would be a model that exhibited year-to-year differences in population growth. Another distribution useful for generating random survival rates (and possibly reproductive rates in some circumstances) is the beta distribution. Random beta variables can be generated in spreadsheets with the BETAINV function. Advantages of the beta distribution are that random values can be constrained to specified intervals. For example, survival rates can be constrained to the interval 0–1. Transformations can also be useful in constraining random variables to specified intervals. The logistic transform, log[$S/(1 - S)$] = f(random variables) is a useful method. The logistic function was used to model density dependence in fawn survival rates earlier. An advantage of using the logistic transform is that correlated random variables can be generated. That is, fawn survival and adult survival logically are not independent. A severe winter with poor survival for adults guarantees a low survival rate for fawns. With the logistic function, two random normal deviates can be used to generate survival rates. The first deviate (D_1) would provide the year-to-year variation, with a large standard deviation. The second deviate (D_2) would provide minor variation of fawn survival from adult survival. The functions might be:

$$\log[S_A/(1 - S_A)] = \beta_A + D_1 \ , \text{ and}$$

$$\log[S_F/(1 - S_F)] = \beta_F + D_1 + D_2 \ .$$

Another approach to incorporating environmental stochasticity into a model is to use the bootstrap technique to sample from observed parameter values for the population of interest. The VLOOKUP or HLOOKUP functions in spreadsheets, combined with a random integer, can be used to bootstrap observations effectively. The correlation of parameters can be maintained by selecting a year-specific set of values (i.e., the fawn and adult survival rates from one year are selected together).

All the models examined so far assume that each animal in the population has exactly the same chance of survival and reproduction, even though these rates are changing with time. What happens if each animal in the population has a different rate of survival and reproduction? Differences between the individuals in the population are termed *individual heterogeneity,* and they create *individual variation.* Many studies have demonstrated individual heterogeneity of individual survival and reproductions. Clutton-Brock and others demonstrated lifetime reproductive success of female red deer varied from 0 to 13 calves reared per female (24). Differences in the frequency of calf mortality between mothers accounted for a larger proportion of variance in success than differences in fecundity. Bartmann and others (17) demonstrated that overwinter survival of mule deer fawns was a function of the fawn's weight at the start of the winter, with larger fawns showing better survival.

Individual variation is caused partially by *genetic variation* (i.e., differences among individuals because of their genome). Individual heterogeneity is the basis of natural selection (i.e., differences between animals is what allows natural selection to operate). However, *phenotypic variation* also is possible, where individual heterogeneity is not a result of genetic variation. Animals that endure poor nutrition during their early development may never be as healthy and robust as animals that are on a higher nutritional plane, even though both are genetically identical. Animals with access to more and better resources have higher reproductive rates, as shown for red deer (24). Thus, individual heterogeneity may result from both genetic and phenotypic variation. Lomnicki developed models of resource partitioning that resulted in phenotypic variation of individuals (25).

Undoubtedly, natural selection plays a role in the genetic variation left in a declining population. Most populations where biologists are concerned about extinction probabilities have suffered a serious decline in numbers. The genotypes remaining after a severe decline are unlikely to be a random sample of the original population. I would expect that the genotypes persisting through a decline are the "survivors," and they would have a much better chance of persisting than would a random sample from the population prior to the decline. Of course, this argument assumes that the processes causing the decline remain in effect, so that the same natural selection forces continue to operate.

The reason that increased individual heterogeneity increases population persistence is that increased variation results in more chance a few animals have exceptionally high reproductive potential and high survival. Therefore, these animals are unlikely to suffer mortality and be removed from the population, and also can be relied on to contribute new births each year. As a result, the population may remain small, but will not go extinct as often. Individual heterogeneity has seldom, if ever, been included in a population viability analysis. Yet, as simple examples show, individual heterogeneity is a very important element in maintaining viability (26).

The combined effects of demographic, temporal, spatial, and individual variation are termed *process variation.* That is, each of these sources of variation affects population processes. Process variation is used as a general term for the inherent stochasticity of changes in the population level. Process variation is in contrast to *sampling variation,* which is the variation contributed when biologists attempt to measure population processes. That is, researchers are unable to measure the exact survival rate of a population. Rather, they observe realizations of the process, but not the exact value. Thus, we observe only estimates of population size, $\hat{N}_t$, instead of true population size N_t. Even if the fate of every animal in the population is observed, the resulting estimate of survival is only an estimate of the true, but unknown, population survival rate, because of demographic variation. We only observe the outcome of the stochastic process and can never measure the underlying parameters.

In reality, we may be fortunate and have a series of survival or reproduction estimates across time that provides information about the temporal variation of the process. However, the variance of this series is not the proper estimate of the temporal variation of the process. This is because each of our estimates includes sampling variation (i.e., we only have an estimate of the true parameter, not its exact value). To estimate properly the temporal variation of the series, the sampling variance of the estimates must be removed. Procedures to remove the sampling variance from a series of estimates and obtain an estimate of the underlying process variation (which might be temporal or spatial variation) are available. Simple methods of moment estimators are explained in Burnham and others and Burnham (27, pp. 260–278). For some kinds of data, where sampling variation is assumed constant, analysis of variance techniques can be used to estimate the underlying process variance. However, these techniques are beyond the scope of this chapter, and are not considered further. White provides a review of methods to estimate process variance and then incorporate the estimate into a population model (26).

Problem 5–5. Add stochasticity to a model.

A. Incorporate demographic variation into the age-structured model of Problem 5–4. Use the CRITBINOM function to produce demographic stochasticity in survival and reproductive rates. Note how you get a different population growth curve each time you recalculate the spreadsheet. Explain why the population is now always an integer value. Change the initial population to 1,000. Note how the population growth function smooths out, demonstrating how demographic variation is only important at small population sizes.

B. Incorporate temporal variation into the demographic stochasticity spreadsheet by making the survival and reproductive rates vary by year. Assume that the standard deviation for each parameter is equal to 10% of its mean value, although White and Bartmann reported that the standard deviation of overwinter fawn survival is closer to 50% of its mean value for a population in northwest Colorado (28). Note how much more variable the resulting population growth curves are than the model with just demographic variation, especially when the initial population size is set at 1,000.

C. For the courageous, modify the age-structured model with density dependence with demographic and temporal variation.

FITTING MODELS TO DATA

Often, data are available from a population, and we desire estimates of the underlying parameters that govern the population. Typically, regression methods are used, but more complex procedures often are required (29). The optimization procedures of spreadsheets allow sophisticated nonlinear parameter estimation, methods that are far beyond the scope of this chapter. However, some exposure to parameter estimation methods is needed to develop useful population models. An example that compares two models is useful in demonstrating the capabilities of these techniques.

To illustrate the possible methods, data from McCorquodale and others and Eberhardt and others will be used (30, 31). These studies report on the growth of an isolated elk population in the shrub-steppe region of southcentral Washington. The elk population colonized the area about 1975. The population estimates are provided in Table 5–2. An interesting question posed by these data is whether the population has shown any decline in its rate of growth, or, in other words, is the population exhibiting density dependence? One approach is to fit the two models, density independence and density dependence, and compare the results. Does the density-dependent model provide an improved fit compared to the density independent-model?

TABLE 5–2

Elk Population Estimates in Southcentral Washington

Year	Estimated Population
1975	8
1976	13
1977	15
1978	
1979	16
1980	
1981	25
1982	27
1983	40
1984	55
1985	71
1986	89
1987	94
1988	95
1989	102
1990	115
1991	133
1992	190
1993	238

To fit the models, I assumed that the estimated population is the true population, that is, that the count is the actual population. Using these values as N_t, I fit the following models:

Density independence: $\hat{N}_{t+1} = N_t(1 + R)$

Density dependence: $\hat{N}_{t+1} = N_t[1 + R(1 - N_t/K)]$.

For each model, the sum of the differences $[\log(\hat{N}_{t+1}) - \log(N_{t+1})]^2$ was minimized. I used logarithms of the observed and predicted values to standardize the variance of the residuals across the nearly 2 orders of magnitude of the observed data. For the years 1979 and 1981, each model was applied twice to the preceding year to obtain the predicted population size.

The result was that the density-dependent model did not improve the fit over the density-independent model ($P = 0.979$). The estimate of K was > 8,000, but was basically not estimable from these data, consistent with no evidence of density dependence in these data. The fit of the two models was so close that the lines cannot be distinguished on a graph (Figure 5–7). The estimate of λ was 1.194, very similar to the values obtained by the original authors. However, a conceptually different model was fit to produce their estimate. The observed population sizes are assumed to have no sampling error, in contrast to the approach used by McCorquodale and others and Eberhardt and others in which all the variation in the data was assumed to be sampling error (30, 31).

Problem 5–6. Use a spreadsheet to fit the models of density independence and density dependence to the elk data.

A. Graph the two models with the observed data to produce a plot similar to that of Figure 5–7.

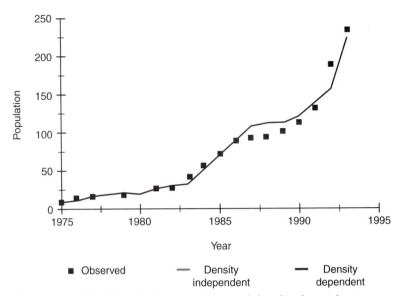

Figure 5–7 Fit of density-independent and density-dependent models to the data of McCorquodale and others and Eberhardt and others from the Arid Lands Ecology Reserve, Washington (30, 31). The two models are indistinguishable on the graph.

Box 5–6: Complex Example of Fitting a Model to Data. Data have been collected on the Piceance mule deer herd in northwestern Colorado (Table 5–3). When a two-age class model (fawns, adults) with sex-specific classes for adults is built from these data, the modeled population crashes. The initial population was computed from the 1981 population estimate assuming that 67% of the animals were counted on the quadrats sampled, based on the work of Bartmann and others (32). Age structure was computed from the 1981 age ratios. Years with missing fawn:female ratios were replaced by the mean of the series. With these inputs, the male:female ratio becomes negative and the population goes to zero. Although the population had been thought to be declining during the 1990s, the decline wasn't observed to be that severe. Details are provided in the spreadsheet available from http://www.cnr.colostate.edu/~gwhite/bgmodel, including graphs of the model and observed values.

The problem identified with this model is that the data are not totally consistent with themselves. That is, sampling variation in the parameter estimates causes the model to crash. We must now fit a model to the observed values that makes the model produce reasonable predictions. The strategy taken was to minimize the weighted least squares between the observed population size and male:female ratios and the predicted values. The weight of each of these observations was taken as the reciprocal of the variance. Thus, for each of these observations, the quantity $[(\theta_i - \hat{\theta}_i)/SE(\theta)]^2$ was computed to create an objective function to minimize. The parameters changed to improve the fit were the fawn and adult survival rates, December fawn:female ratios, and population size. For each of these parameters, a penalty was added to the objective function of the form $[(\theta_i - \hat{\theta}_i)/SE(\theta)]^2$. That is, any change in one of these parameters from its observed value increased the size of the objective function, and thus penalizes the optimization for the change. The resulting fit of the model is a balance between fitting the observed male:female ratios and population sizes corrected for sightability and changing the observed data more than can be justified given its precision. The optimization spreadsheet is included with the spreadsheet mentioned earlier. Results from this procedure were a much-improved fit of the model to the observed data (Figure 5–8). The predicted decline in the population is now consistent with other observations of population size taken as part of a research study (33).

TABLE 5-3
Estimates of Fawn and Adult Survival[a]

Year	Fawns: 100 Female Estimate	SE	Males: 100 Female Estimate	SE	Fawn Survival Estimate	SE	Adult Survival Estimate	SE	Population Size Estimate	SE	Male Harvest Estimate	SE	Female Harvest Estimate	SE
1981	77.7	5.78	13.8	1.95	0.48	0.068	0.86	0.049	21,103	3,592	2,293		19	
1982	75.5	4.34	11.4	1.34	0.36	0.044	0.81	0.048	16,004	2,425	3,072		10	
1983	78.8	4.83	11.4	1.45	0.05	0.021	0.83	0.045	27,309	3,129	3,512		64	
1984	70.2	4.49	7.4	1.16	0.19	0.039	0.88	0.040	21,723	2,387	2,017		12	
1985	72.5	5.57	7.2	1.38	0.41	0.039	0.92	0.038	21,657	2,822	1,849		30	
1986	63.5	4.11	14.0	1.62	0.42	0.038	0.76	0.068			931		21	
1987					0.15	0.033	0.88	0.083			1,326		24	
1988	74.2	4.66	13.9	1.63	0.35	0.064	0.83	0.108	25,248	2,517	1,449	75	585	19
1989	65.7	3.97	12.4	1.42	0.77	0.049	0.90	0.051			2,227	95	1,512	59
1990	61.2	3.92	16.2	1.72	0.32	0.069	0.94	0.035			1,822	92	1,691	48
1991	46.4	2.65	11.9	1.17	0.49	0.072	0.77	0.052			1,917	92	1,238	45
1992	45.5	3.61	10.5	1.51	0.14	0.029	0.71	0.048			1,310	68	1,296	70
1993	42.6	3.50	10.1	1.50	0.65	0.038	0.84	0.038			1,041	63	777	53
1994	46.1	4.05	7.8	1.43	0.76	0.034	0.88	0.035			1,210	65	221	17
1995	47.6	3.81	10.7	1.56	0.70	0.038	0.93	0.029			1,489	68	182	16
1996											1,631	69	206	18
Mean	62.0	4.24	11.3	1.49	0.42	0.045	0.85	0.051	22,174	2,812				
SD	13.6		2.7		0.23		0.07		3,886					

[a]Fawn : female, male : female, and population size from the Piceance mule deer herd, northwestern Colorado, 1981–1995. Missing data are shown as blank entries.

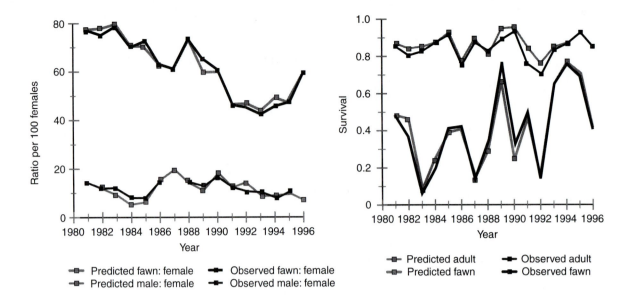

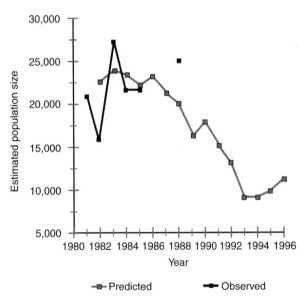

Figure 5–8 Results of fitting a model to the data in Table 5–3 on the Piceance mule deer herd in northwestern Colorado.

PHILOSOPHY ABOUT THE USE OF POPULATION MODELS

There are numerous reasons for constructing models. First, models provide an explicit statement of assumptions about the species of interest. Thus, a realistic model summarizes existing knowledge about a population. Second, models provide predictive capability. This predictive capability is useful for setting harvest levels to meet herd objectives and for resolving controversies. Finally, models allow biologists to play "what if" games. Modeling will highlight your ignorance—parameters in the model that are just guesses become obvious when you are building the model. Models will expose your ignorance as you acknowledge the lack of data available for managing a population.

However, the very reasons that models are useful are also why they are dangerous. One of the most common abuses of models is in using them for purposes for which they were not intended. Too often, a model is constructed with the intention of playing "what if" games, but then is used to make predictions about the population. A second abuse of models is not understanding the limitations of the data when building a model. Examples are using the wrong variance component (e.g., sampling variation in place of temporal variation) and improper sampling frame when data are collected, resulting in biased estimates being used in the model.

I try to adhere to the following criteria when using models. First, models test hypotheses about models, not about actual populations. Inferences made to real populations from models are conditional on the assumptions and data used to build the model. Too often in the literature, we find statements made about model results that get interpreted as statements about real systems. Second, models should be developed from data taken on a population if you are going to use the model to make predictions about that population. Third, models used for prediction should be kept simple, and used only within the breadth of the data used to develop them (i.e., no extrapolation). Building a model from data will generally mean the model is simple because we seldom have enough data to justify a complex model.

The crucial philosophy underlying the last two points just mentioned is that management decisions must be based on data. In other words, the management of a population should not be based on model predictions where the model inputs are not provided from measurements made in the field. Complex models of population dynamics may capture most of our knowledge of this system, but such models do not provide reliable predictions of year-to-year dynamics because of the lack of annual information on required inputs.

The issue of model complexity is better comprehended with an analogy to an auto trip from New York City to Los Angeles. No reasonable driver would start this trip with 7.5-minute USGS topographic quadrangles as his/her model. Certainly the topographic quadrangles contain all the necessary information, but the detail is considerably more than needed. A simpler model will suffice, such as state road maps, and is more likely to result in success. An even simpler model of just a single map of the interstate highways would suffice, but would not provide all the details we might like. Unfortunately, costs usually limit the amount of information available, even though we may desire more. In summary, models are a useful and powerful tool. But like all sharp tools, you can easily cut yourself if you misuse them.

SUMMARY

Population modeling provides a mathematical tool to predict the trajectory of a population or to evaluate potential management strategies. Computer spreadsheets provide a powerful tool for population modeling, including stochastic models and graphical presentations. Density-independent and density-dependent models of populations with single and multiple age classes are developed in this chapter. Difference equation models are employed. Stochastic models can include process variation, such as environmental (temporal and spatial), individual (genetic and phenotypic), and demographic variation, but should not incorporate sampling variation of the estimates used to construct the model into the stochasticity of population growth. The random number generators of modern spreadsheets provide the tools to build reasonable stochastic population models. The numerical optimization capabilities in spreadsheets provide the tools needed to fit complex models to observed data.

LITERATURE CITED

1. LUBOW, B., G. C. WHITE, and D. R. ANDERSON. 1996. Evaluation of a linked-sex harvest strategy for cervid populations. *Journal of Wildlife Management* 60:787–796.
2. GROSS, J. E., J. E. ROELLE, and G. L. WILLIAMS. 1973. Program ONEPOP and information processor: A systems modeling and communications project. Colorado Cooperative Wildlife Research Unit, Colorado State University, Fort Collins.

3. BARTHOLOW, J. 1992. POP-II system documentation. IBM-PC Version 7.00. Fossil Creek Software, Fort Collins, CO.

4. RICKER, W. E. 1954. Stock and recruitment. *Journal of the Fisheries Research Board of Canada* 11:624–651.

5. RICKER, W. E. 1975. Computation and interpretation of biological statistics of fish populations. Bulletin No. 191. Fisheries Research Board of Canada, Ottawa, Canada.

6. HASSEL, M. P. 1975. Density-dependence in single-species populations. *Journal of Animal Ecology* 44:283–295.

7. HASSELL, M. P., J. H. LAWTON, and R. M. MAY. 1976. Patterns of dynamical behaviour in single-species populations. *Journal of Animal Ecology* 45:471–486.

8. MAY, R. M. 1976. Models for single populations. Pages 4–25 *in* R. M. May, ed., *Theoretical Ecology: Principles and Applications.* Blackwell, Oxford, UK.

9. LESLIE, P. H. 1945. On the use of matrices in certain population mathematics. *Biometrika* 33:183–212.

10. CAUGHLEY, G. 1974. Interpretation of age ratios. *Journal of Wildlife Management* 38:557–562.

11. LESLIE, P. H. 1948. Some further notes on the use of matrices in population mathematics. *Biometrika* 35:213–245.

12. LEFKOVITCH, L. P. 1965. The study of population growth in organisms grouped by stages. *Biometrics* 21:1–18.

13. USHER, M. B. 1966. A matrix approach to the management of renewable resources, with special reference to selection forests. *Journal of Applied Ecology* 3:355–367.

14. USHER, M. B. 1969. A matrix model for forest management. *Biometrics* 25:309–315.

15. CASWELL, H. 1989. *Matrix Population Models.* Sinauer, Sunderland, MA.

16. MANLY, B. F. J. 1990. *Stage-Structured Population Sampling, Analysis and Simulation.* Chapman and Hall, London, UK.

17. BARTMANN, R. M., G. C. WHITE, and L. H. CARPENTER. 1992. Compensatory mortality in a Colorado mule deer population. *Wildlife Monograph* 121:1–39.

18. FOWLER, C. W. 1981. Density dependence as related to life history strategy. *Ecology* 62:602–610.

19. NOON, B. R., and J. R. SAUER. 1992. Population models for passerine birds: Structure, parameterization, and analysis. Pages 441–464 *in* D. R. McCullough and R. H. Barrett, eds., *Wildlife 2001: Populations.* Elsevier Applied Science, New York.

20. McCULLOUGH, D. R., ed. 1996. *Metapopulations and Wildlife Conservation.* Island Press, Washington, DC.

21. BOYCE, M. S. 1992. Population viability analysis. *Annual Review of Ecology and Systematics* 23:481–506.

22. KNIGHT, R. R., and L. L. EBERHARDT. 1985. Population dynamics of Yellowstone grizzly bears. *Ecology* 66:323–334.

23. SAS INSTITUTE. 1990. SAS® Language: Reference, Version 6, 1st ed. SAS Institute, Cary, NC.

24. CLUTTON-BROCK, T. H., F. E. GUINNESS, and S. D. ALBON. 1982. *Red Deer Behavior and Ecology of Two Sexes.* University of Chicago Press, Chicago, IL.

25. LOMNICKI, A. 1988. *Population Ecology of Individuals.* Princeton University Press, Princeton, NJ.

26. WHITE, G. C. 1999. Population viability analysis: Data requirements and essential analysis. *In* L. Boitoni and T.K. Fuller, eds., Research Techniques in Animal Ecology. Columbia University Press, New York. In press.

27. BURNHAM, K. P., D. R. ANDERSON, G. C. WHITE, C. BROWNIE, and K. H. POLLOCK. 1987. Design and analysis experiments for fish survival experiments based on capture–recapture. *American Fisheries Society Monograph* 5:260–278.

28. WHITE, G. C., and R. M. BARTMANN. 1998. Mule deer management—what should be monitored? Pages 104–118 *in J.C. deVos, Jr., ed., Proceedings of the 1997 Deer/Elk Workshop,* Rio Rico, Arizona.

29. PASCUAL, M. A., P. KAREIVA, and R. HILBORN. 1997. The influence of model structure on conclusions about the viability and harvesting of Serengeti wildebeest. *Conservation Biology* 11:966–976.

30. McCorquodale, S. M., L. L. Eberhardt, and L. E. Eberhardt. 1988. Dynamics of a colonizing elk population. *Journal of Wildlife Management* 52:309–313.

31. Eberhardt, L. E., L. L. Eberhardt, B. L. Tiller, and L. L. Cadwell. 1996. Growth of an isolated elk population. *Journal of Wildlife Management* 60:369–373.

32. Bartmann, R. M., L. H. Carpenter, R. A. Garrott, and D. C. Bowden. 1986. Accuracy of helicopter counts of mule deer in pinyon-juniper woodland. *Wildlife Society Bulletin* 14:356–363.

33. White, G. C., and R. M. Bartmann. 1998. Effect of density reduction on overwinter survival of free-ranging mule deer fawns. *Journal of Wildlife Management* 62:214–225.

Nutritional Ecology

Donald E. Spalinger

INTRODUCTION

The mechanisms regulating large mammal populations can be divided into two components: those that act to reduce populations and those that act to maintain or enhance them. The former includes predation, hunting, disease, and accidents. Predation and hunting mortality are considered "top-down" regulators of populations, so named for the trophic position that these represent relative to the population of interest. Certainly these play an important role in the management of most large mammal populations today and, accordingly, these factors are addressed elsewhere in this book. Here, we are concerned with the latter mechanisms, those that provide the "capital" for the maintenance and production of large mammal populations.

The mechanisms that maintain or enhance populations are considered "bottom-up" regulators. In this case, the lower trophic level(s) provide the resources necessary for survival, replacement, and/or growth of the population. Although several large mammal species are carnivores or omnivores and therefore consume herbivores (e.g., mountain lion) or a mixture of plants and herbivores (e.g., brown and black bears), most North American large mammals are herbivores. The lower trophic level from which the latter species make their living is the plant community. Because the majority of these species are obligate herbivores (and primarily ruminants), the focus here is on the mechanisms that influence the ability of these mammals to obtain sufficient nutrients from plant communities to sustain productive populations.

All too often when assessing the regulators of populations, wildlife managers fail to consider the effects of nutrition on large herbivore populations. Several reasons may account for this tendency. Nutritional constraints may seem unlikely because most habitats are awash with green plants (1). Hence, plant resource limitations are not apparent. Secondly, the effects of nutritional deficiencies are insidious and commonly disguised as a part of other, more direct effects (e.g., predation and disease). Susceptibility to predation and disease can be greatly increased by malnutrition. Moreover, poor fawning or calving success may be incorrectly diagnosed as predation when, in fact, neonates are born debilitated and weak due to inadequate nutrition of the female (2, 3, 4, 5, 6). Predation in such cases is an obvious proximal cause of death of the neonate, but nutritional factors must be carefully examined and eliminated before assigning an ultimate causal mechanism.

Although a myopic view of the habitats of herbivores may reveal a productive landscape crowded with plants, the abundance of vegetation is not simply a result of a predatory depression of herbivore numbers. In spite of the assistance that large herbivores obtain from their microbial symbionts, making a living from plant communities is a difficult business. Plants are not passive entrées in some sort of herbivore smorgasbord. Rather, they present significant constraints on herbivores, including constraints on the ability of the herbivore to harvest food, the ability to process sufficient quantities of food (gut and toxin constraints), the ability to digest sufficient nutrients at a rate consistent with needs, and the ability to metabolize or convert digested plant nutrients into animal products.

These four constraints imposed by plants on the herbivore must be balanced against the demands of the herbivore. A more precise way of expressing this balancing act that the animal must perform is by equating the nutritional demands of the animal over some arbitrary time scale (e.g., a day) with the ability of the animal to harvest and digest those nutrients within the same time period. In balance equation form,

$$N_i + M_i + S_i = 0 \; , \qquad\qquad (6\text{--}1)$$

where N_i is the nutrient assimilation rate, or the rate at which nutrient i is harvested and digested (e.g., joules/day for energy, grams/day for other nutrients); M_i is the quantity of nutrient i metabolized or "used up" (joules/day or grams/day); and S_i is the quantity of nutrient i stored within the body (joules/day or grams/day). In steady state, $S_i = 0$, although over a sufficiently large time frame (at least for some significant fraction of its reproductive lifetime), S_i must be positive for the animal to remain alive and be productive. However, over a shorter time frame (e.g., a day), most large herbivores are either in an anabolic state, during which they gain nutrients and S_i is positive (e.g., summer), or in a catabolic state, for which S_i is negative (e.g., winter).

The four constraints imposed by plants are imbedded in N_i, the nutrient assimilation rate. Similarly, M_i represents a composite of processes that constitute the maintenance requirements of the animal, while S_i represents the demands of growth, fattening, and reproduction. The processes or constraints associated with nutrient assimilation, and the processes and functions associated with nutrient demands on the animal, can be quantified and compared, much as an accountant would tally the expenses and income of a business to assess its financial health. Hence Eq. (6–1) provides a template for the quantitative assessment of habitat and animal productivity, and measuring or evaluating the components of this balance equation is an important way to understand the linkage between animal productivity and habitat characteristics.

CONSTRAINTS ON FEEDING RATE

Feeding rate is frequently overlooked as a significant factor affecting nutrient assimilation in large herbivores. Often, nutritional ecologists assume that nutrient assimilation is principally a function of digestibility of the foods consumed, and they ignore the importance of time in nutritional interactions (i.e., food acquisition and food passage through the system). Time becomes especially important to very large animals and very small animals, because larger ones must satisfy increasingly larger absolute daily nutritional requirements, while smaller ones must satisfy their more energetically expensive lifestyles. In many cases, the time constraints involved in nutrient assimilation (e.g., daily food intake rate) are considerably more variable than digestibility of the foods, and therefore more critical to the nutritional outcome (7, 8, 9).

Moreover, important life history traits of large herbivores, including body size and reproductive rates, can be correlated to habitat characteristics through their dependence on feeding rate and passage rate (10, 11, 12, 13, 14, 15, 16, 17).

Characteristics of the food being consumed and characteristics of the herbivore govern feeding rate. More specifically, feeding rates of North American large herbivores are related to plant density (plants/meter^2), or the bite size obtainable from the plant (grams/bite; 18, 19, 20, 21, 22). Although feeding rates have often been linked to plant biomass density (grams/meter^2; 22, 23, 24, 25, 26, 27, 28, 29), recent theoretical and empirical examinations suggest that plant biomass density is not likely to be the proximal or driving variable influencing feeding rates, because it is confounded with bite size (22, 30). The relationship between feeding rate and plant density or bite size is curvilinear and asymptotic (Figure 6–1). The relationship is similar to Holling's disc equation or Type II functional response (31). And the asymptotic form is a result of the competition between the forager's need to search for and harvest foods (when food plants are rare in the environment) or harvest and chew its food (when foods are relatively abundant in the environment) (32). Only when plant densities are extremely low (i.e., about 2 meters apart) will feeding rates be limited by plant density (19). In most circumstances, therefore, feeding rates will be governed by the size of bite that the herbivore can obtain from the plant. This relationship is expressed in Eq. (6–2):

$$I = \frac{R_{max}S}{R_{max}h + S},$$
(6–2)

where R_{max} is the maximum rate of food processing (chewing and swallowing food) achievable by the animal (grams/minute), and S is the size of bite cropped (grams/bite).

For browsing herbivores, bite size is commonly a function of plant size (e.g., for small herbs or forbs) or plant architecture (e.g., for shrubs; 18, 22, 33). Moreover, for plants in which bite size can potentially vary, there is generally a negative relationship between bite size and nutritive density of the bite (34, 35, 36, 37). Hence, the forager is

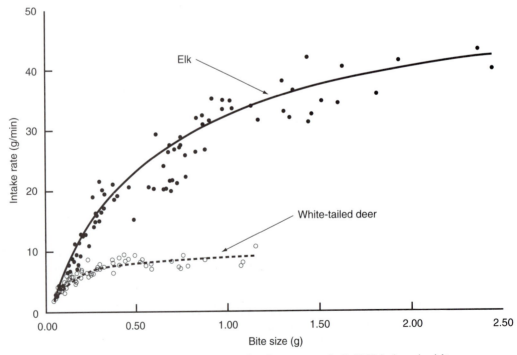

Figure 6–1 Instantaneous feeding rates of elk (266 kg) and white-tailed deer (45 kg) feeding on alfalfa stems in an artificial pasture. (Data from Refs. 18, 20, 40.)

faced with a trade-off between harvesting food more efficiently by taking larger bites, or harvesting higher quality food by sacrificing feeding rate. Current evidence suggests that browsing herbivores select bite sizes that optimize nutrient assimilation rate (i.e., digestible energy intake rate) by taking smaller, but more digestible bites (35, 38).

Bite size, and therefore feeding rate, is also constrained by herbivore morphology. Obviously, mouth size must dictate, at some scale, the size of bite cropped by an herbivore. For example, incisor arcade dimensions (the distance across the incisor row) have been shown to correlate to bite size in large, primarily grazing, herbivores (12, 34, 39, 40). This is not surprising, because grazing on grasslands, especially swards of a continuous (e.g., rhizomatous or stoloniferous) nature, would be facilitated by a wide gape and a broad incisor row that allows the animal to encompass a larger surface area, and hence a larger bite with each cropping motion. The nutritional consequences of large bite size are perhaps best exemplified in the Soay sheep of western Scotland. Overwinter survival of the sheep was highly correlated to incisor arcade width in severe winters, and significant changes in population phenotypes occurred in such winters (41). The clear implication is that incisor arcade width is a variable that influences feeding rate directly through bite size, and that feeding rate in turn directly affects survival in difficult winters.

Feeding rate is also influenced by morphological factors that influence processing rate. Processing rate has been shown in a variety of large herbivores to be related to molar surface area (40). Because molar surface area is greater in larger herbivores, they can potentially consume food at a higher rate than small herbivores. However, if plant morphology restricts bite sizes, then large and small herbivores obtain nearly the same feeding rates (Figure 6–1). In such cases, large herbivores are severely disadvantaged relative to small herbivores. This is why the largest of North America's large mammals are not found in environments where high-quality plants are small, such as in deserts, alpine, temperate rain forests (e.g., the West Coast rain forests from Northern California to the Gulf of Alaska), and high tundra habitats. Although moose extend into southeastern Alaska rain forests and onto the tundra of the North Slope of Alaska, they are restricted, for the most part, to river drainages where willows are abundant and offer relatively large bite sizes.

The role of feeding rate in the nutritional well-being of large herbivores clearly depends on the ecological circumstances. During the growing season, plant quality and density are generally high, and bite sizes and, therefore, feeding rates are sufficient to allow herbivores to fulfill intake requirements with minimal feeding times (e.g., 3 to 6 hours per day). However, in winter or during dormant periods (e.g., late summer in hot, arid climates), plant densities and bite sizes often restrict feeding rates. Under such circumstances, feeding times may approach 10 hours per day or greater, and herbivores may simply not achieve intake rates sufficient to meet energy demands (27). The effects on feeding rates are exacerbated by deep snows that may bury plants (thus reducing plant densities) and elevate energy expenditures for locomotion (42, 43, 44). Dall's sheep in the Brooks Range in winter limit foraging due to snow and severe weather, in spite of losing 22 percent of their body weight over winter. Nevertheless, even at such high latitudes, with little daylight and extremely frigid conditions, they still forage nearly 9 hours per day (45). Although Dall's sheep appear to be well adapted to such rigors, they also select areas where snow depths are moderated significantly by wind, and therefore expend less energy in locomotion while foraging, a luxury that other large mammal species may not enjoy under similar snowfall regimes (46).

CONSTRAINTS ON PROCESSING FORAGE: GUT CAPACITY AND PASSAGE RATE

Feeding rate and foraging time comprise part of a single, more inclusive rate constraint on large herbivores. The time required to process food (i.e., digest and pass) in the digestive tract is the other time constraint. If forage processing were not a constraint, then herbivores should almost never succumb to starvation, but should simply eat more of the ubiquitously abundant, but lower quality, forage that exists. Abundant coarse grasses, weeds, shrubs, and trees occur even on the most severely overgrazed ranges. Theoretically, any nutrient deficiency can be overcome by simply consuming more, as long as the food has a concentration of that nutrient that exceeds the level necessary to offset the costs of consuming and processing the food.

However, it is obvious that herbivores cannot consume unlimited quantities of food to satisfy nutritional requirements. The constraint on food intake is coupled to two causes: (1) the physical restriction of the capacity of the digestive tract (hereafter gut capacity) and the rate at which it empties (passage rate), and (2) the digestive kinetics and toxicity of plants to herbivores. Both constraints are relatively poorly understood at the present time.

The physical restriction of intake due to gut capacity and passage rate depends on the construction of the herbivore. We recognize two distinct forms of digestive systems in vertebrate herbivores, and the distinction between the two revolves around the placement of the fermentation chamber (where the bacterial symbionts reside) relative to the gastric stomach. Those in which the fermentation chamber is in a post-gastric position are termed *hindgut fermenters*. Hence, the fermentation chamber is the caecum and/or large intestine. Because of the morphology of their gut, the hindgut fermenters have few barriers to food passage, and thus have the capacity for high food intake rates (15). High food intake rates, however, lead to lower digestive efficiencies. There are no native North American large mammals that rely on hindgut fermentation, although the horse and wild pig are common species that have been introduced.

Those herbivores that ferment forage in a chamber anterior to the gastric stomach are termed *foregut fermenters*. With few exceptions (i.e., the peccary, with some foregut fermentation capacity) all of the North American foregut fermenters are ruminants (suborder Ruminantia), characterized by their habit of regularly regurgitating the foregut (reticulorumen) contents and rechewing it, a process called *rumination* (47). The remaining North American large mammal species are termed *simple stomached* (single-chambered) animals, and have minimal capacities to ferment plant foods. Although most North American Ursids consume significant quantities of plants, they apparently cannot utilize the complex carbohydrates of the cell wall as energy because of their lack of bacterial digestors (48).

The digestive strategy of the ruminant differs from that of the hindgut fermenter. Their goal is to digest the cell wall components of plants (i.e., cellulose, hemicellulose, and pectin—also referred to as plant "fiber") more efficiently, and this strategy dictates a more complex digestive anatomy and function. The ruminant digestive system consists of a series of four chambers, including, in order from anterior to posterior, the reticulum, the rumen, the omasum, and the abomasum (Figure 6–2). More or less, the reticulum and rumen comprise a single chamber (referred to as the reticulorumen), although they are easily distinguished from each other by their anatomical characteristics. The reticulorumen is the largest chamber of the digestive tract, and the digesta dry matter contained in the rumen typically ranges between 1.6% (browsers) and 2.6% (grazers) of body weight (Figure 6–3), or 6% to 20% of body weight based on digesta wet weight (49). This amounts to about 60% to 77% of the capacity in the entire digestive tract (50, 51, 52, 53, 54). By virtue of its volumetric dominance, the rumen is the site of the majority of fermentation that occurs within the tract (65 to 89% of total cellulosic carbohydrate digestion; (55, 56).

The omasum is a relatively small chamber sandwiched between the reticulorumen and the abomasum. It acts as a filter-pump that, in conjunction with the small papillae surrounding the reticulo-omasal orifice, sorts particles and liquids that enter from the rumen, and passes the smaller particles and liquids to the abomasum. Larger particles trapped by the omasum are probably moved back to the rumen via antiperistaltic contractions (57, 58).

The abomasum is the gastric stomach of the ruminant, and its principal role is to initiate digestion of bacterial protein produced in the rumen (along with small amounts

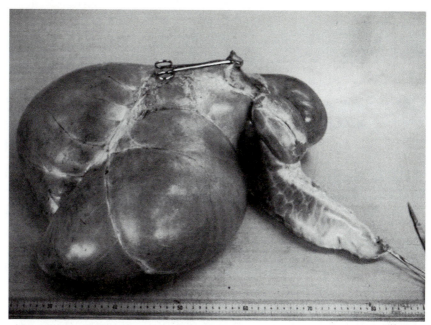

Figure 6–2 The four stomachs of the deer's digestive tract, including the rumen, reticulum, omasum, and abomasum. (Photograph by D.E. Spalinger.)

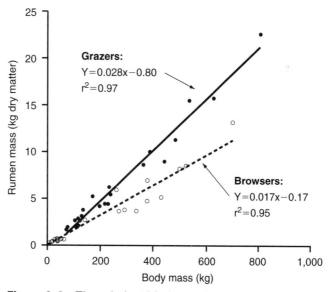

Figure 6–3 The relationship between body weight and rumen dry matter fill in grazers and browsers. The relationships are linear; hence, the slopes of the equations represent the percentages of rumen dry matter as functions of body weight.

of unfermented plant proteins that escape the rumen and are capable of hydrolysis). Following the abomasum is the small intestine, caecum, and large intestine, not significantly modified in form or function from that of nonruminant herbivores.

The passage of materials from the reticulorumen to the lower tract constrains food intake. Passage is restricted by the function of the reticulo-omasal orifice and by the omasum. Particles must be processed to a minimum size (about 0.5 to 1 millimeter in most ruminants) before passing to the lower tract (59). Although fermentation aids

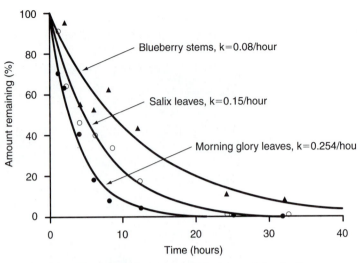

Figure 6–4 The breakdown rate of food particles in the rumen is approximately an exponential decay process. Succulent forages, like spring grasses and herbs (e.g., morning glory), fracture relatively easily, while fibrous and mature foods, like browse stems (e.g., blueberry), break down relatively slowly.

the reduction of particle size, the majority of breakdown is accomplished through rumination (60). Particle breakdown in the rumen follows a negative exponential function, and is, in large part, dependent on the fibrousness of the food eaten (Figure 6–4; 60). High-fiber diets, such as those eaten in winter, require extended rumination time. For example, moose on Isle Royale and in interior Alaska ruminate nearly 12 hours per day in winter when feeding on highly lignified and fibrous browse stems (21, 61). In contrast, succulent spring and summer forages selected by deer and other browsing herbivores (e.g., forbs) are easily fractured and crushed during eating and rumination, and probably are also eroded rapidly by rumen bacteria (Figure 6–4). Hence, animals on such diets spend relatively little time ruminating (Figure 6–3) and have little difficulty achieving intake rates of high-quality foods sufficient to allow them to gain weight (43, 62).

Coupled with breakdown of the food in the rumen is the passage of foods from the rumen through the lower digestive tract. Passage rates can be measured by inoculation of the animal with a bolus of an indigestible dietary marker, either as a solution, in a gelatin capsule, or bound to a food that the animal consumes. The time course of excretion of a marker generally follows the pattern shown in Figure 6–5a. The initial period during which no marker is detected in the feces is called the *transit time*. This corresponds to the time that it takes digesta to transit through the tubular part of the gut (roughly corresponding to the tract starting at the omasum and ending at the rectum in ruminants). The marker concentration in the feces then rises abruptly, followed by a relatively slow exponential decline toward zero. The ascent and decline phases are associated with the mixing and flow from the rumen and caecum/proximal colon. The excretion curve can be used to estimate the indigestible (or rather undigested) fill of the gut and the average or total tract passage rate and total tract mean retention time of the residue (63). The time at which one-half of the undigested fill is excreted is the mean retention time (in hours) of the residue, which is also interpreted as the average time that an individual particle of food remains in the digestive tract. The inverse of mean retention time is called the *rate constant of passage,* or simply *passage rate* (expressed as hour^{-1}). A rough interpretation of this parameter is that it is the proportion of the undigested fill that leaves the gut per hour.

When the preceding measurements are performed in the rumen instead of the feces, the disappearance curve of a marker is somewhat more simplified (Figure 6–5b). There is no transit time, because the marker is inoculated directly into the rumen at the start of the experiment. The decline of marker concentration is a negative exponential function and when plotted on a log–log scale, is linear. The slope of the line fit to this linear pattern of points is the constant rate of passage from the rumen (k_p, hour^{-1}), and the inverse of passage rate is the mean retention time of the marker in the rumen. The marker kinetics shown in Figure 6–5b are those of a soluble marker (Co-EDTA), and represent the passage rate of water from the rumen. Particle marker kinetics are somewhat more complex, because particles must be broken down prior to their passage from the rumen (64).

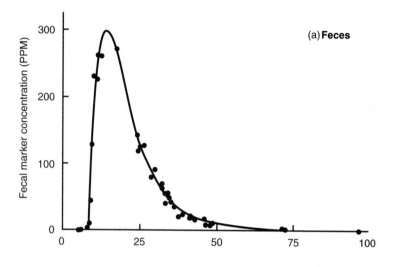

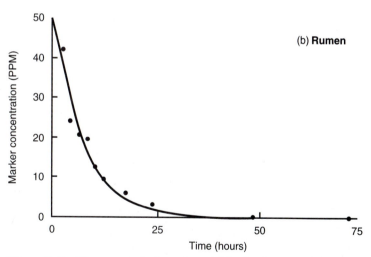

Figure 6–5 The passage of markers (a) in the feces and (b) in the rumen of white-tailed deer. Rumen marker concentration drops exponentially, while the marker in the feces rises abruptly and then drops exponentially. The marker used in this example was Co-EDTA, a soluble marker that traces the passage of water through the digestive system.

Making a living on plants, to a large degree, means digesting the cell walls of these plants. This process requires enzymes not produced by higher animals (e.g., virtually all invertebrate and vertebrate herbivores). Bacteria and some protozoans cultured in the guts of herbivores produce the required enzymes, (i.e., cellulases and hemicellulases). These bacteria live in an anaerobic and chemically reducing environment, and therefore the digestive process is one of fermentation. Measuring and predicting the digestibility of plant foods are extremely important to understanding the adaptations of herbivores for consuming plants.

In the strictest sense, digestibility is the breakdown and absorption of nutrients from the gastrointestinal tract. We measure it by comparing the amount of the nutrient that enters the digestive tract and the amount that escapes to the feces. The digestibility (as a fraction of the amount in the food offered) is then calculated as:

$$D_{N,A} = \frac{IC_{N,I} - OC_{N,O}}{IC_{N,I}} \, , \tag{6-3}$$

where $D_{N,A}$ is the percent apparent digestibility of the nutrient N in the diet, I is the dry mass intake of the diet (grams/day), $C_{N,I}$ is the concentration of nutrient N in the diet (grams/gram of food), O is the dry mass output of feces (grams/day), and $C_{N,O}$ is the concentration of nutrient N in the feces (grams/gram of feces). The entity $D_{N,A}$ is termed the *apparent* digestibility, because for some fractions of the diet, endogenous (i.e., from the animal itself) sources contribute to the feces (65). For example, the endogenous loss of protein from the animal includes unabsorbed enzymes of digestion, bacterial cell wall (considered endogenous because the bacteria are not ingested with the food), and cells abraded from the digestive tract. Because these components deflate the digestion estimate, they can be measured and then mathematically removed to calculate the actual or *true* digestibility of the food. Measurement of the endogenous component is not easy, unless the dietary fraction is uniformly digestible and the endogenous component is a constant amount. When the dietary nutrient is in very low concentrations, it is even possible to obtain negative digestion coefficients. For example, if dietary N is less than endogenous N excreted in the feces, then fecal N will be greater than food N, and according to Eq. (6-3), digestibility will be negative.

The digestible nutrient concentration of a food provides one measure of food quality. However, the true nutrient value to an animal must be measured as a rate. That is, nutritive value must be converted to nutrient assimilation per unit time. An extremely important aspect of digestion, therefore, is the rate at which it occurs. Previous work on digestion kinetics of forage has revealed two important properties that must be considered simultaneously. These include the rate and extent of digestion.

Digestion or fermentation occurs, like any enzymatic process, in a time-dependent fashion. That is, we can follow the digestion of a food over time, and observe that it disappears at a predictable rate over time. For most foods whose fermentations have been measured, they digest according to a negative exponential function (also termed a first-order rate, because it is described by a first-order differential equation). Accordingly, the equation describing digestion of a substance like the cell wall of a plant is:

$$M_t = M_0 e^{-k_d t} \, , \tag{6-4}$$

where M_t is the mass of food item remaining undigested at time t, M_0 is the initial mass of the food item at time 0, and k_d is the rate of digestion (in units time^{-1}). Generally, we express t in hours, and thus the rate constant k is expressed in units "per hour." One way to think of k is in its percentage form. If k is 0.10, then digestion occurs at 10% per hour, or in other words, in each hour, 10% of the remaining undigested food is digested.

This is, however, only an approximation, because biologists are actually dealing with a continuous digestion process acting in a first-order manner. The more accurate approach is simply to substitute the time of interest into Eq. (6–4) to find the amount that remains undigested at time t.

The extent of digestion is obviously related to the true digestibility of the food, but differs in that it is the theoretical limit to digestibility if the food remained in the fermentation chamber for an infinitely long time. In actuality, because fermentative digestion is asymptotic, most plant foods approach the theoretical extent of digestion at about 48 to 72 hours (2 to 3 days), although many foods can take longer than this to digest. Hence, if different herbivores retain the same food for longer or shorter time periods in the fermentation chamber, then these animals will exhibit different true (or apparent) digestibilities for this food.

Plant tissues are an amalgam of many chemical components, and each component has its own characteristic rate and extent of digestion. In spite of this complexity, they can be physically and kinetically subdivided into two fractions: the cell wall and cell contents. The cell wall is composed principally of complex carbohydrates, such as cellulose and hemicelluloses, and other organic and inorganic substances, the most important of which are lignins, pectin, cell wall proteins, and cutins. The cell contents include all of the components of the living plant cell, including starches, sugars, proteins, lipids, and nucleic acids. The extent of digestion and, to some degree, the rate of digestion of both fractions are a relatively predictable function of their chemical composition.

The cell contents are virtually completely digested in the gut (65). Estimates of the extent of true digestibility of cell contents range from 98% to 100%. However, such coefficients overestimate the nutritive value of the cell contents of those plants containing secondary compounds (defensive chemicals) that must be excreted in the urine, at a relatively high energy cost. In addition, some secondary compounds associated with the cell contents (e.g., tannins) bind with proteins of animal and plant origin, making them indigestible. These compounds lower the apparent digestibility of cell contents and protein.

Cell contents digest very rapidly in the rumen. The digestion rate of cell contents is about 40% to 50% per hour (e.g., $k_s = 0.43$; 66). Hence, in just a few hours, nearly all the cell contents have been fermented by microbes (Figure 6–6). Because of their high digestion rate, virtually all cell contents are digested before leaving the rumen.

Digestion rates of cell walls are highly variable and have no apparent affinities to taxonomic groups. For example, digestion rates of grasses may range from 18% per hour for red brome leaves (a C_3 grass) to 4% per hour for ryegrass stems (67). Digestion rates for dicot materials may range from 31% per hour for immature clover leaves down to 5.7% per hour for lignified, woody stems (Figure 6–6; 67, 68).

The extent of digestion of cell walls of plants varies in proportion to the degree of lignification of cell walls and to the extent of silicification. It also depends on the concentration of cutin and suberin in the cell walls. Digestion experiments have revealed that cell wall digestibility decreases as a nonlinear function of lignin/cutin concentration, and at a linear rate with increasing concentration of silica (65). Silica is not a significant component of browse and forb tissues, but is sequestered in grasses, probably as a defense against herbivores.

Measurement of the rate and extent of digestion of foods is complicated by the fact that no simple or reliably accurate methods are available at the present time. The two methods commonly used include the *in vitro* incubation method and the *in situ* nylon bag method. Both have weaknesses, although the nylon bag method, if properly calibrated, is preferable. The *in vitro* method is relatively simple. Samples of the foods to be tested are weighed into test tubes in which a buffer solution and an aliquot of coarsely filtered rumen fluid have been added and flushed with CO_2 (making it anoxic). The tubes are allowed to incubate at rumen temperature (about 39°C) for 48 hours (the standard time for estimating digestibility), or for varying lengths of time for the determination of digestion rate (69). The

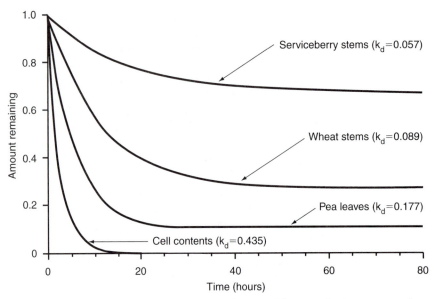

Figure 6–6 The rate of digestion of several forages in the rumen of elk. Included is the relative rate of digestion of cell contents, showing the rapid rate of disappearance of this fraction of the forage in the digestive tract.

digestion is terminated by adding HCl and pepsin to simulate digestion in the abomasum, or by chemically separating the undigested cell wall and the products (e.g., bacteria) by boiling in a detergent solution. With the HCl–pepsin digestion, the mixture is incubated for an additional 48 hours, and then filtered, dried, and weighed. The residues from either method are considered the undigested fraction, and digestion is calculated by difference.

The nylon bag method is similar in principle, but uses a living animal's rumen as the incubation medium. Hence, it alleviates some of the potential problems associated with the *in vitro* method. The procedure requires a fistulated animal, in which a small portion of the dorsal rumen is surgically annealed to the body wall, and a permanent opening is created to the outside (called a fistula). The opening is plugged with a rubber or plastic seal called a cannula, through which digesta or nylon bags can be inserted or removed (Figure 6–7). The foods to be tested are sealed in small nylon bags (similar to tea bags), and placed in the rumen on a cord. The bags are allowed to remain for varying lengths of time, and then removed, washed, and extracted with boiling detergent solution to remove soluble and bacterial residues (70). The fiber in the bag is then dried and weighed to determine extent of removal or digestion.

The relevance of measuring the extent of digestion should be readily apparent, since this tells us how much of the plant is potentially usable by the animal. The importance of digestion rate is more obscure, but is resolved when the "rate of passage" concept is brought to mind. The extent of digestion is, in fact, only significant when foods remain in the gut sufficiently long to achieve this state, and sufficiently long is defined only relative to the rate at which digestion occurs. Hence, if a food digests at a rate of 15% per hour, then [by Eq. (6–4)] in 10 hours, about 61% will remain undigested. If the mean retention time of a food in the rumen is 10 hours, then this food will exit the system with more than half of its potentially digestible components intact. If, instead, the rate of digestion were 10% per hour, then the amount remaining in 10 hours would be around 37%.

Digestion Kinetics

Figure 6–7 White-tailed deer with rumen cannula in a field experiment to examine foraging behavior and rumen function in South Texas. (Photograph by D.E. Spalinger.)

THE EFFICIENCY OF NUTRIENT METABOLISM

Digestion or assimilation of a food is simply the first of many metabolic steps in the conversion of plants into herbivore products (e.g., milk, muscle, fat, or offspring) or useful work for the herbivore. Digestibility is an estimate of the amount of nutrient entering the body of the animal, but this is not the amount of nutrient that can be utilized for maintenance and productive purposes. Many nutrients are excreted in the urine instead of being used for productive purposes. For example, sodium absorbed from the digestive tract can be immediately excreted in the urine, causing one to overestimate the sodium that the animal actually retained and used. The difference between the amount of a nutrient absorbed from the GI tract and that excreted in the urine, or belched from the gut as gas (e.g., methane), is termed the *metabolizable* amount of the nutrient. As one might guess, calculation of the metabolizability of a nutrient requires the measurement or estimate of the loss of the nutrient in the urine or as gas (65).

In addition, when the nutrient of interest is energy, there are other inefficiencies of energy metabolism that must be accounted for. These include the heat or energy lost during fermentation, and the entropy or heat of energetic transformations within the body (called the *heat increment*). The efficiency of energy conversion depends on the metabolic process engaged in by the animal. Generally, energy transformations that provide for the maintenance of animal tissues (e.g., for basal metabolism, maintenance activities, and thermoregulation) are more efficient than those processes that create animal products (e.g., production and storage of fat or protein, growth, and reproduction). To complicate matters further, if the animal is in negative thermal balance, the heat of fermentation or heat increment can be used to offset thermoregulatory energy expenditure, thus increasing the apparent efficiency of energy metabolism. Figure 6–8 provides a generalized diagrammatic representation of the fractionation of energy metabolism for large mammal species.

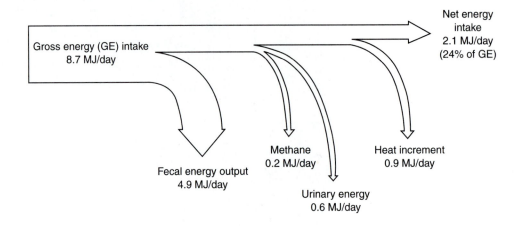

High-quality browse diet
(Hemlock/maple/hobblebush)

Net energy intake 2.1 MJ/day (24% of GE)

Gross energy (GE) intake 8.7 MJ/day

Fecal energy output 4.9 MJ/day

Methane 0.2 MJ/day

Urinary energy 0.6 MJ/day

Heat increment 0.9 MJ/day

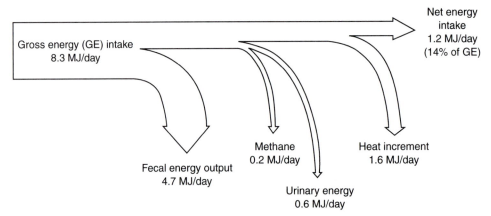

Low-quality browse diet
(Balsam fir/maple/hazelnut)

Net energy intake 1.2 MJ/day (14% of GE)

Gross energy (GE) intake 8.3 MJ/day

Fecal energy output 4.7 MJ/day

Methane 0.2 MJ/day

Urinary energy 0.6 MJ/day

Heat increment 1.6 MJ/day

Figure 6–8 The partitioning of gross energy of two diets fed to white-tailed deer. (Data from Ref. 70a.)

ANTINUTRIENTS: THE DELETERIOUS ROLE OF PLANT DEFENSES

Plants have evolved a complex suite of defenses that serve them adequately (but rarely completely) against the ravages of herbivores. Plant defenses include those that act by reducing the harvest rate of plant parts (mechanical defenses) and those that reduce digestion or induce metabolic or toxic responses in the herbivore (chemical defenses).

Some mechanical defenses of plants are easily recognized. These include the thorns and spines of many temperate and tropical trees, shrubs and herbs, and the hairs and prickles found on stems, leaves, or rachis of leaves of many dicot plants. For the most part, thorns and spines of plant stems rarely inhibit consumption of a plant by an herbivore, although when such defenses are coupled with relatively small leaves, the rate of intake is reduced considerably (Figure 6–9) (71). This defense is probably most

Antinutrients: The Deleterious Role of Plant Defenses

Figure 6–9 Thorns do not protect plants from being consumed by herbivores, such as this white-tailed deer in South Texas. However, combined with small leaves, such defenses slow feeding rates considerably, and thus nutritionally constrain the animal. (Photograph by D.E. Spalinger.)

effective in deterring herbivores from "leaf stripping" the plant, because the thorns located on stems, and commonly at leaf nodes (actually modified stipules) prevent the animal from consuming more than one leaf at a time. Consequently, intake rate on such mechanically defended plants may be depressed considerably. There is some evidence that grazing leads to phenotypic (adaptive) responses that increase the density and length of thorns (72, 73, 74, 75). Such an adaptive response is called an *induced defense,* although this term is traditionally used to describe the chemical responses to herbivory.

While thorns are easily recognized as mechanical defenses, other mechanical defenses exist that are not as easily quantified. The architecture of many plants limits their consumption by herbivores. Many grasses dilute their photosynthetic materials (the high-quality components) with nonphotosynthetic dead or stemmy mass. Unless animals are highly selective, the consequence is to dilute the quality of the food that is removed from the plant, thus lowering its value to the herbivore. Shrubs generally exhibit the same effect when leaves are scattered within a divaricately branching stem canopy. In this case, it is difficult for the herbivore to select leaves without consuming some of the lower quality stems in the process.

The influence that herbivores have on the growth and architecture of plants can be profound. In the case of grasses, the absence of grazing leads to more diluted nutritional quality of the grass to the herbivore. Conversely, consumption by the herbivore, especially during the growing season and after fire, leads to a sward in which tillers proliferate, the density of leaf tissue is greater, and the amount of dead material is less (76, 77, 78, 79, 80). Because such swards offer larger bite sizes and higher quality, they tend to attract herbivores to continue grazing them. Although this appears to be a positive, and therefore unstable, feedback grazing system, it is self-arresting because higher grazing pressures tend to reduce the stature of the grass, eventually resulting in smaller bite sizes, and therefore lower intake rates. When intake rates decline to the average for the landscape, herbivores should become indiscriminate in their selection of such stands. Such grazing systems have been termed *grazing lawns,* and are commonly associated with African grasslands and perhaps the short-grass prairie of North America. Such grazing lawns, however, are unlikely to develop in North American native bunch-grass systems because these grasses do not respond to grazing by tillering, and appear to be less tolerant to high grazing pressure than rhizomatous species (81).

Most browse plants respond in the opposite manner to herbivory, principally because of the difference in location and dominance of the apical meristem. Apical meristems that are nipped by herbivores lose their dominance over other potential meristematic tissues or buds on the stem. These lateral buds subsequently begin to grow, making the plant bushier. These pruned plants produce leaders and leaves that are spatially interwoven with dead and older woody stems, thus making it difficult for the herbivore to select high-quality bites at an efficient rate. In such cases, hedging provides an effective way for the plant to defend itself against further herbivory, decreasing intake rates and diluting the quality of the bites removed.

Chemical defenses of plants are highly diverse and complex, as are their effects on herbivores. Chemical defenses are classified into two broad groups, according to their metabolic or digestive roles. One broad class of defenses acts principally by reducing the digestibility of the plant tissues consumed. These compounds generally reduce digestibility in direct proportion to their concentration in the plant, and are termed *quantitative defenses* (82). Examples of quantitative defenses include cell wall components such as lignin, cutin, and silica. These compounds are indigestible by rumen microorganisms and act to reduce the digestibility of other potentially digestible cell wall components (i.e., cellulose) by binding irreversibly to them.

Tannins are another important quantitative defense of plants. Tannins are a complex group of polyphenolic compounds that bind preferentially to proteins (both plant and animal) and render them indigestible (83). In fact, the name *tannin* refers to the capability of these compounds to tan animal hides—essentially the stabilization of the hide proteins against decomposition. Tannins are common in the leaves of woody plants, but are rare in most grasses. The tannins in plants generally cause a bitter or astringent taste, which is the direct result of the tannin binding with mucoproteins within the mouth. A recognizable human example is the astringency of red wines, resulting from the high tannin concentrations within the skin of red wine grapes. Although humans capitalize on the flavor that this imparts to the wine, this probably represents a defensive deterrent by the grape against herbivorous insects.

Tannins reduce digestibility of protein in proportion to their concentration. However, because of their complex and variable structure, it is difficult to quantify tannins and, more importantly, their biological effects on protein digestibility. Consequently, in lieu of measuring tannin concentrations, nutritionists have developed an assay method that measures relative protein precipitation capabilities. Robbins and others (84) and Hanley and others (85) have developed useful relationships between the precipitating capacity of tannin extracts of plants and the reduction in digestibility of the protein of the plant *in vivo* (84, 85). Browse leaves commonly contain sufficient tannins to lower protein digestibility by 50%. For example, only 36% of the protein in the leaves of red alder was apparently digestible by deer, and nearly half of this reduction was attributable to tannins. Obviously, such reductions in protein digestibility could significantly alter the nitrogen balance of the herbivore, and may be especially important for lactating females whose protein demands for milk production may be nearly six times that of their maintenance protein requirements (65).

The second important group of chemical defenses is comprised of those that act as metabolic toxins to herbivores. Many of these plant compounds are toxic to herbivores at relatively low concentrations. Because their actions are considered relatively dosage independent, they are termed qualitative defenses (82). The list of qualitative defenses is certainly too long to reproduce here, and furthermore, their actions and effectiveness against herbivores are surprisingly species specific. Some of the more important groups include the alkaloids, phenolics, terpenes, and cyanogenic glycosides.

Antinutrients: The Deleterious Role of Plant Defenses

The alkaloids are recognized structurally for their inclusion of nitrogen in a heterocyclic ring. Alkaloids are found commonly in herbs and are rarely found in grasses. Most plants that are defended by bitter-tasting alkaloids are avoided by large herbivores, and generally only support specialist insect herbivores.

Phenolics are a diverse and commonplace group of compounds characterized by the presence of a phenolic ring. Although tannins, discussed earlier as a quantitative defense, are technically phenolics, they are composed of complex chains of phenolic subunits that do not hydrolyze in the digestive tract, and therefore are mostly excreted without metabolic consequences. Most other phenolics, however, are readily hydrolyzed and absorbed from the digestive tract, and can interfere with metabolism or otherwise intoxicate the animal. Robbins and others suggest that these latter phenolics play the more significant role in the nutrition of large herbivores by limiting the animal's daily intake (84). They emphasized that protein precipitating tannins comprised only about 13% of the phenolics of some of their forages, and that the 50% to 60% depression of intake was probably attributable to the abundant soluble phenolics in these foods. Moreover, the subsequent detoxication and excretion of soluble phenolics is energetically expensive. Because these compounds are at relatively high concentrations in browses, understanding phenolic compositions, their absorption in herbivores, and their metabolic consequences and detoxication is extremely critical to a more complete understanding of the nutrition of large herbivores.

Terpenoids are another important plant defensive constituent in browse. Many of the terpenoids are bactericidal and therefore reduce the digestive efficiency of large herbivores. Hence, they act as feeding deterrents, reducing intake of food. Examples include the volatile oils of sagebrush and conifers (86, 87, 88, 89, 90). The terpenes of sagebrush have long been recognized as influential in diet selection and nutrition of mule deer in the intermountain West (91, 92, 93, 94).

A host of additional chemical defenses of plants do not fall into the general categories just listed, but may significantly influence diet selection and behavior of herbivores. Examples include the cyanogenic glycosides, toxic amines, nonprotein amino acids, a few toxic proteins (e.g., ricin, a protein found in castor beans is so toxic that mice are killed by concentrations as small as 0.001 micrograms/gram of body mass), and aflatoxins (actually produced by fungi that live mutualistically with some plants). Cyanogenic glycosides are commonly found in members of the rose family, particularly in the leaves of many cherries (e.g., chokecherry and serviceberry). When digested, the cyanogenic glycoside decomposes into cyanide or prussic acid, causing respiratory failure. Cattle occasionally fall prey to the cyanogenic glycosides of chokecherry, canarygrass, and johnsongrass, but only rarely have wild herbivores been found poisoned. In such instances, it has only occurred where food choice was severely restricted (95). Cyanogenic glycosides can be partially degraded by rumen bacteria, especially when soluble carbohydrates are available (49, 96). In such cases, the nitrogen of the cyanide becomes incorporated into bacterial proteins. Nevertheless, it is important for the herbivore to limit the concentration of cyanogenic glycosides to a tolerable level in the digestive system by careful dietary selection and mixing. This is true as well for all of the potential toxins reviewed earlier. A diverse and well-managed (not overutilized) plant community is crucial for providing such opportunities for large herbivores.

NUTRITIONAL REQUIREMENTS OF HERBIVORES

At the cellular level, the nutritional requirements of large herbivores are essentially the same as those of humans and other omnivores and carnivores. Accordingly, large herbivores require the well-known essential nutrients for life; including protein, energy, water, minerals, vitamins, and fats (49, 65). However, because the conversion of plant constituents to animal products is mediated by fermentation through a symbiont microbial population, the dietary nutritional requirements of herbivores are more general in many respects. In addition, the microbial assistants of herbivores are versatile and effective in digesting plant constituents (e.g., cellulose) that otherwise are refractive to ordinary mammalian digestive enzymes.

As a consequence of this mutualistic herbivore–microbe relationship, large herbivores do not require the array of amino acids that act as building blocks for proteins in carnivores and humans (i.e., the 10 or so essential amino acids). The microbes utilize a variety of plant nitrogenous components, including nonessential amino acids and nitrates, to produce, *de novo,* all of the essential amino acids required by the herbivore. In fact, the bacteria derive some energy from the destruction of plant proteins. In the process, they degrade these to amino acids, and then deaminate the amino acids to produce ammonia in the rumen (97). Other bacteria subsequently consume the ammonia and reconstruct amino acids and proteins for themselves, and ultimately, the herbivore.

For much the same reason, large herbivores do not have the same requirements for dietary vitamins as carnivores and omnivores, because the microbes of the gut are able to manufacture many of these (e.g., the B vitamin complex and vitamin K). The herbivore subsequently "harvests" these vitamins from the bacteria in the small intestine. On the other hand, the bacteria of the gut also pose some unique constraints on large herbivores. For example, they destroy vitamin C and modify plant lipids principally by hydrogenating unsaturated fatty acids (49). Large herbivores, like the cervids, have adapted to these constraints by manufacturing their own vitamin C, and by metabolic adaptations that greatly conserve the relatively small amount of dietary essential fatty acids (i.e., linoleic and linolenic acids) that pass the fermentation process without modification. Consequently, ruminants have a dietary requirement for these fats that is less than half that of other mammals (98).

Maintenance Needs

The animal body is in a constant state of repair and maintenance, and even under the best and most efficient conditions, the animal will excrete, secrete, or slough nutrients and tissues that must ultimately be replaced. Proteins, membrane lipids, minerals, and vitamins are continually being constructed, recycled, and replaced. Like energy, such biochemical transformations are not 100% efficient, leading to the excretion of breakdown products (e.g., urea from protein, creatinine from muscle metabolism) or the loss of the nutrient in urine (e.g., plasma sodium, calcium, and phosphorus) or feces (e.g., unabsorbed bile salts). All essential dietary nutrients are designated as such because they are not perfectly conserved in the animal body. Although there are many nutritional demands on large mammals, including the demand for energy, protein, water, macro- and micronutrients, vitamins, and essential fatty acids, nutritional ecologists generally agree that, for most wildlife populations, energy and protein are the most likely limiting nutrients. Additional needs are recognized for some segments of the population at specific times of the year. These may include the need for sodium in early spring and summer, especially for lactating animals, and the need for phosphorus and calcium for lactation and antler growth in spring and summer.

The maintenance of life is dependent on a constant supply of energy. The maintenance energy needs of an animal are most often defined as the minimal energy required to maintain basic body functioning and posture, to thermoregulate, and to transport and provide nourishment for the animal. As implied by the second law of thermodynamics, all molecular transformations within the body are accompanied by the production of heat, or entropy that, under most conditions, is lost to the animal as unusable energy. The energy lost for the maintenance of basic body functioning is termed the *basal metabolic rate* or *BMR*. More precisely, basal metabolism refers to the minimal energy or metabolic expenditure necessary to maintain the animal, exclusive of the costs of maintaining posture or activity, digestive processes, body temperature (under conditions of thermal stress), or production (e.g., lactation or gestation). One of the important nutritional principles established nearly 50 years ago was that, for a wide

Nutritional Requirements of Herbivores

variety of mammals, ranging in size from the mouse to the elephant, basal metabolism varied directly with body size, expressed in the following equation:

$$E_B = 293\,M^{0.75}\ , \tag{6-5}$$

where E_B is basal metabolic rate (kilojoules/day) and M is body mass (kilogram) (99). Because basal metabolic rate scales as a fractional power of body mass, this means that larger animals have relatively lower energy requirements (per unit of body mass) than do smaller animals. For example, a 45-kilogram pronghorn antelope would require about 113 kilojoules per kilogram of body mass for basal metabolic energy needs, while a 450-kilogram moose would require 64 kilojoules per kilogram of body mass to maintain itself, or 56% of the energy per unit of body mass as the pronghorn. The ecological significance of this is that large animals are more efficient (relative to basal metabolic energy needs) than are small animals.

Many additional measurements of metabolic rate have been added to the list examined by Kleiber, and effects of evolutionary history (e.g., placental versus marsupial mammals), trophic status (carnivores versus herbivores), and feeding behavior (granivores versus folivores versus graminivores) have been observed (65, 100). Nevertheless, basal metabolic rate of large herbivores apparently scales very near the original value expressed in Eq. (6–5). The latest, and most careful, measurements of basal metabolic rate for North American large mammals confirm these results. Hubbert computed the basal metabolic rate of moose to be approximately 288 kj/kg$^{0.75}$/day, while several researchers have measured basal metabolic rate in white-tailed deer and mule deer between 356 and 378 kj/kg$^{0.75}$/day (101, 102, 103, 104). The latter measurements represent metabolic rates that are between 20% and 30% higher than the interspecific mean metabolic rate found by Kleiber (100).

The protein required for maintenance of body tissues is fractionated into the protein required for replacement of abraded intestinal tissues and the protein required to replace the constant loss associated with muscle metabolism. Technically, both losses include many nitrogenous substances other than protein (e.g., creatinine in urine, a product of muscle metabolism), and hence nutritionists generally quantify these losses in terms of nitrogen lost to the animal. Hence, the intestinal erosion losses are termed *metabolic fecal nitrogen,* or *MFN.* Nitrogen losses associated with muscle metabolism are excreted in the urine and termed *endogenous urinary nitrogen* or *EUN.* Metabolic fecal nitrogen losses are a function of the fibrousness of the diet of the animal and intake rate. For large mammals in North America, metabolic fecal nitrogen is a predictable function of intake rate, and ranges between about 5 and 6.4 grams of nitrogen excreted per kilogram of dry matter intake (65). Because endogenous urinary nitrogen is a function of muscle metabolism, it is closely coupled to body size. The protein required to offset EUN losses amounts to between 160 milligrams N/kg$^{0.75}$/day in elk (8.5 grams/day for a 200-kg elk) to about 56 milligrams N/kg$^{0.75}$/day in moose (5.9 grams/day for a 500-kilogram moose; 105).

Nutrient Demands for Reproduction and Growth

Reproduction is one of the most expensive undertakings that an animal will engage in during its lifetime. Both males and females are taxed heavily for this investment, even in social systems in which the male is not actively involved in the rearing of offspring. Male cervids, for example, pay neither the cost of gestation nor lactation, yet their energy expenditure during the breeding season may be so high that they may succumb during or shortly after breeding from exhaustion of energy reserves (106, 107).

The energy costs of reproduction for females are generally separated into the costs associated with gestation, and the costs associated with lactation. Because of the continual and nonlinear growth rates of fetuses and neonates, the costs of gestation are significantly greater in the last trimester of pregnancy than in the first two trimesters, and the costs of lactation are much greater than the costs of gestation (i.e., two to three times more costly) (108). The energy cost for lactation is not constant over the nursing period, and varies with the number of neonates nourished. In general, milk output rises rapidly after birth, approaches a maximum or peak several weeks later, and then falls as the neonate is weaned and begins to feed for itself. The peak of lactation varies with species. For example, in caribou, peak of lactation occurs virtually immediately after birth, while the peak in lactation for a variety of other North American large mammals occurs about 3 to 4 weeks after birth (109, 110, 111, 112). As a result of their early lactational peak, caribou calves exhibit extremely rapid rates of growth very early in life at a time when other neonates grow more slowly (110).

The energy yield in milk at the peak of lactation in ungulates that nurse a single young scales with body mass of the dam:

$$Y = 368M^{0.71} \, , \tag{6-6}$$

where Y is the peak milk energy yield in kilojoules per day (65). Thus the energy demand at the peak of lactation scales nearly identically with basal metabolic rate (i.e., the same exponent), and represents an energy demand that is slightly higher than that of basal metabolic rate. The true energetic cost of lactation, however, is much greater than this, because the energetic cost of producing milk is quite high. The total energy demand at the peak of lactation has been estimated to be as high as four to seven times basal metabolic rate, or between 1,000 and 2,000 kilojoules/kg$^{0.75}$/day.

The protein demands for gestation and lactation are similar in form to those of energy. For example, the daily protein requirement for the production of a white-tailed deer fetus is 0.25 grams per day on day 70 of gestation, but rises exponentially to more than 6 grams per day on day 190 of gestation (108). In comparison, the protein production in milk at the peak of lactation is about 83 g/day (52). Milk protein for caribou and muskoxen can be as high as 150 to 175 g/day at the peak of lactation (110). This protein demand by the female is approximately five times greater than her maintenance protein needs.

Sportsmen and managers have debated the relative roles of nutrition and genetics on antler (or horn) size. There is little doubt that both play a significant role in the development of large antlers or horns (113). With regard to the nutritional needs for antler development, it is often assumed that the nutritional role is dominated by the animal's need for minerals (i.e., calcium and phosphorus). Although calcium is rarely severely deficient in soils or plants, phosphorus availability in forages in some areas of the country may be as low as 0.1% of dry matter. In some instances, where phosphorus demands are high, this may lead to osteophagia ("bone-eating") by cervids (i.e., caribou in tundra ranges). Nevertheless, diets with phosphorus levels as low as 0.14% of dry matter have not been shown to significantly impair antler growth in deer (114).

Antlers, however, are not composed solely of calcium and phosphorus and, in fact, the composition of dried antler is nearly 45% protein. Moreover, the protein requirements during antlerogenesis can be substantially greater than this protein accumulation may indicate, because much of the mass of the growing antler is protein (including the velvet, which is living tissue, not bone). Nevertheless, the protein required for antlerogenesis in white-tailed deer is approximately 10% of the protein required for maintenance (115). Further, because this relatively small added demand comes at a time of year when food is generally abundant and nutritious, it is unlikely that protein deficiencies will constrain antler growth.

A detailed analysis of the incremental energy costs of large animals can lead to a quantitative understanding of their energetic demands and the capability of habitats to supply these energy demands (43, 116). Field measures of energetic costs have also been obtained for many mammals using a doubly labeled water technique (117). This technique integrates all of the above energetic expenditures over a time span of between a few hours to a few weeks (65, 118). Although few North American large mammals have been included in these studies, field metabolic rates of 44 eutherian mammals are related to body mass as:

$$E_F = 894.5M^{0.762} \; , \tag{6-7}$$

where E_F is the field metabolic rate (kilojoules per day; 117). Accordingly, daily energy expenditures of free-ranging animals scale very closely with basal metabolic rate (i.e., have about the same exponent of body mass), but appear to be about three times higher than basal metabolic rates.

ANIMAL BODY CONDITION AND CONDITION INDICES

Two approaches are taken to evaluating habitats of large herbivores for management purposes. The direct approach is to examine the plant communities themselves, monitoring the availability of food, their use by herbivores, and their stability and persistence in the plant community. The approach has several advantages: it provides a means to rank habitat quality, it allows for the assessment of the roles and interactions of all species of herbivores in the community (e.g., competition between livestock and wildlife), and, probably more importantly, it engages and encourages the biologist to think about the lower trophic levels of the ecosystem that, ultimately, determine the density and productivity of herbivores.

The direct approach, however, is not accomplished without difficulty. Plant communities are nearly infinitely variable in their species composition, productivity, and spatial heterogeneity. Moreover, the vagaries of weather juxtapose on this melange a temporal variance that further complicates the assessment of habitat quality. Finally, the nutritional qualities of plants are temporally and spatially variable, and difficult to measure accurately. Herbivores, in turn, are adaptable and resourceful, and consequently their choice of foods is neither static nor highly predictable. As a consequence, it is difficult to quantify efficiently all the variables of plants and plant communities that must be integrated into an accurate habitat assessment. This difficulty has led many biologists to a second, indirect approach to evaluating habitats and herbivore nutritional status. This approach examines the animal itself, either through physiological and physical measures of animal body condition or chemical indices of body condition. In so doing, it is assumed that the physiological condition of the animal directly reflects and integrates over time the quality of the habitat and foods available to the animal.

Body condition is assessed using three general approaches, including the direct and indirect measurement of body stores (fat, protein, ash, and water), the indirect assessment of physiological condition using blood chemistry profiles, and the indirect assessment of diet quality and/or physiological condition using animal products (urine and feces). In addition, chemical methods are used to assess ancillary physiological parameters, including early gestation pregnancy rates (i.e., pregnancy specific protein b), pregnancy-associated glycoprotein and exposure and stress associated with environmental contaminants (119, 120, 121, 122).

Direct and indirect measures of body stores of animals have long been a part of the management of large herbivores. Because lipids represent the animal's principal energy stores, it is not surprising to find that most indirect measurements focus on quantifying fat. Several field methods provide an index to body fat, and these include the assessment of femur marrow fat, kidney fat, and rump fat thickness (123, 124, 125, 126, 127, 128). All three methods assume that the relative quantities of fat in these deposits are related directly to the total body fat of the animal. Although the mass of fat in these deposits has been shown to be related to total body fat, the relationships are generally curvilinear because not all body fat stores are metabolized at the same rate or at the same time (the apparent exception being the rump fat index measured ultrasonically in moose, which appears to be related linearly with total body fat; 129). For example, femur marrow fat is generally one of the last fat stores to be catabolized during starvation, and therefore does not begin to deplete until other body fat is virtually gone (127, 130, 131, 132). Hence, over most of the range in total body fat, measurement of femur fat will be relatively constant and high. Femur marrow fat is therefore only useful as an indicator of starvation and imminent death. Curvilinearity does not preclude the use of such fat indices for monitoring body condition of wild ungulates, although it does limit the usefulness of a single measure as an indicator of body condition. More appropriately, body condition should be assessed from a combination of fat deposit measurements (131, 132, 133).

Recent advances in the measurement of body condition include the use of body water dilution techniques, bioelectric impedance analysis, and ultrasonography. These techniques are advantageous because they are noninvasive and do not require the sacrifice of the animal. Of the three methods, body water dilution is the most precise and therefore the preferred method, when its use is feasible. As the name suggests, the method measures the total body water content of the animal. Because lipids are hydrophobic, adipose tissue contains very little water. Hence, there is a strong inverse relationship between the percentage of body water and the percentage of fat in the animal. Physiologists take advantage of this relationship by measuring the total body water, the simpler of the two parameters to measure. They do so by injecting a known amount of "labeled" water and measuring the concentration of this label over time in the fluids (blood, digesta, feces, or urine) of the animal. The labeled water is generally an isotopically labeled water molecule [^2H_2O (deuterium-labeled water) or ^3H_2O (tritium-labeled water)] that chemically and physiologically mimics "normal" water, and therefore mixes evenly and rapidly in all body water pools (e.g., blood, across the digestive tract wall and into saliva and digesta, and into cells and extracellular fluids). The subsequent dilution of the concentration of the label in the body fluids is directly proportional to the amount of body water into which it was mixed. Because of the time required for equilibration (between 2 and 8 hours), the method is difficult to use on wild animals.

Bioelectrical impedance analysis (BIA) holds promise for rapid field measurement of body composition of anesthetized mammals (134, 135). The method measures the electrical resistance or impedance of an electrical current through the body of the animal. A weak electrical current is applied across electrodes placed at a specified distance apart on the animal, and the resistance between electrodes is measured. The resistance is inversely proportional to the amount and concentration of body water between the electrodes, and hence fat composition is directly related to resistance.

Portable ultrasonography has been used to measure body composition (and pregnancy) in free-ranging herbivores. The portability and resolution of the latest ultrasound machines allow biologists to measure accurately the thickness of subcutaneous (generally scapular or lumbar) fat deposits on tranquilized animals in the field. Stephenson and others have recently compared ultrasonography to quantitative measures of body fat in moose and found that lumbar back fat volumes computed from ultrasonography are linearly related to total body fat (129). This also suggests that in moose the catabolism of back fat is relatively constant over time as the animals deplete body fat stores, which is not apparently true for other species (132).

A variety of hematological parameters have been examined for their usefulness in predicting the nutritional and physiological status of large mammals. Their usefulness depends on satisfying two criteria. First, the blood parameter must have a reliable baseline for comparison within the species. Human blood parameters, such as glucose, have well-established baseline values derived from many subjects under controlled conditions. In contrast, baseline information for most wildlife species is relatively sparse, and therefore it is difficult to determine if hematological values are within the normal range. Secondly, the background conditions, or intraspecific variation in the parameters, must be accounted for. Background conditions refer to the effects of specific environmental or physiological conditions that influence or modify the parameters. For example, age, sex, pregnancy, lactation, and capture or handling stress alter many hematological values (e.g., blood proteins, blood nitrogen, blood pH, and blood enzymes), and therefore may mask a nutritional effect (136). Background conditions for blood parameters must be quantified under controlled conditions with captive animals. For that reason, the usefulness of most blood chemistry parameters for evaluating the nutritional status of free-ranging large mammals is relatively low.

Of the many blood parameters evaluated in the past several decades, perhaps the most reliable for indicating nutritional condition of large mammals are blood urea nitrogen and the ratio of urea and creatinine (136, 137, 138, 139). In general, blood urea nitrogen concentrations and the urea:creatinine ratio increase as the animal depletes its fat reserves and begins to catabolize muscle tissue for energy. As a consequence of protein catabolism, certain amino acids are converted to carbohydrates while the amino group is cleaved and converted into urea. Although in ruminants, much of this urea is recycled to the rumen for use by bacteria to produce new proteins, the rate of bacterial protein production is limited by the rate that carbon skeletons (i.e., carbohydrates) are made available from foods. The concentration of blood urea nitrogen is also affected by protein concentration in the diet. High dietary protein levels are associated with high levels of blood urea nitrogen, because bacteria produce greater amounts of ammonia, which is absorbed and converted by the liver to urea (49). This obviously complicates the use of blood urea nitrogen as a nutritional indicator, because both very high quality diets and very low quality diets may result in similar blood urea nitrogen concentrations. In practice, however, this is seldom a problem, because blood urea nitrogen levels are generally measured on animals under nutritional stress (i.e., winter). Nevertheless, interpretation of blood urea nitrogen concentrations must be made carefully, because they will also reflect the relative stage of fat depletion of the animal, and not necessarily the relative difference in diet quality between populations at the time of sampling. The use of urea:creatinine ratios is also becoming popular for evaluating the nutritional stress of animals, although its measurement is mostly performed on urine samples instead of blood. This parameter is discussed later.

Physiological status may be evaluated by monitoring blood levels of thyroxine, triiodothyronine, and cortisol. Thyroxine levels directly influence metabolic rate, and several researchers have found that concentrations decline during winter. As expected, these correspond to a seasonal depression in metabolic rate and food intake. Hence, the relative concentrations of thyroxine and triiodothyronine of different populations should reflect their relative nutritional status (137, 138). Cortisol concentrations have been proposed as indicators of physiological stress and body condition as well, although most cortisol assays are more easily performed on urine samples than blood (140, 141).

To avoid the difficulty, expense, and stress (to the animals) associated with immobilizing wild animals to determine nutritional condition, several researchers have been exploring the potential nutritional indices that may lie concealed in the excretory products of the animals. One promising method is the measurement of urinary urea:creatinine ratios. As indicated earlier, urea is a consequence of undernutrition and the subsequent catabolization of protein stores. Because urinary urea is a relatively direct reflection of the blood urea concentration, urinary urea excretion should increase with prolonged starvation (49). Creatinine, on the other hand, is a product of muscle metabolism and is excreted in the urine at relatively stable rates (138, 142). Hence, the ratio of urea:creatinine should reflect primarily the dynamics of urea excretion, since creatinine should remain constant (142). The expression of urea as a ratio allows one to collect samples of urine that may be either diluted (by endogenous or exogenous water) or concentrated (by evaporation or dehydration). A simple way to view this is to think of the consequences of drinking many cups of coffee. The urea concentration of urine may drop significantly in such a case, even if your nutritional status remained the same. However, the ratio of urea and creatinine would remain constant throughout, accurately reflecting your nutritional status. This principle has been exploited in wildlife management by collecting snow urine samples and using these to measure the nutritional status of the population (142, 143). A similar approach has been used recently, substituting allantoin for urea in the urinary ratio (144). Allantoin is a purine derivative of dietary and microbial nucleic acids and is principally a product of bacterial decomposition and absorption in the gut. Metabolism of allantoin is limited, and hence it is subsequently excreted in the urine. In theory, the ratio of allantoin:creatinine should reflect dietary digestible dry matter intake, since this promotes the growth and subsequent flow of bacteria to the lower tract.

Although the creatinine ratio techniques show promise for tracking nutritional status of large mammals, the techniques should be used with caution pending further evaluation. Among other potential problems with the ratio technique, it depends heavily on the assumption that creatinine output remains constant over time. This has been shown not to be the case in several instances (e.g., 138, 139). Moreover, because creatinine production is a function of muscle mass and metabolism, it seems unlikely that creatinine levels would remain constant as an animal loses mass and its muscle mass is catabolized. Secondly, it is generally observed that fat stores are preferentially catabolized during the initial and middle stages of starvation (145, 146). Although some protein is catabolized during these stages of depletion, the rates may be relatively steady until most of the fat stores are consumed. Hence, urinary urea:creatinine ratios may plateau for significant periods of time while body condition continues to decline.

Fecal indices of animal body condition and diet quality have also been examined in recent years. In particular, fecal protein and fecal diaminopimelic acid concentrations have been found to be directly related to dietary nitrogen and digestibility (137, 147, 148). In theory, as dietary protein and digestibility increase, the excretion of nitrogen (primarily microbial N that has not been digested by the animal) and diaminopimelic acid (a microbial cell wall component that is indigestible in the vertebrate digestive tract) increases. In practice, fecal nitrogen and diaminopimelic acid concentrations consistently reflect the seasonal patterns of diet quality of free-ranging animals and differences among diets within seasons (137, 147, 149). However, these fecal indices may provide only limited information regarding the physiological status or body condition of herbivores (149). Furthermore, fecal N may be confounded when plants with high tannin concentrations are consumed (150). Binding of dietary and endogenous proteins with tannins causes their precipitation and subsequent excretion, which elevates fecal nitrogen concentrations irrespective of the digestibility or quality of the diet.

NUTRITIONAL BASIS FOR ANIMAL HABITAT ASSESSMENT

The nutritional ecologist has a multitude of techniques with which to assess forage availability and quality, and animal nutritional status. The dilemma is that none provides a foolproof assay of habitat quality, carrying capacity, or an animal's nutritional profile. The question remains, which methods are most appropriate or most pragmatic for measuring or indexing the quality of habitats? Some biologists advocate that animal measures rather than plant community and nutritional measures are more parsimonious and cost effective. Certainly for measuring animal nutritional fitness, and its correlate—productivity—this is true (151). However, body condition indices alone do not reflect habitat carrying capacity or productive potential for at least two reasons. First, individual animal condition reflects the product of both intraspecific competition for resources and the productive potential of the plant community. In other words, poor animal condition may reflect crowding in a productive habitat, or inadequate food quality and quantity in a sparsely populated one. In addition, assessments of animal condition are generally snapshots in time, and it is difficult, if not impossible, to infer anything about rates of change in body condition, which are certainly as important as body condition itself. For example, the observation that an animal or population in midwinter is in poor condition tells us nothing of its past or present nutritional history. Such a condition may result from an animal entering winter in poor shape and remaining in such condition (indicative of poor summer range but good winter range), or it may result from an animal entering winter in good shape and losing body condition rapidly (i.e., good summer range but poor winter range). Without ancillary population data, and serial measures of body condition, such questions will remain unanswered by the wildlife manager.

Secondly, and perhaps more importantly, it is unknown whether animal condition is coupled tightly and in synchrony with the condition of the plant community. The consequences of uncoupled or loosely coupled plant–herbivore systems will be instability. Such uncoupling not only invalidates the assumption that animal condition reflects habitat condition, but has important implications for population management. For example, under the concept of sustained-yield population management, the population is assumed to follow a predictable sigmoidal growth pattern that is asymptotic at K-carrying capacity. The sigmoidal pattern is a result of differential recruitment that approaches zero as the population approaches K (107, 152). A result of uncoupling is that recruitment can be potentially higher than is sustainable by the plant community, and harvest rates that are based on recruitment would therefore fail to maintain long-term stability of plant productivity.

Solid evidence for coupling of plant–herbivore populations is relatively sparse. The coupling between population density and reproduction and recruitment that has been documented (152) strongly suggests a density dependence between animals and their habitats. However, concomitant examinations of the relationship between plant community dynamics and productivity have not been accomplished. The implication is that most wildlife biologists view the plant community as static (i.e., that K remains constant), or that its dynamics are completely and instantaneously reflected by the animal's state. Nevertheless, strong, but anecdotal, evidence implies that herbivores often fail to track plant community dynamics over longer time scales. The argument for this is rather simple: Many herbivore populations are cyclic or prone to catastrophic dynamics (e.g., population buildup followed by dramatic crashes), suggesting that recruitment can exceed zero when the population is at K-carrying capacity, or that the plant community is not stable when the animal population approaches or remains at K.

The productivity and dynamics of large mammal populations are dependent on numerous influences, from trophic levels below and above. Predation, hunting, and plant community characteristics must all be considered simultaneously to understand and manage the population. Although predation and hunting may depress and perhaps control the trajectory of a population, its potential productivity is ultimately a function of the plant community. The nutrition of large animals is the currency that translates plant community productivity into animal productivity and, hence, a nutritional understanding of animal–habitat relationships is the ultimate tool in population management. Although our knowledge of nutritional ecology of large mammals has increased greatly during the past few decades, there is much that remains a mystery, and therefore much remains to be done.

SUMMARY

The productivity of large mammal populations is dependent on the nutritional well-being of its members. Nutritional status is a function of the rates of intake, assimilation, and metabolic processing of nutrients, particularly energy and protein, and the demands for these nutrients for maintenance and production. Hence, nutritional ecologists view the interaction of large mammals and their habitats as being similar to an accounting process using a nutritional currency such as energy or protein.

The rate of nutrient intake is a function of the efficiency of foraging, digestion, and absorption/metabolism of plant foods. Foraging rate is often constrained by the size and architecture of plants, or by the availability of plants in the landscape. As a result of the physical constraints on the foraging animal, feeding rates are an asymptotic function of bite size or plant density.

Once ingested, plant material continues to make life difficult for herbivores. Much of the plant is composed of recalcitrant compounds, such as cellulose, that require an extended digestion time in the gut of the animal. Lengthy retention of food in the rumen of large animals is typical because the more recalcitrant foods, such as mature grasses and browse, are difficult to break down and pass from the rumen. This constrains intake and thereby limits the nutrient assimilation rate. Moreover, the easily digested components often contain toxic defensive substances that the plant uses to discourage consumption. Some chemical defenses, such as tannins, defend the plant by reducing digestibility. Others, such as alkaloids or cyanogenic glycoside, defend by intoxicating the animals. Normally, herbivores avoid poisoning by limiting the consumption of these plants and by mixing their diet to avoid overloading on detoxification mechanisms.

Food fuels the processes that ensure a healthy population: maintenance of the individual, growth, and reproduction. The nutrients most likely to be limiting to large mammals for these processes include energy and protein. Maintenance requirements for energy and protein scale to the three-quarter power of body weight. This means that small animals, such as black-tailed deer and pronghorn, require significantly more energy and protein per unit of body weight than do larger herbivores such as moose or bison. Gestation and lactation greatly increase the nutrient demands of the female. Of the two reproductive processes, lactation is a far greater burden to the female than is gestation. The latter only becomes a significant nutrient demand in the last trimester of pregnancy.

The nutritional status of large mammals can be assessed using several techniques. Among the more direct measures is the assessment of body condition or body fat. Body fat is measured indirectly through water dilution, bioelectric impedance, and ultrasonography. The first two methods rely on the fact that body fat varies inversely with body water content. Hence, measuring the amount of water in the body by water dilution or electrical resistance allows the determination of fat.

Other nutritional indices of animal condition include blood chemistry profiling, particularly the measurement of blood levels of urea nitrogen, thyroid hormones, and cortisol. Measures of the urinary excretion of urea, allantoin, and creatinine allow the estimation of nutritional status as well.

The wildlife manager has many tools available for the assessment of the nutritional well-being of populations. However, none of the methods is foolproof for evaluating habitats, carrying capacity, or the animal's nutritional status. Because large mammal populations are prone to buildup and declines, it is probable that their populations are capable of overshooting carrying capacity. Combinations of techniques are necessary to evaluate the capacity of habitats to support animals, determine the levels of populations that meet management objectives, and monitor populations and habitats to ensure that these objectives are met.

LITERATURE CITED

1. HAIRSTON, N. G., F. E. SMITH, and L. B. SLOBODKIN. 1960. Community structure, population control and competition. *The American Naturalist* 44:421–425.

2. ADAMS, L. G., B. W. DALE, and L. D. MECH. 1996. Wolf predation on caribou calves in Denali National Park, Alaska. Pages 245–260 in L. N. Carbyn, S. H. Fritts, and D. R. Seip, eds., *Ecology and Conservation of Wolves in a Changing World.* Canadian Circumpolar Institute, University of Alberta, Edmonton.

3. CLUTTON-BROCK, T. H., M. MAJOR, S. D. ALBON, and F. E. GUINNESS. 1987. Early development and population dynamics in red deer. Density-dependent effects on juvenile survival. *Journal of Animal Ecology* 56:53–67.

4. GUSTAFSON, L. L., W. L. FRANKLIN, R. J. SARNO, R. L. HUNTER, K. M. YOUNG, W. E. JOHNSON, and M. J. BEHL. 1998. Predicting early mortality of newborn guanacos by birth mass and hematological parameters: A provisional model. *Journal of Wildlife Management* 62:24–35.

5. SINGER, F. J., A. HARTING, K. K. SYMONDS, and M. B. COUGHENOUR. 1997. Density dependence, compensation, and environmental effects on elk calf mortality in Yellowstone National Park. *Journal of Wildlife Management* 61:12–25.

6. THORNE, E. T., R. E. DEAN, and W. G. HEPWORTH. 1976. Nutrition during gestation in relation to successful reproduction in elk. *Journal of Wildlife Management* 40:330–335.

7. BAKER, D. L., and N. T. HOBBS. 1987. Strategies of digestion: Digestive efficiency and retention time of forage diets in montane ungulates. *Canadian Journal of Zoology* 65:1978–1984.

8. ILLIUS, A. W., and M. S. ALLEN. 1994. Assessing forage quality using integrated models of intake and digestion by ruminants. Pages 869–890 *in* G. C. Fahey, ed. *Forage Quality, Evaluation, and Utilization*. University of Nebraska Press, Lincoln.

9. MILNE, J. A., J. C. MACRAE, A. M. SPENCE, and S. WILSON. 1978. A comparison of the voluntary intake and digestion of a range of forages at different times of the year by the sheep and red deer. *British Journal of Nutrition* 40:347–357.

10. CLUTTON-BROCK, T., and P. HARVEY. 1978. Mammals, resources and reproductive strategies. *Nature* 273:191–195.

11. DEMMENT, M., and P. VAN SOEST. 1985. A nutritional explanation of body-size patterns of ruminant and nonruminant herbivores. *The American Naturalist* 125:641–672.

12. GORDON, I. J., and A. W. ILLIUS. 1988. Incisor arcade structure and diet selection in ruminants. *Functional Ecology* 2:15–22.

13. HOFMANN, R. 1989. Evolutionary steps of ecophysiological adaptation and diversification of ruminants: A comparative view of their digestive system. *Oecologia* 78:443–457.

14. ILLIUS, A. W., and I. J. GORDON. 1990. Variation in foraging behaviour in red deer and the consequences for population demography. *Journal of Animal Ecology* 59:89–101.

15. ILLIUS, A. W., and I. J. GORDON. 1992. Modelling the nutritional ecology of ungulate herbivores: Evolution of body size and competitive interactions. *Oecologia* 89:428–434.

16. JANIS, C., and D. EHRHARDT. 1988. Correlation of relative muzzle width and relative incisor width with dietary preference in ungulates. *Zoological Journal of the Linnean Society* 92:267–284.

17. LAWS, R.M. 1981. Large mammal feeding strategies and related overabundance problems. Pages 217–232 *in* P.A. Jewell and S. Holt, eds. *Problems in Management of Locally Abundant Wild Mammals*. St. John's College, Cambridge.

18. GROSS, J. E., N. T. HOBBS, and B. A. WUNDER. 1993. Independent variables for predicting intake rate of mammalian herbivores: Biomass density, plant density, or bite size? *Oikos* 68:75–81.

19. SHIPLEY, L. A., D. E. SPALINGER, J. E. GROSS, N. T. HOBBS, and B. A. WUNDER. 1996. The dynamics and scaling of foraging velocity and encounter rate in mammalian herbivores. *Functional Ecology* 10:234–244.

20. GROSS, J. E., L. A. SHIPLEY, N. T. HOBBS, D. E. SPALINGER, and B. A. WUNDER. 1993. Functional response of herbivores in food-concentrated patches: Tests of a mechanistic model. *Ecology* 74:778–791.

21. RISENHOOVER, K. L. 1987. Winter foraging strategies of moose in subarctic and boreal forest habitats. Ph.D. thesis. Michigan Technological University, Houghton. 108 pages.

22. SPALINGER, D. E., T. HANLEY, and C. ROBBINS. 1988. Analysis of the functional response in foraging in the sitka black-tailed deer. *Ecology* 69:1166–1175.

23. COLLINS, W. B., P. J. URNESS, and D. D. AUSTIN. 1978. Elk diets and activities on different lodgepole pine habitat segments. *Journal of Wildlife Management* 42:799–810.

24. HUDSON, R., and M. NIETFELD. 1985. Effect of foraging depletion on the feeding rate of wapiti. *Journal of Range Management* 38:80–82.

25. HUDSON, R., and W. WATKINS. 1986. Foraging rates of wapiti on green and cured pastures. *Canadian Journal of Zoology* 64:1705–1708.

26. RENECKER, L. A., and R. J. HUDSON. 1986. Seasonal foraging rates of free-ranging moose. *Journal of Wildlife Management* 50:143–147.

27. ROMINGER, E. M., and C. T. ROBBINS. 1996. Winter foraging ecology of woodland caribou in northeastern Washington. *Journal of Wildlife Management* 60:719–728.

28. TRUDELL, J., and R. WHITE. 1981. The effect of forage structure and availability on food intake, biting rate, bite size and daily eating time of reindeer. *Journal of Applied Ecology* 18:63–81.

29. WICKSTROM, M., C. ROBBINS, T. HANLEY, D. SPALINGER, and S. PARISH. 1984. Food intake and foraging energetics of elk and mule deer. *Journal of Wildlife Management* 48:1285–1301.

30. LACA, E. A., E. D. UNGAR, and M. W. DEMMENT. 1994. Mechanisms of handling time and intake rate of a large mammalian grazer. *Applied Animal Behaviour Science* 39:3–19.

31. HOLLING, C. 1965. The functional response of predators to prey density and its role in mimicry and population regulation. *Memoirs of the Entomological Society of Canada* 45:1–60.

32. SPALINGER, D. E., and N. T. HOBBS. 1992. Mechanisms of foraging in mammalian herbivores: New models of functional response. *The American Naturalist* 140:325–348.

33. COOPER, S. M., and N. OWEN-SMITH. 1986. Effects of plant spinescence on large mammalian herbivores. *Oecologia* 68:446–455.

34. ILLIUS, A. W., and I. J. GORDON. 1987. The allometry of food intake in grazing ruminants. *Journal of Animal Ecology* 56:989–999.

35. SHIPLEY, L. A., A. W. ILLIUS, K. DANELL, N. T. HOBBS, and D. E. SPALINGER. 1998. Predicting bite size selection of mammalian herbivores: A test of a general model of diet optimization. *Oecologia* (in press).

36. VIVAS, H. J., and B. E. SAETHER. 1987. Interactions between a generalist herbivore, the moose *Alces alces,* and its food resources: An experimental study of winter forage behaviour in relation to browse availability. *Journal of Animal Ecology* 56:509–520.

37. VIVAS, H. J., B. E. SAETHER, and R. ANDERSEN. 1991. Optimal twig-size selection of a generalist herbivore, the moose *Alces alces:* Implications for plant–herbivore interactions. *Journal of Animal Ecology* 60:395–408.

38. KIELLAND, K., and T. OSBORNE. 1998. Moose browsing on feltleaf willow: Optimal foraging in relation to plant morphology and chemistry. *Alces* 34:149–155.

39. ILLIUS, A. W. 1989. Allometry of food intake and grazing behaviour with body size in cattle. *Journal of Agricultural Science* 113:259–266.

40. SHIPLEY, L. A., J. E. GROSS, D. E. SPALINGER, N. T. HOBBS, and B. A. WUNDER. 1994. The scaling of intake rate in mammalian herbivores. *The American Naturalist* 143:1055–1082.

41. ILLIUS, A. W., S. D. ALBON, J. PEMBERTON, I. J. GORDON, and T. H. CLUTTON-BROCK. 1995. Selection for foraging efficiency during a population crash in Soay sheep. *Journal of Animal Ecology* 64:481–493.

42. SCHWAB, F., M. PITT, and S. SCHWAB. 1987. Browse burial related to snow depth and canopy cover in northcentral British Columbia. *Journal of Wildlife Management* 51:337–342.

43. PARKER, K. L., M. P. GILLINGHAM, T. A. HANLEY, and C. T. ROBBINS. 1996. Foraging efficiency: Energy expenditure versus energy gain in free-ranging black-tailed deer. *Canadian Journal of Zoology* 74:442–450.

44. PARKER, K. L., C. T. ROBBINS, and T. A. HANLEY. 1984. Energy expenditures for locomotion by mule deer and elk. *The Journal of Wildlife Management* 48:474–488.

45. HANSEN, M. C. 1996. Foraging ecology of female Dall's sheep in the Brooks Range, Alaska. Ph.D. thesis, University of Alaska Fairbanks, Fairbanks. 117 pages.

46. HEIMER, W. E., F. J. MAUER, and S. W. KELLER. 1994. The effects of physical geography on Dall sheep habitat quality and home range size. Pages 144–148 *in Biennial Symposium of the North American Wild Sheep and Goat Council.*

47. CARL, G. R., and R. S. BROWN. 1986. Comparative digestive efficiency and feed intake of the collared peccary. *Southwest Naturalist* 31:79–85.

48. PRITCHARD, G. T., and C. T. ROBBINS. 1990. Digestive and metabolic efficiencies of grizzly and black bears. *Canadian Journal of Zoology* 68:1645–1651.

49. VAN SOEST, P. J. 1994. *Nutritional Ecology of the Ruminant.* Cornell University Press, Ithaca, NY.

50. FREUDENBERGER, D. O. 1992. Gut capacity, functional allocation of gut volume and size distributions of digesta particles in two macropodid marsupials (*Macropus robustus robustus* and *M. R. erubescens*) and the feral goat (*Capra hircus*). *Australian Journal of Zoology* 40:551–561.

51. JENKS, J. A., D. M. LESLIE, JR., R. L. LOCHMILLER, and M. A. MELCHIORS. 1994. Variation in gastrointestinal characteristics of male and female white-tailed deer: Implications for resource partitioning. *Journal of Mammalogy* 75:1045–1053.

52. ROBBINS, C., and A. MOEN. 1975. Milk consumption and weight gain of white-tailed deer. *Journal of Wildlife Management* 39:355–360.

53. SIBBALD, A. M., and J. A. MILNE. 1993. Physical characteristics of the alimentary tract in relation to seasonal changes in voluntary food intake by the red deer (*Cervus elaphus*). *Journal of Agricultural Science (Cambridge)* 120:99–102.

54. STAALAND, H., and R. G. WHITE. 1991. Influence of foraging ecology on alimentary tract size and function of Svalbard reindeer. *Canadian Journal of Zoology* 69:1326–1334.

55. ELLIS, W., M. WYLIE, and J. MATIS. 1988. Dietary–digestive interactions determining the feeding value of forages and roughages. *World Animal Science. B. Disciplinary Approach. 4 Feed Science* B.4:177–229.

56. PARRA, R. 1978. Comparison of foregut and hindgut fermentation in herbivores. Pages 205–229 *in* G. G. Montgomery, ed., *The Ecology of Arboreal Folivores.* Smithsonian Institution Press, Washington, DC.

57. STEVENS, C., A. SELLERS, and F. SPURRELL. 1960. Function of the bovine omasum in ingesta transfer. *American Journal of Physiology* 198:449–455.

58. STEVENS, C. E., and I. D. HUME. 1995. *Comparative Physiology of the Vertebrate Digestive System.* Cambridge University Press, Cambridge, UK.

59. MCLEOD, M. N., P. M. KENNEDY, and D. J. MINSON. 1990. Resistance of leaf and stem fractions of tropical forage to chewing and passage in cattle. *British Journal of Nutrition* 63:105–119.

60. SPALINGER, D., C. ROBBINS, and T. HANLEY. 1986. The assessment of handling time in ruminants: The effect of plant chemical and physical structure on the rate of breakdown of plant particles in the rumen of mule deer and elk. *Canadian Journal of Zoology* 64:312–321.

61. RISENHOOVER, K. L. 1986. Winter activity patterns of moose in interior Alaska. *Journal of Wildlife Management* 50:727–734.

62. WATKINS, W. G., R. J. HUDSON, and P. FARGEY. 1991. Compensatory growth of wapiti (*Cervus elaphus*) on spring ranges. *Canadian Journal of Zoology* 69:1682–1688.

63. HOLLEMAN, D. F., and R. G. WHITE. 1989. Determination of digesta fill and passage rate from nonabsorbed particulate phase markers using the single dose method. *Canadian Journal of Zoology* 67:488–494.

64. SPALINGER, D. E., and C. T. ROBBINS. 1992. The dynamics of particle flow in the rumen of mule deer (*Odocoileus hemionus hemionus*) and elk (*Cervus elaphus nelsoni*). *Physiological Zoology* 65:379–402.

65. ROBBINS, C. 1993. *Wildlife Feeding and Nutrition.* Academic Press, San Diego, CA.

66. HUNGATE, R. 1966. *The rumen and its microbes.* Academic Press, New York.

67. FISHER, D., J. BURNS, and K. POND. 1989. Kinetics of *in vitro* cell-wall disappearance and *in vivo* digestion. *Agronomy Journal* 81:25–33.

68. SPALINGER, D. E. 1985. The dynamics of forage digestion and passage in the rumen of mule deer and elk. Ph.D. thesis. Washington State University, Pullman. 176 pages.

69. GRANT, R. J., and D. R. MERTENS. 1992. Impact of *in vitro* fermentation techniques upon kinetics of fiber digestion. *Journal of Dairy Science* 75:1263–1272.

70. UDEN, P., and P. J. VAN SOEST. 1984. Investigations of the *in situ* bag technique and a comparison of the fermentation in heifers, sheep, ponies and rabbits. *Journal of Animal Science* 58:213–221.

70a. MAUTZ, W. W., H. SILVER, J. B. HOLTER, H. H. HAYES, and W. E. URBAN, JR. 1976. Digestibility and related nutritional data for seven northern deer browse species. *The Journal of Wildlife Management* 40:630–638.

71. KOERTH, B. H., and J. W. STUTH. 1991. Instantaneous intake rates of 9 browse species by white-tailed deer. *Journal of Range Management* 44:614–618.

72. BAZELY, D. R., J. H. MYERS, and K. B. DASILVA. 1991. The response of numbers of bramble prickles to herbivory and depressed resource availability. *Oikos* 61:327–336.

73. BELOVSKY, G. E., O. J. SCHMITZ, J. B. SLADE, and T. J. DAWSON. 1991. Effects of spines and thorns on Australian arid zone herbivores of different body masses. *Oecologia* 88:521–528.

74. MILEWSKI, A. V., T. P. YOUNG, and D. MADDEN. 1991. Thorns as induced defenses: Experimental evidence. *Oecologia* 86:70–75.

75. YOUNG, T. P. 1987. Increased thorn length in *Acacia drepanolobium*—An induced response to browsing. *Oecologia (Berlin)* 71:436–438.

76. BAZELY, D. R., and R. L. JEFFERIES. 1989. Leaf and shoot demography of an Arctic stoloniferous grass, *Puccinellia phryganodes,* in response to grazing. *Journal of Ecology* 77:811–822.

77. MCNAUGHTON, S. J. 1979. Grazing as an optimization process: Grass–ungulate relationships in the Serengeti. *The American Naturalist* 113:691–703.

78. McNaughton, S. J. 1984. Grazing lawns: Animals in herds, plant form, and coevolution. *The American Naturalist* 124:863–886.

79. Milchunas, D. G., W. K. Lauerroth, P. L. Chapman, and M. K. Kazempour. 1989. Effects of grazing, topography, and precipitation on the structure of a semiarid grassland. *Vegetatio* 80:11–23.

80. Milchunas, D. G., A. S. Varnamkhasti, W. K. Lauenroth, and H. Goetz. 1995. Forage quality in relation to long-term grazing history, current-year defoliation, and water resource. *Oecologia* 101:366–374.

81. Mack, R. N., and J. N. Thompson. 1982. Evolution in steppe with few large, hooved mammals. *The American Naturalist* 119:757–773.

82. Rhoades, D. F. 1979. Evolution of plant chemical defense against herbivores. Pages 3–54 *in* G. A. Rosenthal and D. H. Janzen, eds. *Herbivores. Their Interaction with Secondary Plant Metabolites.* Academic Press, Orlando, FL.

83. Hagerman, A. E., and L. G. Butler. 1991. Tannins and lignins. Pages 355–388 *in* G. A. Rosenthal and M. R. Berenbaum, eds., *Herbivores: Their Interactions with Secondary Plant Metabolites,* 2nd edition. Academic Press, New York.

84. Robbins, C. T., T. A. Hanley, A. E. Hagerman, O. Hjeljord, D. L. Baker, C. C. Schwartz, and W. W. Mautz. 1987. Role of tannins in defending plants against ruminants: Reduction in protein availability. *Ecology* 68:98–107.

85. Hanley, T. A., C. T. Robbins, A. E. Hagerman, and C. McArthur. 1992. Predicting digestible protein and digestible dry matter in tannin-containing forages consumed by ruminants. *Ecology* 73:537–541.

86. Nagy, J., H. Steinhoff, and G. Ward. 1964. Effects of essential oils of sagebrush on deer rumen microbial function. *Journal of Wildlife Management* 28:785–790.

87. Nagy, J., and R. Tengerdy. 1967. Antibacterial action of essential oils of *Artemisia* as an ecological factor I. Antibacterial action of the volatile oils of *Artemisia tridentata* and *Artemisia nova* on aerobic bacteria. *Applied Microbiology* 15:819–821.

88. Nagy, J., and R. Tengerdy. 1968. Antibacterial action of essential oils of *Artemisia* as an ecological factor II. Antibacterial action of the volatile oils of *Artemisia tridentata* (big sagebrush) on bacteria from the rumen of mule deer. *Applied Microbiology* 16:441–444.

89. Oh, H., M. Jones, and W. Longhurst. 1968. Comparison of rumen microbial inhibition resulting from various essential oils isolated from relatively unpalatable plant species. *Applied Microbiology* 16:39–44.

90. Oh, H., T. Sakai, M. Jones, and W. Longhurst. 1967. Effect of various essential oils isolated from Douglas fir needles upon sheep and deer rumen microbial activity. *Applied Microbiology* 15:777–784.

91. Bray, R. O., C. L. Wambolt, and R. G. Kelsey. 1991. Influence of sagebrush terpenoids on mule deer preference. *Journal of Chemical Ecology* 17:2053–2062.

92. Longhurst, W., H. Oh, M. Jones, and R. Kepner. 1968. A basis for the palatability of deer forage plants. *Transactions of the North American Wildlife Conference* 33:181–192.

93. Personius, T. L., C. L. Wambolt, J. R. Stephens, and R. G. Kelsey. 1987. Crude terpenoid influence on mule deer preference for sagebrush. *Journal of Range Management* 40:84–88.

94. Wallmo, O. C., L. H. Carpenter, W. L. Regelin, R. B. Gill, and D. L. Baker. 1977. Evaluation of deer habitat on a nutritional basis. *Journal of Range Management* 30:122–127.

95. Quinton, D. A. 1985. Saskatoon serviceberry toxic to deer. *Journal of Wildlife Management* 49:362–364.

96. Harborne, J. B. 1988. *Introduction to Ecological Biochemistry.* Academic Press, London, UK.

97. Orskov, E. R. 1982. *Protein Nutrition in Ruminants.* Academic Press, London, UK.

98. Mattos, W., and D. L. Palmquist. 1977. Biohydrogenation and availability of linoleic acid in lactating cows. *Journal of Nutrition* 107:1755–1761.

99. Kleiber, M. 1947. Body size and metabolic rate. *Physiology Review* 27:511–541.

100. McNab, B. K. 1988. Complications inherent in scaling the basal rate of metabolism in mammals. *The Quarterly Review of Biology* 63:25–54.

101. Hubbert, M. E. 1987. The effect of diet on energy partition in moose. Thesis. University of Alaska Fairbanks, Fairbanks. 157 pages.

Literature Cited

102. MAUTZ, W. W., J. KANTER, and P. J. PEKINS. 1992. Seasonal metabolic rhythms of captive female white-tailed deer: A reexamination. *Journal of Wildlife Management* 56:656–661.

103. PARKER, K. L., and C. T. ROBBINS. 1984. Thermoregulation in mule deer and elk. *Canadian Journal of Zoology* 62:1409–1422.

104. PEKINS, P. J., W. W. MAUTZ, and J. J. KANTER. 1992. Reevaluation of the basal metabolic cycle in white-tailed deer. Pages 418–422 *in* R. D. Brown, ed. *The Biology of Deer.* Springer-Verlag, New York.

105. SCHWARTZ, C. C., W. L. REGELIN, and A. W. FRANZMANN. 1987. Protein digestion in moose. *Journal of Wildlife Management* 51:352–357.

106. GAVIN, T. A., L. H. SURING, P. A. VOHS, and E. C. MESLOW. 1984. Population characteristics, spatial organization, and natural mortality in the Columbian white-tailed deer. *Wildlife Monographs* 91:41.

107. VAN BALLENBERGHE, V., and W. B. BALLARD. 1997. Population dynamics. Pages 223–245 *in* A. W. Franzmann and C. C. Schwartz, eds., *Ecology and Management of the North American Moose.* Smithsonian Institute Press, Washington, DC.

108. ROBBINS, C., and A. MOEN. 1975. Uterine composition and growth in pregnant white-tailed deer. *Journal of Wildlife Management* 39:684–691.

109. ALLAYE CHAN-MCLEOD, A. C., R. G. WHITE, and D. F. HOLLEMAN. 1994. Effects of protein and energy intake, body condition, and season on nutrient partitioning and milk production in caribou and reindeer. *Canadian Journal of Zoology* 72:938–947.

110. PARKER, K. L., R. G. WHITE, M. P. GILLINGHAM, and D. F. HOLLEMAN. 1990. Comparison of energy metabolism in relation to daily activity and milk consumption by caribou and muskox neonates. *Canadian Journal of Zoology* 68:106–114.

111. REESE, E. O., and C. T. ROBBINS. 1994. Characteristics of moose lactation and neonatal growth. *Canadian Journal of Zoology* 72:953–957.

112. ROBBINS, C. T., R. S. PODBIELANCIK-NORMAN, D. L. WILSON, and E. D. MOULD. 1981. Growth and nutrient consumption of elk calves compared to other ungulate species. *Journal of Wildlife Management* 45:172–186.

113. LUKEFAHR, S. D., and H. A. JACOBSON. 1998. Variance component analysis and heritability of antler traits in white-tailed deer. *Journal of Wildlife Management* 62:262–268.

114. GRASMAN, B. T., and E. C. HELLGREN. 1993. Phosphorus nutrition in white-tailed deer: Nutrient balance, physiological responses, and antler growth. *Ecology* 74:2279–2296.

115. ASLESON, M. A., E. C. HELLGREN, and L. W. VARNER. 1996. Nitrogen requirements for antler growth and maintenance in white-tailed deer. *Journal of Wildlife Management* 60:744–752.

116. HOBBS, N. T. 1989. Linking energy balance to survival in mule deer: Development and test of a simulation model. *Wildlife Monographs* 101:39.

117. NAGY, K. A. 1994. Field bioenergetics of mammals: What determines field metabolic rates? *Australian Journal of Zoology* 42:43–53.

118. NAGY, K. A. 1989. Field bioenergetics: Accuracy of models and methods. *Physiological Ecology* 62:237–252.

119. HAIGH, J. C., W. J. DALTON, C. A. RUDER, and R. G. SASSER. 1993. Diagnosis of pregnancy in moose using a bovine assay for pregnancy-specific protein B. *Theriogenology* 40:905–911.

120. HOUSTON, D. B., C. T. ROBBINS, C. A. RUDER, and R. G. SASSER. 1986. Pregnancy detection in mountain goats by assay for pregnancy-specific protein b. *Journal of Wildlife Management* 50:740–742.

121. WOOD, A. K., R. E. SHORT, A. E. DARLING, G. L. DUSEK, R. G. SASSER, and C. A. RUDER. 1986. Serum assays for detecting pregnancy in mule and white-tailed deer. *Journal of Wildlife Management* 50:684–687.

122. OSBORN, D. A., J.- F. BECKERS, J. SULON, J. W. GASSETT, L. I. MULLER, B. P. MURPHY, K. V. MILLER, and R. L. MARCHINTON. 1996. Use of glycoprotein assays for pregnancy diagnosis in white-tailed deer. *Journal of Wildlife Management* 60:388–393.

123. NEILAND, K. A. 1970. Weight of dried marrow as indicator of fat in caribou femurs. *Journal of Wildlife Management* 34:904–907.

124. VERME, L. J., and J. C. HOLLAND. 1973. Reagent-dry assay of marrow fat in white-tailed deer. *Journal of Wildlife Management* 37:103–105.

125. ANDERSON, A. E., D. C. BOWDEN, and D. E. MEDIN. 1990. Indexing the annual fat cycle in a mule deer population. *Journal of Wildlife Management* 54:550–556.

126. DAUPHINE, T. 1975. Kidney weight fluctuations affecting the kidney fat index in caribou. *Journal of Wildlife Management* 39:379–386.

127. TORBIT, S. C., L. H. CARPENTER, R. M. BARTMANN, A. W. ALLDREDGE, and G. C. WHITE. 1988. Calibration of carcass fat indices in wintering mule deer. *Journal of Wildlife Management* 52:582–588.

128. STEPHENSON, T. R., K. J. HUNDERTMARK, C. C. SCHWARTZ, and V. VAN BALLENBERGHE. 1993. Ultrasonic fat measurement of captive yearling bull moose. *Alces* 29:115–123.

129. STEPHENSON, T. R., K. J. HUNDERTMARK, C. C. SCHWARTZ, and V. VAN BALLENBERGHE. 1997. Prediction of body fat and body mass in moose using ultrasonography. Alaska Department of Fish and Game, Juneau.

130. ADAMCZEWSKI, J. Z., P. F. FLOOD, and A. GUNN. 1995. Body composition of muskoxen (*Ovibos moschatus*) and its estimation from condition index and mass measurements. *Canadian Journal of Zoology* 73:2021–2034.

131. HOLAND, O. 1992. Fat indices versus ingesta-free body fat in European roe deer. *Journal of Wildlife Management* 56:241–245.

132. WATKINS, B. E., J. H. WITHAM, D. E. ULLREY, D. J. WATKINS, and J. M. JONES. 1991. Body composition and condition evaluation of white-tailed deer fawns. *Journal of Wildlife Management* 55:39–51.

133. ANDERSON, A., D. MEDIN, and D. BOWDEN. 1972. Indices of carcass fat in a Colorado mule deer population. *Journal of Wildlife Management* 36:579–594.

134. FARLEY, S. D., and C. T. ROBBINS. 1994. Development of two methods to estimate body composition of bears. *Canadian Journal of Zoology* 72:220–226.

135. HILDERBRAND, G. V., S. D. FARLEY, and C. T. ROBBINS. 1998. Predicting body condition of bears via two field methods. *Journal of Wildlife Management* 62:406–409.

136. FRANZMANN, A. W. 1985. Assessment of nutritional status. Pages 239–259 *in* R. J. Hudson and R. G. White, eds., *Bioenergetics of Wild Herbivores*. CRC Press, Boca Raton, FL.

137. BROWN, R. D., E. C. HELLGREN, M. ABBOTT, D. C. RUTHVEN III, and R. L. BINGHAM. 1995. Effects of dietary energy and protein restriction on nutritional indices of female white-tailed deer. *Journal of Wildlife Management* 59:595–609.

138. DELGIUDICE, G. D., L. D. MECH, and U.S. SEAL. 1990. Effects of winter undernutrition on body composition and physiological profiles of white-tailed deer. *Journal of Wildlife Management* 54:539–550.

139. WOLKERS, H., T. WENSING, and J. T. SCHONEWILLE. 1994. Effect of undernutrition on haematological and serum biochemical characteristics in red deer (*Cervus elaphus*). *Canadian Journal of Zoology* 72:1291–1296.

140. SALTZ, D., and G. C. WHITE. 1991. Urinary cortisol and urea nitrogen responses to winter stress in mule deer. *Journal of Wildlife Management* 55:1–16.

141. SALTZ, D., G. C. WHITE, and R. M. BARTMANN. 1992. Urinary cortisol, urea nitrogen excretion, and winter survival in mule deer fawns. *Journal of Wildlife Management* 56:640–644.

142 MOEN, R., and G. D. DELGIUDICE. 1997. Simulating nitrogen metabolism and urinary urea nitrogen:creatinine ratios in ruminants. *Journal of Wildlife Management* 61:881–894.

143. DELGIUDICE, G. D., U.S. SEAL, and L. D. MECH. 1991. Indicators of severe undernutrition in urine of free-ranging elk during winter. *Wildlife Society Bulletin* 19:106–110.

144. VAGNONI, D. B., R. A. GARROTT, J. G. COOK, P. J. WHITE, and M. K. CLAYTON. 1996. Urinary allantoin:creatinine ratios as a dietary index for elk. *Journal of Wildlife Management* 60:728–734.

145. PARKER, K. L., G. D. DELGIUDICE, and M. P. GILLINGHAM. 1993. Do urinary urea nitrogen and cortisol ratios of creatinine reflect body-fat reserves in black-tailed deer? *Canadian Journal of Zoology* 71:1841–1848.

146. TORBIT, S. C., L. H. CARPENTER, D. M. SWIFT, and A. W. ALLDREDGE. 1985. Differential loss of fat and protein by mule deer during winter. *Journal of Wildlife Management* 49:80–85.

147. LESLIE, D. M., and E. E. STARKEY. 1985. Fecal indices to dietary quality of cervids in old-growth forests. *Journal of Wildlife Management* 49:142–146.

148. Leslie, D. M., Jr., and E. E. Starkey. 1987. Fecal indices to dietary quality: A reply. *Journal of Wildlife Management* 51:321–325.

149. Kucera, T. E. 1997. Fecal indicators, diet, and population parameters in mule deer. *Journal of Wildlife Management* 61:550–560.

150. Hobbs, N. T. 1987. Fecal indices to dietary quality: A critique. *Journal of Wildlife Management* 51:317–320.

151. Gerhart, K. L., R. G. White, R. D. Cameron, and D. E. Russell. 1996. Estimating fat content of caribou from body condition scores. *Journal of Wildlife Management* 60:713–718.

152. McCullough, D. R. 1979. *The George Reserve Deer Herd.* University of Michigan Press, Ann Arbor.

CHAPTER 7

Carrying Capacity

Karl V. Miller and James M. Wentworth

INTRODUCTION

Perhaps no concept is more central to the practice of wildlife management while at the same time shrouded by cloudy definitions and simplistic interpretations as the concept of carrying capacity (*K*). Often this term is used by biologists as a general concept without considering its exact meaning. Because this term envelops a number of diverse concepts with varied meanings, Macnab (1) included carrying capacity with overpopulation, over-harvesting, and overgrazing in his list of resource management shibboleths (1). Clarification of terminology is required for effective communication of biological concepts.

In this chapter we discuss the historical development of the carrying capacity concept, describe types of carrying capacities (both biologically based and culturally based), and evaluate the utility of the concept in current understandings of large mammal population dynamics. Finally, we discuss methods that have been used by wildlife biologists to estimate carrying capacity.

A CAPACITY OF CARRYING CAPACITIES

Since the concept of carrying capacity was first articulated, varied and conflicting meanings have been assigned to the term. Most definitions of carrying capacity are vague and some are totally meaningless (2). For example, Dasmann (2, 3) listed three ways the term *carrying capacity* is used in the wildlife literature: the number of animals of a species that a habitat does support; the upper limit of population growth, as defined by the logistic curve; and the number of animals that a habitat can maintain in a healthy and vigorous condition (3).

This disparate use of the term is a direct result of its development and application by managers in diverse fields. Range managers generally view carrying capacity as a measure of the productivity of the land. In other words, carrying capacity is regarded as the maximum number of animals that the land can support while providing maximum sustained production and without inducing a downward trend in forage production or forage quality (4). This concept of an "economic" carrying capacity conflicts with a wildlife manager's concept of an "ecological" carrying capacity, which Leopold defined as the maximum density of wild game that a particular range is capable of carrying (5).

Additional confusion has arisen based on differences of opinions regarding whether carrying capacity has meaning only with reference to forage quantity, or whether it should be expanded to include other welfare factors and perhaps even decimating factors such as predation and disease. Many wildlife biologists, particularly those working with ungulates or other primary consumers, generally view carrying capacity exclusively as population responses as a function of forage supplies. As illustration, Caughley defined carrying capacity as an equilibrium between animals and plants (6).

However, population densities of wildlife species, including many large mammal species, clearly are not limited exclusively by food. Other factors, such as cover, interspersion, and social tolerance, can limit populations below the potential of the food resources. For some species, these "security" or "tolerance" carrying capacities are as important as "subsistence" carrying capacities and must be considered in the management of the appropriate species (3).

Since the 1960s, numerous authors have attempted to synthesize the varied meanings proposed for the term *carrying capacity* and present a standardized terminology (1, 3, 6, 7, 8, 9). These varied definitions of carrying capacity can be condensed into ecologically based carrying capacities and culturally based carrying capacities (Table 7–1).

Ecologically Based Carrying Capacities

Although all definitions of carrying capacity are necessarily ecologically based, several definitions are intrinsically based on the response of populations to their environments and not influenced by human requirements or expectations. Each of these definitions describes populations at a stable equilibrium with some environmental influence (i.e., welfare factor) such as forage availability, cover, behavioral restrictions, or others.

K-Carrying Capacity.
The most commonly used definition of carrying capacity, K-carrying capacity, is defined as the maximum number of animals of a given population supportable by the resources of the area (9). Equivalent to subsistence density, potential carrying capacity, and ecological carrying capacity, this concept of carrying capacity is generally applied to ungulate populations that are limited primarily by forage resources (3, 6, 7, 8, 10). Some authors, however, include limitations based on space, cover, or other resources in their definition of ecological carrying capacity. We choose to treat definitions based on these other limitations as unique cases separate from forage-based carrying capacities because removal or manipulation of these other limitations may allow a population to grow until it becomes limited by the availability of food resources.

K-carrying capacity is the basis for most models of population growth and game harvest management. K-carrying capacity corresponds to the upper asymptote (K) of the sigmoid growth curve (Figure 7–1). If we consider a simplistic model of ungulate population growth, density will increase to a point where an equilibrium will be reached between the herbivores and their food supply. The growth rate of this population at equilibrium (K) will, by definition, be zero. Mortality will increase via increased juvenile or adult mortality and natality will decrease due to depressed reproductive rates and advanced age of sexual maturity such that the number of births minus the number of deaths will equal zero.

Because competition for forage is the causative factor in the reduced natality and increased mortality at K-carrying capacity, animal condition will be relatively poor and habitat conditions will be reduced. However, some definitions of carrying capacity suggest that animal quality should be considered such that carrying capacity is ". . . the maximum number of grazing animals . . . that can be maintained in good flesh year after year on a grazing unit without injury to the range . . . or the soil resource" (15, p. 400). We argue that maintenance of animal condition should not, and cannot, be included in definitions of K-carrying capacity because reduction in animal condition must be a direct result of decreased availability of resources as populations pass the inflection point (I) on the sigmoid curve.

As a population grows toward K, the biomass of plant material must necessarily decline (Figure 7–2a). Thus, the carrying capacity of a habitat cannot be viewed as the amount of available biomass in the absence of the herbivore, but rather the biomass

TABLE 7–1
Definitions of Carrying Capacity

Notation	Definition	Synonyms
Ecologically Based CC's		
KCC	K-Carrying capacity—maximum number of animals of a given population supportable by the resources of a specified area. (K of the logistics model)	Subsistence density (3,7) Ecological carrying capacity (6) Potential carrying capacity (10)
BCC	Behavioral carrying capacity—maximum density of a population limited by intraspecific behavior such as territoriality	Saturation point (5) Tolerance density (3)
RCC	Refugium carrying capacity—maximum population density below which animals are relatively invulnerable to predation	Threshold of security (11,12) Security density (3)
ECC	Equilibrium carrying capacity—density of a population at a consistent equilibrium, but the limiting variables are known (may include KCC, BCC, or RCC)	
Culturally Based CCs		
ICC	I-Carrying capacity—population density that yields maximum sustained yield (inflection point of the logistic growth curve)	Optimum density (3) Economic carrying capacity (6) Maximum harvest density (7)
OCC	Optimum carrying capacity—population density that best satisfies human expectations for it (may be equivalent to any of the other definitions of CC)	Relative deer density (13)
MCC	Minimum-impact carrying capacity—population density that minimizes impact on other wildlife, vegetation, or humans without eliminating the population (7)	Wildlife acceptance capacity (14)

Source: Adapted from Ref. 9.

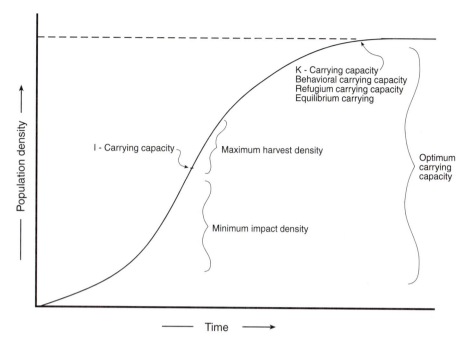

Figure 7–1 Types of ecologically based and culturally based carrying capacities described by various authors and their relation to the sigmoid population model.

of plants available (or the population of animals that it can support) when equilibrium is reached. The theoretical trajectories of plants and herbivores (Figure 7–2b) are the result of the dynamics of the animals and the dynamics of the plants on which they feed (8). An equilibrium is reached because the rate of growth of a plant population, per unit of biomass, tends to rise as plant biomass declines. Concurrently, forage intake per animal declines as plant density is reduced and therefore the rate of increase per animal declines.

Ungulate populations usually overshoot the ultimate K-carrying capacity before an equilibrium is reached. Often there will be substantial impact on the vegetation and K is the residual population size resulting in the maximum defoliation that the vegetation is capable of sustaining (16). This initial irruption and subsequent crash of a population, followed by dampening oscillations of population growth and decline mirrored by decreases and increases in forage availability, are generally viewed as the norm in ungulate populations (1, 10, 17). However, the generality of the irruption followed by oscillations to a stable equilibrium paradigm has been questioned by McCullough (18; see later section on Irruptive Populations and Habitat Resilience).

The degree to which populations overshoot K is related to the reproductive potential of the species (19). Those species with higher reproductive rates are prone to overshoot K more than species with lower reproductive rates. Therefore, it follows that ungulates with high reproductive rates (e.g., white-tailed deer) are more prone to habitat alteration than many other species (Figure 7–3). However, the degree of damage to a habitat is also dependent on the length of time that the population exceeds the habitat capacity.

In natural systems, populations fluctuate around K because of natural fluctuations in environmental conditions. Populations at K fluctuate more than those reduced by hunting and these large fluctuations are more likely to damage vegetation and reduce K (19a). Similarly, in highly variable environments, populations at K fluctuate more widely than those in more stable environments. Therefore, at least theoretically, vegetation damage from unhunted populations is not inevitable because a relatively

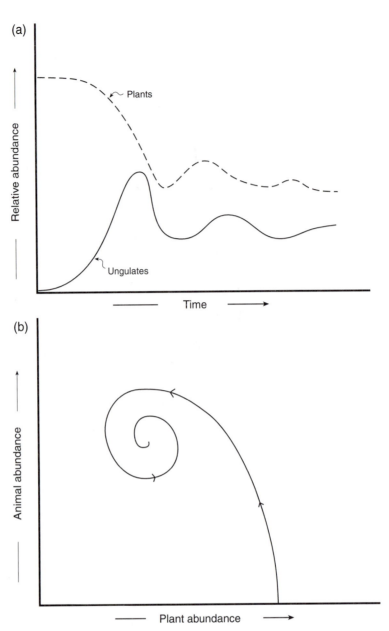

Figure 7–2 The dynamics of ungulates and the plants on which they feed. (a) Expected trajectories of plant and animal relative abundance during a population irruption. (b) Expected changes in relative densities of plants and animals during a population irruption. (Adapted from Ref. 8.)

constant equilibrium may be maintained in stable environments (19). Additionally, in extreme environments (e.g., areas with severe winters or droughts), the population may be intermittently reduced such that it never or rarely reaches K.

Management guidelines for many ungulates are based on an understanding of the K-carrying capacity of the habitat. Habitat management guidelines generally attempt to increase K-carrying capacity and, therefore, reduce the density of the population relative to an increased K. Harvest management guidelines generally project removals to decrease populations relative to current K. Several factors, however, can influence the manager's ability to accurately or precisely predict population responses to harvest or habitat manipulations.

A Capacity of Carrying Capacities

Figure 7–3 The result of population irruptions that overshoot forage resources is often habitat destruction, which may result in dramatic reductions in the carrying capacity of the habitat. (Photograph by B. R. Chapman.)

McCullough discussed environmental variability and its influence on the application of management strategies (19). Relatively stable environments exhibit a leptokurtic (i.e., narrow) distribution of good and bad years (Figure 7–4a). In these stable regions, competition for resources drives populations, and population responses to environmental variations are small. Therefore, management can be relatively precise. Where environmental variation is greater, but still relatively small (Figure 7–4b), management can still be based on density-dependent responses, but responses are less precise. In a highly variable environment (i.e., platykurtic distribution, Figure 7–4c, or random distribution, Figure 7–4e), environmental variations will overshadow density, and management must respond to the uncontrollable environmental variable on an *ad hoc* basis. Deer herds that are dependent on hard mast production may follow this latter distribution. In these herds, K fluctuates widely and irregularly according to acorn production (20). In areas with infrequent poor years (Figure 7–4d), management may be relatively precise, but modifications may be required to adjust for the extreme year. Similarly, management may be relatively precise and reflect density-dependent responses in areas where most years are poor with infrequent good years.

In addition to the direct influence of herbivores and year-to-year environmental variation, K-carrying capacity for many species of ungulates is influenced by successionary stage of the habitat. Carrying capacity may increase or decrease as succession progresses from herbaceous to shrubs to trees depending on the habitat and the ungulate species using that habitat. For example, in forested areas of Pennsylvania, over-winter carrying capacity for white-tailed deer is estimated as 23.1 deer per square kilometer in sapling stands, 7.7 deer per square kilometer in mature hardwood forests, and 1.9 deer per square kilometer in pole timber sized stands (21). Similarly, major disturbances that alter plant communities (i.e., natural and anthropogenic in origin) can shift

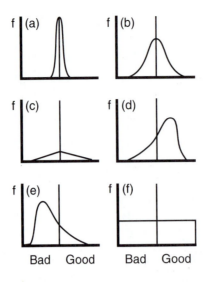

Figure 7–4 Frequency distribution of the occurrence of good and bad years in different environments. See text for explanation. (From Ref. 19.)

plant communities to alternate successionary stages resulting in marked increases in carrying capacity. In turn, herbivory from ungulates may drive successionary patterns. Schmitz and Sinclair suggested two possible successionary pathways in response to the presence or absence of deer herbivory (22). If, for example, herbivore numbers approach carrying capacity during early succession, deer may determine the structure of the plant community by preventing forest regeneration and maintaining the vegetation in an herbaceous state. Alternately, low deer numbers during early succession may allow succession to proceed to a woody vegetation community resulting in a highly different carrying capacity.

The fact that deer can influence plant communities has long been established (23, 24). Recent studies have provided additional details of deer population impacts on plant communities and other vertebrate communities, clearly demonstrating that ungulate populations do not simply respond to vegetative communities and changes in plant biomass (25, 26, 27, 28, 29). Rather, the dynamics of plant and ungulate interactions are highly complex in that ungulate herbivory can have profound influences on the vegetative community and on subsequent future herbivore population trends. There may be at least three mechanisms whereby deer populations can create alternate stable states in vegetation communities (i.e., stable conditions at a different stage than would be predicted based on prevailing ecological and successional conditions) (30). These mechanisms include fire followed by deer browsing, clear-cutting followed by browsing, and sustained long-term suppression of regeneration.

Refugium Carrying Capacity. Refugium carrying capacity, also referred to as threshold of security and security density, is the number of animals that a habitat will support when welfare factors needed to alleviate predation are limiting (3, 7, 9, 11, 12). These factors may include such things as escape cover and habitat interspersion. According to this concept of carrying capacity, animals are relatively immune to predation once population size falls below refugium carrying capacity.

Although Dasmann cites several examples of big game populations being held at or near a security density through hunting or predation, it is unlikely that these populations are in fact at a threshold of security as described by Errington (3, 11, 12). Rather, these examples of populations maintained at an equilibrium below a K-carrying capacity arguably are examples of populations regulated at some sustained yield via natural or human predation. We know of no examples where the concept of refugium carrying capacity is applicable to large mammal populations.

Behavioral Carrying Capacity. Behavioral carrying capacity is synonymous with Dasmann's concept of tolerance density and similar to Leopold's concept of saturation point density (3, 5). Behavioral carrying capacity is the maximum population size that a given area will support when intrinsic behavioral or physiological mechanisms, such as territoriality, are the primary factors controlling animal populations. Although Leopold suggested that saturation density is an intrinsic property of the species unrelated to habitat conditions, he did not state that the underlying mechanisms specifically are behavioral. Rather he proposed that ". . . some internal force or property, which is not subject to large variation . . . is operative, and sets the upper limit beyond which wild populations do not increase." (5)

As with K-carrying capacity, behavioral carrying capacity occurs at the asymptote of the sigmoid model. However, the mechanisms through which population growth rate declines and ultimately stops are different. Resource limitations per se are not resulting in increased mortality or decreased natality, although these may be implicit in the development of territorial behavior. Rather, declines in net recruitment as density approaches behavioral carrying capacity may be due to increased dispersal of juvenile or subordinate individuals or to reduced per capita reproductive rates because the percentage of animals breeding declines as population increases. Because resource availability likely is not limiting at behavioral carrying capacity, at least to those individuals holding territories, the animals and the habitat (i.e., resource availability) should be in good condition.

Equilibrium Carrying Capacity. Equilibrium carrying capacity was proposed by McCullough as the density of a population at a consistent equilibrium for which the limiting variables are unknown (9). Clearly, equilibrium carrying capacity must encompass one of the other types of ecologically based carrying capacities, but the specific mechanism regulating population density has not yet been identified.

Culturally Based Carrying Capacities

Because most populations tend to grow until they reach some form of an ecologically based carrying capacity, any population maintained at a steady density below these carrying capacities by human-induced factors must be viewed as being controlled at a culturally based carrying capacity. Although the population responses to these anthropogenic influences is biologically driven, the level at which they are maintained is derived from some human value placed on the resource. These values may be derived from exploitation of the resource itself (e.g., sustained yield), or from a value placed on some other component of the biotic community on which the animal population being manipulated has a major influence (e.g., timber production, recreation).

I-Carrying Capacity. The population density that yields the maximum sustained yield from a population is referred to as the I-carrying capacity (9). This term is equivalent to the use of carrying capacity in range management and has been variably called optimum density, economic carrying capacity, and maximum harvest density (3, 6, 7, 8). I-carrying capacity equates to the inflection point on the sigmoid curve and is the density where population growth rates are maximal.

Management of most exploited ungulate populations attempts to maintain densities at or near I-carrying capacity. This density allows for maximal removal; however, the young age structure and high turnover rate at I-carrying capacity result in the production of few older trophy-quality individuals. At I-carrying capacity, animals generally will be in good health and habitat conditions will be good.

The simplistic logistics model places I-carrying capacity, or the maximum sustained yield, at 50% of K-carrying capacity. McCullough has argued that the logistics model is not a sound model for management of populations of K-selected mammals such as ungulates (19). In fact, I-carrying capacity when viewed as a residual (post-harvest) population relative to K-carrying capacity varies according to the reproductive potential of the species, including the age of sexual maturity. As an example, maximum sustained yield for white-tailed deer occurs at residual populations that are 56%

of *K*-carrying capacity, whereas for mule deer it occurs at 63% (19). Maximum sustained yield for other big game species will vary, with *I*-carrying capacity being a larger or smaller percentage of *K*, depending on the reproductive potential of the species. As *I*-carrying capacity becomes a larger percentage of *K*, the susceptibility to overexploitation increases.

One counterintuitive result of the management of populations at or near *I*-carrying capacity rather than at *K*-carrying capacity is that fewer deer result in greater recruitment and therefore greater sustained yields. Many biologists have been faced with the dilemma of explaining to the hunting public why seeing fewer deer is good, and how it results in higher harvest rates. This is especially true in areas where a tradition of very high deer densities has produced the expectation of seeing numerous deer while afield (31).

Minimum-Impact Carrying Capacity. Minimum-impact density as proposed by Bailey is the population density that minimizes impact on other wildlife or vegetation without eliminating the population (7). In this culturally defined carrying capacity, greater value is placed on the forage resource itself or on some other wildlife species dependent on that resource. An example of minimum-impact density would be the maintenance of herbivore densities at very low levels to prevent browsing impacts on rare or sensitive plant species. At minimum-impact carrying capacity, animal condition is excellent, reproductive rates are high, and the population's habitat should also be in excellent condition, receiving only minor use from the depressed population (7). Because of high reproductive rates, and the high effort associated with removal of animals from low-density populations, maintenance of populations at minimum-impact carrying capacity is likely to require a great deal of effort. This effort also must include frequent adjustments to harvest goals due to the unstable nature of populations maintained below *I*-carrying capacity.

Analogous to minimum-impact carrying capacity is the concept of a wildlife acceptance capacity (14). This concept entails the maximum population level in an area that is acceptable to people. Wildlife acceptance capacity may or may not be based on some biologically based carrying capacity. Rather, it may be impacted by human acceptance thresholds for damage or nuisance associated with a species, perceived competition with another species of interest to humans, the role of the species in disease transmission to humans or domestic animals, and other economic, aesthetic, ecological, educational, scientific, or intrinsic values (14). Implicit in this concept is that wildlife carrying capacity will vary among human constituents, and also may change over time for the same constituency. For example, deer hunters in New York have a higher wildlife acceptance capacity than do dairy farmers, who in turn have a higher acceptance for deer than fruit growers (32). Because the role of the wildlife manager is to consider these views and determine the appropriate population level, determination of a wildlife acceptance carrying capacity may be as important as, or more important than, determination of a biological carrying capacity. For example, management below a wildlife acceptance carrying capacity could eliminate some benefits associated with the resource, whereas management above this level may exceed public tolerance. Thus, target densities are defined by human expectations, and not by the potential of the habitat, provided wildlife acceptance carrying capacity is below *K*-carrying capacity.

Optimum Carrying Capacity. Optimum carrying capacity as proposed by McCullough should not be confused with optimum density as defined by Dasmann (3,9). Optimum carrying capacity is the population density that best satisfies human expectations (9). Because optimum carrying capacity can vary from total absence of animals to populations at *K*-carrying capacity, it necessarily includes all types of carrying capacity discussed earlier. Optimum carrying capacity may approach 0 on a nature preserve devoted to preservation of an endangered plant species or in highly developed urban/suburban areas. In contrast, optimum carrying capacity in unhunted reserves where natural ecosystem processes are allowed to proceed without human intervention may approach *K*-carrying capacity.

Recently, deCalesta and Stout proposed a similar concept called relative deer density (13). This density is defined as the current deer density as a proportion of deer density at *K*-carrying capacity. Deer densities are reported relative to the capacity of a particular area, and their potential impacts on other ecosystem functions or values. Therefore, relative deer densities can be viewed as a common currency for comparing the interactions among deer and their ecosystem without comparing absolute deer densities.

IRRUPTIVE POPULATIONS AND HABITAT RESILIENCE

Most introductory wildlife texts describe populations growing at an increasing rate until environmental resistance results in a dampening of the growth rate. Ultimately, populations reach an equilibrium at some carrying capacity of the habitat. However, most studies of ungulate populations suggest a strong tendency to overshoot the carrying capacity resulting in population declines from some peak density. McCullough reviewed three paradigms of these population irruptions in ungulates:

1. The predominant paradigm among ungulate biologists is an irruption followed by a recovery to a reduced carrying capacity through a dampened oscillation (6).
2. Leopold proposed an initial irruption followed by a stable equilibrium at a lowered carrying capacity (33).
3. The third paradigm involves repeated irruptions and population crashes with no long-term declines in carrying capacity. In this model, equilibrium is never reached between herbivores and their forage base (18).

Although there is evidence that many ungulate populations follow the dampening oscillation model, other ungulate populations clearly do not stabilize at some equilibrium. In these populations, repeated irruptions may be the norm (18). In particular, repeated irruptions appear to be the normal population response where herbaceous vegetation predominates the forage resources as a result of natural or anthropogenic causes. These herbaceous communities likely have an inherent resilience and the rapid recovery of the plant community following a crash of the herbivore population leads to a repeat of the population irruption and crash.

McCullough sparked the debate over whether one of the paradigms listed is "typical" of ungulate populations (18). As with most biological phenomenon, pigeonholing a process into one model of reality is likely to lead to false management prescriptions based on poorly conceived justifications for management. Although management at a culturally based carrying capacity (i.e., at a goal based on societal needs or preferences) certainly is justifiable, manipulation of populations to achieve some human-conceived notion of ecological balance is debatable, particularly in parks and reserves where natural processes are valued (34, 35). Warren and McCullough argue that if irruption and overpopulation result from past human influences, then intervention may be necessary (18, 34). If, however, repeated irruptions can be viewed as the natural population process on an area, population reduction to maintain an ecological balance is not defensible.

MEASUREMENT OF K-CARRYING CAPACITY

As with the varied use of the concept of carrying capacity, numerous models and methods have been proposed to determine *K*-carrying capacity (see review in Ref. 36). Here we discuss the dominant approaches used by wildlife biologists to determine habitat carrying capacity.

The basis for most estimates of habitat carrying capacity is the assumption that forage quantity and/or quality determines the number of animals a specific area of land can support. Therefore, by definition, these models are directed at the evaluation of K-carrying capacity.

Hobbs provided an overview of the major approaches of carrying capacity estimation for wild ungulates (37). To varying degrees, these models require information on diet, forage availability, forage nutritional quality, and nutritional requirements. Carrying capacity is derived from the relationship between available forage resources and specific animal nutritional requirements.

The most basic form of carrying capacity models is the forage quantity or biomass model (37, 38). In this approach, carrying capacity is determined by dividing the biomass of usable forage by the daily dry-matter intake:

$$k = \frac{A}{B \times \text{days}} ,$$

where:

$$k = \text{carrying capacity}$$

$$A = \text{usable forage (kg/ha)}$$

$$B = \text{average daily dry-matter intake (kg/day), and}$$

$$\text{days} = \text{season length (days of use).}$$

Generally only the dominant food items as determined from diet analyses are considered in the estimation of forage availability. To allow for sustained forage production, usable forage based on an assumed level of use is employed rather than total available forage. Wallmo and others suggested that 50% use of total forage would be appropriate to prevent range degradation (39). Where specific allowable use factors have been determined, these may be incorporated into the model. Dasmann provided a summary of allowable use factors for several key browse species (40). Reported values ranged from 10% to 20% use for doghobble and red maple to 60% to 70% use for aspen, honeysuckle, serviceberry, and wild rose.

Examples of the biomass-based approach include studies of mule deer, white-tailed deer, and moose (39, 41, 42). The key advantage to this model is the limited requirements for data collection. The biomass model provided comparable relative carrying capacity estimates to nutritional-based models in situations where nutritional quality was not limiting (41). However, there may be numerous situations where forage quality may be more limiting than forage quantity. Under these circumstances, the biomass model will greatly overestimate carrying capacity. For example, Biomass and nutritional-based models yielded similar carrying capacity estimates in areas where summer forage quality is high (39). Early winter carrying capacity estimates based on forage quantity were comparable to summer estimates. However, winter forage nutritional quality was insufficient to meet the energy requirements at any population level.

Because of the importance of forage quality, many models incorporate measures of the nutritional content of the forages into carrying capacity estimates. Nutritional-based models have been used with mule deer, elk, mountain sheep, and white-tailed deer (39, 41, 43, 44, 45, 46). A generalized version of the model is:

$$k = \frac{(B_i \times \text{F}_i)}{(R_q \times \text{days})} - E_n .$$

where:

k = number of animals the range can support for the winter period

B_i = consumable biomass of principle forage species i

F_i = nutrient content of principle forage species i

R_q = individual animal requirements, metabolic requirements for daily maintenance

days = number of days animals occupy the winter range

E_n = endogenous reserves of nutrients (44).

As in the biomass model, principal forages are determined from diet studies. Nutrient content, specifically metabolizable nitrogen and/or energy levels, is determined for each forage using standard analysis procedures. Daily energy and nitrogen requirements from food are estimated from total maintenance requirements minus nutrients provided from the catabolism of fat reserves and lean tissues.

Carrying capacity estimates that incorporate forage nutritional quality more closely reflect animal response where forage quality is limiting. However, the nutrition-based approach has much greater data demands than the biomass model, requiring the additional analysis of forage nutritional quality. These models also require a number of assumptions, particularly related to nutritional requirements. Sensitivity analyses show that estimates can vary greatly with relatively small changes in the model assumptions (44, 46).

An additional limitation of these traditional nutrient-based models is that they treat food resources as a single sum of all available forages combined (47). In these models, ranges with large quantities of low-quality forages and ranges with lesser quantities of high-quality forages are not distinguished. Hobbs and Swift described a model that integrates forage quantity and quality of individual foods and selects for a mix of forages that provides the maximum biomass that meets or exceeds a specific nutritional level (47). Using this model, they were able to represent the fundamental differences in the quality of burned and unburned shrub habitats for mule deer and mountain sheep. Burned areas had less overall forage, but had more forage with high nutritional concentrations. As a result, burned areas could support more animals consuming a high-quality diet than could the unburned sites, whereas the unburned areas could carry a larger number of animals on a low plane of nutrition.

Hanley and Rogers described a similar approach that incorporated protein and energy constraints simultaneously (48). The model also includes limitations on proportion of the diet comprised by any single forage species and minimum biomass available for consumption. They provide an example involving black-tailed deer in four forest habitats during summer and winter with three sets of snow conditions and two levels of metabolic requirements.

Carrying capacity estimates using digestible energy consumed by tame deer in enclosures have been used as an alternative to the forage-based approaches discussed earlier (41, 49, 50). Rate of weight change is used to predict digestible energy intake during the trial period. Carrying capacity is estimated by dividing the total digestible energy consumed by the product of daily digestible energy requirements and trial length. The tame-deer technique provides relative carrying capacity estimates similar to those for forage-based techniques although this technique may be six to eight times more costly than forage-based approaches (41).

The foundation of all forage-based approaches is an estimate of resource supplies for the area evaluated. We have previously discussed the influence of environmental variability on carrying capacity. Yearly variation in resource supplies can strongly influence carrying capacity estimates (44). Environmental influences on food availability, such as snow depth, have been incorporated into carrying capacity models (46, 48). However, the failure to cope with environmental stochasticity is a major limitation of these forage-based models (36).

Refinements to forage-based approaches to the measurement of carrying capacity will be possible as knowledge of nutritional requirements and animal–forage interactions improves. However, because of difficulties associated with variable environments, these measurements are likely to continue to be very imprecise. Additionally, the requirements for significant site-specific data collection will limit the use of these forage-based estimates of carrying capacity in most management situations.

INDICES OF CONDITION

Because of the limitations of direct estimates of carrying capacity, managers often rely on data from the animals (e.g., reproductive rates, body condition, antler development, and parasite levels) to provide inferences about habitat carrying capacity and relative population densities. Management decisions based on measurements of animal condition are based on the premise that herd health gradually deteriorates as population density grows from I-carrying capacity toward K-carrying capacity. However, although nutritional indices can provide reliable estimates of overall herd health, they do not always provide an accurate assessment of population density as related to carrying capacity. Condition indices may be influenced by numerous factors including animal age, sex, and reproductive status, parasite burdens, forage quality, soil fertility, season, and latitude. For example, in areas of abundant but low-quality forage, overall ungulate carrying capacity may be high, but animal condition poor regardless of herd density. Thus, condition indicators must be based on site-specific data and used based on experience with the herd under evaluation.

Generally, nutritional indices can be grouped into individual-specific or herd-specific indicators of nutritional condition. Individual-specific indices can provide an assessment of the nutritional condition of individual animals as well as general herd health. Most individual-specific indices are measures of the fat or energy stores of the body. These include kidney fat indices, femur marrow fat, mandibular cavity fat, and tail fat (51, 52, 53, 54, 55). Additionally, blood and urine characteristics can be used as indices of current nutritional status. Blood urea nitrogen, serum cholesterol, nonesterified fatty acids, and ketones have shown promise as energy indices but none is consistent enough to be used across a range of energy intake levels (56). Herd-specific indices include structural measurements (e.g., average yearling beam diameter, age- and sex-specific weights), measures of reproductive performance (e.g., age-specific corpora lutea or fetal counts, lactation rates, and placental scars in some mammals), and evaluation of abomasal parasite burdens (57).

CONCLUSION

Carrying capacity is a nebulous concept that often has been applied with various meanings in the wildlife literature. Standardization of terminology is critical for effective communication among wildlife professionals and with professionals in other disciplines. Similarly, the concept of carrying capacity is variously applicable to wildlife species. Although generally a useful concept for ungulate management, its applicability to other large mammal species, particularly carnivores, is questionable. Because it is most applicable to ungulate population biology and management, measurements of carrying capacity historically have focused on direct evaluations of range capacity as related to forage quantity and quality or on indirect evaluations of herd condition as related to range condition. Neither evaluation has provided managers with a singular tool to assess population density and allow precise management prescriptions. Management of ungulate populations requires integration of all available tools by knowledgeable and experienced professionals, along with adaptive management based on outcomes of previous management prescriptions. As such, there remains a significant amount of "art" in the science of wildlife management.

SUMMARY

Carrying capacity is a well-established term in the vocabulary of wildlife biologists. However, the concept has been variously applied to differing management objectives or population responses to environmental constraints. Effective communication requires precise terminology to avoid semantic confusion. The various definitions of carrying capacity can be condensed into ecologically based and culturally based carrying capacities. Ecological carrying capacities describe populations that are maintained at a relatively stable equilibrium by some intrinsic or extrinsic influence such as forage limitations (*K*-carrying capacity), territoriality (behavioral carrying capacity), or cover (refugium carrying capacity). Alternately, populations at a steady density below a resource limitation carrying capacity due to anthropogenic influences are at a culturally based carrying capacity. These include a maximum sustained yield density (i.e., *I*-carrying capacity), a density that minimizes impacts on other wildlife, vegetation, or humans (i.e., minimum-impact carrying capacity), and densities that satisfy other human expectations (i.e., optimum carrying capacity).

Various methods are used to assess the *K*-carrying capacity of a range. Direct measures include estimating the available biomass and quality of forages or estimating digestible energy consumed by tame deer in enclosures. The requirement of site-specific data and the difficulties associated with variable environments limit the use of direct measures of carrying capacity in most management situations. Therefore, managers often rely on indirect measures based on indicators of herd nutritional status such as fat stores, blood or urine parameters, structural measurements, reproductive performance, and parasite burdens, along with subjective estimates of relative density based on experience with a particular herd.

LITERATURE CITED

1. MACNAB, J. 1985. Carrying capacity and other slippery shibboleths. *Wildlife Society Bulletin* 13:403–410.
2. EDWARDS, R. Y., and C. D. FOWLE. 1955. The concept of carrying capacity. *Transactions of the North American Wildlife Conference* 20:589–602.
3. DASMANN, R. F. 1964. *Wildlife Biology.* John Wiley & Sons, New York.
4. STODDART, L. A., A. D. SMITH, and T. BOX. 1975. *Range Management,* 3rd ed. McGraw-Hill, New York.
5. LEOPOLD, A. 1933. *Game Management.* Charles Scribner's Sons, New York.
6. CAUGHLEY, G. 1976. Wildlife management and the dynamics of ungulate populations. Pages 183–246 *in* T. H. Coaker, ed., *Applied Biology.* Academic Press, London, UK.
7. BAILEY, J. A. 1984. *Principles of Wildlife Management.* John Wiley and Sons, New York.
8. CAUGHLEY, G. 1979. What is this thing called carrying capacity? Pages 2–8 *in* M. S. Boyce and L. D. Hayden-Wing, eds., *North American Elk: Ecology, Behavior and Management.* University of Wyoming Press, Laramie.
9. MCCULLOUGH, D. R. 1992. Concepts of large herbivore population dynamics. Pages 967–984 *in* D. R. McCullough and R. H. Barrett, eds., *Wildlife 2001: Populations.* Elsevier Science Publishers, London, UK.
10. RINEY, T. 1982. *Study and Management of Large Mammals.* John Wiley & Sons, New York.
11. ERRINGTON, P. 1934. Vulnerability of bob-white populations to predation. *Ecology* 15:110–127.
12. ERRINGTON, P. 1956. Factors limiting higher vertebrate populations. *Science* 124:304–307.

13. deCALESTA, D. S., and S. L. STOUT. 1997. Relative deer density and sustainability: A conceptual framework for integrating deer management with ecosystem management. *Wildlife Society Bulletin* 25:252–258.

14. DECKER, D. J., and K. G. PURDY. 1988. Toward a concept of wildlife acceptance capacity in wildlife management. *Wildlife Society Bulletin* 16:53–57.

15. DASMANN, R. F. 1945. A method for estimating carrying capacity of range lands. *Journal of Forestry* 43:400–402.

16. McCULLOUGH, D. R. 1984. Lessons from the George Reserve, Michigan. Pages 211–242 *in* L. K. Halls, ed., *White-Tailed Deer: Ecology and Management.* Stackpole Books, Harrisburg, PA.

17. CAUGHLEY, G. 1970. Erruption of ungulate populations, with emphasis on Himalayan tahr in New Zealand. *Ecology* 51:51–72.

18. McCULLOUGH, D. R. 1997. Irruptive behavior in ungulates. Pages 69–98 *in* W. J. McShea, H. B. Underwood, and J. H. Rappole, eds., *The Science of Overabundance.* Smithsonian Institution Press, Washington, DC.

19. McCULLOUGH, D. R. 1987. The theory and management of *Odocoileus* populations. Pages 535–549 *in* C. M. Wemmer, ed., *Biology and Management of the Cervidae.* Smithsonian Institution Press, Washington, DC.

19a. McCULLOUGH, D. R. 1979. *The George Reserve Deer Herd-Population Ecology of a K-Selected Species.* University of Michigan Press, Ann Arbor.

20. WENTWORTH, J. M., A. S. JOHNSON, P. E. HALE, and K. E. KAMMERMEYER. 1992. Relationships of acorn abundance and deer herd characteristics in the southern Appalachians. *Southeastern Journal of Applied Forestry* 19:5–8.

21. DRAKE, W. E., and W. L. PALMER. 1991. Overwintering deer feeding capacities of mixed oak forests in central Pennsylvania. Federal Aid in Wildlife Restoration final report, Job No. 06218.

22. SCHMITZ, O. J., and A. R. E. SINCLAIR. 1997. Rethinking the role of deer in forest ecosystem dynamics. Pages 201–223 *in* W. J. McShea, H. B. Underwood, and J. H. Rappole, eds., *The Science of Overabundance.* Smithsonian Institution Press, Washington, DC.

23. LEOPOLD, A., K. SOWLS, and D. L. SPENCER. 1947. A survey of overpopulated deer ranges in the U.S. *Journal of Wildlife Management* 11:162–177.

24. WEBB, W. L., R. T. KING, and E. F. PATRIC. 1956. Effect of white-tailed deer on a mature northern hardwood forest. *Journal of Forestry* 54:391–398.

25. MARQUIS, D. A. 1981. Effect of deer browsing on timber in Allegheny hardwood forests of northwestern Pennsylvania. Research Report NE-47, U.S. Department of Agriculture Forest Service, Washington, DC.

26. TILGHMAN, N. G. 1989. Impacts of white-tailed deer on forest regeneration in northwestern Pennsylvania. *Journal of Wildlife Management* 53:424–453.

27. MILLER, D. G., S. P. BRATTON, and J. HADIDIAN. 1992. Impacts of white-tailed deer on endangered plants. *Natural Areas Journal* 12:67–74.

28. WALLER, D. M., and W. S. ALVERSON. 1997. The white-tailed deer: A keystone herbivore. *Wildlife Society Bulletin* 25:217–226.

29. McSHEA, W. J., and J. H. RAPPOLE. 1997. Herbivores and the ecology of forest understory birds. Pages 298–309 *in* W. J. McShea, H. B. Underwood, and J. H. Rappole, eds., *The Science of Overabundance.* Smithsonian Institution Press, Washington, DC.

30. STROMAYER, K. A., and R. J. WARREN. 1997. Are overabundant deer herds in the eastern United States creating alternate stable states in forest plant communities? *Wildlife Society Bulletin* 25:227–234.

31. DIEFENBACH, D. R., W. L. PALMER, and W. K. SHOPE. 1997. Attitudes of Pennsylvania sportsmen towards managing white-tailed deer to protect the ecological integrity of forests. *Wildlife Society Bulletin* 25:244–251.

32. DECKER, D. J., and T. L. BROWN. 1982. Fruit growers' vs. other farmers' attitudes toward deer in New York. *Wildlife Society Bulletin* 10:150–155.

33. LEOPOLD, A. 1943. Deer irruptions. *Wisconsin Conservation Bulletin* 8:3–11.

34. WARREN, R. J. 1991. Ecological justification for controlling deer populations in eastern national parks. *Transactions of the North American Wildlife Natural Resource Conference* 56:56–66.

35. PORTER, W. F. 1992. Burgeoning ungulate populations in national parks: Is intervention warranted? Pages 304–312 *in* D. R. McCullough and R. H. Barrett, eds., *Wildlife 2001: Populations.* Elsevier Science Publishers, London, UK.

36. McLeod, S. R. 1997. Is the concept of carrying capacity useful in variable environments? *Oikos* 79:529–542.

37. Hobbs, N. T. 1988. Estimating habitat carrying capacity: An approach for planning reclamation and mitigation for wild ungulates. Pages 3–7 *in* J. Emerick, S. Q. Foster, L. Hayden-Wing, J. Hodgson, J. M. Monarch, A. Smith, O. Thorne II, and J. Todd, eds., *Proceedings III: Issues and Technology in Management of Impacted Wildlife.* Thorne Ecological Institute, Boulder, Colorado.

38. Harlow, R. F. 1984. Habitat evaluation. Pages 601–628 *in* L. K. Halls, ed., *White-Tailed Deer Ecology and Management.* Stackpole Books, Harrisburg, PA.

39. Wallmo, O. C., L. H. Carpenter, W. L. Reglin, R. B. Gill, and D. L. Baker. 1977. Evaluation of deer habitat on a nutritional basis. *Journal of Range Management* 30:122–127.

40. Dasmann, W. 1981. *Deer Range: Improvement and Management.* McFarland and Company, Jefferson, NC, and London, UK.

41. McCall, T. C., R. D. Brown, and L. C. Brender. 1997. Comparison of techniques for determining the nutritional carrying capacity for white-tailed deer. *Journal of Range Management* 50:33–38.

42. Crete, M. 1989. Approximation of *K* carrying capacity for moose in eastern Quebec. *Canadian Journal of Zoology* 67:373–380.

43. Albert, S. K., and P. R. Krausman. 1993. Desert mule deer and forage resources in southwest Arizona. *Southwestern Naturalist* 38:198–205.

44. Hobbs, N. T., D. L. Baker, J. E. Ellis, D. M. Swift, and R. A. Green. 1982. Energy- and nitrogen-based estimates of elk winter-range carrying capacity. *Journal of Wildlife Management* 46:12–21.

45. Mazaika, R., P. R. Krausman, and R. C. Etchberger. 1992. Forage availability for mountain sheep in Pusch Ridge Wilderness, Arizona. *Southwestern Naturalist* 37:372–378.

46. Potvin, F., and J. Huot. 1983. Estimating carrying capacity of a white-tailed deer wintering area in Quebec. *Journal of Wildlife Management* 47:463–475.

47. Hobbs, N. T., and D. M. Swift. 1985. Estimates of habitat carrying capacity incorporating explicit nutritional constraints. *Journal of Wildlife Management* 49:814–822.

48. Hanley, T. A., and J. J. Rogers. 1989. Estimating carrying capacity with simultaneous nutritional constraints. Research Note PNW 485. U.S. Department of Agriculture Forest Service, Washington, DC.

49. Cowan, R. L., and A. C. Clark. 1981. Nutritional requirements. Pages 72–86 *in* W. R. Davidson, F. A. Hayes, V. F. Nettles, and F. E. Kellogg, eds., *Diseases and Parasites of White-Tailed Deer.* Miscellaneous Publication 7. Tall Timbers Research Station, Tallahassee, FL.

50. Hellickson, M, W. 1991. Predicting carrying capacity using grazeable aboveground biomass: A prerequisite to testing hypotheses about white-tailed deer populations. Thesis. Texas A and I University, Kingsville.

51. Riney, T. 1955. Evaluating condition of free ranging red deer (*Cervus elaphus*), with special reference to New Zealand. *New Zealand Journal of Science and Technology, Section B* 36:429–463.

52. Van Vuren, D., and B. E. Coblentz. 1985. Kidney weight variation and the kidney fat index: An evaluation. *Journal of Wildlife Management* 49:177–179.

53. Verme, L. J., and J. C. Holland. 1973. Reagent-dry assay of marrow fat in white-tailed deer. *Journal of Wildlife Management* 37:103–105.

54. Nichols, R. G., and M. R. Pelton. 1974. Fat in the mandibular cavity as an indicator of condition in deer. *Proceedings of the Annual Conference of Southeastern Association Game and Fish Commissioners* 28:540–548.

55. Cederlund, G. N., R. J. Bergstrom, F. V. Stalfelt, and K. Danell. 1986. Variability in mandible marrow fat in 3 moose populations in Sweden. *Journal of Wildlife Management* 50:719–726.

56. Harder, J. D., and R. L. Kirkpatrick. 1994. Physiological methods in wildlife research. Pages 275–306 *in* T. A. Bookhout, ed. *Research and Management Techniques for Wildlife and Habitats,* 5th edition. The Wildlife Society, Bethesda, MD.

57. Eve, J. H., and F. E. Kellogg. 1977. Management implications of abomasal parasites in Southeastern white-tailed deer. *Journal of Wildlife Management* 41:169–177.

A Dynamic View of Population Regulation

John M. Fryxell and A.R.E. Sinclair

INTRODUCTION

Since the time of Thomas Malthus, humans have recognized that organisms cannot increase without constraint. Early in the development of ecology, there was much debate about how such constraints might come into play: through repeated environmental cataclysms that decimate populations before they overwhelm us or through negative feedbacks that prevent populations from unimpeded growth (1, 2, 3). Although this debate has been largely settled in favor of the negative feedback school, it periodically flares up again in the literature. The concept of natural regulation is of enormous practical interest, because the effective management and conservation of wild species depends, to a considerable degree, on adequate knowledge about sources of population variability and the relative magnitude of stabilizing versus destabilizing processes.

Ecologists are only beginning to appreciate the diversity of different processes shaping animal population dynamics, and research progress with wildlife is appreciably slower than that of shorter-lived organisms more amenable to experimental study. Nonetheless, recently there has been a steady accumulation of theory and data relevant to the issues of population limitation and regulation of large mammals. Our objective is to illuminate and expand on contemporary views of the natural regulation of wildlife populations. We will particularly emphasize the importance of considering wildlife populations as components of larger dynamic systems comprising biotic communities, humans, and the physical milieu in which these interactions occur. We think that a dynamic frame of reference might help shift the emphasis of mammalian population studies from a preoccupation with questions about which factors are regulatory to questions about how limiting factors, regulatory or otherwise, shape the long-term dynamics of wildlife populations.

We define limitation as any natural or anthropogenic process that demonstrably affects the rate of population change (4, 5). By this definition, of course, every source of mortality is limiting, so it is often more useful to discriminate between the relative magnitudes of various limiting factors. This can be particularly instructive if one considers demographic responses to changing population densities of all members of the community.

On the other hand, natural regulation implies that the per capita rate of population growth declines with increasing population in such a way that population density is bounded over time. Note that this definition is somewhat more general than more typical equilibrium-based definitions in which regulation acts to restore populations that are perturbed away from their natural equilibrium, begging the obvious question of how we might identify such equilibria in the first place (4, 5, 6).

In practice, regulation is inferred when mean individual fitness declines with increasing population density through the action of one or more ecological factors. For example, increasing per capita risk of mortality due to starvation would imply a regulatory factor, as would declining reproductive rates. Out of the larger set of factors limiting a given population, not all factors will contribute to natural regulation, nor is there any reason to suppose that regulatory factors will be of larger magnitude than any nonregulatory factors.

We can treat this a little more formally by defining the change in population density from one birth pulse to the next in the following manner:

$$\triangle N = N_{t+1} - N_t = N_t \left[b(N_t) - m(N_t) + \epsilon_t \right] \;,$$

where N_t = population density in year t, $b(N)$ = per capita recruitment rate (birthrate multiplied by postpartum survival), $m(N)$ = per capita mortality rate, and ϵ_t = stochastic variation in environmental conditions, as it affects demographic parameters. The classic model of population regulation is one in which the per capita rate of change ($b - m +$ mean ϵ) declines monotonically with population density. The per capita rate of annual growth is the difference between per capita rates of offspring recruitment and mortality (Figure 8–1). If recruitment rate declines in a monotonic fashion (without local maxima or minima) with density and/or mortality rate increasing with density, then there will be a density at which the recruitment rate matches the mortality rate, a point at which the population can be considered to be at equilibrium. It matters little whether the density dependence stems from effects on reproduction or mortality; the key issue is that some demographic variable must change with density for regulation to occur. At population densities below this equilibrium, recruitment rates exceed mortality rates and the population increases (Figure 8–1), whereas at densities above the equilibrium mortality rates exceed

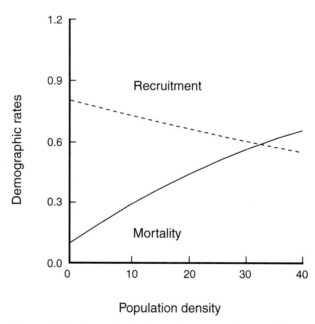

Figure 8–1 Hypothetical recruitment and mortality functions for a naturally regulated population. The point of intersection is the sustainable population equilibrium, commonly termed the *carrying capacity*.

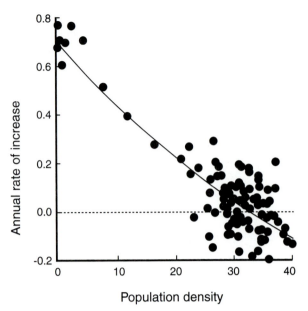

Figure 8–2 Annual rate of increase for a population subject to density-dependent recruitment ($b[N] = 0.8*\exp[-0.01*N]$), density-dependent mortality ($m[N] = 1.1 - \exp[-0.2*N]$), and ϵ is distributed normally with mean = 0 and SD = 0.2. The solid line represents the predicted relationship in the absence of stochastic environmental noise.

recruitment rates and the population declines. Hence, the equilibrium is locally stable, implying that slight perturbations in population density away from this equilibrium would lead to changes in population growth rates restoring the equilibrium once again.

In practice, one might expect that random environmental variation (ϵ_t) would impinge on per capita rates of birth or death. Hence it is perhaps more realistic to think of the density-dependent function for the annual per capita rate of growth as a probability cloud around a regression line (Figure 8–2). Such stochastic variability in demographic variables implies that even tightly regulated populations would be chasing continually shifting equilibria (Figure 8–2). Hence, we would not expect to see population constancy even over long time periods, but rather a distribution of population densities around the mean carrying capacity over the entire study interval (7).

This is nicely demonstrated by the Leirs and others demographic study of the multimammate mouse (8). Monthly fluctuations in mouse abundance were linked to density-dependent and density-independent effects on juvenile maturation rates and adult survival rates, colored by stochastic variation in rainfall. The most compelling models of population dynamics included all three components. This work suggests that populations cannot be fully understood without reference to seasonal and stochastic changes in their environment, showing the utility of mixing approaches based on time series, theory, and experimentation.

One of the clearest demonstrations of environmental variation in large mammals is Owen-Smith's decade-long study of population dynamics of greater kudus in Kruger National Park in South Africa (9). Age- and time-specific variations in demographic variables were obtained from a mix of marked and otherwise-recognizable individuals. African savannah environments are characterized by large year-to-year variability in rainfall. More importantly, rainfall has a direct relationship to annual production of grasses and low-lying forbs and shrubs (10, 11). Subpopulations in two different regions of Kruger National Park went through several years of increase, linked to above-average rainfall, followed by several years of decline during a subsequent drought. Survival rates

Introduction

of both sexes were affected by variations in rainfall, particularly juveniles, but survival was also density dependent. Temporal variation in survival was accordingly best explained by the supply-to-demand ratio of rainfall divided by kudu biomass (9). Interestingly, males older than 3 years had lower survival rates than females of similar age, a difference that was partially attributable to differential vulnerability to malnutrition and carnivores, although no single mortality agent was clearly responsible (12). Similarly, the rate of increase by kangaroos in arid Australian savannahs was closely linked with rainfall over the preceding months (13).

In temperate climates, winter snow conditions can be more important than rainfall effects on resource productivity. There is a strong influence of snow level on rates of recruitment and population growth in moose and white-tailed deer in Minnesota and Isle Royale, with intriguing evidence of a residual effect of snow level on offspring production in subsequent years, which the authors attributed to influences on maternal condition. Similar demonstrations of correlations between demographic rates and precipitation have been shown for other large mammals (14, 15, 16, 17).

We rarely know population density with absolute certainty. Under these circumstances, all too familiar to most wildlife biologists, the density-dependent relationship can be readily obscured by noise along both axes, falsely implying that there is little or no regulatory response to changes in population density (18). For wildlife managers, this can be a disastrous conclusion, because it encourages a belief that at any level of harvesting the constant annual rate of increase can be safely accommodated by the population.

This more complex world view diverges from the idealized population regulation of introductory ecology textbooks. Moreover, it is unclear whether ecological stability is a particularly useful concept in a constantly changing environment (19, 20). It certainly presents a set of new challenges for detecting the underlying density-dependent signal buried in the demographic noise and for interpreting the meaning of population trajectories over short time frames.

TESTING FOR POPULATION REGULATION

The obvious way to test for density dependence is simply to regress per capita rates of increase recorded during a sequence of population censuses against population density. For such time series data, each subsequent change in densities is unavoidably related to densities recorded at earlier points in time. The resulting nonindependence of annual rates of increase can seriously bias conventional regression tests for density dependence. Numerous numerical procedures have been advocated as a means out of this statistical conundrum (7, 21, 22). While the jury is still out on these various alternatives, it would seem that the most robust approach to use, provided that one has an uninterrupted time series, is the parametric bootstrap likelihood ratio test (7). This procedure evaluates the amount of variation in annual rates of increase explainable by past population density and compares that with the explanatory power of a purely density-independent model (sometimes termed a *random walk*). This procedure is repeated on hundreds of simulated data sets that were generated with the density-independent model colored by computer-generated stochasticity of the magnitude found in the real data. If one finds that >5% of the density-independent simulations record similar levels of density dependence by chance as that observed in the real data, then this leads one to reject the regulation hypothesis. However, we do not know of any published case of avowed density dependence that has been reversed when subjected to more appropriate statistical scrutiny. At this stage we do not know how significant the bias problem may be in practice.

The best way around this statistical problem is to directly test for density dependence from independent measurements of per capita demographic rates (i.e., field estimates of birthrates or death rates) and population density (23). Examples of direct field measurements of offspring recruitment rates are relatively common in the wildlife literature, perhaps because it is usually simple to count females during late lactation that have calves or fawns at heel. A typical example derives from long-term studies in Yellowstone National Park. For many years the established policy for managing elk in the park included substantial culling or hunting. Termination of this policy in the late 1960s led to increased elk abundance. Regression analyses demonstrated a concomitant decline in the per capita rate of offspring recruitment during this population increase, indicative of a regulatory effect on offspring recruitment (17, 24, 25). Similar direct tests showing density-dependent recruitment of offspring have been conducted on other populations, including red deer, Soay sheep, wild reindeer, white-tailed deer, elk, moose, greater kudu, and feral donkeys (9, 26, 27, 28, 29, 30, 31, 32, 33, 34, 35).

Direct observational tests of density-dependent survival are somewhat less common in the literature, perhaps because radiotelemetry, high local population densities, and individually known animals are often required to obtain statistically meaningful sample sizes. Good examples include long-term studies of island populations of red deer and Soay sheep, greater kudus, mule deer, African buffalo, (37), wildebeest, and elk (9, 15, 17, 23, 25, 27, 36, 37, 38).

Oddly enough, few studies of density dependence in large mammals actually show evidence that food scarcity is responsible, although many of the documented examples of ungulate density dependence make this presumption. Fragmentary evidence that food availability can be limiting comes from opportunistic studies of demographic responses to year-to-year variation in food availability mediated by weather effects (15, 16, 28). To our knowledge there is only one study that has simultaneously linked vegetation availability, weather, and density-dependent demographic responses by herbivores to changing food availability: the collaborative project on interactions between kangaroos and sheep in rangelands of semiarid Australia (13, 39).

Evidence on regulatory effects of disease is even more fragmentary. Eradication of rinderpest in the late 1950s from cattle herds surrounding Serengeti National Park was associated with rapid growth of buffalo and wildebeest populations in the reserve, ungulates that cannot sustain the disease by themselves (10, 15, 37). This circumstantial evidence is certainly consistent with the notion that rinderpest is a limiting factor. We know of no evidence for wildebeest, buffalo, or indeed any other large terrestrial mammals that mortality due to disease or parasites is density dependent.

The observational approach to testing density dependence, either through time series measurements of population density alone or through direct measurements of demographic rates in relation to variation in population density over time, has severe disadvantages (20). First and foremost, long periods of observation are required before a valid test is possible. Second, it is backward in orientation, incapable by definition of anticipating new limiting factors as they come into play. Hence, it is not well suited to the demands of proactive management. Krebs advocates that we abandon the time series approach in favor of a more experimentally based approach, involving manipulation of a limiting factor and testing whether it has a demonstrable effect on the rate of increase (20). This approach is attractive, rooted in the here-and-now rather than in the entrails of past population patterns. Yet, the so-called "experimental paradigm" itself has some serious flaws. This kind of thinking logically overvalues the importance of limiting factors and devalues understanding of the dynamic nonlinear interactions and time delays that can influence population dynamics. Indeed, in a purely experimental paradigm, temporal patterns become superfluous. To truly understand complex interactions, we must check the goodness of fit of our models against real population trajectories. Long-term population monitoring is therefore just as essential as short-term experimentation.

Perturbation of population densities can indeed accelerate the detection of natural regulation. For example, the historical changes in hunting policy in Yellowstone National Park can be interpreted as a *de facto* management experiment testing natural regulation by temporarily perturbing elk densities away from their assumed prior natural

Testing for Population Regulation

equilibrium via hunting and then restoring the former equilibrium by eliminating the hunt (4, 24). Such perturbations speed detection of density dependence by rapidly producing data from a wide range of population densities.

Experimental manipulation of limiting factors is a common, albeit expensive and logistically difficult, approach to testing regulation in the field. For example, Bartmann and others experimental manipulation of mule deer densities within large enclosures provided a direct test of whether demographic rates are density dependent (36). Another ambitious experiment tested regulatory processes in the boreal forest ecosystem by manipulating mammalian predators, food availability, and nutrient input, alone or in factorial combinations (40). Their results suggest that a number of factors interact synergistically (i.e., more than additively) to affect population densities of not only snowshoe hares, but also other herbivores in the Kluane system (40). Of particular interest, neither exclusion of mammalian predators nor provision of nutritious food supplements was sufficient to prevent the famous 10-year cycle in abundance (40).

In this same tradition, a multitude of carnivore removal experiments have been conducted to test the importance of predation as a limiting factor, if not conclusively test for predation as a regulatory agent (41). Population reductions of wolves and/or bears in numerous jurisdictions in the United States and Canada often have produced similar responses: Predator reduction leads to enhanced survival of calves and sometimes adults, at least in the short term (e.g., 42, 43, 44, 45). Similar experiments have been conducted in recent years in Australia, with similar results on the limiting effect of carnivores (46, 47).

There is little doubt that mammalian predators constitute an important limiting factor (41). The evidence is fragmentary at best, however, that predators can regulate ungulate populations (44, 45, 48). It seems more likely that wolves and bears do not act in an immediately density-dependent fashion (34, 41). Overshadowing this entire debate is the lurking possibility that unharvested populations of ungulates and carnivores have cyclical or more complex patterns of fluctuation (49). Understanding these systems requires a more subtle examination of both positive and negative feedbacks on population growth, a theme that we will address later.

OVERCOMPENSATION AND POPULATION REGULATION

Theoretical models often assume that per capita changes in survival or recruitment are linearly related to population density, for the obvious reason that linear functions are far easier to analyze and solve than more complex functions. There is no reason whatsoever to presume *a priori* that density-dependent patterns in nature will be correspondingly simple. Indeed, there are even reasons to suspect that simple density dependence should be the exception rather than the norm for large terrestrial mammals (50).

Large mammals are unlikely to experience appreciable food scarcity until population density is relatively high, because each individual has a limit on its maximum rate of feeding. Rates of food consumption (i.e., the functional response) by terrestrial mammals in relation to resource abundance have been measured on various occasions for both terrestrial carnivores and herbivores (34, 47, 51, 52, 53, 54). Most of these measurements derive from natural variation in consumer and resource densities, so data often are rather sparse. Nonetheless, the overwhelming pattern is that consumption rates tend to increase and then level off. In most cases, the functional response has a steadily decelerating slope, characteristic of a Type II functional response in Holling's terminology (55). If food availability is inversely related to the population density of consumers, then this consumption pattern implies that increases in population density when consumers are scarce will have less effect on per capita demographic rates than when consumers are common. Many studies of density dependence in large mammals support this pattern, showing nonlinear declines in per capita survival or recruitment with increasing population density (50).

The importance of nonlinear demographic responses is demonstrated by long-term studies of feral Soay sheep and red deer on islands off the coast of Scotland (26, 27, 56, 57, 58). As red deer on the Island of Rum doubled over two decades, changes in recruitment and survival were recorded for individually known animals. Population density of females was used as the measure of abundance, because adult males and females are typically segregated in space. Survival of yearlings and adults changed in mildly accelerating fashion with increasing female density. Per capita recruitment showed a linear decline in relation to population density of females, due to density-dependent mortality of neonates and effects on age at first breeding and overall fecundity. When these density-dependent relationships were incorporated into an age-specific (i.e., Leslie matrix) model, the simulated dynamics closely matched the sigmoid logistic growth observed in the real population.

In contrast, feral sheep in the Saint Kilda archipelago, first studied during the 1960s by Grubb and later by Clutton-Brock and others, exhibit much more dramatic responses to changing population density (27, 59). Overwinter survival of lambs, yearlings, and adults drops precipitously with increase in female density above a threshold of 150 females, well approximated in each case by sigmoid curves fit by logistic regression (Figure 8–3). Sheep reproductive rates were influenced less dramatically by density. Age-specific modeling of these density-dependent responses reproduced the 2- to 3-year cycles in abundance (Figure 8–4) recorded in the long-term time series from Saint Kilda (27, 58).

These unusually detailed studies suggest that the degree of density dependence in the vicinity of equilibrial population densities has a large effect on population dynamics. Strong density dependence in per capita growth rates of the population leads to overcompensation by the population as it approaches its carrying capacity. As a consequence, the population decreases too quickly as it declines from above the equilibrium, increases too rapidly as it grows upward towards the equilibrium, doomed to never converge (58). Thus, such overcompensation implies that too much density dependence is as much a cause of fluctuation as weak density dependence coupled with environmental stochasticity, termed *density vagueness* (59a). Similar examples of overcompensation have been identified in insect and small mammal populations with erratic, cyclic, or even chaotic population dynamics over time suggesting that overcompensation is a real cause of fluctuations in free-living populations (8, 60, 61, 62, 63).

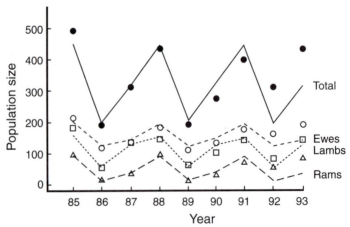

Figure 8–3 Per capita rates of survival, lambing, and recruitment for Soay sheep on the island of Hirta in the Saint Kilda archipelago (27). Data for males is represented by dotted lines and that of females by solid lines. (Reprinted with permission of University of Chicago Press.)

Overcompensation and Population Regulation

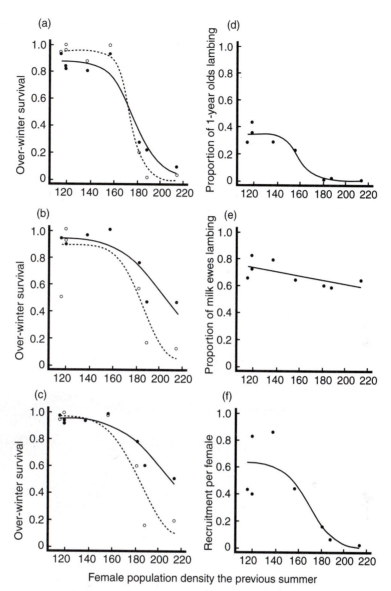

Figure 8–4 Predicted (lines) and observed (points) population densities over time for Soay sheep on the island of Hirta in the Saint Kilda archipelago (27). (Reprinted with permission of University of Chicago Press.)

TIME LAGS AND POPULATION REGULATION

Basic logistic theory starts from the supposition that populations respond immediately to changes in population density. This makes sense for short-lived species, such as insects with univoltine life cycles, rodents, or even passerine birds. It makes less sense for longer-lived organisms.

The negative feedback implicit in density-dependent demographic rates is analogous to the self-regulatory role played by a thermostat attached to a furnace. As house temperature falls below the equilibrium, the thermostat triggers the furnace to come on, bringing house temperature back to equilibrium. When the temperature climbs above the equilibrium, shutdown of the furnace leads to cooling of the house and a return to equilibrium.

Imagine instead what might happen if the thermostat responded to temperature levels recorded 12 hours earlier. In the middle of the night, the ambient temperature would be low, yet the furnace would not be triggered to come on for a further 12 hours, by which point the furnace would be heating an already warm house. The process would be reversed at night. Such a lagged thermostat would produce cyclic swings in house temperature, owing to application of negative feedback at inappropriate times.

Time lags in demographic processes have recently received a great deal of scrutiny in small mammals and insects. Using a flexible procedure for fitting nonlinear models to time series data, Turchin and Taylor claimed that many insect and mammal populations show evidence of delayed density dependence, frequently of a strongly nonlinear form (60). Later work on mammalian time series also suggests that some mammal populations are best described by lagged demographic models incorporating complex density-dependent relationships (61, 63). Countering this view, however, is other work demonstrating that simpler logistic models without lag effects may be sufficient to explain dynamics of other small populations with large degrees of variation (64, 65).

Little empirical work has been directed at testing lagged feedback mechanisms in populations of large mammals. Per capita rates of recruitment by white-tailed deer in Ontario have been shown to be most responsive to population density recorded 3 to 4 years previously, as were growth rates of yearlings (31). The delayed effect of changes of population density on both early growth rates of individuals and per capita rates of recruitment suggests that food availability caused lagged demographic responses following peaks in population density.

It should be possible to test delayed density dependence experimentally in large mammal populations, as has been done with meadow voles (66). By culling within large enclosures, they maintained replicates at high, medium, or low densities characteristic of the species, each with different levels of plant resource quality and biomass. At the end of 2 years, Ostfeld and others removed all individuals and reintroduced colonizers from free-living populations. There was no apparent lag effect of prior vole population densities on plant resources, because each reinoculated population grew at comparable rates. Although such levels of intervention are not often feasible in large mammals, this would be one of the best ways in principle to distinguish between effects of population density per se versus resource depletion.

Age and social structure considerably complicate the long-term dynamics of density-dependent populations (67). The inclusion of age structure, particularly in cases where adult survival is low and less responsive to changes in population density than that of juveniles, can cause delayed population responses to changing environmental conditions. For example, if fecundity is age or size dependent, as is often the case in large herbivores, then changes in age structure itself can affect population recruitment rates, beyond the density-dependent response (68, 69). Demographic characteristics are also shaped to a considerable degree by past patterns of energy gain and body growth. For example, stunted early growth can occur in large herbivores produced at high population densities, a size handicap that persists through to later stages in life, even if environmental circumstances become more favorable (29, 68, 70). Such differences in size can influence fecundity and dominance status within localized kin groups (68, 71). The latter effects may be of particular interest, because asymmetries in competitive ability among individuals in a population can have important stabilizing characteristics (72). On the other hand, persistent cohort-specific effects would tend to destabilize dynamics to some degree (73). The interplay between these opposing time-delayed mechanisms merits further theoretical consideration and field tests.

CONSUMER–RESOURCE INTERACTIONS AND POPULATION REGULATION

Consumer or consumed, there is no mammal population whose temporal dynamics can be completely understood in isolation, because all mammals interact with other trophic levels. If one presumes that resources cannot increase without limit, even without consumers, then a dynamic representation of population interactions of the entire food chain is as follows:

$$\frac{dR}{dt} = rR\left(1 - \frac{R}{K}\right) - \frac{aRN}{1+ahR} + \epsilon_R \quad,$$

$$\frac{dN}{dt} = N\left(\frac{eaR}{1+ahR} - d\right) - \frac{ANP}{1+AHN} + \epsilon_N$$

$$\frac{dP}{dt} = P\left(\frac{EAN}{1+AHN} - D\right) + \epsilon_P \quad,$$

where R, N, and P = local densities of plants, herbivores, and carnivores, respectively; a = the area searched per unit time by herbivores; A = the rate of search by carnivores; h = the handling time for each unit of plant resources consumed; H = the handling time for each herbivore consumed; e = the energy content of each unit of plant tissue consumed; E = the energy content of each herbivore consumed; and the ϵ's represent stochastic environmental variation acting independently on each trophic level.

This dynamic representation of the food chain is faithful to many of the biological characteristics that recur over and over in large mammal communities. We know that plant resources do not grow without bound, even in the absence of any herbivory. The logistic formulation ($rR[1-R/K]$) is one commonly accepted way to model such self-regulation, although it is far from clear from the existing evidence that it is preferable to simpler density-dependent functions (74), such as $r[1-R/K]$. Nonetheless, we will accept its applicability here. We know that consumption rates (i.e., the functional response) of large mammals commonly tend to saturate with increasing prey density. This is reflected by the Type II functional responses for herbivores ($aR/[1+ahR]$) and carnivores ($AN/[1+aHN]$).

The model components that we know least are the numerical responses by herbivores and carnivores to changes in prey density. We can reasonably assume, in the absence of any direct agonistic interactions among conspecifics, that the rate of consumer increase should be linked to the rate of consumption: high feeding rates leading to high rates of population growth. In the food chain model, we assume that the numerical responses are calculated by simply multiplying the energy content of prey (e for plants and E for herbivores) by their respective functional responses.

There is little evidence by which we can assess this assumption. Bayliss related the rate of increase by red kangaroos and grey kangaroos to the standing crop of ground-level vegetation in eastern Australia over the course of 2 years, spanning a wide range of monthly rainfall levels (39). He showed that the kangaroo numerical responses did indeed show a monotonically saturating shape similar to that depicted in our formulation, with negative rates of increase at low levels of vegetation biomass ($-d$ in the limit for our formulation) changing to positive rates of increase when vegetation biomass exceeded the level needed to just match kangaroo birthrates with their rate of mortality.

For carnivore populations, there is at least some field evidence consistent with these results. Messier's compilation of abundance data for wolves and moose from a variety of systems across North America is suggestive of a monotonically saturating curve (34). Assuming that wolves rapidly equilibrate to prey abundance, this is indicative of a saturating numerical response. A good deal of uncertainty is involved in lumping data from widely disparate systems, however, including both hunted and unhunted populations ranging from northern Alaska to the continental United States. Fryxell and others have analyzed long-term dynamics of martens in Ontario in relation to densities of other predators and prey (75). These results suggest that the numerical responses of martens also saturate with increasing prey density, in addition to direct density dependence arising from territorial interactions among carnivores.

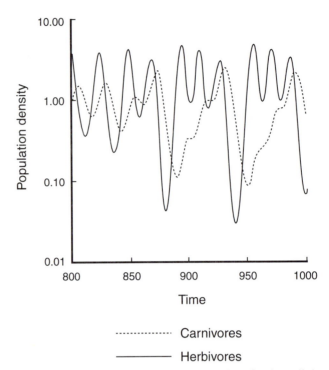

------------------ Carnivores

——————— Herbivores

Figure 8–5 Simulated chaotic dynamics of a three-link food chain (plants, herbivores, and carnivores), as described in the text. The following parameter values were used in the simulation: $a = 0.4$, $A = 0.8$, $h = 1.0$, $H = 2.0$, $r = 1.0$, $d = 0.4$, $D = 0.3$, $e = 1.5$, $E = 1.0$, and $K = 4$. Note the lag in population responses across different trophic levels as well as the inconsistent repetition of cyclic tendency.

The dynamics of such three-link food chains depends on system parameters, ranging from stability to limit cycles to chaotic fluctuations (Figure 8–5) (76, 77, 78, 79, 80). As in most models like that of Figure 8–5, variability of each component link in the chain is positively related to carrying capacity of the lowest link (plants in this case). As carrying capacity increases, the stabilizing effect of intraspecific competition within plants is superceded by the destabilizing effect of consumption by herbivores (77). This destabilizing effect of consumption is due to the monotonically saturating functional response by consumers. The per capita risk of mortality to prey (holding predator density constant) declines with prey density; hence, predation rates are inversely density dependent. The strength of that inverse density dependence is itself contingent on plant density at equilibrium. As a consequence, if a coexisting equilibrium of herbivores, carnivores, and plants occurs at low plant density, then destabilizing processes overwhelm the stabilizing processes and the chain fluctuates in regular or chaotic fashion.

Chaotic dynamics in such plant–herbivore–carnivore systems tend to arise naturally if the pairwise interactions between consumers and their resources are cyclical in isolation and the period of these cycles differs substantially between trophic levels (76, 80). For example, suppose carnivores have much slower rates of prey encounter, longer life spans, and lower mortality rates than do herbivores. Under such circumstances, the period of carnivore–herbivore cycles would be much longer than that of herbivore–plant cycles; in such cases chaos would ensue.

It is also possible for such systems to have more than one domain of attraction, meaning that either cycles or chaos can arise, depending on perturbations to the system that may occur (80). Perhaps even more worrisome from a conservation point of view is that it is possible for a top carnivore population to fluctuate chaotically for an extended period before precipitously crashing to extinction, with no prior warning that something has gone awry (79). Such "blue sky catastrophes" have been discussed in the theoretical physics literature, but are a new concern for ecologists. A crucial question is whether or not the parameter values we find in nature are capable of generating complex food chain dynamics. McCann and Yodzis's analysis suggests that chaotic dynamics could arise for parameters commonly observed in nature (78).

Complex food chain dynamics have been well documented in the lab using flour beetles (81). Mustelid–vole communities in northern Scandinavia also appear to exhibit complex cycles, or perhaps even chaotic time dynamics (62). Field evidence on food chain dynamics for larger mammals is more rare, largely because population biologists usually focus on a single trophic level, no doubt due to logistic constraints. Perhaps the best documented example is the long-term study of moose and wolves in Isle Royale National Park. Population estimates during the past four decades strongly suggest cycles in abundance of herbivores and carnivores (Figure 8–6), slightly out of phase (49, 82). There also is evidence that plant growth rates are affected by herbivore

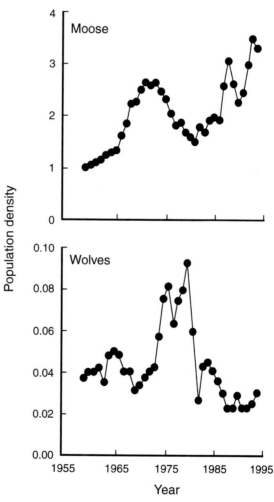

Figure 8–6 Observed population densities of moose and wolves in Isle Royale National Park, Michigan. Note the lagged population responses across different trophic levels and time-varying pattern of population fluctuation. (From 1997 unpublished annual research report of Peterson; reprinted with permission of R. O. Peterson.)

consumption, such that growth rates are cyclical and slightly out of phase with changes in herbivores (49). The overall impression is of an unstable food chain, but even this exemplary study is of insufficient length to test whether fluctuations are of a periodic or more complex nature.

Thus far, we have presumed that a natural predator fills the top position in the food chain. For many systems of interest to wildlife biologists, however, humans fill this role. This does not necessarily have to change the dynamic nature of the system, because societal and managerial processes provide many of the same kind of dynamic feedback responses that occur in natural predators, even though man selects his prey in complex ways that scarcely relate to feeding preferences by natural predators. For example, we can assume that interest in hunting is influenced by hunting success in past years and that managers respond to evidence of a "healthy" increase in harvest success by increasing license sales. This can have the effect of making hunting effort a function of past population densities, analogous to the carnivore population growth built into the food chain model. Such a model is capable of explaining cycles in hunting effort and abundance seen in some harvested populations (31, 69, 83, 84). For example, there is evidence from Ontario that white-tailed deer fluctuate slightly out of phase with hunting effort (Figure 8–7), which could contribute to population cycles.

The simple three-link trophic model we have discussed so far ignores important features that could influence long-term dynamics and the potential for regulation. For one thing, we have assumed that consumers show no behavioral changes in response to changing prey abundance or else the risk of themselves becoming prey for somebody else. A large body of work in behavioral ecology has demonstrated that foragers are responsive to changing environmental circumstances, often in ways that enhance their individual fitness (85). Inclusion of such behavioral mechanisms as optimal diet selection,

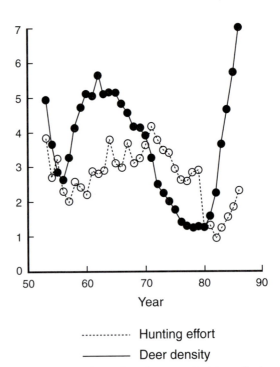

·············· Hunting effort

————— Deer density

Figure 8–7 Annual estimates of white-tailed deer (number per square kilometer) and hunting effort (thousands of hours) in the Canonto district in southern Ontario. Note the lagged population responses across trophic levels similar to that of natural predator–prey systems. (From data in Ref. 31.)

Consumer–Resource Interactions and Population Regulation

optimal habitat selection, or balancing the trade-off between energy gain and predation risk can have important stabilizing or destabilizing effects depending on context (86).

A particularly important behavioral feature that is not routinely incorporated into mammalian trophic models is agonistic social interactions between predators. Many mammalian carnivores, such as wolves, mustelids, large cats, and hyenas, are territorial. Defense of exclusive home ranges against conspecifics can have a stabilizing effect on population dynamics, bounding fluctuations even when not strong enough to actually regulate abundance (86). This can be diagnosed by direct density dependence in carnivore numerical responses above and beyond the usual prey dependence, which has been postulated for wolves feeding on moose (34) and mustelid predators feeding on small mammals (62). It has been argued on theoretical grounds that direct regulation via social mechanisms is unlikely for mammals unless they are territorial, are vulnerable to infanticide, and have males with strong site fidelity, which seems to be consistent with the limited behavioral data available (87). Nonetheless, this assertion demands rigorous testing, particularly in top carnivores.

PATTERNS OF REGULATION IN DYNAMIC SYSTEMS

What does all this have to do with mechanisms of regulation? In principle, population regulation is readily identifiable in systems with simple equilibrium behavior—a point equilibrium to be exact. Perturbations of the focal population away from this equilibrium produce immediate compensatory changes in demographic parameters pushing the population back toward equilibrium.

But what if the focal population fluctuates because of overcompensation, time lags in negative feedback, or complex trophic interactions? The density-dependent response becomes circular, rather than linear, for populations with cyclic dynamics (Figure 8–8). By the time stochastic noise is added to such systems, detection of density dependence is nearly impossible. Boutin found that moose mortality is not density dependent, but

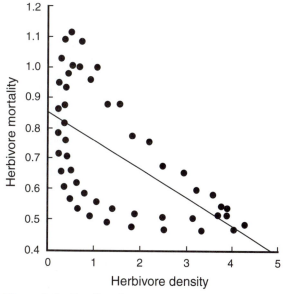

Figure 8–8 Predicted pattern of predation mortality for a herbivore population interacting chaotically with plants and carnivores in a three-link food chain, for time steps 801–849. Parameters as in Figure 8–6. Note the cyclic pattern of density-dependent response to population changes, due to time lags.

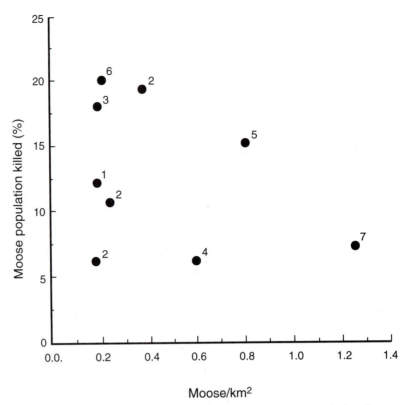

Figure 8–9 Observed moose mortality rates due to predation from a range of field studies (41). Numbers refer to specific study sites. Note the circular distribution of values similar to that of unstable trophic models (e.g., Figure 8–8). (Reprinted with permission of The Wildlife Society.)

rather forms a circular pattern similar to that arising from cyclical food chain models (Figure 8–9) (41). This suggests that some kind of mixture of positive and negative feedbacks, such as that arising from predator–prey interactions, could make it difficult to detect regulatory processes. Indeed, we might argue that regulation is a poor metaphor for describing such dynamic, nonlinear processes.

One method of analyzing such populations is to test for negative feedbacks on demographic parameters in relation to time-lagged density estimates. This works well in cyclic systems with short time lags, but requires long time series data to achieve reasonable statistical power, because there are correspondingly fewer data with which to assess lagged rather than direct demographic responses (60). The arsenal of statistical procedures available for classic time series analysis is of little use, however, because they are by and large organized around strictly linear systems (i.e., those without *any* negative feedback).

SUMMARY

The search for population regulation has long dominated the research agenda in the wildlife research community. There are many recorded instances of negative feedbacks, mostly tracing back to food shortage, but occasionally predation and/or disease. Nonetheless, the regulation concept does not lend itself particularly well to modern interpretations of ecosystem dynamics, because it derives from an equilibrium-based view of nature. If we accept that stochastic factors continually buffet any ecosystem, then it is more practical to think of a probabilistic equilibrium, with an associated probabilistic pattern of natural regulation. This shifts the emphasis away from identifying the regulatory factors to identifying sources of stochastic variability and adequate representation of feedback properties.

Particularly strong feedbacks do not produce demonstrably more stable population dynamics—indeed overcompensation can lead to persistent oscillations. Moreover, many regulatory processes applicable to long-lived terrestrial mammals have built-in time lags due to age structure and physiological responses to crowding. Such lags once again predispose populations to persistent oscillations (68). Inclusion of trophic interactions is even more likely to produce instability in population abundance over time, consistent with the slow variation in abundance seen in several wildlife time series of extended duration.

These biological processes suggest that it may be more appropriate to emphasize the importance of identifying all demographic processes that have limiting effects on populations, including those with both positive and negative feedbacks. These processes will then indicate the potential for complex dynamics over the long term. The literature is especially depauperate with respect to studies at more than one trophic level. We can think of only a handful of wildlife studies that have simultaneously recorded carnivore and herbivore densities or herbivore and plant densities, let alone all three. Yet it is clear that all large mammals are members of larger food webs. How can we pretend to understand wildlife populations, let alone make informed management decisions, without reference to a larger ecological community context? This is ample challenge for the next generation of wildlife biologists!

LITERATURE CITED

1. ANDREWARTHA, H. G., and L. C. BIRCH. 1954. *The Distribution and Abundance of Animals.* University of Chicago Press, Chicago, IL.

2. NICHOLSON, A. J. 1933. The balance of animal populations. *Journal of Animal Ecology* 2:132–178.

3. NICHOLSON, A. J. 1954. An outline of the dynamics of animal populations. *Australian Journal of Zoology* 2:9–65.

4. SINCLAIR, A. R. E. 1989. Population regulation in animals. Pages 197–241 *in* J. M. Cherrett, ed., *Ecological Concepts.* Blackwell Scientific Publications, Oxford, UK.

5. SINCLAIR, A. R. E., and R. P. PECH. 1996. Density dependence, stochasticity, compensation, and predator regulation. *Oikos* 75:164–173.

6. CONNELL, J. H., and W. P. SOUSA. 1983. On the evidence needed to judge ecological stability or persistence. *American Naturalist* 121:789–824.

7. DENNIS, B., and M. L. TAPER. 1994. Density dependence in time series observations of natural populations: Estimation and testing. *Ecological Monographs* 64:205–224.

8. LEIRS, H., N. C. STENSETH, J. D. NICHOLS, J. E. HINES, R. VERHAGEN, and W. VERHEYEN. 1997. Stochastic seasonality and nonlinear density-dependent factors regulate population size in an African rodent. *Nature* 389:176–180.

9. OWEN-SMITH, N. 1990. Demography of a large herbivore, the greater kudu, in relation to rainfall. *Journal of Animal Ecology* 59:893–913.

10. SINCLAIR, A. R. E. 1974. The resource limitation of trophic levels in tropical grassland ecosystems. *Journal of Animal Ecology* 44:497–520.

11. MCNAUGHTON, S. J. 1985. Ecology of a grazing ecosystem: The Serengeti. *Ecological Monographs* 55:259–294.

12. OWEN-SMITH, N. 1993. Comparative mortality rates of male and female kudus: The cost of sexual size dimorphism. *Journal of Animal Ecology* 62:428–440.

13. BAYLISS, P. 1985. The population dynamics of red and western grey kangaroos in arid New South Wales, Australia. I. Population trends and rainfall. *Journal of Animal Ecology* 54:111–125.

14. MECH, L. D., R. E. MCROBERTS, R. O. PETERSON, and R. E. PAGE. 1987. Relationship of deer and moose populations to previous winters' snow. *Journal of Animal Ecology* 56:615–627.

15. SINCLAIR, A. R. E., H. DUBLIN, and M. BORNER. 1985. Population regulation of Serengeti wildebeest: A test of the food hypothesis. *Oecologia* 65:266–268.

16. FRYXELL, J. M. 1987. Food limitation and demography of a migratory antelope, the white-eared kob. *Oecologia* 72:83–91.

17. COUGHENOUR, M. B., and F. J. SINGER. 1996. Elk population processes in Yellowstone National Park under the policy of natural regulation. *Ecological Applications* 6:573–593.

18. WALTERS, C., and D. LUDWIG. 1981. Effects of measurement errors on the assessment of stock-recruitment relationships. *Canadian Journal of Fisheries and Aquatic Sciences* 38:704–710.

19. WOLDA, H. 1989. The equilibrium concept and density dependence tests. What does it all mean? *Oecologia* 81:430–432.

20. KREBS, C. J. 1995. Two paradigms of population regulation. *Wildlife Research* 22:1–10.

21. BULMER, M. G. 1975. The statistical analysis of density dependence. *Biometrics* 31:901–911.

22. POLLARD, E., K. H. LAKHANI, and P. ROTHERY. 1987. The detection of density dependence from a series of annual censuses. *Ecology* 68:2046–2055.

23. WHITE, G., and R. M. BARTMANN. 1997. Density dependence in deer populations. Pages 120–135 *in* H. B. Underwood and J. H. Rappole, eds., *The Science of Overabundance.* Smithsonian Institution Press, Washington, DC.

24. HOUSTON, D. B. 1982. *The Northern Yellowstone Elk.* Macmillan Publishing Company, New York.

25. SINGER, F. J., A. HARTING, K. T. SYMONDS, and M. B. COUGHENOUR. 1997. Density dependence, compensation, and environmental effects on elk calf mortality in Yellowstone National Park. *Journal of Wildlife Management* 61:12–25.

26. CLUTTON-BROCK, T. H., M. MAJOR, and F. E. GUINNESS. 1985. Population regulation in male and female red deer. *Journal of Animal Ecology* 54:831–846.

27. CLUTTON-BROCK, T. H., A. W. ILLIUS, K. WILSON, B. T. GRENFELL, A. D. C. MACCOLL, and S. D. ALBON. 1997. Stability and instability in ungulate populations: An empirical analysis. *American Naturalist* 149:195–219.

28. SKOGLAND, T. 1985. The effects of density-dependent resource limitations on the demography of wild reindeer. *Journal of Animal Ecology* 54:359–374.

29. SKOGLAND, T. 1990. Density dependence in a fluctuating wild reindeer herd: Maternal vs. offspring effects. *Oecologia* 84:442–450.

30. McCULLOUGH, D. R. 1979. *The George Reserve Deer Herd: Population Ecology of a K-Selected Species.* University of Michigan Press, Ann Arbor.

31. FRYXELL, J. M., D. J. T. HUSSELL, A. B. LAMBERT, and P. C. SMITH. 1991. Time lags and population fluctuations in white-tailed deer. *Journal of Wildlife Management* 55:377–385.

32. SAUER, J. R., and M. S. BOYCE. 1979. Time series analysis of the National Elk Refuge census. Pages 9–12 *in* M. S. Boyce, and L. D. Hayden, eds., *North American Elk: Ecology, Behavior, and Management.* University of Wyoming Press, Laramie.

33. BOYCE, M. S. 1989. *The Jackson Elk Herd.* Cambridge University Press, Cambridge, UK.

34. MESSIER, F. 1994. Ungulate population models with predation: A case study with the North American moose. *Ecology* 75:478–488.

35. CHOQUENOT, D. 1991. Density-dependent growth, body condition, and demography in feral donkeys: Testing the food hypothesis. *Ecology* 72:805–813.

36. BARTMANN, R. M., G. C. WHITE, and L. H. CARPENTER. 1992. Compensatory mortality in a Colorado mule deer population. *Wildlife Monographs* 121:1–39.

37. SINCLAIR, A. R. E. 1974. The natural regulation of buffalo populations in East Africa. III. Population trends and mortality. *East African Wildlife Journal* 12:185–200.

38. SAUER, J. R., and M. S. BOYCE. 1983. Density dependence and survival of elk in northwestern Wyoming. *Journal of Wildlife Management* 47:31–37.

39. BAYLISS, P. 1987. Kangaroo dynamics. Pages 119–134 *in* G. Caughley, N. Shepherd, and J. Short, eds., *Kangaroos: Their Ecology and Management in the Sheep Rangelands of Australia* Cambridge University Press, Cambridge, UK.

40. KREBS, C. J., S. BOUTIN, R. BOONSTRA, A. R. E. SINCLAIR, J. N. M. SMITH, M. R. T. DALE, K. MARTIN, and R. TURKINGTON. 1995. Impact of food and predation on the snowshoe hare cycle. *Science* 269:1112–1115.

41. BOUTIN, S. 1992. Predation and moose population dynamics: A critique. *Journal of Wildlife Management* 56:116–127.

42. GASAWAY, W. C., R. O. STEPHENSON, J. L. DAVIS, P. E. K. SHEPHERD, and O. E. BURRIS. 1983. Interrelationships of wolves, prey, and man in interior Alaska. *Wildlife Monographs* 84:1–50.

43. GASAWAY, W. C., R. D. BOERTJE, D. V. GRANGAARD, D. G. KELLEYHOUSE, R. O. STEPHENSON, and D. G. LARSEN. 1992. The role of predation in limiting moose at low densities in Alaska and Yukon and implications for conservation. *Wildlife Monographs* 120:1–59.

44. MESSIER, F., and M. CRÊTE. 1984. Body condition and population regulation by food resources in moose. *Oecologia* 65:44–50.

45. MESSIER, F., and M. CRÊTE. 1985. Moose–wolf dynamics and the natural regulation of moose populations. *Oecologia* 65:503–512.

46. NEWSOME, A. 1990. The control of vertebrate pests by vertebrate predators. *Trends in Ecology and Evolution* 5:187–191.

47. PECH, R. P., A. R. E. SINCLAIR, A. E. NEWSOME, and P. C. CATLING. 1992. Limits to predator regulation of rabbits in Australia: Evidence from predator-removal experiments. *Oecologia* 89:102–112.

48. CRÊTE, M., and R. COURTOIS. 1997. Limiting factors might obscure population regulation of moose (Cervidae: *Alces alces*) in unproductive boreal forests. *Journal of Zoology* 242:765–781.

49. MCLAREN, B. E., and R. O. PETERSON. 1994. Wolves, moose, and tree rings on Isle Royale. *Science* 266:1555–1558.

50. FOWLER, C. W. 1981. Density dependence as related to life history strategy. *Ecology* 62:602–610.

51. SHORT, J. 1985. The functional response of kangaroos, sheep and rabbits in an arid grazing system. *Journal of Applied Ecology* 22:435–447.

52. SPALINGER, D. E., T. A. HANLEY, and C. T. ROBBINS. 1988. Analysis of the functional response in foraging in the Sitka black-tailed deer. *Ecology* 69:1166–11715.

53. LUNDBERG, P., and K. DANELL. 1990. Functional response of browsers: Tree exploitation by moose. *Oikos* 58:378–384.

54. GROSS, J. E., L. A. SHIPLEY, N. T. HOBBS, D. E. SPALINGER, and B. A. WUNDER. 1993. Foraging by herbivores in food-concentrated patches: Tests of a mechanistic model of functional response. *Ecology* 74:778–791.

55. HOLLING, C. S. 1959. Some characteristics of simple types of predation and parasitism. *Canadian Entomologist* 91:385–398.

56. CLUTTON-BROCK, T. H., M. MAJOR, S. D. ALBON, and F. E. GUINNESS. 1987. Early development and population dynamics in red deer. I. Density-dependent effects on juvenile survival. *Journal of Animal Ecology* 56:53–67.

57. CLUTTON-BROCK, T. H., S. D. ALBON, and F. E. GUINNESS. 1987. Interactions between population density and maternal characteristics affecting fecundity and juvenile survival in red deer. *Journal of Animal Ecology* 56:857–871.

58. GRENFELL, B. T., O. F. PRICE, S. D. ALBON, and T. H. CLUTTON-BROCK. 1992. Overcompensation and population cycles in an ungulate. *Nature* 355:823–826.

59. GRUBB, P. J. 1974. Population dynamics of Soay sheep. Pages 360–373 *in* P. A. Jewell, C. Milner, and J. Morton Boyd, eds., *Island Survivors*. Athlone Press, London.

59a. STRONG, D. R. 1986. Density-vague population change. *Trends in Ecology and Evolution* 1:39–42.

60. TURCHIN, P., and A. D. TAYLOR. 1992. Complex dynamics in ecological time series. *Ecology* 73:289–305.

61. TURCHIN, P. 1993. Chaos and stability in rodent population dynamics: Evidence from nonlinear time-series analysis. *Oikos* 68:167–172.

62. HANSKI, I., P. TURCHIN, E. KORPIMAKI, and H. HENTTONEN. 1993. Population oscillations of boreal rodents: Regulation by mustelid predators leads to chaos. *Nature* 364:232–235.

63. SAUCY, F. 1994. Density dependence in time series of the fossorial water vole, *Arvicola terrestris*. *Oikos* 71:381–392.

64. FRYXELL, J. M., J. B. FALLS, E. A. FALLS, and R. J. BROOKS. 1997. Long-term dynamics of small-mammal populations in Ontario. *Ecology* 79:213–225.

65. SAITOH, T., N. CHR. STENSETH, and O. N. BJORNSTAD. 1997. Density dependence in fluctuating grey-sided vole populations. *Journal of Animal Ecology* 66:14–24.

66. OSTFELD, R. S., C. D. CANHAM, and S. R. PUGH. 1993. Intrinsic density-dependent regulation of vole populations. *Nature* 366:259–261.

67. HIGGINS, K., A. HASTINGS, and L. W. BOTSFORD. 1997. Density-dependence and age structure: Nonlinear dynamics and population behavior. *American Naturalist* 149:247–269.

68. SAETHER, B. E. 1997. Environmental stochasticity and population dynamics of large herbivores: A search for mechanisms. *Trends in Ecology and Evolution* 12:143–149.

69. SOLBERG, E. J., B. E. SAETHER, O. STRAND, and A. LOISON. 1999. Dynamics of a harvested moose population in a variable environment. *Journal of Animal Ecology* 68:186–204.

70. ALBON, S. D., T. H. CLUTTON-BROCK, and F. E. GUINNESS. 1987. Early development and population dynamics in red deer. II. Density-dependent effects and cohort variation. *Journal of Animal Ecology* 56:69–81.

71. ALBON, S. D., H. J. STAINES, F. E. GUINNESS, and T. H. CLUTTON-BROCK. 1992. Density-dependent changes in the spacing behaviour of female kin in red deer. *Journal of Animal Ecology* 61:131–137.

72. LOMNICKI, A. 1988. *Population Ecology of Individuals*. Princeton University Press, Princeton, NJ.

73. ALBON, S. D., T. H. CLUTTON-BROCK, and R. LANGVATN. 1992. Cohort variation in reproduction and survival implications for population stability. Pages 15–21 *in* R. Brown, ed., *Biology of Deer*. Springer-Verlag, New York.

74. SCHMITZ, O. S. 1992. Exploitation in model food webs with mechanistic consumer–resource dynamics. *Theoretical Population Biology* 41:161–183.

75. FRYXELL, J. M., J. B. FALLS, E. A. FALLS, R. J. BROOKS, L. DIX, and M. A. STRICKLAND. 1999. Density dependence, prey dependence, and population dynamics of martens in Ontario. *Ecology* 80:1311–1321.

76. HASTINGS, A., and T. POWELL. 1991. Chaos in a three-species food chain. *Ecology* 72:896–903.

77. ABRAMS, P. A., and J. D. ROTH. 1994. The effects of enrichment of three-species food chains with nonlinear functional responses. *Ecology* 75:1118–1130.

78. MCCANN, K., and P. YODZIS. 1994. Biological conditions for chaos in a three-species food chain. *Ecology* 75:561–564.

79. MCCANN, K., and P. YODZIS. 1994. Nonlinear dynamics and population disappearances. *American Naturalist* 144:873–879.

80. MCCANN, K., and P. YODZIS. 1995. Bifurcation structure of a three-species food chain model. *Theoretical Population Biology* 48:93–125.

81. CONSTANTINO, R. F., R. A. DESHARNAIS, J. M. CUSHING, and B. DENNIS. 1997. Chaotic dynamics in an insect population. *Science* 275:389–391.

82. PETERSON, R. O., R. E. PAGE, and K. M. DODGE. 1984. Wolves, moose, and the allometry of population cycles. *Science* 224:1350–1352.

83. BERRYMAN, A. A. 1991. Can economic forces cause ecological chaos? The case of the Northern California Dungeness crab fishery. *Oikos* 62:106–109.

84. MCGARVEY, R. 1994. An age-structured open-access fishery model. *Canadian Journal of Fisheries and Aquatic Sciences* 51:900–912.

85. STEPHENS, D. W., and J. R. KREBS. 1986. *Foraging Theory*. Princeton University Press, Princeton, NJ.

86. FRYXELL, J. M., and P. LUNDBERG. 1997. *Individual Behavior and Community Dynamics*. Chapman and Hall, New York.

87. WOLFF, J. O. 1997. Population regulation in mammals: An evolutionary perspective. *Journal of Animal Ecology* 66:1–13.

Behavioral Ecology

David H. Hirth

INTRODUCTION

The behavioral repertoire of animals has evolved within environmental restraints to maximize those animals' success in the face of intraspecific and interspecific competitors. Each individual is an evolutionary product that represents a successful outcome of this long series of competitions. Among the life-history strategies individuals possess are the ability to capture sufficient food at minimal cost, the ability to avoid predation (more important for herbivores than large carnivores), and the ability to attract mate(s) and produce successful young. The formula for success in these endeavors varies in different parts of a species' range in response to different climatic pressures, cover regimes, and levels of competition from other herbivores or carnivores.

All field-oriented biologists should have some knowledge of the behavior of the animals with which they are working. Behavioral knowledge enables a biologist to know what an animal is doing throughout the year, to understand what pressures animals are living under, and to interpret animal activities to a public that is increasingly interested in animal behavior.

The species under consideration in this volume include native herbivores, omnivores, and carnivores. However, the group is interesting in that it includes predators and prey and species that occupy all habitats from mountain to desert and arctic to subtropics. Regardless of trophic level, all of these species face many of the same problems in trying to stay alive, forage efficiently, and produce successful offspring.

In this chapter I discuss important aspects of behavioral ecology and examine broad trends within and between groups of animals. Authors of individual species chapters will doubtless discuss behavior in greater detail but from a narrower perspective. Being able to look at behavior from a broader perspective allows one to consider "why" questions, rather than just the "how" questions asked at the individual species level (1). Asking the more theoretical "why" questions allows one to make predictions based on general knowledge that can be applied at the individual species level (2, 3). To understand the behavior of an individual species, that species should be placed in a broader context of species that are related, genetically or ecologically.

FORAGING STRATEGIES

Foraging strategies involve the evolution of behavior patterns in which individuals maximize their intake of energy and minimize their output of energy to obtain their food. Individuals that are able to maximize the difference between energy expended while foraging and energy taken in while foraging are likely to be most successful.

All the large mammals of North America display some flexibility in their choice of food and shift their foraging strategy to account for seasonal, and perhaps annual, changes in availability. The four large carnivores (i.e., wolf, mountain lion, jaguar, and polar bear) may have the least flexibility in their choices of prey. The herbivores all exhibit significant seasonal shifts in their diets as some plants become available and others become unavailable (or less desirable) at different seasons. The black bear and the brown bear may show the greatest levels of opportunism, supplementing what is essentially an herbivorous diet with carrion and live prey (e.g., neonatal ungulates) when available.

The mountain lion, jaguar, and polar bear are solitary predators. All use an ambush style of hunting in which the predator uses stealth to approach its prey as close as possible and then makes a short rush to capture the individual. The mountain lion in the North American West is primarily a predator of mule deer and elk, but it supplements its diet with lagamorphs and other small mammals when necessary (4, 5, 6, 7). Lions in the only remaining eastern population in southwest Florida prey predominantly on white-tailed deer and feral hogs (8). Felids as a group have powerful jaws and large powerful front paws with sharp retractable claws, which enable them to kill prey that are as large as or larger than they are (e.g., deer and elk) (9). Canids, on the other hand, lack this strength and are not able to kill large prey alone. Species that prey on large ungulates, such as wolf, dhole, bush dog, and African hunting dog, have evolved a group hunting strategy (3, 9).

Polar bears are the most specialized of the large North American predators. They hunt seals almost exclusively, primarily ringed seals but secondarily bearded seals (10, 11). Polar bears employ several hunting strategies to capture seals. The first is a seasonal one that involves locating birth lairs (i.e., snow caves on the sea ice) of seals in the spring and capturing pups before they can escape to the water (11). When seal pups are not available, bears use a still-hunting technique in which they wait for a seal to appear at a breathing hole or haul-out site (10). Less often bears stalk seals that are basking on the sea ice and approach to within 10 to 30 meters before rushing the seal (10). Seals represent large prey items (ringed seal, 90 kilograms; bearded seal females, 300 kilograms), but polar bears are much larger (females, 300 kilograms; males, 700 kilograms) and can handle seals as solitary predators (11, 12).

The wolf uses an entirely different strategy to hunt its prey. The wolf is a cursorial species, adapted to running down its prey over a long distance (9, 13). It is well known as a social predator that takes advantage of a group or pack to capture ungulates that are frequently much larger than individual wolves. Even though wolves are ungulate specialists, when ungulates are scarce they also opportunistically take smaller prey, such as beaver and lagamorphs (5, 14). It is clear, however, that the social system of the wolf did not evolve to exploit small prey species. Wolf-pack size increases with increasing density of white-tailed deer and moose and wolf density increases with increasing prey density (15, 16). Wolf packs are territorial, and the boundary areas between territories are predictably places with highest intraspecific conflict (17). Perhaps because of this potential for conflict, boundary areas are parts of territories that wolves frequent least, and they appear to serve as buffer zones between territories. In Minnesota, deer densities are highest in these territory buffer zones where predation pressure is least (18, 19, 20). Wolf packs that exploit migratory caribou herds usually maintain their territories, rather than following caribou, and switch to moose as their principle source of food when caribou depart (21). Wolves follow caribou to their wintering grounds only when the density of alternate ungulate prey within their territory is low.

The black bear and brown bear are omnivores, but a large portion of their diet is plant material and insect larvae (22, 23, 24). Adult females travel and forage with their young until the young are yearlings, but otherwise these two types of bears are solitary feeders. All temperate and arctic bears avoid the problems of foraging in winter by hibernating (25). Black bears and brown bears emerge from dens in the North when food is limited. Both species actively search for neonatal ungulates for a short time in late spring. Black bears search out white-tailed and mule deer fawns, and brown bears hunt moose calves and deer (26, 27). Summer diets of both species include mostly grasses, roots, tubers, and soft mast. Their diet in autumn includes significant amounts of hard mast in preparation for winter denning (23, 28, 29).

It is interesting to speculate why black and brown bears have become so large (30). What is the selective advantage of large size to an animal that is primarily an herbivore? There are several potential answers to this question. Large size enables bears to have large foraging ranges, which may be important in view of the low digestive efficiency of these animals in breaking down plant material (30, 31). Also large size allows these animals to store energy as fat for protection against seasonal fluctuations in food supplies. Lastly, large size also enables omnivorous bears to prey on ungulates, even if on an irregular basis (30). Black bears kill adult deer on occasion, and brown bears kill adult moose and caribou (32). Polar bears are larger than their commonest prey, ringed seals, but their large size enables them to kill belugas and walruses occasionally, which weigh 500 to 600 kilograms (33, 34). Large size enables brown bears to exploit more open prairie and tundra habitat and move out from the protection of forests (30, 35). Smaller black bears, and especially their young, are dependent on the forest for cover and for trees to climb for refuge.

The remaining species considered in this volume are ungulates. With the exception of the collared peccary, all of these ungulates are ruminants. However, the peccary has partially separated chambers in its forestomach where some volatile fatty acids are produced, and thus can be considered "ruminant-like" for the sake of this discussion (36).

Ruminants are a diverse group, and with the aid of symbiotic bacteria and protozoans in their foregut they are able to digest cellulose in a variety of forms (37). Although most ruminants specialize to a certain degree, as a group they are able to digest grasses, forbs, and woody browse. Hanley has argued that foraging selection in these animals is based largely on body size, relative rumen size, and mouth size (38). All species are predicted to select the most nutritious and most digestible foods, but the reality is that availability may make most individuals forage on foods that are less than the most desirable. Availability and accompanying time constraints are major factors in determining optimal foraging strategies (39). Large ungulates, such as bison and elk, have greater absolute demand for nutrients than small species, such as white-tailed deer and pronghorns. However, because metabolic rate only increases by the 0.75 power of body weight, large species can save time by selecting commoner, less nutritious food items (40). Small ungulates have less absolute demand for food but spend more time selecting more nutritious food items than do large species (37, 38).

The most important factor determining diet in ruminants is relative rumen size (38). Species with higher rumen volume:body weight ratios retain material in their rumens for a longer time and therefore can break down food with thick cell walls, such as mature grasses. These species are the grass and roughage feeders (41). Species with lower rumen volume:body weight ratios (i.e., concentrate selectors) turn over the contents of their rumens more rapidly to obtain the same amount of energy as a large-rumen species (41). Concentrate selectors must choose food that can be digested quickly, usually with thin cell walls, because of the high rate of rumen turnover (38). Woody browse presents a difficult problem for ruminants because a large portion of the cellulose in the stem has become lignified and is not digestible. Small-rumen species are best adapted to use browse because of their rapid turnover rate, which enables them to pass indigestible material quickly, rather than tie up rumen space for a long time on forage that cannot be digested (38). On this basis, large-rumen species, such as bison, domestic cows, and sheep, can be predicted to specialize on grasses and forbs but avoid browse. Small-rumen species, such as white-tailed and mule deer and pronghorns, can be predicted to select forbs and browse but avoid grasses. Species with rumens that are intermediate in size, such as moose and elk, are able to function as generalist herbivores (41, 42).

Mouth size is another variable that determines degree of selectivity in grazing animals. Species with large mouths, such as bison and domestic cows, probably cannot be selective of either plant parts or even individual plants. However, species with small mouths, such as white-tailed deer, mule deer, and caribou, can select individual plants for feeding and then further select the most nutritious parts of those plants (43, 44). Hence small-rumen species with small mouths can be the most selective about their choice of food, and large-rumen species with large mouths are the least selective.

Foraging strategies of these ungulates are tied to their social behavior and the abundance and distribution of their food. Some species typically occur in small social groups (e.g., white-tailed deer, mule deer, and moose), whereas others typically occur in large social groups (e.g., bison, caribou, and elk). Small-group species are not as likely as large-group species to deplete their food resources, and as a result often have relatively small home ranges. Herds of caribou or bison, by contrast, may rapidly deplete local food resources and thus be forced to move to new feeding areas on a regular basis. These species may be nomadic or migratory because of their social biology.

Temporary depletion of local resources and covering more ground in search of food would seem to be an inherent disadvantage of herding behavior in large grazing animals. However, McNaughton has shown that cropping grass has a number of advantages for herding animals (45). In productive grasslands, herd foraging can create "grazing lawns" in which vegetation height is greatly reduced but vegetative biomass may be concentrated by a factor of 2, resulting in increased bite size and grazing efficiency on the part of the animals. In addition, grazing lawns have higher productivity than ungrazed areas, and regrowth by grasses is stimulated by grazing (46). The net result of herding behavior by plains-dwelling ungulates is an increase in their food supply. In Yellowstone National Park grazing by bison and elk stimulated productivity of grasses by 48% compared to growth on protected control plots (47). However, this grazing-lawn effect does not extend to the less productive semiarid grasslands, where plant regrowth is limited by water availability (48). Although herding behavior under these circumstances may not be beneficial from a foraging standpoint, it might still be advantageous for predator avoidance.

In addition to differences in foraging preferences that are well documented between species, there are differences in foraging preferences between sexes of the same species (49). In some carnivores these differences have been inferred in species where males are much larger than females (e.g., polar and brown bears) even though actual data are scarce (32, 50). In ungulates sexual differences in diet have been documented in white-tailed deer, mule deer, moose, caribou, bighorn sheep, bison, and musk oxen (51, 52, 53, 54, 55, 56, 57, 58, 59, 60, 61, 62, 63, 64). A number of explanations have been offered to explain sex-related foraging differences, but two explanations seem most plausible (49). Differences in body size between sexes in ungulates may create a situation that roughly parallels what is seen between bigger and smaller species (38). Body size differences are the result of sexual selection that, in turn, stems from the polygynous breeding systems seen in most ungulates. Larger males may select a coarser diet with higher cell wall constituents and lower cell contents than that selected by smaller females. Males have a greater absolute need for quantity of food but a lower need for quality of food. Males do not select a coarser diet than females out of a sense of altruism for females and their offspring, but rather because it saves time to be less selective. Females need less food, but the trade-off is that they need to spend time selecting more nutritious food than males. Different energetic demands may result in selection of different food within the same habitat.

Another explanation for sexual differences in diet is that these are the result of different habitat selection with each sex choosing independently the habitat that suites its needs best. Males may select summer habitat with high-energy forage that allows them to maximize growth in preparation for the breeding season. Females in summer may respond to the needs of lactation or protection of offspring and thus select habitat that provides a higher protein diet or proximity to protective cover. Females are smaller and thus more vulnerable to predators than males (61). The result of this explanation is that females may not be able to exploit the high-quality foraging areas that males occupy because of higher risk of predation to themselves and their offspring.

Bleich and others argue that predator avoidance by female bighorn sheep and their young in southern California results in different habitat selection and thus sex-related differences in diet for much of the year (61). Miquelle and others, working with moose in Alaska, again argue that predator avoidance by cows and calves in summer is an important source of habitat segregation and thus sex-related differences in diet (56). However, bulls select areas with highest forage biomass in winter because of their depleted energy reserves following the rut. It is possible that there is no single explanation for sexual differences in diet for all ungulates.

PREDATOR AVOIDANCE

All herbivores are prey for carnivores at some time in their lives. Neonatal ungulates are particularly vulnerable to predation until they are strong enough to elude predators. Ungulates that are small and medium-sized as adults may be vulnerable to a variety of predators, but some species, such as bison, may be large enough to be difficult prey for most predators (63). Many predators key in on newborn fawns and calves and search specifically for neonates when they are available. Black and brown bears are significant predators on young deer, moose, and caribou (24, 26, 27). Wolves also prey on neonatal deer, moose, and caribou, but it is not as clear that wolves hunt young ungulates disproportionately to their abundance (5, 65, 66). Coyotes are probably better known as predators on young deer than wolves (67, 68, 69, 70).

Ungulates use two general behavioral adaptations to reduce predation on their young. These are referred to as "hider" and "follower" strategies (71). Species with hider-type young have neonates that are altricial and remain bedded and concealed in dense cover for 2 to 4 weeks following birth (Figure 9–1). White-tailed deer and mule deer are familiar examples. Species in this category occur in woodland habitat or have access to cover in which their young can be concealed from predators (71, 72). White-tail fawns in their first 2 weeks exhibit an "alarm bradycardia" response to things that

Figure 9–1   White-tailed deer fawn bedded alone, still in "hider" phase. (Photograph by D. M. Hirth.)

frighten them, during which their heart rates decline by about 40 percent (73). Bradycardia and the accompanying prone posture aid in concealing neonates from predators. Other ungulates with hider-type young also use the prone "freezing" posture for concealment, but it is not known whether they exhibit bradycardia (71).

Species with follower-type young have neonates that are precocial and able to follow their mother soon after birth, such as caribou and bison. Species with this strategy typically are plains dwellers (e.g., savannah, prairie, tundra) and do not have access to cover for their young. Faced with the problem of little cover, the best option is to have young that can follow their mother within hours of birth. Both mother and young remain with the main herd and enjoy some protection in that manner (71).

Although these strategies sound different on the surface, there is a lot of variation in the amount of time before hiders join their mothers on a regular basis. Hiders and followers represent extremes in a continuum. Whitetail fawns remain hidden for about 4 weeks before they follow their mothers routinely (72). Moose and elk calves, on the other hand, remain hidden for 2 to 3 weeks, although cow moose apparently remain closer to their neonates during the hider phase than other female cervids (74, 75). Bighorn lambs appear in ewe groups when they are a week old (76). Musk ox and bison calves are incorporated into cow groups almost immediately (77, 78, 79).

Avoidance of predators remains a problem for herbivores even as adults. Small herbivores, such as voles and lagamorphs, can take advantage of their size and remain concealed by vegetation while feeding, but large herbivores, such as ungulates, are generally more visible to predators. Ungulates use two principal strategies to avoid predation, but both involve seeking cover (80). Species that live most commonly in woodland habitat, such as white-tailed and black-tailed deer, typically occur in small social groups and use the woody vegetation around them for cover from predators. Common social groups in these species involve a single adult female and her young or perhaps small groups of males. By contrast, species that typically occur in open prairie or tundra habitat, such as bison, pronghorn, or caribou, do not have physical cover with which to conceal themselves from predators, and therefore they use groups of conspecifics as cover. This is the "selfish herd" concept (81). Although the group as a whole is conspicuous to predators, the probability that any one individual will be singled out for predation is reduced by a factor roughly equal to the number of animals in the group. Each member of the group benefits from the presence of other group members, hence the term "selfish." Openland ungulates (e.g., caribou and elk) typically occur in large herds that include many adults of both sexes. This behavior is widespread in bovids in North America and on African savannahs (82).

Herd membership has other predator-related benefits in addition to the "dilution effect." The collective senses of the group aid in the detection of approaching predators. The eyes, ears, and noses of a herd seem intuitively more effective at predator detection than those of a single individual or a small group, although data do not specifically support this contention. When the group takes flight from an approaching predator, group cohesion makes it more difficult for predators to single out an individual for pursuit. Experiments with schools of small fish demonstrate decreased predator efficiency as group size is increased (83). Finally, membership in a herd allows for greater foraging efficiency. As group size increases, individual vigilance declines. Each member of the group is able to spend a larger portion of its time feeding and a smaller portion of its time scanning for predators (84).

A great deal of research on herding ungulates in North America and Africa in recent years has supported the herd foraging efficiency principle (85, 86, 87, 88, 89). Scanning time decreases in Dall's sheep as group size increases but changes in scanning time also are related to distance from cover (89). Scanning-time benefits of grazing in larger herds of African antelope apply to single-species and mixed-species herds (90). Scanning rates differ among individuals within a herd. Females with young have higher scanning rates than females without young or males, and animals at the periphery of the group have higher scanning rates than animals in the center of the group (87, 91).

The white rump patch has evolved as a common cohesive signal in ungulates that keeps groups together. The function of rump patches in general is to provide a cohesive signal that will help keep the group together in flight situations. By staying within a cohesive group, each individual enjoys group benefits. Almost all ungulates have some white on their rump, but the size of this rump patch and the ability to turn it on and off vary in concert with a species' habitat characteristics (92, 93).

Woodland ungulates have small rump patches where the white signal can be turned on and off. White-tailed deer are named for their very conspicuous rump patch and white underside of tail that are displayed during flight. However, this same signal is turned off when the deer is not in flight, and there is no white to betray the animal's presence. Openland species, such as pronghorn, Rocky Mountain elk, and bighorn sheep, have much larger rump patches that are visible at all times. Because these animals are likely to be conspicuous to predators at all times, group cohesion is also important at all times. Woodland species may resort to flight against predators secondarily because these animals may be able to use a cryptic strategy and avoid detection by a predator entirely. For this reason having a rump patch that can be concealed during normal circumstances has a real adaptive advantage. However, when the cryptic strategy breaks down and a predator detects a woodland ungulate, these animals are able to turn on a white rump-patch signal and use it to help keep their group together during flight (93).

Habitat used by ungulates ranges from woodland to savannah or prairie. Rocky Mountain and desert mule deer occupy open habitat and have large white rump patches that are constantly visible (94). Black-tailed and Sitka subspecies of the mule deer occupy woodland habitat in the Pacific Northwest and have small rump patches that can be turned on and off, much like that of the white-tailed deer (94).

There are a few exceptions to the generalizations linking habitat and cover to rump-patch size. Familiar North American ungulates that have no rump patch at all are the moose, bison, and musk ox. These species are large and stand their ground against predators or even employ an aggressive defense (5). Moose are solitary or a small-group species, whereas bison and musk oxen are herding species.

SOCIAL SYSTEMS

Animal social systems are products of external environmental forces being brought to bear on individuals of each species. Density and distribution of food and foraging strategy are important factors contributing to the evolution of social systems. Predator avoidance is an important factor in the case of prey species. Density of food and demand for food ultimately determine the number of individuals that can be supported per unit area. Small animals do not need as much food as large animals, but they cannot afford to travel long distances to obtain it. Large animals have greater demand for food, but they are better adapted to travel greater distances to obtain it.

The minimum social group for large mammals is a mother–young group that persists until young are old enough to be independent. In most species there is no paternal role in raising young, and males are solitary or live in bachelor groups for most of the year. The mother–young bond usually lasts until the mother has her next litter. In many species this is one year, but in species such as musk oxen and black and brown bears, where young are born every other year, young may stay with their mother for 2 years.

Group living has several inherent disadvantages compared to a solitary life style (95). Groups represent competition for food, and the larger the group, the greater the additional competition and the more widely the group will have to range in search of food. Transmission of diseases and parasites also is facilitated by living in a group. The result is that many species are essentially solitary, except for mothers and their dependent young.

There must be selective advantages that outweigh the disadvantages of group living that promote the evolution of social systems. For ungulates the most important advantage of larger social groups is protection from predators, which has already been discussed. Species that are found in open habitat, ranging in size from collared peccary to bison, live in large social groups that include many adults and often males and females. Another important reason for living in large social groups is that there may be advantages in acquiring food. This advantage is generally thought to benefit social predators and not herbivores (but see the grazing-lawn advantage). Predators working in a group may be able to capture larger units of prey than they would if each individual hunted alone. The wolf is the best known North American example of a social predator, although coyotes may also hunt in this manner in some places (96). However, even among mammalian carnivores social hunting is unusual and the majority of species are solitary hunters. Only 4 of 30 species in the dog family worldwide are social predators, and only 1 of the 36 members of the cat family is a social species (9). The large North American predators and omnivores, except for wolves and sometimes coyotes, have social systems with just mother–young groups, solitary adult males, and solitary subadult males and females.

BREEDING STRATEGIES

A paternal role has been identified only in the wolf for large North American mammals. In all other species, young are reared by their mothers alone. This sets the stage for the evolution of polygynous breeding systems, where a small number of dominant males, presumably the most fit, mate with all the adult females in the population and most younger, less-experienced males are excluded from the breeding population. Where the only contribution that males make to the next generation of young is a genetic one, females are under selective pressure to choose the most fit males as their mates and the fact that other females may have selected the same male as a mate is not a disadvantage.

In a polygynous breeding system males are under selective pressure to breed with as many females as possible. To do this, males need to become dominant in the local male hierarchy. This is accomplished in part by simply living a long time, because male size is correlated with age in ungulates (63, 97, 98, 99). Dominant males also may need to have large antlers, horns, or tusks for success in male–male competition (100, 101). In polygynous species most males probably do not mate at all, because they may not live long enough to achieve necessary social dominance or they never become large enough to rank at the top of the male hierarchy. However, those males that achieve breeding status may be able to leave many offspring in a few rutting seasons (101).

Males communicate their fitness and rank to females using visual or olfactory means (102, 103, 104, 105). Most ungulate breeding system studies have focused on male–male competition for access to females, and to date little attention has been paid to the role of females in selecting a mate, except for lek breeding species (but see Ref. 63). Males may use scent marking to mark themselves, places within their home ranges, or perhaps boundaries of territories (102). White-tailed and mule deer make "rubs" and "scrapes" during the rut, which involve scent from their glandular foreheads, scent from their tarsal glands, and urine (80, 104, 106). Elk and moose create wallows (105, 107, 108, 109). Black and brown bears make scrapes on trees (30). Pronghorns mark the boundaries of territories by rubbing their cheek gland on tall forbs (110).

Breeding systems in many ungulates are largely a function of the number of adult females in a social group. In white-tailed deer, black-tailed deer, and moose, groups are typically small, and there may be only one adult doe per mother–young group. Elk, bighorn sheep, musk oxen, and pronghorns occur in larger social groups, often with 4 to 10 females. Finally, caribou and bison occur in large social groups with many females per group.

In small-group species, males roam widely during the breeding season trying to locate estrous females (80, 106). Whitetail males approach female–fawn groups

Figure 9–2 Flehmen by male white-tailed deer used when identifying females in estrus. (Photograph by D. M. Hirth.)

Figure 9–3 White-tailed buck "tending" a doe, which is bedded. Buck stands and waits for the doe to get up and resume feeding before "testing" her again. (Photograph by D. M. Hirth.)

and investigate adult and yearling females. If no females smell as though they are approaching estrus, the male moves on in search of another group (Figure 9–2). If a female in the group seems close to estrus, the buck stays with her and forms a "tending" bond in which he isolates her from members of her group and waits until she is ready to stand for copulation (Figure 9–3). A tending buck remains about 5 meters behind the doe and makes occasional slow approaches to test the doe's readiness. If the doe is ready for copulation, she stands still as the buck approaches, but if the doe is not ready, she walks away and the buck stops his approach (80).

Ungulates typically found in medium-sized social groups show two types of breeding systems, based largely on density and distribution of foraging resources (111, 112). Where resources are clumped and predictable, dominant males establish territories and attract female–young groups to their territories. This is referred to as *resource-defense polygyny.* Where resources are evenly distributed, sparse, or clumped but unpredictable, it is not profitable for males to defend territories. Under these circumstances males simply take advantage of the fact that females are already social and defend groups of females from other competing males. This is referred to as *female-defense polygyny.*

Elk, bighorn sheep, and musk oxen exhibit female-defense polygyny. During the breeding season dominant males defend mother–young groups from other males and have breeding rights to these females as long as they can keep other males away (74, 76, 78, 113, 114). Resources for these species are sparse, and groups must keep moving to obtain forage, hence males cannot be territorial. Mountain goats also exhibit female-defense polygyny but differ from the above species in that mother–young groups are small and may have only one or two breeding-age females (115). Males of species that use this female-defense strategy spend most of their time keeping females herded into defensible groups and driving away challenging males. The largest males in the population breed most of the females, but later in the rut as these males lose condition, they are replaced by other less dominant males that breed the remaining females.

The only North American ungulate that exhibits resource-defense polygyny is the pronghorn, although this type of breeding behavior occurs in many African antelopes (82, 110). In fact only certain populations of pronghorns are territorial, and others are harem breeders (i.e., female-defense polygyny) (116, 117).

In large-group species, females are part of very large groups or herds, and it is impossible for even dominant males to expel other males from the herd. In the case of caribou and bison, dominant males form tending bonds with estrous females within large herds and stay with that female until she is ready for copulation (77, 118, 119, 120).

The breeding system of the collared peccary does not follow the same pattern seen in other North American ungulates. Peccaries live in mixed-sex social groups throughout the year, whereas sexes are segregated except during the breeding season in other North American ungulates (121, 122). Social groups are territorial throughout the year, and each group may include several breeding-age males (122). The dominant male forms a brief tending bond with estrous females in the group, but subordinate males may do some breeding, especially when several females are in estrus simultaneously (122).

The peccary is the smallest North American ungulate (20 to 30 kilograms) and it is found in relatively open, arid habitat, and this combination may make it unusually vulnerable to predators. Protection from predators afforded by group membership is probably even more important in peccaries than in other ungulates. Young peccaries are especially vulnerable to predation by mountain lions, bobcats, coyotes, and wolves, and single adults could be prey for any of these predators (122).

If group living in peccaries is essential for protection from predators, males may play an unusual role among ungulates in the survival of their own young, which would place a high selective value on mixed-sex social groups. All individuals within a group need protection from predators, but males can provide more effective protection for their young than females because of their larger size. Lack of sexually selected display structures (horns and antlers) might be part of a mechanism to promote cohesion of mixed-sex social groups.

Social behavior and breeding systems are products of a number of extrinsic factors, such as density of food, distribution of food, risk of predation, availability of protective cover, and perhaps others. It should not be surprising, therefore, that breeding systems might differ in different parts of a species' range or even in the same place over time as conditions change (123). Intraspecific variation in breeding systems has been shown in a number of ungulates (117, 124, 125). Group size in white-tailed deer was significantly larger in a south Texas savannah population compared to a woodland population in Michigan (80). Michigan bucks courted does that occurred as singles, whereas courtship in south Texas contained elements of female-defense harem breeding. The breeding system in mule deer generally resembles that described by Hirth for woodland white-tails, but some authors have suggested that mule deer bucks court groups of does, much like savannah white-tails (80, 106, 126; Barnum, 1930, and McLean, 1940, cited in Ref. 106). A similar shift in breeding behavior may occur in moose in open areas in Alaska compared to woodland populations (107, 127, 128, 129). The most dramatic example of a change in breeding system of a North American ungulate was reported in pronghorns by Byers and Kitchen (117). Bromley and Kitchen observed pronghorns at the National Bison Range in western Montana using resource-defense territoriality (110, 130). However, as a result of a sudden decline in the proportion of older males in the population, the breeding system shifted to a female-defense harem system (117).

Male breeding behavior in polygynous species is energetically expensive for males because of the amount of time and effort they spend locating and defending estrous females. The time required for these rutting activities comes at the expense of time that would otherwise be spent feeding. The result is that dominant breeding males in temperate- and arctic-zone species lose approximately 25% of their body weight during the rut and enter the winter in poor condition (76, 97, 107, 131, 132, 133). For breeding males there is a trade-off between maximizing reproductive output in a given breeding season and entering winter in good enough condition to survive (101, 134). The winners of the game are males that maximize their reproductive output over a lifetime and not over a single breeding season. The largest males and probably young of the year are clearly the two most vulnerable segments of the population with respect to winter mortality (56, 74, 135, 136).

The black bear, brown bear, and mountain lion have similar breeding systems. Each is polygynous and their breeding system is based on dominant males with very large territories that overlap the much smaller home ranges or territories of several females (4, 23, 30, 137). Breeding male polar bears are not territorial because females are located on floating ice during the breeding season, which results in an unpredictable and thus undefensible resource (50). In these species females are solitary, as are males, and occur at low density, so male polar bears may have fewer breeding opportunities than male ungulates.

The wolf is the only large North American mammal under consideration in this volume that has a monogamous mating system, although monogamy is the rule within the Canidae (9). As has been discussed earlier, wolves are also unusual in that they are a social predator (5). Usually a single, dominant pair of wolves monopolizes reproduction for the group (pack), which suggests that members of the pack are closely related (138, 139, 140). The driving force for monogamy and a male parental role in reproduction in the canids is not clear because it occurs in species that hunt large prey, such as wolves, hunting dogs, and dholes, and species that hunt small prey, such as coyotes and foxes (3). An explanation of monogamy in wolves and other canids should clarify why monogamy would not have the same selective value for other carnivores (15, 141). Monogamy and male parental care are advantageous because they give nursing females easier access to food, they may provide additional protection for young, and they may enable canids to have larger litter size than if females had to hunt independently for themselves and their young, as do female felids. Litter size in canids ranges from 2 to 13 with a mode of 5, whereas litter size in felids ranges from 1 to 5 with a mode of 3 (9). In the case of wolves, other members of a pack also may help provision the female and dependent young and may provide additional protection (138, 140).

SUMMARY

I examined four aspects of the behavioral ecology of large mammals in North America: foraging behavior, predator avoidance, social systems, and breeding systems. Each of these topics has received a great deal of attention from research biologists in recent years. By examining broad behavioral trends within and between groups of animals, biologists are better able to understand the behavior of individual species.

In this review a few key variables have emerged that have major impacts on the evolution of behavior in large mammals. Size of animal dictates metabolic demand for food, and this is related to size of home range and level of predation pressure on herbivores. Density of food resources imposes a ceiling on the density of animals that can be supported, whether predators or prey. However, distribution of food resources also is important, whether it is evenly distributed or occurs in patches. The young of all but one of these species are reared by their mothers alone, and this has set the stage for the evolution of polygynous breeding systems and thus sexual dimorphism, where males are significantly larger than females. Finally, availability of protective cover has important ramifications for life-history strategies of predators and prey alike and is an important variable in the evolution of social systems.

LITERATURE CITED

1. GAVIN, T. A. 1991. Why ask "why": The importance of evolutionary biology in wildlife science. *Journal of Wildlife Management* 55:760–766.

2. CLUTTON-BROCK, T. H., and P. H. HARVEY. 1978. Mammals, resources and reproductive strategies. *Nature* 273:191–195.

3. MACDONALD, D. W. 1983. The ecology of carnivore social behaviour. *Nature* 301:379–384.

4. HORNOCKER, M. 1970. An analysis of mountain lion predation upon mule deer and elk in the Idaho Primitive Area. *Wildlife Monographs* 21:1–39.

5. MECH, L. D. 1970. *The Wolf: Ecology and Behavior of an Endangered Species.* Natural History Press, Doubleday, New York.

6. SEIDENSTICKER, J. C., M. HORNOCKER, W. WILES, and J. MESSICK. 1973. Mountain lion social organization in the Idaho Primitive Area. *Wildlife Monographs* 35:1–60.

7. KOEHLER, G. M., and M. G. HORNOCKER. 1991. Seasonal resource use among mountain lions, bobcats, and coyotes. *Journal of Mammalogy* 72:391–396.

8. MAEHR, D. S., R. C. BELDEN, E. D. LAND, and L. WILKINS. 1990. Food habits of panthers in southwest Florida. *Journal of Wildlife Management* 54:420–423.

9. KLEIMAN, D. G., and J. F. EISENBERG. 1973. Comparison of canid and felid social systems from an evolutionary perspective. *Animal Behaviour* 21:637–659.

10. STIRLING, I. 1974. Midsummer observations on the behavior of wild polar bears (*Ursus maritimus*). *Canadian Journal of Zoology* 52:1191–1198.

11. SMITH, T. G. 1980. Polar bear predation of ringed and bearded seals in the land-fast sea ice habitat. *Canadian Journal of Zoology* 58:2201–2209.

12. NOWAK, R. M. 1991. *Walker's Mammals of the World,* 5th ed. Johns Hopkins Press, Baltimore, MD.

13. BEKOFF, M., T. J. DANIELS, and J. L. GITTLEMEN. 1984. Life history patterns and the comparative social ecology of carnivores. *Annual Review of Ecology and Systematics* 15:191–232.

14. PETERSON, R. O. 1977. Wolf ecology and prey relationships on Isle Royale. Scientific Monograph Series 11. National Park Service, Washington, DC.

15. NUDDS, T. D. 1978. Convergence of group size strategies by mammalian social carnivores. *The American Naturalist* 112:957–960.

16. MESSIER, F. 1985. Social organization, spatial distribution, and population density of wolves in relation to moose density. *Canadian Journal of Zoology* 63:1068–1077.

17. MECH, L. D. 1994. Buffer zones of territories of gray wolves as regions of intraspecific strife. *Journal of Mammalogy* 75:199–202.

18. HOSKINSON, R. L., and L. D. MECH. 1976. White-tailed deer migration and its role in wolf predation. *Journal of Wildlife Management* 40:429–441.

19. MECH, L. D. 1977. Wolf pack buffer zones as prey reservoirs. *Science* 198:320–321.

20. ROGERS, L. L., L. D. MECH, D. K. DAWSON, J. M. PEEK, and M. KORB. 1980. Deer distribution in relation to wolf pack territory edges. *Journal of Wildlife Management* 44:253–258.

21. BALLARD, W. B., L. A. AYRES, P. R. KRAUSMAN, D. J. REED, and S. G. FANCY. 1997. Ecology of wolves in relation to a migratory caribou herd in northwest Alaska. *Wildlife Monographs* 135:1–47.

22. SERVHEEN, C. 1983. Grizzly bear food habits, movements, and habitat selection in the Mission Mountains, Montana. *Journal of Wildlife Management* 47:1026–1035.

23. ROGERS, L. L. 1987. Effects of food supply and kinship on social behavior, movements, and population growth of black bears in northeastern Minnesota. *Wildlife Monographs* 97:1–72.

24. SCHWARTZ, C. C., and A. W. FRANZMANN. 1991. Interrelationship of black bears to moose and forest succession in the northern coniferous forest. *Wildlife Mongraphs* 113:1–58.

25. FARLEY, S. D., and C. T. ROBBINS. 1995. Lactation, hibernation, and mass dynamics of American black bears and grizzly bears. *Canadian Journal of Zoology* 73:2216–2222.

26. MATHEWS, N. E., and W. F. PORTER. 1988. Black bear predation of white-tailed deer neonates in the central Adirondacks. *Canadian Journal of Zoology* 66:1241–1242.

27. BALLARD, W. B. 1992. Bear predation on moose: A review of recent North American studies and their management implications. *Alces (Suppl)* 1:162–176.

28. ROGERS, L. L. 1976. Effects of mast and berry crop failures on survival, growth, and reproductive success of black bears. *Transactions of the North American Wildlife and Natural Resources Conference* 41:431–438.

29. ELOWE, K. D., and W. E. DODGE. 1989. Factors affecting black bear reproductive success and cub survival. *Journal of Wildlife Management* 53:962–968.

30. STIRLING, I., and A. E. DEROCHER. 1990. Factors affecting the evolution and behavioral ecology of the modern bears. *International Conference on Bear Research and Management* 8:189–204.

31. BUNNELL, F. L., and T. HAMILTON. 1983. Forage digestibility and fitness in grizzly bears. *International Conference on Bear Research and Management* 5:179–185.

32. BOERTJE, R. D., W. C. GASAWAY, D. V. GRANGAARD, and D. G. KELLYHOUSE. 1988. Predation on moose and caribou by radio-collared grizzly bears in east central Alaska. *Canadian Journal of Zoology* 66:2492–2499.

33. LOWERY, L. F., J. J. BURNS, and R. R. NELSON. 1987. Polar bear, *Ursus maritimus,* predation on belugas, *Delphinapterus leucas,* in the Bering and Chukchi Seas. *The Canadian Field-Naturalist* 101:141–146.

34. KILIAAN, H. P. L., and I. STIRLING. 1978. Observations on overwintering walruses in the eastern Canadian High Arctic. *Journal of Mammalogy* 59:197–200.

35. HERRERO, S. 1972. Aspects of evolution and adaptation in American black bears (*Ursus americanus* Pallus) and brown and grizzly bears (*Ursus arctos* Linn.) of North America. Pages 221–231 *in Bears—Their Biology and Management.* Publication No. 23. International Union for Conservation of Nature, Morges, Switzerland.

36. LANGER, P. 1979. Adaptational significance of the forestomach of the collared peccary, *Dicotyles tajacu* (L. 1758) (Mammalia: Artiodactyla). *Mammalia* 43:235–245.

37. VAN SOEST, P. J. 1994. *Nutritional Ecology of the Ruminant,* 2nd ed. Cornell University Press, Ithaca, NY.

38. HANLEY, T. A. 1982. The nutritional basis for food selection by ungulates. *Journal of Range Management* 35:146–151.

39. SCHOENER, T. W. 1971. Theory of feeding strategies. *Annual Review of Ecology and Systematics* 2:369–404.

40. KLEIBER, M. 1961. *The Fire of Life: An Introduction to Animal Energetics.* John Wiley and Sons, New York.

41. HOFMANN, R. R. 1989. Evolutionary steps of ecophysiological adaptation and diversification of ruminants: A comparative view of their digestive system. *Oecologia* 78:443–457.

42. HANLEY, T. A., and K. A. HANLEY. 1982. Food resource partitioning by sympatric ungulates on the Great Basin rangeland. *Journal of Range Management* 35:152–158.

43. JARMAN, P. J. 1974. The social organisation of antelope in relation to their ecology. *Behaviour* 48:215–267.

44. TRUDELL, J., and R. G. WHITE. 1981. The effect of forage structure and availability on food intake, biting rate, bite size and daily eating time in reindeer. *Journal of Applied Ecology* 18:63–81.

45. MCNAUGHTON, S. J. 1984. Grazing lawns: Animals in herds, plant form, and coevolution. *The American Naturalist* 124:863–886.

46. MCNAUGHTON, S. J., L. L. WALLACE, and M. B. COUGHENOUR. 1983. Plant adaptations in an ecosystem context: Effects of defoliation, nitrogen, and water on growth of an African C_4 sedge. *Ecology* 64:307–318.

47. FRANK, D. A., and S. J. MCNAUGHTON. 1993. Evidence for the promotion of above ground grassland production by native large herbivores in Yellowstone National Park. *Oecologia* 96:157–161.

48. HOBBS, N. T., and D. M. SWIFT. 1988. Grazing in herds: When are nutritional benefits realized? *The American Naturalist* 131:760–764.

49. MAIN, M. B., F. W. WECKERLY, and V. C. BLEICH. 1996. Sexual segregation in ungulates: New directions for research. *Journal of Mammalogy* 77:449–461.

50. RAMSAY, M. A., and I. STIRLING. 1986. On the mating system of polar bears. *Canadian Journal of Zoology* 64:2142–2151.

51. BEIER, P. 1987. Sex differences in quality of white-tailed deer diets. *Journal of Mammalogy* 68:323–329.

52. MCCULLOUGH, D. R., D. H. HIRTH, and S. J. NEWHOUSE. 1989. Resource partitioning between sexes in white-tailed deer. *Journal of Wildlife Management* 53:277–283.

53. WECKERLY, F. W., and J. P. NELSON, JR. 1990. Age and sex differences of white-tailed deer diet composition, quality, and calcium. *Journal of Wildlife Management* 54:532–538.

54. BOWYER, R. T. 1984. Sexual segregation in southern mule deer. *Journal of Mammalogy* 65:410–417.

55. WECKERLY, F. W. 1993. Intersexual resource partitioning in black-tailed deer: A test of the body size hypothesis. *Journal of Wildlife Management* 57:475–494.

56. MIQUELLE, D. G., J. M. PEEK, and V. VAN BALLENBERGE. 1992. Sexual segregation in Alaska moose. *Wildlife Monographs* 122:1–57.

57. JAKIMCHUK, R. D., S. H. FERGUSON, and L. G. SOPUCK. 1987. Differential habitat use and sexual segregation in the central Arctic caribou herd. *Canadian Journal of Zoology* 65:534–541.

58. SHANK, C. C. 1982. Age and sex differences in the diets of wintering Rocky Mountain bighorn sheep. *Ecology* 63:627–633.

59. FESTA-BIANCHET, M. 1988. Seasonal range selection in bighorn sheep: Conflicts between forage quality, forage quantity, and predator avoidance. *Oecologia* 75:580–586.

60. KRAUSMAN, P. R., B. D. LEOPOLD, R. F. SEEGMILLER, and S. G. TORRES. 1989. Relationships between desert bighorn sheep and habitat in western Arizona. *Wildlife Monographs* 102:1–66.

61. BLEICH, V. C., R. T. BOWYER, and J. D. WEHAUSEN. 1997. Sexual segregation in mountain sheep: Resources or predation. *Wildlife Monographs* 134:1–50.

62. KOMERS, P. E., F. MESSIER, and C. C. GATES. 1993. Group structure in wood bison: Nutritional and reproductive determinants. *Canadian Journal of Zoology* 71:1367–1371.

63. BERGER, J., and C. CUNNINGHAM. 1994. *Bison: Mating and Conservation in Small Populations.* Columbia University Press, New York.

64. OAKES, E. J., R. HARMSEN, and C. EBERL. 1992. Sex, age, and seasonal differences in the diets and activity budgets of muskoxen (*Ovibos moschatus*). *Canadian Journal of Zoology* 70:605–616.

65. PETERSON, R. O., J. D. WOOLINGTON, and T. N. BAILEY. 1984. Wolves of the Kenai Peninsula, Alaska. *Wildlife Monographs* 88:1–52.

66. BALLARD, W. B., J. S. WHITMAN, and C. L. GARDNER. 1987. Ecology of an exploited wolf population in south-central Alaska. *Wildlife Monographs* 98:1–54.

67. TRUETT, J. C. 1979. Observations of coyote predation on mule deer fawns in Arizona. *Journal of Wildlife Management* 43:956–958.

68. HARRISON, D. J., and J. A. HARRISON. 1984. Foods of adult Maine coyotes and their known-aged pups. *Journal of Wildlife Management* 48:922–926.

69. MACCRACKEN, J. G., and D. W. URESK. 1984. Coyote food habits in the Black Hills, South Dakota. *Journal of Wildlife Management* 48:1420–1423.

70. ANDELT, W. F. 1985. Behavioral ecology of coyotes in south Texas. *Wildlife Monographs* 94:1–45.

71. LENT, P. C. 1974. Mother–infant relationships in ungulates. Pages 14–55 *in* V. Geist and F. Walther, eds., *The Behaviour of Ungulates in Relation to Management.* Publication (New Series) 24. International Union for Conservation of Nature and Natural Resources, Morges, Switzerland.

72. HIRTH, D. H. 1985. Mother–young behavior in white-tailed deer, *Odocoileus virginianus. The Southwestern Naturalist* 30:297–302.

73. JACOBSEN, N. K. 1979. Alarm bradycardia in white-tailed deer fawns (*Odocoileus virginianus*). *Journal of Mammalogy* 60:343–349.

74. McCullough, D. R. 1969. The tule elk: Its history, behavior, and ecology. *University of California Publications in Zoology* 88:1–209.

75. Stringham, S. F. 1974. Mother–infant relations in moose. *Le Naturaliste Canadien* 101:307–323.

76. Geist, V. 1971. *Mountain Sheep: A Study in Behavior and Evolution.* University of Chicago Press, Chicago.

77. McHugh, T. 1958. Social behavior of the American buffalo. *Zoologica* 43:1–41.

78. Tener, J. S. 1965. *Musk oxen in Canada: A Biological and Taxonomic Review.* Monograph Series 2. Canadian Wildlife Service.

79. Lott, D. F., and J. C. Galland. 1985. Parturition in American bison: Precocity and systematic variation in cow isolation. *Zeitschrift für Tierpsychologie* 69:66–71.

80. Hirth, D. H. 1977. Social behavior of white-tailed deer in relation to habitat. *Wildlife Monographs* 53:1–55.

81. Hamilton, W. D. 1971. Geometry for the selfish herd. *Journal of Theoretical Biology* 31:295–311.

82. Estes, R. D. 1974. Social organization of the African Bovidae. Pages 166–205 *in* V. Geist and F. Walther, eds., *The Social Behaviour of Ungulates and Its Relation to Management.* Publication (New Series) No. 24. International Union for Conservation of Nature and Natural Resources, Morges, Switzerland.

83. Landeau, L., and J. Terborgh. 1986. Oddity and the "confusion effect" in predation. *Animal Behaviour* 34:1372–1380.

84. Pulliam, H. R. 1973. On the advantages of flocking. *Journal of Theoretical Biology* 38:419–422.

85. Berger, J. 1978. Group size, foraging, and antipredator ploys: An analysis of bighorn sheep decisions. *Behavioral Ecology and Sociobiology* 4:91–99.

86. Risenhoover, K. L., and J. A. Bailey. 1985. Relationships between group size, feeding time, and agonistic behavior of mountain goats. *Canadian Journal of Zoology* 63:2501–2506.

87. Lipetz, V. E., and M. Bekoff. 1982. Group size and vigilance in pronghorns. *Zeitschrift für Tierpsychologie* 58:203–216.

88. Underwood, R. 1982. Vigilance behaviour in grazing African antelopes. *Behaviour* 79:82–107.

89. Frid, A. 1997. Vigilance by female Dall's sheep: Interactions between predation and risk factors. *Animal Behaviour* 53:799–808.

90. FitzGibbon, C. D. 1990. Mixed-species groups in Thomson's and Grant's gazelles: The antipredator benefit. *Animal Behaviour* 39:1116–1126.

91. Burger, J., and M. Gochfeld. 1994. Vigilance in African mammals: Differences among mothers, other females, and males. *Behaviour* 131:153–169.

92. Guthrie, R. D. 1971. A new theory of mammalian rump patch evolution. *Behaviour* 38:132–145.

93. Hirth, D. H., and D. R. McCullough. 1977. Evolution of alarm signals in ungulates with special reference to white-tailed deer. *The American Naturalist* 111:31–42.

94. Cowan, I. McT. 1956. What and where are the mule and black-tailed deer. Pages 334–359 *in* W. P. Taylor, ed., *The Deer of North America.* Stackpole Books, Harrisburg, PA.

95. Alexander, R. D. 1974. The evolution of social behavior. *Annual Review of Ecology and Systematics* 5:325–383.

96. Brundige, G. C. 1993. Predation ecology of the eastern coyote, *Canis latrans* var., in the Adirondacks, New York. Ph.D. dissertation. State University of New York, College of Environmental Science and Forestry, Syracuse.

97. Wood, A. J., I. McT. Cowan, and H. C. Nordan. 1962. Periodicity of growth in ungulates as shown by deer of the genus *Odocoileus. Canadian Journal of Zoology* 40:593–603.

98. Townsend, T. W., and E. D. Bailey. 1981. Effects of age, sex and weight on social rank in penned white-tailed deer. *The American Midland Naturalist* 101:92–101.

99. Barrette, C., and D. Vandal. 1986. Social rank, dominance, antler size, and access to food in snow-bound wild woodland caribou. *Behaviour* 97:118–146.

100. Geist, V. 1966. The evolution of horn-like organs. *Behaviour* 27:175–214.

101. Clutton-Brock, T. H., S. D. Albon, and F. E. Guiness. 1982. *Red Deer: The Ecology of Two Sexes.* University of Chicago Press, Chicago.

102. Eisenberg, J. F., and D. G. Kleiman. 1972. Olfactory communication in mammals. *Annual Review of Ecology and Systematics* 3:1–32.

103. Coblentz, B. F. 1976. Functions of scent urination in ungulates with special reference to feral goats (*Capra hircus*). *The American Naturalist* 110:549–557.

104. Kile, T. L., and R. L. Marchinton. 1977. White-tailed deer rubs and scrapes: Spatial, temporal and physical characteristics and social role. *The American Midland Naturalist* 97:257–266.

105. Bowyer, R. T., V. van Ballenberghe, and K. R. Rock. 1994. Scent marking by Alaskan moose: Characteristics and spatial distribution of rubbed trees. *Canadian Journal of Zoology* 72:2186–2192.

106. Kucera, T. 1978. Social behavior and breeding system of the desert mule deer. *Journal of Mammalogy* 59:463–476.

107. Lent, P. C. 1974. A review of rutting behavior in moose. *Le Naturaliste Canadien* 101:307–323.

108. Bowyer, R. T., and D. W. Kitchen. 1987. Significance of scent marking by Roosevelt elk. *Journal of Mammalogy* 68:418–423.

109. Miquelle, D. G. 1991. Are moose mice? The function of scent urination in moose. *The American Naturalist* 138:460–477.

110. Kitchen, D. W. 1974. Social behavior and ecology of the pronghorn. *Wildlife Monographs* 38:1–96.

111. Clutton-Brock, T. H. 1989. Mammalian mating systems. *Proceedings of the Royal Society of London* B 236:339–372.

112. Krebs, J. R., and N. B. Davies. 1993. *An Introduction to Behavioural Ecology,* 3rd ed. Blackwell Scientific Publications, London.

113. Struhsaker, T. T. 1967. The behavior of elk (*Cervus canadensis*) during the rut. *Zeitschrift für Tierpsychologie* 24:80–114.

114. Franklin, W. L., A. S. Mossman, and M. Dole. 1975. Social organization and home range of Roosevelt elk. *Journal of Mammalogy* 56:102–118.

115. Geist, V. 1965. On the rutting behavior of the mountain goat. *Journal of Mammalogy* 45:551–568.

116. Buechner, H. K. 1950. Life history, ecology and range use of the pronghorn antelope in Trans-Pecos, Texas. *The American Midland Naturalist* 43:257–354.

117. Byers, J. A., and D. W. Kitchen. 1988. Mating system shifts in a pronghorn population. *Behavioral Ecology and Sociobiology* 22:355–360.

118. Fuller, W. A. 1960. Behaviour and social organization of the wild bison of Wood Buffalo National Park, Canada. *Arctic* 13:3–19.

119. Lent, P. C. 1965. Rutting behaviour in a barren-ground caribou population. *Animal Behaviour* 13:259–264.

120. Kelsall, J. P. 1968. The migratory barren-ground caribou. Monograph Series 3. Canadian Wildlife Service, Ottawa.

121. Sowls, L. K. 1974. Social behavior of the collared peccary, *Dicotyles tajacu.* Pages 144–165 *in* V. Geist and F. Walther, eds., *The Behaviour of Ungulates and Its Relation to Management.* Publication (New Series) 24. International Union for Conservation of Nature and Natural Resources, Morges, Switzerland.

122. Bissonette, J. A. 1982. Ecology and social behavior of the collared peccary in Big Bend National Park, Texas. Scientific Monograph Series No. 16. National Park Service, Washington, DC.

123. Lott, D. F. 1991. *Intraspecific Variation in the Social Systems of Wild Vertebrates.* Cambridge University Press, Cambridge, UK.

124. Langbein, J., and S. J. Thirgood. 1989. Variation in mating systems of fallow deer (*Dama dama*) in relation to ecology. *Ethology* 83:195–214.

125. Hirth, D. H. 1997. Lek breeding in a Texas population of fallow deer, *Dama dama. The American Midland Naturalist* 138:276–289.

126. Geist, V. 1981. Behavior: Adaptive strategies in mule deer. Pages 157–224 *in* O. C. Wallmo, ed. *Mule and Black-Tailed Deer of North America.* University of Nebraska Press, Lincoln.

127. Peek, J. M., R. E. LeResche, and D. R. Stevens. 1974. Dynamics of moose aggregations in Alaska, Minnesota, and Montana. *Journal of Mammalogy* 55:126–137.

128. VAN BALLENBERGHE, V., and D. G. MIQUELLE. 1993. Mating in moose: Timing, behavior, and male access patterns. *Canadian Journal of Zoology* 71:1687–1690.

129. BUBENIK, A. B. 1997. Behavior. Pages 173–221 *in* A. W. Franzmann and C. C. Schwartz, eds. *Ecology and Management of the North American Moose.* Smithsonian Institution Press, Washington, DC.

130. BROMLEY, P. T. 1969. Territoriality in pronghorn bucks on the National Bison Range. *Journal of Mammalogy* 50:181–189.

131. KNOWLTON, F. F., M. WHITE, and J. G. KIE. 1979. Weight patterns of wild white-tailed deer in southern Texas. Pages 55–64 *in* D. L. Drawe, ed., *Proceedings of the 1st Welder Wildlife Foundation Symposium,* Sinton, TX.

132. BOWYER, R. T. 1981. Activity, movement, and distribution of Roosevelt elk during the rut. *Journal of Mammalogy* 62:574–582.

133. MIQUELLE, D. G. 1990. Why don't bull moose eat during the rut? *Behavioral Ecology and Sociobiology* 27:145–151.

134. GUNN, A. E., F. L. MILLER, and B. McLEAN. 1989. Evidence for and possible causes of increased mortality of bull muskoxen during severe winters. *Canadian Journal of Zoology* 67:1106–1111.

135. ROBINETTE, W. L., J. S. GASHWILER, J. B. LOW, and D. A. JONES. 1957. Differential mortality by sex and age among mule deer. *Journal of Wildlife Management* 21:1–16.

136. VAN BALLENBERGHE, V., and W. B. BALLARD. 1997. Population dynamics. Pages 223–245 *in* A. W. Franzmann and C. C. Schwartz, eds., *Ecology and Management of the North American Moose.* Smithsonian Institution Press, Washington, DC.

137. GARSHELIS, D. L., and M. R. PELTON. 1981. Movements of black bears in the Great Smokey Mountains National Park. *Journal of Wildlife Management* 45:912–925.

138. MURIE, A. 1944. *The Wolves of Mt. McKinley. Fauna of the National Parks of the United States.* Fauna Series No. 5. National Park Service, Washington, DC.

139. MECH, L. D. 1975. Population trend and winter deer consumption in a Minnesota wolf pack. Pages 55–83 *in* R. L. Phillips and C. Jonkel, eds., *Proceedings of the 1975 Predator Symposium.* University of Montana, Missoula.

140. HARRINGTON, F. H., L. D. MECH, and S. H. FRITTS. 1983. Pack size and wolf pup survival: Their relationship under varying ecological conditions. *Behavioral Ecology and Sociobiology* 13:19–26.

141. GITTLEMAN, J. L., and P. H. HARVEY. 1982. Carnivore home-range size, metabolic needs and ecology. *Behavioral Ecology and Sociobiology* 10:57–63.

Harvest Management Goals

Len H. Carpenter

INTRODUCTION

As the new millennium approaches, the landscape for big game harvest management is changing rapidly. Gone are days when wildlife managers only considered harvest strategies to optimize herd size and to satisfy the hunting public. Today, the wildlife manager must accommodate a variety of desires and values of interested publics. In some states such as California, the state wildlife agency must prepare environmental assessment documents for all hunting seasons to meet requirements of the California Environmental Protection Act. As a result, processes involved in establishing big game management goals and objectives are increasingly complex. Objectives of this chapter are to provide a discussion on harvest management including goal setting, presentation of key steps that must be addressed in developing harvest objectives, role of carrying capacity in the decision process, selection of season and license types, methods to measure the harvest, and the public-involved regulation setting process.

Importance of Harvest Management

Sport harvest has been the dominant method for managing game populations since modern wildlife management began with the "Doctrine of Wise Use" advanced by Theodore Roosevelt and Gifford Pinchot in 1910 (1). Leopold wrote one of the earliest scientific papers on aspects of game harvest when he discussed the kill factor for deer in the Southwest (2). From 1900 to 1940, the primary goal of harvest management was to prevent overharvest (3, cited in Ref. 1). From 1945 to about 1975, the goal generally became one of controlling increasing big game populations. In many situations the biggest problem was attracting a sufficient number of hunters to obtain harvest goals. Meeting harvest goals was necessary to control big game populations and prevent widespread overuse of their ranges (4).

Since about 1975, harvest management strategies have varied to accommodate a growing demand for big game hunting while maintaining aspects of a quality hunting experience. Big game harvest management in the last decade of the century has become a balancing act of satisfying hunter recreation demands, private landowner concerns about wildlife damage, and broader values of wildlife held by the general population (5, 6). The history of harvesting wildlife is further discussed by Caughley (7).

Recreational hunting in the United States produced nearly $21 billion of expenditures in 1996 (8). More than 11 million hunters sought big game. For most wildlife agencies, money derived from big game hunting pays for the majority of wildlife management programs including nongame and threatened and endangered species management (9). Money spent by hunters each year in local communities has become critical to survival of many small businesses and towns. Economic importance always increases political attention and the same can be seen in wildlife management. This increased interest in hunting has added another dimension to the harvest management equation. Managers must now seriously consider the economic consequences of changes in harvest management strategies.

Contemporary Harvest Systems

Harvest management for big game is the art of melding objectivity of wildlife science and subjectivity of public desire for the attainment of a management goal (1). Scheduled harvests can control population size, alter distribution, and modify behaviors of wild animals. Society's encroachment on wild lands and wildlife habitat is increasing the importance of this aspect of harvest management. However, expanding rural developments by humans also limits harvest options available to the wildlife manager as developed areas are closed to hunting and use of firearms for safety and political reasons.

Recently, economic returns to the private landowner through payment of trespass or service fees by big game hunters have also complicated harvest strategies. Because hunters are willing to pay higher fees for the opportunity to kill a larger animal with less hunting pressure, landowners often severely restrict access to their lands. This limits opportunity for the wildlife manager to achieve adequate harvests (especially of females). When surrounded by public lands with higher hunting pressure as are many areas in the west, these private lands become refuges. The reduced harvest and concentration of animals also frequently contribute to increased private landowner complaints about game damage once the hunting season closes (10, 11, 12, 13).

Big game harvests on private land are typically much more intensively managed than harvests on public lands. Several western states such as California, Colorado, Montana, New Mexico, Oregon, Utah, and Washington administer "ranching for wildlife" programs that allow the private land owner more flexibility in designing their hunting and harvest programs (14). Many of these hunts involve "public" hunters that hunt on private land without paying an access fee. This is the "price" that the private landowner pays to obtain the flexibility desired in their hunting program. To the extent possible, the wildlife manager should incorporate private land harvests into the population database to assess overall herd status.

Big game species fall into two orders: Artiodactyla and Carnivora. These categories facilitate discussions of harvest management because harvest strategies for the two groups are usually different as a result of basic differences in population characteristics and public attitudes toward the groups. Species in order Artiodactyla are characterized by high recruitment rates and rapid population growth, while species in the order Carnivora characteristically have low recruitment and slow population growth. These biological characteristics significantly affect how species respond to harvest regimes.

The different recruitment and growth rate characteristics of each order require use of different harvest management strategies. Deer, sheep, and pronghorn management strategies are designed to obtain maximum yield over time, whereas harvest systems for carnivores are structured for lower yields and to minimize property damage and maximize human safety (15). In addition, public values and attitudes toward the two groups are considerably different. The wildlife manager must consider strongly held protection and preservationist attitudes by segments of the public when designing harvest systems for species of carnivores (6, 16).

Harvest Data

Harvest data are the most common, generally most accurate, and frequently the only information obtained by wildlife managers. Estimates of total kill and kill by age and sex for selected management units typically are obtained. Often harvest data are key inputs to development of population models. Most decisions on big game population management are based on harvest data. This importance requires that harvest measurement systems be statistically based and scientifically valid. Concerns for accuracy and precision of data must be addressed and appropriate methodology designed before data are obtained.

DETERMINING HARVEST MANAGEMENT GOALS

Setting Population Objectives

As with most endeavors, planning is a key first step to determining harvest management goals. It is critical that strategies developed for harvest management fit into a broader framework of agency goals and objectives and be articulated to all publics so that the purpose and role of the harvest program are clearly understood. Most natural resource agencies have comprehensive or strategic plans that describe the general direction of their management actions. Goals are typically subjective while objectives are usually quantifiable and measurable. The planning process includes four basic steps:

1. Determining status of the resource/population (i.e., Where are you?)
2. Defining goals and objectives of the management program (i.e., Where do you want to be?)
3. Establishing management strategies to attain objectives (i.e., How do you get there?)
4. Determining how closely the applied management strategy achieved the resource/population objectives (i.e., Did you get there?) (1).

Development of harvest management strategies should follow these basic steps.

Population Census. To address population status, it is first necessary to conduct a population census to determine number, density, and herd composition (age and sex ratios) of the species of concern, and to define the geographical area that applies to the management action prescribed. The most typical geographical descriptor is a management unit or group of management units that enclose the key habitat requirements of the selected population (Figure 10–1). To be effective for hunter distribution and law enforcement, boundaries of units must be easily recognized. Information on distribution of the population by season of the year is critical in developing management units and in designing inventory efforts. Inventories should be made when conditions are conducive to effective detection of the animals.

Winter and summer habitats of big game species may be separated by considerable distances. Most aerial census systems are designed for times when animals are most concentrated and visible. In mountainous areas of the West the best time to survey is winter. Aerial estimates in habitats with heavy deciduous tree canopies must be conducted after leaf fall. If habitats of the species under study do not lend themselves to aerial reconnaissance, then alternate and more indirect methodology such as measurements of sign, mark–recapture analysis, or modeling must be used. Typically, population inventories for western species like deer, elk, bighorn sheep, mountain goats, and pronghorn are conducted from fixed-wing aircraft or helicopters. Eastern management agencies often do not apply large-scale direct estimates of cervids. Indirect measures, such as hunter success rates, may be involved. Population estimates for carnivores are usually based on indirect methodologies such as harvest trends, damage levels, and modeling.

Sample-based estimate. It is important that the population inventory be a sample-based estimate (17, 18, 19). This allows for calculation of confidence intervals and provides the investigator information on variability (repeatability) of the

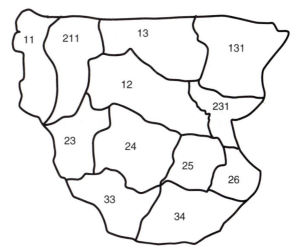

Figure 10–1 Example of a map depicting a group of game management units making up an elk data analysis unit in Colorado.

sample. It also ensures that sampling effort is dispersed across the entire area from which inferences will be drawn. All inventories may not follow strict sampling plans, but it is important that the investigator realize limitations of data obtained from these efforts. In today's world of litigation and conflict concerning management of wildlife resources, it is vital that decisions on harvest management be based on solid information.

Use of computer models. Today most big game managers use computer models to describe populations (20). Models vary from the complex to rather simple spreadsheets. Most models estimate population size for various age and sex categories at various time steps. It is important that necessary model inputs not exceed data availability. Typically, data gaps exist for most of these models. If data are not available, it becomes necessary to guess. The key is to minimize inputs that must be "guesstimated." A main benefit of model construction is to force the systematic and structured organization of data and help identify knowledge gaps. Frequently, harvest data are major components of population models (21, 22).

Computer models allow exploration of various management strategies and options. "What if" scenarios provide the manager with insights into population performance, assuming a harvest option is implemented for a specified time period. The manager can run the model to see if the initial population size selected will sustain the measured harvest over time. If not, it is a good indication that the initial population size was too low. Assuming an initial population size that is too low seems to be a common problem in model building.

Managers face inherent problems when using computer models for harvest decisions. A good example is elk harvest management in Colorado. Colorado historically has managed big game populations from population simulations by data analysis units with the program ONEPOP and later versions such as POP-II (23, 24, 25). State and regional elk population sizes are developed by summing appropriate DAU population estimates from POP-II. Harvest data are important inputs to these models.

Population density objectives are established for each elk population using a data analysis unit public planning process. Objectives are stated for population size and for minimum sex and age ratios and are updated approximately every 5 years. The problem arises in these updates. Over time data for the models change. Frequently, census methodology or intensity of survey changes, model parameters are changed, or new

Figure 10–2 Controlled recreational harvests help maintain a balance between big game populations and their available winter habitat. Deep snows and cold temperatures induce considerable stress on big game populations. (Photograph by L.H. Carpenter.)

models are built. Predictably, model updates result in estimates of higher elk populations. When these updates occur across several large elk populations, the estimate of elk for the state of Colorado increases substantially. Each new and increased population estimate results in substantial concern to the agricultural interests in Colorado having conflicts with elk and their operations. The challenge for the Colorado Division of Wildlife and Wildlife Commission is explaining what these updated model estimates mean with regard to number of elk on the ground. These phenomena became commonly referred to as the "paper elk" syndrome. The risk is to the credibility and trust that various publics have in the elk management program.

Biological Carrying Capacity. Once the first step of making a population census is complete, the next requirement is to determine optimum population level. Many big game populations are managed under the assumption that tenets of K-carrying capacity apply. This means that a population will continue to increase until it approaches limits of the habitat to support it (26, 27, 28, 29). This growth pattern is the logistic growth curve (Figure 7–1, Chapter 7). The inflection point of the growth curve is commonly labeled the I-carrying capacity.

The population size resulting in the maximum increase in the population is the population level that produces the maximum sustainable yield. Sustainable yield is the harvest that can be removed each year. With the logistic model, the optimum strategy for maximum harvest is to maintain the population near the midpoint ($K/2$) of the growth curve. The population must remain below K-carrying capacity to provide for maximum population recruitment and subsequent harvest. Managers can plot population growth and yield curves and extrapolate to the population level that produces maximum sustainable yield (1). This population level would then become the target objective if the goal was to manage the population for maximum sustainable yield.

With many big game populations it is difficult to determine the population level that reflects the K-carrying capacity. In the intermountain west variable annual winter weather can cause considerable change in K-carrying capacity. As a result of severe winters and hunting removal, populations are frequently maintained below

Figure 10-3 Fawns, because of their small body size and lack of fat reserves, suffer the highest mortality rate each winter. Harvests designed to reduce overwintering populations minimize these losses. (Photograph by L.H. Carpenter.)

K-carrying capacity. Deer populations throughout the southeastern United States and Texas regularly reach K and remain at or near this point. Deer harvest levels are not sufficient to reduce the population below K-carrying capacity. Some deer managers have proposed regressing measures of population performance such as net recruitment or antler growth against estimated population size to determine K-carrying capacity (28, 30). Once K-carrying capacity is reached and exceeded, such population parameters should decrease.

Frequently, some measure or description of habitat quality is used to establish numbers of animals the habitat will support. A good example is use of various forest types to describe the relative ability of the environment to support white-tailed deer. In Pennsylvania, white-tailed deer goals are established for different forest stand types (31). Typically, direct measurements of vegetative carrying capacity are too complex and costly to justify. Changing habitat conditions, however, can be critical to achieving harvest objectives. The manager should integrate harvest plans with ongoing habitat and land management activities like road closures and timber harvests. Consideration should also be given to harvest strategies that address multiple species and ecosystem management (32).

Given limits of habitat and climate, the manager must select long-term "average" levels for K-carrying capacity. In environments without extreme weather variation, an average K-carrying capacity level will usually suffice. In environments with extreme weather changes from year to year, selecting for the average K-carrying capacity will result in too low of a population level in mild winters and severe winter losses in bad winters (1) (Figures 10–2 and 10–3). Experience, past histories, and judgment are obviously important to this process.

Generally, species in the order Artiodactyla are managed more for maximum sustainable yield (28). These species are known as K-selected species, and populations will increase until the biological carrying capacity is reached (28, 29). Big game species in the order Carnivora have lower reproductive potentials and population size is typically not habitat limited. Tenets of maximum sustainable yield are not as applicable with carnivores.

Carnivore species are considered more of a trophy big game animal and less as a human food source. Aspects of providing a quality hunting experience and conservative harvest strategies usually dominate in developing harvest goals for carnivores. Carnivore harvest levels are commonly determined by minimum and maximum tolerance levels of interested publics. Public values such as human safety and private property damage usually define population targets and harvest levels for these species.

Cultural Carrying Capacity. The carrying capacity established when maximum wildlife population levels are set by tolerance levels of people has been termed *cultural carrying capacity* or *wildlife acceptance capacity* (33, 35). Frequently, cultural carrying capacity is much lower than *K*-carrying capacity. Examples where cultural carrying capacity are used are elk densities in the West, white-tailed deer populations in eastern urban, park, and farming areas, mule deer in urban areas, deer and motorists, and cougars near outdoor parks and recreation areas in the West.

The wildlife manager must communicate closely with various publics as population objectives are established. It is increasingly important that the manager systematically and scientifically assesses public attitudes toward the managed wildlife species (34, 35). In many situations, these data will be as important as biological data in determining harvest management strategies.

Surplus Concept. Once desired and agreed upon population objectives are determined, the next task is to compare those objectives to estimated current population parameters. If the current population level is greater than the desired level, that difference becomes the harvestable surplus and the harvest goal. In some situations, due to extreme weather impacts or other natural mortality factors, the estimated current population may be less than the long-term population goal. When this occurs, wildlife managers must carefully consider implications of their harvest strategy decisions. Usually, because of tradition and economic considerations, the decision will be to continue with a minimum harvest. It is critical, however, that over the planning horizon (usually 3 to 5 years) progress toward the established population goal be made.

Establishing Hunting Seasons

Once the harvest goal is known, the manager must determine the framework for the hunting season to best accomplish that harvest. Increasingly, due to complexities involved in the planning and setting of hunting seasons and increased communication needs to notify hunters of changes, states prefer to establish big game hunting season frameworks for several years with only minimal changes annually.

Season Framework. Typically, public input and debates are greater when broader hunting season guidelines are established. Once broad guidelines are in place, public comment on the annual hunting season is minimal. Major factors to consider in determining hunting season frameworks include impacts on other activities like livestock grazing and timber management, conflicts with other outdoor recreationists, hunter safety, and predicted weather patterns.

Decisions must also be made on allocation of time to the various hunting interests such as archers, muzzle loaders, and rifle hunters. Season structures usually consist of identification of opening and closing dates, species combinations to be hunted, number of separate or combined seasons for each species, and broad geographical boundaries for the hunts. Season structures are established in most states by wildlife commission or board action in public meetings.

For example, in Colorado there are three separate and/or combined deer and elk hunting seasons. Opening and closing dates, regulations for participation within and among the three seasons, geographical units open for different management strategies such as quality management units (i.e., restricted participation), and decisions on antler point restrictions are approved for the 5-year period. Numerous public meetings and workshops are held to discuss advantages and disadvantages of various alternate season structure strategies. To alleviate hunter crowding in each season, fairly elaborate rules are set for season participation. The goal is to limit participation of each hunter to one of the combined seasons even if the hunter chooses to hunt both deer and elk.

Annual decisions for deer and elk hunting in Colorado are primarily related to determining number of licenses for each sex and changes in license numbers by hunting unit to improve hunter distribution. More significant changes could be made in the hunting season structure on an annual basis, if necessary.

Hunting seasons for carnivores are generally less structured and usually longer than for deer, elk, pronghorn, or mountain sheep seasons. Total harvests for black bear and cougars are more controlled by number of licenses or permits than by season length. In many states the general structure of carnivore seasons is also established for 3 to 5 years. Several states and provinces control cougar harvest with quotas (36). Quota systems require close administrative controls. The season is concluded once the desired kill per management unit is obtained.

License Type.   Another key decision for the wildlife manager is the type of license to be issued for each species hunted. The first category of license types is usually one of two types: limited in number or unlimited. If licenses are limited, some process for distribution to selected hunters is necessary. Today most states use elaborate computer systems that randomly select allocated licenses from a pool of qualified applicants. Some states offer preference systems based on previous unsuccessful attempts at securing a license. Unlimited licenses are distributed at commercial vendors and wildlife agency offices.

Another category of license type is based on sex and/or age of an animal hunted (37). For species with antlers, the category is usually designated antlered or antlerless. Obviously, this is primarily a separation of males and females with exception of young animals where both sexes would be antlerless. For species that do not have antlers, determination of sex is more difficult and license type may not specify sex (e.g., collared peccary). Pronghorn antelope are distinguishable by presence or absence of a cheek patch and horn size, and licenses are issued for both sexes. In species like mountain sheep and mountain goats, horn size and shape are used as sex distinguishing criteria. The manager must be careful not to make the distinction too complex to explain or to determine in the field.

Many states have different license types depending on hunter residence (e.g., resident or nonresident). In all states that allow nonresident hunting there are different fees for purchasing resident and nonresident licenses. In some states, there are significant additional restrictions on nonresidents such as requirements for hiring a guide. The manager must have information on expected numbers of various types of licenses to be issued to predict harvest levels, because each license type will result in varied hunter success and participation rates.

Season Length and Timing.   Impacts of season length and timing are important concerns. Considerations should deal with potential impacts on important biological activities of species to be hunted. Independence of young animals at the time of the hunt, impacts of the hunt on breeding activities, and impacts on other types of mortality must be considered (38, 39, 40, 41).

Hunter desires and preferences for timing of the season must also be evaluated (42). Very short seasons are not acceptable to hunters and should be avoided. Generally, reducing the length of a season is not a good strategy for reducing harvest. If the desired harvest is so minimal as to require only a 2- or 3-day season, the need for a harvest should be questioned. A general effect of shortening the hunting season is simply to crowd more hunters into a shorter time frame with a similar total harvest. To maintain and perpetuate the hunting tradition, it is important that quality outdoor experiences be provided. Longer, rather than shorter, seasons should be the goal of the manager. Obviously, prevailing weather patterns for the hunting areas must be considered to allow for access and safety when deciding length and timing of the season.

Opening Day Impact. No matter what day of the week hunting season opens, opening day will be the most important day of the season with regard to total hunting pressure and harvest (43, 44). Because of standard workweeks, it is traditional for big game seasons to open on Saturday so hunters can hunt both days of the weekend (45). When the season opens early in the week, hunting pressure is spread more evenly throughout the season. When the season opens on or near the weekend, the hunting pressure is more concentrated in the first few days of the season (43).

Many managers are implementing midweek openings to alleviate crowding on opening day. This is especially important for hunting areas that are close to human population centers where workers typically hunt on the weekend, then return home Sunday evening. Hunters prefer longer seasons, but traditionally only hunt for 3 to 4 days (43). The manager must be cognizant of dates of holidays that fall within hunting seasons. Holidays will increase hunter participation, especially when they fall on a Friday or Monday. Inclement weather on opening day can also significantly alter these patterns. Choice of opening day can be effective in distributing hunter pressure and hunter harvest. Daily harvest records over several hunting seasons allow the manager to construct harvest percentages by day to better predict total harvest for a given season length.

Measuring the Harvest

Measuring harvest is essential. It is necessary that sampling systems be designed to reliably estimate hunter participation, total number of harvested animals, composition of harvest by sex and age, and hunter success rates. Over time a variety of methods have been employed (46). Earlier methodology focused on either direct measures from animals checked in the field at check stations or from report cards distributed with licenses to all hunters. These methods were inefficient and costly, but more importantly did not provide a random basis for the sample (47).

Mandatory Reporting. Traditionally, mandatory registration was used as the primary method to obtain hunter harvest information (48). A common approach was a report card attached to the license requiring each hunter to complete the report and mail it to the wildlife agency at completion of the hunt.

Another form of mandatory check requires that all hunters report to a specific site once an animal is harvested. Here officials inspect and tag the animal and record necessary information. Mandatory checks on smaller management units or for units with relatively few hunters and close control of traffic are very successful. If the harvest unit is large, effective coverage of all travel roads becomes more difficult. Costs of mandatory checks for extensive hunting areas are considerable. However, mandatory field checks provide benefits for agency public relations by increasing public confidence gained from the personal contacts.

Drawbacks of mandatory reporting include low compliance rates for mailed report cards and response biases. Mandatory report cards are not based on a random sample and unless total compliance can be reached, it is essentially impossible to determine precision of the estimate. It is also difficult to determine local (i.e., management unit level) hunting pressure from report cards. If harvest estimates are desired over large geographic units (i.e., states or regions within states) mandatory field checks are not advisable. Mandatory checks for species such as mountain sheep, mountain goats, and cougars are usually effective.

Check Stations. Another approach to obtaining information on the harvest is operation of a check station (Figure 10–4). In contrast to mandatory checks, check stations generally operate on the premise of voluntary visits by the hunting public. Check stations historically have been used to gather information on big game harvests (49, 50). Strategically located check stations can provide much useful data on the harvest. They also provide excellent opportunities for law enforcement and public relations. Some check stations are operated primarily for law enforcement and may require stopping

Determining Harvest Management Goals

and checking of all traffic. It is common for law enforcement check stations to operate 24 hours a day and involve several law enforcement jurisdictions such as the sheriff and the highway patrol. If traffic flow is heavy, the logistics of operating these check stations are considerable and require detailed planning and personnel to execute safely and efficiently.

The most common use of check stations is to obtain specialized biological data from harvested animals. Data on sex, age, nutritional condition of the animals, and harvest location can be obtained. With big game animals, age of the harvested animal is commonly recorded. Most big game animals are aged by inspection and removal of the teeth (Figure 10–4). Frequently, specific body measurements or tissue samples are taken at check stations. Selected hunters may be instructed to bring the entire animal to the checkpoint where biologists will eviscerate the animal and collect samples. Interviews generate related information pertinent to the hunt such as harvest location, hunter crowding, and hunter success. In addition, hunters may be able to secure specific biological samples from their harvested animal in the field and return them to the check station if provided with clear and understandable instructions at the beginning of the hunt.

In many states check stations have become traditional places where hunters and agency personnel discuss all factors related to the hunt. This opportunity provides an excellent forum for the agency to communicate important information to the hunting public. It is also a good place to teach the hunter important biological and ecological facts about the species of concern. Check stations also are important for getting key agency personnel from different work units together to work toward a common goal. In this age of specialization and diverging efforts, this is no small accomplishment.

Check stations have been used also to compare data on the harvest obtained from different methods. Data from the check station can be used as "ground truth" information and compared to subsequent mail or phone reports by hunters to estimate sampling biases. Steinert and others compared data from a check station and telephone surveys conducted 2 months later and found that information on total harvest, game management unit of harvest, and day of harvest were reported the same for 60% to 80% of deer and elk hunters (51). The authors concluded that telephone surveys were providing valid estimates.

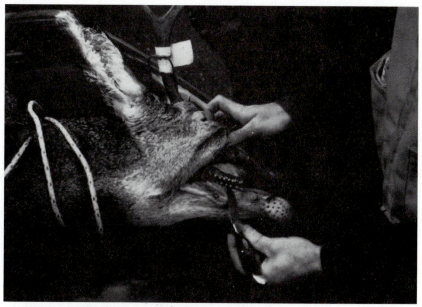

Figure 10–4 Big game check stations provide excellent opportunities for wildlife managers to collect important biological and human dimensions information. Inspection of the teeth of a hunter-killed deer at a check station provides useful information on age of the harvest. (Photograph by L.H. Carpenter.)

Limitations to check stations include absence of a basis for a random sample. This prevents construction of confidence limits on the data. As discussed previously, in some check stations, total checks are obtained so concerns about a valid sample are unnecessary. Increasing human resource costs also limit extensive use of check stations because they are labor intensive.

Mail Surveys. Due to the limitations discussed, other methodologies have been employed to measure harvest. One of the more common sampling approaches employed to estimate harvests is the mail survey (52, 53, 54). The basic requirement for a successful mail survey is a random sample of the hunting population. Samples are drawn from the record of license sales. Often, the population is stratified as to resident and nonresident and license type to minimize variances of responses. Separate surveys are often sent for each species of big game even though the hunter may have hunted more than one species. Development of computer databases has greatly simplified this process. Decisions on precision of harvest estimates necessary for management need to be made early so adequate sample sizes can be secured.

Two sources of bias exist in surveys: nonresponse bias and response bias (55). Nonresponse bias occurs when hunters fail to respond to the survey. Response bias results from a failure to report correctly the results of the hunt (51). It is important to understand source and magnitude of the bias in any survey. A frequent response bias is memory loss. Surveys mailed soon after the hunt period are recommended to alleviate this problem.

A standard mail survey includes questions on species harvested, participation in a hunt, harvest, sex and age of harvest (i.e., young/mature), management unit of harvest (map is provided), and days or dates of hunt effort. Supplementary information may be requested on topics such as rating of quality of the hunt, animals wounded but not retrieved, and total number of animals seen. Mail surveys typically have at least two follow-up contacts to determine biases of the nonrespondent. Later respondents and nonrespondents predictably have lower success rates than do earlier respondents (56). Surveyors typically target total return rates of greater than 70% to ensure adequate response data. The time required for following up contacts is a major limitation of mail surveys because results of the hunt are not known for months after the season (54).

Response to each question must be carefully analyzed and evaluated depending on wording of the questions, completeness of response, and any other biases (57, 58). A number of excellent studies on accuracy and application of mail surveys to estimate hunter harvest have been conducted (59, 60, 61, 62, 63, 64). Although most of these studies were done on waterfowl harvest estimates, many of the findings are pertinent to big game harvest surveys conducted by mail.

Telephone Surveys. Computer technology applied to telephones has ushered in the era of harvest surveys conducted by phone. Telephone surveys have several advantages. Hunters can be contacted immediately after the season, thus greatly decreasing time for results to be known. Phone numbers of nonresident hunters can be programmed to be called at the most appropriate time across multiple time zones to maximize contacts with hunters. The immediate response eliminates concern about biases of late or nonrespondents. Because of time efficiencies and available commercial equipment, costs for phone surveys are very comparable to mail surveys. Hunters missed on the initial phone contact can easily be called a second or third time. The immediacy and flexibility of a phone call also facilitate last-minute changes to questionnaire text and minimize biases due to memory loss.

Telephone surveys do, however, have several disadvantages. Problems include limited competence of contracted telephone surveyors and databases that do not list current phone numbers. Estimates of harvest location may be less accurate. Potential biases result from hunters that either do not have phones or are on schedules that minimize opportunity for contact by phone. Some concerns have also been raised about contributing to more "junk" phone calls. Well-designed and published studies to compare results of phone and mail surveys are surprisingly lacking. However, my experience in Colorado where both types of surveys have been used extensively suggests similar results from the two methodologies.

The manager must evaluate several factors before choosing the survey method. These include fiscal resources available, time constraints, and level of precision desired. Because accuracy (bias) is rarely known, quality of harvest surveys is usually evaluated by measures of precision. Generally, harvest estimates for large geographical areas have an acceptable level of precision. However, precision of harvest estimates can be poor for smaller geographical units where sample sizes are considerably less. The wildlife manager must target the geographical level where management decisions will be made and the level of management precision needed. Then the survey sample is designed accordingly.

Data Analysis.　Once data are obtained, processes must be in place for analysis, summary, and reporting. Obviously, analyses will vary depending on the type of data. However, there are some standard "pitfalls" that the manager must be aware of and avoid.

Reporting biases.　A major concern is with biases inherent in the method used to obtain information. If known, corrections for bias must be made. Nonresponse, response, and survey biases must be suspected. Fortunately, bias with a given method tends to be consistent from year to year and it is feasible to correct the data accordingly. Unfortunately, degree of bias is frequently unknown. Thus, the wildlife manager simply must be aware that harvest data are prone to biases.

Because truth is rarely known, it is important that the manager strive to compare methodologies whenever possible. The manager must be aware, however, that although two methods provide the same answer it does not mean both are correct and without bias. If the answer is significantly different, it does alert the manager to potential problems and frequently highlights areas for further investigation.

Differential vulnerability.　The hunter harvest may not represent the population from which it was removed because of differential vulnerability of individual animals to hunting. The manager must be careful in drawing conclusions from the sex and age composition of the harvest. Factors like hunter behavior, animal behavior, and weather influences may affect vulnerability (65, 66, 67). Specific hunting regulations in place such as antler point restrictions will also affect hunter selectivity (68, 69). Because antler point restrictions for cervids are used to protect specific age cohorts, it should not be surprising that the harvest sample does not represent the population.

Sample size.　A common problem in analysis of harvest data is insufficient sample sizes for the parameter to be estimated. This is especially true for projections of harvest estimates for the smallest management unit. Typically, harvest estimates are generated from statistical regressions that project total harvests from a sample. When survey samples are randomly drawn, management units with the highest density of hunters make up the majority of the samples. Smaller units and units with low hunter density will be poorly represented in the sample and resulting projections will be relatively poor (i.e., have low precision). The problem is exacerbated with estimates of age or sex categories that are not prominent in the sample. The manager must realize these limitations of the data when making harvest decisions.

Unretrieved harvest.　Another factor associated with analysis of harvest data is estimating unretrieved harvest (70). Frequently, this is referred to as wounding loss or death of animals shot, but not retrieved. It may also include an estimate of poaching or illegal kill. Poaching in this sense is defined as kill of the species of concern by people who do not hold appropriate licenses. These deaths are obviously not included in a harvest sampling scheme and must be accounted for. It is known that hunting regulations can impact wounding loss considerably. For instance, antler point restrictions on cervids typically increase unretrieved kills because of greater difficulty in determining a legal animal (69, 71). Estimates of illegal kill are essentially unknown for big game populations but some studies have suggested it may be as large as 50% of legal harvest (70).

Unretrieved loss is typically estimated as a percentage of the legal harvest and added to total harvest. Values of 20% to 30% of the legal harvest are commonly used for deer harvests (72). Values of 10% to 15% would be reasonable for species with lower population densities and lower hunting pressure. Wounding loss will vary by species and sex of animal, hunter density, weapon, and vegetative cover type.

Often when the manager is constructing a big game population model, values for wounding loss and illegal kill are varied widely to get the model to perform. This is indicative of the variable and largely unknown parameter being estimated. Measurements of mortality (other than legal harvest) are desirable but seldom collected because mortality data are difficult to obtain. Natural annual mortality by age and sex class is usually estimated and varied with severity of winter or drought.

Evaluation. Once data are complete and analyzed, managers need to determine if objectives of the harvest were met. Did the total harvest reach numbers necessary to approach the population objective? Were the sex and age components of the harvest near predicted levels? Did the precision of estimates meet expectations? All aspects of the hunt should be evaluated. Factors such as hunter participation, weather during the hunt, animal distribution, and hunter success must be considered. If objectives were met, the manager is well prepared for the next cycle. However, if hunt objectives were not met, the manager must identify reasons why and take corrective steps for future hunts.

Meeting harvest objectives is paramount to keeping a big game population in balance with carrying capacity. If objectives are not met, then typically it becomes harder to meet increased objectives the following year. In many situations, credibility of the wildlife agency rides on the successful conduct of big game hunting seasons. Wildlife managers should be forthright with findings and disclose results to the public.

POPULATION MANIPULATION USING HARVESTS

Harvest strategies provide the manager with tools to manipulate the big game population to achieve stated objectives. Recreational harvest management for big game populations remains a very high profile activity in North America. The high public interest and importance of these hunts require responsible and professional management.

Population Density

The most common objective addressed through hunting is population density. Frequently, the objective is simply to reduce the total population from the estimated pre-season size to the agreed upon post-season size. In some situations, the harvest objective may be minimal, which requires a conservative harvest strategy involving male-only hunts. In situations with a larger harvest objective, it may be necessary to implement additional strategies such as liberal female hunts, additional seasons, and higher bag limits.

Closely controlling hunter harvest is a difficult task (73). If the manager has control over total number and type of licenses issued, predictability of harvest is much greater (74). Harvests of mule deer could be predicted when regulated by either-sex permits but could not be predicted with unrestricted harvests of antlered-only licenses (74). Obviously, the most predictable situation for cervid harvests would be with controlled numbers of antlered and antlerless licenses. Data on total licenses issued, hunter participation rates, and hunter success rates from previous hunts allow accurate prediction of total harvest.

Difficulty in estimating numbers of large carnivores typically results in more conservative hunting seasons and tighter control of license numbers than for ungulates. In addition, variable vulnerability to hunting method by sex and fluctuating population levels complicates predictions of the total harvest (36). Recent research on cougars in New Mexico also points out the importance of understanding the social organization of cougars prior to removal of animals from the population (75).

The most accurate licensing protocol for the large carnivore is the quota system (36). The quota system requires that the season end once the targeted harvest is achieved. However, quota systems require considerably more administration and the manager must be adequately prepared to implement such a system. The wildlife manager for large carnivore populations must also plan for unexpected deaths resulting from animals being killed for wildlife damage reasons. For the program to be successful the wildlife manager should receive all wildlife damage kill records from other state and federal agencies and from local landowners.

Sex and Age Composition

Often, there is no specific objective for sex and age composition of the harvest. The manager typically relies on past harvest records to predict harvest composition. Depending on the hunt and type of licenses offered, however, the resulting sex ratios in the remaining population can be different from those before harvest, and be either a positive or negative result for future management.

In recent years in the West, the typical cervid hunting regime has been heavy hunting pressure on the antlered segment of the population and reasonably controlled harvests on the antlerless population. In some states, this pressure on the male segment has been exacerbated by a policy of unrestricted or very high numbers of antlered licenses. This policy results in tremendous hunting pressure on males and low post-season male sex ratios. In certain game management units and after hard winters in Colorado, it was common for post-season bull elk ratios to be less than 10 bulls per 100 cows and for mule deer ratios to be less than 5 per 100 does (69). The problem was further complicated because typically with low ratios only 1 or 2 of the 10 males were mature (older than 2.5 years old).

Unbalanced sex ratios present several problems and concerns to the wildlife manager. A major concern relates to the reproductive potential of a population with such skewed sex ratios (76). An inadequate number of males may cause females to be bred during second or third estrous cycles, resulting in later births and lower survival of young. Behavior characteristics associated with the male age structure may also affect reproduction. Conception dates of elk in Oregon occurred earlier as bulls in the controlled population within the Starkey enclosure grew older (76) The rut became more synchronous and shortened from 71 days to 41 days when the bulk of breeding changed from yearling bulls to 5-year-old bulls (76). Pregnancy rates also increased as bull age increased. Other workers have discussed importance of age of bulls to breeding and reported conflicting conclusions (39, 77, 78). Impact on reproduction from low sex ratios in mule deer is not well understood. In Colorado, recruitment rate of fawns remained comparable in areas with a wide variance in buck ratios. Factors such as nutrition of the female during the breeding cycle appear to complicate this relationship.

The impact of low buck and bull numbers on hunter satisfaction is clear. Complaints from hunters, guides, and outfitters are common. This concern is soon transmitted to members of the wildlife commission or the state legislature. In typical political fashion these groups promulgate regulations or legislation to "fix" the problem. A typical fix for low deer and elk sex ratios in the West in the 1980s and 1990s has been the implementation of antler point restrictions. Antler point restrictions define legal animals by minimum number of points. For instance, in Colorado, a legal deer under antler point restrictions is a buck with at least three points on one antler. For elk, the minimum is four points. These regulations target bulls and bucks that are at least 2 years old. Even though past experience has demonstrated that antler point restrictions are generally not good fixes, they are a popular initiative (78).

Antler point restrictions increase illegal and unretrieved kill. In many habitats determination of legal animals by counting the number of points is difficult. Without reductions in the total number of hunters, the increased hunting pressure on legal males is intense and typically results in fewer, older animals surviving the hunting season. If sufficient escape cover or large blocks of private land that can serve as refuges are available, antler point restrictions can help populations recover from low sex ratios. Antler point restrictions should be used as are other "tools" in the manager's tool box. They should not be used as blanket prescriptions for treatment of low sex ratios. The best "fix" for low sex ratios is reduced hunting pressure on the male segment of the herd. The responsible manager should recognize and correct low male ratios before they become controversial and difficult political problems.

In carnivore populations, method of take and timing of the hunt influence sex and age of the harvest (36). Commonly, with large carnivores, especially bears, the management objective is to minimize harvest of adult females. To plan for future harvests, the manager should construct databases identifying method of harvest, date of harvest, and sex and age of the kill. Mandatory check initiatives are very helpful in this effort.

Animal Distribution

Big game harvests increasingly address distribution of animals, especially as related to ungulates and private land damage (10, 11). In several states with mixes of public and private lands it is also speculated that proliferation of earlier hunting seasons (i.e., archery and muzzle loading) has contributed to movement of big game animals from public lands onto private land (79).

This movement phenomenon, whether real or imagined, creates problems for the wildlife manager when establishing big game seasons. Private land owners become very vocal about timing of seasons and sportsmen become concerned that populations of wildlife available for hunting on public lands will be reduced. Proponents of the earlier seasons argue they are not the cause of the problem. The wildlife manager and public land manager are caught in the middle.

The growing reality of increased trespass fees for hunting on private lands and the growing closure of other wild lands to hunting by large land owners have contributed to this problem by creating refuges for deer and elk. Animals congregate on these refuges soon after hunting begins on adjacent public lands. The result is a much reduced harvest and inability to reach stated harvest objectives. The responsible manager must continue to maintain and develop hunting access and work aggressively to develop hunting programs on private lands. Consideration of animal distribution and movements should be part of the analysis for planning a big game harvest (80).

MANAGING HARVEST REGULATIONS

There is a paucity of literature on the regulation process, regulation impacts, and regulation evaluation (1). Perhaps this is because of the wide diversity of organization structures, public review processes, agency authorities, wildlife boards, wildlife commissions, and state governments involved in the establishment of wildlife regulations. Denney and Mohler and Toweill discuss various regulations and provide insights into how regulations can achieve stated harvest objectives (81, 82). McCullough discusses the relationship of various regulations to the harvest of white-tailed deer with specific reference to why a bucks-only regulation may not be the best way to achieve maximum sustainable yield (29).

Regulatory Authority and Process

Authority for establishing regulations is given in state and federal statutes. States have statutes that provide authority for establishing hunting regulations. In addition, a number of federal statutes provide legal authority for hunting regulations. All states have detailed legal processes for adoption of regulations. Typically these are described in administrative procedures acts. Administrative acts include guidelines for due process and public review and comment. All adopted regulations must be supported by appropriate state or federal legislation. New regulations are routinely reviewed by the office of the state attorney general to verify this requirement.

Agency Review. To be effective, harvest regulations must have the support of the entire agency. Law enforcement personnel, biologists, and public information specialists must be involved in developing regulations. Regulations must be designed to accomplish a purpose, must be easily understood by the public, and must be enforceable. To accomplish these needs, the agency must have a well-defined regulation development process. Typically, regulations are developed at the grassroots level to address a management problem or opportunity and then forwarded through the process to regional and statewide levels. At each step the regulation should be evaluated as to its purpose and probability of success. During development input should be requested from other state and federal agencies and key publics. Input at earlier stages can prevent delays later in the process.

Public Review. Adoption of big game hunting regulations attracts many players to the table. A common theme facing wildlife agencies as they develop big game harvest regulations is how to "share the wealth." During the past several years, the growth of archery and muzzle loader hunting has intensified discussions on how best to allocate the limited resource among the competing interests. It is helpful if these basic policy issues have been discussed and a decision reached prior to the annual regulation process. These decisions usually are not based on biology and simply reflect public demand or philosophy of the agency or the wildlife commission.

Today, the wildlife manager must be responsive to a wide range of public interest groups including those who do not hunt and those who oppose hunting. Wildlife agencies have broad mandates to represent all wildlife and all publics. Frequently, these broad public interests present challenges to the regulatory process (see Chapter 11).

The wildlife manager should recognize and communicate with all interest groups as hunting regulations are developed (Figure 10–5). A good example of communication occurred a few years ago as Colorado began adopting moose hunting seasons. Thirty-six moose were introduced into North Park and the Laramie River drainage in 1978 and 1979 (83). Moose soon increased their numbers, expanded their range extensively, became a large attraction for tourists and wildlife-watching enthusiasts, and became a harvestable resource. They also had a propensity to browse near roads in one critical area of a moose habitat. This combination created a conflict between hunters and moose watchers.

The Colorado Division of Wildlife working with the State Forest Service agreed to close two hunting areas near main roads in the area of conflict. This solution appears to be working and all interest groups are satisfied. To be prepared for conflicts like this, the wildlife agency should develop systematic and scientific processes to gather information on human dimensions aspects of issues. Interested publics must understand they are welcome to participate in decision making and that they have effective methods for expressing their interests. It is also critical that the agencies be responsive to public desires once they are communicated.

Wildlife Commission or Board Review. Wildlife commissions or wildlife boards establish policy and adopt hunting regulations in most states. These boards serve as buffers between the public and the state wildlife agencies. Board members are

Figure 10–5 All interested publics should be invited and involved in the development of harvest regulations. Forums to encourage public participation are encouraged. (Photograph by D. Guynn, Management Assistance Team, USFWS.)

typically appointed by the governor and serve 2- to 4-year terms. State wildlife agencies formulate necessary regulations to implement stated harvest objectives and bring recommendations to the wildlife commission for adoption.

Wildlife commissions are primarily in place to receive public input. For adoption of many wildlife regulations, states use a two- or three-step process. Typically, the first step is one of introducing the proposed regulation in the draft stage and asking for public input at a wildlife commission meeting. After input is received, the agency prepares a final draft of the regulation for adoption. The wildlife commission then accepts additional public comment at its next meeting before adopting the final regulation.

Wildlife commission meetings dealing with big game regulations are frequently controversial. People care deeply about big game animals and their management. Wildlife commissions should accept input from all interested publics. If people become convinced the commission is not listening to their concerns, they will find other avenues such as ballot initiatives to address their needs. Generally, this approach is not beneficial to sound, scientific wildlife management. The wildlife manger can facilitate this process by developing regulations that are needed, worded well, and supported by most affected parties.

FUTURE CONSIDERATIONS

The future of harvest management for large mammals will depend on how well managers maintain and protect habitats and wild places for animals to live and for hunters to hunt. It will also depend on how well big game managers address the increasingly complex demands that society expects from wildlife management programs. Technological developments have provided the manager with more tools and information than ever before to address biological questions. On the other hand, technological developments as applied to hunter equipment have presented the manager with a dilemma in determining what constitutes fair chase. The challenge will be to openly discuss these issues, make timely decisions, communicate those decisions to all interested people, and develop responsible and scientific management programs. The need to scientifically manage these large mammals and their habitats will only increase in the future.

Finally, wildlife managers should beware of the "trophy" syndrome. Numerous studies have shown that hunting for the primary purpose of obtaining food is supported by a majority of the people. However, hunting designed primarily for purposes of producing trophies has much less support (84, 85, 86). Big game management should be designed with all users in mind.

SUMMARY

Implementing harvest systems for large mammals has become complex. The wildlife manager must consider conflicting values and desires when setting harvest goals. Input from a wide spectrum of publics is necessary as harvest management goals and objectives are formulated. Harvests on private lands should be integrated into the overall harvest plan. Sampling systems to estimate numbers of animals and some measure of habitat carrying capacity are necessary to establish scientifically-defensible harvest objectives. Most harvest systems utilize concepts of biological carrying capacity to arrive at the harvestable surplus. In many situations, tenets of cultural carrying capacity or wildlife acceptance capacity apply and the maximum population level tolerated by people may be lower than the biological carrying capacity.

A variety of season and license types have been used to accomplish harvest objectives. Generally, hunting season approaches for ungulates differ from those for the large carnivores. Ungulate hunting seasons typically employ concepts of maximum sustainable yield and harvest levels are higher (as a percentage of the total population) than are harvests for carnivores. Typically factors such as human property loss, trophy values, and human social values are important in determining carnivore harvest goals. Hunting season length and timing are important considerations in obtaining harvest objectives. Hunter interests and satisfaction are important considerations in season setting decisions. Very short seasons should be avoided to minimize hunter crowding. Longer seasons maximize hunter enjoyment.

Harvest data are some of the most important data gathered on big game populations. Computer models for populations are commonly developed to project harvest impacts. The importance of harvest data to these models requires that data be both accurate and precise. Sound sampling methods must be employed to obtain harvest results. Several methods are used to obtain harvest data. Field check stations are important for obtaining specific biological data and for public relations. Intensive telephone surveys are becoming more popular than surveys by mail because results can be obtained soon after hunting seasons are complete.

Analysis of harvest data is important. The manager must be aware of standard types of biases that frequent harvest data sets. These include reporting biases and differential vulnerability to hunting of various age and sex classes in the population. The level of unretrieved harvest is an important, but difficult to measure, parameter and is usually estimated as a standard percentage of the legal harvest (typically 10% to 15%). Evaluation of the harvest obtained as compared to the harvest objective is critical. If harvest objectives are not met, the harvest system needs to be modified.

Harvest strategies are designed to provide the manager with tools to manipulate populations to achieve stated objectives. The most common objective addressed is population density. Altering the sex and age composition of the population is another common objective. Increasingly, hunting strategies are being designed to alter animal distribution. This is especially important in many areas of the West where big game mammals cause property damage to private lands.

Developing and implementing effective harvest regulations are critical to the successful harvest program. Effective regulations are developed through public involvement processes. Harvest regulations must be understood and enforceable. Wildlife commissions or boards are typically the responsible group for adopting hunting regulations. These boards must accept inputs from all interested publics. The manager designing harvest systems must be sensitive to the concerns that many people have toward the trophy syndrome. Most people strongly support hunting systems designed to provide food for people. Much less support is available for hunting designed to provide trophies.

LITERATURE CITED

1. STRICKLAND, D. M., H. J. HARJU, K. R. McCAFFERY, H. W. MILLER, L. M. SMITH, and R. J. STOLL. 1994. Harvest management. Pages 445–473 *in* T. A. Bookout, ed., *Research and Management Techniques for Wildlife and Habitats,* 5th ed. The Wildlife Society, Bethesda, MD.

2. LEOPOLD, A. 1920. Determining the kill factor for black-tail deer in the Southwest. *Journal of Forestry* 18:131–134.

3. LEOPOLD, A. 1933. *Game Management.* Charles Scribner's Sons, New York.

4. RUSSO, J. P. 1964. The Kaibab North deer herd: Its history, problems and management. Wildlife Bulletin No. 7. Arizona Game and Fish Department, Phoenix, AZ.

5. PURDY, K. G., and D. J. DECKER. 1989. Applying wildlife values information in management: The wildlife attitudes and values scale. *Wildlife Society Bulletin* 17:494–500.

6. LOKER, C. A., and D. J. DECKER. 1995. Colorado black bear hunting referendum: What was behind the vote? *Wildlife Society Bulletin* 23:370–376.

7. CAUGHLEY, G. 1985. Harvesting of wildlife: Past, present, and future. Pages 3–14 *in* S. L. Beasom and S. F. Roberson, eds., *Game Harvest Management.* Caesar Kleberg Wildlife Research Institute, Kingsville, TX.

8. U.S. DEPARTMENT OF THE INTERIOR. FISH AND WILDLIFE SERVICE AND U.S. DEPARTMENT OF COMMERCE, BUREAU OF THE CENSUS. 1996. National survey of fishing, hunting, and wildlife-associated recreation.

9. POTTER, D. R. 1982. Recreational use of elk. Page 529 *in* J. W. Thomas and D. E. Toweill, eds., *Elk of North America.* Stackpole Book, Harrisburg, PA.

10. NIELSEN, D. B., D. D. LYTLE, and F. WAGSTAFF. 1985. Who gains (or loses) when big-game uses private land? *Utah Science* 46:48–51.

11. LACEY, J. R., K. JAMTGAARD, L. RIGGLE, and T. HAYES. 1993. Impacts of big game on private land in southwestern Montana: Landowner perceptions. *Journal of Range Management* 46:31–37.

12. ADKINS, R. J., and L. R. IRBY. 1994. Private land hunting restriction and game damage complaints in Montana. *Wildlife Society Bulletin* 22:520–523.

13. IRBY, L. R., W. E. ZIDACK, J. B. JOHNSON, and J. SALTIEL. 1996. Economic damage to forage crops by native ungulates as perceived by farmers and ranchers in Montana. *Journal of Range Management* 49:375–380.

14. ARHA, K. 1996. Sustaining wildlife values on private lands: A survey of state programs for wildlife management on private lands in California, Colorado, Montana, New Mexico, Oregon, Utah, and Washington. *Transactions of the North American Wildlife and Natural Resources Conference* 61:267–273.

15. WALTERS, C. J., and P. J. BANDY. 1972. Periodic harvest as a method of increasing big game yields. *Journal of Wildlife Management* 36:128–134.

16. LOKER, C. A., D. J. DECKER, R. B. GILL, T. D. I. BECK, and L. H. CARPENTER. 1994. The Colorado black bear hunting controversy: A case study of human dimensions in contemporary wildlife management. HDRU Series No. 94-4. Cornell University Human Dimensions Research Unit, Ithaca, NY.

17. KUFELD, R. C., J. H. OLTERMAN, and D. C. BOWDEN. 1980. A helicopter quadrat census for mule deer on Uncompahgre Plateau, Colorado. *Journal of Wildlife Management* 44:632–630.

18. BOWDEN, D. C., A. E. ANDERSON, and D. E. MEDIN. 1984. Sampling plans for mule deer sex and age ratios. *Journal of Wildlife Management* 48:500–509.

19. NEAL, A. K., G. C. WHITE, R. B. GILL, D. F. REED, and J. H. OLTERMAN. 1993. Evaluation of mark–resight model assumptions for estimating mountain sheep numbers. *Journal of Wildlife Management* 57:436–450.

20. JENSEN, A. L. 1996. Density-dependent matrix yield equation for optimal harvest of age-structured wildlife populations. *Ecological Modeling* 88:125–132.

21. LANCIA, R. A., K. H. POLLOCK, J. W. BISHIR, and M. C. CONNER. 1988. A white-tailed deer harvesting strategy. *Journal of Wildlife Management* 52:589–595.

22. MCCULLOUGH, D. R. 1996. Spatially structured populations and harvest theory. *Journal of Wildlife Management* 60:1–9.

23. GROSS, J. E., J. E. ROELLE, and G. L. WILLIAMS. 1973. Program ONEPOP and information processor: A systems modeling and communications project. Colorado Cooperative Wildlife Research Unit, Colorado State University, Fort Collins, CO.

24. POJAR, T. M. 1977. Use of a population model in big game management. *Proceedings of the Western Association of State Game & Fish Commissioners* 57:82–92.

25. BARTHOLOW, J. 1992. POP-II system documentation. IBM-PC Version 7.00. Fossil Creek Software, Fort Collins, CO.

26. DASMANN, R. F. 1964. *Wildlife Biology.* John Wiley and Sons, New York.

27. CAUGHLEY, G. 1970. What is this thing called carrying capacity? *Ecology* 51:53–72.

28. MCCULLOUGH, D. R. 1979. *The George Reserve Deer Herd: Population Ecology of a K-Selected Species.* University Michigan Press, Ann Arbor.

29. MCCULLOUGH, D. R. 1984. Lessons from the George Reserve, Michigan. Pages 211–242 *in* L. K. Halls, ed., *White-Tailed Deer: Ecology and Management.* Stackpole Books, Harrisburg, PA.

30. MOEN, A. N., C. W. SEVERINGHAUS, and R. A. MOEN. 1986. *Deer CAMP: Computer-Assisted Management Program Operating Manual and Tutorial.* Corner Brook Press, Lansing, NY.

31. DIEFENBACH, D. R., W. L. PALMER, and W. K. SHOPE. 1997. Attitudes of Pennsylvania sportsmen towards managing white-tailed deer to protect the ecological integrity of forests. *Wildlife Society Bulletin* 25:244–251.

32. DECALESTA, D. S., and S. L. STOUT. 1997. Relative deer density and sustainability: A conceptual framework for integrating deer management with ecosystem management. *Wildlife Society Bulletin* 25:252–258.

33. ELLINGWOOD, M. R., and J. V. SPIGNESI. 1986. Management of an urban deer herd and the concept of cultural carrying capacity. *Transactions of the Northeast Deer Technical Committee, Vermont Fish and Game Department* 22:42–45.

34. BROWN, T. L., and D. J. DECKER. 1979. Incorporating farmers' attitudes into management of white-tailed deer in New York. *Journal of Wildlife Management* 43:236–239.

35. DECKER, D. L., and K. G. PURDY. 1988. Toward a concept of wildlife acceptance capacity in wildlife management. *Wildlife Society Bulletin* 16:53–57.

36. ROSS, I. P., M. G. JALKOTZY, and J. R. GUNSON. 1996. The quota system of cougar harvest management in Alberta. *Wildlife Society Bulletin* 24:490–494.

37. STEWART, R. R. 1985. A sex and age-selective harvest strategy for moose management in Saskatchewan. Pages 229–238 *in* S. L. Beasom and S. F. Roberson, eds., *Game Harvest Management.* Caesar Kleberg Wildlife Research Institute, Kingsville, TX.

38. LINDZEY, F. G. 1981. Denning dates and hunting seasons for black bears. *Wildlife Society Bulletin* 9:212–216.

39. SQUIBB, R. C. 1985. Mating success of yearling and older bull elk. *Journal of Wildlife Management* 49:744–750.

40. CASSCLES, K. M. 1992. Effects of harvest strategies on breeding season timing and fecundity of white-tailed deer. Thesis. Mississippi State University, Starkville.

41. DUSEK, G. L., A. K. WOOD, and S. T. STEWART. 1992. Spatial and temporal patterns of mortality among female white-tailed deer. *Journal of Wildlife Management* 56:645–650.

42. BOYLE, K. J., R. L. DRESSLER, A. G. CLARK, and M. F. TEISL. 1993. Moose hunter preferences and setting season timings. *Wildlife Society Bulletin* 21:498–504.

43. MURPHY, D. A. 1965. Effects of various opening days on deer harvest and hunting pressure. *Proceedings of the Annual Conference of Southeastern Association of Game and Fish Commissioners* 19:141–146.

44. CARDOZA, J. E. 1972. Game population trend and harvest survey: Hunter utilization of wildlife management areas. Federal Aid Project W-035-R, Work Plan 5, Job 1. Massachusetts Division of Fisheries and Game, Boston, MA.

45. HUNTER, G. N., 1957. The techniques used in Colorado to obtain hunter distribution. *Transactions of the North American Wildlife and Natural Resources Conference* 22:584–598.

46. BATE, L. J., E. O. GARTON, and R. K. STEINHORST. 1995. Audit of big game harvest surveys in Idaho and the western United States. Report submitted to Idaho Department of Fish and Game. University of Idaho, Moscow.

47. HAWN, L. J., and L. A. RYEL. 1969. Michigan deer harvest estimates: Sample surveys versus a complete count. *Journal of Wildlife Management* 33:871–880.

48. ANEY, W. W. 1974. Estimating fish and wildlife harvest, a survey of methods used. *Proceedings of the Western Association of Game and Fish Commissioners* 54:70–79.

49. ROGERS, G. E. 1953. Function and operation of big game check stations in Colorado. *Journal of Wildlife Management* 17:256–267.

50. SOWLS, L. K. 1961. Hunter-checking stations for collecting data on the collared peccary (*Peccary tajacu*). *Transactions of the North American Wildlife Conference* 26:497–505.

51. STEINERT, S. F., H. D. RIFFEL, and G. C. WHITE. 1994. Comparisons of big game harvest estimates from check station and telephone surveys. *Journal of Wildlife Management* 58:335–340.

52. TURNER, D. W. 1970. Mail surveys of hunting—Precision and sample size. *Proceedings of the Annual Conference of Southeastern Association of Fish and Wildlife Agencies* 24:292–303.

53. KANUK, L., and C. BEVENSON. 1975. Mail surveys and response rates: A literature review. *Journal of Market Research* 12:440–453.

54. GREEN, A. W., and J. C. BARRON. 1980. An adjustment for non-response bias in a mail-out game harvest survey. *Proceedings of the Annual Conference of Southeastern Association of Fish and Wildlife Agencies* 33:123–126.

55. MACDONALD, D., and E. G. DILLMAN. 1968. Techniques for estimating non-statistical bias in big game harvest surveys. *Journal of Wildlife Management* 32:119–129.

56. BARKER, R. J. 1991. Nonresponse bias in New Zealand waterfowl harvest surveys. *Journal of Wildlife Management* 55:126–131.

57. FILLION, F. F. 1981. Importance of question wording and response burden in hunter surveys. *Journal of Wildlife Management* 45:873–882.

58. WHITE, G. C. 1993. Precision of harvest estimates obtained from incomplete responses. *Journal of Wildlife Management* 57:129–134.

59. ATWOOD, E. L. 1956. Validity of mail survey data on bagged waterfowl. *Journal of Wildlife Management* 20:1–16.

60. WRIGHT, V. L. 1978. Causes and effects of biases on waterfowl harvest estimates. *Journal of Wildlife Management* 42:257–262.

61. RASMUSSEN, G. P., and P. P. MARTIN. 1981. Big game management investigations: Feasibility and cost effectiveness of a mail survey to obtain biological data from white-tailed deer taken during the big game hunting season. Federal Aid Project Report W-089-R-25, Study 11, Job 13. New York Division of Fish and Wildlife, Albany, NY.

62. GEISSLER, P. H. 1990. Confidence intervals for the federal waterfowl harvest surveys. *Journal of Wildlife Management* 54:201–205.

63. KUFELD, R. C. 1991. Accuracy of mule and white-tailed deer harvest reported by deer hunters on mailed questionnaire surveys. Federal Aid Project Report W-153-R-4, Work Plan 2, Job 8. Colorado Division of Wildlife, Denver, CO.

64. PENDLETON, G. W. 1992. Nonresponse patterns in the federal waterfowl hunter questionnaire survey. *Journal of Wildlife Management* 56:344–348.

65. ROSEBERRY, J. L., and W. D. KLIMSTRA. 1974. Differential vulnerability during a controlled deer harvest. *Journal of Wildlife Management* 38:499–507.

66. COE, R. J., R. L. DOWNING, and B. S. MCGINNES. 1980. Sex and age bias in hunter-killed white-tailed deer. *Journal of Wildlife Management* 44:245–249.

67. DOWNING, R. L. 1981. Deer harvest sex ratios: A symptom, a prescription, or what? *Wildlife Society Bulletin* 9:8–13.

68. BOYD, R. J., and J. P. LIPSCOMB. 1976. An evaluation of yearling bull elk hunting restrictions in Colorado. *Wildlife Society Bulletin* 4:3–10.

69. CARPENTER, L. H., and R. B. GILL. 1987. Antler point regulations: The good, the bad and the ugly. *Proceedings of the Annual Conference of Western Association Game and Fish Commissioners* 67:94–107.

70. EBERHARDT, L. 1960. Estimation of vital characteristics of Michigan deer herds. Report No. 2282. Michigan Department of Conservation, Game Division, Lansing, MI.

71. WINCHELL, C. S., S. G. BUCK. 1992. Illegal harvest of spike bucks during a regulated mule deer hunt. *California Fish and Game* 78:153–159.

72. STORMER, F. A., C. M. KIRKPATRICK, and T. W. HOEKSTRA. 1979. Hunter-inflicted wounding of white-tailed deer. *Wildlife Society Bulletin* 7:10–16.

73. CONNOLLY, G. E. 1981. Limiting factors and population regulation. Pages 245–285 *in* O. C. Wallmo, ed., *Mule and Black-Tailed Deer of North America.* University of Nebraska Press, Lincoln.

74. FREDDY, D. J. 1982. Predicting mule deer harvest in Middle Park, Colorado. *Journal of Wildlife Management* 46:801–806.

75. LOGAN, K. A., L. L. SWEANOR, T. K. RUTH, and M. G. HORNOCKER. 1996. Cougars of the San Andres Mountains, New Mexico. Project W-128-R Final Report. Hornocker Wildlife Institute, Moscow, ID.

76. NOYES, J. H., B. K. JOHNSON, L. D. BRYANT, and S. L. FINDHOLT. 1996. Effects of bull age on conception dates and pregnancy rates of cow elk. *Journal of Wildlife Management* 60:508–517.

77. FOLLIS, T. B. 1972. Reproduction and hematology of the Cache elk herd. Publication 72–78. Utah Division Wildlife Resources, Salt Lake City, UT.

78. HARPER, J. A. 1985. *Ecology and Management of Roosevelt Elk in Oregon,* rev. ed. Oregon Department of Fish and Wildlife, Portland, OR.

79. CONNER, M. M., 1999. Elk movement in response to early-season hunting in the White River area. Ph.D. dissertation, Colorado State University, Fort Collins, CO.

80. RUDD, W. J., A. L. WARD, and L. L. IRWIN. 1983. Do split hunting seasons influence elk migrations from Yellowstone National Park? *Wildlife Society Bulletin* 11:328–331.

81. DENNEY, R. N. 1978. Managing the harvest. Pages 395–408 *in* J. L. Schmidt and D. L. Gilbert, eds., *Big Game of North America: Ecology and Management.* Stackpole Books, Harrisburg, PA.

82. MOHLER, L. L., and D. E. TOWEILL. 1982. Regulated elk populations and hunter harvests. Pages 561–597 *in* J. W. Thomas and D. E. Toweill, eds., *Elk of North America.* Stackpole Books, Harrisburg, PA.

83. KARNS, P. D. 1997. Population distribution, density and trends. Page 129 *in* A. W. Franzmann and C. C. Swartz, eds., *Ecology and Management of the North American Moose.* Smithsonian Institution Press, Washington, DC.

84. KELLERT, S. D. 1980. Public attitudes toward critical wildlife and natural habitat issues. Phase 1 of United States Fish and Wildlife Service Study. Government Printing Office, Washington, DC.

85. THOMAS, J. K., and C. E. ADAMS. 1989. The public and Texas wildlife. Texas Agriculture Experiment Station, Texas A&M University, College Station.

86. MILLER, S. E. 1992. Public attitude survey. Inservice Document 5 (791). Ohio Department Natural Resources, Division of Wildlife, Columbus, OH.

CHAPTER

11

Human Dimensions and Conflict Resolution

Michael E. Patterson, Dwight E. Guynn, and David C. Guynn, Jr.

INTRODUCTION: PROGNOSIS FOR CONFLICT

Most important natural resource issues ultimately are resolved in the political arena because they center around conflicting value systems rather than more objective, fact-based questions. Wildlife managers have increasingly felt the weight of political pressures due to increasing public interest in issues such as endangered species, urban deer population control, and hunting in general. Since the late 1970s there have been important social changes in North America, but a professional reluctance to deal with these changes has produced a basis for serious conflicts in the stewardship of wildlife resources.

Demographic trends suggest that the general public will be increasingly isolated from traditional uses of wildlife and the outdoors (1). These trends include an increasing segment of the population living in urban areas (35% in 1900 to 78% in 1990). About 25% of households now contain only one parent, up from 9% in 1960. The population is aging and the number of people older than 60 years is expected to double in the next century (2). These changes forewarn a likely decline in wildlife recreation, especially hunting.

These social changes are manifesting themselves at a time when institutions and government agencies are in flux. Downsizing and reorganization are rampant due to public distrust and expectations of incompetence. This phenomenon is particularly important to management of large mammals because the stewardship of wildlife resources in the United States is the charge of federal and state agencies.

A four-stage cycle depicts the current dilemma of many public institutions including those agencies responsible for wildlife resources (3). Most agencies were formed between 1900 and 1950 when government programs were viewed as a good way to serve the needs of the people (4). The inception of an agency is based on perceived need that is in the public interest (stage 1). State game and fish agencies were formed to protect and replenish wildlife primarily to provide recreational opportunities due to the concerns of sportsmen and others interested in protection of wildlife. Their initial goals were to regulate harvest and restore populations to suitable habitats. After its inception, an agency goes through a period of false starts and misfires while it searches for an identity (stage 2). Passage of the Pittman-Robertson Act in 1937 provided funding and defined a process that greatly aided the state agencies in developing programs. From 1940 to about 1970, the agencies grew and matured with a growing body of scientific information and techniques that all but assured desired outcomes (stage 3). With widespread trust and support, professional judgment was sufficient for management decisions. The restoration of large mammals across the United States is testimony to the success of the agencies during this era. Stage 4 begins when the agency matures enough to fulfill its stated mission. Having fulfilled its original charge, the agency must be recharted or disbanded. If left to its own devices, senescence begins and the agency may become dysfunctional.

Agency dysfunction occurs in three instances (3). The first occurs when a person or persons take control and listen only to vested interests at the expense of the agency's mission, its people, and the public at large. In the second instance, an agency created for one purpose is charged with something it was not designed to do and its people are unwilling or unable to change the agency's perceived mission or infrastructure to accommodate the new identity. In the third instance, the agency has outlived its original purpose and becomes self-serving and institutionalized merely to perpetuate its survival. This is basically the situation that many government agencies, including those charged with management of wildlife resources, face today.

Trends confronting wildlife resource agencies seem to be tracking the rest of government (5). In some western states, governors view wildlife agencies as embarrassments, not to be trusted. They are perceived as resisting new priorities (e.g., workforce diversity), generating bad press, and pandering to hunters and anglers to the detriment of other constituents. Agency responsibilities have been pruned and legislative oversight intensified. The public also has lost trust, with ballot initiatives to overturn agency regulations becoming a common occurrence.

Wildlife resource agencies are caught in a paradigm shift that is reshaping government. Most wildlife resource agencies continue to follow the Progressive Era model that prevailed when they were founded (5). The agencies contained professional experts entrusted to make decisions and act as trustees for the public good. They managed the resource and judged the worthiness of recreational pursuits. The public was the beneficiary and user, but had little role in decision making. The Progressive Era model is being replaced by a model based on partnerships and empowerment of the public. Emphasis will shift to trust and relationships and away from regulation and science. Agencies will be challenged to develop leadership skills to allow an informed public and partners (e.g., other agencies) to make decisions to fit local situations rather than function as controllers through regulation and laws.

The prognosis for conflict regarding wildlife resource issues is clear. A changing constituency, expanding management goals, and eroded public confidence will produce an environment rich in conflict. Wildlife agencies will need employees who can understand and respect the values of various publics and who are versed in negotiation and conflict resolution skills.

HUMAN DIMENSIONS RESEARCH

The wildlife profession has responded to these concerns in part by developing human dimensions research. This area of research has been defined in a variety of ways. For example, many in wildlife management equate human dimensions research simply with public polls and surveys (6). However, Manfredo and others argue for a broader definition: ". . . an area of investigation which attempts to describe, predict, understand, and affect human thought and action toward natural environments"(6, p. 54). Decker and others provide an even broader definition, characterizing human dimensions research as an effort to provide wildlife managers with information regarding political, economic, and sociocultural factors, which, when combined with biological and ecological information, comprise the body of knowledge necessary to direct wildlife management (7).

With respect to big game management and conflict resolution, the definition of human dimensions research adopted in this chapter most closely parallels that of Decker and others (7). Specifically, wildlife professionals dealing with conflict resolution need a greater understanding of two aspects of human dimensions research: social science and policy science. Social science refers to research that attempts to understand the beliefs, attitudes, and values that shape human response to wildlife management issues. Policy science is a closely related yet distinct field that encompasses the political component of human dimensions research. Clark describes policy science as the ". . . study of how knowledge is used or not used in decision and policy processes" (8, p. 500). This body of knowledge also addresses the philosophy and tactics of conflict resolution. Because its scope encompasses biological, social, political, and public knowledge, and the processes by which these various sources of knowledge are combined to form and implement policy, this is a highly integrative area of study. The purpose of this chapter is to provide a brief overview of current philosophy, concepts, and approaches to conflict resolution in wildlife management within these two fields.

OVERVIEW OF SOCIAL SCIENCE IN HUMAN DIMENSIONS RESEARCH

Social science is a broad label that encompasses many disciplines including anthropology, human geography, social psychology, and sociology. While examples of research representing each of these disciplines can be found in wildlife management literature, social psychology has dominated human dimensions in wildlife research. In particular, "attitudes" have been the focal point for studies attempting to understand human perception, response, and behavior toward wildlife.

While attitude-based approaches have dominated human dimensions in wildlife literature, research representing a meaning-based approach is becoming increasingly prevalent (9, 10, 11, 12). Research grounded in a meaning-based perspective will continue to grow in importance in the area of conflict resolution and therefore we also provide an overview of this approach.

Attitude-Based Approaches

Conceptual Approach. Attitudes are broadly defined ". . . as a learned predisposition to respond in a consistently favorable or unfavorable manner with respect to a given object" (13, p. 317). This general definition indicates why human dimensions researchers have been so interested in the concept; attitudes are related to the way humans behave with respect to an issue and, just as importantly, they are learned, and therefore may be influenced or changed to help promote desirable behavior. However, there are several distinct approaches to studying attitudes that differ with respect to how attitudes are defined and the intended use of the results. Within human dimensions research, two distinct approaches to understanding attitudes are apparent. The first is illustrated by Kellert's typology of attitudes, the second by Bright and Manfredo (14, 15). The following discussion emphasizes differences in the nature and applicability of insights generated by each approach.

Kellert's research was one of the earliest comprehensive treatments of attitudes in the wildlife literature and identified 10 attitude types that characterize people's perception of wildlife (14). This research provided wildlife managers with a better understanding of how the public perceives wildlife and of differences in perceptions among various stakeholders. However, Kellert clearly emphasizes the distinction between attitudes and behavior, pointing out that attitudes are ". . . broadly integrated feelings, beliefs, and values . . ." that are not necessarily consistent with an individual's behavior (16, p. 31). Thus, the original intent of this approach to studying attitudes was descriptive (i.e., to describe types of attitudes and how they differ across different stakeholders) rather than explanatory or predictive (i.e., to use attitudes as a basis for explaining or predicting behavior).

In contrast, Bright and Manfredo adopted an attitude model from social psychology, the theory of reasoned action, that provides a basis for empirically demonstrating the link between attitudes and behavior (15). This theory also provides a basis for identifying factors that shape attitudes (specifically, attitudes are thought to be determined by underlying beliefs). If the salient beliefs can be identified, they provide an avenue for ultimately changing behavior through persuasive communication (i.e., if attitudes have a strong influence on behavior, the specific beliefs that lead to inappropriate attitudes can be identified and targeted for change). Thus, this approach characterizes attitudes of various stakeholders and provides a basis for understanding the extent to which these attitudes influence behavior, the factors that shape attitudes, and ways for changing or influencing attitudes.

Case Studies. Kellert's typology of attitudes toward animals has been used in a variety of ways (14). For example, Kellert noted a historical trend in the United States representing a decline in utilitarian and negativistic attitudes toward wildlife (17). However, such attitudes remain common among lower socioeconomic groups, elderly, people living in rural areas, and natural resource dependent groups (17). Further, his research indicates that the more "appreciative" attitudes are restricted to a rather limited number of charismatic mega vertebrates and that ecological and scientific attitudes are limited to a small fraction of the general public.

Peyton and Langenau found wildlife biologists had a distinctly different attitude profile than the general public (18). Biologists demonstrated lower negativistic, utilitarian, and moralistic attitudes than the general public, and stronger ecologistic, scientistic, and dominionistic attitudes. Based on the differing profiles, Peyton and Langenau argued for the importance of values training for wildlife professionals because management goals reflect value priorities and failure to understand how their values differ from the public's may lead to conflict. In particular, they expressed concern that biologists' strong ecologistic and scientistic orientation versus the public's greater moralistic orientation may lead biologists to reduce conflicts involving moral issues to objective arguments that ignore the moral component.

Bright and Manfredo's study concerning public response to wolf reintroduction in Colorado illustrates the second approach to studying attitudes (15). Their purposes were to identify attitudes influencing support or opposition to wolf reintroduction; to identify the beliefs, knowledge, and emotions that shape attitudes; and to make recommendations regarding the design of public information campaigns.

The results indicated a strong relationship between intention to support or oppose reintroduction and attitudes (attitudes explained 93% of the variance in the intention measure in a regression model). Additionally, the research was successful in identifying beliefs that were important in shaping attitudes (a regression model with attitude toward reintroduction as the dependent variable explained 93% of the variance). The most important factors explaining variation in attitudes were "symbolic beliefs." Symbolic beliefs refer to beliefs about how wolves fit into society and make it a better or worse place to live. The next most important factors shaping attitudes were emotions and attitudes toward wolves in general. Finally, beliefs about specific outcomes related to wolf reintroduction (e.g., reintroducing wolves would result in large numbers of wolf attacks on livestock) and objective knowledge about wolves explained almost none of the variation in attitudes.

This research reflects the utility and the limitations of this approach. On the positive side, attitudes were closely linked to behavioral intentions. Additionally, the second analysis provides important insights into the nature of beliefs that drive attitudes and intentions. Specifically, deeply held values about the rights of wolves to exist and emotions are most important in explaining the variation in attitudes while the more objective, factual type beliefs appear to be of little importance.

The finding that differences in values and emotions were the dominant factor in explaining variation in support or opposition is, we believe, characteristic of many of the most contentious conflicts likely to face wildlife managers in the future. It also represents the limitations of this approach with respect to conflict resolution. Broad scale public education and information campaigns are most effective at influencing or changing beliefs that deal with factual knowledge or consequences and outcomes that entail questions answerable through research. In contrast, deeply held values are resistant to change, and scientific evidence, though relevant, does not directly resolve the conflict (15). Thus, while this approach may provide useful insights regarding what values the public might respond to in information campaigns, it is not well suited for yielding insights into how to negotiate a resolution to problems where fundamental values are in conflict.

Meaning-Based Approaches

Conceptual Approach. The meaning-based perspective and attitudinal approach differ on fundamental philosophical grounds, especially with respect to questions concerning the nature of reality, how one goes about knowing, and the ultimate goals research seeks to achieve. Some of the key distinctions are highlighted here.

A key distinction between attitude and meaning-based approaches can be characterized by distinguishing between the concepts of information versus meaning. A goal of the attitudinal approach is to gain insight into how to design information campaigns to persuade the public to adopt a particular position or to provide them with the information necessary to make an informed choice (15). Such an approach is highly appropriate when conflicts center around physical, tangible, or factual characteristics. Consider, for example, the use of blaze orange clothing while deer hunting. A salient, factual belief about which people may be mistaken is whether or not deer are more likely to see hunters wearing blaze orange, thereby reducing the likelihood of a successful hunt when blaze orange is worn. This question may be resolved through research on deer vision (e.g., the mixture of rod and cone photoreceptors and their ability to absorb different wavelengths of light) (19). In this case, a definitively right or wrong answer can be found through research, and results presented to a receptive public should resolve the controversy.

However, conflicts often do not center around the facts or information, but rather how people interpret a situation in light of their underlying values. Consider the issue of cruelty to animals. While it would be difficult to find someone in our society who endorses cruelty to animals, what constitutes cruelty is subject to a wide range of interpretations (e.g., different views on the morality of using dogs to hunt deer, death by starvation versus a bullet in situations where mortality is compensatory, or whether changes in deer behavior as a consequence of immunocontraceptives are more or less humane than hunting as a means of managing populations). In situations like this, while the facts are relevant, what is more important is what the facts mean to people. As a consequence, research interest shifts from the concept of information to the concept of meaning.

Meaning does not exist independently in nature, it is socially constructed. With respect to wildlife, people may endow animals with emotionally charged meanings that extend beyond simple objective or physical properties, they interact with these constructed meanings in ways that create highly individualized or culturally bound social realities, and they respond to natural resource controversies on the basis of these constructed realities (20). One implication is that "environmental reality" as constructed by some public constituencies may have little resemblance to the factual, physical, biology-based world as seen by wildlife professionals (21).

Those adopting a meaning-based perspective typically maintain that these different social constructions are not often easily characterized in terms of a small number of attitudinal statements of the type used in traditional surveys, but rather that they represent complex, context dependent, and evolving belief systems that are difficult to capture using standardized surveys. For example, Rolston illustrates this perspective in the following passage describing human ethics:

> An ethic, it may be insisted, has to be formal, general, universal, applicable without regard to time and place. . . . Keep promises. Tell the truth. Do to others as you would have them do to you. Do not cause needless suffering. Respect life. Yes, but ethics is not lived like that. . . . Each person lives in a particular time and place, and such abstractions are not yet a flesh and blood ethic but only a skeleton. An ethic too has an environment, a niche to inhabit. Like a species it is what it is where it is. Ethics evolve as do species. . . (22, p. 342).

In other words, for example, although we all believe in general values like telling the truth, in certain situations we have all told "white" lies in the belief that it was the right thing to do.

As a consequence, proponents of a meaning-based approach maintain that we need to get beyond simply describing or classifying these belief systems in terms of general attitude types at an abstract level. Instead, research needs to begin exploring how individuals interpret values and apply them in specific contexts. To accomplish this goal, meaning-based research typically adopts methodology employing systematic, but nonstandardized in-depth interviews and observations and often collects, represents, and analyzes data in non-numerical form. Such approaches often are treated with skepticism and have not been widely applied in mainstream human dimensions of wildlife research, in part because they produce results that are not statistically generalizable in the traditional sense.

However, statistical generalization, while an important and often desirable criterion in science, is only one way of characterizing representative data. The question "Are the findings representative?" also has a second meaning: "How well do the findings represent the actual subject being studied?" In some situations research is forced to make trade-offs between these two forms of data. For example, at a very general level it is possible to discover insights true of all humans (e.g., we must breathe or die). However, to some degree it is also true that every individual's experiences and perspective are unique. At times conflicts can be usefully explained and dealt with on the basis of broad generalizations, while in other situations dispute resolution requires an in-depth understanding of specific individuals or groups that may not broadly generalize to the public as a whole. Research strategies employed within a meaning-based approach emphasize detailed understanding of specific individuals at the possible expense of statistical generalization.

While the meaning-based perspective is more likely to recognize and accept the importance of individual cases, this does not mean that generalizations are not possible. Instead the approach to collecting and analyzing data differs. As noted earlier, data are collected through in-depth interviews rather than standardized surveys to be more sensitive to individual variation. With respect to analysis, traditionally research that emphasizes statistical generalization begins analysis at the aggregate level (especially using statistics associated in some way with the criterion of squared distance from the mean). In contrast, meaning-based research typically begins analysis by focusing on developing an understanding of individuals first, aggregating across individuals (seeking generalizations) only as a second stage. This emphasis on individuals is consistent with an emerging trend in conflict resolution strategies discussed in the overview of policy science presented later where the goal is to promote understanding and communication between conflicting parties.

Case Studies. The first example of a meaning-based approach is an analysis of a conflict surrounding habitat protection for the endangered golden-cheeked warbler in Texas (23). Noting that the apprehensions of ranchers and the U.S. Fish and Wildlife Service (FWS) regarding the "other" side have inhibited successful collaborative decision making between the two groups, the purpose of the research was to gain an understanding of ranchers' belief systems that may serve as a basis for promoting a constructive dialogue between the two groups.

The analysis emphasizes the ranchers' self-image as stewards of the land. This stewardship ethic is intimately linked to three other aspects of the ranchers' meaning system: common sense, independence, and the human–land connection. Common sense, which ranchers believe can only be achieved through personal experience, provided one foundation for their stewardship ethic. A second foundation was the ranchers' understanding of the human–land connection that defined humans as part of a complex web of life and led to a stewardship ethic in which ranchers saw themselves as responsible for protecting biological diversity on their lands. The third foundation was independence. Ranchers' comments suggested that independence (i.e., the ability to do their own thing) was an aspect central to their very nature, and that the freedom to decide (based on commonsense experience) what land management practices would best nurture life on the ranch was a central part of the stewardship ethic.

This stewardship ethic was threatened by the FWS's approach to the controversy, which was perceived as ignoring personal experience, replacing personal choice with coercion (through increasingly confining regulations), and trivializing ranchers' sense of connectedness to the land (23). The researchers maintain lack of understanding and sensitivity to these core aspects of the ranchers' stewardship ethic unnecessarily exacerbated the conflicts between ranchers and the FWS. They note that in many ways the ranchers' stewardship ethic bears strong similarities to Leopold's (1949) land ethic, which serves as a guide for many in the wildlife profession. For example, paralleling the independence theme in the ranchers' ethic, Leopold repeatedly argued that to be successful, a land ethic must develop within a community rather than be imposed by outside experts (24). Further, acknowledging and legitimizing the ranchers' self-image as land stewards does not require endorsing specific practices (23). What it would do, however, is create an opportunity for a constructive dialogue that currently does not exist. While such understanding does not guarantee conflicts can be resolved, it would ". . . encourage more meaningful participation in conversations needed to achieve a resolution" and provide a basis for encouraging disputants to examine and resolve contradictions between their stewardship ethic and actual land management practices (23, p. 145).

A second example of a meaning-based analysis is Lange's study examining the information campaigns used by environmental and timber groups involved in the spotted owl and old growth forest controversy (25). His purpose was to explore the "logic of interaction" that characterizes communicative strategies used by groups who do not interact directly with each other, but who hope to persuade the public through the mass media.

Lange found the logic of interaction between these groups was a negative downward spiral in which contesting groups mirror or match each other's strategies in ways that become increasingly louder, more shrill, and less informative (25). He found the information campaigns used the same strategies, four of which are reviewed here: frame/reframe, select high/select low, vilify/ennoble, and simplification/dramatization. Frame/reframe refers to the use of facts, explanations, and interpretations to frame or construct a reality favorable to the group's goals. Mirroring this response, opponents try to reframe the situation to create a reality favorable to their position. For example, the timber industry focused on economic issues such as mill shutdowns, loss of jobs, and retraining of displaced workers. Environmental groups in turn tried to reframe the issue by acknowledging the pain of transition, but suggested it is inevitable because old growth is finite. Further they argue that decades of overcutting and increasing technology were the true factors in eliminating jobs while the owl was being used by the industry as a scapegoat.

In select high/select low, contesting groups try to win public support by selectively using studies and "expert opinions" that support their arguments while intentionally ignoring results or experts that seemingly contradict their position. Vilification entails formulating specific adversarial opponents who are cast in an exclusively negative light, attributed sinister motives, and attributed greatly magnified power. At the same time groups ennoble themselves by claiming benevolent motives, arguing they are the ones who have compromised, that science is on their side, and that their side represents morality and the common good. In simplification/dramatization, groups capitalize on the public's tendency to avoid complexity and simplify that which cannot be avoided. Groups seek to create images and slogans that make good "sound bites" for the mass media and capture public attention. The goal is to convince the public that an "injustice which needs their attention" has occurred rather than to educate the public about the issue (25, p. 251).

Lange concludes that this escalating negative downward spiral tends to be self-perpetuating as groups mirror and match each other's strategies (25). "A specific communicative act by one [group] practically 'forces' a predetermined response by the other. Parties become locked into a systematic, self-reinforcing, patterned, and repetitive practice" (25, p. 254). Thus, the challenge is to find a way of creating a logic of interaction or dialog between contesting groups that avoids this negative spiral-like logic. This issue is addressed in the overview of policy science.

OVERVIEW OF POLICY SCIENCE IN HUMAN DIMENSIONS RESEARCH

Alternative Philosophies of Conflict Resolution

Wildlife professionals often express discomfort, frustration, and even agitation when faced with public involvement in wildlife management decisions. We complain about the inaccurate and emotion-filled appeals that seem to sway the public. We note, as do Ehrlich and Ehrlich, "the general lack of scientific knowledge among citizens" decrying not only their meager "knowledge . . . of the findings of science, . . . [but also their] understanding of scientists and the process by which those findings are made" (26, p. 26). As a consequence, researchers and wildlife professionals often express a sense of indignation when nonscientists express opinions about issues we consider to be science based. We often adhere to a belief that the public should leave the decisions to the experts with the technical background and training to understand the issues and resolve wildlife management problems. Or we may adhere to the notion that dealings with the public should focus on education, which certainly will lead the public to accept our decisions.

However, within the policy sciences, this mind-set is seen as an impediment to successfully resolving wildlife management controversies. For example, Clark argues that the approach scientists take often reflects a narrowly bounded rationality that is likely to lead them to misunderstand a problem and incorrectly specify its solution (8). Further, he questions the adequacy of the traditional model of science that he describes as a positivistic perspective on a post-positivist social process.

Similar critiques are evident in a series of articles in *Policy Sciences* where scientific discourse is seen as suffering from two important flaws. First, the scientific community fails to recognize and understand the languages of politics and the public. By failing to understand the discourse of these other two communities, Throgmorton argues that conventional scientists often condemn their research to dusty office shelves (27). Second, scientists often suffer from the misconception that factual and value-laden aspects of science can be neatly separated (28, 29). Such a misconception leads scientists to believe that their work has some privileged status over the emotion-laden discourse of politics and the lay public (27).

Some of the core differences between the attitude of professionalism illustrated in the introductory paragraph to this section and the policy science critiques can be understood as reflecting different philosophies about how "public interest" is defined and achieved. Williams and Matheny's discussion of democracy and environmental disputes outlines these different philosophies in a way that is relevant to the issue of conflict resolution in wildlife management (30). The remainder of this section is a brief overview of the discussion of Williams and Matheny (30).

Wildlife professionals expressing sentiments similar to those described in the introduction to this section are, at least implicitly, subscribing to a philosophy about democracy and dispute resolution (i.e., the managerial model). This perspective on public policy evolved during a period in which industrialization, rapid development of technology, and the emergence of large corporations seemed to threaten the welfare of individuals. It reflects the underlying belief that there is one best policy that can be discovered by scientific experts who, alone, have sufficient expertise and familiarity with the issues and technology to understand the problems and who ". . . would transcend the petty political squabbles of self-interested groups and serve the general public interest" (30, p. 13). Thus, a key assumption of this perspective is that the public interest is an objective reality discoverable by neutral scientific experts. Although this philosophy may at first seem to conflict with democratic ideals such as public participation in decision processes, its supporters respond to these concerns by defining democracy as a means of serving the public interest and argue that if public interest is better served by alternative decision-making processes, participatory procedures should be abandoned. Instead of public participation in decision making, the assumption that science can identify the correct and politically neutral policy solution leads to a belief that interactions with the public should focus on education that increases public faith in science and serves primarily as a means to the end of letting experts make the decisions.

Despite its appeal to a group with a strong sense of professionalism and a recognition of the complexity of its subject matter, the managerial model suffers from several weaknesses. The principal weakness is the belief that it is possible to readily separate the technical facts from value-laden aspects of environmental disputes to reach an objectively best solution. Williams and Matheny, Latour, Malm, and Mishler provide extensive discussions of how science reflects and is shaped by social values (30, 31, 32, 33). Thus, Williams and Matheny argue that ". . . although science may still be the best model we have of human rationality, it is far from being the neutral, objective technique assumed by the managerial model"(30, p. 40).

A second limitation of this model is that it is based on the myth that scientific experts can be expected to agree. The assumption underlying this myth is that, given the same objective information, any disagreement among experts reflects political biases; truly neutral experts, in contrast, will agree about the interpretation of the information. However, even a limited amount of practical experience in research quickly dispels this notion. In fact, as Williams and Matheny point out, disagreement is at the heart of science (30).

The final major weakness of this model revolves around the issue of public education. When conflicts involve scientific and technological uncertainty, the more educated a person is, the less likely he or she is to concede decision-making authority to others. The concept of public education within the managerial model provides no answer for the problem of how to handle an educated public that still disagrees with the experts.

A second model of democratic dispute resolution that seems to be incorporated, in part, in natural resource legislation like the National Environmental Policy Act that requires public participation in planning is the pluralist model. Unlike the managerial model, this philosophy maintains that the public interest cannot be discovered *a priori* by neutral, objective, scientific experts. Instead, the public interest is achieved through an open and balanced political process that allows conflicting groups to define the public interest. In other words, public interest is defined procedurally as a fair and open set of decision-making processes rather than as a specific outcome.

However, in its purest form, this model also has significant weaknesses. First, the scientific and technical complexities of many issues makes it difficult for many individuals to adequately understand their interests. This represents a barrier that this model fails to deal with successfully. Additionally, there are often economic barriers to participation among certain groups that violate the pluralist assumption that absence of participation indicates consensus. Finally, because politics deals with policies affecting large numbers of people, it must aggregate interests and take the group as the unit of analysis. However, this approach does not always adequately deal with the way individuals or specific communities define their interest. As Williams and Matheny point out:

> A central problem . . . is whether citizens see themselves as members of a wider society with a common interest in addressing the negative externalities . . . or simply as members of a local community resisting the specific costs of regulation imposed upon them (30, pp. 45–46).

To overcome the weaknesses apparent in the managerial and pluralist models of conflict resolution, Williams and Matheny present a dialogic model (30). With respect to the issue of defining and achieving the public interest, this model maintains that public interest is not some final end product that can be objectively discovered and defined once and for all through science, but something that is continually evolving and must repeatedly be created through public dialogue.

Second, although this model takes a different perspective on concepts like public interest and science, it does not represent a complete rejection of the managerial and pluralist models. Instead it attempts to incorporate or integrate the strengths of the other models. For example, as with the managerial model, the dialogic model maintains that science provides the best model for structuring debates regarding environmental disputes. However, rather than emphasizing the answers science provides, the dialogic model underscores the importance of understanding science as a social process. Rather than seeing the role of science as predominantly one of producing facts or answers that resolve conflicts, this model emphasizes understanding the way science operates, attempting to incorporate key features of scientific discourse into public debates. For example, Williams and Matheny argue that the deeply held skepticism toward science and experts that is increasingly apparent in environmental disputes is explained by the inability of the public to evaluate scientific claims (30). Yet one of the key features of science as a social process is that research results are presented in a way that makes it possible for others to verify the conclusions through examination of the evidence or independent reproduction of the results. Thus, extending this feature of scientific debates to conflict resolution means that all parties must have the ability to independently evaluate claims of other parties.

As indicated in the discussion of the dialogic model to this point, consistent with the pluralist model, it emphasizes public involvement. However, rather than focusing on dealing with the public only at discrete points in time when conflict over specific issues arises, the dialogic model emphasizes the importance of establishing and maintaining contact across time. Such repeated interactions and long-term relationships with stakeholders are necessary to establish the shared norms and values that are needed to promote cooperation and reduce conflict intensity. Thus, this perspective argues that attempts to educate or unite citizens over specific issues are likely to fail unless some degree of rapport has already been established and maintained over time.

In summary, the dialogic model emphasizes the importance of public dialogue as a way of reducing disagreement and clarifying contested concepts in environmental disputes. In fact, this model maintains the public interest is not something that exists and can be discovered independently of a community, but something that is continually created through dialogue. Williams and Matheny argue that their model provides a blueprint for structuring public dialogue regarding environmental disputes. Central principles from this blueprint include the following:

1. *All parties must have access to information.* This information must be communicated in a way the listener can comprehend. At least in principle, the information must be verifiable by all parties.
2. *All parties must be committed to achieving understanding rather than manipulation.* Seeing citizen participation as an ongoing dialogue rather than viewing conflict as a series of discrete events and restricting involvement to specific issues will facilitate this end.
3. *All parties must have the power to influence the decisions* (30).

The dialogic model has only recently been outlined and has not served directly as a guide for resolving wildlife management conflicts. However, the philosophy underlying this model is consistent with emerging trends in approaches to conflict resolution in wildlife management reviewed later.

In a variety of different dispute contexts, wildlife management agencies are beginning to explore conflict resolution strategies that seek to identify and interact with communities of interest rather than relying primarily on broad-scale public education efforts or public meeting formats that are characteristic of post-NEPA public involvement efforts (34). Examples include the use of citizen task forces to establish deer population objectives and management strategies in New York, the National Park Service's recent use of negotiated rule making (an approach to conflict resolution outlined in the Federal Advisory Committee Act) to resolve a conflict regarding an endangered shorebird and off-road vehicle use at Cape Cod National Seashore, and the use of a citizens' advisory council to resolve big game hunting related conflicts involving hunters, land owners, and outfitters in Montana (35, 36, 37, 38, 39, 40). The remainder of this section discusses specific techniques for conflict resolution used in the Montana big game hunting controversy.

Background and Formation of Governor's Council

Williams and Matheny's blueprint for structuring constructive public dialogue requires that everyone should be committed to achieving understanding (30). However, sometimes one or more participants prefer adversarial conflict as a way of mustering strong bases of support. The concept is, if conflict escalates, then people become polarized and those on your side come to your aid and support you verbally, financially, and in other ways. Increased fund raising and enrollment of members to the lead organization are examples of perceived financial benefits. This strong show of support can lead to a certain degree of fame among one's supporters that in itself can be a strong motivation for some. Regardless of the underlying motivation, the result is a preferred method of dealing with differences through adversarial conflict. Support is marshaled through sheer force of numbers, manipulation, or other means to win the conflict.

This adversarial approach is a win or lose perspective and derives from a scarcity mentality (41). However, the long-term costs of winning through conflict are often overlooked. Acrimony, distrust, and malcontent that result in attempts to get even by those who lost the conflict or make resolving future differences less likely are some of the prices paid when winning through conflict. For some, however, the short-term glory of having led the charge that won the battle is more important than considering the long-term costs. This is especially true when the more spectacular adversarial approach generates strong support and considerable personal visibility.

These factors were only part of the interwoven history in the conflicts over private land and public wildlife issues in Montana. These issues existed for many years prior to formation of the Governor's Council in 1993 (40, 42, 43). There were three principal areas of conflict: the leasing of private lands by outfitters, which closes access to sportsmen, many landowners' desire to be compensated for allowing public hunting fueled by lower profits from traditional agriculture, wildlife damage to crops, and property damage by hunters, and outfitters' desire to receive licenses for all their booked clients instead of having them participate in a drawing (42, 44, 45). Deep-seated values involving private property rights, a tradition of free hunting access to private lands, fairness in hunting license distributions, and government influences on private enterprise (i.e., the outfitting business) were all at stake in the ensuing conflicts.

The conflicts carried over into the 1993 Montana legislature when 12 different bills were introduced by the various interest groups. None of the bills were supported by all the interests, several bills conflicted, and, ultimately, none of the bills were passed. Instead, the legislature passed a joint resolution requesting that the governor form a citizens' council to deal with the conflicts.

The governor's office took into account the long history of conflicts and some constituents' potential preference for adversarial conflict as a means of dealing with issues.

When selecting citizens to serve on the council, the governor defined criteria that included a willingness to seek resolution and a reputation for being reasonable. In addition, no officers of state or national interest groups were permitted to be on the council because of concern that they might have been more susceptible to reverting to a win/lose approach.

Early meetings of council members were tense (Nina Baucus, chair, Montana Governor's Council, personal communication). Eighteen council members were selected from different geographical regions of Montana to represent the interest groups of landowners, outfitters, and sportsmen. Most of the members did not know each other prior to the council formation. Peterson and Horton's observations about apprehensions inhibiting communication was a primary reason for the tense atmosphere of the initial meetings (23).

To address this initial tension, meetings were arranged where overnight stays were required and meals were eaten as a group. This informal, yet planned, part of council meetings provided an opportunity for council members to get to know one another in social settings free from work-related obligations. This was an important factor that reduced council members' apprehensions. In addition, formal team-building exercises were used (46). Burke states that team building is an important intervention if individuals are expected to work together successfully toward mutual goals (47). Past experience suggests that when working with a group where participants do not know each other, three or four meetings managed in the ways mentioned are required to lessen apprehensions for maximum working effectiveness.

The formation of a working group is an important step in setting groups up for success. There are five critical factors of formation:

1. Clarification of objectives for the group and definition of deliverables such as a white paper, other written documents, or verbal reports
2. Clarification of sideboards such as deadlines and financial constraints
3. Clarification of roles of group members (i.e., chair, observer, participant, resource expert)
4. Clarification of the final decision process and who has what roles (clearly stated levels of authority for the group as a whole, for individual members of the group, for recipients of the group's products)
5. Group clarification of the decision process it will use before the group begins to address issues.

The governor gave specific instruction to the Montana Governor's Council at its first meeting. These instructions described issues to be addressed with sideboards. Sideboards included stabilizing the outfitting industry in Montana, recommending means for generating revenues when recommendations would require such funds, and that final council recommendations must be supported by all interest groups. These directions narrowed the scope of council work, focused efforts, and prevented unworkable recommendations that the governor's office could not support.

The governor provided clarification of roles by appointing the chair of the council, Nina Baucus, and informing the council of her role in council meetings. Other members of the council were to serve as participants with responsibilities to report back to their interest groups and bring forth comments and suggestions from them. Authority levels of the group as a whole were specified; the council had authority to provide recommendations in the form of a written report and draft legislation prior to the next legislative session. The governor had final decision authority but promised to support recommendations that all interests could agree upon, although the council members understood that legislative action would be required on any draft legislation.

During the council's first meeting the facilitator initiated a discussion to select a decision process that would be used for all major decisions by the council. Informed consent was selected as the preferred process (48). In effect, each council member had veto power over decisions. This forced work toward win/win decisions that were harder to achieve than the win/lose decision by a majority vote (49). However, the informed consent decision process moved the group more into the dialogic model and ensured that all parties had the power to influence decisions (30). Consequently, comfort levels were raised for interest groups who were in the minority on specific issues.

Management of the Governor's Council

The council held regular meetings usually once a month for 2 to 3 days. As a consequence, successful meeting management was a critical factor. Peyton and Eberhardt list requirements for an effective meeting: Everyone works on the same problem; everyone understands and uses the same process; someone must be able to maintain an open and balanced conversation among participants, protect individuals from personal attack, and maintain an environment of fairness; and everyone's roles and responsibilities are clearly defined and agreed on (50). Interaction Associates' facilitation techniques (49) were used to maximize meeting effectiveness and to establish a cooperative working relationship among all participants (39).

Lange described a trend toward a downward spiral in information campaigns from contesting groups and the tendency for information efforts to become less informative (25). In addition, Fiske and Taylor use cognitive dissonance theory to predict that people are motivated to avoid information inconsistent with their attitudes or choices and that people are likewise biased toward information that reinforces their beliefs (51). These factors were minimized during council meetings through application of a modification of the active listening technique described by Robert (52). In essence, when an issue was raised for discussion, council members were required to describe the concerns of another interest group on the council with regard to that specific issue. Representatives of the interest group then critiqued the description and the process was repeated until all could agree on the accuracy of the description of concerns. For example, a landowner representative might be asked to describe the concerns of sportsmen regarding an issue. After the first attempt, sportsmen representatives were asked to critique the description. The landowner representative then restated the description. This process was continued until both interest groups' representatives agreed on the description of concerns. Each group's concerns were covered in this manner by the council before proceeding into discussions of the issues.

A structured process was required for the council members to effectively begin to address the wide array of private lands and public wildlife issues. Council members were trained on the use of a seven-step problem-solving process (49, 53, 54). This structured approach allowed the council to systematically address selected issues and work collectively toward solutions. This process focused the council members' attention on problem definition and collective understanding of the problems before individuals began generating and defending different solution alternatives.

Public Involvement Tools

Key to leading a public involvement process is the personal attitude and interest of the leader(s). There must be an honest interest in gathering input from all potentially affected interests (PAIs) and in fair and open representation of all PAIs (48). These groups readily recognize shallow interest, unfair representation, or attempts to manipulate and the credibility and trust necessary for successful public support of recommended solutions will be lost if these tactics are used!

All council participants recognized that even if they could agree on recommendations for solving private land and public wildlife issues, statewide support of council recommendations was still required from all interest groups. Following debates and discussions the council adopted a pluralistic model as the citizen participation model most likely to gain public input and support for resolutions to conflicts (30). Over the course of the 2 years of council operation, this model evolved into a dialogic model (30).

Techniques for successful citizen participation included making maximum efforts to involve all PAIs. This took the form of holding all council meetings as open meetings and moving the meeting locations around the state to afford easy access to as many PAIs as possible. In addition, each council member was charged with getting information on recommendation developments to all their interest groups through individual phone contacts and attending their interest groups' individual meetings. A mailing list of more than 800 PAIs was developed, and each time a council meeting was held, a meeting summary was sent to all on the list. Written and telephone comments were taken throughout the process. When such comments were received, the council chair sent written postcard replies to those submitting comments to let them know their comments had been received and would be considered. In addition, a local working group was formed in each of six different geographical areas of the state. These groups were to represent all the PAIs in their geographical region, make suggestions to the council, and evaluate the council's draft recommendations.

The processes mentioned were most useful for involving PAIs in development of initial council recommendations. However, once a draft of recommendations was produced, another means of involving PAIs was added, eight public meetings. Each meeting was held in a different geographical area of the state. The format for these meetings has been described by Guynn:

> A modified open house format (Bleiker and Bleiker 1990) was used. This format provided for individual discussion between Council members and interested persons and reduced the potential for polarization. The format was informal and each person attending an open house meeting was presented with a written list of Council recommendations. Posters were put on the meeting room walls explaining each recommendation. Next to each poster was a blank poster sheet for writing in comments. Also, a Council member wearing a visible name badge was positioned near each poster. These Council representatives were there to receive verbal comments and to clarify any information-related questions. Local group members from the geographical area where the meeting was held attended each meeting in their respective areas and provided the Council with a degree of local credibility at each of these public meetings. Council members were instructed beforehand to listen to all comments, and while they were encouraged to explain the rationale for Council recommendations, they were advised to practice active listening and refrain from defending any of the recommendations. During these open house meetings, Council members often facilitated small, informal, spontaneous discussions about various recommendations and recorded the input for future overall Council consideration. All participants at each public meeting were encouraged to sign up on the Council's mailing list to receive any further drafts or other information from the Council regarding recommendations (39, p. 48).

Two drafts of recommendations were prepared using this public review process, and a 60-day comment period was held. After the second draft's review, a final draft was prepared and a 30-day comment period held. Upon completion of this comment period the council prepared its final recommendations. Twenty recommendations were agreed on by all interests (Table 11–1). These final recommendations were accepted by the governor and submitted to the Montana legislature as House Bill 195. One senator and one house of representatives member (who were on the council) sponsored the bill. There was no testimony against the bill when it was introduced (Montana House of Representatives, 1995 record of testimony). The bill passed the house with an 88–11 vote and the senate with a 46–4 vote. The governor proclaimed the success of the council at resolving conflicts as a miracle (Mark Racicot, Montana Governor, personal communication).

The council recognized that continued dialogue would be needed to evaluate implementation of their recommendations and to address future issues. The council's recommendation package included a request that a standing oversight committee of citizens be established and that all local groups organized by the council be continued. The Montana Department of Fish, Wildlife and Parks accepted responsibility for continuing the local groups, and the governor's office appointed an oversight group of citizens.

In this manner the model used by the Governor's Council completed its evolution from a pluralistic model to a dialogic model of conflict resolution.

The results of this conflict resolution effort culminated in many benefits. One tangible benefit was that approximately $3 million per year in new revenues are generated for the Montana Department of Fish, Wildlife, and Parks to support a hunting access program. Some of the intangible, but no less important, benefits were increased trust and a sound base for continuing dialogue between interest groups. Additional benefits and results are described by Guynn and Landry (40).

TABLE 11–1

Twenty Recommendations Agreed on by All Interests Represented by the Governor's Council on Hunting Access Conflicts in Montana, 1995

1. The Montana Department of Fish, Wildlife and Parks (FWP) provide tangible benefits to landowners as an incentive to allow public hunting. (This included incentives such as up to $8,000 and a free hunting license to the landowner that is nontransferable.)
2. FWP develop enhancements to its current access program (block management) to implement recommendation number 1. Contractual agreements between FWP and landowners regarding development of a hunt management plan, defining lands not eligible for enrollment (such as those on which a trespass fee is charged).
3. FWP develop an enhanced program for hunter education. This program would be similar to a master hunter program where graduates have met more stringent requirements than in the current hunter enrollment program.
4. FWP and other land management agencies encourage the creation of more walk-in areas.
5. Ensure more equitable distribution of "landowner sponsor" class licenses by limiting the number a single landowner can apply for. (A special category of licenses for big game exists in Montana that allows landowners to have hunters apply for these in order to hunt only on the landowner's property. The purpose was initially to allow landowner's relatives and friends that were nonresident Montanans a greater chance to draw a license. However, some landowners began using this class of license as a tool to develop an outfitting business.)
6. The governor initiate an interagency access council to address common access related issues. (The access council would include U.S. Forest Service, Bureau of Land Management, State Lands, FWP, and the Montana Association of Counties.)
7. FWP work with other appropriate agencies to provide increased technical assistance to landowners regarding habitat management.
8. A statement of strong council support of a major habitat program (Habitat Montana) and recommendations that not only purchases of habitat be utilized by FWP but encourage the use of easements, land exchanges, and cooperative projects.
9. Support at the federal, state, and local levels to consolidate isolated parcels of state and federal lands.
10. FWP add an access program coordinator position to its staff. This person would be responsible for handling the block management program and continuing the local working groups established by the council.
11. Legislative moratorium on increasing the number of outfitter licenses beyond the 1994 level. Establish authority for the Montana Board of Outfitters to develop rules and review new operating-area plans and expansion of current plans and for the board to turn down expansions or new outfitting plans that are determined to have an undue conflict with other uses of the area.
12. Legislation requiring nonresident hunters sponsored by an outfitter for licenses in the outfitter set-aside license category to hunt big game with that outfitter. (A pool of nonresident big game licenses is annually available for clients booked by outfitters. Some clients applied for a combination deer/elk license and then hunted only one species with the outfitter and hunted the other species without the outfitter services.)
13. Variable-priced nonresident hunting outfitter licenses be made available in unlimited quantity. The Fish, Wildlife and Parks commission sets the license price annually and is to meet a target of a 5-year average on no more than the current cap on outfitter licenses. This would be done through adjusting the price to reduce applicants for outfitted clients only. No change to be made in the license structure for nonresident applicants that do not book with an outfitter.
14. Funds generated from the variable-priced license to be used for increased public access including the FWP block management program and the tangible benefits described in recommendation number 1. (This currently amounts to approximately $3 million dollars per year in new revenues for FWP.)
15. Reduce the number of nonresident deer/elk combination licenses for outfitted clients from 5,600 to 5,500.
16. Increase the number of nonresident deer licenses for outfitted clients by 300 and increase the number of nonresident deer licenses for nonoutfitted clients by 300.

Tactics for Establishing a Constructive Dialogue in Conflict Resolution

TABLE 11–1 *Continued*

17. Increase the authority, funding, and personnel for the board of outfitters. Add 3.20 full-time equivalents, and add authority to issue citations for infractions of outfitting laws.
18. FWP consider the options of splitting the deer/elk license and report to the governor.
19. Increase the number of game wardens in the field during hunting seasons. Including use of retired or part-time local/county/state law enforcement personnel.
20. Program funding for the previous 19 proposals be derived from the variable-price license and increase the resident hunting licenses by $1.00 per license with the revenues to be designated to the hunting access program.

THE ROLE OF HUNTING IN WILDLIFE MANAGEMENT

Hunting has been the central focus of most controversy related to wildlife management issues. Conflicts over the status of species (i.e., game, nongame, pest), baiting, use of dogs, hunter access, and wildlife damage have been common. The role and justification of hunting as a legitimate management tool have been questioned by referendum (55, 56). As agencies are forced to shift constituents from primarily hunters to the public at large, the role of hunting as a wildlife management tool will come under increasing scrutiny.

The shift in agency constituency will require institutional change and reform (57). Institutional perspective and organizational character will change from emphasis on game and commodity species to emphasis on all wildlife and their ecosystems. A primary interest of managing for utilitarian values must be expanded to include a wide range of wildlife values diversely held by various groups. This will require fundamental changes in agency culture and tradition. The basis for funding wildlife management must be broadened from taxes on consumptive users and license fees to include all citizens who benefit from wildlife resources. The primary focus of management will be the maintenance of the health and vitality of the wildlife resource rather than providing services to meet the interests of consumptive users.

The goals of big game management have expanded dramatically since the 1970s. With completion of restoration programs, primary management goals have shifted from increasing population distribution and abundance to concerns for depredation control, reduction of vehicle collisions, reduction of ecological impacts, maintenance of natural herd structure, and enhancement of the quality of recreational experiences (58). These concerns are particularly acute for white-tailed deer in the eastern United States (59). Hunting is a cost-effective and efficient management technique in most, but not all, situations involving these concerns.

The role of sportsmen has changed dramatically during this century. The wildlife conservation movement of the first 40 years of this century led to a great alliance between sportsmen and the wildlife profession. In addition to financial and political support, sportsmen were directly involved in raising and releasing game, winter feeding, and protection. This alliance was essential to the restoration of many species. Hunting ethics and sportsmanship were paramount. Since that time management techniques have changed greatly with increased knowledge of wildlife biology. With this increase in knowledge, however, has been the general trend of decreased involvement in the management process by sportsmen. In most situations the sportsman has been reduced to a consumer. Kozicky discusses how this trend has led from the sportsmanship and hunting ethics common prior to 1940 to the present emphasis on gadgets and technology by today's sportsmen (60). More recently, Woods and others reported that actual involvement in management had the strongest total effect on hunters' satisfaction with quality deer management (61).

Hunters must be viewed as managers and not just consumers if hunting is to continue as an effective management technique. For this to happen, both hunters and professional wildlife managers must understand and respect the values of nonhunters and communicate their values about hunting. Hunters must demonstrate shooting proficiency that will ensure a quick and relatively painless death, an understanding of wildlife biology and habitats, an ability to police themselves, and a willingness to pay more for wildlife conservation. They must be willing to work with local communities to accommodate the specific needs of the area. They must contribute financially and must be good citizens who demonstrate respect for the wildlife resource and the residents of the areas they hunt.

SUMMARY

The last two decades have produced significant social and professional changes that increase the likelihood and severity of conflicts over the stewardship of wildlife resources. The emerging field of human dimensions research, with its focus on the policy and social sciences, provides a valuable resource for helping wildlife management agencies negotiate this potentially contentious future. Policy science focuses explicit attention on the philosophy and tactics underlying public relations and conflict resolution. Specifically, discussions in this area suggest the need to shift from the Progressive Era management model in which decisions meant to achieve the public interest are left to the experts, to a dialogic model which empowers public stakeholders and attempts to encourage long-term relationships. The Montana Governor's Council on land access and big game hunting issues serves as an illustration of the changes that this shift in philosophy and tactics will entail. Social science research offers insights regarding the different values or belief systems that underlie and drive conflicts and the "logic of interaction" that characterizes different public relations strategies. In addition to the traditional focus on public attitudes, there is a growing emphasis on a more anthropological approach that focuses on how meaning is constructed and communicated. This latter emphasis seems especially consistent with emerging tactics for conflict resolution in the policy sciences in that it seeks to promote understanding necessary to achieve constructive dialogue rather than simply predict behavior or develop persuasive information and education campaigns. As a consequence, research reflecting this perspective will continue to gain in importance.

LITERATURE CITED

1. DiCAMILLO, J. A. 1995. Focus groups as a tool for fish and wildlife management: A case study. *Wildlife Society Bulletin* 23:616–620.

2. U.S. CENSUS BUREAU. 1991. *Statistical Abstract of the U.S. Population.* Government Printing Office, Washington, DC.

3. MASER, C. 1994. *Sustainable Forestry: Philosophy, Science and Economics.* St. Lucie Press, Delray Beach, FL.

4. OSBORNE, D. and T. GAEBLER. 1992. *Reinventing Government.* Addison-Wesly, Reading, MA.

5. HAYS, R. L. 1997. Beyond command and control. *Transactions of the North American Wildlife and Natural Resource Conference* 62:164–169.

6. MANFREDO, M. J., J. J. VASKE, and L. SIKOROWSKI. 1996. Human dimensions of wildlife management. Pages 53–57 *in* A. W. Ewert, ed., *Natural Resource Management: The Human Dimension.* Westview Press, Boulder, CO.

7. DECKER, D. J., T. L. BROWN, and B. A. KNUTH. 1996. Human dimensions research: Its importance in natural resource management. Pages 29–47 *in* A. W. Ewert, ed., *Natural Resource Management: The Human Dimension.* Westview Press, Boulder, CO.

8. CLARK, T. W. 1993. Creating and using knowledge for species and ecosystem conservation: Science, organizations, and policy. *Perspectives in Biology and Medicine* 36:497–525.

9. DIZARD, J. E. 1993. Going wild: The contested terrain of nature. Pages 111–135 *in* J. Bennett and W. Chaloupka, ed., *The Nature of Things: Language, Politics, and the Environment.* University of Minnesota Press, Minneapolis.

10. FELT, L. F. 1994. Two tales of a fish: The social construction of indigenous knowledge among Atlantic Canadian Salmon Fishers. Pages 251–286 *in* C. L. Dyer and J. R. McGoodwin, ed., *Folk Management in the World's Fisheries: Lessons for Modern Fisheries Management.* University Press of Colorado, Niwot.

11. HYMAN, J. B., and K. WERNSTEDT. 1995. A value-informed framework for interdisciplinary analysis: Application to recovery planning for Snake River salmon. *Conservation Biology* 9: 625–635.

12. SUTHERLAND, A., and J. E. NASH. 1994. Animal rights as a new environmental cosmology. *Qualitative Sociology* 17:171–186.

13. LUTZ, R. J. 1990. The role of attitude theory in marketing. Pages 317–339 *in* H. H. Kassarjian and T. S. Roberson, eds., *Perspectives in Consumer Behavior,* 4th ed. Prentice Hall, Englewood Cliffs, NJ.

14. KELLERT, S. R. 1980. Americans' attitudes and knowledge of animals. *Transactions of the North American Wildlife and Natural Resources Conference* 45:111–124.

15. BRIGHT, A. D., and M. J. MANFREDO. 1996. A conceptual model of attitudes toward natural resource issues: A case study of wolf reintroduction. *Human Dimensions of Wildlife* 1:1–21.

16. KELLERT, S. R. 1980. Contemporary values of wildlife in American society. Pages 31–60 *in* W. W. Shaw and E. H. Zube, eds., *Wildlife Values.* Institutional Series Report No. 1. University of Arizona Center for Assessment of Noncommodity Natural Resource Values, Tuscon.

17. KELLERT, S. R. 1987. The contributions of wildlife to human quality of life. Pages 222–229 *in* D. J. Decker and G. R. Goff, eds., *Valuing Wildlife: Economic and Social Perspectives.* Westview Press, Boulder, CO.

18. PEYTON, R. B., and E. E. LANGENAU, JR. 1985. A comparison of attitudes held by BLM biologists and the general public towards animals. *Wildlife Society Bulletin* 13:117–120.

19. MURPHY, B. P., K. V. MILLER, R. L. MARCHINTON, J. DEEGAN, II, J. NEITZ, and G. H. JACOBS. 1993. Photo pigments of white-tailed deer. Page 32 *in Proceedings of 16th Annual Southeast Deer Study Group Meeting,* February 21–24, 1993, Jackson, MS.

20. PALMER, C. E. 1991. Human emotions: An expanding sociological frontier. *Sociological Spectrum* 11:213–229.

21. CANTRILL, J. G. 1992. Understanding environmental advocacy: Interdisciplinary research and the role of cognition. *Journal of Environmental Education* 24:35–42.

22. ROLSTON, III, H. 1988. *Environmental Ethics: Duties To and Values in the Natural World.* Temple University Press, Philadelphia, PA.

23. PETERSON, T. R., and C. C. HORTON. 1995. Rooted in the soil: How understanding the perspectives of landowners can enhance the management of environmental disputes. *Quarterly Journal of Speech* 81:139–166.

24. LEOPOLD, A. 1949. *A Sand County Almanac.* Oxford University Press, New York.

25. LANGE, J. I. 1993. The logic of information campaigns: Conflict over old growth and the spotted owl. *Communication Monographs* 60:239–257.

26. EHRLICH, P. R., and E. H. EHRLICH. 1996. *Betrayal of Science and Reason: How Anti-Environmental Rhetoric Threatens Our Future.* Island Press, Washington DC.

27. THROGMORTON, J. A. 1991. The rhetorics of policy analysis. *Policy Sciences* 24:153–179.

28. DOBUZINSKIS, L. 1992. Modernist and postmodernist metaphors of the policy process: Control and stability vs. chaos and reflexive understanding. *Policy Sciences* 25: 355–380.

29. FISCHER, F. 1993. Citizen participation and the democratization of policy expertise: From theoretical inquiry to practical cases. *Policy Sciences* 25:355–380.

30. WILLIAMS, B. A., and A. R. MATHENY. 1995. *Democracy, Dialogue, and Environmental Disputes: The Contested Languages of Social Regulation.* Yale University Press, New Haven, CT.

31. LATOUR, B. 1990. Postmodern? No, simply amodern! Steps towards and anthropology of science. *Studies in the History and Philosophy of Science* 21:145–171.

32. MALM, L. 1993. The eclipse of meaning in cognitive psychology: Implications for humanistic psychology. *Journal of Humanistic Psychology* 33: 67–87.

33. MISHLER, E. G. 1990. Validation in inquiry-guided research: The role of exemplars in narrative studies. *Harvard Educational Review* 60:415–442.

34. SIRMON, J., W. E. SHANDS, and C. LIGGETT. 1993. Communities of interests and open decision making. *Journal of Forestry* 91:17–21.

35. CURTIS, P. D., R. J. STOUT, and L. A. MYERS. 1995. Citizen task force strategies for suburban deer management: The Rochester experience. Pages 143–149 *in* J. B. McAninch, ed., *Urban Deer: A Manageable Resource? Proceedings of the 1993 Symposium of the North Central Wildlife Society.*

36. CURTIS, P. D., R. J. STOUT, B. A. KNUTH, L. A. MYERS, and T. A. ROCKWELL. 1993. Selecting deer management options in a suburban environment: A case study from Rochester, New York. *Transactions of the North American Wildlife and Natural Resources Conference* 58:102–116.

37. CANZANELLI, L., and M. REYNOLDS. 1996. Negotiated rule making as a resource and visitor management tool: A case study in the use of FACA. *Park Science* 16(2):1, 16–17.

38. BARRY, D. 1997. Cape Cod National Seashore, off road vehicle use: Proposed rule. *Federal Register* 62:24624–24631.

39. GUYNN, D. E. 1997. Miracle in Montana—Managing conflicts over private lands and public wildlife issues. *Transactions of the North American Wildlife and Natural Resources Conference* 62:146–154.

40. GUYNN, D. E., and M. K. LANDRY. 1997. A case study of citizen participation as a success model for innovative solutions for natural resource problems. *Wildlife Society Bulletin* 25:392–398.

41. COVEY, STEVEN R. 1992. *Principle-Centered Leadership.* Simon and Schuster, New York.

42. IRBY, L. R., W. E. ZIDACK, J. B. JOHNSON, and J. SALTIEL. 1996. Economic damage to forage crops by native ungulates as perceived by farmers and ranchers in Montana. *Journal of Range Management* 49:375–380.

43. SWENSSON, E. J. 1996. A survey of Montana hunter/rancher problems and solutions. Thesis. Department of Range Science, Montana State University, Bozeman, MT.

44. GUYNN, D. E., and D. STEINBACH. 1987. Wildlife values in Texas. Pages 117–124 *in* D. Decker and G. R. Goff, eds., *Valuing Wildlife: Economic and Social Perspectives.* Westview Press, Boulder, CO.

45. GUYNN, D. E., and J. L. SCHMIDT. 1984. Managing deer hunters on private lands in Colorado. *Wildlife Society Bulletin* 12:12–19.

46. PFEIFFER, J. W. 1991. *The Encyclopedia of Team-Development Activities.* University Associates, San Diego, CA.

47. BURKE, W. W. 1982. Organization development: Principles and practices. Pages 272–282 *in* W. B. Reddy and K. Jamison, eds. *Team Building: Blueprints for Productivity and Satisfaction.* University Associates, San Diego, CA.

48. BLEIKER, H., and A. BLEIKER. 1990. *Citizen Participation Handbook for Public Officials and Others Serving the Public,* 5th ed. Institute for Participatory Management and Planning, Monterey, CA.

49. DOYLE, M., and D. STRAUSS. 1982. *How to Make Meetings Work.* Jove Publications, New York.

50. PEYTON, R. B., and R. EBERHARDT. 1990. Communication and dispute resolution for fisheries and wildlife managers. Responsive Management Application Package. National Office of Responsive Management, Harrisonburg, VA.

51. FISKE, S. T., and S. E. TAYLOR. 1991. *Social Cognition,* 2nd ed. McGraw-Hill, New York.

52. ROBERT, M. 1982. *Managing Conflict from the Inside Out.* University Associates, San Diego, CA.

53. KOBERG, D., and J. BAGNALL. 1981. *The All New Universal Traveler.* William Kauffman, Los Altos, CA.

54. ARNOLD, J. D. 1992. *The Complete Problem Solver.* John Wiley and Sons, New York.

55. HORN, W. P. 1992. Strings and arrows: Challenges to sport hunting and wildlife management. *Proceedings of the Governor's Symposium on North American Hunting Heritage* 1:108–115.

56. LOKER, C. A., and D. J. DECKER. 1995. Colorado black bear hunting referendum: What was behind the vote? *Wildlife Society Bulletin* 23:370–376.

57. KELLERT, S. R. 1995. Managing for biological and sociological diversity, or "déja vu, all over again." *Wildlife Society Bulletin* 23:274–278.

58. JACOBSON, H. A., and D. C. GUYNN, JR. 1995. A primer. Pages 81–102 *in* K. V. Miller and R. L. Marchinton, eds., *Quality Whitetails—The Why and How of Quality Deer Management.* Stackpole Books, Mechanicsburg, PA.

59. WARREN, R. J. 1997. Deer overabundance. *Wildlife Society Bulletin* 25:213–573.

60. KOZICKY, E. 1977. Tomorrow's hunters—Gadgeteers or sportsmen? *Wildlife Society Bulletin* 5:175–178.

61. WOODS, G. R., D. C. GUYNN, JR., W. E. HAMMITT, and M. E. PATTERSON. 1996. Determinants of participant satisfaction with quality deer management. *Wildlife Society Bulletin* 24:318–324.

CHAPTER 12

Genetic Applications for Large Mammals

Rodney L. Honeycutt

INTRODUCTION

Molecular genetic approaches to the study of wildlife species are becoming widespread because they offer concepts and methodology that can be used in combination with demographic, geographic, and ecologic information for the conservation and management of biodiversity. Some of the primary information provided by genetically based studies includes (1) estimates of genetic variation within species and genetic divergence between species; (2) genealogical relationships among individuals within populations and phylogenetic relationships among populations, species, and higher taxonomic categories; (3) correlations between genetic variation, fitness, and life history traits; (4) relationships between current and future patterns of genetic variation and environmental changes occurring in ecological and geological time; (5) consequences of range fragmentation on existing patterns of genetic variation within and between species; (6) independent tests of hypotheses, derived from behavioral and ecological studies, that define the demography of populations with respect to population structure, breeding patterns, and dispersal; and (7) the development of objective criteria for the recognition of units of conservation and management.

With the advent of new molecular genetic techniques, the examination of genetic variation in wild species is becoming more straightforward, and the range of molecular markers currently available allows biologists to examine patterns of genetic variation at several spatial and temporal scales. I focus on how genetic markers and genetic information can be used in combination with ecological, behavioral, and biogeographical data to obtain a better understanding of the basic biology and management of large mammals in North America. I will address four primary questions in this chapter. First, is genetic variation important to the survival and sustainability of wildlife populations? Second, what processes contribute to the patterns of genetic variation that one observes for wildlife species? Third, how is genetic variation measured, and what are some commonly used molecular markers? Fourth, how have genetic markers been used in studies of large mammals in North America, and what are the management implications of these studies?

Without genetic recombination, populations and species would show a trend toward increasing homozygosity (i.e., lack of genetic variation). Although short-term survival of populations and species is influenced more by demography and ecology, a lack of genetic variation can compromise long-term resilience to environmental change (1, 2). If genetic variation contributes to the overall fitness (i.e., reproductive success) of an individual, then presumably there is a relationship between measures of genetic diversity, survival of genotypes, and the evolutionary potential of a population or species.

Fitness relates to an individual's ability to contribute to the next generation, implying differential survival of individuals. Heterozygosity is a measurement used to determine the extent of variation within an individual or population, and overall heterozygosity for an individual can be estimated by totaling the number of genes (or loci) for which the individual is heterozygous (two different alleles per locus). Several studies have demonstrated a correlation between fitness and genetic diversity as measured by estimates of heterozygosity. Some of the traits related to fitness that seem correlated with heterozygosity include developmental stability, growth and development rates, metabolic efficiency, fertility, survival, and resistance to disease (3).

In bighorn sheep, there appears to be a relationship between horn size in rams and breeding success. Fitzsimmons and others examined four polymorphic allozyme loci in 347 individuals, representing 12 populations of bighorn sheep (4). A positive correlation between heterozygosity and horn growth in rams was observed for older age classes (6, 7, and 8 years), with heterozygosity in 7-year-old rams explaining 21% of variation in horn volume. Therefore, near the onset of breeding at 7 years of age, rams with higher heterozygosity have higher horn volumes, suggesting a higher fitness.

Similar relationships between heterozygosity and antler size (i.e., main beam diameter and length) and body mass have been observed in white-tailed deer (5). As in bighorn sheep, there is some evidence that the correlation between heterozygosity and antlers and body mass is stronger in males 1.5 years of age or older (6).

Hedrick and Miller indicate that evaluating individual or population fitness on the basis of heterozygosity may be risky (7). Nevertheless, one can conclude that genetically depauperate lineages have a higher probability of extinction, and the presence of genetic variation is important to the long-term survival of the population. In terms of conservation, this conclusion means that a conservation plan should consider the amount of heterozygosity in the population under management.

Processes Contributing to Patterns of Genetic Variation

Factors influencing patterns of genetic variation within a species include mutation, selection, genetic drift, mating systems, and gene flow. The ultimate source of genetic variation is mutation, with genetic recombination in bisexual species allowing for the amplification and mixing of that variation. The minimum viable population (MVP) is a term used by conservation biologists to describe the minimum size of a population necessary for maintaining viability and long-term survival, and this size is generally determined on the basis of population dynamics (2). The effective population size (N_e) relates to the number of successfully breeding individuals rather than overall census size (N_c), and there is a relationship between N_e and the overall level of heterozygosity seen in a population (Table 12–1). As N_e decreases in size, the effects from genetic drift, a random process that reduces genetic variation, will become greater, increasing the probability that potentially deleterious alleles will increase in frequency within the population and beneficial alleles will be lost. Therefore, Nunney and Campbell have suggested that MVP be five to 10 times that of N_e (8).

TABLE 12-1

Patterns of Genetic Variation in African Buffalo as a Consequence of Refuge Size, Census Size, and Effective Population Size

Refuge	Area (ha)	Current Population Size	Effective Population Size[a]	Average No. Alleles/Locus[b]	Total Alleles[c] (%)	Heterozygosity[d] (%)
Kruger National Park	1,945,500	35,000	5,250	8.6 (28)	93.8	72.6
Umfolozi-Hluhluwe Complex	47,753	8,400	1,260	4.4 (3)	50.0	54.9
St. Lucia	38,826	175	26	3.1 (3)	50.0	45.0
Addo National Park	9,000	85	13	2.7 (1)	29.7	48.2

Source: Revised from Ref. 9. Reprinted with the permission of Cambridge University Press.

[a]Estimated based on the effective population size (N_e) being 15% of current population size.

[b]Average number of alleles across seven microsatellite loci derived from cattle.

[c]Percentage of total 64 alleles observed for all microsatellite loci examined.

[d]Mean heterozygosity calculated for all microsatellite loci.

Genetic Drift. Many large mammal species reveal effects of genetic drift in small populations, especially as they have become more subdivided and reduced in size. A classic example is the African buffalo, a once widespread species that has been subdivided into small, fragmented populations isolated on game reserves (9). Using microsatellite markers, there was a significant correlation between the amount of genetic variation (i.e., heterozygosity and the number alleles at a locus) and population size on game reserves (Table 12–1). Furthermore, they suggested that current levels of genetic variation are a consequence of genetic drift subsequent to recent habitat fragmentation, with the rate of loss related to N_e. As a consequence of their findings, predictions about the rate at which genetic variation will be lost in the smaller populations can be made, and a management plan for the genetic enhancement of those populations through translocations was developed.

Some other examples of drift in small populations can be seen in comparisons between mainland and island populations (Table 12–2). Black bear from Newfoundland have a heterozygosity of 36% in comparison to 80% for mainland populations (10).

TABLE 12–2
Genetic Diversity (Percent) in Large Mammal Populations

Species	mtDNA[1]	P (A)[2]	H (A)[3]	H (M)[4]	References
Artiodactyla					
Cervidae					
Moose					
Scandinavia		9.4	2.0		47
Sweden		4.0	0.6		46
Europe and N. America	3.9				39
Elk		4.0	1.2		19
White-tailed deer	0.4	32.0	10.0		20, 35
Texas				82.0	27
West Texas	2.2				17
New York		23.4	7.6		37
Georgia		26.3	6.3		50
South Carolina		41.4	8.0	82.0	27, 50
Arkansas			6.1		28, 44
Maryland		20.7	7.6		49
Tennessee		42.0	9.7		29
S.E. USA (mainland)	0.0–0.8		11.3		22
S.E. USA (island)	0.5–0.6		8.3		22
Florida Keys	0.0		6.4		22
Mule deer		10.5	5.3		16
West Texas			3.4		17
Antilocapridae					
Pronghorn					
North America	0.8	13.2	2.4	42.0–73.0	31, 33
Texas		13.2	2.7		30
Bovidae					
Domestic sheep				66.2	24
Bighorn sheep	0.4	12.4	4.1	56.6	23, 24, 34
Dall's sheep		3.7	1.5		48
Mediterranean mouflon				61.0	43
Alpine ibex		6.1	0.7		51
American bison	0.7	5.3	2.3		16, 56
Domestic cow	0.8	10.5	3.5		16, 56

Importance of Genetic Variation to Wildlife Conservation

TABLE 12–2 *Continued*

Species	mtDNA[1]	P (A)[2]	H (A)[3]	H (M)[4]	References
Suiidae					
Collared peccary					
Arizona	0.2–0.8				53
Arizona/Texas	2.0–3.6				53
Domestic pig		14.3	5.0		26
European wild boar		20.0	3.1		26
Carnivora					
Canidae					
Domestic dog	0.4			72.9	25, 27
Gray wolf					
Old World	0.5				59
New World	0.2				59
North America	0.1–0.8	20.0	6.1	62.0	45, 58
Vancouver Island				42.0	45
Isle Royale	0.0	8.0	4.0		58
Coyote	2.3			67.5	11, 32
Ethiopian wolf	0.0			24.1	11, 25
Golden jackal				52.0	11
Black-backed jackal				67.4	11
Ursidae					
Grizzly bear					
Alaska				75.0	21
Brown bear					
Europe (East/West)	7.1				52
Europe (within clade)	0.4–1.5				52
American black bear					
North America	3.6	13.3	1.5	80.0	10, 18, 36
N. America (within clade)	0.1				18
Newfoundland				36.0	10
USA, Tennessee		22.0	8.0		57
Felidae					
Domestic cat		21.0	8.2	77.0	38, 40, 41
Puma					
North America		27.0	1.8–6.7	61.0	38, 45
Texas		9.8	4.10	29.4–46.0	38, 45
Florida		4.9	1.8		45
Bobcat				76.6–81.0	54
Iberian lynx					
Donana National Park				19.7	55
Sierra Morena Mts.				53.2	55
African lion					
Serengetti ecosystem		11.0	3.8		42
Kruger National Park		7.0	2.3		42
Ngorongoro Crater		4.0	1.5		42
Gir Forest		0.0	0.0		42
Cheetah				39.0	38
East Africa		4.0	1.4		45
Southern Africa		2.0	0.0		45
Ocelot		20.8	7.2		41
Mexico				69.8	55
Texas (Willacy County)				55.0	55
Texas (Laguna Atascosa)				36.0	55

[1]Mitochondrial nucleotide diversity estimated from either control region sequences or RFLPs.
[2]Average number of polymorphic allozyme loci.
[3]Average allozyme heterozygosity.
[4]Average heterozygosity for microsatellite loci.

Wolves from Vancouver Island have decreased levels of heterozygosity, and island populations of white-tailed deer from the southeastern United States show a reduction in allozyme heterozygosity and mitochondrial nucleotide diversity (11, 12).

Processes Eroding Variation.   Three other processes that contribute to the loss of genetic variation in populations are population bottlenecks, founder effect, and inbreeding. Population bottlenecks result from a reduction in population size, and experimental studies have verified that bottlenecks can result in loss of genetic variation (13). The extent of a bottleneck's contribution to the loss of genetic variation and reduction in fitness (i.e., fixation of deleterious alleles and inbreeding depression) depends on the magnitude and duration of the bottleneck. For instance, if N_e is low and population growth subsequent to the bottleneck is slow, then more genetic variation (i.e., number of alleles) will be lost, especially with respect to mitochondrial DNA markers (14, 15).

Several species of large animals have experienced recent demographic bottlenecks as a consequence of range fragmentation, decline in population size, or overharvesting. In many cases a loss of genetic variation, resulting from a genetic bottleneck, accompanied the overall reduction in population size. For instance, several large felid species have undergone regional or species wide bottlenecks that have resulted in an overall reduction of genetic variation (Table 12–2). Today there is only one small isolated population of approximately 30 Florida panthers in southern Florida. In comparison to larger populations in the western United States, where bottlenecks have not occurred, this subspecies reveals lower levels of variation for all genetic markers that have been examined (45). African lion from Ngorongoro Crater, East Africa, Gir Forest, and western India represent populations that have undergone recent population bottlenecks as a result of either disease outbreaks or habitat loss (60, 61). In comparison to larger lion populations in the Serengetti ecosystem and Kruger National Park of South Africa, the Ngorongoro Crater and Gir Forest populations have lower levels of genetic variation. Unlike the recent bottlenecks experienced by Florida panther and African lion, the cheetah presumably underwent two population bottlenecks, one dating to historical times in the late Pleistocene and one within the last century (42). According to the authors, these bottlenecks and increased levels of inbreeding have left the cheetah genetically depauperate in comparison to other species of felids. Although the bottleneck hypothesis is somewhat controversial, cheetah reveal evidence of decreased fitness (consistent with inbreeding depression) including high infant mortality, low female fecundity, increased male sterility, and possibly increased susceptibility to disease (62, 40, 41, 63).

In North America, pronghorn antelope and American bison experienced extreme population declines in the 1800s as a consequence of overharvesting. Unlike the examples of large felids, neither species experienced extreme loss of genetic variation for nuclear or mitochondrial DNA markers (Table 12–2) (33, 31, 64). The lack of a genetic bottleneck in response to the decline in population size may be related to the relatively fast increase in population size in only a few generations subsequent to the demographic bottleneck. Additionally, the reduction in population size may not have been severe enough to result in an extreme loss of either nuclear gene variation (i.e., alleles and heterozygosity) or mitochondrial DNA variation (i.e., number of haplotypes and nucleotide diversity among haplotypes).

Unlike pronghorn and bison, several ungulates show low levels of genetic variation throughout their range, possibly as a result of genetic bottlenecks. Moose, in the Old World and New World, show low levels of variation for both nuclear and mitochondrial markers. Swedish populations of moose experienced a severe population bottleneck during the 19th century, and the bottleneck persisted for many generations (46). As a consequence, these populations today reveal extremely low levels of allozyme variation (Table 12–2). Recent studies involving broad-scale surveys of mtDNA variation reveal low variability throughout the Holarctic (65). A similar result was found for the normally polymorphic Mhc class II gene, which revealed low levels of allelic

Importance of Genetic Variation to Wildlife Conservation

variation and low divergence among alleles in moose relative to other mammals (39). These results suggest that the species experienced a recent population bottleneck followed by an expansion.

Elk from Yellowstone National Park show low levels of allozyme heterozygosity, even though there is no evidence of a recent population bottleneck (19). Recent data from our lab indicate low levels of variation in elk for microsatellite loci as well. Sage and Wolfe also found low heterozygosity for Dall's sheep from Alaska (48). These authors suggested that lower levels of genetic variation in large mammals occupying northern latitudes may be the result of historical processes of expansion and contraction of ranges in response to glacial cycles. These cycles of glacial activity and serial recolonizations by small and peripheral founding populations may have resulted in a serial loss of alleles. Such events may very well explain patterns observed in elk and moose.

Founder effect is the result of the establishment of a new population by a small number of individuals. The genetic makeup of the new population depends on variation seen in the founders, with one consequence being a loss of genetic variation in the newly founded population. Generally, island populations are the consequence of founding events resulting from dispersal from mainland populations. Many large mammal populations occupying islands, such as white-tailed deer in the southeastern United States, black bear from Newfoundland, and gray wolves from Isle Royale, show considerably less variation than their mainland populations as a result of founder effect and genetic drift (Table 12–2) (22, 10, 58). In the case of Isle Royale wolves, an estimated 50% loss of allozyme heterozygosity was observed as a consequence of complete isolation from the mainland and small effective population size. There also was a decline in average percent similarity, as assessed with multilocus DNA fingerprints, of 58% in the island population in comparison to mainland forms.

Reintroductions or restocking of wildlife populations can result in founder effects. For instance, ibex populations in Switzerland have undergone four phases of population reduction including (1) an original population bottleneck, (2) establishment of a captive breeding program, (3) founding of new populations from the captive stock, and (4) establishment of new populations from newly founded stocks (51). This process has resulted in low levels of genetic variation within populations and high divergence between populations. At several localities in the Rocky Mountains, bighorn sheep from Wyoming were used to restock declining populations with a small number of founders. As a consequence of these founding events, the new populations lost low-frequency alleles and heterozygosity in comparison to the original source population. A similar pattern can be seen for pronghorn populations from Yellowstone National Park that were used to restock depleted populations. Pronghorn from Yellowstone show more mtDNA haplotypes and higher levels of heterozygosity in comparison to other populations (31).

Finally, inbreeding (i.e., mating between close relatives) alters patterns of genetic variation by increasing homozygosity and is more likely in small subdivided populations. A major result of inbreeding is a decrease in individual fitness as a result of increased homozygosity of deleterious alleles or loss of heterozygosity. Inbreeding is a potentially serious problem for small captive populations, and in normally outcrossing species inbreeding can result in a reduction of viability, growth and development, and fertility. For instance, high levels of inbreeding in captive ungulates and accompanying low levels of heterozygosity are correlated with the production of less viable offspring (66). As shown by Hedrick and Miller, species differ with respect to inbreeding depression (cost of full-sib and parent–offspring mating). Nevertheless, in mammals the cost of inbreeding averages 0.33, suggesting that females should avoid mating with close relatives (7, 67).

Population Subdivision and Reduced Gene Flow. Gene flow tends to homogenize populations in terms of genetics (68). As populations within a species become subdivided and gene flow is reduced, alternate alleles will tend to become fixed as a consequence of genetic drift, with the rate depending on the size of the isolated populations. Several factors can contribute to reduction in gene flow among populations including (1) dispersal distance and sex-biased dispersal, (2) nonrandom mating among groups of individuals, (3) historical geographic barriers, and (4) ecological barriers resulting from recent habitat fragmentation and other alterations to the landscape.

Although one might expect large mammals to reveal little evidence of population subdivision, genetic studies of ungulates have revealed considerable population subdivision over short distances and spatial structuring in relatively small areas. Genetic studies of white-tailed deer have shown evidence of genetic subdivision over short geographical distances, especially when maternally inherited mtDNA markers have been used (69, 70, 37, 12, 22). In white-tailed deer males generally disperse and females demonstrate philopatry and remain in their natal groups, and this pattern of social structuring and sex-biased dispersal contributes to the subdivision seen in deer (71, 72). For instance, Matthews and Porter found genetic subunits related to social groups within populations of white-tailed deer from New York (37). This spatial aggregation of related individuals appears to be the result of matrilineal groups occupying the same winter range. Moose from central Sweden also were found to demonstrate significant differences in allele frequencies over short distances (73). One explanation for this observation is that moose show strong site fidelity and do not disperse long distances to breed.

Patterns of genetic subdivision sometimes can be related to metapopulation structure. Bleich and others found mtDNA haplotype frequency differences among six metapopulations of mountain sheep, suggesting low female dispersal among metapopulations (74). In several cases the appearance of single unique mtDNA haplotypes indicate that the metapopulations may have been founded by single mtDNA lineages. In the case of collared peccary in Arizona, mtDNA markers reveal geographic discontinuities across the state. Theimer and Keim interpret this pattern to be the result of kin-founding of three localities by different maternal lineages from Mexico (53).

Range fragmentation can influence patterns of genetic diversity in large mammals depending on their ability to disperse across the fragmented landscape. Wolves and coyotes differ with respect to their degree of genetic subdivision, with wolves showing a high level of subdivision and coyotes showing little evidence (59, 11). This may be the result of recent fragmentation and population declines in wolves or perhaps the consequences of drift in populations isolated during the Pleistocene.

Finally, major historical barriers to gene flow can result in high levels of genetic divergence among isolated populations of a species. Some examples are as follows (Table 12–2): (1) Collared peccary populations in Arizona and Texas show high levels of mtDNA haplotype diversity of 2% to 3.6% in comparison to values of 0.2% to 0.8% within Arizona (53). (2) Despite the low levels of within-population variation, moose from the Old and New World differ by an mtDNA divergence of 3.9% (39). (3) North American black bear populations can be subdivided into two major mtDNA lineages (i.e., a coastal and continental lineage) separated by 3.6% divergence (18). (4) European brown bear populations can be subdivided into an eastern and western group separated by 7.13% mtDNA divergence, with the western group consisting of two smaller sublineages (52). (5) White-tailed deer populations from the southeastern United States reveal a pattern of genetic and geographic subdivision similar to that observed for unrelated groups of animals (12). This is most likely the result of a common barrier to gene flow resulting from the same historical biogeographic event.

MOLECULAR MARKERS AND MEASURING GENETIC VARIATION

There are two major sources of genetic information. The nuclear genome contains coding (e.g., structural and regulatory genes) and noncoding (e.g., introns, repetitive sequences) sequences of DNA. These sequences are inherited either biparentally, if found on autosomal or X chromosomes, or from the male only, if located on the Y chromosome. The mitochondrial genome is a closed circular DNA molecule housed within mitochondria in the cellular cytoplasm. In mammals the mitochondrial genome is approximately 16,000 bases in length and is composed of coding sequences (i.e., 22 tRNA, 2 rRNA, and 13 protein-encoding genes) and a noncoding control region (i.e., D-loop), responsible for the initiation of transcription and replication. Unlike most nuclear DNA markers, mitochondrial DNA (mtDNA) is inherited only from the mother (75).

Genetic variation within and between individuals is measured at the protein level by estimating mutations involving amino acid replacements and at the DNA level by quantifying mutations resulting from single base substitutions or the addition/deletion of bases. Measurements from protein and DNA can be used to estimate overall genetic similarity among individuals, populations, and species and genetic diversity represented by individual or population heterozygosity (i.e., percentage of loci heterozygous per individual).

Measuring Variation in Proteins. One of the earliest and most cost-effective methods for estimating levels of genetic similarity and heterozygosity in game animals was protein electrophoresis. This technique measures allelic variation at a locus as a consequence of differences in the net charge of a protein. Specific tissue extracts are loaded onto a starch or polyacrylamide gel, and proteins are separated according to their migration in an electric current. Variation at specific loci is identified by histochemical staining, and the genotype (number and frequency of alleles at each locus) is determined from the resultant banding patterns (Figure 12–1a) (for a review see Ref. 76). The major disadvantage of protein electrophoresis is that it underestimates genetic variation because it detects changes in primarily charged amino acids rather than all amino acid differences in a protein (77). This observation is true for large mammals where most species average under 4% heterozygosity. Thus, protein electrophoretic markers have limited usefulness for detailed studies of the genetic structure of populations, especially when compared to the highly variable microsatellite DNA loci, which reveal heterozygosity estimates several times higher than estimates derived from protein electrophoresis for the same species (Table 12–2).

Although protein electrophoresis has limitations, the technique has proven useful in several respects. First, estimates of genetic diversity have been used to examine patterns of geographic variation and gene flow among populations and determine relationships among species of game animals (47, 16, 78, 79, 70, 80, 81, 12, 31). Second, levels of heterozygosity provide a basis for comparisons among diverse species. For instance, mammals have relatively low levels of heterozygosity, averaging 4.1% (82). Many North American large mammals (e.g., moose, elk, pronghorn antelope, Dall's sheep, and bison) have heterozygosity estimates below the average (Table 12–2). One major exception is the white-tailed deer, which averages twice the level of heterozygosity found in other mammal species. Some of the differences observed for these species may be explained by historical fluctuations in population size in response to population expansion, population subdivision, population bottlenecks, founder events, or genetic drift in small populations.

Finally, estimates of heterozygosity derived from protein electrophoresis have been used in studies designed to evaluate the relationship between genetic variation and individual fitness. Some of the most extensive studies have been conducted on white-tailed deer. These studies have revealed a correlation between overall heterozygosity and several traits related to fitness including rates of fetal growth, number of fetuses, body fat, antler size and development, and twinning in females (83, 83a, 5, 6, 84).

Measuring Variation in DNA. The two primary methods of estimating genetic divergence at the level of nucleotide substitutions in DNA are restriction fragment length polymorphisms (RFLP analysis) and direct nucleotide sequencing. The restriction fragment polymorphism analysis uses the digestion of purified DNA with restriction endonucleases (or restriction enzymes) to detect single nucleotide substitutions or

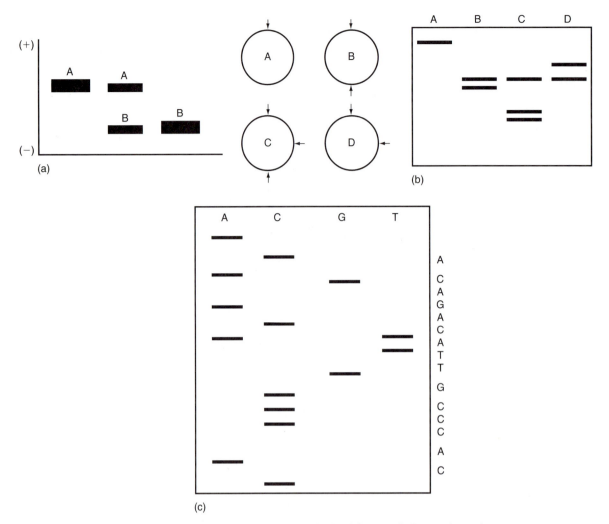

Figure 12–1 (a) Allozyme variants derived from gel electrophoresis. The picture represents patterns of variation for a monomeric protein represented by three genotypes, two of which are homozygotes (AA and BB) and one of which is a heterozygote (AB). (b) Restriction endonuclease digestion of mitochondrial DNA followed by gel electrophoresis. Note that for each restriction endonuclease site a fragment is produced. Four unique haplotypes (A,B,C, and D) result from different patterns of restriction endonuclease digestion. (c) Nucleotide sequencing gel showing four possible bases at each site. The gel is read from bottom to top (see continuous sequence at the right of the figure) with larger fragments shown at the top.

insertion/deletion (indel) events, involving one or more bases. These restriction endonucleases cleave at specific sites (e.g., the restriction enzyme *Eco*RI recognizes GAATTC, cleaving between the G and A) representing combinations of four, five, six, or more bases, with the result being a fragment pattern (RFLP) denoting the number of times a specific restriction site occurs in a piece of DNA (Figure 12–1b). Any base substitutions that alter specific restriction sites will result in a change in the size and number of fragments (RFLP) as a consequence of changes in the number of restriction sites. Most studies of RFLPs within and between populations and species survey restriction site variation for a large number of restriction enzymes. Restriction fragment polymorphism analysis was a common technique employed by many early studies of mtDNA variation either among populations within a species or among closely related species (77).

Restriction fragment polymorphism analysis involves the sorting of DNA fragments, resulting from restriction endonuclease digestion, by size with the use of either agarose or polyacrylamide gel electrophoresis. There are several ways to visualize RFLPs (85): (1) Gels containing digested DNA can be stained with ethidium bromide, a fluorescent stain that binds with DNA, and then exposed to UV light. This is a common procedure for either digestion of total mitochondrial DNA or the digestion of specific DNA fragments amplified with the polymerase chain reaction (PCR). (2) Southern blot hybridization involves the transfer of digested DNAs from agarose gels to nylon membranes followed by hybridization with a probe specific to either the entire mtDNA genome or a region of either the mitochondrial or nuclear genomes. (3) Digested DNA fragments can be labeled with radioisotopes, electrophoresed on agarose or acrylamide gels, and visualized with autoradiography.

The most accurate means of estimating genetic diversity is nucleotide sequencing of specific DNA fragments (86) (Figure 12–1c). At one time this procedure was difficult in that a knowledge of recombinant DNA techniques was required prior to obtaining nucleotide sequence data. With the advent of PCR, fragments can be sequenced directly using either manual or automated sequencing. The combination of PCR and automated sequencing allows for larger databases of genetic variation to be accumulated more cost effectively and at a faster rate.

Genetic variation at the DNA level can be estimated from RFLP and nucleotide sequence data. Nucleotide sequence diversity provides an estimate of sequence divergence between haplotypes or alleles, and haplotype diversity provides an equivalent measure of heterozygosity (87, 88). Both of these measures provide information of genetic variation within and between populations and species.

The polymerase chain reaction has revolutionized the study of genetic variation in natural populations. The method uses enzymatic amplification and thermal cycling to synthesize a specific fragment of DNA in large numbers (89). The procedure requires small amounts of material and provides many useful genetic applications. For instance, feces and hair have been used in analyses employing PCR for the assessment of genetic variation and the determination of size and sex ratios of populations (90, 91, 92, 93, 94, 95). One specific example involves studies of the Pyrenean brown bear, whereby a combination of track data and genetic typing from hair and feces was used to determine the number and sex of remaining individuals of this threatened species (96).

The PCR technique also allows for the genetic analysis of fossil material and specimens housed in museum collections (97, 98). Therefore, one can obtain historical information on genetic variation in populations. Antlers and bone provide a source of DNA that can be extracted and used in genetic analyses. My colleagues and I have routinely used this technique to study microsatellite DNA variation in white-tailed deer and other cervids. Genetic information derived from small quantities of tissue, hair, and bone has direct application for wildlife law enforcement. For instance, genetic markers, diagnostic for individual species of large mammals, provide a direct means of identifying illegally taken animals (99). Microsatellite DNA loci, amplified with PCR, also provide a means of identifying individuals collected illegally.

Nuclear DNA Markers

Y Chromosome Markers. Zinc finger genes (e.g., ZFY) and the SRY gene provide Y-chromosome-specific markers in mammals (100, 101). Unlike the high intraspecific variation seen for mtDNA, these Y-chromosome markers do not show intraspecific variation in the mammal species that have been examined (102, 103, 33). Therefore, their usefulness for detailed population genetic studies of male lineages is limited. Nevertheless, these markers do have other uses in studies of the genetics and population biology of large mammals.

Despite little within-species variation, the ZFY gene does differ between species, thus providing species-specific Y markers that can be used to study the dynamics of interspecific hybridization. For instance, allopatric populations of mule deer, white-tailed deer, and black-tailed deer are fixed for alternate alleles at the albumin locus and unique mtDNA haplotypes (16, 17). In areas of overlap, however, white-tailed deer hybridize with mule deer and black-tailed deer, and wild hybrids of mule deer/white-tailed deer crosses have been found in Texas, the southwestern United States, Montana, and the Pacific Northwest (104, 105, 106, 107, 17). A distinct asymmetry to hybrid crosses between white-tailed and mule deer was observed but the interpretation of that asymmetry differs among researchers. Carr and others and Ballinger and others suggested that based on a maternal marker (i.e., mtDNA) the cross generally involved a white-tailed female and mule deer male, whereas Cronin suggested the opposite interpretation from his studies of mtDNA variation in deer from Montana and Oregon (104, 17, 106). Nucleotide sequence data from the mitochondrial cytochrome *b* gene supported a cross between mule deer females and white-tailed males (20). Recently, Cathey and others, using a ZFY intron as a male specific marker, provided strong support for mule deer females crossing with white-tailed males (103). This interpretation makes biological sense in that habitat changes have favored the expansion of white-tailed deer populations at the expense of mule deer, thus increasing the likelihood of hybridization occurring from crosses suggested by ZFY.

The ZFY locus may prove to be useful for diagnosing phylogenetic relationships among species and families of large mammals. For instance, the determination of relationships among the artiodactyl families Antilocapridae, Bovidae, and Cervidae has been difficult, with different molecular markers providing weak support for alternate phylogenetic trees (108, 109, 110). Nucleotide sequences of the ZFY intron provide strong support for a relationship between the families Antilocapridae and Bovidae, followed by Cervidae (33).

Fragments of DNA containing SRY and ZFY gene sequences can be amplified with the use of PCR. As a result, this method provides a reliable and indirect identification of males, simply by the presence or absence of a PCR amplification product (95). For instance, PCR and SRY-specific markers, were used to determine the sex of free-ranging brown bears from hair or feces collected in the field (92, 94). This same procedure can easily be applied to other large mammals.

Major Histocompatibility Complex (MHC). MHC is a gene family composed of two distinct classes of polymorphic loci, I and II, that are important to immune defense in mammals (111). Most mammals reveal high levels of polymorphism presumably maintained by balancing selection (112, 113). For instance, RFLP analyses of MHC class II loci of bighorn sheep reveal a mean heterozygosity value of 32.5%, even though this species has relatively low levels of allozyme heterozygosity (Table 12–2) (114). Class I genes in many other mammals, such as the domestic cat (H = 28.9%), humans (H=17.4%), and the Palestinian mole-rat (H=51.3%), also reveal high levels of heterozygosity (113). Although it has been suggested that species with low levels of genetic variation at MHC loci may be more susceptible to disease, the relationship between lack of MHC diversity and increased susceptibility to disease is not well understood (63, 45, 8).

Exceptions to the general pattern of high heterozygosity have been observed in populations and species that have presumably experienced demographic reductions or population bottlenecks (113, 115). The European beaver represents a species that underwent extreme population bottlenecks in regions of Scandinavia but has subsequently increased in numbers in more recent time. Restriction fragment polymorphism analyses of MHC class I and class II genes indicate an absence of polymorphism within populations and low levels of divergence among geographic regions (115). The low levels of genetic variation detected with DNA fingerprints, derived from minisatellite loci, confirm that this lack of MHC may be the result of population declines in the 1800s. A similar pattern of reduced variation in MHC class I genes can be seen in other populations and species that have undergone population bottlenecks or population declines, including East African and South African cheetah (H = 6.7% and 5.1%, respectively) and lion from both Ngorongoro Crater (H = 8.0%) and the Gir Forest (H = 0%) in comparison to Serengetti lion (21.8%) (113). Recent nucleotide sequence comparisons of exon 2 of DRB1, an MHC class II gene, in European and North American moose also reveal low levels of genetic diversity, with the frequency of synonymous substitutions between alleles in moose being 0.009 in comparison to 0.087 and 0.046 for humans and cattle, respectively (39). These authors suggest that moose underwent a population bottleneck prior to their divergence in the Old and New Worlds.

Nuclear DNA Satellites. Within the nuclear genome are classes of repetitive sequences that demonstrate high levels of genetic variation that can be used to establish individual-specific markers. The term *DNA fingerprinting* is associated with a class of hypervariable repetitive sequences distributed throughout the nuclear genome. The traditional DNA fingerprints, known as hypervariable minisatellite loci (VNTRs), are characterized by a tandem array of repeating units varying in length. Variation in VNTRs can be detected through RFLP analysis and hybridization with specific VNTR probes known to cross-hybridize with many different organisms (116). The digestion of DNA outside the repetitive array results in differences in the number of repeats detected on the basis of band size. The technique of DNA fingerprinting has been used for estimating relative genetic diversity among populations and for the establishment of parentage and kinship (117, 115, 45, 118). The major difficulty with VNTRs is related to the problem of identifying alleles at a locus. Although the accurate estimation of population genetic statistics such as heterozygosity, relatedness, and gene flow is made difficult by the uncertainty associated with allelic designation, DNA fingerprints do provide an assessment of genetic variation in populations that is similar to estimates derived from other types of markers. For instance, the Florida panther is represented by a small number of individuals in comparison to more widely distributed subspecies. Estimates of heterozygosity from VNTRs are consistent with these findings, with H = 7.5% for the Florida subspecies and 46.8% for the western U.S. subspecies (45).

Microsatellite DNA loci are tandemly repeated sequences composed of a simple repeat unit (2 to 5 bases) flanked by unique sequences that are dispersed throughout the mammalian genome (119, 120, 121). Microsatellite loci vary in length or the number of repeat units, and allelic variation at these loci results from the addition or deletion of individual repeat units, and unlike VNTRs, alleles can be scored and heterozygosity estimated for individual loci. Genetic variation at each locus can be screened with the use of locus-specific flanking primers, PCR amplification, and manual or automated sequencing (Figure 12–2). The primary value of microsatellites as genetic markers is associated with their high level of heterozygosity in comparison to more traditional markers (e.g., protein electrophoresis; Table 12–2). This high level of polymorphism and heterozygosity is the consequence of a mutation rate that exceeds 10^{-3} to 10^{-4} per generation (122, 123).

Microsatellite DNA loci have been characterized for several species of large mammals including the white-tailed deer, caribou, and black bear (124, 125, 10). In addition, the unique primer sites, flanking many DNA microsatellites, are conserved across taxa, thus allowing use of microsatellite loci isolated from one species to study related species (126, 127). As a consequence of the conserved nature of many microsatellite priming sites, large panels isolated from domestic species for detailed gene mapping can be used to study genetic variation in wild species. For instance, hundreds of microsatellite loci have been isolated from the cow and sheep genomes, and many of these have proven useful for studies of wild species including mouflon, bighorn sheep, pronghorn, mule deer, elk, moose, and many other species of artiodactyls (128, 129, 130, 131, 132). Microsatellite loci characterized from the domestic dog

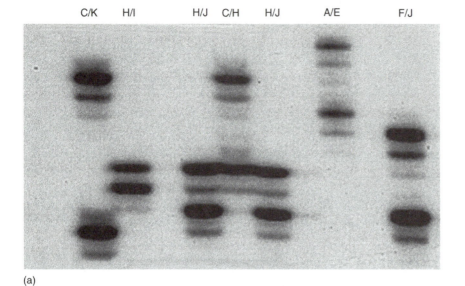

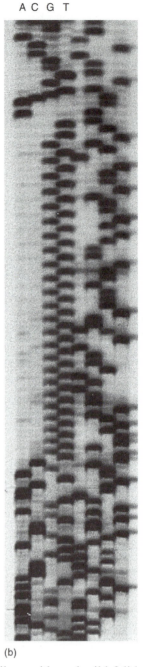

Figure 12–2 Genetic variation can be estimated using DNA extracted from a variety of body parts. Cast deer antlers were sampled: (a) PCR amplified antler DNA and microsatellite genotypes (b) nucleotide sequencing gel showing sequence of DNA microsatellite. (Photograph by R. L. Honeycutt.)

and domestic cat also have been used for studies of wolf-like canids and wild felid species, respectively, and those isolated from black bear work well with more divergent ursid species (133, 38, 10).

High polymorphism and high mutation rates make microsatellites ideal markers for establishing relationships among individuals (e.g., paternity), analyzing genetic variation within and between populations, and determining the genetic structure of populations. If mother and offspring relationships can be identified, then it is possible to exclude potential fathers from a pool of males in the population. Markers demonstrating a high degree of polymorphism increase the overall accuracy of determining relationships among individuals. In the case of microsatellite loci used for grizzly bears, the probability of paternal exclusion ranged between 1×10^{-5} to 4×10^{-9}, thus providing a means of studying reproductive success of males (21). Using microsatellite markers, it was determined that no male sired more than 11% of known offspring, with an overall breeding success of males being 49% (21). In the case of African wild dogs, 14 microsatellite loci allowed for an exclusion probability for father/offspring of 99.6% and for mother/offspring of 99.7% (134). For white-tailed deer the probability of two individuals having the same genotype, derived from six loci, is 2.6×10^{-8}. Therefore, a small panel of highly variable microsatellite loci can provide a powerful means of parental exclusion and the determination of relationships among individuals in a deer population.

The high mutation rates and levels of polymorphism associated with DNA microsatellites make these markers ideal for studying patterns of gene flow, responses of recent population subdivision as a consequence of habitat fragmentation, and population bottlenecks. Examples of how microsatellite variation can be used to assess recent declines in population numbers as a result of habitat fragmentation and restricted gene flow can be seen for the Iberian lynx population in Donana National Park and ocelot populations in Texas (Table 12–2). In both cases recent changes in land use have altered the habitat to an extent that populations have become isolated and numbers of individuals greatly reduced. The smaller isolated populations, such as Donana National Park for the lynx and Laguna Atascosa National Wildlife Refuge for the ocelot, show significant reduction in microsatellite heterozygosity, thus providing a means of estimating the rate at which variation has been lost in these populations as a result of recent isolation. Similar responses to isolation and small populations can be seen including (Table 12–2) (1) black bear from mainland North America relative to Newfoundland; (2) the Ethiopian wolf, a species that has undergone severe population declines, in comparison to other wolf-like canids; and (3) wolves on Vancouver Island in comparison to mainland North America.

Mitochondrial DNA Markers

In mammals, mitochondrial genes that encode proteins and RNAs and noncoding sequences demonstrate a faster rate of sequence divergence relative to nuclear genes. As a consequence of maternal inheritance and the lack of recombination, mtDNA variation becomes fixed at a faster rate within populations and along lineages than the rate seen for nuclear variation (75). Therefore, populations that have restricted gene flow as a result of historical and ecological barriers are more likely to show distinct patterns of mtDNA variation, especially when there is biased female dispersal (12, 22).

The use of mtDNA to study patterns of geographic variation within a species is referred to as *intraspecific phylogeography* (135). A phylogeographic pattern of mtDNA variation reflects the relationship among mtDNA haplotypes (gene tree or phylogeny) obtained within and between populations surveyed over either a subset or the entire range of the species. Phylogeographic studies of codistributed species provide information on the recent history of communities (136). The basic procedure used in studies of intraspecific phylogeography is to relate the gene tree to geography. The degree of subdivision observed with this approach depends on several factors including the vagility of the species, the degree to which females show philopatry, and the age and extent of any potential barriers to dispersal. Within a species, populations may maintain unique mtDNA haplotypes with frequencies and phylogeographic distributions reflecting isolation by distance and the presence of geographic barriers to gene flow. If dispersal distance is high, one would expect little mtDNA subdivision.

Phylogeographic studies of mtDNA have employed RFLP analysis of either total mtDNA or PCR fragments and direct nucleotide sequencing of PCR amplified regions of the mtDNA molecule. One of the major observations from phylogeographic studies of large mammals is the lack of relationship between unique mtDNA lineages and formally recognized subspecies. For instance, there are five recognized subspecies of pronghorn in North America, yet mtDNA markers, allozymes, and microsatellite DNA markers reveal only two major lineages (31, 33). The wood and plains bison subspecies show little difference with mtDNA and nuclear markers (137, 138, 64). The seven subspecies of black bear in western North America were found to represent only two distinct mtDNA lineages (18), and Cronin and Bleich found only one distinct mtDNA haplotype for the five subspecies of mule deer in California (139). Subspecies of white-tailed deer from the

Figure 12–3 A consensus phylogeny of mitochondrial DNA (mtDNA) haplotypes of white-tailed deer populations in the southeastern United States. This genetic study encompassed the range for five recognized subspecies, *Odocoileus virginianus virginianus* (#1), *O. v. osceola* (#2), *O. v. seminolus* (#3), *O. v. clavium* (#4), and *O. v. nigribarbis* (#5). There were only three primary mtDNA haplotype assemblages (shown by circled areas), and these haplotype assemblages do not coincide with the ranges of recognized subspecies. (From Ref. 22.)

southeastern United States also are not characterized by unique mtDNA lineages (12) (Figure 12–3). One of the few cases of agreement between mtDNA phylogeographic patterns and subspecific designations is the collared peccary (53).

Several pairs of North American mammals show evidence of interspecific hybridization as a result of recent range overlap in response to changes in the landscape and demographic changes in each species. Two potential consequences of interspecific hybridization are the loss of fitness in the introgressed genotypes and the genetic swamping of one species by another. Mitochondrial DNA markers have been used to characterize the patterns of interspecific hybridization in several species of large mammals. One of the more controversial studies has involved evidence of hybridization between North American bison and cow (140, 56). Of the 572 bison examined, 5.2% revealed domestic cow mtDNA, suggesting a cross between a male bison and female cow (139). Hybridization between these two species is probably the result of deliberate crossbreeding. Other cases of interspecific hybridization include (1) red wolf and coyote, with individuals of the former species possessing a coyote mtDNA haplotype; (2) coyote and gray wolf; and (3) white-tailed deer and mule deer (141, 32, 104, 103). The common factor shared by these examples is that one species has expanded its range (e.g. coyote and white-tailed deer) and the other has experienced a reduction in population size as a result of range fragmentation (gray wolf, red wolf, and mule deer).

Finally, mtDNA markers have been used in several phylogenetic studies of large mammals. The mitochondrial control region is one of the most popular molecular markers for phylogenetic studies at several levels of divergence. Portions of the control region demonstrate high levels of nucleotide sequence variation, making these regions ideal for population level comparisons. In the family Cervidae the control region averages 1,099 base pairs in length and is subdivided into three major regions, a central conserved region (CCR) and two divergent peripheral domains flanking the CCR at the 5′ and 3′ ends (142). The entire control region sequence has been used to determine relationships among both members of the family Cervidae and populations of the American bison (142, 64). The 5′ variable region also has been used to examine nucleotide sequence variation within many species of large mammals including sheep, cattle, African bovids, deer, bear, and moose (143, 144, 145, 146, 147, 52, 148).

Another mitochondrial gene that has been examined extensively in large mammals is the cytochrome *b* gene. Carr and Hughes sequenced a 359-base pair fragment of the cytochrome *b* gene of the cervid genus *Odocoileus* to examine patterns of introgressive hybridization between mule deer and white-tailed deer (20). Cytochrome *b* gene sequences also have been used to examine phylogenetic relationships among subfamilies of the family Cervidae, among species of peccary, among populations of North

American black bear, and among species of the family Ursidae (149, 150, 18, 151, 152). Two interesting findings from the cytochrome *b* gene are evidence for antlered deer not being monophyletic and the polar bear lineage originating within brown bears (149, 152).

GENETICS AND WILDLIFE MANAGEMENT

Genetic Stocks

Patterns of genetic variation can be used to evaluate the degree to which populations and lineages within a species have been isolated over historical or ecological time (see Chapter 1). These patterns provide information on processes that have influenced species subdivision and the current patterns of genetic variation. In many cases geographic patterns, defined by genetics, are different from those demarcated by traditional taxonomy. The identification of unique genetic stocks is important to management because genetic analysis provides an objective means of identifying units of conservation. Patterns of genetic variation overlaid on geographical distribution allow for the delimitation of either "evolutionary significant units" or "management units" and the identification of unique communities sharing a common biogeographic history (153, 154, 136). Evolutionary significant units are characterized by mtDNA haplotypes, sharing a common ancestry, and significant nuclear gene divergence in terms of allele frequencies (153). Management units represent populations with significant allele frequency differences of either nuclear or mitochondrial loci that reveal low levels of gene flow, yet have alleles that are not phylogenetically distinct (153). In many cases genetically distinct lineages, defined by both nuclear and mitochondrial markers, diagnose units of conservation equivalent to a phylogenetic species (154). Such units require the highest level of conservation consideration. For instance, genetic markers for pronghorn antelope define two major evolutionary significant units that should be managed differently with respect to harvest quotas and translocation efforts. Even within well-defined evolutionary significant units, subgroups, defined by reduced gene flow and either unique nuclear alleles or mtDNA haplotypes, allow one to identify populations or other genetic stocks that might require special management consideration. These genetically defined management units generally reflect more recent changes in populations in response to habitat fragmentation and other processes. In the case of mountain sheep, where mtDNA variation can be partitioned according to meta-population structure, it might be wise to manage these genetically defined populations in a manner that conserves the overall meta-population structure from a genetic standpoint, especially if populations have been founded by unique matrilines.

As large mammal species become more fragmented and local populations begin to either decline or go extinct, translocation of wildlife becomes an essential practice for the enhancement of small populations and restoration of locally extinct species. There are several reasons why a genetically based approach to translocation and reintroduction efforts is important for successful establishment of local populations. First, the source and size of founding stock determine the rate at which genetic variation will be lost as a consequence of random genetic drift and founder effects. Ample evidence suggests, as seen for ibex, that genetic variation can be lost sequentially as populations become founded by stocks with an already depleted store of genetic variation. Restocking from one source of already related individuals also contributes to increased inbreeding and the potential for inbreeding depression. Therefore, one should select founding stock with a high level of variability and low level of relatedness among individuals.

Second, outbreeding depression is a form of hybridization that can result from mixing genetically distinct stocks that are adapted differentially to local environments (155). Such mixing can result in incompatibilities that may contribute to an overall reduction in fitness in the offspring of hybrids. Some of these incompatibilities may be associated with chromosome number, phenotypic differences, and different interacting gene complexes. Small populations are more vulnerable because outbreeding depression can alter the genetic structure and demographic structure of populations. Templeton describes an example of outbreeding depression in populations of ibex (156). In this case ibex from Turkey and the Sinai, rather than Austria, were used to restock populations in Czechoslovakia. The entire population went extinct because the rut was different between local ibex and introduced ibex, resulting in hybrids that bred and produced kids during the coldest time of the year.

Third, genetic markers provide a means of evaluating the success or failure of reintroduction efforts. Studies that have used allozymes and mtDNA to evaluate restocking efforts indicate mixed success of previous efforts. For example, patterns of allozyme variation in southern Arkansas white-tailed deer are consistent with past relocation efforts (28). Just the opposite result was found for white-tailed deer in Tennessee, with relationships among populations depicting similar physiographic regions as opposed to being defined by herd origin (29). Deer restoration efforts were quite common in the southeastern United States, and two studies have evaluated the success of those efforts. On a large geographic scale, the pattern of genetic diversity is consistent with the biogeographic history of the region, reflecting patterns of mtDNA variation similar to those found in unrelated species (12, 22). This suggests that restocking efforts had minimal impact on a more regional scale. A similar geographic study, using allozyme markers, found relationships among geographically divergent populations sharing a history of reintroduction in terms of herd origin (157). This suggests that on a smaller geographic scale restocking efforts may have affected local populations.

Harvest Strategies

Many large mammals, especially moose and white-tailed deer, show considerable genetic subdivision and structure over small distances and within small regions. This structure relates to patterns of dispersal and mating. Some populations also show age-specific differences related to genetic variation and fitness traits. Harvesting has the potential of altering the genetic structure of populations and overall patterns of genetic diversity within and between populations (158). For instance, the selective removal of bighorn sheep males with large horns may contribute to a reduction in overall genetic variation in bighorn sheep and a loss of fitness in the population (4).

Captive Breeding and Artificial Enclosures

Small isolated populations with restricted gene flow are more subject to the loss of variation and increase of potentially deleterious alleles as a consequence of genetic drift, founder effects, and inbreeding. Because genetic variation is important to the long-term survival of a population or species and is related to individual fitness, managed populations in captive breeding programs and artificial enclosures are vulnerable to a loss of genetic variation and viability. Social organization, sex ratios, and asymmetries in the demography of populations influence genetic variation, and in enclosed populations this influence may be increased. To offset the loss of genetic variation, managers must design plans for the replenishment of genetic variation by periodic introduction of new individuals into the gene pool and the removal of individuals exhibiting inferior traits.

Genetic Markers as Ecological Tools

The current molecular genetic tools allow one to test predictions from ecological and behavioral observations that relate to relationships among individuals including general population structure, breeding structure, and population size. It is possible with

PCR to reconstruct the breeding success of individuals through paternity or maternity exclusion with the use of highly polymorphic markers. In addition, these same markers allow for the identification of individuals by genotypic marking. Therefore, genetic markers allow a genetic estimate of population size that can be compared to estimates from other survey methods. Sex ratios also can be obtained with the use of Y-specific markers, and a combination of nuclear and maternal markers can be used to evaluate social structure and asymmetries related to dispersal.

SUMMARY

Genetic variation is important for two primary reasons. First, heterozygosity can be related to traits associated with individual fitness. Second, genetic variation within populations and species is essential for long-term survival, especially as it relates to the ability to respond as the environment changes. The effective population size (N_e), number of successfully breeding individuals, can be related to the level of overall heterozygosity seen in a population and the rate at which variation becomes fixed.

As N_e decreases, processes that erode genetic variation in populations become more active, especially if there is limited gene flow between populations. Genetic drift, a random process that reduces genetic variation, is more active in populations with low N_e. As has been demonstrated for many species with recently fragmented populations, the rate at which variation can be lost as a result of genetic drift can be extensive. Population bottlenecks, as a consequence of decline due to overharvesting, disease, etc., not only reduce the size of a population or species but can drastically reduce levels of genetic variation when the bottleneck has been severe and the population growth subsequent to the bottleneck is slow. Founder events result in the establishment of populations from a small number of individuals. In most cases founder events result in not only a reduction in genetic variation but an increase in the opportunity for extensive inbreeding. Inbreeding, mating between close relatives, can have deleterious effects on species that are normally outcrossed. In many cases this can result in a reduction in overall fitness as can be seen by the large number of cases that relate inbreeding to a reduction in viability, fertility, and growth and development.

Many species of game animals show patterns of genetic subdivision on either a macrogeographic or microgeographic scale. Genetic markers have been used to identify geographic subdivision in large mammals, suggesting evidence of historical geographic subdivision and the development of barriers to gene flow. In most cases the unique genetic stocks do not correspond to subspecies recognized by traditional taxonomy. Therefore, these large-scale geographic patterns of genetic variation provide an objective means of identifying populations requiring special management considerations. Recent range fragmentation and a species' demography and ecology also can contribute to genetic subdivision of populations on a microgeographic scale. Although capable of long-range dispersal, many large mammals show genetic subdivision over short geographic distances. In some cases this microgeographic variation is more pronounced in species with biased dispersal by males, whereby females stay close to the site where they were born.

Currently, there are numerous molecular markers, inherited either biparentally or from one sex. These markers provide a means of determining the amount of genetic variation in a population or species. In addition, genetic markers provide a means of evaluating relationships at several levels of organization including individuals, populations, and species. These markers have several applications to issues pertaining to the management and conservation of large mammals. First, sex-specific markers allow for a separate evaluation of male and female gene flow, which is important if hybridization between species is asymmetrical or there is a bias in which sex disperses. Second, nuclear gene markers associated with functionally important regions (e.g., MHC, which is related to immune response) provide a potential means of assessing the overall "genetic health" of a population or species. In addition, these markers provide a means of determining the level of heterozygosity, which has been related to fitness traits. Third, patterns of mitochondrial DNA divergence allow for both the identification of geographically and genetically distinct lineages and the determination of phylogenetic relationships among species and populations. Fourth, nuclear gene markers help establish relationships among individuals, especially with respect to parentage. Fifth, the advent of PCR and molecular techniques allows one to evaluate hypotheses derived from other sources (e.g., behavior, ecology). The indirect determination of sex ratios, individual animal identification, number of individuals in a population, number of individuals actually breeding, and movement patterns of individuals is possible as a result of genetic techniques. Finally, genetic methods provide an evaluation of effects from range fragmentation, even in cases where populations have been subdivided as a consequence of recent land use practices.

In conclusion, genetic information is relevant to those interested in understanding more about the evolution, ecology, behavior, and survival of populations and species. As genetic technology has become more advanced, the application of genetic approaches to conservation and management questions has become easier. Consequently, genetic studies for most of the North American large mammals are becoming more common.

LITERATURE CITED

1. LANDE, R. 1988. Genetics and demography in biological conservation. *Science* 241:1455–1460.

2. GILPIN, M. E., and M. E. SOULÉ. 1986. Minimum viable populations: Processes of species extinction. Pages 19–34 *in* M. E. Soulé, ed., *Conservation Biology,* Sinauer Press, Sunderland, MA.

3. ALLENDORF, F. W., and R. F. LEARY. 1986. Heterozygosity and fitness in natural populations of animals. Pages 57–76 *in* M. E. Soulé, ed., *Conservation Biology: The Science of Scarcity and Diversity.* Sinauer and Associates, Sunderland, MA.

4. FITZSIMMONS, N. N., S. W. BUSKIRK, and M. H. SMITH. 1995. Population history, genetic variability, and horn growth in bighorn sheep. *Conservation Biology* 9:314–323.

5. SCRIBNER, K. T., M. H. SMITH, and P. E. JOHNS. 1989. Environmental and genetic components of antler growth in white-tailed deer. *Journal of Mammalogy* 70:284–291.

6. SMITH, M. H., R. K. CHESSER, F. G. COTHRAN, and P. E. JOHNS. 1982. Genetic variability and antler growth in a natural population of white-tailed deer. Pages 365–387 *in* R. D. Brown, ed., *Antler Development in Cervidae.* Kleberg Wildlife Research Institute, Kingsville, TX.

7. HEDRICK, P. W., and P. S. MILLER. 1992. Conservation genetics: Techniques and fundamentals. *Ecological Applications* 2:30–46.

8. NUNNEY, L., and K. A. CAMPBELL. 1993. Assessing minimum viable population size: Demography meets population genetics. *Trends in Ecology and Evolution* 8:234–239.

9. O'RYAN, C., E. H. HARLEY, M. W. BRUFORD, M. BEAUMONT, R. K. WAYNE, and M. I. CHERRY. 1998. Microsatellite analysis of genetic diversity in fragmented South African buffalo populations. *Animal Conservation* 1:85–94.

10. PAETKAU, D., and C. STROBECK. 1994. Microsatellite analysis of genetic variation in black bear populations. *Molecular Ecology* 3:489–495.

11. ROY, M. S., E. GEFFEN, D. SMITH, E. A. OSTRANDER, and R. K. WAYNE. 1994. Patterns of differentiation and hybridization in North American wolflike canids, revealed by analysis of microsatellite loci. *Molecular Biology and Evolution* 11:553–570.

12. ELLSWORTH, D. L., R. L. HONEYCUTT, N. J. SILVY, J. W. BICKHAM, and W. D. KLIMSTRA. 1994. White-tailed deer restoration to the southeastern United States: Evaluating mitochondrial DNA and allozyme variation. *Journal of Wildlife Management* 58:586–697.

13. LEBERG, P. L. 1992. Effects of population bottlenecks on genetic diversity as measured by allozyme electrophoresis. *Evolution* 46:477–494.

14. NEI, M., T. MARUYAMA, and R. CHAKRABORTY. 1975. The bottleneck effect and genetic variability in populations. *Evolution* 29:1–10.

15. WILSON, A. C., R. L. CANN, S. M. CARR, M. GEORGE, JR., U. B. GYLLENSTEN, K. M. HELM-BYCHOWSKI, R. HIGUCHI, S. R. PALUMBI, E. M. PRAGER, R. D. SAGE, and M. STONEKING. 1985. Mitochondrial DNA and two perspectives on evolutionary genetics. *Biological Journal of Linnean Society* 26:375–400.

16. BACCUS, R., N. RYMAN, M. H. SMITH, C. REUTERWALL, and D. CAMERON. 1983. Genetic variability and differentiation of large grazing mammals. *Journal of Mammalogy* 64:109–120.

17. BALLINGER, S. W., L. H. BLAKENSHIP, J. W. BICKHAM, and S. M. CARR. 1992. Allozyme and mitochondrial DNA analysis of a hybrid zone between white-tailed deer and mule deer (*Odocoileus*) in West Texas. *Biochemical Genetics* 30:1–11.

18. BYUN, S. A., B. F. KOOP, and T. E. REIMCHEN. 1997. North American black bear mtDNA phylogeography: Implications for morphology and the Haida Gwaii glacial refugium controversy. *Evolution* 51:1647–1653.

19. CAMERON, D. G., and E. R. VYSE. 1978. Heterozygosity in Yellowstone Park elk, *Cervus canadensis*. *Biochemical Genetics* 16:651–657.

20. CARR, S. M., and G. A. HUGHES. 1993. Direction of introgressive hybridization between species of North American deer (*Odocoileus*) as inferred from mitochondrial cytochrome b sequences. *Journal of Mammalogy* 74:331–342.

21. CRAIGHEAD, L., D. PAETKAU, H. V. REYNOLDS, E. R. VYSE, and C. STROBECK. 1995. Microsatellite analysis of paternity and reproduction in Arctic grizzly bears. *Journal of Heredity* 86:255–261.

22. ELLSWORTH, D. L., R. L. HONEYCUTT, N. J. SILVY, J. W. BICKHAM, and W. D. KLIMSTRA. 1994. Historical biogeography and contemporary patterns of mitochondrial DNA variation in white-tailed deer from the southeastern United States. *Evolution* 48:122–136.

23. FITZSIMMONS, N. N., S. W. BUSKIRK, and M. H. SMITH. 1997. Genetic changes in reintroduced Rocky Mountain bighorn sheep populations. *Journal of Wildlife Management* 61:863–872.

24. FORBES, S. H., J. T. HOGG, F. C. BUCHANAN, A. M. CRAWFORD, and F. W. ALLENDORF. 1995. Microsatellite evolution in congeneric mammals: Domestic and bighorn sheep. *Molecular Biology and Evolution* 12:1106–1113.

25. GOTELLI, D., C. SILLERO-ZUBIRI, G. D. APPLEBAUM, M. S. ROY, D. J. GIRMAN, J. GARCIA-MORENO, E. A. OSTRANDER, and R. K. WAYNE. 1994. Molecular genetics of the most endangered canid: The Ethiopian wolf *Canis simensis*. *Molecular Ecology* 3:301–312.

26. HARTL, G. .B., and F. CSAIKL. 1987. Genetic variability and differentiation in wild boars (*Sus scrofa ferus* L.): Comparison of isolated populations. *Journal of Mammalogy* 68:119–125.

27. HONEYCUTT, R. L. 1999. Unpublished data. Texas A&M University, College Station, TX.

28. KARLIN, A. A., G. A. HEIDT, and D. W. SUGG. 1989. Genetic variation and heterozygosity in white-tailed deer in southern Arkansas. *American Midland Naturalist* 121:273–284.

29. KENNEDY, P. K., M. L. KENNEDY, and M. L. BECK. 1987. Genetic variability in white-tailed deer (*Odocoileus virginianus*) and its relationship to environmental parameters and herd origin (Cervidae). *Genetica* 74:189–201.

30. LEE, T. E., J. N. DERR, J. W. BICKHAM, and T. L. CLARK. 1989. Genetic variation in pronghorn from west Texas. *Journal of Wildlife Management* 53:890–896.

31. LEE, T. E., J. W. BICKHAM, and M. D. SCOTT. 1994. Mitochondrial DNA and allozyme analysis of North American pronghorn populations. *Journal of Wildlife Management* 58:307–318.

32. LEHMAN, N., A. EISENHAWER, K. HANSEN, L. D. MECH, R. O. PETERSON, P. J. P. GOGAN, and R. K. WAYNE. 1991. Introgression of coyote mitochondrial DNA into sympatric North American gray wolf populations. *Evolution* 45:104–119.

33. LOU, Y. 1998. Genetic variation of pronghorn (*Antilocapra americana*) populations in North America. Ph.D. dissertation. Texas A&M University, College Station.

34. LUIKART, G., and F. W. ALLENDORF. 1996. Mitochondrial-DNA variation and genetic-population structure in Rocky Mountain bighorn sheep (*Ovis canadensis canadensis*). *Journal of Mammalogy* 77:109–123.

35. MANLOVE, M. N., M. H. SMITH, H. O. HILLSTAD, S. E. FULLER, P. E. JOHNS, and D. O. STRANEY. 1976. Genetic subdivision in a herd of white-tailed deer as demonstrated by spatial shifts in gene frequencies. *Proceedings Annual Conference of Southeastern Association of Game and Fish Commission* 30:487–492.

36. MANLOVE, M. N., R. BACCUS, M. R. PELTON, M. H. SMITH, and D. GRUBER. 1980. Biochemical variation in the black bear. Pages 37–41 *in* C. J. Martinka and K. L. McArthur, eds., *Bears—Their Biology and Management,* Vol. 3 , Bear Biology Association Conference Series. U.S. Government Printing Office, Washington, DC.

37. MATHEWS, N. E., and W. F. PORTER. 1993. Effect of social structure on genetic structure of free-ranging white-tailed deer in the Adirondack Mountains. *Journal of Mammalogy* 74:33–43.

38. MENOTTI-RAYMOND, M. A., and S. J. O'BRIEN. 1995. Evolutionary conservation of ten microsatellite loci in four species of Felidae. *Journal of Heredity* 86:319–322.

39. MIKKO, S., and L. ANDERSSON. 1995. Low major histocompatibility complex class II diversity in European and North American moose. *Proceedings of National Academy of Sciences USA* 92:4259–4263.

40. O'BRIEN, S. J., D. E. WILDT, D. GOLDMAN, C. R. MERRIL, and M. BUSH. 1983. The cheetah is depauperate in genetic variation. *Science* 221:459–462.

41. O'BRIEN, S. J., M. E. ROELKE, L. MARKER, A. NEWMAN, C. A. WINKLER, D. MELTZER, L. COLLY, J. F. EVERMANN, M. BUSH, and D. E. WILDT. 1985. Genetic basis for species vulnerability in the cheetah. *Science* 227:1428–1434.

42. O'BRIEN, S. J., J. S. MARTENSON, C. PACKER, L. HERBST, V. DE VOS, P. JOSLIN, J. OTT-JOSLIN, D. E. WILDT, and M. BUSH. 1987. Biochemical genetic variation in geographic isolates of African and Asiatic lions. *National Geographic Research* 3:114–124.

43. PETIT, E., S. AULAGNIER, R. BON, M. DUBOIS, and B. CROUAU-ROY. 1997. Genetic structure of populations of the Mediterranean mouflon (*Ovis gmelini*). *Journal of Mammalogy* 78:459–467.

44. PRICE, P. K., M. CARTWRIGHT, and M. J. ROGERS. 1979. Genetic variation in white-tailed deer from Arkansas. *Proceedings of Arkansas Academy of Science* 33:64–66.

45. ROELKE, M. E., J. S. MARTENSON, and S. J. O'BRIEN. 1993. The consequences of demographic reduction and genetic depletion in the endangered Florida panther. *Current Biology* 3:340–359.

46. RYMAN, N., G. BECKMAN, G. BRUUN-PETERSEN, and C. REUTERWALL. 1977. Variability of red cell enzymes and genetic implications of management policies in Scandinavian moose (*Alces alces*). *Hereditas* 85:157–162.

47. RYMAN, N., C. REUTERWALL, K. NYGREN, and T. NYGREN. 1980. Genetic variation and differentiation in Scandinavian moose (*Alces alces*): Are large animals monomorphic? *Evolution* 34:1037–1049.

48. SAGE, R. D., and J. O. WOLFE. 1986. Pleistocene glaciations, fluctuating ranges, and low genetic variability in large mammals (*Ovis dalli*). *Evolution* 40:1092–1095.

49. SHEFFIELD, S. R., R. P. MORGAN II, G. A. FELDHAMER, and D. M. HARMAN. 1985. Genetic variation in white-tailed deer (*Odocoileus virginianus*) populations in western Maryland. *Journal of Mammalogy* 66:243–255.

50. SMITH, M. H., R. BACCUS, H. O. HILLESTAD, and M. N. MANLOVE. 1984. Population genetics. Pages 119–128 *in* L. K. Halls, ed., *White-tailed Deer, Ecology, and Management*. Stackpole Books, Harrisburg, PA.

51. STUWE, M., and K. T. SCRIBNER. 1989. Low genetic variability in reintroduced alpine ibex (*Capra ibex ibex*) populations. *Journal of Mammalogy* 70:370–373.

52. TABERLET, P., and J. BOUVET. 1994. Mitochondrial DNA polymorphism, phylogeography, and conservation genetics of the brown bear *Ursus arctos* in Europe. *Proceedings of Royal Society of London Series B* 255:195–200.

53. THEIMER, T. C., and P. KEIM. 1994. Geographic patterns of mitochondrial DNA variation in collared peccaries. *Journal of Mammalogy* 75:121–128.

54. WALKER, C. W. 1998. Patterns of genetic variation in ocelot (*Leopardus pardalis*) populations for south Texas and northern Mexico. Ph.D. thesis. Texas A&M University, College Station.

55. HONEYCUTT, R. L. 1999. Unpublished data. Texas A&M University, College Station, TX.

56. WARD, T. J., J. P. BIELAWSKI, S. K. DAVIS, J. W. TEMPLETON, and J. N. DERR. 1998. Identification of domestic cattle hybrids in wild cattle and bison species: A general approach using mtDNA markers and the parametric bootstrap. *Animal Conservation* 2:51–57.

57. WATHEN, W. G., G. F. MCCRACKEN, and M. R. PELTON. 1985. Genetic variation in black bears from the Great Smoky Mountains National Park. *Journal of Mammalogy* 66:564–567.

58. WAYNE, R. K., N. LEHMAN, D. GIRMAN, P. J. P. GOGAN, D. A. GILBERT, K. HANSEN, R. O. PETERSON, U. S. SEAL, A. EISENHAWER, L. D. MECH, and R. J. KRUMENAKER. 1991. Conservation genetics of the endangered Isle Royale gray wolf. *Conservation Biology* 5:41–51.

59. WAYNE, R. K., N. LEHMAN, M. W. ALLARD, and R. L. HONEYCUTT. 1992. Mitochondrial DNA variability of the gray wolf—Genetic consequences of population decline and habitat fragmentation. *Conservation Biology* 6:559–569.

60. O'BRIEN, S. J., D. E. WILDT, M. BUSH, T. M. CARO, C. FITZGIBBON, I. AGGUNDEY, and R. E. LEAKEY. 1987. East African cheetahs: Evidence for two population bottlenecks. *Proceedings of National Academy of Sciences USA* 84:508–511.

61. PACKER, C., A. E. PUSEY, H. ROWLEY, D. A. GILBERT, J. MARTENSON, and S. J. O'BRIEN. 1991. Case study of a population bottleneck: Lions of the Ngorongoro Crater. *Conservation Biology* 5:219–230.

62. MEROLA, M. 1994. A reassessment of homozygosity and the case for inbreeding depression in the cheetah, *Acinonyx jubatus:* Implications for conservation. *Conservation Biology* 8:961–971.

63. O'BRIEN, S. J., and J. F. EVERMANN. 1988. Interactive influence of infectious disease and genetic diversity in natural populations. *Trends in Ecology and Evolution* 3:254–259.

64. POLZIEHN, R. O., R. BEECH, J. SHERATON, and C. STROBECK. 1996. Genetic relationships among North American bison populations. *Canadian Journal of Zoology* 74:738–749.

65. LISTER, A., and I. VAN PIJLEN. 1998. Molecular and morphological evidence on speciation and subspeciation in Holarctic *Alces* and *Cervus.* European-American Mammal Congress, Santiago de Compostela, Spain, Abstract 55.

66. RALLS, K., and J. BALLOU. 1983. Extinction: Lessons from zoos. Pages 164–184 *in* C. M. Schonewald-Cox, S. M. Chambers, B. MacBryde, and L. Thomas, eds., *Genetics and Conservation: A Reference for Managing Wild Animal and Plant Populations.* Benjamin/Cummings, Menlo Park, CA.

67. RALLS, K., J. D. BALLOU, and A. TEMPLETON. 1988. Estimates of lethal equivalents and the cost of inbreeding in mammals. *Conservation Biology* 2:185–193.

68. SLATKIN, M. 1987. Gene flow and the geographic structure of natural populations. *Science* 236:787–792.

69. CHESSER, R. K., M. H. SMITH., P. E. JOHNS, M. N. MANLOVE, D. O. STRANEY, and R. BACCUS. 1982. Spatial, temporal, and age-dependent heterozygosity of beta-hemoglobin in white-tailed deer. *Journal of Wildlife Management* 46:983–990.

70. CRONIN, M. A., M. E. NELSON, and D. F. PAC. 1991. Spatial heterogeneity of mitochondrial DNA and allozymes among populations of white-tailed deer and mule deer. *Journal of Heredity* 82:118–127.

71. HAWKINS, R. E., and W. D. KLIMSTRA. 1970. A preliminary study of the social organization of white-tailed deer. *Journal of Wildlife Management* 34:407–419.

72. NELSON, M. E. 1993. Natal dispersal and gene flow in white-tailed deer in northeastern Minnesota. *Journal of Mammalogy* 74:316–322.

73. CHESSER, R. K., C. REUTERWALL, and N. RYMAN. 1982. Genetic differentiation of Scandinavian moose *Alces alces* populations over short geographical distances. *Oikos* 39:125–130.

74. BLEICH, V. C., J. D. WEHAUSEN, R. R. RAMEY II, and J. L. RECHEL. 1996. Metapopulation theory and mountain sheep: Implications for conservation. Pages 353–373 *in* D. R. McCullough, ed., *Metapopulation and Wildlife Conservation.* Island Press, Washington, DC.

75. MORITZ, C., T. E. DOWLING, and W. M. BROWN. 1987. Evolution of animal mitochondrial DNA: Relevance for population biology and systematics. *Annual Review of Ecology and Systematics* 18:269–292.

76. MURPHY, R. W., J. W. SITES, JR., D. G. BUTH, and C. H. HAUFLER. 1996. Proteins: Isozyme electrophoresis. Pages 51–120 *in* D. M. Hillis, C. Moritz, and B. K. Marble, eds., *Molecular Systematics,* 2nd ed., Sinauer Associates, Sunderland, MA.

77. AVISE, J. C. 1994. *Molecular Markers, Natural History and Evolution.* Chapman & Hall, New York.

78. WAYNE, R. K., R. E. BENVENSITE, and S. J. O'BRIEN. 1989. Molecular and biochemical evolution of the Carnivora. Pages 465–494 *in* J. L. Gittleman, ed., *Carnivore Behavior, Ecology and Evolution.* Cornell University Press, Ithaca, NY.

79. WAYNE, R. K., and S. J. O'BRIEN. 1987. Allozyme divergence within the Canidae. *Systematic Zoology* 36:339–355.

80. GEORGIADIS, N. J., P. W. KAT, H. OKETCH, and J. PATTON. 1991. Allozyme divergence within the Bovidae. *Evolution* 44:2135–2149.

81. SCRIBNER, K. T., and M. STÜWE. 1993. Genetic relationships among alpine ibex (*Capra ibex*) populations from a common ancestral source. *Biological Conservation* 69:1–7.

82. NEVO, E., A. BEILES, and R. BEN-SHLOMO. 1984. The evolutionary significance of genetic diversity: Ecological, demographic, and life history correlates. Pages 13–213 *in* G. S. Mani, ed., *Lecture Notes in Biomathematics,* Vol. 53. Springer-Verlag, Berlin.

83. COTHRAN, E. G., R. K. CHESSER, M. H. SMITH, and P. E. JOHNS. 1983. Influences of genetic variability and maternal factors on fetal growth rate in white-tailed deer. *Evolution* 37:282–291.

83a. CHESSER, R. K., and M. H. SMITH. 1987. Relationship of genetic variation to growth and reproduction in the white-tailed deer. Pages 168–177 *in* C. M. Wemmer, ed., *Biology and Management of the Cervidae.* Smithsonian Institution Press, Washington, DC.

84. JOHNS, P. E., R. BACCUS, M. N. MANLOVE, J. E. PINDER III, and M. H. SMITH. 1977. Reproductive patterns, productivity and genetic variability in adjacent white-tailed deer populations. *Proceedings Annual Conference Southeastern Association Fish and Wildlife Agencies* 31:167–172.

85. DOWLING, T. E., C. MORITZ, J. D. PALMER, and L. H. RIESEBERG. 1996. Nucleic acids III: Analysis of fragments and restriction sites. Pages 249–320 *in* D. M. Hillis, C. Moritz, and B. K. Marble, eds., *Molecular Systematics,* 2nd ed. Sinauer Associates, Sunderland, MA.

86. SANGER, F., S. NICKLEN, and A. R. COULSON. 1977. DNA sequencing with chain-terminating inhibitors. *Proceedings of National Academy of Sciences USA* 74:5463–5467.

87. NEI, M., and W.-H. LI. 1979. Mathematical model for studying genetic variation in terms of restriction endonucleases. *Proceedings of the National Academy of Sciences USA* 76:5269–5273.

88. NEI, M., and F. TAJIMA. 1981. DNA polymorphism detectable by restriction endonucleases. *Genetics* 97:145–163.

89. SAIKI, R. K., D. H. GELFAND, S. STOFFEL, S. J. SCHARF, R. HIGUCHI, G. T. HORN, K. B. MULLIS, and H. A. ERLICH. 1988. Primer-directed enzymatic amplification of DNA with a thermostable DNA polymerase. *Science* 239:487–491.

90. HOSS, M., M. KOHN, S. PAABO, F. KNAUER, and W. SCHRODER. 1992. Excrement analysis by PCR. *Nature* 359:199.

91. GRIFFITHS, R., and B. TIWARI. 1993. Primers for the differential amplification of the sex-determining region Y gene in a range of mammal species. *Molecular Ecology* 2:405–406.

92. TABERLET, P., H. MATTOCK, C. DUBOIS-PAGANON, and J. BOUVET. 1993. Sexing free-ranging brown bear *Ursus arctos* using hairs found in the field. *Molecular Ecology* 2:399–403.

93. CONSTABLE, J. J., C. PACKER, D. A. COLLINS, and A. E. PUSEY. 1995. Nuclear DNA from primate dung. *Nature* 373:393.

94. KOHN, M., F. KNAUER, A. STOFFELLA, W. SCHRODER, and S. PÄÄBO. 1995. Conservation genetics of the European brown bear—A study using excremental PCR of nuclear and mitochondrial sequences. *Molecular Ecology* 4:95–103.

95. WASSER, S. K., C. S. HOUSTON, G. M. KOEHLER, G. G. CADD, and S. R. FAIN. 1997. Techniques for application of faecal DNA methods to field studies of ursids. *Molecular Ecology* 6:1091–1097.

96. TABERLET, P., J. J. CAMARRA, S. GRIFFIN, E. UHRES, O. HANOTTE, L. P. WAITS, C. DUBOIS-PAGANON, T. BURKE, and J. BOUVET. 1997. Noninvasive genetic tracking of the endangered Pyrenean brown bear population. *Molecular Ecology* 6:869–876.

97. PÄÄBO, S., R. G. HIGUCHI, and A. C. WILSON. 1989. Ancient DNA and the polymerase chain reaction. *Journal of Biological Chemistry* 264:9709–9712.

98. THOMAS, W. K., S. PÄÄBO, F. VILLABLANCA, and A. C. WILSON. 1990. Spatial and temporal continuity of kangaroo rat populations shown by sequencing mitochondrial DNA from museum specimens. *Journal of Molecular Evolution* 31:101–112.

99. CRONIN, M. A., D. A. PALMISCIANO, E. R. VYSE, and D. G. CAMERON. 1991. Mitochondrial DNA in wildlife forensic science: Species identification of tissues. *Wildlife Society Bulletin* 19:94–105.

100. MARDON, G., S-W. LUOH, E. M. SIMPSON, G. GILL, L. G. BROWN, and D. C. PAGE. 1990. Mouse Zfx is similar to Zfy-2: Each contains an acidic activating domain and 13 zinc fingers. *Molecular and Cell Biology* 10:681–688.

101. LUNDRIGAN, B. L., and P. K. TUCKER. 1994. Tracing paternal ancestry in mice, using the Y-linked, sex-determining locus, Sry. *Molecular Biology and Evolution* 11:483–492.

102. BURROWS, W., and O. A. RYDER. 1997. Y-chromosome variation in great apes. *Nature* 385:125–126.

103. CATHEY, J. C., J. W. BICKHAM, and J. C. PATTON. 1998. Introgressive hybridization and non-concordant evolutionary history of maternal and paternal lineages in North American deer. *Evolution* 52:1224–1229.

104. CARR, S. M., S. W. BALLINGER, J. N. DERR, L. H. BLANKENSHIP, and J. W. BICKHAM. 1986. Mitochondrial DNA analysis of hybridization between sympatric white-tailed deer and mule deer in West Texas. *Proceedings of the National Academy of Sciences* 83:9576–9580.

105. GAVIN, T., and B. MAY 1988. Taxonomic status and genetic purity of columbian white-tailed deer. *Journal of Wildlife Management* 52:1–10.

106. CRONIN, M. A. 1991. Mitochondrial and nuclear genetic relationships of deer (*Odocoileus* sp.) in western North America. *Canadian Journal of Zoology* 69:1270–1279.

107. DERR, J. N. 1991. Genetic interactions between white-tailed and mule deer in the southwestern United States. *Journal of Wildlife Management* 55:228–237.

108. KRAUSE, F., and M. M. MIYAMOTO. 1991. Rapid cladogenesis among the pecoran ruminants: Evidence from mitochondrial DNA sequences. *Systematic Zoology* 40:117–130.

109. HONEYCUTT, R. L., M. A. NEDBAL, R. M. ADKINS, and L. L. JANECEK. 1995. Mammalian mitochondrial DNA evolution: A comparison of the cytochrome *b* and cytochrome *c* oxidase II gene. *Journal of Molecular Evolution* 40:260–272.

110. CRONIN, M. A., R. STUART, B. J. PIERSON, and J. C. PATTON. 1996. K-casein gene phylogeny of higher ruminants (Pecora, Artiodactyla). *Molecular Phylogenetics and Evolution* 357:295–311.

111. KLEIN, J. 1986. *Natural History of the Major Histocompatibility Complex.* John Wiley & Sons, New York.

112. HUGHES, A., and M. NEI. 1988. Pattern of nucleotide substitution at the major histocompatibility complex class I loci reveals over-dominant selection. *Nature* 335:167–170.

113. YUHKI, N., and S. J. O'BRIEN. 1990. DNA variation of the mammalian major histocompatibility complex reflects genomic diversity and population history. *Proceedings of National Academy of Sciences USA* 87:836–840.

114. BOYCE, W. M., P. W. HEDRICK, N. E. MUGGLI-COCKETT, S. KALINOWSKI, M. C. T. PENEDO, and R. R. RAMEY II. 1997. Genetic variation of major histocompatibility complex and microsatellite loci: A comparison in bighorn sheep. *Genetics* 145:421–433.

115. ELLEGREN, H., G. HARTMAN, M. JOHANSSON, and L. ANDERSSON. 1993. Major histocompatibility complex monomorphism and low levels of DNA fingerprinting variability in a reintroduced and rapidly expanding population of beavers. *Proceedings of National Academy of Sciences USA* 90:8150–8153.

116. JEFFREYS, A. J., V. WILSON, and S. L. THEIN. 1985. Hypervariable minisatellite regions in human DNA. *Nature* 316:76–79.

117. GILBERT, D. A., N. LEHMAN, S. J. O'BRIEN, and R. K. WAYNE. 1990. Genetic fingerprinting reflects population differentiation in the California Channel Island fox. *Nature* 344:764–766.

118. GILBERT, D. A., C. PACKER, A. E. PUSEY, J. C. STEPHENS, and S. J. O'BRIEN. 1991. Analytical DNA fingerprinting in lions: Parentage, genetic diversity, and kinship. *Journal of Heredity* 82:378–386.

119. TAUTZ, D., and M. RENZ. 1984. Simple sequences are ubiquitous repetitive components of eukaryotic genomes. *Nucleic Acids Research* 12:4127–4138.

120. TAUTZ, D. 1989. Hypervariability of simple sequences as a general source for polymorphic DNA markers. *Nucleic Acids Research* 17:6463–6471.

121. WEBER, J., and P. E. MAY. 1989. Abundant class of human DNA polymorphisms which can be typed using the polymerase chain reaction. *American Journal of Human Genetics* 44:388–396.

122. DIETRICH, W., H. KATZ, S. E. LINCOLN, H. S. SHIN, J. FRIEDHAM, N. C. DRACOPOLI, and E. S. LANDER. 1992. A genetic map of the mouse suitable for typing intraspecific crosses. *Genetics* 131:423–447.

123. WEISSENBACH, J., G. GYAPAY, C. DIB, A. VIGNAL, M. J. P. MILLASSEAU, G. VAYSSIEX, and M. LATHROP. 1992. A second-generation linkage map of the human genome. *Nature* 359:794–801.

124. DEWOODY, J. A., R. L. HONEYCUTT, and L. K. SKOW. 1995. Microsatellite markers in white-tailed deer. *Journal of Heredity* 86:317–319.

125. WILSON, G. A., C. STROBECK, L. WU, and J. W. COFFIN. 1997. Characterization of microsatellite loci in caribou *Rangifer tarandus,* and their use in other artiodactyls. *Molecular Ecology* 6:697–699.

126. MOORE, S. S., L. L. SARGEANT, T. J. KING, J. S. MATTICK, M. GEORGES, and D. J. S. HETZEL. 1991. The conservation of dinucleotide microsatellites among mammalian genomes allows use of heterologous PCR primer pairs in closely related species. *Genomics* 10:654–660.

127. ENGEL, S. R., R. A. LINN, J. F. TAYLOR, and S. K. DAVIS. 1996. Conservation of microsatellite loci across species of artiodactyls: Implications for population studies. *Journal of Mammalogy* 77:504–518.

128. VAIMAN, D., R. OSTA, D. MERCIER, C. GROHS, and H. LEVEZIEL. 1992. Characterization of five new bovine dinucleotide repeats. *Animal Genetics* 23:537–541.

129. VAIMAN, D., D. MERCIER, K. MOAZAMI-GOUDARZI, A. EGGEN, R. CIAMPOLINI, A. LÉPINGLE, R. VELMALA, J. KAUKINEN, S. L. VARVIO, P. MARTIN, H. LEVEZIEL, and G. GUERIN. 1994. A set of 99 cattle microsatellites: Characterization, synteny mapping, and polymorphism. *Mammalian Genome* 5:288–297.

130. VAIMAN, D. M. IMAM-GHALI, K. MOAZAMI-GOUDARIZI, G. GUÉRIN, C. GROHS, H. LEVÉZIEL, and N. SÁIDIMEHTAR. 1994. Conservation of a syntenic group of microsatellite loci between cattle and sheep. *Mammalian Genome* 5:310–314.

131. BUCHANAN, F. C., L. J. ADAMS, R. P. LITTLEJOHN, J. F. MADDOX, and A. M. CRAWFORD. 1994. Determination of evolutionary relationships among sheep breeds using microsatellites. *Genomics* 22:397–403.

132. BUCHANAN, F. C., R. P. LITTLEJOHN, S. M. GALLOWAY, and A. M. CRAWFORD. 1993. Microsatellite and associated repetitive elements in the sheep genome. *Mammalian Genome* 4:258–264.

133. OSTRANDER, E. A., G. F. SPRAGUE, and J. RINE. 1993. Identification and characterization of dinucleotide repeat $(CA)_n$ markers for genetic mapping in dog. *Genomics* 16:207–213.

134. GIRMAN, D. J., M. G. L. MILLS, E. GEFFEN, and R. K. WAYNE. 1997. A molecular genetic analysis of social structure, dispersal, and interpack relationships of the African wild dog (*Lycaon pictus*). *Behavioral Ecology and Sociobiology* 40:187–198.

135. AVISE, J. C., J. ARNOLD, R. M. BALL, E. BERMINGHAM, T. LAMB, J. E. NEIGEL, C. A. REEB, and N. C. SAUNDERS. 1987. Intraspecific phylogeography: The mitochondrial bridge between population genetics and systematics. *Annual Review of Ecology and Systematics* 18:489–522.

136. ZINK, R. M. 1996. Comparative phylogeography in North American birds. *Evolution* 50:308–317.

137. BORK, A. M., C. M. STROBECK, F. C. YEH, R. J. HUDSON, and R. K. SALMON. 1991. Genetic relationship of wood and plains bison based on restriction fragment length polymorphisms. *Canadian Journal of Zoology* 69:43–48.

138. CRONIN, M. A., and N. COCKETT. 1993. Kappa-casein polymorphisms among cattle breeds and bison herds. *Animal Genetics* 24:135–138.

139. CRONIN, M. A., and V. C. BLEICH. 1995. Mitchondrial DNA variation among populations and subspecies of mule deer in California. *California Fish and Game* 81:45–54.

140. POLZIEHN, R. O., C. STROBECK, J. SHERATON, and R. BEECH. 1995. Bovine mtDNA discovered in North American bison populations. *Conservation Biology* 9:1638–1643.

141. WAYNE, R. K., and S. M. JENKS. 1991. Mitochondrial DNA analysis implying extensive hybridization of the endangered red wolf *Canis rufus*. *Nature* 351:565–568.

142. DOUZERY, E., and E. RANDI. 1997. The mitochondrial control region of Cervidae: Evolutionary patterns and phylogenetic content. *Molecular Biology and Evolution* 14:1154–1166.

143. WOOD, N. J., and S. H. PHUA. 1996. Variation in the control region sequence of the sheep mitochondrial genome. *Animal Genetics* 27:25–33.

144. LOFTUS, R. T., D. E. MacHUGH, D. G. BRADLEY, P. M. SHARP, and P. CUNNINGHAM. 1994. Evidence for two independent domestications of cattle. *Proceedings of National Academy of Sciences USA* 91:2757–2761.

145. ARCTANDER, P., P. W. KAT, R. A. AMAN, and H. R. SIEGISMUND. 1996. Extreme genetic differentiation among populations of *Gazella granti*, Grant's gazelle, in Kenya. *Heredity* 76:465–475.

146. ARCTANDER, P., P. W. KAT, B. T. SIMONSEN, and H. R. SIEGISMUND. 1996. Population genetics of Kenyan impalas—consequences for conservation. Pages 399–412 *in* R. K. Wayne and T. B. Smith, eds., *Molecular Genetics in Conservation*. Oxford University Press, Oxford, UK.

147. GONZÁLEZ, S., J. E. MALDONADO, J. A. LEONARD, C. VILA, J. M. BARBANTI-DURATE, M. MERINO, N. BRUM-ZORRILLA, and R. K. WAYNE. 1998. Conservation genetics of the endangered Pampas deer (*Ozotoceros bezoarticus*). *Molecular Ecology* 7:47–56.

148. HUNDERTMARK, K., G. SHIELDS, and R. T. BOWYER. 1998. Phylogeography of North American moose: Evidence for a new theory of colonization. Euro-American Mammal Congress, Santiago de Compostela, Spain, Abstract 57.

149. RANDI, E., N. MUCCI, M. PIERPAOLI, and E. DOUZERY. 1998. New phylogenetic perspectives on the Cervidae (Artiodactyla) are provided by the mitochondrial cytochrome *b* gene. *Proceedings of Royal Society of London, Series B* 265:793–801.

150. THEIMER, T. C., and P. KEIM. 1998. Phylogenetic relationships of peccaries based on mitochondrial cytochrome *b* DNA sequences. *Journal of Mammalogy* 79:566–572.

151. ZHANG, Y.-P., and O. A. RYDER. 1994. Phylogenetic relationships of bears (the Ursidae) inferred from mitochondrial DNA sequences. *Molecular Phylogenetics and Evolution* 3:351–359.

152. TALBOT, S. L., and G. F. SHIELDS. 1996. A phylogeny of the bears (Ursidae) inferred from complete sequences of three mitochondrial genes. *Molecular Phylogenetics and Evolution* 5:567–575.

153. MORITZ, C. 1994. Defining "evolutionary significant units" for conservation. *Trends in Ecology and Evolution* 9:373–375.

154. VOGLER, A. P., and R. DeSALLE. 1994. Diagnosing units of conservation management. *Conservation Biology* 8:354–363.

155. TEMPLETON, A. R. 1986. Coadaptation and outbreeding depression. Pages 105–116 *in* M. E. Soulé, ed., *Conservation Biology: The Science of Scarcity and Diversity.* Sinauer Associates, Sunderland, MA.

156. MEFFE, G. K., and C. R. CARROLL. 1997. Genetics: Conservation of diversity within species. Pages 161–201 *in* G. K. Meffe and C. R. Carroll, eds., *Principles of Conservation Biology,* 2nd ed. Sinauer Associates, Sunderland, MA.

157. LEBERG, P. L., P. W. STANGEL, H. O. HILLESTAD, R. L. MARCHINTON, and M. H. SMITH. 1994. Genetic structure of reintroduced wild turkey and white-tailed deer populations. *Journal of Wildlife Management* 58:698–711.

158. RYMAN, N., R. BACCUS, C. REUTERWALL, and M. H. SMITH. 1981. Effective population size, generation interval, and potential loss of genetic variability in game species under different hunting regimes. *Oikos* 36:257–266.

Big Game Ranching

Ronald J. White

INTRODUCTION

Big game ranching is not a new concept and sport hunting has been a motive for the activity. When hunting emerged as a sport for the ruling classes, game preserves were established with exotic game (1). Marco Polo described a hunting preserve of Kublai Khan's that existed during 1259 to 1294 (2). Père David's deer, believed to have originally inhabited the alluvial plains of northern China, survived due to the herd formerly kept in the Imperial Hunting Park, Beijing, China (3). Currently, most big game ranching enterprises on private lands in North America are an extension of the private game park concept in Europe. In Europe and in North America access for sport hunting is allowed at the discretion of the landowner.

Animals have been domesticated through big game ranching efforts for various reasons, but they were evidently much appreciated for red meat. Records show that wild animals have been raised as a protein source throughout human history. Domestication efforts were not confined to the animals presently considered domestics. Gazelles, oryx, and addax antelopes were domesticated in Egypt more than 4,000 years ago (4). Raising big game animals for commercial meat production is in its initial stages in North America, with most attention focused on raising plains bison, wapiti, and various other species of deer.

Asia is a significant consumption center for deer by-products and will remain so in the foreseeable future (5). Twenty-eight deer by-products are used primarily for medicinal purposes in China (6). Antlers are one profitable by-product of deer farming. The sale of antler products is a multimillion dollar industry (7). Velvet antler (i.e., antler enveloped in soft vascular skin that nourishes the developing antler) has a market in Asia, and antler jewelry is popular in North America (Figure 13–1).

This chapter is intended to acquaint the reader with the concept of big game ranching and farming in North America. Considerable literature exists elaborating on the different aspects of the subject. The references for this chapter provide a beginning point for those interested in additional information.

WHAT IS BIG GAME RANCHING?

The terminology used to describe commercial wildlife raising ventures varies with the animals involved. *Shooting preserve, hunting reserve, game farm,* and *big game ranch* are familiar terms. These terms imply hunters paying landowners for access to their land for sport hunting, that is, fee hunting. The terms, however, can have different meanings

Figure 13–1 Wapiti provide hunting opportunities and by-products, including red meat, antlers, and leather. (Photograph by G.G. White.)

to various people. They might imply fee hunting, or they might be interpreted to mean the husbandry of animals for meat and by-products, for esthetic values, or for other forms of nonconsumptive recreation. Common usage of the term *big game* has often served to identify the large mammalian species. Big game, therefore, usually includes such species as hoofed animals, large cats, and various other species. The term *big game* is restricted here to include only wild ungulates (hoofed mammals), inclusive of exotic and indigenous species. Exotic big game is defined as any nonindigenous species.

Four categories of terms describing emerging systems of commercial ungulate management—ranching, hunting, herding, and farming—were described by Hudson (8). Although each of these four categories contains variations and overlaps to some degree, they serve as the basis for the terminology applicable to this chapter.

Ranching

Ranching refers to the extensive management of wild big game populations that may or may not be confined by a fence. The animals may be harvested, or captured, by different methods for various purposes including meat and by-products, esthetic values, or fee hunting. Hence, big game ranching is the intentional raising, or husbandry, of wildlife ungulates for any purpose.

Hunting

Hunting describes the harvest of essentially wild ungulates. Commercial hunting is distinguished from sport hunting in that it serves the marketplace. For example, ungulates may be commercially harvested from a ranching enterprise to supply meat and by-products. Sport hunting is a recreational activity that contributes meat for human consumption.

Exotic big game animals are usually controlled by physical barriers, but are otherwise essentially wild. The owner of exotic big game can charge a fee for harvesting the animal. The introduction of exotic big game species in North America has been controversial and is expanded on elsewhere in this chapter (9, 10, 11). Management and husbandry aspects for exotics in the United States have been reviewed by White and Mungall and Sheffield (9, 12).

Indigenous animals may not necessarily be confined to a particular piece of private property, but they are considered a product of a big game ranching enterprise if management takes harvest into consideration, as opposed to the incidental harvest of a few animals. The private property owner can usually only charge a fee for access to hunt indigenous ungulates, that is, *fee hunting*. The major indigenous big game species, or subspecies, hunted in North America are black-tailed deer, Rocky Mountain mule deer, desert mule deer, white-tailed deer, caribou, moose, mountain goat, mountain sheep, musk oxen, peccary, pronghorn, and wapiti.

Herding

Herding denotes systems in which animals are managed extensively by taking advantage of their behavior. Various techniques are applied including luring, herding, habituation, and taming. Periodic herd confinement may be involved.

Farming

Farming involves intensive husbandry of wild ungulates under close containment for the production of meat and by-products. Essentially, the animals are raised like domestic livestock, with modifications made to accommodate specific needs, such as harvesting antlers.

NORTH AMERICA'S BIG GAME INDUSTRY

Two basic motives for big game ranching programs are aesthetic or altruistic reasons and various commercial purposes. Four primary commercial strategies for generating income from big game are live animal sales to various markets, big game animal parks where visitors may view and photograph animals, harvesting and sale of meat and by-products, and sport hunting by paying clients (9).

Aesthetic Values or Altruistic Purposes

Some enterprises tolerate or actively raise big game animals solely for aesthetic enjoyment. Endangered species may be added to the collection because of an altruistic desire to contribute personally to the survival of an imperiled species. A major objective of this chapter is to address big game ranching as a commercial industry. Hence, discussions will not dwell on aesthetic or altruistic motives for raising big game animals, except to note here that contributions to wildlife conservation are made by these types of big game ranches.

Commercial Strategies

Most landowners realize that long-term ecological productivity is more important than short-term profits. Traditionally, wildlife has not paid its own way. Landowners have not been able to derive incomes from wildlife that were comparable to those derived from livestock, crops, timber, and other products. This scenario may be changing due to rising human population numbers, the development of a societal ecological conscience, an increased need for outdoor recreational opportunities, and trends in demand for healthier foods. These changes present various alternatives for the wise sustained exploitation of big game animals for profit. The profit motive simultaneously provides one incentive for landowners to maintain wildlife habitat and wildlife populations.

Live Animal Sales. Some big game ranchers raise animals for live sale to various markets. The plains bison is one indigenous species that is popularly used for this purpose, but the largest volume of sales commonly involves exotic species. These animals may be sold to zoos for display, or to other big game ranchers for sport hunting or nonhunting purposes. Popular species such as blackbuck antelope, aoudad (Barbary sheep), and axis deer find ready markets on big game ranches. Expensive rare animals are usually sold to zoos.

Viewing Parks. Regardless of the purposes for which big game animals are maintained, or whether a species is common or rare, an enterprise will often find itself responsible for unmarketable animals. Sometimes, for example, the demand for a species is due to a fad that later fades. Also, the marketing of expensive animals requires time, or some animals cannot be sold immediately due to an injury or pregnancy. A surplus of male animals will usually develop unless they are sold or utilized for hunting purposes. Whatever the reason for their unmarketability, these animals may be a drain on the business. This particular situation has led some enterprises to establish big game animal viewing parks.

During the 1970s, a fad in the zoo world was drive-through areas where animals could be seen in "natural" surroundings. This idea spread to big game ranches, where similar areas were developed. Although a venture of this type is not without problems, admission charges can defray the cost of maintaining surplus animals. Additionally, prospective hunters can observe species available for harvest and gain experience in judging trophy characteristics.

Red Meat and By-Products. A commercial wildlife production industry has evolved to varying degrees in North America. Although the marketplace will ultimately determine their future, new wildlife farms have emerged to meet the increasing demand for specialized meat and various by-products. As part of a more healthy lifestyle, North Americans are showing a preference for meat that is low in fat and extrinsic chemicals. The meat of big game animals meets these criteria. By-products include hides, antlers, horns, skulls, hooves, sausage, jerky, and various items used as novelties and for medicinal purposes (Figure 13–2).

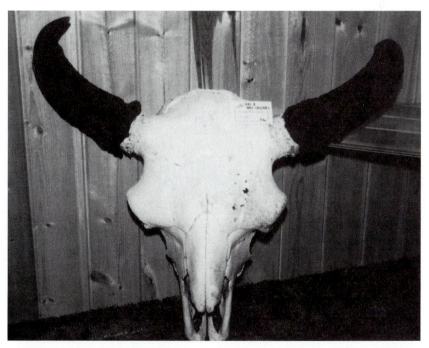

Figure 13–2 Bison skulls are sold as novelties. The price tag on this skull in an art gallery in Montana was $400. (Photograph by G. G. White.)

Canada. Extensive big game ranching formed the initial basis for the commercial production of indigenous big game in Canada. However, the necessity for a large contiguous land base resulted in a trend among private landowners toward smaller, intensively managed big game farms. Since about 1970, interest in big game farming has expanded quickly across Canada, with most private interest occurring in the western provinces (13). Most growth in the new industry has occurred since the mid-1980s (13, 14). Big game production has provided ranching, hunting, herding, and farming opportunities, and leaner meats to supply consumer demand. The industry has grown primarily as a result of the need to diversify conventional agriculture. The major constraint of development has been the availability of breeding stock (15). Establishment of the big game farming industry has not been without controversy. Animal rights advocates, animal welfare groups, and environmental interests strongly opposed the concept (16, 14).

Big game farming legal requirements and standards vary greatly throughout the provinces (15). Several provinces allow only farming of indigenous deer species, while others only permit farming of nonindigenous cervids. Plains bison have been reclassified from a wildlife category to an agricultural species in all provinces except British Columbia. In British Columbia, plains bison remain classified as wildlife, but the species can be raised commercially under permit (13). In Alberta, only indigenous wildlife can be farmed (17). Plains bison, wapiti, and fallow deer dominate the industry in the provinces of Alberta, Saskatchewan, and Manitoba (13). In British Columbia, approval was granted for the farming of fallow deer and plains bison in 1987 and for reindeer in the northeast portion of the province in 1988. Attempts to obtain approval for wapiti farming failed due to opposition (14).

The reindeer (domesticated caribou) industry in North America is unique. Reindeer farming is established in Alaska and northern Canada. Reindeer are owned and husbanded by indigenous people in both countries. Husbandry is often accomplished through various levels of herding. For example, a herd may be loosely herded over an extensive area seasonally, and then be confined twice a year in a smaller area under close management for various handling purposes such as sorting, marking, medical attention, and harvesting of antlers (18).

Reindeer are important to local economies, providing income from the sale of live animals, meat, velvet antlers, and other by-products. The future appears promising for a profitable reindeer industry of moderate size if land use problems such as settlement of native land claims are resolved. Environmental and animal welfare concerns may, however, become issues to be addressed. Improved management through the application of modern technologies is being adopted and may assist the reindeer industry in reaching its potential (19, 18).

Big game farming has a bright future in Canada (13). Increased public scrutiny, however, will require that emphasis be placed on animal welfare. Provincial big game farmer organizations and a Game Farm Advisory Council consisting of several interest groups have developed codes of ethics and procedures manuals (13,14). The documents provide guidelines for the new industry to promote continuing public and government support.

Mexico. Big game farming exists in Mexico, but its present impact and potential is undetermined. The possibility of exporting game meat to Europe or the United States has been suggested. The general lack of financial resources and trained professionals, habitat degradation, indiscriminate exploitation, and weak control over hunting has resulted in a dramatic decline in Mexico's wildlife (20). However, in response to markets for fee-hunting opportunities and meat, the popularity of big game ranching is increasing, especially in northeastern Mexico. Thirty to 40 species of exotic herbivores occur on ranches in northern Mexico, but many species are represented by only a few animals (21). White-tailed deer and mule deer are the important indigenous species, and the exotic aoudad (Barbary sheep) is increasingly popular with hunters. The Rancher's Association of Diversified Producers (Asociación de Ganaderos Diversificados) was established in 1987 to promote big game production. Wildlife enterprises have contributed to the economy of rural producers, and benefitted wildlife in general, through greater protection and habitat restoration (20).

United States. Most development of big game farming in the United States has been in recent years. Some operations function as breeding facilities to produce broodstock for sale to other producers or to zoos. Other enterprises cater to American consumers who are experimenting with a wide variety of "game" meat. Health conscious consumers are increasingly showing a preference for meat that is low in fat and extrinsic chemicals. The concept that big game animals could play a potential role in meat production has been slow to materialize in the United States. However, the potential economic and ecological advantages of cropping big game, instead of domestic livestock, as an alternative meat source have advanced the industry beyond its infancy. Demand for a wider variety of meat and by-products has created the opportunity for the expanding industry. Plains bison, wapiti, red deer, and fallow deer presently dominate the industry, but various species under consideration may increase in popularity.

The North American Deer Farmers Association estimated its members farmed 60,000 deer in 1996, a 10% increase over 1995. Association members represented 144 venison-producing operations with 58 harvesting velvet antlers. The North American deer farming industry is a multimillion dollar industry. Venison sales exceeded $1 million and velvet antler gross sales exceeded $500,000 in 1996. The association lists members among 42 states, five Canadian provinces, two Mexican districts, and three countries outside North America (22).

The indigenous plains bison is a logical candidate for big game farming or ranching operations because it was, and perhaps still is, adapted to many ranges. The National Bison Association, with more than 2,300 members throughout the United States and Canada, claims that plains bison raising is more than a hobby or a passing fad, it is a viable industry. The association estimates the plains bison population to be between 200,000 and 250,000, including 40,000 in public herds with the remainder being privately owned (S. Albrecht, National Bison Association, personal communication, 1996).

The big game farming industry in North America has a receptive market for lean meat and by-products from big game animals. A strong demand for plains bison meat and by-products exceeded industry production in 1993 (23). The demand for venison exceeded the supply in 1987 (24). Any significant expansion of consumer markets will likely require an expanded and dependable supply and distribution system. Restauranteurs and other retailers are reluctant to develop a demand for a product that cannot be consistently supplied. Effective promotion is then required to maintain the market.

Sport Hunting. Sport hunters currently provide a large market for big game ranches. Although many enterprises are primarily in the livestock business, indigenous big game animals are harvested for an access fee simply because they are present. However, the definition of big game ranching stipulates that management of big game animals must be intentional. Many ranchers are realizing that if the ranch is properly managed, the stocking of livestock will be commensurate with the ability of the range resources to support livestock and big game animals. Thus, many landowners intentionally manage indigenous big game animals and livestock as dual crops. Some individuals have introduced exotic big game into the overall hunting scheme of their business.

BIG GAME RANCHING FOR CONSERVATION AND SPORT HUNTING

The American tradition that wildlife should be "freely" available to all citizens is traceable to established practices of royal power over wildlife in England that gradually gave way to parliamentary control. These shared features of English wildlife law were in effect at the time the New World was settled (25).

Colonists who settled America brought the common law of England with them. After the American Revolution, title to wildlife was passed from the Crown to the colonies and then to the newly formed states (26, 27). The inherited English common law has traditionally been retained as wildlife is held in public ownership until it is legally reduced to possession by the individual (28). A potential conflict occurs, however, when wildlife owned by the public at large inhabits land under private ownership and access is controlled by the landowner.

Settlers of the New World found a wealth of natural resources and the literature is replete with accounts of wildlife abundance. Understandably, the early colonists being of European and English descent, viewed wildlife as "a common heritage, not subject to restrictive controls which smacked of Old World class structure" (27, pp. 17–18). Hence, wildlife has played an important role in the lives of immigrant Americans since the United States was colonized. The new Americans, indeed citizens in all North America, used abundant fish and wildlife for domestic and commercial food and fiber products. However, the human population expanded, agriculture flourished, technology developed, and land use patterns changed the complexion of the landscape. Wildlife came to be valued more for recreational purposes than for essential food or economic gain. A growing, urbanized, population now hunts principally for recreation (29).

Disincentives for Big Game Ranching Practices

The burgeoning human population in North America is likely to continue to place heavy demands on natural resources. Competition for use of the land changes the value of the land. Ski areas, dams, summer camps, dude ranches, vacation homes, camping facilities, and access roads, for example, are on the increase throughout much of the mountain and scenic rural areas. All of these factors remove wildlife habitat and increase human use of remaining habitat fragments (9, 30).

Tremendous pressures for economic survival force landowners to use land for the intensified production of marketable food and fiber products. Often, little land remains for wildlife use, and available lands receive minimal attention for wildlife needs. Some landowners have found that development of their land for wildlife resulted in increasing harassment when the public arrived to enjoy the country and the wildlife (31). When the public arrives to view or hunt animals, problems such as trespass, vandalism to landowner property, liability concerns, and disruption of management activities may be exacerbated (27, 32, 33).

Depending on the location and the situation, the income derived from a big game ranching enterprise for fee hunting can be substantial. However, the income is not all profit. Big game animals use forage, a portion of which would otherwise be used for domestic livestock production. They also eat salt and supplemental feed intended for livestock. The presence of big game animals may add to the operational cost of a typical ranch in other ways. Wapiti, for example, tear down fences, eat haystacks, and graze in hay fields and other areas intended for livestock production.

Big game populations are susceptible to affliction by parasites and diseases under free-ranging conditions. The animals' vulnerability is increased when exotic and native species are confined with livestock in pastures surrounded by a game-proof fence. The services of a veterinarian experienced in working with big game should be obtained when problems with diseases or parasites occur or are suspected.

Predation may be a prevalent operational problem. The productivity of indigenous big game species may be reduced and the successful introduction and establishment of exotic big game animals can be affected by predation. The introduced herd may never become established and eventually disappear, or it may only maintain itself at an uneconomic level.

Besides paying for all this, the landowner must interrupt normal business operations to patrol or otherwise attempt to alleviate the problems. Thus, a significant cost is incurred by the landowners whether hunters, or other recreationists, realize it or not.

Figure 13–3 A mature axis deer buck offers outstanding trophy qualities, rufus brown coat profusely dappled with permanent white spots, and excellent venison. The consequences of exotic introductions, however, are often unpredictable and should be carefully considered. (Photograph by H. Swiggett.)

Expenses other than those associated with routine ranching operations may also be incurred to establish a big game ranching enterprise. Appropriate fences may need to be constructed, exotic animals may be purchased, advertising costs may be experienced, and construction of lodging facilities may be necessary depending on the type of enterprise desired.

A variety of concerns exist when exotic animals are considered for introduction. Native species may be displaced by exotics, the ecological status of existing communities may be disturbed, crossbreeding between exotics and indigenous species may occur, exotic big game animals may be difficult to control if they increase rapidly or disperse into areas where they are not desired, exotics can cause depredations to crops and forests, exotics can contribute to land use conflicts, and they may introduce new diseases or parasites to livestock or indigenous wildlife species (34). The consequences of exotic introductions are unpredictable and should not be taken lightly. Caution is advocated when the release of foreign big game animals is under consideration (9) (Figure 13–3).

Legal disincentives also serve to discourage private big game ranching programs. Landowners may have little protection from liability claims under existing laws. Some states are beginning to recognize this problem and are passing protective laws. However, case law interpretation of liability claims requires years of deliberation by the courts. Under these circumstances, insurance costs for protection against liability may be exorbitant. Trespass laws may be another disincentive to those landowners seeking to manage big game commercially. The laws are often vague, poorly enforced, and may favor the trespasser instead of the landowner (9).

These situations and associated problems are complex and have no easy answers. However, the economic and legal concerns effectively hinder incentives to develop and maintain wildlife habitat and wildlife populations on private lands. The trespass and vandalism aspects foster ill will and may prevent the landowners from obtaining an adequate return on property investments. The problems have been exacerbated as our nation's human population has increased and huntable acreage has decreased. Landowner attitudes toward hunters have changed and the amount of private land closed to hunting has increased rapidly in many states. Instead of hunters being welcomed by friendly landowners, outdoor enthusiasts are often greeted by "no trespassing" and "no hunting" signs.

Figure 13–4 Various species or hybrids of exotic sheep, or other exotic big game species, can be hunted throughout the year. (Photograph by H. Swiggett.)

Incentives for Big Game Ranching Practices

The addition of a hunting enterprise to an existing business amounts to diversification of the overall operation. Diversification protects against losses in revenue from normal ranching or farming operations. Because indigenous big game animals compete to some extent with domestic livestock, prudent landowners charge an access fee for hunters to harvest these animals. The wildlife resource is an expense that should also generate its own portion of income. Exotics can add another dimension to the business. Exotics can often be hunted year-round and lodging facilities can thus be utilized when they would otherwise be vacant (Figure 13–4). Both the animals and the facilities thereby provide income when other ranch products are not being marketed.

Other incentives include payment by the state wildlife agency to landowners for indigenous animals harvested on their land, or the issuance of a specified number of landowner permits that can be dispensed by the landowner. Usually the permits are sold to hunters. Some state wildlife agencies also pay landowners for damages caused by indigenous wildlife. For example, big game animals eat or otherwise destroy various truck crops, hay fields, and haystacks, and wapiti tear down pasture fences. Other possible measures to encourage cooperation among interested parties include stricter trespass laws, reduced landowner liability, tax incentives, and technical assistance for wildlife management purposes.

A basic advantage of fee hunting is that it provides an incentive for landowners to produce wildlife from which primary business incomes may be derived. Landowners diversify their risk and opportunity by making wildlife one component of the land-based production aspects of the business. As a component of agricultural production, the commercialization of wildlife has earned respectability, and the concept has been endorsed by the World Conservation Strategy and by the 1987 Report of the World Commission on Environment and Development (8). Benefits listed include influencing land use by redirecting rewards of conservation to landowners, stabilizing markets for wildlife products, and securing the gene pool of rare species. Other opportunities include maintenance of productive landscapes, development and supplying of new markets, and providing livelihoods for local people.

In many cases, industrial expansion, urban sprawl, and other factors have eroded the quality and quantity of wildlife habitat. This situation can be reversed through the collective cooperation of professional resource managers, landowners, concerned conservation organizations, and society in general. Two-thirds of the land in the United States is privately owned (35). Obviously, private lands must be included in successful programs to benefit wildlife and other renewable natural resources. However, incentives for land and wildlife conservation must be provided or landowners cannot afford to make the investments of money and time.

The missing ingredient to address the escalating problem is that consultation and cooperation with the people who effectively control much of the wildlife resources—the private landowners—have not been pursued with the vigor needed for good wildlife policy. Americans have taken the availability of "free" wildlife for granted when, in reality, landowners have been producing much of this public resource at private expense (27). Stated another way, hunters and other recreationists have been subsidized by landowners who produce a product that is common property. But landowners may or may not benefit through their efforts. Author and noted conservationist Aldo Leopold chaired the Committee on American Game Policy that succinctly summarized the issue with the following statement on American Game Policy (36, p. 288):

> *Recognize the landowner as the custodian of public game on all* [private] *land,* protect him from the irresponsible shooter, and compensate him for putting his land in productive condition.
>
> . . . on condition that he preserves the game seed and otherwise safeguards the public interest. In short, make game management a partnership enterprise to which the landholder, the sportsman, and the public each contributes appropriate services, and from which each derives appropriate rewards. (Emphasis in original)

The wisdom of Leopold and the committee has been generally ignored. Wildlife management must be more than a noble crusade. It must include a willingness by the wildlife agencies that hold the wildlife in public trust, and the landowners who control access to the wildlife, to cooperate. Otherwise, the agencies cannot fulfill their administrative mandates and landowners will view wildlife as an intolerable liability.

The Big Game Ranching Issue

The evolution of commercial big game ranching for fee hunting in North America has not been a smooth process. Differences in opinion exist regarding the pros and cons of big game ranching (37, 38, 39, 11, 16, 8, 10). Some critics doubt whether commercial wildlife production is a viable conservation and agriculture alternative. Others question if it will work because of legal, biological, or socioeconomic questions. Philosophical concerns also have been voiced that wildlife should remain "wild" and that illegal activities will result from a social revolt against the elite who "control" the wildlife. This highly charged issue led one critic to write that markets in venison and wildlife parts, fee hunting, and frivolous hunting for fun represent three grim horsemen of destruction bent on changing the North American system of wildlife conservation (40).

Although extreme positions exist, a rational evaluation should prevail because big game ranching is a global fact. Three international wildlife ranching symposiums have been held and the proceedings from each have been published: in the United States (1988), Canada (1990), and South Africa (1992). Attendance has grown to more than 450 participants per event (41). Private resources are being spent worldwide to provide quality hunting and other wildlife experiences for a fee. From the savannas of Africa, to the countrysides of Canada and western Europe, in the forests of the southeastern United States, and on the rangelands of Texas, a successful wildlife recreation industry is evident. Despite benefits to wildlife and sportsmen alike, however, some American opposition to big game ranching persists. Concerns over big game ranching (fee hunting) for state-owned wildlife on private lands can be classified in six categories.

First, a belief exists that the public will lose associated hunting rights and wildlife agencies will forfeit management control (42, 43, 44). This criticism, however, overlooks a significant reality. The right of landowners to control access or trespass to their land is recognized and protected through the Constitution and state trespass laws (27, 45). Although indigenous wildlife is treated as a public trust, the enforcement of trespass laws provides landowners *de facto* control over wildlife by virtue of their landownership (28, 45, 44). Landowners can, therefore, charge a fee for right of entry onto private property for purposes of hunting or other recreational pursuits.

Private landowners involved in the management of indigenous wildlife, due to agreements with state wildlife agencies, do not own the wildlife. However, vesting of some responsibilities and incentives in the landowner leads to better management (46). Under a property rights system, wildlife "damages" to private property become simple production costs to landowners who derive income from wildlife resources on their lands (47).

Second, a concern exists that a market for big game ranchers will reduce wildlife populations through poaching and other forms of overexploitation (48, 49, 40, 37, 38, 39). This concern is based on the belief that wildlife suffers when left to the mercy of individuals seeking private gain. Market hunting and markets for wildlife meat and by-products are cited as an historic cause of the uncontrolled decimation of North America's wildlife in the late 19th and early 20th centuries; similarly, big game ranching is viewed as a current threat to wildlife conservation (48, 49, 40, 37, 38).

Ample evidence exists, however, demonstrating successful conservation results from big game ranching programs despite the fears of its critics. For example, International Paper initiated and developed an extensive lease-hunting and wildlife management program on 668,016 hectares in the southeastern United States because wildlife production became profitable. Hunting leases became one of the company's best products in terms of investment revenue (50). Weyerhaeuser Company also has successful wildlife management and hunting programs across the south. Fees are set to generate marginal profit (51).

The state of Texas contains little public land. A total of 99% of the rangelands and 97% of the forests are under private ownership (29). Under various approaches, the Texas Parks and Wildlife Department exercises legal authority over the indigenous wildlife while landowners provide management stewardship. Record numbers of white-tailed deer occur, and the trophy bucks harvested annually in the "south Texas brush country" are legendary. The successful Texas hunting system results from commercial incentives provided to landowners through various fee-hunting arrangements (29).

Several large private land ranches in the southwestern United States have developed wapiti hunting enterprises (52). They offer quality hunts with high success rates and higher opportunities to harvest trophy bulls, compared to public land hunts. The 193,927-hectare privately owned Vermejo Park Ranch in northern New Mexico exemplifies the attributes for wapiti hunting and several other species as well. In 1989, hunter success was reportedly over 60% for wapiti, 85% for deer, and 100% for pronghorn. Vermejo hosted 1,192 wapiti hunters during 1974–1976, who accounted for 16% of New Mexico's wapiti harvest. A full-time wildlife biologist managed the program for optimum production and trophy animals (53).

The privately owned 80,972-hectare Deseret Land and Livestock ranch in northeastern Utah shows that wildlife ranching and integrated management, including livestock, can work in the far west. In 1977, the ranch owners decided that big game should pay its way. Subsequently, receipts from the wildlife program paid for two full-time biologists, wildlife research, ranch security, and ranch maintenance. Wapiti numbers rose from 350 in 1977 to 1,500 head, mule deer numbers and quality improved, the program expanded into moose and bison, livestock numbers more than doubled, and the range significantly improved (54). Public lands have also benefitted from the private efforts of the ranch. More than 600 wapiti were transplanted to public land in Utah, and others migrated off the ranch onto adjacent public lands (42).

Third, a concern exists that private management of wildlife will cater only to the wealthy, resulting in fewer hunters (48, 49, 40, 37, 38, 42). A concern persists that the typical hunter will not benefit from widespread private management of wildlife. For example, one statement claims that "very slowly, paid hunting concentrates access to wildlife in favor of a smaller and smaller segment of an increasingly affluent, politically powerful people" (40, p. 27). Perhaps the statement holds merit, but the wealthy typically reflect affluence. The rich live in larger homes, drive costlier automobiles, wear nicer clothes, and travel to exotic destinations not affordable to the "average citizen." However, one landowner who provides fee-hunting opportunities in California reported that most hunters on ranches similar to his are in middle-income brackets. The percentages of low- and high-income hunters were similar (32).

Accepted economic theory holds that when a market exists, competition in private industry usually results in increased supplies and lower costs for the commodities and services. Accordingly, the North American economy often results in products and services once available only to the wealthy becoming accessible to the general public because of the workings of the free market. Familiar examples include four-wheel-drive vehicles, snowmobiles, elaborate camping units, well-equipped boats, customized rifles, pistols, and shotguns, and a multitude of other sophisticated gadgets commonly used by today's sportsmen. Fee hunting should be no different. Assuming governmental restrictions do not thwart growth and competition in the private hunting industry, the cost of fee hunting to the average hunter can be expected to decrease while the opportunities increase.

Fourth, a concern exists that replacement of formerly free opportunities by fee hunting reduces hunting opportunities and leads to deterioration of the quality of hunting on public land (42, 43). The rationale for this concern is unclear. Private fee hunting expands the opportunities for hunting. Those who wish, can hunt on private land for a fee. Hunters who would rather hunt for "free" on public land are not precluded from this option. Fee hunting on private land does not replace the opportunity to hunt on public land, as evidenced in California, New Mexico, Utah, and other western states where hunters pay to hunt on private land, often immediately adjacent to accessible public land (53, 54, 32). Private lands in New Mexico, for example, offer some of the best hunting for wapiti and mule deer in the state. The clientele for these private land hunts may thus improve the quality of hunting on public land by reducing the number of hunters on public land. Public land also may benefit by acquiring wildlife that is produced on private land but migrates to adjacent public land or is translocated and released onto public land.

Fifth, criticism exists that big game ranchers who benefit financially from state-owned wildlife should compensate the public for the opportunity (42). It has been argued that landowners who charge a fee, thus benefitting from state-owned wildlife, should be required to pay for this right through a licensing procedure (42). In reality, landowners often do pay the state for annual licenses to operate their businesses. The licensing fee compensates the state for providing the original animals for their operations. The public receives revenues, through the state agency, and the landowner provides wildlife and habitat management that benefits the public with little, if any, public expenditures of financial resources.

Finally, concern exists that private management of wildlife with an economic value will adversely impact other species and their habitat (39). On any defined geographic area under private management, perhaps some species are neglected in favor of others. Nevertheless, numerous landowners managing for wildlife over a large expanse of land will approach their goals in different ways, resulting in a diverse mosaic of habitats and wildlife.

Habitat is the most fundamental management issue now confronting resource management agencies. Habitat loss and habitat degradation were ranked, by state and federal biologists, as the two most important wildlife management issues (30). When landowners can derive benefits from wildlife production under a private property rights

Figure 13–5 A male pronghorn is a popular indigenous big game trophy. Pronghorns and other big game can benefit when their needs for habitat are met through private management practices. (Photograph by G. A. Littauer.)

system, private interest, effort, and investment in habitat and wildlife management will increase (Figure 13–5). Local people with a vested interest in wildlife have a much greater incentive to conserve it (55).

The expanding big game farming industry in Canada has benefitted wildlife. Many large wild ungulate species have reappeared on lands throughout Canada where they were once considered to have disappeared (15). One professional wildlife biologist, who is also a rancher in California, concisely summarized the issue: "I know when I manage for one species, I automatically deprive some other species of some habitat, but most wildlife will benefit from game management. And if fee-hunting allows me to prevent converting the ranch to other uses, it is a big benefit for all species" (32, p. 307).

Solutions to the state-owned wildlife on private lands situation will likely be found "in recognition of local, sectional, and regional differences in perceptions of wildlife values, and in the economic, social, and attitudinal benefits derived therefrom. Here, best results should come from sets of incentives and regulations; tailor-made to fit—realistically—local, sectional, or regional wildlife management needs and goals" (28, p. 275). Several state wildlife agencies have recognized the "win–win" attributes of a public and private partnership approach to wildlife management. The states of California, Colorado, Montana, New Mexico, Oregon, Utah, and Washington have implemented cooperative programs with varying degrees of success in achieving their objectives (56).

The U.S. Department of Agriculture, Natural Resources Conservation Service (57, p. 8) summarized the local approach to land stewardship: "From a national perspective, then, our land will be healthy not because of broad public policies and programs but because each landowner will make his or her own individual place healthy." A positive attitude toward partnerships for stewardship of state-owned wildlife on private lands will augment and strengthen a conservation trend that increasingly stresses the inclusion of local people in the management and protection of natural resources. Economically, individuals will have an incentive to conserve and enhance wildlife and its habitat; ecologically, the natural resources will benefit; and, environmentally, everyone will be winners!

SUMMARY

The long-term future of big game ranching is problematical. Uncontrolled commercial public exploitation took its toll on wildlife resources as America succumbed to European settlement. Protective laws against wanton slaughter and regulated or eliminated markets for game meat and by-products allowed many wildlife species to stage a comeback. Agriculture came into its own as an industry as new technology and technical knowledge evolved. Simultaneously, new consumer tastes and increased leisure time led to wildlife becoming a source for recreation, with meat and by-products a bonus.

The concept that big game animals could play a potential role in commercial meat production has been slow to materialize subsequent to American colonization. Consumers, however, are trying alternative meats for the differing flavors and as a health food. Due to the buying public's unpredictable nature, the consumption of big game meat may prove to be a permanent interest or it might be a passing fad.

The future of big game farming ventures will depend on how successfully the industry deals with a myriad of challenges and opportunities. Minimal governmental restrictions and interference and a favorable economic environment are necessary for expansion. A dependable and adequate supply of products and certified processing plants must be established. Improved distribution, marketing, and promotion of commodities are necessary for development of the industry. Proponents of animal rights and supporters of wildlife for its intrinsic value will present challenges requiring ethical and professional responses and attention to appropriate husbandry. North America has ample agricultural grazing and farmland suited to big game farming. Innovative approaches to solving the existing social, technical, and management obstacles will open new doors to exploit the potential market.

Big game animals are part of the wildlife resources in an area. From a conservation viewpoint, commercial big game ranchers may improve big game management practices and restore habitat. In turn, many of the associated wildlife species benefit from the enhanced habitat. Motives for private wildlife conservation efforts may be aesthetic or economic. Noneconomic motives are largely dependent on landowner philanthropy. Because of the harsh economic pressures facing many landowners, however, attempts are being made to wisely manage private landholdings financially, so that operational expenses are paid. Wildlife resources often must become economically competitive with other land use pressures on private property to encourage economically strapped landowners to commit their time, effort, and money to wildlife enhancement.

An increasing human population is placing a heavy demand on outdoor recreation. However, increased sport hunting opportunities resulting from big game ranching may alleviate existing pressures on private and public lands. Opposition to hunting by animal rights supporters, fluctuating economic conditions, court interpretations of trespass laws, and urbanization of society will present challenges to the hunting fraternity. Cumulatively, effects may be significant.

Big game ranching has its critics. Aspiring big game ranching endeavors, however, occur throughout North America and around the globe. In 1930, the American Game Policy, drafted by a group of wildlife statesmen, indicated, in part, that professional managers of wildlife, and society in general, needed to recognize landowners as the custodians of state-owned wildlife on private land. Further, the landowners should be compensated for their stewardship of the "public interest." Little progress has been made on this front until recently.

Given the significant degradation and loss of wildlife habitat, the increasing demand for outdoor recreation, and the *de facto* control landowners have over state-owned wildlife resources, the time is appropriate for change. Americans should abandon expectations of gratuitous private production efforts of state-owned wildlife and seize the benefits associated with compensation for private conservation efforts. Several states have implemented cooperative wildlife management programs with landowners. The programs differ because they are tailored to local circumstances. However, all involved parties benefit, with the wildlife and its habitat intended as the primary beneficiaries.

LITERATURE CITED

1. RAMSEY, C. W. 1968. Texotics. Bulletin 49. Texas Parks and Wildlife Department, Austin.

2. LEOPOLD, A. 1933. *Game Management.* Charles Scribner's Sons, New York.

3. WALKER, E. P. 1964. *Mammals of the World.* Volume 2. Johns Hopkins Press, Baltimore, MD.

4. ISAAC, E. 1970. *Geography of Domestication.* Prentice Hall, Englewood Cliffs, NJ.

5. LEE, C. H., and T. S. CH'ANG. 1985. Marketing and utilization of deer products in Asia. Pages 307–310 *in* P. F. Fennessy and K. R. Drew, eds., *Biology of Deer Production.* Bulletin 22. The Royal Society of New Zealand, Wellington, New Zealand.

6. KONG, Y. C., and P. P. H. BUT. 1985. Deer—The ultimate medicinal animal (antler and deer parts in medicine). Pages 311–324 *in* P. F. Fennessy, and K. R. Drew, eds., *Biology of Deer Production.* Bulletin 22. The Royal Society of New Zealand, Wellington, New Zealand.

7. BROWN, R. D., ed. 1983. Introduction. Page V *in Antler Development in Cervidae.* Texas A & I University, Kingsville, TX.

8. HUDSON, R. J. 1989. History and technology. Pages 11–27 *in* R. J. Hudson, K. R. Drew, and L. M. Baskin, eds., *Wildlife Production Systems.* Cambridge University Press, New York.

9. WHITE, R. J. 1987. *Big Game Ranching in the United States.* Wild Sheep and Goat International, Las Cruces, NM.

10. DEMARAIS, S., D. A. OSBORN, and J. J. JACKLEY. 1990. Exotic big game: A controversial resource. *Rangelands* 12:121–125.

11. TEER, J. G. 1994. Game harvest systems in Texas. Pages 47–50 *in* W. V. Hoven, H. Ebedes, and A. Conroy, eds. *Proceedings Third International Wildlife Ranching Symposium.* Centre for Wildlife Management, University of Pretoria, Pretoria, South Africa.

12. MUNGALL, E. C., and W. J. SHEFFIELD. 1994. *Exotics on the Range.* Texas A&M University Press, College Station.

13. RENECKER, L. A. 1989. Overview of game ranching in Canada. Pages 47–62 *in* R. Valdez, ed., *Proceedings First International Wildlife Ranching Symposium.* New Mexico Cooperative Extension Service, Las Cruces, NM.

14. IRELAND, D. B., and R. J. LEWIS. 1994. Game farming in British Columbia, Canada. Pages 63–65 *in* W. V. Hoven, H. Ebedes, and A. Conroy, eds., *Proceedings Third International Wildlife Ranching Symposium.* Centre for Wildlife Management, University of Pretoria, Pretoria, South Africa.

15. RENECKER, L. A. 1991. Status of game production in Canada. Pages 70–73 *in* L. A. Renecker and R. J. Hudson, eds., *Wildlife Production: Conservation and Sustainable Development.* Agricultural and Forestry Experiment Station, Fairbanks, Alaska.

16. RENECKER, L. A., C. B. BLYTH, and C. C. GATES. 1989. Game production in western Canada. Pages 248–267 *in* R. J. Hudson, K. R. Drew, and L. M. Baskin, eds., *Wildlife Production Systems.* Cambridge University Press, New York.

17. STEVENSON, R. E. 1991. Big game farming in Alberta. Pages 516–517 *in* L. A. Renecker and R. J. Hudson, eds. *Wildlife Production: Conservation and Sustainable Development.* Agricultural and Forestry Experiment Station, Fairbanks, AK.

18. DIETERICH, R. A. 1991. Reindeer management in Alaska. Pages 213–217 *in* L. A. Renecker and R. J. Hudson, eds., *Wildlife Production: Conservation and Sustainable Development.* Renecker and Hudson, 1991: Agricultural and Forestry Experiment Station, Fairbanks, Alaska.

19. SCOTTER, G. W. 1989. Reindeer husbandry in North America. Pages 223–241 *in* R. J. Hudson, K. R. Drew, and L. M. Baskin, eds., *Wildlife Production Systems.* Cambridge University Press, New York.

20. DIETRICH, U. 1989. Status of the commercialization of Mexican wildlife with special reference to the exploitation of important game species in northeastern Mexico. Pages 65–72 *in* R. Valdez, ed., *Proceedings First International Wildlife Ranching Symposium.* New Mexico Cooperative Extension Service, Las Cruces, NM.

21. MELLINK, E. 1991. Exotic herbivores for the utilization of arid and semiarid rangelands of Mexico. Pages 261–266 *in* L. A. Renecker and R. J. Hudson, eds., *Wildlife Production: Conservation and Sustainable Development.* Agricultural and Forestry Experiment Station, Fairbanks, Alaska.

22. FOX, B. R., ed., 1996. NADeFA census. *The North American Deer Farmer.* Autumn 1996:10.

23. ANDERSON, D. L., ed. 1993. Buffalo are back. *Live Animal Trade & Transport* 3:5–16.

24. VON KERCKERINCK, J. 1987. *Deer Farming in North America.* Phanter Press, New York.

25. BEAN, M. J. 1977. The evolution of national wildlife law. Council on Environmental Quality, U.S. Government Printing Office, Washington, DC.

26. SIGLER, W. F. 1956. *Wildlife Law Enforcement.* Wm. C. Brown Company, Dubuque, IA.

27. TOBER, J. A. 1981. *Who Owns the Wildlife?* Greenwood Press, Westport, CT.

28. BURGER, G. V., and J. G. TEER. 1981. Economic and socioeconomic issues influencing wildlife management on private land. Pages 252–278 *in* R. T. Dumke, G. V. Burger, and J. R. March, eds., *Proceedings Wildlife Management on Private Lands.*

29. PAYNE, C. H. 1989. Sport hunting in North America. Pages 134–146 *in* R. J. Hudson, K. R. Drew, and L. M. Baskin, eds., *Wildlife Production Systems.* Cambridge University Press, New York.

30. FLATHER, C. H., and T. W. HOEKSTRA. 1989. An analysis of the wildlife and fish situation in the United States: 1989–2040. General Technical Report RM-178. U.S. Forest Service, Washington, DC.

31. HYDE, D. O. 1985. Conflicts in the use of range—Can they be resolved? Pages 27–29 *in Proceedings of National Range Conference.* U.S. Department of Agriculture, Washington, DC.

32. FITZHUGH, E. L. 1989. Pros and cons of fee hunting. Pages 303–309 *in* R. Valdez, ed., *Proceedings First International Wildlife Ranching Symposium.*

33. MILLER, J. 1990. Educational perspective: Incentives and disincentives of fee hunting on private lands in the south. Pages 53–60 *in* G. K. Yarrow and D. C. Guynn, Jr., eds., *Proceedings of Symposium on Fee Hunting on Private Lands in the South.* Clemson University Cooperative Extension Service, Clemson, SC.

34. DECKER, E. 1978. Exotics. Pages 249–256 *in* J. L. Schmidt and D. L. Gilbert, eds., *Big Game of North America: Ecology and Management.* Stackpole Books, Harrisburg, PA.

35. LANGNER, L. L. 1987. Hunter participation in fee access hunting. *Transactions of the North American Wildlife and Natural Resources Conference* 52:475–482.

36. LEOPOLD, A. 1930. The American game policy in a nutshell. Pages 281–309 *in Transactions Seventeenth American Game Conference.*

37. GEIST, V. 1988. How markets in wildlife meat and parts, and the sale of hunting privileges, jeopardize wildlife conservation. *Conservation Biology* 2:15–26.

38. GEIST, V. 1989. Legal trafficking and paid hunting threaten conservation. *Transactions of the North American Wildlife and Natural Resources Conference* 54:171–178.

39. TEER, J. G. 1989. Commercial utilization of wildlife resources: Can we afford it? Pages 1–7 *in* R. Valdez, ed. *Proceedings First International Wildlife Ranching Symposium.* New Mexico Cooperative Extension Service, Las Cruces, NM.

40. GEIST, V. 1987. Three threats to wildlife conservation. *Deer and Deer Hunting* 10:22–31.

41. RENECKER, L. A., and R. VALDEZ. 1994. The wildlife ranching symposium: History, structure, achievements, goals, and implications for wildlife management. Pages 1–6 *in* W. V. Hoven, H. Ebedes, and A. Conroy, eds., *Proceedings Third International Wildlife Ranching Symposium.* Centre for Wildlife Management, University of Pretoria, Pretoria, South Africa.

42. KWONG, J. O. 1988. Evolving institutions in wildlife management: The case for fee-hunting. *Western Wildlands* 14:26–31.

43. BENSON, D. E. 1989. What fee hunting means to sportsmen in the USA: A preliminary analysis. Pages 296–302 *in* R. Valdez, ed., *Proceedings First International Wildlife Ranching Symposium.* New Mexico Cooperative Extension Service, Las Cruces, NM.

44. DAVIS, R. K., and E. G. PARSONS. 1989. Fee-hunting on public lands. Pages 310–318 *in* R. Valdez, ed., *Proceedings First International Wildlife Ranching Symposium.*

45. MORRILL, W. I. 1987. Fee access views of a private wildlife management consultant. *Transactions of the North American Wildlife and Natural Resources Conference* 52:530–543.

46. BENSON, D. E. 1989. Changes from free to fee-hunting. *Rangelands* 11:176–180.

47. DANA, A. C., J. BADEN, and T. BLOOD. 1985. Ranching and recreation: Covering costs of wildlife production. Working paper 85–11. Political Economy Research Center, Bozeman, MT.

48. GEIST, V. 1985. Game ranching: Threat to wildlife conservation in North America. *Wildlife Society Bulletin* 13:594–598.

49. GEIST, V. 1987. Three threats to wildlife: Game markets, pay hunting and hunting for "fun." Pages 46–58 *in Proceedings Privatization of Wildlife and Public Lands Access Symposium.* Wyoming Game and Fish Department, Cheyenne.

50. BLOOD, T., and J. BADEN. 1984. Wildlife habitat and economic institutions; feast or famine for hunters and game. *Western Wildlands* 10:8–13.

51. MELCHIORS, M. A. 1990. Wildlife and fee hunting on forest industry lands in the south: A Weyerhaeuser company perspective. Pages 99–106 *in* G. K. Yarrow and D. C. Guynn, Jr., eds., *Proceedings of Symposium on Fee Hunting on Private Lands in the South.* Clemson University Cooperative Extension Service, Clemson, SC.

52. WOLFE, G. J. 1989. Private land elk hunting: Opportunities for landowners and sportsmen. Page 217 *in* R. Valdez, ed., *Proceedings First International Wildlife Ranching Symposium.* New Mexico Cooperative Extension Service, Las Cruces, NM.

53. WOLFE, G. J. 1977. Goals and procedures of wildlife management on a large western ranch. *Transactions of the North American Wildlife and Natural Resources Conference* 42:271–277.

54. SIMONDS, G. 1987. Changing management to meet goals. Pages 43–46 *in* L. D. White, T. R. Troxel, and J. M. Payne, eds. *Proceedings 1987 International Ranchers Roundup.* Texas Agricultural Research and Extension Center, Uvalde, TX.

55. STONE, R. D. 1989. Zambia's innovative approach to conservation. World Wildlife Fund Letter. Number 7.

56. ARHA, K. 1996. Sustaining wildlife values on private lands: A survey of state programs for wildlife management on private lands in California, Colorado, Montana, New Mexico, Oregon, Utah and Washington. *Transactions of the North American Wildlife and Natural Resources Conference* 61:267–273.

57. U.S. DEPARTMENT OF AGRICULTURE, NATURAL RESOURCES CONSERVATION SERVICE. 1996. *America's Private Land, a Geography of Hope.* Program Aid 1548. Washington, DC.

Big Game Management on Tribal Lands

Brian Czech

INTRODUCTION

A lot has happened in "Indian country" (18 United States Code 1151) in recent years. Twenty years ago, Indian tribes in the United States were in the political infancy of self-determination, pursuant to the Indian Self-Determination and Education Assistance Act of 1975 (Public Law 93-638). Tribes were to be gradually dissociated from the federal bureaucracy, and began to govern more of their own affairs. Wildlife jurisdiction, always a primary concern for many tribes, would become a bellwether issue for interpreting the doctrine of self-determination. Political relationships of tribes have changed in Canada, too.

Advances in archaeology and paleoanthropology in the past 20 years have provided new information about ancestral Indian populations and ecology. Meanwhile, tribal traditional knowledge has gained in stature as a complement to Western science. The findings from these diverse fields are pertinent to a widening debate about the prehistoric relationship of Indians and large mammals.

The purposes of this chapter are to summarize prevailing theories on the origins of people in North America and their prehistoric relationship with large mammals, to describe basic North American cultures, demography, and use of large mammals at the dawn of European colonization, to document the political status of North American tribes with regard to large mammal management, to provide an overview of the tribal land base, and to discuss important tribal large mammal management issues.

TRIBAL ORIGIN PERSPECTIVES

Tribal cultures possess traditional knowledge that lies beyond the purview of Western science. Some scientists believe that Indians are descended from Asian people that migrated across Beringia (i.e., the landmass that periodically connected Siberia and Alaska during the Pleistocene epoch) and generally omit any mention of tribal tradition. Yet Deloria noted that "American Indians, as a general rule, have aggressively opposed the Bering Strait migration doctrine because it does not reflect any of the memories or traditions passed down by the ancestors over many generations" (1, p. 97). He also exposed doubt among the scientific community about archaeology methods, especially carbon dating.

Oral traditions of momentous geological events in the Northwest were often disregarded as scientific impossibilities, but in some cases science has advanced only to verify such "myths" (2, 3). Similarly, a common tribal tradition explains that animals and people were once one and the same, but gradually became distinguished (1). Such stories were despised as pagan for centuries; now evolution is the most common explanation for human existence.

Unfortunately, oral traditions are not well documented in the literature, making it difficult for Western scientists to incorporate traditional knowledge in their theories, much less their management implications. There is a trend toward the consolidation and consideration of traditional knowledge, but for now, Western science dominates academic discussions of tribal origins and big game. Furthermore, it is also doubtful that oral traditions contain detail sufficient to satisfy scientists or students. The reality of pre-Columbian North America might best be explained by a combination of Western science and tribal tradition.

After Haven compiled compelling evidence, a consensus gradually formed in the scientific community that Indians are descendants of Asians that immigrated via Beringia (4). The apparent time of native American settlement has been continually pushed back in the literature. At least until the 1928 excavations at Folsum, New Mexico, most anthropologists believed that Beringians first entered North America within the last 6,000 years. The Folsum spear points were thought to be much older. Soon Clovis points were found in association with mammoth kills that outdated the Folsum culture (5). Pre-Clovis artifacts also have been found. By the late 1970s, with the aid of radiocarbon dating, a scientific consensus acknowledged that Beringians had colonized the Americas at least 10,000 years ago, with most scientists opting for a date closer to 13,000 years (6). Many scientists now believe that the earliest migration occurred between 14,000 and 18,000 years ago down an ice-free corridor at the eastern edge of the Rocky Mountains (5).

There is controversial evidence for North American colonization during an earlier emergence of Beringia around 50,000 years ago, including artifacts from Chile and Brazil that have been dated to 28,000 years (7). Some artifacts from Texas may be 37,000 years old (8). MacNeish, with Peruvian artifacts arguably dated at 22,000 years, speculated that North American entry may have occurred 100,000 years ago (9). Stannard asserted that, "the best scientific evidence to date, drawing on coalescent findings from several disciplines, suggests a more prudent estimate would be for an entry date of around 40,000 B.C." (10, p. 265).

OBLIGATE BIG GAME CULTURES

Beringian immigrants at the end of the Pleistocene were nomadic hunters who focused on woolly mammoths (5). As Beringia gradually submersed from 14,000 to 10,000 years ago, most of its steppe tundra fauna went extinct. Since Martin proposed his "overkill hypothesis," it has been popular to concur that the early Indians emanating from Beringia were the primary cause of extinction (11). Surely they played a role, but "Not even the most ardent proponents of the steppe-tundra [vs. a less productive arctic tundra] hypothesis envisage Beringia as a land teeming with vast herds of herbivores. More likely, these animals were scattered through the landscape, concentrated at dozens of special sites, in lowland meadows and near rivers" (5, p. 116).

Meanwhile, the ice-free corridor was widening, and around 11,500 years ago the Clovis culture blossomed on the Great Plains, which then included much of the present Southwest and eastern woodlands (12). Clovis hunters threw spears accelerated with the atlatl. While extant, mammoths remained the favored prey, along with large ancestors of modern plains bison. The Clovis hunters also killed mastodons, camels, horses, tapirs, ground sloths, and bears (5, 6). Clovis populations expanded rapidly, and Clovis artifacts are found throughout much of North America.

In about 500 years, the Clovis culture vanished, along with much of the remaining North American large mammal fauna. While habitat loss was clearly a problem for

Figure 14–1 Plains Indians hunting buffalo. (Engraving by W. W. Rice, 1873; Library of Congress.)

the Beringian fauna, hunting may have played a more important role in the southerly latitudes, especially in Mesoamerica (present-day Mexico, Guatemala, Belize, El Salvador, and Honduras), where big game never was common during the late Pleistocene (13). Martin speculated that human populations may have doubled every 20 to 50 years, and that it took less than 1,000 years for the Americas to become populated at a density of one person per 1.6 square kilometers (14). The fauna was not adapted to efficient human hunters and supposedly would have been easy prey. Although Martin's overkill hypothesis is criticized by many, human and nonhuman factors probably contributed to the Pleistocene extinctions (15).

One finding that tends to refute the overkill hypothesis is the flourishing of bison populations in the midst of the megafauna extinctions (Figure 14–1). Shortgrass plains expanded rapidly throughout much of North America in the wake of the Pleistocene, and bison evolved accordingly (5). Despite an increasing dependence of post-Clovis hunters on plains bison, the bison proliferated. Bison species and plains Indian cultures were keystones in a general economy of nature that lasted almost 10,000 years, until European Americans nearly eliminated the buffalo (16). Elk, pronghorn antelope, and grizzly bear were common dietary supplements for the plains Indians at European contact, as were white-tailed deer in the prairie river bottoms and mule deer in the shortgrass breaks. Lesser amounts of small game and wild plants were eaten (12, 17).

The bison jump hunting strategy, where Indians would stampede a group of bison over a steep ledge, is sometimes cited as evidence of a poor conservation ethic, but it seems impertinent to retroactively apply a conservation ethic to a people who were literally dependent for their survival on an adequate harvest of bison (18). Plains Indians did not develop a bow-hunting culture until 1,400 years ago (5). Stalking and killing bison on the open range with spears was probably a difficult and dangerous task. Driving buffalo over a cliff, where possible, would have been much more efficient, and roughly analogous to a modern-day cattle drive. Waste was not wanton, but incidental to the unpredictable outcome. Furthermore, according to the archaeological record, large bison jumps were relatively infrequent events (5, p. 208).

Meanwhile, coastal and maritime cultures were evolving in the Northwest and in the far North. The Northwest cultures probably were descended from the original (or "Amerind") Beringian migrants, but the Eskimos and Aleuts of the North apparently descended from a distinct migration around 10,000 years ago. They may have even crossed the Bering Strait shortly after the Pleistocene via skin boats. However they

Figure 14–2 Tlingits of southern Alaska with Chilkat blanket woven with mountain goat wool and cedar bark, 1896. (Library of Congress.)

crossed, Eskimos were primarily barren ground caribou and muskox hunters for millennia (5). Moving eastward along the tundra, they reached Greenland no later than 4,000 years ago. Eskimos who settled near the Bering Strait (and Aleuts that settled the archipelago named for them) gradually became sea mammal hunters. Walrus and whales were favored quarry, but seals were probably the staple. Sea lions and polar bears also were taken (19, 20, 5, 21, 6). This "Thule tradition" of sea mammal hunters spread from the Bering Strait to Greenland primarily about 1,000 years ago, replacing to some extent the interior Eskimo hunting culture. Eskimos were (and are) the most dependent of American cultures on large mammals. They were entirely carnivorous, and large mammal parts supplied them with virtually all of their shelter, clothing, and tools (Figure 14–2).

OTHER TRIBAL CULTURES AT EUROPEAN CONTACT

Spear points throughout much of the continent diversified rapidly after 10,000 years ago, suggesting an increasing use of various upland game species, but the basic big game hunting economy dominated the cultural scene throughout the first half of the Holocene. Around 5,000 years ago, culture rapidly diversified (5). By 1492, seven broad cultures existed in addition to plains and arctic. Most corresponded with a distinct biome. In each case, large and small mammals and birds comprised part of the Indian diet. In most cases, fish and wild plants also were important. Herptofauna and invertebrates were of local importance. In a few areas, cultivated plants were staples. These cultures are discussed in approximate decreasing order of their dependence on large mammals.

Boreal Forest

Sometime between the Amerind and Eskimo/Aleut migrations, the other widely acknowledged Beringian race appeared in North America. Some of their descendants eventually found their way to the Southwest, where they diverged into Apache and Navajo cultures (5). Most, however, remained in the northern reaches between the Arctic tundra and the temperate regions. These "Athabascans" were dependent on moose, barren ground caribou, and woodland caribou for their subsistence. The ratio of moose : caribou in the diet generally increased from north to south. In some areas fish

Figure 14–3 Florida Indians hunting deer with deerskin disguises in 1564. (From T. De Bry's *Grandes Voyages,* 1591; Library of Congress.)

were nearly as important, especially salmon in the Alaskan interior. Black bears and grizzly bears were commonly hunted, and polar bears were probably hunted in the Hudson's Bay area (12). Deer and elk were occasionally taken, apparently along the southern margins of Athabascan territory and in the foothills of the Canadian Rockies (17). Waterfowl and a few wild plants supplemented the diet, but agriculture was nonexistent. As with the Eskimos/Aleuts, the Athabascan big game economy remains largely intact, especially in the more northern latitudes.

Woodland

At European contact, bison were widely dispersed in the woodlands of the eastern United States and Canada (16). Indians cleared and maintained large openings with stone axes and fire, and wildfire assisted. However, bison were by no means a staple, and were absent from many areas. At the ecotone with boreal forest, where the woodland cultures abutted Athabascan territory, moose, woodland caribou, and black bear were the most important species, along with wild rice, berries, and waterfowl (22). Eastern elk were distributed and hunted throughout the woodland proper, except for the extreme Northeast and Southeast (23). In the New England states, Florida (Fig. 14–3), and the coastal plains of the Atlantic and Gulf of Mexico, white-tailed deer and turkeys were the most important big game species (22). A great variety of small game and plant foods were obtained, especially in the central and southern regions, and farming was well developed in some areas. Fishing was particularly important in the Great Lakes region and along the Atlantic and Gulf coasts (12).

Desert

As the continent dried following the Pleistocene, the "Desert tradition" arose about 10,500 years ago in the Great Basin and came to dominate Indian life throughout the Southwest, including northeast Mexico (12). The most prominent artifacts of the Desert tradition are baskets and millstones, reflecting the importance of edible wild plants. However, large mammals were fairly common and important subsistence items. During much of the Holocene, the most important big game species in the Desert tradition was the pronghorn, followed by bighorn sheep (20). Mule deer also were taken, but were relatively uncommon until the 1800s, when overgrazing by cattle made much of the Great Basin more favorable to mule deer at the expense of pronghorn (12, 24). Elk and turkeys were important to the Apache and Pueblo tribes in the high country of New Mexico and eastern Arizona, and bison were occasionally taken along the eastern edges of the deserts (17). Farming was important primarily in the valleys of southern Arizona and New Mexico, the tribes of which are sometimes classified as a separate Southwestern culture (6). Reptiles and insects were also important food items, especially in the Great Basin (12, 22).

Columbian Plateau

The Columbian Plateau is a large, two-dimensional ecotone between boreal forest and desert, and between coastal rain forest and Great Plains. It is mountainous, adding to its ecological diversity. In fact, the plateau includes considerable elements of plains (notably the Palouse grasslands), woodlands, deserts, boreal forest, and even tundra on the peaks. Some tribes in the eastern portion of the plateau ventured onto the Great Plains to hunt bison after they had horses, but rarely before then. In fact, there may have been few people in the northern Rockies except for seasonal visitors. On the western portion of the plateau, however, numerous tribes were supported by salmon and roots (6). Mule deer probably were the most important big game species over most of the area, but elk were a staple of the northern Shoshone and Bannock tribes (25). Moose were important in the northern and eastern portions (12). Bighorn sheep were also hunted, as were mountain goats, perhaps primarily for the wool trade with coastal tribes (22, 17, 6). Bears (presumably black and grizzly) and woodland caribou were occasionally hunted, and the latter may have been important to the subsistence of local bands in the northern Cascades and Rockies (17).

Coastal Rainforest

By 1492, there was an incredible diversity of Indian culture along the coast from northern California to Southeast Alaska (6). Large mammals were frequently hunted, especially Columbian black-tailed deer, Columbian white-tailed deer, and Roosevelt elk along the shore (17). Black bears (and perhaps grizzly bears in northern areas) were hunted with dogs (17). Mountain goats commonly were taken for food and wool (17, 6, 22). Underhill reported that moose was a food item, presumably north of Puget Sound (22). Some of the tribes (including Makah, Quileute, and Quinault) specialized in whaling, and most exploited other sea mammals. For the overall culture. however, salmon was the staple (22).

California

Virtually all North American biomes are represented in California, and its borders are contiguous with desert, plateau, and rain forest regions. In the sweeping valleys and mountains within, however, lived by far the greatest number of Indians of any area its size north of Mexico—more than 350,000 at European contact (8). Due to the remarkable variety of California ecosystems, there were a large number of small tribes. The common cultural theme was gathering, especially of acorns. Small game and a wide variety of invertebrates probably were more important food items than big game, but deer, Roosevelt elk, Tule elk, and pronghorn were commonly hunted (12). Bighorn sheep and bears were pursued less frequently (17, 22).

Mexico

Pre-Columbian Mexico was populated by four types of cultures, including the Desert tradition, which was most common in present-day Sonora. Plains culture extended into northern Chihuahua and Coahuila. The loose-knit California culture extended into Baja California, where dependence on shellfish and seafood was greater. Most Mexican culture was

Other Tribal Cultures at European Contact

Mesoamerican, characterized by agricultural villages and large cities with prominent markets, road systems, water diversions, and government, military, and religious institutions (26). Large mammals were uncommon, but wild corn and an incredible diversity of cultivars were native to the Mexican highlands. By the seventh century, agriculture provided about 75% of food items in the Mesoamerican diet (13). By 1492, elaborate trading networks had been established, and freshwater fish and amphibians from highland lakes, shellfish, seafood, small mammals, waterfowl, and upland game birds were important dietary supplements. Domesticated dog may have been the primary source of red meat in crowded areas (13). Of the large mammals, only white-tailed deer was a relatively frequent portion of the diet. Mesoamerican culture was highly urbanized, and even deer must have been uncommon in the urban areas, although the "small Yucatec deer" (probably the brocket deer, which still ranges into eastern Mexico from the south) may have been semi-domesticated in the Mayan villages of the Yucatan Peninsula (13, p. 143). The brocket deer has traits conducive to its capture and is common in bean and corn fields (21).

Indian Populations in 1492

In the 1930s, prevailing opinion was that North America was populated at European contact by less than 3 million people, with less than 1 million living north of Mexico (27). Population estimates have generally risen, with most of the increase resulting from refined estimates for large cities (28). For example, the Aztec capital of Tenochtitlán may have been occupied by 350,000 people, and it was far larger than any European city of the time (10). Dobyns estimated that there were about 11 million people north of Mexico and about 34 million in Mexico (29). In the past two decades, experts have come to agree substantially with the latter (10).

Some have argued that even Dobyns's estimates were low because retrospective estimating is influenced by post-Columbian Indian populations that were decimated by European diseases and genocide. For example, Stannard (10, p. 24) proposed that a liberal population estimate of 700,000 for California, "rather than being excessive—will in time likely turn out to have been an excessively conservative estimate." Dobyns and Swagerty adjusted Dobyns's estimate of the population north of Mexico to 18 million (30, 10).

Several things are clear, however. Regardless of the precise dates of first entry and settlement, Indian populations were not exploding in the late 15th century, nor was their exploitation of large mammals. They had been settled in all North American biomes for millennia, with cultures relatively stabilized (13, 20, 1, 12, 5, 8, 6, 10, 22). The pre-Columbian population of North America was an order of magnitude lower than what it is today, as was the human : large mammal biomass ratio. Finally, the latter two phenomena were most pronounced north of Mexico.

THE POLITICAL STATUS OF TRIBES IN NORTH AMERICA

When France and England explored the Americas, they generally acknowledged the political sovereignty of Indian tribes, pursuant to the international law philosophy of Francisco de Vitoria (31). However, Columbus landed at San Salvador a year before de Vitoria was born, so the same cannot be said for Spain. In fact, de Vitoria's principles were influenced by the cruel treatment that Indians had already received by Spaniards in Mexico. Therefore, Native Americans have achieved various amounts of autonomy and jurisdiction, ranging from virtually none in Mexico to pending territorial status in Canada. The political status of tribes has direct application to their management authority over large mammals.

Mexico

Circumstances particular to Spain and prehistoric Mexico contributed to a rapid loss of Indian sovereignty. Spain needed wealth to stem an economic crisis and found what it was looking for in the Aztec treasures. Spain also had embarked on the religious fervor of the Inquisition, and the Indian population was seen as a rich source of souls to be saved. Those factors combined to drive the conquistadors to a frenetic pace (32). Meanwhile, the Indians were found in concentrated masses and the great majority were accustomed to an organized system of labor and taxation (12). Once the leaders were defeated, revolt was not forthcoming. Furthermore, the peak effects of diseases and genocide were concurrent with the conquest of Mexico, rather than preceding exploration as in many areas to the north (10).

When the pace of conquest slowed, Spain did grant small tracts of land to Indian communities, but after Mexico declared independence in 1821, those tracts were privatized and the Indian owners rapidly lost control to a small minority of *hacendados,* wealthy landowners (12). The Mexican constitution of 1917 protected the concept of communal properties, but few of those were left. From 1910 to 1968, the government established *ejidos,* communal properties for farmers. Many Indians are members of *ejidos.* However, a 1992 reform of the constitution authorized the sale of communal lands, and they probably face the same fate as the Spanish land grants (33).

Tribal sovereignty never has been as controversial an issue in Mexico as in the United States and Canada. Most Mexican citizens are *mestizos* (i.e., of Spanish and Indian ancestry), with Indian genes accounting for about 80% of the gene pool (12). In that sense, Indians are the predominant political force in Mexico, even though most Indian culture is gone. In fact, pure-blooded Indians have historically been major political actors, including two presidents. Nevertheless, a growing number of tribal groups in Mexico are striving for some measure of political autonomy (33).

United States

The most pervasive theme in the history of United States Indian affairs has been inconsistency (34). Tribes were a nebulous political entity until a series of Supreme Court cases from 1823 to 1832 granted tribes a measured sovereignty. That sovereignty eroded during the Indian wars of the 1800s, and from 1885 to 1934, a policy of assimilation evolved. Then came a brief resurgence of sovereignty under the Indian Reorganization Act (IRA) of 1934 (48 Statutes 984), followed abruptly by a 1953 political movement called *termination,* under which tribes were to lose all political status. Of all Indian affairs doctrine, however, termination was the least institutionalized, and in 1975 the aforementioned Indian Self-Determination Act was passed.

As a result of the earlier doctrines, especially the 1934 reorganization, tribes are entwined with the federal government. Congress is the trustee of Indian status, and is responsible for the protection of tribal property, the tribal right to self-government, and tribal welfare in general. This trusteeship has been administered primarily through the Bureau of Indian Affairs. Under self-determination, tribes are relieved of federal bureaucracy via contracts that evolve into tribally managed programs. These "638 contracts" (after Public Law 93-638) have been instrumental in building sound wildlife management programs (35). The most advanced stage of self-determination under the current act is when a tribe establishes a compact with the federal government. Compact tribes receive an annual allocation of funds, which they use in the manner they see fit, as opposed to 638 funds, which are strictly earmarked.

There are 296 federally recognized Indian reservations and about the same number of federally recognized tribes in the lower 48 states (36). About half of the reservations are comprised of 100% tribal land, and the others have fragmented land ownership patterns. The land mass contained within reservation boundaries is more than 20,930,564 hectares, of which more than 18,556,685 hectares are tribally owned. [Reddy's data are incomplete for some small reservations (36).] Nontribal land within reservation boundaries may be owned by individual Indians, non-Indians, utilities, corporations, or government agencies. In the East, 27 tribes are recognized by states, and some of them have state-recognized reservations (37).

Most Alaskan natives (i.e., Eskimos, Aleuts, and Athabascans) are organized in "villages" pursuant to the Alaska Native Claims Settlement Act. The 226 Alaskan native villages are federally recognized tribes, but the state of Alaska does not acknowledge tribal sovereignty. There is only one Indian reservation per se in Alaska, the Metlakatla Reservation (on an island near Ketchikan). In 1996, the Ninth Circuit Court of Appeals upheld the Venetie Tribe's attempt to regain reservation status for 729,000 hectares converted to tribal ownership pursuant to the Alaska Native Claims Settlement Act, but the Supreme Court reversed that decision (38).

Under self-determination doctrine, tribes have authority to manage the large mammal populations on their lands. One of the strongest cases for tribal wildlife jurisdiction (and for self-determination) came in 1983, when the Supreme Court decided *New Mexico v. Mescalero Apache Tribe.* The Court unanimously held that state wildlife jurisdiction and hunting license fees could not be applied on a reservation with an active tribal wildlife management program. Especially since *Mescalero,* tribes have invested heavily in their wildlife management programs. Many have game and fish departments that are similar to state agencies, with a code and commission, director, law enforcement personnel, wildlife biologists, planners, and other specialists (39). Often they are assisted by Bureau of Indian Affairs 638 contracts. A preliminary listing includes more than 1,100 tribal game and fish personnel (G. Rankel, Bureau of Indian Affairs, Office of Trust and Economic Development, Washington, DC, unpublished data, 1997).

Several factors complicate this general scenario. Many tribes entered into treaties with the United States that serve to inflate or deflate their rights to big game harvest and management. Supreme Court decisions reached during periods of other doctrine come back to haunt lawmakers, courts, and tribes. Checkerboard land ownership creates difficult management scenarios. Alaskan natives are denied jurisdiction over wildlife by the state of Alaska, and some villages have entered into cooperative management agreements with the state. The conservative trend of the 1980s and early 1990s, particularly in the executive branch, deterred self-determination (40). Finally, the federal judiciary, much of which was appointed during the presidency of Ronald Reagan, may deter self-determination for decades (41).

Canada

In Canada, the term *Indian* is generally reserved for the numerous bands of Athabascans (or Dene), coastal tribes, and southern tribes from woodland, plains, and plateau. That distinguishes them from Inuits (Eskimos) and Métis (mixed Indian/Europeans common in the plains provinces) (40). In legal parlance, these groups are collectively known as "aboriginals," and 596 "bands" are recognized by the federal government in Ottawa. In popular circles, native political entities are collectively referred to as "First Nations."

In many ways, the history of Canadian natives resembles that of American Indians. First Nations were recognized as distinct political entities, but wards of the government, in the Constitution Act of 1867 that formed the Dominion of Canada. Except for Inuits and Métis, their distinctive status was well defined by the Indian Act of 1876, and elaborated by the Indian Act of 1951 and the Constitution Act of 1982 (40). Treaty making (which implicitly acknowledges political sovereignty) was common in the 1800s, but the general strategy was assimilation from 1867 to 1945. An acceptance of ethnic plurality with a simultaneous focus on integration and equality marked the next period. In 1969 Canada suddenly adopted the infamous "White Paper," which called for a forced version of assimilation that more resembled the termination policy pursued by the United States in the 1950s. The White Paper spawned the most organized political movement in First Nations history, and it was withdrawn in 1970. Since then, the policy of Ottawa has been leaning toward self-determination. In practice, policy has been determined case by case. Canadian natives, however, are united in striving for self-determination via self-government (42).

As of 1990, there were about 500 outstanding land claims involving the First Nations (40). Ottawa's goal in settling these claims is "extinguishment" of constitutionally established but nebulous "aboriginal rights" by replacing them with well-defined settlement rights (43). The few settlements reached are generally favorable to the First Nations, but disputes with provinces are one factor that set a slow pace for settlement.

The legal and political situation is very different in the two territories, over most of which no treaties have ever applied. Native Canadians comprise a 62% majority of the Northwest Territories. In 1992, Northwest Territories voters approved the creation of a self-governing homeland for the 17,500 Inuit living in the eastern part of the Territories. The new territory will be called Nunavut ("Our Land" in Inuit) and will cover an area of about 1,994,000 square kilometers. An Inuit government was to be engaged by 1999 (40).

As of 1985, there were 2,261 Indian reserves in Canada, totaling 2,654,686 hectares (36). More than half of these are in British Columbia, which like the American states of Washington and Oregon, is host to a great variety of coastal tribes. The largest reserves, however, are in Ontario, Saskatchewan, and Alberta.

MAJOR ISSUES IN TRIBAL BIG GAME MANAGEMENT

Indian country tends to be relatively remote, as do big game populations. Therefore, some of the most renowned big game herds in North America live on reservations, and there are many big game issues of far-reaching importance centered on tribal lands. I have selected a few of these to demonstrate the novelty that tribal sovereignty can bring to big game management and the importance of tribal management to large mammal ecology.

Mexico

Much of Mexico is crowded and has relatively little game. Furthermore, because tribes have no political status in Mexico, their managerial impact on large mammals is slight. However, there are exceptions, potentially the greatest being a management scenario involving the Seri Indians and bighorn sheep.

The Seris occupy an *ejido* on the Sonoran shore of the Gulf of California across from the 1,208 square kilometer Tiburón Island. Tiburón Island was declared a wildlife refuge in 1963, and hunting was prohibited. Enforcement is lax, however. In 1975, 20 bighorn sheep were transplanted from the mainland to the island by the New Mexico Department of Game and Fish in collaboration with Sonoran authorities, in an effort to increase the range of the bighorn sheep (44). Because the Seris are the primary users of the island, the National Institute of Ecology, the National University of Mexico, the state of Sonora, and Conservation International have worked with the Seris to develop a management plan for the island that addresses the potential for harvesting bighorn sheep. The Seris may become beneficiaries of hunt revenues, which would provide an incentive to protect the herd from poaching. Meanwhile, the bighorn sheep population has been studied to ascertain potential harvest levels (R. Medellín, National University of Mexico, personal communication). Technically, any plan will require approval from the Mexican government, but the federal authority is highly centralized and often not influential in remote areas (L. Bourillón, University of Arizona, personal communication).

United States

Similar to the major land management agencies, the vast majority of tribal property is in the West. On western reservations, natural resource extraction remains the economic focus (gaming has made recent inroads on that claim). For several of these tribes, big game is the most profitable resource. For instance, on the 728,000-hectare San Carlos Apache Reservation, annual big game license revenues are approximately $435,000, almost all of which is from nontribal members. That is despite the fact that only two (i.e., elk and bighorn) of the eight big game species on the reservation are managed primarily for trophy characteristics, and the most common big game animal (i.e., mule deer) may only be hunted by tribal members (45). Nevertheless, the tribe's Recreation and Wildlife Department is one of the few tribal enterprises that makes a profit for the tribe, and is far more profitable than the tribe's federally subsidized timber program or the seven tribal cattle associations (46).

A similar situation has been noted for other tribes in the Southwest, and is probably common among Western tribes, because they tend to manage large, uncrowded tracts of land (47). Such lands have a high ratio of big game to humans. Tribes therefore feel less political pressure to support masses of hunters like those that crowd state jurisdictions. That allows them to cater their management strategy toward quality, not quantity (48). For example, the San Carlos Apache Tribe manages one of their elk herds for what is probably the highest bull : cow ratio (i.e., 70:100) managed for in the United States, and the largest elk antlers ever recorded were shed by a male from the San Carlos Apache Reservation in 1987 (39). In 1997, a pending world record Boone and Crockett male was taken at San Carlos. Closely rivaling the San Carlos males are those found on the White Mountain Apache Reservation (adjacent to the San Carlos Apache Reservation), the Mescalero and Jicarilla Apache reservations, and the Yakima Reservation—all areas with high male : female ratios and old age structures. Other famous examples include bighorn on the Hualapai Reservation, mule deer on the Jicarilla Apache Reservation and Southern Ute Reservation, Coues deer on the San Carlos Apache Reservation, and pronghorn, black bears, and mountain lions on the White Mountain and San Carlos Apache Reservations.

The trophy status of large animals is very important to the welfare of big game populations on tribal lands. For example, in 1992 the San Carlos Apache Tribe sold three of their special Dry Lake elk tags for $43,000 each. These tags are earmarked by annual commission order for elk habitat improvement projects, which have included numerous watering areas and the establishment of a 2,500-hectare Dry Lake Elk Production Area, from which all cattle were removed and is now managed for elk calving.

Tribes cooperate liberally among themselves and with state and federal agencies in big game reintroductions. The Laguna and Acoma Pueblos have each translocated pronghorn with assistance from the New Mexico Department of Game and Fish, and the pronghorn were obtained from the private lands of Fort Union Ranch (J. Antonio, Bureau of Indian Affairs, personal communication). The Fort Belknap Assiniboin/Gros Ventre Community is planning to translocate bighorns, and the Crow Tribe plans to trade pronghorn for elk from the Uintah/Ouray Reservation (R. Skates, U.S. Fish and Wildlife Service, personal communication). Turkeys have been frequently traded among the tribes of the northern Great Plains, by the Ute tribes, and between the White Mountain Apache Tribe and the Zuni Pueblo. Probably the longest nonaerial translocations of pronghorn ever conducted took place when the San Carlos Apache Tribe translocated pronghorn from Fort Belknap in 1990 and 1991 (49). Most of the operation was conducted by tribal personnel, with technical assistance provided by the U.S. Fish and Wildlife Service and some equipment borrowed from the Arizona Game and Fish Department. On the other hand, a translocation of the endangered wood bison into the Yukon Flats of Alaska has been stalled by a political dispute between the Council of Athabascan Tribal Governments and the state of Alaska.

About $26 million of Bureau of Indian Affairs fish and wildlife program 638 contracts were administered in fiscal year 1997, and another $20 million in fish and wildlife compact funds (G. Rankel, Bureau of Indian Affairs, Office of Trust and Economic Development, personal communication). Federal funding is a result of treaty obligations and trust responsibilities to tribes whose lands were forcefully taken by the United States at great expense to tribal cultures (10). Nevertheless, tribes also spend their own money on large mammal management programs. In some cases, tribal spending far outweighs federal funding. For example, of the $696,000 budget of the San Carlos Apache Tribe's Recreation and Wildlife Department for fiscal year 1993, $576,000 came from tribal hunting and fishing revenues, and $120,000 was administered via 638 contracts (39). These figures did not include the $129,000 generated by the Dry Lake elk hunts.

For many tribal members, the economic value of big game is less important than the spiritual value. Wolves are spiritually important to many tribes. Thus, the Nez Perce Tribe participated with the U.S. Fish and Wildlife Service and the state of Idaho in the environmental impact assessment process for wolf translocation in central Idaho. Once the alternative to release wolves was selected, however, Idaho politics precluded the state from being involved, so the tribe was asked by the Fish and Wildlife Service to accept the responsibility for monitoring the wolves. The tribe accepted, and wolves were released at three national forest sites in 1995 and 1996, accompanied by a group of tribal elders that blessed the wolves and the project (50). Tribal personnel now monitor wolves year-round across the entire Central Idaho Experimental Wolf Population Area. The tribe also participates in public outreach, research, and control efforts when necessary.

Now the Nez Perce are working with federal and state agencies on a potential grizzly bear reintroduction (T. Kaminski, Nez Perce Tribe, personal communication). Bears are revered by tribes in all regions, perhaps because bears are the most anthropoid form in North America (51). Mule deer are spiritually significant to some tribes (52).

For the plains tribes, bison were the source of life and the most important spiritual entity, and today some of the most important controversies involving large mammals and tribes focus on bison (16). The Intertribal Bison Cooperative in Rapid City, South Dakota, is an expanding coalition of 42 tribes whose mission is to restore bison to tribal lands in a manner compatible with cultural beliefs and practices. During the heated controversy over the killing of dispersing Yellowstone bison by the U.S. Park Service in late winter of 1997, the Bison Cooperative protested and proposed that the bison instead be captured and used to reestablish free-ranging herds throughout the West, beginning with the reservations (53). The tribal factor may become pivotal in the debate over returning bison to the Great Plains (54).

In some situations, tribal big game hunting can be more important off the reservation than on. For example, members of the Ute Mountain Tribe are authorized to hunt deer on the western slope of the Colorado Rockies pursuant to the "Brunot agreement" (Act of April 29, 1874, 18 Statutes 36). Treaty rights are most important in the northern plains and Rockies, where treating was a common method of dealing with tribes prior to 1871 (55). Subsistence hunting rights provided to rural Alaskans (including most Alaskan Natives) by the Alaska National Interest Lands Conservation Act of 1980 (16 United States Code 410) apply to nontribal lands, too.

Tribes in the United States have made swift progress in developing professional wildlife management programs in the past two decades. They have gone beyond managing for maximum yield, and instead manage for socially healthy herds and physically healthy individuals. They also manage big game for the spiritual welfare of themselves and the animal world (56). They operate some of the most progressive large mammal programs in the nation. Their challenge will be to continue that trend in the likely face of increasing reservation populations and ever-present fiscal shortages.

Canada

When Nunavut is formed, a great number of caribou, muskox, wolves, bears, and large sea mammals will be managed entirely by the Inuits. The Inuits will be well prepared, as they have been comanagers of big game in the Northwest Territories since passage of the Western Arctic Claims Settlement Act of 1984 (57). Inuits have also served on the Beverly and Kaminuriak Caribou Management Board, which has managed caribou populations from Hudson Bay to Great Slave Lake since 1982 (58).

One of the most endangered large mammal subspecies in North America is the wood bison. The major herd in Wood Buffalo National Park has long been prone to hybridization with introduced plains bison and is infected with brucellosis (59). Three smaller free-ranging herds exist, including one on the reserve of the Waterhen band in Manitoba (60). A population reintroduced to the Yukon Flats in Alaska would probably

Major Issues in Tribal Big Game Management

become larger than all others (Alaska Department of Fish and Game, unpublished data; Council Athabascan Tribal Governments, unpublished data). If more tribal sovereignty is recognized in Alaska and Canada, it is conceivable that the Waterhen band could provide bison directly to the Council of Athabascan Tribal Governments via aerial transport to the Yukon Flats.

SUMMARY

The ecological health of North America prior to European colonization is a controversial topic with important management implications for large mammals. On one hand, some subscribe to the theory that the New World encountered by European explorers was a verdant, vibrant ecosystem teeming with large mammals. They emphasize the naturalness of pre-European North America, where "natural" refers to a relative lack of human influence on ecosystem structure and function (61, 62).

Others emphasize that humans had an impact on ecosystem structure and function for thousands of years before European colonization. Apparently the earliest Americans were nomadic hunters from Beringia. By 1492, when Columbus reached San Salvador, the progeny of those Beringians had hunted, fished, farmed, irrigated, drained, filled, logged, burned, and urbanized vast areas of the continent (10).

These two schools of thought are not entirely contradictory. Few authors deny that native Americans had an impact on pre-European America, and few deny that ecological deterioration is a much greater problem today than it was in 1492. Perhaps it is most important to recognize that most tribal cultures north of Mexico were materially and spiritually dependent on large mammals for thousands of years, and remain so to a varying extent. Especially for tribes in the West and North, they now have the political opportunity and a land base to reclaim their profound relationship with large animals, to demonstrate their traditional knowledge, and to coexist with some of the finest large mammal populations in North America.

LITERATURE CITED

1. DELORIA JR., V. 1995. *Red Earth, White Lies: Native Americans and the Myth of Scientific Fact.* Charles Scribner's Sons, New York.

2. CLARK, E. E. 1952. *Indian Legends of the Pacific Northwest.* University of California Press, Berkeley.

3. CLARK, E. E. 1966. *Indian Legends from the Northern Rockies.* University of Oklahoma Press, Norman.

4. HAVEN, S. 1856. *Archaeology of the United States.* Smithsonian Institution, Washington, DC.

5. FAGAN, B. M. 1987. *The Great Journey: The Peopling of Ancient America.* Thames and Hudson, London, UK.

6. TURNER, G. E. S. 1979. *Indians of North America.* Blandford Press, Poole, UK.

7. WOLKOMIR, R. 1991. New finds could rewrite the start of American history. *Smithsonian* 21:130–144.

8. JOSEPHY, A. M. JR. 1991. *The Indian Heritage of America,* rev. ed. Houghton Mifflin, Boston, MA.

9. MACNEISH, R. S. 1971. Early man in the Andes. *Scientific American* 224:36–46.

10. STANNARD, D. E. 1992. *American Holocaust: The Conquest of the New World.* Oxford University Press, New York.

11. MARTIN, P. S. 1967. Prehistoric overkill. Pages 75–120 *in* P. S. Martin and H. E. Wright, eds., *Pleistocene Extinctions: The Search for a Cause.* Yale University Press, New Haven, CT.

12. DRIVER, H. E. 1969. *Indians of North America.* 2nd ed. University of Chicago Press, IL.

13. ADAMS, R. E. W. 1991. *Prehistoric Mesoamerica,* rev. ed. University of Oklahoma Press, Norman.

14. MARTIN, P. S. 1973. The discovery of America. *Science* 179:969–974.

15. OWEN-SMITH, R. N. 1988. *Megaherbivores: The Influence of Very Large Body Size on Ecology.* Cambridge University Press, Cambridge, UK.

16. McHUGH, T. 1972. *The Time of the Buffalo.* Alfred A. Knopf, New York.

17. McCABE, R. E. BRYANT, L. D., and C. MASER. 1982. Elk and Indians: Historical values and perspectives. Pages 61–123 *in* J. W. Thomas and D. E. Toweill, eds., *Elk of North America: Ecology and Management.* Stackpole Books, Harrisburg, PA.

18. ALLEN, D. L. 1962. *Our Wildlife Legacy.* Funk and Wagnalls, New York.

19. BURT, W. H., and R. P. GROSSENHEIDER. 1976. *A Field Guide to the Mammals,* 3rd ed. Houghton Mifflin, Boston, MA.

20. CRESSMAN, L. S. 1977. *Prehistory of the Far West: Homes of Vanished Peoples.* University of Utah Press, Salt Lake City.

21. NOWAK, R. M. 1991. *Walker's Mammals of the World.* 5th ed. Johns Hopkins University Press, Baltimore, MD.

22. UNDERHILL, R. M. 1971. *Red Man's America: A History of Indians in the United States.* rev. ed. University of Chicago Press, IL.

23. BRYANT, L. D., and C. MASER. 1982. Classification and distribution. Pages 1–60 *in* J. W. Thomas and D. E. Toweill, eds., *Elk of North America: Ecology and Management.* Stackpole Books, Harrisburg, PA.

24. JULANDER, O., and J. B. LOW. 1976. A historical account and present status of the mule deer in the West. Pages 3–11 *in* G. W. Workman and J. B. Low, eds., *Mule Deer Decline in the West: A Symposium.* College of Natural Resources, Utah State University, Logan.

25. CZECH, B. 1996. *Ward vs. Racehorse*—Supreme Court as obviator? *Journal of the West* 35:61–69.

26. GORENSTEIN, S. 1975. *Not Forever on Earth: Prehistory of Mexico.* Charles Scribner's Sons, New York.

27. KROEBER, A. L. 1939. Cultural and natural areas of native North America. University of California Publications in American Archaeology and Ethnology 38.

28. THORNTON, R. 1987. *American Indian Holocaust and Survival: A Population History Since 1492.* University of Oklahoma Press, Norman.

29. DOBYNS, H. F. 1966. Estimating aboriginal American population: An appraisal of techniques with a new hemisphere estimate. *Current Anthropology* 7:395–449.

30. DOBYNS, H. F., and W. R. SWAGERTY. 1983. *Their Number Become Thinned: Native American Population Dynamics in Eastern North America.* University of Tennessee Press, Knoxville.

31. SHATTUCK, P. T., and J. NORGREN. 1993. *Partial Justice: Federal Indian Law in a Liberal Constitutional System.* E. B. Edwards Brothers, Lillington, NC.

32. LIVERMORE, H. V. 1966. *A History of Spain,* 2nd ed. Allen & Unwin, London, UK.

33. BARTOLOMÉ, M. A. 1996. Indians and Afro-Mexicans at the end of the century. Pages 299–306 *in* L. Randall, ed., *Changing Structure of Mexico: Political, Social, and Economic Prospects.* M. E. Sharpe, Armonk, NY.

34. CZECH, B. 1995. American Indians and wildlife conservation. *Wildlife Society Bulletin* 23:568–573.

35. RANKEL, G. L. 1994. Few know extent of outdoor recreation in Indian country. *Recreation Executive Report* 24:1–6.

36. REDDY, M. A. 1993. *Statistical Record of Native North Americans.* Gale Research, Detroit, MI.

37. RUSSELL, G. 1996. *American Indian Reservation Roster.* Russell Publications, Phoenix, AZ.

38. WILKINS, D. 1998. Supreme Court dashes tribes' dreams of sovereignty. *Arizona Daily Star* 157(7 May):17A.

39. CZECH, B. 1993. Statement of Brian Czech (Director, San Carlos Recreation & Wildlife Department, San Carlos Apache Tribe). Pages 183–189 *in* the report of the Oversight Hearing before the Committee on Natural Resources, House of Representatives, One Hundred Third Congress. U.S. Government Printing Office, Serial No. 103–5.

40. FLERAS, A., and J. L. ELLIOTT. 1992. *The 'Nations Within': Aboriginal-State Relations in Canada, the United States, and New Zealand.* Oxford University Press, Toronto, Ontario, Canada.

41. O'BRIEN, D. M. 1993. *Storm Center: The Supreme Court in American Politics.* Norton, New York.

42. COMEAU, P., and A. SANTIN. 1990. *The First Canadians: A Profile of Canada's Native People Today.* James Lorimer, Toronto, Ontario, Canada.

43. ROYAL COMMISSION ON ABORIGINAL PEOPLES. 1995. Treaty making in the spirit of coexistence: An alternative to extinguishment. Canada Communication Group, Ottawa, Canada.

44. MONTOYA, B., and G. GATES. 1975. Bighorn capture and transplant in Mexico. *Desert Bighorn Council Transactions* 28–32.

45. CZECH, B. 1990. Wildlife management on the San Carlos Apache Reservation. Page 258 *in* P. R. Krausman and N. S. Smith, eds, *Managing Wildlife in the Southwest.* Arizona Chapter of The Wildlife Society, Phoenix.

46. CZECH, B., and L. TARANGO. 1998. Wildlife as an economic staple; an example from the San Carlos Apache Reservation. Pages 209–215 *in Proceedings of the 9th U.S./Mexico Border States Conference on Recreation, Parks and Wildlife.* U.S. Department of Agriculture Forrest Service, Rocky Mountain Research Station, Fort Collins, CO. RMRS-P-5.

47. ANTONIO, J. E. 1994. Concerns, values, and approaches being taken in management of fishery and wildlife resources: A Southwest perspective. Pages 156–163 *in* Intertribal Timber Council, compiler, *Proceedings of the Eighteenth Annual National Indian Timber Symposium.* Intertribal Timber Council, Portland, OR.

48. ESKEW, L. 1994. Where the giant bucks roam. *Safari* May/June 1994:52, 94–97.

49. CZECH, B. 1992. Lessons learned from an intertribal, interregional pronghorn antelope reintroduction. *Proceedings of the Biennial Pronghorn Antelope Workshop* 15:100–108.

50. ROBBINS, J. 1997. Return of the wolf. *Wildlife Conservation* 100:44–51.

51. CZECH, B. 1992. The bear in Apache culture. *Ursus* 1:33.

52. STROH, T. L. 1990. Wildlife management on the Zuni Reservation. Pages 252–257 *in* P. R. Krausman and N. S. Smith, eds., *Managing Wildlife in the Southwest.* Arizona Chapter of The Wildlife Society, Phoenix, AZ.

53. ROBBINS, J. 1997. In the West, a matter of the spirit. *New York Times* 146(21 Jan):A8.

54. CALLENBACH, E. 1996. *Bring Back the Buffalo!: A Sustainable Future for America's Great Plains.* Island Press, Washington, DC.

55. BURTON, L. 1991. *American Indian Water Rights and the Limits of Law.* University Press of Kansas, Lawrence.

56. CRUM, R. 1997. Healing the spirit. *Wildlife Conservation* 100:36–43.

57. BAILEY, J., N. B. SNOW, A. CARPENTER, and L. CARPENTER. 1995. Cooperative wildlife management under the Western Arctic Inuvialuit Claim. Pages 11–15 *in* J. A. Bissonette and P. R. Krausman, eds., *Integrating People and Wildlife for a Sustainable Future. Proceedings of the First International Wildlife Management Congress.* The Wildlife Society, Bethesda, MD.

58. SCOTTER, G. W. 1991. The Beverly and Kaminuriak Caribou Management Board: An example of cooperative management. *Transactions of the North American Wildlife and Natural Resources Conference* 56:309–320.

59. CARBYN, L. N., S. M. OOSENBRUG, and D. W. ANIONS. 1993. Wolves, bison and the dynamics related to the Peace-Athabascan Delta in Canada's Wood Buffalo National Park. Circumpolar Res. Series 4. Canadian Circumpolar Institute, University of Alberta, Alberta.

60. TURBAK, G. 1988. Of waterhens and wood bison. *International Wildlife* 18:30–35.

61. GÖTMARK, F. 1992. Naturalness as an evaluation criterion in nature conservation: A response to Anderson. *Conservation Biology* 6:455–458.

62. GRUMBINE, R. E. 1992. *Ghost Bears: Exploring the Biodiversity Crisis.* Island Press, Washington, DC.

History of Management of Large Mammals in North America

Richard J. Mackie

INTRODUCTION

Wildlife management may be among the oldest and most basic human enterprises (1). Early man coexisted with and was dependent on wild animals for food, clothing, and other necessities of life. Because of the close association with animals, these early hunters probably were the world's first naturalists or behaviorists. They had to learn the habits and behavior of species on which their livelihood depended and apply that knowledge when gathering food and other resources. This led to early forms of husbandry and ultimately to domestication of species most amenable to living with humans.

Later, as human populations increased and civilizations developed, practices such as harvest restrictions, predator control, establishment of game reserves, artificial feeding, planting of food and cover, and the use of gamekeepers to protect and care for wild animals were implemented. Ancient Greeks took great interest in natural history and founded game preserves and feeding grounds to maintain stocks of wild animals while variously encouraging and discouraging hunting. Husbandry of game animals for recreation and commerce also was common during Roman times and gained popularity with declining wild populations (2).

The specific animals involved in historical and early Old World management efforts generally are unknown, though most chronicles indicate that large, or big game, mammals were particularly important. The earliest clear record of big game management may be in the narratives of Marco Polo. These relate that game laws of Kublai (A.D. 1259–1294) prohibited ". . . every person throughout all the countries subject to the Great Khan, from daring to kill . . . roebucks, fallow deer, stags, or other animals of that kind . . . between the months of March and October . . . that they may increase and multiply" (3). Translocation and restocking of big game was first recorded in England during the late 17th century when Charles II transplanted deer from Germany to restock royal forest reserves. Habitat improvements for or to attract big game animals may have occurred even earlier. For example, Borland reported observations of Giovanni da Verrazona from explorations along the Atlantic Coast of North America in 1524 that indicated American Indians had created ". . . openings where deer and other wild browsers could feed"(4).

Although many current management policies and practices can be traced to our early ancestors, they did not constitute the organized efforts to protect, wisely use, or deliberately control wildlife populations and habitats for the societal benefits now associated with wildlife conservation and management. As defined and practiced today, *wildlife conservation* is a social process that seeks to attain wise use of wildlife resources and maintain the productivity of wildlife habitats (5). *Wildlife management* is a part of wildlife conservation and entails human effort to control wildlife through specific population or habitat management practices. It has also been defined as the "art of making land produce sustained annual crops of wild game for recreational use" or simply "the art of making land produce wildlife" (3, 6). Conservation may involve a broad spectrum of lay and professional activities. Modern wildlife management evolved as a predominantly professional discipline to help achieve conservation goals through science and education. From a historical perspective these are relatively recent human endeavors that emerged in North America since 1850.

EARLY HISTORY

The histories of wildlife, wildlife conservation, and wildlife management in North America have been recounted in various detail in countless books and other reviews. These are largely chronologies of (1) abundance and exploitation during settlement and spread of European immigrants across the continent; (2) concern, outrage, and stirring of early sportsmen and others to reduce the "butchery" and protect remaining stocks; and (3) the rise and evolution of efforts to restore and scientifically manage at first game animals, especially big game, and later all wildlife and the habitats in which they occur. The chronology of important policies and practices relating to conservation and management of big game mammals in North America has evolved since early colonial times (Table 15–1).

TABLE 15–1

Chronology of Development of Policies and Practices Relating to Conservation and Management of Big Game Mammals in North America

1623	Plymouth Colony: Provided that hunting would be free to all members of the colony—a measure to ensure that people could use wild game animals as a source of food; landholders retained the right to control animals on their property.
1630	Massachusetts: Bounty offered for control of predators (wolves, coyotes, bears, mountain lion).
1639	Rhode Island: Colonial ordinance prohibited killing of deer between May 1 and November 1 each year (the first law relating to harvest of big game in North America).
1646	Rhode Island: Portsmouth colony prohibited shooting of doe deer between May 1 and November 5, with a fine of 5 pounds for violation (first fine for violation of game law).
1694	Massachusetts: First closed season on deer.
1718	Massachusetts: Enacted first closure on deer hunting for period of years.
1739	Massachusetts: Appointed deer wardens (first American game officers).
1750	Delaware: Prohibited hunting on Sunday.
1776	Closed seasons in effect for taking deer in all colonies except Georgia.
1776	Pennsylvania: Legally recognized public right to hunt (state constitution included provision for public hunting on lands not enclosed).
1779	North Carolina: Prohibited "shining" of game animals (deer).
1784	North Carolina: Prohibited hunting on private land without permission.
1788	New York: Prohibited "hounding" of deer.
1789	U.S. Constitution adopted: Established federal authority to make treaties, set federal land policy, and regulate interstate commerce (basis for federal involvement in wildlife conservation and management).
1802	Massachusetts: Permitted taking of deer in private parks at any time (first "shooting preserve" act).
1832	Congress reserved game on Indian lands for Indians (first federal game law).
1836	The Bureau of Indian Affairs established.
1844	New York Sportsmen's Club formed, which later became the New York Association for Protection of Game and played an important role in establishing formal game and fish administration in the United States.

continued

TABLE 15-1

Chronology of Development of Policies and Practices Relating to Conservation and Management of Big Game Mammals in North America—*continued*

1849	U.S. Department of Interior established to administer General Land Office, Bureau of Indian Affairs, and other federal agencies.
1851	Missouri: First law prohibiting sale of game species during closed season.
1852	California: Prohibited hunting of antelope and elk (first for these species).
1852	Maine: Appointed first salaried game wardens in America.
1862	Congress established U.S. Department of Agriculture.
1862	Homestead Act authorized largely unrestricted settlement on 160-acre tracts of public land.
1864	Idaho: Provided first seasonal protection for bison, mountain sheep, and mountain goats as well as protection for deer, elk, and antelope.
1864	New York: Became first state to require license for hunting.
1865	Massachusetts: Established Commission of Fisheries and Game (first state game commission).
1871	Congress established U.S. Fisheries Commission to redistribute, propagate, and introduce desirable game and food fishes (first federal fish and game agency and forerunner to U.S. Fish and Wildlife Service).
1871	*American Sportsmen* was first nationally circulated American sporting journal published, followed by *Forest and Stream* in 1873 and *Field and Stream* in 1887. These helped spread concern and call for protection of dwindling game populations.
1872	Congress established Yellowstone National Park (first national park).
1872	Colorado: Prohibited waste of edible parts of game animals.
1873	New Jersey: First hunting license required for nonresidents.
1876	Colorado: Established bounty on mountain lion.
1877	John D. Caton published *Antelope and Deer of America,* the first comprehensive treatise on the natural history of North American big game animals.
1877	Montana: Territorial government prohibited killing game animals for hides alone.
1878	Vermont: Sportsmen bought and transplanted 17 white-tailed deer from New York to Vermont (first transplant for restoration of big game in America).
1885	Congress authorized Division of Entomology in U.S. Department of Agriculture to study food habits, distribution, and migration of birds in relation to agriculture (first federal entry into wildlife research); became Division of Economic Ornithology and Mammalogy in 1886 and Division of Biological Survey in 1896.
1886	First national "conservation" meeting (National Association for Protection of Game, Birds, and Fish) in St. Louis. Recommended establishment of uniform game laws among states east of the Rocky Mountains and unified state fish and game commissions with wardens appointed by governors and power to prosecute violators of fish and game laws.
1886	Enforcement of federal wildlife regulations delegated to the U.S. Biological Survey, precursor to the federal game warden service.
1886	U.S. Army assumed administration of Yellowstone National Park, marking the beginning of "big game management" in North America by public agency.
1887	Boone and Crockett Club founded by Theodore Roosevelt to promote hunting and "conservation" of big game.
1887	Colorado: Prohibited sale of wild game, fish, or fowl killed in the state.
1891	Forest Reserve Act gave president authority to set aside forest reserves on federal lands.
1892	First elk shipped from Yellowstone National Park for transplanting—beginning a program of trapping and translocation that helped reestablish or establish elk populations over much of North America and other parts of the world.
1893	Michigan, North Dakota: First states to require licenses for all hunters.
1896	U.S. Supreme Court affirmed the concept of state ownership of game (*Geer v. Connecticut*).
1897	First national forest reserve: Yellowstone Park Timberland Reserve.
1900	Congress passed "Lacy Act": First general federal wildlife protection law (prohibited interstate transport of wild game killed in violation of state laws).
1901	Congress created Bureau of Forestry in U.S. Department of Interior. In 1905 it moved to U.S. Department of Agriculture and was renamed U.S. Forest Service to administer the national forests created from national forest reserves.
1902	Federal Reclamation Act established Reclamation Service and set aside money from sale of public lands for building dams and irrigation systems to reclaim arid land in the West.
1902	First "professional" game administrators meeting: National Association of Game and Fish Wardens and Commissioners was convened in Yellowstone National Park (forerunner of International Association of Fish and Wildlife Agencies).
1903	Colorado, Montana: First required guide and outfitter licenses.
1905	Illinois, Massachusetts, Pennsylvania: First recorded occurrence of winter feeding of big game by states. Colorado Game and Fish Department began winter feeding in 1909.

TABLE 15–1

Chronology of Development of Policies and Practices Relating to Conservation and Management of Big Game Mammals in North America—*continued*

1905　First national game preserve: Wichita National Game Preserve, Oklahoma.

1908　U.S. Forest Service set aside a portion of the upper Gallatin River drainage in Montana for use by big game (eliminated livestock grazing), one of the first recorded instances of "habitat management" for big game by resource managers.

1909　First international wildlife conference: North American Game Conference convened by Theodore Roosevelt in Washington, DC. President Roosevelt announced first "American Game Policy" outlining private and public responsibility for "game management."

1911　Weeks Act authorized purchase of lands for national forests in the eastern United States.

1912　Congress appropriated first federal funds for purchase of private lands for creation of the National Elk Refuge, Wyoming. Artificial feeding of elk initiated.

1914　Congress appropriated first federal funds for predator control on public lands.

1916　National Park Service created "to conserve the scenery and the natural and historical objects and the wildlife" of national parks and monuments.

1918　Aldo Leopold formally proposes formation of a new discipline, game management, in *Journal of Forestry* article.

1921　Wyoming: First use of limited permit season in open hunting season on moose.

1923　Colorado: Enacted first antler point restriction ($>2 \times 2$) on mule deer.

1923　Montana classified grizzly bear a game animal (first protection for this species).

1924　California: Established first big game disease control program (to combat outbreak of hoof and mouth disease in deer on the Stanislaus National Forest).

1924　Arizona: First attempt to control erupting big game (mule deer) population on the Kaibab Plateau by driving deer from "overstocked" range across the Grand Canyon to deer-depleted ranges in other parts of the state (attempt by sportsmen failed).

1924　Clarke-McNary Act extended federal authority to acquire lands for a national forest system and provided for private, state, and federal cooperation in forest management.

1924　Upper Mississippi Wildlife and Fish Refuge Act authorized acquisition of public lands for purposes of refuge and protection of game animals, fur animals, wild birds, and fish.

1925　Alaska: Alaska Game Act created an Alaska Territorial Game Commission under the USDI (established protection for game species equivalent to U.S. state laws).

1925　First National Forest Wilderness Area established (Gila National Forest, New Mexico).

1925　Kaibab Plateau, Arizona: First recorded use of exclosures in U.S. Forest Service studies to show effects of browsing by mule deer on forage plants.

1927　Michigan: Game Management Division established in Department of Conservation to inventory the status and collect information on game, especially white-tailed deer, in the state and to recommend plans for management.

1927　University of Michigan offered first formal university course in game management.

1928　Arizona: U.S. Forest Service conducted first program to directly reduce deer numbers on federal land. A total of 1,124 mule deer were killed by government hunters to reduce the deer herd in the Kaibab National Forest.

1928　U.S. Supreme Court: Ruled in the case of Kaibab deer reduction (*Hunt v. United States*) that federal government has the right to protect its lands and property, state law notwithstanding.

1928　Michigan: Hired first biologically trained personnel for game management in a state.

1930　North American Game Policy adopted at North American Game Conference.

1931　Book *Game Survey of the North Central States* (Sporting Arms and Ammunition Manufacturers' Institute, Madison, WI) by Aldo Leopold described existing game conditions and opportunities for game restoration.

1932　Iowa State University established first cooperative wildlife training program.

1932　Iowa: Completed first comprehensive state game survey and management plan.

1932　North Carolina: U.S. Forest Service initiated public hunts to control deer numbers on game preserve within the Pisgah National Forest. Later began to trap and ship deer out of state despite opposition from state Game and Inland Fish Commission.

1933　First textbook in game management: *Game Management* by Aldo Leopold.

1933　Congress created Civilian Conservation Corp (CCC) to work on public development and conservation projects.

1934　Wildlife Restoration Committee established by President Roosevelt to make policy recommendations for rehabilitation of marginal lands for game.

1934　Coordination Act authorized secretaries of agriculture and commerce to cooperate with federal, state, and other agencies to develop a nationwide program of wildlife conservation and rehabilitation in association with federal water projects.

1934　National Forest Refuge Act established fish and game sanctuaries in national forests as approved by states.

continued

TABLE 15–1

Chronology of Development of Policies and Practices Relating to Conservation and Management of Big Game
Mammals in North America—*continued*

1934	Department of Agriculture Regulation G-20-A gave Secretary of Agriculture broad power to regulate hunting and fishing within national forests. State–federal jurisdictional battle that began in Kaibab and Pisgah national forests continues.
1934	Taylor Grazing Act provided for protection of public lands from overgrazing and soil erosion by regulating their use and occupancy.
1934	Model game breeding law proposed at 20th American Game Conference.
1935	U.S. Soil Conservation Service formed in U.S. Department of Agriculture.
1935	American Wildlife Institute formed to promote and help coordinate wildlife conservation, restoration, and management among existing agencies.
1935	Cooperative Wildlife Research Unit Program established to provide academic training in professional wildlife management and conduct wildlife research at nine land-grant colleges and universities.
1936	First North American Wildlife Conference held in Washington, DC, to consider wildlife restoration and conservation issues.
1936	U.S. Forest Service established a Division of Game Management.
1936	The National Wildlife Federation was formed to promote public interests in conservation and restoration of wildlife.
1937	Congress passed the Pittman-Robertson (Federal Aid in Wildlife Restoration) Act providing for a 10% tax on sporting arms and ammunition for distribution to the states for wildlife restoration projects. Required that states dedicate hunting and trapping license fees to wildlife conservation and restoration.
1937	The Wildlife Society organized to support professional wildlife management through science and education, including publication of scientific wildlife management journals and books. New *Journal of Wildlife Management* provided primary outlet for publication of research findings and management techniques for big game.
1937	Missouri: Missouri Conservation Department reorganized to become the most "politically free" state wildlife agency in the United States.
1937	Colorado: Big game hunter check stations first established by U.S. Forest Service and National Park Service officers.
1938	Colorado conducted first aerial counts of big game.
1939	Montana: U.S. Supreme Court ruled (*State v. Rathbone*) that "One who acquires property in Montana does so with notice and knowledge of the presence of wild game and presumably is cognizant of its natural habits." Private landowners had to tolerate a certain amount of wildlife and could not spontaneously and deliberately reduce depredating game populations on their property, which set precedent for future handling of big game damage problems.
1940	First report on the status of wildlife, including big game, in the United States was developed by Biological Division of the U.S. Soil Conservation Service.
1940	U.S. Fish and Wildlife Service established in USDI by consolidation of Bureau of Biological Survey and Bureau of Fisheries.
1940	Federal courts ruled against state in the matter of deer trapping and transplanting from the Pisgah National Forest, ruling (*Chalk v. United States*) that on ceded lands, federal administrators could reduce the deer population to protect the land from injury.
1940	Following the Pisgah ruling, the Secretary of Agriculture met with state game commissioners and the International Association of Fish and Game Agencies and replaced regulation G-20-A with regulation W-2, which recognized state authority and encouraged state–federal cooperation in wildlife management on national forest lands.
1941	Montana first transplanted Rocky Mountain goats.
1941	Pittman-Robertson excise tax increased to current 11%.
1942	Montana first transplanted Rocky Mountain bighorn sheep.
1946	Colorado established the first western state field research station for big game (Little Hills).
1946	Bureau of Land Management created in Department of Interior by combining General Land Office and Grazing Service.
1948	A. S. Einarsen's book, *The Pronghorn Antelope and Its Management,* published as the first modern text on the biology, ecology, and management of a North American big game species; followed by O. J. Murie, *The Elk of North America,* in 1951, and W. P. Taylor, ed., *The Deer of North America: Their History and Management,* in 1956.
1952	Colorado: First use of Game Management Unit concept for managing big game.
1956	Fish and Wildlife Act established Fish and Wildlife Service in Department of Interior
1956	Soil Bank Act conservation reserve program recognized fish and wildlife values.
1958	Engle Act limited military withdrawals of public lands and provided for increased public use and conservation of wildlife on lands within military reservations.
1958	Congress extended and amended Coordination Act of 1934 to Fish and Wildlife Coordination Act authorizing the Secretary of Interior to make surveys and investigations of public lands or other federal lands suitable for wildlife conservation.

Early History

TABLE 15–1

Chronology of Development of Policies and Practices Relating to Conservation and Management of Big Game Mammals in North America—*continued*

1960	National Forest Multiple Use - Sustained Yield Act established policy for multiple use of forest resources and management of forestlands on a sustained yield basis.
1964	Wilderness Act established a National Wilderness Preservation System.
1964	Congress established the Public Land Law Review Commission to suggest improvements in policies, laws, and regulations affecting federal lands, their resources, and their uses.
1965	Land and Water Conservation Fund Act provided funds to plan, acquire, and develop land and water areas, plus facilities, for recreational purposes.
1966	Endangered Species Preservation Act authorized the Secretary of Interior to maintain a list of native endangered species (forerunner to Endangered Species Conservation Act of 1969 that authorized the Secretary of Interior to generate a list of species threatened with extinction).
1968	Wild and Scenic Rivers Act provided for preservation of some rivers with outstanding scenic, recreational, geologic, fish and wildlife, etc., values in free-flowing condition.
1969	National Environmental Policy Act (NEPA) required an environmental impact statement for any federal action affecting the quality of the human environment (assumed wildlife was part of human environment) and formed the Environmental Protection Agency.
1969	Classification and Multiple Use Act of 1969 mandated multiple-use management on lands administered by the Bureau of Land Management.
1969	Federal courts, in the matter of a state–federal dispute (*New Mexico State Game Commission v. Udall*) over control of deer in the Carlsbad Caverns National Park, again ruled that federal administrators had the power, under the Constitution's property clause, to protect federal lands from deer degradation without state interference.
1970	USDI issued a regulatory statement similar to USDA W-2 of 1940 that recognized state authority over resident wildlife and encouraged cordial state–federal relations.
1971	Wild Free-Roaming Horses and Burros Act directed the secretaries of agriculture and interior to protect these animals on public lands.
1971	Mountain lion classified as a game animal in Montana. This generally marked the end of an era of classification and persecution as a predator in western North America.
1972	President Nixon signed Executive Order No. 11643 prohibiting use of compound 1080 (sodium flouroacetate) and other toxicants in federal predator control programs or on federal lands and banning interstate shipment of toxicants for use in predator control.
1973	Endangered Species Act of 1973, extending the ESAs of 1966 and 1969, further prohibited the taking of species listed as "endangered," protected "look-alikes," or species threatened over only part of their range, and designated "critical" habitats.
1973	The United States ratified the Convention on International Trade in Endangered Species (became effective in 1975).
1974	Federal Court ruled (*Humane Society of U.S. v. Morton*) public hunting was an authorized use of a national wildlife refuge and that hunting was a legitimate tool for management of wildlife under state control.
1974	Forest and Rangeland Renewable Resources Planning Act of 1974 called for preparation of land management plans for protection and development of national forests.
1974	Sikes Act of 1974 provided for wildlife, fish, and game conservation and rehabilitation programs on military and other public lands in cooperation with the states.
1976	Supreme Court ruled (*Kleppe v. New Mexico*) that under the property clause of the Constitution Congress exercises complete control over public lands including the power to regulate and protect wildlife on those lands.
1976	Federal Land Policy and Management Act (FLPMA) recognized state authority over fish and resident wildlife on federal lands, reserving specified limited controls to federal administrators. Mandated land use plans by Bureau of Land Management.
1976	National Forest Management Act (NFMA) of 1976 stipulated that national forest management plans must comply with NEPA and that management would maintain viable populations of native vertebrates on national forests.
1976	Symposium on "Mule Deer Decline in the West" failed to explain apparent declines in mule deer numbers across much of the West during the late 1960s and early 1970s.
1977	Soil and Water Resources Conservation Act of 1977.
1978	Supreme Court affirmed (*Baldwin v. Montana, Terk v. Gordon*) a state's right to charge unequal hunting license fees for residents and nonresidents.
1980	Congress passed the Fish and Wildlife Conservation Act of 1980. The act authorized the federal government to provide financial and technical assistance to develop and service state conservation plans for fish and wildlife, specifically nongame species.
1980	Northern Rocky Mountain Wolf Recovery Plan offered strategies for conserving northern gray wolves in three areas of the northern Rockies through natural recolonization and reintroduction. Revision approved 1987.

continued

TABLE 15–1

Chronology of Development of Policies and Practices Relating to Conservation and Management of Big Game Mammals in North America—*continued*

1982	Grizzly Bear Recovery Plan developed for conservation and recovery of grizzly bear in six to seven ecosystems with suitable habitat. Revision approved 1993.
1984	U.S. Fish and Wildlife Service initiated red wolf recovery program in southeastern United States to include captive propagation and reintroduction.
1984	Rocky Mountain Elk Foundation founded to promote sound management of wild, free-ranging elk, other wildlife, and their habitat.
1992	Congress passed the Partnership for Wildlife Act of 1992. This act encouraged a comprehensive conservation program to respond to declining numbers of wildlife before they become endangered, and provided for matching grant proposals for state wildlife conservation programs.
1995	U.S. Fish and Wildlife Service reintroduced gray wolves to northern Idaho and Yellowstone National Park by trapping and transplanting from Alberta.

Sources: 50, 31, 19, 12, 8, 18, 2, 56, 3, 45, 24, 7, 57, 1, 9, 44, 58.

Abundance and Exploitation

The period of colonization has been called the era of abundance and an era of exploitation (7, 8). When European settlers arrived and began spreading across the frontiers of North America, they found a rich array of wildlife resources of great and seemingly unlimited abundance. Some of these resources were important commercially and a driving force for colonization. Others became staples in the diets of settlers or posed threats to human welfare. Thus, similar to earlier episodes of largely savage life, wildlife was either a source of subsistence or in the way of humans seeking homes and livelihood. It was not widely regarded as a source of sport, to be viewed and enjoyed, or to have value unto itself. Moreover, it was generally considered to be a common or "free good" accessible to any or all who needed or chose to pursue it. Free-ranging animals or their parts became private property only when "reduced to possession" through hunting or trapping.

The subsistence hunters were especially fortunate in the variety, numbers, and distribution of large mammals they encountered. Ungulates and carnivores were abundant. There was no reason to restrict use of any wildlife. Species of value were hunted or trapped indiscriminately, while those that presented some threat to humans were pursued relentlessly under bounty and other devices designed to reduce their numbers. Market hunting to supply wild game to settlers, miners, and other frontier builders, for hides or for bounty, became a profession equal to other commercial endeavors with the end often justifying wanton butchery of populations and species (Figure 15–1). Skill in harvesting was paramount, while spoilage was not of particular concern. Thus, the era of abundance associated with colonization gave way to an era of exploitation as human populations increased—a sequence repeated time and again as exploration and settlement produced new frontiers across the continent.

Concern and the Birth of Conservation

Unlimited exploitation combined with widespread transformation of landscapes from wildland habitat to cities, villages, farms, ranches, mines, and other human developments greatly affected the distribution and abundance of North America's wildlife resources. Large mammals such as deer and elk that were most useful to settlers were especially affected. Concern for white-tailed deer and other game near settlements resulted in restriction on hunting and establishment of bounties on wolves, coyotes, bears, mountain lions, and eagles around some colonies as early as 1630 (9, 10). By 1720, all colonies had adopted a prohibition on deer hunting during some part of the year (8). Those efforts did little to check the decline of populations because of the lack of public support and little, if any, enforcement. Moreover, chronicles indicate that the

Figure 15–1 Near the end of an era. Slaughter of bison on the northern plains, 1880. (Photograph by L. A. Huffman. Courtesy of Montana Historical Society, Helena.)

white-tailed deer was the only species to receive protection until well into the 1800s even though elk and wood bison were all but eliminated from the United States east of the Mississippi River by 1850 (4).

It was only after the 1850s, coincident with the latter stages of the pioneer era, settlement of the plains and mountains of western North America, and the growth of sport hunting across the continent, that concern about the future of wildlife, especially large mammals, became widespread. The Louisiana Purchase and explorations of Lewis and Clark and others had opened the West to trappers and traders. The discoveries of gold in California in 1848 and elsewhere in the mountain West during the 1850s and 1860s combined with the Homestead Act of 1862 to bring thousands of settlers to the frontiers. Mining, farming, logging, and grazing spread and within 50 to 75 years dramatically changed the landscape. The rapid decline of vast populations of bison, antelope, elk, and deer on the plains as a result of burgeoning hide and other commercial and subsistence hunting was especially striking. Some species vanished or nearly so; others were greatly reduced in distribution and abundance and persisted as scattered populations in least disturbed or inaccessible areas.

The declines fostered increased concern, outrage, and calls for restrictions on hunting of premier species. Sportsmen's clubs and game protective associations were formed to focus and carry the concerns to the people and territorial, state, and federal governments. Publication of the first American sporting magazines and journals including *American Sportsmen, Forest and Stream,* and *Field and Stream* along with early books on hunting and the natural history of big game animals (e.g., Caton, *The Antelope and Deer of America*) during the 1870s and 1880s attracted further public attention and support for protection of declining populations (11). In the 1880s, the Audubon Society and the Boone and Crockett Club were formed. Led by Theodore Roosevelt, the latter included some of the most influential political and scientific figures of the day and rapidly became a major force in the drive for conservation of big game. Other leaders included G. Grinnell, G. Pinchot, W. Hornaday, and E. T. Seton, whose words, pens, lobbying, and other efforts aroused widespread sympathy for protection and wise use of all wildlife resources. These efforts, combined with governmental measures taken to protect and limit or apportion use of remaining stocks during the late 19th and early 20th centuries, marked the genesis of wildlife conservation and management in America.

Although born of an immediate need for protection of diminishing populations of bison, elk, antelope, and other species, the developing American notion or system of conservation was influenced greatly by the social, economic, and political conditions that existed prior to and during settlement and development of America. Americans lived in a free and democratic society, the first since ancient times. This, together with the pioneering experience, fostered new attitudes and perspectives about wildlife. Most prominent were the concepts of public ownership of all wildlife *ferae naturae* and the right of all to hunt that were recognized from early colonial times. These, plus restrictions on indiscriminate killing and commercial trade in parts and products of wildlife that emerged in reaction to market hunting, collectively provided the foundation for modern conservation theory and practice (12).

Public ownership and the right to hunt gave every citizen a vested interest in wildlife, protecting breeding stocks, and apportioning its use. This was the basis for the widespread public concern and debate about wildlife that led to public control and establishment of publicly funded systems for administering wildlife within state and federal governments. It also provided the basis and support for maintaining and managing national forests, rangelands, and other public lands for wildlife habitat and other uses. Restrictions on killing and commercial trade in wildlife and wildlife parts strengthened the interest and involvement of the public, especially among people in more populated cities and towns, and eliminated any direct, commercial interest of the landowner in wildlife on their land.

CONSERVATION AND THE RISE OF MANAGEMENT

Driven by a near frenzy for protection, large mammal conservation efforts during the pre-1900 period were dominated by development of a framework of laws and administration to limit the ongoing exploitation of wildlife. However, movement was slow and frustrated by the diverse political structures and conditions that existed across the developing nation. Although public debate over how to deal with wildlife abundance or scarcity was established, there was no central authority toward which concerns could be directed. Prior to the late 1800s the federal government did not assume or exercise authority over wildlife and the national forest and rangeland on which it occurred. Calls for federal protection of embattled bison, antelope, and other plains game went unheeded or were swept aside by Congress. While some individuals and interests were concerned with the demise of large mammals, others were encouraging it as a means of controlling Indians and opening up the West to settlement, farming, ranching, mining, and logging. Because of this lack of concern at the federal level, the wildlife conservation initiative passed largely to the states and territories, with highly variable results.

Development of Big Game Conservation in States and Territories

As pioneering gave way to more ordered and cultured life, state and territorial legislatures were pushed to assume increasing interest in and authority over the killing of wildlife. However, because each state and territory controlled its own resources, each moved independently to address game issues and protection in its own way. There was little or no coordination of concerns or conservation efforts among jurisdictions until early in the 20th century. As a result, protective policies and practices unfolded slowly and piecemeal, often accompanied by heated controversy over the best way to conserve the various species.

Generally, state and territorial measures enacted during the late 1800s and early 1900s focused on restricting and apportioning the take of dwindling stocks. California prohibited hunting of antelope and elk in 1852, the first closed seasons on these species. From then through the early 1900s, all states and territories increasingly restricted harvest of large mammals by shortening or closing seasons, limiting the numbers or sex and age of animals that could be taken legally, and licensing hunters, especially non-residents. Gradually, prohibitions developed against wasting of meat and killing for market, and hunting at night with artificial light and from vehicles or roads, over bait, during certain days, or with some kinds of weapons. Fines were provided for violation of game laws and some type of enforcement or government warden service was established to oversee and enforce restrictions. However, wardens were few and poorly equipped and paid. In many areas, local constables, sheriffs, or other county officials were charged with enforcing wildlife laws in which they held little interest.

Neither the increased restrictions nor law enforcement efforts were successful in stemming exploitation. The new laws, if known, were widely ignored and market hunting of large mammals persisted openly until seasons on some species were closed, limits on the numbers of others that could be taken were implemented, and interstate traffic in illegally killed game was prohibited around the turn of the century. Thereafter, commercial hunting for hides, meat, and other animal products was reduced or restricted to illicit traffic and more remote regions, especially the northern frontiers of Canada and Alaska. Subsistence hunting or the killing of game for food by individual settlers, frontiersmen, and backwoodsmen continued on a broad scale into the early 1900s, when it became increasingly controlled but never eliminated.

The first state game and fish agencies were founded during the mid- to late 1870s and by about 1915 this concept was gradually adopted in different forms by all states. The *Geer* decision of the U.S. Supreme Court in 1896 formally established wild game and fish as the property of the state and provided the basis for each to administer or manage its wildlife resources. Initially, however, at least from a large mammal perspective, the state agencies were concerned primarily with game law enforcement and had little or no involvement in conservation policy. The power to establish hunting regulations and other game preservation or conservation measures remained largely with legislatures.

State agencies were poorly funded. Although licenses were required of nonresident and/or resident hunters in some states beginning in the early 1860s, they did not become a general requirement for hunting until after the turn of the century. It was only after the U.S. Congress passed the Federal Aid in Wildlife Restoration (Pittman-Robertson, or P-R) Act of 1937 and individual state legislatures ratified and agreed to provisions of the act that state hunting license money was earmarked for use by state game agencies for wildlife conservation and management.

State wildlife employees remained predominantly game wardens and political appointees with no formal training in dealing with wildlife resources or conservation issues until the late 1930s or later. Their role was simply to enforce the law, to ensure that license requirements and other restrictions were obeyed, to conduct or administer predator control programs, including payment of bounties, and to maintain the sanctity of game preserves and refuges established by legislative decree. In some cases, agency personnel also became involved in early efforts by sportsmen to transplant deer, elk, and other big game animals and winter feeding programs that developed as conservation programs around the turn of the century.

Efforts to control predators had been ongoing since settlement. With declining game populations, predation was increasingly viewed as a major drain on wildlife and competition with hunters. Predators were viewed as major threats to settlers and the range livestock industry that spread rapidly across the West during the 1870s and 1880s.

Bounties, which were first applied to help protect deer in the colonies, were effected throughout the West beginning in the 1870s. Under the bounty system, individual hunters and trappers combined in cooperative efforts with stockmen and game officers to bring predator control to a commanding place in wildlife preservation during

Figure 15–2 This grizzly bear was shot on the northern plains, 1881. Grizzly were not protected as a game animal in Montana until 1923. (Photograph by L. A. Huffman. Courtesy Montana Historical Society, Helena.)

the early 1900s. Establishment of the federal predatory bird and mammal control program in 1915 and introduction of poisons further intensified control efforts. As a result, the wolf was eliminated from most of the contiguous 48 United States; the numbers and distribution of mountain lion were greatly reduced; and the grizzly bear became restricted to Canada, Alaska, and remnant populations in the Yellowstone and northern Rocky Mountain areas of Montana. Although the grizzly bear was protected as a game animal by Montana in 1923, similar protection for mountain lion and wolves would not occur for another 50 years or more (Figure 15–2).

The third major thrust of early state game preservation and conservation efforts, the establishment of statutory game preserves and refuges, did not come fully into vogue until the early 20th century when continental big game populations had declined to all time lows. The concept may have originated in the U.S. Congress with creation of Yellowstone National Park in 1872. Establishment of the Wichita National Game Reservation in Oklahoma in 1905 and the Grand Canyon National Game Preserve in Arizona in 1906 triggered widespread efforts by federal and state governments to rebuild game populations by allowing animals to increase and flow out from protected areas. Montana, for example, saw its first legislative game preserves, to protect pronghorn, deer, and game birds, in 1910; by 1936, a maximum of 46 preserves protected some members of all big game species in the state. Deficiencies of the concept would become apparent only years later as states attempted to implement population management and control increasing numbers of deer and other species.

Efforts to restock depleted ranges by translocation apparently began in 1878 when a group of sportsmen brought 17 white-tailed deer from New York to Vermont. Other states followed. For example, Pennsylvania imported and released 1,192 deer from other states between 1905 and 1925 (13). Trapping and transplanting elk from Yellowstone National Park to poorly stocked or uninhabited ranges throughout much of North America began about 1892. It became important in conservation and management of elk in the park and across the West from about 1910, when elk were first transplanted to Montana, through the early 1970s. Translocation as a management tool or practice for big game would come into greatest use after 1940, when biologists and game managers employed trapping and transplanting to enhance or restore populations of nearly all-native species.

Winter feeding first became a tool for conservation of big game with feeding of white-tailed deer around 1905. The Wyoming legislature first authorized feeding of elk in Jackson Hole during 1909, while Congress appropriated funds for the U.S. Biological Survey to begin winter feeding on the National Elk Refuge in 1912 (14, 15). By the 1920s, supplemental feeding to augment natural forage became a standard tool in deer and elk conservation and early management programs. This continued until studies evaluating the effectiveness of winter feeding during the 1930s and 1940s showed winter feeding of hay to be ineffective in reducing starvation and created problems associated with crowding, disease, and overuse of natural forage in the vicinity of feeding grounds (16, 17).

Development of Big Game Conservation Programs at the Federal Level

Involvement of the federal government in fish and wildlife conservation and management began in 1871 with establishment of the U.S. Fishery Commission to study and propagate fish. Authorization of the Division of Entomology in the Department of Agriculture to study the diet, distribution, and migration of birds in relation to agriculture similarly marked the entry of the federal government into wildlife science in 1885. This agency later became the U.S. Biological Survey, which did much to coordinate and increase public interest in wildlife in America through the turn of the century (1). In 1900, the agency was also given responsibility for enforcing the Lacy Act passed by Congress to prohibit interstate transport of game killed or shipped in violation of state laws and restrict importation of foreign game. This greatly increased federal involvement in wildlife administration and gave "conservation" a major boost in efforts to control market hunting and trade in meat and other products of game animals. However, other developments through the same time period were also important.

Establishment of Yellowstone National Park in 1872 marked the first federal act to set aside and protect lands for their natural values, including wildlife. Although the park constituted the first big game refuge, substantial effort to control hunting and protect big game in Yellowstone did not occur until the U.S. Army assumed administration in 1886 (18). As observed by Geist, this move was particularly significant and probably marked the beginning of big game management in North America (12).

Under military jurisdiction many of the practices that came to characterize early big game management, especially for elk, in North America were implemented. The park attracted natural historians and mammalogists to study animals and habitats, and brought science into wildlife protection. Surveys to determine big game numbers on winter range and elk migrations were conducted and reported (18). Illegal hunting was suppressed. Predator control and supplemental winter feeding programs were developed to help restore diminished populations. Beginning in the 1890s, elk were trapped for relocation to vacant and underpopulated ranges throughout much of North America. From the 1920s to the 1960s, this program, which resulted in relocation of more than 13,500 elk to 36 states and several countries, was keystone to restoration and management of elk across the West (15). It also was a basic part of the increasingly intensive elk population management program that developed in and around the park after 1930. During this same period, the national parks in the United States and Canada became primary sites for pioneering research on big game mammals and habitats.

A second significant step toward management of big game and other wildlife in North America at the federal level was passage of the National Forest Reserves Act of 1897 providing administration, conservation, and use of large areas of federal lands that became the national forests in 1905. Most of these lands were important in sustaining the nation's big game resources; and the U.S. Forest Service, organized from the old Bureau of Forestry, rapidly became the lead organization in the development of big game management in the United States. Leadership was provided by people like Gifford Pinchot, the first chief of the Forest Service, who defined conservation in terms of development and use of resources, and Aldo Leopold, a forester who promoted conservation in the field and would later be recognized as the "father of game management" (19). Leopold, in particular, promoted game conservation for recreational hunting, a concept central to development of management through the 1950s.

In the early 1900s foresters were the only body of scientifically trained professionals acquainted with game conditions on the ground and capable of dealing with the problem of saving game for public recreation (20). Their efforts between 1905 and 1940 contributed significantly to developing a framework of theory and practice that would characterize management of deer and other big game into modern times.

Although formal game censuses were not initiated until 1914, forest officers began to survey and evaluate large mammal numbers and conditions in the field soon after the national forests were organized. They found these resources severely depleted and nonproductive. In response, they initially pushed efforts to remove predators (21, 22). However, recognizing needs for winter range and possible competition from livestock grazing for forage, they also moved to restrict the largely uncontrolled grazing that was occurring, at least on some important game ranges. For example, in 1908 a large area adjacent to Yellowstone National Park in the Gallatin National Forest, Montana, was set aside solely for use by big game. Cattle and sheep grazing was also reduced in the Kaibab National Forest in Arizona. These measures may have constituted the first recorded instances of "habitat management" for big game by resource managers.

Beginning in 1914, the U.S. Forest Service began to census large mammals on the national forests. In the early 1920s surveys of deer and elk food and range use habits in relation to forestry, livestock grazing, and range conditions were initiated. Reports of these surveys provided the basis for the first efforts by game agents to break away from the preservationist policies of the late 1800s and early 1900s and to initiate more flexible management of big game, including liberalized hunting and hunter harvest in some areas. In this respect, no single event was more important to future philosophy and practice of big game management than the classic case of mule deer on the Kaibab National Forest.

The efforts to protect big game through restriction of hunting, increased law enforcement, predator control, establishment of refuges, and translocation combined with favorable habitat conditions to influence recovery of some populations, especially deer, by the early 1900s. On the Kaibab Plateau, creation of the Grand Canyon Game Preserve in 1906, predator control, and extensive reductions in numbers of cattle and sheep grazed led to dramatic increases in mule deer numbers through the early 1920s. Extensive winter losses and reduced fawn survival during 1924 to 1925 resulted in a population "crash" (23, 24). Surveys suggested that "overbrowsing" by the rapidly growing deer population had resulted in deterioration and loss of desirable browse and forbs and replacement by less desirable and unpalatable species (Figure 15–3). It appeared that mule deer were being forced to exist on a starvation diet resulting in increased mortality and reduced fawn production. As a result, the U.S. Forest Service began to advocate and initiate measures to control and reduce deer herds as much as possible by conducting public hunts, shooting, and trapping and shipping deer to other areas. Range studies, including the first use of exclosures to document effects of overbrowsing, were established to objectively evaluate problems and needs and support control efforts (24).

The high populations and deteriorating habitat conditions on the Kaibab had generated concerns about overabundance of deer and damage to forest resources, including rangeland, timber, and game resources as early as 1918. Paradoxically, perhaps because of a lack of experience with increasing deer populations, managers misread or ignored the signs of impending problems. Differences of opinion among professional foresters, state game administrators, politicians, sportsmen, livestock operators, and others with respect to numbers of deer, health of the animals, the significance of range problems, and what if anything could or should be done about them, also hampered attempts to seriously address the apparent problems even after the "dieoffs" began. Measures that were taken constituted "too little too late." They also generated legal controversy and actions that first brought big game management and the issue of state versus federal authority to the courtroom.

State Challenges to Early Federal Management Programs

Sportsmen and others, including the Arizona Department of Conservation, reacted strongly against the U.S. Forest Service control programs in the Kaibab. Sportsmen had

Figure 15–3   Heavily browsed and deteriorating shrubs and "high-lined" trees from deer over-population on winter range in the northern Rocky Mountains. (Photograph by R. J. Mackie.)

embraced restricted hunting and other protectionist policies as necessary conservation tools and were reluctant to accept liberalized management. The infant state game agency had not developed the capability to evaluate and respond to problems associated with increasing big game populations. State game administrators and politicians were concerned about their constituents' views, the need for control, and federal encroachment on state ownership and their right to administer game populations. They challenged the U.S. Forest Service control program that sanctioned the killing of Kaibab deer by federal hunters all the way to the U. S. Supreme Court (*Hunt, Governor of Arizona, and others vs. U.S. 278 U.S. 96*).

While attempts to manage deer in the Kaibab continued to flounder during the late 1920s and early 1930s, the controversy about deer, deer management, and state versus federal imperatives increased and spread. Recovery and growth of deer populations in other areas brought further concerns about overabundance, overgrazing, overbrowsing, and damage to forage plants and other resources. In 1932, in another attempt to manage increasing deer numbers, the U.S. Forest Service initiated efforts to control white-tailed deer numbers on a game preserve within the Pisgah National Forest in North Carolina—first by conducting public hunts and later by trapping and shipping deer out of state. This highly controversial move also resulted in court action relating to the right of the federal agency to control deer and remove them from the state. Again the U.S. Supreme Court ruled that federal administrators could, upon evidence of injury to lands from overpopulation, act to protect their property and reduce big game populations by hunting or trapping and transplanting to other areas (*Chalk, Commissioner of Game and Inland Fisheries and others vs. U.S. 114F. 2d 297*).

The Kaibab and Pisgah rulings could have made it possible for the U.S. Forest Service to take complete control of management of deer and other big game populations and their habitats in states or areas with large amounts of national forest and other federal lands. However, by the 1930s, the concept of game management had become established in state game and fish agencies; and representatives of the agencies, the U. S. Department of Agriculture, and the U.S. Forest Service met and negotiated a new federal regulation (G-20-A) calling for state–federal cooperation in establishment and maintenance of game populations (19). The regulation returned administration of resident game animals and population control to the states; the federal agency was a partner and cooperator in the sense of controlling and managing use of the habitat game animals required. This action, together with the emergence of game management as a professional endeavor and passage of the Federal Aid in Wildlife Restoration Act in 1937, paved the way for emergence of modern big game management systems across the country by the 1940s.

BIRTH AND DEVELOPMENT OF SCIENCE-BASED MANAGEMENT, 1920 TO 1940

By 1920 there was growing recognition among foresters, some state game administrators, and other conservationists that wildlife must be produced and otherwise managed, and protected, to be successfully conserved. Leopold made a strong case for new efforts in game conservation and development of a new science of game management (20). Modeled after the science of forestry, the new discipline would entail knowledge of game conditions, including census and information on distribution, habitat conditions, predation, and hunter kill. In addition to protection, it would involve provision for hunting based on the concept of a *sustained annual yield* determined by numbers of breeding stock, annual production of young, and losses in areas capable of sustaining harvest. Although refined in later articles and his text to recognize variation in environmental conditions, animal numbers, and other factors, Leopold's concepts became widely accepted and central to emerging big game management programs (3).

Although the 1920s and 1930s marked a period of transition from preservation to restoration and management of game populations, a framework of scientific knowledge, theory, and practice to support and apply Leopold's concepts to big game was largely lacking. Early attempts to define theory and practice in wildlife conservation were strongly preservationist and lacked understanding of the role of animal biology and behavior, habitat characteristics, and environmental controls in distribution and abundance of animals (e.g., 25, 26). The theory that the only thing needed to conserve big game animals was protection of animals and their habitat worked as long as populations were low and increasing. However, it failed and became a handicap as deer continued to increase and overpopulate ranges on the Kaibab Plateau and elsewhere across the continent. A new and better understanding of how populations operated was needed.

Natural scientists, conservationists, and foresters of the late 1800s and early 1900s were keen observers with great interest in big game and often recorded and reported their views and concerns in considerable detail. Leopold noted that "some of the important characteristics of various game species are already so well known that they need no particular comment. They have for years been the subject of investigation by naturalists and sportsmen, and are quite thoroughly recorded in the literature" (3).

Importantly, however, no scientific studies relating to conservation of North American big game animals or their habitats had been conducted or published prior to 1920. Articles in sporting magazine articles and books reporting observations of hunters or the general occurrence, characteristics, and habits of deer and other large animals constituted nearly all of the published information available for assessing management problems and needs (11, 27, 28). The few "management" publications available dealt with raising deer and other large game animals in captivity, or deer farming; and at least some conservationists and early game managers apparently viewed elk and other wild ungulates as "nature's cattle" that could be managed as such (29, 30).

New Science, Theory, and Practice

More or less "scientific" studies for big game management first appeared in the 1920s. By that time, it had been apparent ". . . that the amount of food in winter yarding areas was one of the most important factors limiting the number of deer . . ." (31). When, where, and how this concept was first recognized and established as fact is unknown. It was generally consistent with events on the Kaibab National Forest and elsewhere and became widely accepted as the basis for *carrying capacity* or the number of deer or other animals a unit of range could sustain (3). As a result, most early studies were either inventories of winter range and forage conditions or investigations of animal food and range use habits to help evaluate those conditions.

By the late 1920s and early 1930s, the U.S. Forest Service had developed standardized game and range surveys to determine big game numbers and range conditions on the national forests. Studies of elk and mule deer migrations and feeding habits and forage requirements in national parks and national forests were also being conducted to help assess management problems and needs (32, 33, 34, 35, 36).

Some of the earliest state deer management studies were conducted in Michigan where a Game Division, employing biologically trained personnel, was formed in the Department of Conservation in 1927 to determine the status of game, inventory winter deer yards, and recommend plans for management. The studies included surveys to estimate deer numbers, herd composition, and annual harvest and experiments to determine carrying capacity of deer yards, preferences and palatability of forages, changes in condition of deer on various foods, and the role of parasites and disease in winter starvation. A 1938 bulletin summarizing Michigan's deer problem, their studies, and a suggested general plan for deer management in the state was the first substantial publication on management theory and practice for a big game animal in North America (31).

Growing game populations and perplexing problems with overabundant deer led to establishment of professional game management programs during the late 1920s through early 1940s. By the mid-1930s many colleges and universities had curricula to train professionals. These programs, together with Cooperative Wildlife Research Units established at land grant colleges and universities in several states, were initiating graduate training and research important to emerging wildlife science. Passage of the Federal Aid in Wildlife Restoration Act of 1937 provided new funding for state wildlife agencies to restructure, employ trained game biologists and managers, and begin studies on the status and needs of big game.

Despite these developments, it was not until 1945 to 1955 that substantive literature began to accumulate from new state and university programs. The few studies relating to biology and management of big game species were published largely by U.S. Forest Service personnel during the late 1930s (37, 38, 39, 40, 41, 42).

The first modern texts summarizing management concepts or other information for big game species were published during the 1940s and 1950s (13, 43, 44, 45, 46). By then, big game management systems were in place throughout most of North America and deer and other big game managers, scientists, and educators had begun to frame management dogma generalized from observations and early studies of overabundant and problem deer populations. Rasmussen's account of mule deer on the Kaibab Plateau became the center of theory for understanding population ecology, habitat relationships, and management needs (23).

Generally the new theory held that deer populations, released from hunting, predation, or other limitation on favorable habitat, were highly productive, inherently irruptive, and capable of overpopulating and overbrowsing their ranges. Winter range and winter forage were the primary factors limiting populations. Predation, competition, diseases and parasites, and winter weather were usually secondary factors related to nutrition. *Carrying capacity* was the number of deer a range could sustain in balance with the winter forage supply. As deer numbers reached or exceeded carrying capacity, the amount and quality of winter browse available to individual deer declined, resulting in malnutrition and starvation, decreased reproduction, smaller body size, and reduced antler growth (Figure 15–4).

From a habitat perspective, overpopulation and overuse of important forage plants resulted in smaller, less productive plants that would ultimately die and be replaced by less nutritious and palatable plants. Thus, deer and other big game animals were "major influents" that could change the entire character of deer ranges and reduce carrying capacity if not checked by some extrinsic factor.

Figure 15–4 Burgeoning mule deer populations on western rangelands resulted in extensive starvation on winter ranges and calls for more liberal hunting regulations to bring deer numbers in balance with winter forage supplies. (Photograph by R. G. Janson.)

In this scenario, similar to livestock grazing and range management theory, grazing and browsing were the primary factors affecting the kinds, amount, and quality of forage available. Proper levels of forage use sustained healthy, productive winter forage plants which, in turn, sustained healthy productive game populations. The way to maintain healthy, productive big game populations and habitat carrying capacity was population control through harvest. Because recreational hunting was a primary goal of game management as defined by Leopold and the primary source of funding for management under the federal aid program, theory and practice were complementary.

Under this ideology, successful restoration of big game populations was achieved when numbers were sufficient to sustain recreational hunting, at least under limited entry or permit seasons. By the late 1950s and 1960s, excellence in deer management in particular was equated by many with either-sex hunting seasons that provided for maximal recreational opportunity, hunter success, and total harvest (Figure 15–5).

Managing big game species and their habitats under the new concepts required additional science and education. As suggested initially by Leopold, a base of knowledge or facts about animals, including their habits and requirements, their habitats, and their habitat relationships, was essential to assessment of management opportunities and needs (20). One of the first tasks of new management programs was to establish or expand research and other information gathering programs. Another was to convince sportsmen, politicians, and others, including some game managers and administrators, of the need for management that often liberalized hunting seasons and harvests, including the killing of females and young.

In spite of new science, rapidly changing official policies, and support in some conservation circles, the new programs were not easily implemented. Protective theories and practices had become traditional and set in the minds of sportsmen, and in state laws and regulations that had to be changed or amended. The application of biology and ecology, including accepted management principles, and the need to achieve a reasonable annual harvest that included females and young, was not easily understood and accepted, even within some conservation agencies and institutions of higher learning.

Figure 15–5 Reducing overabundant deer populations and implementing sustained-yield harvest management required that early closed seasons and buck-only hunting be replaced by more liberal hunting of animals of both sexes and all ages. (Photograph by R. J. Mackie.)

Moreover, the new game managers and biologists employed by state agencies were considered invaders into a field dominated by law enforcement and other personnel. In some cases, the need for research and information or education programs was viewed as self-serving, to promote and expand management rather than benefit game populations.

In many areas, the road to big game management based on facts became the "deer wars," exemplified by almost continuous controversy and conflict between efforts to implement new management philosophy and practices on one hand and those associated with disbelief and opposition to change on the other. The conflicts often became extremely bitter and personal, marked by professional and political recriminations. Perhaps the best description of all aspects and considerations inherent in the conflict can be found in *Doe Day: The Antlerless Deer Controversy in New Jersey,* a case study of the effort to implement deer management in New Jersey (47).

Despite this tumultuous early history, the new management concepts and efforts endured to become generally accepted and respected functions of wildlife administration by the 1960s and 1970s. Even with this acceptance, however, social, political, economic, and evolving biological and ecological considerations often precluded implementation of sustained yield management and achieving a balance between large herbivores and their forage supplies.

INTENSIVE MANAGEMENT, 1940 TO 1970

With rapidly accumulating field data and research studies, North American large mammal management became organized around three major objectives: (1) to develop and sustain maximum game populations consistent with available habitat and other uses of the land, (2) to ensure maximum production and utilization of annual game surpluses, and (3) to provide the maximum possible amount of recreational opportunities for sportsmen. The manner in which these objectives were addressed by management agencies and units varied greatly depending on species or mix of species that occurred, their distribution and abundance, and other factors. Sportsmen, ranchers, loggers, and resort operators were the segments of society concerned or influenced most by the occurrence and abundance of large mammal populations. Management programs were directed heavily toward important game species and habitats, or problem situations, and increased in relation to distribution, abundance, and opportunities for hunting.

Early efforts to apply science to management of large mammals were directed largely at white-tailed deer, mule deer, and elk. Deer were most widely distributed and abundant and increasing rapidly across most of their geographic ranges from the 1920s through the 1950s. Problems associated with this abundance demanded management attention, generating public and professional controversy. Thus, deer management programs, policies, and practices developed concomitant with and in response to increasing populations and problems.

Elk, though slower to recover, were highly prized and in great demand for restoration in vacant and poorly stocked habitats throughout the West. As a result, elk populations increased gradually, but steadily, roughly paralleling the development of big game management through the 1950s and 1960s. Controversy regarding sustained yield management policies and the best way to manage elk centered largely on historic ranges around Yellowstone National Park and other northern Rocky Mountain areas where large populations occurred.

Attention directed to other species varied across the continent and over time in relation to abundance and public values or concerns. Moose and caribou were most common and target species for early management primarily in Canada and Alaska, where both were widely distributed and exploited for subsistence by resident humans and for sport. In the northern United States, caribou were absent or very rare. Huntable or potentially huntable populations of moose occurred only in the northern Rocky Mountains, northern Minnesota, and Maine.

Pronghorn were relatively slow to recover to huntable numbers in most plains environments where agriculture, livestock production, fences, and other developments had significantly reduced the amount and continuity of habitat. Early protective management policies had led to gradually increasing populations on rangelands across the northern and southern plains, the Great Basin, and elsewhere in historic open, shrub-grassland habitats from the 1940s through the 1960s. Where they occurred in abundance in huntable populations, pronghorn offered excellent opportunity to apply controlled sustained yield management policies by the mid-1950s.

Bighorn sheep and mountain goats also generated relatively minor concern and only moderate, though gradually increasing, management attention because of their limited distribution and abundance across North America. Protection, trapping, and transplanting beginning in the 1940s led to expanded populations and controlled recreational hunting under sustained yield concepts. Musk ox were limited to treeless arctic tundra habitat where they drew relatively minor management concern. Bison persisted only in Yellowstone National Park and fenced refuges such as the National Bison Range in the United States and the Woods Bison National Park in Canada. For most of these species, and for black bear, management continued under traditional protective or limited harvest programs designed to encourage recovery until new management programs for deer and elk were established.

Big game management programs became established and expanded in most states and federal agencies during the 1950s and 1960s. Management personnel typically recommended programs, including hunting regulations, to a citizens' game board or commission, and implemented programs or policies that were adopted. They collected, analyzed, and interpreted biological and ecological data from management surveys as a basis for those recommendations. They also conducted basic and applied research, initiated habitat acquisition and development projects, provided colleagues and the public with information, and cooperated with other agencies and institutions. However, the mix of species and habitats, environmental conditions, and social, political, and economic forces that existed in each state, agency, or other unit often influenced differences in programs. In some jurisdictions, managers saw little need for research or other programs, or were restricted by law or limited funds from involvement in them. As a result, what constituted big game management at one time or in one agency, state, or part of the country did not necessarily follow in another. Both the kinds and intensity of effort varied greatly.

To implement sustained yield management, most states adopted some form of herd or habitat units or hunting districts to distribute recreational opportunities and harvests in accord with the distribution and abundance of game animals. The units also served to evaluate management needs and determine effects of management practices. In many areas, especially in the West, studies evaluating big game range conditions and forage production, use, and condition trends became standard practice for evaluating population status relative to carrying capacity and for harvest management recommendations. Some of these practices dated back to the late 1920s and early 1930s when winter range and forage had become accepted as the primary factors limiting deer and other large herbivores and the U.S. Forest Service began developing wildlife handbooks that outlined procedures for sustained yield management of wildlife resources on the national forests. Developing and testing methods for measuring range conditions and relationships between forage use and range conditions became a primary focus of research in state and federal agencies during the 1940s and early 1950s.

In Montana, game biologists established and annually measured up to 1,035 permanent browse use and condition trend transects during 1955 to 1971 (Figure 15–6) (48, 49). Numerous exclosures and other vegetation sampling units were also established and measured periodically by federal and state personnel to assess vegetation characteristics on winter ranges of five to seven big game species. Key grass use and condition trend transects and other measures of forage or range conditions were employed on foothill grassland winter ranges grazed by elk and bighorn sheep (48). Many other western states followed similar procedures.

Figure 15–6 Measurements of forage production, utilization, and condition trends were a primary database for management of deer and other big game during the 1950s and 1960s. (Photograph courtesy Montana Fish, Wildlife and Parks, Helena.)

Efforts to estimate species population size or trend and sex and age composition, productivity, mortality, and animal condition in management units or using individual winter ranges or yards also became important. Attempts to estimate big game numbers and trends for management probably began with early surveys in Yellowstone National Park during the late 1800s, though the first use of population estimates in relation to deer management apparently occurred on the Kaibab Plateau, Arizona, where the Kaibab National Forest supervisor enumerated mule deer population trends from 1906 through the 1920s. These and other estimates from surveys by the U.S. Biological Survey, naturalist George Shiras III, and others, combined with surveys of deer range conditions, provided the basis for U.S. Forest Service recommendations to limit deer numbers on the forest during the mid-1920s (23).

Most early attempts to enumerate big game populations and trends were simply direct counts by individuals or groups organized for the purpose. Aerial surveys were first applied in big game management in Colorado in 1937 and became widely applied during the late 1940s and 1950s (50). By the 1960s, aerial surveys were standard in managing pronghorn and widely used in surveys that employed various sampling procedures to determine population size or trend and composition for nearly all North American big game. In environments where overhead cover and other factors precluded aerial surveys, pellet group and various direct observation methods and other techniques were applied. Where counts or other population estimates were impractical or unnecessary, classification counts to determine population composition, reproduction, and mortality provided supplementary data. The difficulty or costs of obtaining population estimates, combined with the opportunity for sportsmen and others to dispute numbers or management strategies related to numbers, led some biologists to avoid population studies. Others believed them unnecessary and, instead, relied largely on trends in range conditions and harvest information to determine management needs.

Harvest estimates and trends were the third major data set available to managers seeking sustained yield. Knowledge of total harvest and composition related to forage use and condition trends and, where available, data on population trend, reproduction, and mortality provided bases for adjustment of hunting regulations and subsequent harvest goals. Harvests were determined by hunter check stations, hunter report cards, and mail and/or telephone questionnaire surveys that first came into extensive use in the late 1940s and became increasingly sophisticated and statistically based during the 1960s. Hunter checking stations were first used to determine harvest trends and to collect biological information on harvested animals during the 1930s. By the 1950s, when techniques for determining the age of deer and elk by tooth replacement and wear were developed, checking stations became widely used to measure population age structure and animal condition and harvests in local areas.

While big game management focused on harvest management, many programs entailed efforts to produce and maintain maximum populations, especially breeding stocks, in all suitable habitats. State agencies that controlled big game populations rarely owned or controlled more than a small percentage, if any, of the lands that sustained those populations. Thus, opportunities for habitat management were often limited to population control, to acquisition and development of key winter ranges, and to cooperative efforts with forest and range managers and private landowners.

Whether public or private, the lands occupied by big game usually were used primarily for other purposes that had greatly and often repeatedly altered landscapes and vegetation across much of the continent from pioneer to current times. Public rangelands, federal game ranges and refuges, and the national forests where early management concepts and practices were developed had been managed primarily for other values, including livestock grazing and timber production, during the early to mid-20th century. Wildlife production and recreation remained secondary uses or by-products to other uses of public lands until the 1960s and 1970s when public pressures pushed Congress and agencies to enact new laws and regulations influencing greater consideration for wildlife in public land use and management programs (Table 15–1).

Intensive Management, 1940 to 1970

Figure 15–7 Clear-cutting of this timber stand for regeneration and enhancement of browse supplies adjacent to winter range in northern Michigan was conducted during the mid-1950s. (Photograph by R. J. Mackie.)

Early in developing big game management programs, wildlife biologists recognized the importance of clarifying relationships between big game animals and prevailing land uses and management practices. Concerns about game–livestock grazing relationships and competition on rangeland became some of the earliest issues in efforts to recover and manage big game, especially elk, in the West. They also were the basis for numerous administrative and research studies throughout the 1960s.

In other areas, forestry and timber management practices were or became of equal or greater concern to big game managers (Figure 15–7). Good forestry did not always represent good habitat management for big game, and practices that seemed beneficial in one area, at one time, or to one species could be detrimental in other areas, at other times, or for other species. Close coordination between game and land management agencies was essential.

Cooperative interagency agreements relating to research and other information gathering and development of management objectives, policies, and practices were often developed to foster better communications and working relationships at field and administrative levels. Under these agreements, state biologists spent considerable time evaluating the possible impacts of various grazing, logging, road building, drainage, and habitat enhancement projects for big game on federal lands. Federal agencies conducted research important to big game management, developed various habitat enhancement projects, participated in gathering management information, and assisted in implementing management practices. Occasionally federal agencies transferred certain lands to states for management as winter game ranges, or set aside lands from grazing and other uses to enhance big game management.

Many developing big game populations occupied private lands. The existence and productivity of game on these lands are often determined by the owner's land use practices. Hunting and harvest to control game populations or meet other management objectives were also subject to landowner decision. Because of this, game personnel in many areas found it essential to work closely and cooperatively with private landowners and with agencies such as the Soil Conservation Service, which provided technical and cost-sharing assistance to landowners for land, water, and wildlife conservation projects.

During early efforts to restore big game populations, especially elk, it became evident that suitable winter range was necessary in certain foothill and valley areas. These areas were predominantly private lands where opportunities were high for game damage and game–livestock competition. One solution was acquisition and development of key rangelands offering potential for wintering elk and other migratory species. Although some tracts (e.g., the National Elk Refuge near Jackson, Wyoming) were acquired as early as 1912, state efforts to acquire winter game range for elk began during the 1940s with federal aid funds. The kinds and amounts of land acquired and the developments following purchase varied greatly, but extensive programs were established in some states. In Montana, for example, more than 60,000 hectares were purchased or leased between 1940 and 1970 for development as winter game ranges by eliminating livestock grazing and other uses (51).

As deer, elk, and other big game populations increased to fill nearly all their habitats, management concepts, policies, and practices changed. Many early policies and practices were reevaluated and deemphasized or discarded. Closed, short, and male-only hunting seasons for most species were dropped in favor of regulations tailored to management needs and opportunity that included hunting animals of both sexes and all ages. The value of preserves and refuges in big game management declined, and many were eliminated or their functions were changed. The need for predator control declined and bounties were dropped in most areas. Other techniques such as salting, baiting, and driving used occasionally to influence distribution and range use were also discarded for general use.

EXPANDING HORIZONS AND NEW DIMENSIONS, 1970 TO PRESENT

By the early 1970s, big game management programs were firmly in place and becoming increasingly challenging and complex. Restoration of nearly all native ungulate species to areas capable of sustaining populations was under way or complete. Some species, notably deer, probably were more widely distributed and abundant than at the time of settlement. Others, including new populations and subspecies of elk, bighorn sheep, and occasionally other species, were still increasing. Although the black bear had gradually achieved status and management as an important big game animal throughout North America, other large carnivores, including the grizzly and polar bear, mountain lion, and gray and red wolves, were only beginning to attract attention. Hunting, controls imposed due to depredation, and habitat loss limited numbers and distribution, raising concern about the survival of viable populations in many areas. Passage of the Endangered Species Preservation Acts in 1966 and 1973 gave impetus to efforts to protect and ultimately restore species and subspecies, the polar bear in the north, the grizzly bear in and around Yellowstone National Park and other northern Rocky Mountain habitats, red wolves in the Southeast, and gray wolves in the northern Rocky Mountains, northern Minnesota, and the Southwest.

In eastern, midwestern, and southern states, white-tailed deer were not only abundant in traditional forest and rural environments, but were rapidly increasing in urban areas and in and around parks and preserves where traditional hunting was not possible. In most of these states, black bear that had achieved new status as an important big game animal during the 1950s and 1960s were the only other large animals of management concern. On the other hand, in most western states, Alaska, and Canadian provinces, 3 to 10 or more big game species had to be considered. Some big game ranges in the western United States and Canada could accommodate elk, mule deer, white-tailed deer, bighorn sheep, and moose. Occasionally mountain goats, woodland caribou, black bear, mountain lion, grizzly bear, and wolves could also occur. In some other environments, especially in the south and southwest, one or more exotic big game animals could occur; in others, feral burros and horses had also become sufficiently abundant to merit concern and management attention in terms of interspecific relations.

Because of this, management in many areas had begun to trend from individual species toward management of species complexes or at landscape and ecosystem levels.

Although state and provincial agencies had matured and become reasonably effective in dealing with known biological and ecological problems and established management goals, new challenges arose. Knowledge applicable to management had increased greatly as research expanded and evolved from basic natural history and ecological studies of individual species and populations to analysis of interspecific relationships and the impacts of land use and management practices on individual species. As early as the 1960s, biologists in some areas began to question existing knowledge of factors limiting deer and other big game populations. Declines of mule deer populations across the West during the late 1960s and early 1970s were not explained by current knowledge (52). Neither were extensive fluctuations and declines in caribou herds in Alaska and Canada during the 1970s (53). Various reviews indicated reasonably accurate data on population ecology of deer and other large mammals were rare, and at least one authority believed that most of the studies done on population ecology of big game ". . . coalesce into an amorphous mass of nothing much" (54). Once again, theory and practice that evolved under one set of conditions (rapidly increasing, overabundant, and problem populations) were not generally applicable as conditions changed.

"Modern" theory that winter range or winter forage supplies were the primary factors controlling big game populations and the main tool needed to sustain abundance was adequate harvest of animals of both sexes and all ages did not always work after populations became established or declined. Questions about the ability of existing knowledge of deer–habitat relationships to explain population trends, plus the lack of close or consistent correlation between population characteristics and trend and measures of browse utilization and condition, resulted in reduced emphasis on browse and other range surveys as bases for estimating carrying capacity and determining harvest and other management needs by the late 1970s and early 1980s (55).

New needs and challenges for research and management information and for understanding population ecology and animal–habitat relationships were greatly abetted by two major technological developments during the 1960s and 1970s. The advent of miniature radio transmitters and portable receivers allowed researchers to accurately measure animal movements and other behavior, to accurately identify population units, and to monitor reproduction, mortality, and other aspects of population dynamics (Figure 15–8). Secondly, the development and adaptation of computers, especially personal computers, to biology and management during the late 1970s and 1980s greatly expanded research and management capability to analyze field data with precision and speed. Early data sets could be reexamined and reanalyzed for accuracy and result, and new hypotheses could be developed and tested with almost immediate results. Further, complex data sets and interactions among diverse factors could be examined, and models could be developed to explain population phenomena, determine alternative management strategies, and evaluate or predict management impacts. This has led managers in some areas, for some big game species, to adapt field surveys and population data collections specifically for population management models that project appropriate harvests for populations and other management units.

New challenges and horizons for big game management also developed as a result of evolving social, political, and economic factors. Beginning during the 1960s, public perceptions of wildlife and how it is treated or how animals are produced, allocated, and utilized began to change rapidly. From an early focus on sportsmen's concerns, information and education efforts funded by hunting license fees were expanded to reach nonhunters, seeking broad public appreciation for wildlife and for understanding and supporting habitat preservation and other management needs. Other groups developed broad-scale media and other efforts concerning nongame wildlife, preservation of threatened and endangered species and habitats, animal rights and welfare, effects of logging, grazing, and other land uses on wildlife habitats, and the human environment and related issues. All helped to increase public interest, concern, and involvement in wildlife conservation and management. Surveys

Figure 15–8 Development of miniature radio transmitters and portable receivers during the 1960s and 1970s led to new studies and understanding of the behavior, biology, and ecology of large mammals. (Photograph by R. J. Mackie.)

to determine sportsmen and public involvement, interests, and concerns and public hearings to seek input on management proposals were gradually incorporated into management programs.

Passage of the National Environmental Policy Act in 1969 and subsequent enactment of similar legislation in many states also increased public involvement in big game management. This act required assessment or an environmental impact statement for any federal action significantly affecting the quality of the human environment, including any significant actions involving management of public lands. Wildlife was assumed to be part of the human environment, and major wildlife and habitat management issues, including the impacts of changes in grazing management, logging and road development, and mineral development, were subjected to intense public and professional management scrutiny. Many sportsmen's groups and others used the act to become more involved in management decision making and to challenge programs or data supporting programs. Analysis and reporting of big game habitat and population impacts or the reviewing of environmental impact statements rapidly became a major task of biologists, especially in federal agencies.

The Endangered Species Preservation Acts, the National Environmental Protection Act, and other congressional acts and actions through the years served to increase the direct involvement of federal agencies and personnel in big game management. Although the Federal Land Policy and Management Act of 1976 recognized state authority over fish and resident wildlife on national resource lands, other acts and actions involving management of national forest and public rangeland resources required increasing staffs and funding directed to wildlife habitat and populations issues. Efforts to develop new management and recovery plans for grizzly bear beginning in the early 1970s and gray wolves during the late 1970s and early 1980s were established as cooperative state–federal efforts steered by interagency committees (Figure 15–9). Funding, leadership, and other aspects of the programs came under administration by the U.S. Fish and Wildlife Service. During the 1990s, these programs and management of bison in and adjacent to Yellowstone National Park have also generated questions about management authority and responsibility for control of possibly disease-carrying animals migrating into adjoining states. Issues of federal–state authority and relationships

Figure 15–9 Translocated gray wolves were released in northern Idaho. (Photograph by L. Parker. Courtesy of U.S. Fish and Wildlife Service, Helena, Montana.)

have been important historically, since the advent of conservation in the 1800s. However, it remains that governments and programs in different agencies at both levels were instrumental in bringing conservation and management to where they are at the dawn of a new millennium.

In concluding this chapter on the history of big game management, it is important to note that the vast majority of scientific knowledge about species, their habitats, and management in North America was generated during the past 50 years—concomitant with development of management structures, policies, and practices across the continent. With due recognition to the diverse programs and practices that exist elsewhere, that knowledge and the system it serves provide models for expansion of conservation and management of wildlife worldwide. The American system, based on public ownership and individual rights to hunt and otherwise enjoy its benefits, demonstrates the capability of greatly diminished populations of a diversity of species and need to recover under concerned management, even while sustaining limited harvests and on lands used intensively for other uses. Generally prosperous national and regional economies and growing human generosity toward living with wildlife, including large mammals, probably contributed significantly to this recovery. Unfortunately, history cannot precisely determine or document the roles of management policies and practices as compared with social, economic, and other environmental factors that also contributed.

SUMMARY

Chapter 15 presents an overview of the development of conservation and management theory, policy, and practice in North America. Both conservation and management emerged during the past century to promote wise use of wildlife resources and to control and maintain productivity of wildlife habitats and populations for recreational use and other human benefits. Their histories chronicle abundance and exploitation of all species through about 1850; concern, outrage, and stirring of early sportsmen and others to protect remaining stocks through the late 1800s and early 1900s; and the rise and evolution of efforts to restore and scientifically manage big game mammals through the mid- to late 20th century. Early conservation policies were largely protective, to reduce exploitation and apportion use of remaining stocks. In the early 1900s it became evident that protection alone would not restore depleted populations. New national parks and the national forest system were established, game preserves were created, predator control programs were implemented, and trapping and transplanting, winter feeding, and other practices were applied to further conservation. At the same time, foresters began promoting conservation in terms of development and use of game resources. In 1918, Aldo Leopold proposed a new professional discipline, game management, to use science to enhance production of wild game for recreational use based on sustained yield management as practiced in forestry. By then, deer populations were increasing rapidly in North America, including some national forests where efforts to reduce overbrowsing and starvation through population control were initiated. Scientific studies, including deer range and population surveys and habitat relationships of deer and elk, were founded during the late 1920s and 1930s, providing new knowledge and support for development of state programs to manage big game populations throughout North America on the basis of winter range carrying capacity and sustained yield theory. These programs grew rapidly through the 1950s and 1960s when deer populations had recovered to historical highs in distribution and abundance and most other species populations had increased to levels providing significant recreational hunting and other management opportunities. The recovery and subsequent population trends, together with changing social, economic, and political conditions, especially heightened public interest and involvement in wildlife conservation, resulted in management of big game populations becoming highly complex at the close of the 20th century. New techniques including radio telemetry, personal computers, and modeling provided new knowledge and understanding of population ecology for management; old ideas and practices were questioned and some had to be discarded. New dimensions included greater consideration for all wildlife, for more biological, ecological, and human values in management at ecosystem and landscape levels, and recovery of grizzly bear and wolves in ecosystems capable of sustaining those species.

LITERATURE CITED

1. TREFATHEN, J. B. 1964. *Wildlife Conservation and Management.* D.C. Heath and Company, Boston, MA.

2. HUDSON, 1993. Origins of wildlife management in the western world. Pages 5–21 *in* A. W. L. Hawley, ed., *Commercialization and Wildlife Management: Dancing with the Devil.* Krieger, Malabar, FL.

3. LEOPOLD, A. 1933. *Game Management.* Charles Scribner's Sons, New York.

4. BORLAND, H. 1975. *The History of Wildlife in America.* National Wildlife Federation, Washington, D.C.

5. BAILEY, J. A. 1984. *Principles of Wildlife Management.* John Wiley & Sons, New York.

6. PEEK, J. M. 1986. *A Review of Wildlife Management.* Prentice-Hall, Englewood Cliffs, NJ.

7. SHAW, J. 1985. *Introduction to Wildlife Management.* McGraw-Hill, New York.

8. GRAY, G. G. 1993. *Wildlife and People: The Human Dimensions of Wildlife Ecology.* University of Illinois Press, Urbana.

9. TREFATHEN, J. B. 1966. Wildlife regulation and restoration. Pages 16–37 *in Origins of American Conservation.* The Ronald Press, New York.

10. CAIN, S. A. 1978. Predator and pest control. Pages 379–398 *in* H. P. Brokaw, ed., *Wildlife in America.* U.S. Government Printing Office, Washington, D.C.

11. CATON, J. D. 1887. The antelope and deer of America. *Hurd Houghton,* New York.

12. GEIST, V. 1993. Great achievements, great expectations: Successes of North American wildlife management. Pages 47–72 *in* A. W. L. Hawley, ed. *Commercialization and Wildlife Management: Dancing with the Devil.* Krieger, Malabar, FL.

13. GABRIELSON, I. N. 1941. *Wildlife Conservation.* Macmillan, New York.

14. ANDERSON, C. C. 1958. The elk of Jackson Hole. Bulletin No. 10. Wyoming Game and Fish Commission, Laramie, Wyoming.

15. ROBBINS, R. L., D. E. REDFEARN, and C. P. STONE. 1982. Refuges and elk management. Pages 479–507 in J. W. Thomas and D. E. Toweill, eds., *Elk of North America: Ecology and Management.* Stackpole Books, Harrisburg, PA.

16. CARHART, A. H. 1943. Fallacies in winter feeding of deer. *Transactions of the North American Wildlife Conference* 8:333–337.

17. DOMAN, E. R., and D. I. RASMUSSEN. 1944. Supplemental winter feeding of mule deer in northern Utah. *Journal of Wildlife Management* 8: 317–338.

18. HOUSTON, D. B. 1982. *The Northern Yellowstone Elk: Ecology and Management.* Macmillan, New York.

19. BELANGER, D. O. 1988. *Managing American Wildlife: A History of the International Association of Fish and Wildlife Agencies.* University of Massachusetts Press, Amherst.

20. LEOPOLD, A. 1918. Forestry and game conservation. *Journal of Forestry* 16:404–411.

21. LEOPOLD, A. 1915. The varmint question [1915]. Pages 47–48 in S. L. Flader and J. B. Callicot, eds., 1991. *The River of the Mother of God and Other Essays by Aldo Leopold.* University of Wisconsin Press, Madison.

22. FLADER, S. L., and J. B. CALLICOTT, eds. 1991. *The River of the Mother of God and Other Essays by Aldo Leopold.* University of Wisconsin Press, Madison.

23. RASMUSSEN, D. I. 1941. Biotic communities of Kaibab Plateau, Arizona. *Ecological Monographs* 3:229–275.

24. RUSSO, J. P. 1964. The Kaibab North deer herd - Its history, problems and management. Wildlife Bulletin No. 7. Arizona Game and Fish Department, Phoenix.

25. HORNADAY, W. T. 1914. *Wildlife Conservation in Theory and Practice.* Yale University Press, New Haven, CT.

26. GRAVES, H. S., and E. W. NELSON. 1919. Our national elk herds: A program for conserving the elk on national forests about the Yellowstone National Park. Department Circulation 51. U.S. Department of Agriculture, Washington, D.C.

27. ROOSEVELT, T., T. S. VAN DYKE, D. G. ELLIOT, and A. J. STONE. 1902. *The Deer Family.* Macmillan, New York.

28. MERRILL, S. 1916. *The Moose Book.* E. P. Dutton, New York.

29. LANTZ, D. E. 1910. Raising deer and other large game animals in the United States. U.S. Bureau of Biological Survey Bulletin No. 36.

30. LANTZ, D. E. 1916. Deer farming in the United States. U.S. Department of Agriculture, Farmers Bulletin No. 330.

31. BARTLETT, I. H. 1938. Whitetails: Presenting Michigan's deer problem. Game Division Bulletin, Michigan Department of Conservation, Lansing.

32. SKINNER, M. P. 1925. Migration routes in Yellowstone Park. *Journal of Mammalogy* 6:184–192.

33. RUSH, W. M. 1932. Northern Yellowstone elk study. Montana Fish and Game Commission Bulletin, Helena.

34. RUSSELL, C. P. 1932. Seasonal migration of mule deer. *Ecological Monographs* 2:1–41.

35. DIXON, J. S. 1934. A study of the life history and food habits of mule deer in California. *California Fish and Game* 20(3 and 4):1–146.

36. ROBINSON, C. S. 1931. Feeding habits and forage requirements of Rocky Mountain mule deer in the Sierra Nevada Mountains. *Journal of Forestry* 29:557–564.

37. CASE, G. W. 1938. The influence of elk on deer populations. *Proceedings of the First and Second Idaho Game Management Conference, University of Idaho Bulletin* 33:25–27.

38. CASE, G. W. 1938. The use of salt in controlling the distribution of game. *Journal of Wildlife Management* 2:79–81.

39. DENIO, R. M. 1938. Elk and deer foods and feeding habits. *Transactions of the North American Wildlife Conference* 3:421–427.

40. GREFFENIUS, R. J. 1938. Results of the Copper Ridge Basin elk study. U.S. Department of Agriculture Forest Service, *Rocky Mountain Bulletin* 21:14–15.

41. CLIFF, E. P. 1939. Relationship between elk and mule deer in the Blue Mountains of Oregon. *Transactions of the North American Wildlife Conference* 4:560–569.

42. YOUNG, V. A., and W. L. ROBINETTE. 1939. A study of the range habits of elk on the Selway Game Preserve. University of Idaho Bulletin No. 34 (16).

43. EINARSEN, A. S. 1948. The pronghorn antelope and its management. Stackpole Books, Harrisburg, PA.

44. TRIPPENSEE, R. E. 1948. *Wildlife management: Upland Game and General Principles.* McGraw-Hill, New York.

45. MURIE, A. 1951. *The Elk of North America.* Stackpole Books, Harrisburg, PA.

46. TAYLOR, W. P., ed. 1956. *The Deer of North America: Their History and Management.* Stackpole Books, Harrisburg, PA.

47. TILLETT, P. 1963. *Doe Day: The Antlerless Deer Controversy in New Jersey.* Rutgers University Press, New Brunswick, NJ.

48. COLE, G. F. 1958. *Range Survey Guide.* Booklet. Montana Fish and Game Department, Helena.

49. MACKIE, R. J. 1974. Application of the key browse utilization-condition trend survey method in Montana, 1957–1971, and description of existing transects. Unpublished Job Progress Report, Federal Aid Project No. W-120-R-5, Job B6-201. Montana Fish and Game Department, Helena.

50. BARROWS, P., AND J. HOLMES. 1990. *Colorado's Wildlife Story.* Colorado Division of Wildlife, Denver.

51. MUSSEHL, T. W., and F. W. HOWELL. 1971. *Game Management in Montana.* Montana Fish and Game Department, Helena.

52. WORKMAN, G. W., and J. B. LOW, eds., 1976. *Mule Deer Decline in the West: A Symposium.* Utah State University College of Natural Resources, Logan.

53. KLEIN, D. R. 1987. Caribou: Alaska's wilderness nomads. Pages 191–207 *in* H. Kallman, ed., *Restoring America's Wildlife* 1937–1987. U.S. Department of the Interior, Fish and Wildlife Service, Washington, DC.

54. CAUGHLEY, G. 1980. A population of mammals. A review of *The George Reserve Deer Herd: Population Ecology of a K-Selected Species. Science* 207:1338–1339.

55. MACKIE, R. J. 1975. Evaluation of the key browse survey method. Fifth Mule Deer Workshop, Albuquerque, NM.

56. LEOPOLD, A. 1931. *Game Survey of the North Central States.* Sporting Arms and Ammunition Manufacturers' Institute, Madison, WI.

57. SIGLER, W. F. 1956. Wildlife law enforcement. 4th ed. Wm. C. Brown, Dubuque, IA.

58. USDI BUREAU OF LAND MANAGEMENT. 1962. *Historical Highlights of Public Land Management.* U.S. Government Printing Office, Washington, DC.

Wolf

Warren B. Ballard and Philip S. Gipson

INTRODUCTION

No other carnivore has caused as much controversy as wolves, the largest member of the family Canidae. Wolves have been feared by humans for centuries. When Europeans arrived in North America wolves were distributed throughout most of the continent. After decades of persecution, gray wolves were eliminated from the eastern half of the United States (except for the Great Lakes region) by about 1900, and from much of the western United States by about 1930 (1). By the 1950s gray wolves had reached their lowest numbers in North America, and within the contiguous United States, gray wolves occurred only on Isle Royale, Michigan, in northern Minnesota, and when individuals occasionally crossed into the southwest United States from Mexico (2, 3).

Gray wolf populations declined in Canada also; they were exterminated from the eastern provinces by about 1900 (4). Numbers were lowest in western Canada around 1930, but populations expanded from 1940 to 1960, as did predator control programs (4,5).

Red wolves still occurred in eastern Oklahoma, southern Arkansas, Texas, Louisiana, and southern Mississippi into Alabama in 1940 (6). Their numbers declined rapidly through the 1960s primarily due to animal damage control activities and hybridization with coyotes (7). By 1970 only isolated red wolves and red wolf × coyote hybrids occurred along the coast of southeastern Texas and possibly southwestern Louisiana.

Wolves have been sought by hunters and trappers for the value of their pelts. Until recently, pelts may have been the only value attributed to wolves by humans except for a very small minority of individuals and native peoples (8). During the 1960s human perceptions about wolves evolved with the environmental movement to include the wolf's natural intrinsic value, and the wolf was one of the first species listed under the United States Endangered Species Act (2). Wolves had value beyond pelts; scientists began to view them as important components of natural ecosystems, while others believed wolves had aesthetic, spiritual, and medical values (9, 10). Wolves became a symbol of wilderness during the 1970s, even though on a worldwide basis they were numerous relative to other carnivores (11, 12).

During the early 1980s, as translocation plans were being developed, wolves began to naturally recolonize areas where they had not existed for decades (13). Also, biologists determined that wolves did not require "wilderness" to survive, but only abundant prey and some measure of protection from human persecution (2). As we approach the 21st century the dilemma for managers will be when, where, and how wolves should be managed.

The wolf is the largest undomesticated member of the dog family (Canidae), and in North America is represented by the gray wolf and the red wolf. Wolves range in color from white to black to gray to brown with numerous shades such as "cinnamon" and "tawny" (14). There is much variation in coloration and pelage among individuals and by geographic areas. Males are generally larger than females for each age class. Adult females weigh from 18 to 55 kilograms and males from 20 to 70 kilograms depending on species or subspecies. Heavier wolves have been reported (i.e., 80 kilograms), although differences in types of scales and not knowing when wolves had eaten make such reports questionable (1, p. 69; 4). No recent wild wolves have approached 80 kilograms. The Mexican wolf is the smallest subspecies of gray wolves; females weigh from 25 to 34 kilograms and males from 31 to 41 kilograms (14, 3, 15, 16). Red wolf characteristics are intermediate between gray wolves and coyotes, and were suggested as being the only wolf species to have evolved entirely in North America (6). Wolves have 42 teeth. The upper/lower dental formula is incisors, 3/3; canines, 1/1; premolars, 4/4; and molars 2/3.

Historically, gray wolves were distributed throughout most of North America above 20 degrees North latitude (e.g., above mid-Mexico) in all habitats that supported ungulate populations (17). Currently, their distribution is limited primarily to the northern portions of the continent (Figure 16–1). Former distribution of the Mexican wolf included central and northern Mexico, west Texas, southern New Mexico, southern Arizona, and southeastern and central Mexico (3, 15). The Mexican wolf was extirpated from the wild in the United States and less than 50 breeding pairs were estimated in Mexico in 1978, but it is doubtful if any live there today (16, 15). The red wolf was distributed in the southeastern United States (Figure 16–1) but was extinct in the wild by 1980 (18). Between 1987 and 1992 captive red wolves had been released in North Carolina and Tennessee (18).

Subspecies designation of gray wolves has been controversial. Hall and Young and Goldman (based on morphological studies by Goldman and Hall and Kelson) recognized 24 subspecies of gray wolf in North America (19, 17, 20, 14, 21). However, many of these classifications were based on relatively few type specimens and without the aid of modern statistical treatment.

Wolf species relationships have been clarified with new techniques, such as molecular genetics, which examine genetic variations among individuals and populations. Wayne and others concluded there was little genetic differentiation among gray wolf populations and that gene flow was high (22). However, the Mexican wolf was slightly distinct genetically, and hybridization between gray wolves and coyotes was widespread in southeastern Canada and Minnesota. In addition, Mech, Brewster and Fritts, and Wayne and others concluded that there is very little difference in natural history or behavior among the various gray wolf subspecies and that specific species designations refer more to where the individual came from rather than differences among populations (23, 24, 22). Alternatively, Theberge proposed that the recognition of wolves in Canada be defined by prey-based or geographic-forest-based ecotypes, resulting in 10 wolf ecotypes (25). Brewster and Fritts suggested that future taxonomists may consider recognizing no subspecies of gray wolf in North America (24, p. 372). Nowak used 10 standardized measurements of 580 adult male wolf skulls to conclude there were five modern subspecies of gray wolves (Figure 16–1): Arctic Island wolf, Mexican wolf, Algonquin park wolf, Minnesota wolf, and Alaskan wolf (26).

Controversy exists on whether the red wolf is a separate species, a subspecies of gray wolf, or a gray wolf : coyote hybrid (27). Nowak and others analyzed skull morphological features and compared behavior between coyotes and red wolves (27). They concluded that the red wolf was probably a full species. Wayne and Jenks analyzed mitochondrial DNA and suggested the red wolf had obtained its genotypes from hybridization of coyotes and gray wolves, and that their results were consistent with the hybrid origin hypothesis (28). They also maintained that existing morphological data

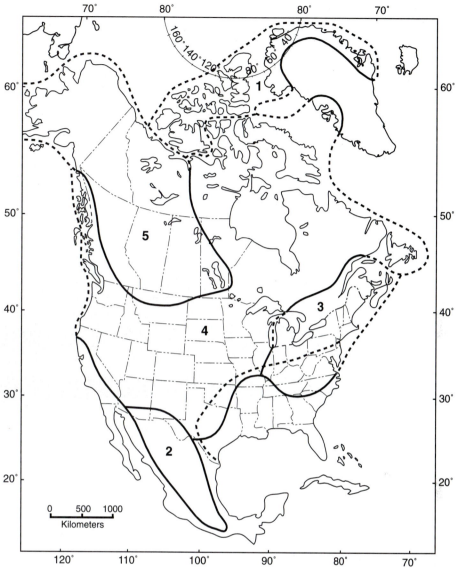

Figure 16–1 Original distribution of wolves in North America, showing the five recognized subspecies of gray wolves: (1) Arctic Island wolf, (2) Mexican wolf, (3) Algonquin Park wolf, (4) Minnesota wolf, and (5) Alaskan wolf. Red wolf historical distribution delineated by dashed line in southeastern United States. (Modified from Ref. 26.)

were consistent with a species of hybrid origin. Nowak and others indicated that the taxonomic identity of the red wolf could not be "unequivocally defined," and they supported additional studies (27, p. 415). Furthermore, because samples cannot be obtained from more than half of the red wolf's range as a result of extinction, the true genetic origin of the red wolf may never be known (29).

LIFE HISTORY

Diet

Wolves are opportunistic feeders, but their principal prey are ungulates. Deer, elk, caribou, moose, bison, and muskox are the principal species preyed on year-round by wolves, although Dall's sheep can be an important prey item (17, 9, 30, 31, 32). Hares and beavers may be important food items for some populations. Where several species of ungulates occur in the same area, wolves exhibit a preference for certain species (33). For example, Carbyn found that mule deer were preferred over elk in Jasper National Park, Canada, even though elk were more abundant (34). Similarly, white-tailed deer were taken disproportionately over elk in Riding Mountain National Park, whereas in Alaska wolves prefer caribou over moose when both are available (35, 36, 31, 37).

Wolf diets in summer are more diverse than winter diets, and smaller prey items are taken more often than during winter, but ungulates still compose the largest proportion (i.e., greater than 75%) of biomass eaten (32, 38, 39, 36, 30, 40). Although white-tailed deer are the most important prey (i.e., 39% biomass) for translocated red wolves, raccoons and marsh rabbits also are important (i.e., 39% and 11% biomass, respectively) (18) (Table 16–1).

Ungulate kill and consumption rates of wolves vary greatly among areas, seasons, and year based on prey vulnerability. Schmidt and Mech reported that gray wolf kill rates vary from 0.5 to 45.2 kilograms per wolf per day (41). Normally, however, winter kill rates range from 2.0 to 7.2 kilograms per wolf per day (Table 16–1). Generally larger wolf packs kill more prey than smaller packs, but consumption per individual wolf is lower for larger packs (42, 36, 31, 43, 41). Single wolves can and do kill large ungulates (9,44). Wolves sometimes kill in excess of what they can consume, and in such cases, prey are usually vulnerable because of age or snow conditions (45, 46, 47, 48). When prey are less abundant or less vulnerable, wolf consumption rates decline and wolves conserve energy by sleeping more and traveling less or they switch prey (49, 34, 50, 31). Ungulates most often preyed on by wolves generally are young or old individuals, those in poor condition, or those made vulnerable by deep snow conditions or disease (32, 9, 51, 52, 53, 50). Wolves prey heavily on calves or fawns, sometimes as neonates, but particularly during winter when snow is present (54, 36, 30). Ballard and Van Ballenberghe summarized several moose calf mortality studies and found that wolves killed 1.4% to 17.9% of neonates annually (55). Wolf predation can

TABLE 16–1

Ungulate Biomass Index per Wolf Versus Estimated Winter Consumption Rates of Ungulates by Wolves During Winter in North America[a]

Location	Major Prey Species	Ungulate Biomass Index/Wolf	Estimated $\bar{X}$ Mass of Wolf (kg)	Consumption Rate in Winter		Reference
				kg/wolf/day	kg/kg of wolf/day	
East-central Ontario	Deer	112	30	2.9	0.10	197
Northern Alberta	Bison	152	38	5.3	0.14	109
West-central Yukon	Sheep	153	37	3.0	0.08	198
North-central Minnesota	Deer	161	34	2.0	0.06	30
Southwest Quebec	Moose	162	30	2.2	0.07	126
Northeast Minnesota	Deer	178	30	2.9	0.10	112
Isle Royale, Michigan	Moose	225	33	7.2	0.22	54
Northeast Alberta	Moose	231	38	5.5	0.14	173
Isle Royale, Michigan	Moose	264	33	4.9	0.15	199
Northwest Alaska	Caribou, moose	267	43	5.3	0.12	31
Denali Park, Alaska	Moose	334	40	4.5	0.11	200
Southwest Manitoba	Elk	336	32	6.8	0.21	35
Kenai Peninsula, Alaska	Moose	345	40	4.8	0.12	51
Northwest Minnesota	Deer	400	32	2.9	0.09	42
South-central Alaska	Moose	659	40	7.1	0.18	36

Source: From Ref. 31 as modified from Ref. 30a.

[a]Bison, moose, elk, bighorn sheep, mountain goat, mule deer, white-tailed deer assigned relative biomass values of 8, 6, 3, 2, 1, 1, 1, respectively.

be the greatest cause of mortality to neonate caribou (56, 57, 58). Wolf predation also has resulted in significant losses to elk and deer neonates (59, 9, 34, 60). Interestingly, because the eastern wolf subspecies is relatively small, white-tailed deer are their preferred prey species and moose are not regularly killed. The few moose that are preyed on may have been predisposed due to winter tick infestations or infections with meningeal worm (61, 62). However, eastern wolves frequently scavenge moose carcasses (62).

The degree to which wolves prey on ungulates that are predisposed to die from various causes has been widely debated. Biologists have frequently relied on bone marrow fat to assess the physical condition of ungulate prey at death (63). However, marrow fat does not appear to be a very reliable indicator of condition and may be only a good predictor that an animal was close to starvation (64, 65). Mech suggested that a high proportion of prey, if not all, possess some trait that puts them at a disadvantage and that these traits are subtle and not easily measured (66). Little is known about what traits predispose neonates to wolf predation, but Adams and others indicated that caribou calves born following severe winters were lighter than normal and suffered higher mortality rates from wolf predation (58). Mech and others maintain that there is a "grandmother" effect on ungulate survival whereby the nutritional health of an individual's grandmother affects survival (67). Further, Mech and others and McRoberts and others suggested that the cumulative effect of three consecutive severe winters resulted in adult females that were in relatively poor condition and produced calves in poor condition with lower survival (68, 69). Messier challenged their hypothesis and suggested there was no carryover effect (70, 71).

Reproductive Strategy

The pack is the basic social unit of wolves and generally consists of a breeding pair (i.e., alpha male and female) and subordinates (i.e., primarily offspring from previous years) (9). Social standing influences which animals breed in a wolf pack. Lower ranking adults usually do not reproduce due to social suppression rather than lack of reproductive development (72, 73). Wolves become sexually mature at 22 months of age but reproduction has been documented in 9- to 10-month-old captive wolves (74). Although wolves exhibit monogamous and polygamous mating systems, monogamy is more common in wild wolves (75).

Estrus lasts 5 to 7 days or longer, and occurs during January through April depending on latitude; in southern latitudes breeding is initiated earlier (76). Gestation lasts about 63 days and litters are usually composed of 4 to 6 (with a range of 1 to 11) pups (9, 77). Usually only one litter is produced per pack but multiple litters are not uncommon (Figure 16–2). Harrington and others estimated that the frequency of multiple litters in wild wolves may range from 20% to 40% (75). However, in four Alaskan studies, the frequency of multiple litters varied by population from 0% to about 13%, and Ballard and others suggested that Harrington's estimate was too high (51; 36, p. 27; 78; 79). Ballard and others also suggested that the frequency of multiple litters may have been a compensatory mechanism for low survival rates, but Hayes hypothesized that multiple litters were the result of social disruption caused by human harvest (36, 43). However, multiple litters also occurred at a rate of about 13% in Denali National Park where wolves were not exploited by humans, and Mech and others attributed multiple litters to favorable food conditions (79, 80). Wolf productivity appears to be highly correlated with available ungulate biomass per wolf (81) (Table 16–1).

Wolf summer activities are centered around den and rendezvous sites (i.e., resting areas) and the raising of pups. Both adult and subadult wolves participate in the care and feeding of wolf pups. Weaning usually occurs about 2 weeks after den emergence (9, p. 143). Pups are fed by regurgitation of prey items while at den and rendezvous sites. Den site attendance by pack members is highly variable (82, 83).

Harrington and others suggested that pup survival was positively correlated with pack size (84). However, several other studies have found no such relationship (51, 36, 43). In fact, accounts of single adults raising wolf litters have been reported (83, 85).

Figure 16–2 Wolf summer activities are centered around den and rendezvous sites. Usually only one litter is produced per pack, but multiple litters do occur. Pups first emerge from dens at 16 to 21 days of age. (Photograph by J. Foster.)

Wolf pairs usually are unrelated (86). Because wolf packs have been thought to be composed mostly of family members, concerns have been expressed that genetic inbreeding may occur (87). However, high dispersal rates, acceptance of foreign wolves into existing packs, formation of new packs, and pack splitting all suggest that inbreeding is not a problem except in isolated populations such as Isle Royale (88, 42, 89, 36, 31, 79, 90). However, even at Isle Royale the wolf population has persisted for 48 years and is again increasing (R. O. Peterson, School of Forestry and Wood Products, Michigan Technological University, personal communication). Lehman and others found that wolves from different packs had high levels of genetic similarity and that the degree of relatedness varied by population (91). Also, some packs contained members that were not offspring of the breeding pair.

Sex Ratios

Fuller reported that sex ratios of wild adult (greater than 1 year old) wolves often were skewed in favor of males with the percentage of females ranging from 11% to 58% (77). Hayes and others reported that sex ratios favored females in a Yukon study area while Mech reported a higher proportion of male pups in a high-density Minnesota population with declining prey (53, 92). Fuller suggested that pup sex ratios were less skewed at birth, but that the mechanism by which initial or subsequent sex ratios occurred was unknown (77).

Longevity

Little is known about the longevity of wild wolves. Mech reported one wolf that lived for 13 years and suggested that wolves remained reproductively active until at least 11 years of age (93). W. B. Ballard (unpublished data) reported two Alaskan wolves that lived to 14 years of age. The oldest known captive wolf lived 16 years (1).

Physiology and Growth

Pups are born blind and deaf and weigh about 500 grams (9, p. 123; 94). Eyes open at 11 to 15 days. Pups first emerge from dens at 16 to 21 days of age and weaning occurs at about 35 days of age (9, p. 143). At 16 to 26 weeks deciduous teeth are replaced and winter pelage becomes apparent (95). Mech summarized tooth development and replacement in captive wolves (9, p. 140). In Alaska, male and female wolves attain most of their weight by 11 and 23 months of age, respectively (31). Sensory organs (e.g., vision, auditory system, taste, smell) were reviewed by Asa and Mech, but most information was derived from the study of dogs (96).

BEHAVIOR

Social Structure and Communication

Wolf packs have an elaborate social structure that usually includes an alpha male and female and their subordinate offspring (e.g., beta males and females and other lower ranking members). Social interactions among pack members result in a hierarchy. A wolf's position in the pack is determined by sex, age, and social interactions (97). The alpha pair is the nucleus of the pack, and often they are the only wolves that breed, but as mentioned earlier, multiple litters are more common than previously thought.

Packs have a wide range of ceremonies that are used to reaffirm dominance relationships. Most studies of dominance relationships have been conducted on captive packs. Zimen identified 48 different social behavioral patterns (97). Tail posture, facial movements, body posture, baring of teeth, head and ear movements, and other behaviors convey the intentions, status, and mood of individual pack members (98).

Howling is the type of communication for which wolves are most known, although several other types of vocalizations occur such as whimpering, growling, and barking (99). There is considerable variation in the strength and intensity of howls by individual wolves, and howling can be heard throughout the year. The main purposes of howling are to assemble pack members and to advertise the presence of a territory (9). Wolf howls can be heard by other wolves from 6.4 to 9.6 kilometers (100). Scent marking also is an important method of advertising territorial boundaries (101).

Den Sites and Activity Patterns

Wolves den in a wide variety of sites including hollow logs, rock caves, dug out beaver lodges, earthen dens, and dens of other species (32, 17, 99, 9, 102, 54, 103, 104). Most often, wolves enlarge red fox dens or dig their own. Most den site preparation is done by alpha females (9). Den sites frequently occur on elevated knolls or on sides of hills and are near water. Entrance tunnels consist of one or more holes from 35 to 63 centimeters in width (9). Many pack territories contain several alternate den sites that are used as rendezvous sites also.

A pack occupies its den site for several weeks until midsummer when adults move pups to a rendezvous site. Pups are left unattended at den and rendezvous sites for varying periods of time (105, 82, 83). By mid- to late autumn pups begin traveling with adults on a regular basis.

Den sites are frequently reused during many years (32, 9, 104, 78). Mech and Packard reported one case on Ellesmere Island where radiocarbon dating bones suggested that one den site had possibly been used for 700 years or more (106).

Movements

Wolves are capable of moving long distances (Figure 16–3). Movements can be classified as daily movements, extraterritorial forays, migrations, and dispersal. Daily movement rates range from 1.6 to 72 kilometers (107, 38, 108, 109). In northwest Alaska, wolves were more active during summer when they attended den and rendezvous sites than during winter (110).

Extraterritorial forays typically consist of movements of single wolves temporarily leaving territories for varying periods, directions, and distances, but returning to the original territory (42, 111, 31). All sexes and age classes make extraterritorial

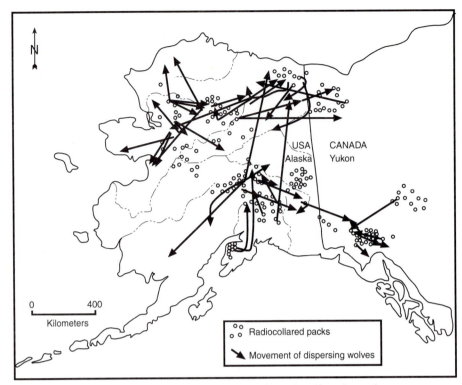

Figure 16–3 General location of radiocollared wolf packs and dispersal movements within Alaska and Yukon Territory from 1975 to 1991. (Data sources include Refs. 203, 113, 36, 31, 51, 201, 202, 67, 53.)

movements that may occur at any time of the year and occur in most, if not all, wolf populations. Messier found that rates of extraterritorial movements were higher for adults while Fuller found higher rates for yearlings (111, 30). Extraterritorial forays may be stimulated by interpack conflicts, low prey availability, and breeding activity, but most appear to be related to dispersal (30, 112, 51, 42, 111, 36, 30).

Dispersal occurs year-round, in all directions, and by all sex and age classes. Highest dispersal rates occur among yearling, single wolves, but dispersal of groups and family groups has been observed (113, 31, 80). Dispersal distances are highly variable and have been documented up to 917 kilometers (114). Fuller summarized results from several areas of North America and found that 16% to 83% of yearlings dispersed (77).

Rate of dispersal depends on many factors including wolf and prey densities, pack size and composition, territory sizes, social stresses, and behavior (Figure 16–4) (89, 115, 30, 116, 31). Dispersal is the primary mechanism by which wolves repopulate areas from which they have been extirpated, and this movement results in genetic exchange among packs and regions (79, 117). The success of dispersal depends on the availability of vacant territories and competition for prey resources (42, 111, 115, 30, 116).

Migratory movements usually consist of entire packs leaving their original territory to follow migratory prey to wintering areas, and then returning to their original territory for denning. Depending on resident prey availability, wolves will follow migratory caribou or deer to wintering areas outside their home territories (115, 118, 31). Ballard and others found that 11% of the wolf packs in northwest Alaska migrated from 64 to 272 kilometers each year, but the packs that migrated varied each year (31). Forbes and Theberge found that wolves traveled up to 35 kilometers from resident territories to follow yarding white-tailed deer (118). Spatial and social relationships of wolves while on migratory winter range are poorly understood.

Behavior

Figure 16–4 Dispersal occurs year-round, in all directions, and by all sex and age classes. Highest dispersal rates occur among yearling, single wolves, but dispersal of groups and family groups has been observed. (Photograph by W. B. Ballard.)

Territoriality

Wolf packs usually maintain year-round territories, although overlaps frequently occur along territorial boundaries. Where pack territories are relatively small, overlap is slight to nonexistent (30). In areas with relatively large territories, overlaps appear to be the result of spatial and temporal movement patterns, and the methods used to plot territory boundaries (36, 80).

Territory sizes (excluding extraterritorial and migratory movements) generally range from 125 to 4,312 square kilometers (Figure 16–5) (112, 31). Nondenning male packs occupy larger areas, but it is unclear whether these areas are defended. Generally, wolf packs that depend on deer have relatively small territories, whereas those that depend on larger migratory prey have larger territories (4). Winter territories are usually larger than summer territories, presumably because denning and prey availability restrict wolf movements, although much variation exists (49, 31, 30, 53). Annual territory sizes are positively correlated with pack size and number of animal relocations (36, 30, 53). Generally, 30 to 60 relocations of radiomarked wolf packs are required per year to adequately assess territory sizes (42, 35, 115, 36, 119).

POPULATION DYNAMICS

Population Trends and Current Status

Wolf population trends in North America up to 1992 (excluding Mexico where few, if any, wolves continue to survive) were provided by Miller, Hayes and Gunson, Stephenson and others, and Theil and Ream (120, 121, 117, 13). Approximately 59,500 to 69,070 wild gray wolves occurred in North America by the early 1990s. Gray wolf densities ranged from 3 to 43 wolves per 1,000 square kilometers (77). Approximately 19 red wolves were free ranging as of 1992 (18).

Figure 16–5 Wolf packs occupy territories ranging in size from 125 to 4,312 square kilometers. Territory sizes are correlated with pack size. (Photograph by W. B. Ballard.)

Wolves are still distributed over much of their historical range in Alaska. During winter 1989–1990 the Alaska population was estimated at 5,900 to 7,200 wolves within 700 to 900 packs (117). Alaska's wolf population is expected to remain stable or increase in future years.

Hayes and Gunson estimated Canadian gray wolf numbers in 1992 at 52,000 to 60,000 (121). They concluded that the wolf's future in Canada was secure because of limited control programs to reduce livestock depredations, and a reluctance of the government to initiate wolf control programs to increase native ungulates. These developments have allowed wolves to increase and they continue to reoccupy vacant habitats. Excluding provinces where wolves are already extinct, no management agencies reported that wolves were in danger of extinction.

By 1995, 1,740 to 2,030 wolves existed in Minnesota, and an additional 100 to 120 occurred in small breeding populations in Michigan, Montana, Wisconsin, and possibly Washington (122, 13). Smaller numbers of wolves also were reported in the Dakotas, Idaho, and Wyoming. As of August 1996, there were at least 76 wolves in Yellowstone National Park and Idaho as a result of translocation (123). Mexican wolves have been released within the Blue Range Primitive Area of eastern Arizona with a population goal of 100 free-ranging wolves. In addition, Theil and Ream have proposed that, despite many problems, wolves will continue to naturally reoccupy areas from which they were previously exterminated (13).

Rates of Increase and Survival

Wolf populations have been thought to exhibit numerical (i.e., changes in wolf density in response to changes in prey density) and functional (i.e., changes in the number and kind of prey items eaten per predator at different prey densities) responses. Acting in conjunction they constitute the predation rate that was thought to regulate the numbers of wolves and number of prey (124, 125, 126, 30, 127).

Wolves have tremendous potential to increase depending on availability of prey. Reported annual finite rates of increase (λ) have ranged from 0.40 to 2.53 with the highest rates occurring where wolf populations were recovering from control programs and prey availability was relatively high (77, 43, 36, 43).

Wolves recolonize areas primarily through dispersal, pair-bonding with another dispersing wolf, and subsequent reproduction. Pack splitting at high pack numbers, pack budding, and high reproductive rates allow wolf populations to increase quickly (43, 79).

Annual survival rates vary by population, with pups having lower survival rates than adults. Ranges of annual survival rates by age class are as follows: pups, 0.13 to 0.70; yearlings, 0.26 to 1.00; and adults, 0.32 to 0.89 (51, 36, 31, 30, 43). Dispersing wolves had lower survival rates than residents in some studies, whereas others could not detect such differences (51, 115, 30, 53, 31).

Mortality Factors

Most mortality in wild wolf populations occurs during winter or periods of snow cover. Little is known about the causes of early pup mortality. Pups raised to 3 weeks old, when they first appear at the den entrance, tend to survive at least through summer, except in part of Minnesota, where canine parvovirus is a problem (51, 36, 2, 80, 112, 128). Excluding populations that have deliberately been reduced, the largest cause of mortality to most unprotected wolf populations is hunting and trapping (Table 16–2). In these populations hunting and trapping account for 69% to 92% of mortalities to radiocollared wolves greater than 4 months old. In areas where wolves receive partial protection such as parks (i.e., protected while in the park, but seasonally migrate out), human-caused mortality is still the primary cause of death. Intraspecific strife and other natural mortality factors are most important in parks where packs remain within the area and in remote areas that receive little human use. For example, Meier and others attributed 52% of radiocollared wolf mortalities to other wolves in Denali National Park, Alaska (79). They concluded that such intraspecific strife was a normal consequence of wolf territoriality in the absence of human intervention. Other sources of mortality to North American wolves include predation by brown and black bears, injuries from prey, disease, starvation, automobile collisions, old age, and other miscellaneous factors.

The effects of diseases and parasites on wolf population dynamics are poorly understood and largely unstudied (129). Mortality of wolves from rabies, canine distemper, canine parvovirus, blastomycosis, tuberculosis, and mange have been reported, and in some cases, epizootics were associated with population declines (129). Rabies was found to be a limiting factor for wolves in northwest Alaska where it caused a population decline and greatly diminished the capacity of the wolf population to rebound following winter mortality (31, 130). Rabies was an important mortality factor in the Great Lakes region also (131). Canine parvovirus has influenced wolf numbers in Minnesota, Wisconsin, and Montana (132, 128, 133, 134). In addition, other diseases and parasites such as infectious canine hepatitis, oral papillomatosis, brucellosis, leptospirosis, lyme disease, tularemia, dog heartworm, dog hookworm, liver fluke, hydatid tapeworm, dog lice, and mites affect wolf populations and may pose a risk to their prey species and humans (129).

Mech originally proposed that harvest " . . . seems to stimulate both reproduction and pup survival . . . " (9, p. 61). Mech proposed that wolf populations could withstand losses of 50% annually (9, p. 64). Peterson and others pointed out that managers interpreted this as meaning annual harvests could approach 50 percent of the autumn populations (51). Since that time considerable effort has been expended to determine a sustainable harvest level. Keith estimated that sustainable exploitation could vary from 23% to 38% of early winter numbers (125). Fuller summarized data from nine studies across North America and concluded that, on average, a wolf population could sustain an overall overwinter mortality rate of 35% (30, p. 25). Wolf populations could be sustained if human-caused mortality did not exceed 28%. However, the amount of mortality that a wolf population can withstand is variable and depends on factors such as pack sex and age composition, productivity, multiple litters, and density.

TABLE 16–2
Cause of Mortality (%) to Radiocollared or Tagged Gray Wolves from Several North American Study Areas, 1969–1994

Area	Status[a]	Period	Human-Caused Mortality				Wolves	Rabies	Starved	Other			n	Reference
			Auto	Trap	Shot	Snare				Prey	Natural	Unknown		
Denali, Alaska	P	1986–1992			9.7[b]		51.6				38.7		31	79
Algonquin, Ontario	P[c]	1987–1992	3.4	20.7	24.1		3.4	20.7		3.4		24.1	29	118
NE Minnesota	C	1969–1972	21.6		72.5[b]		2.0					3.9	51	39, 112
NW Minnesota	H	1972–1974			80.0[b]		20.0						5	42
NW Minnesota	P	1974–1977			66.7						33.3[d]		9	42
Kenai, Alaska	H	1976–1981		2.6	44.7	44.7	1.9		2.6		5.3[e]		38	51
NW Alaska	H	1987–1992			69.2[f]		1.9	21.2	5.8		1.9		52	31
SC Alaska	H	1975–1982	1.4	9.9	70.4[b]		11.3		1.4	1.4		4.2	71	36
Yukon	P[g]	1990–1994		3.7	11.1						85.2		27	43
SE Quebec	H	1980–1984	18.2	31.8	4.5		18.2		27.3				22	115

[a]P, protected; C, partially protected; H, open hunting and trapping seasons.
[b]Exact cause not reported but definitely human-caused and not auto.
[c]Wolves followed migratory deer out of park.
[d]Unknown, probably natural.
[e]Canine distemper.
[f]Thirty-four of 36 shot from snowmobile; 2 with aid of aircraft.
[g]Open to hunting and trapping, but remote.

Estimates of allowable exploitation based on the relationship between exponential rate of increase of individual packs and mortality rate also should incorporate sex and age composition of the pack. Ballard and others determined that small packs (i.e., fewer than six individuals) could sustain higher rates of mortality than large packs (i.e., more than seven wolves) (31). Pairs of adults have high reproductive potential, and unless large packs have multiple litters, they can not compensate for similar mortality rates. Wolf packs remained stable when total overwinter mortality was equivalent to about 53% in northwest Alaska, similar to Mech's original estimate (31, 9).

Relationships with Other Predators

Peterson reviewed interspecific competition among wolves and small canids, particularly coyotes and red foxes (135). Wolves kill coyotes and foxes but usually do not eat them. More often, coyotes and foxes appear to avoid contact with wolves if possible. When wolves are exterminated coyote populations increase. Although coyotes and wolves are sympatric in many areas, coyotes are often excluded from wolf pack territories (135). There is no evidence of fox population declines due to wolves. Peterson suggested that food was the principal factor that influenced competition among wolves, coyotes, and foxes (135).

Wolves frequently interact with black bears and brown bears at kill and den sites. Wolves sometime kill bears and bears sometimes kill wolves, but these deaths are not important to the population dynamics of either species (136, 137, 138). However, brown bears frequently usurp wolf kills and this could increase wolf kill rates. It is likely that wolves occasionally interact with other predator species, but such occurrences are probably rare and insignificant.

Wolf-Prey Relationships

Considerable research during the past 25 years has documented the effects of wolves on prey populations (Figure 16–6). Whether wolves regulate or limit prey population growth has been debated in the literature. However, misuse of terminology, different interpretations of data [see Bergerud and others and Bergerud and Snider versus Thompson and

Figure 16–6 Most wolf packs occupy year-round territories, but some packs that prey largely on migratory caribou may migrate with the caribou. (Photograph by W. B. Ballard.)

Peterson; Van Ballenberghe, and Eberhardt and Pitcher versus Bergerud and Ballard; Keith versus Theberge; Mech and others and McRoberts and others versus Messier (1991)], and lack of appropriate hypothesis testing have limited the benefits of this debate (70, 139, 140, 141, 142, 55, 143, 145, 146, 147, 148, 149, 150, 125, 151, 68, 69, 70, 141).

Several terms have consistently been misused and appropriate definitions are required to understand predator–prey relationships. First, any factor that reduces the growth rate of a prey species is a limiting factor, which includes density-dependent and density-independent processes (152, 139, 70). Regulating factors (i.e., factors that keep a prey population in equilibrium) are a subset of limiting factors and are solely density dependent. Thus, by definition, any factor that reduces a prey population's rate of growth is a limiting factor. Secondly, biologists have frequently referred to predation rate when they were actually discussing killing rate. Predation rate is the product of both numerical and functional responses (71).

Several prey models have been proposed to describe predator–prey relationships. Some models have focused primarily on moose–wolf relationships using data from Isle Royale to test hypotheses. Although Isle Royale data sets have provided a wealth of valuable information, this island is a relatively simple predator–prey system and is not representative of predator–prey systems on mainland areas. More typical predator–prey systems involve wolves, one or both species of bears, and multiple species of ungulates and alternate prey (55). Most efforts to understand and quantify predator–prey relationships focused on moose–predator relationships primarily because both species inhabit environments that have received less impact from humans, and are probably more reflective of natural ecosystems. Boutin in a review of Ballard and Larsen and Van Ballenberghe concluded that the evidence that predation by wolves and bears was acting as a major limiting factor in most moose populations was less than convincing, and there was little in the way of experimental evidence that predation regulated population growth (141, 153, 154). Four basic models have been proposed to explain moose–predator relationships: low-density equilibria, multiple stable states, stable-limit cycles, and recurrent fluctuations (154, 141, 142, 55). The low-density equilibria hypothesis predicts that ungulate populations are regulated to low levels by density-dependent predation. The multiple stable states model predicts that ungulate populations are regulated by density-dependent predation at low ungulate densities but that food competition regulates ungulates at high densities. The stable-limit cycle hypothesis predicts that predation is density dependent during ungulate population increases and inversely density dependent during ungulate population declines, while the recurrent fluctuations model predicts that predation is inversely density dependent at high moose densities and that predation is not regulatory.

Messier reviewed data from 27 studies of wolf predation and concluded that wolf predation regulated moose populations at low moose densities (i.e., from 200 to 400 per 1,000 square kilometers) but was not regulatory above those densities (127). Hayes added several data points to Messier's analyses and concluded that if wolves regulated moose population growth, it occurred at moose densities of 70 to 120 moose per 1,000 square kilometers (43). Both analyses assumed that there was a strong wolf population functional response. Recently, Ballard and Van Ballenberghe questioned whether the traditional functional response is appropriate for wolf predation, and argued that the density of vulnerable prey, rather than absolute prey numbers, may be more important (155). Although the evidence that predation regulates population growth is equivocal, all authors concur that additional hypothesis testing is necessary. The low-density equilibria and the recurrent fluctuations models appear to be supported by existing empirical, experimental, and predator reduction studies (156, 157, 142, 158, 55, 155).

When managing ungulates it may not matter whether predation regulates prey population growth if reductions in limiting factors result in the desired increases in prey population yields (142). Gasaway and others demonstrated that where predator populations were naturally regulated, prey population densities and their harvest yields were substantially lower than areas where populations of wolves and bears were manipulated by humans (157). Because of differences in predator and prey densities, differences in weather patterns, and most importantly, differences in the magnitude of human impacts on habitats, predation may significantly limit prey population growth and prey populations may mimic any of the proposed population models (142).

Wolf predation on domestic livestock was the primary justification for passage of the 1931 Animal Damage Control Act authorizing the "eradication, suppression, or bringing under control . . . of wolves, coyotes . . . and other animals" (Public Law 776, 71st Congress). Ironically, wolves were already gone from many western states by the time the act was passed, and the emphasis of the predator control programs shifted from wolves to coyotes (3, 159, 160, 161). Wolves no doubt preyed on domestic livestock during the early 1900s, but only a few brief field studies were conducted, and most of the evidence against wolves consisted of reports by ranchers who perceived that wolves caused the damage (162, 163, 160). Recent studies in Minnesota and the western provinces of Canada suggest that livestock made up a relatively small portion of the diets of wolves, even in areas where the ranges of wolves overlap farms with cattle and other domestic animals (164, 165, 166, 167). In Alberta the summer diet of wolves consisted of cervids in preference to cattle, and during the winter cervids constituted 93% of prey occurrences (165). Wolves in Minnesota averaged killing only 32 cattle and 41 sheep per year from 1987 to 1989. During this period there were approximately 1,200 wolves and 6,800 farms in the wolf range that contained about 230,000 cattle and 90,000 sheep. White-tailed deer were the primary prey, but wolves also consumed moose, beaver, snowshoe hares, and occasionally livestock (164).

Seton and Young provided numerous accounts of famous wolves and their purported damage (168, 1, 160). However, Gipson and Ballard suggested that the accuracy of many accounts should be questioned and the importance of domestic livestock to wolves at that time was not known (169).

Population and Habitat Management

The density of wolf populations is often estimated by researchers and managers on an annual or occasional basis using a variety of methods. Harvest statistics, howling surveys, hunter observations to assess trends, a variety of aerial reconnaissance surveys, radiotelemetry, ground surveys, past studies, and, recently, line-intercept track sampling have been used to survey wolves (170, 82, 171, 157, 171, 36, 119, 121, 121, 172). However, there is no consistency in methods used among agencies (121). All of the methods have advantages and disadvantages, but most estimates are not objective or contain no measure of precision (172). Aerial reconnaissance surveys and line-intercept track sampling from aircraft when snow conditions are adequate appear to be the best methods when large, relatively open areas must be surveyed (172).

Ages of gray wolves have been assessed by a number of different methods including tooth wear and replacement, epiphyseal closure of longbones, dentine width and root closure, and counts of cementum annuli (39, 173, 42, 36, 170, 174, 175, 176). Of those methods, counts of cementum annuli probably offer the most accurate and cost-effective method available (176).

Current management of wolves consists of regulation of legal harvest, protection from harvest, translocation, and wolf population reduction. Exports and imports of wolves to the United States are governed by the Convention on International Trade of Endangered Species. Within the United States, Alaska is the only state that permits public harvest of wolves. Since the wolf was declared a game species in 1960, Alaska has maintained an elaborate system for monitoring harvest. Wolf hunting and trapping seasons are set annually for each of 26 game management units by the Alaska Board of Game. Current regulations require that each harvested wolf be inspected by Alaska Department of Game and Fish officials or their designee. At the time of examination, color, sex, age class (i.e., pup or adult), method of take (i.e., trapped, shot, snared), and date and location of harvest are recorded and each hide is fitted with a metal tag. Export of wolves from the state requires an export permit. Compliance with these regulations is good in many areas of the state, but compliance by rural subsistence users is poor (31).

Wolf hunting and trapping regulations vary by game management unit, but generally they are liberal because harvests are relatively low and well within sustained yield levels. Annual harvests fluctuate depending on snow conditions and other weather factors, and range from about 700 to 1,600. Until eliminated by public referendum in 1996, Alaska allowed an individual using an aircraft for access to shoot wolves the same day they were airborne provided they were at least 100 meters from the aircraft when a wolf was taken.

Hunting and trapping seasons in Canada vary by province and territory and were summarized by Hayes and Gunson (121). Indigenous people are allowed to hunt wolves without restriction. Wolves are not hunted in Saskatchewan or the Okanagan region of British Columbia. Trapping harvests are generally better recorded than hunting harvests. Yukon and Ontario require that wolf hides be presented to officials for "sealing" with a tag, while in other areas harvests are monitored by questionnaire and auction sale records. The wolf is the only big game animal in Canada that is hunted year-round, has no bag limits in many areas, and does not require a hunting license (121). A minimum of 4,053 wolves are killed annually in Canada (121).

Wolves receive partial or full protection from hunting and trapping within various parks and special use areas in Canada and the United States. Although none of these areas was established specifically for the protection of wolves, wolves receive protection along with other species. However, some parks may be too small to harbor viable populations of wolves, and their seasonal movements from these areas subject wolves to human-caused mortality. In such cases, management may consist of an addition of wolf management buffer zones that offer further protection (177).

Within park and translocation areas, managers sometimes place human-use restrictions on activities surrounding den sites out of fear that these activities would disrupt denning activities (178, 179, 180). Restrictions range from 100 meters to 1.6 kilometers and often are controversial (181). However, recent observations of wolves adapting to close, disruptive human presence indicate that many of these restrictions may be conservative (181).

Several studies indicated a negative relationship between road densities and the presence of wolves (30, 182, 183, 184, 185). However, Mech suggested that wolves would occupy roaded areas if protected from human persecution (2). Mladenoff and others and Person and others suggested that roaded areas are only occupied after roadless areas are saturated (186, 185). Person and others recommended controlling access on roads following logging operations to increase protection for wolves (185).

Wherever wolf and livestock distributions overlap, wolf control is a part of management (187). Annual wolf depredations in North America range from 0.23 to 3.0 per 1,000 cattle, and 2.66 per 1,000 sheep (167, 188, 164, 189). Management includes the removal of damaging wolves, education of livestock owners, and compensation to owners (167, 164, 187). Bangs and others suggested that the wolf depredations feared by ranchers in the western United States (e.g., in areas where there are no wolves) may have been due to local cultural values, traditions, legends concerning wolves in the early 1900s, and dislike of the Endangered Species Act (i.e., government intervention and outside influence) (187). Historical and current beliefs about depredating wolves were largely influenced by the accounts of individual wolves by Seton and Young (168, 1, 160). Gipson and Ballard suggested that many of these early accounts were inaccurate and, in some cases, fabricated (169). Nevertheless, current wolf management efforts appear to be contributing to the recovery of gray wolves in the western United States (187).

Wolf control to enhance native ungulate populations has become extremely controversial among some parts of society. Wolf control consists of lethal and nonlethal methods that were summarized by Cluff and Murray and Boertje and others (190, 191). Within recent years Alaska, Yukon, and British Columbia have conducted wolf control programs to benefit ungulates (157, 156, 192, 36, 53, 158). All of these programs have increased ungulate survival, increased human harvest of prey species, and maintained viable wolf populations. In spite of these management successes, wolf control remains controversial, and objectionable to some (193, 117).

In an effort to reach consensus on wolf management among interested groups, a number of state, provincial, and territorial agencies have initiated planning strategies centered on establishing predator and prey management goals and objectives, and the establishment of wolf management zones (194, 195, 196). Management in these proposed zones would range from the complete protection of wolves, to wolf control that would benefit ungulate populations. In spite of these efforts, segments of society remain polarized on this issue, and continued debate and litigation are likely.

With good planning and proper management, gray wolf populations will continue to expand and thrive in areas where they once had been extirpated. Translocations of Mexican and red wolves likely will be successful. Successful wolf translocations will allow the wolf to be removed from the endangered species list. However, once wolf populations recover they will need to be managed and, in some cases, controlled.

SUMMARY

When, where, and how gray wolves and red wolves should be managed are complex issues facing natural resource managers. Gray wolves occurred over most of North America, except the present southeastern United States, until the late 1800s. By 1950, wolves were extirpated over the contiguous United States except for northern Minnesota, Isle Royale, and isolated areas in the southwestern states. Red wolves survived in southeastern Texas and adjacent Louisiana until about 1970. Gray and red wolves have been translocated to limited areas within their former ranges and recovery plans call for additional translocations. Gray wolves have naturally recolonized parts of some northern states.

Molecular biology studies have helped clarify the taxonomic relationships of wolves. Genetically, there appears to be little difference between gray wolves from across North America, although Mexican wolves appear to be distinct from other gray wolves. Recent studies based on morphology suggest that there are five modern subspecies of gray wolves. Controversy exists on whether the red wolf is a separate species, a subspecies of gray wolf, or a gray wolf × coyote hybrid. The true genetic origin of the red wolf may never be known because no specimens were preserved before red wolves became extinct in more than half of their range.

Most studies of wolf and prey relationships have focused on responses of moose populations to predation by wolves. Predation may regulate moose when moose densities are low. Prey densities and their harvest yields may be substantially lower in areas where wolves are naturally regulated than in similar areas where wolves are controlled. Fear of wolf predation on domestic livestock has been the primary justification for control of wolves. Early accounts of damaging wolves contributed to fears of serious losses of livestock to wolves. Researchers question the accuracy of many early accounts of damaging wolves and suggest that losses of livestock to wolves may be lower than previously reported.

Wolf populations have potential for rapid increase, depending on mortality and availability of prey. The largest cause of mortality in unprotected wolf populations is hunting and trapping. Territorial disputes are important causes of mortality in areas where wolves have little interaction with humans. Other sources of mortality include predation by bears, injuries from prey, disease, and starvation.

LITERATURE CITED

1. YOUNG, S. P. 1944. Their history, life habits, economic status, and control. Part 1. Pages 1–386 *in* S. P. Young and E. A. Goldman, eds., *The Wolves of North America.* Dover, New York.
2. MECH, L. D. 1995. The challenge and opportunity of recovering wolf populations. *Conservation Biology* 9:270–278.

3. BROWN, D. E. 1983. *The Wolf in the Southwest: The Making of an Endangered Species.* University of Arizona Press, Tucson.

4. CARBYN, L. N. 1987. Gray wolf and red wolf. Pages 359–376 *in* M. Novak, J. A. Baker, M. E. Obbard, and B. MALLOCH, eds., *Wild Furbearer Management and Conservation in North America.* Ontario Ministry of Natural Resources, Toronto.

5. NOWAK, R. M. 1983. A perspective on the taxonomy of wolves in North America. Pages 10–19 *in* L. N. Carbyn, ed., *Wolves in Canada and Alaska: Their Status, Biology, and Management.* Report Serial No. 45. Canadian Wildlife Service. Ottawa, Canada.

6. NOWAK, R. M. 1979. *North American Quaternary Canis.* Monograph 6. University of Kansas Museum of Natural History, Lawrence, KS.

7. McCARLEY, H. 1979. Recent changes in distribution and status of wild red wolves (*Canis rufus*). Endangered Species Report No. 4. U.S. Fish and Wildlife Service. Washington, DC.

8. STEPHENSON, R. O. 1982. Nunamiut Eskimos, wildlife biologists and wolves. Pages 434–440 *in* F. H. Harrington and P. C. Paquet, eds., *Wolves of the World: Perspectives of Behavior, Ecology, and Conservation.* Noyes Publications, Park Ridge, NJ.

9. MECH, L. D. 1970. *The Wolf: The Ecology and Behavior of an Endangered Species.* Natural History Press, Garden City, NY.

10. VEST, J. H. C. 1988. The medicine wolf returns: Traditional Blackfeet concepts of *Canis lupus. Western Wildlands* 14:28–33.

11. THEBERGE, J. B. 1975. *Wolves and Wilderness.* J. M. Dent and Sons, Toronto, Ontario, Canada.

12. HARRINGTON, F. H., and P. C. PAQUET, eds. 1982. *Wolves of the World: Perspectives of Behavior, Ecology, and Conservation.* Noyes Publications, Park Ridge, NJ.

13. THEIL, R. P., and R. R. REAM. 1995. Status of the gray wolf in the lower 48 United States to 1992. Pages 59–62 *in* L. N. Carbyn, S. H. Fritts, and D. R. Seip, eds., *Ecology and Conservation of Wolves in a Changing World.* Canadian Circumpolar Institute, University of Alberta, Edmonton.

14. GOLDMAN, E. A. 1944. The wolves of North America. Part 2. Classification of wolves. Pages 389–636 *in* S. P. Young and E. A. Goldman, eds., *The Wolves of North America.* Dover, New York.

15. PARSONS, D. R., and J. E. NICHOLOPOULOS. 1995. Status of the Mexican wolf recovery program in the United States. Pages 141–146 *in* L. N. Carbyn, S. H. Fritts, and D. R. Seip, eds., *Ecology and Conservation of Wolves in a Changing World.* Canadian Circumpolar Institute, University of Alberta, Edmonton.

16. McBRIDE, R. T. 1980. The Mexican wolf (*Canis lupus baileyi*): A historical review and observations on its status and distribution. Endangered Species Report No. 8. U.S. Fish and Wildlife Service.

17. YOUNG, S. P., and E. A. GOLDMAN. 1944. *The Wolves of North America.* Dover, New York.

18. PHILLIPS, M. K., R. SMITH, V. G. HENRY, and C. LUCASH. 1995. Red wolf reintroduction program. Pages 157–168 *in* L. N. Carbyn, S. H. Fritts, and D. R. Seip, eds., *Ecology and Conservation of Wolves in a Changing World.* Canadian Circumpolar Institute, University of Alberta, Edmonton.

19. HALL, E. R. 1981. *The Mammals of North America.* 2nd ed. John Wiley and Sons, New York.

20. GOLDMAN, E. A. 1937. The wolves of North America. *Journal of Mammalogy* 18:37–45.

21. HALL, E. R., and K. R. KELSON. 1959. *The Mammals of North America.* Vol. 2. The Ronald Press, New York.

22. WAYNE, R. K., N. LEHMAN, and T. K. FULLER. 1995. Conservation genetics of the gray wolf. Pages 399–407 *in* L. N. Carbyn, S. H. Fritts, and D. R. Seip, eds., *Ecology and Conservation of Wolves in a Changing World.* Canadian Circumpolar Institute, University of Alberta, Edmonton.

23. MECH, L. D. 1991. Returning the wolf to Yellowstone. Pages 309–322 *in* R. B. Keiter and M. S. Boyce, eds., *The Greater Yellowstone Ecosystem: Redefining America's Wilderness Heritage.* Yale University Press, New Haven, CT.

24. BREWSTER, W. G., and S. H. FRITTS. 1995. Taxonomy and genetics of the gray wolf in western North America: A review. Pages 353–373 *in* L. N. Carbyn, S. H. Fritts, and D. R. Seip, eds., *Ecology and Conservation of Wolves in a Changing World.* Canadian Circumpolar Institute, University of Alberta, Edmonton.

25. THEBERGE, J. B. 1991. Ecological classification, status, and management of the gray wolf, *Canis lupus,* in Canada. *Canadian Field-Naturalist* 105:459–463.

26. NOWAK, R. M. 1995. Another look at wolf taxonomy. Pages 375–397 *in* L. N. Carbyn, S. H. Fritts, and D. R. Seip, eds., *Ecology and Conservation of Wolves in a Changing World.* Canadian Circumpolar Institute, University of Alberta, Edmonton.

27. Nowak, R. M., M. K. Phillips, V. G. Henry, W. C. Hunter, and R. Smith. 1995. The origin and fate of the red wolf. Pages 409–415 in L. N. Carbyn, S. H. Fritts, and D. R. Seip, eds., *Ecology and Conservation of Wolves in a Changing World.* Canadian Circumpolar Institute, University of Alberta, Edmonton.

28. Wayne, R. K., and S. M. Jenks. 1991. Mitochondrial DNA analysis implying extensive hybridization of the endangered red wolf *Canis rufus. Nature (London)* 351:565–568.

29. Henry, V. G. 1992. Finding on a petition to delist the red wolf (*Canis rufus*). *Federal Register* 57:1246–1250.

30. Fuller, T. K. 1989. Population dynamics of wolves in north-central Minnesota. *Wildlife Monograph* 105.

31. Ballard, W. B., L. A. Ayres, P. R. Krausman, D. J. Reed, and S. G. Fancy. 1997. Ecology of wolves in relation to a migratory caribou herd in northwest Alaska. *Wildlife Monograph* 135.

32. Murie, A. 1944. *The Wolves of Mount McKinley.* Fauna Serial No. 5. U.S. National Park Service, Washington, D.C.

33. Bergerud, A. T. 1990. Rareness as an antipredator strategy to reduce predation risk. *Congress of International Game Biologist* 19:15–25.

34. Carbyn, L. N. 1975. Wolf predation and behavioral interactions with elk and other ungulates in an area of high prey diversity. Ph.D. thesis. University of Toronto, Ontario, Canada.

35. Carbyn, L. N. 1983. Wolf predation on elk in Riding Mountain National Park, Manitoba. *Journal of Wildlife Management* 47:963–976.

36. Ballard, W. B., J. S. Whitman, and C. L. Gardner. 1987. Ecology of an exploited wolf population in south-central Alaska. *Wildlife Monograph* 98.

37. Dale, B. W., L. G. Adams, and R. T. Bowyer. 1995. Winter wolf predation in a multiple prey system. Pages 223–230 in L. N. Carbyn, S. H. Fritts, and D. R. Seip, eds., *Ecology and Conservation of Wolves in a Changing World.* Canadian Circumpolar Institute, University of Alberta, Edmonton.

38. Mech, L. D. 1966. *The Wolves of Isle Royale. Fauna of the National Parks of the United States.* Fauna Serial No. 7. U.S. National Park Service, Washington, D.C.

39. Van Ballenberghe, V., A. W. Erickson, and D. Byman. 1975. Ecology of the timber wolf in northeastern Minnesota. *Wildlife Monograph* 43.

40. Weaver, J. L. 1993. Refining the equation for interpreting prey occurrence in gray wolf scats. *Journal of Wildlife Management.* 57:534–538.

41. Schmidt, P. A., and L. D. Mech. 1997. Wolf pack size and food acquisition. *American Naturalist* 150:513–517.

42. Fritts, S. H., and L. D. Mech. 1981. Dynamic, movements, and feeding ecology of a newly protected wolf population in northwestern Minnesota. *Wildlife Monograph* 80.

43. Hayes, R. D. 1995. Numerical and functional responses of wolves, and regulation of moose in the Yukon. M.S. thesis. Simon Fraser University, Vancouver, British Columbia.

44. Smith, T. G. 1980. Hunting, kill, and utilization of a caribou by a single gray wolf. *Canadian Field-Naturalist* 94:175–177.

45. Mech, L. D., and L. D. Frenzel, Jr. 1971. Ecological studies of the timber wolf in northeastern Minnesota. Research Paper No. NC-52. U.S. Department of Agriculture Forest Service NCFES, St. Paul, MN.

46. Peterson, R. O., and D. L. Allen. 1974. Snow conditions as a parameter in moose–wolf relationships. *Naturalist Canadian* 101:481–492.

47. Eide, S. H., and W. B. Ballard. 1982. Apparent case of surplus killing of caribou by gray wolves. *Canadian Field-Naturalist* 96:87–88.

48. Miller, F. L., A. Gunn, and E. Broughton. 1985. Surplus killing as exemplified by wolf predation on newborn caribou. *Canadian Journal of Zoology* 63:295–300.

49. Mech, L. D. 1977. Population trend and winter deer consumption in a Minnesota wolf pack. Pages 55–83 in R. L. Phillips and C. Jonkel, eds., *Proceedings of the 1975 Predator Symposium.* Montana Forest and Conservation Experiment Station, University of Montana, Missoula.

50. Mech, L. D., T. J. Meier, J. W. Burch, and L. G. Adams. 1995. Patterns of prey selection by wolves in Denali National Park, Alaska. Pages 231–249 in L. N. Carbyn, S. H. Fritts, and D. R. Seip, eds., *Ecology and Conservation of Wolves in a Changing World.* Canadian Circumpolar Institute, University of Alberta, Edmonton.

51. PETERSON, R. O., J. D. WOOLINGTON, and T. N. BAILEY. 1984. Wolves of the Kenai Peninsula, Alaska. *Wildlife Monograph* 88.

52. LANKESTER, M. W. 1987. Pests, parasites and diseases of moose (*Alces alces*) in North America. *Swedish Wildlife Research Supplement* 1:461–489.

53. HAYES, R. D., A. M. BAYER, and D. G. LARSON. 1991. Population dynamics and prey relationships of an exploited and recovering wolf population in southern Yukon. Final Report. No. TR-91-1. Yukon Fish and Wildlife Branch.

54. PETERSON, R. O. 1977. Wolf ecology and prey relationships on Isle Royale. Science Monograph Serial No. 11. U.S. National Park Service, Washington, DC.

55. BALLARD, W. B., and V. VAN BALLENBERGHE. 1997. Predator–prey relationships. Pages 247–274 *in* C. C. Schwartz and A. W. Franzmann, eds., *Ecology and Management of the North American Moose.* Smithsonian Institution Press, Washington, DC.

56. MILLER, F. L., and E. BROUGHTON. 1974. Calf mortality on the calving grounds of the Kaminuriak caribou. Report Serial No. 26. Canadian Wildlife Service. Ottawa, Canada.

57. MILLER, F. L., E. BROUGHTON, and A. GUNN. 1983. Mortality of newborn migratory barren-ground caribou calves, Northwest Territories, Canada. *Acta Zoologica Fenn.* 175:155–156.

58. ADAMS, L. G., B. W. DALE, and L. D. MECH. 1995. Wolf predation on caribou calves in Denali National Park, Alaska. Pages 245–260 *in* L. N. Carbyn, S. H. Fritts, and D. R. Seip, ed., *Ecology and Conservation of Wolves in a Changing World.* Canadian Circumpolar Institute, University of Alberta, Edmonton.

59. PIMLOTT, D. H. 1967. Wolf predation and ungulate populations. *American Zoologist* 7:267–278.

60. KUNKEL, K. E., AND L. D. MECH. 1995. Wolf and bear predation on white-tailed deer fawns in northeastern Minnesota. *Canadian Journal of Zoology* 72:1557–1565.

61. WHITLAW, H. A., AND M. W. LANKESTER. 1994. The co-occurrence of moose, white-tailed deer, and *Parelaphostrongylus tenuis* in Ontario. *Canadian Journal of Zoology* 72:819–825.

62. FORBES, G. J., and J. B. THEBERGE. 1996. Response by wolves to prey variation in central Ontario. *Canadian Journal of Zoology* 74:1511–1520.

63. BALLARD, W. B. 1995. Bone marrow fat as an indicator of ungulate condition—how good is it? *Alces.* 31:105–109.

64. MECH, L. D., and G. D. DELGIUDICE. 1985. Limitations of the marrow fat technique as an indicator of body condition. *Wildlife Society Bulletin* 13:204–206.

65. WATKINS, B. E., J. H. WHITMAN, D. E. ULLREY, D. J. WATKINS, and J. M. JONES. 1991. Body composition and condition evaluation of white-tailed deer fawns. *Journal of Wildlife Management* 55:39–51.

66. MECH, L. D. 1996. A new era for carnivore conservation. *Wildlife Society Bulletin* 24:397–401.

67. MECH, L. D., M. E. NELSON, and R. E. MCROBERTS. 1991. Effects of maternal and grandmaternal nutrition on deer mass and vulnerability to wolf predation. *Journal of Mammalogy* 72:146–151.

68. MECH, L. D., R. E. MCROBERTS, R. O. PETERSON, and R. E. PAGE. 1987. Relationships of deer and moose populations to previous winters' snow. *Journal of Animal Ecology* 56:615–627.

69. MCROBERTS, R. E., L. D. MECH, and R. O. PETERSON. 1995. The cumulative effect of consecutive winters' snow depth on moose and deer populations: A defense. *Journal of Animal Ecology* 64:131–135.

70. MESSIER, F. 1991. The significance of limiting and regulating factors on the demography of moose and white-tailed deer. *Journal of Animal Ecology* 60:377–393.

71. MESSIER, F. 1995. On the functional and numerical responses of wolves to changing prey density. Pages 187–198 *in* L. N. Carbyn, S. H. Fritts, and D. R. Seip, eds., *Ecology and Conservation of Wolves in a Changing World.* Canadian Circumpolar Institute, University of Alberta, Edmonton.

72. PACKARD, J. M., L. D. MECH, and U. S. SEAL. 1983. Population regulation in wolves. Pages 43–53 *in* L. N. Carbyn, ed., *Wolves in Canada and Alaska: Their Status, Biology, and Management.* Report Serial No. 45. Canadian Wildlife Service, Quebec.

73. PACKARD, J. M., U. S. SEAL, L. D. MECH, and E. D. PLOTKA. 1985. Causes of reproductive failure in two family groups of wolves (*Canis lupus*). *Z. Tierpsychologhy* 68:24–40.

74. MEDJO, D. C., and L. D. MECH. 1976. Reproductive activity in nine- and ten-month-old wolves. *Journal of Mammalogy* 57:406–408.

75. HARRINGTON, F. H., P. C. PAQUET, J. RYON, and J. C. FENTRESS. 1982. Monogamy in wolves: A review of the evidence. Pages 209–222 *in* F. H. Harrington and P. C. Paquet, eds., *Wolves of the World: Perspectives of Behavior, Ecology, and Conservation.* Noyes Publications, Park Ridge, NJ.

76. MECH, L. D. 1974. *Canis lupus.* Mammalian Species 37:1–6.

77. FULLER, T. K. 1995. Comparative population dynamics of North American wolves and African wild dogs. Pages 325–328 *in* L. N. Carbyn, S. H. Fritts, and D. R. Seip, eds., *Ecology and Conservation of Wolves in a Changing World.* Canadian Circumpolar Institute, University of Alberta, Edmonton.

78. BALLARD, W. B. 1993. Demography, movements, and predation rates of wolves in northwest Alaska. Ph.D. dissertation. University of Arizona, Tucson.

79. MEIER, T. J., J. W. BURCH, L. D. MECH, and L. G. ADAMS. 1995. Pack structure and genetic relatedness among wolf packs in a naturally-regulated population. Pages 293–302 *in* L. N. Carbyn, S. H. Fritts, and D. R. Seip, eds., *Ecology and Conservation of Wolves in a Changing World.* Canadian Circumpolar Institute, University of Alberta, Edmonton.

80. MECH, L. D., L. G. ADAMS, T. J. MEIER, J. W. BURCH, and B. W. DALE. 1998. *The Wolves of Denali.* University of Minnesota Press, Minneapolis.

81. BOERTJE, R. D., and R. O. STEPHENSON. 1992. Effects of ungulate availability on wolf reproductive potential in Alaska. *Canadian Journal of Zoology* 70:2441–2443.

82. HARRINGTON, F. H., and L. D. MECH. 1982. Patterns of den attendance in two Minnesota wolf packs. Pages 81–105 *in* F. H. Harrington and P. C. Paquet, eds., *Wolves of the World: Perspectives of Behavior, Ecology, and Conservation.* Noyes Publications, Park Ridge, NJ.

83. BALLARD, W. B., L. A. AYRES, C. L. GARDNER, and J. W. FOSTER. 1991. Den site activity patterns of gray wolves, *Canis lupus,* in southcentral Alaska. *Canadian Field-Naturalist* 105:497–504.

84. HARRINGTON, F. H., L. D. MECH, and S. H. FRITTS. 1983. Pack size and wolf pup survival: Their relationship under varying ecological conditions. *Behavioral Ecology and Sociobiology* 13:19–26.

85. BOYD, D. K., and M. D. JIMENEZ. 1994. Successful rearing of young by wild wolves without mates. *Journal of Mammalogy* 75:14–17.

86. SMITH, D., T. J. MEIER, E. GEFFEN, L. D. MECH, L. G. ADAMS, and J. W. BURCH. 1997. Is incest common in gray wolf packs? *Behavioral Ecology* 8:384–391.

87. WOOLPY, J. H., and I. ECKSTRAND. 1979. Wolf pack genetics, a computer simulation with theory. Pages 206–224 *in* E. Klinghammer, ed., *The Behavior and Ecology of Wolves.* Garland Publications, New York.

88. ROTHMAN, R. J., and L. D. MECH. 1979. Scent-marking in lone wolves and newly-formed pairs. *Animal Behavior* 27:750–760.

89. VAN BALLENBERGHE, V. 1983. Extraterritorial movements and dispersal of wolves in southcentral Alaska. *Journal of Mammalogy* 64:168–171.

90. WAYNE, R. K., D. A. GILBERT, N. LEHMAN, K. HANSEN, A. EISENHAWER, D. GIRMAN, R. O. PETERSON, L. D. MECH, P. J. GOGAN, U. S. SEAL, and R.. J. KRUMENAKER. 1991. Conservation genetics of the endangered Isle Royale gray wolf. *Conservation Biology* 5:41–51.

91. LEHMAN, N. A., P. CLARKSON, L. D. MECH, T. J. MEIER, and R. K. WAYNE. 1992. A study of the genetic relationships within and among wolf packs using DNA fingerprinting and mitochondrial DNA. *Behavior Ecology and Sociobiology* 30:83–94.

92. MECH, L. D. 1975. Disproportionate sex ratios in wolf pups. *Journal of Wildlife Management* 39:737–740.

93. MECH, L. D. 1988. Longevity in wild wolves. *Journal of Mammalogy* 69:197–198.

94. RUTTER, R. J., and D. H. PIMLOTT. 1968. *The World of the Wolf.* J. B. Lippincott Co., Philadelphia, PA.

95. VAN BALLENBERGHE, V., and L. D. MECH. 1975. Weights, growth, and survival of timber wolf pups in Minnesota. *Journal of Mammalogy* 56:44–63.

96. ASA, C. S., and L. D. MECH. 1995. A review of the sensory organs in wolves and their importance to life history. Pages 287–291 *in* L. N. Carbyn, S. H. Fritts, and D. R. Seip, eds., *Ecology and Conservation of Wolves in a Changing World.* Canadian Circumpolar Institute, University of Alberta, Edmonton.

97. ZIMEN, E. 1982. A wolf pack sociogram. Pages 282–322 *in* F. H. Harrington and P. C. Paquet, eds., *Wolves of the World: Perspectives of Behavior, Ecology, and Conservation.* Noyes Publications, Park Ridge, NJ.

98. SCHENKEL, R. 1947. Expression studies of wolves. *Behaviour* 1:81–129.

99. JOSLIN, P. W. B. 1966. Summer activities of two timber wolf (*Canis lupus*) packs in Algonquin Park. M.S. thesis. University of Toronto, Ontario, Canada.

100. HARRINGTON, F. H., and L. D. MECH. 1979. Wolf howling and its role in territory maintenance. *Behaviour* 68:207–249.

101. PETERS, R. P., and L. D. MECH. 1975. Scent-marking in wolves: A field study. *American Scientist* 63:628–637.

102. PETERSON, R. O. 1974. Wolf ecology and prey relationships on Isle Royal. Thesis, Purdue University, Lafayette, IN.

103. ALLEN, D. C. 1979. *Wolves of Minong.* Houghton Mifflin Company, Boston, MA.

104. BALLARD, W. B., and J. R. DAU. 1983. Characteristics of gray wolf, *Canis lupus,* den, and rendezvous sites in southcentral Alaska. *Canadian Field-Naturalist* 97:299–302.

105. CHAPMAN, R. C. 1977. The effects of human disturbance on wolves (*Canis lupus* L.). Master's thesis. University of Alaska, Fairbanks.

106. MECH, L. D., and J. M. PACKARD. 1990. Possible use of wolf, *Canis lupus,* den over several centuries. *Canadian Field-Naturalist* 104:484–485.

107. BURKHOLDER, B. L. 1959. Movements and behavior of a wolf pack in Alaska. *Journal of Wildlife Management* 23:1–11.

108. MECH, L. D., L. D. FRENZEL, JR., R. R. REAM, and J. W. WINSHIP. 1971. Movements, behavior, and ecology of timber wolves in northeastern Minnesota. Pages 1–34 *in* L. D. Mech, and L. D. Frenzel, Jr., eds., *Ecological Studies of the Timber Wolf in Northeastern Minnesota.* Forest Service Research Paper No. NC–52 . U.S. Department of Agriculture, NCFES, St. Paul, MN.

109. OOSENBRUG, S. M., and L. N. CARBYN. 1982. Winter predation on bison and activity patterns of a wolf pack in Wood Buffalo National Park. Pages 43–53 *in* F. H. Harrington and P. C. Paquet, eds., *Wolves of the World: Perspectives of Behavior, Ecology, and Conservation.* Noyes Publications, Park Ridge, NJ.

110. FANCY, S. G., and W. B. BALLARD. 1995. Monitoring wolf activity by satellite. Pages 329–333 *in* L. N. Carbyn, S. H. Fritts, and D. R. Seip, eds., *Ecology and Conservation of Wolves in a Changing World.* Canadian Circumpolar Institute, University of Alberta, Edmonton.

111. MESSIER, F. 1985. Solitary living and extraterritorial movements of wolves in relation to social status and prey abundance. *Canadian Journal of Zoology* 63:503–512.

112. MECH, L. D. 1977. Productivity, mortality, and population trends of wolves in northeastern Minnesota. *Journal of Mammalogy* 58:559–574.

113. BALLARD, W. B., R. FARNELL, and R. O. STEPHENSON. 1983. Long distance movement by gray wolves (*Canis lupus*). *Canadian Field-Naturalist* 97:333.

114. FRITTS, S. H. 1983. Record dispersal by a wolf from Minnesota. *Journal of Mammalogy* 64:166–167.

115. MESSIER, F. 1985. Social organization, spatial distribution, and population density of wolves in relation to moose density. *Canadian Journal of Zoology* 63:1068–1077.

116. GESE, E. M., and L. D. MECH. 1991. Dispersal of wolves (*Canis lupus*) in northeastern Minnesota, 1969–1989. *Canadian Journal of Zoology* 69:2946–2955.

117. STEPHENSON, R. O., W. B. BALLARD, C. A. SMITH, and K. RICHARDSON. 1995. Pages 43–54 *in* L. N. Carbyn, S. H. Fritts, and D. R. Seip, eds., *Ecology and Conservation of Wolves in a Changing World.* Canadian Circumpolar Institute, University of Alberta, Edmonton.

118. FORBES, G. J., and J. B. THEBERGE. 1995. Influences of a migratory deer herd on wolf movements and mortality in and near Algonquin Park, Ontario. Pages 303–313 *in* L. N. Carbyn, S. H. Fritts, and D. R. Seip, eds., *Ecology and Conservation of Wolves in a Changing World.* Canadian Circumpolar Institute, University of Alberta, Edmonton.

119. FULLER, T. K., and W. J. SNOW. 1988. Estimating wolf densities from radiotelemetry data. *Wildlife Society Bulletin* 16:367–370.

120. MILLER, F. L. 1995. Status of wolves on the Canadian Arctic Islands. Pages 35–42 *in* L. N. Carbyn, S. H. Fritts, and D. R. Seip, eds., *Ecology and Conservation of Wolves in a Changing World.* Canadian Circumpolar Institute, University of Alberta, Edmonton.

121. HAYES, R. D., and J. R. GUNSON. 1995. Status and management of wolves in Canada. Pages 21–34 *in* L. N. Carbyn, S. H. Fritts, and D. R. Seip, eds., *Ecology and Conservation of Wolves in a Changing World.* Canadian Circumpolar Institute, University of Alberta, Edmonton.

122. MECH, L. D., D. H. PLETSCHER, and C. J. MARTINKA. 1995. Gray wolves. Pages 98–100 *in* E. T. LaRoe, G. S. Farris, C. E. Puckett, P. D. Doran, and M. J. Mac, eds., *Our Living Resources: A Report to the Nation on the Distribution, Abundance, and Health of U.S. Plants, Animals, and Ecosystems.* U.S. Department of the Interior, National Biological Service, Washington, DC.

123. BANGS, E. E., and S. H. FRITTS. 1996. Reintroducing the gray wolf to central Idaho and Yellowstone National Park. *Wildlife Society Bulletin* 24:402–413.

124. PACKARD, J. M., and L. D. MECH. 1980. Population regulation in wolves. Pages 135–150 *in* M. N. Cohen, R. S. Malpass, and H. G. Klein, eds., *Biosocial Mechanisms of Population Regulation.* Yale University Press, New Haven, CT.

125. KEITH, L. B. 1983. Population dynamics of wolves. Pages 66–77 *in* L. N. Carbyn, ed., *Wolves in Canada and Alaska: Their Status, Biology, and Management.* Report Serial No. 45. Canadian Wildlife Service, Quebec.

126. MESSIER, F., and M. CRÊTE. 1985. Moose–wolf dynamics and the natural regulation of moose populations. *Oecologia (Berlin)* 65:503–512.

127. MESSIER, F. 1994. Ungulate population models with predation: A case study with the North American moose. *Ecology* 75:478–488.

128. MECH, L. D., and S. M. GOYAL. 1993. Canine parvovirus effect on wolf population change and pup survival. *Journal of Wildlife Diseases* 29:330–333.

129. BRAND, C. J., M. J. PYBUS, W. B. BALLARD, and R. O. PETERSON. 1995. Infectious and parasitic diseases of the gray wolf and their potential effects on wolf populations in North America. Pages 419–430 *in* L. N. Carbyn, S. H. Fritts, and D. R. Seip, eds., *Ecology and Conservation of Wolves in a Changing World.* Canadian Circumpolar Institute, University of Alberta, Edmonton.

130. BALLARD, W. B., and P. R. KRAUSMAN. 1997. Occurrence of rabies in wolves of Alaska. *Journal of Wildlife Diseases* 33:242–245.

131. THEBERGE, J. B., G. J. FORBES, I. K. BARKER, and T. BOLLINGER. 1994. Rabies in wolves of the Great Lakes region. *Journal of Wildlife Diseases* 30:563–566.

132. MECH, L. D., S. M. GOYAL, C. N. BOTA, and U. S. SEAL. 1986. Canine parvovirus infection in wolves (*Canis lupus*) from Minnesota. *Journal of Wildlife Diseases* 22:104–106.

133. WYDEVEN, A. P., R. N. SCHULTZ, and R. P. THIEL. 1995. Monitoring of a recovering gray wolf population in Wisconsin, 1979–1991. Pages 147–156 *in* L. N. Carbyn, S. H. Fritts, and D. R. Seip, eds., *Ecology and Conservation of Wolves in a Changing World.* Canadian Circumpolar Institute, University of Alberta, Edmonton.

134. JOHNSON, M. R., D. K. BOYD, and D. H. PLETSCHER. 1994. Serology of canine parvovirus and canine distemper in relation to wolf (*Canis lupus*) pup mortalities. *Journal of Wildlife Diseases* 30:270–273.

135. PETERSON, R. O. 1995. Wolves as interspecific competitors in canid ecology. Pages 315–324 *in* L. N. Carbyn, S. H. Fritts, and D. R. Seip, eds., *Ecology and Conservation of Wolves in a Changing World.* Canadian Circumpolar Institute, University of Alberta, Edmonton.

136. BALLARD, W. B. 1980. Brown bear kills gray wolf. *Canadian Field-Naturalist* 94:91.

137. HOREJSI, B. L., G. E. HORNBECK, and R. M. RAINE. 1984. Wolves, *Canis lupus,* kill female black bear, *Ursus americanus,* in Alberta. *Canadian Field-Naturalist* 98:368–369.

138. BALLARD, W. B. 1982. Gray wolf–brown bear relationships in the Nelchina Basin of south-central Alaska. Pages 71–80 *in* F. H. Harrington and P. C. Paquet, eds., *Wolves of the World: Perspectives of Behavior, Ecology, and Conservation.* Noyes Publications, Park Ridge, NJ.

139. SINCLAIR, A. R. E. 1991. Science and the practice of wildife management. *Journal of Wildlife Management* 55:767–773.

140. SKOGLAND, T. 1991. What are the effects of predators on large ungulate populations? *Oikos* 61:401–411.

141. BOUTIN, S. 1992. Predation and moose population dynamics: A critique. *Journal of Wildlife Management* 56:116–127.

142. VAN BALLENBERGHE, V., and W. B. BALLARD. 1994. Limitation and regulation of moose populations: The role of predation. *Canadian Journal of Zoology* 72:2071–2077.

143. BERGERUD, A. T., W. WYETT, and J. B. SNIDER. 1983. The role of wolf predation in limiting a moose population. *Journal of Wildlife Management* 47:977–988.

144. BERGERUD, A. T., and J. B. SNIDER. 1988. Predation in the dynamics of moose populations: A reply. *Journal of Wildlife Management* 52:559–564.

145. THOMPSON, I. D., and R. O. PETERSON. 1988. Does wolf predation alone limit the moose population in Pukaskwa Park?: A comment. *Journal of Wildlife Management* 52:556–559.

146. VAN BALLENBERGHE, V. 1985. Wolf predation on caribou: The Nelchina herd case history. *Journal of Wildlife Management* 49:711–720.

147. VAN BALLENBERGHE, V. 1989. Wolf predation on the Nelchina caribou herd: A comment. *Journal of Wildlife Management* 53:243–250.

148. EBERHARDT, L. L., and K. W. PITCHER. 1992. A further analysis of the Nelchina caribou and wolf data. *Wildlife Society Bulletin* 20:385–395.

149. BERGERUD, A. T., and W. B. BALLARD. 1988. Wolf predation on caribou: The Nelchina herd case history: A different interpretation. *Journal of Wildlife Management* 52:344–357.

150. BERGERUD, A. T., and W. B. BALLARD. 1989. Predation on the Nelchina caribou herd: A reply. *Journal of Wildlife Management* 53:251–259.

151. THEBERGE, J. B. 1990. Potentials for misinterpreting impacts of wolf predation through prey:predator ratios. *Wildlife Society Bulletin* 18:188–192.

152. SINCLAIR, A. R. E. 1989. Population regulation in animals. Pages 197–241 *in* J. M. Cherrett, ed., *Ecological Concepts: The Contribution of Ecology to an Understanding of the Natural World.* Blackwell Science Publications, Oxford, U.K.

153. BALLARD, W. B., and D. G. LARSEN. 1987. Implications of predator–prey relationships to moose management. *Swedish Wildlife Research Supplement* 1:581–602.

154. VAN BALLENBERGHE, V. 1987. Effects of predation on moose numbers: A review of recent North American studies. *Swedish Wildlife Research Supplement* 1:431–460.

155. BALLARD, W. B., and V. VAN BALLENBERGHE. 1998. Moose–predator relationships: Research and management needs. *Alces* 34:91–105.

156. GASAWAY, W. C., R. D. BOERTJE, D. V. GRANDGAARD, D. G. KELLYHOUSE, R. O. STEPHENSON, and D. G. LARSEN. 1992. The role of predation in limiting moose at low densities in Alaska and Yukon and implications for conservation. *Wildlife Monograph* 120.

157. GASAWAY, W. C., R. O. STEPHENSON, J. L. DAVIS, P. E. K. SHEPHERD, and O. E. BURRIS. 1983. Interrelationships of wolves, prey, and man in interior Alaska. *Wildlife Monograph* 84.

158. BOERTJE, R. D., P. VALKENBERG, and M. E. MCNAY. 1996. Increases in moose, caribou, and wolves following wolf control in Alaska. *Journal of Wildlife Management* 60:474–489.

159. DAY, A. M., and A. P. NELSON. 1928. Wildlife conservation and control in Wyoming under the leadership of the United States Biological Survey. *U.S. Biological Survey, Predatory Animal and Rodent Control Office,* Cheyenne, WY.

160. YOUNG, S. P. 1970. *The Last of the Loners.* Macmillan Press, New York.

161. DUNLAP, T. R. 1984. Values for varmits. *Pacific Historical Review* LlII:141–161.

162. BAILEY, V. 1907. Wolves in relation to stock, game, and the national forest reserves. Forest Service Bulletin 72:1–31. U.S. Department of Agriculture, Washington, DC.

163. BELL, W. B. 1921. Hunting down stock killers. Pages 289–300 *in* L. C. Everard, ed., *Yearbook of Agriculture 1920.* U.S. Department of Agriculture, Washington, DC.

164. FRITTS, S. H., W. J. PAUL, L. D. MECH, and D. P. SCOTT. 1992. Trends and management of wolf-livestock conflicts in Minnesota. Resource Publication No. 181. U.S. Fish and Wildlife Service, Washington, DC.

165. BJORGE, R. R., and J. R. GUNSON. 1983. Wolf predation of cattle on the Simonette River pastures in northwestern Alberta. Pages 106–111 *in* L. N. Carbyn, ed., *Wolves in Canada and Alaska: Their Status, Biology, and Management.* Report Serial No. 45. Canadian Wildlife Service.

166. BJORGE, R. R., and J. R. GUNSON. 1985. Evaluation of wolf control to reduce cattle predation in Alberta. *Journal of Range Management* 38:483–487.

167. GUNSON, J. R. 1983. Wolf predation of livestock in western Canada. Pages 102–105 *in* L. N. Carbyn, ed., *Wolves in Canada and Alaska: Their Status, Biology, and Management.* Report Serial No. 45. Canadian Wildlife Service, Quebec.

168. SETON, E. T. 1929. *Lives of Game Animals* Vol. 1, Part 1, pp. 313–316. Charles T. Branford, Boston, MA.

169. GIPSON, P. S., and W. B. BALLARD. 1998. Accounts of famous North American wolves. *Canadian Field-Naturalist* 112:724–739.

170. RAUSCH, R. A. 1967. Some aspects of the population ecology of wolves, Alaska. *American Zoologist* 7:253–265.

171. Crête, M., and F. Messier. 1987. Evaluation of indices of gray wolf, *Canis lupus,* density in hardwood-conifer forests of southwestern Quebec. *Canadian Field-Naturalist* 101:147–152.

172. Ballard, W. B., M. E. McNay, C. L. Gardner, and D. J. Reed. 1995. Use of line-intercept track sampling for estimating wolf densities. Pages 469–480 *in* L. N. Carbyn, S. H. Fritts, and D. R. Seip, eds., *Ecology and Conservation of Wolves in a Changing World.* Canadian Circumpolar Institute, University of Alberta, Edmonton.

173. Fuller, T. K., and L. B. Keith. 1980. Wolf population dynamics and prey relationships in northeastern Alberta. *Journal of Wildlife Management* 44:583–602.

174. Parker, G. R., and J. W. Maxwell. 1986. Identification of pups and yearling wolves by dentine width in the canine. *Arctic* 39:180–181.

175. Goodwin, E. A., and W. B. Ballard. 1985. Use of tooth cementum for age determination of gray wolves. *Journal of Wildlife Management* 49:313–316.

176. Ballard, W. B., G. M. Matson, and P. R. Krausman. 1995. Comparison of two methods to age gray wolf teeth. Pages 455–460 *in* L. N. Carbyn, S. H. Fritts, and D. R. Seip, eds., *Ecology and Conservation of Wolves in a Changing World.* Canadian Circumpolar Institute, University of Alberta, Edmonton.

177. Forbes, G. J., and J. B. Theberge. 1996. Cross-boundary management of Algonquin Park wolves. *Conservation Biology* 10:1091–1097.

178. Chapman, R. C. 1979. Human disturbance at wolf dens—A management problem. Pages 323–328 *in* R. M. Linn, ed., *Proceedings of the 1st Conference Scientific Research in the National Parks.* Proceedings Series No. 5, Vol. 1. U.S. National Park Service, Washington, D.C.

179. Wydeven, A., and R. N. Schultz. 1992. Management policy for wolf den and rendezvous sites. Unpublished report. Wisconsin Department of Natural Resources.

180. Fritts, S. H., E. E. Bangs, and J. F. Gore. 1994. The relationship of wolf recovery to habitat conservation and biodiversity in the northwestern United States. *Landscape Urban Planning* 28:23–32.

181. Thiel, R. P., S. Merrill, and L. D. Mech. 1998. Tolerance by denning wolves, *Canis lupus,* to human disturbance. *Canadian Field-Naturalist* 112 (in press).

182. Jensen, W. F., T. K. Fuller, and W. L. Robinson. 1986. Wolf (*Canis lupus*) distribution on the Ontario to Michigan border near Sault Ste. Marie. *Canadian Field-Naturalist* 100:363–366.

183. Mech, L. D., S. H. Fritts, G. L. Radde, and W. J. Paul. 1988. Wolf distribution and road density in Minnesota. *Wildlife Society Bulletin* 16:85–87.

184. Thiel, R. P. 1985. The relationship between road densities and wolf habitat suitability in Wisconsin. *American Midland Naturalist* 113:404–407.

185. Person, D. K., M. Kirchhoff, V. Van Ballenberghe, G. C. Iverson, and E. Grossman. 1996. The Alexander Archipelago Wolf: A conservation assessment. General Technical Report No. PNW-GTR-384. U.S. Department of Agriculture, Forest Service, Pacific Northwest Research Station, Portland, OR.

186. Mladenoff, D. J., T. A. Sickley, R. G. Haight, and A. P. Wydeven. 1995. A regional landscape analysis and prediction of favorable gray wolf habitat in the northern Great Lakes region. *Conservation Biology* 9:279–294.

187. Bangs, E. E., S. H. Fritts, D. R. Harms, J. A. Fontaine, M. D. Jimenez, W. G. Brewster, and C. C. Niemeyer. 1995. Control of endangered gray wolves in Montana. Pages 127–134 *in* L. N. Carbyn, S. H. Fritts, and D. R. Seip, eds., *Ecology and Conservation of Wolves in a Changing World.* Canadian Circumpolar Institute, University of Alberta, Edmonton.

188. Tompa, F. S. 1983. Status and management of wolves in British Columbia. Pages 20–24 *in* L. N. Carbyn, ed., *Wolves in Canada and Alaska: Their Status, Biology, and Management.* Canadian Wildlife Service Report Serial No. 45.

189. Mack, J. A., W. G. Brewster, and S. H. Fritts. 1992. A review of wolf depredation on livestock and implications for the Yellowstone area. Pages 5-21–5-44 *in* J. D. Varley and W. G. Brewster, eds., *Wolves for Yellowstone? A report to the United States Congress,* Vol. IV. Research and Analysis, U.S. National Park Service, Yellowstone National Park, Mammoth, WY.

190. Cluff, H. D., and D. L. Murray. 1995. Review of wolf control methods in North America. Pages 491–504 *in* L. N. Carbyn, S. H. Fritts, and D. R. Seip, eds., *Ecology and Conservation of Wolves in a Changing World.* Canadian Circumpolar Institute, University of Alberta, Edmonton.

Here

the

191. BOERTJE, R. D, D. G. KELLEYHOUSE, and R. D. HAYES. 1995. Methods for reducing natural predation on moose in Alaska and Yukon: An evaluation. Pages 505–514 *in* L. N. Carbyn, S. H. Fritts, and D. R. Seip, eds., *Ecology and Conservation of Wolves in a Changing World*. Canadian Circumpolar Institute, University of Alberta, Edmonton.

192. BERGERUD, A. T., and J. P. ELLIOT. 1986. Dynamics of caribou and wolves in northern British Columbia. *Canadian Journal of Zoology* 64:1515–1529.

193. GILBERT, F. F. 1995. Historical perspectives on wolf management in North America with special reference to humane treatments in capture methods. Pages 13–17 *in* L. N. Carbyn, S. H. Fritts, and D. R. Seip, eds., *Ecology and Conservation of Wolves in a Changing World*. Canadian Circumpolar Institute, University of Alberta, Edmonton.

194. CLARKSON, P. L. 1995. Recommendations for more effective wolf management. Pages 537–545 *in* L. N. Carbyn, S. H. Fritts, and D. R. Seip, eds., *Ecology and Conservation of Wolves in a Changing World*. Canadian Circumpolar Institute, University of Alberta, Edmonton.

195. HAGGSTROM, D. A., A. K. RUGGLES, C. M. HARMS, and R. O. STEPHENSON. 1995. Citizen participation in developing a wolf management plan for Alaska: An attempt to resolve conflicting human values and perceptions. Pages 481–487 *in* L. N. Carbyn, S. H. Fritts, and D. R. Seip, eds., *Ecology and Conservation of Wolves in a Changing World*. Canadian Circumpolar Institute, University of Alberta, Edmonton.

196. MECH, L. D., T. J. MEIER, J. W. BURCH, and L. G. ADAMS. 1991. Demography and distribution of wolves, Denali National Park and Preserve, Alaska. Natural Resources Report No. AR-91/01. U.S. National Park Service, Anchorage, AK.

197. KOLENOSKY, G. B. 1972. Wolf predation on wintering-deer in east central Ontario. *Journal of Wildlife Management* 36:357–369.

198. SUMANIK, R. S. 1987. Wolf ecology in the Kluane Region, Yukon Territory. Thesis, Michigan Technical University, Houghton.

199. PETERSON, R. O., and R. E. PAGE. 1988. The rise and fall of Isle Royale wolves, 1975–1986. *Journal of Mammalogy* 69:89–99.

200. HABER, G. C. 1977. Socio-ecological dynamics of wolves and prey in a subarctic ecosystem. Ph.D. thesis, University of British Columbia, Vancouver, Canada.

201. WEILER, G. J., and G. W. GARNER. 1987. Wolves of the Arctic National Wildlife Refuge: Their seasonal movements and prey relationships. Pages 691–742 *in* G. W. Garner and P. E. Reynolds, eds., *1985 Update Report Baseline Study of the Fish, Wildlife, and their Habitats*. U.S. Fish and Wildlife Service, Anchorage, AK.

202. ADAMS, L. G., R. O. STEPHENSON, B. W. DALE, and B. SHULTS. 1989. Population ecology of wolves in Gates of the Arctic National Park and Preserve, Alaska, 1988 Progress Report. Natural Resources Report No. AR-89/15. U.S. National Park Service, Anchorage, AK.

203. STEPHENSON, R. O., and D. D. JAMES. 1982. Wolf movements and food habits in northwest Alaska. Pages 26–42 *in* F. H. Harrington and P. C. Paquet, eds., *Wolves of the World: Perspectives of Behavior, Ecology, and Conservation*. Noyes Publications, Park Ridge, NJ.

Puma

Kenneth A. Logan and Linda L. Sweanor

INTRODUCTION

Pumas once had the broadest geographic distribution of any terrestrial mammal in the Western Hemisphere, ranging from northern British Columbia to the tip of South America and from coast to coast (1). But settlement of the Americas by European immigrants since the 1500s brought persecution and habitat loss. In North America, pumas were killed principally to protect livestock and wild ungulates and because they were perceived as "bad" animals. Development for settlements, agriculture, and industry transformed habitats, making them unsuitable for pumas. By the late 1800s, eastern populations were extinct or severely reduced; western populations were diminished by the early 1900s (2). The result was a reduction in puma distribution to about one-half of its historical geographic range on the continent (Figure 17–1).

Only recently have people supported laws to protect pumas. Since 1965, regulations on hunting by all of the western states and provinces, excluding Texas, have enabled populations to recover from historical low levels. Furthermore, the passage of the Endangered Species Act in 1973 gave eastern subspecies full protection and may have prevented the extinction of the Florida panther. Most western states and provinces presently report increasing trends in puma populations (Table 17–1). Although sightings of pumas are reported every year in many eastern states and provinces, there is not sufficient evidence to indicate that any breeding populations exist in eastern North America except in southern Florida (34).

Human perspectives toward the puma contribute to the difficulty of managing the species. Economic values are a principal guiding force, such as when pumas prey on livestock and bring financial loss to ranchers, or where pumas are hunted as trophies and, thus, contribute directly to the economies of outfitters, guides, and rural communities. Yet, others believe that trophy hunting or control of pumas is unethical. Some people, recognizing potential dangers, value human safety above conserving pumas. To wildlife professionals, biological and ecological values are also important (5, 6, 7, 8). Moreover, the growing human population and attendant loss of habitat complicate management. The challenge for wildlife managers is to consider all of these issues and to balance the needs of people with the conservation of pumas (9).

Figure 17–1 Distribution of the puma in North America. (Adapted from Refs. 41, 13, 142, 143, 144, 4, 50, 145.) Precolonial range in the United States and Canada comprised about 8,900,000 square kilometers (1, 146, 147). Pumas presently occupy about 2,906,000 square kilometers (compiled in part from Ref. 105). The Florida panther occupies a minimum area of 10,360 square kilometers (143). Confirmed records include a puma shot on Wrangell Island, Alaska, in 1989, a male captured in southwestern Minnesota in 1991, and a puma shot near Lake Abitibi, Ontario, in 1992 (4).

TAXONOMY

The puma may have more common names than any other American mammal: catamount, cougar, *leon,* mountain lion, panther, and several other names given by local native American tribes (1, 10). The name *puma* apparently was originally given to the animal by the Incas of South America (1).

The evolution of the puma is obscure. *Puma concolor* may have evolved during the late Pleistocene from a larger ancestor, *Felis inexpectata,* that was present in the late Pliocene and early Pleistocene (circa 2 to 7 million years) (11). Recently, O'Brien used genetic and morphological characters to estimate that *Puma concolor* emerged 5 to 6 million years ago (12). Whether the puma originated in North or South America is not known, but its closest known relative is thought to be the cheetah (13, 12).

TABLE 17–1
Puma Status and Approaches to Puma Management in the United States and Canada in 1997[a]

State or Province	Percent Area Puma Habitat	Population (Numbers Trend)[b]	Legal Status	Trapping Season	Hunting Season Length	Hunting Restrictions and Hunter Kill[c]	Depredation Control[d]	Public Safety[e]	Trans-Location[f]
Alberta	20	650–750: S-I	Big game	No	Dec. 1–Feb. 28	BL = 1; Q; FSQ; FCP; 52(62)	R; DP; 28/2(0)	6/2(6)	Yes (1–2/yr)
Arizona	77	2,500; S	Big game	No	Year-round	BL = 1; Q of 1 puma in SW region; no FCP; 203(225)	R; 50/31(35)	2–5/1	No
British Columbia	60	>4,000; S-I	Big game	No	Sep. 16–Jun. 31, Nov. 15–Mar. 31, Nov. 15–Feb. 28	BL = 2; PS; Q in area with shortest season; no FCP; 277(508)	R; ?/105(168)	Combined with depredation	Yes
California	60	4,000–6,000; S–D	Specially protected mammal	No	None	N/A 0	R; 252/92(103)	300*/5–10	Yes (<10/year)
Colorado	No estimate	?; I	Big game	No	Nov. 15–Mar. 31	BL = 1; Q; FCP; 291(327)	R; DP; 79/13(6)	<12/0–3(0)	Yes
Florida	13	30–50 adults; S–D	Federal endangered	No	None	N/A 0	None	0	Yes
Idaho	95	5,000+; I	Big game	No	Sep. 15–Mar. 31 most areas	BL = 1; PS; FQ in south hunt units; FCP; 419(632)	R; DP; 7/3(17)	<5/<1(0)	Yes (4–5 in 6 yrs)
Montana	50	?; I	Big game	No	Dec. 1–Feb. 15 most areas	BL = 1; PS; MQ with FQ or FSQ in some hunt units; FCP; 425(567)	R; 23/4(?)	21/11	No* (0 since 1993)
Nevada	45	2,500–4,000 D	Big game	No	Oct. 1–Apr. 30 most areas	BL = 2; Q; FCP; 149(143)	R; 52/43(23)	<5/2	Yes (<5/yr)
New Mexico	59	?; ?	Big game	No	Dec. 1–Mar. 31	BL = 1; FCP; 123(168)	R; PC; 30/9(11)	1/1 in 12 years (0)	Yes
Oregon	75	3,000–3,300; I	Game mammal	No	Aug. 30–Mar. 31; 4 high-damage areas open year-round	BL = 1; Q; no hunting with dogs; FCP; 137(47)*	R; 402/39(94)	Included under depredation	No
South Dakota	?	?; I	State threatened	No	None	N/A; 0	No depredation	Increased sightings	No
Texas	40	?; I	Nongame	No	Year-round	None; 155(163)*	R; No restrictions	<1/<1	Yes

continued

TABLE 17-1
Puma Status and Approaches to Puma Management in the United States and Canada in 1997[a]—continued

State or Province	Percent Area Puma Habitat	Population (Numbers Trend)[b]	Legal Status	Trapping Season	Hunting Season Length	Hunting Restrictions and Hunter Kill[c]	Depredation Control[d]	Public Safety[e]	Trans Location[f]
Utah	60	2,000–3,000; D	Game	No	Dec. 17–Jun. 7 most areas; Nov. 15–Jun. 7 in 4 hunt units	BL = 1; PS; LE in units subjected to intense hunt pressure; Q other units; FSQ in 1 unit; 1 unit closed; FCP; 332(576)	R; DP up to 1/2 of value of loss. 75/36(?)*	5–10/4–5	Yes
Washington	51	2,400 min.; I	Game	No	Aug. 15–Mar. 15	BL = 1; general permit; no hunting with dogs; FCP; 168(178)	R; ?/?(43)	Included under depredation	Yes (11 in 1996)
Wyoming	No estimate	?; ?	Trophy game	No	Sep. 1–Mar. 31 (most of state): year long in 2 units with high livestock losses	BL = 1; Q; FSQ in 7 of 27 units; FCP; 78/140	R; DP; ?/0.4(0)	?/1.2(4)	Yes; (0 in 1996)

[a]As reported by Padley (105). Information was updated from telephone interviews conducted in fall 1997 by L. L. Sweanor with personnel from responsible management agencies in each of the western states and provinces.

[b]All puma population numbers reported are best guesses based on a variety of factors, including estimated available habitat and puma densities; Oregon and Nevada apply population models. Trends are estimated based on trends in hunter kill, depredation, sightings, nuisance incidences, and age structure of hunter kill. Key: S, stable; D, decreasing; I, increasing; ?, no estimate available.

[c]Texas requires a general hunt license; all other states and provinces that allow puma hunting (n=12) require a specific puma hunt license. Seven states charge a trophy/tag fee. Key: BL, bag limit; PS, separate pursuit season; Q, quotas or (in Arizona and Nevada) harvest objectives; FSQ = female subquota; LE, limited entry; FCP, females accompanied by cubs and spotted cubs protected. Numbers indicate average annual number of pumas killed by hunters, 1991–1995; 1996 hunter kill is in parentheses. Texas puma kill includes all reported mortalities (hunter kill, depredation, and accidents). Hunter kill in Oregon and Washington declined when hunters were prohibited from using dogs beginning in 1995 and 1996, respectively.

[d]In all states and provinces, an individual has the right to kill a puma if it is about to harm him or his property. R: Where puma depredation is suspected, the puma(s) can be captured and killed, either by landowner, landowner agent or Wildlife Services (WS) after (before in Oregon) confirmation of depredation by state game agency or WS, but kill must be reported to state game agency. DP: State or provincial agency pays damage claims for confirmed depredations. PC: Preventive control allowed in localized areas where high incidences of depredation have been documented. Numbers generally indicate mean annual number of puma depredation incidents/pumas killed from 1991–1995. Pumas killed in 1996 is in parentheses. However, Utah reports number of incidents each year and 10-year average in number of kills. A ? means no numbers were available.

[e]Numbers indicate annual range or mean annual number of public safety incidents/mean annual number of pumas killed, 1991–1995. The number of pumas killed in 1996 is in parentheses when available. Incidents in California include sightings.

[f]Most management agencies do not have specific policies on translocating problem pumas. Most euthanize pumas that are causing depredations or directly threatening human safety. The decision to move pumas is usually made on a case-by-case basis, and most are not tagged or monitored. Montana can potentially translocate pumas, but only if vacant habitat is available. Washington has some specific guidelines for translocation: All individuals must be marked, and they must be moved a minimum of 120 kilometers from their capture sites. Pumas in British Columbia must be eartagged. Only Florida has attempted translocation for reintroduction purposes (134). Numbers of pumas translocated are reported in parentheses if available.

Linneaus classified the puma as *Felis concolor* in 1771; Jardine reclassified the genus as *Puma* in 1834 (14). Recent taxonomic frameworks have placed as many as 29 species, including the puma, within the genus *Felis* (15). Wozencraft, however, upgraded many of the subgenera within *Felis* to full generic status (14). He placed the puma back in the monotypic genus *Puma,* in support of Jardine's 1834 recognition of the unique evolution of the species (4).

Goldman recognized 13 subspecies of puma in the United States and Canada, principally from cranial features (16). However, recent genetic studies (M. Culver, Laboratory of Genomic Diversity, National Cancer Institute, personal communication) have shown no genetic variation in North American pumas at the subspecies level, unless the most rapidly evolving genetic markers are applied. Because South American pumas show abundant genetic variation, even using moderately evolving markers, Culver speculates that the differentiation observed in North American pumas is more on the population level, while South American pumas differ more on the subspecies level. Consequently, fewer puma subspecies should be recognized in North America. For example, McIvor and others could not find any supporting evidence for the subspecific designation of the Yuma puma (17).

DESCRIPTION

Pumas are the second largest cat in North America. Jaguars, which are extremely rare in the Southwest, are slightly larger. In general, body size of pumas increases with latitude (18). Information on puma morphometry gathered by Anderson clearly shows that males and females are sexually dimorphic (13). Male pumas weighed about 1.4 times more than females in five out of six North American subspecies for which data were available; insufficient sample size probably precluded the sixth. Mean masses for adult specimens (at least 24 months old) ranged from 52.8 to 68.0 kilograms for males and from 34.4 to 48.0 kilograms for females (Figure 17–2). The heaviest recorded mass of 125.2 kilograms was for a viscerated male in Arizona (19). Adult male pumas

Figure 17–2 This 37-kilogram adult female puma was captured with a foot-hold snare in southern New Mexico. (Photograph by K. Logan.)

Figure 17–3 Dark facial markings, spots on the body, and rings on the tail are clearly visible on this 3-month-old puma cub in the Big Horn Mountains of Wyoming. (Photograph by K. Logan.)

reached total lengths (body plus tail) of 202.2 to 230.8 centimeters and adult females measured 183.6 to 201.7 centimeters. Shoulder height measurements for adults ranged from 56.0 to 78.7 centimeters for males and 53.4 to 76.2 centimeters for females.

The puma has a slender, muscular, cylinder-shaped body, a relatively compact head, heavy forelimbs, and a long tail that is slightly over one-third of the animal's total length. The appearance of the exposed parts of the adult puma is largely of one color with varying hues of tawny, reddish brown, and grayish-brown; these variations can occur within the same population. Black patches are on the backs of the ears and the tip of the tail. A black "mustache" accents the sides of the muzzle and contrasts sharply against the white hair around the mouth. The underparts of the neck, chest, and abdomen and all the way to the base of the tail are covered with whitish hair.

Cubs are born fully furred with black spots on reddish- to gray-brown coats and with black rings on their tails. Birth weights average 508.3 grams (13). Within 2 weeks of birth, cubs' eyes are fully open (1). When cubs are as young as 5 months old, their eyes turn from blue to the brown or amber color of adult pumas (Figure 17–3). Spots and rings fade as cubs grow older, appearing as light brown dapples by the time cubs reach 9 months of age. Dapples on the pelage disappear within about 24 months, but some may persist on the forelimbs and hind limbs for up to 30 months (K. Logan and L. Sweanor, unpublished data).

Gender of adult pumas can be determined by characteristics of the external sex organs. Males have a spot of black hair (2.5-centimeter diameter) encircling the opening of the penis sheath, which is about 12 centimeters anterior-ventral to the anus. The scrotum is between the anus and black spot. The female's vulva is directly below the anus. When treed, a puma's sex can be distinguished with the naked eye or with binoculars. The male sex organs are visually evident, but those of the female are often hidden beneath the base of the tail.

The puma has 30 teeth. The upper/lower dental formula is incisors, 3/3; canines, 1/1; premolars, 3/2; molars, 1/1. The shape of the teeth reflects the puma's strict carnivory. Large conical canines are effective in grabbing prey and separating skeletal joints, while the robust carnassials are used for shearing apart tissues and breaking bones. The small, closely spaced incisors are used for stripping tissues from bone and for grooming. The relatively small number of teeth and vestigial teeth (P^2, M^1) probably are partly the result of selection for shorter jaws that delivered a more effective biting action.

A puma's feet have retractile claws used to grapple with prey, for self-defense, and to climb trees for escape. The fore feet have five toes, four of which support the puma on the ground. Although the pollex is nonsupporting, it has the largest claw and is used to grasp prey and in combat. The hind feet are smaller than the fore feet and have four toes. Toe pads are oval and the heel pad has three distinct rear lobes. Generally, width measurements of heel pads can be used to distinguish tracks made by adult male and female pumas within the same population; however, overlap in track heel pad width between small males and large females can sometimes cause confusion. For example, hind heel pad widths for male and female pumas in northern Wyoming ranged from 49 to 62 and 36 to 54 millimeters, respectively (20).

LIFE HISTORY

Diet

Pumas in North America consume a variety of prey. Although most food consists of animals that they kill, pumas will opportunistically scavenge wild and domestic animals that die of other causes (21, 22). Pumas sometimes ingest green grass or other vegetation, perhaps as an emetic to help expel ingested hair, as roughage to hasten expulsion of gut parasites, or as a source of micronutrients (23, 13). Plant material is also ingested accidentally when it adheres to consumed animal tissues. Hydration is maintained by drinking surface water and consuming prey tissues. Cubs are entirely dependent on their mother's milk for the first 6 to 8 weeks of life; thereafter, they are carnivorous and follow their mother to her kills (K. Logan and L. Sweanor, unpublished data).

Deer are the puma's most important food, although other species of ungulates are eaten depending on local abundance and vulnerability (Table 17–2) (13). In the Southwest and along the Pacific Coast, pumas rely principally on mule deer and black-tailed deer (24, 25, 26, 27, 28, 29, 22, 30). In the Northwest, elk, moose, and white-tailed deer are important components of the puma's diet, along with mule deer (5, 31, 32, 33, 32, 34). In central British Columbia, bighorn sheep, which occur with few other mammalian species, made up 58% of the puma's diet (35). Mule deer declines may cause pumas to shift their diets to other available prey, such as collared peccaries (27). Wild pigs are the most important food item for the Florida panther (Table 17–2), and they are an important component in the year-round diet of pumas in California (Table 17–2) (36, 30). Pronghorn become prey when they use rugged terrain that makes them vulnerable to pumas (37).

Small prey are important in certain locales, comprising from 14% to 21% of the diet. Examples include porcupines in Utah and Nevada, opossums on the Santa Ana Mountains, California, lagamorphs in south-central Utah, snowshoe hares and porcupines in south-central British Columbia, and raccoons in Florida (23, 38, 39, 40, 36). Many other small prey are taken opportunistically but usually comprise less than 10% of the puma's diet (Table 17–2).

Pumas are sometimes cannibalistic. Male pumas have killed and cannibalized cubs, adult females, and other adult males (1, 41, 42, 43, 38, 43, 44). Pumas have also scavenged other pumas (including direct relatives) that died of nonpredation causes (1, 45, 38, 43).

Livestock in the puma's diet may vary from zero to as high as 34% (22, 46). Only in Arizona have puma preyed heavily on cattle; there, cattle contributed slightly more biomass to the diet of pumas than did deer (44.2% versus 40.1%) (46). Other small livestock such as sheep and goats are particularly vulnerable to pumas. In general, pumas infrequently prey on horses, apparently favoring foals, but rarely adults (47, 1, 48, 23). Pumas may occasionally prey on domestic dogs and cats where residential areas overlap puma habitat (49, 38, 50).

TABLE 17–2

Percentage Occurrence of Puma Food Items in Selected Areas of North America

Food item	Idaho[a]	Alberta[b]	British Columbia[c]	Utah[d]	New Mexico[e]	Arizona[f]	Arizona[g]	California[h]	California[i]	Florida[j]
Mule deer		38	13.7	87.5 (87.5)	91.2(85.7)	58	39	82 (74)	58.6 (53.4)	
Elk		22	68.6	3.6 (0.4)						
Mule deer and elk	70									
White-tailed deer		4	3.9							28
Moose		30								
Bighorn sheep		6	3.9		1.9(0.6)		7			
Pronghorn					1.0(0.7)	2				
Peccary					0(0.1)	2	25			
Wild pig								4 (20)	2.8 (3.1)	42
Cattle				0.9 (0.4)		30	13	0 (4)		2
Sheep/goat									5.6 (t)	
Opossum									13.8 (10.4)	1
Small rodents				0 (12.1)	0(2.2)		8	0 (8)		
Lagomorphs	5.5		2.0	2.7 (17.2)	0.2(4.2)	2	8	0 (2)	0.7 (4.7)	4
Porcupine			2.0	0 (0.8)	1.0(3.2)	6	2			
Mustelids[k]				0.9 (0.4)	0.8(2.6)		9	4 (2)	2.8 (t)	
Raccoon							2		41. (t)	12
Puma			3.9	0.9 (0)	2.0(2.0)		1	0 (33)	2.1 (0)	1
Coyote			2.0	1.8 (0)	0.6(0.1)			2 (0)	7.6 (9.3)	
Miscellaneous	24.5		5.9	1.8 (16.9)	1.0(1.3)		5	6 (17)	2.1 (23.9)	9

[a]Ref. 5. Percent frequency of occurrence in scats (n = 235 prey items in 198 scats). Miscellaneous = various small mammals and grass.

[b]Ref. 32. Biomass consumed in winter (n = 368 puma kills). Nonungulate prey comprised <1%.

[c]Ref. 31. Confirmed puma kills that were subsequently consumed and three gastrointestinal samples (n = 51).

[d]Ref. 67. Confirmed puma kills and contents of 1 stomach (n = 112). Percent frequency of occurrence in scats (n = 316 prey items in 239 scats) reported in parentheses. Miscellaneous in kills = carrion; miscellaneous in scats= gray fox, bobcat, marmot, beaver, bird, unknown animal, and vegetation.

[e]Ref. 22. Prey killed and eaten by pumas (n = 525). Miscellaneous = oryx, ringtail, and golden eagle; each comprised <1%. Percent frequency of occurrence in 847 scats and four stomachs in parentheses. Miscellaneous = oryx, ringtail, unknown birds and box turtle, comprising 1.3% in aggregate (135).

[f]Ref. 26. Percent frequency of occurrence in scats (n = 54 prey items in 50 scats).

[g]Ref. 29. Percent frequency of occurrence in scats (n = 159 scats). Miscellaneous = beetle, bobcat, canidae, Gila monster, and chuckwalla; each comprised less than 2%.

[h]Ref. 30. Prey killed by pumas (n = 45); all but one bobcat and one coyote were eaten. Percent frequency of occurrence in scats (n = 160 prey items in 46 scats) reported in parentheses. Miscellaneous in kills = bobcat, turkey vulture; in scats = bobcat, aves, ingested grass.

[i]Ref. 38. Prey killed by pumas (n = 145). Deer comprised about 78% of prey biomass. Percent frequency of occurrence (n = 193 prey items in 178 scats) reported in parentheses; t <2%. Miscellaneous in kills = bobcat; in scats = fox, unidentified canids, domestic cats, beavers, unidentified rodents, voles, and moles.

[j]Ref. 36. Percent frequency of occurrence of prey in scats (n = 281 prey items in 270 scats). Miscellaneous includes armadillo: 8%, and black bear, mustelidae, wild turkey, alligator, and unknown mammal (all ≤1%). One horse was included in with cattle.

[k]Mustelids include badgers and/or skunks.

Consumption Patterns and Prey Selection

The frequency with which pumas kill prey and their rates of consumption largely depend on the energy requirements of the individual population units and the biomass of the prey. Ackerman studied puma predation and ecological energetics in south-central Utah and developed a predictive model for prey utilization (21). Single adult females and males consumed approximately 2.2 to 2.6 kilograms and 3.4 to 4.3 kilograms per day, respectively. Calculated intervals between deer kills were 14 to 17 days for females and 8 to 11 days for males. Consumption by females with cubs varied with litter size and age of the cubs. For example, a family with 3-month-old cubs would consume about 4 kilograms of deer meat per day, with kills occurring about every 10 days. In contrast, a family with 2 to 3 yearling cubs could consume up to 12 kilograms of meat per day and would kill a deer about every 7 days. Based on the energy requirements of females with young, Ackerman concluded that deer-sized ungulates probably were essential for the maintenance of viable puma populations (21). Other estimates of kill intervals of deer-size prey by single pumas range from 7 days during winter to 10 days (23, 51). Hornocker estimated that an adult puma requires 14 to 20 deer or 5 to 7 elk each year (5). But Murphy and others estimated that nonmaternal adult females and family groups would kill 34 and 52 mule deer and elk (combined) per annum, respectively (33). The actual frequency with which pumas kill prey also is influenced by hot seasons or climates where spoilage is higher, and by other carnivores, including coyotes, bears, and wolves, that usurp puma kills (26, 35, 52).

Adult male deer generally appear to be more vulnerable to puma predation than other segments of the population, especially during winter (23, 5, 21, 30). In New Mexico, similar predation rates were found for males and females during a period of relatively good habitat conditions. However, when a drought occurred, pumas killed males at higher rates than females (22). Pumas may kill females in proportions smaller than, or equal to, their occurrence in populations, while fawns may be killed in proportions equal to or greater than their occurrence (23, 5, 21, 30, 23, 23, 30, 5, 53).

Pumas kill elk calves in proportions greater than their availability in the population and may even select calves above all other classes of prey (5, 33). But, females are killed indiscriminately and mature males appear to be avoided (5).

In multi-ungulate-prey environments, prey selection may differ with puma sex, mass, or prey species. In Alberta, males killed moose more frequently than females, and males almost never killed elk (54). Yet, elk were the second most frequent prey item for female pumas after deer. Deer comprised half of all prey occurrences for females, but less than one-third of all prey for males. In the northern Yellowstone ecosystem, there was a positive relationship between ungulate prey mass and puma mass. The largest pumas (i.e., adult males) killed the greatest percentage of adult elk, and the smallest percentage of mule deer, while the opposite was true for the smallest self-sufficient pumas (i.e., subadult females) (33).

Reproductive Strategy

Pumas are polygamous and promiscuous (6, 13, 55). An adult male may breed with a number of different females that range within his territory, and a female may breed with more than one male during individual or subsequent estrus cycles (56). Male and female pumas become sexually mature at about 24 and 21 months old, respectively; however, females as young as 18.5 and 19 months old have successfully conceived in the wild (Figure 17–4) (43, 57, 43). Based on captive females, estrus periods average 6.5 to 8.1 days long, and estrus cycles average 37.6 days (13). Sweanor and others documented wild pumas in breeding associations that lasted 1 to 6 days with apparent estrus cycles that averaged 21 days (56). Pumas probably are induced ovulators and exhibit copulation rates as high as 50 to 70 per day (58, 59).

Gestation lengths for pumas vary from 82 to 103 days (13, 43). The mean gestation period of 91.3 days found for 31 wild litters in New Mexico closely matches the mean of 91.9 days for 42 captive litters from 11 sources (43, 13). Litters may be

Figure 17–4 Female pumas, such as this 45-kilogram adult in the Big Horn Mountains of Wyoming, generally reach sexual maturity at about 2 years of age. (Photograph by K. Logan.)

comprised of 1 to 6 cubs, however, the average size is about 3 (Figure 17–5) (60, 42, 43). The smallest mean litter size reported for pumas was 2.3 cubs for four litters of the endangered Florida panther (61). In contrast, the largest mean litter size of 3.4 cubs was reported for eight first-time mothers in New Mexico (43).

Puma cubs can be born at any time of the year (60, 47, 43). However, birth pulses have been noted during June to September in Utah and Nevada, July to September in New Mexico, September to November in south-central Utah, August to November in Wyoming, and July and August in Alberta (60, 47, 43, 53, 20, 62). In Wyoming and New Mexico, birth pulses coincided with periods when prey were relatively more abundant and vulnerable (20, 43).

Mean birth intervals for wild mothers that successfully raised cubs of the first litter to independence or at least 12 months of age ranged from 17 to 24 months (62, 63, 43). Mothers whose litters die before independence take 24 to 308 days to successfully rebreed. Successful females may produce as many as five litters in their lifetimes (43).

Individual reproductive success is highly variable in male and female pumas (43, 64). Reproductive success apparently favors long-lived territorial males and the adult females residing within their territories (64).

BEHAVIOR

Social Structure

Pumas live a relatively solitary existence. The only enduring social unit is that of a mother and her cubs. Breeding associations are brief, and newly independent siblings only associate for short durations, if at all (6, 55, 62, 56). Mothers provide for and rear offspring by themselves, while males defend territory in which they pursue other potential mates. Most breeding is done by the resident adults. Independent subadults (i.e., young pumas that are independent of their mothers but not yet old enough to breed) may be recruited into the population of their birth or they may emigrate to other subpopulations where they can establish adult residency (56). Most *transients,* a term originally used to describe certain pumas with an ephemeral presence in particular locales or populations, probably are dispersing subadults (65, 6, 55).

Figure 17–5 Puma litters generally contain three cubs. These 21-day-old cubs were at a nursery in the Chihuahua Desert in southern New Mexico. (Photograph by L. Sweanor.)

The social organization of pumas has been described as a land tenure system where dominance over an area is held initially by the resident adult pumas occupying the area (6). Recruits establish residency in areas where there is adequate space and resources, or where a previously occupied area is made available by puma deaths, home range shifting, or by direct competition for space (55). Adult males are territorial. They advertise their presence, patrol their territory, and compete directly with other males for territory and mates. Fights may occur between neighboring territorial males or a territorial male and a new male immigrant attempting to establish a territory. Fights generally result in a subsequent home range shift by a resident, expulsion of the immigrant, or the death of a combatant and either retention or subsequent takeover of the area by the victor (56). Adult females are not territorial and they rarely fight other pumas, except to protect cubs or food. Females generally avoid other pumas, possibly to minimize interference while bearing and raising offspring, and consequently maximize reproductive success (65, 6, 56). Pumas inhabiting an area communicate with one another in time and space using visual and olfactory cues and vocalizations (6, 55, 56). These cues help pumas space themselves in the habitat, and aid them in finding each other to breed, or to defend or contest territory (see Communication).

Home ranges of pumas have been studied extensively using radiotelemetry (Table 17–3). In general, home ranges of males are larger than females within the same population by factors of about 1.5 to 5. Home ranges may be as small as 18 square kilometers for an adult female during an entire reproductive cycle or as large as 792 square kilometers for an adult male (56, 66). The large variation in reported home-range size may result in part from the differing home-range measurement techniques, including number of locations per animal, duration of monitoring, season of monitoring, and method used to estimate area (13). However, the social and reproductive status of animals monitored, the quantity and quality of the habitat, elevational migrations of prey, and puma population density are important ecological influences (31, 56).

TABLE 17–3
Home Range (HR) Estimates (Square Kilometers) for Pumas in North America

Location	Time Interval	Male	n^a	Female	n^a	No. Locations/ Puma/Period	Method[b]	Reference
Alberta	Annual	334	6	140	21	19–56	MCP	62
Arizona	Annual	196	5	109	2	34–59	MCP	46
	Annual	249	5	164	2	34–59	0.90 HM	
British Columbia	Winter			29	4	31–41	MMA	31
	Spring–Fall			40	3	38–99		
	Annual	151	2	55	5	36–140		
California	Annual	199	4/7	84	7/17	31–113	MCP	30
	Annual	252	4/7	114	7/17	31–113	0.95 HM	
	Wet (Nov.–Apr.)	523	2/4	109	12/26	≥45	MCP	38
	Dry (May–Oct.)	447	2/4	118	10/21	≥45		
Colorado	Annual	256	6/12	309	7/27	27–53	MCP	75
	Annual	456	6/12	309	7/27	27–53	0.95 HM	
Florida	Annual	558	8/16	191	10/20	NR[c]	MCP	72
Idaho	Winter–Spring (Dec.–May)	126	3/4	90	6/8	NR	MCP	6
	Summer–Fall (June–Nov.)	293	1	148	4			
	Annual	453	1	233	3			
Montana	Annual	462	2	202	5	29–35	MCP	68
New Mexico	Annual	187	23/69	74	29/69	32–52	MCP	56
	Annual	192	23/69	72	29/69	32–52	0.90 ADK	
Texas	Annual	792	1	159	5	102–361	MCP	66
Utah	Nonwinter	573	1	336	4/5	24–96	MCP	69
	Winter			207	3/4	20–43		
	Total	826	1	685	4	31–188		
Wyoming	Annual	320	2	73	2	31–34	MCP	20

[a]When provided by sources, the total number of individual pumas monitored/number of time intervals is noted (e.g., 4/7 = 4 different pumas monitored over 7 different years). Consequently, HR sizes are means of means.
[b]Method: MCP, minimum convex polygon, (136); HM, harmonic mean, (137,138); MMA, modified minimum area, (139); ADK adaptive kernal, (140).
[c]NR, not recorded.

The arrangement of territories and home ranges of this solitary felid probably is defined by its mating system, its energy requirements, and habitat quality. The larger home-range size of the territorial male probably is not related to his larger body size or to greater energy demands, but instead results from his attempts to encounter as many receptive females as possible (30). This increases his reproductive success. An adult male's home range can overlap home ranges of as many as five adult females (56). Territories of adult males are dynamic. They include and exclude area over time or shift to entirely new areas, depending on the dominance of neighboring and immigrant males, deaths or shifts of other males, and the presence of receptive females (56).

Because females are most concerned with successfully raising cubs, female home-range sizes are keyed to the availability and distribution of food. Female pumas raise cubs without the aid of the male; consequently, their overall energy requirements are probably equal to or greater than the male's (67). Adult female home ranges are the smallest when activities focus around nurseries with suckling cubs, then expand as the cubs grow and become carnivorous (55). In contrast to males, females have relatively stable home ranges (55, 56). This behavior enables the female to develop an intimate knowledge of the resources (e.g., food, nurseries) required to successfully raise offspring.

Home-range dynamics are influenced by seasonal changes in the environment. In the Rocky Mountains, pumas have relatively small winter home ranges where deep snow restricts them and their prey to valley bottoms. But, in summer, home ranges expand as pumas follow prey to higher elevations (6, 68, 69, 70). Some pumas have discrete winter and summer home ranges, but not in environments where prey do not make seasonal elevational movements (62, 30, 55).

Adult male pumas in California, Idaho, Montana, and Wyoming had exclusive home ranges (30, 6, 68, 71). Yet other populations in Alberta, California, Florida, Montana , and New Mexico had varying degrees of overlap between male home ranges (62, 38, 72, 44, 56). Some of these differences may have to do with the sample size of males monitored concurrently (i.e., reports on nonoverlap used two to four males, reports on overlap used at least four males) (38). Female adult pumas exhibit overlapping home ranges in practically every population that has been studied. Although home ranges of pumas of the same sex overlap, pumas use core areas and move about their home ranges in ways that reduce the chances of encountering conspecifics, thus minimizing competition and potentially dangerous associations (55, 56).

Activity and Movements

The diel activity pattern of pumas is mostly crepuscular, although they are sometimes active during midday and at night (21, 30, 55, 73). Activity patterns probably are largely influenced by the activity of the puma's main prey.

Females with nursing cubs tend to travel circuitous daily routes that may cover up to 3 kilometers per day (55). Because of the highly dependent and relatively immobile offspring, mothers must hunt or feed on cached kills and regularly return to cubs to nurse and protect them from predators. By the time cubs are weaned at about 2 months, they are capable of traversing rugged terrain. In addition, because energy requirements of the family unit increase as cubs grow, mothers must hunt over ever broader areas (21). If they are successful, mothers return to their cubs at previous caches or rendezvous sites and bring the cubs to the new kills. Sometimes cubs accompany mothers on hunts. Adult females without cubs tend to have larger daily movements than females with cubs. The larger movements of the lone female may enable her to search for mates and to exploit areas with less competition. Territorial males have the longest daily movements to patrol and defend their large territories and locate estrus females (53, 55, 73).

Philopatry and Dispersal

Puma offspring become independent of mothers when they are 9 to 21 months old (62, 43). Independent subadults either establish their adult home ranges adjacent to or overlapping their natal areas (i.e., philopatry), or they disperse (62, 56). Dispersers leave their natal areas when they are 10 to 33 months old and usually before sexual maturity (69, 74, 62, 38, 43, 75, 43).

Philopatry is generally a characteristic of female pumas (75, 62, 56). When habitat conditions are favorable, females that remain philopatric may enhance their reproductive success. Philopatric females already are in habitat where they are familiar with food sources and potential nurseries. Moreover, philopatric females may already be familiar with adult males (future mates) in the area, and thus minimize the chances of aggressive encounters with unfamiliar males. Although philopatric females risk breeding with fathers, incestuous inbreeding probably occurs too infrequently in pumas to produce deleterious genetic effects in the population (56). Male philopatry has only been documented in Florida where severe habitat restrictions apparently force males to return to the vicinity of their natal areas after failed dispersal attempts (76).

Dispersal of puma offspring from natal areas and even origin populations has been described across North America (6, 71, 62, 75, 38, 63, 43). Dispersal directions of pumas appear to be random; however, pumas tend to use patches of favorable habitat to link their dispersal movements (56, 7). Almost all subadult male pumas disperse from natal areas and emigrate from the subpopulations in which they were born (75, 62, 63, 56). In contrast, less than half of female progeny may disperse from natal areas. Some females establish home ranges in the subpopulation of origin, while others emigrate (56). In general, the dispersal distances of males are two to four times longer than females (75, 63, 43). Anderson and others summarized the dispersal distance for 65 North American pumas; the means were 85.0 kilometers (range = 29–274 kilometers) for 33 males and 31.4 kilometers (range = 9–140 kilometers) for 32 females (75). The longest known dispersal movement is 483 kilometers for a male tagged as a cub in the Big Horn Mountains of northern Wyoming and killed west of Denver, Colorado, at 30 months of age (K. A. Logan, unpublished data). Dispersal distances probably are partially influenced by habitat conditions. Fragmented habitat may impede dispersal, particularly where human development creates barriers (77). However, dispersers from fragmented habitats may have to travel longer distances to find suitable environments in which to establish residency.

Dispersal of puma progeny is adaptive for several reasons: (1) extreme inbreeding is avoided; (2) out-crossing is enhanced; (3) competition for food, space, and mates between siblings and parents is minimized; (4) unoccupied habitat will have a greater likelihood of colonization; and (5) small, isolated populations can be rescued from extinction risks (55, 56; also citing work from 78, 79, 80, 81).

Communication

Communication through visual, tactual, auditory, and olfactory mechanisms may assist pumas to space themselves in the habitat or to find one another (6, 55, 56). Owing to the highly secretive nature of pumas, very little is known about postures, gestures, and tactile responses exhibited by associating wild pumas. However, the results of such communications sometimes are observed (e.g., scarring, mortal injuries, avoidance, pregnancies).

A broad range of vocalizations has been heard from wild pumas, including low gargling growls, spits and hisses, chirps, squeaks and mews, purrs, bird-like whistles, ouch calls, and throaty yowls and caterwauls (K. Logan and L. Sweanor, unpublished data). These vocalizations are important methods of communication and can signal threat or distress, provide comfort, indicate location, and advertise breeding condition (82, 83, 56).

Indirect methods of communication include olfactory cues given by chemicals from the body or glands that persist in tracks, urine, and scats. Often these are in combination with visible cues like scrapes, scratches on trees, scat mounds, and kill caches. Scrapes, which are usually made by territorial male pumas, consist of small mounds of soil, and/or dried vegetation scraped into a pile with the hind feet (6, 56). Sometimes feces and/or urine is deposited on the mound of debris at the end of the scrape. Scrapes are found at prominent locations and along travel ways throughout the male's territory, although their frequency may increase in areas of territorial overlap or around kills (55). Several scrapes may be clustered at particular sites that may be reused at irregular intervals by more than one territorial male. Female pumas, some known to be in estrus, visit scrapes made by males and will sometimes urinate and defecate on them (56). However, females and cubs rarely make scrapes. Scrape sites probably function as "bulletin boards" that pumas use to glean information on the temporal and spatial presence of other pumas in the population (55).

Scratching sites, typically large trees with leaning trunks or exposed roots, are sometimes revisited at irregular intervals, and are used to sharpen the claws (lamella from the claws is shed at the site); however, scent from the feet probably also persists. Pumas may also rub their faces on the trees or other nearby vegetation and rocks to deposit scent from facial glands.

POPULATION DYNAMICS

Density

Puma populations studied in Florida, 12 western states, and two western provinces have generated estimates of population density with a wide range of reliability. This is partially influenced by the different methods used and intensity of the field research

TABLE 17-4

Estimates of Puma Population Composition and Density in North American Studies that Employed Intensive Capture–Recapture and Radiotelemetry Techniques.

Location and Reference	Study Area Size (km²)	Resident		Transients or Subadults	Dependent Offspring	Desity (Pumas/100 km²)	
		Males	Females			Resident Adults	Total
Alberta (62)	780	4–5	8–12	3–10	6–18	1.5–2.2[a]	2.7–4.7
British Columbia (31)	540	1–2	4	2–4	11	0.93–1.1[a]	3.5–3.7
Idaho (6)	520	3	2–6	0–5	1–7	0.96–1.7[a]	1.7–3.5
New Mexico[b] (43): TA	703	4–6	3–9	0–5	3–15	0.84–2.1[c]	2.0–4.3
RA	1,356	7–11	6–16	1–5	6–23	0.94–2.0[c]	1.7–3.9
Utah (63)	1,900	0–4	5–8	0–10	3–14	0.32–0.63[a]	0.58–1.4
Wyoming (71)	741	3	7–9	1–7	13–17	1.4–1.5[a]	3.5–4.6

[a]Density of pumas was estimated when pumas and prey were on winter range.

[b]Study area was divided into two parts. In the reference area (RA) puma numbers were protected over a 10-year period. In the treatment area (TA), pumas were protected for 5 years, then the numbers of adult and subadult pumas were experimentally reduced by 53% and 100%, respectively. After the one-time reduction, the TA was again protected and monitored for the remainder of the 10-year period.

[c]Density was estimated during January of each year in an environment where neither pumas nor prey migrated.

efforts. The most reliable estimates come from six studies that employed intensive search, capture, marking, recapture and radiotelemetry techniques, and defined analytical methods (Table 17–4). In those studies, low densities of 0.3 to 1.5 resident adults or 0.6 to 3.5 total pumas per 100 square kilometers occurred when pumas were hunted during or just prior to study, had significant accidental human-caused mortality, or were experimentally reduced (Table 17–4). Maximum densities reached 0.6 to 2.2 resident adults or 1.4 to 4.7 total pumas per 100 square kilometers. All of the populations (Table 17–4), except New Mexico, were quantified when pumas were restricted to their winter ranges.

Although data are limited, the density estimate of 0.9 adult Florida panthers (resident and transient) per 100 square kilometers falls at the low range of densities for resident adult pumas (Table 17–4). The total population estimated for 5,040 square kilometers of occupied habitat in Florida was about 74 panthers, including adults, subadults, and cubs (72).

Sex and Age Structure

The sex ratio of pumas at birth approximates 1:1 (60, 43). Once pumas become independent subadults, however, fewer males than females appear to be present; this probably occurs because males tend to have greater mortality and emigration rates (62, 43, 43). In the resident adult population, there continue to be fewer males (6, 26, 71, 30, 57, 31, 62, 38, 63, 43). This is expected because only males are territorial, male territories are larger than female home ranges, and females are recruited at higher rates (43). In addition, mortality of adult males may be higher in hunted populations where hunters select males (62).

Little information exists on the age structure of puma populations. A population in New Mexico that was protected from hunting averaged 61% adults, 6% subadults, and 33% cubs. An adjacent population from which pumas were experimentally removed averaged 56% adults, 10% subadults, and 34% cubs (43). A hunted puma population in southwestern Alberta was comprised of 43% to 60% adults, 5% to 37% subadults, and 25% to 49% cubs (62). In a hunted population in Wyoming, population composition for pumas greater than 24 months old, 13 to 23 months old, and 0 to 12 months old averaged 50%, 5%, and 46%, respectively (20). For a hunted population in Idaho, composition of those same age classes averaged 62%, 16%, and 22%, respectively (revision by Hornocker in 13, p. 61). Although these puma populations were subjected to different levels of exploitation, their statistics represent self-sustaining populations.

The oldest reported ages for wild pumas were about 13 years for males and 12 years for females in a protected population in New Mexico (43). In that population, adult males generally lived longer than adult females, reflecting the higher natural survival rates of adult males (43). Sport hunting can reduce the age structure of a population. For example, in a hunted population in Wyoming, the oldest puma was about 7 years old (71). In contrast, captive pumas have lived at least 18 years (13).

Survival and Mortality

Sex- and age-specific survival rates on pumas are scarce. In protected populations in Utah and New Mexico, survival rates of cubs from birth to independence were estimated at 0.67 and 0.71, respectively; however actual survival rates probably were somewhat lower (84, 43). Beier and Barrett, who studied a protected puma population in fragmented habitat in California, estimated that, from birth until establishment of a stable home range, a puma's survival rate was between 0.45 to 0.52 (38). Although a high survival rate (>0.97) was reported for cubs in a hunted population in Alberta, mortality probably was underestimated. Many of the cubs were not detected until they were 0.6 to 1.5 years old (62). Consequently, some cubs could have died at earlier ages when they were more susceptible to mortality. In New Mexico, male and female cubs died at approximately an equal ratio (20 : 18) (43). Sample sizes from other studies are smaller, reporting male : female deaths of 0 : 5, 4 : 2, and 9 : 6 (84, 31, 38).

Limited data from New Mexico suggest that subadult males have a lower survival rate (0.56, $n = 9$) than subadult females (0.88, $n = 16$) (43). The researchers explained that males faced higher risks because they disperse at greater frequencies, travel longer distances from natal areas, and have to compete directly with other males for territory. In contrast, if female subadults disperse at all, distances are shorter, and they generally avoid other pumas. Anderson and others calculated an annual survival rate of 0.64 for pumas (males and females combined) in the 12- to 24-month age class (75). However, they did not indicate the status of those pumas (e.g., dependent, independent, dispersers, or resident adults).

The largest data set on survival rates of adult pumas comes from a protected population in New Mexico where 34 males and 51 females were monitored with radiotelemetry (9–20 males per year, 7–24 females per year). Mean annual survival rates were 0.90 for males and 0.81 for females (43). In protected populations in Utah and California, the annual survival rate for adults (males and females combined) averaged 0.72 and 0.75, respectively (85, 38). Annual survival rates for pumas in a protected Colorado population were 0.69, 0.92, and 0.80 in age classes 24 to 36 months, 36 to 48 months, and 48 to 60 months, respectively (75). Sample sizes in each class were not given. In an Arizona population subjected to heavy predator control and sport hunting pressure, the mean annual survival rates were 0.44 ($n = 4$ per year) for males and 0.78 ($n = 2$ to 5 per year) for females (46). The higher male mortality rate was associated with the greater number of males involved in depredation control (16 males, 6 females killed).

Causes of Mortality

Humans were the major cause of death in puma populations subjected to legal sport hunting (50% to 100%) or control actions to protect livestock (57% to 92%) (5, 68, 71, 62, 28, 46). Humans also were the major cause of death in three puma populations protected from sport hunting for research purposes (26, 42, 75). Mortalities resulted from illegal and legal kills inside and outside of the study areas, predator control, and collisions with vehicles. In California and Florida, where pumas were protected from sport hunting statewide, but where habitats were severely fragmented by human development, vehicle collisions were the first and second (respectively) most important cause of puma deaths (38, 76).

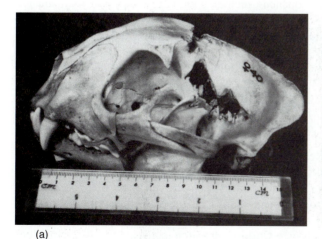
(a)

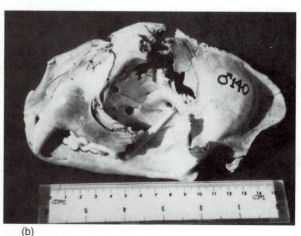
(b)

Figure 17–6 In unhunted puma populations, the most common cause of death may be from other pumas. (a) This 2-year-old female was killed and eaten by a territorial male in the San Andres Mountains of New Mexico. (b) This 14-month-old dispersing male was killed by a territorial male in the San Andres Mountains of New Mexico. (Photographs by K. Logan.)

Infanticide and cannibalism of cubs and intraspecies strife have been documented across the geographic range of pumas in North America (1, 5, 45, 53, 20, 74, 75, 62, 44, 38, 46, 43, 42). In puma populations with minimal human influence, the major cause of death may be pumas themselves (Figure 17–6). For example, in New Mexico, infanticide and cannibalism caused 44% of cub deaths, and intraspecies strife caused 100% and 52% of deaths in subadults and adults, respectively. Male pumas did all of the killing (43). Considering only natural causes of mortality, intraspecies strife was the single greatest identifiable cause of death in Florida panthers, in pumas in California, and in adult pumas in Utah (72, 38, 85).

Although reports are rare, adult, subadult, and cub pumas have been killed by other carnivores, including gray wolves, bears (T. Ruth, Hornocker Wildlife Institute, personal communication), and coyotes (86, 87, 43). Some of the wolf-related mortalities occurred near puma ungulate kills and may indicate interspecific competition. Such incidents may increase as wolves reoccupy more of their former habitat.

Pumas have died from injuries sustained while attempting to kill large prey, such as mule deer, elk, and bighorn sheep (88, 89, 43, 5, 89, 89). Mortality of this type may be more prevalent in pumas that are female, young and inexperienced, or old and infirm (43, 89). Cubs have died after being impaled by porcupine quills (60). Other accidental types of mortality include drowning and rattlesnake bite (46, 43).

Relatively little is known about the effects of parasites and disease in pumas. Anderson tabulated 40 parasites and 23 diseases and pathological conditions reported in pumas and opined "that whatever parasites and diseases are endemic in wild puma none are apt to be population limiting" (13, pp. 38–42). Waid found eight helminth species in pumas in southwestern Texas, which were not reported by Anderson, but concluded the helminths collected from individual pumas had little effect on their condition (66, 13). Rabies, congenital heart defect, mercury poisoning, bacterial infection, and plague have killed pumas (72, 90, 62, 43). Foley concluded that western puma populations are not in imminent danger of catastrophe due to infectious disease, including rabies, canine distemper, feline panleucopenia, feline leukemia, feline immunodeficiency virus, and feline infectious peritonitis (91).

Emigration and Immigration

Emigration and immigration are important components of puma population dynamics. Most progeny from a subpopulation emigrate, and males emigrate at higher rates than females; consequently, most male recruits are immigrants (62, 63, 43, 43). Emigration and the subsequent immigration and recruitment of pumas into other subpopulations and habitats have vital adaptive consequences, including numeric and genetic maintenance and/or enhancement, and colonization of unoccupied habitat (43). Immigrants may comprise as much as one-half of the total recruits into a specific subpopulation. Moreover, emigration and immigration are vital behavioral components that maintain metapopulation dynamics regionwide, particularly where puma habitats are fragmented (43).

Population Increase

Quantitative information on rates of puma population increase is scarce. Logan and others calculated observed rates of increase of 0.24 for adult pumas in a protected population in Utah (using data from Ref. 63) and 0.04 to 0.08 for adult pumas in a hunted population in Alberta (using data from Ref. 62) (7). Long-term research in New Mexico indicates the adult segment of a depressed puma population can increase at observed exponential rates of 0.17 to 0.28 after protection (43). However, the rate of increase slows considerably (in New Mexico the rate declined to 0.05 or less) as adult density approaches or exceeds carrying capacity. During population increases, female recruitment may occur more rapidly than male recruitment because female recruits are mainly comprised of progeny from within the local population, whereas male recruits are mostly immigrants (43). Some populations exhibit density stability over time. For example, a puma population in Utah was stable during the first 7 years of study, and a lightly hunted population in Idaho was stable for 8 years (63, 6).

Population Dynamics

Pumas have adapted to a wide range of habitats in North America, but they appear to favor habitats with vegetation or topographic cover that provide advantages while hunting ungulate prey, refuge for newborn cubs, and landscape linkages for dispersers (70, 92, 77, 43). Just as the species of vegetation are important for the attraction of herbivore prey, vegetation structural characteristics also are important for determining habitat suitability. In north-central Wyoming, pumas preferred mixed conifer and curl leaf mountain mahogany vegetation and steep and rugged topography (greater than 50% slope). Besides providing stalking cover for pumas (increasing prey vulnerability), rugged terrain also provided ungulates with exposed south- and west-facing slopes where food and moderated weather were available in winter (greater prey availability). Douglas fir, juniper/sagebrush-grass, lodgepole pine, and riparian zones and moderate topography (20% to 40% slope) were used in proportion to availability. Sagebrush-grasslands and gentle topography (less than 20% slope) were clearly avoided. Puma kill caches were primarily found in the curl leaf mountain mahogany and mixed conifer vegetation types (70).

Similarly, pumas in south-central Utah preferred pinyon-juniper/lava rock, ponderosa pine/oakbrush, mixed aspen/spruce-fir, and spruce-fir associations. They avoided sagebrush bottomlands, agricultural/pasture lands, slickrock sandstone canyons, and open meadows. Puma kills and caches occurred in disproportionately high numbers in pinyon-juniper/lava rock habitat (92). In Wyoming and Utah, pumas exhibited seasonal elevational movements that followed movements of deer as they sought habitats with more favorable food and weather conditions (70, 92).

Florida panthers preferred dense hardwood hammock communities (93, 72). Pine flatwoods was the second most important vegetation community. Both of these communities commonly had thickets of saw palmetto, which were the most important resting and denning cover for panthers. Vegetation communities that lacked cover or were flooded were avoided (72).

When raising small cubs, females in Utah preferred mixed aspen/spruce-fir and spruce-fir associations with dense understories that contained large lava boulders (92). All 53 puma nurseries examined in the Chihuahua Desert of southern New Mexico occurred in boulder piles, rock outcrops, undercut cliffs, and dense patches of vegetation that provided effective lateral and overhead cover from inclement weather, and most had hiding spaces that cubs could use to escape predators. Examination of 26 birth nurseries indicated females do not prepare a site for their cubs (K. A. Logan and L. L. Sweanor, unpublished data).

EFFECTS ON PREY POPULATIONS

Effects of puma predation on prey populations may vary with ecological conditions. In Idaho, California, Utah, and the northern Yellowstone ecosystem, pumas did not limit elk and deer populations, the major prey in those areas (5, 30, 63, 33). In New Mexico, effects of puma predation on desert mule deer changed as drought modified habitat conditions. During years of adequate precipitation and relatively high fawn production, puma predation rates on deer 1 year old or older were relatively constant despite an increase in puma density, and the deer population increased. However, during a drought that reduced ecological carrying capacity and fawn production, puma predation rates on deer 1 year old or older increased significantly and were related to puma density. Although pumas contributed directly to the deer population decline as the major proximate cause of mortality, habitat quality probably was the ultimate cause for the decline due to increased vulnerability of deer to mortality factors, including predation, malnutrition, and disease (22).

Effects pumas have on small populations of ungulates vary. Pumas may have caused a small population of less than 60 bighorn sheep in the Sierra Nevada Mountains to abandon winter range. This behavioral response may have contributed to reduced nutrient intake, later lambing, and lower lamb survival, which, together with predation, led to an overall population decline (94). On the other hand, puma predation had an insignificant effect on a population of about 125 bighorn sheep in central Idaho wilderness (5). While studying a remnant population of about 40 desert bighorn sheep in southern New Mexico, Logan and others found that the density of adult pumas was independent of sheep mortality rates and the rate of puma predation on sheep (95). Yet, a single puma with a predilection to killing bighorn sheep can cause sporadically high rates of predation in small populations (95, 96). Puma predation on female pronghorn in Arizona was high enough to stabilize or decrease the pronghorn population in rugged, brushy terrain, but puma predation is probably insignificant in rolling grassland habitat where little cover exists (37). A feral horse population on the California–Nevada border was stabilized by high puma predation on foals (48).

In habitats where other large carnivores are present, those carnivores may scavenge on or displace pumas from their kills, and potentially force pumas to increase their killing frequency. Coyotes, bears, and wolves scavenge on and displace pumas from their kills (35, 97, 52, 34). Murphy and others estimated that pumas displaced from kills by bears lost 17% to 26% of their daily energy requirements (52). Given the potential for wolves and pumas to depress prey populations, gauging the effects of interactions between these carnivores will become more important as wolves recolonize the intermountain West and the Southwest (98, 99, 37, 22).

MANAGEMENT AND CONSERVATION

Issues

Major issues that drive puma management today include public safety, protection of private property, predation on large mammals, sport hunting effects and ethics, and species conservation. Although dangerous encounters between humans and pumas are rare (i.e., 15 documented deaths and 59 nonfatal injuries in the United States and Canada between 1890 and 1996), public safety is an important concern (100). Records suggest that encounters and attacks have increased during the last 30 years, concomitant with increases in puma and human densities and human use of puma habitats (49, 101, 102, 50). Subadult and/or underweight pumas are most frequently involved in attacks (49, 101, 103, 101). This may be expected, because young, inexperienced animals are most likely to be dispersing or attempting to establish home ranges in marginal habitats where they face encroaching human populations and potentially dangerous encounters with humans (49).

The loss of private property in the form of livestock or pets is problematic in certain areas. Tully reported that 15 western states and provinces had suffered puma predation of livestock (i.e., depredation) (104). The most significant losses generally occurred where domestic sheep grazed on open range within puma habitat (104, 50). The value of livestock lost to pumas was highly variable. Although some states and provinces, including Alberta and Idaho, have relatively low rates of livestock depredation and attendant monetary losses, other areas have reported significant losses, with increasing trends in depredation complaints and/or losses (104, 105). Colorado and Utah reported livestock losses of more than $100,000 per year from 1994 through 1996, with increases in losses since the 1980s (T. Beck, Colorado Division of Wildlife, and B. Blackwell, Utah Division of Wildlife Resources, personal communication). Confirmed depredation incidents in California increased from 4 in 1972 to 331 in 1995. Livestock depredation was increasing in conjunction with an apparent increase in the puma population, whereas depredation on pets was related to increased human development in puma habitat (50).

Big game hunters and wildlife managers are concerned about potential depressive effects of puma predation on ungulates. Some hunters believe they compete with pumas for game, while managers wonder what effects predation has on ungulate populations, particularly during downtrends that may have been initiated by other events, such as severe winter or drought (B. Blackwell, Utah Division of Wildlife Resources; M. Austin, British Columbia Ministry of the Environment, Wildlife Branch, personal communication; K. Logan and L. Sweanor, records on public meetings in New Mexico). Moreover, puma predation may jeopardize rare ungulates, such as desert bighorn sheep in southern California (94).

Sport hunting of pumas is an issue that has sociopolitical and biological ramifications. Hunting provides recreational opportunity and a direct income for outfitters. Although hunters want continued opportunities to pursue pumas, other publics are concerned about the ethical justification of hunting. Public sentiment led to a prohibition on sport hunting pumas in California since 1972 and to a ban on the use of dogs to hunt pumas in Oregon and Washington in 1994 and 1996, respectively.

Biological effects of hunting on puma populations may be relatively benign if offtake is compensatory. But more aggressive prescriptions can depress puma populations if offtake is additive, and especially if females and progeny are impacted (63, 43, 106). Females are critical because they produce future recruits for local and adjacent populations. Moreover, orphaned cubs left behind by hunter-killed females have poor survival and may become nuisance animals (53, 107, 43). A hunter's dogs may also find and kill cubs during a chase (65, 26, 20, 107, 75). Special pursuit seasons, where hunters are allowed to use dogs to chase pumas but not kill them, may not be without costs. Accidental mauling of cubs and even adults may be significant in high-use areas, and repeated pursuits of the same animal may cause deleterious physiological effects or death (108, 109, 107).

Habitat loss and fragmentation are the greatest threats to long-term puma conservation. The West is the fastest growing region in the United States (population is increasing at a rate of 1.8% per year), resulting in ever increasing fragmentation and loss of puma habitat (110). These changes are critical, because large expanses of habitat and wild landscape linkages are essential for the maintenance of individual puma populations and especially for the regionwide metapopulations they may form (111, 112, 95). Immediate effects of habitat degradation and human development include increased incidents of cougar–human encounters, pet depredation, killing of nuisance pumas, puma deaths from vehicle collisions, and disruption of natural dispersal patterns (111, 95, 50). Ultimate detrimental effects may result in population isolation and depression that could lead to genetic deterioration and finally extinction (111, 61, 113). For example, the extreme isolation of the Florida panther may have resulted in reduced fitness that threatens to cause extinction of the subspecies. Semen quality in Florida panthers is the worst recorded in felids, with more than 90% developmental abnormalities in spermatozoa and 44% incidence of cryptorchidism (113, 114, 115). Moreover, the subspecies has a high incidence of heart defects (116).

Strategies

Wildlife managers face the daunting task of responding to dynamic needs of people while pursuing the long-term goal of conservation. To adequately adapt to those needs, management agencies need a repertoire of management strategies anchored in a reliable understanding of the biology and ecology of the species. To develop the repertoire, puma management should be approached as experiments that involve setting specific objectives, executing management prescriptions, and monitoring programs that provide reliable data to test management performance. Without an experimental framework, puma management will largely be an activity of reactionism with a high probability of mistakes and failure. Such situations jeopardize wildlife and the professional credibility of agencies.

Seven main strategies used by managers to address specific issues identified by publics are control, sustained harvest, translocation, protection, monitoring, research, and education. Efforts to control pumas may occur in areas where livestock, public safety, or, in some cases, other wild species, are threatened. All western states and provinces allow the killing of individual pumas that threaten public safety or private property (117). Most states and provinces attempt to control puma numbers through sport hunting. Utah reported a stabilized depredation trend during 1995 and 1996, in conjunction with an increasing puma harvest and apparent downturning puma population (B. Blackwell, Utah Division of Wildlife Resources, personal communication). Based on that state's crude population estimates, Utah hunters may have killed between 19% and 29% of the statewide puma population during the 1996–1997 season. Puma population depression is achieved when kill rates exceed growth rates, which, depending on population demography, may range from 5% to 28% of adult pumas (43).

Most western states and provinces list the puma as a game animal, and have established regulations that contributed to today's viable populations (Table 17–1). Quotas are used by most agencies to regulate and direct hunting pressure. Sex-based quotas have also been used to give females more protection. Quotas may improve hunt quality by regulating hunting pressure and by sustaining puma population levels and harvest (118). When the objective is sustained harvest, Logan and others recommended that initial annual harvest should not exceed 8% of the adult males (i.e., observed rate of increase in a protected population), and hunting of females should be restricted (43). As empirical field data on population trends are collected, harvest rates can be adjusted to fulfill specific objectives.

Translocation has been used to deal with nuisance pumas (Table 17–1), but rarely has the success of this option been monitored. In the only intensive study on puma translocation of wild animals with known life histories, Ruth and others found that independent pumas of dispersal age (less than 27 months old) were the best candidates for success (119). Furthermore, they recommended when translocation versus euthanasia should be used, the size and character of release areas, and characteristics of pumas for augmenting or reestablishing populations. However, the need for nuisance puma management and future augmentation and reintroduction efforts could be minimized with a proactive approach where large expanses of puma habitat with connecting corridors are maintained to accommodate natural dispersal mechanisms (119).

Protection is a strategy that recognizes critical risks and uncertainties in puma management and the need for habitat. Protected populations function as robust biological savings accounts that contribute to population resilience by countering management-related mistakes in exploited subpopulations and increased mortality in adjacent fragmented habitat (120, 43, 121). Human exploitation through hunting and control probably disrupts the traditional patterns of natural selection in pumas, the long-term effects of which are unknown (43, 106). Moreover, areas subjected to control, heavy sport hunting, or other human-caused mortality may impact other puma subpopulations in the region by functioning as population sinks, where pumas have high rates of mortality. High mortality in dispersing pumas, the potential recruits of subpopulations, probably would lower metapopulation resiliency. Protected areas enable subpopulations to evolve relatively naturally, and thus provide diversity of genotypes in a time when selection is heavily influenced by humans. Dispersal of pumas from protected areas ensures that those genotypes will be carried to human-impacted subpopulations (43).

Self-sustaining puma populations are dependent on large expanses of habitat and wild landscape linkages. Because public lands presently provide the bulk of puma habitat, management of those lands should consider the vital needs of puma populations (i.e., prey, cover, security, linkages). Private lands are important too, particularly those that provide seasonal habitats for ungulate prey (e.g., low elevation winter ranges, migration corridors). Habitat on private lands is essential to the conservation of the endangered Florida panther (112). Thus, for long-term conservation, innovative economic incentives may be needed to compel private landowners to conserve puma habitat (see Ref. 112).

Monitoring is essential if managers are to know if management prescriptions are effective. Puma population monitoring has relied principally on harvest statistics. But such indirect methods make data interpretation difficult because their relationships with puma population dynamics are poorly understood. For example, sex ratios in the harvest may not represent the population because of male-biased hunter selection. In addition, techniques for aging pumas, including tooth eruption, coloration, and wear patterns, premolar cementum annuli counts, pulp cavity to canine tooth diameter ratios, and skull suture characteristics (K. Greer, Montana Fish, Wildlife and Parks, unpublished data), have unknown estimates of accuracy, precision, or bias (47, 122, 123).

Management and Conservation

Although, reliable estimates of local puma population dynamics can be achieved with intensive capture, mark, recapture and radiotelemetry techniques, such efforts are too expensive and time consuming to apply to large regions (5, 13, 7). Hence, track counts are an alternative. Track counts to obtain population trends could focus on areas of prime interest (e.g., areas of control, high harvest, high puma–human conflicts). In addition, trends in protected areas could function as reference points to be compared with trends in human impacted areas. Useful techniques and guidelines have been developed to index puma abundance in snow-covered and dry terrain (124, 125, 126). Track counts to determine puma population trends have been applied at the statewide level, but the relationship between the number of track sets and the population was poorly defined (127, 128).

Reliable research findings should be assimilated into management. Focal topics for puma research include population dynamics, genetics, interactions with prey, habitat use, interactions with humans, population monitoring techniques, and methods of livestock husbandry. Studies of puma population dynamics could help managers develop population objectives, assess impacts of harvest and control, and provide an understanding of how puma metapopulations function. Genetic studies could illuminate puma taxonomy, population structure, and potential human impacts (i.e., population isolation, reduction) on populations and subspecies. Long-term (greater than 10 years) experimental research will be needed to determine to what extent puma predation limits or regulates prey populations (129). To date, studies have not been long enough to include all of the phases of fluctuation of a population of pumas and its ungulate prey. Knowing how pumas use habitat should assist land managers to identify potential degradation and fragmentation, locations of dispersal and migration corridors, and areas of puma–human conflict. Behavioral studies of pumas and people may shed light on how population or behavioral modification may minimize dangerous encounters. Because of the difficulty in censusing pumas, reliable, cost-effective methods are needed by managers to track population trends. Livestock growers in puma habitat may benefit from alternate livestock types (e.g., cattle instead of sheep) or husbandry practices (e.g., guard dogs, herders) that are practical and cost effective. Particularly for population and puma–prey interaction studies, *a priori* multifactorial hypotheses should be tested and empirical data should be collected for inclusion into models. Habitat models could be developed using Geographic Information System technology. In all research, key patterns should be identified to develop theory and to guide future research.

Education, key to the success of any puma management approach, should include both the interested public and wildlife managers. Information on how people should behave while living or recreating in puma habitat has been effective in reducing the probability of dangerous encounters (T. Beck, Colorado Division of Wildlife, personal communication) (100). Publics informed about puma biology also take a more active or understanding role in how pumas are managed by agencies. Likewise, educated agency personnel may have greater success identifying management objectives, devising realistic strategies, and designing relevant research. Furthermore, managers need to be informed about the needs and values that the public places on pumas to assist them in setting management goals and objectives. Integrated participation by educated publics and managers is essential to maintain broad flexibility (i.e., options) in puma management and conservation.

All of these strategies could be incorporated into an adaptive landscape level management plan like the one recently developed for New Mexico (130, 131, 132). The plan provides for the needs of people while also considering the biological needs for puma conservation. This is accomplished using a zone management approach that includes experimental puma control in localized problem areas to protect private property, humans, and endangered species; sport hunting sustained by harvest quotas and protection of females and cubs; and a long-term conservation strategy that uses large protected areas or refuges in excess of 2,400 square kilometers. Monitoring the effects of imposed management prescriptions on representative zones is fundamental to this plan to determine if objectives are being met.

The puma is the last large obligate carnivore that still occurs in viable populations in the wilds of North America, except for the occupied wolf ranges of Canada, Alaska, and Minnesota. The puma's broad geographic distribution, solitary nature, and presence in some of the most rugged and remote habitats helped it escape the regional extinctions that befell the other large carnivores. The recovery of the puma in the West has occurred only during the last three decades, the equivalent of three puma lifetimes. Not only is this recovery an indication of our management successes, but also of the resiliency of the puma. Today, puma populations are "high" relative to our collective memories. What they were historically is conjecture. But pumas today also are facing a completely different world than they would have encountered even 100 years ago. Because of our growing human population, puma habitats and landscape linkages are continually shrinking. Consequently, the recent increases in puma populations may not be sustainable. The ecological role of pumas, including their ability to help dampen oscillations in prey populations, structure biological communities, and direct the evolution of their prey are all reasons why pumas should be conserved (5, 22). Moreover, they can be used as umbrella species to define minimum areas required to preserve ecologically intact ecosystems (8, 133). In the long run, if humans are to successfully conserve pumas in self-sustaining populations, then people living in or impacting their wild environments will have to be educated and caring. Furthermore, wildlife managers will require a thorough understanding of the animal and potential methods for achieving success in dealing with short-term problems and long-term conservation goals.

SUMMARY

Pumas in North America were eliminated from most of the East and severely reduced in the West by the early 1900s. However, protective measures since 1965 resulted in the healthy puma populations observed throughout much of the West today. Although as many as 13 subspecies of puma have been described in North America, recent genetic studies contest this degree of differentiation.

Among the largest obligate carnivores in North America, pumas have specialized morphology and physiology enabling them to individually subdue the largest ungulates on the continent. Deer are their most important prey, though other wild and domestic ungulates and small animals can be important where they are abundant and vulnerable. Kill and consumption rates vary depending on size, reproductive status, likelihood of spoilage, and presence of competing carnivores.

Pumas are polygamous and promiscuous. Reproductive maturity occurs at about 2 years. Impregnated females give birth to an average of three cubs after about 91 days of gestation. Cubs can be born at any time of the year, but birth pulses may correspond to periods with higher prey abundance and vulnerability. Successful females can give birth every 1.5 to 2 years.

Except for family groups, pumas are generally solitary. Females singly care for young and avoid other pumas, while males defend large territories that support multiple mates. Home-range size and shape vary depending on the social and reproductive status of the animal, the quality and quantity of the habitat, migrations of prey, and puma population density. Most young disperse from their natal areas after independence, with males generally dispersing farther than females.

Low puma population densities are typical, and range from 0.6 to 4.7 pumas per 100 square kilometers. The sex ratio at birth is even, but tends to favor females in the subadult and adult age classes. Humans are the major source of mortality in hunted populations, whereas pumas themselves may be responsible for most mortality in protected populations. Population growth occurs by recruitment of progeny from the local population and of young pumas dispersing from regional populations.

Pumas occupy a wide range of habitats, but favor areas that provide sufficient prey and cover for hunting and protecting young cubs. The effect pumas have on prey populations may vary with ecological conditions. Although puma predation may not limit ungulate populations in some areas, in others, pumas may depress ungulate numbers. Puma predation may also dampen prey population oscillations.

Puma management is driven by major issues, including public safety, protection of private property, predation on big game, sport hunting effects and ethics, and species conservation. Strategies that managers use to address these issues include control, sustained harvest, translocation, protection, monitoring, research, and education. Although puma populations are presently doing well in the West, the growing human population threatens their future principally by altering habitat. Ecological reasons for conserving the puma include its role in shaping biological communities and its importance as an umbrella species. Management and the long-term conservation of the puma will require informed wildlife managers engaged with informed and caring publics.

LITERATURE CITED

1. YOUNG, S. P. 1946. History, life habits, economic status, and control, Part 1. Pages 1–173 *in* S. P. Young and E. A. Goldman, eds. *The Puma, Mysterious American Cat.* The American Wildlife Institute, Washington, DC.

2. NOWAK, R. M. 1976. *The Cougar in the United States and Canada.* U.S. Department of the Interior, Fish and Wildlife Service, Washington, DC, and New York Zoological Society, New York.

3. MCBRIDE, R. T., R. M. MCBRIDE, J. L. CASHMAN, and D. S. MAEHR. 1993. Do mountain lions exist in Arkansas? *Proceedings Annual Conference of Southeast Fish and Wildlife Agencies* 47:394–402.

4. NOWELL, K., and P. JACKSON, eds. 1996. *Wild Cats: Status and Conservation Action Plan.* IUCN/SSC Cat Specialist Group, Gland, Switzerland.

5. HORNOCKER, M. G. 1970. An analysis of mountain lion predation upon mule deer and elk in the Idaho Primitive Area. *Wildlife Monograph* 21.

6. SEIDENSTICKER, J. C. IV., M. G. HORNOCKER, W. V. WILES, and J. P. MESSICK. 1973. Mountain lion social organization in the Idaho Primitive Area. *Wildlife Monograph* 35.

7. LOGAN, K. A., L. L. SWEANOR, T. K. RUTH, and M. G. HORNOCKER. 1996. *Cougars of the San Andres Mountains, New Mexico.* Final Report, Federal Aid in Wildlife Restoration Project W-128-R. New Mexico Department of Game and Fish, Santa Fe.

8. NOSS, R. F. , H. B. QUIGLEY, M. G. HORNOCKER, T. MERRILL, and P. C. PAQUET. 1996. Conservation biology and carnivore conservation in the Rocky Mountains. *Conservation Biology* 10:949–963.

9. LOGAN, K. A., and L. L. SWEANOR. 1997. The cougars of New Mexico: Balancing the needs of cougars and people. *New Mexico Wildlife* 42:4–9.

10. BARNES, C. T. 1960. *The Cougar or Mountain Lion.* The Ralton Company, Salt Lake City, UT.

11. KURTEN, B. 1976. Fossil puma (Mammalia: Felidae) in North America. *Netherlands Journal of Zoology* 26:502–534.

12. O'BRIEN, S. J. 1996. Molecular genetics and phylogenetics of the Felidae. Pages xiii–xxiv *in* K. Nowell and P. Jackson, eds., *Wild Cats: Status and Conservation Action Plan.* IUCN/SSC Cat Specialist Group, Gland, Switzerland.

13. ANDERSON, A. E. 1983. A critical review of literature on puma (*Felis concolor*). Special Report No. 54. Colorado Division of Wildlife, Fort Collins.

14. WOZENCRAFT, W. C. 1993. Order Carnivora. Pages 286–346 *in* D. E. Wilson and D. M. Reeder, eds., *Mammal Species of the World: A Taxonomic and Geographic Reference,* 2nd ed. Smithsonian Institution Press, Washington DC.

15. NOWAK, R. M., and J. L. PARADISO. 1983. *Walker's Mammals of the World, Volume II.* Johns Hopkins University Press, Baltimore, MD.

16. GOLDMAN, E. A. 1946. Classification of the races of the puma, Part 2. Pages 177–302 *in* S. P. Young and E. A. Goldman, eds., *The Puma, Mysterious American Cat.* The American Wildlife Institute, Washington, DC.

17. McIvor, D. E., J. A. Bissonette, and G. S. Drew. 1995. Taxonomic and conservation status of the Yuma mountain lion. *Conservation Biology* 9:1033–1040.

18. Kurten, B. 1973. Geographic variation in size in the puma (*Felis concolor*). *Commentationes Biologicae* 63:3–8.

19. Musgrave, M. E. 1926. Some habits of mountain lions in Arizona. *Journal of Mammalogy* 7:282–285.

20. Logan, K. A. 1983. Mountain lion population and habitat characteristics in the Big Horn Mountains of Wyoming. Thesis. University of Wyoming, Laramie.

21. Ackerman, B. B. 1982. Cougar predation and ecological energetics in southern Utah. M.S. thesis. Utah State University, Logan.

22. Logan, K. A., L. L. Sweanor, and M. G. Hornocker. 1996. Cougars and desert mule deer. Pages 193–229 *in* K. A. Logan, L. L. Sweanor, T. K. Ruth, and M. G. Hornocker, eds., *Cougars of the San Andres Mountains, New Mexico*. Final Report, Federal Aid in Wildlife Restoration Project W-128-R. New Mexico Department of Game and Fish, Santa Fe.

23. Robinette, W. L., J. S. Gashwiler, and O. W. Morris. 1959. Food habits of the cougar in Utah and Nevada. *Journal of Wildlife Management* 23:261–273.

24. Hibben, F. C. 1937. A preliminary study of the mountain lion (*Felis oregonensis* ssp.). *University of New Mexico Bulletin, Biology Series* 5:1–59.

25. Donaldson, B. 1975. Mountain lion research. Job Progress Report, P-R Project W-93-R-17, Work Plan 15, Job 1. New Mexico Game and Fish Department.

26. Shaw, H. G. 1977. Impact of mountain lion on mule deer and cattle in northwestern Arizona. Pages 17–32 *in* R. L. Phillips and C. Jonkel, eds., *Proceedings of the 1975 Predator Symposium*. University of Montana, Missoula.

27. Leopold, B. D., and P. R. Krausman. 1986. Diets of 3 predators in Big Bend National Park, Texas. *Journal of Wildlife Management* 50:290–295.

28. Smith, T. E., R. R. Duke, M. J. Kutilek, and H. T. Harvey. 1986. Mountain lions (*Felis concolor*) in the vicinity of Carlsbad Caverns National Park, New Mexico, and Guadalupe Mountains National Park, Texas. Final Report. National Park Service, Washington, DC.

29. Cashman, J. L., M. Peirce, and P. R. Krausman. 1992. Diets of mountain lions in southwestern Arizona. *Southwestern Naturalist* 37:324–326.

30. Hopkins, R. A. 1989. Ecology of the puma in the Diablo Range, California. Thesis. University of California, Berkeley.

31. Spreadbury, B. 1989. Cougar ecology and related management implications and strategies in southeastern British Columbia. Master of Environmental Design project. University of Calgary, Alberta.

32. Ross, P. I., and M. G. Jalkotzy. 1996. Cougar predation on moose in southwestern Alberta. *Alces* 32:1–8.

33. Murphy, K. M., G. S. Felzien, M. G. Hornocker, and T. Lemke. 1998. Cougar food habits, prey selection, and predation rates in the northern Yellowstone ecosystem. Pages 2–62 *in* K. M. Murphy, The ecology of the cougar (*Puma concolor*) in the northern Yellowstone ecosystem: Interactions with prey, bears, and humans. Ph.D. thesis. University of Idaho, Moscow.

34. Kunkel, K. E. 1997. Predation by wolves and other large carnivores in northwestern Montana and southeastern British Columbia. Ph.D. thesis. University of Montana, Missoula.

35. Harrison, S. 1990. Cougar predation on bighorn sheep in the Junction Wildlife Management Area, British Columbia. M.S. thesis. University of British Columbia, Vancouver.

36. Maehr, D. S., R. C. Belden, E. D. Land, and L. Wilkins. 1990. Food habits of panthers in southwest Florida. *Journal of Wildlife Management* 54:420–423.

37. Ockenfels, R. A. 1994. Factors affecting adult pronghorn mortality rates in central Arizona. Arizona Game and Fish Department, *Wildlife Digest* 16.

38. Beier, P. and R. H. Barrett. 1993. *The Cougar in the Santa Ana Mountain Range, California*. Final Report. Orange County Cooperative Mountain Lion Study, CA.

39. Ackerman, B. B., F. G. Lindzey, and T. P. Hemker. 1984. Cougar food habits in southern Utah. *Journal of Wildlife Management* 48:147–155.

40. Spalding, D. J., and J. Lesowski. 1971. Winter food of the cougar in south-central British Columbia. *Journal of Wildlife Management* 35:378–381.

41. Hemker, T. P., F. G. Lindzey, B. B. Ackerman, and A. J. Button. 1982. Survival of cougar cubs in a non-hunted population. Pages 327–332 *in* S. D. Miller and D. D. Everett, eds., *Cats of the World: Biology, Conservation and Management.* National Wildlife Federation, Washington, DC.

42. Spreadbury, B. R., K. Musil, J. Musil, C. Kaisner, and J. Kovak. 1996. Cougar population characteristics in southeastern British Columbia. *Journal of Wildlife Management* 60:962–969.

43. Logan, K. A., L. L. Sweanor, and M. G. Hornocker. 1996. Cougar population dynamics. Pages 22–113 *in* K. A. Logan, L. L. Sweanor, T. K. Ruth, and M. G. Hornocker, eds., *Cougars of the San Andres Mountains, New Mexico.* Final Report, Federal Aid in Wildlife Restoration Project W-128-R. New Mexico Department of Game and Fish, Santa Fe.

44. Williams, J. S. 1992. Population characteristics and habitat use of mountain lions in the Sun River area of northern Montana. M.S. thesis. Montana State University, Bozeman.

45. McBride, R. T. 1976. The status and ecology of the mountain lion *Felis concolor stanleyana* of the Texas–Mexico border. M.S. thesis. Sul Ross University, Alpine.

46. Cunningham, S. C., L. A. Haynes, C. Gustavson, and D. D. Haywood. 1995. Evaluation of the interaction between mountain lions and cattle in the Aravaipa-Klondyke area of southeast Arizona. Technical Report 17. Arizona Game and Fish Department, Phoenix.

47. Ashman, D., G. C. Christensen, M. L. Hess, G. K. Tsukamoto, and M. S. Wickersham. 1983. The mountain lion in Nevada. Federal Aid Wildlife Restoration Final Report, Project No. W-48-15. Nevada Fish and Game Department.

48. Turner, J. W., M. L. Wolfe, and J. F. Kirkpatrick. 1992. Seasonal mountain lion predation on a feral horse population. *Canadian Journal of Zoology* 70:929–934.

49. Aune, K. E. 1991. Increasing mountain lion populations and human-mountain lion interactions in Montana. Pages 86–94 *in* C. E. Braun, ed., *Mountain Lion–Human Interaction: Symposium and Workshop,* Denver, CO, April 24–26, 1991. Colorado Division of Wildlife, Denver.

50. Torres, S. G., T. M. Mansfield, J. E. Foley, T. Lupo, and A. Brinkhaus. 1996. Mountain lion and human activity in California: testing speculations. *Wildlife Society Bulletin* 24:451–460.

51. Connolly, E. J., Jr. 1949. Food habits and life history of the mountain lion (*Felis concolor hippolestes*). M.S. thesis. University of Utah, Salt Lake City.

52. Murphy, K. M., G. S. Felzien, M. G. Hornocker, and T. K. Ruth. 1997. Encounter competition between bears and cougars: some ecological implications. *Ursus* 10:55–60.

53. Hemker, T. P. 1982. Population characteristics and movement patterns of cougars in southern Utah. M.S. thesis. Utah State University, Logan.

54. Jalkotzy, M. G., I. Ross, and J. R. Gunson. 1992. Management plan for cougars in Alberta. Wildlife Management Planning Series, No. 5. Alberta Forestry, Lands, and Wildlife. Fish and Wildlife Division, Edmonton.

55. Sweanor, L. L. 1990. Mountain lion social organization in a desert environment. Thesis. University of Idaho, Moscow.

56. Sweanor, L. L., K. A. Logan, and M. G. Hornocker. 1996. Cougar social organization. Pages 114–192 *in* K. A. Logan, L. L. Sweanor, T. K. Ruth, and M. G. Hornocker, eds., *Cougars of the San Andres Mountains,* New Mexico. Final Report, Federal Aid in Wildlife Restoration Project W-128-R. New Mexico Department of Game and Fish, Santa Fe.

57. Maehr, D., J. C. Roof, E. D. Land, and J. W. McCowan. 1989. First reproduction of a panther (*Felis concolor coryi*) in southwestern Florida, U.S.A. *Mammalia* 53:129–131.

58. Bonney, R. C., H. D. M. Moore, and D. M. Jones. 1981. Plasma concentrations of oestradiol-17β and progesterone, and laparoscopic observations of the ovary in the puma (*Felis concolor*) during oestrus, pseudopregnancy and pregnancy. *Journal of Reproductive Fertilization* 63:523–531.

59. Eaton, R. L. 1976. Why some felids copulate so much. *World's Cats* 3:73–94.

60. Robinette, W. L., J. S. Gashwiler, and O. W. Morris. 1961. Notes on cougar productivity and life history. *Journal of Mammalogy* 42:204–217.

61. Maehr, D. and G. B. Caddick. 1995. Demographics and genetic introgression in the Florida panther. *Conservation Biology* 9:1295–1298.

62. Ross, P. I. and M. G. Jalkotzy. 1992. Characteristics of a hunted population of cougars in southwestern Alberta. *Journal of Wildlife Management* 56:417–426.

63. LINDZEY, F. G., W. D. VAN SICKLE, B. B. ACKERMAN, D. BARNHURST, T. P. HEMKER, and S. P. LAING. 1994. Cougar population dynamics in southern Utah. *Journal of Wildlife Management* 58:619–624.

64. MURPHY, K. M., M. CULVER, M. MENOTTI-RAYMOND, V. DAVID, M. G. HORNOCKER, and S. J. O'BRIEN. 1998. Cougar reproductive success in the northern Yellowstone ecosystem. Pages 78–112 *in* K. M. Murphy, The ecology of the cougar (*Puma concolor*) in the northern Yellowstone ecosystem: Interactions with prey, bears, and humans. Ph.D. thesis. University of Idaho, Moscow.

65. HORNOCKER, M. G. 1969. Winter territoriality in mountain lions. *Journal of Wildlife Management* 33:457–464.

66. WAID, D. D. 1990. Movements, food habits, and helminth parasites of mountain lions in southwestern Texas. Ph.D. thesis. Texas Tech University, Lubbock.

67. ACKERMAN, B. B., F. G. LINDZEY, and T. P. HEMKER. 1986. Predictive energetics model of cougars. Pages 333–352 *in* S. D. Miller and D. Everett, eds., *Proceedings of the International Cat Symposium,* Kingsville, TX.

68. MURPHY, K. M. 1983. Characteristics of a hunted population of mountain lions in western Montana. M.S. thesis. University of Montana, Missoula.

69. HEMKER, T. P., F. G. LINDZEY, and B. B. ACKERMAN. 1984. Population characteristics and movement patterns of cougars in southern Utah. *Journal of Wildlife Management* 48:1275–1284.

70. LOGAN, K. A. and L. L. IRWIN. 1985. Mountain lion habitats in the Big Horn Mountains, Wyoming. *Wildlife Society Bulletin* 13:257–262.

71. LOGAN, K. A., L. L. IRWIN, and R. SKINNER. 1986. Characteristics of a hunted mountain lion population in Wyoming. *Journal of Wildlife Management* 50:648–654.

72. MAEHR, D., E. D. LAND, and J. C. ROOF. 1991. Social ecology of Florida panthers. *National Geographic Research and Exploration* 7:414–431.

73. BEIER, P., D. CHOATE, and R. H. BARRETT. 1995. Movement patterns of mountain lions during different behaviors. *Journal of Mammalogy* 76:1056–1070.

74. MAEHR, D. 1990. *Florida Panther Movements, Social Organization, and Habitat Utilization.* Final Report. Bureau of Wildlife Research, Florida Game and Fresh Water Fish Commission.

75. ANDERSON, A. E., D. C. BOWDEN, D. M. KATTNER. 1992. The puma on Uncompahgre Plateau, Colorado. Technical Publication No. 40. Colorado Division of Wildlife.

76. MAEHR, D. S. 1997. The comparative ecology of bobcat, black bear, and Florida panther in South Florida. *Bulletin of the Florida Museum of Natural History* 40(1):1–176.

77. BEIER, P. 1995. Dispersal of juvenile cougars in fragmented habitat. *Journal of Wildlife Management* 59:228–237.

78. GREENWOOD, P. J. 1980. Mating systems, philopatry and dispersal in birds and mammals. *Animal Behavior* 28:1140–1162.

79. DOBSON, F. S. 1982. Competition for mates and predominant juvenile male dispersal in mammals. *Animal Behavior* 30:1183–1192.

80. MOORE, J., and R. ALI. 1984. Are dispersal and inbreeding avoidance related? *Animal Behavior* 32:94–112.

81. HORNOCKER, M. G., and T. BAILEY. 1986. Natural regulation in three species of felids. Pages 211–220 *in* S. D. Miller and D. D. Everett, eds., *Cats of the World: Biology, Conservation and Management.* National Wildlife Federation, Washington, DC.

82. PETERS, G., and W. C. WOZENCRAFT. 1989. Acoustic communication by fissiped carnivores. Pages 14–56 *in* G. L. Gittleman, ed., *Carnivore Behavior, Ecology and Evolution.* Cornell University Press, Ithaca, NY.

83. PADLEY, W. D. 1997. Mountain lion (*Felis concolor*) vocalizations in the Santa Ana Mountains, California. Page 89 *in* W. D. Padley ed., *Proceedings of the Fifth Mountain Lion Workshop,* San Diego, CA.

84. LINDZEY, F. G., B. B. ACKERMAN, D. BARNHURST, T. BECKER, T. P. HEMKER, S. P. LAING, C. MECHAM, and W. D. VANSICKLE. 1989. *Boulder-Escalante Cougar Project: Final Report.* Utah Division of Wildlife Research, Salt Lake City.

85. LINDZEY, F. G., B. B. ACKERMAN, D. BARNHURST, and T. P. HEMKER. 1988. Survival rates of mountain lions in southern Utah. *Journal of Wildlife Management* 52:664–667.

86. WHITE, P. A., and D. K. BOYD. 1989. A cougar, *Felis concolor,* killed and eaten by gray wolves, *Canis lupus,* in Glacier National Park, Montana. *Canadian Field-Naturalist* 103(3):408–409.

87. BOYD, D. K., and G. K. NEALE. 1992. An adult cougar (*Felis concolor*) killed by gray wolves (*Canis lupus*) in Glacier National Park, Montana. *Canadian Field-Naturalist* 106:524–525.

88. GASHWILER, J. S. and W. L. ROBINETTE. 1957. Accidental fatalities of the Utah cougar. *Journal of Mammalogy* 38:123–126.

89. ROSS, P. I., M. G. JALKOTZY, and P. DAOUST. 1995. Fatal trauma sustained by cougars, *Felis concolor,* while attacking prey in southern Alberta. *Canadian Field-Naturalist* 109:261–263.

90. BASS, O. L., and D. S. MAEHR. 1991. Do recent panther deaths in Everglades National Park suggest an ephemeral population? *National Geographic Research and Exploration* 7:427.

91. FOLEY, J. E. 1997. The potential for catastrophic infectious disease outbreaks in populations of mountain lions in the western United States. Pages 29–36 *in* W. D. Padley, ed., *Proceedings of the Fifth Mountain Lion Workshop.* San Diego, CA.

92. LAING, S. P. and F. G. LINDZEY. 1991. Cougar habitat selection in south-central Utah. Pages 27–37 *in* C. E. Braun, ed., *Proceedings: Mountain Lion–Human Interaction: Symposium and Workshop,* Denver, CO, April 24–26, 1991. Colorado Division of Wildlife, Denver.

93. BELDEN, R. C., W. B. FRANKENBERGER, R. T. MCBRIDE, S. T. SCHWIKERT. 1988. Panther habitat use in southern Florida. *Journal of Wildlife Management* 52:660–663.

94. WEHAUSEN, J. D. 1996. Effects of mountain lion predation on bighorn sheep in the Sierra Nevada and Granite Mountains of California. *Wildlife Society Bulletin* 24:471–479.

95. LOGAN, K. A., L. L. SWEANOR, and M. G. HORNOCKER. 1996. Cougars and desert bighorn sheep. Pages 230–249 *in* K. A. Logan, L. L. Sweanor, T. K. Ruth, and M. G. Hornocker, eds. *Cougars of the San Andres Mountains, New Mexico.* Final Report, Federal Aid in Wildlife Restoration Project W-128-R. New Mexico Department of Game and Fish, Santa Fe.

96. ROSS, P. I., M. G. JALKOTZY, and M. FESTA-BIANCHET. 1997. Cougar predation on bighorn sheep in southwestern Alberta during winter. *Canadian Journal of Zoology* 74:771–775.

97. KOEHLER, G. M., and M. G. HORNOCKER. 1991. Seasonal resource use among mountain lions, bobcats and coyotes. *Journal of Mammalogy* 72:391–396.

98. NELSON, M. E., and D. L. MECH. 1981. Deer social organization and wolf predation in northeastern Minnesota. *Wildlife Monograph* 77.

99. MESSIER, F. 1991. The significance of limiting and regulating factors on the demography of moose and white-tailed deer. *Journal of Animal Ecology* 60:377–393.

100. TORRES, S. 1997. *Mountain Lion Alert.* Falcon Publishing Company, Inc., Helena, MT.

101. BEIER, P. 1991. Cougar attacks on humans in the United States and Canada. *Wildlife Society Bulletin* 19:403–412.

102. GREEN, K. A . 1991. Summary: Mountain lion–human interaction questionnaires, 1991. Pages 4–9 *in* C. E. Braun, ed., *Mountain Lion–Human Interaction: Symposium and Workshop.* Colorado Division of Wildlife, Denver.

103. RUTH, T. K. 1991. Mountain lion use of an area of high recreational development in Big Bend National Park, Texas. M.S. thesis. Texas A&M University, College Station.

104. TULLY, R. J. 1991. Results, 1991 questionnaire on damage to livestock by mountain lion. Pages 68–74 *in* C. E. Braun, eds., *Mountain Lion–Human Interaction: Symposium and Workshop,* Denver, CO, April 24–26, 1991. Colorado Division of Wildlife, Denver.

105. PADLEY, W. D., ed. 1997. *Proceedings of the Fifth Mountain Lion Workshop,* San Diego, CA.

106. MURPHY, K. M., P. I. ROSS, and M. G. HORNOCKER. 1998. The ecology of anthropogenic influences on cougars. Pages 113–147 *in* K. M. Murphy, The ecology of the cougar (*Puma concolor*) in the northern Yellowstone ecosystem: Interactions with prey, bears, and humans. Ph.D. thesis. University of Idaho, Moscow.

107. BARNHURST, D. 1986. Vulnerability of cougars to hunting. Thesis. Utah State University, Logan.

108. HARLOW, H. J., F. G. LINDZEY, W. D. VAN SICKEL, and W. A . GERN. 1992. Stress response of cougars to nonlethal pursuit by hunters. *Canadian Journal of Zoology* 70:136–139.

109. ROBERSON, J., and F. LINDZEY, eds. 1984. *Proceedings of the Second Mountain Lion Workshop.* Utah Division of Wildlife Research and Utah Cooperative Wildlife Research Unit, Logan.

110. U.S. Bureau of the Census, 1996. *Statistical Abstract of the United States.* Washington, DC.

111. Beier, P. 1993. Determining minimum habitat areas for cougars. *Conservation Biology* 6:94–108.

112. Maehr, D. S., and J. A. Cox. 1995. Landscape features and panthers in Florida. *Conservation Biology* 9:1008–1019.

113. Hedrick, P. W. 1995. Gene flow and genetic restoration: the Florida panther as a case study. *Conservation Biology* 9:996–1007.

114. O'Brien, S. J., M. E. Roelke, N. Yuhi, K. W. Richards, W. E. Johnson, W. L. Franklin, A. E. Anderson, O. L. Bass, Jr., R. C. Belden, and J. S Martenson. 1990. Genetic introgression within the Florida panther, *Felis concolor coryi. National Geographic Research* 6:485–494.

115. Barone, M. A., M. E. Roelke, J. Howard, J. L. Brown, A. E. Anderson, and D. E. Wildt. 1994. Reproductive characteristics of male Florida panthers: Comparative studies from Florida, Texas, Colorado, Latin America, and North American zoos. *Journal of Mammalogy* 75:150–162.

116. Roelke, M. E., J. S. Martenson, and S. J. O'Brien. 1993. The consequences of demographic reduction and genetic depletion in the endangered Florida panther. *Current Biology* 3: 344–350.

117. Braun, C. E., editor. 1991. *Mountain Lion–Human Interaction: Symposium and Workshop,* Colorado Division of Wildlife, Denver.

118. Ross, P. I., M. G. Jalkotzy, and J. R. Gunson. 1996. The quota system of cougar harvest management in Alberta. *Wildlife Society Bulletin* 24:490–494.

119. Ruth, T. K., K. A. Logan, L. L. Sweanor, M. G. Hornocker, and L. L. Temple. 1998. Evaluating cougar translocation in New Mexico. *Journal of Wildlife Management* 62:1264–1275.

120. Weaver, J. L., P. C. Paquet, and L. F. Ruggiero. 1996. Resilience and conservation of large carnivores in the Rocky Mountains. *Conservation Biology* 10:964–976.

121. McCullough, D. R. 1996. Spatially structured populations and harvest theory. *Journal of Wildlife Management* 60:1–9.

122. Trainer, C. E., and G. Matson. 1988. Age determination in cougar from cementum annuli counts of tooth sections. Page 71 *in* R. Smith, ed., *Proceedings of the Third Mountain Lion Workshop.* The Wildlife Society and Arizona Game and Fish Department, Prescott.

123. Aune, K. E., and P. Schladweiler. 1995. Statewide wildlife laboratory services. Annual Report, P-R Project W-120-R-15. Montana Fish, Wildlife and Parks, Helena.

124. Van Sickle, W. D., and F. Lindzey. 1991. Evaluation of a cougar population estimator based on probability sampling. *Journal of Wildlife Management* 55:738–743.

125. Van Dyke, F. G., R. H. Brocke, and H. G. Shaw. 1986. Use of road and track counts as indices of mountain lion presence. *Journal of Wildlife Management* 50:102–109.

126. Beier, P., and S. C. Cunningham. 1996. Power of track surveys to detect changes in cougar populations. *Wildlife Society Bulletin* 24:540–546.

127. Smallwood, S. 1994. Trends in California mountain lion populations. *Southwestern Naturalist* 39:67–72.

128. Smallwood, K. S. and E. L. Fitzhugh. 1995. A track count for estimating mountain lion *Felis concolor californica* population trend. *Biological Conservation* 71:251–259.

129. Sinclair, A. R. E. 1989. Population regulation in animals. Pages 197–241 *in* J. M. Cherett, ed., *Ecological Concepts.* Blackwell Scientific Publications, Oxford, U.K.

130. Logan, K. A., and L. L. Sweanor. 1997. Using biology and ecology in cougar management in the West—New Mexico as a template. Paper presented at the Arizona and New Mexico Chapters of the Wildlife Society Annual Joint Meeting, Gallup, NM, February 6–8, 1997.

131. New Mexico Department of Game and Fish. 1997. Long range plan for the management of cougar in New Mexico, 1997–2004. Federal Aid in Wildlife Restoration, W-93-R38, Project 1, Job 5.5. New Mexico Department of Game and Fish, Santa Fe.

132. New Mexico Department of Game and Fish. 1997. Action plan for the management of cougar in New Mexico, 1997–2004. Federal Aid in Wildlife Restoration, W-93-R38, Project 1, Job 5.5. New Mexico Department of Game and Fish, Santa Fe.

133. CLARK, T. W., P. C. PAQUET, and A. P. CURLEE. 1996. Introduction: Large carnivore conservation in the Rocky Mountains of the United States and Canada. *Conservation Biology* 10:940–948.

134. BELDEN, R. C., and B. W. HAGEDORN. 1993. Feasibility of translocating panthers into northern Florida. *Journal of Wildlife Management* 57:388–397.

135. ELMER, M. 1997. Cougar food habitat dynamics in the San Andres Mountains, New Mexico. M.S. thesis, University of Idaho, Moscow.

136. MOHR, C. O. 1947. Table of equivalent populations of North America small mammals. *American Midland Naturalist* 37:223–249.

137. TIMOSSI, I. C., and R. H. BARRETT. 1985. Mico-dixon user's manual. Version 10-1-85, Department of Forestry and Resource Management, University of California, Berkeley.

138. ACKERMAN, B. B., F. A. LEBAN, E. O. GARTON, M. D. SAMUEL. 1989. User's manual for program HOME RANGE. Contribution No. 259. Forestry, Wildlife and Range Experiment Station, University of Idaho, Moscow.

139. HARVEY, M. J., and R. W. BARBOUR. 1965. Home range of *Microtus ochrogaster* as determined by the modified minimum area method. *Journal of Mammalogy* 46:398–402.

140. WORTON, B. J. 1987. A review of models of home range for animal movement. *Ecological Modeling* 38:277–298.

141. BRITISH COLUMBIA FISH AND WILDLIFE BRANCH AND RESEARCH ANALYSIS BRANCH. 1978. *Cougar.* Wildlife Distribution Mapping Generalized Big Game Series. Department of Recreation and Conservation, Victoria, British Columbia.

142. ALBERTA FISH AND WILDLIFE DIVISION. 1992. Management plan for cougar in Alberta. Wildlife Management Planning Series No. 5. Alberta Forestry, Lands and Wildlife, Edmonton.

143. BELDEN, R. C., and T. H. LOGAN. 1996. Status of the mountain lion in Florida: 1996. Pages 14–16 *in* W. D. Padley, ed., *Proceedings of the Fifth Mountain Lion Workshop.* San Diego, CA.

144. McCARTHY, J. 1996. *A Final Environmental Impact Statement: Management of Mountain Lions in Montana.* Montana Fish Wildlife and Parks, Helena.

145. GROVES, C. R., B. BUTTERFIELD, A. LIPPINCOTT, B. CSUTI, and J. M. SCOTT. 1997. *Atlas of Idaho's Wildlife; Integrating Gap Analysis and Natural Heritage Information.* Idaho Department of Fish and Game, Boise.

146. DIXON, K. R. 1982. Mountain lion (*Felis concolor*). Pages 711–727 *in* J. A. Chapman and G. A. Feldhamer, eds., *Wild Mammals of North America: Biology, Management and Economics.* Johns Hopkins University Press, Baltimore, MD.

147. LINDZEY, F. 1987. Mountain lion. Pages 657–668 *in* M. J. Novak, J. A. Baker, M. E. Obbard, and B. Malloch, eds., *Wild Furbearer Management and Conservation in North America.* Ontario Ministry of Natural Resources, Toronto, Canada.

148. KEISTER, G. P., and W. A. VAN DYKE. 1997. Modeling cougar populations in Oregon. Unpublished paper. Oregon Department of Fish and Wildlife, Baker City.

149. SHERRIFF, S. L. 1978. Computer model for mountain lion populations. Thesis, Colorado State University, Fort Collins.

Jaguar

Raul Valdez

INTRODUCTION

To many Americans, jaguars conjure an image of a large spotted cat stealthily stalking prey animals in the midst of a lush, humid South American jungle. This stereotypic notion is incongruous with recent sightings of jaguars in temperate pine-oak forests in southern Arizona and New Mexico. However, jaguars are not a new faunal element in temperate North America. A jaguar photographed in extreme southwestern New Mexico in 1996 and another photographed months later in the Baboquivari Mountains some 320 kilometers to the west in Arizona were a great surprise to most of the American wildlife conservation community (1). In fact, 4 of 10 jaguars listed in the first edition of the Boone and Crockett North American big game trophy records book were collected in the 1920s in Arizona (2). Eleven of the 71 jaguars listed in the sixth edition are from the United States (10 from Arizona and 1 from Texas) (3). Brown cautioned that jaguars collected by hunters between 1958 and 1973 in the United States could have been trapped elsewhere and released by unethical guides (4). Nonetheless, jaguars killed by hunters prior to 1958 and other past records indicate the existence of a historical residence in the United States and northern Mexico.

Jaguars are the most spectacular of the spotted cats of the Americas. Their large size and boldly spotted coats make them a megafaunal charismatic species that immediately attracts the attention of the general public. Their skins would command high prices if sustainable hunting could be initiated, and indeed the illegal skin trade has been a major menace to their survival. They are also in high demand by big game hunters and hunting permits would probably sell in excess of $25,000. However, their greatest worth is as a focal conservation species for obtaining public support of wildlife conservation projects and for stimulating ecotourism income for rural economic development. Their existence value cannot be overestimated. In addition, jaguars are a priority umbrella species because their habitats are rich in biodiversity. Sustainable use of jaguar habitat protects a myriad of plant and animal species, communities, and ecosystems.

Jaguars are probably the least studied large cats in the world. There are 5 published field studies of jaguars in North America, three in Mexico and one each in Belize and Costa Rica, Central America (5, 6, 7, 8, 9). I have relied on these and several other studies conducted in South America to provide an overview of jaguar life histories.

The jaguar is the largest indigenous American felid and third largest felid in the world. Only the lion and the tiger are larger. It is the only species of the genus *Panthera* in the New World. The other species in the genus are the lion and the leopard in Africa and Asia, and the tiger, which is endemic to Asia. Jaguars have been successfully bred in captivity with pumas, lions, and leopards (10).

Jaguars originally occurred from the Rio Santa Cruz and Rio Negro in Argentina with a continuous northern distribution through Paraguay, Bolivia, Peru, Brazil, northern South America, Central America, southern Mexico and north along eastern and western Mexico including northern Baja California, exclusive of the central Mexican highlands (Figure 18–1). In the United States, jaguars have been recorded in California, Arizona, New Mexico, Texas, and Louisiana (11, 12, 13, 14). The distribution of mountain lions overlaps that of jaguars (12, 15).

Jaguars have compact, robust bodies with a large head, short rounded ears, large feet, and a relatively short tail. Females are 10% to 20% smaller than males (16). Weights of 15 males and 10 females from Venezuela averaged 96 and 56 kilograms, respectively (13). Almeida recorded the weights of 43 male and 21 female jaguars killed by hunters he guided over 20 years in the Matto Grosso of western Brazil; males weighed 67 to 119 kilograms (mean = 96 kilograms) and females weighed 62 to 93 kilograms (mean = 86 kilograms) (17). The heaviest of 7 male puma from the same area weighed 64 kilograms (mean = 53 kilograms) and the heaviest of 2 female puma weighed 33 kilograms. Weights up to 149 kilograms for male jaguars have been recorded in Venezuela (18). Six adult male jaguars in Belize, Central America, averaged 57.2 kilograms and were significantly smaller than males from the Pantanal of Brazil (8).

The background coloration of jaguars is a pale yellow to reddish yellow superimposed on which are black rosettes (roselike circular markings) on the back and sides

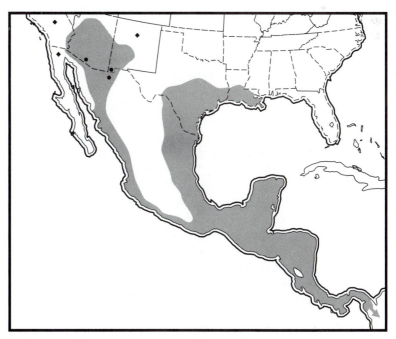

Figure 18–1 Approximate original distribution of the jaguar in North America. Circles represent jaguars recorded between 1996 and 1998. Squares represent distributional extremities.

Figure 18–2 Jaguar photographed in Belize showing typical coloration. (Photograph courtesy of K. Young.)

(Figure 18–2). In most individuals, some of the rosettes enclose one or more black spots of varying size. The background color within the rosettes may be darker than the general body color. Spots tend to merge on the mid-dorsum forming a black, irregular line. The insides of the ears, the jaws, undersides from the chin to the base of the scrotum, and inner portions of the legs are whitish interspersed with black spots or blotches of various sizes. The face, neck, and most of the tail are spotted and the lower third of the tail is ringed (19, 16, 13). Melanistic individuals are more common than in other felids and seem to be more frequent in hot, humid, tropical areas in South America (13) (Figure 18–3).

Five of the eight subspecies of jaguar occurred in North America, following the classification of Pocock (20, 21, 22, 12). *Arizonensis* occurred from northeastern Sonora and northwestern Chihuahua, Mexico, north to northern New Mexico and west to Arizona, northern Baja California, Mexico, and southern California. The type specimen of *arizonesis* is from Arizona. *Veraecrucis* occurred from the coastal region of Tabasco in southeastern Mexico north along the gulf slope of eastern Veracruz and Tamaulipas to southern, central, and eastern Texas and southern Louisiana. *Goldmani* occurred in the Yucatan Peninsula of Mexico south to northern Guatemala and British Honduras. *Hernandesii* occurred from southern Sonora through Sinaloa south to the Isthmus of Tehuantepec. *Centralis* occurred from eastern Chiapas south through southern Guatemala to Colombia, South America. Three other subspecies are recognized in South America.

The validity of the eight recognized subspecies has been questioned because the skull and body color criteria used for designation are too variable (23). There appears to be a cline in body size with the largest animals in the extreme south and smallest animals in the extreme north. Larson recommended that captive jaguars be managed as a single taxon without recognizing subspecies until diagnostic criteria can be determined (23). Too few museum study specimens exist to reach definitive taxonomic conclusions.

Taxonomy and Distribution

Figure 18–3   Melanistic jaguars. (Photograph by H. Montes courtesy of the El Paso Zoo.)

OCCURRENCE IN THE UNITED STATES

The United States probably had a viable jaguar population early in the 20th century. Jaguars were probably breeding residents in Arizona. A female and two cubs were observed in the Grand Canyon, Arizona, in 1890 and a female and its young were harvested in Coconino County, Arizona in 1900 (24, 4). Most of the approximately 70 jaguars killed or reported in the United States in the last 100 years date before 1950 and were males, indicating they may have been dispersing transients from Mexico. However, the number of killed in Arizona between 1900 and 1979 shows a decline characteristic of an overexploited population and not an erratic or irregular pattern indicative of immigrants from Mexico (25, 26). In Texas, jaguars occurred in south, central (east and south of San Antonio), and eastern Texas north to the Red River (27). The last jaguar recorded in Texas was killed in 1948 (25). Jaguar cubs were recorded in California north of Los Angeles in 1855 and the last recorded jaguar in California was killed at Palm Springs in 1860 (19). The occurrence of jaguars in Louisiana was based on two newspaper accounts in the 1880s of a specimen killed south of New Orleans (28, 29). Lowery included it as having occurred in Louisiana based on a specimen collected in east Texas within 42 km of the Louisiana border (30). Most jaguars recorded in the United States in the last 25 years have been in Arizona, including one killed in the Dos Cabezas Mountains, Cochise County, in 1986, two sightings in Pima County in 1988 and 1993, the two recorded in 1996, and one sighting in 1997 in the Cerro Colorado. The last female jaguar recorded in the United States was shot in 1963 in the White Mountains in east-central Arizona (26).

Jaguar populations in the United States declined due to state and federal predator control programs and private livestock producers who indiscriminately killed pumas and other large predators including an occasional jaguar. Jaguar populations were never large enough in the United States since 1900 to maintain viable populations under prolonged, sustained predator control programs or sport hunting. Other major factors probably contributing to their extirpation in the United States were the decline of wildlife prey populations and the widespread degradation and alteration of habitats such as riparian areas in the late 1880s and early 1900s.

LIFE HISTORY

Because they are solitary, secretive cats usually inhabiting areas with dense tropical vegetation, visual observations of jaguars are rare and less is known of their life histories than any of the other large cats in the genus *Panthera*. Most of the present knowledge is based on research conducted in South America (31, 32, 18, 33, 34). Jaguar life histories have been summarized by Perry, Guggisberg, Tewes and Schmidly, Seymour, Leyhausen and others, Emmons , Kitchener, Hoogesteijn and Mondolfi, Oliveira, and Nowell and Jackson (35, 19, 36, 16, 37, 38, 39, 13, 40).

Habitats

Jaguars in the neotropics are usually associated with closed canopy forest and permanent water sources below 1,200 meters, but also occur in arid areas such as the Gran Chaco of southern South America (34, 15). They are excellent swimmers and even inhabit seasonally flooded areas. In the Pantanal region of South America, jaguars used riparian forested areas and forest patches more than expected based on availability, and open forests and grasslands were used less than expected (41). Soto and others reported that the northern limit of the jaguar's present distribution in northeastern Mexico was in the state of Nuevo Leon (42). There, jaguars occupied habitats composed of scrub, chaparral, oak, and oak-pine at altitudes from 500 to 2,700 meters (43). In southern Sonora, Mexico, they occupy tropical deciduous thorn forest and range into montane oak and oak-pine forest in southern Chihuahua and northern Sonora, Mexico, and southern Arizona and New Mexico. At lower elevations in the Southwest, they have been recorded in Sonoran Desert scrub, shrub-invaded semidesert, and riparian areas of river bottoms (25, 26).

Diets

Jaguars are the "big bully" cats of the Americas. Their large, powerful bodies and strong jaw apparatus afford jaguars the ability to overpower all native and introduced domestic animals. They seem to be most active from dusk to dawn, but hunting activity probably varies depending on prey species availability and human disturbance (13). They ambush or more commonly stalk their prey and kill by biting the nape at the base of the skull or puncturing the braincase, the latter being the usual method of killing capybaras, the world's largest rodents, which weigh up to 79 kg (31). They infrequently kill by biting the throat. Jaguars prey on a wide variety of vertebrates ranging in size from fish, turtle eggs, birds, and iguanas that weigh less than 1 kilogram to tapirs that can attain mass exceeding 200 kilograms. They seem particularly adapted to preying on slower, larger mammals such as peccaries, while pumas, which are smaller and more agile than jaguars, are adapted to feeding on quicker, large mammals such as deer. More than 85 species are reported to be prey items of jaguars (16). Their opportunistic diets are an adaptive response to the rich biodiversity of their neotropical habitats.

In Venezuela, prey species include fishes, crocodilians, turtles, snakes, wading birds, peccaries, howler monkeys, giant anteaters, tapirs, brocket deer, sloths, armadillos, giant armadillos, and large rodents such as agoutis and capybaras. Individuals can become depredators of livestock including cattle, horses, donkeys, and pigs, particularly feral pigs (13). In rain forests in southeastern Peru, jaguars hunted along waterside habitats preying on crocodilians and river turtles (33, 44). In Belize, 54% of prey identified in feces were armadillo. The remaining 17 prey species comprised less than 10% of the identified remains. Armadillos were abundant and uniformly distributed and vulnerable to predation by jaguars because of their vocalizations during foraging, limited mobility, and lack of defenses (8).

In the Calakmul Biosphere Reserve in the Yucatan Peninsula of Mexico, seven mammalian species comprised 86% of jaguar diets, followed by birds (10%) and reptiles (4%). The principal mammalian prey species were collared peccaries (42%), coatimundis (18%), and nine-banded armadillos (12%). Less important mammalian prey consisted of red brocket and white-tailed deer, pacas (a large rodent), and tamanduas (5). In Belize, mammals comprised 96% of jaguar diets, while in the Amazon region of Peru, mammals and reptiles comprised 48.7% and 33.3%, respectively, of jaguar diets (8, 33). The large component of crocodilians and turtles probably was due to the abundance of aquatic habitats. In the Chaco region of Paraguay, gray brocket, Brazilian rabbits, armadillos, collared, white-lipped and Chacoan peccaries, salt desert cavies, tapirs, and white-eared opossums comprised 89% of the biomass and 71% of the prey items taken (45).

Collared and white-lipped peccaries comprise a significant component of jaguar diets. Peccaries comprised a higher percentage of jaguar diets in the Calakmul Biosphere Reserve, Mexico, and Amazon region of Peru than expected based on availability and were significant in their diets in Costa Rica (5, 33, 9). Aranda hypothesized that the jaguar's robust, compact body evolved as an adaptation for preying on peccaries (5). Indeed, the jaguar's historical distribution in the United States coincided closely with the distribution of the collared peccary (19). Based on food preference studies in the Calakmul Biosphere Reserve, Aranda and Sanchez-Cordero hypothesized that jaguars and pumas avoided direct competition by feeding on slightly different prey spectra, hunting in different biotypes, and avoiding synchronous use of habitats (7). Jaguars exhibited a preference for collared peccaries, and pumas exhibited a preference for white-tailed and brocket deer.

Jaguars, even though they are primarily a tropical species, can be expected to survive and reproduce in temperate habitats as long as a prey base is available. Indeed, jaguars originated in the Holarctic (46). Lions, leopards, and cheetahs had a widespread distribution in temperate Asia, as did the tiger, where abundant prey species such as gazelles, wild pigs, wild sheep, deer, and other mammalian prey provided a nutritional base. The large body size of jaguars would enable them to withstand subzero temperatures.

Reproduction

Females reach sexual maturity at 2 to 3 years and males at 3 to 4 years after establishing adult home ranges. Jaguars are polyestrous and births have been recorded in every month of the year in captivity (16). Wild females also probably breed throughout the year but with breeding peaks in the midrain season in South America. Births are more common in Mexico from July to September, in Belize from May to September, and in Venezuela from December to February (11, 47). Births are timed to occur when prey is most available.

After an average gestation period of 101 days, females give birth to singletons or up to four cubs (16). In Belize, 52% of females observed with young had twins, 35% had singletons, and 13% had three cubs (47). Births occur in secluded dens in dense vegetation, caves, hollow trees, among and under overhanging rocks, and beneath fallen trees (13). At birth, young weigh 850 to 865 grams and measure about 40 centimeters (16). In the wild, young are weaned at 1.5 to 2 years although they are known to remain for a time in the mother's home range. In Belize, two cubs were weaned at 17 months (47). Jaguars in captivity have lived 23 years but in the wild they probably rarely live more than 11 years (47).

BEHAVIOR

Jaguars are solitary, except for the mating season when females usually associate with one male; estrous females may be followed by more than one male. Jaguars communicate with vocalizations, scratching and scraping of tree trunks, and urinating on trees and other vegetation. They also mark areas by trampling grasses and other soft vegetation by rolling over them with their body and biting surrounding brush and twigs. These sites can occur at intervals of several hundred meters along their trails. They also scrape a prominent spot on the ground with their hind feet, defecating or urinating on the scrape (13). These behaviors can communicate location, area occupancy such as a home range, social status, gender and age, and reproductive condition. Social behavior studies in a zoo environment of a captive-born adult male and female jaguar revealed that several complex grooming and wrestling behaviors were the most common acts and were initiated by both animals. Play, grooming, and sexual behavior patterns were performed in the same bout. Contact behavior was dominated by play and complex grooming activities. Adult jaguars in the wild probably do not engage in prolonged or complex social behavior because of their solitary nature (48).

Home Range and Density

In the Pantanal region of Brazil, the largest seasonally flooded area in the world (greater than 100,000 square kilometers), the mean home range of one male and four females was 142 square kilometers. The home ranges of females overlapped 42%, but only during the dry season. Home-range size averaged 12.8 square kilometers during the wet season and 54.3 square kilometers during the dry season. The smaller home ranges reflect the concentration of prey on islands during the wet season. Of animals located on several consecutive days, the mean distance traveled by one male (3.3 kilometers) was greater than that of four females (1.8 kilometers). Mean distance traveled during the dry season was greater than during the wet season, when prey were concentrated. A dispersing male moved 30 kilometers from its natal area and a female moved 8 kilometers from its natal area (41).

In Belize, Central America, the mean home range of four adult males was 33.4 , with a range of 28 to 40 square kilometers, and two females moved within minimum areas of 10 to 11 square kilometers. Males often remained within 2 to 3 square kilometers an average of 7 days before moving to other areas in a single night. This probably reflected high prey densities in minimally disturbed areas. Two adult males made average linear movements of 0.7 kilometer per hour. Home ranges of males overlapped those of adjacent males, but home ranges of the two females did not overlap. Movements of females were restricted within the home ranges of individual adult males. The home range of one male was overlapped 80% by at least four male jaguars, one female jaguar, and a puma. Ranges of adult males overlapped probably because roads were used for travel and hunting (8).

In Chiapas, southern Mexico, Aranda reported densities of one jaguar per 23 to 35 square kilometers in tropical forests and a density of one jaguar per 13 to 20 square kilometers in coastal areas (6). Within five separate areas encompassing 8,800 square kilometers, there were an estimated 241 to 343 jaguars in 1988. Jaguar density variations are probably due to prey species availability, prey density, habitat structure, and human alterations and disturbance.

POPULATION MANAGEMENT AND CONSERVATION

The jaguar's geographic distribution has continually contracted since the European invasion of the New World. Jaguars are listed as endangered in the United States and throughout its range by the U.S. Department of the Interior and is in Appendix I of the Convention on International Trade in Endangered Species (CITES) (49, 50). Population numbers have decreased considerably in Argentina, Costa Rica, and Panama and it is probably extirpated in El Salvador, Uruguay, and Chile. In Mexico and Central America, jaguar populations were estimated to exist in only 33% of their original historical range, and in areas currently occupied by jaguars, 75% of these populations exist in reduced numbers. In South America, jaguars were estimated to exist in 67% of their original range and 36% of populations currently exist in reduced numbers. The pelt trade in spotted cats, once a major decimating factor, ceased in 1970 when international controls came into effect (51). Strategies for jaguar conservation and management include larger and more extensive protected areas, education and management programs for local ranchers, government assistance to individuals with jaguar problems, and stricter penalties and enforcement for the illegal killing of jaguars (52).

The high rate of deforestation throughout their range is the major threat to jaguar survival. Logging of extensive areas and land clearing followed by human settlement and cattle ranching are recurrent patterns throughout tropical America. Cattle production in jaguar habitat made available a readily accessible prey species. Subsistence hunting of the jaguar's natural prey species exacerbates the problem (13). In some areas, ranching is practiced with minimal husbandry and in these areas livestock constitute a high percentage of jaguar diets (14). In livestock-producing areas, jaguars are usually killed indiscriminately. Translocating jaguars that are livestock depredators has not been successful because translocated animals continue to prey on livestock (41, 13).

Most of the present jaguar habitats are in privately owned property, as in the Pantanal region where 95% of the land is privately owned. Jaguar management will require the cooperative involvement of private landowners. Management programs that provide for the selective removal of individual livestock-depredating jaguars or compensating ranchers for livestock killed will have to be instituted. Fencing cattle to restrict herds in grasslands and planted pastures or minimize access to preferred jaguar habitats such as forested lands is another livestock management option. Enterprise diversification, incorporating ecotourism, traditional livestock, and game ranching of native species, is essential to conservation efforts to develop economic incentives to incorporate wildlife into land management schemes. Particularly urgent is the development of economically viable integrated natural resources management programs that maintain the diversity and integrity of natural ecosystems in tropical and subtropical Latin America (53). Sustainable sport hunting of jaguars will require close supervision by the private and governmental sectors to minimize abuses. Quigley and Crawshaw emphasized the need to integrate ecological requirements with the socioeconomic realities of the region when designing a conservation plan for jaguars (34).

The creation of nature reserves is an essential element in jaguar habitat management. In the Pantanal region, a reserve size of 2,000 to 3,000 square kilometers was recommended to maintain 30 to 50 adult jaguars. Only two reserves exceeding 3,000 square kilometers have been established in Mexico (14). Concomitant with reserves is the need to establish travel corridors between reserves to minimize the extinction potential of isolated populations due to demographic and genetic uncertainties. Riparian areas are preferable corridors where still extant such as in the Pantanal region (34).

Jaguars recorded in 1996 in Arizona and New Mexico probably originated from dispersing animals in Sonora. Determining jaguar distribution and numbers in northern Sonora should be prioritized. Occupied and potential habitats in northern Mexico should be identified, distribution and abundance of potential prey species determined, and currently utilized and potential corridors mapped. Reestablishment of jaguar populations in the American Southwest and northern Sonora, Mexico, will require long-term coordinated research and management programs among American and Mexican private, state, and federal agencies. Above all, jaguar conservation programs will require the support of rural communities and the incorporation of community-based socioeconomic development programs.

Although legally protected throughout their range, jaguars face a pessimistic future due to habitat destruction, habitat fragmentation, decrease in native prey species, and conflicts with ranching enterprises because of jaguar predation on livestock. Their fate has been similar to that of wolves and grizzly bears in the United States. The remaining fragmented jaguar populations could readily be extirpated by trapping, hunting, and poisoning. The sightings of jaguars in New Mexico and Arizona in 1996 are an indication that jaguars could become reestablished in the United States. Every effort should be made to ensure that their recovery in the United States becomes a reality and remaining populations throughout their range not be extirpated.

SUMMARY

In North America, Jaguars originally occurred in Central America, southern, eastern, and western Mexico exclusive of the central Mexican highlands, and northern Baja California and in the United States in portions of California, Arizona, New Mexico, Texas, and Louisiana. Five subspecies occur in North America; however, no diagnostic criteria exist. Jaguar populations declined or were extirpated due to predator control programs, decline of wild prey populations, and the widespread degredation and alteration of habitats such as riparian areas. Most of the approximately 70 jaguars reported in the United States in the last 100 years have been males and date before 1950, indicating they may have been from Mexico. Jaguar numbers killed in Arizona between 1900 and 1979 showed a decline characteristic of an overexploited population and not an erratic or irregular pattern indicative of immigrants. Recent jaguars sighted in the southwestern United States probably originated from dispersing animals in Sonora, Mexico.

In the northern limits of their distribution, jaguars have been recorded in desert scrub, chaparral, scrub-invaded semidesert, and montane oak and oak-pine forests. Jaguars appear to be adapted to prey on slower, larger mammals, particularly peccaries, but prey include more than 85 species. Jaguars are solitary except during mating season. Females, which are polyestrus and usually twin, reach sexual maturity at 2 to 3 years and males at 3 to 4 years.

In Chiapas, Mexico, jaguar densities were one per 13 to 25 square kilometers. Jaguars throughout their range are listed in Appendix I of CITES. Continuing detrimental impacts include habitat destruction, habitat fragmentation, decrease in native prey species, and conflicts with ranching enterprises. Determining jaguar distribution and status in northern Mexico should be a priority. Occupied and potential habitats should be identified, distribution and abundance of prey species determined, and currently utilized and potential corridors identified. Jaguar conservation programs will require the support of rural, community-based socioenomic development programs.

LITERATURE CITED

1. GLENN, W. 1996. *Eyes of Fire: Encounter with a Borderlands Jaguar.* Printing Corner Press, El Paso, TX.

2. GRAY, P. N. 1932. *Records of North American Big Game.* The Derrydale Press, New York.

3. ALBERTS, R. C., ed. 1971. *North American Big Game.* The Boone and Crockett Club, Pittsburgh, PA.

4. BROWN, D. E. 1991. Revival for el tigre? *Defenders* 66:27–35.

5. ARANDA, M. 1994. Importancia de los pecaries (*Tayassu* spp.) en la alimentacion del jaguar (*Panthera onca*). *Acta Zoologica Mexicana* 62:11–22.

6. ARANDA, M. 1996. Distribucion y abundancia del jaguar, *Panthera onca* (Carnivora; Felidae) en el estado de Chiapas, Mexico. *Acta Zoologica Mexicana* 68:45–52.

7. ARANDA, M., and V. SANCHEZ-CORDERO. 1996. Prey spectra of jaguar (*Panthera onca*) and puma (*Puma concolor*) in the tropical forests of Mexico. *Studies on Tropical Fauna and Environment* 31:65–67.

8. RABINOWITZ, A., and B. NOTTINGHAM. 1986. Ecology and behavior of the jaguar (*Panthera onca*) in Belize, Central America. *Journal of Zoology, London (A)* 210:149–159.

9. CHINCHILLA, F. A. 1997. La dieta del jaguar (*Panthera onca*), el puma (*Felis concolor*) y el manigordo (*Felis pardalis*)(Carnivora:Felidae) en el Parque Nacional Concovado, Costa Rica. *Revista de Biologia Tropical* 45:1223–1229.

10. GRAY, A. P. 1972. *Mammalian Hybrids.* Commonwealth Agricultural Bureaux, Farnham Royal, Slough, England.

11. LEOPOLD, A. S. 1959. *Wildlife of Mexico, the Game Birds and Mammals.* University of California Press, Berkeley.

12. HALL, E. R. 1981. *The Mammals of North America,* 2nd ed. John Wiley and Sons, New York.

13. HOOGESTEIJN, R., and E. MONDOLFI. 1993. *The Jaguar.* Armitani Editores, Caracas, Venezuela.

14. NOWELL, K., and P. JACKSON, eds. 1996. The Americas. Pages 114–148 *in Wild Cats: Status, Survey, and Conservation Plan.* IUCN, Gland, Switzerland.

15. EISENBERG, J. F. 1989. *Mammals of the Neotropics. Volume 1. The Northern Neotropics.* University of Chicago Press, Chicago.

16. SEYMOUR, K. L. 1989. Panthera onca. *Mammalian Species* 340:1–9.

17. ALMEIDA, A. 1990. *Jaguar Hunting in the Mato Grosso and Bolivia.* Safari Press, Long Beach, CA.

18. MONDOLFI, E., and R. HOOGESTEIJN. 1986. Notes on the biology and status of the jaguar in Venezuela. Pages 85–123 *in* S. D. Miller and D. D. Everett, eds., *Cats of the World: Biology, Conservation, and Management.* National Wildlife Federation, Washington, DC.

19. GUGGISBERG, C. A. W. 1975. *Wild Cats of the World.* Taplinger Publishing, New York.

20. POCOCK, R. I. 1939. The races of jaguar (*Panthera onca*). *Novitates Zoologiae* 41:406–422.

21. NELSON, E. W., and E. A. GOLDMAN. 1933. Revision of the jaguars. *Journal of Mammalogy* 14:221–240.

22. MILLER, G. S., JR., and R. KELLOGG. 1955. List of North American recent mammals. *United States National Museum Bulletin* 205:1–954.

23. LARSON, S. E. 1997. Taxonomic re-evaluation of the jaguar. *Zoo Biology* 16:107–120.

24. HOFFMEISTER, D. F. 1986. *Mammals of Arizona.* University of Arizona Press, Tucson.

25. BROWN, D. E. 1983. On the status of the jaguar in the Southwest. *Southwestern Naturalist* 28:459–460.

26. BROWN, D. E. 1997. Return of el tigre. *Defenders* 72:13–20.

27. TAYLOR, W. P. and W. B. DAVIS. 1947. The mammals of Texas. *Game, Fish, and Oyster Commission Bulletin* 27.

28. NOWAK, R. M. 1975. Retreat of the jaguar. *National Parks Conservation Magazine* 49:10–13.

29. NOWAK, R. M. 1994. Jaguars in the United States. *Endangered Species Technical Bulletin* 19:6.

30. LOWERY, G. H., JR. 1974. *The Mammals of Louisiana and its Adjacent Waters.* Louisiana State University Press, Baton Rouge.

31. SCHALLER, G. B., and J. M. C. VASCONCELOS. 1978. Jaguar predation on capybara. *Zeitschrift fur Saugetierkunde* 43:296–301.

32. SCHALLER, G. B., and P. G. CRAWSHAW. 1980. Movement patterns of jaguar. *Biotropica* 12:161–168.

33. EMMONS, L. H. 1987. Comparative feeding ecology of felids in a neotropical rainforest. *Behavioral Ecology and Sociobiology* 20:271–283.

34. QUIGLEY, H. B., and P. G. CRAWSHAW, JR. 1992. A conservation plan for jaguar *Panthera onca* in the Pantanal region in Brazil. *Biological Conservation* 61:149–157.

35. PERRY, R. 1970. *The World of the Jaguar.* Taplinger Publishing, New York.

36. TEWES, M. E., and D. J. SCHMIDLY. 1987. The neotropical felids: Jaguar, ocelot, margay, and jaguarundi. Pages 697–712 *in* M. Novak, J. Baker, M. E. Obbard, and B. Malloch, eds., *Wild Furbearer Management and Conservation in North America.* Ministry of Resources, Ontario, Canada.

37. LEYHAUSEN, P., B. GRZIMEK, and V. ZHIWOTSCHENKO. 1990. Panther-like cats and their relatives. Pages 1–48 *in Grzimek's Encyclopedia of Mammals,* Vol. 4. McGraw-Hill, New York.

38. EMMONS, L. H. 1991. Jaguars. Pages 116–123 *in* J. Seidensticker and S. Lumpkin, eds., *Great Cats.* Rodale Press, Emmaus, PA.

39. KITCHENER, A. 1991. *The Natural History of Wild Cats.* Comstock Publishing Associates, Ithaca, NY.

40. OLIVEIRA, T. G. DE. 1994. *Neotropical Cats: Ecology and Conservation.* Presa do Universidade Federal de Maranhao, Maranhao, Brazil.

41. CRAWSHAW, P. G., and H. B. QUIGLEY. 1991. Jaguar spacing, activity and habitat use in a seasonally flooded environment in Brazil. *Journal of Zoology, London* 223:357–370.

42. SOTO, J. H. L., O. C. R. ROSAS, and J. A. RAMIREZ. 1997. El jaguar (*Panthera onca veraecrucis*) en Nuevo Leon, Mexico. *Revista Mexicana de Mastozoologia* 2:126–128.

43. ROSAS, O. C. R. 1996. Distribucion y aspectos ecologicos del jaguar *Panthera onca veraecrucis*, en Nuevo Leon, Mexico. Bachelor's thesis. Universidad Autonoma de Nuevo Leon, Monterrey, Mexico.

44. EMMONS, L. H. 1989. Jaguar predation on chelonians. *Journal of Herpetology* 23:311–314.

45. TABER, A. B., A. J. NOVARRO, N. NERIS, and F. H. COLMAN. 1997. The food habits of sympatric jaguar and puma in the Paraguayan Chaco. *Biotropica* 29:204–213.

46. KURTEN, B., and E. ANDERSON. 1980. *Pleistocene Mammals of North America*. Columbia University Press, New York.

47. RABINOWITZ, A. R. 1986. *Jaguar*. Arbor House, New York.

48. GRITTINGER, T. F., and D. L. SCHULTZ. 1994. Social behavior of adult jaguars (*Panthera onca* L.) at the Milwaukee County Zoo. *Transactions Wisconsin Academy of Sciences, Arts and Letters* 82:73–81.

49. FEDERAL REGISTER. 1994. Endangered and threatened wildlife and plants; proposed endangered status for the jaguar in the United States, 59:35674–35679.

50. FEDERAL REGISTER. 1997. Final rule to extend endangered status for the jaguar in the United States, 62:39147–39157.

51. SWANK, W. G. and J. G. TEER. 1987. Status of the jaguar—1987. *Oryx* 23:14–21.

52. RABINOWITZ, A. 1995. Jaguar conflict and conservation, a strategy for the future. Pages 394–397 *in* J. A. Bissonette and P. R. Krausman, eds., *Integrating People and Wildlife for a Sustainable Future*. The Wildlife Society, Bethesda, MD.

53. FREESE, C. H., and C. J. SAAVEDRA. 1991. Prospects for wildlife management in Latin America and the Caribbean. Pages 430–444 *in* J. G. Robinson and K. H. Redford, eds., *Neotropical Wildlife Use and Conservation*. University of Chicago Press, Chicago.

Black Bear

Michael R. Pelton

INTRODUCTION

Black bears are the most common of the world's eight bear species (1). Compared to the other seven, they have done well in the face of expanding human populations. However, their current status in North America varies from that of pest to being threatened (2, 3).

Black bears also were the most common Ursid in the late Pleistocene in North America (4). Their body size was small early in their evolutionary history but increased over time. In recent history the species was distributed throughout North America with the exception of the more arid regions of the Southwest and treeless areas of northern Canada (5, 6) (Figure 19–1).

Black bears were second only to white-tailed deer in importance to Native Americans (7). Black bears were treated with respect and often incorporated into rituals as objects of worship. Many tribes accorded human attributes to them. However, the species was used for practical purposes as well, such as meat for food, fur for clothing, bones as tools, and fat for cooking, tanning hides, and waterproofing (8). Overhunting likely occurred in some traditional hunting territories but the results of tribal warfare periodically allowed game populations to recover (9).

The advent of white settlers and firearms brought about rapid changes to black bear populations in North America. The first white explorers established a trade system with Native Americans and black bear skins and oil became a common trade item in exchange for firearms, ammunition, cloth, and liquor (10, 11). Technology and land use patterns also changed dramatically and black bear populations suffered the consequences. Any semblance of a "balance" between the Native Americans and black bear populations was soon destroyed by unregulated trade, timber harvest, agriculture, railroads, and overhunting (11). By the early 1900s extensive human settlement, land clearing, the ravages of the Civil War, and unregulated hunting left large areas of previously occupied range devoid of black bears.

The alarm call was sounded early in the 20th century and a network of national reserves (parks, forests, and refuges) and assorted state systems were created. Concomitant with this event was the establishment of state wildlife agencies and laws and regulations governing the harvest of black bears. These two major events eventually enabled black bears to recover, leaving only the intensively settled and cleared areas of the Southeast and Midwest devoid of the species (12).

Today black bears are a highly regarded and popular game species throughout most of their range and populations are regulated under state and provincial laws. However, to society as a whole a dichotomy exists regarding their perceived importance. Some regard black bears as an alluring attraction accounting for tremendous economic value from tourists around parks and other protected areas. Still others deem the species

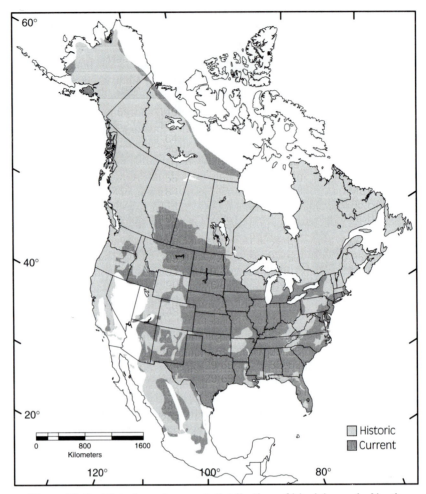

Figure 19–1 Historic and current distribution of black bears in North America (6, 12 Reprinted by permission of John Wiley & Sons, Inc.)

as dangerous or a pest and want to be rid of it (2). For example, the Louisiana subspecies is on the federal endangered species list and a major recovery effort is under way (3). However, in Canada and parts of the northwestern United States, though officially a game species, they are often regarded as a pest. Resource agencies are caught in the middle and try to maintain a balance with their management strategies. Ecologically, bears may serve as keystone species, as valuable seed dispersers, for cycling of nutrients (feces deposits), and soil disturbance (feeding activities) (13). The species is sometimes used as a landscape-level indicator species, reflecting changes taking place across large regions, thus serving as a valuable environmental monitor.

The demand for bear parts in Asia has increased the value and trade of North American black bear gall bladders and stimulated an illegal market (14). International efforts are currently under way to control this traffic. There are no populations of North American black bears known to be currently in jeopardy due to the demands for gall bladders or other body parts.

TAXONOMY AND DISTRIBUTION

Hall lists 16 subspecies of black bears; this contrasts with 18 subspecies previously listed (6, 15). Modern advances in DNA analysis may alter our current understanding of the subspecies of black bears in North America. The integrity of the three predominant subspecies in eastern North America is under evaluation (16).

Black bears have a relatively heavy, compact body structure, short tail (8–15 cm), and massive legs and feet; they are plantigrade. In contrast to brown/grizzly bears, black bears have strong, recurved claws, a straight facial profile, and no shoulder hump. They also have movable lips and a long prehensile-like tongue that add to their dexterity in gathering small food items. Cubs have blue eyes, but they change to dark brown as adults. The massive skull has prominent zygomatic arches, flat-crowned molars for crushing, and strong canines (17, 18). The dental formula is 3/3, 1/1, 2–4/2–4, 2/3 = 34–42 teeth (15).

Black bears exhibit considerable phenotypic plasticity in body size depending on location and food supplies (19). Adult males (100–300 kilograms) are larger than adult females (40–200 kilograms). Females normally reach full skeletal size at 5 years of age, whereas males normally do not reach maximum size until 7 or 8 years of age (20). Both sexes may increase in body mass for another 3 to 4 years of age (18).

The predominant color of this species is black with a brownish muzzle and an occasional white chest blaze. Cinnamon, brown, and blond colored pelages are more common in the western range (21, 22). Bluish-gray and white pelages occur in coastal areas of Alaska and British Columbia.

Black bears have a keen sense of smell. Their ability to stand on their hind legs gives them a significant advantage seeing, hearing, and smelling for long distances. They exhibit some level of color vision and detailed near vision (23). Black bears are excellent runners and can run at more than 50 kilometers per hour for short distances. They also are good swimmers and have been observed crossing bodies of water 10 to 15 kilometers wide.

Black bears presently are found in portions of 39 states, all the Canadian provinces, and mountainous regions of northern Mexico (Figure 19–1) (12). The current range is continuous throughout most of New England but exhibits increasing levels of fragmentation southward through the mid-Atlantic and southeastern states. The range of the species in the western United States is largely restricted to public lands in the mountains. The current estimates of the species range represent approximately 62% of the historic range (12). More than 600,000 black bears exist in North America with most states and provinces reporting stable to increasing numbers (24, 25).

LIFE HISTORY AND POPULATION DYNAMICS

Diet Selection

Although classified taxonomically in the order Carnivora, black bears are functionally omnivores; in reality, the species is predominantly vegetarian, more specifically fructivorous and granivorous. They consume a wide variety of both plant and animal foods. Animal foods range from insect grubs, beetles, ants, and yellowjackets and their larvae to the active capture and consumption of live prey such as white-tailed deer fawns and moose or elk calves (26). They also eat tubers and roots as well as carrion. Black bears have been known to forage more than 30 meters high in trees for acorns and beechnuts (Figure 19–2).

Seasonal diets shift to take advantage of whatever is available and of nutritional benefit. After den emergence in spring, black bears typically graze on fresh young herbaceous material such as squawroot, skunk cabbage, and horsetails. Spring can be a season of nutritional stress and insect larvae, beetles, carrion, and ants are important sources of energy. Summer food habits reflect a predominance of soft mast such as blueberry, huckleberry, blackberry, and serviceberry. Hard mast (nuts) becomes a focus of black bears during fall when they must accumulate substantial amounts of fat. An abundance of nuts or fall berries (pokeweed, grape, cherry, blueberry) is vital to the survival of black bears because they must exist entirely off fat stores while hibernating for 2 to 5 months in winter.

This mobile and adaptable omnivore can take advantage of a wide variety of foods, whether it is prickly pear cactus in the southwestern chaparral, black gum fruits

Figure 19–2 Black bears are excellent tree climbers and take full advantage of their short, strong recurved claws to gain access to tree-borne fruits and nuts. (Photograph by M. Pelton.)

and alligator eggs in the southeastern swamps, abandoned apple orchards in the Northeast, or horsetail in the Northwest. Because black bears have a simple stomach and no caecum, they have a difficult time digesting plants. Consequently, they must consume large quantities of food to compensate for their less efficient digestive system. Often they must expend considerable energy to get the food they need such as climbing oak, beech, or cherry trees. Consequently, black bears are opportunistic feeders, and spend considerable time gorging themselves on a single food item such as a berry patch, productive oak tree, or colony of social insects, when it is locally available. Human food and garbage also represent available and concentrated food sources for black bears. Likewise, beeyards, orchards, various grain crops, garbage dumps, and bird feeders are vulnerable to this omnivore. Thus, conflicts with humans are commonplace within or near occupied habitat (2).

Reproduction

Female black bears reach sexual maturity at 2 to 7 years of age depending on their nutritional intake (27, 28). Nutrition also has an effect on fecundity of females (29). Years of poor berry or nut production can result in decreased litter sizes and an increased incidence of barren females (30, 31). Males are normally sexually mature at 3 to 4 years of age.

Black bears are polygynous, so one dominant male is capable of breeding with several females during the breeding season. Adult males seek estrous females by following their scent. A mated pair may stay together from a few hours to a few days (18). Males commonly compete for estrous females and fighting among males is common but normally brief. Older males may possess battle scars around their shoulders and heads as testament to fighting.

Black bears have a seasonally constant estrus. Females remain in estrus until bred or until the ovarian follicles begin to degenerate. Breeding occurs in summer with females coming into estrus as early as late May and as late as August with the peak of breeding normally in July (32, 33, 34, 35). Black bears are induced ovulators with ovulation occurring only as a result of coital stimulation (36, 37, 38). The mobility of sexually mature males, prolonged estrus, and induced ovulation in females tend to ensure successful reproduction for the species.

Black bear females exhibit delayed implantation. Fertilized eggs (blastocysts) float in the uterus and do not implant until late November or early December. Therefore, it is only during the last 6 to 8 weeks of the 6- to 8-month gestation period that

Life History and Population Dynamics

Figure 19–3 Adult females normally breed every 2 years, producing two to three cubs that stay with her for 15 to 17 months. (Photograph by M. Pelton.)

fetal development occurs. Minimal cell differentiation occurs prior to implantation (36). Poor nutrition resulting in low fall body mass of females (less than 50 kilograms) may result in no implantation of the blastocysts, resorptions of implanted fetuses, or early death of neonates (39).

Cubs are born in winter dens during January or early February. They are small (0.2–0.3 kilograms), lack hair, and eyes are closed. Females have six functional mammae; four in the pectoral region and two in the inguinal area. Litter size is normally two or three but four cubs are not uncommon; litters of five are reported occasionally (40). Sex ratios at birth are normally 50:50. Females nurse their young on milk high in fat and protein. At den emergence in late April or early May, cubs may weigh 3 to 5 kilograms. Cubs continue to nurse through the summer.

Females normally breed every 2 years (Figure 19–3). Placental scars on the uterus and counts of corpora lutea on the ovaries are the traditional methods of evaluating reproduction. Reproductive histories of female bears also may be determined from dental cementum; narrower annulations occur during years of cub production and broader annulations in years of no production (41).

Poor food years can cause a population-wide reproductive failure. Thus, most adult females become pregnant the next summer resulting in breeding synchrony, a cyclic process that will sometimes repeat itself for many years or until another food shortage occurs during a normally high year of cub production (42). If a female loses her cubs early in the first year, she may come into estrus and produce cubs again, thus disrupting breeding synchrony over time.

Maternal females become more protective of young over time and more aggressive toward intruders; they exhibit increasing levels of fidelity to the young. A strong bond has formed between mother and cubs by the time of den emergence. The process of teaching the young begins intensively at that time. The young are seldom far from her side and learn the meanings of various vocalizations and behaviors of the mother. Nursing often occurs at the base of large trees or other types of escape cover. Cubs den with their mother the next winter but they have been known to separate and occupy separate dens close by, or even separate altogether. Cubs also have been known to survive when separated from their mother as early as August. Normally, however, the family unit breaks up after 15 to 17 months, just prior to the female coming into estrus again in May or June.

Although the sex ratio at birth is generally 50:50, the older segment of a population is usually skewed toward females. Males are more likely to come in contact with hunters, campgrounds, or garbage dumps and are killed more frequently than females (43).

Longevity

Annulations in the cementum of sectioned and stained premolar teeth yield a fairly accurate age for black bears (44). Other techniques such as body or skull measurements, long bone or skull suture closure, cheek width, and degree of root canal closure have proven less accurate (45, 46, 47). The average age of males in a healthy bear population ranges from 3 to 5 years of age, whereas females average 5 to 8 years old. Females exhibit more restricted movements, smaller home ranges, and less aggressive behavior than males; this contributes to their longer life span (48). Because they constitute the productive units of the population, it is advantageous for the females to live longer. Although captive black bears may live more than 25 years and some wild bears have attained 20 or more years, few ever reach 10 to 12 years of age in the wild. Survival of cubs depends on the condition, age, and experience of their mother. Normally mothers are very attentive and careful of their young and survival of offspring is high. Cubs of females in poor physical condition are more likely to die than cubs of mothers in well-nourished condition (49, 29).

Mortality

The highest mortality in a black bear population occurs from the time of family breakup at 16 to 17 months until sexual maturity and establishment of a relatively defined home range at 3 to 5 years of age (50, 51). Yearlings and subadults may starve during years of severe natural food shortages. Young black bear must find adequate nutrition for both growth and maintenance, avoid larger bears, and begin to establish a home range. Because of these nutritional demands and the presence of larger and older males in the population, the younger animals, particularly males, are forced to occupy less suitable habitat. Subadult mortality rates likely exceed 35% (51). Mortality rates decline with an increase in age; the mortality rate of males is nearly double that of females, 26% versus 17%, respectively (52). Most mortality is human induced and includes hunting, poaching, depredation control, and accidents with vehicles. Some mortality is attributed to the practice of infanticide by males toward unrelated cubs (53).

Parasites and Diseases

Although a number of neoplastic, rickettsial, viral, bacterial, and traumatic diseases have been reported for black bears, few seem to be major mortality factors (54, 55, 56, 57, 58, 59). Compared to other mammals, black bears have relatively few internal and external parasites. Many parasites of black bears also are found in other hosts occupying sympatric ranges. Other than trichina, there seem to be no serious public health threats (55). Severe demodicosis has been reported in a small percentage of bears (60, 61). Although heartworms can occur in a high percentage of black bears, this parasite apparently presents no serious health hazard (57).

Density and Indices

Abundance estimates of black bears are difficult to obtain but are important to wildlife managers. Good estimates are necessary to detect population trends caused by various natural or anthropogenic events to monitor threatened or declining populations, to understand basic dynamics of a population, and to enable managers to establish appropriate harvest levels (62).

Most efforts to calculate actual densities of black bears are hampered by a failure of one or more of the basic assumptions necessary for accurate and precise estimates

TABLE 19-1

Estimates of Density of Selected Black Bear Populations in North America

State/Province	Per 10 km²	Source
Alaska	0.9 to 2.9	62
Arkansas	0.8	101
Arizona	1.1 to 3.6	113, 114
Colorado	1.1	115
Minnesota	2.0	43
Tennessee	3.3	116
Virginia	5.9 to 8.3	117, 118
Ontario	5.0	119
Idaho	6.3	120
Maine	2.0	42

and an inability to determine the area inhabited by the estimated population (62). Recent advances in the use of radiotelemetry, remote cameras, and DNA technology show promise in resolving many of the historic sampling problems for species such as black bears that exhibit low densities, wide-ranging movements, and cryptic behavior (63, 64, 65, 66). Estimated densities of black bears exhibit wide variation ranging from 1 to 12.4 square kilometers per bear (Table 19-1). Some of this variation is explained by the variety of procedures used to derive the estimates. However, habitat quality accounts for some of the variation as well (62).

Where actual population estimates are not feasible or needed, population indices have been used. The first efforts focused on track counts, direct counts at roadsides, and incidence of scats along prescribed index routes (67, 68, 69). Intrepretation of harvest data continues to be the most common index used by resource management agencies (70). Unfortunately, population trend indicators are subject to numerous environmental, sampling, and interpretation biases and have not been adequately tested against measured changes in the population (71, 72, 66, 73, 70). Bait-station indices in some regions seem to offer a long-term, cost-efficient, and relatively accurate method for monitoring black bear populations (74). Whatever the situation, responsible wildlife resource agencies should use as many estimators and indices as possible to monitor bear populations.

Winter Denning

Black bears circumvent food shortages by becoming dormant for varying periods during winter months. The degree of lethargy and length of inactivity are related to latitude. However, even in the southern part of their range, individuals, particularly females, den for 60 or more days (75, 76). In the northern part of their range they may stay in dens for more than 5 months. Secure dens are important for females because birth and early maternal care of cubs occur during the denning period. Black bears may enter dens as early as October or as late as January. Some individuals in more southern populations, particularly males, never settle into one spot for a long period of time; they may move continuously, build a shallow nest and stay a few days, and then move again (76). Timing of onset and cessation of denning depends on latitude, available food, sex, age, and local weather (77). Adult females den first, then subadults and adult males. Den emergence occurs between mid-March and early May. Males emerge first and females with cubs emerge last. Foot pads may be shed during denning and new pads formed during the post-denning recovery period (78). At the time of emergence, a fecal plug that has blocked the gastrointestinal tract is defecated.

Because black bears can be sedentary for up to 5 months, some remarkable physiological and metabolic processes occur. Although their metabolism decreases slightly, they continue to use a large amount of energy. At the same time there is no food or water intake and no urination or defecation (79). In addition, no net loss of calcium or bone deterioration occurs after months of recumbency (80). These phenomena may have significant implications and applications to human medicine (e.g., osteoporosis, anorexia nervosa).

BEHAVIOR

Movements

Black bears move in response to seasonal availability of food, dispersal as subadults, breeding opportunities, and before and after denning. Concentrations of soft mast, hard mast, or artificial food sources provide the stimulus for extensive movements and range expansion (81, 27). These movements occur primarily during late summer and fall when the physiological need to increase fat levels is high. However, black bears are capable of moving long distances in response to food availability at any time of year outside the denning period. Subadults in the population, particularly males, disperse considerable distances from their natal home range. Female subadults tend to be more sedentary and establish home ranges closer to their previous natal range (48). Adult males move considerably during summer in search of estrous females and often encompass the home ranges of several females. Regardless of where adult males or females go during the period of hyperphagia, they tend to return to their more familiar spring–summer range to den (82). Post-denning movements tend to be limited but this depends on nutritional needs; an animal in poor condition may be forced to make greater than usual movements to obtain necessary amounts of food.

Because of the factors mentioned, the size of home ranges varies depending on sex, age, season, population density, and food resources. Annual home range sizes vary from 5 to 49 square kilometers for females and 22 to 1,721 square kilometers for males (Table 19–2). The home ranges of males are typically many times larger than those of females.

Activity

Black bears are normally solitary except for family groups (adult female with cubs), breeding pairs in summer, and congregations at feeding sites. Male black bears are more temporally than spatially territorial. Home ranges of males overlap considerably

TABLE 19–2

Estimates of Mean Home Range Sizes of Selected Black Bear Populations in North America[a]

| Location | Home Range Size (km²) | | Reference |
	Females	Males	
Arkansas	34.7	89.7	101
California	17.1	22.4	121
Idaho	48.9	112.1	122
Maine	43.0	1,721.0	123
Massachusetts	28.0	318.0	124
North Carolina	12.0	39.0	125
Tennessee	6.9	51.2	126
Virginia	38.0	95.0	105
Virginia	27.0	111.7	127
Washington	5.3	51.6	46

[a]All estimates are from annual 100% convex polygon home ranges.

(temporal spacing), whereas females, particularly those with cubs, tend to exclude other females from their home ranges in summer (spatial spacing). This strategy allows dominant males to spread their genes among a number of females and adult females to protect the resources necessary to more adequately ensure survival of their young. Temporal spacing is likely maintained by a hierarchial system in which the large, older males are dominant.

As with many other mammalian species, black bears tend to be more active after passage of a low-pressure weather front (83). Black bear activities are depressed at temperatures above 25°C or below freezing. After 30 or more days of recovery from denning, black bears typically adopt a crepuscular pattern of activity through the spring and summer. However, they are capable of altering this pattern (e.g., nocturnal or diurnal), depending on the influences of human activities. Where human food or garbage is available, individuals may become distinctly diurnal (on roadsides) or nocturnal (in campgrounds). With the advent of fall, activities increase and, depending on food availability, black bears may be active as much as 24 hours a day, foraging for energy-rich foods (83). This frenzy of activity may continue until denning and depends on satiation and fat deposition. Prior to actual denning, there may be a period of 30 or more days of declining movements and activities (84).

Depending on the level of disturbance (natural or otherwise) winter activities are minimal. Black bears exhibit increasing levels of activity and shorter inactive denning periods with decreasing latitude; this may be due to food availability as well as temperature (75, 76). Males tend to be more active during the denning period than females; there is less necessity for them to stay in one place than for adult females with cubs or yearlings.

Mark Trees

Black bears create marked trees by clawing and biting them (Figure 19–4). This marking may function in the same way as the urinary signposts of canids. Marked trees of black bears do not occur randomly but are located along defined trails, particularly ridge lines, with the marked portions of the tree facing the trail. Marks are usually 1.5 to 2 meters high on the tree and are conspicuous. Marked trees are usually beside a trail. The occasional marked trees located away from the trail (usually less than 5 meters) have paths

Figure 19–4 Tree marking with their claws or teeth serves an unknown but likely social function in the social structure of black bears. (Photograph by M. Pelton.)

leading to them, worn into the forest soil from many years of repeated visits by bears (85). Some trees are marked only once, while others receive repeated markings year after year, sometimes with such frequency that the scarring eventually kills the tree. Often bear hair can be collected from the crevices in the bark below the mark.

Only a relatively small number of bears have been seen biting or clawing trees. Most observations have been of males making such marking before and during the breeding season, which is in summer, but females have been seen marking trees toward the end of summer and in the fall.

The four prevailing hypotheses to explain marking remain untested. The earliest idea was that markings are related to the dominance hierarchy system among the males in the population, particularly during the summer when males compete for females in estrus, and subadult males begin to establish home ranges. A second hypothesis is related to breeding and suggested that marks may serve as a means of communication to ensure that males and females are synchronized properly for successful mating. The third idea involves territorial defense among adult females. Because it is known that some adult females are territorial, marking may serve as a means of communicating with other adult females in the area, enabling them to better avoid one another. Finally, given that marking increases when a bear enters a new area, it has been suggested that marking may be used for orientation by animals when they are visiting unknown or rarely visited areas.

Marking is a highly visual activity and likely serves an important social function within a bear population. It seems likely that there may be several explanations for bears marking trees. Recent advances in DNA analysis of hair and saliva left on marked trees may help answer questions about the role of tree marking in black bear populations.

HABITAT REQUIREMENTS

Black bears thrive in a wide array of habitat types with mixtures of forest, shrub, and openings. They live in dry, shrubby chaparrel in the Southwest and feed on a variety of berry crops, acorns in shaded wet ravines, and prickly pear cactus on drier sites. In coastal North Carolina bears use a mosaic of large, intensively managed pine plantations and agricultural crops (corn and wheat). They use oak-dominated habitats in the Appalachian mountains where the forest canopy is more than 90% closed. In the Northeast bears can be found among second home developments and scattered small white cedar swamps, agricultural fields (corn), abandoned apple orchards, and fragmented oak and beech forests. In the upper Great Lakes region expansive spruce-fir forests dominate the habitat of black bears. However, many other habitat types also are used, for example, upland openings and various types of lowland swamps.

Black bears require five basic habitat components: escape cover, fall sources of hard or soft mast, spring and summer feeding areas, movement corridors, and winter denning habitat.

Escape cover that provides secure refugia is essential. The extent that this cover is necessary depends on the level of human disturbance and intrusion. White cedar thickets in the Northeast; laurel "slicks" in the Appalachian mountains; pocosin, titi, and bay swamps in the South; and chaparral scrub in the Southwest are examples of important types of escape cover. Some type of relatively impenetrable shrub cover seems to be a necessary element for good black bear habitat throughout the range of the species.

Because of the necessity to deposit substantial levels of fat prior to denning, fall sources of high-energy foods such as acorns, beechnuts, or abundant fall berry crops are vital. The presence or absence of these foods in fall can dictate the dynamics of a population; without it females may fail to reproduce and high mortality can occur because of malnutrition, road kills, and nuisance control (86).

Spring and summer feeding areas are important in the recovery of black bears from winter dormancy, particularly females with cubs. Abundant and varied berry

Figure 19–5   Secure winter dens are important for female black bears. Cavities in large old trees provide a high degree of protection during birth and postnatal care. (Photograph by F. T. van Manen.)

crops and colonial insects (ants, bees, yellow jackets) must provide the necessary nutrients to sustain both the female and her growing cubs or young from a previous year.

Seasonal foods are not always juxtaposed and large secure habitats have become fragmented with increased human developments. This logistics problem can be overcome for the relatively mobile black bear by providing secure movement corridors to connect important habitats. Corridors generally provide thick cover for concealment and are located along distinct topographic features such as creek or river drainages, ridgetops, and gaps.

Perhaps the most vulnerable time for black bears is winter dormancy. During this period females are giving birth to cubs, both sexes are sedentary and lethargic, and energy conservation is critical. In the northern parts of the species range, snow acts as a form of cover and ground-level dens in brush piles and thickets provide suitable protection. However, at any latitude female bears prefer cavities in large, standing, dead or live trees. Rock crevices also are used extensively throughout the range of the species. Ground dens, particularly for females, are located in thick, understory cover. High levels of winter disturbance can be sustained by black bears if their winter dens are located in thick, impenetrable cover, high up in cavities of old trees, or deep into rock crevices (Figure 19–5).

POPULATION MANAGEMENT

Harvest

All Canadian provinces and 28 of 39 states with black bear populations conduct legal hunts. More than 40,000 black bears are harvested annually by more than 500,000 hunters (25). States or provinces with the largest populations have more liberal hunting seasons, bag limits, and hunting methods than those with smaller populations. Fall is the traditional hunting season, is common to all states and provinces, and begins as early as 1 August and closes as late as 1 January. Spring hunts are presently conducted in all Canadian provinces but restricted to only a few western states in the United

States. Within the framework of these seasons, states and provinces have a variety of regulations governing harvest. These options include use of dogs, still hunting, use of bait, protection of females with cubs, minimum weight criteria, quota hunts, bag limits, timing of hunts (early season versus late season), use of primitive weapons, and areas with limited access (bear sanctuaries). This latter technique is referred to as spatial management, as opposed to traditional numerical management, and has proven successful as a method to manage black bear populations. By manipulating the factors mentioned, management agencies can regulate the level of harvest and consequently its effect on the population size.

By using various population indicators such as level of nuisance complaints, bait stations, incidence of road kills, intensive census techniques on sample areas, and computer models based on harvest or capture data, appropriate actions can be taken to liberalize or reduce hunting pressure. In areas where population increase or expansion is desired, females may be afforded greater protection than males. This protection can be accomplished by adjusting the hunting season toward the end of fall after most females have denned, total protection of females when cubs are at their side, harvesting around the periphery of large zones of protection such as designated bear sanctuaries or national parks, and conducting short, early spring hunts before females have a chance to emerge from dens.

Sex and age composition of the harvest and hunting success have proven, by themselves, to be unreliable indicators of bear population status because behavioral differences may bias vulnerability (66, 73, 87). Hunting methods, timing of season, and hunter numbers and selectivity may change (51, 88). Black bear populations that exhibit similar sex and age compositions may have altogether differing population growth rates depending on reproductive and mortality rates and patterns (89, 90). Finally, changes in weather and food availability can have dramatic effects on vulnerability to harvest from year to year (91, 72, 92). As a result, management agencies must use as many population, habitat, and environmental parameters as possible in monitoring their black bear populations.

Some methods used to harvest bears (baiting, dogs, spring hunts, and high-tech innovations, such as radio-collared hunting dogs, CB radios, four-wheel-drive and all-terrain vehicles), are deemed too efficient, unsportsmanlike, or inhumane by some constituents of the public. Recent state referenda have resulted in elimination or restriction of some of these practices in some states, whereas other states have voted to retain these practices. Management agencies are sometimes forced to alter their harvest strategies because of public sentiment; this may place limits on the agencies' flexibility to manage black bear populations. These constraints come at a time when virtually all states and provinces report stable and increasing populations of black bears. Because of the high mobility of this species, crossing of political boundaries often necessitates multi-agency cooperation to deal with management issues presented at the landscape level (93). As human populations expand into occupied bear habitat or along its periphery, dealing with the public and managing increasingly healthy black bear populations will become a difficult challenge for resource agencies into the next century. Educating the public about black bears and proactive harvest strategies must be a high priority.

Nuisance Bear Management

Black bears can cause damage to a wide array of agricultural commodities (128). State wildlife agencies consider the damage done by black bears of primary importance in their management plans for this species (94). Row crops such as corn, wheat, oats, sugarcane, soybeans, and peanuts are attractive to black bears (95). Serious economic damage can be incurred quickly on small, concentrated areas such as orchards or beeyards. Electric fencing has proven effective in repelling bears from beeyards (96). However, state wildlife agencies use depredation permits, spring hunts, and more liberal hunting regulations as methods to deal with problems of a more extensive nature (2).

Figure 19–6 As humans move close to occupied black bear habitat, conflicts arise over easily accessible concentrations of human food and garbage. (Photograph by M. Pelton.)

Black bears are a serious problem for resource managers where bears receive complete protection; individuals are sometimes attracted to human food or garbage. The historic method to deal with such "panhandlers" has been translocation, but the technique has had varied success. Black bears have a strong homing ability and are capable of traveling many kilometers to return to their resident area and continue their nuisance activity (97). A few of those that do not return may renew their nuisance activity in their new areas. Many others disappear, likely succumbing to mortality during their wanderings via accidents or illegal kills (97).

Separating people and their food or garbage from bears has been a particularly challenging task for managers of protected black bear populations. Closing traditional roadside pulloffs, bear-proofing garbage cans, enforcing laws prohibiting feeding, more frequent garbage pickup, more aggressive and quicker action toward first-time offender bears, and better information and education programs for the public are techniques that must be used to cope with nuisance bears in protected populations (98, 99, 2).

As human populations push into the periphery of occupied bear habitat (or vice versa), dealing with nuisance activity around commercial (e.g., hotels, motels, and restaurants) and residential developments becomes an issue (Figure 19–6). Black bears are more apt to search for attractive sources of anthropogenic foods during years of natural food scarcity as they wander to the fringes of their habitat. Poorly managed dumpsters at hotels or restaurants, dog food on carports, bird feeders, and garbage cans in residential areas are examples of easily available food for black bears. Natural resource agencies in the 21st century will have to deal with expanding human populations that are adjacent to occupied black bear habitat.

Natural resource agencies must develop programs to educate an increasingly urbanized public about how to live compatibly with black bears. At the same time, they must develop more innovative and comprehensive approaches when dealing with nuisance animals because some traditional tools are inappropriate, ineffective, inefficient, or unacceptable to the public. Recent techniques being tested include use of trained dogs to harass nuisance bears, capture and release on-site rather than translocation of animals, and more creative ways to separate bears and human food or garbage in remote areas (e.g., cable devices for hanging food) and near developed areas (e.g., new lid and latch designs on garbage containers) (100).

HABITAT MANAGEMENT

Any forest management activity that promotes production of summer and fall soft and hard mast benefits black bears. Hardwood cutting and timber rotation that ensure peak production of fall nut crops are important. The production of a given species of nut is highly variable from year to year. Encouraging a variety and mixture of nut-producing species increases the availability of fall foods (86). Normal forest management activities that create small forest openings and thus encourage soft mast production in summer and early fall are important; black bears benefit as long as this activity does not diminish important sources of fall foods. Black bears, particularly females, are less likely to use large, clearcut openings (101).

Maintaining large, contiguous forest tracts with minimal human disturbance is a recommended strategy (102). Concomitant with stabilizing large contiguous forest tracts is the need to control or restrict human use and access to some areas. Roads without controlled access can influence bears in several ways: disturbance by noise and presence of vehicles, displacement of bears to poorer quality habitat, reduction of bear use of habitats that have been modified, social disruption of bears away from roads, and increased human-induced mortality (103). Conversely, roads that receive limited human use may have positive benefits to black bears as travel corridors or feeding sites (104, 105, 106, 107).

In regions where snow does not serve as protective cover in winter, large old trees should be encouraged as prime denning sites for pregnant females. Maintaining old growth on a variety of ecological sites is an important management strategy to perpetuate prime den sites. Slash from cutover areas also should be piled and left as potential ground den sites.

Corridors (e.g., ridgetops, gaps, and drainages) between seasonal feeding areas should be maintained with thick, continuous cover for concealment (108). Maintenance of such movement corridors, particularly as human populations expand, is an important management strategy (109). Depending on the surrounding habitats and vulnerability of the bear population, corridors may need to be only a few meters wide (110). However where human activities impinge on the population, corridors may need to be many kilometers wide (109). Providing road overpasses or underpasses at critical locations can reduce highway mortality and the negative effects of habitat fragmentation (111).

SUMMARY

Black bears have proven to be the most adaptable of the world's bear species. They currently thrive in a wide variety of habitats, and stable and expanding populations are reported throughout most of their range in North America. Consequently, during the past 30 years, natural resource agencies have found themselves dealing with increasing numbers of nuisance complaints as human developments impinge on existing, occupied black bear habitat. This increasing level of conflict between black bears and people necessitates new and innovative methods for coping with such interactions and more creative and intensive educational programs, particularly in an increasingly urbanized society. In some locations there is growing evidence that black bears also can be restored to former range (112). At the same time natural resource agencies must recognize that this species is only adaptable to a point and they must guard against deterioration of existing habitats by stabilizing escape cover, winter denning habitat, summer and fall foods, and movement corridors. Advances in field methods for capturing, handling, marking, and monitoring black bears have greatly facilitated our current knowledge base for this species (59). Emerging technologies such as the use of DNA, global positioning systems, and geographic information systems will contribute to that growing knowledge base and allow us to better manage the species into the 21st century. Besides economic, aesthetic, and biological values, black bears are an important landscape indicator species. Because of their relative abundance they also can serve as a useful surrogate for technology transfer of information to assist in the management of more threatened or endangered bears species and other carnivores.

LITERATURE CITED

1. SERVHEEN, C. 1989. *The Status and Conservation of the Bears of the World.* International Conference Bear Research and Management. Monograph Series No. 2.

2. WARBURTON, G. S., and R. C. MADDREY. 1994. Survey of nuisance bear programs in eastern North America. *Proceedings Eastern Workshop on Black Bear Research and Management* 12:115–123.

3. NEAL, W. A. 1990. Proposed threatened status for the Louisiana black bear. *Federal Register* 5(120):25341–25345.

4. KURTEN, B., and E. ANDERSON. 1980. *Pleistocene Mammals of North America.* Columbia University Press, New York.

5. SETON, E. T. 1929. *Lives of Game Animals.* Vol. 2. Doubleday, Doran and Co., New York.

6. HALL, E. R. *The Mammals of North America,* 2nd ed. John Wiley and Sons, New York.

7. SWANTON, J. R. 1979. *The Indians of the Southeastern United States.* Smithsonian Institution Press, Washington, DC.

8. RAYBOURNE, J. W. 1987. The black bear: Home in the highlands. Pages 105–117 *in* H. Kallman, C. P. Agee, W. R. Goforth, and J. P. Linduska, eds., *Restoring America's Wildlife: 1937–1987.* U.S. Department Interior, Fish and Wildlife Service. U.S. Government Printing Office, Washington, DC.

9. WHITE, R. 1983. *The Roots of Dependency.* University of Nebraska Press, Lincoln.

10. WILLIAMS, S. C. 1930. *Adair's History of the American Indians.* Promontory Press, New York.

11. SHROPSHIRE, C. C. 1996. History, status, and habitat components of black bears in Mississippi. Dissertation. Mississippi State University, Starkville, MS.

12. PELTON, M. R., and F. VAN MANEN. 1994. Distribution of black bears in North America. *Proceedings Eastern Workshop on Black Bear Research and Management* 12:133–138.

13. ROGERS, L. L., and R. D. APPLEGATE. 1983. Dispersal of fruit seeds by black bears. *Journal of Mammalogy* 64:310–311.

14. GASKI, A. L. 1997. An update on the bear trade. Pages 62–76 *in* A. L. Gaski and D. F. Williamson, eds., *Second International Symposium on the Trade of Bear Parts.* Seattle, WA.

15. HALL, E. R., and K. R. KELSON. 1959. *The Mammals of North America.* Vol. 2. The Ronald Press Company, New York.

16. VAUGHAN, M. R., E. M. HOLLERMAN, M. R. PELTON, M. L. KENNEDY, M. A. BOGAN, S. R. FAIN, D. L. MILLER, T. H. EASON, J. W. KASBOHM, and P. K. KENNEDY. 1998. *Systematics of Black Bears in the Southeastern United States. Final Report.* U.S. Fish and Wildlife Service.

17. BANFIELD, A. W. F. 1974. *The Mammals of Canada.* University of Toronto Press, Toronto, Canada.

18. KOLENOSKY, G. B., and S. M. STRATHEARN. 1987. Black bear. Pages 443–454. *in* M. Kovak, J. A. Baker, M. E. Obbard, and B. Malloch, eds., *Wild Furbearer Management and Conservation in North America.* Ministry of Natural Resources, Ontario, Canada.

19. EASON, T. H. 1995. Weights and morphometrics of black bears in the southeastern United States. Thesis. University of Tennessee, Knoxville.

20. RAUSCH, R. L. 1961. Notes on black bears (*Ursus americanus* Pallas) in Alaska, with particular reference to dentition and growth. *Zeitschrift fur Saugetierkunde* 26:65–128.

21. COWAN, I. M. 1938. Geographic distribution of color phases of the red fox and black bear in the Pacific Northwest. *Journal of Mammalogy* 19:202–206.

22. ROUNDS, R. C. 1987. Distribution and analysis of colourmorphs of the black bear (*Ursus americanus*). *Journal Biogeography* 14(6):521–538.

23. BACON, E. S., and G. M. BURGHARDT. 1976. Learning and color discrimination in the American black bear. *International Conference Bear Research and Management* 3:27–36.

24. PELTON, M. R., A. B. COLEY, T. H. EASON, D. L. D. MARTINEZ, J. A. PEDERSON, F. T. VAN MANEN, and K. M. WEAVER. 1999. American black bear conservation action plan. Pages 144–156 *in* C. Servheen, S. Heirero, and B. Peyton, eds., *Bears—Status Survey and Conservation Action Plan.* International Union Conservation Nature Species Survival Commission Bear Specialist Group, IUCN SSC Polar Bears Specialist Group, IUCN, Gland, Switzerland and Cambridge, UK.

25. VAUGHAN, M. R., and M. R. PELTON. 1995. Black bears in North America. Pages 100–103 *in* E. T. La Roe, G. S. Farris, C. E. Puckett, P. D. Doran, and M. J. Mac, eds., *Our Living Resources.* U.S. Department Interior, National Biological Service, Washington, DC.

26. FRANZMANN, A. W., and C. C. SCHWARTZ. 1986. Black bear predation on moose calves in highly productive versus marginal moose habitats on the Kenai Peninsula, Alaska. *Alces* 22:139–154.

27. ROGERS, L. L. 1977. Movements and social relationships of black bears in northeastern Minnesota. Dissertation, University of Minnesota, St. Paul.

28. HAMILTON, R. J. 1978. Ecology of black bears in southeastern North Carolina. Thesis. University of Georgia, Athens.

29. NOYCE, K. V., and D. L. GARSHELIS. 1994. Body size and blood characteristics as indicators of condition and reproductive performance in black bears. *International Conference Bear Research and Management* 9:481–496.

30. EILER, J. H., W. G. WATHEN, and M. R. PELTON. 1989. Reproduction in black bears in the southern Appalachian mountains. *Journal of Wildlife Management* 53:353–360.

31. ELOWE, K. D., and W. E. DODGE. 1989. Factors affecting black bear reproductive success and cub survival. *Journal of Wildlife Management* 53:962–968.

32. KNUDSEN, G. J. 1961. We learn about bears. *Wisconsin Conservation Bulletin* 27:13–15.

33. JONKEL, C. J., and I. McT. COWAN. 1971. The black bear in the spruce-fir forest. *Wildlife Monographs* 27:1–57.

34. BEEMAN, L. E., and M. R. PELTON. 1980. Seasonal foods and feeding ecology of black bears in the Smoky Mountains. *International Conference Bear Research and Management* 4:141–147.

35. REYNOLDS, D. G., and J. BEECHAM. 1980. Home range activities and reproduction of black bears in west-central Idaho. *International Conference Bear Research and Management* 3:181–190.

36. WIMSATT, W. A. 1963. Delayed implantation in the ursidae, with particular reference to the black bear. Pages 49–76 *in* A. C. Ender, ed., *Delayed Implantation.* University Chicago Press, Chicago, IL.

37. ERICKSON, A. W., and J. E. NELLOR. 1964. Breeding biology of the black bear. Part 1. Pages 1–45 *in* A. W. Erickson, J. Nellor, and G. A. Petrides, eds., *The Black Bear in Michigan.* Michigan State Agricultural Experiment Station, Research Bulletin No. 4.

38. BOONE, W. R., J. C. CATLIN, K. G. CASEY, E. T. BOONE, P. S. DYE, R. J. SCHUETT, J. O. ROSENBURG, T. TSUBOTA, and J. M. BAHR. 1998. Bears as induced ovulators—A preliminary study. *Ursus* 10:503–505.

39. HELLGREN, E. C., M. R. VAUGHAN, F. C. GWAZKAUSKA, B. WILLIAMS, P. F. SCANLON, and R. L. KIRKPATRICK. 1990. Endocrine and electrophoretic profiles during pregnancy and nonpregnancy in captive female black bears. *Canadian Journal of Zoology* 69:892–898.

40. ALT, G. L. 1981. Reproductive biology of black bears of Northeastern Pennsylvania. *Transactions Northeast Section The Wildlife Society* 38:88–89.

41. CARREL, W. K. 1994. Reproductive history of female black bears from dental cementum. *International Conference Bear Research and Management* 9:205–212.

42. McLAUGHLIN, C. R., G. J. MATULA, JR., and R. J. O'CONNOR. 1994. Synchronous reproduction by Maine black bears. *International Conference Bear Research and Management* 9:471–479.

43. GARSHELIS, D. L. 1994. Density-dependent population regulation of black bears. Pages 3–14 *in* M. Taylor, ed., *Density-Dependent Population Regulation in Black, Brown, and Polar Bears.* International Conference Bear Research and Management. Monograph Series No. 3.

44. WILLEY, C. H. 1974. Aging black bears from premolar tooth sections. *Journal of Wildlife Management* 38:97–100.

45. MARKS, S. A., and A. W. ERICKSON. 1966. Age determination in the black bear. *Journal of Wildlife Management* 30:389–410.

46. POELKER, R. J., and H. D. HARTWELL. 1973. The black bear in Washington. Washington State Game Department. Biology Bulletin No. 14.

47. SAUER, P. R. 1975. Relationship of growth characteristics to sex and age for black bears from the Adirondack region of New York. *New York Fish and Game Journal* 22:81–113.

48. ROGERS, L. L. 1987. Effects of food supply and kinship on social behavior, movements, and population growth of black bears in northeastern Minnesota. *Wildlife Monographs* 97.

49. ROGERS, L. L. 1976. Effects of mast and berry crop failures on survival, growth, and reproductive success of black bears. *Transactions North American Wildlife and Natural Resources Conference* 41:431–438.

50. BUNNELL, F. L., and D. E. N. TAIT. 1980. Bears in models and in reality—Implications to management. *International Conference Bear Research and Management* 4:15–23.

51. KOLENOSKY, G. B. 1986. The effects of hunting on an Ontario black bear population. *International Conference Bear Research and Management* 6:45–55.

52. BUNNELL, F. L., and D. E. N. TAIT. 1985. Mortality rates of North American bears. *Arctic* 38:316–323.

53. LECOUNT, A. L. 1987. Causes of black bear cub mortality. *International Conference Bear Research and Management* 7:75–82.

54. BOWMER, E. J. 1973. Ursine trichinosis in British Columbia. *Canadian Journal Public Health* 64:84.

55. CRUM, J. M., V. F. NETTLES, and W. R. DAVIDSON. 1978. Studies on endoparasites of the black bear (*Ursus americanus*) in the southeastern United States. *Journal of Wildlife Diseases* 14:178–186.

56. JUNIPER, I. 1978. Morphology, diet and parasitism in Quebec black bears. *Canadian Field-Naturalist* 92:186–189.

57. COOK, W. J., and M. R. PELTON. 1979. Selected infectious and parasitic diseases of black bears in the Great Smoky Mountains National Park. *Proceedings Eastern Workshop on Black Bear Research and Management* 4:120–124.

58. LECOUNT, A. L. 1981. A survey of trichinosis among black bears of Arizona. *Journal of Wildlife Diseases* 17:349–351.

59. PELTON, M. R. 1980. Black bear. Pages 504–514 *in* J. A. Chapman and G. A. Feldhamer, eds., *Wild Mammals of North America: Biology, Management, Economics.* Johns Hopkins University Press, Baltimore, MD.

60. MANVILLE, A. M. 1978. Ecto-and endoparasites of the black bear in northern Wisconsin. *Journal of Wildlife Diseases* 14:97–101.

61. FORRESTER, D. J., M. G. SPALDING, and J. B. WOODING. 1993. Demodicosis in black bears (*Ursus americanus*) from Florida. *Journal of Wildlife Diseases* 29:136–138.

62. MILLER, S. D., G. C. WHITE, R. A. SELLERS, H. V. REYNOLDS, J. W. SCHOEN, K. TITUS, V. G. BARNES, JR., R. B. SMITH, R. R. NELSON, W. B. BOLLARD, and C. C. SCHWARTZ. 1997. Brown and black bear density estimation in Alaska using radiotelemetry and replicated mark-resight techniques. *Wildlife Monographs* 133.

63. EBERHARDT, L. L. 1990. Using radio-telemetry for mark-recapture studies with edge effects. *Journal of Applied Ecology* 27:259–271.

64. GARSHELIS, D. L. 1992. Mark-recapture estimation for animals with large home ranges. Pages 1098–1111 *in* D. R. McCullough and R. H. Barrett, eds., *Wildlife 2001: Populations.* Elsevier Applied Science, London, U.K.

65. PELTON, M. R., and L. C. MARCUM. 1977. The potential use of radio-isotopes for determining densities of black bears and other carnivores. Pages 221–236 *in* R. L. Phillips and C. Jonkel, eds., *Proceedings Predator Symposium,* Montana Forest and Conservation Experiment Station, University of Montana, Missoula.

66. MILLER, S. D. 1990. Population management of bears in North America. International Conference Bear Research and Management 8:357–373.

67. SPENCER, H. E., JR. 1955. The black bear and its status in Maine. Department of Inland Fish and Game. Bulletin No. 4. Augusta, ME.

68. BARNES, V. G., JR., and O. G. BRAY. 1967. *Population Characteristics and Activities of Black Bears in Yellowstone National Park.* Colorado Cooperative Wildlife Research Unit, Colorado State University, Fort Collins.

69. PELTON, M. R. 1972. Use of foot trail travellers in the Great Smoky Mountains National Park to estimate black bear (*Ursus americanus*) activity. *International Conference Bear Research and Management* 2:36–43.

70. GARSHELIS, D. L. 1993. Monitoring black bear populations: Pitfalls and recommendation. *Proceedings Western Black Bear Workshop on Research and Management* 4:123–144.

71. PELTON, M. R., J. CARDOZA, R. CONLEY, C. DUBROCK, and J. LINDZEY. 1978. Census techniques and population indices. *Proceedings Eastern Workshop Black Bear Research and Management* 4:242–252.

72. PELTON, M. R., C. BENNETT, J. CLARK, and K. JOHNSON. 1986. Assessment of bear population census techniques. *Proceedings Eastern Workshop on Black Bear Research and Management* 8:208–223.

73. GARSHELIS, D. L. 1991. Monitoring effects of harvest on black bear populations in North America: A review and evaluation of techniques. *Eastern Workshop on Black Bear Research and Management* 10:120–144.

74. PELTON, M. R. 1984. Bait station indices: Where we are and where we need to go. *Proceedings Eastern Workshop on Black Bear Research and Management* 7:43–47.

75. WOODING, J. B., and T. S. HARDISKY. 1992. Denning by black bears in northcentral Florida. *Journal of Mammalogy* 73:895–898.

76. WEAVER, K. M., and M. R. PELTON. 1994. Denning ecology of black bears in the Tensas River Basin of Louisiana. *International Conference Bear Research and Management* 9:427–433.

77. JOHNSON, K. G., and M. R. PELTON. 1980. Environmental relationships and the denning period of black bears in Tennessee. *Journal of Mammalogy* 61:653–660.

78. ROGERS, L. L. 1974. Shedding of foot pads by black bears during denning. *Journal of Mammalogy* 55:672–674.

79. NELSON, R. A. 1973. Winter sleep in the black bear: A physiologic and metabolic marvel. *Mayo Clinic Proceedings* 48:733–737.

80. FLOYD, T., and R. A. NELSON. 1990. Bone metabolism in black bears. *International Conference Bear Research and Management* 8:135–137.

81. PIEKIELEK , W., and T. S. BURTON. 1975. A black bear population study in northern California. *California Fish and Game Journal* 61:4–25.

82. GARSHELIS, D. L., and M. R. PELTON. 1981. Movements of black bears in the Great Smoky Mountains National Park. *Journal of Wildlife Management* 45:912–925.

83. GARSHELIS, D. L., and M. R. PELTON. 1980. Activity of black bears in the Great Smoky Mountains National Park. *Journal of Mammalogy* 61:8–19.

84. JOHNSON, K. G., and M. R. PELTON. 1979. Denning behavior of black bears in the Great Smoky Mountains National Park. *Proceedings Annual Conference Southeastern Association Fish and Wildlife Agencies* 33:239–249.

85. BURST, T. L., and M. R. PELTON. 1980. Black bear mark trees in the Smoky Mountains. *International Conference Bear Research and Management* 5:45–53.

86. PELTON, M. R. 1989. The impacts of oak mast on black bears in the southern Appalachians. *Proceedings Southern Appalachian Mast Management Workshop*. U.S. Forest Service and Department of Forestry, Wildlife and Fisheries, University of Tennessee.

87. BUNNELL, F. L., and D. E. N. TAIT. 1980. Bears in models and in reality—implications to management. *International Conference Bear Research and Management* 4:15–23.

88. LITVAITIS, J. A., and D. M. KANE. 1994. Relationship of hunting techniques and hunter selectivity to composition of black bear harvest. *Wildlife Society Bulletin* 22:604–606.

89. CAUGHLEY, G. 1974. Interpretation of age ratios. *Journal of Wildlife Management* 38:557–562.

90. HARRIS, R. B., and L. H. METZGAR. 1987. Harvest age structures as indicators of decline in small populations of grizzly bears. *International Conference Bear Research and Management* 7:109–116.

91. LINDZEY, J. S., G. L. ALT, C. R. MCLAUGHLIN, and W. S. KORDEK. 1983. Population response of Pennsylvania black bears to hunting. *International Conference Bear Research and Management* 5:34–39.

92. NOYCE, K. V., and D. L. GARSHELIS. 1997. Influence of natural food abundance on black bear harvests in Minnesota. *Journal of Wildlife Management* 61:1067–1074.

93. CLARK, J. D., and M. R. PELTON. 1998. Management of a large carnivore: Black bear. Pages 209–223. *in* J. D. Peine, ed., *Ecosystem Management for Sustainability*. Lewis Publishers, Boca Raton, FL.

94. CLARK, J. D., D. L. CLAPP, K. G. SMITH, and T. B. WIGLEY. 1991. Black bear damage and landowner attitudes toward bears in Arkansas. *Proceedings Annual Conference Southeastern Association Fish and Wildlife Agencies* 45:208–217.

95. DAVENPORT, L. B., JR. 1955. Agricultural depredation by the black bear in Virginia. *Journal of Wildlife Management* 17:331–340.

96. BRADY, J. R., and D. S. MAEHR. 1982. A new method for dealing with apiary-raiding black bears. *Proceedings Annual Conference of Southeastern Association Fish and Wildlife Agencies* 36:571–577.

97. FIES, M. L., D. D. MARTIN, and G. T. BLANK, JR. 1986. Movements and rates of return of translocated black bears in Virginia. *International Conference Bear Research and Management* 7:369–373.

98. CALVERT, R., D. SLATE, and P. DEBOW. 1992. An integrated approach to bear damage management in New Hampshire. *Proceedings Eastern Workshop on Black Bear Research and Management* 11:96–107.

99. MARTIN, D. D., D. KOCKA, K. DELOZIER, and D. M. CARLOCK. 1994. Protocols for handling nuisance black bears. *Proceedings Eastern Workshop on Black Bear Research and Management* 12:99–106.

100. WOODING, J. B., N. L. HUNTER, and T. S. HARDISKY. 1988. Trap and release apiary-raiding black bears. *Proceedings Annual Conference Southeastern Association Fish and Wildlife Agencies* 42:333–336.

101. CLARK, J. D. 1991. Ecology of two black bear (*Ursus americanus*) populations in the interior highlands of Arkansas. Dissertation. University of Arkansas, Fayetteville.

102. HELLGREN, E. C., and M. R. VAUGHAN. 1994. Conservation and management of isolated black bear populations in the southeastern coastal plain of the United States. *Proceedings Annual Conference Southeastern Association Fish and Wildlife Agencies* 48:276–285.

103. MCLELLAN, B. D. 1990. Relationship between human industrial activity and grizzly bears. *International Conference Bear Research and Management* 8:57–64.

104. CARR, P. C., and M. R. PELTON. 1984. Proximity of adult female black bears to limited access roads. *Proceedings Annual Conference Southeastern Association Fish and Wildlife Agencies* 38:70–77.

105. GARNER, N. P. 1986. Seasonal movements, habitat selection, and food habits of black bears (*Ursus americanus*) in Shenandoah National Park, Virginia. Thesis. Virginia Tech University, Blacksburg.

106. HELLGREN, E. C., and M. R. VAUGHAN. 1988. Seasonal food habits of black bears in Great Dismal Swamp, Virginia - North Carolina. *Proceedings Annual Conference Southeastern Association Fish and Wildlife Agencies* 42:295–305.

107. LOMBARDO, C. A. 1993. Population ecology of black bears on Camp Lejeune, North Carolina. Thesis, University of Tennessee, Knoxville.

108. MYKYTKA, J. M., and M. R. PELTON. 1990. Management strategies for Florida black bears based on home range habitat composition. *International Conference Bear Research and Management* 8:161–167.

109. HARRIS, L. D. 1988. Landscape linkages: The dispersal corridor approach to wildlife conservation. *Transactions North American Wildlife and Natural Resources Conference* 53:595–607.

110. WEAVER, K. M., D. K. TABBERER, L. U. MOORE, JR., G. A. CHANDLER, J. C. POSEY, and M. R. PELTON. 1990. Bottomland hardwood forest management for black bears in Louisiana. *Proceedings Annual Conference Southeastern Association Fish and Wildlife Agencies* 44:342–350.

111. FOSTER, M. L., and S. R. HUMPHREY. 1995. Use of highway underpasses by Florida panthers and other wildlife. *Wildlife Society Bulletin* 23:95–100.

112. EASTRIDGE, R., and J. D. CLARK. 1998. *An Experimental Repatriation of Black Bears into the Big South Fork Area of Kentucky and Tennessee. Final Report.* Tennessee Wildlife Resources Agency, Nashville.

113. LECOUNT, A. L. 1987. Characteristics of a northern Arizona black bear population. Federal Aid Wildlife Restoration Project No. W-76-R. Final Report.

114. LECOUNT, A. L. 1982. Characteristics of a central Arizona black bear population. *Journal of Wildlife Management* 46:861–868.

115. BECK, T. D. I. 1991. Black bears of west-central Colorado. Colorado Division Wildlife, Technical Publication No. 39.

116. MCLEAN, P. K., and M. R. PELTON. 1994. Estimates of population density and growth of black bears in the Smoky Mountains. *International Conference Bear Research and Management* 9:253–261.

117. HELLGREN, E. C., and M. R. VAUGHAN. 1989. Demographic analysis of a black bear population in the Great Dismal Swamp. *Journal of Wildlife Management* 53:969–977.

118. CARNEY, D. W. 1985. Population dynamics and denning ecology of black bears in Shenandoah National Park, Virginia. Thesis. Virginia Tech University, Blacksburg.

119. KOLENOSKY, G. B. 1990. Reproductive biology of black bears in east-central Ontario. *International Conference Bear Research and Management* 8:385–392.

120. BEECHAM, J. J. 1980. Some population characteristics of two black bear populations in Idaho. *International Conference Bear Research and Management* 4:201–204.

121. NOVICK, H. J., and G. R. STEWART. 1982. Home range and habitat preferences of black bears in the San Bernardino Mountains of southern California. *California Fish and Game Journal* 67:21–35.

122. AMSTRUP, S. C., and J. J. BEECHAM. 1976. Activity patterns of radio-collared black bears in Idaho. *Journal of Wildlife Management* 40:340–348.

123. HUGIE, R. D. 1982. Black bear ecology and management in the northern conifer-deciduous forest of Maine. Dissertation. University of Montana, Missoula.

124. ELOWE, K. D. 1984. Home range movements, and habitat preferences of black bear (*Ursus americanus*) in western Massachusetts. Thesis. University of Massachusetts, Amherst.

125. SEIBERT, S. G. 1989. Black bear habitat use and response to roads on Pisgah National Forest, North Carolina. Thesis. University of Tennessee, Knoxville.

126. VAN MANEN, F. 1994. Black bear habitat use in Great Smoky Mountains National Park. Dissertation. University of Tennessee, Knoxville.

127. HELLGREN, E. C., and M. R. VAUGHAN. 1990. Range dynamics of black bears in Great Dismal Swamp, Virginia - North Carolina. *Proceedings Annual Conference Southeastern Association Fish and Wildlife Agencies* 44:268–278.

128. VAUGHAN, M. R., P. F. SCANLON, S. E. P. MERSMANN, and D. D. MARTIN. 1989. Black bear damage in Virginia. *Proceedings Eastern Wildlife Damage Control Conference* 4:147–151.

CHAPTER 20

Brown (Grizzly) and Polar Bears

Maria Pasitschniak-Arts and François Messier

INTRODUCTION

Grizzly or brown bears and polar bears are the two largest carnivores in North America. The grizzly bear is a symbol of the pristine and untamed wilderness of North America's mountains, forests, and plains. The polar bear symbolizes the vast Arctic. Both species are equally impressive in their massive size, rugged beauty, power, and potential danger. Not surprisingly, humans have always been fascinated with bears, and fact and folklore are inseparably intertwined in the history of man and bear. To native people, bears have a spiritual and cultural significance.

Historically, human impact on bears was largely related to direct killing. Humans' uncompromising intolerance of any form of competition with grizzly bears resulted in widespread use of firearms, traps, and toxins to effectively eradicate entire populations (1). A rise in the monetary value of polar bear hides, increased use of oversnow machines, and interest in guided hunting resulted in escalated killing during the 1950s and 1960s (2, 3). More recent impacts on bears have been largely related to habitat loss, especially for grizzlies (4, 5). For example, human encroachment and fragmentation of natural ecosystems are extensive over much of North America; numerous species of fauna and flora are already extinct, and large predators such as bears are frequently in a vulnerable or threatened predicament. More then ever, survival of bears depends on preservation of biological diversity and assurance of resource availability for these species in the remaining wilderness regions.

TAXONOMY AND DISTRIBUTION

Grizzly or brown bears and polar bears belong to the order Carnivora, family Ursidae, subfamily Ursinae. The genus *Ursus* includes brown bears of Eurasia and grizzly bears of North America, polar bears, and the American and Asian black bears. Other species in this family include the sun, sloth, and spectacled bears, and the giant panda (6).

Grizzly and polar bears are larger and more heavily built than other ursids. Grizzly bears are characterized by a massive head with a distinctive dished facial profile, small eyes and ears, a powerful body with a prominent shoulder hump, and long curved claws on the forefeet, about twice as long as on the hind feet (Figure 20–1). The pelage encompasses various shades of tan, blond, gold, gray, silver, and brown. Grizzly bears are sexually dimorphic and show extensive variation in size and color depending on locality. Adult males are on average larger and heavier than adult females, and the heaviest individuals are found along coastal Alaska and on Kodiak Island (7). Extensive

Figure 20–1 A grizzly bear, showing the characteristic dished facial profile, prominent shoulder hump, and long, curved claws. (Photograph by F. Messier.)

variation in skull and skeletal characteristics, body size, color, and other morphological traits has resulted in a formidable and confusing synonymy (8). Currently, all brown or grizzly bears are considered as one Holarctic species (9, 6). Nine subspecies of the grizzly bear are recognized in North America (10).

Polar bears are slightly larger than grizzly bears in size and weight. The species is easily identified by its white fur overlaying a mottled black and pink skin. Compared to grizzly bears, polar bears have smaller and more elongated heads, longer necks, and lack a shoulder hump (Figure 20–2). The claws of polar bears are shorter and sharper than those of the grizzly. The forepaws are large, making them well adapted for swimming and digging through snow. Sexual dimorphism is also prevalent in polar bears, with males being larger and heavier than females. Size and weight vary somewhat geographically, with a slight cline of increasing skull size from Spitzbergen to the Bering Strait, but such variation is quite small compared to grizzlies. Polar bears are not divided into subspecies (11, 2, 6).

Fossil evidence suggests that polar bears arose from coastal grizzly bears (12). Molecular genetics supports the fossil record and places the divergence of the two species at approximately 300,000 to 400,000 years ago (13). The distinguishing morphological features of polar bears likely evolved rapidly in response to selective pressure of the new marine environment. Interbreeding and production of fertile offspring between grizzly and polar bears have been recorded in captivity, suggesting a close phylogenetic relationship between the two species (14) (See Chapter 2).

Historically, grizzly bears inhabited most of western and central North America from the Arctic Ocean to central Mexico, but the species is now restricted to approximately half of its former range (Figure 20–3). In Canada, grizzly bears currently are found in the Yukon, Northwest Territories, British Columbia, and Alberta. In the United States, grizzly bears inhabit most of Alaska, but south of the Canada–United States border they have been eliminated from all but approximately 2 percent of their original

Figure 20–2 Polar bears in the Canadian Arctic. (Photograph by F. Messier.)

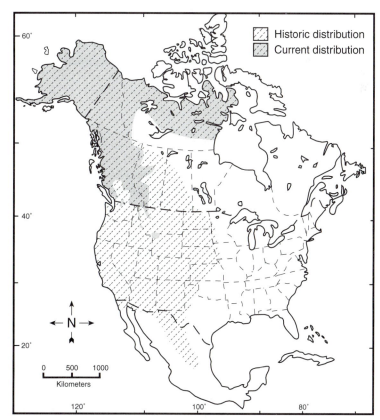

Figure 20–3 Historic and present-day distribution of grizzly bears in North America.

Figure 20–4 Distribution of polar bears throughout the circumpolar basin. In North America, 14 distinct populations are recognized: Southern Beaufort (SB), Northern Beaufort (NB), McClintock Channel (MC), Gulf of Boothia (GB), Viscount Melville (VM), Lancaster Sound (LS), Baffin Bay (BB), Foxe Basin (FB), Davis Strait (DS), Western Hudson Bay (WH), Southern Hudson Bay (SH), Norwegian Bay (NW), Kane Basin (KB), and Queen Elizabeth Islands (QE).

range. In the contiguous United States, grizzlies exist as six disparate subpopulations: (1) Yellowstone ecosystem of northwestern Wyoming, eastern Idaho, and southwestern Montana, (2) Northern Continental Divide ecosystem in northcentral Montana, (3) Cabinet/Yaak ecosystem in northwestern Montana and northeastern Idaho, (4) Selkirk ecosystem in northern Idaho and northeastern Washington, (5) North Cascades ecosystem in northern Washington, and (6) Bitterroot ecosystem in central Idaho (4, 15, 16).

In contrast to grizzly bears, polar bears are almost always associated with sea ice, are circumpolar in their distribution (Figure 20–4), and presently occupy close to all of their original habitat (4). In North America, polar bears range from Alaska on the West Coast to Baffin Island, and Newfoundland and Labrador on the East Coast. The species is distributed throughout the coastal habitat and on the Arctic sea ice. Currently, 14 distinct populations are recognized (17, 18).

LIFE HISTORY

Diet Selection

Grizzly and polar bears are highly opportunistic and respond rapidly to new and available food sources. Grizzly bears are highly efficient omnivores and a diverse range of foods is important for maintaining adequate nutritional levels (5). Polar bears are primarily carnivorous and well adapted to exploit the Arctic marine environment (11).

On the North American continent, grizzly bears use a variety of foods that can be divided into animal matter (i.e., mammals, birds, fish, insects), vegetal matter (i.e., stems and leaves), underground vegetation (i.e., bulbs, corms, and roots), and fruits and nuts. Different food types are available in different regions and bears have adapted to local availability of specific food items (7). Meat is a source of high protein and digestible energy for bears. In early spring, grizzly bears feed extensively on carrion of elk, bison, moose, caribou, mule deer, antelope, mountain sheep, and mountain goats. They are also opportunistic predators of ungulates. During spring elk-calving season, grizzly bears concentrate on catching vulnerable newborn calves (19, 20, 21). Muskox calves and adults are hunted (22, 23). In southcentral Alaska, grizzly bears are the largest cause of moose calf mortality (24, 25). Calves and adult caribou are also preyed on or killed incidentally (26). Killing rates among individual bears are highly variable (27). Hunting techniques include searching, chasing, and ambushing, but a combination of strategies may be used. Predation on black bears has also been documented (28, 29). The spring meat diet is supplemented with immature green vegetation such as grasses, sedges, and forbs.

Agility and excellent paw coordination enable grizzly bears to capture a variety of small mammals and fish (7, 5). Spawning and migrating salmonids are an important food source in coastal regions (30, 31). In Yellowstone National Park, cutthroat trout are eaten from May through July (32). Alpine insect aggregations of army cutworm moths, whitebark pine seeds, and various berries are an important high-fat food source in the Yellowstone ecosystem (33, 7, 34). Depending on location and availability, a wide assortment of plant species, including nutritious tubers, bulbs, lush grasses, flowers, leaves, roots, fungi, eggs, nuts, insects and fruits are consumed (7).

Unlike grizzly bears, polar bears are almost exclusively carnivorous. Ringed seals are the primary food source. During spring, about half of the ringed seals killed are newborn pups, but weaned pups and postparturient females are exploited also. In addition, bearded, harp, and hooded seals are preyed on. Seals are captured primarily by still hunting or breaking through snow lairs, but bears also stalk seals by crawling on ice or swimming. Being opportunistic hunters, polar bears will kill beluga whales, narwhals, and walruses (35, 36, 37). Seal fat is preferred and first to be consumed. Polar bears scavenge on seal, whale, and walrus carcasses, but rarely on carcasses of terrestrial mammals. Occasionally, small mammals, birds, eggs, and vegetation are eaten (38, 39).

Reproductive Strategy

Grizzly and polar bears display some of the lowest reproductive rates among terrestrial mammals. Both species exhibit late sexual maturity and protracted reproductive cycles. First parturition is usually around 5 years of age, but a large variation (4–10 years) is observed among different populations (7, 40, 41, 5). Parturition may be delayed several years as a consequence of limited food supply (42). Grizzly and polar bears typically breed from mid-May to July, and from March to May, respectively. Both species are polygamous and several males may follow a female, resulting in fights between competitors. Intense fighting between males for breeding opportunities suggests that large body size results in reproductive advantage (43, 5). Grizzly bears may choose isolated summit ridges for courtship activities (44, 45). Herding of estrous females by males is a reproductive strategy observed in grizzly and polar bears (44, 43). This herding behaviour has been interpreted as males sequestering estrous females from other adult males. Sequestering in isolated areas of low food abundance suggests reduced food intake by the breeding animals and likely minimizes the probability of encountering other males seeking estrous females (43, 44). In habitats where grizzly bears congregate, or in years of high food availability, females may mate with two males in one day, or a number of different males during the breeding season (46, 5, 47). DNA analyses of Arctic grizzly bears lends support to a flexible reproductive strategy where offspring from one litter may be sired by different males (47). Female polar bears have been observed with two or more males in the same breeding season (43).

Both species exhibit delayed implantation. Following fertilization, embryos develop to the blastocyst stage but remain free in the uterus. This embryonic arrest stage continues until females enter their maternity dens by September or October. In autumn, the blastocysts implant, and the active gestation period of approximately 2 months begins. Grizzly and polar bear cubs are born from January to March and November to January, respectively. Litter size varies from one to four, but two cubs are most typical. Twinning is dizygotic and is consistent with multiple paternity (40). Sex ratio at birth is 1 : 1. Newborn cubs are sightless, helpless, and small, and weigh between 0.5 and 1.0 kilogram. Lactation lasts 1.5 to 2.5 years, which is exceptionally long for carnivores. Cubs remain with their mothers for 2 to 3 years. The time interval between litters for both species is in the range of 2 to 4 years. As a result, more males than females are available for breeding in any given year (43, 40, 2, 48, 5).

Physiology and Longevity

There is controversy on the topic of hibernation and dormancy in bears, but in the literature both terms are used to describe denning bears (49, 50). Because there is a difference between winter sleep of bears and the deep hibernation of small mammals, the term *carnivorean lethargy* has been proposed to describe the former condition (51). Bears in winter sleep are unique compared to other hibernating mammals. While in their dens, grizzly and polar bears do not eat, drink, urinate, or defecate, the drop in their body temperature is slight, they are easily aroused into an active state, and they give birth to cubs and nurse them under these stressful conditions (52, 49, 53, 7).

For grizzly bears, denning is a strategy used to reduce energy expenditure during winter months when food is unavailable. For polar bears, the reduced access to food in winter is a matter of degree. This fundamental difference results in the obligate use of dens by grizzly bears, and the facultative use of dens by polar bears (54). For grizzly bears, the denning period is continuous for 3 to 7 months, while polar bears, with the exception of pregnant females, use temporary shelters during the fall ice-free period and during midwinter. Pregnant polar bears follow the same denning strategy as grizzly bears. Maternity dens serve a dual purpose: They enable pregnant grizzly and polar bears to conserve energy, and provide protection for newborn cubs incapable of thermoregulation (55, 56).

Winter sleep and birth of offspring while in a den require a number of physiological adaptations. While in the den, body temperature of bears is reduced by only a few degrees, but reduction in metabolism is comparable to hibernating mammals. Lean body mass is conserved and mechanisms of protein conservation are well developed. Adipose tissue is used almost exclusively to meet metabolic, thermoregulatory, and lactation demands (55, 57).

Bears give birth to small litters of tiny immature neonates that are much smaller than predicted for females of comparable size. This may be a result of physiological constraints on the ability of pregnant females to meet the requirements of fetal metabolism while fasting. Because mammalian fetuses require glucose as the principal oxidative substrate, fasting bears must provide glucose to developing fetuses through gluconeogenesis by using body protein, which may eventually jeopardize the female's survival. A short gestation period and birth of tiny neonates enables the female to change from transplacental to mammary nourishment of the cubs. Fatty acids from maternal lipid stores are incorporated into milk, sparing maternal body protein (53). Protein content of milk of both species is similar; however, lipid content of polar bear milk is approximately double that of grizzly bear milk (58, 59). Overall, reproductive success appears to be largely dependent on maternal fat reserves to meet the demands of gestation and lactation. Females with large fat stores tend to produce heavier cubs and heavier cubs are more likely to survive the first year of life (57). Food shortages, insufficient energy reserves, and poor body condition of adult females may result in failure of embryonic implantation, fetus reabsorption, inability to maximize cub growth, or cub abandonment (55).

Cub survival during the first year of life depends on rapid growth and weight gain. Protein content of post-denning milk increases with the higher protein needs of growing cubs and the higher protein metabolism of the nonfasting mother. Farley and Robbins found that during the denning period grizzly bear cubs grow at a rate of 98 grams per day and upon emergence from the den weigh approximately 5.1 kilograms (59). During summer, cubs can gain up to 351 grams per day. Upon weaning, grizzlies weigh approximately 102 kilograms. Polar bear cubs weigh on average 11 kilograms upon emergence from dens, 56 kilograms by summer, and 126 kilograms as yearlings (40).

Both grizzly and polar bears show extreme sexual dimorphism, which is largely the result of the longer growth period of males relative to females. In general, female and male grizzly bears attain their maximum size by 4 to 5 and 6 to 8 years of age, respectively. Cranial measurements indicate that rates of growth of males are twice that of females between the ages of 6.5 and ≥15.5 years (7). Similarly, in polar bears, adult body length and mass is reached at approximately 5 years for females, and 8 to 10 years of age for males (11, 40).

In the wild, potential longevity of grizzly and polar bears is about 20 to 30 years, and in captivity 30 years. For both species, reproductive longevity appears to approximate physical longevity (7, 2).

BEHAVIOR

Social Structure

Adult grizzly and polar bears are solitary except during the breeding season. In both species, parental care is provided exclusively by females. Cub adoption occurs occasionally (46, 60, 61). Females have a prolonged association with their cubs, nursing, teaching, and protecting them.

Cubs or females are sometimes preyed on by large males, so females with cubs tend to avoid such encounters (62). The threat of cannibalism may result in differential distribution of bears by age and sex class. Female grizzly bears with cubs tend to avoid males and male-occupied habitats by moving into habitats with fewer resources (63). Similarly, adult female polar bears avoid habitats where males are most abundant (2). Both species of bears, however, tend to congregate at abundant food sources (64, 35). When in proximity to one another, a social hierarchy is maintained within populations. Being more social than polar bears, grizzly bear females and their cubs will occasionally bond and the cooperative families travel, feed, and defend themselves as one cohesive unit. Similarly, weaned juveniles occasionally band together with littermates or same-age nonrelated individuals. In both cases, members of bonded groups attain a higher status in the social hierarchy and have an advantage at food sources and in defense against larger bears (5). Females with cubs may attack and sometimes kill a conspecific that approaches too closely (7, 2).

Differences in feeding behavior are evident between the two species. Grizzly bears often eat a carcass over several days, and may cache it between feedings. A bear guarding a cache may chase or attack any approaching intruder (16). In contrast, polar bears commonly leave considerable portions of prey uneaten and uncovered, and these remains may be a valuable food source for scavengers and subadult bears (65, 66). Differences in the length of time that the two species remain with their kill may be related to prey size and the rate at which prey are captured (67). In both species, larger and older bears displace subadults from carrion and kills, and sometimes kill them. Thus, successful foraging and subsequent survival of independent juveniles is influenced by social factors and by food availability.

Communication

Bears communicate by smell, sight, and sound, but like other fissipeds and many solitary animals, they rely primarily on olfaction (68). Male polar bears, for example, instantly recognize the path of an estrous female, suggesting that a chemical signal is involved (67). Grizzly and polar bears vocalize little. Cubs and juveniles may utter distress calls when hungry, hurt, or separated from their mother and siblings. Females call to their cubs and reassure them with nasal contact. Apprehensive animals may huff and snort, while offensive threats can be accompanied by growling and roaring (2, 67). Tails of bears are conspicuously small and do not serve an important function in communication. Bears tend to approach conspecifics head on, thus enhancing recognition and communication. Dominant or aggressive animals use frontal orientation and approach with ears back, neck outstretched, and muzzle displaying canines. Subordinate animals assume a lateral orientation, turn away, sit, or back away. Claws and canine teeth are important weapons in intraspecific fighting. In fights, opponents use their paws to strike one another in the shoulder or neck region and biting is often directed toward the head, neck, and throat (30, 2, 67). Two opponents may lock jaws in an attempt to assess dominance, which may result in broken canines (43). During the breeding season, dominant male grizzly bears often fight to attain alpha status. Individuals that attain this status are generally characterized by large body size, a high degree of aggressiveness, and superior fighting skills. The winner would likely maintain his hierarchical status for the remainder of the season. Toward the end of the breeding season, threat behaviors and vocalizations replace attacks as a means of communicating status (5). Male polar bears also engage in physical combat. In autumn, polar bears engage in ritualized "play fighting," at which time they may be practicing for the "real fighting" of spring. During the spring breeding season, male polar bears fight aggressively with one another, which may result in considerable tooth breakage, scarring, and wounding (43).

Movements

Distribution and movements of grizzly and polar bears are influenced by food resources, habitat selection, denning areas, avoidance of conspecifics, and philopatry (69, 16, 70). Seasonal movements of grizzly bears are related to major food sources such as emerging spring vegetation, carrion, spawning salmon or trout, high berry or pine nut production, and ant or moth aggregations. Seasonal use of some habitats results in sexual segregation where arrival of males displaces females (63).

Polar bears are closely tied to sea ice and activity is affected by seasons. The period of highest activity occurs from May through July, coinciding with availability of newborn seal pups and moulting adults. Hyperphagia and the large accumulation of body fat during this period ensures the best chance of survival through winter. Activity is lowest during winter, and is likely due to inclement weather, limited accessibility to seals, and energy conservation during the coldest months (56, 54). Land areas are used during times of open water and provide living space and shelter for bears as they wait out the summer period of low food availability. In western Hudson Bay, adult males tend to occupy coastal areas, while pregnant females and families are found further inland. Movement by bears on land is less than on sea ice, and animals spend most of their time resting (70). Little food is consumed, and energy is derived from fat reserves accumulated during the previous spring (69). Although polar bears may travel over large distances, they exhibit seasonal fidelity to geographic areas. A possible advantage of philopatry may be familiarity with conspecifics and distribution of food resources. Philopatry may also be related to navigational patterns learned from the mother (69).

For grizzly bears, accumulation of energy reserves, particularly for pregnant females, and availability of suitable den sites are crucial for successful denning. Most dens of grizzly bears are located in remote, mountainous and timbered regions, far from human disturbance. Individuals tend to dig their own dens, but may occupy natural caves or rocksplits. Dens do not appear to be randomly distributed; bears show habitat preferences and dens tend to be concentrated in areas that possess appropriate environmental conditions. Individual fidelity to specific denning sites has also been observed (7). In spring, males are first to emerge, while females with young cubs are last (52, 71).

Maternity dens of polar bears are located on land, on offshore islands, and on pack ice (72, 73, 54). Females show high fidelity to maternity denning areas and are faithful to denning substrates (i.e., land, pack ice, land-fast ice) and general geographic areas, but appear not to reuse the same den (73). Depending on latitude, grizzly and polar bears enter dens anytime between September and November, and spring arousal tends to occur between February and May (52, 72, 56, 54).

Polar bears have access to their prey throughout winter, although hunting success may be lower than in spring. Thus, with the exception of pregnant females, polar bears of all other ages and reproductive classes may remain active throughout winter. Individuals may build shelters, which are usually occupied in midwinter (54, 70). The use of shelters appears to be related to energy conservation when seal-hunting conditions are unfavorable (56, 54).

For grizzly bears, the home range defines the habitat essential to the animal's survival. Home ranges of grizzly bears overlap, and there is little evidence of territorial defense. Females have smaller home ranges than males, which may decrease females' risks of encountering nonresident, aggressive males. Home ranges of mature males may encompass home ranges of several adult females, thus increasing the probability of breeding. Home range size is influenced by age, sex, social status, distribution and availability of food and cover, location of dens, and proximity to mates (64, 7). Aggressive behavior may be observed at important food locations where numerous individuals aggregate, such as at fish spawning areas, berry patches, and garbage dumps (30, 5).

Polar bears rarely establish stationary home ranges (defined as the area over which an animal normally travels in search of food (74). This is likely due to yearly variations in ice conditions and the subsequent changes in food availability. Female polar bears move to areas of sea ice where hunting success is greatest, and males follow females. In response to female distribution, males are not able to defend specific territories (43).

POPULATION DYNAMICS

Population Trends and Current Status

In Canada, an estimated 25,000 (range = 22,000 to 28,000) grizzly bears occur in Alberta, British Columbia, the Yukon, and Northwest Territories. The plains grizzly bear, which inhabited the prairie regions of Alberta, Saskatchewan, and Manitoba, was eliminated by the 1880s (75). In Alberta, grizzly bears are declining in the Kananaskis Provincial Park and Bow Crow Forest of southwestern Alberta (41). The current strategy is to increase the population of about 790 to 1,000 bears through regulation of human-caused mortality and harvest management. Up to 13,000 grizzly bears occur in British Columbia, but numbers are declining in the south (76, 75). An estimated 6,300 grizzly bears inhabit the Yukon. In the Northwest Territories, the estimated population of 5,050 appears to be stable (75). The range of grizzly bears in Canada has been classified into 14 bear zones differing in climate, landform, vegetation, and land use activity (Table 20–1) (75).

Grizzly bear populations south of the Canadian border were designated as threatened on July 28, 1975, under the United States Endangered Species Act. The estimated population is in the range of 700 to 900 animals in Montana, Wyoming, Idaho, and Washington (4). The minimum population estimate in the Yellowstone ecosystem is approximately 240 and the number appears to be increasing (16). At the same time, human use of Yellowstone National Park is increasing and human infringement on grizzly bear habitat continues to cause direct and indirect mortality of bears (77). In the Selkirk Mountains grizzly bear ecosystem of Idaho, Washington, and British Columbia, some 10 to 15 bears reside in the United States and 16 to 21 in Canada. The population appears to be stable, but illegal mortalities appear to limit population growth (78). The Northern Continental Divide ecosystem is adjacent to British Columbia and Alberta, and links grizzly populations from the United States with those of Canada. This linkage zone in southern Canada is crucial for the long-term survival of grizzly

TABLE 20–1

Population Size and Status (Risk Categories Defined by the Committee on the Status of Endangered Wildlife in Canada) of Grizzly Bears in 14 Zones in Canada

Zone	Population size	Status
Arctic coastal plains	2,860	Not at risk
Taiga shield	790	Vulnerable
Taiga plains	1,520	Vulnerable
Subarctic mountains	2,540	Vulnerable
Subarctic mountains and plains	5,680	Not at risk
Cold boreal plains	960	Vulnerable
Cold moist mountains	2,940	Not at risk
Temperate wet mountains	3,310	Vulnerable
Cool moist plateaus	1,100	Vulnerable
Cool moist mountains	2,540	Vulnerable
Hot dry plateaus	140	Threatened
Cool dry mountains	930	Not at risk
Nonmountainous boreal plains	—	Extirpated
Glaciated prairies	—	Extirpated

Source: From Ref. 100.

bears in the continental United States. From 1979 to 1994, grizzly bears in the North Fork of the Flathead River Drainage of British Columbia and Montana appeared to be increasing (79). The recent minimum population estimate for this ecosystem is greater than 300 bears (16). In the Cabinet/Yaak ecosystem, fewer than 15 bears inhabit the Cabinet Mountains, and a small but unknown number live in the Yaak portion (16). Currently, it is unclear whether grizzly bears in the Selway/Bitterroot ecosystem are permanent residents. Verified grizzly bear tracks were documented in the North Cascades ecosystem, but the current status of the population is unknown (16).

Grizzly bears are a game species in Alaska and numbers are estimated between 32,000 to 43,000 bears. Populations in most regions of the state are stable; however, high harvest rates in some areas are largely due to improved road access related to the timber industry (4).

The current status of polar bears is better than it was in the 1960s due to international cooperation on management and research by the Polar Bear Specialist Group. The total population (worldwide) is estimated between 21,470 and 28,370 bears (18). The sizes of 11 North American populations range from 200 to 3,470 (Table 20–2).

Overall, the species is stable, but potentially vulnerable to human encroachment and activities such as oil and gas exploration. As development continues to expand in the Arctic, the potential for mortality associated with oil spills and contamination of the bears' food supply increases (4). Because polar bears are at the highest trophic level in the Arctic food chain they are most vulnerable to the accumulation of environmental toxins, which may negatively affect their reproduction (80, 81). Higher human density will also result in more frequent human–bear contact (82). Similar to grizzly bears, polar bear populations are fragile when subjected to human-induced mortality. Low reproductive rate and slow response in compensating for population declines do not allow either species to sustain high mortality rates, especially of females (83, 2, 4).

Productivity and Recruitment

Factors affecting the dynamics and productivity of grizzly and polar bear populations are complex. For grizzly bears, because overall densities show considerable variation among areas and appear to be related to food productivity, food may conceivably be the ultimate regulating factor (84). Similarly for polar bears, food appears to be a critical factor influencing survival and recruitment (65). Recruitment is also influenced by the number of adult females, because survival rate of adult females is a major factor affecting population growth rate. Recruitment depends on the number of females available for

TABLE 20–2

Population Size and Status of 11 North American Polar Bear Populations

Population	Size	Status
Western Hudson Bay	1,200	Stable
Southern Hudson Bay	1,000	Stable
Foxe Basin	2,000	Stable
Lancaster Sound/Baffin Bay	2,470	Declining
Davis Strait	950	Stable
Gulf of Boothia	900	Stable
M'Clintock Channel	700	Stable
Viscount Melville Sound	230	Stable
Northern Beauford Sea	1,200	Stable
Southern Beauford Sea	1,800	Stable
Queen Elizabeth Islands	200	—

Source: From Refs. 17, 18, and 101.

breeding and the number of females that produce offspring. Survival of cubs is strongly correlated with the condition of the mother, which is ultimately related to the availability of adequate and nutritious food. A reduction in food supply may result in a decline in cub survival and recruitment. Nutrition and social behavior, including cannibalism, are not mutually exclusive, and the two factors may be important in influencing recruitment. In both species, social interactions between dominant and subordinate individuals may result in decreased foraging efficiency of subordinate animals as a result of intraspecific competition, displacement from feeding sites, increased vigilance, and higher levels of social stress. Intraspecific predation may also be important (62, 65, 84, 63).

Although all populations are ultimately regulated by density-dependent processes, the mechanisms by which density influences population dynamics of grizzly and polar bears are poorly understood. Given the sparse and inconclusive data, wildlife managers should not assume that a reduction in density will result in an increase in reproduction, recruitment or survival of bear populations, particularly populations that are already subjected to some form of human-induced mortality (65, 84).

Mortality Factors

With the exception of man, grizzly and polar bears have no natural enemies (64, 85, 86). Human presence in bear habitats often leads to human–bear conflicts, often with fatal consequences for bears (76). During poor-food years, human-habituated grizzly bears may use native and non-native foods near human developments and are killed more often than nonhabituated bears. Adult females frequently fall into this category, presumably as a result of energetic demands associated with providing food for dependent young (34). Defense kills of grizzly and polar bears, involving potentially dangerous or nuisance bears and protection of property, occur more frequently in areas with higher human populations. Death by gunshot is the major cause of mortality for both species. Sport hunting of grizzly bears tends to be biased for larger males over smaller females (87). Harvest of polar bears by the Inuit also tends to be sex-biased toward males (2, 88).

Cubs-of-the-year sustain the highest mortality rate (66). Deaths may occur during denning or shortly after emergence from dens. Some cubs may not be strong enough to leave the den with their mothers, or are too small to compete with their siblings. Low maternal fat reserves for lactation leads to cub starvation; therefore, malnourishment of the mother appears to be a critical factor in cub survival (7, 40, 89).

Mortality of subadults may be related to the breakup of family groups, which subjects juveniles to stressful and potentially dangerous situations. Subadults are inexperienced at foraging and hunting and they have a greater chance of being displaced from food sources by larger and more dominant individuals. Competition for carrion may decrease

Figure 20–5 Capturing a polar bear using a ski-doo and dogs. (Photograph by F. Messier.)

subadult survival (65). Subadults may thus go into winter in poorer condition than adults, making survival between the time of weaning and adulthood a difficult one.

Intraspecific killing and infanticide by grizzly and polar bears are rarely observed. Females with cubs tend to avoid males and male-occupied habitats because of the likelihood that males may kill cubs. Evidence tends to support the hypothesis that adult males are responsible for most predation events. Cubs-of-the-year and yearlings appear most vulnerable, and infanticide and cannibalism appear to occur more frequently at food aggregations (62, 28, 90, 63). Other natural causes of mortality may be related to malnutrition, disease, senility, and injury (7, 2).

Emigration and Immigration

Studies on density effects of grizzly and polar bear populations are difficult to conduct because most populations inhabit remote areas, making research logistically difficult and expensive (Figure 20–5; 65). In both species of bears, adult males appear to have evolved an aggressive disposition toward subadult males. This aggression results in the emigration or dispersal of subadult males and a low immigration of males into the population. Upon mating, young females tend to remain near the maternal home range. Little information has been published on dispersal of grizzly bears. As a result, dispersal behavior and its relationship to population density are poorly understood (84).

Polar bears of both sexes tend to show fidelity to geographic areas (70). Bears feeding on ice in spring return to the same general feeding areas for many years. Bears tend to disperse from commonly used areas when the availability of seals declines, showing closely linked bear–seal interactions. Immigration and emigration are limited and genetic exchange among distant populations occurs occasionally, but is generally infrequent (91, 69, 17).

POPULATION MANAGEMENT

In Canada grizzly bears are considered a game species. Legal harvest by sport hunters is allowed, but hunting is prohibited in designated national parks, provincial wilderness areas, wildlife sanctuaries, and preserves. In Alberta, British Columbia, Yukon, and Northwest Territories, females with cubs are protected. Additionally, in Alberta and

British Columbia, all cubs and yearlings are protected, while in the Yukon, cubs but not yearlings are protected (75). In the United States, grizzly bears are hunted as a game species in Alaska, but protected south of the Canadian border (4).

Polar bears are hunted by native people in Canada and Alaska. In Canada, sport hunting is allowed, but sport hunters must hire an Inuit guide and hunting must be done using a dog team. Females with offspring and bears in dens are protected. Although sport hunting is not allowed in Alaska, harvest by native people is unrestricted, and there is no protection for bears in dens or females with cubs (2, 3).

Management of bear populations is required wherever bears and people coexist. Activities of people in bear habitats often attract bears, which ultimately results in some form of human–bear conflict. As development and recreational activities expand into previously pristine wilderness areas and as bears are exposed to greater numbers of people, human injuries will inevitably increase. Wildlife managers have the dual responsibility of maintaining bear populations and at the same time taking necessary precautions to ensure human safety (76).

Management of grizzly bears involves dealing with a variety of human-related activities such as recreation, grazing, road building, vehicle use, timber harvest, and oil, gas, and mineral exploration and development. All of these activities can potentially result in human–grizzly bear conflicts, and management strategies should be designed to reduce bear mortality whenever possible (15). In small populations, the loss of even a few females may result in population declines (42). In areas designated for recovery of grizzly bears, priority should be given to bears with respect to human activities and resources. Creation of bear management areas in Yellowstone National Park, where humans are seasonally or permanently excluded, provides refuges for bears and decreases human–bear contact (86).

Human–grizzly bear encounters have increased in and around national and provincial parks in Canada and the United States (92). Several situations associated with grizzly bear-inflicted injuries have been identified. The most common cause is due to a sudden encounter between a bear and person. Unfortunately, as more people use off-trail areas, chances of sudden encounters are likely to increase. A second situation results from a bear's past history. Bears that have fed on people's garbage (i.e., food-conditioned) or been repeatedly exposed to people (i.e., habituated) may potentially view people as prey and thus present a major management challenge. Cases of provoked attacks usually occur when people such as photographers or hunters approach bears too closely, triggering an attack (76).

In places where grizzly bears and humans coexist, human–bear conflict situations cannot always be avoided. For management agencies that have a mandate to protect threatened animals, a relocation program is one means of resolving this conflict (93). Survival rates of transported grizzly bears are largely affected by whether the individuals return to the capture site. Bears should be moved distances at least 100 kilometers to reduce return rates (94, 93). Because subadult females appear least likely to return, transplanting of female grizzly bears may be a useful management tool to help augment small populations, such as in the Cabinet Mountains of Montana. Success of relocated animals depends on age, sex, human-related history of the animal, similarity of food resources between capture and release sites, timing of release, and public support. The ultimate success of this type of conservation effort is the reproduction of transplanted individuals (95). Relocating females is considered a viable management option, because a number of transplanted individuals have successfully reproduced (93). Relocation programs should be matched with extensive management efforts (i.e., incineration of garbage, fencing, and deterrent actions) to make recreational or industrial sites less attractive to bears.

In the far North, polar bear–human encounters have increased as a result of oil and mineral exploration, and subsequent higher human density (92). In human–polar bear confrontations, most bears perceived as aggressive toward people have been killed. Analyses of aggressive encounters reveal two distinct patterns: Young male polar bears are largely responsible for predatory attacks on humans, and females with cubs occasionally attack humans, apparently in defense of cubs. Most aggressive encounters with females have occurred in denning areas and are extremely rare, suggesting that female polar bears are less aggressive than female grizzly bears. In Canadian national parks, subadult males appear most likely to be involved in aggressive interactions with humans. In some cases, food stress may be a contributing variable. The hunting skills of subadults are not yet perfected, and in addition, these subadults are likely to have been outcompeted from natural food sources by larger, more aggressive adults. Polar bears are also attracted to garbage, animal carcasses, and other potentially edible items, so many incidents could potentially be avoided if these attractions were removed (76).

A successful management approach, referred to as the Polar Bear Alert Program, was instigated at Churchill, Manitoba, Canada. The objectives of the program were to ensure the safety of people and protection of property from damage by polar bears and to ensure the conservation of polar bears and avoid undue harassment and killing of bears. Nuisance bears or bears entering the garbage dump are captured and kept in a holding facility until the ice freezes, at which time the bears are released. In some cases, bears are relocated immediately after capture. The program has resulted in a significant decrease in bear deaths. Despite a large increase in tourism at Churchill, the number of polar bears killed by human intervention has dropped from 30 bears per year in the 1970s to 8 bears per year in the 1990s, and the last human mauling occurred in 1983 (C. Elliot, Manitoba province, personal communication).

In national parks, injury rates resulting from human confrontations with grizzly and polar bears are low with respect to total visitation (76). Various management approaches aimed at reducing brown and polar bear conflicts include removing all garbage from access by bears, educating the public on bear behavior and how to act in bear country, enforcement of regulations involving people, vigilance and patrols by management officers, area or campground closures, bear-proof shelters, compulsory guides, use of firearm-propelled deterrents, and research and evaluation of aversive conditioning (92, 15, 96, 97).

HABITAT REQUIREMENTS AND MANAGEMENT

Grizzly and polar bears present a challenging management situation because of their large spatial requirements and aggressive nature. Wilderness and solitude are essential for the bears' survival. The main objectives of bear management should therefore include habitat conservation, preservation of genetic diversity, minimization of human–bear contact, limitation of resource exploitation in critical bear habitats, and educating the public on bear behavior and ecology.

The spacial, temporal, and ecological requirements of the grizzly bear are diverse and complex. The habitat of the grizzly must meet the species' needs for food, space, cover, travel corridors, winter denning, and refugias. Because grizzly bears are wide ranging and omnivorous, they require large home ranges of essential habitat that encompass all their biological needs. While some portions of their home ranges may be used seasonally, or less often than others, large expanses of wilderness habitat encompassing a diversity of habitats are critical to support self-sustaining, healthy populations. To ensure successful management of remaining grizzly bear populations, long-term studies of their home range requirements and detailed habitat use are needed for all zones in which the species live (7, 5).

With continuing demands for natural resources, many remote areas occupied by bears will be developed. This is a major concern in habitat management because development results in habitat loss and fragmentation, increased risk of human–bear confrontations, and easy access into bear habitat. Resource extraction industries often rely on an extensive network of roads, which negatively influences bear populations. In grizzly bear habitats, roads are often constructed along valley bottoms and pass through riparian habitats that contain a variety of important food sources. If grizzly bears use areas near roads less than expected, these areas represent habitat loss for the bears. Roads not only increase the probability of vehicle–bear collisions, but also the bears' vulnerability to legal hunters and poachers, which is often a major source of bear mortality. Management policies must minimize road-related effects on grizzly bears by controlling vehicle access, people with firearms, and human settlement in bear habitats (98).

Despite its relative isolation from humans, protection and proper management of polar bear habitat is of utmost importance for the species' long-term survival. Population boundaries, movement patterns, and sizes of several polar bear populations are inadequately known (3). Long-term studies of habitat use are needed to provide a better understanding of the species' movements and use of resources on ice and land. Some highly productive and critical feeding areas overlap with areas of potential offshore oil production. In addition, the productive shore leads and polynyas used by bears are also used as shipping routes during winter (3).

An increase in industrial development could negatively impact polar bears, especially in the vicinity of denning areas. An increase in tourism and recreational activities certainly will result in higher incidence of human–bear contact in areas where polar bears are forced on shore during the ice-free season (73, 3). Toxic chemicals in the oceans threaten many marine species, including polar bears, which are at the top of the marine food chain. Polychlorinated biphenyls (PCBs), other organochlorines, cadmium, mercury, lead, and radioactive waste contamination are potential threats to polar bears, and their long-term ecological consequences to bears are unknown. Increasing petroleum activity in the Arctic will elevate the potential for oil spills, whose harmful effects may be carried over large areas and last for many months (3). An experiment designed to simulate an Arctic oil spill clearly demonstrated the detrimental effects of such an occurrence on polar bears. Bears coated with oil developed thermoregulatory and metabolic stresses. By licking oil from their fur, polar bears ingested petroleum hydrocarbons, which led to behavioral abnormalities, anorexia, and tissue damage (99).

Although some aspects of bear biology are well understood, other areas urgently require further study. Minimum viable population size, habitat quality, impact of contaminants on immune systems and hence the risk of epizootics, and effects of global change are some issues that have yet to be addressed.

Bears evolved in large expanses of wilderness. For the grizzly bear, much of the wilderness in southern Canada and south of the Canadian border has already been destroyed and can never be reclaimed by bears. The polar bear has fared better. Because most of the original habitat of the polar bear is uninhabited by humans, it is one of the few remaining large carnivores that still occupies most of its historic range. Nevertheless, both species are threatened by increased development and continuous human expansion. Conserving the existing wild and partially wild habitats is the only way to ensure the survival of remaining bear populations. Both species are a vital reminder of healthy ecosystems and of the beauty of pristine wilderness. Human beings now have the obligation to manage wilderness habitats in a manner that will maintain their biodiversity, which is essential for both bear and human survival.

To date, a large wealth of biological knowledge exists on grizzly and polar bears, and we continue to acquire a greater understanding of the species and their management requirements. Unlike the first settlers to North America, we are well aware of the ramifications of wilderness destruction, population declines, and extinctions; unfortunately, we have limited time to ensure the survival of large carnivores that compete with humans for space and resources. It is essential that governments and an educated public support the preservation of remaining wilderness areas and the creatures that inhabit them. Grizzly and polar bears have unique and important aesthetic, recreational, economic, scientific, cultural, and philosophical merits. The value that human beings place on bears and future land use practices will ultimately decide the fate of these magnificent carnivores.

SUMMARY

Grizzly bears and polar bears are the two largest carnivores in North America. Grizzly bears inhabit forests, mountains, and plains, whereas polar bears are found in the Arctic. Despite their similarity in size, grizzly bears are omnivorous and exploit a variety of animal and plant foods; polar bears are almost entirely carnivorous. Unavailability of food during winter results in the obligate use of dens by all grizzlies. Availability of winter prey results in the facultative use of dens by polar bears (with the exception of pregnant females). Bears differ from other hibernating mammals because they do not eat, drink, urinate, or defecate in their dens. In addition, their drop in body temperature is slight, they are easily aroused, and they give birth during winter sleep. Both species have low reproductive rates, exhibit late sexual maturity, and give birth to small litters of immature neonates. Reproductive success appears to be dependent on maternal fat reserves; females with large fat stores tend to produce heavier cubs, and heavier cubs are more likely to survive the first year of life. Parturition may be delayed as a consequence of limited food supply. Recruitment depends on the number of females available for breeding and the number that produce offspring. In the wild, both species may live to 30 years, and reproductive longevity appears to approximate physical longevity.

Distribution and movements of grizzly and polar bears are influenced by food resources, habitat selection, denning areas, social interactions, and philopatry. The grizzly bear is currently restricted to approximately half of its former range, and south of the Canadian border, it is on the threatened list. The polar bear still occupies most of its historic range. Both species are vulnerable to development and continuous human expansion. Management of grizzly bears involves dealing with activities such as recreation, grazing, road construction, vehicle use, timber harvest, and oil, gas, and mineral exploration. Polar bears are threatened by an increase in industrial development, the release of toxic chemicals, and oil spills. The present and future survival of bears is dependent on human beings; conserving and protecting wilderness areas is the only way to safeguard the remaining populations.

LITERATURE CITED

1. BROWN, D. E. 1985. *The Grizzly in the Southwest: Documentary of an Extinction.* University of Oklahoma Press, Norman.

2. STIRLING, I. 1988. *Polar Bears.* The University of Michigan Press, Ann Arbor.

3. PRESTRUD, P., and I. STIRLING. 1994. The International Polar Bear Agreement and the current status polar bear conservation. *Aquatic Mammals* 20.3:113–124.

4. SERVHEEN, C. 1990. The status and conservation of bears of the world. *International Conference on Bear Research and Management* 2:1–32.

5. CRAIGHEAD, J. J., J. S. SUMNER, and J. A. MITCHELL. 1995. *The Grizzly Bears of Yellowstone.* Island Press, Washington, DC.

6. WILSON, D. E., and D. M. REEDER. 1993. *Mammal Species of the World.* Smithsonian Institution Press, Washington, DC.

7. LeFRANC, M. N. JR., M. B. MOSS, K. A. PATNODE, and W. C. SUGG III (eds.). 1987. *Grizzly Bear Compendium.* International Grizzly Bear Committee, Washington, DC.

8. PASITSCHNIAK-ARTS, M. 1993. *Ursus arctos. Mammalian Species* 439:1–10.

9. RAUSCH, R. L. 1963. Geographic variation in size in North American brown bears, *Ursus arctos* L., as indicated by condylobasal length. *Canadian Journal of Zoology* 41:33–45.

10. HALL, E. R. 1984. Geographic variation among brown and grizzly bears (*Ursus arctos*) in North America. Special Publication. University of Kansas, Museum of Natural History 13:1–16.

11. DeMASTER, D. P., and I. STIRLING. 1981. *Ursus maritimus. Mammalian Species* 145:1–7.

12. KURTÉN, B. 1964. The evolution of the polar bear, *U. maritimus* Phipps. *Acta Zoologica Fennica* 108:3–30.

13. TALBOT, S. L., and G. F. SHIELDS. 1996. A phylogeny of the bears (Ursidae) inferred from complete sequences of three mitochondrial genes. *Molecular Phylogenetics and Evolution* 5:567–575.

14. MALYOV, A. V. 1991. Reproduction and sexual behaviour of polar bears (*Ursus maritimus* Phipps) in the Kazan Zoobotanical Garden. Pages 86–89 *in* S. C. Amstrup and O. Wiig, eds., *Polar Bears: Proceedings of the Tenth Working Meeting of the IUCN/SSC Polar Bear Specialist Group*. IUCN, Gland, Switzerland.

15. STRICKLAND, M. D. 1990. Grizzly bear recovery in the contiguous United States. *International Conference on Bear Research and Management* 8:5–9.

16. UNITED STATES FISH AND WILDLIFE SERVICE. 1993. Grizzly bear recovery plan. Missoula, MT.

17. TAYLOR, M. K., and J. LEE. 1995. Distribution and abundance of Canadian polar bear populations: A management perspective. *Arctic* 48:147–154.

18. WIIG, O., E. W. BORN, and G. W. GARNER (eds.). 1995. *Polar Bears: Proceedings of the Eleventh Working Meeting of the IUCN/SSC Polar Bear Specialist Group*. IUCN, Cambridge.

19. FRENCH, S. P., and M. G. FRENCH. 1990. Predatory behaviour of grizzly bears feeding on elk calves in Yellowstone National Park, 1986–88. *International Conference on Bear Research and Management* 8:335–341.

20. GUNTHER, K. A., and R. A. RENKIN. 1990. Grizzly bear predation on elk calves and other fauna of Yellowstone National Park. *International Conference on Bear Research and Management* 8:329–334.

21. HAMER, D., and S. HERRERO. 1991. Elk, *Cervus elaphus,* calves as food for grizzly bears, *Ursus arctos,* in Banff National Park, Alberta. *Canadian Field-Naturalist* 105:101–103.

22. CASE, R., and J. STEVENSON. 1991. Observation of barren-ground grizzly bear, *Ursus arctos,* predation on muskoxen, *Ovibos moschatus,* in the Northwest Territories. *Canadian Field-Naturalist* 105:105–106.

23. CLARKSON, P. L., and I. S. LIEPINS. 1993. Grizzly bear, *Ursus arctos,* predation on muskox, *Ovibos moschatus,* calves near the Horton River, Northwest Territories. *Canadian Field-Naturalist* 107:100–102.

24. BALLARD, W. B., J. S. WHITMAN, and D. J. REED. 1991. Population dynamics of moose in south-central Alaska. *Wildlife Monographs* 114:1–49.

25. BALLARD, W. B. 1992. Bear predation on moose: A review of recent North American studies and their management implications. Third International Moose Symposium, Syktyukar, USSR. *Alces Suppl.* 1:162–176.

26. REYNOLDS, H. V., and G. W. GARNER. 1987. Patterns of grizzly bear predation on caribou in northern Alaska. *International Conference on Bear Research and Management* 7:59–67.

27. BALLARD, W. B., D. M. STIRLING, and J. S. WHITMAN. 1990. Brown and black bear predation on moose in southcentral Alaska. *Alces* 26:1–8.

28. MATTSON, D. J., R. R. KNIGHT, and B. M. BLANCHARD. 1992. Cannibalism and predation on black bears by grizzly bears in the Yellowstone ecosystem, 1975–1990. *Journal of Mammalogy* 73:422–425.

29. SMITH, M. E., and E. H. FOLLMANN. 1993. Grizzly bear, *Ursus arctos,* predation of a denned adult black bear, *U. americanus. Canadian Field-Naturalist* 107:97–99.

30. EGBERT, A. L., and A. W. STOKES. 1976. The social behaviour of brown bears on an Alaskan salmon stream. *International Conference on Bear Research and Management* 3:41–56.

31. GLENN, L. P., and L. H. MILLER. 1980. Seasonal movements of an Alaska Peninsula brown bear population. *International Conference on Bear Research and Management* 4:307–312.

32. REINHART, D. P., and D. J. MATTSON. 1990. Bear use of cutthroat trout spawning streams in Yellowstone National Park. *International Conference on Bear Research and Management* 8:343–350.

33. MATTSON, D. J., C. M. GILLIN, S. A. BENSON, and R. R. KNIGHT. 1991. Bear feeding activity at alpine insect aggregation sites in the Yellowstone Ecosystem. *Canadian Journal of Zoology* 69:2430–2435.

34. MATTSON, D. J., B. M. BLANCHARD, and R. R. KNIGHT. 1992. Yellowstone grizzly bear mortality, human habituation, and whitebark pine seed crops. *Journal of Wildlife Management* 56:432–444.

35. LOWRY, L. F., J. J. BURNS, and R. R. NELSON. 1987. Polar bear, *Ursus maritimus,* predation on belugas, *Delphinapterus leucas,* in the Bering and Chukchi seas. *Canadian Field-Naturalist* 101:141–146.

36. SMITH, T. G., and B. SJARE. 1990. Predation of odontocete whales by polar bears in nearshore areas of the Canadian High Arctic. *Arctic* 43:99–102.

37. CALVERT, W., and I. STIRLING. 1990. Interactions among polar bears and overwintering walruses in the central Canadian high Arctic. *International Conference on Bear Research and Management* 8:351–356.

38. RUSSELL, R. H. 1975. The food habits of polar bears of James Bay and Southwest Hudson Bay in summer and autumn. *Arctic* 28:117–129.

39. SMITH, A. E., and M. R. J. HILL. 1996. Polar bear, *Ursus maritimus,* depredation of Canada goose, *Branta canadensis,* nests. *Canadian Field-Naturalist* 110:339–340.

40. RAMSAY, M. A., and I. STIRLING. 1988. Reproductive biology and ecology of female polar bears (*Ursus maritimus*). *Journal of Zoology (London)* 214:601–634.

41. WIELGUS, R. B., and F. L. BUNNEL. 1994. Dynamics of a small, hunted brown bear *Ursus arctos* population in southwestern Alberta, Canada. *Biological Conservation* 67:161–166.

42. EBERHARDT, L. L. 1990. Survival rates required to sustain bear populations. *Journal of Wildlife Management* 54:587–590.

43. RAMSAY, M. A., and I. STIRLING. 1986. On the mating system of polar bears. *Canadian Journal of Zoology* 64:2142–2151.

44. HAMER, D., and S. HERRERO. 1990. Courtship and use of mating areas by grizzly bears in the Front Ranges of Banff National Park, Alberta. *Canadian Journal of Zoology* 68:2695–2697.

45. BRADY, K. S., and D. HAMER. 1992. Use of a summit mating area by a pair of courting grizzly bears, *Ursus arctos,* in Waterton Lakes National Park, Alberta. *Canadian Field-Naturalist* 106:519–520.

46. CRAIGHEAD, J. J., M. G. HORNOCKER, and F. C. CRAIGHEAD, JR. 1969. Reproductive biology of young female grizzly bears. *Journal of Reproduction and Fertility* 6:447–475.

47. CRAIGHEAD, L., D. PAETKAU, H. V. REYNOLDS, E. R. VYSE, and C. STROBECK. 1995. Microsatellite analysis of paternity and reproduction in Arctic grizzly bears. *Journal of Heredity* 86:255–261.

48. AUNE, K. E., R. D. MACE, and D. W. CARNEY. 1994. The reproductive biology of female grizzly bears in the Northern Continental Divide ecosystem with supplemental data from the Yellowstone ecosystem. *International Conference on Bear Research and Management* 9:451–458.

49. LYMAN, C. P, J. S. WILLIS, A. MALAN, and L. C. H. WANG. 1982. Hibernation and torpor in mammals and birds. Pages 12–36 *in* T. T. Kozlowski, ed. *Physiological Ecology.* Academic Press, New York.

50. HOCHACHKA, P. W., and G. N. SOMERO. 1984. *Biochemical Adaptation.* Princeton University Press, Princeton, NJ.

51. HOCK, R. J. 1960. Seasonal variations in physiologic functions of arctic ground squirrels and black bears. Bulletin. *Museum of Comparative Zoology* 124:155–171.

52. CRAIGHEAD, F. C., JR., and J. J. CRAIGHEAD. 1972. Data on grizzly bear denning activities and behaviour obtained by using wildlife telemetry. *International Conference on Bear Research and Management* 2:84–106.

53. RAMSAY, M. A., and R. L. DUNBRACK. 1986. Physiological constraints on life history phenomena: The example of small bear cubs at birth. *The American Naturalist* 127:735–743.

54. MESSIER, F., M. K. TAYLOR, and M. A. RAMSAY. 1994. Denning ecology of polar bears in the Canadian Arctic Archipelago. *Journal of Mammalogy* 75:420–430.

55. WATTS, P. D., and S. E. HANSEN. 1987. Cyclic starvation as a reproductive strategy in the polar bear. *Symposium of the Zoological Society of London* 57:305–318.

56. MESSIER, F., M. K. TAYLOR, and M. A. RAMSAY. 1992. Seasonal activity patterns of female polar bears (*Ursus maritimus*) in the Canadian Arctic as revealed by satellite telemetry. *Journal of Zoology (London)* 226:219–229.

57. ATKINSON, S. N., and M. A. RAMSAY. 1995. The effect of prolonged fasting on the body composition and reproductive success of female polar bears (*Ursus maritimus*). *Functional Ecology* 9:559–567.

58. DEROCHER, A. E., D. ANDRIASHEK, and J. P. Y. ARNOULD. 1993. Aspects of milk composition and lactation in polar bears. *Canadian Journal of Zoology* 71:561–567.

59. FARLEY, S. D., and C. T. ROBBINS. 1995. Lactation, hibernation, and mass dynamics of American black bears and grizzly bears. *Canadian Journal of Zoology* 73:2216–2222.

60. BARNES, V. G., JR., and R. B. SMITH. 1993. Cub adoption by brown bears, *Ursus arctos middendorffi*, on Kodiak Island, Alaska. *Canadian Field-Naturalist.* 107:365–367.

61. ATKINSON, S. N., M. R. L. CATTET, S. C. POLISCHUK, and M. A. RAMSAY. 1996. A case of offspring adoption in free-ranging polar bears (*Ursus maritimus*). *Arctic* 49:94–96.

62. TAYLOR, M. K., T. LARSEN, and R. E. SCHWEINSBURG. 1985. Observations of intraspecific aggression and cannibalism in polar bears (*Ursus maritimus*). *Arctic* 38:303–309.

63. WIELGUS, R. B., and F. L. BUNNEL. 1994. Sexual segregation and female grizzly bear avoidance of males. *Journal of Wildlife Management* 58:405–413.

64. CRAIGHEAD, F. C., JR. 1979. *Track of the Grizzly.* Sierra Club Books, San Francisco, CA.

65. DEROCHER, A. E., and M. TAYLOR. 1994. Density-dependent population regulation of polar bears. *International Conference on Bear Research and Management* 3:1–43.

66. AMSTRUP, S. C., and G. M. DURNER. 1995. Survival rates of radio-collared female polar bears and their dependent young. *Canadian Journal of Zoology* 73:1312–1322.

67. STIRLING, I., and A. E. DEROCHER. 1990. Factors affecting the evolution and behavioral ecology of the modern bears. *International Conference on Bear Research and Management* 8:189–204.

68. PETERS, G., and W. C. WOZENCRAFT. 1989. Acoustic communication by fissiped carnivores. Pages 14–56 *in* J. L. Gittleman, ed., *Carnivore Behaviour, Ecology, and Evolution.* Cornell University Press, New York.

69. DEROCHER, A. E., and I. STIRLING. 1990. Distribution of polar bears (*Ursus maritimus*) during the ice-free period in western Hudson Bay. *Canadian Journal of Zoology* 68:1395–1403.

70. FERGUSON, S. H., M. K. TAYLOR, and F. MESSIER. 1998. Space use of polar bears in and around Auyuittuq National Park, Northwest Territories, during the ice-free period. *Canadian Journal of Zoology* 75:1585–1594.

71. SCHOEN, J. W., L. R. BEIER, J. W. LENTFER, and L. J. JOHNSON. 1987. Denning ecology of brown bears on Admiralty and Chichagof Islands. *International Conference on Bear Research and Management* 7:293–304.

72. RAMSAY, M. A., and I. STIRLING. 1990. Fidelity of female polar bears to winter-den sites. *Journal of Mammalogy* 71:233–236.

73. AMSTRUP, S. C., and C. GARDNER. 1994. Polar bear maternity denning in the Beauford Sea. *Journal of Wildlife Management* 58:1–10.

74. BURT, W. H. 1943. Territoriality and home range concepts as applied to mammals. *Journal of Mammalogy* 24:346–352.

75. BANCI, V. 1991. The status of the grizzly bear in Canada in 1990. A COSEWIC status report commissioned by British Columbia Ministry of the Environment, Wildlife Branch, Alberta Forestry, Lands and Wildlife, Fish and Wildlife Division, and Yukon Government, Department of Renewable Resources.

76. HERRERO, S., and S. FLECK. 1990. Injury to people inflicted by black, grizzly or polar bears: Recent trends and new insights. *International Conference on Bear Research and Management* 8:25–32.

77. EBERHARDT, L. L., B. M. BLANCHARD, and R. R. KNIGHT. 1994. Population trend of the Yellowstone grizzly bear as estimated from reproductive and survival rates. *Canadian Journal of Zoology* 72:360–363.

78. WIELGUS, R. B., W. L. WAKKINEN, and P. E. ZAGER. 1994. Population dynamics of Selkirk Mountain grizzly bears. *Journal of Wildlife Management* 58:266–272.

79. HOVEY, F. W., and B. N. MCLELLAN. 1996. Estimating population growth of grizzly bears from the Flathead River drainage using computer simulations of reproduction and survival rates. *Canadian Journal of Zoology* 74:1409–1416.

80. CALVERT, W., I. STIRLING, M. TAYLOR, M. A. RAMSAY, G. B. KOLENOSKY, M. CRÊTE, S. KEARNEY, and S. LUTTICH. 1995. Research on polar bears in Canada 1988–92. Pages 33–59 *in* O. Wiig, E. W. Born, and G. W. Garner, eds. *Polar Bears: Proceedings of the Eleventh Working Meeting of the IUCN/SSC Polar Bear Specialist Group.* IUCN, Cambridge, UK.

81. NORSTROM, R. J. 1995. Chlorinated hydrocarbons in polar bears from North America, Greenland and Svalbard. Pages 185–186 *in* O. Wiig, E. W. Born, and G. W. Garner, eds., *Polar Bears: Proceedings of the Eleventh Working Meeting of the IUCN/SSC Polar Bear Specialist Group.* IUCN, Cambridge.

82. AMSTRUP, S. C. 1993. Human disturbances of denning polar bears in Alaska. *Arctic* 46:246–250.

83. TAYLOR, M. K., D. P. DEMASTER, F. L. BUNNELL, R. E. SCHWEINSBURG. 1987. Modelling the sustainable harvest of female polar bears. *Journal of Wildlife Management* 51:811–820.

84. MCLELLAN, B. 1994. Density-dependent population regulation of brown bears. *International Conference on Bear Research and Management* 3:1–43.

85. KNIGHT, R. R., and L. L. EBERHARDT. 1985. Population dynamics of Yellowstone grizzly bears. *Ecology* 66:323–334.

86. MATTSON, D. J., and M. M. REID. 1991. Conservation of the Yellowstone grizzly bear. *Conservation Biology* 5:364–372.

87. MILLER, S. D., and M. A. CHIHULY. 1987. Characteristics of nonsport brown bear deaths in Alaska. *International Conference on Bear Research and Management* 7:51–58.

88. DEROCHER, A. E., and I. STIRLING. 1995. Estimation of polar bear population size and survival in western Hudson Bay. *Journal of Wildlife Management* 59:215–221.

89. DEROCHER, A. E., and I. STIRLING. 1996. Aspects of survival in juvenile polar bears. *Canadian Journal of Zoology* 74:1246–1252.

90. OLSON, T. L. 1993. Infanticide in brown bears, *Ursus arctos,* at Brooks River, Alaska. *Canadian Field-Naturalist* 107:92–94.

91. DURNER, G. M., and S. C. AMSTRUP. 1995. Movements of a polar bear from Northern Alaska to northern Greenland. *Arctic* 48:338–341.

92. MILLER, G. D. 1987. Field tests of potential polar bear repellents. *International Conference on Bear Research and Management* 7:383–390.

93. BLANCHARD B. M., and R. R. KNIGHT. 1995. Biological consequences of relocating grizzly bears in the Yellowstone ecosystem. *Journal of Wildlife Management* 59:560–565.

94. MILLER, S. D., and W. B. BALLARD. 1982. Homing of transplanted Alaskan brown bears. *Journal of Wildlife Management* 46:869–876.

95. SERVHEEN, C., W. F. KASWORM, and T. J. THIER. 1995. Transplanting grizzly bears *Ursus arctos horribilis* as a management tool—results from the Cabinet Mountains, Montana, USA. *Biological Conservation* 71:261–268.

96. CALVERT, W., M. TAYLOR, I. STIRLING, G. B. KOLENOSKY, S. KEARNEY, M. CRÊTE, and S. LUTTICH. 1995. Polar bear management in Canada 1988–92. Pages 61–76 *in* O. Wiig, E. W. Born, and G. W. Garner, eds. *Polar Bears: Proceedings of the Eleventh Working Meeting of the IUCN/SSC Polar Bear Specialist Group.* IUCN, Cambridge, UK.

97. SCHLIEBE, S. L., and T. J. EVANS. 1995. Polar bear management in Alaska 1988–92. Pages 139–144 *in* O. Wiig, E. W. Born, and G. W. Garner, eds., *Polar Bears: Proceedings of the Eleventh Working Meeting of the IUCN/SSC Polar Bear Specialist Group.* IUCN, Cambridge, UK.

98. MCLELLAN, B. N., and D. M. SHACKLETON. 1988. Grizzly bears and resource extraction industries: Effects of roads on behaviour, habitat use and demography. *Journal of Applied Ecology* 25:451–460.

99. ØRITSLAND, N. A., F. R. ENGELHARDT, F. A. JUCK, R. J. HURST, and P. D. WATTS. 1981. Effects of crude oil on polar bears. Environmental Studies No. 24. Indian and Northern Affairs, Ottawa, Canada.

100. BANCI, V., D. A. DEMARCHI, and W. R. ARCHIBALD. 1994. Evaluation of the population status of grizzly bears in Canada. *International Conference on Bear Research and Management* 9:129–142.

101. LUNN, N. J., I. STIRLING, D. ANDRIASHEK, and G. B. KOLENOSKY. 1997. Re-estimating the size of the polar bear population in Western Hudson Bay. *Arctic* 50:234–240.

CHAPTER 21

Collared Peccary

Eric C. Hellgren and Robert L. Lochmiller

INTRODUCTION

The collared peccary in North America is a recent ungulate arrival from the Neotropics that inhabits arid to semiarid environments in the American Southwest. Because of its tropical origins and limited distribution in North America, many sportsmen and biologists know little of its ecology, conservation, and management. It has legal protection by virtue of its classification as a game species in Arizona, Texas, and New Mexico, and thus provides sporting opportunity to hunters and income to state governments, private landowners, and Indian tribes. In Central and South America, the peccary serves a more essential role. It is a main source of protein and plays an important cultural role for some indigenous peoples. Sowls described peccary–people interactions, especially in Latin America (1). For a detailed discussion of the economic, social, and cultural roles of the peccary in human society, we refer the reader to Donkin (2). In this chapter, we focus on the ecology and management of the collared peccary in the American Southwest.

TAXONOMY AND EVOLUTION

The collared peccary is the only native, North American member of the suborder Suiformes. It is one of the members of the family Tayassuidae, which also includes the white-lipped peccary and the Chacoan peccary, which are restricted to the Neotropics. The peccaries apparently diverged from the pigs (family Suidae) about 35 to 40 million years ago and reached South America during the Pliocene period about 7 million years ago (3, 4). The genus *Tayassu* is known from fossil records in South America from the Pleistocene–Recent boundary (4). Only within the Recent epoch (about 10,000 years ago) have collared peccaries radiated into North America.

Collared peccaries have been placed in at least five different genera, but are commonly referred to as *Tayassu* today (see Woodburne for an argument for *Dicotyles* and Theimer and Keim for *Pecari;* 4, 5). As many as 10 subspecies have been recognized, but only 2 purported subspecies occur in North America: one in Texas and northeastern Mexico, and the other in Arizona, southwestern New Mexico, and northwestern Mexico (Figure 21–1) (1).

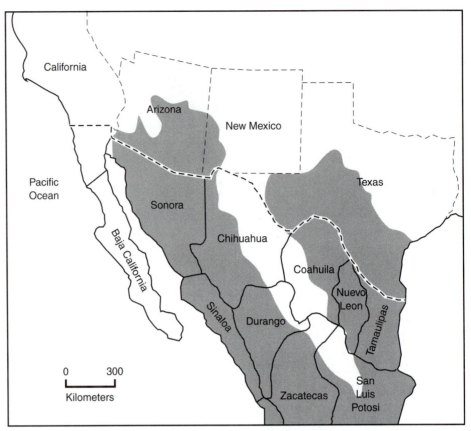

Figure 21–1 Distribution of collared peccary in the United States and northern Mexico. *Tayassu tajacu sonoriensis* occurs in Arizona, New Mexico, and northwestern Mexico and *T.t. angulatus* occurs in Texas and northeastern Mexico (29).

SPECIES DESCRIPTION

Collared peccaries are small ungulates with a pig-like appearance (e.g., short legs, long snout, rounded rostrum, large canines). Indeed, many outdoor enthusiasts consider them pigs. We have even seen brochures that place peccaries in the Perissodactyla! Among the external characters that distinguish peccaries from pigs include the presence of straight canines in peccaries (versus curved in pigs), a short tail (versus long curly tails in pigs), a dorsal scent gland in peccaries, and only three toes (two hooves, one dewclaw) on the hind feet (versus four in pigs: two hooves, two dewclaws) (Figure 21–2). Peccaries have coarse, grizzled fur that ranges in color from light gray to brown to almost black. The hairs have alternating dark and light bands, giving the animal a salt-and-pepper appearance. The collar, which gives the animal its common name, is composed of whitish hairs. It extends dorsally and posteriorly from the ventral side of the neck (6, 1). Peccaries less than 2 to 3 months old are reddish brown with a dark dorsal stripe (6).

Collared peccaries are the least sexually dimorphic of all North American ungulates. Unless the scrotum or nursing is observed, it is very difficult to sex individuals. Body mass, most external measurements, and internal organ weights do not vary by sex (7, 8, 9). Exceptions include the tail (longer in females), the scent gland (larger in males), and skeletal mass of the forequarters (heavier in males). Although these differences are biologically and statistically significant, they usually are very subtle. Sexual dimorphism has been observed in the canines and several skull measurements (8, 10).

Figure 21–2 Collared peccary share a general appearance with pigs, but differ in their canines, tail, dorsal scent gland, and number of toes on the hind foot. (Photograph courtesy of L. Ilse.)

Peccary skulls have been studied in detail (11, 12, 13). The presence of straight, interlocking canines has led to alternative views on their functional role. Herring postulated that canines evolved for use as weapons, whereas Kiltie countered that the canines evolved, along with other skull characters, to assist in breaking extremely hard tropical nuts (11, 13). Slight dimorphisms in canine breadth and robustness also suggest a sexual selection component (4, 8).

The adult peccary has the following dental formula: I 2/3, C 1/1, P 3/3, M 3/3. Permanent canines erupt at 8 to 9 months of age, and permanent dentition is completed at 21 months (14). Adult peccaries have traditionally been placed in five wear classes, which vary based on the extent (i.e., which teeth are worn?) and the level (i.e., how worn are the teeth?) of wear (15). Estimation of age in years with cementum annuli analysis has been used to calibrate wear classes, but is problematic because North American peccaries live in relatively aseasonal environments and deposition of annuli has not been clarified (16).

LIFE HISTORY

Foraging and Nutrition

The peccary does not fit either side of the browsing–grazing dichotomy, though it may be classed as a concentrate selector (*sensu* 17). In Central and South America, peccaries primarily consume large nuts and seeds, fruits, and underground plant parts (18, 19, 20). Peccary crania have architectural features (e.g., interlocking canines, restricted craniomandibular joint) designed to assist in the mastication of hard tropical nuts (13, 21). Underground plant parts (e.g., roots, tubers) contribute to the diet in arid and semiarid tropical areas (22, 20). In the semiarid, subtropical regions of North America inhabited by peccaries, however, peccaries feed heavily on vegetative and reproductive parts of succulents (e.g., prickly pear cactus, yucca, century plant), forbs, and browse (23, 24, 25). Animal matter occurs as a trace or not at all in most diet studies so peccaries can be considered herbivores (reviewed in 1, pp. 73–74).

Figure 21–3 Photograph of the stomach of a collared peccary. Note the two blind-sacs framing the central forestomach and the glandular hindstomach. (Photograph by E. Hellgren.)

Collared peccaries have a complex, pouched stomach with a high pH (about 6.0), which likely assists in digestion and metabolism of their herbivorous diet, especially cactus pads and browse (26, 27). Protozoa have been identified in large concentrations in inocula from peccary stomachs (28). Volatile fatty acids, indicative of microbial (i.e., bacterial) fermentation, occur at high concentrations in the forestomach and glandular stomach (29, 30). Digestibilities for neutral detergent and acid detergent fiber in a rice bran–tropical hay diet averaged 62.0% and 55.7%, respectively (31). Mean retention time averaged 45 hours on a rice–hay diet. Peccaries fed a pelleted deer ration (47.5% neutral detergent fiber) had apparent digestibilities of 53.4% and 36.1% for neutral detergent and acid detergent fiber, respectively. Mean retention time was 52.5 hours. These latter two studies suggest that the fiber digestive efficiency of peccaries is equal to that of ruminants (32). The peccary stomach contains two blind sacs, a forestomach, and a glandular stomach. Although anatomical characteristics of the stomach suggest that separation between compartments (Figure 21–3) is not adequate to reduce passage rate sufficiently for high cellulose digestion, mean retention time of peccaries is longer than in white-tailed deer and may facilitate high fiber digestion (33, 32).

The diet of the collared peccary in North America is very diverse, although dominated in several regions by succulents. In southern Texas, prickly pear cactus pads are a staple, composing 30% to 100% of the diet seasonally (34, 24, 35, 25, 36). Other major food items important on a seasonal basis include prickly pear fruits, mesquite beans, and forbs (e.g., shaggy purslane, pepperweed, sida, zexmania, Russian thistle, ragweed, false mallows, sunflowers). Woody browse, roots, and grass compose a consistently small (less than 12%) portion of the diet. In western Texas, prickly pear dominated the diet (29% to 75% on a monthly basis), but roots and leaves of another succulent, lechuguilla, also occurred in large amounts (up to 41% of a monthly diet; 34, 23). Other major items were forbs and seeds of woody plants. Eddy studied the feeding ecology of peccaries in three Arizona ranges, and found that diets varied by habitat (37). In desert habitat, cactus pads and fruits again dominated

Life History

the diet, with beans of mesquite, paloverde, and catclaw acacia also being consumed. In a desert-grassland, mesquite savanna, cactus and mesquite beans were joined by oak acorns and herbaceous tissue of century plant. In an oak-grassland area, major foods were tubers of morning glory, tissue of century plant, berries, acorns, and several rhizomes and tubers. Succulent grasses, forbs, and tubers composed the majority of the winter–spring (January–April) diet of peccaries in another Arizona study area, with cactus and fruits of woody plants dominating summer diets (38). A fall flush of forbs supplemented cactus in the diet.

The overwhelming prevalence of cactus, especially prickly pear cactus, in the diet of collared peccaries in north America has stimulated considerable research. Prickly pear is very low in digestible energy and protein (4% to 6% crude protein), but high in water. It also contains low levels of several vitamins and minerals, but abundant amounts of oxalic acid, a plant secondary compound that disrupts calcium metabolism (1). Peccaries fed a diet of solely prickly pear cactus developed renal disease consistent with oxalate poisoning (29). Prickly pear herbivory by peccaries is influenced positively by water concentration and negatively by spinescence, calcium oxalate crystals, and neutral detergent fiber content (39). Cactus foraging may represent a trade-off between nutritional value (e.g., water, nitrogen) and plant defenses (chemical and structural), as selected pads are not the highest in nitrogen (39).

It is likely that peccaries eat cactus and other succulents to meet water requirements. Free-living peccaries had water turnover rates of 1.35 liters per day in summer and 1.17 liters per day in winter (40). Captive peccaries can readily consume enough cactus to obtain these amounts of water, yet retain gut space to consume foods of higher nutritional value and thus meet requirements for other nutrients (29). A diet of cactus alone provides below-maintenance levels of digestible energy and protein; therefore, other foods, such as forbs and beans or fruits of woody plants, are necessary to make up the deficit. Nitrogen requirements of 0.8 grams of nitrogen per kilogram metabolic body weight per day correspond to 6.8% crude protein at measured dry matter intake rates (35, 41). Daily energy requirements are 376 and 435 kilojoules per kilogram metabolic body weight per day during summer and winter, respectively (42). The added requirement in winter is associated with thermoregulation and an apparent seasonal change in basal metabolism.

Reproduction

In recent years, we have learned a considerable amount about the basic reproductive biology and nutrition of collared peccaries in the southwestern United States. The necessity of such information has become increasingly clear to wildlife managers tasked with the harvest management of this unique animal. Although similar to wild boar or pigs in their gross appearance, the reproductive biology and fecundity of collared peccaries differ in many ways. Given the arid characteristics of their habitat, collared peccaries are accustomed to dramatic fluctuations in nutritional conditions from season to season and year to year. Nutritional fluctuations impact peccary reproduction (16, 43, 44). Although nutrition, as influenced by precipitation, can account for a sizable portion of the annual variation in juvenile recruitment in wild peccary herds, there is evidence that intrinsic, density-dependent regulatory factors also play an important role in their reproductive life history (45, 9).

Although captive female peccaries can breed as early as 33 weeks of age, most do not breed before 12 months of age (46). The excellent nutritional conditions in captivity do not provide us with a realistic picture of what occurs in wild peccary herds where periods of nutritional stress are common. Although the consequences of malnutrition on sexual development have not been adequately explored in the collared peccary under controlled conditions, examination of breeding performance of females from wild herds has shed some light on what those consequences might be relative to herd dynamics (16, 9).

In southern Texas, the minimum live masses of pregnant adult and yearling peccaries were 15.9 and 17.3 kilograms, respectively, with no pregnant yearling younger than 17 months (primiparity at 20 months of age) (9). Pregnancy rates of 7% to 35% have been reported in southern Texas (16, 9).

The collared peccary is a polyestrous breeder with an estrous cycle extending from 22 to 26 days (46, 16, 47). Conception rates are generally improved with increased copulation frequency and duration of estrus, which lasts from 1 to 5 days (48). Ovulation of 2.0 to 2.2 ova occurs 12 to 24 hours after the onset of estrus (16, 9, 47). Ovulation in female peccaries is sensitive to nutritional condition, with fewer ova and higher incidence of anovulatory estrus reported in malnourished females (43).

Mean gestation length is 145 days, with a range of 143 to 147 days for successful pregnancies (46, 48, 44). Moderate restrictions in the quality of diets consumed by pregnant females do not influence the length of gestation (44). Premature and stillborn births have been observed in this species, but are uncommon under normal nutritional conditions (48). Overall intrauterine losses are about 10% (49, 43, 9).

Endocrine changes in the gravid female peccary have been documented (50, 51). Serum progesterone concentrations rise rapidly during the initial 10 days of gestation and remain elevated from gestation days 11 to 140. A prepartum decline in serum progesterone is an early indication of parturition within the next 8 days. Estradiol-17β levels remain low until day 90 of gestation, after which they increase fourfold by parturition (50).

Most females give birth to twins, although singleton and triplet litters are common. The incidence of twinning in wild herds averages from 70% to 80% (46, 23). Litters of more than three are rare, but can occur (52, 53). Litter size has a tendency to be slightly smaller in yearling (1.7 young) females compared to adults (1.9 young) in wild herds, and also can be reduced in size during periods of nutritional stress (9, 43).

Lactation duration varies from 6 to 8 weeks and probably is influenced by maternal nutrition (46, 54). Peccaries in good nutritional condition will generally express a postpartum heat under captive conditions, but the occurrence of pregnant, lactating females in the wild is confined to good rangelands (16).

Puberty in the male peccary has not been reliably established; however, electroejaculation studies have indicated 10 to 15 months as a suitable range under captive conditions (55). Fertilization of receptive females by males at 10 or 11 months of age has been documented in captivity (46, 16).

Electroejaculation techniques used to characterize the semen in the male peccary produce three fractions (i.e., clear accessory fluid, sperm-rich liquid, and a gelatinous accessory gland component; 55). Although electroejaculation yields about 4 milliliters of semen, the true ejaculate during copulation can exceed 80 milliliters (55). Total sperm count in ejaculates of captive peccaries averages 371×10^6 sperm and is not influenced by season of collection (55).

Low failed to document a circannual rhythm in male reproductive activity based on testicular histology or size (16). However, measures of testosterone concentrations, testicular size, and scrotal circumference suggest that a low-amplitude rhythm is present, with maximal activity from October to March (55). This male rhythm in captivity coincides with the peak of female breeding activity in much of their geographical range in the southern United States. Despite such a rhythm, pregnant females are commonly observed throughout the year, indicating that at least some males remain fertile in any given month of the year (16). Low levels of serum testosterone during summer months may be a reflection of heat stress and could account for observed nadirs in reproductive activity at this time in wild populations (56). Hellgren and others documented a relationship between ambient temperature 1 week prior to sampling and serum testosterone concentrations in adult male peccaries (55). Nutrition probably does not play a significant role in regulating the amount of male reproductive activity in the wild (57).

Figure 21–4 The dorsal scent gland of a male collared peccary. Scent glands are larger in males than females, and are used in the marking of territories and interindividual communication. (Photograph by E. Hellgren.)

Among adults, dominant males are reproductively more active and potentially more aggressive than subordinate males. Sowls thought that reproductive-related aggressive encounters between males occur infrequently in the wild where dominance hierarchies are established (29, 23, but see 58). Males under captive conditions readily confront subordinate males attempting to breed an estrous female (48).

Growth and Development

Morphological characteristics of the collared peccary have been well described. Allometric relationships between internal organ mass and body mass are described for nursling collared peccaries (birth to 6 weeks of age) and adults (59, 7). Although sexual dimorphism in internal organs is limited in peccaries as it is in swine, selected organs differ between males and females in both absolute and relative mass. For example, the prominent dorsal scent gland (Figure 21–4) is much larger in males and its activity is dependent on age, sex, and physiological status (7, 60).

Fetal development has been described for the collared peccary, and morphometric traits such as length of the ear and hind foot can be used to estimate gestational ages of fetuses collected from wild females (49). Physical development after birth was described by Lochmiller and others (8). Postnatal age during the suckling period can be accurately predicted from measures of hind foot length. Chest circumference is the best predictor of body mass when mass cannot be measured directly in the field.

The average body mass of adult peccaries in the wild is highly variable and shows considerable seasonal, geographical, and interherd variation. Mass of adults may range from 15 to 24 kilograms, depending on the season and location, and undoubtedly reflects the recent nutritional history of the rangeland (61, 16, 62). Body size differences exist among subspecies, with the Texas subspecies being slightly larger than the Arizona subspecies (62). Indices of lipolysis such as visceral fat depots and specialized indices of bone marrow and kidney fat can provide a useful means of assessing the nutritional condition of animals in the wild (63, 64).

The collared peccary is a highly social, territorial species that lives year-round in mixed-sex herds. These traits place strong constraints on the ecology of the species, and lead to several considerations relative to peccary management. Peccary herds are cohesive and seemingly closed units with minimal apparent male–male competition, a trait consistent with the lack of sexual dimorphism exhibited by the species (58). Dispersal is poorly documented. Byers and Bekoff hypothesized that kin selection was important in the evolution of group living in collared peccaries, predicting that intraherd, male–male competition would be low if herd members were related (58). Only Bissonette has documented a linear dominance hierarchy among peccaries in a free-ranging situation, with dominant males alone successful in copulating with adult females (23). However, dominance hierarchies have been reported in small, captive herds (65, 55).

In North America, mean herd sizes are 7 to 12 in hunted areas and 10 to 18 in un-hunted areas (29, 38, 66, 58, 23). Herds appear to be generally larger in Texas than Arizona. Herds of over 50 individuals are rare. Herds in Central and South America are similar in size to those in North America (67, 68).

Herds vary in size on a daily and seasonal basis, primarily on the basis of sub-grouping or fragmentation. In Texas, larger groups are observed in winter and smaller groups in spring and summer, especially after precipitation events (23, 69, 70). Bissonette postulated that group dynamics and changes in group size were affected by reproductive events, habitat density, and seasonal changes in temperature regimes and food patchiness (23). Although herds appear to be cohesive units, variation in group size can lead to underestimates of herd and population size, especially during conditions that favor smaller subgroups.

Peccary territory size ranges from 0.2 to 8.0 square kilometers, although most estimates are between 1 and 3 square kilometers, regardless of technique used to collect data (e.g., observations of marked individuals, radiotelemetry) or to determine territorial size (e.g., convex polygon, subjective drawing of boundaries, harmonic mean, kernel) (66, 71, 72, 73, 23, 38, 74, 75). Variables that appear to affect territory size include precipitation, resource availability, and presence of feral pigs (66).

Territorial overlap between adjacent herds has been reported by several workers (56, 73, 38, 74, 75). In general, the overlap zone is narrow (100–200 meters), often includes important features such as water or preferred bedding sites, and is not used simultaneously by adjacent herds. The dorsal scent gland (discussed later) is used to mark territorial boundaries. Although rare, defense of territorial boundaries has been observed (38). Aggressive behavior, such as tooth clacking, jaw popping, and chasing, is often seen when two herds occur in the overlap zone at the same time (38).

Individual behaviors in the collared peccary have been the focus of much study (76, 65, 58, 23, 77, 78, 79). We refer the reader to these original citations for detailed descriptions of aggressive, friendly, display, or other specific behaviors. We will focus briefly on behaviors associated with communication, especially by olfactory means, because of their importance in intraherd and interherd interactions of peccaries.

The collared peccary has a large scent gland (8 × 10 centimeters) located dorsally, anterior to the tail (Figure 21–4). This gland extrudes an odoriferous secretion with numerous volatile components (80). It was noted by European explorers and travelers, has been described morphologically and histologically, and is one of the few sexually dimorphic characters found in peccaries (81, 82, 83) (male glands are larger than female glands; 7, 80). Several functions have been attributed to the scent gland, including territorial marking, herd recognition and cohesion, and conveyance of reproductive information (76, 58, 65, 60). The odor and composition of the glandular secretions may vary with sex, age, reproductive status, and social rank (60). Byers and Bekoff postulated that the mutual rub (also termed reciprocal rub), a behavior in which peccaries rub dorsal gland secretions on the backs and heads of the rubbing participants, functions to orient individuals within their herd (58). Low grunts provide an auditory supplement to maintain herd spacing (58). It is likely that scent gland secretions and the mutual rub lead to unique herd scents, or a "plume of odor," that maintain herd integrity and territorial boundaries (1). It also may contribute to the apparent difficulty that dispersing peccaries have in joining new herds (38).

Dispersal of peccaries from natal herds and interherd movements remain perplexing issues. In Arizona, Day reported adult interchange rates of 10% to 12% between adjacent herds (38). In southern Texas, Ellisor and Harwell reported male interchange to range from 22% to 41%, while Gabor reported that adult peccaries occasionally moved to adjacent herds after adulthood (56, 66). These observations call into question the level of integrity of an individual herd. Gabor reported that four of five males and none of three females that were captured as subadults and recaptured as adults were found in adjacent herds (66). He also noted an apparent dispersal of 8 kilometers from a natal range.

Formation of new herds and colonization of new areas is believed to be by herd fission (84, 6, 38). Feeding subgroups may form the nucleus for new, territorial herds. Supplee found that reduction of a herd led to rapid dispersal of a mixed-sex subgroup from an adjacent herd into the newly available habitat (84). Similarly, Day reported the formation of three new herds from subgroups splintering from large herds and occupying areas abandoned or rarely used by other herds (38). He hypothesized that matrilines were especially important in this process. Fission and dispersal are important in the recolonization of locally extirpated herds. Their role in herd and population dynamics should be addressed by combining an intense mark–recapture effort with modern molecular techniques.

POPULATION DYNAMICS

Dynamics of peccary populations are poorly studied relative to other large ungulates. The lack of a synchronized breeding period, with a subsequent, easily identifiable birth cohort, has complicated demographic analysis (1). In addition, the relative roles of density-dependent and density-independent factors on peccary population growth are unknown. In southern Texas, where precipitation is highly variable, Low hypothesized that it is unlikely that peccary densities reach levels at which density-dependent factors operate (16). He posited that the effects of rainfall on primary productivity drove rapid fluctuations in peccary numbers. Indeed, rainfall and forage production have impacted peccary reproduction and recruitment in several studies (16, 45, 9). Conversely, density-dependent recruitment in arid western Texas and Arizona, possible compensatory responses in juvenile survival to hunting mortality, high adult survival rates (see later discussion), and the high degree of sociality exhibited by peccary herds argue for density-dependent regulation of peccary herds (23, 38, 9). Obviously, relationships among reproduction, survival, and density among peccaries need further investigation.

Demography

For adequate analysis of the population dynamics of any species, data are needed on age-specific natality and survival rates for that population. In a winter sample of 345 female peccaries in Texas collected over 10 years, fecundity of adults ranged from 0.57 to 0.81 female young per female (Table 21–1) (9). Maximal age-specific productivity occurred in 3- to 7-year-old females, while subadults had lower productivity (0.16 female young per female). Using assumptions based on seasonal patterns of parturition, Hellgren and others estimated that annual adult fecundity could range from 0.7 to 1.4 female young per female, depending on rainfall and primary productivity (9). If the assumptions hold true, peccaries in Texas typically only produce one litter per year, which simplifies demographic analysis. Also, yearling productivity is only about 30% to 40% of adult productivity, probably as a result of nutrient limitation (9). Adult collared peccaries in the Peruvian Amazon produce 1.4 to 1.8 litters per year (85).

TABLE 21–1

Age-Specific Reproductive Performance of Female Collared Peccaries Collected in Winter at the Chaparral Wildlife Management Area, Dimmit and LaSalle Counties, Texas, 1989–1998

Age class (yr)[a]	n	n ovulating	Ovulation rate $\bar{x}$	Ovulation rate SE	n pregnant	Litter size $\bar{x}$	Litter size SE	n young	Gross fecundity[c] (F young/F)
Subadults (1)	41	11	1.82	0.12	8	1.75	0.16	14	0.16
Adults									
2–3	111	96	2.04	0.05	79	1.71	0.05	135	0.57
4–5	91	91	2.09	0.04	81	1.94	0.04	157	0.81
6–7	49	48	2.15	0.06	38	2.00	0.07	76	0.73
>7	60	57	2.16	0.05	43	1.91	0.06	82	0.64
All adults[c]	345	324	2.10	0.02	265	1.86	0.03	493	0.67

[a]Age classes based on recategorization of tooth wear classes (15, 16).
[b]Assuming in utero sex ratio of 113M:100F.
[c]Includes adults not aged.
Source: Updated from Ref. 9.

Survival rates of peccaries have been determined from life tables of hunted populations and telemetry data. Using life tables created from ages of hunter-harvested animals in southern and western Texas, Low reported annual survival rates of 0.79 and 0.73, respectively. Similar work in Arizona found annual survival rates to average 0.78 (16, 38). A heavily harvested population in southern Texas (1.9 animals per square kilometer per year) yielded survival rate estimates of 0.65 and 0.73 for males and females, respectively (9). Survival rates from telemetered animals in unharvested herds averaged 0.90 for males and 0.87 to 0.90 for females (9). Gabor reported annual rates of female survival to average 0.85 on the same study area as Hellgren and others, but with a reduced harvest rate. Although harvest rates were not known in all cases, it appears that hunting mortality is additive to natural mortality, and that annual survival of adult females in unharvested populations approaches 0.90 (66, 9).

Sex ratios reported for several populations vary considerably. In Arizona, 2 studies reported a female bias in sex ratio *in utero* (44% to 48% male), whereas fetal sex ratio in Texas was male biased (53% to 56%) (29, 38, 54, 9). In neither state did sex ratio deviate significantly from 50:50. However, large data sets from harvested populations deviate from parity (53% male) (29, 9). Harvest may be biased toward males, which is unlikely because of the lack of sexual dimorphism in peccaries, or there may be observer error in sexing fetuses or harvested adults in one or more of the previous studies.

Juvenile survival rates remain the unknown parameter in peccary population dynamics. However, population modeling efforts, either from individually based or population-based models, suggest that juvenile survival rates range between 0.3 and 0.4 to maintain a stationary population (Figure 21–5) (86, 9). Research efforts to understand population dynamics should focus on this parameter, which is logistically difficult to determine.

Population Estimation and Density

Density estimates are available from throughout the geographic range of the peccary. However, no census technique for peccaries has been tested systematically or validated. Techniques used to arrive at population and density estimates include sightings, hunter reports, annual harvests, strip censuses, and long-term observations of herds (1). Reported densities range from 0.8 animal per square kilometer in Brazil to 16.0 animals per square kilometer on Barro Colorado Island (reviewed in 1, p. 128). Of 14 estimates provided in Sowls, 10 were between 3.0 and 12.0 animals per square

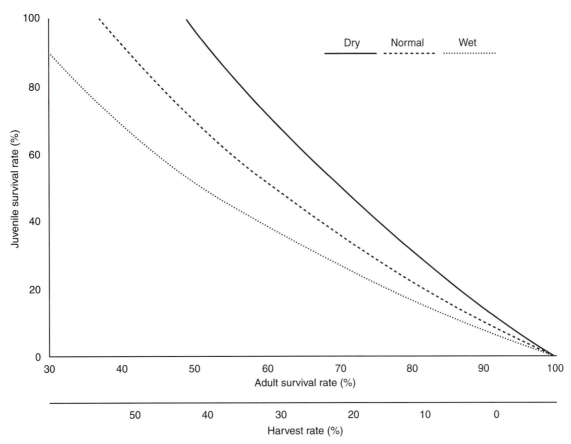

Figure 21–5 Modeled relationships among annual adult survival, annual juvenile survival, and annual harvest rate to maintain a stationary population ($r = 0$) of collared peccaries. We used three fecundity schedules that were representative of dry, normal, and wet years, respectively. Adult survival was kept constant across adult age classes at a specified survival rate. We calculated harvest rate using an additive model of total mortality, with mortality in the absence of hunting set at 0.1. (Adapted from Ref. 9 with the permission of The Wildlife Society.)

kilometer (1). Recently, Gabor used mark–recapture estimates (animals were marked during trapping and recaptured during the hunting season by harvest) to show a six-fold difference in peccary density between adjacent areas with (4 to 5 peccaries per square kilometer) or without (20 to 28 peccaries per square kilometer) feral pigs (66).

POPULATION MANAGEMENT

Annual surveys, often conducted simultaneously with deer surveys, have been conducted for decades in Arizona and Texas to monitor population trends (38). Personnel conducting these surveys, whether on foot, horseback, vehicle, or aircraft, record number of herds seen, average herd size, and adult:juvenile ratio. These three population statistics are used by wildlife managers and administrators to set harvest quotas and make harvest recommendations. In Arizona, populations have been expanding north-westerly and into urban areas, while in Texas, the populations have not shown a distributional shift (although a disjunct population of peccaries now occurs in northern Texas near the Red River as a result of translocation) (38, 87, 88). In southern Texas, helicopter counts of peccaries during deer surveys have declined during the past 25 years (R. B. Taylor, Texas Parks and Wildlife Department, unpublished data).

Harvest is the primary tool of population management in the American Southwest and peccaries provide unique hunting experiences. The seasons are typically long and during the winter months (year-round in some Texas counties) to take advantage of the extended breeding season. The poor vision and hearing of this species permit stalking of herds to close range for archery or handgun use. Indeed, archery equipment, handguns, and muzzle-loaders have become popular for the hunting of peccaries in North America (1). Archers have increased their success rates to about 20%, nearly as high as firearm hunters, because of the development of modern compound bows, better shooting skills, and improved stalking techniques (38).

Overharvest is a management concern at the local herd level and the population level. Accessible herds may be extirpated by heavy localized harvest, which is facilitated by the herd structure of peccary populations. To reduce local extinctions and prevent overharvest, the Arizona Game and Fish Department instituted a harvest quota system in 1972, which allowed a limited number of javelina permits per management area. The general harvest goal is 20% of the accessible population, or 15% of the population (38). A 20% harvest over a 5-year period on a study area in southern Texas did not appear to affect population size (9). However, in Arizona, 16% harvest of a population was associated with a population decline, while another population remained stationary after 7 years of estimated 12% harvest (38). Day reported that average herd size stabilizes at a lower level on hunted areas than on unhunted areas (38).

Collared peccaries may exhibit density-dependent recruitment to harvest mortality (Figure 21–5), with increased recruitment compensating for higher harvest (9). Under the assumption that hunting is an additive factor of mortality (see earlier discussion), recruitment (i.e., juvenile survival) must increase with harvest rate to maintain a stationary population. The range over which compensation can occur probably varies by region and habitat. In southern Texas, Ellisor and Harwell suggested that herds in dense brush could withstand 20% to 30% harvest, whereas herds in open mesquite savannas could only withstand a 15% reduction (56). The latter value may apply to more open, arid Arizona habitats. Relationships among harvest rate, recruitment, and population stability require further investigation.

Sowls concluded that the conservation of the collared peccary in the southwestern United States is secure, because of legal status as a game species and management by regulated and monitored hunting (1). He contended that habitat loss is the main threat to peccary populations and maintenance of suitable habitat should be ensured.

Conservation and management of collared and other peccaries in Central and South America is another story. The Pigs and Peccaries Specialist Group of the International Union for the Conservation of Nature provided a list of priorities for conservation and management of Neotropical peccaries (89). Clearing of land for timber harvest and agriculture is reducing available peccary habitat, and unregulated hunting has unknown effects on peccary populations. A recent study of peccary status in hunted ($N=19$) and unhunted ($N=4$) Amazonian forests highlighted the differential susceptibility of white-lipped and collared peccaries (68). Collared peccaries appear to persist in even the most heavily hunted areas, while white-lipped peccaries are very susceptible to hunting pressure, even in remote forests unaltered by habitat disturbance (68). A full discussion of the future of Neotropical peccaries is beyond the scope of this chapter, but see Sowls and references contained therein for the current status of this issue (1, pp. 248–278).

HABITAT REQUIREMENTS

Landscape and habitat requirements have not been identified for the collared peccary. The species occurs in a wide variety of habitats, from Sonoran and Chihuahuan desert scrub and oak woodland in Arizona and western Texas, to mixed thornscrub (or chaparral) in southern Texas, to a variety of tropical forests and woodlands in Central and South America (1). Because of intense study at its northern range, the collared peccary is sometimes considered an arid-adapted species (6). However, its physiology reflects its tropical origins and restricted range in temperate areas (90). The peccary is well adapted to thermal loads, by virtue of behavioral thermoregulation, pelage characteristics (coloration, lack of insulation), panting, and mild hyperthermia (42). However, it does not produce a highly concentrated urine and has a narrow thermoneutral zone, not characteristics of a desert mammal (42, 90). It inhabits arid and semiarid areas in North America because of a favorable thermal regime, consuming succulent plants to meet water requirements. Peccaries are nocturnal during hot temperatures and require dense bedding cover (thick brush, caves, abandoned mines) (37, 91, 36, 38). They become more diurnal during the mild winters characterizing their range in the American Southwest.

Landscape and habitat selection in the peccary have been studied recently (66). These findings indicate general preferences for dense woody cover (but see 6) and abundant food (e.g., cactus, forbs) to meet foraging and cover requirements and to address thermal constraints. In the Gulf Coast Prairies of southern Texas, peccaries were very selective, preferring dense communities of chaparral-mixed grass and live oak-chaparral and avoiding more open types such as open mesquite-grass savannas and bunchgrass-forb communities (36). In the western Rio Grande Plains of southern Texas, dense brush cover and abundant prickly pear supported the highest peccary populations (56, 66). Peccaries in this region selected high canopy cover (>60%) of brush such as mesquite, whitebrush, and cactus (66). Limited data on habitat selection in Arizona indicated similar results, with peccary herds preferring mesquite-dominated associations for bedding and avoiding several, relatively open types while foraging (87). In arid western Texas, Bissonette reported seasonal changes in habitat use by peccaries (23). In the cooler winter months, peccaries selected open, flat bajadas, but they switched their activities to deep, wooded draws in the summer. These activities were consistent with seasonal changes in forage availability and bedding requirements (i.e., shade; 23). Peccary group size was positively related to cover composition of forbs and succulents but inversely related to woody cover (23). Note that areas of woody cover on the western Texas site were dominated by creosotebush and tarbush, which are not peccary browse, and the landscapes had few favorable characteristics, such as bedding sites.

HABITAT MANAGEMENT

The effects of landscape and habitat manipulation on populations of collared peccaries are unknown. Consequently, recommended practices for habitat management need testing. Ellisor and Harwell suggested that 30% to 50% of brushland could be cleared in peccary habitat without detriment to peccary populations, if thickets of dense brush were retained and interconnected (56). If, as opined by Gabor, peccaries require relatively homogeneous landscapes with abundant succulents, the former recommendation may not be a good one (66). Sympatric populations of feral pigs in southern Texas are more likely to use open areas for grazing and may exert a negative competitive influence on peccaries in landscapes with more savanna and grassland vegetation (36, 66). Pigs may be less of an influence in more arid areas (e.g., western Texas, Arizona) because of their greater dependence on free water than peccaries.

Habitat management has been discussed and research questions developed relative to peccaries (92). For example, at what level of grazing pressure are peccary populations impacted? Conversion of shrub- and brushlands to grasslands are not favorable to peccaries, because of the loss of preferred forages, thermal cover, and escape cover. On the other hand, shrub invasion of desert grasslands resulting from grazing may increase peccary habitat. Timber harvest is a minor issue in peccary range in North America, but is an important consideration in the forested regions in Central and South America. The effects of brush manipulation (burning, chemical, and mechanical) on peccary populations are not understood, although it is likely that removal of woody vegetation is not beneficial. Manipulation practices that increase forb production, such as seasonal burns, may increase forage availability for peccaries.

Water developments are a common habitat management strategy for wildlife, including peccaries, in the southwestern United States. However, population responses of peccaries to water developments have never been measured. Data on water requirements of peccaries and availability of preformed water in peccary foods suggest that free water is not required (40). Nevertheless, peccaries will use available sources of free water, such as tinajas, ponds, stock tanks, and guzzlers (40;93;1, pp. 135–137). Therefore, water developments can serve management purposes other than population enhancement, namely, affecting local distributions of peccaries for consumptive or nonconsumptive use.

SUMMARY

The collared peccary holds a unique place among the large mammals of North America. It is the only tropical ungulate native to the United States and has an interesting life history, with mixed-sex herds and sexual monomorphism. One of three species of peccaries, the collared peccary is a small (20–kilogram) pig-like animal with a whitish collar. Peccaries consume a variety of plant foods, such as large nuts, seeds, fruits, forbs, browse, succulents (e.g., cactus), and underground plant parts. Neither ruminants nor monogastrics, they have a complex, pouched stomach to assist in digestion of their herbivorous diet.

Collared peccaries are highly social, territorial animals that live in mixed-sex herds of 5 to 25 throughout the year. These herds subgroup, or fragment, at different times of the year. Territories are defended from conspecifics, and the large dorsal scent gland possessed by this species serves to mark territories and identify individuals within a herd. Colonization of new areas is believed to be by herd fission.

Reproduction in peccaries differs from other native ungulates in North America. Young may be born at any time of the year, though births are concentrated in spring. Females do not normally produce their first litter until 2 years of age. The gestation period lasts 145 days and average litter size is 2 young. Male peccaries are fertile year-round, though their reproductive activity peaks in October to March. Dominance hierarchies exist in peccary herds.

Harvest is the primary tool for management of collared peccaries in North America. Hunting mortality is additive to natural mortality; however, within a range of harvest rates, juvenile survival may be compensatory to adult hunting mortality. Density-dependent recruitment in unhunted herds and the high degree of sociality argue for density-dependent regulation of peccary populations. Overharvest, particularly at the local level, is a management concern.

Habitat management has lagged behind harvest as a tool to manage peccary populations. Preferred habitats for collared peccaries in North America are desert scrub, oak woodland, and mixed thornscrub. They are highly selective of vegetation types, preferring communities of dense woody cover with abundant succulents. Understanding the effects of land use manipulations on peccary populations is a major management need.

LITERATURE CITED

1. Sowls, L. K. 1997. *Javelinas and Other Peccaries: Their Biology, Management, and Use.* Texas A&M University Press, College Station.

2. Donkin, R. A. 1985. The peccary—with observations on the introductions of pigs to the New World. *Transactions of the American Philosophical Society* 75:1–152.

3. Romer, A. S. 1966. Vertebrate paleontology. 3rd ed., University of Chicago Press. 468pp.

4. Woodburne, M. O. 1968. The cranial myology and osteology of *Dicotyles tajacu,* and its bearing on classification. *Memoirs of the California Academy of Science* 7:1–48.

5. THEIMER, T. C., and P. KEIM. 1998. Phylogenetic relationships of peccaries based on mitochondrial cytochrome *B* DNA sequences. *Journal of Mammalogy* 79:566–572.

6. BISSONETTE, J. A. 1982. Collared peccary. Pages 841–850 *in* J. A. Chapman and G. D. Feldhamer, eds., *Wild Mammals of North America: Biology, Economics, and Management.* Johns Hopkins University Press, Baltimore, MD.

7. LOCHMILLER, R. L., E. C. HELLGREN, and W. E. GRANT. 1986. Absolute and allometric relationships between internal morphology and body mass in the adult collared peccary, *Tayassu tajacu* (Tayassuidae). *Growth* 50:296–316.

8. LOCHMILLER, R. L., E. C. HELLGREN, and W. E. GRANT. 1987. Physical characteristics of neonate, juvenile, and adult collared peccaries (*Tayassu tajacu angulatus*) from south Texas. *Journal of Mammalogy* 68:188–194.

9. HELLGREN, E. C., D. R. SYNATSZKE, P. W. OLDENBURG, and F. S. GUTHERY. 1995. Demography of a collared peccary population in south Texas. *Journal of Wildlife Management* 59:153–163.

10. WRIGHT, D. B. 1993. Evolution of sexually dimorphic characters in peccaries (Mammalia, Tayassuidae). *Paleobiology* 19:52–70.

11. HERRING, S. W. 1972. The role of canine morphology in the evolutionary divergence of pigs and peccaries. *Journal of Mammalogy* 53:500–512.

12. HERRING, S. W. 1974. A biometric study of suture fusion and skull growth in peccaries. *Anatomy and Embryology* 146:167–180.

13. KILTIE, R. A. 1981. The function of interlocking canines in rain forest peccaries (Tayassuidae). *Journal of Mammalogy* 62:459–469.

14. KIRKPATRICK, R. D., and L. K. SOWLS. 1962. Age determination of the collared peccary by the tooth replacement pattern. *Journal of Wildlife Management* 26:214–217.

15. SOWLS, L. K. 1961. Hunter-checking stations for collecting data on the collared peccary (*Pecari tajacu*). *Transactions of the North American Wildlife Conference* 26:496–505.

16. LOW, W. A. 1970. The influence of aridity on reproduction of the collared peccary (*Dicotyles tajacu*) (Linn.)) in Texas. Dissertation. University of British Columbia, Vancouver, Canada.

17. HOFMANN, R. R. 1989. Evolutionary steps of ecophysiological adaptation and diversification of ruminants: A comparative view of their digestive system. *Oecologia* 78:443–457.

18. KILTIE, R. A. 1981. Stomach contents of rain forest peccaries (*Tayassu tajacu* and *T. pecari*). *Biotropica* 13:234–236.

19. BODMER, R. E. 1991. Influence of digestive morphology on resource partitioning in Amazonian ungulates. *Oecologia* 85:361–365.

20. BARRETO, G. R., O. E. HERNANDEZ, and J. OJASTI. 1997. Diet of peccaries (*Tayassu tajacu* and *T. pecari*) in a dry forest of Venezuela. *Journal of Zoology* (London) 241:279–294.

21. KILTIE, R. A. 1985. Craniomandibular differences between rain forest and desert collared peccaries. *American Midland Naturalist* 113:384–387.

22. OLMOS, F. 1993. Diet of sympatric caatinga peccaries (*Tayassu tajacu* and *T. pecari*). *Journal of Tropical Ecology* 9:255–258.

23. BISSONETTE, J. A. 1982. Social behavior and ecology of the collared peccary in Big Bend National Park. Science Monograph No. 16. National Park Service, Washington, DC.

24. EVERITT, J. H., C. L. GONZALEZ, M. A. ALANIZ, and G. V. LATIGO. 1981. Food habits of the collared peccary on south Texas rangeland. *Journal of Range Management* 34:141–144.

25. CORN, J. L., and R. J. WARREN. 1985. Seasonal food habits of the collared peccary in south Texas. *Journal of Mammalogy* 66:155–159.

26. LANGER, P. 1974. Stomach evolution in the Artiodactyla. *Mammalia* 38:295–314.

27. LANGER, P. 1979. Adaptational significance of the forestomach of the collared peccary, *Dicotyles tajacu* (L. 1758) (Mammalia: Artiodactyla). *Mammalia* 43:235–245.

28. CARL, G. R., and R. D. BROWN. 1983. Protozoa in the forestomach of the collared peccary (*Tayassu tajacu*). *Journal of Mammalogy* 64:709.

29. SOWLS, L. K. 1984. *The Peccaries.* University of Arizona Press, Tucson.

30. LOCHMILLER, R. L., E. C. HELLGREN, J. F. GALLAGHER, L. W. VARBER, and W. E. GRANT. 1989. Volatile fatty acids in the gastrointestinal tract of the collared peccary (*Tayassu tajacu*). *Journal of Mammalogy* 70:189–191.

31. COMMIZZOLI, P., J. PEINIAU, C. DUTERTRE, P. PLANQUETTE, and A. AUMAITRE. 1997. Digestive utilization of concentrated and fibrous diets (*Tayassu peccari*[sic], *Tayassu tajacu*) raised in French Guyana. *Animal Feed Science Technology* 64:215–226.

32. CARL, G. R., and R. D. BROWN. 1986. Comparative digestive efficiency and feed intake of the collared peccary. *The Southwestern Naturalist* 31:79–85.

33. LANGER, P. 1978. Anatomy of the stomach of the collared peccary, *Dicotyles tajacu* (L. 1798) (Mammalia, Artiodactyla). *Zeitschrift fur Saugetierkunde* 43:42–59.

34. JENNINGS, W. S., and J. T. HARRIS. 1953. The collared peccary in Texas. Federal Aid Report Series No. 12. Texas Game and Fish Commission. Austin, TX.

35. GALLAGHER, J. F., L. W. VARNER, and W. E. GRANT. 1984. Nutrition of the collared peccary in south Texas. *Journal of Wildlife Management* 48:749–761.

36. ILSE, L. M., and E. C. HELLGREN. 1995. Resource partitioning in sympatric populations of collared peccaries and feral hogs in southern Texas. *Journal of Mammalogy* 76:784–799.

37. EDDY, T. A. 1961. Foods and feeding patterns of the collared peccary in southern Arizona. *Journal of Wildlife Management* 25:248–257.

38. DAY, G. I. 1985. Javelina research and management in Arizona. Arizona Game and Fish Department, Phoenix.

39. THEIMER, T. C., and G. C. BATEMAN. 1992. Patterns of prickly pear herbivory by collared peccaries. *Journal of Wildlife Management* 36:234–240.

40. ZERVANOS, S. M., and G. I. DAY. 1977. Water and energy requirements of captive and free-living collared peccaries. *Journal of Wildlife Management* 41:527–532.

41. CARL, G. R., and R. D. BROWN. 1985. Protein requirements of adult collared peccaries. *Journal of Wildlife Management* 49:351–355.

42. ZERVANOS, S. M., and N. F. HADLEY. 1973. Adaptational biology and energy relationships of the collared peccary (*Tayassu tajacu*). *Ecology* 54:759–774.

43. LOCHMILLER, R. L., E. C. HELLGREN, and W. E. GRANT. 1986. Reproductive responses to nutritional stress in adult female collared peccaries. *Journal of Wildlife Management* 50:295–300.

44. LOCHMILLER, R. L., E. C. HELLGREN, and W. E. GRANT. 1987. Influences of moderate nutritional stress during gestation on reproduction of collared peccaries. *Journal of Zoology (London)* 211:321–316.

45. DAY, G. I. 1977. Climate and its relationship to survival of young javelina. P-R Final Report W-78-R. Arizona Game and Fish Department, Phoenix.

46. SOWLS, L. K. 1966. Reproduction in the collared peccary (*Tayassu tajacu*). Pages 155–172 *in* I. W. Rowlands, ed. *Comparative Biology of Reproduction in Mammals*. Zoological Society of London, Academic Press, UK.

47. COSCARELLI, K. P. 1985. Ovulation in the collared peccary as documented by laparoscopy. Thesis. Texas A&M University, College Station.

48. LOCHMILLER, R. L., E. C. HELLGREN, and W. E. GRANT. 1984. Selected aspects of collared peccary (*Dicotyles tajacu*) reproductive biology in a captive Texas herd. *Zoo Biology* 3:145–149.

49. SMITH, N. S., and L. K. SOWLS. 1975. Fetal development of the collared peccary. *Journal of Mammalogy* 56:619–625.

50. HELLGREN, E. C., R. L. LOCHMILLER, M. S. AMOSS, and W. E. GRANT. 1985. Serum progesterone, estradiol-17, and glucocorticoids in the collared peccary during gestation and lactation as influenced by dietary protein and energy. *General and Comparative Endocrinology* 59:358–368.

51. SOWLS, L. K., N. S. SMITH, D. W. HOLTAN, G. E. MOSS, and V. L. ESTERGREEN. 1976. Hormone levels and corpora lutea cell characteristics during gestation in the collared peccary. *Biology of Reproduction* 14:572–578.

52. HALLORAN, A. 1945. Five fetuses of *Pecari angulatus* from Arizona. *Journal of Mammalogy* 26:434.

53. KNIPE, T. 1957. The javelina in Arizona. Wildlife Bulletin No. 2. Arizona Game and Fish Department, Phoenix.

54. LOCHMILLER, R. L. 1984. Nutritional influences on growth and reproduction and physiological assessment of nutritional status in the collared peccary. Dissertation. Texas A&M University, College Station.

55. HELLGREN, E. C., R. L. LOCHMILLER, M. S. AMOSS, JR., S. W. J. SEAGER, S. J. MAGYAR, K. P. COSCARELLI, and W. E. GRANT. 1989. Seasonal variation in serum testosterone, testicular measurements, and semen characteristics in the collared peccary (*Tayassu tajacu*). *Journal of Reproduction and Fertility* 85:677–686.

56. ELLISOR, J. E., and W. F. HARWELL. 1979. Ecology and management of javelina in south Texas. Federal Aid Report Series No. 16. Texas Parks and Wildlife Department. Austin, TX.

57. LOCHMILLER, R. L., E. C. HELLGREN, L. W. VARNER, L. W. GREENE, M. S. AMOSS, S. W. J. SEAGER, and W. E. GRANT. 1985. Physiological responses of the adult male collared peccary to severe nutritional restriction. *Comparative Biochemistry and Physiology* 82A:49–58.

58. BYERS, J. A., and M. BEKOFF. 1981. Social, spacing, and cooperative behavior of the collared peccary. *Journal of Mammalogy* 62:767–785.

59. LOCHMILLER, R. L., E. C. HELLGREN, and W. E. GRANT. 1985. Relationships between internal morphology and body mass in the developing, nursling collared peccary, *Tayassu tajacu* (Tayassuidae). *Growth* 49:154–166.

60. HANNON, P. G., D. M. DOWDELL, R. L. LOCHMILLER, and W. E. GRANT. 1991. Dorsal-gland activity in peccaries at several physiological states. *Journal of Mammalogy* 72:825–827.

61. SCHWEINSBURG, R. E. 1969. Social behavior of the collared peccary (*Pecari tajacu*) in the Tuscon Mountains. Dissertation, University of Arizona, Tucson.

62. LOCHMILLER, R. L., E. C. HELLGREN, L. W. VARNER, K. MCBEE, and W. E. GRANT. 1989. Body condition indices of malnourished collared peccaries. *Journal of Wildlife Management* 53:205–209.

63. CORN, J. L., and R. L. WARREN. 1985. Seasonal variation in nutritional indices of collared peccaries in south Texas. *Journal of Wildlife Management* 49:57–65.

64. LOCHMILLER, R. L., E. C. HELLGREN, W. E. GRANT, and L. W. VARNER. 1985. Bone marrow fat and kidney fat indices of condition in collared peccaries. *Journal of Mammalogy* 66:790–795.

65. SOWLS, L. K. 1974. Social behavior of the collared peccary *Dicotyles tajacu* (L.). Pages 144–165 *in* V. Geist, and F. Walther, eds., *The Behavior of Ungulates and Its Relation to Management*. International Union for the Conservation of Nature, Morges, Switzerland.

66. GABOR, T. M. 1997. Ecology and interactions of sympatric collared peccaries and feral pigs. Dissertation. Texas A&M University-Kingsville/Texas A&M University, Kingsville and College Station (cooperative program).

67. ROBINSON, J. G., and J. F. EISENBERG. 1985. Group size and foraging habits of the collared peccary *Tayassu tajacu*. *Journal of Mammalogy* 66:153–155.

68. PERES, C. A. 1996. Population status of white-lipped *Tayassu pecari* and collared peccaries (*T. tajacu* in hunted and unhunted Amazonian forests. *Biological Conservation* 77:115–123.

69. GREEN, G. E., W. E. GRANT, and E. DAVIS. 1984. Variability of observed group sizes within collared peccary herds. *Journal of Wildlife Management* 48:244–248.

70. OLDENBURG, P. W., P. J. ETTESTAD, W. E. GRANT, and E. DAVIS. 1985. Structure of collared peccary herds in south Texas: Spatial and temporal dispersion of herd members. *Journal of Mammalogy* 66:764–770.

71. BIGLER, W. A. 1974. Seasonal movements and activity patterns of the collared peccary. *Journal of Mammalogy* 55:851–855.

72. ELLISOR, J. E., and W. F. HARWELL. 1969. Mobility and home range of collared peccary in southern Texas. *Journal of Wildlife Management* 33:425–427.

73. SCHWEINSBURG, R. E. 1971. Home range, movements, and herd integrity of the collared peccary. *Journal of Wildlife Management* 35:455–460.

74. OLDENBURG, P. W., P. J. ETTESTAD, W. E. GRANT, and E. DAVIS. 1985. Size, overlap, and temporal shifts of collared peccary herd territories in south Texas. *Journal of Mammalogy* 66:378–380.

75. ILSE, L. M., and E. C. HELLGREN. 1995. Spatial use and group dynamics in sympatric populations of collared peccaries and feral hogs in southern Texas. *Journal of Mammalogy* 76:993–1002.

76. SCHWEINSBURG, R. E., and L. K. SOWLS. 1972. Aggressive behavior and related phenomena in the collared peccary. *Zeitschrift für Tierpsychologie* 30:132–145.

77. BYERS, J. A. 1985. Social interactions of juvenile collared peccaries, *Tayassu tajacu* (Mammalia: Artiodactyla). *Journal of Zoology (London)* 201:83–96.

78. BABBITT, K. J., and J. M. PACKARD. 1990. Suckling behavior of the collared peccary (*Tayassu tajacu*). *Ethology* 86:102–115.

79. BABBITT, K. J., and J. M. PACKARD. 1990. Parent–offspring conflict relative to phase of gestation. *Animal Behaviour* 40:765–773.

80. WATERHOUSE, J. S., J. KE, J. A. PICKETT, and P. J. WELDON. 1996. Volatile components in dorsal gland secretions of the collared peccary, *Tayassu tajacu* (Tayssuidae, Mammalia). *Journal of Chemical Ecology* 22:1307–1314.

81. TYSON, E. 1683. Anatomy of the Mexico musk hog. *Philosophical Transactions of the Royal Society of London* 13:(153):359–385.

82. EPLING, G. P. 1956. Morphology of the scent gland of the javelina. *Journal of Mammalogy* 37:246–248.

83. WERNER, H. J., W. W. DALQUEST, and J. H. ROBERTS. 1952. Histology of the scent gland of the peccaries. *Anatomical Record* 113:71–80.

84. SUPPLEE, V. C. 1981. The dynamics of collared peccary dispersion into available range. Thesis. University of Arizona, Tucson.

85. GOTTDENKER, N., and R. E. BODMER. 1998. Reproduction and productivity of white-lipped and collared peccaries in the Peruvian Amazon. *Journal of Zoology (London)* 245: 423–430.

86. WILBUR, J. P., P. G. HANNON, and W. E. GRANT. 1991. Effects of varying dietary quality on collared peccary population dynamics—a simulation study. *Ecological Modelling* 53:109–129.

87. BELLANTONI, E. S., and P. R. KRAUSMAN. 1993. Habitat use by collared peccaries in an urban environment. *The Southwestern Naturalist* 38:345–351.

88. DAVIS, W. B., and D. J. SCHMIDLY. 1994. *The Mammals of Texas.* Texas Parks and Wildlife Press, Austin.

89. OLIVER, W. L. R., ed. 1993. *Pigs, Peccaries, and Hippos.* International Union for the Conservation of Nature, Gland, Switzerland.

90. ZERVANOS, S. M. 1975. Seasonal effects of temperature on the respiratory metabolism of the collared peccary (*Tayassu tajacu*). *Comparative Biochemistry and Physiology* 50A:365–371.

91. BISSONETTE, J. A. 1978. The influence of extremes of temperature on activity patterns of peccaries. *The Southwestern Naturalist* 23:339–346.

92. OCKENFELS, R. A., G. I. DAY, and V. C. SUPPLEE. 1985. *Proceedings of the Peccary Workshop.* Arizona Game and Fish Department, Phoenix.

93. ELDER, J. B. 1956. Watering patterns of some desert game animals. *Journal of Wildlife Management* 20:368–378.

Bison

James H. Shaw and Mary Meagher

INTRODUCTION

No other large mammal in North American can match the American bison in relation to presettlement numbers, effects on the landscape, and historical importance. Adapting to the climatic extremes and low-quality forage of the Great Plains, bison practiced large-scale nomadism and, under the right environmental conditions, probably long-distance migration. The behavioral characteristics necessary for survival on the plains persist today and become increasingly evident as their numbers rapidly increase. The objective of this chapter is to review the biology, management, and status of American bison. Historical accounts are included to understand bison on a larger scale than they occur today.

HISTORICAL ACCOUNTS

Numbers

An Army unit under the command of Colonel Richard Dodge encountered a large herd of bison along the Arkansas River in southern Kansas in 1871. In a letter to William Hornaday in 1887, Dodge estimated that the herd was 40 kilometers across and recalled that scouts and hunters estimated its length at roughly 80 kilometers. Most importantly, Dodge included an estimate of density, reporting that the average was no less than 25 to 38 per hectare (1). Taken at face value, these estimates result in 12 to 18 million bison in a single herd (2).

Yet Hornaday arbitrarily reduced Dodge's estimate to 4 million, having decided that it was wedge shaped rather than rectangular. Perhaps Hornaday's reduction was motivated by his desire to align Dodge's herd more closely to his estimate of 3.7 million bison in the southern plains herd, a conservative and highly questionable assessment derived from hide shipping records of one of the region's three railroads, augmented by some arbitrary assessments of losses to subsistence hunting (1).

Besides observations such as Dodge's and shipping records such as Hornaday's, other presettlement estimates of numbers have been derived through estimates of carrying capacity. McHugh calculated that the grasslands of America's Great Plains could have supported an average of about 10 bison per square kilometer or roughly 32 million for the 3.24 million square kilometer region (3).

A similar estimate was developed from a 1910 U.S. Department of Agriculture survey on livestock on the Great Plains, modified to the presumably higher carrying capacity for mobile free-ranging bison, and reduced to account for competition from feral horses and Indian ponies (4). Through such deductions, Flores estimated that the carrying capacity for bison on the southern plains would have been about 6 million during the pre-slaughter period of 1825 to 1850 and about 28 to 30 million for the Great Plains region (4).

The most widely cited estimate for preslaughter bison numbers was Seton's (5). He first accepted Hornaday's revision of Dodge's observation at 4 million. He then estimated (by unknown criteria) that the area required to support such a herd would be about 518,000 square kilometers. The geographic region inhabited by bison was, according to Seton's reckoning, roughly 15 times the area needed by a herd of 4 million, resulting in the famous figure of 60 million (5, 1, 2).

Seton's figure of 60 million, along with his other estimate of 75 million bison for the whole continent (itself based loosely on the same 1910 U.S. Department of Agriculture livestock survey used by Flores) is unacceptable by modern standards of wildlife science (4). The figures from McHugh and Flores seem more reliable, but could be in error because of extensive changes in carrying capacity between 19th century vegetation subject to nomadic bison and extensive wild fires and the 20th century conditions on which the estimates were based (3, 4). Finally, numbers derived from estimates of carrying capacity represent maximum numbers that could be supported, not observed population sizes that actually occurred.

In the early 19th century, there were millions of bison between the Mississippi River and the Rocky Mountains. Just how many millions we cannot know with any scientific certainty. There is no reason to assume that populations of bison were stable, particularly given the species' mobility, human encroachment, and the region's between-year variation in rainfall and resulting differences in primary productivity.

Value

Paleoindians hunted bison for more than 10,000 years by running them over cliffs or "jumps," or by running them into ravines or canyons and spearing them from above. Most of these kills, though, were probably seasonal and opportunistic, given the limitations on Indian mobility and weaponry.

Once the Indians learned to use horses, beginning in the late 17th century, their capabilities for hunting bison greatly expanded (6, 3, 7, 8). By the 19th century roughly 35 tribes or ethnic units became Plains Indians, a mobile, nomadic culture largely specialized for hunting bison (4).

Americans of European descent tend to underestimate the impact of American Indians on wildlife and the landscape. We either presume that Indians were incapable of extensive effects or assume that they possessed an ecological sensitivity that effectively prevented excessive harvest. Neither presumption is consistent with the evidence. Flores, for example, evaluated accounts of Comanche bands experiencing starvation due to shortages of bison between 1825 and 1850, decades before hide hunters exterminated the free-ranging herds (4). The actual extent to which Indian hunting pressures influenced these shortages is not clear, but a combination of Indian hunting for subsistence and for commercially traded robes, and with increased forage competition that bison may have encountered with feral and Indian horses, were likely major forces (4).

The skins of male bison were too heavy to be useful in tepee covers or as robes, so the commercial hunting pressure before the Civil War was directed differentially at females. Although estimates of robes shipped along the Missouri River have been published, the extensive trade network that emerged in the 19th century makes it impossible to determine their origins and thus our abilities to estimate their effects. One classic account, though, suggests that by 1846, bison sex ratios may have been skewed heavily in favor of males (9).

Extermination and Recovery

In 1871, the Mooar brothers helped develop a technique for turning field-dried hides into industrial-grade leather (10). The markets opened by that technological change doomed the free-ranging herds. Bison disappeared from the southern plains by 1875 and from the northern plains by 1882 (11).

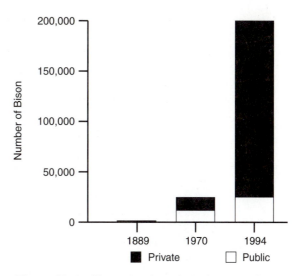

Figure 22–1 Bison numbers have increased greatly since the 1800s, primarily due to private ownership.

C. J. "Buffalo" Jones, Charles Goodnight, and other private individuals undertook the task of saving the bison from extinction (12). They successfully captured about 75 bison, mostly calves, to form the five "foundation" herds from which most of the public herds, and later private ones, were established. Just under 80% of these founders came from the southern plains, while the remainder were taken from Montana and perhaps the Dakotas (13).

The only bison population in the United States that survived in the wild was the one in Yellowstone National Park. Meagher estimated that the population was down to about 22 at its lowest point (14).

By the time Hornaday censused the remaining bison, he could account for only 541, nearly all of which were in captivity (11). Their numbers slowly increased and by 1907, 15 of the descendants of the five foundation herds were shipped from the Bronx Zoo in New York to the Wichita Mountains Forest and Game Reserve (now the Wichita Mountains Wildlife Refuge) in Oklahoma (3). The following year 36 founders were transported to the National Bison Range in Montana.

Recovery was under way by 1970 when McHugh estimated that there were 30,000 American bison, roughly half of which were in the public herds and the other half in private ownership (3). The number of bison in the public herds in the United States and Canada has not appreciably changed since 1970, but the number in private ownership has increased more than 10-fold and is still rising exponentially (Figure 22–1).

TAXONOMY AND DISTRIBUTION

Description

American bison are dark brown as adults, though the coat color may become lighter as summer progresses and the sun bleaches the hair. Newborn calves are reddish-brown or "buff" colored at birth and their coats gradually darken until age 3 or 4 months by which time they are the same color as adults.

Adult males are the largest land mammals native to the Western Hemisphere. They may stand more than 2 meters at the shoulder and weigh up to 900 kilograms. Females are usually about half the body mass of males, weighing up to about 450 kilograms (15). Compared with most other bovids, bison tend to have relatively massive

Figure 22–2 Calf, female, and male bison during summer rut. The male is "roaring" to warn rivals while he "tends" a female in estrus. (Photograph by J. H. Shaw.)

forequarters, particularly in males. Both sexes have horns, though female horns are substantially less massive than those of males, and have a greater tendency to curve inward toward the tips (Figure 22–2).

Taxonomy

Linnaeus placed American bison in the same genus as that of domestic cattle (16, 17, 18). By the late 19th century, however, virtually all authorities recognized sufficient anatomical distinctiveness between bison and other bovines to assign it to its own genus (19). The first "modern" taxonomic treatment for American bison retained the separate genus, as do most other authorities (20).

In recent years though, an increasing number of zoologists have begun placing bison in the same genus as cattle. Bison are largely interfertile with cattle, a fact established over more than two centuries of interbreeding that produced the mostly fertile hybrids known as "cattalo" or "beefalo." The genus debate is likely to flourish in the next few years and may need reevaluation.

Another long-standing taxonomic debate concerns the legitimacy of the wood bison as a distinct subspecies from the plains bison. Differences in pelage and morphometric comparisons seemed to support subspecific distinctiveness (21, 22). Subsequent studies, however, have shown that the pelage differences are strongly influenced by diet (23). Comparisons of DNA concluded that the wood and plains bison did not represent distinct subspecies (24, 25).

Distribution

American bison occurred west of the Appalachian Mountains to Northern Mexico and in to Idaho and the Northwest Territories and probably beyond those limits intermittently (26, 27, 17). The highest numbers seemed to have been between the 98th meridian and the Front Range of the Rocky Mountains. Historical accounts suggest that bison were relatively rare within the tallgrass prairies even as early as the first decade of the 19th century (28, 29) (Figure 22–3).

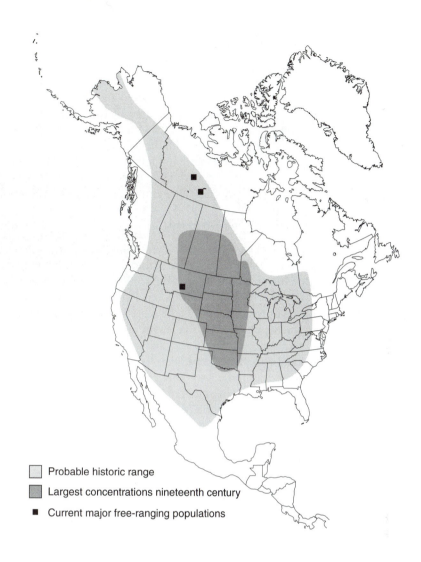

Figure 22–3 Geographic range of American bison (modified from Ref. 11 and Ref. 3.)

Legend:

☐ Probable historic range

■ (dark) Largest concentrations nineteenth century

■ Current major free-ranging populations

Bison have been restored throughout much of their historical range and even beyond, but most herds are now fenced and managed as isolated units. Large, free-ranging herds persist in the MacKenzie Sanctuary, Slave River Lowlands, and Wood Buffalo National Park in Canada and in the Henry Mountains, and Yellowstone and Grand Teton National Parks in the United States.

LIFE HISTORY

From roughly the second half of the Holocene until the hide hunts of the 1870s, herds of American bison had access to an entire continent. Bison, like all heterotrophs, survive only if they maintain an adequate energy balance. But the combination of large body size and strong drive to aggregate meant that the scale over which the energy balance had to be met was especially large. Moreover, the Great Plains region, where historical bison attained the greatest numbers, is climatically variable with hot summers, cold winters, and precipitation. The relentless search for forage over vast areas of unpredictable productivity made the capacity for long-distance travel an integral part of bison life history.

Diet Selection

Bison eat grasses and sedges almost exclusively. On shortgrass prairies, grasses constituted 93% to 98% of bison diets during the growing season and roughly 90% to 99% grasses year-round (30, 31). Plumb and Dodd reported between 83% and 98% grasses in summer diets of bison on a northern mixed grass prairie (32). On tallgrass prairie, bison consumed 11% to 15% sedges in summer and fall, and 20% to 39% sedges during winter and spring, with the balance of the diet consisting almost entirely of grasses (33). Sedges are more common than grasses in bison diets in the mountain meadows of Yellowstone and in the boreal meadows of Canada (14, 34). Bison consumption of browse and forbs has been almost negligible in nearly all diet studies.

Reproductive Strategy

Most female bison first breed during their third summer, with rut reaching a peak between early July at lower latitudes and early September at higher latitudes (26). Following an average gestation of 285 days, females give birth to a single calf in spring (35). At Wind Cave National Park, birth synchrony in bison has been reported as weaker than generally anticipated in a follower species (36). In contrast, Berger presented evidence that the length of gestation was actually shortened in females in good condition, apparently to achieve a greater degree of birth synchrony in Badlands National Park, South Dakota (37). Calves born earlier in the year achieve socially dominance over those born later (36).

The calving rate among sexually mature females ranges from 88.2% at the National Bison Range in Montana to 50% in the Slave River Lowlands (38, 39). These rates seem to be influenced by latitude, growing season, population density in relation to carrying capacity, and age structure of the herd (12). Calving rates decline in females older than 12 years (40, 41). Males are reproductively viable at 2 years of age, but most are inhibited from reproducing by older males until they reach full size at 6 to 8 years (42, 43, 14). In Badlands National Park, 48% of the successful matings were attributed to 11% of males in the 9- to 15-year-old age group (44). Males at Fort Niobrara National Wildlife Refuge, Nebraska, began to participate in rut at age 5 or 6 and reached a peak at ages 8 to 11 (45).

Sex Ratios

The primary (*in utero*) sex ratio in bison is skewed slightly toward males, usually in the 51% to 56% range (46, 35, 17). Higher differential mortality among males tends to push secondary sex ratios more toward equity. Managers opting for higher productivity retain a higher proportion of adult females in the more intensively managed herds. Less intensively managed populations tend to have a more balanced sex ratio and lower calf production.

Longevity

No life table has been published for a bison population, so we know little about "natural" life spans. Although McHugh reported known-aged female bison reaching 40 or more years, the average life span is surely less (43). Meagher doubted that many bison in the wild reach 20 years of age (14). Managers typically remove older individuals before they reach the ends of their lives. Unmanaged herds typically lack handling facilities needed for accurate aging of individual bison.

BEHAVIOR

Bison typically form two types of social groups: mixed groups and male groups. Mixed groups consist of females and their offspring, along with a few males. The male groups are all male and usually range from two to six individuals older than 2 years (43, 47, 48).

The two types of groups largely mix during rut, when males compete for dominance and breeding and mingle among the larger mixed groups. In at least some populations, bison tend to associate by age classes (38). A possible variation in the mixed group may occur, in which females without calves associate with a relatively high proportion of males (adult groups), while those females with calves associate more commonly with one another in mixed groups that contained a substantially lower proportion of males (49). Such differences might be related to different nutritional requirements of females with calves as opposed to those without them (49).

The bison's drive to aggregate is powerful, reaching maximum levels during summer. Private bison breeders quickly learn how difficult it is to keep adjacent herds separated by internal fencing (N. Anderson, President, Oklahoma Bison Association, personal communication). Bison force their way through standard barbed-wire cattle fences to join other bison. Managers at the Wichita Mountains Wildlife Refuge, Oklahoma, keep herd units separated only with heavy welded-wire big game fences, using railroad rails for fence posts.

Historical accounts also emphasize the aggregative tendencies of bison. Besides written accounts, such as the one described earlier by Colonel Dodge, several 19th century artists put their observations of free-ranging herds on canvas. For example, William J. Hays's 1862 painting titled "Herd on the Move" depicted thousands of bison converging in one direction, with smaller units merging into the main body in a manner analogous to creeks flowing into a river (50, p. 61).

Group Stability

Shaw and Carter confirmed long-term associations between female bison and their offspring through the male offsprings' second summer and the female offsprings' third summer, with the sex-bias similar to that reported for female bison and offspring at Santa Catalina Island, California (48, 51). Thus, the most stable social unit in American bison is the female–calf unit, particularly when it involves female calves. As a general rule, composition of large groups is quite fluid (51, 52, 38).

Post-weaning associations offer increased protection from displacement by older bison. Yearling daughters accompanying their mothers have been shown to spend more time in the center of social groups and to be less frequently displaced than yearling females not accompanying their mothers (53). At the Tallgrass Prairie Preserve, Oklahoma, 43 ten-month-old bison calves were introduced to augment the established herd of nearly 300. Other bison were more aggressive to these "orphan" calves than they were to any of the resident age classes (54).

Migratory or Nomadic?

Among living ungulates, some populations of barren-ground caribou and wildebeest make seasonal and directional movements that characterize true migration. Modern bison populations, at greatly reduced scales, are capable of doing the same in Yellowstone National Park, at the Wichita Mountains Wildlife Refuge, Oklahoma, and at least in portions of Wood Buffalo National Park, Northwest Territories (55, 56, 57, 58).

Even the free-ranging bison herds of the 19th century were probably not migratory. In the most thorough review of historical accounts of bison, Roe concluded that their movements in the 19th century were virtually unpredictable. Further doubt about the regularity of bison movements was raised by McHugh (1, 3).

We concur with Roe and McHugh that 19th century bison were more nomadic than migratory (1, 3). For seasonal and directional movement to occur on a regular basis, forage must be available predictably at each end of the migration route. Such conditions are met in today's Yellowstone region, but they were unlikely to exist on the Great Plains in the 19th century. Human encroachment, primarily through hunting pressure and an increase in fires, would have destabilized bison populations and forage production to such an extent as to render predictability impossible. Under such conditions, nomadism would have been more likely than migration.

Figure 22–4 Bison prefer to graze on sites that are burned earlier in the year. (Photograph by J. H. Shaw.)

Fire and "Patch Dynamics"

Roe presumed that fire was destructive to bison, citing accounts of herds fleeing desperately from fires and those overtaken left singed and even blinded by the flames (1). Nine bison perished in the extensive wild fires in Yellowstone National Park in 1988 (59).

Despite the physical dangers of fire, it is a beneficial and even essential force in maintaining prairie. Prairie grasses are fire tolerant and in many cases fire dependent (33). When recovering from fires, grasses put forth lush, green aboveground growth that large grazers find attractive. For example, a 1,860-hectare prescribed burn in the Wichita Mountains, Oklahoma, increased use by bison from 3.9% in June the year before the burn to 58.0% during the June following the burn (57). Bison attraction to recovering burns has been reported in all major types of prairies (60, 61, 62) (Figure 22–4).

Coppedge and Shaw examined bison grazing patterns in relation to prairie sites burned at various seasons (63). Mixed groups of bison grazed on recovering burns significantly more often than expected by chance during the first postfire growing season, regardless of season of burn. Mixed groups used larger burns more than smaller ones. Male groups showed less of a tendency to select recovering burn sites than did mixed groups.

Simulation modeling helps interpret the implications of larger scale burns on bison ecology. Wallace and others, for instance, simulated various patterns of fire for Yellowstone National Park (64). They concluded that bison survival would not be adversely affected until fires affected more than 60% of the landscape. They further noted that for their analysis, there was no biologically important difference between clumped versus random patterns of fires.

Obviously, then, patterns of fires affect subsequent patterns of grazing. As bison preferentially graze sites that are recently burned, they reduce the fuel load on those sites and decrease the likelihood of subsequent fires on those grazed areas. Unburned, and therefore less grazed, sites experience more fuel accumulation and thereby increase the chances of burning during the next round of fires. This combination of fire and grazing creates a shifting mosaic of prairie sites (63).

Productivity and Recruitment

The rate of increase for any bison population depends on the population's age structure, sex ratio, latitude, and presumably forage quantity and quality, and various forces that affect attrition. Managers can most easily manipulate sex ratios and age structures to vary overall calf production. In a relatively unmanaged herd such as Yellowstone, one can expect sex ratios to approach 1 : 1, age structures to be relatively old, and the population to vary in response to winter severity and changing ecological conditions. Such conditions result in relatively low productivity rates such as the 20% (i.e., 20 calves per 100 yearling and older animals) reported for Yellowstone during the 1960s (14).

In contrast, the herd on the Tallgrass Prairie Preserve is managed for high productivity. Managers skew sex ratios toward greater proportions of females, keep age structures relatively young, and maintain numbers below carrying capacity. Such populations may experience productivity rates of 50% or 50 calves per 100 older bison (R. Hamilton, Biologist, Tallgrass Prairie Preserve, personal communication).

Mortality and Attrition

Mortality in more intensively managed herds is low and the populations are regulated through periodic round-up and sale of surplus bison. As long as ownership of bison increases in popularity, this artificial means of attrition will remain widespread.

The least managed herds occur in relatively remote regions with harsh climates. Winter severity and predation are the primary causes of mortality in such bison populations. In Canada's Wood Buffalo National Park, for instance, snow depth impairs bison movements and inhibits their abilities to forage. Carbyn and others concluded that snow depths of 55 centimeters or more became critical for calves, and depths of 65 centimeters or more were critical for adult females (58).

The effects of wolf predation on bison populations have been studied in detail in Wood Buffalo National Park. Calf production was relatively low, roughly 28 : 100 females, and wolf predation rates were calculated in excess of the population's recruitment (58). Simulation models showed that if wolf predation on bison ceased, the bison population would quickly and substantially increase (58).

Diseases and Parasites

From the standpoint of conservation and management, understanding of diseases in bison requires recognition of at least five issues.

1. Presence of or exposure to a pathogen does not always mean disease (pathology). Likewise, presence of a parasite may not result in effects on the host.
2. Evaluation of the host–organism relationship must involve the ecology and behavior of the host, and the ecology of the disease or parasite.
3. Disease effects on an individual may not have consequences for the population.
4. Herd management practices can affect host–organism relationships.
5. Host–organism relationships may differ among host species harboring the same organism (i.e., bison are not cattle).

Three contagious diseases dominate the biology, politics, and management issues for several important free-ranging bison populations: anthrax, brucellosis, and tuberculosis. Of these, anthrax is locally important but does not generate the attention and controversy that has focused primarily on brucellosis, and secondarily on tuberculosis.

Anthrax. Anthrax occurs in the MacKenzie Bison Sanctuary and Wood Buffalo National Park and environs in northern Canada (65). It is not reported in bison in the United States. Outbreaks are associated geographically with neutral or alkaline and calcareous soils where the spores revert to a vegetative form for long periods and multiply with optimal environmental conditions. Occurrence may be seasonal with soil-borne infection, when minimum daily temperatures are 16°C. Outbreaks occur with marked environmental changes such as flooding or drought and may be highly sporadic but repetitive (66).

Necropsies should not be done, but blood, swab, and tissue (ear) samples collected from carcasses can provide diagnosis (65). During an epizootic scavengers may contract the disease or serve as mechanical vectors.

Wood Buffalo National Park and environs experienced eight outbreaks between 1962 and 1991 that killed up to 281 bison (65). Outbreaks in the park in the past resulted in additional mortality as efforts were made to contain the epizootic by eliminating local bison populations.

The first known outbreak of anthrax occurred in the Mackenzie Bison Sanctuary in July and August, 1993. At the time the population exceeded 2,000. Of the 170 deaths, 135 were of adult male bison. The outbreak was associated with several months of drought preceded by several years of flooding of bison habitat. The best explanation for this outbreak appeared to be that the spores are concentrated mechanically by water movement and evaporation, to survive where calcareous soils foster long-term viability (65). Although the Mackenzie Bison Sanctuary attempted to minimize the outbreak and resulting environmental contamination by burning carcasses or treating them with antiscavenger chemicals (i.e., formaldehyde), the sporadic nature of anthrax outbreaks suggests that long-term population consequences need not be of concern unless other decimating factors are operative at the same time.

Brucellosis. This bacterial agent occurs in bison of Yellowstone, Grand Teton, and Wood Buffalo National Parks. Brucellosis was more common in bison in the past, but fenced herds now are maintained brucellosis free. The disease was introduced to North America by cattle (67). It also occurs in elk of the Yellowstone–Grand Teton–northwest Wyoming area, which complicates the prospect of eradication.

The bacterium in bison appears to be the same organism as is found in cattle. Recent research indicates there are also major host response differences between bison and cattle and even between captive herds of bison (68, 69, 70, 71, 72).

Compared with its occurrence in cattle, brucellosis in bison is poorly understood. In bison, abortions happen, but apparently much less commonly than in cattle (68). Seroprevalence does not correlate as well with infection in bison as it does in cattle. In contrast to cattle, seroprevalence and culture are higher in calves and subadults, and in more males than females of all ages (68).

Transmission potential from free-ranging bison to cattle appears to be quite low, although not zero. Transmission has been demonstrated in captive studies, and is assumed to have more potential, provided brucellosis is present, where bison and cattle are maintained together as fenced populations (69). No transmission cases in the wild have been documented. The topic is quite contentious because livestock economics are perceived to be at risk, and because free-ranging bison make difficult neighbors regardless of disease issues (Figure 22–5).

Tuberculosis. Bovine tuberculosis has only been reported in bison within and adjacent to Wood Buffalo National Park and from private bison ranches (73). Tuberculosis appears to be potentially more dangerous to bison populations than anthrax and brucellosis. Game farming of elk and ranching of bison increase the risk by concentrating the animals at higher densities and in transporting them over long distances.

The tuberculosis host–organism relationship is similar between host species, including humans (66). The organism forms a primary focus, often in the lung, with caseous lesions in adjacent lymph nodes as a result of drainage from the primary focus. Additional lesions form a tumor-like, granulomatous tubercle that gives the disease its name.

Figure 22–5 Bison taken in control operations north of Yellowstone National Park. (Photograph by M. Maegher.)

This infectious disease usually occurs in a progressive form, but sometimes quite slowly. It is debilitating for the host and is commonly fatal. The pathogen may have a role in the population decline of the bison of Wood Buffalo National Park, acting as one agent among a complex of environmental change, wolf predation, and possibly in synergy with brucellosis. Tessaro reported tuberculosis in 15 (21%) of 72 bison sampled in Wood Buffalo National Park (73). Most workers regard tuberculosis as an important threat for wildlife and a valid public health concern.

Other Pathogens. Pasturella (i.e., hemorrhagic septicemia) was first reported from bison at the old Buffalo Ranch in Yellowstone National Park in 1911, with recurrences in 1919 and 1922 (14). In 1992, two forms of pasturella were isolated from Yellowstone bison, but there was no evidence of disease, and prevalence was comparatively low (74). *Pasturella multicoida* has been identified as the causative organism elsewhere. In 1965, four young bison died at the National Bison Range, and a 1966–1967 serosurvey detected significant antibody levels in several public herds (73). In sum, the organism may be of concern occasionally on preserves and ranches, but does not appear to be a threat of much consequence to bison maintained as wildlife.

A serosurvey of bison in Yellowstone National Park for exposure to infectious organisms commonly found in cattle showed that exposure to some pathogens had occurred, but antibody titers were low and there was no evidence of active infections. Titers were detected for anaplasmosis, bovine respiratory syncytial virus, bovine viral diarrhea, bovine herpes virus 1, several serovars of leptospirosis, and parainfluenza 3 virus (75). Parainfluenza 3 virus was reported from the Delta bison of Alaska but no clinical disease occurred (76). Tessaro cited the occurrence of several other infectious pathogens, mostly in captive bison, and of no population importance with the exception of malignant catarrhal fever, which has caused fatalities in some ranched herds (73).

Parasites. Several dozen endoparasites have been reported from bison (26, 77, 73). Most are recorded from fenced populations, some very small and closely confined. Although sometimes of importance to an individual host animal, the records of occurrence will be of interest mainly to commercial producers or keepers of small show herds. In the absence of feedgrounds or other artificially induced concentrations, none of these endoparasites appears to be particularly prevalent in or have population consequences for the bison inhabiting the larger refuges and preserves.

Van Vuren and Scott compared internal parasites of sympatric bison and cattle using the Henry Mountains, Utah (78). Prevalences differed between the two host species for the five taxa recorded, particularly for those parasites associated with water. Differences in host suitability and behavioral ecology were identified as probable factors. Neither effects on the host individuals nor their populations were identified.

Various ectoparasites have been reported; most appeared to be unimportant to bison populations (26, 77, 73). Exceptions to this likely occur where organisms such as biting flies may affect summer range distributions, at least locally (14). Additionally, introduced pests such as the face fly have a potential for becoming more important in the future (79).

Other Conditions. Fluoride toxicosis is chronic in Yellowstone National Park wildlife as a feature of the chemical environment associated with geothermal activity (80). It causes jaw deformities and other characteristic bone lesions, affecting individuals, but prevalence and contribution to population consequences are unknown. Periodically harsh environmental conditions, particularly severe winters, may predominate for bison (M. M. Meagher, unpublished data).

Tessaro and Reynolds and others reviewed miscellaneous pathological conditions reported from individuals, none with population consequences (73, 26). Such occurrences could be important to a commercial producer, but are not infectious and have a minimal role on the various refuges and fenced preserves.

POPULATION MANAGEMENT

Bison management is a paradox. On the one hand, perpetuation of bison as a species is a conservation success story after the near-extermination in the wild of the last century. Yet the success has taken place mostly on fenced public preserves of varying sizes and management objectives, and on commercial ranches. Recovery is real numerically, but is removed from the ecologically generated pressures and large-scale variability under which the species evolved. In particular, landscape scale influences are rarely possible. The topic of bison management is compounded further by a very strong spiritual mystique, derived from a Native American and frontier heritage. And, like the California grizzly, bison make a more comfortable icon than a wildlife reality.

Bison management is unique in that this is the only large wildlife species maintained primarily as fenced populations. In turn, management problems appear to be underscored by a perception (sometimes unconscious) that bison are not wildlife, but rather a breed of cattle. This may derive from their stolid temperament; people commonly think of wildlife as fleeing from humans and do not recognize that a running animal is a frightened animal, and that this behavior is not necessarily an indication of "wildness." The second influence toward the human attitude likely comes from the history of the species: near-extermination in the wild coupled with establishment of herds on ranches, initially as something of a hobby. The attitude that bison are cattle is compounded by the many interbreeding efforts since the 1700s (17).

Here, we are concerned mainly with public preserves, where initially the goal was preservation of the species. Now, public opportunities for viewing and enjoyment in natural-appearing conditions, and education and research, are commonly expressed objectives. We recognize the increasing interest in commercial ranching, and that a few commercial enterprises are large and located in traditional bison habitat. But commercial management commonly entails more animal husbandry practices to maximize production, such as forced weaning of calves, supplemental feeding, and a selected age and sex structure, with specifics tailored to the individual operation and lands involved. We also recognize a growing interest on the part of Native Americans in establishing bison herds on tribal lands, for cultural and economic reasons, but these herds are usually managed more intensively than most public preserves.

However, management of public preserves also entails a range of management intensity and activities, depending on lands available and preserve objectives. Most public preserves were established for and are managed primarily for bison, and secondarily for other species and broad ecological relationships such as the U.S. Fish and Wildlife administered National Bison Range, Montana, Fort Niobrara National Wildlife Refuge, Nebraska, and the Wichita Mountains Wildlife Refuge, Oklahoma. The fenced national parks, such as Elk Island in Canada, and Badlands, Theodore Roosevelt, and Wind Cave in the United States, place more emphasis on broad ecological concerns, of which bison are an important, but not necessarily dominant, component. This perspective is apparent in the management plans being developed by the Nature Conservancy for newly established herds such as the Tallgrass Prairie Preserve, Oklahoma. In contrast, Custer State Park, South Dakota, manages its herd for the highest practical yield (81). Each preserve must tailor management to its legislative mandates, agency objectives, and specific needs, some of which may vary over time. When more than one federal or state agency is involved, as is true for the bison of Grand Teton National Park in Jackson Hole, Wyoming, and the Henry Mountains, Utah, interagency cooperation must be maintained.

For all the fenced preserves and the smaller free-ranging public herds such as those in Alaska, the Henry Mountains of Utah, and Jackson Hole, Wyoming, population size is regulated. Live removals, regulated sport hunting, or some combination is commonly used to regulate populations (Figure 22–6).

Finally, several large preserves, unfenced, with thousands of bison, represent the closest attempt now to management of bison as a wildlife species, theoretically with minimal interference by humans. Wood Buffalo National Park, Canada, the MacKenzie Bison Sanctuary, Canada, and Yellowstone National Park in the United States are examples.

Wood Buffalo National Park. The population declined from around 12,000 in the 1960s to less than 4,000 by 1997. The causes appear to be complex, but include cattle diseases (i.e., tuberculosis, brucellosis, anthrax), wolf predation, and environmental change generated partly by the construction of a large dam on the Peace River

Figure 22–6 Live auction sales are a common means of population control for confined bison populations. (Photograph by J. H. Shaw.)

upstream from the park (58). A 5-year research program is under way with an advisory board representing the various stakeholders, including the aboriginal groups (82). The hope is to focus management actions on disease eradication and population restoration.

Yellowstone National Park. The population roughly doubled from about 2,000 to about 4,000 between 1980 and 1994 (82a). With the exceptionally severe snows in the winter of 1996–1997, the population dropped to about 2,000. Nearly 1,100 were removed by the state of Montana after they exited the park; winterkill inside the park accounted for the rest. Although more research is needed, it appears probable that the population will continue to decline. Habitat changes appear to be directional, with functional ability to support bison decreasing within the park. Long-term climatic changes have a role, but changes in bison energetics related to use of the snow plowed roads appears to be a major factor (82a). As bison leave the park, they are unwelcome and cannot shift to new winter ranges in Montana to any extent. The situation is complex and unprecedented for large ungulates. The future is uncertain for bison in Yellowstone National Park.

MacKenzie Bison Sanctuary. This population represents a major management success story as part of the effort by Canada to reestablish bison herds with the genetic complex of the wood (northern) bison strain. The population has increased slightly in recent years, with quite limited hunting removals. The overriding concern here is to keep the population disease free from the tuberculosis and brucellosis present in the bison from Wood Buffalo National Park.

Finally, bison management must consider human management, particularly where populations are maintained on preserves and parks where tourism is promoted. Experience in Yellowstone National Park indicates that visitors often mistake stolidness for tameness, and approach bison, camera in hand, or are simply careless in getting too close. The park has recorded two fatal gorings, and in recent years there have been more bison-caused injuries (nine in 1985) than those inflicted by bears (Yellowstone National Park, unpublished data). Near-misses are often unreported, but of those reported all but one record involved a male bison. Custer State Park reported three goring injuries involving male bison between 1987 and 1993 (81).

Preserves make considerable effort to warn visitors, or physically separate them with fencing, as at the National Bison Range, Montana (81). In Yellowstone with the agency mandate for visitor use, a two-part philosophy has evolved. Developed areas such as campgrounds are managed for human safety, and occasionally a temperamental male is killed. Most of the rest of the park is considered bison country, and if people are injured by bison in those areas, the policy is to consider that the injury was provoked by the person. Each incident is evaluated to improve management decisions and to refine the safety information given to the public.

Regardless of the present, future management poses concerns for bison. A large, truly nomadic mammal is not particularly welcome in much of the modern world. Adult male bison approach the 1,000-kilogram criterion for megaherbivores and present increasing conflicts when managed as large, free-ranging populations (83). Proposals such as that of Popper and Popper for a Buffalo Commons are appealing and have some ecological and economic merit, but must recognize the management conflicts arising from the nomadic and gregarious behavior of bison (84).

HABITAT REQUIREMENTS

The widespread original geographic range of American bison suggests that the species can adapt to a wide range of habitat conditions. Basically, bison need space, sufficient graminoids, and surface water. Especially during summer, bison use vertical structures to rub, although fence posts and utility poles seem to work as well as trees and shrubs (85). As described in the previous section, space is the most difficult component to provide. An adequate amount and diversity of prairie grasses must be managed through prescribed fires as described in the following section.

HABITAT MANAGEMENT

Fire

Bison are dependent on grasses and grasses in turn require periodic fires to reduce competition from invading trees and shrubs. Thus, in most habitats, bison managers will develop, maintain, and evaluate prescribed fires. Season of burns and the size and spatial distributions of burns are important variables that managers will have to decide in each specific case. Livestock managers traditionally burn in late winter or early spring to maximize grass production for the growing season. Bison managers may follow this tradition, or they may opt for a mix of spring, summer, and fall burns that more closely mimic presettlement fire occurrences. At the Tallgrass Prairie Preserve, Oklahoma, bison showed preferences of burned over unburned sites regardless of season of burn, but favored summer burns more than fall burns and fall burns more than spring burns (63).

Size and spatial distribution of prescribed burns are important in influencing bison movements and grazing patterns. Bison use of burns at the Tallgrass Prairie Preserve correlated with burn size, but not with interburn distances (63). However, the limited scale of the bison area at the time of the study along with the extensive movements of bison presumably rendered interburn distances irrelevant. At larger scales, interburn distances might become important.

Prairie Dogs

The 19th century bison concentrations mapped by Hornaday overlapped the geographic range of the black-tailed prairie dog (11, 86). Such strong association between presumed competitors suggests that the two species may benefit one another more than they compete.

Modern field studies, primarily at Wind Cave National Park, South Dakota, have provided an understanding of the ecological interactions between bison and prairie dogs. Prairie dog colonies or "towns" attract bison (87). In additive fashion, the combined effects of bison and prairie dog grazing remove most of the aboveground portions of grasses from colony sites (88). Moreover, both species of herbivore differentially make greater use of the edges of colonies than the center (89).

Bison may be attracted to prairie dog colonies for a variety of reasons. During the early growing season, the lack of dead, standing vegetation on the colonies creates favorable conditions for early emergence of grasses. In addition, bison may favor the more open conditions of the colony sites as a means of detecting predators or perhaps because they offer better sites for wallowing (89). The grazing imposed by bison helps keep prairie grasses low enough to prevent most predators from sneaking close enough to burrows to capture prairie dogs.

Other Ungulates

American elk originally had a huge geographic range, much of which overlapped with bison. Early in the 19th century, elk were encountered in the tallgrass regions of the southern plains more frequently than on either mixed grass or shortgrass regions farther west. Bison were encountered least often in the tallgrass region, suggesting some degree of large-scale habitat segregation between the two species (29). More localized habitat segregation between bison and elk has been reported in Elk Island National Park, Alberta, during summer (90). Elk are far less specialized than bison, shifting from grasses to forbs to browse as seasons and local conditions allow (91).

The pronghorn is, in dietary terms, the antithesis of the American bison. While bison consume almost exclusively graminoids, pronghorns eat mainly forbs and browse (91, 92). Thus the two species presumably complement one another under most conditions. Early 19th century travelers on the southern plains encountered pronghorn most commonly in the shortgrass regions and least frequently in tallgrass regions (29).

In terms of diet and habitat preference, white-tailed deer and mule deer do not overlap significantly with bison. Competition between deer and bison appears to be insignificant.

Fencing

The species' physical strength and drive to move require that most modern bison populations be fenced securely. Effective fences range from simple barbed wire augmented with post extenders and an offset hot wire to more elaborate high-tensile fences with alternating strands carrying up to 10,000 volts. The latter style is used at the Konza Prairie Preserve, Kansas, where the bison herd lives close to an interstate highway. Some bison owners reported that a simple high-tensile wire fence was sufficient at small scales even without electrification (93). Most bison breeders, though, favor a 9-gauge woven wire fence at least 1.9 meters high supported by substantial posts (93). Hot wires are recommended for herds living close to population centers or heavily traveled roads.

SUMMARY

The hide hunts of the 1870s and 1880s reduced the American bison population from millions to a few hundred. Restoration efforts began early in the 20th century and by century's end the species exceeded a quarter million with most of those in private ownership.

Bison are gregarious, active, and efficient at finding the grasses and sedges on which they feed. Calving rates range from 50% to 90% of females 3 years of age and older, depending on latitude, habitat conditions, and herd composition.

For most bison populations, active management consists of prescribed burns, containment, vaccination, and population control. Prescribed fire is an increasingly common tool for improving and maintaining bison habitat. Sturdy fences limit movements for most herds. Unconfined bison can create controversy in cattle country, with most of the concern over disease transmission. The most controversial disease, brucellosis, can be transmitted between species, although an actual occurrence has yet to be documented.

Most private herd owners manage bison as they would cattle. If continued, such practices may eventually convert a wild species into another form of livestock. Truly wild bison require large areas and high numbers, conditions that are currently met only in a few public herds.

LITERATURE CITED

1. ROE, F. G. 1970. *The North American Buffalo: A Critical Study of the Species in Its Wild State.* University of Toronto Press, Toronto, Canada.

2. SHAW, J. H. 1995. How many bison originally populated western rangelands? *Rangelands* 17:148–150.

3. MCHUGH, T. (with the assistance of Victoria Hobson) 1972. *The Time of the Buffalo.* © 1972 by Tom McHugh and Victoria Hobson. Reprinted by permission of Alfred A. Knopf, Inc., New York.

4. FLORES, D. 1991. Bison ecology and bison diplomacy: The southern plains from 1800 to 1850. *Journal of American History* 78:465–485.

5. SETON, E. T. 1929. *Lives of Game Animals,* 4 vol. Doubleday, Doran and Company, Garden City, NJ.

6. HAINES, F. 1970. *The Buffalo: The Story of American Bison and Their Hunters from Prehistoric Times to the Present.* Thomas Y. Crowell Co., New York.

7. DARY, D. A. 1974. *The Buffalo Book: The Full Saga of the American Animal.* Swallow Press, Chicago, IL.

8. GEIST, V. 1996. *Buffalo Nation: History and Legend of the North American Bison.* Voyageur Press, Stillwater, MN.

9. PARKMAN, F. 1945. *The Oregon Trail.* Little, Brown & Company, Boston.

10. MARTIN, C. 1973. *The Saga of the Buffalo.* Hart Publishing Co., New York.

11. HORNADAY, W. T. 1989. The extermination of the American bison, with a sketch of its discovery and life history. Annual Report (1887). Smithsonian Institution, Washington, DC.

12. SHAW, J. H. 1996. Bison. Pages 227–236 *in* P. R. Krausman, ed., *Rangeland Wildlife.* Society for Range Management, Denver, CO.

13. CODER, B. 1975. The national movement to preserve the American buffalo in the United States and Canada between 1880 and 1920. Dissertation. Ohio State University, Columbus.

14. MEAGHER, M. M. 1973. *The Bison of Yellowstone National Park.* Science Monograph Series 1. National Park Service, Washington, DC.

15. MEAGHER, M. M. 1978. Bison. Pages 123–133 *in* J. L. Schmidt and D. L. Gilbert, eds., *Big Game of North America: Ecology and Management.* Stackpole Books, Harrisburg, PA.

16. LINNAEUS, C. 1758. *Systema Naturae per Regna Tria Naturae, Secundum Classes, Ordines, Genera, Species, cum Characteribus, Differentiis, Synonymis,* 10th ed. Laurentii Salvii, Stockholm, Sweden.

17. MEAGHER, M. M. 1986. *Bison bison. Mammalian Species* 266:1–8.

18. WILSON, D. E., and D. M. REEDER. 1993. *Mammal Species of the World: A Taxonomic and Geographic Reference.* Smithsonian Institution Press, Washington, DC.

19. ALLEN, J. A. 1876. The American bisons, living and extinct. Memoirs Museum Comparative Zoology IV, No. 10. Cambridge, MA.

20. SKINNER, M. F., and O. C. KAISEN. 1947. The fossil *Bison* of Alaska and preliminary revision of the genus. *Bulletin of the American Museum of Natural History* 89:123–256.

21. GEIST, V. and P. KARSTEN. 1977. The wood bison (*Bison bison athabasca* Rhoads) in relation to hypotheses on the origin of the American bison (*Bison bison* Linnaeus). *Zeitdschrift fuer Saugetierkunde* 42:119–127.

22. VAN ZYLL DE JONG, C. G. 1986. A systematic study of recent bison with particular consideration of wood bison. Publications in Natural Science Number 6. National Museum of Canada, Ottawa.

23. GEIST, V. 1991. Phantom subspecies: The wood bison *Bison bison "athabasca"* Rhoads 1897 is not a valid taxon, but an ecotype. *Arctic* 44:283–300.

24. STROBECK, C., R. O. POLZIEHN, and R. BEECH. 1993. Genetic relationship between wood and plains bison assayed using mitochondrial DNA sequence. Pages 209–227 *in* R. Walker, ed., *Proceedings of the North American Public Bison Herds Symposium.* Custer State Park, Custer, SD.

25. POLZIEHN, R. O., R. BEACH, J. SHERATON, and C. STROBECK. 1996. Genetic relationships among North American bison populations. *Canadian Journal of Zoology* 74:738–749.

26. REYNOLDS, H. W., R. D. GLAHOLT, and A. W. L. HAWLEY. 1982. Bison. Pages 972–1007 *in* J. A. Chapman and G. A. Feldhamer, eds., *Wild Mammals of North America.* The Johns Hopkins University Press, Baltimore, MD.

27. MCDONALD, J. N. 1981. *North American Bison: Their Classification and Evolution.* University of California Press, Berkeley and Los Angeles.

28. BOTKIN, D. B. 1994. *Our Natural History: The Lessons of Lewis and Clark.* G. P. Putnam's Sons, New York.

29. SHAW, J. H., and M. LEE. 1997. Relative abundance of bison, elk, and pronghorn on the southern plains, 1806–1857 (Memoir 29). *Plains Anthropologist* 42:163–172.

30. SCHWARTZ, C. C., and J. E. ELLIS. 1981. Feeding ecology and niche separation in some native and domestic ungulates on the shortgrass prairie. *Journal of Applied Ecology* 18:343–353.

31. PEDEN, D. G., G. M. VAN DYNE, R. W. RICE, and R. M. HANSEN. 1974. The trophic ecology of *Bison bison* L. On shortgrass plains. *Journal of Applied Ecology* 11:489–497.

32. PLUMB, G. E., and J. L. DODD. 1993. Foraging ecology of bison and cattle on a mixed prairie: Implications for natural area management. *Ecological Applications* 3:631–643.

33. COPPEDGE, B. R. 1996. Range ecology of bison on tallgrass prairie in Oklahoma. Dissertation. Oklahoma State University, Stillwater.

34. REYNOLDS, H. W., and D. G. PEDEN. 1987. Vegetation, bison diets, and snow cover. Pages 39–44 *in* H. W. Reynolds and A. W. L. Hawley, eds., *Bison Ecology in Relation to Agricultural Development in the Slave River Lowlands, NWT.* Occasional Papers Number 63. Canadian Wildlife Service, Edmonton, Alberta.

35. HAUGEN, A. O. 1974. Reproduction in the plains bison. *Iowa State Journal of Research* 49:108.

36. GREEN, W. C. H., and A. ROTHSTEIN. 1993. Persistent influences of birth date on dominance, growth and reproductive success in bison. *Journal of Zoology* 230:177–186.

37. BERGER, J. 1992. Facilitation of reproductive synchrony by gestation adjustment in gregarious mammals: A new hypothesis. *Ecology* 73:323–329.

38. RUTBERG, A. T. 1986. Lactation and fetal sex ratios in American bison. *American Naturalist* 127:89–94.

39. VAN CAMP, J., and G. W. CALEF. 1987. Population dynamics of bison. Pages 21–24 *in* H. W. Reynolds and A. W. L. Hawley, eds., *Bison Ecology in Relation to Agricultural Development in the Slave River Lowlands, NWT.* Occasional Paper Number 63. Canadian Wildlife Service, Edmonton, Alberta.

40. FULLER, W. A. 1961. The ecology and management of bison the American bison. *Extrait de la Terre et la Vie* 2:286–304.

41. SHAW, J. H., and T. S. CARTER. 1989. Calving patterns among American bison. *Journal of Wildlife Management* 53:896–898.

42. HALLORAN, A. F. 1968. Bison (Bovidae) productivity on the Wichita Mountains Wildlife Refuge. *Southwestern Naturalist* 13:23–26.

43. McHUGH, T. 1958. Social behavior of the American buffalo (*Bison bison bison*). *Zoologica* 43:1–40.

44. BERGER, J., and C. CUNNINGHAM. 1994. *Bison: Mating and Conservation in Small Populations.* Methods and Cases in Conservation Science Series. Columbia University Press, New York.

45. MAHER, C. R., and J. A. BYERS. 1987. Age-related changes in reproductive effort of male bison. *Behavioral Ecology and Sociobiology* 21:91–96.

46. FULLER, W. A. 1966. The biology and management of bison of Wood Buffalo National Park. *Wildlife Management Bulletin Series* 1(16):1–52.

47. CALEF, G. W., and J. VAN CAMP. 1987. Seasonal distribution, group size and structure, and movements in bison herds. Pages 15–20 *in* H. W. Reynolds and A. W. L. Hawley, eds., *Bison Ecology in Relation to Agricultural Development in the Slave River Lowlands, NWT.* Occasional Paper Number 63. Canadian Wildlife Service, Edmonton, Alberta.

48. SHAW, J. H., and T. S. CARTER. 1988. Long-term associations between bison cows and their offspring: Implications for the management of closed gene pools. Pages 50–55 *in* J. Malcomb, ed., *North American Bison Workshop.* U.S. Fish and Wildlife Service and Glacier National Historical Assoc., Missoula, MT.

49. KOMERS, P. E., F. MESSIER, and C. C. GATES. 1993. Group structure in wood bison: Nutritional and reproductive determinants. Canadian. *Journal of Zoology* 71:1367–1371.

50. BARSNESS, L. 1977. *The Bison in Art: A Graphic Chronicle of the American Bison.* Northland Press and the Amon Carter Museum of Western Art, Fort Worth, TX.

51. LOTT, D. F. and S. C. MINTA. 1983. Random individual association and social group instability in American bison (*Bison bison*). *Zietschrift Tierpschololgia* 43:153–172.

52. VAN VOREN, D. 1983. Group dynamics and summer home range of bison in southern Utah. *Journal of Mammology* 63:329–332.

53. GREEN, W. C. H., J. G. GRISWOLD, and A. ROTHSTEIN. 1989. Post-weaning associations among bison mothers and daughters. *Animal Behavior* 38:847–858.

54. COPPEDGE, B. R., T. S. CARTER, J. H. SHAW, and R. G. HAMILTON. 1997. Agonistic behaviour associated with orphan bison (*Bison bison* L.) calves released into a mixed resident population. *Applied Animal Behavioral Science* 55:1–10.

55. MEAGHER, M. M. 1989. Range expansion by bison of Yellowstone National Park. *Journal of Mammalogy* 70:670–674.

56. MEAGHER, M. M. 1989. Evaluation of boundary control for bison of Yellowstone National Park. *Wildlife Society Bulletin* 17:15–19.

57. SHAW, J. H., and T. S. CARTER. 1990. Bison movements in relations to fire and seasonality. *Wildlife Society Bulletin* 18:426–430.

58. CARBYN, L. N., S. M. OOSENBRUG, and D. W. ANIONS. 1993. Wolves, bison and the dynamics related to the Peace-Athabasca Delta in Canada's Wood Buffalo national Park. Circumpolar Research Series Number 4. Canadian Circumpolar Institute, University of Alberta, Edmonton.

59. SINGER, F. J., W. Schreier, J. Oppenheim, and E.O. Garton. 1989. Drought, fires, and large mammals. *BioScience* 39:716–722.

60. COPPOCK, D. L., and J. K. DETLING. 1986. Alteration of bison and black-tailed prairie dog grazing interactions by prescribed burning. *Journal of Wildlife Management* 50:452–455.

61. VINTON, M. A., D. C. HARTNETT, E. J. FINCK, and J. M. BRIGGS. 1993. Interactive effects of fire, bison (*Bison bison*) grazing, and plant community composition in tallgrass prairie. *American Midland Naturalist* 129:10–18.

62. PEARSON, S. M., M.G. Turner, L. L. Wallace, and W. H. Romme. 1995. Winter habitat use by large ungulates following fire in northern Yellowstone National Park. *Ecological Applications* 5:744–755.

63. COPPEDGE, B. R. and J. H. SHAW. 1998. Bison grazing patterns on seasonally burned tallgrass prairie. *Journal of Range Management* 51:258–264.

64. WALLACE, L. L., M. G. TURNER, W. H. ROMME, and Y. WU. 1993. Bison and fire: Landscape analysis of ungulate response to Yellowstone's fires. Pages 79–120 *in* R. Walker, ed., *Proceedings North American Public Bison Herd Symposium.* Custer State Park, Custer, SD.

65. GATES, C. C., B. T. ELKIN, and D. C. DRAGON. 1994. Investigation, control and epizootiology of anthrax in a geographically isolated, free-roaming bison population in northern Canada. *Canadian Journal of Veterinary Research* 59:256–264.

66. FRASER, C. M., ed. 1991. *The Merck Veterinary Manual.* Merck and Co. Rahway, NJ.

67. MEAGHER, M., and M. E. MEYER. 1994. On the origin of brucellosis in bison of Yellowstone National Park: A review. *Conservation Biology* 8:645–653.

68. MEYER, M. E., and M. MEAGHER. 1995. Brucellosis in free-ranging bison (*Bison bison*) in Yellowstone, Grand Teton, and Wood Buffalo National Parks: A review. *Journal of Wildlife Diseases* 31:579–598.

69. DAVIS, D. S., J. W. TEMPLETON, T. A. FICHT, J. D. WILLIAMS, J. D. KOPEC, and L. G. ADAMS. 1990. *Brucella abortus* in captive bison. I. Serology, bacteriology, pathogenesis, and transmission to cattle. *Journal of Wildlife Diseases* 26:360–371.

70. DAVIS, D. S., J. W. TEMPLETON, T. A. FICHT, J. D. HUBER, R. D. ANGUS, and L. G. ADAMS. 1991. *Brucella abortus* in bison. II. Evaluation of Strain 19 vaccination of pregnant cows. *Journal of Wildlife Diseases* 27:258–264.

71. DAVIS, D. S. 1993. Summary of bison/brucellosis research conducted at Texas A and M University 1985–1993. Pages 347–359 *in Proceeding of the North American Public Bison Herds Symposium.* Custer State Park, Custer, SD.

72. MEYER, M. E., and M. MEAGHER 1995. Brucellosis in captive bison. *Journal of Wildlife Diseases* 31:106–110.

73. TESSARO, S. V. 1989. Review of the diseases, parasites and miscellaneous pathological conditions of North American bison. *Canadian Veterinary Journal* 30:416–422.

74. TAYLOR, S. K., A. C. S. WARD, D. L. HUNTER, K. GUNTHER, and L. KORTGE. 1996. Isolation of *Pasturella* spp. from free-ranging American bison. *Journal of Wildlife Diseases* 32:322–325.

75. TAYLOR, S. K., V. M. LANE, D. L. HUNTER, K. G. EYRE, S. KAUFMAN, S. FRYE, and M. R. JOHNSON. 1997. Serologic survey for infectious pathogens in free-ranging American bison. *Journal of Wildlife Diseases* 33:308–311.

76. ZARNKE, R. L., and G. A. ERICKSON. 1990. Serum antibody prevalence of parainfluenza 3 virus in a free-ranging bison (*Bison bison*) herd from Alaska. *Journal of Wildlife Diseases* 26:416–418.

77. THORNE, E. T., N. KINGSTON, W. R. JOLLEY, and R. C. BERSTROM. 1982. *Diseases of Wildlife in Wyoming,* 2nd ed. Wyoming Game and Fish Department, Cheyenne.

78. VAN VUREN, D., and C. A. SCOTT. 1995. Internal parasites of sympatric bison, *Bison bison,* and cattle, *Bos taurus. Canadian Field Naturalist* 109:467–469.

79. BURGER, J. F. and J. R. ANDERSON. 1970. Association of the face fly, *Musca autumnalis,* with bison in western North America. *Annals Entomological Society of America* 63(3):635–639.

80. SHUPE, J. L., A. E. OLSON, H. B. PETERSON, and J. B. LOW. 1984. Fluoride toxicosis in wild ungulates. *Journal of the American Veterinary Medical Association* 185:1295–1300.

81. WALKER, R. 1993. An update on the bison management program in Custer State Park. Pages 200–205 *in* R. Walker, ed., *Proceedings of the North American Public Bison Herds Symposium,* Custer State Park. Custer, SD.

82. GELLATLY, S., J. CHISHOLM, and D. HUFF. 1997. Consensus based research to assist with bison management in Wood Buffalo. Abstract *in Symposium on Bison Ecology and Management in North America.* Bozeman, MT.

82a. MEYER, M. E., and M. MEAGHER. 1998. Recent changes in bison numbers and distribution. Pages *in* L. Irby, ed., *Proceedings of the Symposium on Bison Ecology and Management in North America,* Bozeman, MT.

83. OWEN-SMITH, R. N. 1988. *Megaherbivores: The Influence of Very Large Body Size on Ecology.* Cambridge Studies in Ecology Series, Cambridge University Press, Cambridge.

84. POPPER, F. J., and D. E. POPPER. 1993. The future of the Great Plains. Pages 313–343 *in* R. E. Walker, ed., *Proceedings of the North American Public Bison Herds Symposium.* Custer State Park, Custer, SD.

85. COPPEDGE, B. R. and J. H. SHAW. 1997. Effects of horning and rubbing behavior by bison (*Bison bison*) on woody vegetation in a tallgrass prairie landscape. *American Midland Naturalist* 138:189–196.

86. KNOPF, F. L. 1996. Prairie legacies—birds. Pages 135–148 *in* F. B. Sampson and F. L. Knopf eds., *Prairie Conservation: Preserving North America's Most Endangered Ecosystem.* Island Press, Washington, DC.

87. COPPOCK, D. L., J. E. ELLIS, J. K. DETLIN, and M. I. DYER. 1983. Plant–herbivore interactions in a North American mixed grass prairie. II. Responses of bison to modification of vegetation by prairie dogs. *Oecologia* 56:10–15.

88. CID, M. S., J. K. DETLING, A. D. WHICKLER, and M. A. BRIZUELA. 1991. Vegetational responses of a mixed-grass prairie site following exclusion of prairie dogs and bison. *Journal of Range Management* 44:100–105.

89. KRUEGER, K. 1986. Feeding relationships among bison, pronghorn, and prairie dogs: An experimental analysis. *Ecology* 67:76–770.

90. TELFER, E. S., and A. CAIRNS. 1979. Bison-wapiti interrelationships in Elk Island National Park, Alberta. Pages 114–121 *in* M. S. Boyce and L. D. Hayden-Wing, eds., *North American Elk: Ecology, Behavior, and Management.* University of Wyoming, Laramie.

91. WAGNER, F. H. 1978. Livestock grazing and the livestock industry. Pages 121–145 *in* H. P. Brokaw, ed., *Wildlife and America.* Council on Environmental Quality, Washington, DC.

92. HARTNETT, , D. C., A. A. STEUTER, and K. R. HICKMAN. 1996. A comparative ecology of native and introduced ungulates. Pages 72–101 *in* F. L. Knopf and F. B. Samson, eds., *Ecology and Conservation of Great Plains Vertebrates.* Ecological Studies 125. Springer, New York.

93. JENNINGS, D. C., and J. HEBBRING. 1983. *Buffalo Management and Marketing.* National Buffalo Association. Custer, SD.

CHAPTER

23

Mountain Goat

James M. Peek

INTRODUCTION

Mountain goats fill the most precipitous habitat niche of the North American ungulates: the steep, rugged mountains and cliffs. While mule deer and bighorn sheep will occupy the steep crags, the steepest slopes and cliffs provide the mountain goat with habitat that the other species do not use consistently through the annual cycle of seasons. However, mountain goats successfully occupy less precipitous terrain, such as the Suckling Hills east of the Copper River in Alaska. And they can occupy lower elevations, such as the introduced population on the cliffs south of Pend O'Reille Lake in northern Idaho, the steep canyon walls along the Salmon River in central Idaho, and the Black Hills in South Dakota. Steep terrain, regardless of elevation, characterizes the habitat for this species.

Mountain goats tend to remain in steep terrain in winter, although they may move to lower elevations, especially in periods with deep snow. Cliffs were the primary factor determining mountain goat habitat use in the northcentral Cascades in Washington (1); this is the case across their range. For this reason, goats tend to remain isolated in winter and are not as amenable to study as their mule deer and bighorn sheep associates. Aerial search is less regular in the high mountains in winter, and opportunities to study goats from the ground during the colder months are reduced. Even in summer, goats are relatively inaccessible to the ground observer and extensive planning and logistical support are necessary to conduct any sustained investigations. Nevertheless, information has slowly been accumulating since Rideout and Wigal and Coggins summarized the available work, and Chadwick produced a popularized review of this uniquely North American species (2, 3, 4).

TAXONOMY AND DISTRIBUTION

The genus *Oreamnos* is found in the mountainous portions of western North America and is now represented by the mountain goat (5). There is an extinct species, Harrington's mountain goat, whose fossil remains have been found in caves in the Southwest. It was smaller than the extant species (6). Mountain goats are rupicaprids, or goat-antelopes, which means that its nearest relatives are the chamois of the European Alps, and the goral, serow, takin, and tahr of Asia (7, 8). Rupicaprids are paleontologically older and ancestral to Pleistocene caprids, the sheep and goats (9).

The distribution of mountain goats (Figure 23–1) includes southeastern Alaska, southern Yukon, southwestern Northwest Territory to northern Washington, central Idaho and Montana, with introductions to Kodiak, Chichagof, and Baranof Islands, Alaska; the

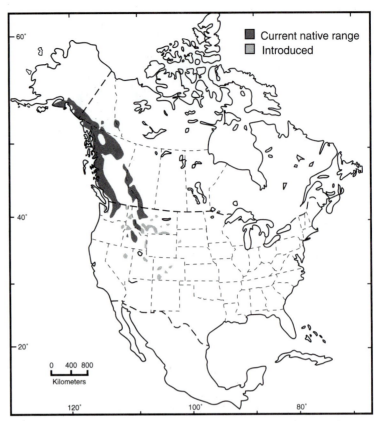

Figure 23–1 Distribution of the mountain goat. (After Ref. 3.)

Olympic Peninsula, Washington; southcentral Montana; the Ruby Mountains, Nevada; northeastern Oregon; the Black Hills, South Dakota; and Colorado (10). Allen recognized differences in size and skull width among individuals in British Columbia, Washington, Idaho, and Montana but Cowan and McCrory concluded that there were no valid reasons for recognizing subspecies among the current populations (11, 12).

LIFE HISTORY

The sexes are monomorphic, but individual animals may be recognized through their unique facial characteristics and horns (13). Horns are up to 30.5 centimeters long in both sexes, black and slightly curved, with males tending to have thicker but not necessarily longer horns than females (14). Males and females exhibit similar patterns of horn growth (15, 16) (Figure 23–2). Annual rings on the horns can be identified sufficiently to be used to indicate age up to approximately 6 years, after which underestimates are likely (17). Cote and others reported that most of the growth of horns occurs in the first 3 to 4 years of age (16).

The hoofs of the mountain goat have a heavy sponge-like convex pad protruding outside the horny exterior walls, which would facilitate maneuverability on the smooth surfaces of cliffs (18). The points are relatively sharp, which also helps footing.

Pelage is a white, thick wool with guard hairs that are shed during summer. Adults, particularly lactating females, may retain vestiges of the previous year's coat into fall, but most shedding is completed by mid-July for males and mid-August for females (19). The pelage provides protection down to −20°C at which temperature

Life History

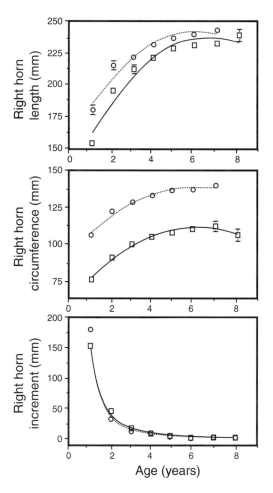

Figure 23–2 Relationship of horn length, circumference, and growth increment (mm) to age of mountain goats. Squares and continuous lines represent females and circles and broken lines represent males. (After Ref. 16.)

oxygen consumption begins to increase (20). At lower temperatures, goats may reduce activity and lie down in the insulating snow to reduce heat loss. Beards, dorsal ruffs, and pantaloons facilitate various forms of agonistic behaviors. The skin has dermal shields, or thicker hide, on the haunches and rump where horn wounds caused by conspecifics most often occur (21).

Males tend to be 10% to 30% larger than females (Table 23–1), with records of more than 136 kilograms (7). Races in the northern part of the range show greater differences in size between sexes than more southerly races (12). Rideout and Worthen presented a useful equation to estimate mass of mountain goats as log mass=1.2574(log length + log girth) −2.1531, with length and girth measurement in inches and mass in pounds (22). Body mass may be predicted from horn length and basal horn circumference for both sexes (15, 16) (Figure 23–3). Lentfer reported mass of 12 kids in June ranging from 2.9 to 7.3 kilograms in the Crazy Mountains of Montana (23). Hutchins and others reported a birth mass of 3.2 kilograms for a captive kid (24). Kids follow their dams shortly after birth and are not hidden and left. The growth patterns of mountain goats in captivity presented by Houston and others reflect the natural situation (25). Steep increases in body mass occur from July to December, then decline through winter (Figure 23–4). Body mass does not change appreciably for males after 5 years and for females after 4 years of age. Cote and others concluded that since the horn length of males reaches an asymptote by 3 to 4 years of age and body mass continues to increase beyond that, body mass may be more important than horn size in determining social rank and reproductive status (16).

TABLE 23–1
Weights and Total Length of Mountain Goats

Location	Sex	Age	No.	Length (cm)	Mass (kg)	Source and Comments
Alaska	Both (Jun)	1	3		38.4	44a
(Kenai)	Both (Aug)	1	2		50.4	Mass only
	F (Jun)	2+	10		52.9	
	F (Aug)	2+	12		73.1	
	M (Jun)	2+	2		70.3	
	M (Aug)	2+	3		97.8	
Alberta	Both (Jul)	4+	135		110.0	16
(Caw Ridge)						Mass only
Idaho	M	Birth	1	55.9	2.9	19
	M	1	2	95.3	19.9	
	F	1	1	99.1	19.1	
	M	4+	5	153.7	69.9	88.0 max. mass
	F	4+	6	141.0	53.1	58.1 max. mass
Montana	M	Kid	1	90.8	16.0	2
	F	Kid	1	85.1	15.0	
	M	1	2	119.4	33.6	
	F	1	2	118.2	32.4	
	M	4+	3	153.9	68.6	
	F	4+	3	141.0	56.5	
Montana	F	13 months	3	114.3	30.4	23
(Crazy	M	13 months	5	120.1	34.9	
Mountains)	F	2	2	142.2	44.5	
	M	2	2	143.5	46.7	
	F	4+	5	154.8	71.4	
	M	4+	4	178.6	81.9	95.3 max. mass
Washington	M	3 weeks	1	13.6		64
(Chopaka	F	3 weeks	1	6.4		Mass only
Mountain)	M	6	2	93.4	109.8	Max. mass
	F	6–10	3	72.1	75.7	Max. mass

Brandborg reported breeding in November through early December, with parturition dates ranging from 15 May to 27 June (19). The age at which breeding is initiated for both sexes depends on body size, which is a function of forage and weather conditions. Lentfer reported presence of a corpus luteum in a 27-month-old female, suggesting a minimum breeding age of 2.5 years for females in Montana (23). Houston and Stevens reported yearling breeding when goats were still colonizing the Olympic Range in Washington, but subsequently Houston and others reported that breeding was delayed until 2 years of age (26, 27). Houston and others reported that goats weighing more than 50 kilograms were lactating while goats below that mass consistently were not (25). The minimum breeding age was 3.5 years at Caw Ridge, Alberta, and in the Black Hills, South Dakota (28, 29). Houston and others reported that measurement of bovine pregnancy-specific protein with bovine antisera provided a reliable diagnosis of pregnancy detection in mountain goats (30). Presence of spermatozoa was confirmed in yearling males by Henderson and O'Gara, but Geist found males participating in rutting activities initially at 2.5 years (31, 13). Twinning is common when conditions permit. Zoo records suggest a gestation period of approximately 186 days (24).

Mountain goats have 32 teeth. The upper/lower dental formula is incisors, 0/3; canines, 0/1; premolars, 3/3; and molars, 3/3. Smith considered the longevity of mountain goats to be related to tooth wear and loss (32). A collection of skulls from the Bitterroot Mountains revealed the oldest male was 15.5 years and the oldest female was 13.5 years. An examination of 165 skulls by Cowan and McCrory revealed 14- and 18-year-old females (12). Brandborg

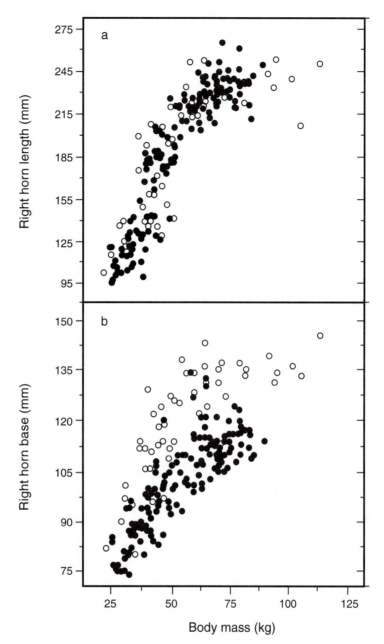

Figure 23–3 Relationship of horn length (mm) and basal circumference (mm) with body mass (kg) of mountain goats at Caw Ridge, Alberta (16). Open circles represent males and filled circles represent females. The equations are: male weight = 0.34 horn length −9.19; female weight = 0.30 horn length −1.68; male weight = 0.85 horn base circumference −45.4; female weight = 1.13 horn base circumference −58.99.

and Geist reported 13-year-old males (19, 13). Chadwick concluded that mountain goats living beyond 10 years were nearing the end of life expectancy (4).

Criteria used to determine sex of mountain goats in the field include observation of genitalia, urination posture, and horn morphology (33). In summer, male genitalia are visible but the longer winter pelage obscures the scrotum. Male goats stretch out when urinating while female goats squat. Male horns are gradually curved through their length while the female horn tends to be straight.

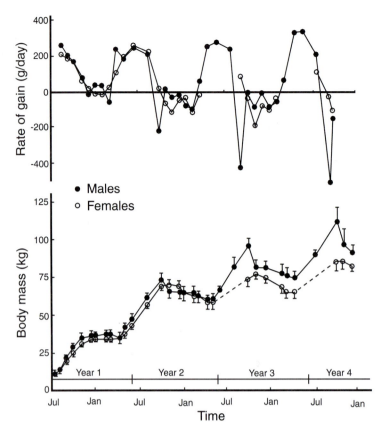

Figure 23–4 Growth of mountain goats in captivity. Weights are means ± 1 standard deviation for four males and six females. (After Ref. 25.)

BEHAVIOR

The mountain goat retains behavior patterns that Geist considered primitive, related to its rupicaprid ancestry and to the habitat it evolved in (13, 34). Rupicaprids, being an ancient group of bovines, have relatively simple horn configurations, less sexual dimorphism, smaller group sizes, and primitive behaviors that relate to occupancy of difficult terrain with naturally fragmented resources in terms of distribution of suitable cover and forage. More social species tend to occur in more open habitats with less patchy distributions of forage and less severe climate, where the herd tends to replace terrain and vegetation as cover from predators, and interactions between individuals are facilitated. Goats tend to occur in small groups ordered along dominance hierarchies. The mean group size was 6.3 goats in summer in British Columbia (35). Mature males tended to spend much of the year alone or with another male. Holmes found that the larger groups of females had positive or neutral effects on foraging time for females but negative effects on yearling males due to increased alertness of the younger animals (35a). This was thought to be a cause of segregation of males from females. Houston and others reported mean group sizes of 2.1 to 2.8 goats, with a range of 1 to 32 in Olympic National Park (27). Groups that Lentfer observed ranged up to 40 in late August (23). Most goats were observed in smaller groups that tended to be larger in winter when range was limited. Casebeer and others reported most groups of goats on native range along the continental divide and Bitterroot Mountains in Montana were less than 5, with highest groups in winter (18). About 50% of all mountain goats Chadwick observed in Glacier Park, Montana, were solitary animals, and an additional 25% were pairs (36). The larger groups occurred on more moderate slopes, which would facilitate the coordination of interactions between conspecifics.

Masteller and Bailey reported that groups appeared to be persistent, dominated by older females that received priority access to resources (37). Group sizes are expected to vary between populations, seasons, habitat, and population density, and generally appear to be loose-knit, except for the female–kid association.

Geist postulated that over evolutionary time, selection favored males with thick hides, great tendency to display before other males, and an inhibition to strike, which favors a social system that reduces potential for injury to conspecifics by males, and creates a matriarchal dominance hierarchy (13). Adult females are considered dominant to all other sex–age classes, and aggression among adult females is more intense than for bighorn sheep and chamois (38). Dominance hierarchies were nonlinear, and reversals in rank occurred from one year to another and within years for 12% to 14% of interactions. In some cases, younger females defeated older ones and in others the reverse occurred. The best cue for explaining dominance was horn length in one year and body mass in another year. Age was correlated with dominance rank, with about two-thirds of interactions between goats of different ages being won by older animals. Reproductive status did not influence rank. There was also no relationship between dominance rank and foraging efficiency or amount of time goats were alert. Fournier and Festa-Bianchet concluded that dominance was worth fighting over, which suggests that benefits of aggressive behavior may be greater for females than for other ungulates where fights among females are infrequent (38). The reasons for this are unclear, but may be related to disposition of forage resources and security from predation (39).

Males attempt to minimize injury to conspecifics, using elaborate behavioral displays. Four threat displays have been described for the mountain goat (13). The rush threat is poorly developed, but involves a walk or trot toward a potential opponent with upward swipes of the horns, usually toward the side of the opponent. The attempt to engage an opponent's flanks is considered a more primitive form of behavior than the head-on engagements typical of deer, mountain sheep, and the African antelopes (40). The weapon threat involves lowering the head and aiming the horns at the opponent, with a sweep of the horns in a half-circle upward. The present threat is used by all sexes and ages, and consists of presenting the broadside to the opponent, back arched, head drawn down, and legs stretched straight. Females typically attack conspecifics with a rush, a horn-threat, and sometimes a plunge into the opponent. Males move more slowly, then typically arch into a present-threat posture and approach on a tangent to the opponent. Usually opponents go into a present-threat posture when confronted and occasionally will withdraw. Goats mark vegetation with secretions from their supraoccipital glands. Geist reported that marking was most obvious during agonistic encounters and may serve to intimidate conspecifics (13).

Courtship displays involve the male pawing a pit, then approaching a female from a low crouch, neck and head extended, tail raised, ears forward and nostrils expanded, tongue flickering, in a low-stretch posture. The male may lick the flank or below the tail of the female and tap her on the flank or between the haunches, before mounting. Females that are not receptive may turn and plunge their horns into the male, so the deliberate courtship is an attempt to reduce risk of injury and aggressiveness by the female.

The presence of horns on females, relative lack of sexual differentiation in appearance, intensely agonistic social system, aggressiveness of females, prominence of displays directed toward the body rather than the head, and generally small group sizes probably are all fundamental adaptations to this species' rugged habitat. Other species that evolved in more open habitat or less precipitous habitat where resources were more evenly distributed show less antagonism in their social systems, cephalization of displays, reduced aggression of females, and increased group sizes. The rupicaprid ancestry predicts retention of primitive patterns, facilitated by occupation of this especially precipitous habitat.

Diurnal activity of mountain goats includes combinations of resting, feeding, vigilance, moving, and social activities (41). Feeding and resting were the main activities of adult females, subadults, and kids in summer (Figure 23–5). Feeding occurs throughout

No. individuals (%)

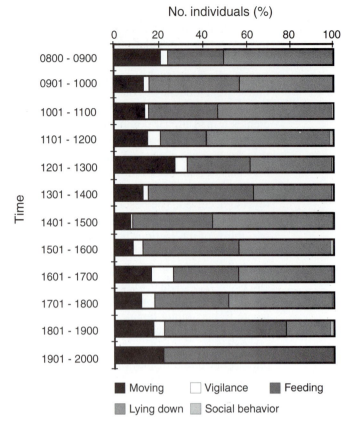

Figure 23–5 Diurnal activity rhythms of adult female mountain goats and subadults of both sexes. (After Ref. 41.)

the daylight hours, but appears to increase in late afternoon. Movement occurs primarily in early morning, midday, and late afternoon. During summer, social interactions between adults and subadults are rare but kids are particularly active at play in late afternoon.

Movements and Home Range

Mountain goats, especially females, show strong site fidelity to established seasonal ranges. In Olympic National Park, 84% of 125 marked goats returned to the same summer ranges each year, some for 5 to 10 years (27). Smith reported home range size did not change much after 2 years of monitoring (42). In some instances seasonal ranges are separated from each other and in other instances goats remain in the same area year-round. Schoen and Kirchoff reported that two males used different seasonal ranges, two males used the same range one year and not the next, and one male used overlapping seasonal ranges in a southeast Alaska study (43). Home ranges tend to be relatively small, but small sample sizes and methods of calculation that tend to underestimate size probably provide minimum estimates (Table 23–2). The extremely high relief and steep topography that add to size are difficult to calculate in most situations, and further produce minimal estimates (44).

Nichols (44a) observed variation in movements and fidelity to summer and winter ranges of goats using the same range among 34 radiocollared goats (45). Some goats used small home ranges year-round, whereas others migrated more than 24 kilometers between winter and summer ranges. In winter when deep snow restricted movements, goats were mostly found on lower south-facing slopes near timberline. One high-quality winter range consisted of heavy timber on broken

Behavior

TABLE 23–2

Home Range Comparisons for Adult Mountain Goats

Location	Sex	Season	No.	Size	Reference
East Front, Montana	M	Jun–Nov	2	18-22 km^2	100
	F		1	14 km^2	
Sapphires, Montana	M	Summer–Fall	1	17.6 km^2	100a
	F	Summer–Fall	5	19.8 km^2	
Bitterroots, Montana	M	Winter	1	22 ha	100b
	F	Winter	5	42-64 ha	
Glacier Park, Montana	M	Annual		6.3 km^2	101
	F	Annual		8.9 km^2	
Southeast Alaska	M	Annual	7	44.9 km^2	42
	F	Annual	13	11.7 km^2	

cliffs that provided snow-free forage and escape terrain. Adult males were least predictable in movements, with one moving more than 68 kilometers through deep snow. Another adult male shifted its entire annual range each of the 4 years it was monitored. Goats, especially females, were most consistent in range use from mid-July to late August, while greatest movement and lowest fidelity to an area used the previous season was from mid-September to early November.

Goats in the northern Montana ranges wintered on the same areas used in summer, but on south and west exposures where the prevailing wind exposes vegetation that goats can use (18). In deep snow areas on the Crazy Mountains in Montana, Lentfer reported that goats occurred at higher elevations than during summer, but that wintering areas were typically in drainages with lower snow depths (23). Brandborg observed very limited movements on winter ranges, with one band of approximately 10 animals remaining on about 81 hectares from February to May in the Salmon River area (19). Summer trips ranged from a few hundred meters to a half of a kilometer.

Home-range sizes of an introduced Colorado population did not differ between two summers, but summer home ranges were eight times greater than winter home ranges. Summer mean distances were 6.4 kilometers with no differences between sexes, while winter distances were 1.3 kilometers (46). Of significance, winter range size was not limited by deep snow, which suggested that an energy conservation strategy was being used.

Diet

Diet of mountain goats may be expected to vary with season and locale. Saunders described them as "snip-feeders," grazing for short times on any one site (47). Grassy slide-rock areas and ridgetops were primary feeding areas, with alpine meadows receiving intensive use in late summer and early fall when they were free of snow. Houston and others generalized mountain goat diets in five ways (Table 23–3):

1. A remarkable variety of plant species are consumed across the geographic range, including mosses, lichens, ferns, grasses, sedges, rushes, forbs, conifers, deciduous trees, and shrubs.
2. Spring and summer diets usually are dominated by grasses or forbs.
3. Winter diets frequently are dominated by browse, including conifers.
4. The proportions of browse in diets increase sharply during severe winters.
5. Conifers often are important components of winter diets, even during mild conditions when other forages are seemingly available (27).

TABLE 23–3

Important Forage Species in the Diets of Mountain Goats from Across the Range

Location	Season	Species	Reference
Crazy Mountains, Montana	Summer	Tufted hair grass, bluegrasses	47
	Winter	Subalpine fir	
Central Idaho	Winter	Bluebunch wheatgrass, mountain mahogany	19
Colorado	Winter	Sagebrush, sedges, fescues	101a
Olympic Mountains, Washington	Summer	Fescues, sedges	27
	Winter	Phlox, Pacific yew	
Bitterroots, Montana	Winter	Bluebunch wheatgrass	100b
Sawtooth Range, Montana	Summer	Fescues, buffaloberry	100
SE Alaska	Winter	Yellow cedar, western hemlock	52
Kodiak, Alaska	Summer	Sagebrushes, fleabane	102
	Winter	Rough fescue	
Flathead, Montana	Winter	Mountain maple, juniper, sedges,	102a
	Summer	Serviceberry, bromes	
Washington	Winter	Wheatgrass,	103
	Summer	Huckleberry	
Alberta	Summer	Wheatgrass	104
	Winter	Willow, conifers	
British Columbia	Summer	Huckleberry	83
South Dakota	Winter	Lichens, conifers, kinnickinnick	105
Idaho	Winter	Mountain mahogany	105a
	Summer	Cinquefoils	
Montana	Summer	Bluebells, clover	106
South Dakota	Winter	Lichens, bearberry, chokecherry	107
Montana	Summer	Wheatgrass, fescue	108
Colorado	Summer	Wheatgrass, clover	109

Forage species used vary depending on what is available (Table 23–4). On ranges where windblown slopes and ridgetops are available for goats in winter, grasses and sedges provide a high component of the diet. On ranges such as those near the coast, conifers and shrubs that protrude above snow level and occur in conifer stands where goats winter provide most of the diet. Fox and Smith showed that forbs and cryptogams constituted more than 68% of the diet when snow depths were less than 50 centimeters, while trees and arboreal cryptogams constituted 93% of the diet when snow depths were more than 100 centimeters in southeast Alaska (48). Summer foods will consist of succulent forbs and shrubs, with individual species again varying with local situations.

Alpine plants provide high-quality forage, which translates into higher nitrogen content of rumen materials in deer (49). This suggests that mountain goats are feeding on an exceptionally high-quality forage base in summer when high-elevation plants are available. Plant growth at high latitudes and elevations is rapid, sometimes being initiated under snow cover. The short and cool summers with long exposure to solar radiation promote rapidly maturing plants that have higher digestibility and nitrogen levels, and remain higher for long periods due to the longer exposure to the sun (50). Underground parts contain large amounts of carbohydrate reserves, which are rapidly translated into production of new tissue, within a week of initiation of spring growth (51). The heavy precipitation and abundant alpine habitat produce a highly productive summer range for goats in southeastern Alaska that contributes to their high growth rates and consistent late summer attainment of prime body condition (52). These same conditions prevail at high elevations elsewhere, when compared to drier habitats at lower elevations.

TABLE 23–4
Estimates of Forage Class Composition of Mountain Goat Diets

Area (Authority)	Season[a]	Gram inoids	Forbs[b]	Browse[c]	Cryptograms	Diet determination method[d]
Alaska						
(52)	W[e]	4	33	28	35	F
	W	tr	2	40	57	F
Alberta						
(104)	S	63	14			R
British Columbia						
(83)	S[f]	24	23	52		R
	S[g]	74	1	25		R
Colorado						
(110)	W	42	5	51		F
	S, F	43	15	33		R
(111)						
(109)	W	88		12		D, S
Idaho						
(19)	F	54		42		R
	W	23		52	24	D
(105a)	Sp	43	1	56		S
	S	21	79			S
	F	11	89			S
	W	15	10	75		S
Montana						
(18)	S	3	1	96		U
	W	63	2	35		U
(102a)	Sp	97				S
	S	13	54	33		S
	W	75	4	12	9	S
(108)	F	90	6	4		R
	W	25		75		S
(47)	Sp	70	13	14	1	R
	W	59	16	25		R
South Dakota						
(105)	W			30	60	D
Washington						
	W	31.3	3.2		1.2	F
(58)	S	43.3	20.0	64.1	0.6	F
	W	17.7	6.2		8.7	F
(1)	S	60.0	40.0	37.0		F
(27)				66.7		

Source: Modified from Ref. 54.

[a]S, summer; Sp, spring; F, fall; W, winter.

[b]Includes ferns.

[c]Includes shrubs and trees.

[d]D, direct observation of wild animals; S, feeding site examination; R, rumen content analysis; U, plant utilization transects; F, fecal analysis.

[e]First row diet is for snow depths < 50 cm, second row is for snow depths > 100 cm.

[f]Coastal sample.

[g]Interior sample.

Fox and others estimated the daily intake of digestible energy of an adult mountain goat to be 4,980 kilocalories in summer, or approximately 39% more energy than required for daily activities (52). The winter estimate of 1,590 kilocalories reflected a 12% loss in net energy per day. This suggests that the mountain goat is adapted to withstand severe winter conditions with poor nutritional conditions, which may be countered by foraging on high-quality plants found in the highest elevations.

Nutrient content and digestibility of forage species in alpine habitat had 50% more crude protein and 100% more phosphorus than grasses in the fescue association at lower elevations in Alberta (53). Protein contents of forbs at high elevations were similar to that of lower elevation, but again phosphorus content was double that of the lower elevation plants. This area along the southeastern Canadian Rockies is characterized by a 60- to 70-day growing season with killing frosts occurring each month except July. Mean maximum temperatures range from 12° to 15°C and mean minimum temperatures average 5°C, which are probably characteristic of mountain goat habitat. The variation in nutrient content among grasses and grass-likes, forbs, and shrubs is high but indicates that goats can obtain high-quality forage from all three forage categories (Table 23–5). Dailey suggested high percentages of crude protein, with high digestibility even during winter of some shrubs and forbs (54). In southeastern Alaska, yellow cedar produces a digestible forage in winter, and, when available, bunchberry, an evergreen forb, is highly digestible (Table 23–5).

Mountain goats occupying coastal mountain ranges occur in a low selenium environment, selenium being an integral part of the blood enzyme glutathione peroxidase (55). Low selenium levels are linked to a variety of pathologies and nutrition problems, so this nutrient is of interest whenever mountain goats and other species decline and causes are not readily identifiable (56). High activity of this enzyme per unit of selenium in goats reduces their dietary selenium requirement relative to domestic cows and horses, an adaptation to low selenium environments (55).

Concern over potential competition for forage between mountain goats and mountain sheep led Dailey and others to investigate the interactions on Colorado alpine habitat (57). Both species used similar diets during summer but not in winter. Mountain sheep chose more graminoids, whereas mountain goats chose more dicots in winter. Dietary overlap was greatest between mountain goats and mule deer on Chopaka Moun-

TABLE 23–5

Nutrient Content and Digestibility of Mountain Goat Forages in Selected Areas of the Range

Location (Authority)	Season	Species	% Digestibility	% Crude Protein
Southeast Alaska (52)	Winter	Yellow cedar	47.3	6.25
		Blueberry	36.1	9.38
		Bunchberry	75.6	10.6
Alberta (53)	Spring	Grasses	33.9-69.5	12.1-24.4
		Forbs	20.7-68.5	13.5-27.3
		Shrubs	7.6-34.7	9.6-20.7
	Summer	Grasses	37.7-60.8	10.1-18.9
		Forbs	19.5-62.5	8.3-18.8
		Shrubs	12.2-19.8	8.9-15.7
	Fall	Grasses	34.9-56.3	5.4-9.1
Colorado (54)	August	Grasses	59±2.7	12±1.1
		Forbs	59±2.2	16±1.2
		Shrubs	36±14.0	16±7.2
	Winter 1	Grasses	55±1.7	6.7±0.6
		Forbs	45±3.6	4.2±0.8
		Shrubs	26±6.6	8.3±1.0
	Winter 2	Grasses	56±2.2	7.0±0.7
		Forbs	50±3.0	7.2±0.5
		Shrubs	33±3.5	7.0±0.9
Washington (27)[a]	Winter	Grass		3.0
		Forb		7.0
		Conifer		7.5
	Spring	Grass		18.0
		Forb		12.5
		Conifer		15.0

[a]Estimated from Ref. 25's figure 30, grass was Idaho fescue, forb was Phlox, conifer was Pacific yew.

Behavior

tain, Washington, where graminoids constituted a major portion of the year-round diet, but diets of goats, cattle, and mule deer varied considerably between and within seasons (58). Dailey and others concluded that the mixtures of dicots and graminoids used by both species can reduce potential for competition for food between these species (57).

Population Trends

Surveys of mountain goat populations by the managing agencies suggest rather dramatic variations in numbers during the past several decades. In Idaho, goat numbers declined during the late 1950s and early 1960s, again in the 1970s and early 1980s, and again in the early 1990s. These declines were attributed to harvests exceeding recruitment by Kuck, and have brought about reductions in harvests (59). More recently, populations in Idaho generally have been stable or increasing (60). Reductions of up to 90% in southeast Alaska were noted in the 1960–1970 period that were attributed to cyclic weather influences and hunter harvest (61). In Alberta, severe winters plus hunter harvest brought about declines in the Willmore Wilderness in the 1980s, whereas unhunted populations were able to increase or decrease only slightly in other portions of westcentral Alberta (62). The Black Hills population increased from the original transplant of 6 animals from Alberta in 1924 to 300 to 400 goats by 1940, stabilized through the 1970s, and then declined in the 1980s and has fluctuated since (29). British Columbia Fish and Wildlife Branch, Cranbrook, B. C. (unpublished report), documented the decline in goats in the Kootenay region of British Columbia, attributing it to increased access and unrestricted hunting. Subsequently, Hebert and Smith reported population estimates for British Columbia between 30,000 and 65,000 goats, with high production and survival in the 1980s (63). Populations in Washington were hunted at levels higher than recruitment prior to 1981, but surveys have since indicated higher survival and recruitment (64). Goat harvests over the 1955–1984 period declined from 55 goats in 1963 to 3 to 5 goats during 1977–1984 on the Rocky Mountain East Front in Montana, even as permits were reduced (65). Trends in Rocky Mountain goat populations on U.S. national forest lands, projected to be relatively stable at 35,000 to 40,000 over the 1964–1984 period, are questionable: Harvests have been declining during that same period (66).

Population surveys are sufficiently variable in accuracy and precision over space and time for this species to make general population estimates and trends questionable. Populations that receive relatively intensive study and become reasonably well known are likely tracked with reasonable accuracy. Efforts to improve the reliability of surveys are available, but logistics and fiscal considerations affect the effort (45). Reintroductions to previously occupied but vacant habitat, augmentations of small populations, and translocation to new habitat have been used to increase populations.

POPULATION DYNAMICS

Geist predicted that species with relatively small home ranges that remain constant between years should be able to vary their reproductive output opportunistically from one year to the next, depending on environmental conditions; he used the mountain goat as an extreme example (67). Under deep snow conditions, females may become territorial and drive off subordinate goats, tolerating only their own offspring. Low production of young and low survival of subadults result and population declines will occur. When snow conditions moderate, females become more tolerant and production and survival increase. However, individuals remain tied to a specific unit of land.

Survival of goats at Caw Ridge, Alberta, was estimated to be 88% from birth to 3 months, with most deaths occurring in fall (28). Survival to 4 years of age from birth was estimated at 29% for males and 34% for females in that Alberta population. Smith reported that prime-age goats 2 to 8 years old were relatively invulnerable to natural mortality factors during mild Alaskan winters (42). Annual mortality rate was 29% for seven yearlings and 32% for goats greater than 9 years old in southeastern Alaska.

Mortality rates for kids reviewed by Adams and Bailey ranged from 15% to 80% (68). Snow depth on 1 May was negatively correlated with kid : adult ratios observed in the Sawatch Range of Colorado and in southcentral Montana (68, 69). Kid : adult ratios from 71 : 100 to 18 : 100 for 14 years in Colorado encompass the variation observed by Houston and Stevens for 10 years in Olympic National Park, Washington, and probably represent a typical range (68, 26).

Mortality factors include accidents such as falls, and predation by cougars, wolves, and grizzly bears. Goat carcasses occasionally are found at the base of cliffs and snowslides, suggesting accidental death. However, predation may be the major mortality factor, especially where wolves occur with goats (62, 70). Festa-Bianchet and others reported that predation was the known or probable cause of death of goats up to 4 years of age (28). Fox and Streveler reported goat remains in 53% of 78 wolf scats collected within goat range in southeastern Alaska, suggesting that wolf predation was a major mortality factor for mountain goats (71). Goats react to the presence of a predator by seeking out a vertical cliff face or confronting the predator with their horns. While prime escape terrain may provide forage because it sheds snow in winter, at other times goats appear vulnerable enough to cause frequent wolf visits to goat habitat (71). Holroyd, Casebeer and others, and Brandborg reported instances of eagles harassing mountain goats, particularly kids (35, 18, 19).

While mountain goat population declines have been related primarily to overharvest, diseases and pathogens also have been implicated. Hebert and Cowan reported that white muscle disease, resulting from selenium deficiency, was present in goats inhabiting the East Kootenay region of British Columbia (72). These goats were subject to stress attributable to capture activities that likely induced the paralytic symptoms typical of this disease. Because it is known that mountain goats live in a selenium-deficient environment, the question arises as to how they respond to the vigorous exertion that may occur in natural situations as from predators. One may assume that the proximity of escape cover that allows the goat to rapidly escape the predator is one means of reducing exercise to tolerable levels. If this is the case, then in habitats that include predators, or where hunting occurs, cliffs that provide the escape cover for mountain goats should be considered a habitat requirement.

Protozoan parasites in the genus Sarcosystis inhabit muscle tissue. The parasite requires a second host to complete its life cycle; in mountain goat habitat the second host is probably a larger mammalian carnivore. The carnivore eats infected muscle tissue from the goat and contracts the parasite. Foreyt reported this parasite in 24 of 56 mountain goats in Washington (73). Mahrt and Colwell found sarcocyst infections in 11 of 15 goats in Alberta (74). The potential for mortality attributable to sarcocyst infections is unknown.

At least 10 species of endoparasites have been reported from mountain goats. Boddicker and others reported that certain nematodes were serious pathogens in goats in South Dakota (75). Intestinal nematodes have been reported in goats in South Dakota and Idaho, and may be common parasites in Alberta and Montana (76, 77, 78). A mountain goat in emaciated condition necropsied by Foreyt and Leathers was determined to have starved due to a large oral fibroma (79).

Other pathogens have been implicated as important proximate factors causing mortality when adverse winter conditions occur or when goats reach high densities relative to forage resources and nutritional levels in segments of the population decline. There are circumstances where the pathogens may be direct causes of mortality, but the proximate factor alternative likely portrays the most common situation where pathogens are associated with mortality. Dunbar and others reported respiratory syncytial virus, which predisposes a number of respiratory diseases, was isolated from mountain goats in Washington (80). Parainfluenza-3 virus, pasteurella, and bovine viral diarrhea also were found in goats (19). Samuel and others found 17 helminths (12 nematodes and 5 cestodes) and one species of tick in 53 goats from the Willmore Wilderness in Alberta (81). Mortality attributable to diseases and parasites, whether the proximate or ultimate cause, is difficult to identify.

Population Dynamics

Mountain goat management typically has assumed that hunters were not sufficiently able to differentiate between sexes, so sex-specific regulations were impractical. However, the skilled observer usually can distinguish between males and females, and females and kids have been successfully protected in the Yukon where guided trophy hunting was the primary harvest (82). Regulations that do not specify sex must be more conservative than regulations that are sex specific, because goats are polygamous and a higher proportion of the males may be harvested than females. Harvest levels between 4% and 7% of British Columbia populations were considered allowable if population maintenance was the goal (83).

During the 1950s and 1960s, the prevailing assumption was that mountain goats inhabited difficult and isolated terrain and hunter harvest was not likely to be limiting, especially in the more remote wilderness. In more accessible mountain ranges, mountain goats were typically hunted with limited entry permits, at levels considered conservative. Investigations of the population declines in the 1960s and 1970s revealed that hunters were taking sufficiently high proportions of some goat populations to cause reductions. Combinations of increased access to mountain goat habitat as mining and logging activities encroached and severe winter weather associated with predation in some areas contributed to population reductions. In British Columbia, a decline from 1969 to 1975 was accompanied by progressively younger ages of harvested goats, and a shift in distribution of harvest from predominantly the southern half of the province to the northern half (84).

Major investigations in Idaho and British Columbia revealed that as populations declined, production and survival of young did not increase as would be expected. Kuck reported a noncompensatory response of goats to significant population reduction from 217 goats to 95 in 6 years attributable to hunter harvest in Idaho (85). He discovered that when adult females were harvested, the areas on which they wintered would be taken over by younger animals. These areas, on the southern and dry end of the native range in Idaho, shed snow during winter and were heavily browsed. Goats that wintered on these least vulnerable areas were not producing young at the same levels as areas where security was lower but forage conditions were better. Recruitment and productivity declined as hunting pressure increased in British Columbia (86). Season lengths and bag limits across most of the province were considered excessively liberal. Recruitment rates in coastal populations were very low and contributed to the negative effects of hunting. For populations with low production and survival of young, conservative hunting strategies must be conducted under the assumption that no compensation in survival or production will occur that is attributable to harvest, or that harvest mortality is additive to natural mortality. Thus harvest management must consider the inability of many hunters to distinguish age and sex classes and the difficulty of distributing harvest away from most accessible portions of the population.

Population performance of introduced goats can differ from those of long-established populations that occupy the natural range. Adams and Bailey calculated that 7.5% of a Colorado population could be harvested. Reproductive success was highly correlated with lower snow depths that allowed more available forage through winter and maintained higher nutrition levels (68). This population grew at a rate of 10% per year for 4 years when 4% of the population was harvested each year, then increased to 16% per year for 2 years when the harvest was stopped. When harvest rates increased to 6% per year, the population grew during periods when production and survival of kids were high and declined when they were low. A simulation predicted that a harvest of 10 to 11 goats from a population of 145, or about 7% of the late-summer population, would stabilize the herd.

Swenson evaluated population trends in an introduced Montana herd, showing wide fluctuations related mostly to hunter harvest (69). The population was established in 1956 to 1958, or 18 years prior to the study, and was increasing rapidly. Again, harvest rate was the most important mortality factor affecting population trend, and in this case snow depths and kid : adult ratios were not correlated with the trend. Males constituted between 54% and 64% of the harvest over the 18 years. This goat population was considered to be in a high-increase phase. Heavy harvests of more than 16% of the population that occurred in one 4-year period and caused declines were changed to increases when decreased to less than 11% of the observed population. The decline in harvest was accompanied by an increase in reproduction, which was not observed by Kuck or Hebert and Turnbull in the native populations (85, 83).

Introduced mountain goats may be a good example of the irruptive sequence that Caughley demonstrated for Himalayan thar in New Zealand (87). An initial irruptive increase is followed by a stabilization, then a major decline, and a subsequent dynamic fluctuation at a lower level (87). Caughley related this sequence to forage supplies relative to numbers of thar (87). This sequence was partially observed in a Montana population where goats introduced in 1941 to 1943 were studied by Lentfer (23, 88). Adult : kid ratios were 100 : 37 in August and 100 : 54 in April and 202 adults were counted in September 1953. Adult goats in summer weighed 47 to 96 kilograms for females and 69 to 95 kilograms for males. Lentfer reported that the goat population had probably started to stabilize and could have been in the initial high stabilization or decline phase of Caughley's population sequence (88, 87). By 1961, kid : adult ratios were 100 : 24, 6 adult females weighed between 47 to 56 kilograms, and horn sizes had declined from 1952 and 1953 (89). Habitat surveys suggested that subalpine fir and two grasses, alpine bluegrass and sheep fescue, showed excessive use (90). While sample sizes were small, the implication was that the goat population had declined in size and productivity 20 years after introduction.

Mountain goat populations in the Olympic Peninsula showed population density responses in production and survival of young (26). At a high population level of approximately 229, kids : female ratio was 20 : 100 (Figure 23–6). The figure suggests a

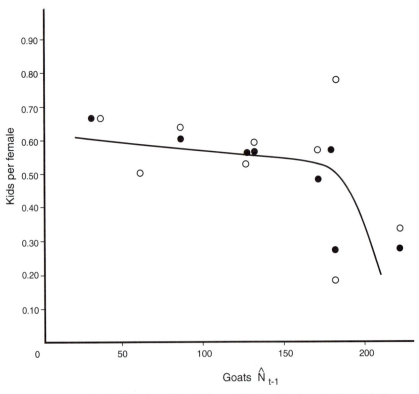

Figure 23–6 Relationship of June production/survival and fall numbers of mountain goat kids to population size, Klahhane Ridge, Olympic National Park, Washington. (After Ref. 26.)

Population Management

threshold response at about 175 goats, or approximately 75% of the highest level, with the kids : female ratio increasing to 50 : 100, and increasing slightly at lower levels. Houston and others documented the results from a sustained effort to extirpate the introduced population (27). Live weights and total horn lengths were negatively associated with goat population size the previous year. Since this population had stabilized after being introduced in the 1920s, the conclusions may be applicable to either native or introduced populations whenever numbers are dropped significantly below the food-based carrying capacity level and the remaining goat population has access to more forage per individual. In cases where this does not happen, the implication is that population declines are merely adjusting to K by several possible means and the available forage per individual is not changing enough to cause compensations in reproduction or survival.

Subsequent work in Alaska revealed that adult mortality was not constant, which meant that use of models assuming constant mortality could underestimate survival (42). Mortality of goats aged 2 to 8 years old was exclusively due to hunting in the Alaska population, which was also the case in British Columbia and Idaho (83, 85). The Alaskan population was increasing at a time when 38% of the total mortality was attributable to hunting, and nonhunting mortality was at a minimum (42).

Bailey synthesized problems in managing mountain goat harvest by concluding that a conservative strategy was justified because the species was easily overharvested (91). Population monitoring may not adequately detect declines, which exacerbates the problem. Conservatively harvested or unharvested populations should be assumed to be at or near ecological or food-based carrying capacity, and no compensatory responses in production and survival attributable to harvest should be anticipated. If maximizing harvest is the goal, then populations should be monitored carefully to ascertain distribution, forage condition, and population performance to detect compensatory responses. Most wildlife management situations will be conservative in nature and such attempts likely are not warranted. Recovery of mountain goat populations from excessive harvest can be prolonged and often confounded by adverse weather or predation.

HABITAT RELATIONSHIPS AND MANAGEMENT

Mountain goat habitat in the interior regions would ordinarily be considered relatively secure because it is typically steep and inaccessible. However, mining and logging activities increase access into habitat or are sufficiently close to habitat to warrant concern. Use of forest types can be important for wintering goats in coastal regions. Dense overlapping canopies that intercept snow and facilitate access to forage may be important to habitats. These forests may exist on steep slopes that goats use. More than 75% of observed goats were on range with greater than 70% slope (64, 65, 42, 92). Exposures vary in their use by region, and Adams and Bailey reported that southern aspects on subalpine range were preferred in winter in Colorado, Schoen and Kirchoff reported a preference for southwest slopes during winter in southeastern Alaska, and Johnson observed goats on east or southwest slopes depending on local snow accumulation in Washington (93, 43, 64). Characteristics of goat winter habitat vary across the range (Table 23–6).

Mountain goat populations increased dramatically following wildfire on Chopaka Mountain, Washington (64). This suggested use of prescribed fire to improve grass and shrub cover by removing dense forest stands. Because burns in these higher elevation forests reduce conifer cover for substantial periods, herbaceous and woody plants important as forage may last for extensive periods. Use of prescribed fire would have to be tailored to the sites and integrated with other resource values.

Assessment of habitat importance relative to animal use is risky because of the constant shifts in distribution relative to weather pattern that also cause fluctuations in population size (52). However, goat winter habitat may be related to commercial forest in coastal ranges (Figure 23–7). Recommendations are to keep facilities away from prime winter habitat, and at least 400 meters from escape terrain. Sequential cuts should emanate away from escape terrain to ensure access to old-growth forest.

TABLE 23–6
Selected Characteristics of Rocky Mountain Goat Winter Ranges

Parameter	Condition	Location and References
Slope	>67%	Montana, Alaska, Washington, Wyoming (65, 42, 64, 92)
Aspect	SE-SW	Alaska, Colorado (subalpine),
	E, SW	Washington (52, 93, 64)
	N	Colorado (alpine) (93, 68)
Distance to cliffs	<400 m	Alberta, Alaska, Wyoming (112, 52, 92)
	<800 m	Washington (64)
Community type	Nonforest	Alaska interior, Colorado, B. C. interior, Montana. (102, 93, 83, 23, 100b)
	Dense forest	Alaska, B. C. coastal, (52, 83)
	Open forest	Washington, Colorado, B. C. interior (64, 93, 83)

Source: Modified from Ref. 94.

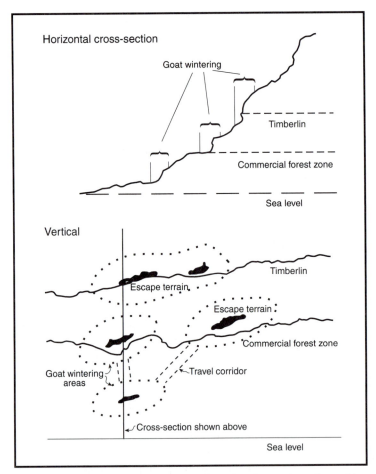

Figure 23–7 Mountain goat wintering sites in conjunction with commercial forest in southeast Alaska. (After Ref. 52.)

Rotations of cuts should be long enough to reestablish dense canopy and understory. Forested travel corridors between wintering sites should be kept intact. In areas where wolf predation is significant, wintering habitat should not be fragmented or goats will be concentrated in the remaining habitats where they may be more vulnerable.

Arnett and Irwin and Joslin provided guidelines for human activities within goat habitat (94, 65). Critical winter ranges, mineral licks, and kidding areas should be identified and protected in planning timber operations and mining activities. Human-related

Habitat Relationships and Management

disturbances within 800 meters of mountain goat winter range should be restricted to the snow-free periods. Helicopter flight paths should be kept above 300 meters from all ground surfaces, including canyon walls, and restricted to corridors of 800 meters in width. In low-quality timber areas, timber harvest should be avoided. Road closure and obliteration of access to minimize disturbance, overharvest, and poaching are important activities in and near goat habitat. Nonmotorized recreation should be emphasized.

Natural salt licks are resources where mountain goats often concentrate activities (95). Although goats have relatively high net values for hunting when compared with other species, their nonconsumptive values around licks like those in Glacier National Park, Montana, and Banff National Park, Alberta, are likely much higher (96). Sodium may be the element that is sought at licks because blood serum levels drop abruptly in spring when use of licks is at its highest (97). Singer reported that goats visited the Walton lick in Glacier from mid-April to mid-September with peaks in late June and early July (98). Goat use of the lick was hindered by heavy vehicle traffic along the road adjacent to the lick. Construction of an underpass that allowed goats to cross under the road was a satisfactory solution to the problem (99). High use of the underpasses by goats in this area was attributed to placement on traditional crossing routes, their large size, fencing and walls along the approaches, and redistribution of visitors to a viewing area removed from the bridges.

SUMMARY

The mountain goat has high recreational and esthetic value and, as an important part of the steep mountain complex with the capability of influencing vegetative composition, may be considered a keystone species. Efforts to conserve the species will center around protecting habitats and conservatively managing populations, where hunting is practiced. It has unique behavior patterns and morphological attributes that are adaptations to the severe terrain in which it evolved. It is a uniquely North American species, with distant relatives on other continents, and serves as a reminder of how past events on earth influence our current flora and fauna.

LITERATURE CITED

1. GILBERT, B. A., and K. J. RAEDEKE. 1992. Winter habitat selection of mountain goats in the North Tolt and Mine Creek Drainages of the north central Cascades. *Proceedings of the Biennial Symposium of the Northern Wild Sheep and Goat Council* 8:305–324.

2. RIDEOUT, C. B. 1978. Mountain goat. Pages 149–159 *in* J. L. Schmidt and D. L. Gilbert, eds. *Big Game of North America.* Wildlife Management Institute, Washington, DC.

3. WIGAL, R. A., and V. L. COGGINS. 1982. Mountain goat. Pages 1008–1020 *in* J. A. Chapman and G. A. Feldhamer, eds., *Wild Mammals of North America.* Johns Hopkins University Press, Baltimore, MD.

4. CHADWICK, D. H. 1983. *A Beast the Color of Winter.* Sierra Club Books, San Francisco.

5. BLAINVILLE, H. M. D. DE. 1816. On several species of mammalian animals from the order of ruminants. *Bulletin Société Philomathique de Paris,* 3(3).

6. MEAD, J. I., P. S. MARTIN, R. C. EULER, A. LONG, A. J. T. JULL, L. J. TOOLIN, D. J. DONAHUE, and T. W. LINICK. 1986. Extinction of Harrington's mountain goat. *Proceedings National Academy of Science* 83:836–839.

7. RIDEOUT, C. R., and R. S. HOFFMANN. 1975. *Oreamnos americanus. Mammalian Species* 63.

8. NOWAK, R. M. 1991. *Walker's Mammals of the World,* 5th ed., Vol. II. Johns Hopkins University Press, Baltimore, MD.

9. GEIST, V. 1983. On the evolution of ice age mammals and its significance to an understanding of specifications. *ASB Bulletin* 30:109–133.

10. HONECKI, J. H., K. E. KINMAN, and J. W. KOEPPL. 1982. *Mammal Species of the World.* Allen Press, Lawrence, KS.

11. ALLEN, J. A. 1904. New forms of the mountain goat (*Oreamnos*). *Bulletin of the American Museum of Natural History* 20:19–21.

12. COWAN, I. McT., and W. McCRORY. 1970. Variation in the mountain goat, *Oreamnos americanus*. *Journal of Mammalogy* 51:60–73.

13. GEIST, V. 1964. On the rutting behavior of the mountain goat. *Journal of Mammalogy* 45:551–568.

14. COWAN, I. McT., and C. J. GUIGET. 1960. *The Mammals of British Columbia Handbook*. British Columbia Provincial Museum 11:1–413.

15. BUNNELL, F. L. 1980. Weight estimation of Dall's sheep and mountain goats. *Wildlife Society Bulletin* 8:291–297.

16. COTE, S. D., M. FESTA-BIANCHET, and K. G. SMITH. 1998. Horn growth in mountain goats (*Oreamnos americanus*). *Journal of Mammalogy* 79:406–414.

17. STEVENS, V., and D. B. HOUSTON. 1989. Reliability of age determination of mountain goats. *Wildlife Society Bulletin* 17:72–74.

18. CASEBEER, R. L., M. J. ROGNRUD, and S. BRANDBORG. 1950. The Rocky Mountain goat in Montana. Wildlife Bulletin Number 5. Montana Department of Fish, Wildlife and Parks. Helena, MT.

19. BRANDBORG, S. M. 1955. Life history and management of the mountain goat in Idaho. *Idaho Department of Fish and Game Wildlife Bulletin* 2.

20. KROG, H., and M. MONSON. 1954. Notes on the metabolism of a mountain goat. *American Journal of Physiology* 178:515–516.

21. GEIST, V. 1967. On fighting injuries and dermal shields of mountain goats. *Journal of Wildlife Management* 31:192–194.

22. RIDEOUT, C. R., and G. L. WORTHEN. 1975. Use of girth measurement for estimating weight of mountain goats. *Journal of Wildlife Management* 39:705–708.

23. LENTFER, J. W. 1955. A two-year study of the Rocky Mountain goat in the Crazy Mountains, Montana. *Journal of Wildlife Management* 19:417–429.

24. HUTCHINS, M., G. THOMPSON, B. SLEEPER, and J. W. FOSTER. 1987. Management and breeding of the Rocky Mountain goat *Oreamnos americanus* at Woodland Park Zoo. *International Zoo Yearbook* 26:297–308

25. HOUSTON, D. B., C. T. ROBBINS, and V. STEVENS. 1989. Growth in wild and captive mountain goats. *Journal of Mammalogy* 70:412–416.

26. HOUSTON, D. B., and V. STEVENS. 1988. Resource limitation in mountain goats: A test by experimental cropping. *Canadian Journal of Zoology* 66:228–238.

27. HOUSTON, D. B., E. G. SCHREINER, and B. B. MOORHEAD. 1994. Mountain goats in Olympic National Park: Biology and management of an introduced species. Scientific Monograph NPS/NROLYM/NRSM-94-25. U.S. National Park Service, Washington, DC.

28. FESTA-BIANCHET, M., M. URQUHART, and K. G. SMITH. 1994. Mountain goat recruitment: Kid production and survival to breeding age. *Canadian Journal of Zoology* 72:22–27.

29. BENZON, T. A., and L. A. RICE. 1988. Mountain goat status, Black Hills, South Dakota. *Proceedings of the Biennial Symposium of the Northern Wild Sheep and Goat Council* 6:168–183.

30. HOUSTON, D. B., C. T. ROBBINS, C. A. RUDER, and R. G. SASSER. 1986. Pregnancy detection in mountain goats by assay for pregnancy-specific protein B. *Journal of Wildlife Management* 50:740–742.

31. HENDERSON, R. E., and B. W. O'GARA. 1978. Testicular development of the mountain goat. *Journal of Wildlife Management* 42:921–922.

32. SMITH, B. L. 1986. Longevity of mountain goats. *Proceedings of the Biennial Symposium of the Northern Wild Sheep and Goat Council* 5:341–346.

33. SMITH, B. L. 1988. Criteria for determining age and sex of American mountain goats in the field. *Journal of Mammalogy* 69:395–402.

34. GEIST, V. 1966. The evolution of horn-like organs. *Behavior* 27:175–214.

35. HOLROYD, J. C. 1967. Observations of Rocky Mountain goats on Mount Wardle, Kootenay National Park, British Columbia. *The Canadian Field-Naturalist* 81:1–22.

35a. HOLMES, E. 1988. Foraging behavior among different age and sex classes of Rocky Mountain goats. *Proceedings of the Biennial Symposium of the Northern Wild Sheep and Goat Council* 6:13–25.

36. CHADWICK, D. H. 1977. The influence of mountain goat social relationships on population size and distribution. *Proceedings of the International Mountain Goat Symposium* 1:74–91.

37. MASTELLER, M. A., and J. A. BAILEY. 1988. Do persisting matrilineal groups partition resources on mountain goat winter ranges? *Proceedings of the Biennial Symposium of the Northern Wild Sheep and Goat Council* 6:26–38.

38. FOURNIER, F., and M. FESTA-BIANCHET. 1995. Social dominance in adult female mountain goats. *Animal Behavior* 49:1441–1459.

39. ALEXANDER, R. D. 1974. The evolution of social behavior. *Annual Review of Ecology and Systematics* 5:325–383.

40. LUNDRIGAN, B. 1996. Morphology of horns and fighting behavior in the family Bovidae. *Journal of Mammalogy* 77:462–475.

41. ROMEO, G., and S. LOVARI. 1996. Summer activity rhythms of the mountain goat *Oreamnos americanus* (de Blainville, 1816). *Mammalia* 60:496–499.

42. SMITH, C. A. 1986. Rates and causes of mortality in mountain goats in southeast Alaska. *Journal of Wildlife Management* 50:743–746.

43. SCHOEN, J. W., and M. D. KIRCHOFF. 1981. Habitat use by mountain goats in southeast Alaska. Final Report Project W-17-R, Alaska Department of Fish and Game, Juneau.

44. NAYLOR, K. S. 1988. Distribution, habitat use and population characteristics of introduced mountain goats at Pend Oreille Lake, Idaho. Thesis. University of Idaho, Moscow.

44a. NICHOLS, L. 1983. Mountain goat movements study. Alaska Department of Fish and Game Progress Report Project W-21-R. Juneau.

45. NICHOLS, L. 1980. Mountain goat management technique studies. Final Report, Project W-17-R, Alaska Department of Fish and Game, Juneau.

46. ADAMS, L. G., M. A. MASTELLER, and J. A. BAILEY. 1982. Movements and home range of mountain goats. *Proceedings of the Biennial Symposium of the Northern Wild Sheep and Goat Council* 3:391–405.

47. SAUNDERS, J. K. 1955. Food habits and range use of the Rocky Mountain goat in the Crazy Mountains, Montana. *Journal of Wildlife Management* 19:429–437.

48. FOX, J. L., and C. A. SMITH. 1988. Winter mountain goat diets in southeast Alaska. *Journal of Wildlife Management* 52:362–365.

49. KLEIN, D. R. 1965. Ecology of deer range in Alaska. *Ecological Monographs* 35:259–284.

50. KLEIN, D. R. 1970. Food selection by North American deer and their response to overutilization of preferred plant species. *British Ecological Society Symposium* 10:25–46.

51. MOONEY, H. A., and W. D. BILLINGS. 1960. The annual carbohydrate cycle of alpine plants as related to growth. *American Journal of Botany* 47:594–598.

52. FOX, J. L., C. A. SMITH, and J. W. SCHOEN. 1989. Relation between mountain goats and their habitat in southeastern Alaska. General Technical Report PNW-GTR-246. U.S. Forest Service, Washington, DC.

53. JOHNSTON, A., L. M. BEZEAU, and S. SMOLIAK. 1968. Chemical composition and *in vitro* digestibility of alpine tundra plants. *Journal of Wildlife Management* 32:773–777.

54. DAILEY, T. V. 1981. Composition and quality of mountain goat diets in alpine tundra, Colorado. Thesis. Colorado State University, Fort Collins.

55. ROBBINS, C. T., S. M. PARISH, and B. L. ROBBINS. 1985. Selenium and glutathione peroxidase activity in mountain goats. *Canadian Journal of Zoology* 63:1544–1547.

56. FIELDER, P. C. 1986. Implications of selenium levels in Washington mountain goats, mule deer and Rocky Mountain elk. *Northwest Science* 60:15–20.

57. DAILEY, T. V., N. T. HOBBS, and T. N. WOODARD. 1984. Experimental comparisons of diet selection by mountain goats and mountain sheep in Colorado. *Journal of Wildlife Management* 48:799–806.

58. CAMPBELL, E. G., and R. L. JOHNSON. 1983. Food habits of mountain goats, mule deer, and cattle on Chopaka Mountain, Washington 1977–1980. *Journal of Range Management* 36:488–491.

59. KUCK, L. 1986. Mountain goat hunting in Idaho. *Proceedings of the Biennial Symposium of the Northern Wild Sheep and Goat Council* 5:63–67.

60. HAYDEN, J. A., G. GADWA, G. MCNEILL, J. ROHLMAN, and R. SHEA. 1990. *Mountain Goat Management Plan 1991–1995.* Idaho Department of Fish and Game, Boise.

61. SMITH, C. A. 1984. Evaluation and management implications of long term trends in coastal mountain goat populations in southeast Alaska. *Proceedings of the Biennial Symposium of the Northern Wild Sheep and Goat Council* 4:395–424.

62. SMITH, K. G. 1988. Factors affecting population dynamics of mountain goats in west-central Alberta. *Proceedings of the Biennial Symposium of the Northern Wild Sheep and Goat Council* 6:308–329.

63. HEBERT, D., and T. SMITH. 1986. Mountain goat management in British Columbia. *Proceedings of the Biennial Symposium of the Northern Wild Sheep and Goat Council* 5:48–59.

64. JOHNSON, R. L. 1983. Mountain goats and mountain sheep of Washington. Biological Bulletin 18. Washington Department of Game, Olympia, WA.

65. JOSLIN, G. 1986. Montana mountain goat investigations, Rocky Mountain Front. Montana Department of Fish, Wildlife, and Parks, Helena.

66. FLATHER, C. H., and T. W. HOEKSTRA. 1989. An analysis of the wildlife and fish situation in the United States: 1989–2040. General Technical Report RM-178. U.S. Forest Service, Washington, DC.

67. GEIST, V. 1981. On the reproductive strategies in ungulates and some problems of adaptation. *Proceedings of the Second International Congress on Systematics and Evolutionary Biology* 2:111–132.

68. ADAMS, L. G., and J. A. BAILEY. 1982. Population dynamics of mountain goats in the Sawatch Range, Colorado. *Journal of Wildlife Management* 46:1003–1009.

69. SWENSON, J. E. 1985. Compensatory reproduction in an introduced mountain goat population in the Absaroka Mountains, Montana. *Journal of Wildlife Management* 49:837–843.

70. BALLARD, W. 1977. Status and management of the mountain goat in Alaska. *Proceedings of the International Mountain Goat Symposium* 1:5–23.

71. FOX, J. L., and G. P. STREVELER. 1986. Wolf predation on mountain goats in southeastern Alaska. *Journal of Mammalogy* 67:192–195.

72. HEBERT, D. M., and I. McT. COWAN. 1971. White muscle disease in the mountain goat. *Journal of Wildlife Management* 35:752–756.

73. FOREYT, W. J. 1989. *Sarcocystis* sp. in mountain goat (*Oreamnos americanus*) in Washington: Prevalence and search for the definitive host. *Journal of Wildlife Diseases* 25:619–622.

74. MAHRT, J. L., and D. D. COLWELL. 1980. Sarcocystis in wild ungulates in Alberta. *Journal of Wildlife Diseases* 16:571–576.

75. BODDICKER, M. L., E. J. HUGGHINS, and A. H. RICHARDSON. 1971. Parasites and pesticide residues of mountain goats in South Dakota. *Journal of Wildlife Management* 35:95–103.

76. BODDICKER, M. L., and E. J. HUGGHINS. 1969. Helminths of big game mammals in South Dakota. *Journal of Parasitology* 55:1067–1074.

77. DIKMANS, G. 1942. New host–parasite records. *Proceedings of the Helminthological Society of Washington* 9:65.

78. BECKLUND, W. W. 1965. *Nematodirus maculosus,* sp. n. (Nematoda: Trichostrongylidae) from the mountain goat, *Oreamnos americanus,* in North America. *Journal of Parasitology* 51:945–947.

79. FOREYT, W. J., and LEATHERS. 1985. Starvation secondary to an oral fibroma in a wild mountain goat (*Oreamnos americanus*). *Journal of Wildlife Diseases* 21:184–185.

80. DUNBAR, M. R., W. J. FOREYT, and J. F. EVERMANN. 1986. Serologic evidence of respiratory syncytial virus infection in free-ranging mountain goats (*Oreamnos americanus*). *Journal of Wildlife Diseases* 22:415–416.

81. SAMUEL, W. M., W. K. HALL, J. G. STELFOX, and W. D. WISHART. 1977. Parasites of mountain goat, *Oreamnos americanus* (Blainville), of west central Alberta with a comparison of the helminths of mountain goat and Rocky Mountain bighorn sheep, *Ovis c. canadensis* (Shaw). *Proceedings of the International Mountain Goat Symposium* 1:212–225.

82. JOHNSON, R. L. 1977. Distribution, abundance and management status of mountain goats in North America. *Proceedings of the International Mountain Goat Symposium* 1:1–7.

83. HEBERT, D. M., and W. G. TURNBULL. 1977. A description of southern interior and coastal mountain goat ecotypes in British Columbia. *Proceedings of the International Mountain Goat Symposium* 1:126–146.

84. FOSTER, B. R. 1977. Horn growth and quality management for mountain goats. *Proceedings of the International Symposium on Mountain Goats* 1:200–226.

85. KUCK, L. 1977. The impacts of hunting on Idaho's Pahsimeroi mountain goat herd. *Proceedings of the International Mountain Goat Symposium* 1:114–125.

86. HEBERT, D. M. 1978. A systems approach to mountain goat management. *Proceedings of the Biennial Symposium of the Northern Wild Sheep and Goat Council* 2:227–243.

87. CAUGHLEY, G. 1970. Eruption of ungulate populations with emphasis on Himalyan thar in New Zealand. *Ecology* 51:53–72.

88. LENTFER, J. W. 1957. Rocky Mountain goat investigations. Job Completion Report W-73-R-2. Montana Department of Fish Wildlife and Parks, Helena.

89. FOSS, A. J. 1962. A study of the Rocky Mountain goat in Montana. Thesis. Montana State University, Bozeman.

90. PEEK, J. M. 1962. Rocky mountain goat investigations (range phase). Project W-98-R-2 Job B-9. Montana Department of Fish, Wildlife and Parks, Helena.

91. BAILEY, J. A. 1986. Harvesting mountain goats: Strategies, assumptions, and needs for management and research. *Proceedings of the Biennial Symposium of the Northern Wild Sheep and Goat Council* 5:37–47.

92. HAYNES, L. A. 1992. Mountain goat habitat of Wyoming's Beartooth Plateau: Implications for management. *Proceedings of the Biennial Symposium of the Northern Wild Sheep and Goat Council* 8:325–339.

93. ADAMS, L. G., 1981 and J. A. BAILEY. 1980. Winter habitat selection and group size of mountain goats, Sheep Mountain-Gladstone Ridge, Colorado. *Proceedings Biennial Symposium of the Northern Wild Sheep and Goat Council* 2:465–481.

94. ARNETT, E. B., and L. L. IRWIN. 1989. Mountain goat/forest management relationships: A review. Technical Bulletin 562. National Council of the Paper Industry for Air and Stream Improvement, Incorporated. Corvallis, Oregon.

95. COWAN, I. McT., and V. C. BRINK. 1949. Natural game licks in the Rocky Mountain national parks of Canada. *Journal of Mammalogy* 30:379–387.

96. LOOMIS, J. B., D. M. DONNELLY, C. F. SORG, and L. OLDENBURG. 1985. Net economic value of hunting unique species in Idaho: Bighorn sheep, mountain goat, moose, and antelope. Resource Bulletin RM-10. U.S. Forest Service, Washington, DC.

97. HEBERT, D., and I. McT. COWAN. 1971. Natural salt licks as a part of the ecology of the mountain goat. *Canadian Journal of Zoology* 49:605–610.

98. SINGER, F. J. 1978. Behavior of mountain goats in relation to U.S. highway 2, Glacier National Park, Montana. *Journal of Wildlife Management* 42:591–597.

99. SINGER, F. J., and J. L. DOHERTY. 1985. Managing mountain goats at a highway crossing. *Wildlife Society Bulletin* 13:469–477.

100. THOMPSON, M. J. 1981. Mountain goat distribution, population characteristics and habitat use in the Sawtooth Range, Montana. Thesis. Montana State University, Bozeman.

100a. RIDEOUT, C. B. 1977. Mountain goat home range in the Sapphire Mountains of Montana. *Proceedings of the First International Mountain Goat Symposium* 1:201–211.

100b. SMITH, B. L. 1976. Ecology of Rocky Mountain Goats in the Bitterroot Mountains, Montana, Thesis. University of Montana, Missoula, Montana.

101. SINGER, F. J., and J. L. DOHERTY. 1985a. Movements & habitat use in an unhunted population of mountain goats, *Oreamnos americanus.* Canadian Field-Naturalist 99:205–217.

101a. ADAMS, L. G., and J. A. BAILEY. 1983. Winter Forages of Mountain Goats in Central Colorado. *Journal of Wildlife Management* 47:1237–1243.

102. HJELJORD, O. 1973. Mountain goat forage and habitat preference in Alaska. *Journal of Wildlife Management* 37:353–362.

102a. CHADWICK, D. H. 1973. Mountain goat ecology-logging relationships in the Bunker Creek drainage of western Montana. Montana Department of Fish, Wildlife and Parks Final Report Project W-120-R-3,4. Helena.

103. ANDERSON, N. A. 1940. Mountain goat study/progress report. Washington Department of Game Biological Bulletin 2. Olympia.

104. COWAN, I. McT. 1944. Report of wildlife studies in Jaspea, Banff, and Yoho National Parks in 1944. Canadian Wildlife Service, Department of Mines and Resources. Ottawa, Canada.

105. HARMON, W. 1944. Notes on mountain goats in the Black Hills. *Journal of Mammalogy* 25:149–151.

105a. KUCK, L. 1973. Rocky Mountain goat ecology. Idaho Department of Fish and Game Job Progress Report Project 144R3. Boise.

106. PALLISTER, G. L. 1974. The seasonal distribution and range use of bighorn sheep in the Beartooth Mountains with special reference to the West Rosebud and Stillwater herds. Thesis. Montana State University, Bozeman.

107. RICHARDSON, A. H. 1971. The Rocky Mountain goat in the Black Hills. South Dakota Department of Game, Fish and Parks Bulletin 2. Pierre.

108. PEEK, S. V. 1972. The ecology of the Rocky Mountain goat in the Spanish Peaks area of southwestern Montana. Thesis. Montana State University, Bozeman.

109. HIBBS, L. D. 1967. Food habits of the mountain goat in Colorado. *Journal of Mammalogy.* 48:242–248.

110. ADAMS, L. G. 1981. Ecology and population dynamics of mountain goats, Sheep Mountain - Gladstone Ridge Colorado. Thesis, Colorado State University, Fort Collins.

111. BAILEY, J. A., and B. K. JOHNSON. 1977. Status of introduced mountain goats in the Sawatch Range of Colorado. *Proceedings of the First International Mountain Goat Symposium* 1:54–63.

112. McFETRIDGE, R. JK. 1977. Strategy of resource use by mountain goat nursery groups. *Proceeding of the First International Mountain Goat Symposium* 1:169–173.

CHAPTER 24

Dall's and Stone's Sheep

R. Terry Bowyer, David M. Leslie, Jr., and Janet L. Rachlow

INTRODUCTION

Dall's sheep inhabit rugged and precipitous mountain ranges in Alaska and western Canada, and are the northernmost species of mountain sheep in the New World (1) (Figure 24–1). The severity and variability of arctic and subarctic environments have placed unique constraints on the ecology, behavior, and evolution of these mountain ungulates. Dall's sheep and Stone's sheep were named for W. H. Dall and A. J. Stone, respectively (2). Following Bowyer and Leslie, we use Dall's sheep as the common name for the species and use subspecific designations as appropriate (1).

(a)

(b)

Figure 24–1 (a) Dall's sheep are the northernmost member of their genus in the New World. (b) Their stocky body conformation is well adapted for inhabiting extremely rugged and precipitous terrain. (Photographs by J. L. Rachlow.)

491

Because Dall's sheep inhabit remote northern ranges, our historical perspectives on population trends and value to native peoples are relatively short and spotty. These perspectives begin with exploration and settlement of northwestern Canada and Alaska in the mid-1800s. As early as 1881, Nelson observed 100 skins of Dall's sheep that were harvested by native peoples that lived along the arctic coast near Point Barrow, Alaska (3). Those sheep were hunted with early Winchester rifles in spring, probably over several years (3). Although accounts are limited, native peoples probably did not affect extant populations of Dall's sheep negatively to the extent that the activities of early European setters did.

Early mining activities and railroad construction throughout the range of Dall's sheep clearly influenced populations of wildlife because of demands for provisions to sustain laborers. Early accounts, notably on the Kenai Peninsula and around Mount McKinley (in present-day Denali National Park and Preserve) in Alaska, noted that populations of Dall's sheep and other large mammals were decimated due to extensive market hunting and poaching (3, 4, 5, 6, 7). Those settlers that remained in remote locations through winter, without access to more sophisticated means of food distribution, placed increased consumptive demands on wildlife for themselves and their sled dogs (4). Early naturalists often located abandoned winter camps with piles of skulls of Dall's sheep (more than 70 males in one instance) (6, 7). When game was abundant, harvest probably was wasteful (4).

The growing rarity of Dall's sheep in the 1930s, particularly in the area of Mount McKinley, caused W. C. Sheldon and conservation leaders of the time to pressure the U.S. Congress to establish Mount McKinley National Park in 1927 (3). Harsh winters in the 1930s and, at times, wolf predation also contributed to declining populations of Dall's sheep throughout its range (8, 6, 7).

Mountain sheep throughout North America—the "gallant mountaineers"—are valued as much for their nonconsumptive and aesthetic appeal as for their consumptive uses (5, 9). Wright and others stated that Dall's sheep were the greatest attraction to visitors to Mount McKinley, likely because of their high visibility, social habits, and occurrence in areas of scenic beauty (4). Harsh and remote conditions of native ranges of Dall's sheep contribute to their allure (10). Some populations are habituated to humans, providing enhanced viewing opportunities and enjoyment to recreationalists (11, 9).

Dall's sheep currently are hunted throughout their range and are highly prized by sportsmen. Typically, hunting of mature rams with at least three-quarter-curl horns is practiced. Subsistence use by native peoples is uncommon, relative to other ungulates (12). Many unhunted populations of Dall's sheep now occur in national parks and other protected areas in Canada and Alaska.

TAXONOMY AND DISTRIBUTION

Ancestors of mountain sheep arose in Asia about 2.5 million years ago during the Villafranchian age (11, 51). *Ovis* was a Palearctic immigrant to the New World and a member of the Rancholabrean fauna of the late Pleistocene (13). The oldest (about 100,000 years ago) remains of *Ovis* in North America were discovered near Fairbanks, Alaska, in the penultimate deposits from Illonian glaciers (14). By the Wisconsin Glacial (about 23,000–40,000 years ago), fossils from *Ovis* were common (15, 16, 17, 18). The fossil record for *Ovis,* however, is not complete, in part, because these bovids inhabited mountainous terrain that was not conducive to the formation or discovery of fossils (11). Indeed, fossils have not resolved the origin or evolution of contemporary species of mountain sheep (19, 20, 21, 22, 23). Nonetheless, ancestral Dall's sheep, which were larger than their modern counterparts, occurred in North America by the late Pleistocene (24).

Schaller reviewed the taxonomy of the Caprinae; he recognized four tribes including the Caprini to which *Ovis* belongs (25). *Ovis* is composed of five species (argali, bighorn sheep, Dall's sheep, snow sheep, urial) in addition to the domestic sheep. Bighorn sheep, Dall's sheep, and snow sheep of Siberia are extremely stocky ungulates that are well adapted to life in rugged, mountainous terrain with precipitous cliffs, whereas the remaining species of *Ovis* have a lithe body conformation better suited for more open, rolling landscapes (26, 25). Although debate over the taxonomy of *Ovis* exists, most researchers recognize three subspecies of snow sheep from Siberia, seven subspecies of bighorn sheep, which extend in distribution from southern Canada through the United States and into Mexico, and three subspecies of Dall's sheep that occur from Alaska into northern Canada (27, 20, 11, 25).

Analyses of chromosome morphology, likewise, have failed to resolve the taxonomy of *Ovis*. Dall's sheep possess a chromosome number (2n) of 54, with a karyotype characterized by three pairs of large metacentric autosomes and 24 pairs of acrocentric chromosomes; the largest pair are the X chromosomes (28, 28a). Nevertheless, Dall's sheep, bighorn sheep, and argalis have an identical chromosome number and similar chromosome morphology (29, 23). Further, the snow sheep, which is thought to share common ancestry with Dall's sheep and bighorn sheep based on chromosomal morphology of the three largest autosomes, has a chromosome number of 2n = 52 (30, 23). Clearly, chromosome number is too variable to be a reliable taxonomic marker for *Ovis*. Likewise, sera proteins and antigens differ only slightly among the species of *Ovis* and offer little promise of discriminating among taxa (31). Groves and Shields reported a close phylogenetic relationship between Dall's sheep and bighorn sheep based on the cytochrorome *b* sequence of mitochrondial DNA, but had no data for snow sheep (32). Mitochrondial DNA from some subspecies of bighorn sheep indicated great variability for that species, with particular subspecies of bighorn sheep showing closer affinities to Dall's sheep than to other subspecies of bighorn sheep (33). The taxonomy of *Ovis* is an area in dire need of more study using modern molecular methods.

Despite these taxonomic difficulties, the Dall's sheep generally is recognized as a valid species with three extant subspecies (Dall's sheep, Stone's sheep and *O. d. kenaiensis*) (1). *Ovis d. kenaiensis* may be synonymous with *Dall's sheep* (2). Although Dall's sheep on the Kenai Peninsula, Alaska, are smaller in size than other populations, this difference may represent clinal variation in Dall's sheep (20, 2). Bowyer and Leslie provide a more complete description of the taxonomy of Dall's sheep including synonymies and localities for type specimens (1).

In general, the Dall's sheep is smaller in body size than bighorn sheep but larger than snow sheep (34, 35, 1). For instance, the body mass of six adult males and eight adult females from the Yukon ranged from 72.8 to 82.3 kilograms and 46.4 to 50.4 kilograms, respectively; maximum mass for males in autumn ranged from 80 to 110 kilograms (36, 37, 2). In comparison, weights for the largest subspecies of bighorn sheep ranged from 72.6 to 143.3 kilograms for adult males and from 53.1 to 90.7 kilograms for adult females (38).

The body lengths of male Dall's sheep range from 1,300 to 1,780 millimeters, whereas females range from 1,324 to 1,620 millimeters in length; height at the shoulder for males ranged from 916 to 1,090 millimeters, but data for females were unavailable (20, 36, 39). Bowyer and Leslie reviewed additional measurements of other body and skull characteristics for Dall's sheep and concluded there was little variation among subspecies (1). The dental formula for adult sheep is I 0/3, c 0/1, p 3/3, m 3/3 = 32 total teeth; all permanent teeth are in place by 48 months of age (40).

The horns of Dall's sheep are intermediate in size between the larger more massive horns of bighorn sheep and the smaller horns of snow sheep; maximum length and basal circumference, respectively, for horn sheaths of adult males were Dall's sheep (130; 38 centimeters); bighorn sheep (124; 47 centimeters); and snow sheep (111; 36 centimeters) (26, 11, 39). There is extreme sexual dimorphism in the size and conformation of horns in mountain sheep; males possess heavier horns that are longer and

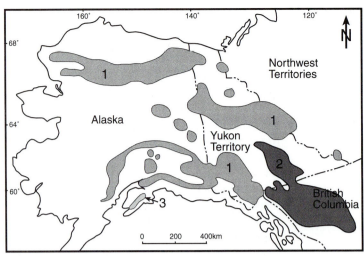

Figure 24–2 Distribution of Dall's sheep: (1) *Ovis dalli dalli,* (2) *O. d. stonei,* and (3) *O. d. kenaiensis.* (From Ref. 1.)

flare more widely than those of females (11). Horns of male Dall's sheep also flare more widely (88 centimeters tip to tip) than do those of bighorn sheep (66 centimeters). Horns of snow sheep are smaller but similar in shape to those of Dall's sheep. Dall's sheep exhibit two distinct color morphs: *O. d. dalli* and *O. d. kenaiensis* are white (or off-white), whereas Stone's sheep is much darker and possesses a rump patch and other markings more typical of bighorn sheep and snow sheep (1). Some *O. d. dalli* may possess a black tail, and Fannin's morph, which is intermediate in coloration between Dall's sheep and Stone's sheep occurs occasionally, where the ranges of the subspecies are close (41, 42) (Figure 24–2).

The distribution of mountain sheep in the New World has been influenced, in part, by the activities of humans; however, Dall's sheep have been little affected compared with bighorn sheep (43, 44, 45, 46, 47). This difference most likely reflects the harsh, remote locations inhabited by Dall's sheep and the relative lack of human developments, including grazing of livestock, at far northern latitudes. Indeed, the Dall's sheep ranges through rugged and steep mountain ranges in Alaska, Northwest Territories, Yukon, and British Columbia from 69°40′ to 59°30′N latitude.

The northernmost subspecies is *O. d. dalli,* but these animals occur in Alaska as well as the Yukon, Northwest Territories, and British Columbia (Figure 24–2). *Ovis d. kenaiensis* is confined to the Kenai Peninsula in Alaska, whereas *O. d. stonei* is distributed the farthest southward with populations occurring in British Columbia and the Yukon. In the southern part of its distribution, Dall's sheep may be restricted to isolated patches of alpine habitat, but the contiguous areas inhabited by Dall's sheep generally are larger than those occupied by bighorn sheep (48, 49, 1). Hoefs reported about 13,000 Stone's sheep and 9,000 Dall's sheep from the Yukon, but now suggests a total estimate of 22,000 animals (personal communication) (50). I. Hatter (Wildlife Branch, British Columbia Environment personal communication) estimates about 14,000 Stone's sheep and 500 Dall's sheep from British Columbia. Hoefs suggested a population of about 7,000 Dall's sheep from the Northwest Territories, but A. Veitch (personal communication) believes there are at least 14,000 animals (50). Hoefs estimated 72,650 Dall's sheep occurred in Alaska including 1,500 *O. d. kenaiensis* (50). Valdez and Krausman estimated a total of 113,750 Dall's sheep (99,750 in Alaska and 14,500 in Canada) (51). Despite some variation in population estimates and the need for more quantitative survey methods, there were probably more than 110,000 Dall's sheep in 1998. Dall's sheep are the most abundant wild sheep in the world and still occupy about 90% of their former distribution (50). More detailed information on taxonomy and distribution of Dall's sheep is available in several other reviews (2, 39, 52, 1, 51a).

The ecology and behavior of mountain sheep should be viewed in an evolutionary perspective to understand how these ungulates have adapted to the harsh environments they occupy. This understanding has far-reaching consequences for the conservation and management of Dall's sheep. As with the evolution of *Ovis,* ideas for how these mountain ungulates became adapted to the landscapes they inhabit are clouded. Two related ideas, in particular, that have contributed to our existing view of mountain sheep need closer examination.

The first of these is the glaciation hypothesis proposed by Geist and its role in the evolution of mountain sheep (11). In general, this hypothesis suggested that sheep evolving at the face of retreating glaciers were subjected to differing selective forces than those inhabiting glacial refugia. Consequently, sheep from glaciated areas should possess larger horns and rump patches than those that evolved in refugia. In addition, Geist proposed that glaciation, in Asia and the New World, led to a pattern of colonization (and evolution) that placed aoudad-like ancestors at the base of an evolutionary tree leading to New World mountain sheep (i.e., the argali cline) (11). Geist further refined his ideas concerning glaciation by formulating his dispersal theory, which attempted to explain rapid evolution of sheep in relation to withdrawal of the glaciers (53). Geist postulated that abundant forage at the face of retreating glaciers would promote increased body size, larger horns, rump patches, increased productivity, and early maturation in mountain sheep (53). These conditions, then, promoted the evolution of neoteny in sheep. This brief overview does not do justice to the entire scope of ideas proposed by Geist but is sufficient to examine their relevance to mountain sheep (11, 53).

Schaller raised serious questions concerning the pattern of evolution for mountain sheep proposed by Geist, especially whether the aoudad was closely related or ancestral to modern-day mountain sheep (25, 11). Indeed, modern genetic techniques indicate that the phylogenetic relationships proposed by Geist were not correct and that morphological similarities he used as support for his hypotheses likely were the result of convergent evolution (11, 54). Of course, Geist did not have access to this information at the time he formulated his hypotheses (11, 53). Nonetheless, the aoudad is not closely related to mountain sheep and, consequently, the argali cline should now be set aside as an explanation for the evolution of mountain sheep (54).

There is little doubt that Geist was correct in assuming that glaciation helped shape the evolution and distribution of mountain sheep (11, 53). Likewise, retreating glaciers unquestionably altered the landscape and the habitats available to sheep. How much new habitat would have been created relative to the ability of nearby populations of sheep to reproduce and expand, however, is the critical question. Glaciers that withdrew at a rate of even several meters each year would likely have adjacent populations of sheep that capitalized on the production of new forage. The dispersal distances of caprids and ovids far outdistance any new habitat created by receding glaciers (25, 47). Similarly, the reproductive potential of mountain sheep would have rapidly filled vacant habitats and made the long periods necessary for the evolution of characteristics postulated by Geist unlikely (2, 11). Given the new information on phylogenetic relationships, dispersal, and population dynamics, there is no longer a need to hypothesize neoteny to explain why male sheep attain sizes larger than females; sexual dimorphism is best explained by degree of polygyny in ungulates (55, 56, 57, 58). Schaller criticized Geist's hypotheses for giving too much attention to social selection and too little to natural selection. Indeed, neither the glaciation nor dispersal hypotheses will explain the evolution of sexual dimorphism or rump patches in ruminants; both of these characteristics are widespread in species that do not have close phylogenetic ties and did not evolve in areas subject to heavy glaciation (25, 11). Moreover, Dall's sheep undoubtedly were exposed to retreating glaciers in the mountainous areas of Beringia, yet sheep in interior Alaska lack a rump patch.

More parsimonious hypotheses now exist to explain the evolution of rump patches, sexual dimorphism, and the evolution of horn-like structures (59, 60, 55, 56,

58, 61, 62, 63). We concur with Schaller that the evolution of life history traits of mountain sheep needs to be viewed in a broad perspective that includes a multitude of factors such as steepness and ruggedness of terrain, snow depth and other aspects of climate, predation, and phylogenetic constraints (25). An approach that emphasizes the role of natural selection in understanding adaptation offers the best explanation for the life-history characteristics of mountain sheep, which include age at first reproduction, litter size, longevity, and many other aspects we discuss later (64).

DIET SELECTION IN RELATION TO HABITAT

Studies on the diet of Dall's sheep are not numerous, but there is general agreement that these ungulates feed mostly on graminoids (7, 65, 2, 66, 67, 68, 69, 70). Nonetheless, Hoefs and Cowan recorded 110 different plant species in the diet of these herbivores (66). In the southwest Yukon, Canada, Hoefs and Cowan noted that reedgrass was the most common grass in the diet of sheep, whereas fescue was eaten most often in Alaska, (66, 2, 70). Other important genera of grasses eaten by Dall's sheep include bromegrass, bluegrass, and wheatgrass. The occurrence of sedges and rushes in the diet of Dall's sheep may increase with north latitude: Hoefs and Cowan reported about 18% of this forage eaten by sheep in the southern Yukon, Nichols noted about 25% from the Kenai Peninsula, and Hansen indicated about 30% from the Brooks Range (66, 2, 70). These northern-most populations appear to consume the common sedge most often.

Dall's sheep also eat a variety of forbs, especially during late spring and summer, including saxifrage, cinquefoil, willow-herb, pussytoes, and many others (66, 70). Browse consumed by Dall's sheep was principally sagebrush in the southern part of their range and willow, dryas, and blueberry in the north (66, 70). Sheep sometimes move to lower elevations in early spring to obtain browse and do so again in early autumn when frost curtails growth in forbs and graminoids at higher elevations; note the peaks in use of browse by Dall's sheep during those periods (Figure 24–3). These sheep also consume limited amounts of lichens and moss.

Hansen reported strong selection (use greater than availability) for grasses during all seasons, selection for forbs in spring and summer, and for sedges in winter; browse was avoided (use less than availability) throughout the year in the Brooks Range, Alaska (70). Hansen cautioned, however, that interannual patterns of

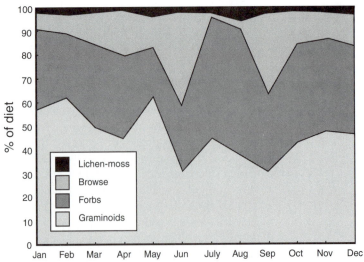

Figure 24–3 The year-round diet of Dall's sheep, Kluane, Yukon, Canada. (Modified from Ref. 66.)

Diet Selection in Relation to Habitat

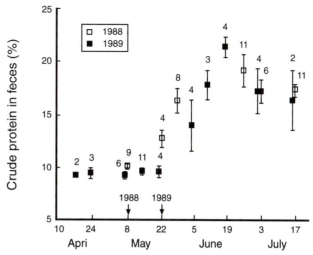

Figure 24–4 Mean (± *SE*) percent crude protein in the feces of female Dall's sheep in interior Alaska, USA. Sample sizes are provided above means; arrows indicate onset of lambing in 1988 and 1989. (From Ref. 73.)

diet selection could be affected by 10 to 15 centimeters of snow cover. Likewise, Rachlow and Bowyer noted that a cool summer that limited growth of forage resulted in differences in selection of habitat by Dall's sheep in interior Alaska (71). Mineral licks also can be important seasonally to Dall's sheep (6, 72, 7, 11).

Diet quality, as indexed by fecal crude protein, increased rapidly during early spring, peaked in June, and then began declining by July (Figure 24–4). Hansen reported a continued decline in fecal crude protein to pre-spring levels by October and that levels remained low throughout winter (70). This same general pattern was evident for digestible nitrogen, digestible energy, and *in vitro* dry matter digestibility of sheep forages (70). The growing season in arctic and subarctic environments is extremely reduced. The number of days between the last freeze in spring and the first freeze in autumn was 79 days in 1988 but only 33 days in 1989 for Dall's sheep in interior Alaska (73). Likewise, degree days greater than 5°C, an index to the length of the growing season in arctic plants, were 576 and 486, respectively, for these same 2 years (74, 73). The cool, short growing season in 1989 resulted in a marked decrease in grasses and dryas available to Dall's sheep (71). Thus, Dall's sheep must acquire the resources necessary to meet their need for reproduction and to cope with harsh winter conditions in a narrow window that may vary markedly between years (73). In addition, forage generally was more plentiful as distance from steep, precipitous terrain preferred by Dall's sheep increases (75, 71). Hence, diet selection was affected by risk of predation in these mountain ungulates (71).

REPRODUCTIVE STRATEGY

Dall's sheep are thought to have an estrous cycle of about 17 days (76). Photoperiod is likely an important cue in timing of reproduction, and the presence of an adult male and physical condition of the female are proximal stimuli that also affect onset of estrus, which is thought to last for 1 day (11, 2). Timing reproduction so that young can be provisioned successfully regulates when parturition occurs in ungulates (77, 78, 73, 79). Dall's sheep are monestrus with rut typically occurring in November and December (11, 80). One young weighing 3 to 4 kilograms normally is born after about 171 days of gestation; twins are rare (81, 80, 82). There is increasing evidence that adjustment in gestation length may be under proximal control of the female in Dall's sheep and other ungulates (78, 83, 79). For instance, Dall's sheep in interior Alaska

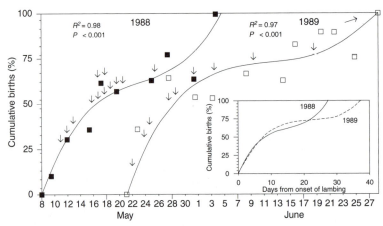

Figure 24–5 Cumulative percent of births determined from young to female female-like ratios showing differences in timing and synchrony (inset) of lambing in interior Alaska 1988 and 1989. Arrows indicate date of birth observed or inferred. (From Ref. 78.)

delayed onset of lambing by 14 days when a spring storm deposited 25 centimeters of fresh snow during the peak lambing period of the previous year (Figure 24–5) (78). There is a trend for date of parturition to be earlier for populations of mountain sheep with increasing north latitude (84, 85). Nonetheless, marked interannual differences in the date of birth can occur. Rachlow and Bowyer reported that median date of birth in a normal year was 18 May, but was 27 May in the year with the late snowstorm. Synchrony of births also differed in these 2 disparate years (78). Evidence that birth synchrony in Dall's sheep is related to predation is lacking; instead, synchronous births result from a limited time in which offspring can develop to a sufficient size in spring and summer to withstand harsh conditions in winter (78).

Reproduction has not been reported for lambs, but young yearlings may become pregnant in highly productive populations (80). Females usually do not begin reproducing, however, until 30 months of age (11). Indeed, young : adult ratios at birth for Dall's sheep in interior Alaska varied from 0.4:1 to 0.6:1 indicating not all adult females reproduced each year, a pattern reported for other arctic ungulates (86, 78, 87). Males can become sexually mature at 18 months, but because of the polygynous mating system they seldom gain an opportunity to breed until 5 to 7 years old (11).

Fetal sex ratios, although skewed slightly toward males in free-ranging populations, do not depart significantly from parity (11, 80). Captive females kept on a high nutritional plain, however, produced proportionally more daughters than sons (88). Thus, nutritional condition of the female likely affects the rate of reproduction and the sex of her offspring.

Females seek steep, rugged terrain where they seclude themselves from other sheep for 1 to 2 days to give birth (89, 78, 73, 71). Neonates are exceptionally precocial and have been observed standing within 30 minutes following parturition; young have been observed traveling with their mothers within 24 hours of birth (89). Weaning generally is completed within 3 to 5 months (36).

In a cool summer with reduced availability of forage, females markedly curtailed the amount of time they spent nursing young (Figure 24–6). These females also attempted to compensate for a reduction in forage biomass by spending more time nursing young early in 1989 (73). Such differences in maternal care between years also were reflected in the proportion of unsuccessful suckles initiated by young and the proportion of suckling bouts terminated by adult females. Although the presence of a neonate may be a strong maternal stimulus, the amount of maternal care delivered clearly is under control of the adult female (73). Moreover, differences in maternal investment between years did not simply track environmental conditions, but represented a female strategy for coping with a harsh, unpredictable environment (73).

Reproductive Strategy

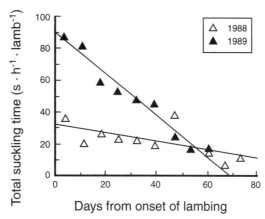

Figure 24–6 Comparison of total time spent suckling per active hour by young during poor (1989) and normal (1988) summers in interior Alaska, USA. Note that female Dall's sheep attempted to compensate by nursing longer in 1989 but also weaned lambs more quickly in that poor year. (From Ref. 73.)

Growth of well-provisioned young is rapid; body mass may attain 27 to 30 kilograms by 9 months of age (36). Indeed, milk of Dall's sheep is rich in protein (70%), and composed of about 12% fat by 10 days postpartum (90, 91). Losses of young may be 40% to 50% during their first winter, and the nutrition obtained during their first short summer of life may be crucial to their overwinter survival (2, 73). In addition to affecting survival, poor nutrition may cause females to forego reproduction in subsequent years and can slow growth of horns in males for 5 years following birth (78, 73, 92).

Males ultimately attain greater (about 40%) body mass than do females, and sustain growth for a longer period of time (6 years in males, 4 years in females) (36). Rates of growth (kilograms body mass/month) for the first 5 years have been estimated at 0.49 for males and 0.48 for females (85).

Maximum longevity of Dall's sheep in the wild appears to be about 14 years for males, and 19 years for females, although there is a slight tendency to underestimate the age of females by counting horn annuli (7, 92, 2, 93). Simply examining life expectancy for populations without considering other aspects of their demography, however, can lead to misinterpretation of data and the dynamics of populations (86).

Like other arctic ungulates, Dall's sheep have a suite of adaptations that help them cope with a severe climate. They possess an undercoat of fine wool, and long, hollow guard hairs that provide insulation against the long, cold winters (20, 11). The winter pelage of Dall's sheep may exceed 5 centimeters and is shed in a single molt that occurs from March through July; mature males molt earlier than females, young, or animals in poor physical condition (2, 11, 20). Another adaptation of these sheep relates to the concentrations of fatty acids in the marrow of leg bones, especially the radius. These fatty acids decrease in concentration distally, which likely helps to reduce heat loss via thermal conduction; hair covering the legs is comparatively short and probably does not provide sufficient insulation to prevent heat loss in winter (94).

Hematology and blood chemistry of Dall's sheep are similar to those reported for bighorn sheep (95, 96, 97, 35). Such data, however, may vary with sex, age, physical condition, season, handling, and numerous other factors (35, 98). Although some blood variables may correlate with physical condition, direct comparisons between populations using these values should be made with appropriate caution (98).

BEHAVIOR

Dall's sheep are an extremely gregarious species that exhibit a high degree of polygyny, with large, dominant males mating most often (11). Dall's sheep possess a tending-bond mating system, in which a dominant male guards, tends, and courts an estrous female (99, 11). Once the male has copulated with a receptive female, he soon leaves in search of additional mates. Competition among males for mates has lead to the evolution of marked dimorphism in body mass and size of horns between the sexes (1).

Rutting groups of Dall's sheep were composed of as many as 21 sheep, with an overall mean of 3.7 for 166 groups; however, group size declined for both males and females by midwinter (11). The mean size of 139 maternal bands ranged from about 4 to 10 individuals from prelambing through postlambing in spring and early summer (71). Group size is related to foraging efficiency with larger groups spending more of their active time feeding than smaller ones, and spending less time in vigilance or alarm behaviors than smaller groups (75, 71). Likewise, group size increases with increasing distance to escape terrain, ostensibly an adaptation to lower the risk of predation (100, 75, 71).

The sexes of Dall's sheep spatially segregate from one another during spring and summer (71). The causes of sexual segregation in bighorn sheep have been more thoroughly studied than in Dall's sheep, but the explanation is likely similar (101). Females are confined to steep, rugged terrain where they and their neonates can reduce the risk of predation, but where forage is less abundant and of lower quality than on areas inhabited by males. Larger males are less susceptible to predation than females with young, forage in areas with more predators, and obtain the forage required to attain large body size necessary to be effective in combat for mates. The hypothesis that this behavior resulted from males avoiding competition with potential mates and offspring (102) is no longer considered viable for most populations of mountain sheep (101).

The sexes begin to associate as rut approaches. By that time males already have formed groups and begun to sort out dominance interactions with relatively low-intensity behaviors that include foreleg kicks, horn displays, and more rarely jump-threats and clashes (11). Males also mount one another in dominance interactions, but the notion that dominant males treat all subordinates as if they were females may not be correct (11). For instance, dominate males seldom direct courtship behaviors such as tongue flicks toward subordinate males or lick their perianal region, and do not flehmen in response to smaller males or their urine. Dominance mounting is simply a common form of aggression among ungulates.

Aggressive interactions during rut are vigorous and tend to lack ritualization. Such behaviors include low-stretch postures, vigorous kicking, jump-threats, and forceful clashing of horns. The skulls of males have undergone pneumation to help absorb such forceful horn clashes (11). Males often display their horns, which serve as indicators of social rank (11). Aggressive interactions determine which males mate, and such behavior between males over estrous females can be fierce; large males may strike opponents in the side of the body with their horns and even push rivals off cliffs (11). Vigorous rutting activities by these large males exhaust their energy reserves, and survivorship drops markedly in older age classes of males (11).

Geist provided excellent descriptions of courtship behaviors in mountain sheep (11). Dominant males approach females in a low-stretch posture while flicking their tongues. Males ascertain the reproductive status of females by licking the perianal region of the female or lapping urine from the ground where a female has urinated. Males then flehmen to determine if the female is in estrus and receptive (103). Courting males may be distracted by females inducing them to investigate an area where a female has urinated, and then moving away while the male flehmens (11). Estrous females sometimes elicit courtship from a male by butting and rubbing against him (11). Males may kick a female with a stiff foreleg during courtship, ostensibly to determine if she is willing to stand for mounting. Copulation cannot occur if a female moves forward, which prevents the male from mating successfully (11). Males may tend females for 2 to 3 days prior to copulation (11). As rut concludes, large males become less gregarious, and aggressive behaviors occur less frequently (11). Some smaller males may continue to direct courtship behaviors toward females, but females typically are not receptive to such advances (11). No evidence of territoriality in Dall's sheep or bighorn sheep exists.

POPULATION DYNAMICS

Densities of Dall's sheep normally range between one and two animals per square kilometer; densities on winter range usually are higher than on summer range (66, 104). Densities of three to six sheep per square kilometer have been reported from Kluane Park, Yukon, and Dry Creek, Alaska (66). Comparisons of densities among populations, however, can reflect differences in habitat quality only if populations are at the carrying capacity (K) of their respective habitats.

Hoefs and Cowan believed that the adult sex ratio for Dall's sheep did not differ from parity (66). Their data, however, were gathered in spring when spatial segregation of the sexes can complicate sampling designs. Murphy and Whitten provided the best and most unbiased estimate of the adult sex ratio for an unhunted population: 66 adult males to 100 adult females (86). Adult sex ratios as low as 13 males to 100 females have been reported for a hunted population, although hunting alone probably was not the only explanation for that skewed ratio (66). A sex ratio favoring adult females is typical of polygynous ungulates and likely results from increased mortality of males engaging in strenuous rutting activities (11, 105, 77).

Several life table approaches have been used to examine the survivorship of male and female Dall's sheep (66, 7, 104). Time-specific life tables, however, do not provide accurate descriptions of the dynamics of Dall's sheep or other large mammals through time because population parameters such as survival and fecundity are fixed at the time of the sample and cannot reflect the changing dynamics of the population (106). Death-series life tables (i.e., reconstructed from ages of skulls) presented by Murie and Hoefs and Cowan are likewise problematical (7, 66). These tables assume a stable-age distribution, and Murphy and Whitten clearly demonstrated that this was not the case for data presented by Murie, because the partial survivorship curves varied markedly among cohorts (106, 86, 7). The restrictive assumptions of life tables and their inability to cope with density dependence reduce their usefulness for modeling the dynamics of populations of Dall's sheep (106).

Another concept that has limited the understanding of population dynamics of mountain sheep is the notion of "population quality" (11, 66). Population of sheep that evolved at the force of retreating glaciers were hypothesized to be more productive than those that evolved in glacial refugia (11). Although it is possible that some populations may be genetically superior to others, this has not been demonstrated for Dall's sheep. Small, bottle-necked or isolated populations might differ from others genetically, but "low-quality" populations have been postulated for areas in Alaska where this is not likely (66). Populations of Dall's sheep in interior Alaska exhibited little genetic variability based on allozymes (107).

More importantly, the concept of "population quality" as applied to Dall's sheep has no theoretical underpinnings in population ecology; indeed, this concept completely ignores density-dependent mechanisms or weather-related effects on productivity, survivorship, behavior, and other population parameters. In addition, no mechanism has been proposed to explain what makes some populations "high quality" or others "low quality," except the area (e.g., glaciated or refugia) in which they presumably evolved, which ignores millennia of evolution since the end of the Pleistocene.

What then regulates population dynamics of Dall's sheep? These mountain ungulates occupy habitats at high elevations that are subjected to extreme climatic conditions during winter, which can be highly variable among years (79). Moreover, summers are short and forage likewise may vary in abundance and quality among years (73, 71). Murphy and Whitten reported a strong inverse relationship between snowfall in the previous winter, and the ratio of young to females the following spring (Figure 24–7) (86).

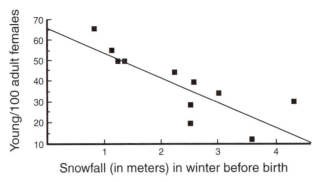

Figure 24-7 Relationship between young to female ratios and total snowfall during the previous winter before births, interior Alaska, USA. (From Ref. 86.)

Rachlow and Bowyer demonstrated that females curtailed investment in offspring more rapidly in a cool summer with lower quality and availability of forage than a normal year (73). Weather clearly affects both productivity and behavior in Dall's sheep.

Both Nichols and Murphy and Whitten argued that density-dependent mechanisms also were important in regulating populations of Dall's sheep; population density affects the per capita availability of forage to individuals (2, 86). Forage available in winter likely affects overwinter survival and perhaps the ability of pregnant females to carry a fetus to term (2, 108). Similarly, forage in spring and summer influences maternal provisioning of young and the likelihood of future reproduction by females (73, 71). Because young invest more resources in body growth than in fat deposition than do older age classes, undernourished young are less likely to survive harsh winter conditions, and overwinter mortality among young can be 40% to 50% (2).

Thus, both weather and density dependence interact to effect the dynamics of Dall's sheep populations. Maximal rates of increase in unhunted populations are probably between 11% and 18% (2). The year-round ability of habitat to support the needs of the sexes of Dall's sheep relative to the number of animals influences the per capita forage available to individuals and, thereby, their body size, age at first reproduction, fecundity, survivorship, and many aspects of their social behavior. This is one reason why comparing densities of sheep among populations may be inappropriate; the carrying capacity of the range in relation to the number of animals influences per capita availability of forage and, hence, productivity for the population. Variation in habitat quality makes comparisons of absolute density among populations meaningless for assessing density-dependent processes. These life-history traits also are strongly influenced by climatic conditions, which interact with population density to regulate populations of this mountain ungulate. Note the high variability in young to female ratios for a single population of Dall's sheep through time (Figure 24-7). The concept of "population quality" is not sufficient to understand this process.

We offer a simple conceptual model to illustrate our point and to help clarify interactions between density-dependent and density-independent phenomena in regulating populations of Dall's sheep (Figure 24-8). Although our model is focused on overwinter mortality, the same general ideas hold for other life-history characteristics (e.g., fecundity, age at first reproduction) related to population demography.

First, consider a population that is well below the carrying capacity (*K*) of the habitat (Figure 24-8a); such a circumstance might occur in a newly colonizing population from a reintroduction, from catastrophic mortality during a severe winter, or from an overharvest of females. Intraspecific competition in such populations is lax, and per capita availability of forage is high. Individuals reach a high nutritional plain, and body condition, including fat reserves, is excellent; reproductive rate and survivorship would be high (109). Winters of mild to moderate severity would have little effect on the overwinter survivorship of these individuals because they are well

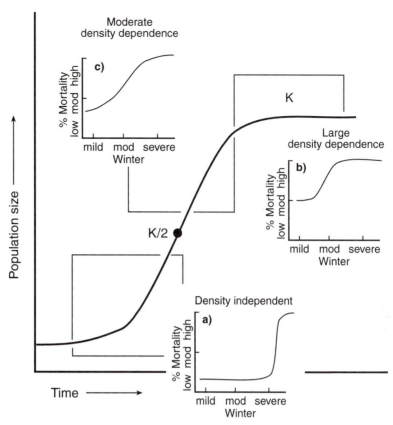

Figure 24–8 A conceptual model showing the relationships between population density, winter severity, and rate of mortality. Representative curves are provided for (a) density independent, (b) large, and (c) moderate density-dependent effects interacting with winter severity. The lines around the inset graphs show the area on the growth curve to which each inset corresponds. Note that the shape of the population-growth curve need not be symmetrical for the proposed relationships to hold.

buffered against such events by substantial body reserves. A sufficiently severe winter, however, might cause high rates of mortality even among animals in excellent physical condition.

Now consider the opposite extreme: a population at or near the *K* of the habitat (Figure 24–8b). Under this circumstance, intraspecific competition for forage is intense and, consequently, the average physical condition of individuals is poor. These animals have limited body reserves to withstand winter conditions, and even a mild winter is capable of causing some mortality. Moderate winter severity can produce high rates of death because these animals are poorly buffered against climatic extremes.

Finally, consider a population between *K*/2 and *K* (Figure 24–8c). Animals in this population would be in better physical condition than those at *K*, but in poorer condition than those at low density because of moderate levels of intraspecific competition for resources. Under those conditions, overwinter mortality would be higher than at low density, but lower than at high density. Indeed, this interaction between population density and winter severity produces a near-linear relationship between winter severity and overwinter rate of mortality (Figure 24–8c). Such linear relationships have been used to infer density-independent regulation of ungulate populations, but other interpretations are possible. Correlation should not be used to infer cause and effect, especially in population ecology (110). As an ungulate population approaches *K* and body reserves are diminished, climatic variables are likely to

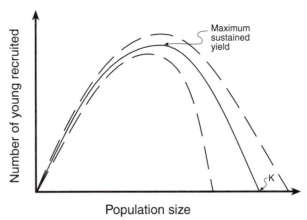

Figure 24–9 Conceptual relationship between the number of young successfully recruited into a population and population size (or density). The solid line shows the long-term mean, and the dashed lines indicate the effects of variable climatic conditions. Variability in recruitment becomes more pronounced at large population size because animals are in poor physical condition and more easily affected by climatic variability.

become increasingly important in predicting population parameters, even where such a population exhibits strong density dependence. Climate plays a more important role as populations approach *K* (Figure 24–9) because physical condition of individuals can be affected more easily by climatic variability. Moreover, our model indicates that individuals in good condition may survive all but the most severe winters, whereas animals in poor condition may succumb even during winters of mild or moderate severity (Figure 24–8), a conclusion that has been reached for other northern ungulates (111).

Predation also might affect the population dynamics of Dall's sheep. These mountain ungulates inhabit steep, rugged terrain that they readily negotiate to elude, avoid, and outdistance predators. Nonetheless wolves have been reported to prey on Dall's sheep, and diets of this canid contained 2% to 25% of Dall's sheep (112, 66, 113, 7, 114). Predation by wolves and coyotes may increase during periods of deep snow (108). Other predators of Dall's sheep include grizzly bears, black bears, wolverines, lynx, and golden eagles (7, 115, 89, 66, 116). Dall's sheep also may perish from accidental falls and be killed by avalanches (66, 108).

We concur with Nichols that predation does not exert an important influence on populations of Dall's sheep under most circumstances (2). For instance, Nichols noted that rates of increase in interior Alaska averaged 11% even though wolves were abundant (2, 117). Likewise, Rachlow and Bowyer noted that losses of young to predators in this same area were slight from May through July (78). Nichols reported that wolves may kill Dall's sheep more frequently when population size exceeds *K,* and sheep are weakened or forced by competition and food scarcity to forage far from escape terrain (2). These outcomes, however, likely involve substantial amounts of compensatory mortality, and it is questionable if predators typically regulate populations of Dall's sheep. This outcome is in stark contrast to the effects of predators on other northern ungulates that cannot seek protection in steep, rugged terrain (118, 119, 120, 79).

Dall's sheep also are infected or infested with numerous disease organisms and parasites, some of which may be capable of limiting or regulating their populations. Lungworm has been reported from Dall's sheep, but there have not been massive die-offs of Dall's sheep from this parasite and its associated pneumonias or hemorrhagic septicemia as reported for bighorn sheep (121, 122, 35). As with other sources of mor-

Population Dynamics

tality, this disease complex may be more prevalent in populations on a low nutritional plain (69). Other nematodes have been reported in Dall's sheep but caused no obvious pathology (69, 123). Perhaps low temperatures in winter or a short summer growing season at high latitudes helps limit direct transmission of parasite larvae or does not provide suitable conditions for intermediate hosts of pathogenic parasites that debilitate populations of bighorn sheep. Additionally, domestic sheep and cattle are uncommon throughout the range of Dall's sheep. Parasitic coccidia, however, have been isolated from feces of Dall's sheep (124). Further, positive titers in blood sera have been reported for several arboviruses, contagious ecthyma, parainfluenza III, epizootic hemorrhagic disease, and Q fever (125, 126, 127, 128, 97, 129, 130). Serology also indicated that Dall's sheep were exposed to the bacteria *Brucella* spp., and *Campylobacter feti* (97, 127). *Mycoplasma ovipneumoniae* has killed captive Dall's sheep, but was not reported for free-ranging sheep (131, 132).

Necrosis of horn cores and the mandible are prevalent in Dall's sheep; *Corynebacterium pyogenes, Fusobacterium necropitorum, Proteus* spp., *Micrococci* spp., and *Escherichia* spp. have been associated with infections of lumpy jaw (133, 7, 134, 135, 66, 136, 137). Causes of horn aberrations in Dall's sheep are uncertain but may involve *Actinomyces* spp. (137). Skull asymmetry was caused by osteoporosis in bighorn sheep (138). Dall's sheep may be predisposed to mandibular infections from excessive tooth wear because of wind-blown silt deposited on forage (139). At present, there is no evidence that diseases play a major role in regulating or limiting populations of Dall's sheep.

Competition between Dall's sheep and other large herbivores with sympatric distributions has not been documented (1). Caribou, moose, and mountain goats use different habitats or select diets differently than Dall's sheep (140, 141, 142).

POPULATION MANAGEMENT

Currently, there is no reason to attempt to manage populations of Dall's sheep using predator control or by intervening in disease processes; predators or diseases do not regulate populations of this mountain ungulate often enough to be of consequence. Effective management, then, is restricted largely to regulating sport and subsistence harvest or providing viewing opportunities for aesthetic and other nonconsumptive uses (1).

Dall's sheep are managed by different state, federal, and provincial government agencies within their distribution. In general, Dall's sheep are managed to furnish trophies for hunters; subsistence take is thought to be small (2, 143). Although some harvesting of females was permitted in the past, harvests now are restricted mostly to three-fourths or full-curl males (143, 144, 145, 146). The total harvest relative to the total size of populations is typically less than 2% (1). Most large males are harvested by nonresident hunters, who are required to follow a set of regulations that includes hiring guides or outfitters, payment of special fees, and hunting only in specified areas (1).

Nichols concluded that the harvest of large males did not influence reproduction in populations of Dall's sheep, and this practice was not sufficient to reduce herd numbers (2). This is not surprising because large males only use the same ranges as females during rut, and one consequence of sexual segregation is a reduction in competition of large males with females and young for much of the year. Thus, only the harvest of females can affect populations in a density-dependent manner, and their density interacting with climatic conditions is the primary factor regulating productivity of populations of Dall's sheep.

This interaction between density and climate also plays an important role in determining horn size in males and, hence, the quality of trophies available to harvest. For instance, Bunnell reported a strong positive relationship between precipitation (an index to primary productivity) and horn growth for male Dall's sheep (92). Moreover, years with depressed recruitment were years in which growth of horns also was diminished (92). The quality (i.e., growth) of horns of young males is mediated through the physical condition of the females, and this effect can be observed for up to 5 years of age (92). Annual growth of horns is much greater in males than females, and there is marked interannual variation in growth. Annual growth of horns and body mass were not correlated significantly for males (92). This analysis, however, was complicated by the horns composing 8% to 12% of body mass in large males and that some broming (i.e., wear) of horn tips occurred (11). Bunnell interpreted this lack of significance to mean that a male could recover from a period of nutritional deprivation and reattain its body mass, but that this period of hardship was recorded in its horns (92). That the cause of differences in horn size was related to nutrition was demonstrated by greater growth of horns in a male held on a game farm than for individuals in a free-ranging population from which the captive male was obtained (92). Finally, Bunnell calculated an index to horn quality that varied markedly among years (92). Clearly, there are not populations of high and low quality, but variable environmental conditions that cause changes in quality within a population that may vary through time and are dependent on nutrition.

One manner in which the harvest of large males might affect population demography was proposed by Geist (11). He argued that young males (i.e., less than a three-quarter curl) would experience high rates of mortality if large males were harvested heavily because younger individuals would begin participating in rutting activities that lead to high rates of mortality in larger males. Heimer and others offered some support for Geist's hypothesis (127, 11). Murphy and others, however, provided the only critical test of this hypothesis (147). They reported no relationship between the ratios of old or younger males to females across a number of populations throughout Alaska, and concluded that a reduction in older males via hunting did not affect survivorship of younger males.

One final way in which the harvest of large males might affect the demographics of a population is via the role of large males in initiation of estrus in females. The presence of a rutting male can hasten onset of estrus in bovids and cervids (148). Whether young but sexually mature males can fulfill this role in Dall's sheep is unknown. Even if estrus is delayed in populations without a sufficient number of large males, whether this would markedly affect timing or synchrony of parturition is uncertain because of the apparent ability of females to adjust the length of gestation (78, 83). Moreover, whether there might be a cost (e.g., low weight of a neonate) from adopting such a strategy requires more study. Indeed, this is one of the least known aspects of the biology of Dall's sheep.

HABITAT REQUIREMENTS

Dall's sheep generally inhabit windswept, dry, steep, and rugged mountains characterized by subalpine-grass and low-shrub communities typical of high elevations and high latitudes (7, 11, 48, 149, 71). Most populations of Dall's sheep are migratory and occupy different ranges in summer and winter, although a few populations are relatively sedentary (6, 7, 11, 66). Typical of other polygynous and sexually dimorphic ruminants, the sexes of adult Dall's sheep spatially segregate around the time of parturition (150, 101, 58, 71). Movements of Dall's sheep between seasonal ranges have been related to plant phenology, temperature, and depth of snow (66). Seasonal movements from 8 to 48 kilometers have been reported (66). Because summers are short at northern latitudes, Dall's sheep spend most of the year on winter range (males 271 to 303 days, females 240 to 263 days; 11). Windswept areas with sufficient forage and suitable escape terrain to elude predators are likely the key elements of winter habitat for Dall's sheep. For instance, Dall's sheep in Kluane National Park, Yukon, spent 70% of their time foraging in areas with little or no snow (less than 5 centimeters

Figure 24–10 Lambing habitat for Dall's sheep in interior Alaska was characterized by steep, rugged terrain intermixed with forage including grasses and dryas. (Photograph by J. L. Rachlow.)

deep), and less than 10% of their time in areas with snow depths greater than 15 centimeters (66). Primary productivity of plants on winter range (29–120 grams per square meter) is an important component of overwinter survival and for production of young (149, 139).

Adult males may occupy a variety of ranges throughout the year including areas inhabited during pre-rut, rut, early to midwinter, late winter, spring, and summer; these sheep also may move to areas with salt licks (11). Geist reported that such ranges were smallest in midwinter (about 0.8 kilometers in diameter) and largest in spring and summer (6 kilometers) (11). Adult females were reported to inhabit seasonal ranges in spring, for lambing, during summer, and in winter (11). Estimates of home-range size from modern, quantitative methods, however, are unavailable (151, 152). Additional information on habitat use outside the lambing period is provided by Hoefs and Cowan and Burles and Hoefs (66, 108).

Because of the severity of winters in the Arctic and sub-Arctic, growth and development of young Dall's sheep and replenishment of female body reserves must occur during the short summer (85, 78, 73). Moreover, maternal females are likely constrained in their selection of habitat because of the vulnerability of young to predators and to exposure and hypothermia from severe climatic conditions (153, 71, 75). Consequently, suitable lambing habitat may be a crucial component affecting the productivity of sheep populations.

Lambing habitat for Dall's sheep in interior Alaska was characterized by steep, rugged terrain intermixed with forage including grasses and dryas (Figure 24–10) (71). Lambing sites typically occurred at high elevation (above 1,180 meters) and were free of snow. Rachlow and Bowyer noted that a suite of variables was useful in discriminating lambing sites from random ones including distance to escape terrain, cover of grasses, cover of dryas, slope aspect, slope brokenness, slope steepness, and presence of snow (Figure 24–11) (71). Moreover, maternal females altered selection of habitat with the chronology of lambing, with additional variables entering at peak lambing that related to climate (windchill and cover from wind provided by browse). Additionally, females selected terrain features more strongly in a year with adequate food, but selected forages in a year with reduced availability of food (71).

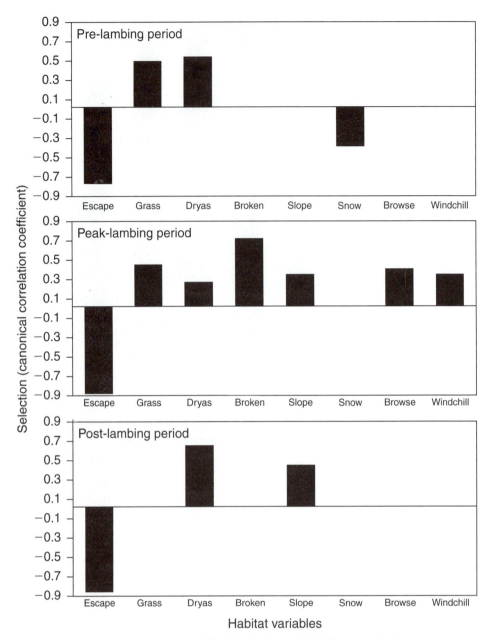

Figure 24–11 Selection of habitat by female Dall's sheep during pre, peak, and post lambing periods in interior Alaska, USA, 1988–1989. (Escape = distance to escape terrain, grass = percent cover of grasses, dryas = percent cover of dryas, broken = brokenness of terrain, slope = percent slope, snow = presence of absence of snow, browse = percent cover of browse, and windchill). Canonical correlations indicate the strength and direction of selection. For instance, the negative correlation for distance to escape terrain indicates females used sites closer to such terrain than randomly located sites. The positive correlation for grass indicates there was more of this forage at sites used by females than at random sites. (From Ref. 71.)

Young Dall's sheep apparently acquire home ranges from adults. Females typically have ranges that are similar to their mother (or maternal group), whereas males gradually disassociate from their mothers and begin associating with groups of mature males; as with many large mammals, males are the initial dispersers (11).

Habitat Requirements

Nonetheless, Dall's sheep exhibit a high degree of fidelity to seasonal ranges (males, 88%; females, 90%) (11).

HABITAT MANAGEMENT

Because of the remote and rugged terrain occupied by Dall's sheep, manipulation of habitat is rare. Subalpine ranges in British Columbia, Canada, were burned in spring to improve habitat (144, 69). Elliott noted that such burning was capable of slowing population declines and enhancing horn size in males (144). Nutrient quality of burned areas was not superior to that of unburned ranges in spring and summer, but burned areas had a greater quantity of forage than unburned areas in winter (69).

Mineral exploration and extraction, road construction, harassment by low-flying aircraft, and other human disturbances of Dall's sheep, especially on lambing grounds, have the potential to affect populations of sheep (12, 145, 146). Nonetheless, most of the range of Dall's sheep remains remote, pristine, and relatively unaffected by human developments or other activities.

Compared with other large mammals in North America, our understanding of the biology of Dall's sheep is incomplete. Much remains to be learned about these unique ungulates that inhabit mountainous areas of the far north.

SUMMARY

Dall's sheep occur in windswept, precipitous mountain ranges in Alaska and western Canada and are the northernmost species of mountain sheep in the New World. Three subspecies of Dall's sheep are recognized; Stone's sheep possess dark pelage and a rump patch, whereas the other subspecies are white or off-white and lack rump patches. Dall's sheep are stocky ungulates that are well suited to the steep, rocky terrain they inhabit. Males are larger in body mass than females and also possess much larger horns.

Dall's sheep feed mostly on graminoids but consume a wide variety of plant species. Forage is only of high quality during the brief summer growing season when sheep must accumulate sufficient body reserves to survive the long and often severe winter.

Dall's sheep are polygynous with a tending-bond mating system and are highly gregarious, especially during rut, which occurs from November to December. No evidence of territoriality has been observed. A single young typically is born in May after a gestation period of about 171 days. Births are highly synchronized, but marked interannual variation in both timing and synchrony of parturition has been reported. The sexes of Dall's sheep tend to be spatially separated outside the mating season.

The population dynamics of Dall's sheep are affected by a combination of density-independent and density-dependent processes. Young to female ratios are negatively correlated with snowfall in the winter prior to the birth of young, and intraspecific competition between females likely exacerbates the influence of such climatic extremes. Females also alter maternal investment in young in relation to environmental conditions in spring and summer. The growth of horns for males is dependent on their nutrition early in life when most horn growth occurs and influences the number of trophy males available for harvest.

Dall's sheep are preyed on by a variety of predators, but the rugged terrain inhabited by these mountain ungulates reduces the effects of predation. Predation, disease, or hunting of large males currently has few negative effects on populations of Dall's sheep.

Management of habitat for Dall's sheep is uncommon; burning of subalpine areas to improve range quality has occurred in British Columbia. Human disturbance holds the potential to affect populations of Dall's sheep, but most areas inhabited by these northern ungulates remain pristine and undisturbed.

LITERATURE CITED

1. BOWYER, R. T., and D. M. LESLIE, JR. 1992. Ovis dalli. *Mammalian Species* 393:1–7.

2. NICHOLS, L., JR. 1978. Dall's sheep. Pages 173–189 *in* J. L. Schmidt and D. L. Gilberd, eds., *Big Game of North America: Ecology and Management.* Wildlife Management Institute, and Stackpole Books, Harrisburg, PA.

3. NELSON, E. W. 1930. *Wild Animals of North America: Intimate Studies of Big and Little Creatures of the Mammal Kingdom.* National Geographic Society, Washington, DC.

4. WRIGHT, G. M., J. S. DIXON, and B. H. THOMPSON. 1933. Fauna of the national parks: A preliminary survey of faunal relations in national parks. *U.S. National Park Service Fauna Series* 1:1–157.

5. HORNADAY, W. T. 1935. *Hornaday's American Natural History: A Foundation of Useful Knowledge of the Higher Animals of North America,* 16th ed. Charles Scribner's Sons, New York.

6. DIXON, J. S. 1938. Birds and mammals of Mount McKinley National Park, Alaska. *U.S. National Park Service Fauna Series* 3:1–236.

7. MURIE, A. 1944. The wolves of Mount McKinley. *U.S. National Park Service Fauna Series* 5:1–238.

8. WRIGHT, R. G. 1992. *Wildlife Research and Management in the National Parks.* University of Illinois Press, Urbana.

9. GEIST, V. 1975. On the management of mountain sheep: Theoretical considerations. Pages 177–105 *in* J. B. Trefethen, ed., *The Wild Sheep in Modern North America.* Boone and Crockett Club and Winchester Press, New York.

10. RUSSELL, A. 1960. *Horns in the High Country.* Alfred A. Knopf, New York.

11. GEIST, V. 1971. *Mountain Sheep: A Study in Behavior and Evolution.* University of Chicago Press, Chicago.

12. NICHOLS, L. 1975. Report and recommendations of the Dall and Stone sheep workshop group. Pages 208–266 *in* J. B. Trefethen, ed., *The Wild Sheep in Modern North America.* Boone and Crockett Club and Winchester Press, New York.

13. KURTÉN, B., and E. ANDERSON. 1980. *Pleistocene Mammals of North America.* Columbia University Press, Columbia, NY.

14. PÉWÉ, T. L., and D. M. HOPKINS. 1967. Mammal remains in pre-Wisconsin age in Alaska. Pages 266–270 *in* D. M. Hopkins, ed., *The Bering Land Bridge.* Stanford University Press, Palo Alto, CA.

15. GUTHRIE, R. D. 1968. Paleoecology of the large-mammal community in interior Alaska during the late Pleistocene. *American Midland Naturalist* 79:346–363.

16. HARINGTON, C. R. 1978. Quaternary vertebrate faunas of Canada and Alaska and their suggested chronological sequence. *Sylogeus Series* 15:1–105.

17. HARINGTON, C. R. 1980. Pleistocene mammals from Lost Chicken Creek, Alaska, USA. *Canadian Journal of Earth Science* 17:168–198.

18. WEBER, F. R., T. D. HAMILTON, D. M. HOPKINS, C. A. REPENNING, and H. HAAS. 1981. Canyon Creek: A late Pleistocene vertebrate locality in interior Alaska, USA. *Quaternary Research* 16:167–180.

19. SUSHKIN, N. 1925. The wild sheep of the Old World and their distribution. *Journal of Mammalogy* 6:145–157.

20. COWAN, I. McT. 1940. Distribution and variation in the native sheep of North America. *American Midland Naturalist* 24:505–580.

21. STOKES, W. L., and K. C. CONDIE. 1961. Pleistocene bighorn sheep from the Great Basin. *Journal of Paleontology* 35:598–609.

22. STOCK, A. D., and W. L. STOKES. 1969. A re-evaluation of Pleistocene bighorn sheep from the Great Basin and their relationship to living members of the genus *Ovis. Journal of Mammalogy* 50:805–807.

23. KOROBITSYNA, K. V., C. F. NADLER, N. N. VORONTSOV, and R. S. HOFFMANN. 1974. Chromosomes of the Siberian snow sheep, *Ovis nivicola,* and implications concerning the origin of Ampniberingian wild sheep (subgenus Pachyceros). *Quaternary Research* 4:235–245.

24. GUTHRIE, R. D. 1984. Alaskan megabucks, megabulls, and megarams: The issue of Pleistocene gigantism. *Special Publication of The Carnegie Museum of Natural History* 8:482–509.

25. SCHALLER, G. B. 1977. *Mountain Monarchs: Wild Sheep and Goats of the Himalaya.* University of Chicago Press, Chicago.

26. CLARK, J. L. 1964. *The Great Arc of Wild Sheep.* University of Oklahoma Press, Norman.

27. HEPTNER, V., A. NASIMOVIC, and A. BANNIKOV. 1966. *Die Säugetiere der Sowjetunion.* Vol. 1, Paarhufer und Unpaarhufer, Gustav Fischer Verlag.

28. NADLER, C. F. 1971. Chromosomes of the Dall sheep, *Ovis dalli dalli. Journal of Mammalogy* 52:461–463.

28a. BUNCH, T. D., R. S. HOFFMAN, and C. F. NADLER. 1999. Appendix: Cytogenetics and genetics. Pages 263–276 *in* R. Valdez and P. R. Krausman, eds. *Mountain Sheep of North America.* University of Arizona Press, Tucson.

29. WURSTER, D. H., and K. BENIRSCHKE. 1968. Chromosome studies in the superfamily Bovidae. *Chromosoma* 25:152–171.

30. NADLER, C. F., R. S. HOFFMANN, and A. WOOLF. 1973. G-band patterns as chromosomal markers, and the interpretation of chromosomal evolution in wild sheep (*Ovis*). *Experientia* 29:117–119.

31. HIGHT, M. E., and C. F. NADLER. 1976. Relationships between wild sheep and goats and the aoudad (Caprini) studied by immuno diffusion. *Comparative Biochemistry and Physiology* 54B:265–269.

32. GROVES, P., and G. F. SHIELDS. 1997. Cytochrome *b* sequence suggests convergent evolution of the Asian takin and Arctic muskox. *Molecular Evolution and Cytogenetics* 8:363–374.

33. RAMEY, R. R., II. 1993. Evolutionary genetics and systematics of North American mountain sheep: Implications for conservation. Dissertation. Cornell University, Ithaca, NY.

34. CHERNIAVSKI, F. B. 1962. On the reproduction and growth of snow sheep (*Ovis nivicola* Esch.). *Zoologicheskii Zhurnal* 41:1556–1566.

35. SHACKLETON, D. M. 1985. *Ovis canadensis. Mammalian Species* 230:1–9.

36. BUNNELL, F. L., and N. A. OLSEN. 1976. Weights and growth of Dall sheep in Kluane Park Reserve, Yukon Territory, Canada. *Canadian Field-Naturalist* 90:157–162.

37. ULMER, F. A. 1941. On the weights of Dall's sheep. *Journal of Mammalogy* 22:448–449.

38. BLOOD, D. A., D. W. WISHART, and D. R. FLOOK. 1970. Weights and growth of Rocky Mountain bighorn sheep in western Alberta. *Journal of Wildlife Management* 34:451–455.

39. HALL, E. R. 1981. *The Mammals of North America,* 2nd ed. John Wiley and Sons, New York.

40. HEMMING, J. E. 1969. Cementum deposition, tooth succession and horn development: A criteria of age in Dall sheep. *Journal of Wildlife Management* 33:552–558.

41. GUTHRIE, R. D. 1972. Fannin's color variation of the Dall sheep, *Ovis dalli,* in the Mentasta Mountains of eastern Alaska. *Canadian Field-Naturalist* 86:288–289.

42. SCOTTER, G. W. 1980. Observation of a dark phase ram, *Ovis dalli,* District of MacKenzi, Northwest Territories, Canada. *Canadian Field-Naturalist* 94:464–465.

43. BUECHNER, H. K. 1960. The bighorn sheep in the United States: Its past, present and future. *Wildlife Monographs* 4:1–174.

44. LESLIE, D. M., JR., and C. L. DOUGLAS. 1980. Human disturbance at water sources of desert bighorn sheep. *Wildlife Society Bulletin* 8:284–290.

45. BLEICH, V. C., R. T. BOWYER, A. M. PAULI, R. L. VERNOY, and R. W. ANTHES. 1990. Responses of mountain sheep to helicopter surveys. *California Fish and Game* 76:197–204.

46. BLEICH, V. C., R. T. BOWYER, A. M. PAULI, M. C. NICHOLSON, and R. W. ANTHES. 1994. Mountain sheep *Ovis canadensis* and helicopter surveys: Ramifications for the conservation of large mammals. *Biological Conservation* 45:1–7.

47. BLEICH, V. C., J. D. WEHAUSEN, R. R. RAMEY, II, and J. L. RECHEL. 1996. Metapopulation theory and mountain sheep: Implications for conservation. Pages 353–373 *in* D. R. McCullough, ed. *Metapopulations and Wildlife Conservation.* Island Press, Coveco, CA.

48. LORD, T. M., and A. J. LUCKHURST. 1974. Alpine soils and plant communities of a Stone sheep habitat in Northeastern British Columbia. *Northwest Science* 48:38–51.

49. HOEFS, M. 1975. Estimation of numbers and description of wild sheep in the Yukon Territory. Pages 17–23 *in* J. B. Trefethen, ed. *The Wild Sheep in Modern North America.* Boone and Crockett Club, Winchester Press, New York.

50. HOEFS, M. 1989. Thinhorn sheep (*Ovis dalli*)—Distribution, abundance and management. Pages 105–137 *in Symposium on Wild Sheep of the World,* Prague, Czechoslovakia.

51. VALDEZ, R., and P. R. KRAUSMAN. 1999. Description, distribution, & abundance of mountain sheep in North America. Pages 3–22 *in* R. Valdez and P. R. Krausman, eds., *Mountain Sheep in North America.* University of Arizona Press, Tucson.

51a. NICHOLS, L., and F. L. BUNNELL. 1999. Natural history of thin horn sheep. Pages 23–77 *in* R. Valdez and R. R. Krausman, eds., *Mountain Sheep of North America.* University of Arizona Press, Tucson.

52. LAWSON, B., and R. JOHNSON. 1982. Mountain sheep. Pages 1036–1055 *in* J. A. Chapman and G. A. Feldhamer, eds. *Wild Mammals of North America: Biology, Management, and Economics.* Johns Hopkins University Press, Baltimore, MD.

53. GEIST, V. 1974. On the relationship of social evolution and ecology in ungulates. *American Zoologist* 14:205–220.

54. HARTL, G. B., H. BURGER, R. WILLING, and F. SUCHENTRUNK. 1990. On the biochemical systematics of the Caprini and Rupicaprini. *Biochemical Systematics and Ecology* 18:175–182.

55. RALLS, K. 1977. Sexual dimorphism in mammals: Avian models and unanswered questions. *The American Naturalist* 122:917–938.

56. ALEXANDER, P. D., J. L. HOOGLAND, R. D. HOWARD, M. NOONAN, and P. W. SHERMAN. 1979. Sexual dimorphism and breeding system in pinnipeds, ungulates, primates, and humans. Pages 402–434 *in* N. A. Chaonon and W. Irons, eds., *Evolutionary Biology and Human Social Behavior: An Anthropological Perspective.* Duxbury Press, North Scitvate, MA.

57. BERGER, J., and C. CUNNINGHAM. 1994. *Bison: Mating and Conservation in Small Populations.* Columbia University Press, NY.

58. WECKERLY, F. L. 1998. Sexual size dimorphism: Influence of mass and mating systems in the most dimorphic mammals. *Journal of Mammalogy* 79:33–52.

59. HIRTH, D. H., and D. R. MCCULLOUGH. 1977. Evolution of alarm signals in ungulates with special reference to white-tailed deer. *American Naturalist* 111:31–42.

60. BOWYER, R. T., J. L. RACHLOW, V. VAN BALLENBERGHE, and R. D. GUTHRIE. 1991. Evolution of a rump patch in Alaskan moose: An hypothesis. *Alces* 27:12–23.

61. GEIST, V. 1966. The evolution of horn-like organs. *Behaviour* 27:175–214.

62. LINCOLN, G. A. 1994. Teeth, horns and antlers. Pages 131–158 *in* R. V. Short and E. Balaban, eds., *The Differences Between the Sexes.* Cambridge University Press, Cambridge, UK.

63. LUNDGREN, B. 1996. Morphology of horns and fighting behavior in the family Bovidae. *Journal of Mammalogy* 77:462–475.

64. WILLIAMS, G. C. 1966. *Adaptation and Natural Selection: A Critique of Some Current Evolutionary Thought.* Princeton University Press, Princeton, NJ.

65. LUCKHURST, A. J. 1973. Stone sheep and their habitat. Thesis. Department of Plant Science, University of British Columbia, Vancouver, Canada.

66. HOEFS, M., and I. MCT. COWAN. 1979. Ecological investigation of a population of Dall sheep (*Ovis dalli dalli* Nelson). *Syesis* 12 (Suppl. 1):1–81.

67. WHITTEN, K. R. 1975. Habitat relationships and population dynamics of Dall sheep (*Ovis dalli dalli*) in Mt. McKinley National Park, Alaska. Thesis. University of Alaska, Fairbanks.

68. WINTERS, J. F., JR. 1980. Summer habitat and food utilization by Dall's sheep and their relation to body and horn size. Thesis. University of Alaska, Fairbanks.

69. SEIP, D. R., and F. L. BUNNELL. 1985. Nutrition of Stone's sheep on burned and unburned ranges. *Journal of Wildlife Management* 49:397–405.

70. HANSEN, M. C. 1996. Foraging ecology of female Dall's sheep in the Brooks Range, Alaska. Dissertation. University of Alaska, Fairbanks.

71. RACHLOW, J. L., and R. T. BOWYER. 1998. Habitat selection by Dall's sheep (*Ovis dalli*): Maternal trade-offs. *Journal of Zoology (London)* 245:457–465.

72. JONES, R. L., and H. C. HANSON. 1985. *Mineral Licks: Geography and Biochemistry of North American Ungulates.* Iowa State University Press, Ames.

73. RACHLOW, J. L., and R. T. BOWYER. 1994. Variability in maternal behavior by Dall's sheep: Environmental tracking or adaptive strategy? *Journal of Mammalogy* 75:328–337.

74. CHAPIN, F. S., III, 1983. Direct and indirect effects of temperature on Arctic plants. *Polar Biology* 2:47–052.

75. FRID, A. 1997. Vigilance by female Dall's sheep: Interactions between predation risk factors. *Animal Behaviour* 53:799–808.

76. ASDELL, S. A. 1964. *Patterns of Mammalian Reproduction.* Cornell University Press, Ithaca, NY.

77. BOWYER, R. T. 1991. Timing of parturition and lactation in southern mule deer. *Journal of Mammalogy* 72:138–145.

78. RACHLOW, J. L., and R. T. BOWYER. 1991. Interannual variation in timing and synchrony of parturition in Dall's sheep. *Journal of Mammalogy* 72:487–492.

79. BOWYER, R. T., V. VAN BALLENBERGHE, and J. G. KIE. 1998. Timing and synchrony of parturition in Alaskan moose: Long-term versus proximal effects of climate. *Journal of Mammalogy* 79:1332–1344.

80. NICHOLS, L., JR. 1978. Dall sheep reproduction. *Journal of Wildlife Management* 42:570–580.

81. BUNNELL, F. L. 1980. Weight estimation of Dall's sheep, *Ovis dalli dalli,* and mountain goats, *Oreamnos americanus. Wildlife Society Bulletin* 8:291–297.

82. HOEFS, M. 1978. Twinning in Dall sheep. *Canadian Field-Naturalist* 92:292–293.

83. BERGER, J. 1992. Facilitation of reproductive synchrony by gestation adjustment in gregarious mammals: A new hypothesis. *Ecology* 73:323–329.

84. BUNNELL, F. L. 1980. Factors controlling lambing period of Dall's sheep. *Canadian Journal of Zoology* 58:1027–1031.

85. BUNNELL, F. L. 1982. The lambing period of mountain sheep: Synthesis, hypotheses, and tests. *Canadian Journal of Zoology* 60:1–14.

86. MURPHY, E. C., and K. R. WHITTEN. 1976. Dall sheep demography in McKinley Park and a reevaluation of Murie's data. *Journal of Wildlife Management* 40:597–609.

87. CAMERON, R. D., and J. M. VER HOEF. 1994. Predicting parturition rate of caribou from autumn body mass. *Canadian Journal of Zoology* 71:480–486.

88. HOEFS, M., and U. NOWLAN. 1994. Distorted sex ratios in young ungulates: The role of nutrition. *Journal of Mammalogy* 75:631–636.

89. PITZMAN, M. S. 1970. Birth behavior and lamb survival in mountain sheep in Alaska. Thesis. University of Alaska, Fairbanks.

90. LAUER, B. H., and B. E. BAKER. 1977. Amino acid composition of casein isolated from the milks of different species. *Canadian Journal of Zoology* 55:231–236.

91. COOK, H. W., H. A. PERSON, N. M. SIMMONS, and B. E. BAKER. 1970. Dall sheep, *Ovis dalli dalli,* Milk: Part 1. Effects of stages of lactation on the composition of milk. *Canadian Journal of Zoology* 48:629–633.

92. BUNNELL, F. L. 1978. Horn growth and population quality in Dall sheep. *Journal of Wildlife Management* 42:764–775.

93. HOEFS, M. 1984. Reliability of aging old Dall sheep ewes by horn annuli technique. *Journal of Wildlife Management* 48:980–982.

94. WEST, G. C., and D. L. SHAW. 1975. Fatty acid composition of Dall sheep bone marrow. *Comparative Biochemistry and Physiology* 50B:599–602.

95. FRANZMANN, A. E. 1971. Physiologic values of Stone sheep. *Journal of Wildlife Diseases* 7:139–141.

96. BUTCHER, P. D., and C. M. HAWKEY. 1979. The nature of erythrocyte sickling in sheep. *Comparative Biochemistry and Physiology* 64A:411–418.

97. FOREYT, W. J., T. C. SMITH, J. F. EVERMANN, and W. E. HEIMER. 1983. Hematologic serum chemistry and serologic values of Dall's sheep, *Ovis dalli dalli,* in Alaska, USA. *Journal of Wildlife Diseases* 19:136–139.

98. KEECH, M. A., T. R. STEPHENSON, R. T. BOWYER, V. VAN BALLENBERGHE, and J. VER HOEF. 1998. Relationships between blood-serum variables and depth of rump fat in Alaskan moose. *Alces* 34:173–179.

99. HIRTH, D. H. 1977. Social behavior of white-tailed deer in relation to habitat. *Wildlife Monographs* 53:1–55.

100. HAMILTON, W. D. 1971. Geometry for the selfish herd. *Journal of Theoretical Biology* 31:295–311.

101. BLEICH, V. C., R. T. BOWYER, and J. D. WEHAUSEN. 1997. Sexual segregation in mountain sheep: Resources or predation? *Wildlife Monographs* 134:1–50.

102. GEIST, V., and R. G. PETOCZ. 1977. Bighorn sheep in winter: Do rams maximize reproductive fitness by spatial and habitat segregation from ewes? *Canadian Journal of Zoology* 55:1802–1810.

103. ESTES, R. D. 1973. The role of the vomeronasal organ in mammalian reproduction. *Mammalia* 36:315–341.

104. SIMMONS, N. M., M. B. BAYER, and L. O. SINKEY. 1984. Demography of Dall's sheep in the MacKenzie Mountains, Northwest Territories. *Journal of Wildlife Management* 48:156–162.

105. BOWYER, R. T. 1981. Activity, movement, and distribution of Roosevelt elk during rut. *Journal of Mammalogy* 62:572–584.

106. CAUGHLEY, G. 1977. *Analysis of Vertebrate Populations.* John Wiley & Sons, New York.

107. SAGE, R. D., and J. O. WOLFF. 1986. Pleistocene glaciations, fluctuating ranges, and low genetic variability in a large mammal (*Ovis dalli*). *Evolution* 40:1092–1095.

108. BURLES, D. W., and M. HOEFS. 1984. Winter mortality of Dall sheep, *Ovis dalli dalli,* in Kluane National Park, Yukon. *Canadian Field-Naturalist* 98:479–484.

109. MCCULLOUGH, D. R. 1979. *The George Reserve Deer Herd: Population Ecology of a K-Selected Species.* University of Michigan Press, Ann Arbor.

110. BOWYER, R. T., S. C. AMSTRUP, J. G. STAHMAN, P. REYNOLDS, and F. BURRIS. 1988. Multiple regression methods for modeling caribou populations. *Proceedings of the North American Caribou Workshop* 3:89–118.

111. WEIXELMAN, D. A., R. T. BOWYER, and V. VAN BALLENBERGHE. 1998. Diet selection by Alaskan moose during winter: Effects of fire and forest succession. *Alces* 34:213–238.

112. CHILD, K. N., K. K. FUJINO, and M. W. WARREN. 1978. Gray wolf, *Canis lupus columbianus,* and Stone sheep, *Ovis dalli stonei,* fatal predator–prey encounter. *Canadian Field-Naturalist* 92:399–401.

113. HOEFS, M., H. HOEFS, and D. BURLES. 1986. Gray wolf, *Canis lupus pambasilens,* in Kluane Lake Area, Yukon. *Canadian Field-Naturalist* 100:78–84.

114. GASAWAY, W. C., R. O. STEPHENSON, J. L. DAVIS, P. E. K. SHEPHERD, and O. E. BURRIS. 1984. Interrelationships of wolves (*Canis lupus*), prey and man in interior Alaska. *Wildlife Monographs* 84:1–50.

115. MURIE, A. 1981. The grizzlies of Mount McKinley. *U.S. National Park Service Fauna Series* 14:1–251.

116. NETTE, T., D. BURLES, and M. HOEFS. 1984. Observations of golden eagles, *Aquila chrysaetos,* predation on Dall sheep, *Ovis dalli dalli,* lambs. *Canadian Field-Naturalist* 98:252–254.

117. MURPHY, E. C. 1974. An age structure and reevaluation of the population dynamics of Dall sheep (*Ovis dalli dalli*). Thesis. University of Alaska, Fairbanks.

118. GASAWAY, W. C., R. D. BOERTJE, D. V. GRANGAARD, D. G. KELLEYHOUSE, R. O. STEPHENSON, and D. G. LARSEN. 1992. The role of predation in limiting moose at low densities in Alaska and Yukon and implications for conservation. *Wildlife Monographs* 120.

119. DALE, B. W., L. G. ADAMS, and R. T. BOWYER. 1994. Functional response of wolves preying on barren ground caribou in a multiple-prey system. *Journal of Animal Ecology* 63:644–652.

120. VAN BALLENBERGHE, V., and W. B. BALLARD. 1994. Limitation and regulation of moose populations: The role of predation. *Canadian Journal of Zoology* 72:2071–2077.

121. GOBLE, F. C., and A. MURIE. 1942. A record of lungworms in *Ovis dalli* (Nelson). *Journal of Mammalogy* 23:220–221.

122. STELFOX, J. G. 1971. Bighorn sheep in the Canadian Rockies: A history 1800–1970. *Canadian Field-Naturalist* 85:101–102.

123. GIBBS, H. C., and W. A. FULLER. 1959. Record of *Wyominia tetoni* Scoot, 1941, from *Ovis dalli* in the Yukon Territory. *Canadian Journal of Zoology* 37:815.

124. CLARK, G. W., and D. A. COLWELL. 1974. *Elmeria dalli*: A new species of protozoan (Elmeriidae) from Dall sheep, *Ovis dalli. Journal of Protozoology* 21:197–199.

125. ZARNKE, R. L., C. L. CALISHER, and J. KERSCHNER. 1983. Serologic evidence of arbovirus infections in humans and wild animals in Alaska, USA. *Journal of Wildlife Diseases* 19:175–179.

126. DIETERICH, R. A., G. R. SPENCER, D. BURGER, A. M. GALLINA, and J. VANDERSCHALIE. 1981. Contagious ecthyma in Alaskan muskoxen and Dall sheep. *Journal of the American Veterinary Medical Association* 179:1140–1143.

127. HEIMER, W. E., R. L. ZARNKE, and D. J. PRESTON. 1982. Disease surveys in Dall sheep in Alaska. *Symposium of Northern Wild Sheep and Goat Council* 3:188–197.

128. SMITH, T. C., W. E. HEIMER, and W. J. FOREYT. 1982. Contagious ecthyma in an adult Dall sheep, *Ovis dalli dalli,* in Alaska, USA. *Journal of Wildlife Diseases* 18:111–112.

129. ZARNKE, R. L. 1983. Serologic survey for selected microbial pathogens in Alaska (USA) Wildlife. *Journal of Wildlife Diseases* 19:324–329.

130. ZARNKE, R. L., R. A. DIETERICH, K. A. NEILAND, and G. RANGLACK. 1983. Serologic and experimental investigations of contagious ecthyma in Alaska. *Journal of Wildlife Diseases* 19:170–174.

131. BLACK, S. R., I. K. BARKER, K. G. MEHREN, G. J. CRAWSHAW, S. ROSENDAL, L. RUHNKE, J. THORSEN, and P. S. CARMAN. 1988. An epizootic of *Mycoplasma ovipneumoniae* infection in captive Dall's sheep (*Ovis dalli dalli*). *Journal of Wildlife Diseases* 24:627–635.

132. ZARNKE, R. L., and S. ROSENDAL. 1989. Serologic survey for *Mycoplasma ovipneumoniae* in freeranging Dall sheep (*Ovis dalli*) in Alaska. *Journal of Wildlife Diseases* 25:612–613.

133. SHELDON, W. C. 1932. Mammals collected or observed in the vicinity of Laurier Pass, B.C. *Journal of Mammalogy* 13:196–203.

134. HOEFS, M., T. D. BUNCH, R. L. GLAZE, and H. S. ELLSWORTH. 1982. Horn aberrations in Dall's sheep (*Ovis dalli*) from Yukon Territory, Canada. *Journal of Wildlife Diseases* 18:297–304.

135. BUNCH, T. D., M. HOEFS, R. L. GLAZE, and H. S. ELLSWORTH. 1984. Further studies on horn aberration in Dall's sheep (*Ovis dalli dalli*) from Yukon Territory, Canada. *Journal of Wildlife Diseases* 20:125–133.

136. GLAZE, R. L., M. HOEFS, and T. D. BUNCH. 1982. Aberrations of the tooth arcade and mandible in Dall's sheep from southwestern Yukon, Canada. *Journal of Wildlife Diseases* 18:305–310.

137. HOEFS, M., and T. D. BUNCH. 1992. Cranial asymmetry in a Dall ram (*Ovis dalli dalli*). *Journal of Wildlife Diseases* 28:330–332.

138. BLEICH, V. C., J. G. STAHMANN, R. T. BOWYER, and J. E. BLAKE. 1990. Osteoporosis and cranial asymmetry in a mountain sheep (*Ovis canadensis*). *Journal of Wildlife Diseases* 26:372–376.

139. HOEFS, M., and M. BAYER. 1983. Demographic characteristics of an unhunted Dall sheep, *Ovis dalli dalli*, population in southwest Yukon, Canada. *Canadian Journal of Zoology* 61:1346–1357.

140. KLEIN, D. R. 1953. A reconnaissance study of the mountain goat in Alaska. Thesis. University of Alaska, Fairbanks.

141. HENSHAW, J. 1970. Conflict between Dall's sheep and caribou. *Canadian Field-Naturalist* 84:388–390.

142. MIQUELLE, D. G., J. M. PEEK, and V. VAN BALLENBERGHE. 1992. Sexual segregation in Alaskan moose. *Wildlife Monographs* 122:1–57.

143. HEIMER, W. E. 1985. Population status and management of Dall sheep in Alaska. Pages 1–15 *in* M. Hoefs, editor. *Wild Sheep: Distribution, Abundance, Management and Conservation of the Sheep of the World and Closely Related Mountain Ungulates.* Special Report. Northern Wild Sheep and Goat Council, Whitehorse, Yukon, Canada.

144. ELLIOT, J. P. 1985. The status of thinhorn sheep (*Ovis dalli*) in British Columbia. Pages 43–47 *in* M. Hoefs, ed., *Wild Sheep: Distribution, Abundance, Management and Conservation of the Sheep of the World and Closely Related Mountain Ungulates.* Special Report. Northern Wild Sheep and Goat Council, Whitehorse, Yukon, Canada.

145. HOEFS, M., and N. BARICHELLO. 1985. Distribution, abundance and management of wild sheep in Yukon. Pages 16–34 *in* M. Hoefs, ed., *Wild Sheep: Distribution, Abundance, Management and Conservation of the Sheep of the World and Closely Related Mountain Ungulates.* Special Report. Northern Wild Sheep and Goat Council, Whitehorse, Yukon, Canada.

146. POOLE, K. G., and R. P. GRAF. 1985. Status of Dall's sheep in the Northwest Territories, Canada. Pages 35–42 *in* M. Hoefs, ed., *Wild Sheep: Distribution, Abundance, Management and Conservation of the Sheep of the World and Closely Related Mountain Ungulates.* Special Report. Northern Wild Sheep and Goat Council, Whitehorse, Yukon, Canada.

147. MURPHY, E. C., F. J. SINGER, and L. NICHOLS. 1990. Effects of hunting on survival and productivity of Dall sheep. *Journal of Wildlife Management* 54:284–290.

148. COBLENTZ, B. E. 1976. Functions of scent urination in ungulates with special reference to feral goats (*Capra hircus*). *American Naturalist* 110:549–557.

149. HOEFS, M. 1984. Productivity and carrying capacity of a subarctic sheep winter range. *Arctic* 37:141–147.

150. BOWYER, R. T. 1984. Sexual segregation in southern mule deer. *Journal of Mammalogy* 65:410–417.

151. Worton, B. J. 1989. Kernel methods for estimating the utilization of distribution in home-range studies. *Ecology* 70:164–168.

152. Kie, J. G., J. A. Baldwin, and C. J. Evans. 1996. CALHOME: A program for estimating animal home ranges. *Wildlife Society Bulletin* 24:342–344.

153. Berger, J. 1991. Pregnancy incentives and predation constraints in habitat shifts: Experimental and field evidence for wild bighorn sheep. *Animal Behavior* 41:66–77.

CHAPTER 25

Bighorn Sheep

Paul R. Krausman and David M. Shackleton

INTRODUCTION

A soldier in Coronado's army saw bighorn sheep at the confluence of the Gila and San Francisco Rivers in 1540, and described them "as big as a horse, with very large horns and little tails. I have seen some of their horns, the size of which was something to marvel at" (1). Many others have "marveled" at bighorns, as they became a part of Western folklore. Today, from western Canada and the United States to Mexico, bighorns are the subject of a myriad of electronic and written media. Not surprisingly, bighorn sheep are also one of the most highly sought big game species in North America. With most bighorn habitat on public lands, they are a resource that everyone can enjoy. Few species anywhere generate as much concern and effort to ensure their survival. Besides the various federal, state, and provincial wildlife agencies, the goals of large, professional organizations like the Foundation for North American Wild Sheep, the Arizona Desert Bighorn Sheep Society, the Desert Bighorn Council, the Fraternity of the Desert Bighorn, the Northern Wild Sheep and Goat Council, and the Society for the Conservation of Bighorn Sheep, emphasize conserving and otherwise benefiting bighorn sheep.

TAXONOMY AND DISTRIBUTION

The ancestral home of wild sheep is most likely Asia, and the fossil record indicates a large argaliform sheep occurred there during the early Pleistocene epoch about 2 million years ago (2). While the oldest known fossil sheep in North America come from Yukon deposits, zoogeographers suggest that the ancestors of North American wild sheep became isolated from their Asiatic progenitors during the middle Pleistocene (3). During this period, massive continental ice sheets periodically covered much of northern North America from coast to coast, causing sea levels to drop, creating a connection (i.e., the Bering Land Bridge) between eastern Siberia and western Alaska. This land bridge provided passage between the two continents for many animals and plants, although most moved from the Old to the New World (4). Once established in North America, mountain sheep probably evolved their unique characteristics in isolation from their Asiatic ancestors, although the snow sheep of Siberia share many similar traits and are included with North American species in the Pachyceriform sheep (2).

Modern mountain sheep in North America are divided into two basic groups: thinhorn (Dall's and Stone's) of Alaska and northwestern Canada, and the bighorn of the western mountain systems and deserts (Figure 25–1). The characteristics used to classify different races of bighorn sheep depend on relatively small differences in

Figure 25–1 Adult male and female Rocky Mountain bighorn sheep. (Photograph by D. M. Shackleton.)

cranial dimensions and it is unlikely that all currently recognized subspecies will be maintained when bighorn taxonomy is revised (5, 6, 7). For now, the bighorn group includes the "type" species—the Rocky Mountain bighorn, the California bighorn, and the desert races. Further details can be found in the historical treatments of the nomenclature of North American mountain sheep, given by Allen, Osgood, and Cowan, and Wehausen and Ramey (8, 9, 10, 5, 6).

Rocky Mountain and California bighorn sheep are widely distributed in localized populations throughout the mountain regions of western North America (Figure 25–2). Not surprisingly, the Rocky Mountain bighorn sheep's distribution closely follows the Rocky Mountains, and extends from about 55°N in Alberta and British Columbia, south through Montana, Idaho, Utah, Wyoming, Colorado, and into northern New Mexico at around 36°N (11, 12). California bighorn historically ranged from the eastern slopes of the Coast Mountains in central British Columbia (51°N), south into Washington, Oregon, and Idaho as far as the Sierra Nevadas in California (37°N) (5). However, from 1900 until 1954, this subspecies became extinct over much of its distribution, especially in the United States. Subsequently it has been translocated, mainly from British Columbia, to restock and reestablish populations in California, Oregon, Washington, Idaho, Nevada, and North Dakota (13). Desert bighorn sheep formerly occupied ranges from Nevada (40°N) to Baja California, Mexico (24°N), and from western Texas, southern New Mexico and Arizona, western Colorado, and Utah to California (14). This vast region encompasses several ecologically distinct life zones and indicates the desert bighorns' affinity for rugged, extremely arid, and sparsely vegetated desert environments.

LIFE HISTORY

Diets

Bighorn sheep diets have been well studied, and research has focused on their nutritional requirements, on foraging impacts on vegetation, and on competition and dietary overlap with native and exotic herbivores. Various methods were used to study bighorn diets including analysis of rumen contents of collected animals, direct observations of

Life History

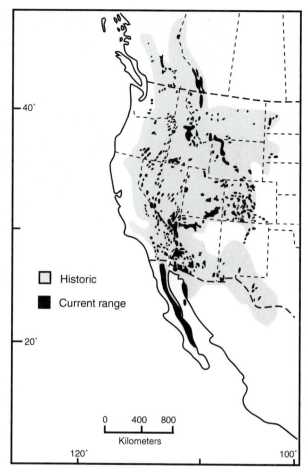

Figure 25–2 Historic and current ranges of bighorn sheep in North America.

feeding, laboratory studies of captive animals, and detailed analyses of monthly and seasonal diets (1, 15, 16, 17, 18, 19, 20, 21, 22, 23, 24, 25, 26, 27, 28, 29, 30, 31, 32, 33). Despite all this work, it is difficult to make more than the broadest generalizations about bighorn diets because of the differing methodologies employed (34).

The general consensus is that bighorn are relatively opportunistic in their diet selection, using whatever palatable species are available to them (35, 27, 36, 34). For example, Wikeem found that more than 267 plant species were eaten by California bighorn, while Browning and Monson's review of desert bighorn diets showed they used more than 470 species throughout their distribution (37, 36). Forbs are most frequently eaten by Rocky Mountain and California bighorn, followed by grasses and lastly browse, whereas browse dominates desert bighorn diets, followed by grasses or forbs depending on precipitation patterns (34, 24, 32, 33). Perhaps not surprisingly, the relative amounts of these three forage classes vary significantly in bighorn diets, among populations, and within subspecies and among age-sex classes and individuals (38, 35, 39, 40, 41, 42, 43, 21, 20, 22, 32, 33, 44, 45).

Besides geographic, taxonomic, and individual variation, bighorn diets show strong seasonal variation. Most probably this reflects changes in the availability and palatability of forage species and in the nutritional requirements of the animals, although forage nutrient quality does not correlate well with what bighorn eat (34). In desert regions, seasonal use of different forages has been attributed to precipitation patterns and the effects of soil moisture on vegetation classes (20, 21, 32). However, observed declines in their use of browse could be due to declines in palatability when twigs become coarse and woody, and possibly when they exceed an acceptable diameter (46).

Figure 25–3 Although critically important to desert bighorns, water availability cannot be used to predict habitat use. (Photograph by R. C. Etchberger.)

Water Requirements

All bighorn sheep seem adapted to arid conditions whether they live in hot or cold climates, and they can survive for long periods without free standing water. They can meet most of their water requirements from vegetation in summer and from snow or ice in winter (47, 48, 49, 50, 51). Desert bighorn are particularly well adapted to survive on preformed water in their food, and on metabolic water formed as a result of oxidative metabolism. They require a minimum of 4% to 5% of their body mass in water per day during summer and 1% to 2% of their body mass during winter, although individual consumption levels vary. Increased day-length, high ambient temperatures, reduced forage moisture content, and mating activities require additional water intake and can result in their dependence on surface water (52, 53). During summer, desert bighorn may go without water for more than 15 days, resulting in a loss of more than 20% of the hydrated body mass (30% of their total body water) (52). Not surprisingly, there is a close relationship between the distributions of desert sheep and water in the Southwest (18, 54, 46). In the Cabeza Prieta National Wildlife Refuge, Arizona, the location of water sources was the major factor influencing elevational distribution of bighorn, particularly during summer. When temperatures dropped, however, sheep became less dependent on surface water and dispersed to the cooler summits of local mountains (55). Despite this need for water, a habitat model for desert bighorn based on vegetation, topography, and water availability indicated that each component was important, but none could be used to predict bighorn habitat use (56)(Figure 25–3).

Reproduction

Ovulation and spermatogenesis usually begin around 18 months of age or earlier, but wild bighorn do not become fully sexually active until later (57, 58, 59). Most females mate first when at least 2.5 years old, and male Rocky Mountain sheep usually do not begin to participate fully in the rut until 7 to 8 years old, well after puberty (58). However, in expanding populations or in rare cases, female bighorn have given birth to their first lamb at 18 months (57, 60, 61, 62, 49, 63, 64). Sexual activity can occur much earlier in captivity; captive desert bighorn male lambs began spermatogenesis at 26 to 28 weeks and exhibited a seasonal spermatogenetic cycle after 21 months of age, while in other studies captive yearling desert bighorn males inseminated all females living in their enclosure (65, 66, 62). Irvine found no apparent decrease in spermatogenesis with increasing age and concluded that even the oldest males were capable of breeding (18).

The estrous cycle in bighorns lasts 28 days, with a receptive period of around 48 hours (67). Gestation lasts 173 to 185 days in desert sheep, 173 to 175 days in California bighorn, and 173 to 176 days in Rocky Mountain sheep (67, 68, 69, 59, 70). Females in all races usually produce a single lamb each year until old age or death overtakes them. Most males, on the other hand, actively mate for only a few years, during which time they may inseminate many females. Successful mating is not entirely a function of sexual maturation, as social hierarchy and behavior play major roles (58, 61).

Turner and Hansen defined the rutting or mating season as that period in which mating activities result in 70% of the lamb production for the following season (67). The duration of the mating season is longer at lower elevations and southern latitudes, and shorter at higher elevations and more northern latitudes (71, 72). Rocky Mountain and northern populations of California bighorn mate in late autumn and early winter, with the mating season beginning as early as late October, or as is typical for most Rocky Mountain populations, in early November (73, 74, 13, 47, 58). Sometimes mating extends into late December and even occasionally into early January, but usually peaks between mid-November and mid-December (75, 76, 77, 73, 35, 74, 78, 58, 60). By contrast, desert bighorn exhibit an extended mating period, and births have been documented in the Sonoran Desert in all months except October (25, 54, 79, 80). This variation in mating seasons appears related to environmental conditions at the time of birth.

Average climatic and forage conditions vary relatively predictably and seasonally in most areas inhabited by California and Rocky Mountain bighorn. Variation in these two factors is extremely important for reproduction, especially for the timing of the birth season. Climatic conditions can affect the survival of newborn lambs and also the forage quality and quantity that are important for lactation (58, 81, 82). However, lambs must also grow large enough to survive their first winter (83). So for bighorns, as with most other northern ungulates, the birth season is a trade-off between young being born early enough for adequate prewinter growth and being born late enough to avoid the thermal stress and poor forage conditions of late winter (84).

Thompson and Turner assessed the temporal variation in parturition seasons for 22 populations of bighorn sheep (72). In those from northern latitudes, parturition seasons were shorter, later, and cued to brief, relatively predictable periods of vegetation growth (71, 72).

The birth period of Rocky Mountain and California bighorn begins in early spring (late April or May) usually coinciding with initiation of spring vegetation growth and ameliorating climatic conditions. Few lambs are born after June (34). However, the same factors may operate differently for desert bighorn sheep due to low population density (85). More importantly, Thompson and Turner found poor correlation between the inception and duration of the vegetation growing season and the lambing period in desert bighorns (72). They, together with Lenarz and Conley, concluded that an extended lambing season was a result of unpredictable precipitation patterns, and consequently plant regrowth, both of which appear essential for maternal and neonatal survival (79). For desert bighorns, seasonal fluctuations in resources are not as predictable from year to year. Throughout much of the range of desert bighorn, plant productivity is related directly to temporal and spatial precipitation patterns, and these vary considerably and unpredictably. Nonseasonal reproductive behavior may be an adaptive strategy of desert bighorn that ensures lamb survival during periods of varying and unpredictable forage production. An extended lambing period would increase the probability that late gestation and early lactation would coincide with a period of adequate precipitation and forage availability (54, 46, 72). Such reproductive responses to unpredictable resources are not uncommon in desert vertebrates (84).

Figure 25–4 Male Rocky Mountain bighorns, like this mature 10-year-old, typically lose mass during rut and over winter. (Photograph by D. M. Shackleton.)

Growth and Development

From birth weights of 3 to 5.5 kilograms, bighorns grow rapidly during their first 6 months of life (35, 58, 59, 86, 87, 88). Thereafter, growth continues each spring through to early fall, until about 3 or 4 years in females and around 6 years in males. As a result of faster growth rates and a longer growing period, male bighorns quickly become larger and heavier than females. Sex differences in mass are noticeable by the first year, and by 2 years of age, males can weigh 18% more than females and 65% more by age 6 years (89). The heaviest mass recorded for a Rocky Mountain male was 137 kilograms, with females reaching a maximum of around 90 kilograms (90, 91). Desert bighorn are smaller, with average weights for adult males and females of 82 and 48 kilograms, respectively (92).

Growth rate varies between and within populations and can be depressed by factors such as parasite loads (88, 90, 93). All age classes of Rocky Mountain, except sometimes lambs, and many Californian bighorn usually lose mass due to rutting expenditures, overwinter maintenance costs, or both (89, 87) (Figure 25–4). For most age classes, these overwinter mass changes probably reflect the use of fat reserves, and reported losses range from 7% in yearlings to more than 20% in adults (90, 87, 88, 89).

Bighorn Sheep have 32 teeth. The upper/lower dental formula is I, 0/3; C, 0/1; P, 3/3; and M, 3/3.

Female horns are much smaller and shorter than male horns. Female horns are relatively thin and gently curved, while those of adult males are massive at the base, tapered, and curled in a spiral as they grow. Horn growth each year occurs when a new keratin horn sheath develops over the underlying bone horn-core attached to the top of the skull. Each year's horn sheath grows beneath the preceding year's horn, so that the horn sheaths are grown one inside the other, and so except for the first or lamb horn, only a part of the annual horn sheath is exposed (94, 34). When horn growth stops, probably in fall or early winter, and is then followed by regrowth of a new sheath the following spring, a distinct break or annual ring is formed.

Horns and horn growth can be of value to wildlife biologists and managers for several reasons. First, individuals can be aged by counting the horn rings or annuli that develop each year. However, with long-lived individuals, it is usually easier to age males than females. After about 4 or 5 years of age female horns grow very little, probably because of reproductive costs, so their horn rings become very crowded and hard to distinguish. In males, the first two years or more of horn growth can be lost to "brooming" or breakage of the horn tips during fighting, making precise aging problematic in older animals (95). Although only relative measures of annual horn length are possible because all but the first year's (lamb) horn are partially hidden by preceding ones, measuring annual horn growth can provide insights into an individual's and a population's status. When the average lengths of the exposed annual horn sheaths are plotted (for males or females), the generalized pattern of horn growth is relatively consistent among bighorn sheep populations. The longest visible horn sheath almost always is the one grown in the second year of life, after which exposed horn length decreases rapidly. Both the maximum annual horn growth in the second year and the rate of "decline" in annual horn length vary among populations reflecting environmental rather than genetic differences (58, 60, 96). Relative annual horn growth is initially greater in high-quality populations (i.e., rapid population growth, fast individual growth rates, early maturation, early mortality, intense social interaction, and high milk production) though it declines faster than in stable or declining, low-quality populations (58, 60, 96, 97). Thus, managers may be able to use the average pattern of annual horn sheath growth and body growth to compare populations and evaluate management options (58, 98, 60, 96, 99, 97). This can be especially valuable for wildlife managers because size limits for trophy hunting are often described in terms of horn size (e.g., three-quarter curl and full curl) and hence are influenced by horn growth.

BEHAVIOR

Social Organization

Bighorn are highly social animals that spend their life in groups with other sheep (Figure 25–5), although group integrity remains flexible to some degree throughout the year (58, 54). In general, group composition consists of spatially and sexually segregated units of all-male groups, and female–juvenile groups comprised of adult females, lambs, and offspring from the previous 1 to 2 years (58). During the mating

Figure 25–5 Band of male Rocky Mountain bighorns. (Photograph by D. M. Shackleton.)

season, adult male and female–juvenile groups join for the duration of the rut, which usually takes place within the females' home range. Sandoval described a fourth group type in desert bighorn—groups of barren adult females, yearlings, and socially mature males (46).

Group sizes vary by group type, seasonally, and geographically, and range from 2 to more than 100 individuals (34, 77, 75, 35, 74, 100, 101, 78, 58, 60, 102, 103). Lone bighorns, most often males, are uncommon and probably only transitory (77, 75, 74, 58, 60, 103). Groups appear to provide two main advantages over individuals in many species: improved foraging efficiency and predator avoidance (104). In bighorns, as in other ungulates, predation is probably a major selective force in the formation and maintenance of groups (105, 106, 107).

Males and females in most bighorn populations occupy separate, seasonal ranges, although spatial and temporal overlap does occur (58, 108, 109, 103, 32, 33). Several explanations have been offered for sexual segregation in bighorn (108, 110, 44, 103, 107). The most probable, which also applies to other ungulates, is that these differences in habitat use are most likely due to females selecting secure areas for raising their young, and to males choosing areas for maximizing body condition (111, 112). Bleich and others supported this and concluded that sexual segregation in mountain sheep most probably results from different reproductive strategies of the sexes (33). To optimize their fitness, male bighorn need to grow large bodies and horns to improve their chances in intrasexual competition and hence mating success. As a result, they need high-quality food resources, and because of their larger size can use areas with a higher predation risk. Females on the other hand are smaller than males and are accompanied by highly vulnerable young, so they cannot afford to use areas with high predation risks. Thus, females may forgo foraging quality for higher security (see also 81).

All but young males are socially dominant over adult females. Among males, it seems that age, horn size, and probably body size and fighting behavior are positively correlated with dominance status (113, 58). Large horned, older males are the most dominant and also those most likely to mate with females (58, 114, 115, 61). Females do not show any clear relationships between physical attributes and social status; in fact, their dominance hierarchies appear less linear and more subtle than those of males (116, 117, 118). Only age and nursing rate have been related to social status in female bighorn, but there is no evidence so far that a female's status is related either to her reproductive fitness or to her differential investment in male and female offspring (118, 116, 117).

Movements

With few exceptions, males and females in most populations migrate between different areas at different seasons of the year, and there are usually at least two seasonal ranges for Rocky Mountain and California bighorn (i.e., winter and summer) and three for desert bighorn (summer, fall–winter, and spring) (77, 35, 58, 119, 120, 121, 54, 122, 123, 32). Bighorn seasonal movements are not attributable to any one specific factor, probably because animal movements are dictated by an individual's response to a variety of stimuli (124). Behavioral, physiologic, and environmental factors seem to influence these migrations, including home range knowledge, water and forage availability, lambing and mating activities, season, topography, and age and sex class.

Female desert bighorns have a relatively small home range, especially compared to those occupied by adult male groups (32, 125). Most desert bighorn populations are restricted to small areas during hot summer months due to suboptimal distribution of resources. As a result, local rainfall patterns significantly affect seasonal range size and movement patterns of desert bighorn. Summer showers that fill natural water catchments allow sheep to use areas not normally available during dry periods.

For most Rocky Mountain and some California bighorn populations, seasonal movements are not just changes in location but also involve vertical (altitudinal) migrations. These annual elevational movements between ranges have been examined from a functional and adaptive viewpoint. Hebert concluded that upward movements in spring and summer allowed bighorn to maintain or prolong a diet composed mainly of new, growing vegetation because the onset of plant phenology is retarded by increased elevations (126). New growing vegetation provides high-quality, readily digestible forage; thus sheep prolong the period of high forage quality by making vertical migrations from low to high elevations in spring and summer. Differences between bighorn populations in the elevation ranges used may affect development of individuals and populations (60, 96, 127).

Although vegetation growth is probably a major factor stimulating the upward vertical migrations of bighorn sheep in spring and summer, other factors are also involved. If suitable lambing grounds are separate from the winter range, the primary stimulus for pregnant females is to move to more secure habitat just before birth to avoid predators. This drive for security from predators can even override selection of forage conditions. Festa-Bianchet showed that females left areas with good forage and gave birth in areas of lower forage quality but with greater security cover (81). Similar trade-offs have been documented for desert bighorn sheep (107, 33, 128). Etchberger and Krausman also concluded that the traditional use of parturition sites in rugged terrain outweighed the increased need for water by lactating females (128). Climatic factors such as snow accumulation are also believed to stimulate Rocky Mountain and California bighorns to return to lower elevations in late summer and early fall (77, 35, 74). Avoidance of biting insects may cause movements to more exposed areas (74). Availability and distribution of mineral licks may also affect space use by bighorn (77).

In general, most Rocky Mountain and many California bighorn populations return to winter ranges in October and November (77, 101, 129, 58, 130). Similarly, most sheep leave their winter range in May and June, which for females is most likely a response to lambing and foraging requirements. Perhaps not surprisingly, distances sheep move between seasonal ranges vary (some move up to 70 kilometers), most probably in relation to the availability and distribution of suitable ranges (74, 100, 101, 131, 132). Desert bighorn migratory patterns also vary considerably, with most populations remaining in isolated areas, while others move between mountain ranges. Migratory patterns of desert bighorn may be classified as seasonal elevational movements within the same range, long-distance annual migration between mountain ranges that include elevational movements and dispersal from water sources, and movement from and to seasonally available water sources (133).

Social Behavior

Social relations begin at birth when a female bighorn and her young must learn to recognize each other. Lactation is costly and females cannot afford to squander scarce resources on strange offspring, while the young needs to be able to find its mother for food and protection. This mutual recognition, or mother–young bond, seems to develop during the first 1 to 2 days following birth when the female is isolated from other group members (58, 134). Females rely more on their neonate's smell for recognition, invariably sniffing its rear when it begins to suckle. Lambs by contrast, quickly learn the sound of their mother's voice and it is interesting to watch when a female gives an alarm bleat. Her young invariably runs straight to her, often briefly suckling, while the female sniffs its rear. The suckle probably reinforces the young's response to the alarm call (134).

Researchers have measured several variables related to nursing-suckling in bighorn and found they change with the lamb's age. General trends are similar, but significant differences occur within and between populations (135, 136, 99, 134, 82, 137, 58, 60, 138, 139,). Length of suckling bouts, suckling rates, and total time spent suckling decrease with age, as the lamb spends time grazing (58, 135, 136, 60, 139, 134, 82). Times spent nursing are also affected by whether they follow bedding periods, and by maternal age, condition, and lungworm loads, but not by the sex of the offspring (134, 82).

Figure 25–6 Clashing horns is a spectacular behavior used to determine social dominance. (Photograph by D. M. Shackleton.)

Adult male bighorn have a diverse repertoire of social behavior patterns they perform to females and other males (58). Fighting is their most spectacular and well-known behavior. Just before the rut, and sometimes in spring, males gather to interact and fight with each other, presumably to determine relative dominance status. Fights are usually between pairs of males, but sometimes small groups or "huddles" of three or more males will interact (58). When fighting, males rear up on their hind legs and run bipedally toward each other, lunging headfirst at the last moment to clash horns (Figure 25–6). These echoing clashes can be repeated many times before one of the males is defeated. Dominant males then treat the subordinates like estrous females, using the same behavior patterns as in courtship (58).

The social behavior pattern most often used by bighorns is the "low stretch." An animal, usually a male, approaches or passes another with head low, neck extended, and nose pointed upward (58). This pattern is used in many social situations, and especially in the rut when males use it to approach and test females. Females usually respond by squatting and urinating, then the male sniffs the urine and lip curls (flehmen) testing whether the female is coming into estrus (Figure 25–7). Once a female is found coming into heat, she is guarded and courted by a dominant male, though the pair is often surrounded by eager, but subordinate, males of various ages.

Rocky Mountain bighorn use as many as three mating tactics. The more typical is the tending pair, less frequent is "coursing" or rape-chase, and an even rarer tactic is "blocking" (58, 61, 114, 115). During actual courtship of an estrous female, the tending male uses a variety of behavior patterns that include nosing the female's flank and rear, usually while twisting the top of the head to one side and accompanied by flicking the tongue and vocalizing, gently kicking her with his foreleg, pushing his chest against her rump, and finally copulation attempts (58, 61).

POPULATION DYNAMICS

Population Trends and Current Status

Seton estimated that in pristine times there were about 2 million mountain sheep (thinhorn and bighorn) in the contiguous United States, and another 2 million in Canada and Alaska combined (1). Seton's estimate of 4 million sheep is often cited

Figure 25–7 A male bighorn does a lip curl or flehmen to determine if a female is coming into estrus. (Photograph by D. M. Shackleton.)

as a reliable approximation of mountain sheep numbers (1). However, Valdez doubted wild sheep numbers ever exceeded 500,000 for all North America (140). Mountain sheep are highly selective in their habitat preferences, and it is probably a misconception that sheep were uniformly distributed throughout the mountains of western North America.

The present distribution ranges of all races of bighorn sheep seem considerably veduced, and they may be occupying habitats in the most remote portions of their historical distribution (77, 35). Bighorn sheep were quite well adapted to habitats far from the rugged terrain more often considered typical for the species (5, 77, 75, 73, 12, 141, 142). Valdez and Krausman and Krausman estimated the number of bighorns in North America in 1991 (143, 144). At that time, Canada supported 12,700 Rocky Mountain bighorn, and 3,000 California bighorn sheep. Mexico supported 3,500 desert bighorns and in the contiguous United States there were 25,219 Rocky Mountain bighorn sheep, 4,901 California bighorns, and 17,450 desert bighorns. The total number of bighorn sheep in North America has not changed significantly since the 1991 survey (144).

Fecundity and Sex Ratios

Bighorn sheep are characterized by low fecundity (i.e., the mean number of female live births in each female age class) (145). Single lambs are the rule, and twinning is rare (73, 146, 58, 86, 67). Females typically first mate at 2.5 years of age, but under very favorable environments they can produce lambs on their second birthday (57, 88, 147). In desert populations, even yearling females have given birth to lambs (133, 148, 54, 64). Fecundity in bighorn populations has been reported to increase up to at least 5 years of age and then may decline after 8 years of age (99, 147). The few studies of bighorn pregnancy rates show they can be very high, ranging from more than 90% to 100% (149, 150, 151). Factors potentially affecting conception and pregnancy rates have been discussed for bighorn, but there are few data to support any general conclusions (see review in 34).

Buechner reported that sex ratios in bighorn sheep populations varied considerably (73). Some of this variability may be attributed partially to biases in the census techniques used, to season of survey, to behavioral differences between males and females, and whether the population is hunted. More recent studies of unmanipulated populations suggest that adult ratios are typically near unity, although Russo found a male:female ratio of 123:100 in Arizona, and others reported sex ratios skewed toward females in hunted populations (152, 25, 153, 58, 46, 154, 155, 133, 122, 156, 54, 157, 158, 159). By contrast, McQuivey found that in Nevada, the average male:female ratio for unhunted populations was 57:100 compared to a statewide mean ratio of 60:100 (133). There are few data on sex ratios at birth. Woodgerd reported a 50:50 sex ratio of young lambs, and Geist assumed a unitary sex ratio at birth (57, 58).

Mortality

Bighorn sheep can suffer a high rate of mortality during their first year of life, and reported mortality can be as high as 40% to 90% (160, 78, 161). Most lambs are lost during their first few weeks, often due to predation (162, 49, 163, 164, 150, 161, 128, 83). Coyotes are usually cited as the main predator, but other causes of lamb mortality are also found and seem to be interrelated (164, 165, 150). They may include pneumonia, weather, inbreeding depression, poor maternal nutrition, poor mothering, human disturbance, and predation (161, 150). The impact of these mortality agents may also be influenced by underlying factors such as birth date, range condition, population density, and quality of security cover (34).

Published mortality data are limited for subadult bighorn. For males, Geist reported no mortality among yearlings and only 3% among 2-year-olds (58). However, others have found yearling mortality rates of 33% in males and 41% in females, and 16% to 18% and 41% for 2-year-old males and females, respectively (163, 166).

In their recent review of Rocky Mountain and California bighorn data, Shackleton and others concluded that for males in most populations, death rates between 3 and 5 years old were similar, ranging between 3% and 14% (34). For older males, age-specific death rates were higher but the slope of the increase depended on whether the population was hunted or not. Without hunting, age-specific mortality rates in stable populations increased slowest, reaching more than 60% by age 15 years, and life expectancy was high. In hunted populations living on high-quality ranges and increasing in numbers, male mortality rates increased more rapidly with increasing age, and life expectancy was shorter.

Only limited mortality data are available for females, partly because it is difficult to age females older than 5 years by horn annuli counts. Adult female mortality rates seem to be higher than those of males, and have been estimated to be around 11% (35, 59, 167, 163, 132).

In desert bighorn sheep, Hansen found that the highest adult mortality of 47% occurred in age classes 10 through 12 years (167). By contrast, McQuivey reported that the average age structure of males in Nevada declined gradually through successive years of life, indicating relatively constant mortality rates (133). Also, he found no single mortality factor to be affecting any age group for males older than 1 year. These results closely agree with those of Leslie and Douglas, who found relatively constant mortality rates for males from the River Mountains, Nevada (54). They hypothesized that increased mortality of young males could be due to involvement in rutting activities, especially during the pre-rut period. High levels of pre-rut activity of socially immature males were also observed by Sandoval (46).

Survivorship of desert bighorn females did not differ between the Little Harquahala and Harquahala Mountains, Arizona (32). However, the probability of survival for males was consistently higher in the Harquahala Mountains. Mortality factors were similar between the study areas except that mountain lion predation was higher for female sheep in the Harquahala Mountains.

Population Regulation

The amount and distribution of available resources such as food and water, the effects of predation, and physical factors such as climate, cover, and disease are some of the factors that may limit population growth within a given habitat. Although no hard data are available, current indirect evidence indicates that food may be the most frequent factor limiting bighorn sheep, but only in winter (34). Similarly, there is little published evidence of the role of interspecific competition in population dynamics of bighorns, although the bighorn's diet can overlap considerably with elk, cattle, and to a lesser degree with mountain goats, mule deer, and horses (168, 169, 170). The availability of food might set only an upper limit to population growth. If food were the only factor regulating population size, one would expect population control to take place exclusively by malnutrition and diseases accompanying starvation. This is usually not the case as a number of decimating factors begin to slow and eventually stop the growth of an expanding population (171).

Sinclair found no evidence that predation regulates ungulate populations, but bighorn with limited escape terrain are very vulnerable and predation levels can be high (172, 164, 173, 150, 174). However, Wehausen reported that mountain lion predation caused two populations of bighorn sheep to decline to low densities (175). Although most bighorn live in precipitous habitats, Festa-Bianchet recorded only two individuals having died in falls (176). By contrast, disease can be a major limiting factor. There are well-documented examples since the end of the last century of "die-offs" of bighorn populations suffering more than 50% mortality within 1 year (73, 12, 177). Populations that are stressed seem most predisposed to a die-off, with lungworm typically involved, followed by the development of acute bronchial pneumonia and death (178, 177, 179). Another problem disease is *Pasteurella* bacteria, which causes acute bronchial pneumonia and eventually kills bighorn. Domestic sheep are often implicated in the transmission of these diseases, especially the latter, to which bighorn seem to have no immunity. In southwestern British Columbia, such disease-related die-offs may occur about every 20 years (180). A less problematic disease is mange, but even this has caused die-offs. Despite their obvious and catastrophic effect, we still do not know if die-offs involve primarily density-dependent (e.g., range condition) or density-independent (e.g., presence of domestic sheep) factors (181, 179, 182, 183).

In two Rocky Mountain bighorn populations with different demographic histories, juvenile female survival in the population that increased to carrying capacity was negatively affected by density, but not survival of all other classes (184). In the population that suffered a die-off, Jorgensen and others were unable to detect any density effects on survival because the population never returned to numbers at which density limitations would have an effect (184). Current available evidence suggests that at least some bighorn populations are regulated by density-dependent feedback on fecundity and lamb survival, but density dependence only operates at intermediate population levels (34, 184). In the absence of die-offs, bighorn populations seem to be relatively stable, though the density-dependent mechanism(s) involved are uncertain. Geist postulated that they are nutritionally regulated, and suggested classifying bighorn populations as either increasing with high recruitment and high adult mortality (i.e., "high-quality" or "dispersal" phenotype), or stable and food limited with low recruitment rates and low adult mortality (i.e., "low-quality" or "maintenance" phenotype) (58, 185). However, there are as yet no data that directly support food as limiting or regulating bighorn populations.

HABITAT REQUIREMENTS

Topography

The general name of "mountain sheep" reflects the bighorn's preference for steep, rugged topography typically found in mountains, though they can meet their requirements in other areas as well (186, 58, 187, 141). The range of habitat components they can use varies from steep or gentle slopes, broken cliffs, rock outcrops, canyons and adjacent river benches to mesa tops (73, 35, 25, 153, 188, 189, 42, 133, 49, 190, 191, 32, 192). Alpine and subalpine slopes and river benches are mainly used for feeding, while cliffs and other precipitous ground supply bighorn with escape terrain. These steep habitats provide them with their best defense against predators, and bighorn are rarely found far from such security cover (193, 194, 46, 41, 110, 195, 196, 197). Though males may move further away than female–juvenile groups, escape terrain is especially important for females when giving birth (110, 195, 160, 188, 48, 198, 161, 81). While specific characteristics of birth sites for Rocky Mountain and California bighorn have not been studied, Etchberger and Krausman compared microsites used for parturition to those used by females at other times of the year (128). Parturition microsites were comprised of steep rugged terrain and had less barrel cacti for up to 8 weeks following parturition than sites used by females without lambs. Parturition site fidelity was strong between years with a mean distance between parturition sites for individual females of 450 meters.

The range of elevations that bighorn occupy is extreme. Desert sheep are found from 78 meters below sea level in Death Valley, California, to more than 4,267 meters above sea level in the White Mountains, California, while other subspecies are found on ranges from 450 meters to more than 3,300 meters above sea level (152, 199, 34). Sheep in most populations make seasonal altitudinal movements, but a few remain at similar elevations year-round (77, 35, 188, 200). Because of these variations, it is difficult to generalize about the elevational preferences of bighorn. A wide variety of exogenous and endogenous factors may influence elevational use of a particular range, and together reflect the animals' environmental, physiologic, and behavioral preferences.

Climate

Areas inhabited by Rocky Mountain and California bighorn are relatively arid. Summers are warm or even hot with highs up to 35°C, and winters are usually cold with temperatures sometimes reaching minus 40°C (77, 201, 188, 5, 38, 77). Extreme conditions also prevail in desert habitats where temperatures as low as minus 29°C have been recorded in winter and more than 49°C in summer (242, 202).

Desert bighorn activities are affected by temperature (203). In the Cabeza Prieta National Wildlife Refuge, Arizona, desert bighorn bedded in the shade an average of 7 hours each day when wet bulb temperatures were above 18°C (55). During the hot, dry summer months, heat stress is a serious obstacle to desert bighorn survival; they minimize its effect by limiting their activities, bedding in the shade during the hottest part of the day, and feeding and watering on shaded slopes. Wind, cold temperatures, and high precipitation levels, on the other hand, are more limiting to Rocky Mountain bighorn.

Snow accumulation in winter limits habitat use by California and especially Rocky Mountain bighorn. Generally, areas with snow deeper than 30 centimeters are avoided, because deep snow increases the costs of foraging and traveling (42). These activities also can be hampered by snow crusts that develop during freeze–thaw cycles (35, 204). Consequently, most Rocky Mountain and California bighorn winter ranges are located in areas with low snow accumulation. Low snow depths can occur as a result of low precipitation, heat gains on south-facing aspects, high winds blowing snow away, or some mix of these three factors.

Precipitation in desert ranges is low and unpredictable, ranging from less than 2.5 centimeters in the Mojave Desert to more than 251 centimeters in the White and San Gabriel Mountains, California (202). Erratic rainfall patterns can reduce the significance of any precipitation measurements taken at a single locality. However, rainfall can determine the distribution and production of forage and is an important physical requisite for desert sheep survival in the Southwest (154).

Vegetation Cover

Climate, elevation, and latitude vary widely throughout the geographic distribution of bighorn sheep. Such variation is naturally reflected in the structure and floristic composition of the vegetation components of their habitats (160, 205, 206, 207, 13, 41, 194, 208, 49, 195, 48, 187, 32, 33, 34). Different habitats can meet specific requirements of bighorn for activities such as foraging, resting, mating, lambing, thermal regulation, and predator avoidance, so habitat use can vary daily and seasonally as the sheep's requirements change (206, 208, 195, 48, 193, 49, 187).

Generally, bighorn use open habitats such as grasslands and shrub-steppe communities located at various elevations and on different slope gradients. Such open habitats provide bighorn with good visibility, improving their chances of detecting predators, but the sheep are usually not far from cliffs or other precipitous habitat. Open grasslands are used primarily for foraging and may contain grasses such as fescues, wheatgrasses, and ricegrasses, along with forbs and often shrubs at lower elevations, and sedges at the higher elevations used in summer (209, 210, 133, 199, 34).

Throughout much of the desert bighorn's habitat, the vegetation associations are predominantly adapted to dry, rocky, or sandy soils, and plants characteristically have a thickened epidermis and a reduction of their leaf surface (210). In general, the vegetation is uniformly sparse, and the plants widely spaced by the demands of their root systems in the shallow soils, while a rounded canopy results from equal exposure to solar radiation from all sides (211, 25, 133, 46, 212, 54, 213, 214, 32, 128). Needless to say, such plants must withstand severe drought lasting up to several years.

For most bighorn habitat in the Great Basin Desert, sagebrush, shadscale, blackbrush, and cliffrose comprise the major browse species. The major grasses include wild rye, Indian rice, galleta, bluegrass, and fescue (17, 202). In the lower elevations of the White, San Gabriel, San Jacinto, and Santa Rosa Mountains, California, their habitat is characterized by lowland browse types dominated by sagebrush, redberry, and chaparral white-thorn. Bighorn habitat extends through piñon-juniper and ponderosa pine associations. The summer range includes the subalpine and alpine biotic communities (210, 133, 199, 122). In the Painted Desert, sagebrush, blackbrush, shadscale, galleta, and piñon-juniper comprise the major vegetation associations used by desert bighorn (25). Characteristic plants of desert bighorn habitat in the Sonoran Desert include paloverde, ironwood, saguaro, and organpipe cactus. The major grass species include grama, galleta, and sacaton (215, 216). Bighorn habitat throughout the Chihuahuan Desert is characterized by few trees, agave, yucca, small cacti, and numerous spiny shrubs (217, 46, 212). In general, volcanic soils support relatively homogeneous grasslands, and sedimentary parent material produces creosote, mesquite, tarbrush, cactus savannas, and agave thickets. Much of the bighorn habitat in Baja California and Sonora in Mexico is characterized by agave, ocotillo, ironwood, cholla, acacia, and numerous cacti (217a, 218, 155).

Bighorn rarely use densely forested areas, probably because forage and visibility are more limited, although trees may be used for shade when bedding or during cold days with high winds (47, 201, 58, 206). Similarly, desert bighorn in the San Gabriel Mountains of California use heavy cover for thermal cover (122). In the San Andres Mountains, New Mexico, the piñon-juniper community received little use by desert bighorn, except for a few occasions where they were traveling along established trails close to escape terrain (46).

Visibility

For bighorn, the structure of the vegetation has been suggested to be more important than the type of plant species present. Studies have shown that open habitats, with high visibility, were used most by Rocky Mountain bighorn to facilitate detection of predators (187, 219). However, open habitat may not be used if escape terrain is not readily available (47). Visibility is an important habitat feature for bighorn sheep because their predator-evasion strategy involves foraging diurnally in relatively large dispersed groups on open habitat close to escape terrain. Predators are detected visually, and a larger dispersed group of sheep may be more alert to potential predators over a relatively large area. Risenhoover and Bailey compared foraging efficiency of bighorns relative to group size and habitat visibility; they concluded that foraging efficiency was higher when sheep were in large groups and in habitats with greater visibility (187).

Krausman and Leopold concluded that some areas of the Harquahala Mountains, Arizona, were not used, in part, because large boulders obstructed vision (196). Desert bighorn in Arizona have also been found to abandon areas because fire suppression allowed vegetation to grow and obstruct visibility (191, 220, 221). Similarly, DeForge observed that reduced visibility in maturing chaparral lowered its suitability for bighorn, thus decreasing carrying capacity potential, and eventually resulting in the total loss of bighorn range (122).

MANAGEMENT

Populations

Bighorn sheep populations have declined significantly since 1900 with a subsequent reduction in their geographic distribution (154, 73, 222, 167). Declines in number and distribution have been attributed to a wide range of human-induced factors that include overgrazing by livestock and feral burros, diseases transmitted by livestock, road construction, housing developments, canals, fire suppression, and recreation activities (223, 216, 224, 225, 178, 180, 177, 54, 32, 197, 226, 191, 227). As a result, population management has been directed at reducing human-caused decimating factors and at learning more about other population-limiting factors. Understanding limiting factors is especially important for resource managers charged with managing bighorns because many sheep populations are small and susceptible to extinction (228, 229, 230, 221).

Management efforts have been directed at eliminating domestic stock from bighorn habitat, understanding the effects of diseases on populations, eliminating competition with species such as feral burros, and minimizing human disturbance. However, one of the most active management programs for bighorn sheep has been the translocation of animals to various locations within their historic ranges. In general, translocations have been successful and many populations have been returned to historical habitats. However, human disturbance to translocated and other herds needs to be closely monitored. In general, bighorn are intolerant of human activities, especially outside protected areas, and can have difficulty adjusting to human encroachment (Figure 25–8). They are animals with traditional range use and movement patterns that are probably passed from adults to young; consequently bighorn do not adjust well when these patterns are disrupted (58). Disturbances, whether specifically directed toward bighorn or not, are known to have adverse effects on sheep populations, and they have been known to abandon the use of ranges following increased levels of human activity (124, 133, 199, 122, 231, 216, 232, 121, 142, 226, 233).

Habitat

Because bighorn sheep exist in arid environments, habitat-related management generally has been limited to prescribed burns and the provision of water. Prescribed burns are used to control forest and heavy shrub encroachment on bighorn ranges in an attempt to mimic natural fires and maintain open grasslands for foraging and visibility (234, 194, 58). Water sources have been developed in many areas, but their value to bighorn sheep has been questioned because there is no empirical evidence demonstrating the benefit of such artificial water (235, 236, 233, 237).

The relationship between food supply and population density is recognized as a basic concept of wildlife management. Management of renewable resources is based on the knowledge that there is a limit to the intensity of harvest that each species can tolerate. If this critical level is exceeded, the species will decline and the future annual harvest will diminish. Notwithstanding lower forage production in desert mountain ranges, it is probable that sufficient food supply exists to support more sheep than are

Figure 25–8 Bighorn sheep in general, and these desert bighorns in particular, are relatively intolerant of human activities outside of protected areas. (Photograph by R. C. Etchberger.)

currently present. When a certain population has a number of different requirements, the one in shortest supply relative to the demand obviously will be the limiting factor (171). The apparent excess of one life requirement, such as rough topography, cannot make up for the scarcity of another such as suitable interspersion of food and water.

Until recently, bighorn sheep have often been managed on a case-by-case basis, usually within a mountain range. However, the flatter areas between mountains may act as corridors for sheep to gain access to other ranges for lambing and foraging (238, 239, 240). Bleich and others proposed a model for the conservation of desert bighorn sheep, and Schwartz and others were among the first to suggest a management strategy based at the landscape level (238, 241). Overall, they suggested that management needs to address intermountain travel corridors for bighorn and, where necessary, take steps to minimize potential barriers. Habitat for bighorn sheep still exists in the west, but managers (and the public) have to ensure that sufficient habitat is protected, that movement corridors remain open, that human disturbance is reduced or kept to a minimum, and that transmission of diseases from livestock is eliminated. Only if these are accomplished will efforts to enhance viable populations of bighorn have a chance to be successful.

SUMMARY

Bighorn sheep are an important renewable natural resource that humans have a keen interest in managing and conserving. Modern mountain sheep are divided into thinhorn (i.e., Dall's and Stone's) and bighorns. Bighorn extend from Canada through the Rocky Mountains and other western ranges, west into Mexico.

Bighorn are relatively opportunistic in their selection of diet. There is a strong seasonal dietary variation that probably reflects changes in the availability and palatability of forage. Bighorns can survive for long periods without free standing water by obtaining moisture from vegetation, snow, or ice.

Most females mate when at least 2.5 years old but males mate well after puberty. Estrus lasts 28 days with a receptive period around 48 hours. Gestation is from 173 to

176 days and females usually produce a single lamb each year. Complex, aggressive, and courtship behaviors play major roles in reproduction.

The optimum period for birth likely determines the timing of the season, but mating occurs for longer periods in southern ranges. Rocky Mountain bighorns mate in autumn and early winter, whereas desert races can mate over a longer period and in most months. Lambs weigh up to 5.5 kilograms at birth and grow rapidly during their first 6 months of life. Adult males are larger than adult females in horn and body dimensions.

Bighorn are highly social animals, but males and females usually occupy separate seasonal ranges with limited temporal or geographic overlap. Several explanations for sexual segregation in bighorn have been made. The most likely is that females select secure areas for raising young, whereas males choose areas for maximizing growth and body condition.

Mortality rates are variable among populations. Lamb mortality can be up to 90% with most mortality occurring 1 to 3 weeks after birth, often due to predation. Yearling mortality rates are less and adult mortality is 50% to 60%.

The amount and distribution of available resources, the effect of predation, and climate, cover, and disease are some of the factors that may limit population growth within a given habitat. This complex area deserves more attention.

Bighorn prefer open habitats adjacent to steep, rugged topography that provide security cover. The range of elevations that bighorn can use is extreme. Climate, elevation, and latitude vary widely throughout the geographic distribution of bighorn. This variation is naturally reflected in the structure and floristic composition of the vegetation in their habitats and in their diets.

Populations have declined significantly since 1900 due to overgrazing by livestock and in desert areas by feral burros, diseases transmitted by livestock, road construction, housing developments, and other anthropogenic factors. Recent management has effectively minimized disturbances in most areas and translocated animals into their historic ranges to supplement or reestablish populations in many areas. Managers must ensure that habitat is protected, movement corridors remain open, livestock are separated from bighorn, and human disturbance is minimized to maintain and enhance viable populations of bighorn sheep in North America.

LITERATURE CITED

1. SETON, E. T. 1929. The bighorn. Pages 517–574 *in Lives of the Game Animals,* Vol. 3, Part 2. Doubleday, Doran Company, Garden City, NY.

2. VALDEZ, R. ed. 1982. *The Wild Sheep of the World.* Wild Sheep and Goat International, Mesilla, NM.

3. KURTEN, B., and E. ANDERSON. 1980. *Pleistocene Mammals of North America.* Columbia University Press, New York.

4. PIELOU, E. C. 1991. *After the Ice Age.* University of Chicago Press, Chicago, IL.

5. COWAN, I. M. 1940. Distribution and variation in the native sheep of North America. *American Midland Naturalist* 24:505–580.

6. WEHAUSEN, J. D., and R. R. RAMEY II. 1993. A morphometric reevaluation of the peninsular bighorn subspecies. *Desert Bighorn Council Transactions* 37:1–10.

7. JESSUP, D. A., and R. R. RAMEY II. 1995. Genetic variation of bighorn sheep as measured by blood protein electrophoresis. *Desert Bighorn Council Transactions* 39:17–25.

8. ALLEN, J. A. 1912. Historical and nomenclature notes on North American sheep. *Bulletin American Museum Natural History* 31:1–29.

9. OSGOOD, W. H. 1913. The name of the Rocky Mountain sheep. *Proceedings of the Biological Society Washington* 26:57–62.

10. OSGOOD, W. H. 1914. Dates for *Ovis canadensis, Ovis cervina, and Ovis montana. Proceedings of the Biological Society Washington* 27:1–4.

11. CLARK, J. L. 1978.$*The Great Arc of the Wild Sheep.* University of Oklahoma Press, Norman.

12. STELFOX, J. G. 1971. Bighorn sheep in the Canadian Rockies: A history. 1800–1970. *Canadian Field-Naturalist* 85:101–122.

13. DEMARCHI, R. A., and H. B. MITCHELL. 1973. The Chilcotin River bighorn population. *Canadian Field-Naturalist* 87:433–454.

14. MONSON, G. 1980. Distribution and abundance. Pages 40–51 *in* G. Monson and L. Sumner, eds., *The Desert Bighorn.* University of Arizona Press, Tucson.

15. BARRET, R. H. 1964. Seasonal food habits of the bighorn at the Desert Game Range, Nevada. *Desert Bighorn Council Transactions* 8:85–93.

16. YOAKUM, J. 1964. Bighorn food habit-range relationships in the Silver Peak Range, Nevada. *Desert Bighorn Council Transactions* 8:95–102.

17. BRADLEY, W. G. 1964. The vegetation of the Desert Game Range with special reference to the desert bighorn. *Desert Bighorn Council Transactions* 8:43–67.

18. IRVINE, C. A. 1969. Factors affecting the desert bighorn in southeastern Utah. *Desert Bighorn Council Transactions* 13:6–13.

19. BROWN, B. W., D. D. SMITH, D. E. BERNHARDT, K. R. GILES, and J. B. HELVIE. 1975. Food habits and radionuclide tissue concentrations of Nevada desert bighorn, 1972–1973. *Desert Bighorn Council Transactions* 19:61–68.

20. BROWN, B. W., D. D. SMITH, and R. P. McQUIVEY. 1977. Food habits of desert bighorn sheep in Nevada 1956–1976. *Desert Bighorn Council Transactions* 21:32–61.

21. SANCHEZ, D. R. 1976. Analysis of stomach contents of bighorn sheep in Baja, California. *Desert Bighorn Council Transactions* 20:21–22.

22. HICKEY, W. O. 1978. Bighorn sheep ecology. Federal Aid Wildlife Restoration Project W-160-R-5. Idaho Department of Fish and Game. Boise.

23. WILSON, L. O. 1976. Biases in bighorn research relating to food preferences and determining competition between bighorn and other herbivores. *Desert Bighorn Council Transactions* 20:46–48.

24. KRAUSMAN, P. R., B. BOBEK, F. WHITING, and W. BROWN. 1988. Dry matter and energy intake in relation to digestibility in desert bighorn sheep. *Acta Theriolgical* 33:121–130.

25. WILSON, L. O. 1968. Distribution and ecology of desert bighorn sheep in southeastern Utah. Publication Number 68-5. Utah Department Natural Resources and Division of Fish and Game, Salt Lake City.

26. CONSTAN, K. J. 1972. Winter foods and range use of 3 species of ungulates. *Journal of Wildlife Management* 36:1068–1075.

27. TODD, J. W. 1972. A literature review of bighorn sheep food habits. Special Report 27. Colorado Department Game, Fish and Parks Cooperative Wildlife Research Unit, Fort Collins.

28. JOHNSON, J. D. 1975. An evaluation of the summer range of bighorn sheep (*Ovis canadensis canadensis* Shaw) on Ram Mountain, Alberta. Thesis. University of Calgary, Alberta, Canada.

29. STEWART, S. T. 1975. Ecology of the West Rosebud and Stillwater bighorn sheep herds, Beartooth Mountains, Montana. Federal Aid Wildlife Restoration Projects W-120-R-6 and R-7. Montana Fish and Game Department, Helena.

30. WIKEEM, B. M., and M. D. PITT. 1979. Interpreting diet preference of California bighorn sheep on native rangeland in south-central British Columbia. *Rangelands* 1:200–202.

31. WIKEEM, B. M., and M. D. PITT. 1992. Diet of California bighorn sheep: Assessing optimal foraging habitat. *Canadian Field-Naturalist* 106:327–335.

32. KRAUSMAN, P. R., G. LONG, R. F. SEEGMILLER, and S. G. TORRES. 1989. Relationships between desert bighorn sheep and habitat in western Arizona. *Wildlife Monograph* 102:1–66.

33. BLEICH, V. C., R. T. BOWYER, and J. D. WEHAUSEN. 1997. Sexual segregation in mountain sheep: Resources or predation? *Wildlife Monograph* 134:1–50.

34. SHACKLETON, D. M., C. C. SHANK, and B. M. WIKEEM. 1999. Natural history of Rocky Mountain and California bighorn sheep. Chapter 3 *in* R. Valdez and P. R. Krausman eds., *Mountain Sheep of North America.* University of Arizona Press, Tucson.

35. SUGDEN, L. G. 1961. The California bighorn in British Columbia with particular reference to the Churn Creek herd. British Columbia Department Recreation and Conservation, British Columbia, Canada.

36. BROWNING, B. M., and G. MONSON. 1980. Food. Pages 80–99 *in* G. Monson, and L. Sumner, eds., *The Desert Bighorn: Its Life History, Ecology and Management*. University of Arizona Press, Tucson.

37. WIKEEM, B. M. 1984. Forage selection by California bighorn sheep and the effects of grazing on an *Artemisia-Agropyron* community in southern British Columbia. Dissertation. University of British Columbia, Vancouver, Canada.

38. JONES, F. L. 1950. A survey of the Sierra Nevada bighorn. *Sierra Club Bulletin* 35:29–76.

39. BLOOD, D. A. 1967. Food habits of the Ashnola bighorn sheep herd. *Canadian Field-Naturalist* 81:23–29.

40. BROWN, B. W. 1974. Distribution and population characteristics of bighorn sheep near Thompson Falls in northwestern Montana. Thesis. University of Montana, Missoula.

41. PALLISTER, G. L. 1974. The seasonal distribution and range use of bighorn sheep in the Beartooth Mountains, with special reference to the West Rosebud and Stillwater herds. Federal Aid Wildlife Restoration Project W-120-R-5. Montana Fish and Game Department, Helena.

42. STELFOX, J. G. 1975. Range ecology of Rocky Mountain bighorn sheep in Canadian National Parks. Dissertation. University of Montana, Missoula.

43. BEAR, G. D. 1978. Evaluation of fertilizer and herbicide applications on two Colorado bighorn sheep winter ranges. No. 10, Federal Aid Wildlife Restoration Project W-41-R. Colorado Division of Wildlife, Division Representative.

44. SHANK, C. C. 1982. Age-sex differences in the diets of wintering Rocky Mountain bighorn sheep. *Ecology* 63:627–633.

45. HICKEY, W. O. 1975. Bighorn sheep ecology. Federal Aid Wildlife Restoration Project W-160-R-2. Idaho Department of Fish and Game. Boise.

46. SANDOVAL, A. V. 1979. Preferred habitat of desert bighorn sheep in the San Andres Mountains, New Mexico. Thesis. Colorado State University, Fort Collins.

47. McCANN, J. L. 1956. Ecology of mountain sheep. *American Midland Naturalist* 56:297–324.

48. KORNET, C. A. 1978. Status and habitat use of California bighorn sheep on Hart Mountain, Oregon. Thesis. Oregon State University, Corvallis.

49. VAN DYKE, W. A. 1978. Population characteristics and habitat utilization of bighorn sheep, Steens Mountain, Oregon. Thesis. Oregon State University, Corvallis.

50. KRAUSMAN, P. R., S. TORRES, L. L. ORDWAY, J. J. HERVERT, and M. BROWN. 1985. Duel activity of ewes in the Little Harquahala Mountains, Arizona. *Desert Bighorn Council Transactions* 29:24–26.

51. WARRICK, G. D., and P. R. KRAUSMAN. 1989. Barrel cacti consumption by desert bighorn sheep. *Southwestern Naturalist* 34:483–486.

52. TURNER, J. C. 1979. Osmotic fragility of desert bighorn sheep red blood cells. *Comparative Biochemistry Physiology* 64A:167–175.

53. TURNER, J. C., and R. A. WEAVER. 1980. Water. Pages 100–112 *in* G. Monson and L. Sumner, eds., *The Desert Bighorn*. University of Arizona Press, Tucson.

54. LESLIE, D. M., and C. L. DOUGLAS. 1979. Desert bighorn of the River Mountains, Nevada. *Wildlife Monograph* 66:1–56.

55. SIMMONS, N. M. 1969. Heat stress and bighorn behavior in the Cabeza Prieta Game Range, Arizona. *Desert Bighorn Council Transactions* 13:56–63.

56. BLEICH, V. C., M. C. NICHOLSON, A. T. LOBARD, and P. V. AUGUST. 1992. Using a geographic information system to test mountain sheep habitat models. *Proceedings of the Biennial Symposium on Northern Wild Sheep and Goat Council* 8:256–263.

57. WOODGERD, W. 1964. Population dynamics of bighorn sheep on Wildhorse Island. *Journal of Wildlife Management* 28:381–391.

58. GEIST, V. 1971. *Mountain Sheep: A Study in Behavior and Evolution*. University of Chicago Press, Chicago.

59. BLUNT, M. H., H. A. DAWSON, and E. T. THORNE. 1972. The birth weights and gestation in captive Rocky Mountain bighorn sheep. *Journal of Mammalogy* 58:106.

60. SHACKLETON, D. M. 1973. Population quality and bighorn sheep (*Ovis canadensis canadensis* Shaw). Dissertation. University of Calgary, Alberta, Canada.

61. SHACKLETON, D. M. 1991. Social maturation and productivity in bighorn sheep: Are young males incompetent? *Applied Animal Behavior Science* 29:173–184.

Literature Cited

62. McCutchen, H. E. 1976. Status of Zion National Park desert bighorn restoration project 1975. *Desert Bighorn Council Transactions* 20:52–54.

63. Sandoval, A. V. 1981. New Mexico bighorn sheep status report. *Desert Bighorn Council Transactions* 25:66–68.

64. Morgart, J. R., and P. R. Krausman. 1983. Early breeding in bighorn sheep. *Southwestern Naturalist* 28:460–461.

65. Turner, J. C. 1976. Initial investigations into the reproductive biology of the desert bighorn ram, *Ovis canadensis nelsoni, O. c. cremnobates. Proceedings of the Biennial Symposium on Northern Wild Sheep and Goat Council* 4:22–25.

66. Blaisdell, J. A. 1976. The Lava Beds bighorn—So who worries? *Desert Bighorn Council Transactions* 20:50.

67. Turner, J. C., and C. G. Hansen. 1980. Reproduction. Pages 145–151 *in* G. Monson and L. Sumner, eds., *The Desert Bighorn.* University of Arizona Press, Tucson.

68. Sandoval, A. V., R. G. Peterson, J. Haywood, and A. Bottrell. 1984. Gestation period in *Ovis canadensis. Journal of Mammalogy* 65:337–338.

69. Shackleton, D. M., R. G. Peterson, J. Haywood, and A. Botrell. 1984. Gestation period in *Ovis canadensis. Journal of Mammology* 65:337–338.

70. Whitehead, P. E., and E. H. McEwen. 1980. Progesterone levels in peripheral plasma of Rocky Mountain bighorn ewes (*Ovis canadensis*) during the oestrous cycle and pregnancy. *Canadian Journal of Zoology* 58:1105–1108.

71. Bunnell, F. L. 1982. The lambing period of mountain sheep: Synthesis, hypothesis, and tests. *Canadian Journal of Zoology* 60:1–14.

72. Thompson, R. W., and J. C. Turner. 1982. Temporal geographic variation in the lambing season of bighorn sheep. *Canadian Journal of Zoology* 60:1781–1793.

73. Buechner, H. K. 1960. The bighorn sheep in the United States, its past, present, and future. *Wildlife Monograph* 4.

74. Blood, D. A. 1963. Some aspects of behavior of a bighorn herd. *Canadian Field-Naturalist* 77:77–94.

75. Wishart, W. D. 1958. The bighorn sheep of the Sheep River Valley. Thesis. University of Alberta, Edmonton, Canada.

76. Honess, R. F., and N. M. Frost. 1942. A Wyoming bighorn sheep study. Bulletin 1. Wyoming Game and Fish Department, Laramie.

77. Smith, D. R. 1954. The bighorn sheep in Idaho. Its status, life history and management. Wildlife Bulletin 1. Idaho Game and Fish Department, Boise.

78. Morgan, J. K. 1970. Ecology of the Morgan Creek and East Fork of the Salmon River bighorn sheep herds and management of bighorn sheep in Idaho. Thesis. Utah State University, Logan.

79. Lenarz, M. S., and W. Conley. 1982. Reproductive gambling in bighorn sheep (*Ovis*): A simulation. *Journal of Theoretical Biology* 98:1–7.

80. Witham, J. H. 1983. Desert bighorn sheep in southwestern Arizona. Dissertation. Colorado State University, Fort Collins.

81. Festa-Bianchet, M. 1988. Seasonal range selection in bighorn sheep conflicts between forage quality, forage quantity, and predator avoidance. *Oecologia* 75:580–586.

82. Festa-Bianchet, M. 1988. Nursing behavior of bighorn sheep: Correlates of ewe age, parasitism, lamb age, birthdate and sex. *Animal Behavior* 36:1445–1454.

83. Festa-Bianchet, M. 1988. Birthdate and survival in bighorn lambs (*Ovis canadensis*). *Journal of Zoology (London)* 214:653–661.

84. Sadleir, R. M. 1987. Reproduction in female cervids. Pages 123–144 *in* C. M. Wemmer, ed., *Biology and Management of the Cervidae.* Smithsonian Institute Press, Washington, DC.

85. Lenarz, M. S. 1979. Social structure and reproductive strategy in desert bighorn sheep (*Ovis canadensis mexicana*). *Journal of Mammalogy* 60:671–678.

86. Eccles, T. R., and D. M. Shackleton. 1979. Recent records of twinning in mountain sheep. *Journal of Wildlife Management* 43:974–976.

87. McEwan, E. H. 1975. The adaptive significance of the growth patterns in cervids compared with other ungulate species. *Zool. Zh.* 54:1221–1232.

88. Jorgensen, J. T., and W. D. Wishart. 1984. Growth rates of Rocky Mountain Bighorn sheep on Ram Mountain, Alberta. *Northern Wild Sheep and Goat Council Proceedings* 4:270–284.

89. FESTA-BIANCHET, M., J. T. JORGENSON, W. J. KING, K. G. SMITH, and W. D. WISHART. 1996. The development of sexual dimorphism: Seasonal and lifetime mass changes in bighorn sheep. *Canadian Journal of Zoology* 74:330–342.

90. STELFOX, J. G., and J. MCGILLIS. 1970. Seasonal growth patterns of bighorns correlated with range condition and endoparasite loads. *Transactions of Northern Wild Sheep Council* 1:35–38.

91. BLOOD, D. A., D. R. FLOOK, and W. D. WISHART. 1970. Weights and growth of Rocky Mountain bighorn sheep in western Alberta. *Journal of Wildlife Management* 34:451–455.

92. HANSEN, C. G., and O. V. DEMING. 1980. Growth and development. Pages 152–171 *in* G. Monson and L. Sumner, eds., *The Desert Bighorn.* University of Arizona Press, Tucson.

93. WOODARD, T. N., C. HIBLER, and W. RUTHERFORD. 1972. Bighorn lamb mortality investigations in Colorado. *Transactions of Northern Wild Sheep Council* 2:44–47.

94. TAYLOR, R. A. 1962. Characteristics of horn growth in bighorn rams. Thesis. University of Montana, Missoula.

95. SHACKLETON, D. M., and D. A. HUTTON. 1971. An analysis of the mechanisms of brooming in mountain sheep horns. *Z. Säugetierk.* 36:342–350.

96. SHACKLETON, D. M. 1976. Variability in physical and social maturation between bighorn sheep populations. *Transactions of Northern Wild Sheep Council* 4:1–8.

97. WISHART, W. D., and BROCHU, D. 1982. An evaluation of horn and skull characteristics as a measure of population quality in Alberta bighorns. *Proceedings of the Biennial Symposium on Northern Wild Sheep and Goat Council* 3:127–142.

98. GILCHRIST, D. 1992. Why is Montana the land of giant rams? *Proceedings of the Biennial Symposium on Northern Wild Sheep and Goat Council* 8:8–13.

99. SMITH, K. G., and W. D. WISHART. 1978. Further observations of bighorn sheep non-trophy seasons in Alberta and their management implications. *Proceedings of the Biennial Symposium on Northern Wild Sheep and Goat Council* 1:52–74.

100. MCCULLOUGH, D. R., and E. R. SCHNEEGAS. 1966. Winter observations on the Sierra Nevada bighorn sheep. *California Fish and Game Department* 52:68–84.

101. BERWICK, S. H. 1968. Observations on the decline of the Rock Creek, Montana, population of bighorn sheep. Thesis. University of Montana, Missoula.

102. BAUMEN, T. G., and D. R. STEVENS. 1978. Winter habitat preferences of bighorn sheep in the Mummy Range, Colorado. *Proceedings of the Biennial Symposium on Northern Wild Sheep and Goat Council* 1:320–330.

103. ASHCROFT, G. E. W. 1986. Sexual segregation and group sizes in California bighorn sheep. M.S. thesis. University of British Columbia, Vancouver, Canada.

104. PULLIAM, H. R., and T. CARACO. 1984. Living in groups: Is there an optimal group size? Pages 122–147 *in* J. R. Krebs, and N. B. Davies, eds., *Behavioral Ecology: An Evolutionary Approach,* 2nd ed. Blackwell Scientific Publication, Oxford, UK.

105. JARMAN, P. J. 1974. The social organization of antelope in relation to their ecology. *Behavior* 48:215–267.

106. JARMAN, P. J., and M. V. JARMAN. 1979. The dynamics of ungulate social organization. Pages 185–220 *in* A. R. E. Sinclair and M. Norton-Griffiths, eds., *Serengeti: Dynamics of an Ecosystem.* University of Chicago Press, Chicago.

107. BERGER, J. 1991. Pregnancy incentives, predation constraints and habitat shifts: Experimental and field evidence for wild bighorn sheep. *Animal Behavior* 41:61–77.

108. GEIST, V., and R. G. PETOCZ. 1977. Bighorn sheep in winter: Do rams maximize reproductive fitness by spatial separation and habitat segregation from ewes? *Canadian Journal of Zoology* 55:1802–1810.

109. MORGANTINI, L. E., and R. J. HUDSON. 1981. Sex differential in use of the physical environment by bighorn sheep (*Ovis canadensis*). *Canadian Field-Naturalist* 95:69–74.

110. SHANK, C. C. 1979. Sexual dimorphism and the ecological niche of wintering Rocky Mountain bighorn sheep. Dissertation. University of Calgary, Alberta, Canada.

111. MAIN, M. B., and B. E. COBLENTZ. 1990. Sexual segregation among ungulates: A critique. *Wildlife Society Bulletin* 18:204–210.

112. MAIN, M. B., F. W. WECKERLY, and V. C. BLEICH. 1996. Sexual segregation in ungulates: New directions for research. *Journal of Mammalogy* 77:449–461.

113. HASS, C. C., and D. A. JENNI. 1991. Structure and ontogeny of dominance relationships among bighorn rams. *Canadian Journal of Zoology* 69:471–476.

114. HOGG, J. T. 1984. Mating in bighorn sheep: Multiple creative male strategies. *Science* 225:526–529.

115. HOGG, J. T. 1987. Intrasexual competition and mate choice in Rocky Mountain bighorn sheep. *Ethology* 75:119–144.

116. ECCLES, T. R., and D. M. SHACKLETON. 1986. Correlates and consequences of social status in female bighorn sheep. *Animal Behavior* 34:1392–1401.

117. FESTA-BIANCHET, M. 1991. The social system of bighorn sheep: Grouping patterns, kinship and female dominance rank. *Animal Behavior* 42:71–82.

118. HASS, C. C. 1991. Social status in female bighorn sheep (*Ovis canadensis*): Expression, development and reproductive correlates. *Journal of Zoology (London).* 225:509–523.

119. EUSTIS, G. P. 1962. Winter lamb surveys on the Kofa Game Range. *Desert Bighorn Council Transactions* 6:83–86.

120. BATES, J. W., JR., J. C. PEDERSON, and S. C. AMSTROP. 1976. Bighorn sheep range, population trend and movement. *Desert Bighorn Council Transactions* 20:11–12.

121. KING, M. M., and G. W. WORKMAN. 1982. Desert bighorn on BLM lands in southeastern Utah. *Desert Bighorn Council Transactions* 26:104–106.

122. DEFORGE, J. R. 1980. Population biology of desert bighorn sheep in the San Gabriel Mountains of California. *Desert Bighorn Council Transactions* 24:29–32.

123. ELENOWITZ, A. S. 1983. Habitat use and population dynamics of transplanted desert bighorn sheep in the Peloncillo Mountains, New Mexico. Thesis. New Mexico State University, Las Cruces.

124. LESLIE, D. M., JR. 1977. Home range, group size, and group integrity of the desert bighorn sheep in the River Mountains, Nevada. *Desert Bighorn Council Transactions* 21:25–28.

125. SCOTT, J. E., R. R. REMINGTON, and J. C. DEVOS, JR. 1990. Numbers, movements, and disease status of bighorn in southwestern Arizona. *Desert Bighorn Council Transactions* 34:9–13.

126. HEBERT, D. M. 1973. Altitudinal migration as a factor in the nutrition of bighorn sheep. Thesis. University of British Columbia, Vancouver, Canada.

127. KLEIN, D. R. 1965. Ecology of deer range in Alaska. *Ecological Monograph* 35:259–284.

128. ETCHBERGER, R. C., and P. R. KRAUSMAN. 1999. Parturition of desert bighorn sheep in western Arizona. *Southwestern Naturalist.* In press.

129. WOOLF, A., T. O'SHEA, and D. L. GILBERT. 1970. Movements and behavior of bighorn sheep on summer ranges in Yellowstone National Park. *Journal of Wildlife Management* 34:446–450.

130. BECKER, K., T. VARCALLI, E. T. THORNE, and G. B. BUTLER. 1978. Seasonal distribution patterns of Whiskey Mountain bighorn sheep. *Proceedings of the Biennial Symposium on Northern Wild Sheep and Goat Council* 1:1–16.

131. FESTA-BIANCHET, M. 1986. Site fidelity and seasonal range use by bighorn rams. *Canadian Journal of Zoology* 64:2126–2132.

132. HENGEL, D. A., S. H. ANDERSON, and W. G. HEPWORTH. 1992. Population dynamics, seasonal distribution and movement patterns of the Laramie Peak bighorn sheep herd. *Proceedings of the Biennial Symposium on Northern Wild Sheep and Goat Council* 8:83–96.

133. MCQUIVEY, R. P. 1978. The desert bighorn sheep of Nevada. Bulletin No. 6. Nevada Department of Wildlife Biology, Las Vegas.

134. SHACKLETON, D. M., and J. HAYWOOD. 1985. Early mother–young interactions in California bighorn sheep, *Ovis canadensis californiana. Canadian Journal of Zoology* 63:868–875.

135. HOREJSI, B. L. 1972. Behavioral differences in bighorn lambs (*Ovis canadensis canadensis* Shaw) during years of high and low survival. *Transactions Northern Wild Sheep Council* 2:51–73.

136. HOREJSI, B. L. 1976. Suckling and feeding behavior in relation to lamb survival in bighorn sheep (*Ovis canadensis*). Dissertation. University of Calgary, Alberta, Canada.

137. HASS, C. C. 1990. Alternative maternal-care patterns in two herds of bighorn sheep. *Journal of Mammalogy* 71:24–35.

138. BERGER, J. 1979. Weaning conflict in desert and mountain sheep (*Ovis canadensis*): An ecological interpretation. *Z. Tierpsychology* 50:188–200.

139. BERGER, J. 1979. Social ontogeny and behavioural diversity: Consequences for bighorn sheep *Ovis canadensis* inhabiting desert and mountain environments. *Journal of Zoology (London)* 192:251–266.

140. VALDEZ, R. 1988. *Wild Sheep and Wild Sheep Hunters of the New World.* Wild Sheep and Goat International, Mesilla, NM.

141. SHACKLETON, D. M. 1985. *Ovis canadensis. Mammalian Species* 230.

142. KRAUSMAN, P. R. 1993. The exit of the last wild mountain sheep. Pages 242–250 *in* G. P. Nabhan, ed., *Counting Sheep.* University of Arizona Press, Tucson.

143. VALDEZ, R., and P. R. KRAUSMAN. 1999. Distribution and abundance of mountain sheep. *In* R. Valdez and P. R. Krausman, eds., *Mountain Sheep of North America.* University of Arizona Press, Tucson.

144. KRAUSMAN, P. R. 1997. 10.4—Regional summary. Pages 316–317 *in* D. M. Schackleton, ed., *Conservation of Wild Sheep and Goats and Their Relatives: Status Survey and Conservation Action Plan for Caprinae.* International Union for the Conservation of Nature, Gland, Switzerland.

145. CAUGHLEY, G. 1977. *Analysis of Vertebrate Populations.* John Wiley and Sons, New York.

146. SPALDING, D. J. 1966. Twinning in bighorn sheep. *Journal of Wildlife Management* 30:207.

147. FESTA-BIANCHET, M. 1988. Age-specific reproduction of bighorn ewes in Alberta, Canada. *Canadian Journal of Zoology* 69:157–160.

148. BERGER, J. 1982. Female breeding age and lamb survival in desert bighorn sheep (*Ovis canadensis*). *Mammalia* 46:183–190.

149. HARPER, W. L., and R. D. H. COHEN. 1985. Accuracy of Doppler ultrasound in diagnosing pregnancy in bighorn sheep. *Journal of Wildlife Management* 49:793–796.

150. HASS, C. C. 1989. Bighorn lamb mortality: Predation, inbreeding, and population effects. *Canadian Journal of Zoology* 67:699–705.

151. JORGENSON, J. T. 1992. Seasonal changes in lamb: Ewe ratios. *Proceedings of the Biennial Symposium on Northern Wild Sheep and Goat Council* 8:219–226.

152. WELLES, R. E., and F. B. WELLES. 1961. The bighorn of Death Valley. Fauna Series 6. U.S. National Park Service, Washington, DC.

153. WELCH, R. D. 1969. Behavioral patterns of desert bighorn sheep in south-central New Mexico. *Desert Bighorn Council Transactions* 13:114–129.

154. RUSSO, J. P. 1956. The desert bighorn in Arizona. Bulletin No. 1. Arizona Game and Fish Department, Phoenix.

155. ALVAREZ, T. 1976. Status of desert bighorns in Baja California. *Desert Bighorn Council Transaction* 20:18–21.

156. DeFORGE, J. R. 1984. Population estimate of Peninsular desert bighorn sheep in the Santa Rosa Mountains, California. *Desert Bighorn Council Transactions* 28:41–43.

157. REMINGTON, R. R. 1983. Arizona bighorn sheep status report 1983. *Desert Bighorn Council Transactions* 27:39–41.

158. GUYMON, J. G., and J. W. BATES. 1984. Utah's desert bighorn sheep status report, 1984. *Desert Bighorn Council Transactions* 28:49–50.

159. OLDING, R. J. 1984. Arizona bighorn sheep status report—1984. *Desert Bighorn Council Transactions* 28:51–53.

160. BLOOD, D. A. 1961. An ecological study of California bighorn sheep (*Ovis canadensis californiana* Douglas) in southern British Columbia. Thesis. University of British Columbia, Vancouver, Canada.

161. AKESON, J. J., and H. A. AKESON. 1992. Bighorn sheep movements and summer lamb mortality in central Idaho. *Proceedings of the Biennial Symposium on Northern Wild Sheep and Goat Council* 8:14–27.

162. SPRAKER, T. R. 1974. Lamb mortality. *Transactions of Northern Wild Sheep Council* 3:102–103.

163. STEWART, S. T. 1980. Mortality patterns in a bighorn sheep population. *Proceedings of the Biennial Symposium on Northern Wild Sheep and Goat Council* 2:313–330.

164. HARPER, W. L. 1984. Pregnancy rates and early lamb survival of California bighorn sheep (*Ovis canadensis californiana,* Douglas 1871) in the Ashnola Watershed, British Columbia. Thesis. University of British Columbia, Vancouver, Canada.

165. HEBERT, D. M., and S. HARRISON. 1988. The impact of coyote predation on lamb mortality patterns at the Junction Wildlife Management Area. *Proceedings of the Biennial Symposium on Northern Wild Sheep and Goat Council* 5:283–291.

166. FESTA-BIANCHET, M. 1989. Survival of male bighorn sheep in southwestern Alberta. *Journal of Wildlife Management* 53:259–263.

167. HANSEN, C. G. 1980. Population dynamics. Pages 217–235 *in* G. Monson and L. Sumner, eds., *The Desert Bighorn*. University of Arizona Press, Tucson.

168. STREETER, R. G. 1969. A literature review on bighorn sheep population dynamics. Special Report No. 24. Colorado Division of Game, Fish and Parks, Fort Collins.

169. PICTON, H. D. 1984. Climate and the prediction of reproduction of 3 ungulates. *Journal of Applied Ecology* 21:869–879.

170. BERGER, J. 1986. *Wild Horses of the Great Basin: Social Competition and Population Size.* University of Chicago Press, Chicago, IL.

171. LEOPOLD, A. 1933. *Game Management.* Charles Scribner's Sons, New York.

172. SINCLAIR, A. R. E. 1989. Population regulation in animals. Pages 197–241 *in* J. M. Cherrett, ed., *Ecological Concepts: The Contribution of Ecology to an Understanding of the Natural World.* Blockwell Scientific Publishing, Oxford, UK.

173. HARRISON, S., and D. A. HEBERT. 1988. Selective cougar predation on the Junction Wildlife Management Area. *Proceedings of the Biennial Symposium on Northern Wild Sheep and Goat Council* 5:292–306.

174. HARRISON, S. 1990. Cougar predation on bighorn sheep in the Junction Wildlife Management Area, British Columbia. Thesis. University of British Columbia, Vancouver, Canada.

175. WEHAUSEN, J. D. 1996. Effects of mountain lion predation on bighorn sheep in the Sierra Nevada and Granite Mountains of California. *Wildlife Society Bulletin* 24:471–479.

176. FESTA-BIANCHET, M. 1987. Bighorn sheep, climbing accidents, and implications for mating. *Mammalia* 51:618–620.

177. RYDER, T. J., E. S. WILLIAMS, K. W. MILLS, K. H. BOWLES, and E. T. THORNE. 1992. Effect of pneumonia on population size and lamb recruitment and recruitment in Whiskey Mountain bighorn sheep. *Proceedings of the Biennial Symposium on Northern Wild Sheep and Goat Council* 8:136–146.

178. SPRAKER, T. R., C. P. HIBLER, G. G. SCHOONVELD, and W. S. ADNEY. 1984. Pathological changes and microorganisms found in bighorn sheep during stress-related die-off. *Journal of Wildlife Disease* 20:319–327.

179. DUNBAR, M. R. 1996. Theoretical concepts of disease versus nutrition as primary factors in population regulation of wild sheep. *Proceedings of the Biennial Symposium on Northern Wild Sheep and Goat Council* 8:174–192.

180. SCHWANTJE, H. M. 1988. Causes of bighorn sheep mortality and dieoff—Literature review. Wildl. Working Rep. No. WR-35, Wildlife Branch, Ministry Environment, Government of British Columbia, Victoria, Canada.

181. DUNBAR, M. R. 1992. Theoretical concepts of disease versus nutrition as primary factors in population regulation of wild sheep. *Proceedings of the Biennial Symposium on Northern Wild Sheep and Goat Council* 8:174–192.

182. ONDERKA, D. K., and W. D. WISHART. 1988. Experimental contact transmission of *Pasteurella haemolytica* from clinically normal domestic sheep causing pneumonia in Rocky Mountain bighorn sheep. *Journal of Wildlife Diseases* 24:663–667.

183. COGGINS, V. L., and P. E. MATTHEWS. 1992. Lamb survival and herd status of the Lostine bighorn herd following a *Pasteurella* die-off. *Proceedings of the Biennial Symposium on Northern Wild Sheep and Goat Council* 8:147–154.

184. JORGENSEN, J. T., M. FESTA-BIANCHET, J. M. GAILLAD, and W. D. WISHART. 1997. Effects of age, sex, disease, and density on survival of bighorn sheep. *Ecology* 78:1019–1032.

185. GEIST, V. 1985. On Pleistocene bighorn sheep: Some problems of adaptation and relevance to today's American megafauna. *Wildlife Society Bulletin* 13:351–359.

186. ADAMS, L. G., K. L. RISENHOOVER, and J. A. BAILEY. 1982. Ecological relationship of mountain goat and Rocky Mountain bighorn sheep. *Proceedings of the Biennial Symposium on Northern Wild Sheep and Goat Council* 3:9–22.

187. RISENHOOVER, K. L., and J. A. BAILEY. 1985. Foraging ecology of mountain sheep: Implications for habitat management. *Journal of Wildlife Management* 49:797–804.

188. DREWEK, J. R. 1970. Population characteristics and behavior of introduced bighorn sheep in Owyhee County, Idaho. Thesis. University of Idaho, Moscow.

189. MERRITT, M. F. 1974. Measurement of utilization of bighorn sheep habitat in the Santa Rosa Mountains. *Desert Bighorn Council Transactions* 18:4–17.

190. HOLL, S. A., and V. C. BLEICH. 1983. San Gabriel mountain sheep: Biological and management considerations. San Bernardino National Forest Administration Report. U.S. Forest Service, San Bernardo, CA.

191. ETCHBERGER, R. C., P. R. KRAUSMAN, and R. MAZAIKA. 1989. Mountain sheep habitat characteristics in the Pusch Ridge Wilderness, Arizona. *Journal of Wildlife Management* 53:902–907.

192. WAKELING, B. F., and W. H. MILLER. 1989. Bedsite characteristics of desert bighorn sheep in the Superstition Mountains, Arizona. *Desert Bighorn Council Transactions* 33:6–8.

193. OLDEMEYER, J. L., W. J. BARMORE, and D. L. GILBERT. 1971. Winter ecology of bighorn sheep in Yellowstone National Park. *Journal of Wildlife Management* 35:257–269.

194. ERICKSON, G. L. 1972. The ecology of Rocky Mountain bighorn sheep in the Sun River area of Montana with special reference to summer food habits and range movements. Federal Aid and Wildlife Restoration Project W-120-R-2 and R-3. Montana Fish and Game Department, Helena.

195. HANSEN, M. C. 1982. Status and habitat preferences of California bighorn sheep on Sheldon National Wildlife Refuge, Nevada. Thesis. Oregon State University, Corvallis.

196. KRAUSMAN, P. R., and B. D. LEOPOLD. 1986. The importance of small populations of desert bighorn sheep. *Transactions of North American Wildlife Natural Resource Conference* 51:52–61.

197. GIONFRIDDO, J. P., and P. R. KRAUSMAN. 1986. Summer habitat use by mountain sheep. *Journal of Wildlife Management* 50:331–336.

198. HALL, E. R. 1981. *The Mammals of North America,* Vol. 2, 2nd ed. Ronald Press, New York.

199. KOVACH, S. D. 1979. An ecological survey of the White Mountain Peak bighorn. *Desert Bighorn Council Transactions* 23:57–61.

200. SPALDING, D. J., and J. N. BONE. 1970. The California bighorn sheep of the south Okanagan Valley, British Columbia. Wildlife Management Publication 3. Fish and Wildlife Branch, Victoria, British Columbia, Canada.

201. SCHALLENBERGER, A. D. 1966. Food habits, range use and interspecific relationships of bighorn sheep in the Sun River area, west-central Montana. Thesis. Montana State University, Bozeman.

202. HANSEN, C. G. 1980. Habitat. Pages 64–79 *in* G. Monson and L. Sumner, eds., *The Desert Bighorn: Its Life History, Ecology and Management.* University of Arizona Press, Tucson.

203. CHILELLI, M. E., and P. R. KRAUSMAN. 1981. Group organization and activity patterns of desert bighorn sheep. *Desert Bighorn Council Transactions* 25:17–24.

204. PETOCZ, R. G. 1973. The effect of snow cover on the social behavior of bighorn rams and mountain goats. *Canadian Journal of Zoology* 51:987–993.

205. TODD, J. W. 1972. Foods of Rocky Mountain bighorn sheep in southern Colorado. Thesis. Colorado State University, Fort Collins.

206. DALE, A. R. 1987. Ecology and behavior of bighorn sheep, Waterton Canyon, Colorado, 1981–1982. Thesis. Colorado State University, Fort Collins.

207. DEMARCHI, R. A. 1965. An ecological study of the Ashnola bighorn winter ranges. Thesis. University of British Columbia, Vancouver, Canada.

208. GOODSON, N. J. 1978. Status of bighorn sheep in Rocky Mountain National Park. Thesis. Colorado State University, Fort Collins.

209. JORGENSEN, M. C., and R. E. TURNER. 1975. Desert bighorn of the Anza-Borrego Desert State Park. *Desert Bighorn Council Transactions* 19:51–53.

210. JAEGER, E. C. 1957. *The North American Deserts.* Stanford University Press, Palo Alto, CA.

210a. WEAVER, R., and J. L. MENSCH. 1970. Desert bighorn sheep in northern Inyo and southern Mono counties. Federal Aid Restoration Project W-51-R. California Department Fish and Game, Sacramento, CA.

211. BRADLEY, W. G., and J. E. DEACON. 1965. The biotic communities of southern Nevada. Reprint Series 9. Desert Research Institute, Nevada Southern University, Las Vegas.

212. WATTS, T. J. 1979. Detrimental movement patterns in a remnant population of bighorn sheep (*Ovis canadensis mexicana*). Thesis. New Mexico State University, Las Cruces.

213. KELLY, W. E. 1979. A comparison of 3 bighorn areas on the Humboldt National Forest. *Desert Bighorn Council Transactions* 23:37–39.

214. DOUGLAS, C. L., and L. D. WHITE. 1979. Movements of desert bighorn sheep in the Stubbe Spring Area, Joshua Tree National Monument. *Desert Bighorn Council Transaction* 23:71–77.

215. MENDOZA, V. J. 1976. The bighorn sheep of the state of Sonora. *Desert Bighorn Council Transactions* 20:25–26.

216. SEEGMILLER, R. E., and R. D. OHMART. 1981. Ecological relationships of feral burros and desert bighorn sheep. *Wildlife Monographs* 78:1–58.

217. MOORE, T. D. 1958. Transplanting and observation of transplanted bighorn sheep. *Desert Bighorn Council Transactions* 2:43–46.

217a. GUZMAN, G., JR. 1961. Vegetation zones of the territory of Baja California in relation to wildlife. *Desert Bighorn Council Transactions* 5:68–74.

218. FLORES, M. G., L. J. JIMENEZ, S. X. MADRIGAL, R. F. MONCAYO, and T. F. TAKAKI. 1972. Tipos de vegetacion de la Republica Mexicana, Subsecretaria de Planeacion. Dirección General de Estudios. Dirección Agricola. S.R.H.

219. WAKELYN, L. A. 1987. Changing habitat conditions on bighorn sheep ranges in Colorado. *Journal of Wildlife Management* 51:904–912.

220. ETCHBERGER, R. C., P. R. KRAUSMAN, and R. MAZAIKA. 1990. Effects of fire on desert bighorn sheep habitat. Pages 53–57 *in* P. R. Krausman and N. S. Smith, eds. *Managing Wildlife in the Southwest*. Arizona Chapter Wildlife Society, Phoenix.

221. KRAUSMAN, P. R., G. LONG, and L. TARANGO. 1996. Desert bighorn sheep and fire, Santa Catalina Mountains, Arizona. Pages 162–168 *in* P. F. Ffolliott, L. F. DeBano, M. B. Baker, Jr., G. J. Gottfried, G. Sols-Garza, C. B. Edminster, D. G. Neary, L. S. Allen, and R. H. Hamre, tech. coords. *Effects of Fire on the Madrean Province ecosystems*. RM-GTR-289. U.S. Forest Service. Fort Collins, CO.

222. BAILEY, J. A. 1980. Desert bighorn, forage competition and zoogeography. *Wildlife Society Bulletin* 8:208–216.

223. GALLIZIOLI, S. 1977. Overgrazing on desert bighorn ranges. *Desert Bighorn Council Transactions* 21:21–23.

224. BUNCH, T. D. 1980. A survey of chronic sinusitis in the bighorn of California. *Desert Bighorn Council Transactions* 24:14–18.

225. KRAUSMAN, P. R. 1996. Problems facing bighorn sheep in and near domestic sheep allotments. Pages 59–64 *in* W. D. Edge, ed., *Sustaining Rangeland Ecosystem Symposium*. Oregon State University, SR 953.

226. KRAUSMAN, P. R. 1985. Impacts of the Central Arizona Project on desert mule deer and desert bighorn sheep. Final Report 9-730-X069. U. S. Bureau of Land Reclamation, Phoenix, AZ.

227. HARRIS, L. K. 1992. Recreation in mountain sheep habitat. Dissertation. University of Arizona, Tucson.

228. BERGER, J. 1990. Persistence of different sized populations: An empirical assessment of rapid extinctions in bighorn sheep. *Conservation Biology* 4:91–98.

229. FITZSIMMONS, N. N., and S. W. BUSKIRK. 1992. Effective population sizes for bighorn sheep. *Proceedings of the Biennial Symposium on Northern Wild Sheep and Goat Council* 8:1–7.

230. KRAUSMAN, P. R., G. LONG, and R. M. LEE. 1993. Mountain sheep population persistence in Arizona. *Conservation Biology* 7:219.

231. PURDY, K. G., and W. W. SHAW. 1981. An analysis of recreational use patterns in desert bighorn habitat: The Pusch Ridge Wilderness case. *Desert Bighorn Council Transactions* 25:1–5.

232. HAMILTON, K. S., S. A. HOLL, and C. L. DOUGLAS. 1982. An evaluation of the effects of recreational activity on bighorn sheep in the San Gabriel Mountains, California. *Desert Bighorn Council Transactions* 26:50–55.

233. KRAUSMAN, P. R., and R. C. ETCHBERGER. 1995. Response of desert ungulates to a water project in Arizona. *Journal of Wildlife Management* 59:292–300.

234. McWHIRTER, D., A. SMITH, E. MERRILL, and L. IRWIN. 1992. Foraging behavior and vegetation responses to prescribed burning on bighorn sheep winter range. *Proceedings of the Biennial Symposium on Northern Wild Sheep and Goat Council* 8:264–278.

235. BURKETT, D. W., and B. C. THOMPSON. 1994. Wildlife association with human-altered water sources in semi-arid vegetation communities. *Conservation Biology* 8:682–690.

236. BROYLES, B. 1995. Desert wildlife water developments: Questioning use in the Southwest. *Wildlife Society Bulletin* 23:663–675.

237. KRAUSMAN, P. R., and B. CZECH. 1998. Water developments and desert ungulates. *In Symposium on Environmental, Economic and Legal Issues Related to Rangeland Water Developments*. The Center for the Study of Law, Science, and Technology, Arizona State University, Tempe.

238. BLEICH, V. C., J. D. WEHAUSEN, and S. A. HOLL. 1990. Desert-dwelling mountain sheep: Conservation implications of a naturally fragmented distribution. *Conservation Biology* 4:383–390.

239. BLEICH, V. C., J. D. WEHAUSEN, R. R. RAMEY II, and J. L. RECHEL. 1996. Metapopulation theory and mountain sheep: Implications for conservation. Pages 353–373 *in* D. R. McCullough, ed., *Metapopulations and Wildlife Conservation*. Island Press, Washington, DC.

240. KRAUSMAN, P. R. 1997. The influence of scale on the management of desert bighorn sheep. Pages 349–367 *in* J. A. Bissonette, ed., *Primer in Landscape Ecology*. Springer-Verlag, New York.

241. SCHWARTZ, O. A., V. C. BLEICH, and S. A. HOLL. 1986. Genetics and the conservation of mountain sheep. *Biology Conservation* 37:179–190.

242. SANDOVAL, A. V. 1980. Management of a psoroptic scabies epizootic in bighorn sheep (*Ovis canadensis mexicana*) in New Mexico. *Desert Bighorn Council Transactions* 24:21–28.

The Muskox

David R. Klein

INTRODUCTION

The muskox, among the large herbivorous mammals of the world, is the most highly adapted for life in the Arctic. Although widely distributed throughout the Arctic following the last glacial period, it became extirpated in Siberia over 2,000 years ago, and in Alaska and much of mainland Canada early in the 20th century. Its population recovery, through legal protection and reestablishment by translocation in its former range areas, stands out as one of the great successes in wildlife restoration efforts (Figure 26–1).

Historical Perspective

Archaeological evidence and historical records indicate that muskoxen have been an important food source for indigenous peoples, especially when other major food sources, such as caribou, fish, and marine mammals, were not available (1, 2, 3, 4). The relatively sedentary nature of muskoxen and their habit of grouping together to face predators have made them particularly vulnerable to being overhunted locally. With the aid of dogs, native peoples were able to effectively kill entire groups of muskoxen with spears or bows and arrows. The availability of firearms and the purchase of hides and meat by white traders, explorers, and overwintering whalers increased the pressure on muskoxen in historical times (5). Furthermore, the demise of the plains bison in the last century led to the demand for muskox hides as a replacement for buffalo robes. The extirpation of muskoxen at the southern limits of their distribution in North America and Russia appears associated with their proximity to humans.

Alaska's last muskoxen were reported to have been shot by a native hunter west of Point Barrow around 1858; however, there are unsubstantiated reports of the killing of the final two groups of muskoxen on the south side of the eastern Brooks Range in the winters of 1882 to 1883 and 1896 to 1897 or 1897 to 1898, respectively (3, 6). Muskoxen were returned to Alaska in 1930 when 34 were brought to Alaska from Greenland, held in a captive experimental herd at College, Alaska, until 1935 and 1936, when they were released on Nunivak Island in the Bering Sea. Much of Nunivak Island had been established as a National Wildlife Refuge and was without wolves or bears (7). Nunivak Island, however, is south of the known past distribution of muskoxen in Alaska, and from 1967 through 1981, some muskoxen from the Nunivak Island population were translocated to the mainland. First, they were introduced to adjacent Nelson Island, a complex of low mountains surrounded by the ancient delta of the Yukon and Kuskokwim Rivers, then to historical muskox range in northeastern Alaska in 1969 and 1970

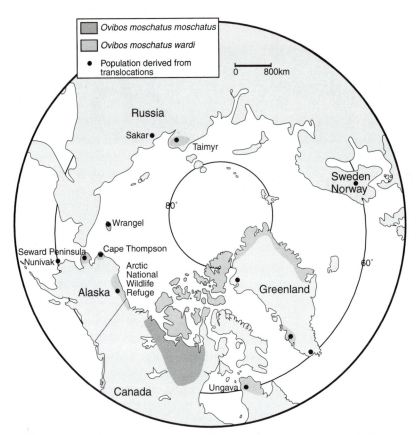

Figure 26–1 World distribution of free-ranging muskox populations. Populations established through translocation of animals into range areas occupied in historical times and to areas with no historical record of muskoxen are shown by cross hatching or, where they have remained in very localized areas, by an open circle.

and northwestern Alaska in 1970 and 1977. Muskoxen were also introduced to the Seward Peninsula, which is south of the known historical distribution of muskoxen, in 1970 and 1981 (7). All of these translocations of muskoxen have resulted in established populations, and dispersal from the northeastern Alaska population has resulted in the establishment of breeding populations westward to the central Alaskan Arctic and in the northern Yukon Territory of Canada and adjacent Northwest Territories.

Muskoxen from Canada and Alaska have been translocated to Russia to establish populations on the Taimyr Peninsula, where the most recent fossil record is dated approximately 2,800 years ago, to Wrangel Island in the Chuckchi Sea, and in 1996 and 1997 50 muskoxen from the Taimyr were released in the northern Sakar Republic (Yakutia) (8). Muskox populations have also been successfully established outside of their historical range in western Greenland, northern Quebec and adjacent Labrador, and Norway and adjacent Sweden.

The alarming decline of muskoxen on the North American mainland during the 19th and beginning of the 20th centuries led the Canadian government to grant legal protection for the muskox in 1917. The Thelon Game Sanctuary in the Northwest Territories was established in 1927, primarily to protect one of the few small populations of muskoxen remaining on the mainland (8). These actions were probably responsible for preventing the extirpation of muskoxen on the North American mainland. In the Canadian Arctic Archipelago and in Greenland, remote muskox populations have not been heavily exploited by humans.

Value

Muskoxen apparently have not been the primary food resource of indigenous peoples of the Arctic. However, archaeological evidence from Alaska, Canada, and Greenland indicates that they have served as an important seasonal subsistence food for some groups of arctic peoples (1, 9, 2, 10). With the reestablishment and increase of muskoxen in recent years throughout the Arctic, indigenous peoples are increasingly harvesting muskoxen as a subsistence food. Some use has been made by native peoples of the bones, horn, skin, and underwool of muskoxen for cultural items and, more recently, for sale to tourists.

Commercial exploitation of the large muskox population on Banks Island for export of meat to largely Canadian urban markets has been under way since 1981 under a harvest quota system. Annual harvests of a few hundred animals have been well below the annual quotas of more than 2,000 (11, 12). Indigenous peoples in Canada, primarily on Banks and Victoria Islands, in Greenland, and in Alaska, primarily on Nunivak Island, derive income from guiding, outfitting, and provision of other services to nonresident trophy hunters and ecotourists. It is clear that the muskox is viewed by those living outside of the Arctic as a unique species emblematic of the Arctic and acts as a drawing card for tourists and for sport and trophy hunters.

Efforts to develop a cottage industry in the Arctic based on harvesting of the extremely fine underwool of the muskox, marketed under its Inuit name [i.e., *qiviut,* and using captive herds, date from the 1930s (13, 14)]. Most of these efforts that took place in Alaska, Quebec, and Norway have been abandoned, although a muskox farm with approximately 40 animals continues to operate in Palmer, Alaska. There, income is derived from tourist visits and the production of *qiviut,* which is marketed through a cooperative that contracts with native women in rural villages to weave the *qiviut* yarn into fashionable women's garments. Captive herds for research purposes are maintained by the Universities of Alaska, Saskatchewan, and Tromsø.

Muskoxen, while valued by many as a unique and interesting component of Arctic ecosystems, also play an important role, along with caribou, as the only large herbivores involved in the transfer of primary production to higher tropic levels, which include predators and scavengers.

TAXONOMY AND DISTRIBUTION

The muskox is a member of the family Bovidae, which includes bison, cattle, sheep, and goats, but its relationship is closest to the North American mountain goat and to the goral and serow of Asia (15). It is included along with sheep and goats in the subfamily Caprinae. Although its closest relative was thought to be the takin, a similar appearing mountain-dwelling ungulate found in China and the adjacent Himalayan mountains, analysis of mitochondrial DNA from the muskox and takin does not support this suggestion, and their similarity appears to result instead from convergent evolution (15).

Three subspecies of muskoxen have been recognized on the basis of slight differences in body size, horn characteristics, and color of pelage. These are the barren ground muskoxen on the northern central mainland of Canada and originally extending across northern Alaska; the Hudson Bay muskoxen occurring on the Canadian mainland west of Hudson Bay; and the white-faced muskoxen present in northeastern and northern Greenland and the Canadian Arctic Archipelago. However, recently completed studies of the mitochondrial DNA from all three subspecies show little genetic diversity within the species and do not support the previous subspecific separations (16).

Muskoxen are blocky and short legged (Figure 26–2). Adult bulls weigh about 340 kilograms and stand about 135 centimeters at the shoulder, whereas females are about 90 kilograms less and are about 110 centimeters at the shoulder. However, there is considerable variation in size in relation to forage conditions, and body size decreases into

Figure 26–2 Muskoxen have a compact blocky body with short legs. The horns, with sharp upturned tips, are effective weapons against wolves. (Photograph by D. R. Klein.)

the High Arctic. The thick underwool and very long guard hairs of muskoxen give them a massive appearance. Both sexes have similar dark brown pelage on the body, with pale muzzles and nearly white saddles and lower legs. The amber-colored horns with black tips extend down along the side of the head in both sexes before curving outward and upward to sharp tips, forming effective weapons against wolves. The horns of adult males expand into broad bases that nearly meet at the top of the head, forming a thick bos that absorbs the shock of head-on horn clashes during rut. Boone and Crockett Club records of muskox trophies include several bulls with individual horn lengths exceeding 63.5 centimeters, bos widths of more than 25 centimeters, and tip-to-tip spreads exceeding 76.2 centimeters; however, these are exceptional individuals (17).

On the basis of the fossil record, muskoxen first appeared in the late Pliocene or early Pleistocene in Asia (18). They probably reached North America via the Bering Land Bridge in the mid-Pleistocene about 90,000 years ago, and were present from Europe across Siberia and into Alaska and the adjacent Yukon during the last glacial period, which lasted until about 10,000 years ago (19). Following recession of the continental ice sheets and the northern advance of the treeline, muskoxen persisted only in arctic regions north of the boreal forests. Muskoxen were present in northern Siberia until about 2,800 years ago (8).

LIFE HISTORY

The rutting period of muskoxen extends for nearly 2 months beginning in late summer. Gestation has been documented among captive muskoxen to vary from 244 to 252 days (20). There is an extended calving period, which may begin in early April and continue through mid-June, with no pronounced synchrony of calving such as that characteristic of caribou and many other northern ungulates. Normally a single calf is born, although a few cases of twinning have been recorded. Under conditions of adequate quantity and quality of available forage during the summer growth period, female

Life History

muskoxen generally breed for the first time in their second year, giving birth when 3 years old. They may then bear a calf annually in subsequent years, although if stressed nutritionally or energetically by extreme weather conditions that limit forage production or availability, they may fail to accumulate sufficient body reserves to allow them to come into estrus at the time of the rut and thus will not bear a calf for 1 or more years (21). Young male muskoxen, although sexually mature by 3 years of age, are usually prevented from breeding by the dominant older herd males. Older males that have not achieved or have lost herd male status are present throughout the distribution of muskoxen as singles or in small all-male groups. They are therefore more vulnerable to predation than adult animals in the larger mixed sex groups. As a consequence, the sex ratio among adults favors females. Muskoxen of both sexes can live for more than 20 years in captivity (W. Hauer, Institute of Arctic Biology, University of Alaska Fairbanks, personal communication) and there is a record from the wild of a female living for more than 30 years (H. Thing, Danish Polar Center, personal communication).

Muskoxen are endowed with a variety of morphologic and physiologic adaptations to life in the Arctic (22). Their compact body, with short legs, neck, ears, and tail, results in a low surface area in relation to their body size and minimizes heat loss in winter. The extremely fine and dense underwool of muskoxen, which is replaced annually, is thought to provide better insulation than the pelage of other mammals and further reduces winter heat loss. Prior to winter, muskoxen accumulate extremely large amounts of body fat, primarily around the viscera and subcutaneously, which provide a reserve for winter when forage is of low quality and its availability is restricted. Female muskoxen also draw on these fat reserves during early lactation following the birth of their young, which occurs prior to the initiation of plant growth of high forage quality. Young muskoxen, like the young of other northern ungulates, grow rapidly during the summer period of high forage quality and abundance, and then enter a growth plateau period during winter that reduces their metabolic requirements (23). Additionally, muskoxen further reduce their metabolic requirements by greatly reducing activity during winter, and in this regard can be considered among the most energetically conservative of ungulates (22).

BEHAVIOR

Muskoxen are highly social animals. They are generally found in groups of a few to 30 or more animals (Figure 26–3). Sociality among muskoxen, as with most other ungulates living in open terrain, is thought to have evolved as a strategy to minimize predation (24). The increased surveillance provided by the multiple eyes within a group means that any single animal will need to spend less time watching for the approach of predators and therefore will have more time to devote to foraging and other activities. In addition, the muskox relies on a group defense against the wolf, its major predator, and other possible predators (25). When approached by wolves, muskoxen generally run to high ground where they closely aggregate, with the adults facing the attackers and the young animals behind the adults or surrounded by them (Figure 26–4). In this defensive formation the group stands its ground, and if the wolves move in too closely an adult will charge out trying to hook and throw a wolf with its horns before spinning around and returning to the protection of the group. This group defense against wolves is generally effective, and wolves are therefore more frequently successful in making kills when they can surprise the muskoxen and attack calves or yearlings while the group is fleeing to a defensive location, or when encountering single male muskoxen.

Most muskoxen occur in groups of females and young animals, with a dominant herd male; occasionally a second male is tolerated by the herd male provided that it does not challenge the dominance status (26). Group composition is not fixed over time and there is exchange of individuals of both sexes between groups.

Figure 26–3 Muskoxen are social ungulates usually found in groups composed of females and young with a dominant herd male associated with most groups. Adult males that have not been successful in joining mixed sex groups exist as single animals or in small all-male groups. (Photograph by D. R. Klein.)

Figure 26–4 When attacked by wolves muskoxen generally come together in tight groups with young animals in the center and adults facing the attackers. The group shown here is responding in a similar fashion to Eskimo sled dogs used to hold the muskoxen so that they can be darted and marked during an investigation of their movements and social structure in northeastern Greenland. (Photograph by D. R. Klein.)

Female muskoxen do not isolate themselves from the group when giving birth as caribou do. Although a bond between the female and her calf is developed, it does not appear to be as strong as the one formed by caribou, thus muskox calves are vulnerable to abandonment if mixed sex groups are subjected to disturbance, such as from low-flying aircraft or other human activities, shortly after calving (27).

Muskoxen do not make long seasonal migratory movements and are generally present in the same areas year-round; however, shorter seasonal movements are not uncommon, particularly in the High Arctic (28). Adult male muskoxen may occasionally wander long distances from traditional areas. Movements are usually related to seasonal forage availability and quality, as influenced by variation in winter snow depth. Home ranges may include overlapping summer and winter ranges or seasonal ranges that are separate from one another (29). Muskoxen are not territorial.

As muskoxen have become reestablished in historical range areas or in new areas not previously occupied, indigenous peoples living or hunting within these areas have expressed concern that muskoxen may displace caribou or reindeer. However, aggressive interaction between these species has not been observed to occur when they are together in the same area (30; C. Ihl, Department of Biology and Wildlife, University of Alaska Fairbanks, personal communication).

POPULATION DYNAMICS

Muskox populations have recovered from alarmingly low numbers in remote residual populations in the High Arctic early in the 20th century to numbers of more than 150,000 at the end of the 20th century. This includes their reestablishment in formerly occupied habitats through range expansion and translocation of animals, and their establishment in several areas where they were not previously present, at least in historical times (Table 26–1).

Partly because of their past precipitous decline in numbers, coupled with their remoteness that has limited investigation into their biology, it was previously thought that muskoxen were characterized by an extremely low reproductive rate (8). Age of sexual maturation was thought to be 3 or 4 years in females, with females producing calves only in alternate years. With observation of the rapid increase of muskox populations following their legal protection and introductions into new habitats, it became apparent that earlier assumptions needed revision. Under favorable habitat and weather conditions, most female muskoxen will give birth at 3 years of age and a few at 2 years, and may produce a single young annually (31, 32). Exponential rates of increase may exceed 25% annually. Under poor habitat conditions, however, female muskoxen may not come into estrus at the time of the rut and, therefore, fail to bear a calf for 1 or more years.

Productivity varies primarily in response to weather conditions. Extremely deep snows and ice layers at ground level or in the snow may restrict access to forage with resulting heavy mortality, especially of very young and old animals. Icing has been a major factor in the decline of muskox populations in the Canadian High Arctic (33). Occasionally, extreme summer weather conditions, including delayed snow melt and low temperatures, may so severely restrict plant growth that forage production may not be adequate to meet the nutritional requirements of muskoxen in summer and the subsequent winter. Weather extremes appear to be more frequently involved in muskox population declines with increasing latitude into the High Arctic.

Predation, primarily by wolves, can be an important source of mortality. This has been the case in northern Greenland where local muskox populations are often partially isolated from one another by glaciers, high mountains, and broad fjords. Wolves have periodically reached these areas of Greenland from Ellesmere Island, established breeding populations, and preyed heavily on the muskoxen, the only ungulate prey available (34). The wolves declined to extinction in the 1930s, presumably after muskox densities had reached low levels and following intensive efforts by trappers to

TABLE 26–1
Population Numbers, Trends, and Legal Status of Muskoxen

Location	Year(s) of Introduction	Population Estimate	Legal Status	Source (Personal Communication or Reference)
Alaska				
Nunivak Island	1935, 1936	634 (1998)[a]	Permit hunts (1975+)	R. Seavoy, Alaska Department of Fish and Game
Nelson Island	1967, 1968	293 (1998)[a]	Permit hunts (1981+)	R. Seavoy, Alaska Department of Fish and Game
Adjacent mainland		57 (1997)	No open season	U.S. Fish and Wildlife Service
Seward Peninsula	1970, 1980	>1,432 (1998)[b]	Permit hunts (1997)	J. Dau, Alaska Department of Fish and Game
Northeast Alaska	1969, 1970	508 (1998)[b]	Permit hunts (1983+)	P. Reynolds, U.S. Fish and Wildlife Service
Northwest Alaska	1970, 1977	>387 (1998)[a]	No open season	J. Dau, Alaska Department of Fish and Game
Total		**3,311**		
Canada				
Northwest Territories Mainland	Indigenous	19,121± (1986–97)	Hunting quotas	B. Fournier and A. Gunn, Government, Northwest Territories
Banks Island	Indigenous	45,833± 1,938 (calves); 10,614± 488 (calves)	Hunting quotas	J. Nagy and M. Branian, Government, Northwest Territories
Total		56,447± 2,426 (1998)		
Victoria Island	Indigenous	46,502± (1994, 1998)	Hunting quotas	B. Fournier, A. Gunn, J. Nagy, and M. Branian, Government, Northwest Territories
Other arctic islands	Indigenous	11,492± (1990–97)	Hunting quotas	B. Fournier and A. Gunn, Government, Northwest Territories
Yukon Territory	From NE Alaska 1973–1978, 1983	160± (1998)	No open season	D. Russell, Canadian Wildlife Service
Quebec		600 (1997)	No open season	J. Hout, Laval University
Total		**134,352**		
Greenland				
North and northeast	Indigenous	20,000 (1983)	Hunting quota	44
Northwest (Thule)	1986	150 (1995)	No open season	P. Aastrup, Greenland Environmental Institute
Western	1962–65, 1991–92	3,000+(1995)	Hunting quota	P. Aastrup, Greenland Environmental Institute
Southwest	1987	200 (1997)	Hunting quota	H. Thing, Danish Polar Center
Total		**23,350**		
Russia				
Taimyr	1974, 1975	1,000 (1995)	No open season	G. Yakushkin, Extreme North Agricultural Research Institute
Wrangel Island	1975	298 (1995)	No open season	F. Chernyavsky and D. Kovalev, Institute of Biological Problems of the North
Sakar (Yakutia)	1996, 1997	50 (1997)	No open season	G. Yakushkin, Extreme North Agricultural Research Institute
Total		**1,348**		
Norway	1947–53	50–80	No open season	M. Forchhammer, University of Oslo
Sweden	1971 (Norway)	ca. 10	No open season	A. Danell, University of Alaska Fairbanks
World Total		**162,436±**		

[a]Post-calving.
[b]Pre-calving.

552

take wolves through trapping, shooting, and poisoning, but returned again in the 1970s. Grizzly bears prey on muskoxen where their distribution overlaps that of muskoxen on the North American mainland. In the absence of wolves, such as in western Greenland and on Nunivak Island, and where wolf densities have been low, such as on the Seward Peninsula and Banks Island, other sources of mortality, such as malnutrition and winter starvation, that are more directly related to food limitations may become important.

Yersiniosis, a bacterial disease, has been implicated in numerous deaths of muskoxen under high-density conditions on Banks Island (35). No other diseases or parasites have been reported as important causes of mortality in wild populations. Adult male muskoxen frequently injure themselves in fights during the rut and these may result in mortality. Accidental deaths by wandering onto the sea ice or from falls from cliffs have also been documented (36, 37).

POPULATION MANAGEMENT

Management of muskox populations has in the past often taken the form of complete legal protection to allow recovery of overexploited populations or establishment of introduced populations. Following total protection from hunting imposed by the Canadian government in 1917, it was not until 1969 that hunting was again permitted through a limited quota system on Ellesmere Island (11). Quotas have now been established for many arctic communities in Canada and total more than 2,000 animals annually, although quotas in several areas are generally not filled due to the remoteness of muskoxen from communities. In the case of Banks Island, the very high quota greatly exceeds the potential for their harvest for subsistence, trophy hunting, and commercial harvest.

In Alaska, reintroduced muskox populations have been protected from hunting until populations reach sufficient size to permit limited annual harvests (Table 26–1). Efforts in the late 1960s and early 1970s by the Fish and Wildlife Service and the Alaska Department of Fish and Game to allow hunting of muskoxen on Nunivak Island to reduce the population within the range carrying capacity were opposed by segments of the public on an ethical basis (38). Efforts to control the population by translocating animals from the island, while successful in establishing new populations, were not effective in reducing the Nunivak Island population to the desired level.

On Nunivak Island, where the introduced muskox population has been hunted since 1975, permits are usually limited to males, hunted primarily by trophy and sport hunters from outside of the area. Harvest permits for a quota of females favor local residents who hunt for subsistence use of the meat. The hunting season is generally during September and from February through March 15. The local Yupik Eskimos favor this system, which enables them to derive income from guiding, transporting, and providing other services to trophy and sport hunters and engage in some subsistence hunting. The meat from all animals killed is required to be salvaged and local residents are usually given the meat from muskoxen killed by hunters. On Nelson Island, however, hunting permits for either sex animals are issued on a first come first served basis in the local village, a strategy that favors subsistence use by local residents. The hunting season is from February 1 through March 25. The Yupik Eskimos of Nelson Island have not expressed an interest in deriving income from encouraging hunting of the muskoxen by sport or trophy hunters from outside of their area. Muskoxen are the only ungulate available as a subsistence food to the Nelson Island people, in contrast to Nunivak Island where there are reindeer. The increasing numbers of muskoxen in Alaska and their expansion into new habitats should offer increased opportunities for muskox hunting for subsistence, sport, and trophy in the future.

The permit hunt system to limit the annual harvest to an allowable quota has been a relatively effective tool for management of muskoxen in Alaska. However, in northern Alaska and on the Seward Peninsula there has been disagreement between sport and subsistence hunters about how permits are allocated.

In Jameson Land, eastern Greenland, a 40-year tradition exists among the local Inuit for hunting muskoxen with the aid of sled dogs. Under a quota system, hunting takes place between August 20 and September 20. A party of hunters accompanied by two or more experienced dogs approach a group of muskoxen (usually 5 to 10 animals), the dogs are released to hold the muskoxen in a compact defensive formation, and the entire group is shot. This method of hunting causes minimal disturbance and stress to other muskoxen in the area and avoids breaking up the social structure of mixed sex and age groups, which is the case if individual animals are selectively shot from these groups. If adult males are to be the focus of the hunt, however, single males or small male groups can usually be found.

In Canada, with the rapid increase of muskoxen in recent decades on Banks Island, and more recently on Victoria Island, it has become apparent that control of muskox numbers to levels below carrying capacity of the habitat may not be possible where human densities are low and animals are remote from communities. Attempts to reduce the extremely large muskox population on Banks Island through liberal quotas and commercial harvests, because of concerns for overgrazing and competition with the depressed caribou population, have failed to bring about the desired reduction in muskox numbers. Some increase in the annual harvest may be possible through encouragement of increased trophy hunting and improvements in handling, transportation, and marketing of muskoxen killed in the commercial harvest. However, even given these incentives, it is doubtful that the Banks Island population can be reduced through human harvest to a level considered compatible with the carrying capacity of the habitat. The long-term consequences of deterioration of the habitat and a reduced population of muskoxen, however, may be buffered by heavy mortality during severe winters and increasing wolf predation.

HABITAT REQUIREMENTS

Muskoxen have broad muzzles and a large rumen capacity, which equips them for rapid forage ingestion and digestion of moderately low-quality forage. Their diet is dominated by graminoids, predominantly sedges, although willow leaves are important dietary components in early summer (Figure 26–5) (39). Forbs, which are of high nutritive value in early summer, are also selectively grazed. The diet of muskoxen varies seasonally and regionally. In winter, sedges make up the major part of the diet although in their southern distribution shrubby willows in riparian communities are browsed in early winter. The quality of forage consumed also varies widely from summer to winter. Muskoxen, like caribou, require a high quality diet during the brief arctic summer to provide for milk production, growth of the young, fat deposition prior to winter, and energy expenditure by males during the rut. During winter, the dried sedges and other graminoids available to muskoxen are of poor quality, but the slow rate of passage through their gut increases its digestibility. Muskoxen conserve energy in winter by greatly reducing their activity, thereby lowering their metabolic and forage requirements.

Muskoxen are restricted to treeless habitats. Their predator avoidance strategy is dependent on existence in open terrain where approaching predators can be seen in time for the muskoxen to form their defensive grouping. However, their short legs and small hooves make them poorly adapted for existence in areas of deep snow. Therefore, they seek areas with low winter snowfall and windy conditions that expose the vegetation. In the High Arctic under polar desert conditions, the low precipitation results in little winter snow, most of which is blown into drifts, thus causing little impedance to muskox movements and foraging. An exception occurs in winters of deep snows or when storms from the south in early winter bring rain that can create an ice layer on the frozen ground or in the snow, thus making the engulfed vegetation unavailable to the muskoxen as forage (40). In summer in the High Arctic, the dry conditions result in

Figure 26–5 Willow leaves are an important component of the diet of muskoxen during early summer. In the more southern regions of their distribution, where willows grow as shrubs along stream drainages, these riparian habitats are often favored locations for muskoxen. (Photograph by C. Ihl.)

most plant growth being restricted to areas where water from melting snow, accumulated in drifts during winter, provides moisture.

Snow plays an important role in the ecology of muskoxen, both through its direct effect on the animals and its influence on the distribution and composition of plant communities. The amount of snowfall, wind patterns, and terrain variability determine the amount of snow and where it is distributed on the landscape, and thus are major variables influencing habitat suitability for muskoxen (41).

Like other northern ungulates, muskoxen use natural mineral licks; however, availability of mineral licks does not appear to limit habitat occupancy over most of their distribution. The mineral found to be common to several licks used by muskoxen in Jameson Land, in eastern Greenland, was sodium, which has also been found in mineral licks used by other ungulates (42, 43). Mineral licks are visited most frequently in early summer when high potassium levels in newly growing plants inhibit retention of dietary sodium, and lactation and rapid growth of body tissues increase physiologic requirements for sodium.

HABITAT MANAGEMENT

The remoteness of most muskox habitat from human habitation and industrial development activities has spared muskoxen from the loss and degradation of habitat that has affected many of the other North American large mammal species. The industrialized world, however, is increasingly focusing on the Arctic as a source of fossil fuels and other minerals. To ensure the protection of critical habitats for muskoxen, it is essential that management agencies acquire detailed information on the seasonal habitat requirements and patterns of habitat use by muskoxen. This information is essential in planning land use activities throughout the present and potential future distribution of muskoxen, whether it be for establishment of parks and wildlife reserves or for planning for mining and petroleum exploration, development, and location of associated transportation corridors. The primary criteria that define the suitability of habitats for muskoxen include plant community composition and associated production of forage species, characteristics of winter snow cover as it relates to access to forage, proximity of seasonal range components and movement corridors between them, and density and species of other herbivores and predators.

The distribution of muskoxen has expanded greatly in recent decades through legal protection from overhunting, reestablishment in historical range areas, and introduction to areas outside of their original range. This increase in the overall distribution of muskoxen has occurred mainly throughout the southern portions of their distribution in the low Arctic and Subarctic. In the High Arctic, factors associated with limitations on muskox populations include the limited and widely scattered forage, and weather extremes that further limit forage production and availability. In their more southern distribution, populations are more likely regulated by muskoxen density in relation to the forage resource.

Little is known of the response of tundra vegetation to grazing pressure by muskoxen and associated carrying capacity of muskox habitats. A further complication exists in areas where populations have only recently become established. In such areas, plant communities that existed in the absence of muskoxen are now undergoing changes in their structure, species composition, and productivity as they adjust to the impact of the muskox grazing pressure. Until these plant communities have reached some degree of stability in relation to the muskoxen, it will be extremely difficult to assess carrying capacity of these ranges. It is apparent that management of muskox populations in the Arctic and Subarctic that have been increasing in number and distribution will require an increased emphasis on understanding habitat relationships and development of techniques for assessment of range status.

SUMMARY

Muskoxen, although well adapted for life under the extreme conditions of the Arctic, are poorly adapted for defense against humans as predators. As a consequence, they persisted only in areas remote from humans and were extirpated from southern portions of their native ranges during prehistoric and early historic times. Through legal protection and translocation efforts their populations have recovered throughout much of their original distribution and they number about 160,000.

Muskoxen have adapted to arctic extremes physiologically through energy conservation and the ability to accumulate large stores of body fat prior to winter, morphologically through their extremely effective insulative pelage, and behaviorally through conservation of energy by minimizing activity and movements. Muskoxen are restricted to treeless tundra habitats and localize in winter in areas that are often windy and with limited snow cover. Their diet consists mainly of sedges in winter, whereas in summer willow leaves, sedges, grasses, and forbes are important. They have a potentially high rate of reproduction and under favorable conditions females bear their first young at 3 years of age and may produce a single calf each year. Such conditions are more frequently realized in their southern distribution. Where populations are well established and productive, hunting is permitted under quota systems. Most hunting is by indigenous people for subsistence purposes, although some trophy hunting is possible in Alaska, Canada, and Greenland, and a limited commercial harvest is carried out on Banks Island in Canada.

LITERATURE CITED

1. ALLEN, E. 1913. Ontogenetic and other variations in muskoxen, with a systematic review of the muskox group, recent and extinct. *Memoirs of the American Museum of Natural History, New Series* 1:101–226.

2. HAHN, J. 1977. Besiedlung und sedimentation der Pra-Dorset-Statio Umingmak I D, Banks Island, N.W.T. *Polarforschung* 47:26–37.

3. HONE, E. 1934. The present status of the muskox in arctic North America and Greenland. American Committee for International Wildlife Protection, Publication No. 5.

4. BURCH, E. S. 1977. Muskox and man in the central Canadian Subarctic, 1689–1974. *Arctic* 30:135–154.

5. BARR, W. 1991. Back from the brink: The road to muskox conservation in the Northwest Territories. Komatic Series, No. 3. Arctic Institute of North America, Calgary, Alberta, Canada.

6. REED, I. M. 1946. Notes on the extinction of the muskox in Alaska. Appendix C to a letter to the Hon. Julius A. King, Secretary of the U.S. Department of Interior (Nov. 18, 1946).

7. KLEIN, D. R. 1988. The establishment of muskox populations by translocation. Pages 298–318 *in* L. Nielsen and R. D. Brown, eds., *Translocation of Wild Animals.* Wisconsin Humane Society and Caesar Kleberg Wildlife Research Institute, Milwaukee, WI, and Kingsville, TX.

8. TENER, J. 1965. *Muskoxen in Canada: A Biological and Taxonomic Review.* Queen's Printer, Ottawa, Ontario, Canada.

9. KNUTH E. 1967. Archaeology of the musk-ox way. Publication No. 5. Contributions du Centre d'Etudes Arctiques et Finno-Scandinaves, Sorbonne.

10. WILL, R. T. 1984. Muskox procurement and use on Banks Island by nineteenth century Copper Inuit. *Biological Papers of the University of Alaska, Special Report* 4:153–161.

11. URQUHART, D. R. 1982. Muskox: Life history and current status of muskoxen in the NWT. Northwest Territories Renewable Resources, Wildlife Service.

12. GUNN, A., C. SHANK, and B. MCLEAN. 1991. The history, status and management of muskoxen on Banks Island. *Arctic* 44:188–195.

13. PALMER, L. J., and C. H. ROUSE. 1963. Muskoxen investigations in Alaska 1930–1935. Department of the Interior, Bureau of Sport Fisheries and Wildlife, Juneau, AK.

14. WILKINSON, P. F. 1971. The domestication of the muskox. *Polar Record* 15:683–690.

15. GROVES, P., and G. F. SHIELDS. 1996. Phylogenetics of the Caprinae based on cytochrome *b* sequence. *Molecular Phylogenetics and Evolution* 5:467–476.

16. GROVES, P. 1997. Intraspecific variation of mitochondrial DNA of muskoxen based on control region sequences. *Canadian Journal of Zoology* 75:568–575.

17. NESBITT, W. H., and J. RENEAU. 1988. *Records of North American Big Game.* The Boone and Crockett Club, Dumfries, VA.

18. HARINGTON, C. 1961. History, distribution and ecology of the muskoxen. Thesis. McGill University, Montreal, Quebec, Canada.

19. MCDONALD, H., and R. DAVIS. 1989. Fossil muskoxen in Ohio. *Canadian Journal of Zoology* 67:1159–1166.

20. LENT, P. C. 1988. *Ovibos moschatus. Mammalian Species* 302:1–9.

21. SCHULMAN, A. B., and R. G. WHITE. 1997. Nursing behavior as a predictor of alternate-year reproduction in muskoxen. *Rangifer* 17:31–35.

22. KLEIN, D. R. 1996. Arctic ungulates at the northern edge of terrestrial life. *Rangifer* 16:51–56.

23. WHITE, R. G., F. L. BUNNELL, E. GAARE, T. SKOGLAND, and B. HUBERT. 1981. Ungulates on arctic ranges. Pages 397–483 *in* L. C. Bliss, O. W. Heal, and J. J. Moore, eds., *Tundra Ecosystems: A Comparative Analysis.* Cambridge University Press, Cambridge, UK.

24. KLEIN, D. K. 1992. Comparative ecological and behavioral adaptations of *Ovibos moschatus* and *Rangifer tarandus.* Rangifer 12:47–55.

25. GRAY, D. R. 1974. The defense formation of the muskox. *Musk-ox* 14:25–29.

26. GRAY, D. R. 1987. *The Muskoxen of Polar Bear Pass.* Fitzhenry and Whiteside, Ontario, Canada.

27. MILLER, F. L., and A. GUNN. 1979. Responses of Peary caribou and muskoxen to turbo-helicopter harassment, Prince of Wales Island, Northwest Territories, 1976–77. Occasional Paper No. 40. Canadian Wildlife Service, Ottawa.

28. RAILLARD, M. 1992. Influence of muskox grazing on plant communities of Sverdrup Pass (70 °N), Ellesmere Island, N.W.T., Canada. Doctoral dissertation. University of Toronto, Ontario, Canada.

29. MILLER, F. L., R. H. RUSSELL, and A. GUNN. 1977. Peary's caribou and muskoxen on western Queen Elizabeth Islands, N.W.T., 1972–4. Canadian Wildlife Service Report, Series No. 40, Ottawa: Queen's Printer.

30. BIDDLECOMB, M. E. 1992. Comparative patterns of winter habitat use by muskoxen and caribou in northeastern Alaska. M.Sc. thesis. University of Alaska, Fairbanks.

31. JINGFORS, K. T., and D. R. KLEIN. 1982. Productivity in recently established muskox populations in Alaska. *Journal of Wildlife Management* 46:1092–1096.

32. ROWELL, J. 1989. Survey of reproductive tracts from female muskoxen harvested on Banks Island, NWT. Page A57 *in* P. F. Flood, ed., *Proceedings of the Second International Muskox Symposium,* Saskatoon, Saskatchewan, Canada. National Research Council of Canada, Ottawa, Ontario, Canada.

33. PARKER G. R., D. C. THOMAS, E. BROUGHTON, and D. R. GRAY. 1975. Crashes of muskox and Peary caribou populations in 1973–74 on the Parry Islands, Arctic Canada. *Canadian Wildlife Service Progress Notes* 56:1–10.

34. DAWES, P. R., M. ELANDER, and M. ERICSON. 1986. The wolf (*Canis lupus*) in Greenland: A historical review and present status. *Arctic* 39:119–132.

35. BLAKE, J. E., B. D. McLEAN, and A. GUNN. 1989. Yersiniosis in free-ranging muskoxen on Banks Island, NWT, Canada. Page A58 *in* P. E. Flood, ed., *Proceedings of the Second International Muskox Symposium,* Saskatoon, Saskatchewan, Canada. National Research Council, Ottawa.

36. SPENCER, D. L., and C. J. LENSINK. 1970. The muskox of Nunivak Island. *Journal of Wildlife Management* 34:1–15.

37. KLEIN, D. R., and H. STAALAND. 1984. Extinction of Svalbard muskoxen through competitive exclusion: An hypothesis. *Biological Papers of the University of Alaska, Special Report* 4:26–31.

38. LENT, P. C. 1971. Muskox management controversies in North America. *Biological Conservation* 3:255–263.

39. KLEIN, D. R., and C. BAY. 1994. Resource partitioning by mammalian herbivores in the high Arctic. *Oecologia* 97:439–450.

40. GUNN, A., F. L. MILLER, and B. McLEAN. 1989. Evidence for possible causes of increased mortality of bull muskoxen during severe winters. *Canadian Journal of Zoology* 67:1106–1111.

41. NELLEMANN, C., and M. G. THOMSEN. 1994. Terrain ruggedness and caribou forage availability during snowmelt on the Arctic Coastal Plain, Alaska. *Arctic* 47:361–367.

42. KLEIN, D. R., and H. THING. 1989. Chemical elements in mineral licks and associated muskoxen feces in Jameson Land, northeast Greenland. *Canadian Journal of Zoology* 67:1092–1095.

43. JONES, R. L., and H. C. HANSON. 1985. *Mineral Licks, Geophagy, and Biochemistry of North American Ungulates.* Iowa State University Press, Ames.

44. THING, H., P. HENRICHSEN, and P. LASSEN. 1983. Status of muskox in Greenland. Pages 1–6 *in* D. R. Klein, R. G. White, and S. Keller, eds., *Proceedings of the First International Muskox Symposium.* Biological Papers of the University of Alaska, Special Report No. 4.

Pronghorn

J. D. Yoakum and B. W. O'Gara

INTRODUCTION

Pronghorn are resilient wild ruminants endemic to North America, where they have lived for more than 10 million years. Nelson speculated that populations numbered more than 30 million during the early 1800s, but were nearly extirpated (1). However, the 20th century witnessed populations rebounding to become the second most abundant big game in the United States (2).

When Lewis and Clark crossed the continent, they remarked that these small speedy ungulates reminded the explorers of Old World gazelles; consequently, the explorers called them antelope (3). However, authors of contemporary technical literature use pronghorn in recognition of the animals' unique horns (Figure 27–1). This is even more appropriate in modern times as Old World antelope have been translocated and established on pronghorn habitats. In Mexico they are called *berrendo* and French-Canadians refer to them as *cabrie*. Native Americans have given pronghorn more than 20 different names.

Figure 27–1 The pronghorn, so named because of the prongs on the annual deciduous horns, is common on much of the open grasslands and shrub steppes of North America. (Photograph by B.W. O'Gara.)

Historical Perspective

Prior to the 19th century, Native Americans harvested pronghorn for food, tanned hides for clothing, and used the animals' parts as artifacts and in spiritual ceremonies. The harvest had little effect on populations. Pronghorn were apparently first seen and hunted by arriving Euroamericans in the 1500s in Mexico. However, it was not until 250 years later that the species was collected for science by the Lewis and Clark expedition. The explorers obtained skins and skeletons, which became the basis for the first scientific description of the species (4). Ord used the term *pronghorn* to differentiate the North American species from the antelope of Asia and Africa (4).

The arrival of homesteaders on the western prairies had a devastating effect on pronghorn, resulting in the diversion of waters, plowing of native plants, construction of barrier fences, and relentless yearlong hunting for consumption and marketing. Thus, pronghorn were reduced to less than 1% of their pristine numbers at the end of the 19th century.

The beginning of the 20th century witnessed isolated populations slowly increasing as protective laws were enacted. Beginning in 1930, wildlife management practices such as translocation projects and water developments accelerated recovery. Populations increased by 300 times in less than 50 years. Pronghorn populations were well on the road to recovery and became the epitome of a restored big game species in North America (2).

As pronghorn increased and hunting seasons reopened, they became an asset to local communities because of their beauty and observability. Hunters and tourists were attracted and enriched entrepreneurs in the area. Evaluations of the monetary value of pronghorn are lacking. However, the value of pronghorn to Alberta—a province with about 22,300 pronghorn in 1989—was estimated at nearly $7 million annually at that time (5).

Another factor promoting public interest and support was the development of conservation and management organizations dedicated to the welfare of pronghorn (e.g., the Order of the Antelope in Oregon and the Arizona Pronghorn Foundation). The newest conservation organization (established in 1991) dedicated to pronghorn is the North American Pronghorn Foundation headquartered in Casper, Wyoming. The foundation is concerned with the welfare of pronghorn throughout North America.

Two technical organizations function to benefit pronghorn: the Interstate Antelope Conference (California, Nevada, and Oregon) and the Pronghorn Antelope Workshop with representatives from Canada, Mexico, and the United States. Both groups have published guides for management of pronghorn and their habitat (6, 7).

TAXONOMY AND DISTRIBUTION

Physical Description

According to O'Gara, pronghorn are the smallest North American big game ruminant, averaging 48 kilograms for adult females and 54 kilograms for adult males in Montana, while newborn fawns average 3.5 kilograms (8). Mean measurements in centimeters are as follows: total length 140, shoulder height 89, and tail 13. Body color is contrasting white and rusty brown to tan with black or dark brown markings about the head and a black patch below the ears on males, but absent on females. Eyes average 36 centimeters and are the largest in relation to body size for any North American wild ungulate. Dentition is I 0/3, C 0/1, P 3/3, M 3/3 = 32 that includes 12 upper (M, PM) and 20 lower (I, C, P,M). All permanent teeth are present at 36 months of age (9). Males have the one median, two subauricular, two rump, and four interdigital skin glands. The median and subauricular are absent on females.

The most unique physical characteristic of pronghorn is their horns. They possess a true horn core that is part of the skull, covered with a black sheath. This sheath is grown from cells similar to skin and not hair, as is sometimes alluded to in literature (10). Horn sheaths are cast each fall and new ones immediately start to grow. Pronghorn are the only ungulate in the world that annually cast the horn sheath and it is this biological characteristic that contributed to their classification as a separate family. Males have a prong of about 7.5 centimeters midway on the horn sheath protruding forward. Males have horns averaging 25 to 38 centimeters, while approximately 70% of the females have horns 2.5 to 10 centimeters long, generally lacking the prongs. Lateral toes (dew claws) are absent. Front feet are larger than rear.

Taxonomy

The scientific classification of pronghorn has been a challenging endeavor resulting in the recognition of one family, genus, and species. Therefore, most taxonomists place the pronghorn in the order Artiodaclyla, family Antilocapridae, genus *Antilocapra,* and species *americana* (11). Five subspecies generally are recognized including American pronghorn, Oregon pronghorn, Mexican pronghorn, Sonoran pronghorn, and peninsular pronghorn.

The original geographic delineation of subspecies was based on small samples and questionable assumptions. Certain presumed subspecies have been mixed due to numerous translocations. Recent studies by Lee and others supported (but did not prove) subspecies status for Mexican pronghorn and questioned (but did not disprove) subspecific status for Oregon pronghorn (12).

Distribution

Pronghorn historically ranged from south-central Canada through all western states in the United States, south to San Luis Potosi, Mexico (Figure 27–2). Herds occupied environments from sea level in Mexico to more than 3,535 meters in Oregon and Wyoming. Most

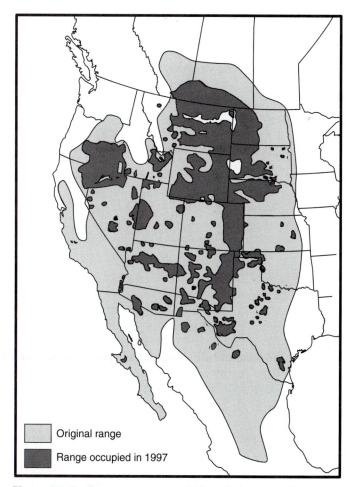

Figure 27–2 Contemporary and historical range of pronghorn. "Original range" modified from Nelson (1). Not all areas within these boundaries occupied. Range occupied in 1997. (From Ref. 35.)

ancestral habitats are occupied today; however, herds are much smaller and many are isolated. All translocation attempts outside of ancestral habitats (e.g., Florida, Hawaii, Washington) failed to establish sustaining populations for more than several decades.

LIFE HISTORY

Reproduction

Pronghorn females usually breed for the first time at 16 to 17 months of age, but fawns occasionally conceive a month or more later than adults and yearlings (8). Although pronghorn on southern rangelands have breeding seasons as long as 3 months, those in most of the species' range have a comparatively short (about 3 weeks) and intense rut. If not bred, females have recurrent estrus (13, 14). Differences in breeding dates are related to rainfall patterns and elevation that contribute to the timing of spring green-up (15).

Known gestation periods for captive pronghorn averaged 250 days at the Sybille Wildlife Research Station, Wyoming (W. Hepworth, Wyoming Game and Fish Department, personal communication). However, gestation periods were longer following dry summers than those following wet summers on the National Bison Range, Montana (16).

The gestation period is relatively long for such a small ruminant, apparently because pronghorn females have the highest level of maternal investment in their offspring of any North American ruminant (17). The percent litter birth mass and daily fawn growth rate compared to maternal mass is 18 and 0.07, respectively. By comparison, the same values for bighorn sheep are 5 and 0.04 (16). For this reason, female pronghorn are especially sensitive to nutritional stress during spring.

Twins were the rule during first pregnancies and 98% of subsequent pregnancies, but singles were three times as common as triplets in Wyoming (18). Three to nine embryos commonly are conceived, but intrauterine mortality ordinarily reduces the litter size to two (8, 19). Based on a sample size of 645 fetuses collected in six states, the sex ratio was 112 males per 100 females (20).

Longevity

It is common for pronghorn to live to 9 years of age (21). Failing teeth coupled with the stress of winters or droughts weaken the animals. Then they may die of pneumonia, predation, or other causes. However, during favorable environmental conditions, pronghorn have lived to 15 years of age or more (22).

Physiology

The pronghorn is well adapted to its open, arid environment. Physiologic characteristics and adaptations such as selective cooling of the brain during heat stress, large respiratory organs and airways, high buffering capacity and hemoglobin levels of the blood, and high blood volumes cannot be seen (23, 24). However, these attributes complement the pronghorn's long legs with heavy proximal muscles and lightweight feet that indicate a swift and enduring runner (25). Physiologic adaptations to help the animals cope with semiarid environments include conserving water through countercurrent exchange of respiratory air in the long nose and lowering urine output by reducing food intake during periods of dehydration (26, 27). These and other physiologic adaptations, many of which need further study, combine with behavior to make pronghorn extraordinarily unique animals.

Fawn growth is extremely rapid. In Alberta, fawns gained about 27.2 kilograms during their first summer and autumn (19). Mitchell reported that the mass of yearling pronghorn during their second winter corresponded to that of adults (19). In contrast, Roseberry and Klimstra indicated that white-tailed deer reached full mass at 4.5 years of age (28).

Nutrient Requirements

Recent diet studies provide new insights regarding how much grasses, forbs, and shrubs contribute to pronghorn diets in differing biomes (29). Pronghorn are opportunistic foragers and shift use of forage classes depending on availability, succulence, and nutritional value. Forbs are the dominant forage class ingested (Figure 27–3). The

GRASSLANDS BIOME

SHRUB-STEPPE BIOME

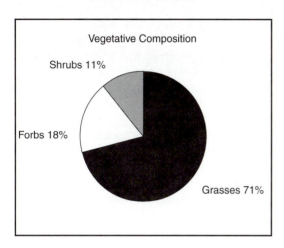

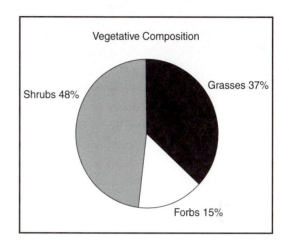

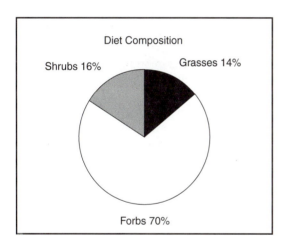

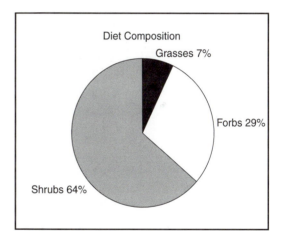

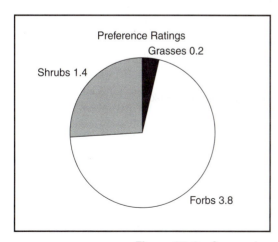

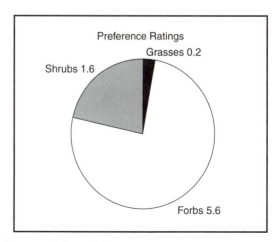

Figure 27–3 Comparison of vegetational composition, diet selection, and preference ratings for pronghorn on grassland and shrub-steppe biomes. (Modified from Ref. 29.)

true value of forbs may be even greater when correction factors for differential digestion rates are applied. Forbs are highly nutritious and provide water needed for survival and reproduction (30, 31). This is most important during spring and summer when quality forage is needed for fetus growth and lactation.

Foraging on shrubs varies according to habitat, season, and vegetative community (32). Next to forbs, shrubs are the forage class most frequently consumed by pronghorn (Figure 27–3). Protein values of shrubs are high during winter when the reverse is true for herbaceous vegetation (33, 29). Martinka documented pronghorn mortality from malnutrition on grasslands and survival on shrub/grasslands during a severe winter in Montana (34).

The role of grasses in the diet of pronghorn has been poorly understood in the past. Pronghorn often graze grasses heavily during spring and fall "green-up" when new growth is highly nutritious and succulent, but total use remains around 10% of annual diets (Figure 27–3) throughout all pronghorn habitats. Pronghorn consume less than 1% of graminoids on western rangelands (2).

Habitat managers should strive to sustain native vegetative communities with a mixture of grasses, forbs, and shrubs. Such communities should likewise have high numbers of forage taxa. Additional diet selection studies are needed because no studies could be located that used the technique of assessing differential digestion rates.

Pronghorn obtain moisture by drinking and from preformed water (obtained from forage) with lesser quantities produced by metabolic processes (27). Snow may replace free water. Pronghorn drink from natural sources such as springs, seeps, streams, rivers, lakes, and ephemeral pools. They likewise consume water from man-made reservoirs, troughs, and precipitation catchments or guzzlers. Populations occupying northern rangelands drink water frequently when it is available. When pronghorn are deprived of water, they exhibit dehydration stress, but those on arid rangelands apparently have adapted to living with low quantities and infrequent access to drinking water (35).

Whisler reported that summer water requirements for laboratory animals in Wyoming were 3.6 liters per day per 45-kilogram animal (27). The water needs for free-ranging animals may be greater. Habitat managers developing water sources should provide an average of 3.8 liters per day per animal for pronghorn accustomed to drinking during summers (7). Drinking water in quantities of approximately one-fourth the summer rates should be available during winters when free water is not available.

The availability of drinking water appears to be related to pronghorn densities. Herds occupying northern rangelands where drinking water usually is available maintain higher densities than populations in deserts (35).

BEHAVIOR

Social Structure

Pronghorn are gregarious, but group sizes vary throughout the year with different environmental conditions. Herds in Montana are fairly discrete and include the resident pronghorn on a particular land area that have routine social interactions and often winter as a single group (36). Combinations of large streams or other natural barriers, highways, railroads, fences, and livestock-use areas may define herd ranges. Bands are subdivisions of herds. Female bands are associations of females that have nearly exclusive use of traditional summer rangelands. Bachelor bands associate during the interval between spring dispersal and the onset of rut. Herds are separated primarily by environmental barriers, while bands within herds are isolated by behavioral mechanisms.

Studies have indicated nearly equal numbers of herds with harem-type breeding systems (in which dominant males controlled and defended females without regard to specific locations) and territorial systems in which males defended particular areas (37). The differences in social organization are not caused by major differences in specific behavior or displays. Rather, social organization is influenced by differences in density

Behavior

Figure 27–4 Young males in bachelor bands spar extensively while establishing dominance hierarchies. (Drawing by E. Fichter.)

and spatial dispersion. In other words, the same behavioral repertoire might result, under different environmental conditions, in different social systems (Figure 27–4).

In territorial populations, territorial males do almost all of the breeding. The territorial system appears beneficial because it ensures that the largest, most aggressive males do most of the breeding, that there are havens where females can escape the overzealous courtship of bachelor males during parturition, lactation, and the rut, and that bachelors are excluded from competing with pregnant or lactating females on the best rangelands (8).

The spatial distribution of pronghorn in any given area is partly a function of departure, return, exchange and coalescence among bands (38). Sizes of pronghorn home (20–1,144 square kilometers) and seasonal (0.03–64 square kilometers) ranges and distances moved daily (less than 1 kilometer to more than 10 kilometers)—as reported in the literature—vary so much with weather and habitat conditions that results from studies in one area seldom have application to another area, or even another year.

Movements

The timing and length of seasonal movements vary with altitude, latitude, weather, and rangeland conditions. Today, snow forces some herds to move 320 kilometers or more between summer and winter rangelands (39). Before humans constructed barriers to pronghorn travel, the animals apparently made longer movements.

Animals form large wintering herds, usually composed of both sexes and all age classes, and disperse during spring (40). In territorial populations, dominant males establish territories, usually during March. Females form bands of about 5 to 12 animals and bachelor males form larger bands that travel extensively.

Communication

In bright sunshine, the flash of erected rump patches can be seen over 1.5 kilometers. These visual signals undoubtedly alert other pronghorn at even greater distances. When the rump patch is erected, rump glands discharge a pungent airborne scent that alerts nearby pronghorn that have not observed the white flash. Males produce scent in subauricular glands that sexually arouses females and is used to mark territories (41). Females also are attracted by excretions of the median scent glands of males (42).

POPULATION DYNAMICS

Productivity

Two hundred and ninety pregnant females, collected in seven states, carried 450 fetuses for a ratio of 190 fetuses per 100 females (20). Thus, most yearlings and adults bear twins, and a few fawns also breed.

Pronghorn's high productivity allows them to recover rapidly after population crashes resulting from climatic conditions. For example, North Dakota had about 15,000 pronghorn during winter 1964 to 1965. Between then and 1980, the state experienced five major winter kills, and hunting seasons were closed during 1978 through 1981, but numbers dropped to 1,800. The winters from 1980 through 1985 were comparatively mild, and the population increased to 5,000 to 6,000 despite an annual harvest of about 1,000 animals during 1982 through 1985 (43).

The most variable, and probably the most important, factor in pronghorn productivity is fawn survival, measured as fawns per 100 females in midsummer. An example of variability was documented on a single fawning ground in California by Ellis (30). There, ratios ranged from 7 to 82 fawns per 100 females over a 13-year period.

Mortality Factors

Droughts and cold winters with deep crusted snow probably have held pronghorn to the limits of their habitat for eons. However, man-made barriers to movements, heavy grazing by livestock, and other modifications of habitat magnify the problems of climatic conditions. Conversely, favorable weather usually results in population increases despite hunting and other predation (44).

Malnutrition of adults usually is associated with severe weather. In addition, fawn survival is closely related to availability of palatable forbs and grasses during spring (30). Climatic and vegetative conditions can hardly be separated from levels of nutrition.

Human predation takes a large number of adult pronghorn annually. Hunters have harvested more than 4 million pronghorn since 1935 (45). However, such harvests are carefully regulated and generally help to maintain stable populations. Many other predators take pronghorn (mostly fawns), but predation that limits population stability or growth usually is associated with poor habitat conditions or small pronghorn populations and many predators. Of 882 radiocollared fawns monitored during 15 studies in 10 provinces and states, predators killed 459 (52%). Of these, 200 were killed by coyotes, 79 by bobcats, and 49 by golden eagles. The remaining 24 were killed by mountain lions, badgers, and a small raptor (44). Most of these studies were conducted because of poor fawn survival and they should not be considered typical for all pronghorn populations.

Extensive epizootics controlling pronghorn populations are uncommon. However, 14 viruses or viral diseases, 21 genera of bacteria, 8 species of protozoa, 1 fluke, 5 species of tapeworms, 33 species of roundworms, 4 species of ticks, and the louse fly have been reported in or on pronghorn (46). The impact of most of these agents on free-ranging populations is unknown.

Bluetongue, carried by cattle, probably is the most serious disease affecting pronghorn, and at least 3,200 died from the disease in eastern Wyoming during 1976 (47). Bever presented evidence that rangelands heavily grazed by sheep resulted in high parasite loads in pronghorn (48).

Poisonous plants, road and railroad accidents, parturitional problems, drowning, miring in mud, being stranded on ice, hail, fighting, falls and large ungulates all kill a few pronghorn (44). Although a train or a fall over a cliff may kill more than 100 animals at once, none of these mortality factors seriously affects populations over long periods of time.

Population Trends and Current Status

Pronghorn apparently numbered 30 to 40 million prior to the arrival of Euroamericans. Herds were heavily exploited during early western settlement and decreased to around 15,000 by 1915. An aroused citizenry, together with the enactment of protective laws, resulted in herds increasing to more than 30,000 within a decade according to the first survey conducted by Nelson (1). Wildlife management practices such as translocation

TABLE 27–1

Estimated Population of Pronghorn in Canada, Mexico, and the United States During the 1980s and 1990s

Country: Providence or State	1984	1995	1997
Canada:			
Alberta	21,500	17,300	13,200
Saskatchewan	10,000	22,000	10,000
Subtotal	31,500	39,300	23,200
Mexico:			
Chihuahua	300	200	
Coahuila	12	65	
Sonora	100	300	
South Lower California	100	200	
Subtotal	512	765	
United States:			
Arizona	9,000	11,000	10,200
California	6,800	6,500	5,600
Colorado	57,500	68,000	72,000
Idaho	23,500	12,500	12,000
Kansas	1,200	1,800	2,500
Montana	161,000	225,000	57,500
Nebraska	9,000	4,100	6,500
Nevada	9,800	15,600	14,500
New Mexico	30,000	36,000	35,000
North Dakota	5,700	10,000	5,200
Oklahoma	400	1,500	1,500
Oregon	14,000	17,100	15,000
South Dakota	67,000	47,900	20,000
Texas	12,000	12,000	9,600
Utah	6,000	11,000	10,500
Wyoming	608,000	441,500	370,300
Subtotal	1,020,900	921,500	649,000
Total	1,052,912	961,600	672,200

Source: From Ref. 49.

All figures (except Coahuila, Mexico) rounded to 100.

to extirpated areas, development of watering sites, and return of many habitats to native vegetation following abandonment of homesteads all contributed to accelerated population growth. In 1964, a survey indicated there were 386,000 pronghorn (2). Subsequent surveys document a continued increase to 1,052,912 in 1984 (2). Since then, the population has decreased approximately 36% (Table 27–1) (49).

The population during the 20th century of around 1 million animals is a tribute to the adaptability of the species and to mankind's willingness to share western rangelands with pronghorn. Numbers probably will never return to historic records where every pronghorn alive today represents 30 to 40 alive two centuries ago.

Threatened Classification

Three subspecies of pronghorn are classified as threatened: the Sonoran and the Mexican in Mexico and the United States, and the peninsular in Mexico. Less than 1% of the total North American population is endangered. The desert ecosystems that the Sonoran and peninsular subspecies historically and presently occupy have been and are areas of low densities. The three subspecies in Mexico appear to be experiencing the greatest survival problems (50).

Management Plans

Management plans are valuable practices for managing pronghorn and their habitat. They should list objectives, goals, and procedures to best manage the animals such as maintenance or improvement of forage, water, and space, and coordination of pronghorn management into holistic land use plans (7).

Wildlife management plans usually are initiated by provincial or state wildlife agencies. They typically emphasize practices to protect, reduce, maintain, or enhance populations, and spell out methods to census herds, control limiting factors, and harvest or translocate surplus animals.

Habitat management plans generally are prepared by government land agencies or private landowners. They emphasize the maintenance or improvement of forage, water, and space for pronghorn.

In the United States, management plans or recovery plans for threatened species are mandatory in accordance with the Endangered Species Act of 1973. Such plans identify procedures to increase the number of a species or subspecies to a population level adequate to justify removal from the list. The various guides suggesting techniques and practices to manage pronghorn and their habitat are aids to biologists developing management plans (51, 52, 52a, 6, 7).

Capture and Translocation

Capture and translocation have been integral parts of pronghorn management since the birth of restoration efforts early in the 20th century. Although restoration of this species has been phenomenally successful, trapping and translocation still are important in some provinces and states.

Pronghorn are captured by a wide variety of nets, traps, drugs, and, under certain conditions, by hand (7). Amstrup and others suggested parameters to consider when selecting a capture method (53). Number, age and sex of animals needed, density of animals in the trapping area, terrain and proximity to roads, whether the animals are accustomed to fences, how wary they are, the possibility and acceptability of mortalities, and the cost (in time and money) per captured or marked animal are important considerations. Lee and others give details of capture methods (7).

Translocation should be considered only after it has been determined that pronghorn can survive in historic rangelands possessing sufficient quantity and quality of forage, water, and space without conflict with other major environmental issues. Translocation should be preceded by a feasibility study and a management plan that documents the objectives, translocation procedures, and post-release monitoring of the animals and their habitat (7).

Predator Control

Pronghorn, although they have made an impressive comeback, sometimes are restricted in their movements by agricultural areas, highways, and fences. Thus, some herds are localized and relatively small. Under such artificial circumstances, predators may keep pronghorn populations from increasing or even eliminate them (7). Control of predators to benefit a big game population often involves a large area. Even if desirable, predator control seldom is economically feasible. However, Smith and others indicated that selective, time-specific application of aerial gunning in areas of high coyote density was an economically beneficial means of increasing fawn survival on Anderson Mesa in north-central Arizona (54).

In treating a problem situation where pronghorn populations are reduced and predators are prevalent, Lee and others recommended the following (7):

1. Determine the pronghorn herd parameters that are desirable in terms of total number, rate of recruitment to the herd, and age classes.

2. Determine the year-round distribution of pronghorn and the habitats involved.

3. Consider other population influences including, but not limited to, mortality factors.

4. If a factor such as predation is determined to be a significant inhibitor of a pronghorn herd, the cost of actually controlling predators in the short term must then be balanced against the long-term return.

5. If it is determined that the increase in pronghorn justifies the cost, predator control should be done on those herd units where documentation dictates predator reduction would be beneficial in meeting management objectives.

Crop Depredations

Although pronghorn generate considerable income, they also damage agricultural crops. Losses appear modest when viewed on a provincial or statewide basis, but such losses can be important to individual landowners. In nearly all cases, depredations are brought about by environmental changes made by humans. Results of a survey sent to conservation agencies of 18 western states and provinces in 1991 indicated that depredations were stable in 9 and increasing in 8. Only in Nebraska were depredation complaints decreasing, apparently because pronghorn numbers were depressed. Almost every agency reported some damage to alfalfa. Most state agencies did not pay compensation for crop damage. Those that did paid about $85,000 in 1990 (7).

The most practical solution to crop depredation is to reduce herds in the general area by issuing sufficient numbers of females/fawn permits during the regular hunting season. If problems persist, depredation permits issued at the target fields and at the time of depredations may solve the problem.

Harvest Strategies

Given the pronghorn's high reproductive rate, human population growth and sprawl, and the vagaries of weather and habitat conditions, managing the harvest is one of the most practical ways of ensuring sustainability and overall well-being of pronghorn populations. Nearly all pronghorn hunting is via limited-entry licenses. This reflects the low number of animals in some states and the need to distribute the harvest in others.

Pronghorn harvests should be carried out according to clear objectives. To do this, the manager needs sound biological data concerning the number of animals, habitat conditions, male:female ratios, reproductive potential, pronghorn behavior and movements, hunter access, and competing land uses. Setting population objectives is a balance between the species' needs and human interests. Simulation modeling has proven a valuable tool for determining harvest strategies suitable for achieving management objectives (45).

Pronghorn harvest regulations manipulate the type and level of harvest and "manage" the hunter. Regulating the harvest includes setting bag limits, season lengths, number of permits, and other strategies to ensure that a specific number and sex of pronghorn are harvested. Regulating the hunter involves restrictions to ensure that hunts are conducted legally, safely, and ethically and maximize opportunities for participation and harvest within the sustained yield.

HABITAT CHARACTERISTICS AND REQUIREMENTS

Pronghorn habitats exhibit biotic and abiotic components in varying degrees of quality and quantity that relate to population densities. Knowledge of habitat characteristics and requirements for a site often becomes the basic information needed to develop strategies to enhance populations and habitats. Approximately 68% of populations occur in grasslands (Figure 27–5) and 31% in shrub steppes (Figure 27–6) with fewer than 1% in deserts. They occupy large expanses of flat or low

Figure 27-5 Grasslands sustain the highest density of pronghorn per biome. Although grasses are dominant, the rangelands also support forbs and shrubs. Pronghorn and bison have grazed grasslands for centuries. (Photograph by R. Hitchcock.)

Figure 27-6 The shrub steppe is the second highest producing landscape for pronghorn. Such communities produce a mixture of grasses, forbs, and shrubs all consumed in various quantities and preference by pronghorn. (Photograph by R. Hitchcock.)

rolling terrain lacking major barriers to seasonal movements and a mixed vegetative community of grasses, forbs, and shrubs (Table 27–2).

Ellis compared populations for the shrub steppes of the Great Basin with the grasslands of the Great Plains, noting that fecundity was 190 fawns per 100 producing females for both ecosystems (30). However, fawn survival was twice as high and grass and forb production was greater on the Great Plains than in the Great Basin. Nutritive values (particularly protein) of grasses and forbs were greater than shrubs during late spring and early summer. Ellis concluded that fawn survival was twice as high on the Great Plains because of abundant, nutritious grasses, and forbs during gestation and lactation.

Habitat Characteristics and Requirements

TABLE 27–2
Habitat Requirements for Pronghorn in Grassland and Shrub-Steppe Biomes

	Pronghorn Requirements	
Habitat Factor[a]	Grassland Biome	Shrub-Steppe Biome
Abiotic		
1. Physiography: Low rolling terrain No major physical barriers Slopes less than 30%	Large, expansive area (40/km minimum)	
2. Climate		
Precipitation	25–40 cm	20–30 cm
Snow depth	25–40 cm	25–40 cm
Temperature	Not a major problem, as herds inhabit hot, semiarid areas to cold alpine steppes.	
3. Soils	Not a determining factor except in relation to soil/site vegetation production.	
4. Water (drinking)		
Quantity	1.0–5.5 L/day	1.0–5. 5 L/day
Distribution	1.5–6.5 km	1.6–6.5 km
Biotic		
1. Vegetation		
Forage consumption	1.0–3.5 kg air-dry	1.0–3.5 kg air-dry
Ground cover	60–80% vegetation	30–50% vegetation
Plant spp. composition	50–80% grasses 10–20% forbs <5% shrubs	5–15% grasses 5–10% forbs 10–35% shrubs
Plant spp. diversity	10–20 grasses 20–60 forbs 5–10 shrubs	5–10 grasses 10–70 forbs 5–10 shrubs
Height	25–45 cm >65 cm unfavorable for most habitats	25–45 cm
Succulence	The more availability year-round the better for all forage classes.	
Communities	Greater variety and diversity important (meadows, playas, wildfire burns).	
2. Animal		
Wild ungulates	Few competition compatibility problems.	
Predators	Pronghorn may be reduced or limited in areas of low densities or isolated herds.	
Mankind	Suitable habitat being usurped. Increased construction of barriers (mainly fences). Predator control, water developments, and alfalfa plantings beneficial. Livestock commensal in grasslands; competitive for forage in shrub-steppe and desert biomes.	

Source: From Ref. 35.

[a]These requirements should be available in the right combination. Too little or too much of any component may become a limiting factor to pronghorn production and/or survival. Populations exist in areas with factor parameters beyond these requirements; however, high densities are generally associated with these criteria.

Rangelands with drinking waters within a radius of 3.2 kilometers sustained higher pronghorn numbers than landscapes with a paucity or complete lack of waters (35). Habitat assessment models in Arizona indicated that fences, divided highways, and fenced railroads contributed to habitat fragmentation, thereby decreasing pronghorn habitat (55).

HABITAT MANAGEMENT—THE FOUR CARDINAL STEPS

Inventory and Monitoring

The foundation for habitat management is a base inventory of the quality and quantity of abiotic and biotic components. Periodic monitoring should determine if habitat conditions are stable, increasing, or decreasing. How often monitoring should be accomplished varies with the degree of change in the habitat. However, 5 years appears adequate for relatively stable landscapes. Rangelands experiencing rapid environmental changes should be monitored more frequently. Biotic and abiotic factors (Table 27–2) should be included in inventory and monitoring programs.

Evaluate Habitat Quality

Determining habitat quality is difficult, but can be accomplished using 10 habitat assessment models developed for pronghorn (50a). Examples of such models for shrub steppes in the Great Basin are provided by the U.S. Bureau of Land Management and Kindschy and others (56, 51). For the shrub steppe of the Rocky Mountains and grasslands, a habitat suitability index model was developed by Allen and others (57).

Managers may find habitat assessment models of assistance in making resource decisions. This is especially true when developing plans to translocate herds, evaluating the effects of vegetation manipulation projects, or completing environmental impact studies to determine the relationship between livestock and pronghorn.

Maintain Quality Habitats

When ecosystems have the right combination of habitat factors, they have the potential to produce optimum numbers. Therefore, recognizing habitats in good ecological condition and maintaining them, by objective, is paramount. This is especially true when the lands are managed for multiple resources, and land management decisions favoring other resource objectives may damage pronghorn habitat conditions.

Enhance Low-Quality Landscapes

Two situations call for improving pronghorn habitats. One such case is when a site has a limiting component (e.g., vegetation, water) with the potential for improvement. A second case is when a component is limited in number or distribution but can be enhanced through improvement practices. Yoakum and others, Valentine, Kie and others, and Payne and Bryant review habitat improvement practices (58, 59, 60, 61).

Vegetation Enhancement

Disturbance of plant communities generally serves to accelerate or retard plant succession. Rangelands with vegetation in low ecological status or late plant succession can be enhanced by fire, mechanical, or chemical treatment and artificial seedings.

Wild fires ignited by lightning are a natural alteration that favors herbaceous plants. Prescribed fires are now being used to simulate wildfires on pronghorn habitats to improve forage conditions (7). Prescribed fires are one of the most economical and ecologically feasible disturbances to maintain low growing, mixed plant communities favorable for pronghorn forage and fawn cover.

Mechanical and chemical treatments also can change plant communities. Apparently, chaining retains more herbaceous plants than does plowing. When sufficient undercover of native herbaceous plants is present, follow-up artificial seedings may not be necessary.

When treated sites (e.g., prescribed fires, mechanical, and chemical projects) lack adequate herbaceous plants, artificial seedings can be beneficial by providing a mixture of grasses, forbs, and shrubs. A minimum of six species each of grasses, forbs, and shrubs should be seeded to simulate natural plant communities and provide a variety of preferred forage species.

Vegetative manipulation projects that change native vegetation to large monoculture grasslands of coarse, alien perennial grasses usually produce limited habitat for pronghorn. Treatments of large areas require pronghorn to travel long distances to obtain preferred shrubs. Grass monocultures frequently provide low densities and diversity of forbs and shrubs vital to pronghorn during various seasons.

Habitat Management—The Four Cardinal Steps

Although fertilizing western uplands can increase plant production, treatments are most feasible in areas receiving more than 38 centimeters of annual precipitation. Treatments may be most effective on spring and summer rangelands needing increases in nutrients for fawn production and survival.

Water Improvements

Pronghorn readily drink from various water developments including reservoirs, springs and seeps, horizontal wells, or windmills with troughs. Beale and Smith suggested that water developments supported pronghorn where natural water sources were limited, particularly during dry seasons or drought years (62). Such developments should be placed every 2.4 to 4.0 kilometers. Hundreds of small reservoirs have been constructed on western rangelands that are natural in appearance and serve pronghorn well, especially during late summer and hot weather when vegetation becomes desiccated and physiologic requirements for water increase.

RESOURCE ISSUES AND CHALLENGES

Fences

Fences can be major obstacles limiting or restricting pronghorn movements to food and water or to escape from deep snow. Most fences on western rangelands are built to control livestock. Wire fences with the lower wire smooth and 40 centimeters above the ground, with minimum use of stays, and a top wire 91 centimeters can best be negotiated by pronghorn (Figure 27–7).

Let-down and adjustable fence designs have mitigating benefits, especially in deep snow. However, experience indicates that weather, manpower limitations, and other factors at times intercede and limit the effectiveness of these fence designs. Electric fences are gaining popularity and can help limit pronghorn movements, especially to agriculture crops. Wolf-type, anticoyote, and woven wire fences are effective barriers for all age classes of pronghorn, and are inappropriate on rangelands occupied by pronghorn. Although structures such as antelope passes have been designed to allow pronghorn access through fences, these devices have limited application and, at times, cause injury and mortality.

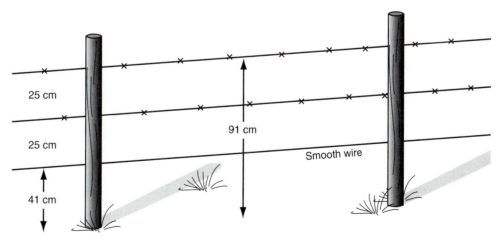

25 cm

25 cm

41 cm

91 cm

Smooth wire

Figure 27–7 Suggested specifications for barbed-wire fences on rangelands occupied by pronghorn and cattle. (Adapted from Ref. 54.)

Livestock Relations

Rangeland vegetation can be altered rapidly by livestock and these changes can affect the quality and quantity of preferred forage needed to sustain thrifty pronghorn herds (30, 63, 35). McNay and O'Gara reported that livestock displaced parturient females in Nevada (64).

At times, livestock and pronghorn have commensal relationships. Rangelands with an abundance of grasses can be heavily grazed by livestock, resulting in increased forbs and shrubs preferred by pronghorn. Then too, pronghorn consume many plants known to be noxious or poisonous to livestock. Predator control programs for livestock sometimes provide benefits to pronghorn.

Competition with cattle for forage does not appear to be a serious problem on healthy ecosystems because of the consumption of different forage classes by the two species. However, Ellis found serious competition for grasses and forbs on heavily grazed Great Basin rangelands during spring and early summer, resulting in low fawn survival (30). For domestic sheep, competition is more prevalent due primarily to both animals consuming large quantities of forbs and shrubs.

Industrial Development

The greatest potential for impacts from oil and gas development and production on pronghorn are the fragmentation and loss of habitat. New oil and gas wells often intrude into previously undisturbed roadless areas, causing increased human activity as well as direct loss of habitat. Winter rangelands, seasonal movement corridors, and fawning areas require special management attention to reduce stress from extractive activities.

The development of energy and mineral resources and post-development land use practices will continue to alter pronghorn habitats. Impacts of these developments will be compounding and cumulative. In regions with potential for energy extraction, actions to protect and preserve pronghorn habitat should be planned, coordinated, and applied to the entire area of expected impact. Suggestions to accomplish these recommendations are provided in Lee and others (7).

Urban Development

Pronghorn are experiencing habitat fragmentation and loss with accelerating human occupation of western North America, especially since World War II. Ockenfels and others reported that pronghorn in Arizona often avoided urban development, though they tolerated certain high human activity levels associated with development (55). Ockenfels and others recommended working with local planning commissions to create travel corridors through planned urban developments. Fences and human-related disturbances increase as urban development proceeds. These may cause pronghorn to vacate ancestral habitats.

SUMMARY

Pronghorn are endemic to North America, inhabiting arid, rolling rangelands from central Canada to mid-Mexico. Herds prior to 1800 numbered 30 to 40 million; however, during Euroamerican settlement they were severely exploited and around 13,000 remained by 1915. Now, an estimated 1 million graze western grasslands, shrub steppes, and deserts.

Pronghorn obtain their name from the prongs on the annually deciduous horns of males. Females usually produce two fawns annually. Adults generally live up to 9 years. Mortality is mainly due to malnutrition, climate, and predation. They consume 3.6 kilograms of forage per day, mainly forbs and shrubs with few grasses. Herds have higher densities where a mixture of preferred forage classes is abundant and drinking waters are readily available. Although water requirements are obtained mainly from vegetation, they can drink up to 3.8 liters per day.

Pronghorn are gregarious and herd size varies throughout the year. Nearly equal numbers of herds have harem-type or territorial breeding systems. Animals form large wintering herds, usually of both sexes and all age classes.

For centuries, pronghorn coinhabited the vast western rangelands with mankind. During the past 75 years, they have increased more than 3,000%. This indicates there are ample natural resources to sustain a million pronghorn and more than 200 million humans. In the 21st century, it will be a challenge for humans to continue their modern lifestyle and allow pronghorn to prosper. The keys to pronghorn and human compatibility are understanding pronghorn habitat requirements and human tolerance in allowing survival of pronghorn herds in an accelerating human civilization.

LITERATURE CITED

1. NELSON, E. W. 1925. Status of the pronghorned antelope, 1922–24. Bulletin 1346. U.S. Department of Agriculture, Washington, DC.

2. YOAKUM, J. D., B. W. O'GARA, and V. W. HOWARD. 1995. Pronghorn on western rangelands. Pages 211–226 in P. Krausman, ed., *Rangeland Wildlife.* Society for Range Management, Denver, CO.

3. THWAITES, R. G., ed. 1904. *Original journals of the Lewis and Clark expedition 1804–1806,* eight volumes. Dodd, Mead and Company, New York.

4. ORD, G. 1818. Sur plusieurs animaux de l'Amerique septentrionale, et entre autres sur le *Rupicapra americana,* l'*Antilocapra americana,* le *Cervus major* on wapiti, etc. *Journal de Physigui de Chimie, d'histroire naturellie el des arts* 87:146–155.

5. ALBERTA FISH AND WILDLIFE DIVISION. 1990. Management plan for pronghorn antelope in Alberta. Wildlife Management Planning Series, No. 3. Edmonton, Alberta, Canada.

6. SALWASSER, H. 1980. Pronghorn antelope population and habitat management in the northeastern Great Basin environments. Interstate Antelope Conference, Alturas, CA.

7. LEE, R. M., J. D. YOAKUM, B. W. O'GARA, T. M. POJAR, and R. OCKENFELS, eds. 1998. Pronghorn management guides. 18th Pronghorn Antelope Workshop, Prescott, AZ.

8. O'GARA, B. W. 1978. *Antilocapra americana.* Mammalian Species 90. American Society of Mammalogy.

9. DOW, S. A., and A. L. WRIGHT. 1962. Changes in mandibular dentition associated with age in pronghorn antelope. *Journal of Wildlife Management* 26:1–18.

10. SETON, E. T. 1927. *Lives of Game Animals,* Vol. 3: *Hooved Animals.* Doubleday, Doran and Company, Garden City, NY.

11. WILSON, D. E., and D. M. REEDER. 1993. *Mammal Species of the World,* 2nd ed. Smithsonian Institution, Washington, DC.

12. LEE, T. E. JR., J. W. BICKHAM, and M. D. SCOTT. 1994. Mitochondrial DNA and allozyme analysis of North American pronghorn populations. *Journal of Wildlife Management* 58:307–318.

13. LEHMAN, V. W., and J. B. DAVIS. 1942. Experimental wildlife management in south Texas chaparral. Quarterly Progress Report, Federal Aid Project 1-R, Unit 5, Section 1. Texas Game, Fish, and Oyster Commission, Austin.

14. POJAR, T. M., and L. L. MILLER. 1984. Recurrent estrus and cycle length in pronghorn. *Journal of Wildlife Management* 48:973–979.

15. TICER, C. L. D., R. A. OCKENFELS, and J. C. DE VOS, JR. 1996. Pronghorn fawning dates in Arizona. *Pronghorn Antelope Workshop Proceedings* 17: in press.

16. BYERS, J. A., and J. T. HOGG. 1995. Environmental effects on prenatal growth rate in pronghorn and bighorn: Further evidence of energy constraint on sex-biased maternal expenditure. *Behavioral Ecology* 6:451–457.

17. BYERS, J. A., and J. D. MOODIE. 1990. Sex-specific maternal investment in pronghorn, and the question of a limit on differential provisioning in ungulates. *Behavioral Ecology and Sociobiology* 26:157–164.

18. HEPWORTH, W. G., and F. BLUNT. 1966. Research findings on Wyoming antelope. *Wyoming Wildlife* 30(6):24–29.

19. MITCHELL, G. J. 1980. The pronghorn antelope in Alberta. University of Regina, Saskatchewan, Canada.

20. O'GARA, B. W. Reproduction. Chapter *in* B. W. O'Gara and J. D. Yoakum, eds., *The Pronghorn: Ecology and Management.* Wildlife Management Institute, Washington, DC.

21. HEPWORTH, W. G. 1965. Investigations of pronghorn antelope in Wyoming. *Antelope States Workshop Proceedings* 1:1–12.

22. KERWIN, M. L., and G. J. MITCHELL. 1971. The validity of the wear-age technique for Alberta pronghorns. *Journal of Wildlife Management* 35:473–747.

23. CARLTON, C., and T. McKEAN. 1977. The carotid and orbital retina of the pronghorn, deer and elk. *Anatomical Record* 189(1):91–108.

24. McKEAN, T., and B. WALKER. 1974. Comparison of selected cardiopulmonary parameters between pronghorn and the goat. *Respiratory Physiology* 21:365–370.

25. HILDEBRAND, M., and J. P. HURLEY. 1985. Energy of the oscillating legs of a fast-moving cheetah, pronghorn, jackrabbit, and elephant. *Journal of Morphology* 184:23–31.

26. BROMLEY, P. T. 1977. Aspects of the behavioral ecology and sociology of the pronghorn (*Antilocapra americana*). Dissertation. University of Calgary, Alberta, Canada.

27. WHISLER, S. 1984. Seasonal adaptations of pronghorn antelope to water deprivation. Dissertation. University of Wyoming, Laramie.

28. ROSEBERRY, J. L., and W. D. KLIMSTRA. 1975. Some morphological characteristics of the Crab Orchard deer herd. *Journal of Wildlife Management* 39:48–59.

29. YOAKUM, J. D. 1990. Food habits of the pronghorn. *Pronghorn Antelope Workshop Proceedings* 14:102–111.

30. ELLIS, J. E. 1970. A computer analysis of fawn survival in pronghorn antelope. Dissertation. University of California, Davis.

31. SMITH, A. D., and D. M. BEALE. 1980. Pronghorn antelope in Utah: Some research and observations. Publication 80-13. Utah Division of Wildlife Resources, Salt Lake City.

32. SUNDSTROM, C., W. G. HEPWORTH, and K. L. DIEM. 1973. Abundance, distribution, and food habits of the pronghorn. Bulletin 12. Wyoming Game and Fish Commission, Cheyenne.

33. BARRETT, M. W. 1974. Importance, utilization, and quality of *Artemesia cana* on pronghorn winter ranges in Alberta. *Antelope States Workshop Proceedings* 6:26–57.

34. MARTINKA, C. 1967. Mortality of northern Montana pronghorn antelope in a severe winter. *Journal of Wildlife Management* 31:159–164.

35. YOAKUM, J. D. 1994. Water requirements of pronghorn. *Pronghorn Antelope Workshop Proceedings* 17:143–157.

36. PYRAH, D. B. 1987. American pronghorn antelope in the Yellow Water Triangle, Montana. Montana Department of Fish, Wildlife and Parks, Helena.

37. MAHER, C. R. 1991. Activity budgets and mating system of male pronghorn at Sheldon National Wildlife Refuge, Nevada. *Journal of Mammalogy* 72:739–744.

38. FICHTER, E. 1972. On the nature of pronghorn groups. *Antelope States Workshop Proceedings* 5:222–232.

39. RIDDLE, P. 1990. Wyoming antelope status report—1990. *Pronghorn Antelope Workshop Proceedings* 14:24.

40. AUTENRIETH, R. E., and E. FICHTER. 1975. On the behavior and socialization of pronghorn fawns. *Wildlife Monograph* 42:111.

41. MOY, R. F. 1970. Histology of the subauricular and rump glands of the pronghorn (*Antilocapra americana* Ord). *American Journal of Anatomy* 129:665–87.

42. MULLER-SCHWARZE, D., and C. MULLER-SCHWARZE. 1972. Social scents in hand reared pronghorn (*Antilocapra americana*). *Zoologica Africana* 7:251–271.

43. McKENZIE, J. V. 1986. North Dakota antelope status report. *Pronghorn Antelope Workshop Proceedings* 12:25–28.

44. O'GARA, B. W. In preparation. Mortality factors. Chapter *in* B. W. O'Gara and J. D. Yoakum, eds., *The Pronghorn: Ecology and Management.* Wildlife Management Institute, Washington, DC.

45. O'GARA, B. W., and B. MORRISON. In preparation. Managing the harvest. Chapter *in* B. W. O'Gara and J. D. Yoakum, eds., *The Pronghorn: Ecology and Management.* Wildlife Management Institute, Washington, DC.

46. O'GARA, B. W. In preparation. Diseases and parasites. Chapter *in* B. W. O'Gara and J. D. Yoakum, eds., *The Pronghorn: Ecology and Management.* Wildlife Management Institute, Washington, DC.

47. THORNE, E. T., E. S. WILLIAMS, T. R. SPRAKER, W. HELMS, and T. SEGERSTROM. 1988. Bluetongue in free-ranging pronghorn antelope (*Antilocapra americana*) in Wyoming: 1976 and 1984. *Journal of Wildlife Disease* 24:113–119.

48. BEVER, W. 1957. The incidence and degree of the parasite load among antelope and the development of field techniques to measure such parasitism. Pittman-Robertson Project 12-R-14. Job Outlook A-52. South Dakota Fish and Game Department, Pierre.

49. YOAKUM, J. D., B. W. O'GARA, R. MCCABE. 1998. 1996–1997—Survey of pronghorn for North America. *Proceedings of Pronghorn Antelope Workshop* 18: in press.

50. O'GARA, B. W., and J. D. YOAKUM. In preparation. *The Pronghorn: Ecology and Management.* Wildlife Management Institute, Washington, DC.

50a. YOAKUM, J. D., B. W. O'GARA, and V. W. HOWARD. 1995. Pronghorn on western rangelands. Pages 211–226 *in* P. Krausman, ed., *Rangeland Wildlife.* Society for Range Management, Denver, CO.

51. KINDSCHY, R. R., C. SUNDSTROM, and J. YOAKUM. 1982. Wildlife habitats in managed rangelands—The Great Basin of southeastern Oregon: Pronghorns. General Technical Report PNW-145. U.S. Department of Agriculture, Pacific Northwest Forest and Range Experiment Station, Portland, OR.

52. RIPLEY, T. H. 1980. Planning wildlife management investigations and projects. Pages 1–6 *in* S. D. Schemnitz, ed., *Wildlife Management Techniques Manual.* The Wildlife Society, Washington, DC.

52a. INTERNATIONAL UNION FOR CONSERVATION OF NATURE. 1996. Red list of threatened animals. Gland, Switzerland, and Cambridge, UK.

53. AMSTRUP, S. C., J. MEEKER, B. W. O'GARA, and J. MCLUCAS. 1980. Capture methods for free-ranging pronghorns. *Pronghorn Antelope Workshop Proceedings* 9:98–131.

54. SMITH, R. H., D. J. NEFF, and N. G. WOOLSEY. 1986. Pronghorn response to coyote control—A benefit cost analysis. *Wildlife Society Bulletin* 14:226–231.

55. OCKENFELS, R. A., A. ALEXANDER, C. L. DOROTHY TICER, and W. CARREL. 1994. Home ranges, movement patterns, and habitat selection of pronghorn in central Arizona. Technical Report 13, Arizona Game and Fish Department, Phoenix.

56. U.S. BUREAU OF LAND MANAGEMENT. 1980. Manual supplement 6630—Big Game studies: Guidelines for the evaluation of pronghorn antelope habitats. Reno, NV.

57. ALLEN, A. W., J. C. COOK, and M. J. ARMBRUSTER. 1984. Habitat suitability index models: Pronghorn. FWS/OBS-82/10.65. United States Fish and Wildlife, Fort Collins, CO.

58. YOAKUM, J. D., W. P. DASMANN, R. SANDERSON, C. NIXON, and H. CRAWFORD. 1980. Habitat improvement techniques. Pages 329–403 *in* S. Schemnitz, ed., *Wildlife Management Techniques Manual.* The Wildlife Society, Washington, DC.

59. VALENTINE, J. F. 1989. *Range Development and Improvements,* 3rd ed. Academic Press, San Diego, CA.

60. KIE, J. G., V. C. BLEICH, A. L. MEDINA, J. D. YOAKUM, and J. W. THOMAS. 1994. Managing rangelands for wildlife. Pages 663–688 *in* T. A. Bookhout, ed., *Research and Management Techniques for Wildlife and Habitats.* The Wildlife Society, Bethesda, MD.

61. PAYNE, N. F., and F. C. BRYANT. 1994. *Techniques for Wildlife Management of Uplands.* McGraw-Hill, New York.

62. BEALE, D. M., and A. D. SMITH. 1978. Birth rate and fawn mortality among pronghorn antelope in western Utah. *Pronghorn Antelope Workshop Proceedings* 8:445–448.

63. HOWARD, V. W., J. L. HOLECHEK, R. D. PIEPER, L. GREEN-HAMMOND, M CARDENAS, and S. L. BEASOM. 1990. Habitat requirements for pronghorn on rangelands impacted by livestock and net wire in east central New Mexico. Agriculture Experiment Station Bulletin 750. New Mexico State University, Las Cruces.

64. MCNAY, M. E., and B. W. O'GARA. 1982. Cattle-pronghorn interactions during the fawning season in northwestern Nevada. Pages 593–606 *in* J. M. Peek and P. D. Dalke, eds., *Wildlife–Livestock Relationships Symposium.* Proceedings 10. University of Idaho, Forest, Wildlife and Range Experiment Station, Moscow.

65. O'GARA, B. W. 1996. Estimates of pronghorn numbers. *Pronghorn Antelope Workshop Proceedings* 17: in press.

CHAPTER
28

Moose

Albert W. Franzmann

INTRODUCTION

Moose are the largest living member of the deer family and inhabit the northern spruce and boreal forests of North America. Native Americans respected this magnificent creature and were dependent on it for their survival in many areas. Early European explorers were fascinated by moose and many accounts appear in writings regarding their importance as a source of food, clothing, tools, and recreation. Moose continue to provide the basic staples for people who live off the land and for others that depend on moose meat as a supplementary food source. Many still enjoy the traditional hunt and their activity provides an economic base for many northern forest communities. Nonhunters appreciate the opportunity to view and photograph this denizen of the north.

Historical Perspective

Moose crossed the Bering Land Bridge to North America from Siberia during the Wisconsin glaciation into unglaciated refugia in Alaska (1). Peterson theorized that moose from Alaska then invaded three distinct refugia south of the ice fields providing isolation for subspeciation (6). However, there are other theories as to the timing and routes of invasion and dispersal (2, 3, 4, 5). Their past distribution was determined by the dynamics of glacial epochs and associated boreal forests. Moose once ranged over the then-forested Great Plains and eastern United States. They spread northward with the retreat of the glaciers and extension of the boreal forests (6).

Reeves and McCabe provided an extensive review of how moose and native Americans interacted (7). They opened their treatise with a quote from Seton: "What the Buffalo was to the Plains, the White-tailed Deer to the Southern woods, and the Caribou to the Barrens, the Moose is to this great Northern belt of swamp and timberland. . . . It is the creature that enables the natives to live" (8, p. 189). The following five paragraphs of historical perspective come from Reeves and McCabe unless otherwise cited (7).

Humans arrived in North America via the Bering Land Bridge or by boat about the same time as the moose. These meat eaters preferred moose because of the high return per unit of time and energy invested. Natives used a variety of weapons (e.g, clubs, spears, knives, alatl), devices (e.g., snares, traps, canoes, calls, flares), and strategies (e.g., ambush, stalk, pursuit, drive, scavenge, fire, dogs) to procure moose during all seasons. Peak hunting periods included the rut, during deep and crusting snow, and during summer on open rivers and lakes.

For the boreal meat eater, use of moose defined subsistence. They used everything except the call. Even dung was used for sprains. Moose meat and organs that were not consumed fresh were frozen, dried, or smoked. The hide was used for clothing, moccasins, shields, blankets, packs, sled runners, lashing, skin boats, snowshoe webbing, lodging, and as food during severe environmental conditions. Bones and antlers were used for tools, clubs, hide fleshers, fish hooks, and gaming dice. Tendons were used to sew clothing, wrap arrowheads and fletching, and make glue, snares, and fishnets. Moose hair was used for decorative embroidery, burdenstraps, trumplines, and roaches. Various body parts were culturally important and used widely in medicine.

The physical importance of moose to sustain natives imparted a tremendous cultural value. Many legends, myths, taboos, and beliefs were associated with moose. Calendars were related to activities of moose. The moose was an important food and gift item during feasts or potlatch and a trade item between family groups and tribes. Some tribes tied the beginning of manhood to a boy's first moose kill. The lack of a written language leaves our understanding of the cultural importance of moose to oral legends and myths passed through early European immigrants, leaving a potential for misinterpretation.

The earliest Europeans often were provisioned by natives with moose meat and fat. Moose hides often were used as a trade item. Moose provided a valuable source of nourishment and became a primary subsistence item for European immigrants.

Moose populations were depleted as European settlements spread across North America because of their food value and the availability of firearms. Peterson noted: "This pattern of depletion extended across most of the settled areas with the advance of white populations in eastern Canada and northern United States so that by the beginning of the nineteenth century, the future of moose seemed uncertain in many areas" (6, p. 21). Over time, this led first to protection and then to today's active management. Some of the settlers' farming and forestry practices created seral vegetation that benefited moose. Therefore, depending on the time and place, the settlers' impact could be described on a continuum from depletion to expansion.

Value

The cultural importance of moose to natives was not harbored by early European immigrants. Their view was based on utilitarian value. Moose was the most important food item for these early immigrants. They were the "fuel" for the Hudson Bay Company and the Northwest Fur Company (7). However, when humans depend on an animal for their basic needs, it becomes culturally important. Today, in remote areas of North America dependence on moose is the focus of the fall season and a cultural event for native and non-native people.

From the time of the first European immigrants to the present, the utilitarian value of moose decreased as humans evolved culturally to the point where they could "afford" to experience moose from an aesthetic viewpoint. Very few people today are dependent on moose for subsistence and the aesthetic value associated with the hunt has taken on a different meaning. The experience in the field becomes as important or more important than the harvest. Today, many who do not hunt place great value on programs such as "Watchable Wildlife." Many people may never hunt or see a moose, but they appreciate the fact that they are there and are eager to have information about them (9).

The difficulty of quantifying the economic value of moose has much to do with their aesthetic value. How much is one's first harvest or first view of a moose worth? One can calculate the value of the meat from harvest in relation to beef prices, but most who have consumed moose believe the moose are more valuable and healthy than beef. Crichton noted that lean beef has eight times more saturated fat than moose meat (9). The economy generated by hunting or viewing moose (travel, housing, licenses, equipment, guides, etc.) also can be calculated. Regelin and Franzmann estimated that in Alaska activities associated with moose hunting annually generate about $31,000,000 into the economy (10).

Systematic Characteristics

Moose are in the deer family Cervidae, order Artiodactyla (even-toed), infraorder Pecora, suborder Ruminantia (11). The systematics of deer are based on anatomic features including retention of certain bones of the lower foreleg and characteristics of the skull. Moose have the following skull characteristics: upper jaw without incisors; upper canine teeth virtually always absent; lower canines modified; the first premolar lost; diastema between the incisors and second premolar; second and third premolar tightly spaced; maxillary bone articulates with the nasals; pedicles short and the first antlers appear 4 to 6 months after birth (3). The major postcranial skeletal characteristics include the third and fourth metacarpal bones fused to form the "cannon bone"; the third and fourth digits formed hooves; the second and fifth digits formed dewclaws; and the second and fifth metacarpal bones regressed.

General Characteristics and Description

Moose are the largest member of the deer family, with a massive body on long slender legs, a shoulder hump, a long nose with a large, flexible overhanging upper lip, a "bell" or pendulous dewlap at the throat, and a very short tail (Figure 28–1).

Newborn moose pelage is light red to reddish brown in color with shades of gray to black on the lower abdomen, chest, legs, muzzle, hooves, eye rings, and ears. This juvenile coat is replaced in 2 to 3 months by a darker coat. The coat is shed in spring and is replaced by short, fine, nearly black hair. As summer progresses, these scale-like guard hairs grow rapidly, up to 25 centimeters long on the hump, and attain a lustrous red-brown appearance with shading toward black, particularly in the lower extremities (12). Some moose in prime coat, particularly males, are nearly black. White hair around the vulva of females is used to identify females during aerial sexing (13). Under the guard hairs there is an undercoat of grayish, fine wool-like hair that provides insulation for survival in cold regions. Albinism and a white colorphase occur rarely (14).

Figure 28–1 Bull moose demonstrating unique characteristics of North America's largest member of the deer family. Note the "bell" or pendulous dewlap at the throat, the massive body with the shoulder hump, the long nose with a large, flexible overhanging upper lip, the massive antlers, and the short tail. (Photograph by J. Faro, Alaska Department of Fish and Game, Sitka, Alaska.)

Adult moose have 32 teeth (0/3, 0/1, 3/3, 3/3) (6). Tooth eruption, tooth wear, and incisor cross sectioning are used to age moose (6, 15, 16). Incisor cementum layer age determinations have become the accepted method for moose with mean age only overestimating known age by 0.5 years with errors ranging from between minus 1 and plus 3 years (17).

Moose lack metatarsal and interdigital external glands, but have lachrymal glands below each eye and small tarsal glands on the medial side of each hind leg. Four mammae in the perineal area produce milk composed of solids (24.5%), lipids (5.8%), ash (2%), protein (10.3%), lactose (6.8%), and gross energy (1.4 kilocalories/gram), with a specific gravity of 1.048, and pH of 6.8 (18).

Males develop large palmicorn antlers during summer that are nourished by an extensive vascular system in its skin (i.e, velvet cover) (Figure 28–2). The velvet is shed as the rut begins in August or September, assisted by rubbing and thrashing on shrubs and trees that stain antlers from white to dark tan. Antlers are shed by large bulls generally by November and as late as April in younger bulls. Antlers from mature bulls may weigh up to 35 kilograms and have a spread up to 205 centimeters (19, 20). Growth rates and size of antlers are a function of age, genetic potential, and nutrition.

Moose may increase their late winter weights by 21% to 55% to their fall maximum (21). Maximum live weights of moose at the Moose Research Center (MRC), Alaska, were 771 kilograms for a male with antlers just prior to the rut and 573 kilograms for a mature female in November (K. Hundertmark, MRC, personal communication). Mean skeletal measurements of 23 adult females from a highly productive, expanding moose population in Alaska were total body length, 302 centimeters; chest girth, 201 centimeters; hind foot length, 82 centimeters; and shoulder height, 186 centimeters. Total body length is useful in predicting total body mass (21). Measurements reported for adults from other populations and subspecies of moose are generally smaller, but within 3% to 8% of these values (22, 23, 24).

Figure 28–2 Bull moose shedding the antler "velvet" at the onset of the rut or breeding season. The extensive vascular system in the velvet that nourishes antler growth ceases, predisposing the shed. Bulls thrash shrubs and trees during the process that stains the antlers. (Photograph by C. C. Schwartz, Alaska Department of Fish and Game, Soldotna, Alaska.)

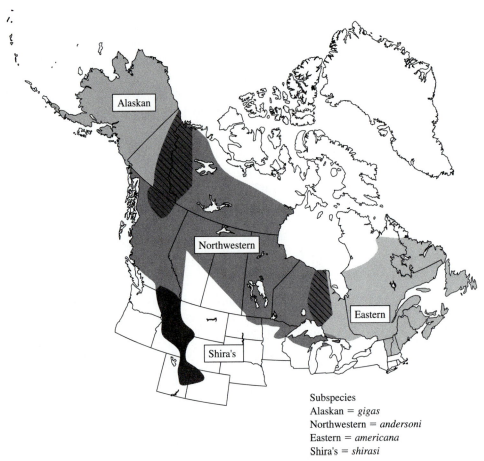

Figure 28–3 Distribution of moose in North America by subspecies (From Ref. 26.)

Subspecies
Alaskan = *gigas*
Northwestern = *andersoni*
Eastern = *americana*
Shira's = *shirasi*

Distribution

Moose have a unique association with the boreal forests in the Northern Hemisphere that provide the proper climate and vegetation. Present distribution is limited in the north by suitable food on the tundra (not cold temperature) and in the south by high ambient temperature (25). Moose are relative newcomers to the Rocky Mountain chain in the western United States, areas north of Lake Superior, and extended ranges in Labrador and Alaska. They continue to occupy new territory (26).

The subspecies of North American moose are Shira's or Wyoming, eastern or taiga, northwestern, and Alaska/Yukon (Figure 28–3). Subspeciation of the genus *Alces* in North America is based primarily on skull differences attributed to post-glacial geographical isolation (27). Recent range expansion has reduced their isolation and differences are not definitive. Subspeciation of North American moose should be reevaluated.

LIFE HISTORY

Diet Selection

Moose are primarily browsers with a variable diet that is characterized by use of early successional woody browse regrowth following disturbances such as fire, floods, and logging. They select twigs from shrubs in winter and leaves from deciduous species in summer. Renecker and Schwartz reported that moose sampled 164 different plant species in North America, but only aspen, birch, and willow were eaten in large quantities throughout their range (28). Moose are a pioneering-type animal and adapt to available forage, but their reproductive performance peaks when preferred species are abundant and available (12). Other species may take on an important role in their diet depending on season, snow depth, locality, and availability. Examples include balsam fir in their eastern range, maples and beaked hazel in the midcontinent, and aquatics wherever lakes occur (29).

Reproductive Strategy

The seasonally polyestrous breeding season or rut of moose is from early September to late October, peaking in late September and early October with calving occurring from mid-May to mid-June. Schwartz and Hundertmark reported the receptive period for females was 15 to 26 hours, the cycle interval was 24 to 25 days, and mean gestation length was 231 days (30). Cows that bred during their first estrus gave birth from May 18 to June 7 with 67% of births within 1 week. Yearling females can conceive and calve when bred by yearling or older males; however, young males are generally excluded from breeding by mature males (31, 32, 33). Single calves generally are produced, but twins are not uncommon. Twinning in populations varies from 5% in a static Jackson Hole, Wyoming, population during the 1960s to 72% in an expanding population on the Kenai Peninsula during the 1980s (34, 35). The twinning rate is a function of nutrition and condition of the population and the rates may be used to assess the status of a population (35). Triplets are rare (36).

The breeding strategy is characterized by harem breeding by tundra moose and by pair breeding by the closed forest moose (32, 37). Bubenik described the former as an assemblage and a strategy where the male dominates the female; conversely, in the taiga the female dominates the male (33). The tundra harem strategy is characterized by congregations of moose in large rutting groups in open habitats. A dominant bull forms the core, surrounded by females and then subdominant and young males. The taiga breeding is more monogamous, or paired.

Functional sex ratios for harem breeding moose can be skewed toward females, but may vary depending on breeding success, distribution, and density. Sex ratios commonly range from 20 to 30 bulls per 100 cows, which ensures synchronous breeding of all females during first estrus (38). The pair breeding strategy of taiga moose requires a higher proportion of bulls in the population because a single bull can fertilize only a few females during a short, synchronous rut. The ideal ratio varies based on density and distribution of moose. The ratio should not fall below 50 bulls per 100 cows if the density is above 0.3 moose per square kilometer and should be close to parity when density is below 0.3 moose per square kilometer (39, 40).

Longevity

Female moose can reproduce yearly until age 18, but their maximum reproductive potential is between ages 4 and 12 (12). Longevity data for male moose are limited, but one bull at the MRC lived to age 11.5 and another to age 14 (K. Hundertmark, MRC, personal communication). The peak reproductive age of males is approximately 4 to 8 years.

Physiology

Growth of moose is continuous and consists of three phases: prenatal, suckling, and maturity. The first two phases are wholly or partially dependent on the cow for sustenance (38). Estimates of the incidence of delayed breeding can be made using hind foot length of the fetus (30).

Single calves are larger than twins at birth, but there is no mass difference by sex. Birth mass averaged 16.2 kilograms for 26 single calves, and 13.5 kilograms for 50 twins (30). Calves grow rapidly, gaining approximately 1 kilogram per day to October when they attain a mean mass of 181.4 kilograms (21). From October to May calves generally lose mass, but gain rapidly as forage becomes available.

Alaskan female moose reach their maximum mass at age 4 (500 kilograms) and males at 7 to 9 (700 kilograms). Moose gain and lose weight seasonally based on the abundance and quality of forage and their reproductive status. Pregnant females reach their minimum annual mass postpartum when males are gaining mass. Males lose from 12% to 18% mass during the rut when they cease eating (21, 41). Older moose lose body mass.

Schwartz summarized the physiologic and nutritional adaptations of moose to northern environments (42, p. 139):

> Moose exploit the boreal forest where food resources have high nutritive value during brief summers, and low quality and availability during long winters. To accommodate this fluctuating environment, moose store large quantities of fat during summer and fall which helps to offset their winter energy deficit. Annual rhythms are keyed to this cycle. Intake rates vary seasonally and correspond with nutrient quality and forage availability. Moose are hyperphagic in summer and reduce food intake in winter. Activity budgets vary among environments and seasons with foraging and resting/ruminating occupying most of their time. Metabolism follows a circannual cycle that peaks in mid-summer with a nadir in late winter; peak metabolism corresponds to maximum energy intake and storage. Moose are classified as seasonally adaptable concentrate selectors that choose a diet primarily of browse foliage and twigs. This diet is high in lignin as well as readily digestible nutrients. Energy and protein requirements are similar to other cervidae. Body composition, like metabolism and intake, is dynamic seasonally. Nutritional adaptations stabilize energy balance and allow moose to withstand energy shortages in a fluctuating environment.

Physiologic changes in blood parameters reflect the seasonality of moose nutrition. Blood parameters (e.g., packed cell volume, hemoglobin, total serum protein, albumin, phosphorus, calcium) respond to condition extremes in moose; higher with excellent condition and lower with poor condition (12). Other blood parameters reflect intake levels of protein and energy, stress and excitability, and starvation (43). Mineral levels of calcium, copper, iron, potassium, magnesium, manganese, sodium, and zinc in hair showed peak values in late summer related to increased mineral intake (44). Conversely, low hair values indicated a copper deficiency in a subpopulation of moose (45).

BEHAVIOR

Social Structure

The behavioral strategy of the harem-breeding tundra and pair-breeding taiga moose was discussed previously. The complex rituals of moose during rut have been described and are summarized (46, 47, 32, 12, 33, 48, 30). Males begin their challenge with a swaying gait dipping their antlers from side to side with stiff, long strides generally circling the rival bull. Their eyeballs are rolled back displaying white (Figure 28–4). Thrashing shrubs, displacement feeding, and exaggerated movements may proceed a mock battle.

A reinforcement of drive may follow where the bull returns to the cow displaying a low-stretch posture and flehmen (Figure 28–5) to detect pheromones. Activation of a wallow established earlier may occur characterized by squat urination, digging, pawing, and rolling. Cows may use the wallow also.

For bulls, there is then the decision to yield or fight. The fight is generally of short duration consisting of a quick clash and some intense pushing. This may be repeated until one moves off or a stalemate occurs that results in a repeat of this behavior. Rarely, a fight may become extremely aggressive and cause the death of one or both bulls. The victor drives the cow and mounts her. He may breed her repeatedly through the 15- to 26-hour receptive period of estrus.

Figure 28–4 Bulls during the rut displaying their antlers to one another assessing relative rank. They may do this for some time when rank is even and for a short time when the differential is great. The fight, if it occurs, generally consists of a quick clash and some intense pushing. This may occur only once or be repeated until one moves off. Rarely, a fight may become extremely aggressive, resulting in death of one or both bulls. (Photograph by C. C. Schwartz, Alaska Department of Fish and Game, Soldotna, Alaska.)

Figure 28–5 Bull moose doing lip curl or "flehmen." This behavior exposes the vomeronasal or Jacobson's organ so the bull can better detect the pheromones produced by the estrous cow. (Photograph by C. C. Schwartz, Alaska Department of Fish and Game, Soldotna, Alaska.)

Tundra females generally move into the area, often traditional, selected by the bull; taiga females appear to select the breeding area. Females are very aggressive toward one another during rut. Tundra females are highly mobile and vocalize extensively in their search for males.

Bubenik described the social world of moose as not territorial in the classical sense and preferred the term *individualistic* (48, 49). Houston called moose *quasisolitary* and noted that large groups are uncommon (34). Females with calves almost always occur alone (50).

A firm cow–calf bond requires constant contact for the first week. During the first days of a calf's birth, the cow remains within visual or vocal distance (51, 52). Moose calves display a "lying-out" behavior until they gain coordination (about 4 days), in contrast to the hiding behavior of some cervids. The female is very protective of her calf and challenges to her space will generally result in an aggressive attack. This protective bond increases the chances for the calf's survival the first year of its life. The female provides defense against predation and provides experience in habitat selection (12). However, there is evidence that the bond may not be totally essential because several moose introductions were accomplished in Alaska using only calves (53).

Individualism of moose during winter with deep snows and limited food resources is not as evident as during summer. Large aggregations of moose may come together where a food resource is relatively more abundant (54). "Yarding" type behavior may occur when snow conditions limit movements (6).

Interspecies Relationships

Moose share habitat spatially and temporally with caribou, mule deer, white-tailed deer, bison, elk, musk oxen, Rocky Mountain sheep, Dall's sheep, snowshoe hares, and beaver. Boer described four forms of interactions between moose and other species (55):

1. Direct competition for food or other resources occurs between white-tailed deer and moose during winter, and generally between elk and moose and Snowshoe hares and moose. (56, 57, 58, 59, 60, 61, 62). Moose browsing of regenerating aspen and birch may inhibit their growth and negatively impact beaver (63, 64).

2. Parasite-mediated competition occurs with the presence of a shared parasite that differentially affects competitors such as the meningeal worm of white-tailed deer infecting moose and causing neurologic disease (65).

3. In multiprey systems moose may influence, or be influenced by, the number of other prey species. Moose increases likely resulted from increased availability of caribou as alternative prey in parts of Denali National Park and in northern Alaska (66, 67). Sheep and musk oxen numbers may be interactively linked to moose numbers in multiprey–predator systems (55).

4. Commensalism occurs when one species benefits while the other is unaffected. Beaver ponds provide aquatic habitat for moose, and beaver cutting aspen promotes suckering and sprout growth for moose (63, 55). Moose browsing stabilizes the forest edge to maintain grassy openings for bison (68).

These interactions occur spatially and temporally, but their population-level significance is difficult to assess. One must be cautious when evaluating their effects and realize that native species evolved mechanisms to reduce competition. This is best exemplified by African ungulates on the Serengeti Plain that coexist through a system of resource sharing (69). In Elk Island National Park, Alberta, sharing of similar resources occurs among moose, white-tailed deer, elk, and bison (55).

Movements, Migration, and Home Range

Moose move within seasonal ranges, migrate between seasonal ranges, and disperse to new ranges. Within a seasonal range, moose establish home ranges where normal activities occur. Moose may migrate from one seasonal range to another or may remain year-round residents in an area. Movements, migrations, and home range were poorly documented until the advent of immobilizing drugs and radiotelemetry. Biologists can now document these activities, but our understanding of the "why" is often incomplete (70).

Home range varied from 2 to 60 square kilometers in Maine and up to 92 square kilometers in Alaska (71, 72). These great variances in home range sizes are attributed to differences in season, geography, demography, weather, habitat, sex, and age (70).

Migration is a regular annual movement with return to at least one common area each year (73). Migration distances vary seasonally on a continuum from 2 kilometers to 170 kilometers and are generally mediated by snow conditions or changes in quantity or phenology of forage (74, 75, 73). Migrations have a traditional pattern that may persist for generations (74, 73, 76). A population of moose occupying the same range for a period of time may consist of several "demes" or segments of that population that may seasonally migrate; one to one area, one to another, and some remaining year-round (76). Migration of moose is an adaptation to optimize survival by using the most favorable environments (77).

Moose dispersal generally is limited to short distances from their natal range, but long dispersals up to 250 kilometers were documented (78). Colonization of new habitats by moose are accomplished by dispersal (34, 29, 67, 79).

Communication

Humans are conceptionalists and as logical and abstract thinkers have difficulty respecting the horizons in which animals communicate and thereby understanding their dialogues (48). Animals are perceptionists and respond to species-specific signals as single or complex codes (80, 48). These signals may be olfactory, visual, or vocal and the more potent signals are known as key-releasers (81). Some odors are so potent they are called super-releasers, but may go undetected by humans due to poor olfactory senses (48).

Olfactory cues originate from tarsal glands, urine, feces, saliva, lachrymal glands, and velvet. Pheromones and vomerones are produced and are intense indicators of the stage of estrus in cows. The vomeronal components from the cow help create a scent aura that sexually stimulates both partners (48).

The visual distance capabilities of moose are unknown and are probably less important to distance assessment than stereophonic hearing and possibly smell (48). Red deer and white-tailed deer have rods and cones in the retina and distinguish six colors and it is likely that moose see colors as well (82, 83, 48). Visual discrimination of antler rank by females using antlered dummies of male moose was confirmed by Bubenik (33). Other visual signals include guard hair erection, rolling eyes, ear movement, tongue "smack," head posture, flehmen, and torso posture (48).

The excellent hearing and echolocation abilities of moose are aided by the large reflective surface of the external ear, the distance between ears, and independent movability of each ear (48). Lent described male vocalizations during the rut as a "croak" while cows emit a long, quavering "moan" (32). A year-round seeking call used to attract other moose, described by Lent as a "whine" and by Bubenik as a soft nasal "i-i-e-h-n," becomes a distress call when made hard and loud (32, 48). Under extreme stress moose may emit a loud "roar bellow" (12). A "snort" or "bark" is a low degree vocal threat (48, 32).

POPULATION DYNAMICS

VanBallenberghe and Ballard reviewed moose population dynamics and stated: "Understanding the dynamics of a population requires knowledge of how many animals it contains, how fast it is increasing or decreasing, its rate of production of young, and its rate of loss through mortality" (84, p. 223). McCullough concluded that the slow progress in the field of population dynamics of large wild herbivores relates to the complexity of forces working continuously on a population (85).

Population Trends and Status

Moose population fluctuations are the norm. They are naturally regulated primarily by predation and carrying capacity with food being the ultimate component of carrying capacity (86, 87, 88). Alteration of these forces affects trends. Increased carrying capacity with limited or no predation generally results in moose population irruptions (89, 88, 90). Conversely, habitat decline or loss with high predation may result in severe population declines (90, 91). In between these extremes, there is a continuum of moose–predator–carrying capacity relationships, including equilibrium in rare instances (67). In a simple wolf–moose system, predators can maintain moose at low densities for decades (92).

Population estimates, trends, and status in moose numbers vary by state and province (Table 28–1). Population variations result from reintroduction, fire, expansion of range, change in habitat, snow conditions, and forestry practices (26). However, the total estimated North American moose population during the last three decades varied little: 938,981 in 1960 and 975,241 in 1990.

Productivity and Recruitment

Productivity of a population is the rate at which new individuals of breeding age are produced based on natality rates and survival to maturity. The natality rate of moose is the number of calves born in a population each year and generally expressed as

TABLE 28–1

Moose Population Estimates, Status, and Trends by Jurisdiction in North America, Early 1990s

Jurisdiction	Population	Status/Trend
Alaska	144,000–166,000	Varies regionally
Alberta	100,680	Below carrying capacity (CC)
British Columbia	175,000	Close to CC in south and central; below CC in north
Colorado	700	Recently reintroduced; range expanding
Connecticut	A few transients	Considered undesirable
Idaho	5,100	Range expanding
Labrador	3,000–5,000	Recent influx from Quebec
Maine	21,000–24,000	Whole state occupied
Manitoba	27,100	Presently low densities, but increasing
Massachusetts	12–45	Recent range extension
Michigan	300–400	Reintroduced on mainland in 1980s, expansion under way; 1996–1997 crash on Isle Royale
Minnesota	12,000–15,000	Expanding range, primarily in northwest
Montana	6,000	At desired level
Newfoundland	97,000	Occupy entire island
New Brunswick	20,000	Stable; most of mainland occupied, some islands not
New Hampshire	2,500–5,500	Recent range extension
New York	15–30	Very recent range extension
North Dakota	700	Below CC; expanding
Northwest Territories		Well below CC
Nova Scotia	3,700	Stable
Ontario	103,000	Population and distribution increasing
Oregon	A few transients	Reintroductions have failed
Quebec	69,000	Increasing range in south
Saskatchewan	60,000–70,000	Below CC
Utah	3,500–4,000	Range expanding
Vermont	1,300	Recent range extension
Washington	185	Range expanding
Wisconsin	30–40	Occur sporadically
Wyoming	13,505	Range expanding
Yukon Territory	60,000	Pristine habitat below CC

Source: Adapted from Ref. 26.

calves/cow/year. Natality is often difficult to measure because of early mortality of calves, particularly from predation. Moose under ideal conditions have a potential natality rate exceeding one calf per cow per year due to twinning rates up to 72%, yearling breeding, and conception rates approaching 100% (35, 31, 91). A maximum natality rate and a minimum mortality rate produce a great, but seldom achieved, biotic and recruitment potential.

Moose managers evaluate recruitment by estimating calves per 100 cows in the fall. In the Nelchina Basin, Alaska, fall ratios varied from 13 : 100 in 1972 to 90 : 100 in 1953 (90). The 13 : 100 ratio was attributed to calf mortality and would not sustain the population.

Mortality Factors

Moose populations in good habitat experience high productivity and recruitment rates, and they generally can absorb mortality from predation, severe winters, hunting, poaching, disease, and accidental deaths. Conversely, mortality factors can be devastating in areas with poor productivity.

Predation by wolves, brown or grizzly bears, and black bears can have a significant impact on moose populations, as can mountain lions and coyotes under certain conditions. Most ecosystems in North America have alternative prey species and more than one predator (93).

Connolly listed 31 studies where predators limited or had a regulatory influence on prey and 27 that did not (94). No moose studies were available. Since then, with advances in capture and telemetry technology, we know that wolves, bears, and mountain lion can reduce the abundance of moose (95, 96, 97, 72, 98, 99, 91, 100, 101, 102, 103, 104). Other studies indicate that when major predators were removed or controlled, prey numbers increased (105, 106, 100, 107, 108, 72, 91, 109).

Many factors work in conjunction with predation to influence the impact on moose, but three direct factors are notable: alternate prey, predator–prey ratios, and multipredator systems (94). When alternative prey are available the effects of predation may be ameliorated or accentuated depending on conditions and seasons (94). Where moose are the primary prey of wolves and the ratio is less than 20 moose per wolf, predation can cause a decline in moose (110). The combination of wolf (11 moose per wolf) and grizzly bear (6 moose per bear) predation limited moose to low densities and placed them in a "predator-pit" in eastcentral Alaska even when environmental factors favored their expansion (91).

Boertje and others expressed concern that most predator–moose studies were not long term (109). They examined 20-year effects of wolf control on the abundance of wolves and their primary prey, moose and caribou. They concluded that controlling wolf populations, combined with favorable weather, enhances long-term abundance of wolves, moose, and caribou. Wolf control recently has become more contentious, making it nearly impossible to manage wolves as a primary predator on moose, even with extensive planning, because of animal rights activists (111).

Winter mortality results from an array of factors related primarily to snow depth, density, hardness, and persistence over time. These factors affect mobility of moose, the availability of food, and their resultant energy balance. Adverse snow conditions on the Kenai Peninsula, Alaska, were responsible for nearly 100% calf mortality during winter 1971 to 1972 and approximately 85% during winters of 1989 to 1990 and 1994 to 1995 (12, T. Spraker, Alaska Department of Fish and Game, personal communication) (Figure 28–6).

Unregulated hunting and unchecked poaching can have detrimental impacts on moose populations, particularly one in decline or in a "predator-pit." Conversely, moose populations can benefit from the culling influence of regulated hunting, particularly by selective harvest (40).

Many agents of disease infect moose, but few have the potential to impact moose populations. Included are the meningeal worm in eastern North America, winter ticks, the arterial worm, and possibly *Brucella* (112). Diseases transmissible to humans or that

Figure 28–6 Snow depth is an important factor influencing movements of moose. A depth of 70 centimeters impedes movements, but snow density, hardness, and crusting are contributing factors. These moose have found it necessary to track one another in the deep snow. Deep snow conditions also may result in moose "yarding" or grouping together as these moose have. (Alaska Department of Fish and Game photograph.)

may make moose inedible include toxoplasmosis and brucellosis. The hydadid cyst tapeworm is a human risk, but is only indirectly associated with moose that serve as an intermediate host (112). Perhaps the greatest threat to moose may result from contact with domestic and game farm animals. Brucellosis and tuberculosis are obvious threats, but others may surface. Therefore, it is advantageous to maintain a blood serum bank of populations to detect past and present disease status by presence of antibodies (18).

Motor vehicles and trains are the primary sources of accidental mortality for moose. Annual moose fatalities in North America approached 3,500 from vehicles and 1,500 from trains (113). A 70% decline of a local moose population occurred along a road/railroad transportation corridor in Alaska (114).

Many essential habitat requirements for moose are available within transportation corridors that function as an "ecotonal trap" and create a high collision risk. Potential solutions include speed limits, signs, lighting, vegetation clearing, fencing, and media awareness (113). McDonald reported that moose-proof fencing, one-way gates, underpasses, and lighting on a section of the Glenn Highway, Alaska, reduced collision fatalities by 65% (115). Other incidental moose mortality reported by Child includes moose–aircraft collisions, fighting, calf abandonment, drownings, falls, protection of life and property, forest fire, avalanches, and fence entanglements (113).

Emigration and Immigration

Emigration and immigration of moose affect population dynamics because a given population can gain or lose numbers by these movements. In a rapidly growing moose population in Alberta, early increases could not be explained by reproduction. Later, the population grew more slowly. During the early phase of population growth, moose immigrated into the population and later emigrated when they exceeded carrying capacity (116). This scenario is not unusual for moose and is most notable for introduced populations (117, 118, 70).

POPULATION MANAGEMENT

The attitude that moose populations need not be managed ignores the impact humans have on this resource. Positive management rather than benign neglect is necessary to perpetuate moose habitats and populations (77). To manage moose, an information base provided by research is needed to attain the goals for populations that the public process provides (119). For example, goals outlined for management of moose in one area of Alaska include the following: (1) Maintain, protect, and enhance moose habitat; (2) maintain viable populations in their historic range; (3) manage moose on a sustained yield basis; (4) manage moose consistent with interests and desires of the public; (5) manage primarily for meat hunting; (6) manage for greatest hunter participation possible, consistent with other goals; (7) provide opportunity to view and photograph moose; and (8) develop and maintain a database for making management decisions (120). These goals are not exclusionary and are predicated by Alaska law that gives subsistence use top priority. Goals for other jurisdictions may vary from these and more specific subgoals may evolve through the public process for specific areas or populations.

Population and harvest management require reliable estimates of major population parameters such as population size, annual rate of change, recruitment, sex composition, and mortality. These factors are estimated primarily by aerial surveys, but ground surveys such as pellet counts, direct observation, and harvest statistics are useful supplements (40). Aerial surveys vary with objectives and their complexity and applications were outlined by Timmermann and Buss (40).

Harvest management is employed to stabilize moose populations within carrying capacity and to provide excess animals for human use. Strategies detailed by Timmermann and Buss include seasons, bag limits, access control, hunter education, weapon control, and selective harvest (aimed at certain sex and age segments) (40). Predator management, under certain conditions, can provide greater numbers of moose and their predators, thereby providing more opportunities for human use (91, 109).

The indicator animal concept is based on the premise that conditions in the environment such as quality and quantity of food available, stress, and weather affect the physiologic and morphometric status of the animal living in that environment (121, 122). Considering this, various measurements can be obtained from animals in a population to quantify the environmental changes. Strategies used include morphometric data and condition indices, blood parameters, body fat measurements, urine analysis, and mineral uptake in hair (123, 124, 125, 126, 127).

HABITAT REQUIREMENTS

Habitat is the primary limiting factor for moose. Large quantities of forage intake are needed to maintain moose whose rumen content weights varied seasonally from 51 to 81 kilograms (128). Intake of forage varies seasonally with sex and age and is a function of availability, digestibility, and rate of passage (129). Schwartz and Renecker extensively reviewed quality and quantity of nutritional requirements of moose and how they utilize them (130).

Moose primarily have been associated with boreal forests that are fire-dependent ecosystems (131). Peek characterized moose habitat as fire influenced, non or limited fire influenced, and aquatic (132). Examples of how fire positively affects moose exist from Alaska to Nova Scotia (133, 134). The optimum successional stage of burned boreal forests occurs from 11 to 30 years following fire, but peaks near 15 years (135, 136). Non or limited fire-influenced habitats include those above timberline, river delta systems, extreme northern shrub communities, coastal forests, and riparian willow communities (137, 138, 67, 139, 34). Aquatic habitats are used by moose from spring to fall, in some areas, and provide highly digestible and nutritious forage (140). Moose also use aquatic habitats to escape insect pests and may use them to cool themselves (141, 132) (Figure 28–7).

Figure 28–7 Moose use aquatic habitats to help them thermoregulate and stay cool, escape insects, drink water, and feed on aquatic vegetation that has a high mineral content. (Photograph by C. C. Schwartz, Alaska Department of Fish and Game, Soldotna, Alaska.)

Depth, density, and hardness of snow may limit availability of certain habitats or vegetation during winter. Moose activity in winter often is regulated by snow conditions. Snow depths of 71 to 97 centimeters limit travel (142). Snow conditions may alter the energy balance of moose by increasing metabolic demands for locomotion and decreasing energy intake. Moose winter in the coldest regions of the world when adequate shelter from wind is available. Conversely, regions where temperatures exceed 27°C for long periods without shade or access to lakes and streams do not support moose (25). Moose adapt to winter conditions by selecting habitats that favor conservation of energy and by reducing their metabolic rate (132, 143).

HABITAT MANAGEMENT

Components needed for moose habitat include abundant high-quality winter food, shelter near food, isolated calving sites, aquatic feeding areas, seral forest stands with deciduous shrubs and forbs in summer, mature forests for shelter from snow and heat, travel corridors, and mineral licks. Habitat management protects, enhances, or creates these components, and should be based on objectives; some on a short-term (10–20 years) and others on a long-term (80–150 years) basis (144).

Assessment of the quality of moose habitat may be based on moose productivity (29). Franzmann and Schwartz suggest using twinning rates and number of females without calves (35). Methods using key browse species or twig biomass are labor intensive and limited to relatively small areas (29, 145, 146, 144). Habitat suitability index models are valuable for making and testing predictions, but are static and cannot be applied on a broad scale (147, 148, 149, 144).

Thompson and Stewart questioned whether habitat management can be conducted to minimize the negative influences of population-limiting factors, particularly predation (144). The answer lies in a sound information base on the population and the forces acting on it.

Habitat Management

TABLE 28–2

Cutblock Size Recommendations in Moose Habitat Guidelines from North American Game Management Agencies

Location	Cutblock Size (ha)		Comments
	Conifer	Hardwood	
Alberta	24–32	100	
Saskatchewan	40	120–140	
Manitoba	None	None	Distance to thermal cover <200 m, and line of sight <400 m
Ontario	80–130	N/A	Distance to cover <200 m
Quebec	250		4% of the cut must remain in islands >3 ha
New Brunswick	125		
Nova Scotia	50		Includes deer range
Minnesota	81		

Source: Adapted from Ref. 144.

Wildfire is the ultimate habitat creator for moose, but logging also has historically benefited moose (6). Both create seral vegetation preferred by moose. Timber harvest creates habitat in some areas, but can be destructive if components of habitat are lost, access by logging roads is not controlled, or large areas are converted to monotypic stands (150). The recommended size of forest cuts varies from 24 to 250 hectares depending on the productivity of the sites (Table 28–2). Cuts should be made with consideration for the greatest vegetation regeneration benefit and the components of good habitat previously listed. Coordination with foresters for mutual benefit is necessary.

Nonforestry practices that create seral vegetation include prescribed burning, selected land clearing, abandonment of farmsteads, and mechanical rehabilitation (146). These methods are limited to relatively small areas and may be very costly. However, when rehabilitation areas are selected based on their regenerative ability, the cost:benefit ratio often favors rehabilitation. This is becoming more feasible as biologists begin to recognize that the economic and aesthetic value of the moose resource has been extremely undervalued in the past. Also, fire management plans, which in the past focused solely on fighting fires, are now being reevaluated to allow for wildfires to burn under certain conditions for the benefit of moose and other wildlife.

SUMMARY

Moose, the largest member of the deer family, have great aesthetic, ecological, cultural, and commercial value. Native Americans depended on moose for survival and early Europeans considered moose the fuel of expansion and exploration. Moose came to North America over the Bering Land Bridge and in time occupied the northern boreal forests. Four subspecies have been identified, but recent range expansion has reduced their isolation and differences are not definitive. Moose are primarily browsers of early successional deciduous leaves and twigs. They breed during the fall rut and calve in spring. Breeding strategy is characterized by harem breeding on the tundra and pair breeding in the closed forest. Most calves are born as singles, but twinning is common. Calves grow rapidly during their first summer. Female moose reach their maximum mass at age 4 (500 kilograms) and males at 7 to 9 (700 kilograms). They are well adapted to northern climates and limited by expansion to the south by high ambient temperatures. Moose move and migrate with seasons, snow depth, and food resources. Some migrations are traditional. Dispersal occurs to colonize new habitats.

Moose population fluctuations are primarily regulated by predation and food availability. Other mortality factors include adverse snow conditions, unregulated hunting and poaching, disease, motor vehicles, and trains. Positive management is necessary to perpetuate moose habitats and populations based on sound research. Goals for moose management include habitat protection and enhancement, maintenance of viable populations in their historic range, sustained yield management, public interest input, meat hunting, quality hunting experience, opportunities for viewing and photographing, and subsistence needs. These goals are not exclusionary. Population and harvest management require reliable estimates of population size, annual rate of change, recruitment, sex composition, and mortality. These population parameters are primarily obtained by aerial surveys. Habitat is the primary limiting factor for moose because they need large quantities of forage. Wildfire is the ultimate habitat creator for moose, but others include habitats above timberline, river delta systems, extreme northern shrub communities, coastal forests, riparian willow communities, and aquatics.

LITERATURE CITED

1. KURTEN, B., and E. ANDERSON. 1980. *Pleistocene Mammals of North America.* Columbia University Press, New York.

2. BUBENIK, A. B. 1986. Taxonomic position of *Alcinae* Jerdon, 1974, and the history of the genus *Alces* Gray, 1821. *Alces* 22:1–67.

3. BUBENIK, A. B. 1998. Evolution, taxonomy, and morphophysiology. Pages 77–123 *in* A. W. Franzmann and C. C. Schwartz, eds., *Ecology and Management of North American Moose.* Smithsonian Institution Press, Washington, DC.

4. GEIST, V. 1987. On the evolution and adaptations of *Alces. Swedish Wildlife Research Supplement* 1:11–24.

5. LISTER, A. M. 1993. Evolution of mammoth and moose: The Holarctic perspective. Pages 178–204 *in* R. A. Martin and A. D. Barnaby, eds., *Quaternary Mammals of North America.* Cambridge University Press, Cambridge, England.

6. PETERSON, R. L. 1955. *North American Moose.* University of Toronto Press, Toronto, Ontario, Canada.

7. REEVES, H. M. and R. E. McCABE. 1998. Of moose and man. Pages 1–75 *in* A. W. Franzmann and C. C. Schwartz, eds., *Ecology and Management of North America Moose.* Smithsonian Institution Press, Washington, DC.

8. SETON, E. T. 1929. *Lives of Game Animals,* Vol. III, 2. Doubleday Doran, New York.

9. CRICHTON, V. F. J. 1998. Hunting. Pages 617–653 *in* A. W. Franzmann and C. C. Schwartz, eds., *Ecology and Management of North American Moose.* Smithsonian Institution Press, Washington, DC.

10. REGELIN, W. L. and A. W. FRANZMANN. 1998. Past, present, and future management and research in Alaska. *Alces* 34(2):279–286.

11. SIMPSON, G. G. 1946. The principles of classification and a classification of mammals. *Bulletin of the American Museum of Natural History,* Vol. 85.

12. FRANZMANN, A. W. 1978. Moose. Pages 67–81 *in* J. L. Schmidt and D. L. Gilbert, eds., *Big Game of North America.* Stackpole Books, Harrisburg, PA.

13. MITCHELL, H. 1970. Rapid aerial sexing of antlerless moose in British Columbia. *Journal of Wildlife Management* 34:645–646.

14. TROYER, W. A. 1980. White moose of McKinley. *Alaska Magazine* September:76–77.

15. PASSMORE, R. C., R. L. PETERSON, and A. T. CRINGAN. 1955. A study of mandibular tooth-wear as an index of age in moose. Pages 223–238 *in* R. L. Peterson, ed., *North American Moose.* University of Toronto Press, Toronto, Canada.

16. SERGEANT, D. E., and D. H. PIMLOTT. 1959. Age determination in moose from sectioned incisor teeth. *Journal Wildlife Management* 23:315–321.

17. GASAWAY, W. C., H. B. HARKNESS, and R. A. RAUSCH. 1978. Accuracy of moose age determination from incisor cementum layers. *Journal of Wildlife Management* 42:558–563.

18. FRANZMANN, A. W., P. D. ARNESON, and D. E. ULLREY. 1975. Composition of milk from Alaskan moose in relation to other North American wild ruminants. *Journal of Zoo Animal Medicine* 6:12–14.

19. GASAWAY, W. C. 1975. Moose antlers: How fast do they grow? Miscellaneous Publication. Alaska Department of Fish and Game, Fairbanks.

20. RENEAU, J., and S. C. RENEAU, eds. 1993. *Records of North American Big Game,* 10th ed. The Boone and Crockett Club, Missoula, MT.

21. FRANZMANN, A. W., R. E. LeRESCHE, R. A. RAUSCH, and J. L. OLDEMEYER. 1978. Alaskan moose measurements and weights and measurement–weight relationships. *Canadian Journal of Zoology* 56:298–306.

22. BLOOD, D. A., J. R. McGILLIS, and A. L. LOVASS. 1967. Weights and measurements of moose in Elk Island National Park, Alberta. *Canadian Field-Naturalist* 81:263–269.

23. DOUTT, J. K. 1970. Weights and measurements of moose, *Alces alces shirasi. Journal of Mammalogy* 51:808.

24. LYNCH, G. M., B. LAJEUNESSE, J. WILLMAN, and E. S. TELFER. 1995. Moose weights and measurements from Elk Island National Park, Alberta. *Alces* 31:199–207.

25. KELSALL, J. P., and E. S. TELFER. 1974. Biogeography of moose with particular reference to western North America. *Naturaliste Canadien* 101:117–130.

26. KARNS, P. D. 1998. Population distribution, density and trends. Pages 125–139 *in* A. W. Franzmann and C. C. Schwartz, eds., *Ecology and Management of North American Moose.* Smithsonian Institution Press, Washington, DC.

27. PETERSON, R. L. 1952. A review of the living representatives of the genus *Alces.* Contribution to the Royal Ontario Museum of Zoology and Paleontology Number 34. Toronto, Ontario, Canada.

28. RENECKER, L. A., and C. C. SCHWARTZ. 1998. Food habits and feeding behavior. Pages 403–439 *in* A. W. Franzmann and C. C. Schwartz, eds., *Ecology and Management of North American Moose.* Smithsonian Institution Press, Washington, DC.

29. PEEK, J. M. 1974. A review of moose food habits studies in North America. *Naturaliste Canadien* 101:307–323.

30. SCHWARTZ, C. C., and K. J. HUNDERTMARK. 1993. Reproductive characteristics of Alaskan moose. *Journal of Wildlife Management* 57:454–468.

31. SCHWARTZ, C. C., W. L. REGELIN, and A. W. FRANZMANN. 1982. Male moose successfully breed as yearlings. *Journal of Mammalogy* 63:334–335.

32. LENT, P. C. 1974. A review of the rutting behavior in moose. *Naturaliste Canadien* 101:307–323.

33. BUBENIK, A. B. 1987. Behavior of moose (*Alces alces*) of North America. *Swedish Wildlife Research Supplement* 1:333–366.

34. HOUSTON, D. B. 1968. The Shira's moose of Jackson Hole, Wyoming. Technical Bulletin Number 1. Grand Teton Natural History Association, Jackson, WY.

35. FRANZMANN, A. W., and C. C. SCHWARTZ. 1985. Moose twinning rates: A possible population condition assessment? *Journal of Wildlife Management* 49:394–396.

36. FRANZMANN, A. W. 1981. *Alces alces.* Mammalian Species, No. 154. American Society of Mammalogists.

37. KNORRE, E. P. 1959. Ekologiya locya [Moose ecology]. *Trudy Pechoro-Ilychskohgo Gos. Zapov.* 9:5–113.

38. SCHWARTZ, C. C. 1998. Reproduction, natality and growth. Pages 141–171 *in* A. W. Franzmann and C. C. Schwartz, eds., *Ecology and Management of North American Moose.* Smithsonian Institution Press, Washington, DC.

39. CRETE, M., R. J. TAYLOR, and P. A. JORDAN. 1981. Optimization of moose harvest in southwestern Quebec. *Journal of Wildlife Management* 45:598–611.

40. TIMMERMANN, H. R., and M. E. BUSS. 1998. Population and harvest management. Pages 559–615 *in* A. W. Franzmann and C. C. Schwartz, eds., *Ecology and Management of North American Moose.* Smithsonian Institution Press, Washington, DC.

41. SCHWARTZ, C. C., W. L. REGELIN, and A. W. FRANZMANN. 1987. Seasonal weight dynamics of moose. *Swedish Wildlife Research Supplement* 1:301–310.

42. SCHWARTZ, C. C. 1992. Physiological and nutritional adaptations of moose to northern environments. *Alces Supplement* 1:139–155.

43. FRANZMANN, A. W. 1985. Assessment of nutritional status. Pages 239–259 *in* R. J. Hudson and R. G. White, eds., *Bioenergetics of Wild Herbivores.* CRC Press, Boca Raton, FL.

44. FLYNN, A., and A. W. FRANZMANN. 1987. Mineral element studies in North American moose. *Swedish Wildlife Research Supplement* 1:289–299.

45. FLYNN, A., A. W. FRANZMANN, P. D. ARNESON, and J. L. OLDEMEYER. 1977. Indications of a copper deficiency on a subpopulation of Alaskan moose. *Journal of Nutrition* 107:1182–1188.

46. ALTMANN, M. 1959. Group dynamics of Wyoming moose during the rutting period. *Journal of Mammalogy* 40:420–424.

47. HOUSTON, D. B. 1974. Aspects of social organization of moose. Pages 690–696 *in* V. Geist and F. Walther, eds., *The Behavior of Ungulates and Its Relation to Management,* Vol. II. Publication Number 24. International Union for the Conservation of Nature and Natural Resources, Morges, Switzerland.

48. BUBENIK, A. B. 1998. Behavior. Pages 173–221 *in* A. W. Franzmann and C. C. Schwartz, eds., *Ecology and Management of North American Moose.* Smithsonian Institution Press, Washington, DC.

49. WECKERLY, F. W. 1992. Territoriality in North American deer: A call for a common definition. *Wildlife Society Bulletin* 20:228–231.

50. PEEK, J. M., R. E. LeRESCHE, and D. R. STEVENS. 1974. Dynamics of moose aggregations in Alaska, Minnesota, and Montana. *Journal of Mammalogy* 55:126–137.

51. STRINGHAM, S. F. 1974. Mother–infant relations in moose. *Naturaliste Canadian* 101:325–369.

52. CEDERLAND, B-M. 1987. Parturition and early development of moose (*Alces alces L.*) *Swedish Wildlife Research Supplement* 1:399–422.

53. BURRIS, O. E., and D. E. McKNIGHT. 1973. Game transplants in Alaska. Wildlife Technical Bulletin 4. Alaska Department of Fish and Game, Juneau.

54. SIGMAN, M. 1977. A hypothesis concerning the nature and importance of the over-winter cow–calf bond in moose. *Proceedings of the North American Moose Conference and Workshop* 13:71–90.

55. BOER, A. H. 1998. Interspecific relationships. Pages 337–349 *in* A. W. Franzmann and C. C. Schwartz, eds., *Ecology and Management of North American Moose.* Smithsonian Institution Press, Washington, DC.

56. LUDWIG, H. A., and R. T. BOWYER 1985. Overlap of winter diets of sympatric moose and white-tailed deer. *Journal of Mammalogy* 55:390–392.

57. FLOOK, D. R. 1964. Range relationships of some ungulates native to Banff and Jasper National Parks, Alberta. Pages 119–128 *in* D. J. Crisp, ed., *Grazing in Terrestrial and Marine Environments.* Blackwell Science Publications, Oxford, England.

58. MARTINKA, C. J. 1969. Population ecology of summer resident elk in Jackson Hole, Wyoming. *Journal of Wildlife Management* 33:465–481.

59. TELFER, E. S., and A. L. CAIRNS. 1986. Resource use by moose versus sympatric deer, wapiti and bison. *Alces* 22:113–137.

60. BERGERUD, A. T., and F. MANUEL. 1968. Moose damage to balsam fir-white birch forests in central Newfoundland. *Journal of Wildlife Management* 32:729–746.

61. TELFER, E. S. 1972. Forage yield and browse utilization on logged areas in new Brunswick. *Canadian Journal of Forest Research* 2:346–350.

62. WOLFE, J. O. 1980. Moose–snowshoe hare competition during peak hare densities. *Procedures of the North American Moose Conference and Workshop* 16:238–254.

63. WOLFE, M. L. 1974. An overview of moose coactions with other animals. *Naturaliste Canadian* 101:437–474.

64. SHELTON, P. C., and R. O. PETERSON. 1983. Beaver, wolf and moose interactions in Isle Royale National Park, USA. *Acta Zoologica Fennica* 174:265–266.

65. ANDERSON, R. C. 1965. An examination of wild moose exhibiting neurological signs in Ontario. *Canadian Journal of Zoology* 43:635–639.

66. SINGER, F. J., and J. DALLE-MOLLE. 1985. The Denali ungulate-predator system. *Alces* 21:339–358.

67. COADY, J. W. 1980. History of moose in northern Alaska and adjacent regions. *Canadian Field-Naturalist* 94:61–68.

68. CAIRNS, A. L., and E. S. TELFER. 1980. Habitat use by 4 sympatric ungulates in boreal mixed wood forest. *Journal of Wildlife Management* 44:849–857.

69. JARMAN, P. J., and A. R. E. SINCLAIR. 1979. Feeding strategy and patterns of resource partitioning in ungulates. Pages 130–163 *in* A. R. E. Sinclair and M. Norton-Griffiths, eds., *Serengeti—Dynamics of an Ecosystem.* University of Chicago Press, Chicago.

70. HUNDERTMARK, K. J. 1998. Home range, dispersal, and migration. Pages 303–335 *in* A. W. Franzmann and C. C. Schwartz, eds., *Ecology and Management of North American Moose.* Smithsonian Institution Press, Washington, DC.

71. LEPTICH, D. J., and J. R. GILBERT. 1989. Summer home range and habitat use by moose in northern Maine. *Journal of Wildlife Management* 53:880–885.

72. BALLARD, W. B., J. S. WHITMAN, and D. J. REED. 1991. Population dynamics of moose in southcentral Alaska. *Wildlife Monograph* 114:1–49.

73. LeRESCHE, R. E. 1974. Moose migrations in North America. *Naturaliste Canadian* 101:393–415.

74. BERG, W. E. 1971. Habitat use, movements, and activity patterns of moose in northwestern Minnesota. M.Sc. thesis. University of Minnesota, St. Paul.

75. BARRY, T. W. 1961. Some observations of moose at Wood Bay and Bathurst Peninsula, N.W.T., Canada. *Canadian Field-Naturalist* 75:164–165.

76. ANDERSEN, R. 1991. Habitat deterioration and the migratory behavior of moose (*Alces alces* L.) in Norway. *Journal of Applied Ecology* 28:102–108.

77. COADY, J. W. 1982. Moose. Pages 902–922 *in* J. A. Chapman and G. A. Feldhammer, eds., *Wild Mammals of North America.* The Johns Hopkins University Press, Baltimore, MD.

78. MYTTON, W. R., and L. B. KEITH. 1981. Dynamics of moose populations near Rochester, Alberta, 1975–1978. *Canadian Field-Naturalist* 95:39–49.

79. GASAWAY, W. C., S. D. DuBOIS, and K. L. BRINK. 1980. Dispersal of subadult moose from a low density population in interior Alaska. *Proceedings of the North American Moose Conference and Workshop* 16:314–337.

80. CARR, D. E. 1972. *The Forgotten Senses.* Doubleday, Garden City, NY.

81. LORENZ, K. 1954. Das angeborene Erkennen. *Natur und Volk* 84:285–295.

82. BACKHAUS, D. 1959. Experimentelle Untersuchungen uber die Sehscharfe und das Farbsehen eineger Huftiere. *Zoologisch Tierphychologie* 16:445–467.

83. WITZEL, D. A., M. D. SPRINGER, and H. H. MOLLENHAUER. 1978. Cone and rod photoreceptors in white-tailed deer (*Odocoileus virgianus*). *American Journal of Veterinary Research* 39:699–701.

84. VANBALLENBERGHE, V., and W. B. BALLARD. 1998. Population dynamics. Pages 223–245 *in* A. W. Franzmann and C. C. Schwartz, eds., *Ecology and Management of North American Moose.* Smithsonian Institution Press, Washington, DC.

85. McCULLOUGH, D. R. 1979. *The George Reserve Deer Herd.* University of Michigan Press, Ann Arbor.

86. KEITH, L. B. 1974. Population dynamics of mammals. *International Congress of Game Biologists* 11:2–58.

87. CAUGHLEY, G. 1970. Eruptions of ungulate populations, with emphasis on the Himalayan thar in New Zealand. *Ecology* 51:53–72.

88. PETERSON, R. O. 1977. Wolf ecology and prey relationships on Isle Royale. Scientific Monograph Number 11. National Park Service, Washington, DC.

89. PIMLOTT, D. H. 1953. Newfoundland moose. *Transactions of the North American Wildlife Conference* 18:563–581.

90. BISHOP, R. H., and R. A. RAUSCH. 1974. Moose population fluctuations in Alaska. *Naturalists Canadian* 101:559–593.

91. GASAWAY, W. C., R. D. BOERTJE, D. V. GRANGARD, D. G. KELLEYHOUSE, R. O. STEPHENSON, and D. G. LARSEN. 1992. The role of predation in limiting moose at low densities in Alaska and Yukon and implications for conservation. *Wildlife Monograph* 120:1–59.

92. BALLARD, W. B., and D. G. LARSEN. 1987. Implications of predator–prey relations to moose management. *Swedish Wildlife Research Supplement* 1:581–602.

93. BALLARD, W. B., and V. VANBALLENBERGHE. 1998. Predator–prey relationships. Pages 247–273 *in* A. W. Franzmann and C. C. Schwartz, ed., *Ecology and Management of North American Moose.* Smithsonian Institution Press, Washington, DC.

94. CONNOLLY, G. E. 1978. Predators and predator control. Pages 369–394 *in* J. L. Schmidt and D. L. Gilbert, eds., *Big Game of North America.* Stackpole Books, Harrisburg, PA.

95. FRANZMANN, A. W. 1998. Restraint, translocation, and husbandry. Pages 519–557 *in* A. W. Franzmann and C. C. Schwartz, eds., *Ecology and Management of North American Moose.* Smithsonian Institution Press, Washington, DC.

96. FRANZMANN, A. W., C. C. SCHWARTZ, and R. O. PETERSON. 1980. Moose calf mortality in summer on the Kenai Peninsula, Alaska. *Journal of Wildlife Management* 44:764–768.

97. BALLARD, W. B., T. H. SPRAKER, and K. P. TAYLOR. 1981. Causes of neonatal moose calf mortality in southcentral Alaska. *Journal of Wildlife Management* 45:335–342.

98. HAUGE, T. M. and L. B. KEITH. 1981. Dynamics of moose populations in northeastern Alberta. *Journal of Wildlife Management* 45:573–597.

99. GASAWAY, W. C., R. O. STEPHENSON, J. L. DAVIS, P. E. K. SHEPHERD, and O. E. BURRIS. 1983. Interrelationships of wolves, prey, and man in interior Alaska. *Wildlife Monograph* 84:1–50.

100. BOERTJE, R. D., W. C. GASAWAY, D. V. GRANGAARD, and D. G. KELLEYHOUSE. 1988. Predation on moose and caribou by radio-collared grizzly bears in eastcentral Alaska. *Canadian Journal of Zoology* 66:492–499.

101. LARSEN, D. G., D. A. GAUTHIER, and R. L. MARKEL. 1989. Causes and rate of moose mortality in the southwest Yukon. *Journal of Wildlife Management* 53:548–557.

102. SCHWARTZ, C. C., and A. W. FRANZMANN. 1991. Interrelationship of black bears to moose and forest succession in the northern coniferous forest. *Wildlife Monograph* 113:1–58.

103. OSBORNE, T. O., T. F. PARAGI, J. L. BODKIN, A. J. LORANGER, and W. N. JOHNSON. 1991. Extent, cause, and timing of moose calf mortality in western interior Alaska. *Alces* 27:24–30.

104. ROSS, I. P., and M. G. JALKOTZY. 1996. Cougar predation on moose in southwestern Alberta. *Alces* 32:1–8.

105. CRETE, M., and F. MESSIER. 1984. Responses of moose to wolf removal in southwestern Quebec. *Alces* 20:107–128.

106. STEWART, R. R., E. W. KOWAL, R. BEAULIEU, and T. W. ROCK. 1985. The impact of black bear removal on moose calf survival in eastcentral Saskatchewan. *Alces* 21:403–418.

107. BALLARD, W. B., and S. M. MILLER. 1990. Effects of reducing brown bear density on moose calf survival in southcentral Alaska. *Alces* 26:9–13.

108. BALLARD, W. B., J. S. WHITMAN, and C. L. GARDNER. 1987. Ecology of an exploited wolf population in southcentral Alaska. *Wildlife Monograph* 98:1–54.

109. BOERTJE, R. D., P. VALKENBURG, and M. E. MCNAY. 1996. Increases in moose, caribou, and wolves following wolf control in Alaska. *Journal of Wildlife Management* 60:474–489.

110. PETERSON, R. O., and R. PAGE. 1983. Wolf-moose fluctuations at Isle Royale National Park, Michigan, USA. *Acta Zoologica Fennica* 174:251–253.

111. FRANZMANN, A. W. 1993. Biopolitics of wolf management in Alaska. *Alces* 29:9–26.

112. LANKESTER, M. W., and W. M. SAMUEL. 1998. Pests, parasites and diseases. Pages 479–517 *in* A. W. Franzmann and C. C. Schwartz, eds., *Ecology and Management of North American Moose.* Smithsonian Institution Press, Washington, DC.

113. CHILD, K. N. 1998. Incidental mortality. Pages 275–301 *in* A. W. Franzmann and C. C. Schwartz, eds., *Ecology and Management of North American Moose.* Smithsonian Institution Press, Washington, DC.

114. SCHWARTZ, C. C., and B. BARTLEY. 1991. Reducing incidental moose mortality: Considerations for management. *Alces* 27:227–231.

115. MCDONALD, M. G. 1991. Moose movement and mortality associated with the Glenn Highway expansion, Anchorage, Alaska. *Alces* 27:208–219.

116. ROLLEY, R. E., and L. B. KEITH. 1980. Moose population dynamics and winter habitat use at Rochester, Alberta, 1965–1979. *Canadian Field-Naturalist* 94:9–18.

117. MERCER, W. E., and D. A. KITCHEN. 1968. A preliminary report on the extension of moose range in the labrador Peninsula. *Procedures of the North American Moose Conference and Workshop* 5:62–81.

118. NOWLIN, R. A. 1985. Distribution of moose during occupation of vacant habitat in northcentral Colorado. Ph.D. thesis. Colorado State University, Fort Collins.

119. FRANZMANN, A. W., and C. C. SCHWARTZ. 1987. Management of North American moose populations. Pages 517–522 *in* C. M. Wemmer, ed., *Biology and Management of the Cervidae.* Smithsonian Institution Press, Washington, DC.

120. ALASKA DEPARTMENT OF FISH AND GAME. 1990. Strategic plan for management of moose in Region 1, southeast Alaska 1990–1994. Alaska Department of Fish and Game, Juneau.

121. FRANZMANN, A. W. 1970. Ecology and the veterinarian. *Journal of the American Veterinary Medical Association* 157:1981–1982.

122. FRANZMANN, A. W., R. COOK, C. M. SINGH, and J. V. CHEERAN. 1995. Health and condition evaluation of wild animal populations: The animal indicator concept. Pages 365–399 *in* S. H. Berwick and V. B. Saharia, eds., *The Development of International Principles and Practices of Wildlife Research and Management.* Oxford University Press, Delhi, India.

123. FRANZMANN, A. W. 1977. Condition assessment of Alaskan moose. *Proceedings of the North American Moose Conference and Workshop* 13:119–127.

124. FRANZMANN, A. W., and R. E. LeResche. 1978. Alaskan moose blood studies with emphasis on condition evaluation. *Journal of Wildlife Management* 42:334–351.

125. STEPHENSON, T. R., K. J. HUNDERTMARK, C. C. SCHWARTZ, and V. VANBALLENBERGHE. 1993. Ultrasonic fat measurement of captive yearling bull moose. *Alces* 29:115–124.

126. DelGUIDICE, G. D., R. O. PETERSON, and U. S. SEAL. 1991. Differences in urinary chemistry profiles of moose on Isle Royale during winter. *Journal of Wildlife Management.* 27:407–416.

127. FLYNN, A., A. W. FRANZMANN, and P. D. ARNESON. 1975. Sequential hair shaft as an indicator of prior mineralization in the Alaskan moose. *Journal of Animal Science* 41:906–910.

128. GASAWAY, W. C., and J. W. COADY. 1974. Review of energy requirements and rumen fermentation in moose and other ruminants. *Naturaliste Canadien* 101:227–262.

129. SCHWARTZ, C. C., W. L. REGELIN, A. W. FRANZMANN, and M. HUBBERT. 1987. Nutritional energetics of moose. *Swedish Wildlife Research Supplement* 1:265–280.

130. SCHWARTZ, C. C., and L. A. RENECKER. 1998. Nutrition and energetics. Pages 441–478 *in* A. W. Franzmann and C. C. Schwartz, eds., *Ecology and Management of North American Moose.* Smithsonian Institution Press, Washington, DC.

131. ROWE, J. S., and G. W. SCOTTER. 1973. Fire in the boreal forest. *Quaternary Research* 3:444–464.

132. PEEK, J. M. 1998. Habitat relationships. Pages 351–375 *in* A. W. Franzmann and C. C. Schwartz, eds., *Ecology and Management of North American Moose.* Smithsonian Institution Press, Washington, DC.

133. SPENCER, D. L., and E. F. CHATELAIN. 1953. Progress in the management of moose in south-central Alaska. *Transactions of the North American Wildlife Conference* 18:539–552.

134. ROWE, J. S. 1982. Forest regions of Canada. Forestry Branch Bulletin 123. Canadian Department of Northern Affairs and Natural Resources, Ottawa, Ontario.

135. KELSALL, J. P., E. S. TELFER, and T. D. WRIGHT. 1977. The effects of fire on the ecology of the boreal forest, with particular reference to the Canadian north: A review and selected bibliography. Occasional Paper 32. Canadian Wildlife Service, Ottawa, Ontario.

136. SCHWARTZ, C. C., and A. W. FRANZMANN. 1989. Bears, wolves, moose, and forest succession: Some management considerations on the Kenai Peninsula, Alaska. *Alces* 25:1–10.

137. RISENHOOVER, K. L. 1989. Composition and quality of moose winter diets in interior Alaska. *Journal of Wildlife Management* 53:569–576.

138. MacCRACKEN, M. R., V. VANBALLENBERGHE, and J. M. PEEK. 1997. Habitat relationships of moose on the Copper River Delta in coastal south-central Alaska. *Wildlife Monograph* 136:1–52.

139. TELFER, E. S. 1984. Circumpolar distribution and habitat requirements of moose (*Alces alces*). Pages 145–182 *in* R. Olsen, R. Hastings, and F. Geddes, eds., *Northern Ecology and Resource Management.* University of Alberta Press, Edmonton, Alberta, Canada.

140. JORDAN, P. A. 1987. Aquatic foraging and the sodium ecology of moose: A review. *Swedish Wildlife Research Supplement* 1:119–137.

141. RITCEY, R. W., and N. A. M. VERBEEK. 1969. Observations of moose feeding on aquatics in Bowron Lake Park, British Columbia. *Canadian Field-Naturalist* 83:339–343.

142. KELSALL, J. P. 1969. Structural adaptations of moose and deer for snow. *Journal of Mammalogy* 50:302–310.

143. REGELIN, W. L., C. C. SCHWARTZ, and A. W. FRANZMANN. 1985. Seasonal energy metabolism of adult moose. *Journal of Wildlife Management* 49:394–396.

144. THOMPSON, I. D. and R. W. STEWART. 1998. Management of moose habitat. Pages 377–401 *in* A. W. Franzmann and C. C. Schwartz, eds., *Ecology and Management of North American Moose.* Smithsonian Institution Press, Washington, DC.

145. TELFER, E. S. 1974. A trend survey for browse ranges using the Shafe twig count technique. *Proceedings of the North American Moose Conference and Workshop* 10:160–171.

146. OLDEMEYER, J. L., and W. L. REGELIN. 1987. Forest succession, habitat management, and moose on the Kenai National Wildlife Refuge. *Swedish Wildlife Research Supplement* 1:163–179.

147. ALLEN, A. W., P. A. JORDAN, and J. W. TERRELL. 1987. Habitat suitability index models: Moose. Lake Superior region. Biological Report 82. United States Fish and Wildlife Service, Washington, DC.

148. ALLEN, A. W., J. W. TERRELL, W. L. MANGUS, and E. L. LINDQUIST. 1991. Application and partial validation of a habitat model for moose in the Lake Superior region. *Alces* 27:50–64.

149. ADAIR, W. P., P. JORDAN, and J. TILLMA. 1991. Aquatic forage ratings according to wetland type: Modifications for the Lake Superior HSI. *Alces* 27:140–149.

150. THOMPSON, I. D., and D. L. EULER. 1987. Moose habitat in Ontario: A decade of change in perception. *Swedish Wildlife Research Supplement* 1:181–193.

CHAPTER 29

White-tailed Deer

Stephen Demarais, Karl V. Miller, and Harry A. Jacobson

INTRODUCTION

Human history in North America has been linked closely with native deer species. The most abundant archaeological remains uncovered in association with the early human inhabitants were those of white-tailed deer (1, 2). As much as 25% of the diet of the early woodland Indian cultures consisted of venison, and much like the boundaries of wolf packs, tribal boundaries were associated with hunting grounds for white-tailed deer (2). Early Americans used virtually every part of the animal. Antlers and long bones were used for utensils, tools, and spear and arrow points; tendons became thread, fishing line, and bow strings; fat became preservative, soap, and lamp light; brains were used for tanning; and hides became clothing, blankets, lodging, and in later years currency.

White-tailed deer were associated with European colonial expansion in North America. Early concern over dwindling deer populations sparked the first closed season on deer in 1639 (3). However, early concern did not prevent overexploitation. As late as the 1750s, deer hides provided greater returns than other commodities shipped from ports along the Southeastern coast of the United States (4). From 1755 through 1773 hides from about 600,000 deer were shipped from Savannah, Georgia (5). Market hunting of deer and other game from 1800 to 1850 served as an inexpensive source of protein and helped fuel the westward expansion. During this period, deer were extirpated from much of their range, including many states east of the Mississippi (6). Population estimates at the end of the 19th century ranged from 300,000 in 1890 to 500,000 in 1900 (7).

Market hunting and unregulated harvest for subsistence use were not effectively stopped until after 1900. The Lacey Act and the sheer scarcity of deer were the impetus that halted commercial harvest (2). Restoration efforts in some states began as early as the late 1800s, but were relatively small and usually consisted of a single restocking. Pittman-Robertson Act funds allowed larger, organized restoration efforts from the 1930s through the 1960s, resulting in the restoration of deer populations over much of their former range (8).

Protective game laws and ideal habitat conditions resulting from the regeneration of cut-over forests allowed white-tailed deer to recover quickly and become abundant on the American landscape. Sport hunting became the dominant use of this resource and has become a major economic asset. In 1991, an estimated $1.8 billion was attributed to nonconsumptive use of the white-tailed deer resource, and 10.3 million people hunted white-tailed deer spending an estimated $5.1 billion in pursuit of their sport (9). Numbers of deer are estimated to exceed those of precolonial days with estimates exceeding 26 million animals. The adaptability and recruitment powers of the white-tailed deer are evidenced by the increase in annual harvest from 2 million deer in 1978 to more than 5.3 million deer in 1994 (10).

Throughout human history in North America, the white-tailed deer has been considered a valued resource. However, current high populations of this animal have economic costs, as well as benefits. Conover estimated that deer annually cause $100 million dollars in crop damage, $750 million in forest damage, more than $1 billion in vehicle damage, and $251 million in household damage (9). Vehicle–deer collisions result in about 29,000 human injuries and 211 lives lost annually (11).

In addition to economic issues, abundant white-tailed deer populations can alter ecological communities. As a keystone herbivore, overabundant populations impact forest regeneration and alter plant successional patterns (12, 13). White-tailed deer are now recognized as agents of ecological change, with the ability to influence area and regional biodiversity (14, 15). Deer herbivory in the central and northeastern United States can essentially remove Eastern hemlock from the understory (16). Negative herbivory effects from white-tailed deer on ground-nesting and shrub-nesting songbirds have been noted (17, 18).

Concurrent with burgeoning deer populations throughout much of its range, societal values have changed toward use of this species. An increasingly urban society, divorced from the reality of life in a natural setting, has placed increasing stigma on sport hunting. The future management of white-tailed deer rests with our ability to understand the ecological and social interactions of deer with humans and the environment. How well we understand these interactions will determine how well we address the management challenges of the 21st century.

TAXONOMY, DISTRIBUTION, AND GENETIC VARIATION

The white-tailed deer belongs to the order Artiodactyla, suborder Ruminantia, infraorder Pecora, superfamily Cervoidea, family Cervidae, subfamily Odocoileinae, genus *Odocoileus,* species *virginianus* (19). The first fossil record appears in North America about 3.5 million years ago (20).

White-tailed deer are among the most widely distributed mammals, occurring from the Northwest territories and southern provinces of Canada to Peru and Brazil in South America (Figure 29–1). They occur in all 48 coterminous United States except Utah, although their range is restricted in Nevada and California. They have been introduced into the British Isles, Bulgaria, the Czech Republic, Finland, Yugoslavia, New Zealand, Cuba, Virgin Islands, and other Caribbean Islands (19, 21, 22, 23). Few mammals other than deer and humans have a geographic distribution ranging from the boreal forests to the tropics.

Thirty-eight subspecies of white-tailed deer have been described (19). However, translocations and reintroductions to areas from which they had been extirpated have clouded the issue of current subspecies distribution. At least 25 states obtained deer from outside their state, with Texas, Michigan, North Carolina, and Wisconsin contributing the most deer (24). This movement of deer from state to state often introduced subspecies from disparate areas of North America, and thus raises questions regarding the integrity of the 17 subspecies described in North America north of Mexico.

Regardless of subspecies distinction, there is considerable geographic variation in phenotypic characteristics of this animal. Body sizes range from the small Florida key deer, where a mature buck may weigh 27 kilograms, to large northern variants, which may weigh 136 kilograms or more. The largest known white-tailed deer weighed 181 kilograms dressed weight and would have had a live weight in excess of 225 kilograms (25).

Antlers of white-tailed deer are highly variable and, as with body mass, depend on age, genetics, nutrition, and other site conditions. They range in size from smaller antlers of Coues and tropical subspecies to the more massive antlers of the northern subspecies. However, large antlers are not necessarily restricted to the northern states.

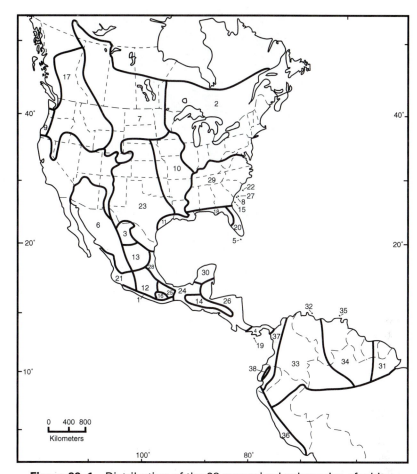

Figure 29–1 Distribution of the 38 recognized subspecies of white-tailed deer in North, Central, and South America. Subspecies as follows: 1, Acapulco; 2, northern woodland; 3, Carmen Mountains; 4, Chiriqui; 5, Florida Key; 6, Coues; 7, Dakota; 8, Hilton Head Island; 9, Columbian; 10, Kansas; 11, Avery Island; 12, Mexican tableland; 13, Miquihuana; 14, Chiapas; 15, Blackbeard Island; 16, Oaxaca; 17, northwest; 18, Florida coastal; 19, Coiba Island; 20, Florida; 21, Sinaloa; 22, Bulls Island; 23, Texas; 24, Mexican lowland; 25, rain forest; 26, Nicaragua; 27, Huntington Island; 28, northern Veracruz; 29, Virginia; 30, Yucatan; 31, cariacou; 32, curassavicus; 33, gudodotii; 34, gymnotis; 35, margaritae; 36, peruvianus; 37, tropicalis; 38, ustus. (Adapted from Refs. 215 and 19.)

Antler size generally increases as males grow to maturity at 5 to 7 years of age (Figure 29–2). Average antler mass of captive males increased fourfold from yearling to 2.5 years of age and ninefold from yearling to 5.5 years of age (26). Antler points, main beam length, and antler spread generally increase with age. Males with 8 points (i.e., 4×4) occur with the highest frequency. However, some males may never obtain more than 6 or 7 total points and others may grow 30 or more points. Antler points are classified as typical (points generally symmetric and arising from the top of the main beams) or atypical (points arising from other than the top of the main beam). The frequency of atypical points increases in older males. White-tailed deer antlers usually have more prominent brow tines than do mule deer antlers and lack bifurcated branching of the antler tines. However, exceptions occur for both species and the Sitka black-tailed deer resembles white-tailed deer (27).

Figure 29–2 Antler size and complexity generally increase with age up to about 5 to 7 years. (Photograph by R. E. Zaiglin.)

Other physical characteristics of white-tailed deer are body height ranging from 61 to 102 centimeters; length 122 to 242 centimeters; and a distinctive tail about 28 centimeters in length. The tail is broad at the base and brown above, although often with a dusky tip, and with a prominent white fringe (28). The white underside of the tail is displayed when the deer is alarmed or threatened. The forward curving antlers, smaller ears, narrower snout, and smaller metatarsal glands (20–50 millimeters) distinguish white-tailed deer from the closely related mule and black-tailed deer (27). They have 32 teeth (i 0/3, c 0/1, p 3/3, m 3/3) (29). Tooth replacement and wear patterns and cementum annuli can be used to estimate age, although accuracy above 2.5 years can be limited (30, 31).

The potential for antler development is regulated by an animal's DNA, although debate continues over the relative heritability of specific antler features and the age at which ultimate antler size can be predicted (31a, 32, 33). Geographic origin and antler quality have been linked in a Georgia study, where areas restocked with deer from Wisconsin had larger antlered deer than areas restocked from other geographic sources (34). Pelage traits also may be linked to geographic origin and genetics, as are physiologic traits such as digestive and tannin-binding enzymes (35). Genetics and geographic origin also affect the timing of antler growth and casting (36).

LIFE HISTORY

Diet Selection

Understanding the diet selection process of white-tailed deer and applying this to management is best accomplished by first knowing what whitetails "should" eat. Hoffman provides an intriguingly simple basis for understanding the characteristics of forages animals can consume and digest successfully (37). Whitetails are concentrate-selectors, which means they selectively forage for high-quality, easily digestible plant parts. Their anatomic, behavioral, and physiologic adaptations allow them to select nutritious forage and obtain the easily digestible nutrients quickly. Their large salivary glands produce enzymes that help deactivate secondary plant compounds, such as tannins, commonly found in browse. Although high-fiber, low-quality forages such as mature grasses may provide adequate nutrition to animals such as elk, white-tailed deer have a quicker digestive process and require highly digestible forages to fulfill their energy and protein requirements. On severely overpopulated ranges, white-tailed deer have starved to death with their stomachs full of low-quality forages.

A list of plants consumed by white-tailed deer would be quite long (38). However, the majority of the diet comes from a relatively small number of forages. Although deer consumed 113 species and 41 species groups in southcentral Oklahoma, 47 species or species groups accounted for 93% of the overall diet (39). Regular sampling from a wide range of species allows deer to continually evaluate new sources of nutrition.

Diet selection changes in response to seasonal changes in forage abundance, quality, and metabolic needs of the animal. For example, varying the amounts of forbs, browse, mast, and grasses selected during spring and summer allows deer to satisfy changing nutritional needs. Highly digestible forbs and grasses address protein and phosphorous requirements, browse provides rapidly fermented cell solubles, and mast provides a concentrated energy source (40). Throughout the range of white-tailed deer, browse, mast, and forbs combine in varying amounts as the predominant forage classes (typically greater than 85% of the overall diet) (41, 42, 43). Soft mast, including grapes, blackberries, persimmons, and crab apples, are used heavily during summer, whereas hard mast (primarily acorns) are important fall forages. In some areas, evergreen browse is an important source of protein throughout the year, but is particularly important during winter when other forages are limited (44, 45). A wide variety of agricultural crops also are used readily.

Land management decisions affect forage availability and thus diet quality for white-tailed deer. Rangelands in poor condition have lower species richness and abundance of forbs than rangelands in excellent condition (46). Poor range conditions cause deer to forage almost exclusively on browse, instead of taking advantage of the typically higher protein content and digestibility of forbs.

Timber management decisions also impact diet selection. Young clear-cuts provide abundant forbs, vines, and woody browse. Soft mast production in young clear-cuts may average 107 or more kilograms per hectare (47). However, hard mast production may be severely reduced or eliminated. Thill and others concluded that deer may be able to obtain nutritionally superior diets from recent clear-cuts compared to older stands (43).

Reproductive Strategy

The seasonally polyestrous breeding activity of white-tailed deer is synchronized by photoperiod, but allows considerable temporal variation in breeding dates among regions and even populations within an area. Seasonal timing of reproduction by temperate species allows the most energy-demanding phases of the reproductive cycle (i.e., late pregnancy and lactation) to coincide with the period of maximum food abundance and quality (48). The adaptive significance of breeding date varies from very high in northern, temperate populations, where most breeding takes place within a short time (e.g., 2 weeks in southern Michigan), to low in southern, tropical populations, where breeding activity may extend throughout the year (e.g., Virgin Islands) (49, 50).

Within the seasonal constraints set by photoperiod, endogenous rhythms, and genetics, the exact timing of the majority of breeding activity, or the rut, is affected primarily by nutrition and deer herd demographics. Poor nutrition, caused by factors such as poor-quality habitat, overpopulation, and drought, can delay breeding. Improving habitat quality and reducing deer and livestock density can shift peak breeding to an earlier date. An inadequate number of males (i.e., unbalanced sex ratio) can keep females from being bred during their first or second estrous period, which extends and delays the breeding season. A common misconception among recreational hunters is that cold weather causes deer to "go into rut," when in fact the weather change probably just increases general deer activity.

Adult females may exhibit three to seven estrous cycles separated by 21 to 29 days if not bred (51, 52, 53, 54). Breeding can occur several times and involve more than one male within a 24-hour heat or estrous period (55). Reported gestation lengths vary from 187 to 221 days due perhaps to nutritional, regional, or subspecies differences (54, 56, 57, 55).

Female fawns may breed their first autumn if they reach a critical mass (e.g., 36 kilograms in the northern United States), which is related to nutritional condition and birth dates (58). Prevalence of fawn breeding is a sensitive indicator of habitat quality, and is assessed by measuring lactation prevalence in 1.5-year-old harvested females. First-time breeders typically produce one fawn, followed by twins and sometimes triplets in subsequent years. The number of eggs, fetuses, or fawns produced varies with nutrient intake. Nutrient deficiency is one explanation for the low productivity of Key deer (0.76 fetuses per yearling or older female and only 61% of 2- to 4-year-old females breeding) (59). Verme showed that autumn nutritional quality impacted number of fawns in females aged 1.5 years (0.1 on low diet versus 1.2 on high diet), 2.5 years (0.5 on low diet versus 1.5 on high diet), and 3.5 plus years (1.3 on low diet versus 1.8 on high diet) (60). Deer populations with adequate nutrient intake produce 1.6 to 1.8 fetuses per adult female (61, 62).

Sex ratio at birth may vary from the theoretical 50% males and 50% females due to nutrient deficiency. Verme reviewed available literature and concluded that undernourished females tend to produce a surplus of males and females with good nutrition at estrus produce more females (63). Caley and Nudds concluded that competition for food on the maternal home range (i.e., local resource competition) caused observed facultative adjustments in deer sex ratios at birth (64). Their assumption is that male fawns, which have a high tendency to disperse, will be less competitive for local food resources than female fawns, which tend to establish home ranges that overlap with their dam.

The polygynous breeding behavior of white-tailed deer allows a range of biologically acceptable sex ratios. However, heavy annual harvest of adult males, or significant harvest of males before the rutting period, may result in skewed sex ratios during the rut. In some populations, an insufficient supply of males may not allow breeding of all females during their first estrous cycle, and thereby prolong the rutting period. A simulation study showed that an 81% male harvest rate resulted in 24% of adult females not conceiving during their first estrous cycle (65). Prolonged breeding seasons may have several negative effects (66). Male survival may be reduced as rutting behavior continues as long as females continue to enter estrus (67). Prolonged breeding seasons also may result in greater predation on fawns (68). In southern herds, body mass and antler development of yearling males is related directly to date of birth. Therefore, in some populations, male reproductive fitness may be related more to date of birth than to genetic background.

Longevity

Life expectancy for white-tailed deer varies dramatically with human-induced mortality factors. Few wild whitetails live beyond 10 years, although some have been documented to live more than 20 years (28). Wild known-aged females lived in excess of 14 years in New York and Michigan (69, 70). In heavily exploited populations, life expectancy is much lower. Life expectancy was 2 to 3 years for deer in heavily exploited populations in Illinois and Pennsylvania (71, 72). A selective buck harvest strategy will increase the prevalence of older aged males and increase their average longevity. Poor habitat quality combined with severe winter weather may decrease longevity in northern areas.

Physiologic Patterns

The physiology of any wild animal adapted to temperate regions will be keyed to optimizing survival and short-term influence of the individual with positive effects on the long-term vigor of the population. A review of the literature on white-tailed deer physiology provides support for this conclusion in the areas of growth, metabolism, fat deposition, and the seasonal timing of nutritional requirements. In sum, metabolism must support behavior, behavior must provide for metabolism, and all processes must take place within the appropriate time frame.

An intricate set of physiologic processes results in seasonal patterns of growth, metabolism, and food intake. Fawns take advantage of seasonally favorable weather and forage supplies during their first 3 months, growing at 0.2 kilogram per day (73). Fawn growth essentially halted by mid-November while on a high-quality diet in Michigan and during January and February in South Texas (74, 75). After their first winter, forage quality and availability, sex, age, and reproductive physiology affect growth patterns. In southern Texas, body mass increased annually through 3 to 4 years in females and 5 to 6 years in males (75). Mature deer showed cyclic mass changes related to reproductive phenology and forage supplies. Males lost 27% of their mass during November through February due to breeding activities and lower quality forage. Females gained mass during gestation and lost mass during parturition and lactation. However, seasonal mass patterns may differ among regions reflecting responses to climate, weather, and reproductive phenology, as body mass of adult females reached a low in May in Minnesota (76). Overpopulation and resultant damage to the forage supply decreases body mass.

Adaptations to extreme northern winters include a highly insulative coat, behavioral and habitat choices to conserve energy, and reduced thyroid function (i.e., reduced metabolic rate) resulting in deer that are "geared down to a relatively torpid, almost semihibernating state" (77, p. 110). Voluntary restriction of food intake during winter has been reported in other areas, including New Hampshire, Virginia, and Texas (78, 79, 80). A reduced metabolic rate during winter allows conservation of limited energy resources (81). However, Mautz and others questioned if the previously reported winter metabolic depression was an inherent seasonal adaptation to conserve resources or an artifact of experimental design (82, 78).

Fat deposition during autumn is cued by photoperiod to ensure that animals deposit energy stores prior to the occurrence of resource restriction (83). Because it occurs in autumn, Verme and Ozoga suggested that fat deposition was an obligatory physiologic event that proceeds until a serious negative energy balance occurs (74). The importance of fat supplies to survival during forage-limited winter conditions is evident in results from Worden and Pekins; deer with 20% body fat can balance their daily energy requirements for about 3 months, even with reduced intake rates (84).

Accurate determination of white-tailed deer nutrient requirements has been limited because of the animal's physiologic and behavioral adaptations to minimize their nutrient requirements in a potentially limiting environment. For example, understanding protein requirements depends on understanding the deer's ability to recycle urea in increasingly greater amounts to account for declining protein supplies. As dietary protein content decreased from 26% to 5%, the percent of urea recycled increased from 41% to 92% (85).

Protein requirements depend on the physiologic demands of the animal and vary with age and season. Weaned fawns require 13% to 20% protein for adequate growth (86). Adult requirements are limited primarily to maintenance and antler growth in males and gestation and lactation in females. Adult maintenance estimates for crude protein vary from a high of 12% to a low of 4% (87, 88). Protein requirements would be expected to be lower during winter than spring or summer due to the animal's resource conservation strategy. Lactation increases protein requirements because deer milk is 8.2% protein on a wet basis (89). A diet of 16% protein is generally considered optimum for antler development, although as low as 10% may be adequate (88).

Energy requirements also depend on the animal's physiologic demands and vary with age and season. Gross energy, such as that supplied by acorns, is important for fat deposition in autumn. A 10% decrease in energy intake stunted growth and lipogenesis of fawns (74, 77). Energy requirements for reproductively active females vary across the seasons. A northern female in winter requires about 3,200 kilocalories per day of digestible energy for maintenance, which does not include other energetic costs, such as for foraging and walking through deep snow (90). Energy required for peak lactation with two fawns is about 2.3 times the basal metabolic rate (i.e., energy needed for maintenance at rest in a thermal-neutral environment) (58, p. 362).

Very little is known about the mineral and vitamin requirements of white-tailed deer. Research on macronutrients, such as calcium and phosphorus, has provided varied results. Early work indicated that a diet of 0.64% calcium and 0.56% phosphorus was required for antler growth (91). Phosphorous requirements were recently estimated to vary between 0.12% and 0.29% seasonally (92). Magnesium requirements likely are similar to the 0.20% required by domestic livestock (93). Little is known about the requirements for microminerals, such as aluminum and cobalt, and for fat-soluble vitamins.

BEHAVIOR

Social Structure

Adult deer are sexually segregated during much of the year. This sexual segregation is most pronounced during spring and summer. At this time, adults reside in small, single-sex groups that remain separate spatially or temporally (94). However, in open habitats or areas with highly concentrated food resources, large mixed-sex aggregations can occur (95). These groupings are temporary and fragment into subgroups to return to cover after feeding.

The female social unit is the matriarchal family unit consisting of an adult female, her fawns from the current year, and often female offspring from previous years (Figure 29–3) (96). Matriarchal group membership varies seasonally. Prior to parturition and for the initial 4 to 6 weeks of fawn rearing, matriarchal females defend exclusive fawning territories (97). Young females usually establish fawning territories on the periphery of their mothers' defended areas.

During the first few days after birth, females rarely venture more than 100 meters from their offspring (98) (Figure 29–4). Fawns begin to accompany their dam at 3 to 4 weeks of age. As the female's home range sizes increase during summer, they become increasingly tolerant of related females and their offspring. In forested areas, matriarchal groups remain small, except during winter. In the more open habitats of the western portion of the whitetail's range, larger groups of females with their fawns are common.

Figure 29–3 The typical matriarchal family unit consists of an adult doe, her most recent fawns, and one or more yearling does. (Photograph by R. E. Zaiglin.)

Figure 29–4 Fawn survival during their first month is aided by their cryptic coloration and behavior. (Photograph by S. Demarais.)

In intensively farmed areas of the Midwest where forested cover is limited, young females often disperse long distances to locate suitable fawning habitat. In an Illinois study, 50% of female fawns and 21% of yearling females dispersed (99). In Nebraska, 16% of all aged females emigrated and in Minnesota, 20% of yearling females dispersed (100, 101). Under some circumstances, female dispersal rates may be less than 5%, regardless of deer density (102).

Adult males are gregarious and form bachelor groups from late winter through summer. In forested habitats, these groups often vary in size from two to four, while larger groups of five or more individuals often form in open habitats (103, 95). Bachelor groups are transitory with membership changing frequently (49). Social interactions are limited, although mutual grooming and dominant–subordinate interactions occur commonly (Figure 29–5) (104, 105).

Yearling males disperse from their natal areas and seek association with bachelor groups (106). Dispersal distances typically range from 3 to 10 kilometers, but distances of more than 150 kilometers have been reported. Although antagonism by mature males and competition for breeding privileges may be partially responsible for yearling dispersal, social aggression from the yearling's dam and related females may be the primary impetus for dispersal (105, 107, 108).

Movement Patterns

Although movement patterns are crepuscular, activity cycles are flexible and may correspond to a variety of environmental and anthropogenic factors (105). During winter, deer may tend to be more diurnal, with a single, late afternoon activity peak, whereas in other seasons deer are crepuscular with activity peaks at dusk and just after dawn (109). Deer activity is depressed by heavy rain, strong or gusty winds, and above-average summer temperatures (96, 110). In northern regions, temperature and snow depth exert the strongest effects on deer activity (109). Although associations between deer movements and lunar phase have been proposed, no published scientific studies have demonstrated clear relationships.

Figure 29–5 Aggression-related mortality during the breeding season is a rarity in buck-exploited populations due to lack of older aged bucks, but it increases as populations develop older age structure. (Photograph by R. E. Zaiglin.)

The total distance that whitetails move during a 24-hour period generally ranges from 1.5 to 4.5 kilometers, although in an Illinois study, distances between radiolocations over a 24-hour period averaged about 5.7 kilometers (105, 99). Individual distances vary greatly according to the sex and age of the individual, season, habitat, and weather. Daily movements are greatest during the breeding season, particularly for dominant, mature males (111).

In Alabama, females increased activity during the rut, but total distance moved decreased (112). However, estrous females may undertake brief breeding excursions outside of their normal range, apparently in search of a suitable breeding partner (113).

Home range size likewise varies by sex and age of the individual, habitat, and season. Reported home range sizes vary greatly over regions of the country (Table 29–1). Annual home range size of adult females is about 50% that of adult males.

Seasonal Migrations

In the northern portion of the whitetail's range, there are often pronounced seasonal migrations in response to snow depth and cold weather (114, 115). Migrations to wintering habitat generally are less than 16 kilometers, but migrations in excess of 90 kilometers have been reported from the Upper Peninsula of Michigan (116). Because related females may band together during autumn, fawns learn migration routes from female relatives (117). In individual deer yards, deer tend to represent distinct subpopulations or demes with deer from each yard occupying different summer ranges (118).

During winter, deer congregate in areas of coniferous forest that provide thermal cover, reduced wind velocities, and decreased snow depths compared to adjacent areas. These winter yards may support from a few deer to several hundred. The Mead Deer Yard in Upper Michigan supported an estimated 43,000 deer on 69,300 hectares during winter 1987 (119). Travel within yards often is confined to heavily used trails that minimize energy drain. Mature northern white cedar is the preferred cover species in deer yards because it provides a narrow thermal range, firm snow pack, and significantly reduced wind speeds. In addition, it is a preferred and nutritious winter browse (119). Unfortunately, the thick coniferous cover and heavy deer use combine to limit available browse in deer yards.

TABLE 29-1

Reported Home Range Size for White-Tailed Deer in Various Parts of Its Range by Sex and Age Class

Location	Animal (sex/age)	No.	Home Range	Reference
Mississippi	Female	4	747 ha ± 219 SD	204
	Male	5	1511 ha ± 571 SD	
Nebraska	Female	14	170 ha (C.I. = 38)	205
Florida	Male	10	7.0 km^2 ± 1.4 SE	206
		23	29 km^2 ± 0.4 SE	
Florida	Female (yearling)	7	2458 ha	207
	Female (adult)	6	344 ha	
	Male	5	701 ha	
Washington	Female	18	103.6 ha ± 16.5 SE	208
	Male	7	208.6 ha ± 24.6 SE	
Kentucky	Female	6	642 ha ± 132 SE	209
Montana	Female (adult)	11	71 ha ± 18.4 SE	210
	Female (yearling)	5	91 ha ± 30.4 SE	
Illinois	Female	7	50.8 ha ± 23 SE	211
Michigan	Female		45 ha	109
	Male		142 ha	
New York	Female (summer)	64	221 ha ± 19.0 SE	212
	Female (winter)	45	132 ha ± 18.3 SE	
	Male (summer)	34	233 ha ± 23.4 SE	
	Male (winter)	12	150 ha ± 31.6 SE	
Texas	Female	27	84 ha ± SE	213
		14	139 ha + 37 SE	
Wisconsin	Female	15	178 ha ± 102 SD	214

Deer leave yarding areas for summer ranges as soon as snowmelt advances sufficiently to allow ease of travel and daytime temperatures are consistently above freezing (120). Spring migration is more direct and faster than fall migration (121). Fawns must be led back to their ancestral grounds by their dam or other adult female, as with the fall migration. Fawns orphaned overwinter wander aimlessly during spring and summer (120).

Although migrations are most pronounced on northern ranges, seasonal shifts in activity centers have been reported in other regions. In an Illinois study, 20% of adult females during winter migrated an average of 13 kilometers away in spring and returned in fall. In Nebraska, 13% of females migrated seasonally (100). Kammermeyer and Marchinton reported seasonal shifts onto a refuge in northwestern Georgia in response to habitat conditions and hunting pressure (122). Movements in response to seasonal flooding are common in southern river swamps (123, 124).

Communication

White-tailed deer use a variety of vocalizations, visual displays, and olfactory cues to communicate. Several vocalizations have been identified and grouped into alarm/distress calls, agonistic calls, maternal/neonatal calls, and mating calls (125). Alarm calls include the characteristic snort produced by an alerted deer, particularly adults, and the bawl produced by distressed deer of both sexes and all ages. Agonistic calls include the grunt, the grunt-snort, and the grunt-snort-wheeze. These are emitted during progressively intense agonistic interactions between males. Various other grunts include maternal grunts directed at fawns, tending grunts given by males tending an estrous female, and a cohesive call given by females when separated from a group (126). Care-soliciting calls produced by fawns include the nursing whine, mew, and bleat.

Subtle visual cues are used to communicate dominance status, aggressive intent, and alarm. Low-level threats are signaled by an ear drop in which the ears are laid back along the neck coupled with a direct stare by the aggressor. The head-high and head-low threats are more aggressive and signal intent to strike or chase (49). Dominance displays by a rutting male include the ear drop, erection of body hairs, and flaring of the preorbital gland and nostrils (127). Upon detection of potential danger, a deer often will stomp one or alternating forefeet. When fleeing, whitetails expose the white underside of their tails to maintain group cohesiveness (128).

Olfaction likely is the most important communicative route in white-tailed deer. Whitetails possess a variety of specialized skin glands including the nasal, forehead, preorbital, tarsal, metatarsal, interdigital, and preputial glands. Although the forehead, tarsal, and interdigital glands are known to produce socially significant odors, little is known of the communicative significance of the other glands.

Compounds arising from the interdigital gland presumably mark a trail while a deer walks and convey home range familiarity, although chemical analyses of interdigital secretions suggest that dominance status likewise may be conveyed (129). The forehead gland is most active during the breeding season and is used in marking rubs and overhanging branches at scrape sites (130, 131). Whitetails obtain information on individual identity, dominance status, physical condition, and reproductive status from the tarsal glands. Odors arise from interactions among glandular secretions, urine deposited during a behavior called rub-urination, and bacteria residing on the tarsal hairs (132, 133).

Reproductive Behavior

The breeding season consists of several phases. Following velvet shed in late summer or early fall, males engage in sparring contests in preparation for rut. Rutting activity is mediated by changes in photoperiod, although there is considerable geographic variation in the timing of breeding, particularly in southerly regions (134, 135). In most regions, shortening day length triggers a chain of hormonal events initiated by the pineal gland that ensures conception occurs in the fall. In northern regions, photoperiod changes more abruptly than in southern latitudes. A short, precisely timed breeding season is necessary in northern climates so fawns have optimal nutrition and ample time for growth prior to harsh winter weather. In southern regions, seasonal climatic fluctuations are less severe and reproductive timing is less critical. In these regions, changing photoperiod apparently opens a reproductive window when breeding can occur. Exactly when breeding occurs varies according to a variety of factors, including genetics, herd demographics, nutrition, and likely traditional physiologic cues passed from mother to daughter (36, 66). In southern states, peak breeding may occur in September in portions of South Carolina, November in Georgia, and December to January in portions of Alabama and Mississippi. In peninsular Florida, breeding occurs during mid-summer, and in Central and South America, deer reproduce year-round (136).

Deer use a variety of signposts to communicate throughout the year. During spring and summer, males communicate via communal licking branches located near trails or at field borders (137). Following velvet shedding and throughout the breeding season, males make antler rubs by debarking small trees and shrubs and anointing them with secretions from the forehead gland. Other males and females respond to antler rubs less than 3 days old by smelling, licking, or rubbing (138). In producing a scrape, a male marks an overhanging branch, paws a shallow depression underneath, and then rub-urinates over the pawed depression (139).

Peak scraping activity typically occurs 2 weeks in advance of peak breeding (139). Although scrape sites may be defended in the presence of a dominant male, scrapes may be visited and remarked by several males during the breeding season (140, 141).

Males detect estrus via visual and olfactory cues. Volatile compounds produced in the female's reproductive tract apparently are a source of sexually attractive odors (142). Urinary compounds of low volatility likely are analyzed by the vomeronasal system with a behavior called flehmen, although this olfactory route does not appear to detect reproductive attractiveness (143). Anosmic males can detect estrous females by behavioral cues, whereas males with occluded vomeronasal organs are attracted to estrous females even in the absence of behavioral cues (144). Vomerofaction may mediate physiologic changes that ensure males and females are simultaneously in peak reproductive condition, but this concept is speculative (127).

POPULATION DYNAMICS

Population Trends and Current Status

The dynamics of white-tailed deer populations have changed since European colonization of North America. McCabe and McCabe theorized four major stages of population status in the United States and Canada from 1500 through 1980 (2). The first stage, lasting from 1500 to 1800, was characterized by massive harvest but minimal habitat modifications. This stage saw a 50% to 65% reduction in a population estimated to be 23 to 34 million deer in 1500. Deer numbers rebounded modestly during the second stage, 1800 to 1865, due to their resurgence in abandoned farmland in the eastern United States and their expansion into new habitats elsewhere. The third stage, the latter half of the 1800s known as the exploitation era, saw a decline from about 13 million animals to perhaps as low as 350,000. Deer populations were devastated during this stage because intensive harvest by market hunters was magnified by a lack of sanctuaries and negative land use impacts on habitat. Protection and restoration efforts during the fourth stage, 1900 to 1980s, have proven highly successful. McCabe and McCabe theorized that white-tailed deer populations have returned to their pre-exploitation era levels (2).

It was not long after intensive restoration efforts began in the 1930s that concern was expressed about the size of deer populations and the need to inhibit population growth rates (145). White-tailed deer have proven themselves to be highly successful at surviving and propagating under a wide range of habitat conditions. Estimates for 45 states indicate a United States population exceeding 26 million deer in 1993, which rivals estimates of the pre-1500 population (10).

Regulatory agencies are faced with increasingly complex management scenarios as they look for effective methods to control this expanding resource. Recreational harvest remains the most cost-effective means, although alternative strategies involving fertility control, predator reintroduction, fencing and repellents, and use of professional shooters are being evaluated (10). Declining public involvement in recreational hunting indicates additional problems for hunting-based solutions in the future. Managing our burgeoning urban deer populations will be particularly challenging as regulatory agencies and stakeholders grapple to develop acceptable management strategies (146).

Population Productivity

Population productivity is the result of the difference between potential productivity and mortality. Population growth occurs when reproduction exceeds mortality. Although conceptually simple, myriad influences affect potential productivity and mortality.

Reproductive parameters including timing of the breeding season, fertility rates, conception rates, age at first breeding, and sex ratio of offspring can be altered by population density, habitat conditions, and perhaps genetics (147). Females over 2.5 years of age on good range typically have pregnancy rates greater than 90% and reproductive rates of 1.3 to 2.0 embryos per female (148). Typical reproductive rates (fetuses per adult female) include 1.7 in Mississippi and South Carolina, 1.8 to 2.0 in Missouri, and 1.9 in Illinois (149, 62, 150, 151).

On high-quality habitats, many females will breed at 6 months of age. Reported fawn breeding rates range from 0% in Montana, to 27% to 41% in North Carolina, Illinois, South Carolina, and Missouri, and up to 65% in Michigan (152, 153, 151, 154, 150, 155). Conversely, on inherently poorer ranges such as the Southern Coastal Plain, or in overpopulated herds, some females may not breed until 2.5 years of age (147).

As population density increases toward the carrying capacity of the habitat, pregnancy rates and reproductive rates of adults decline. In a Michigan study, productivity of females on poor nutrition was only one-third that of females on good nutrition (60). Stillbirth and postnatal mortality, possibly due to fawn abandonment and imprinting failure by malnourished females, likewise increase as herd density increases (97).

In poor-quality habitats, females may have substantially lower *in utero* conception rates. For example, number of fetuses per pregnant adult female was 1.3 in various regions of Florida and 1.1 on some southeastern coastal islands (135, 156). In some of these poor-quality habitats, forage quality is limited and reproductive parameters apparently are not affected by changes in herd density (157).

In most habitats, reproduction decreases and mortality increases as the population density increases toward the carrying capacity. However, density-dependent effects often lag behind relative density, allowing populations to exceed the carrying capacity (158). These population irruptions result in significant mortality and habitat damage.

Mortality Factors

Predation. Historically, wolves and mountain lions were the primary predators of white-tailed deer; bobcats, black bears, and coyotes likely were incidental predators concentrating on fawns (20). Where they occur, wolves kill whitetails year-round and constitute an important mortality factor (117, 159). Although an average annual kill rate is 20 adult deer per wolf, the degree to which wolves limit deer populations is not clear (160, 161). In summer, wolves prey on fawns, and in winter they kill primarily fawns, adults greater than 5 years old, deer with abnormalities, and adult males weakened by rutting activities (159, 160).

Although there is little information on predation of whitetails by mountain lions, bobcats, and black bears, where their ranges overlap, whitetails are consumed. Black bear predation likely is limited to opportunistic consumption of young fawns and sick or debilitated animals. Similarly, bobcat predation is limited, although high incidences of predation have been reported (162).

Extirpation of the mountain lion and gray wolf from much of their former range has allowed the coyote to expand its range eastward to New Brunswick and Maine and south to the Southeastern states. The coyote can be an important predator of whitetails. In northern regions, coyote predation increases during winter when deep snow increases prey vulnerability. When whitetails are disadvantaged by winter conditions, adult coyotes may feed almost exclusively on fresh deer kills and do not restrict predation to malnourished or injured animals (163, 164). Significant coyote predation on adult white-tailed deer occurs in South Texas (165).

Diseases and Parasites. White-tailed deer are susceptible to several diseases and are hosts to numerous parasites. Deer tolerate most of these diseases and parasites well, and in most populations, disease-related mortality is limited. A major exception is hemorrhagic disease caused by the epizootic hemorrhagic disease or bluetongue virus. Five serotypes of bluetongue and two serotypes of epizootic hemorrhagic disease have been detected in the United States. Mortality from this disease occurs in the South every year. Mortality rates typically are less than 15%, although rates greater than 50% have been reported (166). Virus transmission occurs in late summer and fall and coincides with peak populations of biting midges that serve as the vector. Viral exposure and serotype diversity is greater in southern states, whereas disease severity increases with latitude (166). Genetics may influence resistance to hemorrhagic disease. In a Mississippi study, deer with northern origins were less resistant to hemorrhagic disease than deer of southern origin (36).

More than 100 species of internal or external parasites, including protozoans, flukes, nematodes, tapeworms, and arthropods have been reported (167). Highest parasite diversity occurs in the southeastern states. However, heavy parasite loads commonly are associated with nutritionally stressed herds regardless of whether the stress results from inherently poor-quality habitat or habitat degraded by overbrowsing.

The internal nematodes and the external arthropods tend to be the most important groups of parasites. The large stomach worm and the large lungworm are the two most pathogenic helminths. The large lungworm is prevalent across the southeastern United States, with young deer and males most susceptible (168). Other helminth parasites of importance in this region are the liver fluke, the meningeal worm, the muscle worm, the arterial worm, and stomach worms.

Parasitic arthropods include ticks, deer keds, lice, and bot flies (167). Tick infestation is rarely a major cause of mortality in deer, although in some areas, significant fawn mortality has been attributed to lone-star tick infestations. The black-legged tick is a vector of the agents of Lyme disease and human granulocytic ehrlichiosis, whereas the lone-star tick is the vector of the rickettsial disease, human monocytic ehrlichiosis. Other external arthropods are seldom of significant health consequences to whitetails.

Automobile Accidents. Vehicle collisions can be a significant source of mortality in some regions, and a major human safety concern. The national deer road-kill for 1991 was at least 500,000 deer (169). Average damage cost per collision was estimated at $1,600 in Michigan in 1986, $1,881 in Vermont during the 1980s, and $1,415 in New York in late 1980s (169). Most accidents occur at dawn, dusk, or after dark with peaks at sunrise and 2 hours after sunset (170). Seasonal peaks coincided with rutting activities and hunter disturbance (170).

Winter Mortality. Overt starvation or chronic influences of inadequate nutrition can be responsible for mortality in many regions where population density exceeds habitat carrying capacity (Figure 29–6). In Texas, starvation was a significant mortality factor, resulting in 16% natural mortality of deer during a 6-year period (171). However, winter mortality is greatest in northern ranges. In northern Michigan, winter mortality was estimated at 117,000 deer in 1994 and 125,000 in 1986 (119). Starvation

Figure 29–6 Physical development is retarded and mortality rates increase in most habitats when population density exceeds the habitat's ability to fulfill nutritional requirements. (Photograph by J. J. Jackley.)

rates are influenced by population density relative to carrying capacity, snow depth, and duration of snow cover (172, 173, 119). A winter severity index that incorporates wind chill, snow depth, and snow support ability is highly correlated with spring counts of winter-killed deer (174, 175). Approximately two-thirds of winter-killed deer in the Adirondack region of New York are fawns (176). Severe winter weather in December is a major determinant of fawn survival overwinter, whereas prolonged winter conditions into April affect newborn fawn survival (177).

POPULATION MANAGEMENT

Population management of white-tailed deer involves manipulating harvest rates to adjust density, sex ratio, and/or age structure to fulfill management goals. Management goals vary depending on the landowners (e.g., private lands versus public lands), conflicting land uses (e.g., urban–rural interface, agricultural crops depredated by deer), and human perspectives (e.g., recreational hunters versus nonhunters).

Increased involvement from a variety of interest groups has complicated deer population management in most locations. Gone are the days when management objectives were as simple as increasing populations using translocation to generate a surplus for recreational harvest.

White-tailed deer are keystone herbivores, with significant impacts on forest ecology (12). Their ecosystem-level impact guarantees involvement of varied stakeholders in the development of population management goals. Overabundance at the rural–urban interface has caused many problems because traditional hunting-based deer management strategies may not be acceptable in many suburban areas (146). There are widely divergent opinions on appropriate management approaches among the public sectors and even within natural resource professionals, the latter divergence being of fairly recent origin (178).

White-tailed deer populations have wide flexibility inherent within their basic biology relative to how their population characteristics should be manipulated. For example, their polygynous breeding strategy allows flexibility as to the sex ratio. The sex ratio goal of management thus becomes a choice of the decision makers, geared to the preferred population composition. Managers attempting to maximize the prevalence of mature males in the population typically target one male per one to two adult females.

Population management of white-tailed deer is based on the population growth models, such as those developed with data from the George Reserve (94). Density-dependent responses to population reduction are expected within much of their range, but exceptions do occur. Soil fertility is so low in some areas of Florida that abundant, low-quality forage limits population productivity equally within a wide range of density (157).

Humans manipulate deer population density to fulfill one or more human-oriented management objectives. Most often these management approaches involve harvest by recreational hunters. Nonlethal methods using immunocontraception are not yet applicable to population-level management, although remotely delivered, abortion-producing prostaglandin has potential for localized areas (179, 180). For the vast majority of deer management cases, recreational harvest remains the only practical and the most desirable option.

Selective manipulation of harvest rates among age classes and sexes allows managers to alter sex ratio and age structure. There is a growing realization within the hunting public that they are *de facto* managers, and their harvest selection impacts herd structure and social behavior. Male-only harvest strategies were appropriate during the restoration phase of deer management. Occasional male-only harvest regulations still may be "politically appropriate" in areas, such as northern Wisconsin, with periodic (about every 3–4 years) high natural winter mortality; protection of females from harvest allows a rapid recovery to prescribed population densities (K. McCaffery,

Wisconsin Department of Natural Resources, personal communication). A selective harvest strategy designed to balance sex ratio and increase the prevalence of older males has grown from its origin in Texas and is becoming increasingly accepted (181, 182).

The future of recreational hunting may lie in the ability of the hunting public to convince the nonhunting majority that they are willing and able managers of this resource. Proactive involvement in the setting of population management objectives followed by ethically appropriate actions in the field should ensure that recreational harvest remains a viable management option for deer population management.

HABITAT REQUIREMENTS

White-tailed deer have amazingly flexible habitat requirements that allow them to live successfully within a wide range of habitat types. Their basic requirements include adequate nutrition, escape cover, thermal cover in extreme environments, and water during periods when moisture in foods is insufficient. In general, the distribution of the white-tailed deer in North America is limited only by deep snow, extreme cold, desert, large open spaces devoid of escape cover, and mountainous terrain. In the West, the white-tailed deer is found most often along riverine corridors. Thermal protection and shelter from deep snow provided by mature conifers with dense canopies are important habitat components in its north-central and northeastern range. In dry environments, white-tailed deer characteristically inhabit brush country and riparian habitat. The white-tailed deer thrives in agricultural and forested areas that have adequate amounts of early successional stage habitats consisting of an abundance of woody and herbaceous forage. Specific and detailed descriptions of habitat for white-tailed deer throughout their range are provided by Halls (183).

HABITAT MANAGEMENT

Population control is an essential component of habitat management. In many regions, unchecked deer populations may destroy their own habitat and that of other wildlife species. Not only are the nutritional resources of deer removed by overpopulation, but also important escape and thermal cover. Thus, regardless of other management practices, it is important to maintain deer populations below the K-carrying capacity of the environment to prevent range damage. The amount below K-carrying capacity that should be targeted depends on relationships with human land use decisions. When management decisions are based primarily on human-based values, the target density is considered "optimum carrying capacity."

Aside from population control, forest management has the most important influence on deer populations over much of their range. The production of understory vegetation is related directly to a number of factors including site quality, forest type, and stand age and structure (184). Abundant forage is produced in early succession stands after cutting. Timber harvest allows sunlight to reach the forest floor, creates soil disturbance, and stimulates the production of desirable herbaceous vegetation, woody vines, and other browse species that are of high nutritional value to deer and provide escape cover. Post-harvest regeneration treatments also can significantly alter deer habitat. A selective herbicide can eliminate undesirable trees, grasses, and woody vegetation and promote desirable legumes and vines (185). Fertilizer can be used to promote growth and increase the nutritional qualities of native vegetation and agricultural plantings.

Production of understory vegetation decreases as stands close although production typically will increase slightly as stands reach sawtimber size and natural mortality thins the canopy. Silvicultural practices, which emphasize interspersion and juxtaposition of different age forest stands, generally optimize deer habitat.

Where they occur, acorns can be an important component of the fall and winter diet of deer. Kammermeyer and Thackston recommend maintaining at least 20% of an area in mast-producing hardwoods (186). Because most oaks do not produce substantial amounts of acorns until they are at least 40 years old, rotation lengths of 100 years or more may be appropriate for some stands (187). Selection cuts that promote desirable mast production and crown growth of white and red oaks, persimmons, pecans, and other mast also are desirable.

In northern areas providing critical winter cover, vegetation should include dense stands of mature conifer cover to interrupt snowfall and provide yarding areas and timber stands with lighter cuts on shorter intervals for hardwood browse production (188). Northern white cedar and hemlock are the preferred winter cover type for yarding deer. Ideally these will have about 70% canopy closure and cover at least 50% or more of an area greater than 3 square kilometers (188, 189). Effective management of critical winter habitat involves development of cooperative timber management agreements (190).

In certain forest types, fire can be an important management tool. For example, in some Southern pine types, fire increases browse production and palatability, soft mast production, nutritional value of some forages, and the abundance of some important forbs (191, 192, 193, 194, 195). Controlled burning also may reduce parasite abundance, particularly immature stages of ticks and helminths. Most pine species and some oaks, such as post oak, have thick bark or cambium layers that allow them to survive mild understory fires without damage. Prescribed fire has also been shown to benefit deer habitats in Minnesota, Wisconsin, western states, and elsewhere (196, 197, 184).

Controlled fire also may promote establishment of less desirable vegetative types. For example, hot fires in post oak vegetation can lead to grassland savanna rather than a brush understory. In the pine flatlands of the Southern Piedmont, repeated prescribed fire can result in a dense understory of saw palmetto and loss of desirable browse species. Addition of mechanical disturbances, such as roller chopping, in these pine flatlands can prevent the understories from being dominated by palmetto and promote more desirable plant species.

In dry or drought-prone environments, addition of water holes or catchments is assumed to expand the usable habitat for deer, although tests of this assumption are lacking. One water source for each 150 to 200 hectares likely is sufficient.

Agricultural plantings (i.e., food plots) can increase deer numbers, harvest rates, condition indices, and reproductive performance, and also alter movements, core areas, and home ranges (198, 199, 200, 201). Because well-managed food plots may produce up to 12,800 kilograms of dry forage per hectare, and forage consumption may exceed 5,300 kilograms per hectare, as little as 1% of an area in high-quality plots may provide substantial benefits to a deer herd (202, 203, 198).

The species selected for food plots should be adaptable to the site and region, easy to establish, affordable, and provide abundant forage at the appropriate time of the year. Fall plantings of winter wheat, rye, oats, or ryegrass along with clovers or alfalfa help meet protein requirements and provide a highly digestible diet of high caloric content. Corn, grain sorghum, or plantings of soybeans, iron and clay peas, and peanuts also provide highly nutritious supplements. Kammermeyer and Thackston provide a detailed discussion of agricultural plantings for deer in various regions of the whitetailed deer's range (186).

SUMMARY

From exploitation to restoration, deer populations have been a direct function of human use of the resource. Currently, deer have tremendous but conflicting economic and ecological values. Future deer management will require an understanding of the ecological and social interactions of deer with humans and the environment.

Whitetails are one of the most widely distributed mammals, with 38 subspecies. There is wide geographic variation in phenotypic characteristics. Whitetails selectively forage for high-quality, easily digestible plant parts. Forage selection changes seasonally as forage quality changes, but browse, mast, and forbs will combine in varying amounts to be the predominant forage classes in their diets. Agricultural crops are used heavily when available.

The seasonally polyestrous breeding activity of white-tailed deer is synchronized by photoperiod, but allows considerable temporal variation in breeding dates. The most energy-demanding phases of the reproductive cycle coincide with the period of maximum food abundance and quality, but the exact timing is affected primarily by nutrition and sex ratio. Prevalence of fawn breeding is a sensitive indicator of habitat quality. Managers estimate prevalence of fawn breeding by measuring lactation prevalence in 1.5-year-old harvested females.

Physiologic patterns are timed to optimize survival. Adaptations to temperate climate include a highly insulative coat, behavioral and habitat choices to conserve energy, and reduced thyroid function (i.e., reduced metabolic rate). Fat deposition during autumn ensures that animals deposit energy stores prior to the occurrence of resource restriction.

Home range size varies by sex and age, habitat, and season. Annual home range size of adult females is approximately 50% that of adult males. Although migrations are most pronounced on northern ranges, seasonal shifts in activity centers have been reported in other regions. During northern winters, deer congregate in areas of coniferous forest that provide thermal cover, reduced wind velocities, and decreased snow depths compared to adjacent areas.

Population productivity is simply the potential productivity minus mortality. Population growth occurs when reproduction exceeds mortality. Although conceptually simple, myriad influences affect both potential productivity and mortality. As population density increases toward the carrying capacity of the habitat, pregnancy rates and reproductive rates of adults decline. In most habitats, reproduction decreases and mortality increases as the population density increases toward the carrying capacity.

Coyotes can be an important predator of whitetails. In most populations disease-related mortality is limited; a major exception is hemorrhagic disease. Vehicle collisions cause significant mortality and are a major human safety concern. Overt starvation or chronic influences of inadequate nutrition cause mortality, especially in northern areas.

Population management involves manipulating harvest rates to adjust density, sex ratio, and/or age structure to fulfill management goals. Management goals vary widely depending on the landowners, conflicting land uses, and human perspectives. Density-dependent responses to population reduction generally are expected within much of their range, but exceptions occur.

The habitat requirements of white-tailed deer are few. This animal is amazingly flexible in its habitat needs. Habitat management concepts revolve around provision of cover and nutrition.

LITERATURE CITED

1. EMERSON, T. E. 1980. A stable white-tailed deer population model and its implications for interpreting prehistoric hunting patterns. *Mid-continental Journal of Archeology.* 5:117–127.

2. McCABE, R. E., and T. R. McCABE. 1984. Of slings and arrows: A historical retrospection. Pages 19–72. *in* L. K. Halls, ed., *White-tailed Deer: Ecology and Management.* Stackpole Books, Harrisburg, PA.

3. LEOPOLD, A. 1933. *Game Management.* Charles Scribner's Sons, New York.

4. HARLOW, R. F., and F. K. JONES, Jr. 1965. Reproduction. Pages 108–124 *in* R. F. Harlow and F. K. Jones, Jr., eds., The white-tailed deer in Florida. Technical Bulletin Number 9. Florida Game and Fresh Water Fish Commission, Tallahassee, FL.

5. YOUNG, S. P. 1956. The deer, the Indians and the American pioneers. Pages 1–27 *in* W. P. Taylor, ed., *Deer of North America.* Stackpole Books, Harrisburg, PA.

6. TREFETHEN, J. B. 1975. *An American Crusade for Wildlife*. Boone and Crockett Club. Winchester Press, New York.

7. DOWNING, R. L. 1987. Success story: White-tailed deer. Pages 45–57 *in Restoring America's Wildlife*. U.S. Department of Interior, Fish and Wildlife Service, Washington, DC.

8. WILLIAMSON, L. L. 1987. Evolution of a landmark law. Pages 1–17 *in Restoring America's Wildlife*. U.S. Department of Interior, Fish and Wildlife Service, Washington, DC.

9. CONOVER, M. R. 1997. Monetary and intangible valuation of deer in the United States. *Wildlife Society Bulletin* 25:298–305.

10. JACOBSON, H. A., and J. C. KROLL. 1994. The white-tailed deer *(Odocoileus virginianus)*—"The most managed and mismanaged species." Pages 25–35 *in* J. A. Milne, ed. *Recent Developments in Deer Biology. Proceedings of the Third International Congress on the Biology of Deer*. Macaulay Land Use Research Institute, Craigiebuckler, Aberdeen, UK, and Moredun Research Institute, Edinburgh, United Kingdom.

11. CONOVER, M. R., W. C. PITT, K. K. KESSLER, T. J. DuBow, and W. A. SANBORN. 1995. Review of human injuries, illnesses, and economic losses caused by wildlife in the United States. *Wildlife Society Bulletin* 25:407–414.

12. WALLER, D. M., and W. S. ALVERSON. 1997. The white-tailed deer: A keystone herbivore. *Wildlife Society Bulletin*. 25:217–226.

13. STROMAYER, K. A., and R. J. WARREN. 1997. Are overabundant deer herds in the eastern United States creating alternate stable states in forest plant communities? *Wildlife Society Bulletin* 25:227–234.

14. WITMER, G. W., and D. S. DeCALESTA. 1992. The need and difficulty of bringing the Pennsylvania deer herd under control. *Proceedings of the Eastern Wildlife Damage Control Conference* 5:130–137.

15. SCHMITZ, O. J., and A. R. E. SINCLAIR. 1997. Rethinking the role of deer in forest ecosystem dynamics. Pages 201–223 *in* W. J. McShea, H. B. Underwood, and J. H. Rappole, eds., *The Science of Overabundance: Deer Ecology and Population Management*. Smithsonian Institution Press, Washington, DC.

16. ANDERSON, R. C., and O. L. LOUCKS. 1979. White-tailed deer *(Odocoileus virginianus)* influence on structure and composition *of Tsuga canadensis* forests. *Journal of Applied Ecology* 16:855–861.

17. DeCALESTA, D. S. 1994. Effect of white-tailed deer on songbirds within managed forests in Pennsylvania. *Journal of Wildlife Management* 58:711–718.

18. McSHEA, W. J., and J. H. RAPPOLE. 1997. Herbivores and the ecology of forest understory birds. Pages 298–309 *in* W. J. McShea, H. B. Underwood, and J. H. Rappole, eds., *The Science of Overabundance: Deer Ecology and Population Management*. Smithsonian Institution Press, Washington, DC.

19. BAKER, R. W. 1984. Origin, classification and distribution. Pages 1–18 *in* L. K. Halls, ed., *White-tailed Deer: Ecology and Management*. Stackpole Books, Harrisburg, PA.

20. SMITH, W. P. 1991. *Odocoileus virginianus*. *Mammalian Species*. No. 388.

21. HARRIS, L. H. 1984. New Zealand. 1984. Pages 547–556 *in* Halls, L. K. ed. *White-tailed Deer: Ecology and Management*. Stackpole Books, Harrisburg, PA.

22. BOJOVIC, D., and L. K. HALLS. 1984. Central Europe. Pages 557–560 *in* L. K. Halls, ed., *White-tailed Deer: Ecology and Management*. Stackpole Books, Harrisburg, PA.

23. NYGREN, K. F. A. 1984. Finland. Pages 561–570 *in* Halls, L. K. ed., *White-tailed Deer: Ecology and Management*. Stackpole Books, Harrisburg, PA.

24. McDONALD, J. S., and K. V. MILLER. 1993. A history of white-tailed deer restocking in the United States—1878 to 1992. Research Publication 93-1. Quality Deer Management Association. Greenwood, SC.

25. FASHINGBAUER, B. A., J. M. IDSTROM, C. KINSEY, W. H. PETRABORG, D. W. BURCAPOW, and F. B. Lee. 1965. Big game in Minnesota. Technical Bulletin Number 9. Minnesota Department of Conservation, St. Paul, MN.

26. JACOBSON, H. A. 1995. Age and quality relationships. Pages 103–111 *in* K. V. Miller and R. L. Marchinton, eds., *Quality Whitetails*. Stackpole Books. Mechanicsburg, PA.

27. GEIST, V. 1994. Origin of the species. Pages 2–16 *in* D. Gerlach, S. Atwater, and J. Schnell, eds., *Deer*. Stackpole Books, Mechanicsburg, PA.

28. HALLS, L. K. 1978. White-tailed deer. Pages 43–66 *in* J. L. Schmidt, and D. L. Gilbert, eds., *Big Game of North America: Ecology and Management*. Stackpole Books, Harrisburg, PA.

29. SAUER, P. R. 1984. Physical characteristics. Pages 73–90 *in* L. K. Halls, ed., *White-Tailed Deer: Ecology and Management*. Stackpole Books, Harrisburg, PA.

30. SEVERINGHAUS, C. W. 1949. Tooth development and wear as criteria of age in white-tailed deer. *Journal of Wildlife Management* 13:195–216.

31. GILBERT, F. F. 1966. Aging white-tailed deer by annuli in the cementum of the first incisor. *Journal of Wildlife Management* 30:200–202.

31a. DEMARAIS, S. 1998. Managing for antler production: Understanding the age–nutrition–genetic interaction. Pages 33–36 *in* D. Rollins, ed., *The Role of Genetics in White-Tailed Management*. Texas A&M University, College Station.

32. WILLIAMS, J. D., W. F. KRUEGER, and D. H. HARMEL. 1994. Heritabilities for antler characteristics and body weight in yearling white-tailed deer. *Heredity* 73:78–83.

33. LUKEFAHR, S. D., and H. A. JACOBSON. 1998. Variance component analysis and heritability of antler traits in white-tailed deer. *Journal of Wildlife Management* 62:262–268.

34. MARCHINTON, R. L., K. V. MILLER, and J. S. MCDONALD. 1995. Genetics. Pages 169–192 *in* K. V. Miller, and R. L. Marchington, eds., *Quality Whitetails*. Stackpole Books, Harrisburg, PA.

35. WHITE, A. K. 1986. A comparison of phenotypic and genotypic characteristics of captive white-tailed deer *(Odocoileus virginianus)* and their offspring. Master's Thesis, Mississippi State University, Starkville.

36. JACOBSON, H. A., and S. D. LUKEFAHR. 1998. Genetics research on captive deer at Mississippi State University. Pages 46–50 *in* D. Rollins, ed., *The Role of Genetics in White-Tailed Deer Management*. Texas A&M University, College Station.

37. HOFFMAN, R. R. 1985. Digestive physiology of the deer—Their morphophysiological specialisation and adaptation. Pages 393–407 *in* P. F. Fennessy and K. R. Drew, eds., *Biology of Deer Production*. Bulletin 22. Royal Society of New Zealand.

38. NUDDS, T. D. 1980. Forage "preference": Theoretical considerations of diet selection by deer. *Journal of Wildlife Management* 44:735–740.

39. GEE, K. L., M. D. PORTER, S. DEMARAIS, F. C. BRYANT, and G. VAN VREEDE. 1994. *White-Tailed Deer: Their Foods and Management in the Cross Timbers*, 2nd ed. NF-WF-94-01, Samuel Roberts Noble Foundation, Ardmore, OK.

40. VANGILDER, L. D., O. TORGERSON, and W. R. PORATH. 1982. Factors influencing diet selection by white-tailed deer. *Journal of Wildlife Management* 46:711–718.

41. STORMER, F. A., and W. A. BAUER. 1980. Summer forage use by tame deer in northern Michigan. *Journal of Wildlife Management* 44:98–106.

42. CRAWFORD, H. S. 1982. Seasonal food selection and digestibility by tame white-tailed deer in central Maine. *Journal of Wildlife Management* 46:974–982.

43. THILL, R. E., H. F. MORRIS, Jr., and A. T. HARREL. 1990. Nutritional quality of deer diets from southern pine-hardwood forests. *American Midland Naturalist* 124:413–417.

44. BLAIR, R. M., R. ALCANIZ, and A. HARRELL. 1983. Shade intensity influences the nutrient quality and digestibility of southern deer browse leaves. *Journal of Range Management* 36:257–264.

45. WENTWORTH, J. M., A. S. JOHNSON, and P. E. HALE. 1990. Influence of acorn use on deer in the Southern Appalachians. *Proceedings Annual Conference of the Southeastern Association of Fish and Wildlife Agencies*. 44:142–154.

46. BRYANT, F. C., C. A. TAYLOR, and L. B. MERRILL. 1981. White-tailed deer diets from pastures in excellent and poor range condition. *Journal of Range Management* 34:193–200.

47. STRANSKY, J. J., and L. K. HALLS. 1980. Fruiting of woody plants affected by site preparation and prior land use. *Journal of Wildlife Management* 44:258–263.

48. LINCOLN, G. A. 1992. Biology of seasonal breeding in deer. Pages 565–574 *in* R. D. Brown, ed., *The Biology of Deer*. Springer-Verlag, New York.

49. HIRTH, D. H. 1973. Social behavior of white-tailed deer in relation to habitat. Dissertation. University of Michigan, Ann Arbor.

50. WEBB, J. W., and D. W. NELLIS. 1981. Reproductive cycle of white-tailed deer of St. Croix, Virgin Islands. *Journal of Wildlife Management* 45:253–258.

51. CHEATUM, E. L., and G. H. MORTON. 1946. Breeding season of white-tailed deer in New York. *Journal of Wildlife Management* 10:249–263.

52. KNOX, W. M., K. V. MILLER, and R. L. MARCHINTON. 1988. Recurrent estrous cycles in white-tailed deer. *Journal of Mammalogy* 69:384–386.

53. SEVERINGHAUS, C. W., and E. L. CHEATUM. 1956. Life and times of the white-tailed deer. Pages 57–187 *in* W. P. Taylor, ed., *The Deer of North America*. Stackpole Books, Harrisburg, PA.

54. HAUGEN, A. O. 1959. Breeding records of captive white-tailed deer in Alabama. *Journal of Mammalogy* 40:108–113.

55. VERME, L. J. 1965. Reproduction studies on penned white-tailed deer. *Journal of Wildlife Management* 29:74–79.

56. HAUGEN, A. O., and L. A. DAVENPORT. 1950. Breeding records of white-tailed deer in the Upper Peninsula of Michigan. *Journal of Wildlife Management* 14:290–295.

57. ADAMS, W. H., Jr. 1960. Population ecology of white-tailed deer in northeastern Alabama. *Ecology* 41:706–715.

58. MOEN, A. N. 1973. *Wildlife Ecology: An Analytic Approach*. W. H. Freeman and Company, San Francisco, CA.

59. FOLK, M. J., and W. D. KLIMSTRA. 1991. Reproductive performance of female Key deer. *Journal of Wildlife Management* 55:386–390.

60. VERME, L. J. 1967. Influence of experimental diets on white-tailed deer reproduction. *Transactions North American Wildlife and Natural Resources Conference* 28:431–443.

61. BARRON, J. C., and W. F. HARWELL. 1973. Fertilization rates of South Texas deer. *Journal of Wildlife Management* 37:179–182.

62. RHODES, O. E., JR., K. T. SCRIBNER, M. H. SMITH, and P. E. JOHNS. 1985. Factors affecting the number of fetuses in a white-tailed deer herd. *Proceedings of the Annual Conference of the Southeastern Association of Fish and Wildlife Agencies* 39:380–388.

63. VERME, L. J. 1983. Sex ratio variation in *Odocoileus*: A critical review. *Journal of Wildlife Management* 47:573–582.

64. CALEY, M. J., and T. D. NUDDS. 1987. Sex ratio adjustment in *Odocoileus*: Does local resource competition play a role. *The American Naturalist* 129:452–457.

65. GRUVER, B. J., D. C. GUYNN, Jr., and H. A. JACOBSON. 1984. Simulated effects of harvest strategy on reproduction in white-tailed deer. *Journal of Wildlife Management* 48:535–541.

66. MILLER, K. V., and J. J. OZOGA. 1997. Density effects on deer sociobiology. Pages 136–150 *in* W. J. McShea, H. B. Underwood, and J. H. Rappole, eds., *The Science of Overabundance: Deer Ecology and Population Management*. Smithsonian Institution Press, Washington, DC.

67. BUBENIK, G. A., A. B. BUBENIK, G. M. BROWN, and D. A. WILSON. 1977. Sexual stimulation and variations of plasma testosterone in normal, antiandrogen and antiestrogen treated white-tailed deer *(Odocoileus virginianus)* during the annual cycle. *Proceedings of the International Congress of Game Biologists* 13:377–386.

68. RUTBERG, A. T. 1987. Adaptive hypotheses of birth synchrony in ruminants: An interspecific test. *American Naturalist* 130:692–710.

69. OZOGA, J. J. 1969. Some longevity records for female white-tailed deer in northern Michigan. *Journal of Wildlife Management.* 33:1027.

70. TULLAR, B. F., Jr. 1983. A long-lived white-tailed deer. *New York Fish and Game Journal* 30:119.

71. FORBES, S. E., L. M. LANG, S. A. LISCINSKY, and H. A. ROBERTS. 1971. The white-tailed deer in Pennsylvania. Pennsylvania Game Commission, Harrisburg.

72. CALHOUN, J., and F. LOOMIS. 1974. *Prairie Whitetails*. Illinois Department of Conservation, Springfield, IL.

73. RAWSON, R. E., G. D. DELGIUDICE, H. E. DZIUK, and L. D. MECH. 1992. Energy metabolism and hematology of white-tailed deer fawns. *Journal of Wildlife Disease* 28:91–94.

74. VERME, L. J., and J. J. OZOGA. 1980. Effects of diet on growth and lipogenesis in deer fawns. *Journal of Wildlife Management* 44:315–324.

75. KNOWLTON, F. F., M. WHITE, and J. G. KIE. 1979. Weight patterns of wild white-tailed deer in southern Texas. Pages 55–64 *in* Drawe, D. L., ed., *Proceedings First Welder Wildlife Foundation Symposium*. Welder Wildlife Foundation, Sinton, TX.

76. DELGUIDICE, G. D., L. D. MECH, K. E. KUNKEL, E. M. GESE, and U. S. SEAL. 1992. Seasonal patterns of weight, hematology, and serum characteristics of free-ranging female white-tailed deer in Minnesota. *Canadian Journal of Zoology* 70:974–983.

77. VERME, L. J., and D. E. ULLREY. 1984. Physiology and nutrition. Pages 91–118 *in* L. K. Halls, ed., *White-tailed Deer: Ecology and Management.* Stackpole Books, Harrisburg, PA.

78. SILVER, H., N. F. COLOVOS, J. B. HOLTER, and H. H. HAYES. 1969. Fasting metabolism of white-tailed deer. *Journal of Wildlife Management.* 33:490–498.

79. WARREN, R. J., R. L. KIRKPATRICK, A. OELSCHLAEGER, P. F. SCANLON, and F. C. GWAZDAUSKAS. 1981. Dietary and seasonal influences on nutritional indices of adult male white-tailed deer. *Journal of Wildlife Management* 45:926–936.

80. WHEATON, C., and R. B. BROWN. 1983. Feed intake and digestive efficiency of white-tailed deer. *Journal of Wildlife Management* 47:442–450.

81. SEAL, U.S., L. J. VERME, J. J. OZAGA, and A. W. ERICKSON. 1972. Nutritional effects on thyroid activity and blood of white-tailed deer. *Journal of Wildlife Management.* 36: 1041–1052.

82. MAUTZ, W. W., J. KANTER, and P. J. PEKINS. 1992. Seasonal metabolic rhythms of captive female white-tailed deer: A re-examination. *Journal of Wildlife Management* 56:656–660.

83. ABBOTT, M. J., D. E. ULLREY, P. K. KU, S. M. SCHMITT, D. R. ROMSOS, and H. A. TUCKER. 1984. Effect of photoperiod on growth and fat accretion in white-tailed doe fawns. *Journal of Wildlife Management* 48:776–787.

84. WORDEN, K. A., and P. J. PEKINS. 1995. Seasonal change in feed intake, body composition, and metabolic rate of white-tailed deer. *Canadian Journal of Zoology* 73:452–457.

85. ROBBINS, C. T., R. L. PRIOR, A. N. MOEN, and W. J. VISEK. 1974. Nitrogen metabolism of white-tailed deer. *Journal of Animal Science* 38:186–191.

86. ULLREY, D. E., W. G. YOUATT, H. E. JOHNSON, L. D. FAY, and B. L. BRADLEY. 1967. Protein requirements of white-tailed deer fawns. *Journal of Wildlife Management* 31:679–685.

87. HOLTER, J. B., H. H. HAYES, and S. H. SMITH. 1979. Protein requirements of yearling white-tailed deer. *Journal of Wildlife Management* 43:872–879.

88. ASLESON, M. A., E. C. HELLGREN, and L. W. VARNER. 1996. Nitrogen requirements for antler growth and maintenance in white-tailed deer. *Journal of Wildlife Management* 60:744–752.

89. OFTEDAL, Q. T. 1981. Milk, protein, and energy intakes of suckling mammalian young: A comparative study. Doctoral dissertation. Cornell University, Ithaca, NY.

90. ULLREY, D. E., W. G. YOUATT, H. E. JOHNSON, L. D. FAY, and B. L. BRADLEY. 1970. Digestible and metabolizable energy requirements for winter maintenance of Michigan white-tailed does. *Journal of Wildlife Management* 34:863–869.

91. MCEWEN, L. C., C. E. FRENCH, N. D. MAGRUDER, R. W. SWIFT, and R. H. INGRAM. 1957. Nutrient requirements of the white-tailed deer. *Transactions North American Wildlife Conference* 22:119–132.

92. GRASMAN, B. T., and E. C. HELLGREN. 1993. Phosphorus nutrition in white-tailed deer: Nutrient balance, physiological responses, and antler growth. *Ecology* 74:2279–2296.

93. JONES, R. L., and H. P. WEEKS, Jr. 1985. Ca, Mg, and P in the annual diet of deer in south-central Indiana. *Journal of Wildlife Management* 49:129–133.

94. MCCULLOUGH, D. R. 1979. *The George Reserve Deer Herd: Population Ecology of a K-Selected Species.* University of Michigan Press, Ann Arbor.

95. HIRTH, D. H. 1977. Social behavior of white-tailed deer in relation to habitat. *Wildlife Monograph* 53:1–55.

96. HAWKINS, R. E., and W. D. KLIMSTRA. 1970. A preliminary study of the social organization of the white-tailed deer. *Journal of Wildlife Management* 34:460–464.

97. OZOGA, J. J., L. J. VERME, and C. S. BIENZ. 1982. Parturition behavior and territoriality in white-tailed deer; Impact on neonatal mortality. *Journal of Wildlife Management* 46:1–11.

98. HIRTH, D. H. 1994. Does and their young. Pages 129–134 *in* D. Gerlach, S. Atwater, and J. Schnell, eds., *Deer.* Stackpole Books, Mechanicsburg, PA.

99. NIXON, C. M., L. P. HANSEN, P. A. BREWER, and J. E. CHELSVIG. 1991. Ecology of white-tailed deer in an intensively farmed region of Illinois. *Wildlife Monographs* 118:1–77.

100. VERCAUTEREN, K. C. 1998. Dispersal, home range fidelity, and vulnerability of white-tailed deer in the Missouri River Valley. Ph.D. dissertation. University of Nebraska, Lincoln.

101. NELSON, M. E. 1993. Natal dispersal and gene flow in white-tailed deer in northeastern Minnesota. *Journal of Mammalogy* 74:316–322.

102. PORTER, W. F., N. E. MATHEWS, H. B. UNDERWOOD, R. W. SAGE, Jr., and D. F. BEHREND. 1991. Social organization in deer: Implications for localized management. *Environmental Management* 15:809–814.

103. BROWN, B. A., Jr. 1974. Social organization in male groups of white-tailed deer. Pages 436–446 *in* V. Geist and F. Walther, eds., *The Behaviour of Ungulates and Its Relation to Management.* New Series, Publication 24. International Union for the Conservation of Nature, Morges, Switzerland.

104. FORAND, K. J., and R. L. MARCHINTON. 1989. Patterns of social grooming in adult white-tailed deer. *American Midland Naturalist* 122:357–364.

105. MARCHINTON, R. L., and D. H. HIRTH. 1984. Behavior. Pages 129–168 *in* L. K. Halls, ed., *White-tailed Deer: Ecology and Management.* Stackpole Books, Harrisburg, PA.

106. BROWN, B. A., and D. H. HIRTH. 1979. Breeding behavior in white-tailed deer. *Proceedings of the First Welder Wildlife Foundation Symposium* 1:83–95.

107. HOLZENBIEN, S., and R. L. MARCHINTON. 1992. Emigration and mortality of orphaned male white-tailed deer. *Journal of Wildlife Management* 56:147–153.

108. HOLZENBIEN, S., and R. L. MARCHINTON. 1992. Spatial integration of maturing male white-tailed deer into the adult population. *Journal of Mammalogy* 73:326–334.

109. BEIER, P. and D. R. McCULLOUGH. 1990. Factors influencing white-tailed deer activity patterns and habitat use. *Wildlife Monographs* 109:1–51.

110. MICHAEL, E. D. 1970. Activity patterns of white-tailed deer in South Texas. *Texas Journal of Science* 21:417–428.

111. PLEDGER, J. M. 1975. Activity, home range, and habitat utilization of white-tailed deer *(Odocoileus virginianus)* in southeastern Arkansas. Master's thesis. University of Arkansas, Fayetteville.

112. IVEY, T. L., and M. K. CAUSEY. 1988. Social organization among white-tailed deer during the rut. *Proceedings of the Annual Conference of the Southeastern Association of Fish and Wildlife Agencies* 42:266–271.

113. SAWYER, T. G. 1981. Behavior of female white-tailed deer with emphasis on pheromonal communication. Master's thesis. University of Georgia, Athens.

114. VERME, L. J., and J. J. OZOGA. 1971. Influence of winter weather on white-tailed deer in Upper Michigan. Pages 16–28 *in* A. O. Haugen, ed., *Snow and Ice in Relation to Wildlife and Recreation Symposium.* Iowa State University, Ames.

115. HOSKINSON, R. L., and L. D. MECH. 1976. White-tailed deer migration and its role in wolf predation. *Journal of Wildlife Management* 40:429–441.

116. DOEPKER, R. V., and J. J. OZOGA. 1991. Wildlife value of northern white-cedar. Pages 15–34 *in* D. O. Lantagne, ed., *Northern White-Cedar in Michigan.* Report 512, Michigan State University, Ann Arbor.

117. NELSON, M. E., and L. D. MECH. 1981. Deer social organization and wolf predation in northeastern Minnesota. *Wildlife Monographs* 77:1–53.

118. NELSON, M. E., and L. D. MECH. 1987. Demes within a northeastern Minnesota deer population. Pages 27–40 *in* B. D. Chepko-Sade and Z. T. Halpin, eds., *Mammalian Dispersal Patterns.* University of Chicago Press, Chicago.

119. OZOGA, J. J. 1995. *Whitetail Winter.* Willow Creek Press, Minocqua, WI.

120. NELSON, M. E., and L. D. MECH. 1986. Relationship between snow depth and gray wolf predation on white-tailed deer. *Journal of Wildlife Management* 50:471–474.

121. RONGSTAD, O. J., and J. R. TESTER. 1969. Movements and habitat use of white-tailed deer. *Journal of Wildlife Management* 33:366–379.

122. KAMMERMEYER, K. E., and R. L. MARCHINTON. 1976. The dynamic aspects of deer populations utilizing a refuge. *Proceedings Annual Conference of the Southeastern Association of Game and Fish Commissioners* 29:466–475.

123. BYFORD, J. L. 1970. Movements and ecology of white-tailed deer in a logged floodplain habitat. Dissertation. Auburn University, Auburn, AL.

124. JOANEN, T., L. McNEASE, and D. RICHARD. 1985. The effects of winter flooding on white-tailed deer in southwestern Louisiana. *Proceedings Louisiana Academy of Sciences* 48:109–115.

125. ATKESON, T. D., R. L. MARCHINTON, and K. V. MILLER. 1988. Vocalizations of white-tailed deer. *American Midland Naturalist* 120:194–200.

126. RICHARDSON, L. W., H. A. JACOBSON, R. J. MUNCY, and C. J. PERKINS. 1983. Acoustics of white-tailed deer *(Odocoileus virginianus). Journal of Mammalogy* 64:245–252.

127. MILLER, K. V., and R. L. MARCHINTON. 1994. Deer talk: Sounds, smells, and postures. Pages 158–168 *in* D. Gerlach, S. Atwater, and J. Schnell, eds., *Deer*. Stackpole Books, Mechanicsburg, PA.

128. HIRTH, D. H., and D. R. MCCULLOUGH. 1977. Evolution of alarm signals in ungulates with special reference to white-tailed deer. *American Midland Naturalist* 111:31–42.

129. GASSETT, J. W., D. P. WIESLER, A. G. BAKER, D. A. OSBORN, K. V. MILLER, R. L. MARCHINTON, and M. NOVOTNY. 1996. Volatile compounds from interdigital gland of male white-tailed deer *(Odocoileus virginianus)*. *Journal of Chemical Ecology* 22:1689–1696.

130. ATKESON, T. D., and R. L. MARCHINTON. 1982. Forehead glands in white-tailed deer. *Journal of Mammalogy* 63:613–617.

131. GASSETT, J. W., D. P. WIESLER, A. G. BAKER, D. A. OSBORN, K. V. MILLER, R. L. MARCHINTON, and M. NOVOTNY. 1997. Volatile compounds from forehead region of male white-tailed deer *(Odocoileus virginianus)*. *Journal of Chemical Ecology* 23:569–578.

132. MILLER, K. V., B. JEMIOLO, J. W. GASSETT, I. JELINEK, D. WIESLER, and M. NOVOTNY. 1998. Putative chemical signals from white-tailed deer *(Odocoileus virginianus)*: Social and seasonal effects on urinary volatile excretion in males. *Journal of Chemical Ecology* 24: 673–683.

133. GASSETT, J. W. and K. V. MILLER. 1997. Odor production from the tarsal glands of male white-tailed deer. *International Congress of the International Society for Applied Ethology, Prague, Czech Republic* 31:149.

134. GOSS, R. J. 1983. *Deer Antlers: Regeneration, Function, and Evolution*. Academic Press, New York.

135. RICHTER, A. R., and R. F. LABISKY. 1985. Reproductive dynamics among disjunct white-tailed deer herds in Florida. *Journal of Wildlife Management* 49:964–971.

136. JACOBSON, H. A. 1994. Reproduction. Pages 98–108 *in* D. Gerlach, S. Atwater, and J. Schnell, eds. *Deer*. Stackpole Books, Mechanicsburg, PA.

137. MARCHINTON, R. L., K. L. JOHANSEN, and K. V. MILLER. 1990. Behavioural components of white-tailed deer scent marking: Social and seasonal effects. Pages 295–310 *in* D. W. Macdonald, D. Muller-Schwarze, and S. E. Natynczuk, eds., *Chemical Signals in Vertebrates* V. Oxford University Press, Oxford, UK.

138. SAWYER, T. G., R. L. MARCHINTON, and K. V. MILLER. 1989. Response of female white-tailed deer to scrapes and antler rubs. *Journal of Mammalogy* 70:431–433.

139. MOORE, W. G., and R. L. MARCHINTON. 1974. Marking behavior and its social function in white-tailed deer. Pages 447–456 *in* V. Geist and F. Walther, eds., *The Behaviour of Ungulates and Its Relation to Management*. New Series, Publication 24. International Union for the Conservation of Nature, Morges, Switzerland.

140. MARCHINTON, R. L., and T. D. ATKESON. 1985. Plasticity of socio-spatial behaviour of white-tailed deer and the concept of facultative territoriality. Pages 375–377 *in* P. F. Fennessy and K. R. Drew, eds., *The Biology of Deer Production*. Bulletin 22. The Royal Society of New Zealand, Wellington.

141. DASHER, K. A., J. W. GASSETT, D. A. OSBORN, and K. V. MILLER. 1998. Remote monitoring of scraping behaviour in a wild population of white-tailed deer. *Biology of Deer, Kaposvar, Hungary* 4:34.

142. WHITNEY, M. D., D. L. FORSTER, K. V. MILLER, and R. L. MARCHINTON. 1991. Sexual attraction in white-tailed deer. Pages 327–333 *in* R. D. Brown, ed., *The Biology of Deer*. Springer-Verlag, New York.

143. MULLER-SCHWARZE, D. 1994. The senses of deer. Pages 58–65 *in* D. Gerlach, S. Atwater, and J. Schnell, eds., *Deer*. Stackpole Books, Mechanicsburg, PA.

144. GASSETT, J. W., K. A. DASHER, D. A. OSBORN, and K. V. MILLER. 1998. What the nose knows: Detection of oestrus by male white-tailed deer. *Biology of Deer, Kaposvar, Hungary* 4:35.

145. LEOPOLD, A, L. K. SOWLS, and D. L. SPENCER. 1947. A survey of over-populated deer ranges in the United States. *Journal of Wildlife Management* 11:162–177.

146. MESSMER, T. A., L. CORNICELLI, D. J. DECKER, and D. G. HEWITT. 1997. Stakeholder acceptance of urban deer management techniques. *Wildlife Society Bulletin* 25:360–366.

147. JACOBSON, H. A., and D. C. GUYNN, Jr. 1995. A primer. Pages 81–102 *in* K. V. Miller and R. L. Marchinton, eds., *Quality Whitetails*. Stackpole Books, Mechanicsburg, PA.

148. MATTFELD, G. F. 1984. Northeastern hardwood and spruce/fir forests. Pages 305–330 *in* L. K. Halls, ed., *White-tailed Deer: Ecology and Management*. Stackpole Books, Harrisburg, PA.

149. JACOBSON, H. A., D. C. GUYNN, JR., R. N. GRIFFIN, and D. LEWIS. 1979. Fecundity of white-tailed deer in Mississippi and periodicity of corpora lutea and lactation. *Proceedings Southeastern Association of Fish and Wildlife Agencies* 33:30–35.

150. HANSEN, L. P., J. BERINGER, and J. H. SCHULZ. 1996. Reproductive characteristics of female white-tailed deer in Missouri. *Proceedings Southeastern Association of Fish and Wildlife Agencies* 50:357–366.

151. ROSEBERRY, J. L., and W. D. KLIMSTRA. 1970. Productivity of white-tailed deer on Crab Orchard National Wildlife Refuge. *Journal of Wildlife Management* 34:23–28.

152. MUNDINGER, J. G. 1981. White-tailed deer reproductive biology in the Swan Valley, Montana. *Journal of Wildlife Management* 45:132–139.

153. CHIAVETTA, K. J. 1958. Harvest antlerless deer! *North Carolina Wildlife* 22:16–19.

154. RHODES, O. E., JR., J. M. NOVAK, M. H. SMITH, and P. E. JOHNS. 1986. Assessment of fawn breeding in a South Carolina deer herd. *Proceedings Southeastern Association of Fish and Wildlife Agencies* 40:430–437.

155. HAUGEN, A. O. 1975. Reproductive performance of white-tailed deer in Iowa. *Journal of Mammalogy* 56:151–159.

156. OSBORNE, J. S., A. S. JOHNSON, P. E. HALE, R. L. MARCHINTON, C. V. VANSANT, and J. M. WENTWORTH. 1992. Population ecology of the Blackbeard Island white-tailed deer herd. Bulletin 26. Tall Timbers Research Station, Tallahassee, FL.

157. SHEA, S. M., T. A. BREAULT, and M. L. RICHARDSON. 1992. Herd density and physiological condition of white-tailed deer in Florida flatwoods. *Journal of Wildlife Management* 56: 262–267.

158. MCCULLOUGH, D. R. 1997. Irruptive behavior in ungulates. Pages 69–98 *in* W. J. McShea, H. B. Underwood, and J. H. Rappole, eds., *The Science of Overabundance: Deer Ecology and Population Management.* Smithsonian Institution Press, Washington, DC.

159. FRITTS, S., and L. D. MECH. 1981. Dynamics, movements, and feeding ecology of a newly protected wolf population in northwestern Minnesota. *Wildlife Monographs* 80:1–79.

160. MECH, L. D., and L. D. FRENZEL, JR. 1971. An analysis of the age, sex, and condition of deer killed by wolves in northeastern Minnesota. Pages 35–51 *in* L. D. Mech and L. D. Frenzel, eds., *Ecological Studies of the Timber Wolf in Northeastern Minnesota.* Research Paper NC-52. U.S. Department of Agriculture, Forest Service. Washington, DC.

161. MECH, L. D. 1984. Predators and predation. Pages 189–200 *in* L. K. Halls, ed., *White-tailed Deer: Ecology and Management.* Stackpole Books, Harrisburg, PA.

162. MCCORD, C. M. 1974. Selection of winter habitat by Bobcats *(Lynx rufus)* on the Quabbin Reservation, Massachusetts. *Journal of Mammalogy* 55:428–437.

163. HUEGEL, C. N., and O. J. RONGSTAD. 1985. Winter foraging patterns and consumption rates of northern Wisconsin coyotes. *American Midland Naturalist* 113:203–207.

164. PARKER, G. R., and J. W. MAXWELL. 1989. Seasonal movements and winter ecology of the coyote, *Canis latrans*, in northern New Brunswick. *Canadian Field-Naturalist* 103:1–11.

165. DEYOUNG, C. A. 1989. Mortality of adult male white-tailed deer in south Texas. *Journal of Wildlife Management* 53:513–518.

166. DAVIDSON, W. R., and G. L. DOSTER. 1997. Health characteristics and white-tailed deer population density in the Southeastern United States. Pages 164–184 *in* W. J. McShea, H. B. Underwood, and J. H. Rappole, eds., *The Science of Overabundance: Deer Ecology and Population Management.* Smithsonian Institution Press, Washington, DC.

167. SAMUEL, W. M. 1994. The parasites and diseases of whitetails. Pages 233–235 *in* D. Gerlach, S. Atwater, and J. Schnell, eds., *Deer.* Stackpole Books, Mechanicsburg, PA.

168. PRESTWOOD, A. K., J. F. SMITH, and J. BROWN. 1971. Lungworms in white-tailed deer of the southeastern United States. *Journal of Wildlife Diseases* 7:149–154.

169. ROMIN, L. A., and J. A. BISSONETTE. 1996. Deer-vehicle collisions: Status of state monitoring activities and mitigation efforts. *Wildlife Society Bulletin* 24:276–283.

170. ALLEN, R. E., and D. R. MCCULLOUGH. 1976. Deer-car accidents in southern Michigan. *Journal of Wildlife Management* 40:317–325.

171. TEER, J. G., J. W. THOMAS, and E. A. WALKER. 1965. Ecology and management of white-tailed deer in the Llano Basin of Texas. *Wildlife Monographs* 15:1–62.

172. SEVERINGHAUS, C. W. 1972. Weather and the deer population. *The Conservationist* 28:28–31.

173. BOYER, R. T., M. E. SHEA, and S. A. McKENNA. 1986. The role of winter severity and population density in regulating northern populations of deer. Miscellaneous Publication Number 689. Maine Agricultural Experiment Station, Orons, ME.

174. VERME, L. H. 1968. An index of winter weather severity for northern deer. *Journal of Wildlife Management* 32:566–574.

175. KARNS, P. D. 1980. Winter—The grim reaper. Pages 47–53 *in* R. L. Hine and S. Nehls, eds., *White-tailed Deer Population Management in the North Central States. Proceedings of a Symposium.* 41st Midwest Fish and Wildlife Conference, Urbana, IL.

176. SEVERINGHAUS, C. W. 1982. Sex and age composition of winter-killed deer in the central Adirondack region of New York. *New York Fish and Game Journal* 29:199–203.

177. VERME, L. J. 1977. Assessment of natal mortality in upper Michigan deer. *Journal of Wildlife Management* 41:700–708.

178. SWIHART, R. K., and A. J. DeNICOLA. 1997. Public involvement, science, management, and the overabundance of deer: Can we avoid a hostage crisis? *Wildlife Society Bulletin* 25:382–387.

179. MULLER, L. I., R. J. WARREN, and D. L. EVANS. 1997. Theory and practice of immunocontraception in wild mammals. *Wildlife Society Bulletin* 25:504–514.

180. DeNICOLA, A. J., D. J. KESLER, and R. K. SWIHART. 1997. Remotely delivered prostaglandin F_{2a} implants terminate pregnancy in white-tailed deer. *Wildlife Society Bulletin* 25:527–531.

181. BROTHERS, A., and M. E. RAY, Jr. 1975. *Producing Quality Whitetails.* Fiesta Publication Company, Laredo, TX.

182. MILLER, K. V., and R. L. MARCHINTON, eds., 1995. *Quality Whitetails: The How and Why of Quality Deer Management.* Stackpole Books, Mechanicsburg, PA.

183. HALLS, L. K., ed. 1984. *White-tailed Deer: Ecology and Management.* Wildlife Management Institute, Stackpole Books, Harrisburg, PA.

184. CRAWFORD, H. S. 1984. Habitat management. Pages 629–646 *in* L. K. Halls, ed., *White-tailed Deer: Ecology and Management.* Stackpole Books, Harrisburg, PA.

185. HURST, G. A., and R. C. WARREN. 1981. Enhancing white-tailed deer habitat of pine plantations by intensive management. Technical Bulletin 107. Mississippi Agricultural and Forestry Experiment Station, Mississippi State University, Starkville.

186. KAMMERMEYER, K. E., and R. THACKSTON. 1995. Habitat management and supplemental feeding. Pages 129–154 *in* K. V. Miller and R. L. Marchinton, eds., *Quality Whitetails.* Stackpole Books, Mechanicsburg, PA.

187. GOODRUM, P. D., V. H. REID, and C. E. BOYD. 1971. Acorn yields, characteristics and management criteria of oaks for wildlife. *Journal of Wildlife Management* 35:520–532.

188. MARSTON, D. L. 1977. Deer wintering area management in Maine: A progress report. *Transactions, Northeast Deer Study Group* 11:103–110.

189. OZOGA, J. J., R. V. DOEPKER, and M. S. SARGENT. 1994. Ecology & management of white-tailed deer in Michigan. Wildlife Division Report 3209. Michigan Department of Natural Resources, Lansing, MI.

190. WILEY, J. E., III, K. F. STRONG, J. W. LANIER, and B. J. HILL. 1978. Winter habitat management for white-tailed deer in New Hampshire—A case history. *Transactions, Northeast Deer Study Group* 14:63–72.

191. LAY, D. W. 1967. Browse palatability and the effects of prescribed burning in southern pine forests. *Journal of Forestry* 65:826–828.

192. JOHNSON, A. S., and J. L. LANDERS. 1978. Fruit production in slash pine plantations in Georgia. *Journal of Wildlife Management* 42:606–613.

193. STRANSKY, J. J., and L. K. HALLS. 1976. Nutrient content and yield of burned vs. mowed Japanese honeysuckle. *Proceedings Southeastern Association of Game and Fish Commissioners* 29:403–406.

194. LANDERS, J. L. 1987. Prescribed burning for managing wildlife in Southeastern pine forests. Pages 19–27 *in* J. G. Dickson and O. E. Maughan, eds., *Managing Southern Forests for Wildlife and Fish: A Proceedings.* General Technical Report SO-65. U.S. Department of Agriculture, Forest Service. Washington, DC.

195. CUSHWA, C. T., E. V. BENDER, and R. W. COOPER. 1966. The response of herbaceous vegetation to prescribed burning. Research Note SE-53. U.S. Department of Agriculture, Forest Service, Washington, DC.

196. RUTSKE, L. H. 1969. A Minnesota guide to forest game habitat improvement. Technical Bulletin 10. Minnesota Department of Conservation, Division of Game and Fish, St. Paul, MN.

197. VOGL, R. J., and A. M. BECK. 1970. Response of white-tailed deer to a Wisconsin wildfire. *American Midland Naturalist* 84:270–273.

198. JOHNSON, M. K., B. W. DELANEY, S. P. LYNCH, J. A. ZENO, S. R. SCHULTZ, T. W. KEEGAN, and B. D. NELSON. 1987. Effects of cool-season agronomic forages on white-tailed deer. *Wildlife Society Bulletin* 15:330–339.

199. KAMMERMEYER, K. E., and E. B. MOSER. 1990. The effect of food plots, roads, and other variables on deer harvest in Northeastern Georgia. *Proceedings Southeastern Association of Game and Fish Commissioners* 44:364–373.

200. VANDERHOOF, R. E. 1995. Production and utilization of agronomic forages and their effects on deer movement and hunter behavior. Dissertation. Mississippi State University. Starkville, MS.

201. SCANLON, J. J., and M. R. VAUGHN. 1985. Movements of white-tailed deer in Shenandoah National Park, Virginia. *Proceedings Southeastern Association of Fish and Wildlife Commissioners* 39:397–402.

202. BALL, D. M., C. B. HOVELAND, and G. D. LACEFIELD. 1991. *Southern forages.* Potash and Phosphate Institute, Norcross, GA.

203. KAMMERMEYER, K. E., E. A. PADGETT, W. M. LENTZ, and R. L. MARCHINTON. 1992. Production, utilization, and quality of three ladino clovers in Northeastern Georgia. *Southeast Deer Study Group* 15:15.

204. MOTT, S. E., R. L. TUCKER, D. C. GUYNN, and H. A. JACOBSON, 1985. Use of Mississippi bottomland hardwoods by white-tailed deer. *Proceedings Southeastern Association of Fish and Wildlife Agencies* 39:403–411.

205. VERCAUTEREN, K. C., and S. E. HYGNSTROM. 1998. Effects of agricultural activities and hunting on home ranges of female white-tailed deer. *Journal of Wildlife Management* 62:280–285.

206. SARGENT, R. A., Jr. 1992. Movement ecology of adult male white-tailed deer in hunted and non-hunted populations in the wet prairie of the everglades. Thesis. University of Florida, Gainesville.

207. LABISKY, R. F., D. E. FRITZEN, and J. C. KILGO. 1991. Population ecology and management of white-tailed deer in the Osceola National Forest, Florida. Final Report to the Florida Game and Fresh Water Fish Commission. Department of Wildlife and Range Science, University of Florida, Gainesville.

208. GAVIN, T. A., L. H. SURING, P. A. VOHS, Jr., and E. C. MESLOW. 1984. Population characteristics, spatial organization, and natural mortality in the Columbian white-tailed deer. *Wildlife Monographs* 91:1–41.

209. PAIS, R. C., W. C. MCCOMB, and J. PHILLIPS. 1991. Habitat associated with home ranges of female *Odocoileus virginianus* (Mammalia: Cervidae) in eastern Kentucky. *Brimleyana* 17:57–66.

210. LEACH, R. H., and W. D. EDGE. 1994. Summer home range and habitat selection by white-tailed deer in the Swan Valley, Montana. *Northwest Science* 68:31–36.

211. CORNICELLI, L., A. WOOLF, and J. L. ROSEBERRY. 1996. White-tailed deer use of a suburban environment in southern Illinois. *Transactions of the Illinois State Academy of Science* 89:93–103.

212. TIERSON, W. C., G. F. MATTFELD, R. W. SAGE, Jr., and D. F. BEHREND. 1985. Seasonal movements and home ranges of white-tailed deer in the Adirondacks. *Journal of Wildlife Management* 49:760–769.

213. INGLIS, J. M., R. E. HOOD, B. A. BROWN, and C. A. DEYOUNG. 1979. Home range of white-tailed deer in Texas coastal prairie brushland. *Journal of Mammalogy* 60:377–389.

214. LARSON, T. J., O. J. RONGSTAD, and F. W. TERBILCOX. 1978. Movement and habitat use of white-tailed deer in southcentral Wisconsin. *Journal of Wildlife Management* 42:113–117.

215. WHITEHEAD, G. K. 1972. *Deer of the World.* Constable, London.

CHAPTER
30

Mule and Black-tailed Deer

John G. Kie and Brian Czech

INTRODUCTION

Mule and black-tailed deer, considered different subspecies of a single species within the deer family, evolved in North America before the presence of humans. They have played a role in myth, human subsistence, recreation, and important roles in ecosystem structure and function. Their prevalence throughout western Canada, Mexico, and the United States make them an important big game species.

Mule and black-tailed deer provide recreational hunting opportunities over most of their range, and have nonconsumptive aesthetic values. The average value of a deer-hunting trip (net willingness to pay) to a hunter in California was estimated at $191, or $115 per recreation-visitor day (1). Comparable data for Alaska were $155 per deer-hunting trip by residents, and $101 per deer-hunting trip by nonresidents (2). In addition, the general public in California derived an average value of $11 per trip on outdoor trips where they saw deer and $15 per trip on trips taken primarily to view deer (1). In 1987, deer hunters in California spent $184 million and derived an additional $230 million above and beyond the money they spent. Nonconsumptive use of deer added another $43 million in value (1).

Mule and black-tailed deer can have negative economic value in other situations. Deer can cause damage to gardens and other landscaping, to agricultural crops, and to new tree seedlings (3, 4, 5, 6). Damage to deer, humans, and property from deer–vehicle collisions can also be substantial (7).

More importantly, mule and black-tailed deer play valuable roles in healthy ecosystems (Figure 30–1). There are important reasons to consider mule and black-tailed deer as ecological indicators (8). The biology of deer is well known compared to that of other species of wildlife. They have relatively large home ranges, and are often seasonally migratory, requiring resource managers to consider entire landscapes rather than isolated patches of habitat. They require temporally and spatial diverse habitat elements such as food and cover. Finally, mule and black-tailed deer can have significant effects on vegetation composition and basic ecosystem processes such as nutrient cycling, thereby acting as keystone species (9, 8, 10).

TAXONOMY AND DESCRIPTION

Mule deer and black-tailed deer occur over much of western North America from 23°N to 60°N (Figure 30–2). The most southerly records are from northern San Luis Potosi, Mexico, and the most northerly from southern Yukon Territory, Canada (11, 12). Current

Figure 30–1 Mule and black-tailed deer provide economic, esthetic, and subsistence values to many people and also play important roles in the functioning of healthy ecosystems. (Photograph by H. Ritter.)

taxonomy places mule and black-tailed deer in a single species (order Artiodactyla, suborder Ruminantia, and family Cervidae) and recognizes 7 to 9 subspecies (13, 14). Of the five species of cervids that occur in North America, only mule and black-tailed deer were characterized by mitochondrial DNA exhibiting high divergence between and low divergence within subspecies (15). This suggests limited gene flow between populations of mule and black-tailed deer since the Pleistocene and supports the validity of the subspecies concept for this species.

Rocky Mountain mule deer range from the Yukon to Arizona and New Mexico. Desert mule deer extend from Arizona and New Mexico south into Mexico as far as San Luis Potosi. Peninsula mule deer occupy Baja California del Sur. Southern mule deer extend from Baja California del Norte northward into southern California. California mule deer are found throughout central and southern California. Columbian black-tailed deer range from central California to central British Columbia. A small population of Columbian black-tailed deer has become established on the island of Kaua'i, Hawaii, following several introductions from western Oregon in the 1960s (16, 17). Sitka black-tailed deer occur in northern British Columbia, southeastern Alaska, and in disjunct populations elsewhere in Alaska (Figure 30–2).

Two subspecies previously described, burro deer and Inyo mule deer, were not recognized as distinct subspecies by Mackie and others and Anderson and Wallmo, and are not discussed here (13, 14). Tiburón Island mule deer and Cedros Island mule deer are found off the west coast of Sonora, Mexico, and off the west coast of Baja California, respectively. Consistent with Bergman's rule, mule deer from Cedros and Tiburón Islands and Peninsula mule deer exhibit the smallest body sizes of any subspecies of mule deer (18). However, Wallmo and Anderson and Wallmo have questioned the validity of these two insular populations as distinct subspecies, and they were not recognized by Mackie and others (12, 14, 13).

Taxonomy and Description

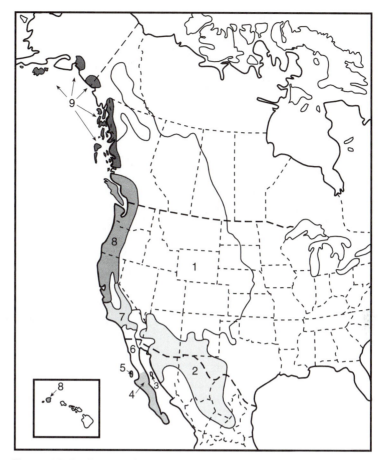

Figure 30–2 Range of subspecies of mule and black tailed deer: (1) Rocky Mountain mule deer, (2) desert mule deer, (3) Tiburón Island mule deer, (4) peninsula mule deer, (5) Cedros Island mule deer, (6) southern mule deer, (7) California mule deer, (8) Columbian black-tailed deer, and (9) Sitka black-tailed deer. (After Refs. 12, 13, 14, and 16.)

LIFE HISTORY

Diet Selection

Deer possess a stomach with four chambers, the first and largest of which is the rumen (18). Food is brought into the rumen and recticulum where microbial action aids in the breakdown of plant cell-wall constituents. Rumination, the periodic regurgitation and chewing of the bolus or cud, hastens the process by additional mechanical action. Large-bodied ruminants such as domestic cattle and elk are characterized by relatively large rumen : body size ratios, long intestinal lengths, and high small intestine : large intestine ratios (19, 20, 21). In addition, they have lower metabolic rates than smaller ruminants because metabolic rate increases as some fractional power of body mass, often estimated at 0.75 (22). As such, these bulk-feeders have the luxury of ingesting fairly coarse forages of low digestibility such as dried grasses, and slowing down the rate of food passage to allow microbial digestion to break down the cell-wall constituents.

Small-bodied ruminants like mule and black-tailed deer must feed on forages in which the nutrients are more concentrated and more digestible. Even large-bodied ruminants such as moose that are adapted to feeding on highly concentrated forages exhibit digestive morphology similar to their smaller bodied relatives and are restricted to those forage species (20, 21).

Selection of specific dietary items by mule and black-tailed deer is varied and complex (23, 24, 25). Rocky Mountain mule deer eat at least 673 species of plants (26, 27). Diets vary by subspecies and location, season, sex, and other factors (26, 27, 28). Mule and black-tailed deer traditionally have been thought of as browsers, particularly during winter, relying primarily on twigs and other vegetative parts of woody plants. Their requirement for forages with concentrated nutrients precludes them from making heavy use of forages of low digestibility such as dried grasses, and may at times result in their preferential use of woody plant material. However, the concept of deer as browsers by choice is perhaps one of the oldest and most persistent myths in deer ecology and management (29). Given access to seasonally abundant, nutritious, herbaceous plants of high digestibility, deer will tend to select these species in preference to browse species of lower digestibility.

For example, diets of Sitka black-tailed deer on Admirality Island, Alaska, ranged from 57% to 79% nonbrowse items such as forbs, ferns, grasses, sedges, lichens, algae, and mosses depending on season, although during a winter with deep snows browse consumption peaked at 87% (30). Diets of Columbian black-tailed deer on oak woodland-annual grass ranges in northern California included as much as 62% newly germinated annual grasses during winter when those species are green and highly digestible (31). Finally, black-tailed deer on the island of Kaua'i, Hawaii, are primarily frugivorous in the fall, with diets in September and October consisting of 58% passion fruit and guava, both fruits of non-native plant species (17). Other seasonally important forage classes for mule and black-tailed deer include acorns, mistletoe, lichens, mushrooms, and succulents (32, 33, 30, 32, 34). Many of these forages are often underrepresented in diet studies because of their high digestibility or deficiencies in sampling and analytical methodology (32).

The average, year-round food intake rate of mule deer has been estimated at 22 grams of dry matter per kilogram of body mass (bled carcass mass) per day (35). This amounts to 1.5 kilograms per day for a 68-kilogram deer (14). Penned black-tailed deer voluntarily reduce their forage intake during late fall and winter, even when fed *ad libitum* levels (36, 37, 38). Mean intake rate of dry matter by nonlactating black-tailed females was estimated at 67 grams per kilogram$^{0.75}$ body mass per day, with lactating females with single and twin fawns at 135% and 170% of that rate, respectively (39). These rates are equivalent to 1.3, 1.7, and 2.1 kilograms per day for a 50-kilogram female black-tailed deer, respectively.

Reproductive Strategy

Breeding. Mule and black-tailed deer are polygynous with a tending-bond breeding system (40). Some cervids such as elk exhibit true harem formation characterized by females that collect around dominant males, who in turn aggressively advertise their social status to maintain dominance over subordinate males. In contrast, mule and black-tailed deer males stay with and tend a single female until she is bred, thereby exhibiting serial polygyny (40). To minimize the time spent with any single female, it is advantageous for a male to stimulate urination in a female so that he can assess her state of estrus and readiness to breed (41). One way he can do this is with a rush courtship display, similar to that used by white-tailed deer, where he charges the female, exciting her, and stimulating her to urinate (40). Mule deer males, unlike whitetails, also mimic juvenile behavior and vocalizations to allow them to remain close to the female until she finally urinates and he can determine if she is receptive (40).

Gestation. Most mule deer females breed for the first time at 1.5 years of age and bear their first fawns when they turn 2 (13, 14). Reproduction in fawns has been noted in penned situations, particularly for black-tailed deer, but appears unusual in the wild (42, 43). Female mule and black-tailed deer appear to remain capable of conceiving throughout most of their lives, although some very old individuals may cease to breed, presumably because of inadequate nutritional intake (44).

Breeding among mule deer occurs in mid- to late autumn and early winter, with exact dates varying by subspecies and location (13). Desert mule deer breed somewhat later, in December and January, while black-tailed deer may breed as early as September and October (27). The estrous cycle in female mule deer is 22 to 28 days, while the period of estrus itself when the female is receptive is 24 to 36 hours (13). If the female is not bred on first estrus, she may go through additional cycles (27). Timing and synchrony of reproduction are adaptations to long-term climatic patterns that help ensure females have adequate nutrition during late gestation and parturition and that fawns are born at an optimal time of year (45, 27, 46).

The mean length of gestation in mule deer is 200 to 208 days, with a range of 183 to 218 days (47). It is not known whether mule and black-tailed deer can facultatively vary gestation length in response to short-term variations in weather and forage conditions as has been shown for Dall's sheep and bison (48, 49).

The average number of fetuses per yearling female ranged from less than 0.66 to 1.71, and the average for adult females was 0.92 to 1.96 based on a summary of data from 12 sites (45). Mature females commonly produce two young when nutritional levels are adequate, while primiparous females will often only conceive a single fawn (14). A maximum of five fetuses was reported by Nellis and others (43). Ovulation rates are closely related to nutritional intake, and where deer are existing at or above carrying capacity, reproductive rates decline in a density-dependent manner (13, 50).

Sex Ratio. Fetal sex ratios in mule deer show a preponderance of males. Robinette and others reported an average of 110 males : 100 females (51). In white-tailed deer, male fetuses predominate among females on lower planes of nutrition, when delayed breeding occurs (i.e., when females do not conceive at first estrus), or when populations are at or near carrying capacity (52, 53, 54, 55). Conversely, Trivers and Willard suggested that adult females in poor condition should maximize their reproductive fitness by bearing more female offspring because they are less expensive to produce (56). In addition, a female offspring in poor condition is still capable of breeding whereas a male in the same condition cannot successfully compete against bigger, stronger rivals. In red deer, male offspring were more expensive to produce, but dominant females also produced more male than female fetuses (57, 58). Similar contradictory data can be found for other species of cervids. Female offspring predominate when females are on good diets, male offspring predominate when females are on good diets indicating no apparent effect of diet on sex ratio of offspring (59, 60, 61). Fetal sex ratios in mule and black-tailed deer are probably influenced by several factors including familial ties, social rank, habitat quality, and nutritional levels (62).

Adult females outnumber adult males in most populations by more than 2:1 (13). Selective removal of adult males by sport hunting obviously influences adult sex ratios. Heavily hunted populations may average only 5 to 25 males:100 females, while lightly hunted populations average 20 to 40 males:100 females (13). The adult sex ratio among a lightly hunted population at San Carlos Apache Reservation, Arizona, was estimated at 53 males:100 females (B. Czech, unpublished data). Even populations of mule deer that are not subject to sport hunting can show an abundance of adult females. For example, mule deer in the national parks of western Canada averaged 53 males:100 females, while those in Glacier National Park, Montana, averaged about 25 males:100 females (63, 64).

Adult female cervids are usually in their poorest condition after late gestation and lactation because of the nutritional stresses associated with the production of offspring. However, forage conditions are often good at that time of year. Conversely, adult males are in poorest condition following rut when winter forage conditions prevail, and therefore suffer greater overwinter mortality (65, 55, 66).

Longevity

Anderson and Wallmo reviewed the following accounts of maximum longevity in mule and black-tailed deer (14). Robinette and others reported an estimated age of 19 years for a male mule deer living in the wild (45). The estimate was based on tooth-cementum layers, however, and the authors acknowledged that they misjudged the age on 51% of the deer examined (45). Ross reported an estimated age of 20 years for a female mule deer in the wild (67). A semi-tame female Columbian black-tailed deer reached an estimated 22 years of age, based on an assumed age of 2 years when she was first seen (68). Estimates of maximum longevity in each of these cases is subjective.

Average life spans of mule and black-tailed deer in the wild are considerably shorter than the potential maximums. Female mule deer in Montana seldom lived beyond 10 to 12 years, and males seldom beyond 8 years (69). In Utah, only about 1% of the legally killed male mule deer were estimated to be 8 years of age or older (45).

Physiology

Growth. In one study, male black-tailed deer averaged 3.1 kilograms at birth, and gained an average of 0.20 kilograms/day for the first 5 months of life (70). Robinette and others reported an average birth mass of 3.76 kilograms for male mule deer and 3.62 kilograms for females, with mass gains over the next 5 months of 0.22 kilograms/day for males and 0.21 kilograms/day for females (71). Mass gains were higher when the amount of dietary supplements was increased (71).

The rate of growth in mule and black-tailed deer slows after 5 to 6 months, and individuals begin to exhibit annual mass cycles, increasing from spring to fall and declining from fall to spring (Figure 30–3) (37, 12). Growth in female mule deer continues until around 2 years of age when they reach about 60 to 75 kilograms, while males may show additional increases in body mass until they reach 9 to 10 years of age and about 90 to 115 kg (72).

Fat Deposition and Mobilization. Fat deposition and subsequent mobilization occur at different locations and at different rates in mule and black-tailed deer (73). Blood serum lipid levels such as cholesterol and tryglycerides fluctuate as a function of time since last meal, and are of little use in assessing body condition in deer (74). Subcutaneous fat stores such as those on the back and brisket and abdominal fat around the kidney and heart are deposited and mobilized more slowly, and high amounts indicate an animal in good condition. Marrow fat in bones such as the femurs, metatarsals, and mandibles is the first to be deposited and the last to be mobilized, and an animal on a poor plane of nutrition will use up most of its kidney and back fat before it starts to draw on marrow fat as a source of energy (75).

In general, mule and black-tailed deer store fat rapidly during spring and summer and deplete it during fall and winter (75, 12). The low point in the fat storage cycle in males occurs following rut and into winter and early spring (75). Females undergo an annual cycle of fat deposition and loss less pronounced than that of males (Figure 30–4), and they reach a low point in their fat storage cycle during late gestation and lactation because of the high energetic demands of producing milk for young fawns (75). Females that do not produce a fawn or who lose their fawns soon after giving birth are able to start replenishing their fat reserves earlier and enter the following winter in better condition. Adult females are generally able to maintain higher fat reserves during the critical winter months, hence their survival rates are usually higher than that of adult males.

Water Requirements. Various studies have reported daily water intake rates for mule and black-tailed deer from 24 to 104 milliliters/kilogram body mass, depending on season (14). Where mule deer live in arid and semiarid environments they are adapted to a scarcity of free water. Desert mule deer in Arizona on average visited sources of water once

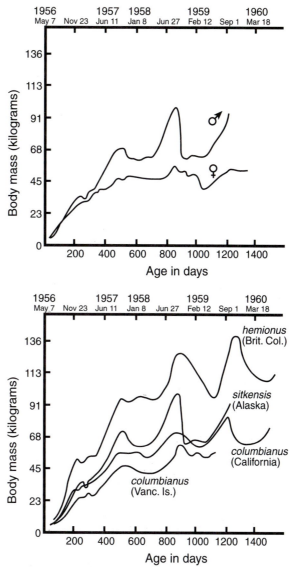

Figure 30–3 Growth patterns in male and female Sitka black-tailed deer (top) and male mule and black-tailed deer (bottom), all maintained on a high plane of nutrition (After Ref. 37, cited in Ref. 27.)

a day and consumed 5 to 6 liters of water per visit during the hot summer months, while visitation rates and amount of water consumed per visit declined during cooler seasons of the year (76, 77). Mule deer also obtain water from succulent plant material, dew on the surface of plants, and from metabolic processes (47). Feeding at night in hot, arid environments provides not only relief from thermal stress but may also be timed to take advantage of diel cycles in plant water content (78). Whether mule deer actually require free surface water has been debated (79). However, when access to free water is severely restricted in penned white-tailed deer, they reduce their consumption of forage (80). Therefore, although deer in the wild may exist for some periods of time without access to standing water this likely represents a marginal, survival situation (79).

The abundance and spacing of water sources does influence the distribution of mule deer in arid environments. In Arizona and New Mexico, mule deer are usually found within 2.4 kilometers of free water (81, 82, 83). Mule deer in northern

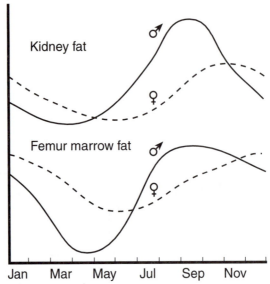

Figure 30–4 Generalized annual cycle of fat storage and depletion for mule deer in Colorado. (After Ref. 75, cited in Ref. 27.)

California averaged 1.19 to 1.55 kilometers away from water sources, with a mean greatest distance of 2.46 kilometers (84). Female mule deer drink more water than males during late summer (77). Females are often found closer to sources of water than males, presumably because of the demands of lactation, although in one study females remained close to water sources year-round and not just during lactation (85, 86, 84, 87). Recommendations for the spacing of free water for mule deer have included a maximum of 4.0 to 4.8 kilometers in New Mexico, and a preferred spacing in northern California of less than 3.2 kilometers with a maximum of 4.6 kilometers (83, 84). However, artificial water developments may have some negative ecological consequences in arid regions, and their use should be carefully considered (88, 89).

Thermal Relationships.   Mean body temperature in mule deer has been reported at 38.0 to 38.2°C, and at 37.5 to 39.7°C (90, 91). Yet mule deer exist across a wide range of temperature gradients averaging from −15°C or less during winter in the Rocky Mountains to over 30°C in the summer in the southern end of its range (12). Thermoregulation is accomplished by shivering, changing posture, erecting hair, and by making use of environmental temperature variations afforded by different habitats and topographic features (13). During winter in Colorado, mule deer seek out a preferred thermal zone between −9° and 7°C (92).

Deer and other cervids are adapted to cold winter temperatures because of the thickness of their winter coats. The lower boundary of the thermal neutral zone for black-tailed deer in winter pelage was measured at −10°C, while the lower and upper boundaries for deer in summer pelage were measured at 10° and 27°C (93). Rainfall occurring during summer may pose significant energetic costs of thermoregulation to black-tailed deer as a result of the differences in the lower boundary (93).

Snow depth has a great influence on the movements of mule and black-tailed deer (13). Snow depths of 25 to 30 centimeters may impede movements of mule deer in Colorado, and depths exceeding 50 centimeters may completely prevent their use of areas (92). During the heaviest snow accumulations in the Bridger Mountains, Montana, mule deer may be restricted to only 20% to 50% of their winter range (13).

Social Structure

Social structure in mule and black-tailed deer consists of "female clans related by maternal descent that are facultative resource defenders and males dispersed as individuals or groups of unrelated individuals" (40, p. 213). Adult females are often seen with their offspring, particularly females, from the previous year. Once their fawns are born, they are relatively intolerant of yearlings and other deer while nursing but associate with them between nursing bouts. Neonatal mule and black-tailed deer are often considered classical hider-type ungulates, in which the young fawns remain hidden for most of the day while the dam may be as much as 500 meters away foraging or resting (94, 40). The dam will return to her fawns to nurse them three to four times each day. Where abundant forage resources are spatially concentrated, however, fawns as young as 2 to 3 days have been observed accompanying adult females out into open meadows with groups of mixed-sex adults as the adults feed (95). Indeed, the distinction between hider and follower types of neonates is indistinct.

As with many other polygynous ruminants, mule and black-tailed deer exhibit strong patterns of sexual segregation between adult females and males in all seasons of the year except during rut, although detection and measurement of this phenomenon is scale dependent (85, 86, 96). Numerous reasons for such segregation have been suggested but only two hypotheses are currently considered viable (28, 97). The first hypothesis is that because of their smaller size and smaller mouthparts, females can better use closely cropped forages and thereby competitively displace larger bodied males. The second hypothesis is that females with young are restricted to habitats that provide adequate cover for survival of their young (98).

Movements

Unidirectional movements greater than 5 kilometers were considered dispersal among black-tailed deer, and most occurred between 1 and 2 years of age (99). Dispersal among young males averaged 15.2 kilometers and among females 12.2 kilometers. Dispersal movements greater than 12 kilometers were rare, although one yearling male moved 16.8 kilometers to establish a new home range just before its first birthday (100). One young male mule deer in Arizona dispersed over 44 kilometers (101).

Where resources are abundant and climate is mild, mule and black-tailed deer can be year-round residents (31). However, deer often exhibit migratory behavior, moving to high elevation montane ranges during summer to take advantage of seasonally abundant herbaceous forages and retreating to lower elevation ranges in winter when snow starts to fall (102). Migratory mule deer can concentrate on winter ranges in large numbers (Figure 30–5).

In some instances, herds of mule deer may be comprised of migratory and non-migratory individuals (102). Migratory deer may be at greater risk of predation than resident deer in such cases, but may be selected for during years when winter weather is particularly severe. During mild years, resident deer are favored. Climate variability in such a system likely allows the maintenance of both behaviors in a single population of deer (102).

Mule and black-tailed deer are strong swimmers and cross open bodies of water during migration (40, 103, 68). In southeast Alaska, black-tailed deer swim across Frederick Sound for distances greater than 22 kilometers (103). In northern California, groups of black-tailed deer have been regularly observed swimming across a reservoir for periods as long as 23 minutes and distances greater than 1 kilometer (104). Although deer commonly swim open bodies of water during migration without ill effect, unusually cold weather may allow a partial sheet of ice to build up

Figure 30–5 Migratory mule deer often concentrate on winter ranges in large numbers. (Photograph by D. J. Freddy.)

around the margins of freshwater reservoirs that would otherwise be ice-free. In such cases, deer may become trapped and suffer increased rates of winter mortality (105).

Home-range size in mule and black-tailed deer is related to sex, age, body mass, season, race, habitat, computational method, and other factors (13, 14). Reported home-range sizes have ranged from 39 hectares for a yearling female black-tailed deer in western Oregon to 3,379 hectares for an adult female mule deer in the open plains of South Dakota, reflecting general trends between closed and open habitats (106, 107, 13). Mule deer generally have larger home ranges than black-tailed deer, but again, this may reflect differences in the structure of typically occupied habitats rather than inherent interspecific differences (100).

The reasons for variations in the sizes of home ranges in deer are not well understood. Riley and Dood reported decreases in home-range size among mule deer fawns with increasing population size in Montana (108). Hervert and Krausman observed that when denied access to water, female mule deer in Arizona searched outside their known home ranges for alternate sources (76). In California, cattle grazing led to increases in home-range size among mule deer on summer range, and in 3 of 4 years of a study of black-tailed deer on winter range (109, 110).

Measures of landscape heterogeneity (e.g., number of different types of habitat patches, habitat patch shape and spatial arrangement, and contrast between adjacent patches) may also play important roles in determining home-range size in deer. Landscape metrics such as patch richness density, mean nearest neighbor, double log fractal dimension, and mean edge contrast index accounted for more than half the variation in home-range size in 80 female mule and black-tailed deer from five sites in California (J. Kie, unpublished data).

Communication

Sebaceous and sudoriferous secretory cells located in various integumentary glands in mule and black-tailed deer produce specific pheromones that elicit reactions of conspecifics (111, 112, 14). Secretions from the metatarsal glands on the outer hind foot act as alarm pheromones, those from the tarsal glands on the inner hind foot aid in recognition, and those from the interdigit glands may leave a scent trail (111). Urine is often deposited on the tufts of hair around the tarsal glands, and this rub-urinating behavior is a dominance display in adult mule deer (111, 40, 14).

Behavior

Territoriality

The question of whether deer in North America are territorial has resulted in considerable debate (113). Those authors that have argued that white-tailed deer, black-tailed deer, and mule deer exhibit territoriality commonly have done so on the basis of adult females excluding others from areas near their fawns for up to 4 weeks postpartum (114, 115, 116, 117, 113). Because adult females may show high fidelity to fawning sites from one year to the next, some have suggested that selection of sites may indeed be related to forage resources and where such resources are abundant, mule and black-tailed deer exhibit facultative territoriality (40). The counterargument is that no firm data exist to show that deer defend exclusive areas solely to sequester forage resources, and even during parturition adult female home ranges overlap to some degree (54, 118). Indeed, where herbaceous forage is abundant on summer ranges, black-tailed deer may reduce agonistic behavior toward other deer following parturition (95). Whether or not mule and black-tailed deer exhibit true territoriality can only be resolved by the acceptance of a common definition followed by the careful construction and testing of appropriate hypotheses (113).

POPULATION DYNAMICS

Population Trends and Current Status

Seton estimated the mule and black-tailed deer population in North America at 10 million at the time of contact with European explorers, but firsthand accounts of the explorers invariably indicated a scarcity of deer in the west relative to bison, elk, and pronghorn (119, 120). As with most North American large mammal populations, overhunting precipitated declines in the late 1800s, but habitat loss and deterioration caused by human economic activity were becoming a more important factor (121, 122, 123). By 1908, there were only about a half million mule and black-tailed deer remaining (124).

Populations increased from the 1920s to the 1960s, when there were about 4.7 million mule and black-tailed deer in the United States (124). Much of the increase was attributable to irruptions in Montana, Wyoming, Idaho, Utah, and Nevada (125). Four major hypotheses were presented to explain these irruptions: (1) succession from grassland to shrubland, (2) conversion of forests to shrubland, (3) predator control, and (4) reduction in livestock numbers. Gruell concluded that data best supported the first hypothesis (125). A period of decline in deer numbers began as early as 1950 in some areas and lasted until about 1980 (120). The primary cause of the decline was habitat loss or degradation (124). Inadequate census methodology, insufficient ecological knowledge, and a tradition of uncritical acceptance of new ideas all cast doubt on the magnitude and even existence of the decline (29). Those factors may have caused an inaccurate portrayal of deer population trends in many areas, but the extent and impact of habitat loss resulting from economic development clearly played a role. For example, reservoirs and roads each replaced millions of hectares of deer habitat in the west during the latter half of the 20th century (126).

In many areas of the intermountain West, the succession from grassland to shrubland that had benefited mule deer populations earlier in the century continued to a less than optimal stage with excessive ratios of woody : herbaceous vegetation (125). Continued overgrazing by cattle removed ground cover needed to carry wildfire that had been instrumental in checking succession. Fire suppression by the federal government also played a role in advancing succession (127).

State game agency figures suggest that mule and black-tailed deer populations have been relatively stable since about 1980 (128). While habitat loss has continued, expertise in census methodology has increased, confounding interpretations of long-term trends. Under pressure to produce more deer, managers may be obtaining numbers that are "acceptably precise, although inaccurate" (50, p. 7). Despite the state agency compilations, there is a general impression among managers that the range-wide mule deer population has been declining in recent years (128).

Productivity and Recruitment

An example of a mule deer population approaching its biotic potential comes from the Kaibab Plateau, Arizona, where, the population went from about 3,500 in 1906 to 100,000 in 1924 (129). These figures correspond to an average exponential rate of increase of 0.19 [or $\log_e(100,000/3,500)/18$ year] for an average annual percentage increase of 20% (130). However, the estimates of deer population size in 1924 range from 30,000 to 100,000, casting doubt on the accuracy of the population growth estimate (131, 132).

Another irruption in mule deer occurred between 1917 and 1939, when the Cache, Utah, population grew from approximately 250 to 6,000, corresponding to an average exponential rate of increase of 0.14 [or $\log_e(6,000/250)/22$ year] for an average annual percentage increase of 16% (133, 130). In the Kaibab and Cache populations the figures account for relatively long-term trends, in which individual years of higher growth rates have been embedded.

The concept of biotic potential is useful in establishing a ceiling for discussions about population growth rates. Normally, however, populations tend to fluctuate in relative equilibrium with dampened dynamics. The influence of nutritional levels on adult female reproductive rates has been discussed. Other factors that operate to prevent a deer population from achieving its biotic potential at a specific location are mortality and emigration.

Mortality Factors

Mortality factors include hunting, nonhuman predation, disease, starvation, or accidents. Periodically, one of these sources of mortality is readily identified as the limiting factor of a population, offering managers relative certainty regarding the potential effects of management practices. Usually, however, it is difficult to identify a single limiting factor. Moreover, mortality in deer is often difficult to detect and its cause difficult to determine. Finally, the net effect on population size is sometimes unclear, because of compensatory mortality factors (134, 135).

In addition to providing a source of recreation, hunting may be used to regulate mule and black-tailed deer populations (136). In such cases, harvest of females is necessary to affect population size (54). Conversely, hunting can be compensatory to other mortality factors in many instances, offering managers the opportunity to increase hunting opportunities with little effect on deer populations.

The most common predators of mule and black-tailed deer are coyotes, which overlap the geographic distribution of mule and black-tailed deer (137, 138, 139, 45, 140). Mountain lions prey heavily on deer and are more efficient than coyotes at taking adults (141, 142, 45, 143). Black bears and bobcats are common predators of fawns (144, 141, 45). Grizzly bears and wolves will also kill deer where their ranges overlap (145, 146). Golden eagles also kill fawns, and dogs kill deer on occasion (45, 126).

Predation should not be considered as a regulating factor without simultaneously considering habitat conditions. Better forage conditions enable deer to spend less time feeding, lowering the chances of predation, (147, 147a). There is an interaction between foraging and predation, and it is meaningless to consider these factors in isolation (148, 149). Finally, lethal control of predators is socially unacceptable in many cases, thus lowering predation may be more efficiently achieved by improving habitat conditions such as hiding cover (150).

Viral diseases known or suspected to infect mule and black-tailed deer include bluetongue and epizootic hemorrhagic disease, foot-and-mouth disease, malignant catarrhal fever, rabies, and bovine virus diarrhea/mucosal disease complex (151, 152). Few of these have been documented at epizootic levels, but an outbreak of foot-and-mouth disease among mule deer in California resulted in the shooting of more than 22,000 deer between 1924 and 1926 (144).

Bacterial diseases known or suspected to infect mule and black-tailed deer include pasteurellosis, brucellosis, necrobacillosis, actinomycosis, blackleg, caseous lymphadenitis, and anthrax. Necrobacillosis has been noted at epizootic levels (153, 154, 151, 155).

There are numerous parasitic diseases of mule and black-tailed deer, including elaeophorosis and setaria (152). Parelaphostrongylosis could become problematic as white-tailed deer continue to expand their range in the West (156). Numerous gastrointestinal parasites, lungworms, foot worms, eye worms, tapeworms, legworms, trematodes, and botfly larvae subcutaneously parasitize mule and black-tailed deer, while black flies, houseflies, horseflies, blowflies, stableflies, botflies, gnats, mosquitoes, lice, mites, ticks, and fleas parasitize mule and black-tailed deer cutaneously (31, 157, 152, 151, 152). Neoplastic disease (tumors) also occurs, especially among black-tailed deer in California (158).

Severe weather can cause mortality directly or can lead to nutritional stress and therefore susceptibility to predation, accidents, and diseases. Probably the worst weather combination for mule and black-tailed deer is heavy snow, high winds, and very low temperatures. High winds and cold temperatures cause rapid convective heat loss, creating higher energy and therefore foraging demands. Meanwhile, deep snows prevent deer from obtaining cured herbaceous plants and low shrubbery, and make locomotion a more energy-intensive activity (159). High winds can pack snow tightly, making it difficult to paw through, and can pile it into high drifts that act as barriers to travel (160). In summer, high temperature in combination with drought lowers the production of forage and limits the distribution of water. As with winter extremes, the lowered nutritional status that results makes deer weaker and more easily preyed upon and prone to accidents.

Robinette and others observed six mule deer fawns with broken limbs over the course of 15 years in Utah (45). They noted that fawns occasionally fell into abandoned mine shafts, which are abundant in many parts of the West. Deer frequently get caught in fences during all seasons, and have been trapped in steel water tanks and concrete water canals (161, 162, 163).

The number of deer killed annually by automobiles along highways has exceeded 1,000 in Arizona, Colorado, Kansas, Montana, Nebraska, South Dakota, Texas, and Wyoming (126). Given that mule deer are the most common deer species in several of these states, mule deer probably accounted for much of this mortality.

POPULATION MANAGEMENT

The long-term prospects for mule and black-tailed deer is a function of land development in the West, and maintaining the base of relatively undeveloped public lands is especially important. Maintenance of deer populations will eventually require coordination with the economic development of lands that provide habitat, and deer managers might be most effective by helping to educate the public about this fundamental principle.

On lands that remain undeveloped or are used in a manner compatible to simultaneous deer production, the emphasis should be on maintaining good habitat or improving poor habitat, while judiciously considering the needs of other species and other natural resources managers.

For the sake of efficient breeding, genetic integrity, and hunter satisfaction, it is also desirable to maintain a significant cohort of older males (164). Maximizing deer production has resulted in habitat deterioration in some areas, low male : female ratios, young age structures, poor productivity and recruitment, and a lack of large-antlered males, as has also been demonstrated with white-tailed deer (165, 166). The perception of managers that total numbers of mule deer are declining may simply reflect the presence of fewer and smaller males (L. Carpenter, Wildlife Management Institute, personal communication). The social structure of cervid populations on state and federal lands is one of the primary reasons for the high prices obtained for hunting permits on the Indian reservations and large private landholdings of the West, where less intensive hunting pressure in the presence of intact predator populations has resulted in the maintenance of high male:female ratios, old males, and large antlers.

As Giles pointed out, wildlife management includes habitat management, animal management, and human management (167). The challenge, then, is not to produce high numbers of deer using intensive management techniques, but to use educational approaches to reduce demand for low-quality hunting opportunities, and to increase demand for healthy social structures and habitat quality. However, habitat management and animal management are themselves an increasingly social enterprise. By their nature, deer managers may prefer to work directly with animals and habitat, but their efforts will meet with futility if they neglect their human management responsibilities.

HABITAT REQUIREMENTS

Mule and black-tailed deer are adapted to a wide variety of habitats throughout western North America (Figure 30–2). They occur in all life zones except the arctic, tropics, and most xeric deserts (13). Typical habitats can be found throughout species strongholds from northern Arizona and New Mexico northward to British Columbia and Alberta along the Rocky Mountains and associated basins (13). Mule deer in these habitats are typically migratory, spending summers at high elevations and retreating to foothill and basin ranges in winter. Montane and subalpine forest dominate summer ranges, along with wet meadows, riparian areas, aspen stands, and other high-value habitats. Winter ranges include shrub dominated habitats, where mule deer often prefer open ridges and south-facing slopes where snow does not accumulate (13).

Mule deer are also found on open plains and prairies at the eastern edge of their geographic range. These deer are usually nonmigratory and make extensive use of topographic cover associated with breaks along river drainages (13). In these instances, habitat use may be patchy and involve large movements by individuals (13). Mule deer also occur in semidesert habitats in the southwestern United States and Mexico. Here they are often found in low numbers and most commonly near sources of water. They are also found in the northern desert ranges in the Great Basin of Nevada, western Utah, and southeast Oregon (13). Black-tailed deer can be found in shrub-dominated habitats in the southern portion of their range in California. Vegetation consisting of mixed species of shrubs along with oak woodlands provides higher quality habitat than do dense stands of chamise chaparral, and deer densities are correspondingly higher (31).

Mule and black-tailed deer have long been considered a species that thrives in early to midseral vegetation following natural or human-induced disturbance (27). Wildfire, prescribed burning, and clear-cut logging have resulted in increased numbers of deer in many areas of the West in the past. Increases in forage production following such disturbances generally benefit mule and black-tailed deer. For example, in spruce forests of the northern Pacific Coast, deer forage is most abundant between about 5 and 10 years following disturbance such as logging (Figure 30–6). As forest stands start to mature and trees shade out desirable forage species, the value of habitat to deer declines. In old-growth forests, openings created when large trees fall over allow sunlight to reach the forest floor and forage plants for deer to become reestablished. In coniferous forests where snow accumulation is heavy, such as in northern British Columbia and southeast Alaska, new forage created by disturbance is largely unavailable to deer during winter. In such situations, resident deer are dependent on older forests where tree canopy cover is sufficient to intercept snow and provide access to understory forage supplies (168). In these areas, black-tailed deer are climax-associated species rather than successional species.

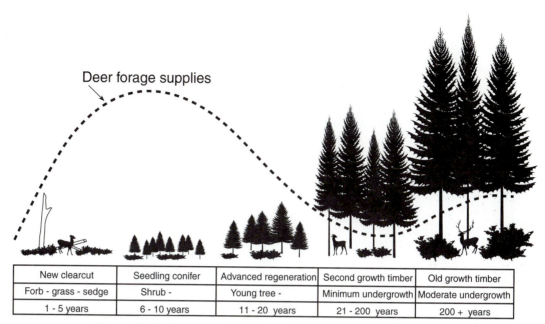

Deer forage supplies

New clearcut	Seedling conifer	Advanced regeneration	Second growth timber	Old growth timber
Forb - grass - sedge	Shrub -	Young tree -	Minimum undergrowth	Moderate undergrowth
1 - 5 years	6 - 10 years	11 - 20 years	21 - 200 years	200 + years

Figure 30–6 Generalized pattern of deer forage supplies during secondary succession in coniferous forest habitat. Duration of stages represents averages for western hemlock/Sitka spruce forests of the northern Pacific Coast. (After Ref. 25.)

HABITAT MANAGEMENT

Wallmo (1978) stated three general axioms applicable to habitat management for mule deer: (1) Early stages of plant succession are more beneficial than climax vegetation, (2) a mixture of plant communities provides better habitat than any single community, and (3) more browse is preferable to less browse. As with any broad generalities, these axioms need to be considered on a case-by-case basis. With respect to the first axiom, fire and logging have been considered favorable for producing high-quality habitat for mule and black-tailed deer (27). However, the value of old-growth, mature forest for black-tailed deer in coastal British Columbia and southeast Alaska has already been discussed. Furthermore, it is unlikely that large forest clear-cuts will be a silvicultural option on public lands in the future. Finally, the use of prescribed fire at a large enough spatial scale to have a significant effect on deer populations is expensive and may not see widespread use, while "let burn" policy is hampered by a formidable set of social and political factors (169). In reality, the extent of early and midsuccessional vegetation seen in the early 20th century as a result of wildfire, logging, and livestock grazing practices is unlikely to reoccur in the foreseeable future. As a result, numbers of mule and black-tailed deer in the western United States will remain lower than those seen during the mid-1900s. As a corollary, wildlife managers will need to develop tools and techniques for maximizing habitat benefits for deer in mid- and late successional habitats.

Wallmo's second axiom is nearly universally true (27). Mixes of plant communities that provide forage and cover for deer are preferable to monotypic blocks of single habitat types. How those mixes of habitats occur with respect to patch shape, spatial arrangement, and contrast between adjacent patches are important variables in determining their value to deer (147a).

Finally, the third axiom may apply only in limited situations. The best foraging habitats for mule and black-tailed deer contain abundant, nutritious plant species spatially arranged such that deer have access to those forages in a way that minimizes the risk of predation (147a). Those forage species are often highly digestible forbs and even grasses in early growth stages.

Chapter 30 / Mule and Black-tailed Deer

Direct Habitat Improvement

Seeding, fertilization, and prescribed burning have all been used as tools to improve habitat for mule and black-tailed deer. Such activities are expensive, difficult to accomplish on a scale large enough to substantially increase deer numbers, and may conflict with other land management goals.

Although direct habitat improvement projects may provide an illusion of actively benefiting deer, careful cost–benefit analyses should be performed using realistic estimates of anticipated population responses of deer prior to initiating such projects. In most cases, efforts directed at coordinating deer habitat management with other land uses such as livestock grazing, timber management, roads and recreation, and home site development have greater potential to provide benefits to mule and black-tailed deer.

Coordination with Livestock Management

The impacts of livestock grazing on mule and black-tailed deer can be classified as direct negative, indirect negative, operational, and beneficial (170, 171). Competition between cattle and deer for food would be a direct negative impact (170, 171). Factors affecting competition between deer and livestock include diet similarity, forage availability, animal distribution patterns, season of use, and behavioral interactions (79, 161). Indirect negative impacts of cattle grazing include (1) gradual reductions in the vigor of some plants and in the amount and quality of forage produced; (2) elimination or reduction of the ability of forage plants to reproduce; (3) reduction or elimination of locally important cover types and replacement by less favorable types or communities, either by direct actions over time or by changing the rate of natural successional processes; and (4) general alterations and reduction in the kinds, qualities, and amounts of preferred or otherwise important plants through selective grazing or browsing or other activities (170). Operational impacts include fence construction, water development, brush control, and livestock herding activities (170, 172, 173, 174, 175, 176). In some cases, prescribed livestock grazing can be used as a tool to manage habitat for mule and black-tailed deer (177, 178, 176, 179, 180, 181, 182, 183, 184).

Livestock use often affects mule deer in a variety of ways. On summer ranges in California, cattle stocking rates affected mule deer hiding cover, habitat use, home-range size, and activity patterns. Hiding cover for mule deer declined over the summer as a result of natural processes, even in the absence of cattle grazing (150). However, hiding cover is most important early in summer when fawns are young. Cattle grazing accelerated the decline in hiding cover for deer early in summer, and the effects were nonlinear, with heavy grazing having a more pronounced effect. Deer home ranges were smallest in the absence of cattle, and were centered primarily in creek-bottom, meadow-riparian habitats (109). Under heavy cattle grazing, centers of activity did not shift, but home ranges became larger and included areas with steeper slopes. Patterns of habitat use also changed, and deer spent more time feeding with higher levels of cattle grazing (185, 147).

Livestock management practices that can affect deer include livestock numbers, timing and duration of grazing, animal distribution, livestock types, and specialized grazing systems (186). The timing and duration of grazing, along with livestock distribution, are the bases for various specialized grazing systems, including continuous grazing, deferred-rotation, and rest-rotation grazing. Timing of livestock grazing is critical to mule and black-tailed deer, and they are particularly susceptible to adverse effects during fawning season (150, 185, 147, 109). Therefore, carefully designed deferred-rotation and rest-rotation grazing systems have great potential benefit to deer (Figure 30–7) (186).

Livestock congregate around sources of water, supplemental feed, and mineral blocks, and their impacts are most pronounced in those areas. Riparian zones, because of their abundant forage and water, are good examples of livestock concentration areas. Cross-fencing, developing alternative water sources, and providing mineral and

Figure 30-7 Overgrazed ranges provide few benefits to mule deer (Left). However, carefully designed grazing programs can provide benefits to deer similar to those in this ungrazed meadow (Right). (Photograhs by J. Kie and H. Ritter.)

nutritional supplements on upland sites away from riparian areas more evenly distribute livestock and often benefit deer (186).

Deer often crawl under fences when not hurried, but jump them when startled or chased (161). When a deer jumps a fence, its feet can become entangled between the top wires, resulting in death. This problem can be reduced by limiting total fence height to 96 centimeters (186). The top wire should be separated from the next wire by 30 centimeters. The bottom wire should be smooth and placed 41 centimeters above ground level. Wire stays should be placed every 2.5 meters between posts to keep the top wires from twisting around the leg of a deer. Net-wire fences no higher than 91 centimeters allow movement of adult deer but prevent passage of fawns. They should be avoided on summer and fall migration routes used by deer.

Coordination with Timber Management

Timber management programs treat large areas of forested lands each year and provide an important source for indirect management of deer habitats. The array of activities associated with timber management changes the structural characteristics of forest stands and strongly influences the amount and distribution of forage and cover on deer ranges. Timber management offers an opportunity for restoring or improving early seral stage forest habitats that are critical for herd productivity in many areas. Well-planned timber management can enhance deer habitat and poorly planned projects can seriously degrade important areas.

With the exception of black-tailed deer in British Columbia and southeast Alaska, which are dependent on canopy cover in mature forests, mule deer generally benefit from the successional vegetation that becomes established following timber harvest (Figure 30-6). Silvicultural systems affect deer habitats by influencing vegetation structure and species composition within managed stands. While individual site responses are highly variable, the amount of understory response generally varies inversely with the amount of total overstory tree canopy remaining after harvest (187).

The length of time that successional vegetation is of value to deer and other early successional animal species is of course dependent on the specific forest type, soil quality, climate, and other factors. In general, however, truncating the successional sequence by use of herbicides, planting of seedlings, and other artificial regeneration practices will shorten the time period over which benefits to deer will occur (188).

The sizes and shapes of even-age, or clear-cut, harvest units can affect deer use by influencing the distance between forage and cover. Reynolds demonstrated that most deer activity occurs near the edge between forage and cover and use dramatically declines at distances exceeding 180 meters from edges (189). Thomas and others suggested that (1) circular harvest units that exceed about 10 hectares in size will receive very low foraging use, and (2) openings smaller than about 2 hactares will receive the most use (190).

Armleder and others provided examples of how black-tailed deer make use of microhabitat features in forests of British Columbia and how timber management practices can be used to improve deer habitat values (191). In those forests, black-tailed deer make extensive use of ridge tops, knolls, and topographic breaks, and proportionally less use of gully bottoms and dense stands of young trees (Figure 30–8). Deer may be prevented from foraging in open areas when snow is deep, but rely on stands of older trees with interlocking crowns to intercept snow as do black-tailed deer in southeast Alaska (168, 191, 30).

Timber harvest prescriptions must be written on a site-specific basis, but some general principles apply in incorporating deer habitat values. Leaving unlogged strips as screening cover next to roads will benefit deer by minimizing human disturbance. It may also be possible to design forest road intersections in such a way as to minimize the distances at which deer can be seen (Figure 30–8).

Coordination with Other Uses

Information regarding the response of mule and black-tailed deer to roads and vehicular traffic is surprisingly scarce and imprecise. Perry and Overly found that main roads had the greatest impact on mule deer and elk, and primitive roads the least impact (192). Roads through meadow habitats reduced deer and elk use, while roads through open forest habitat reduced habitat use by elk but not by mule deer. Rost and Bailey reported reductions in habitat use by mule deer and elk within 200 meters of roads (193). Deer use was reduced most near roads in shrubland habitats and less in pine and juniper habitats. Livezey also found that black-tailed deer whose home ranges were located within 200 meters of secondary roads were displaced by traffic during hunting seasons (194).

Where large numbers of mule deer are being killed in collisions with motor vehicles along paved highways, a variety of techniques are available to reduce deer mortality and damage to vehicles. Clearing screening cover from beside highways and reducing deer forage availability in the right-of-way can minimize deer–vehicle collisions. Simply providing a fenced crosswalk system that forces deer to cross highways at well-marked points where motorists can anticipate them can reduce deer mortality between 37% and 43% (195).

Mule and black-tailed deer respond to different forms of recreational use in a variety of ways. Adverse impacts can include the direct death of an animal or the predisposition to death at a later time (indirect mortality). Other more subtle effects include lowered productivity, reduced use of refugia, reduced use of preferred habitats, aberrant behavior, and stress (196). These have the potential to lead to reductions in reproductive success and increases in mortality. Recreational use by humans on foot and on motorized vehicles such as motorcycles, four-wheelers, and snowmobiles can result in behavioral changes, and in some cases, documented reductions in reproductive success in mule deer (197).

In the southern Sierra Nevada of California, Cornett and others found that habitat use by mule deer was reduced by about 30% within 45 meters of trails (198). Deer became habituated to the presence of humans hiking directly on trails, but retreated much sooner when humans were encountered away from well-used trails. Yarmoloy and others used all-terrain vehicles to follow three female mule deer for 9 minutes per day for 15 days during fall (197). The following spring, those three deer collectively raised a single fawn, although they exhibited normal reproductive outputs in the years preceding and following the experiment. Freddy and others found that mule deer in Colorado exhibited mild alert responses to people on snowmobiles at greater distances than they did to people on foot, but that they actively fled people on foot at a greater threshold distance (199). They suggested that minimizing all levels of response by deer to people on foot and on snowmobiles required

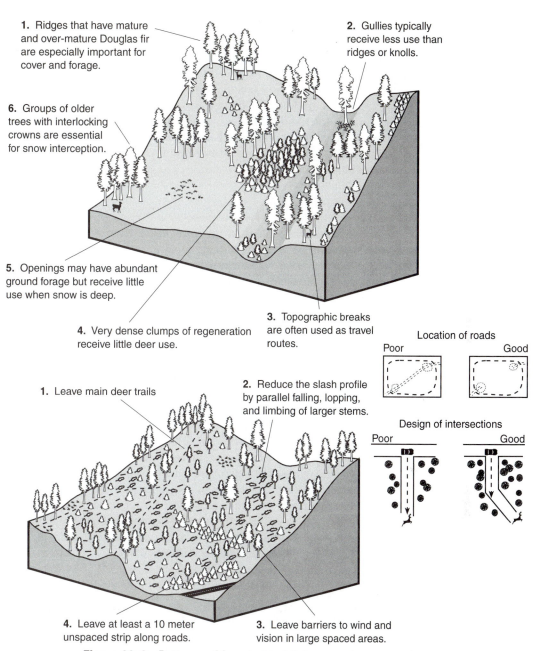

1. Ridges that have mature and over-mature Douglas fir are especially important for cover and forage.

2. Gullies typically receive less use than ridges or knolls.

6. Groups of older trees with interlocking crowns are essential for snow interception.

5. Openings may have abundant ground forage but receive little use when snow is deep.

4. Very dense clumps of regeneration receive little deer use.

3. Topographic breaks are often used as travel routes.

Location of roads
Poor   Good

1. Leave main deer trails

2. Reduce the slash profile by parallel falling, lopping, and limbing of larger stems.

Design of intersections
Poor   Good

4. Leave at least a 10 meter unspaced strip along roads.

3. Leave barriers to wind and vision in large spaced areas.

Figure 30–8 Patterns of forested habitat use and suggested management options to improve habitat for black-tailed deer on winter ranges in British Columbia. (After Ref. 192.)

buffers of greater than 334 meters and greater than 470 meters, respectively. Minimizing flight responses by deer would require buffers of at least 191 and 133 meters for people on foot and on snowmobiles, respectively (199). Flight distances in mule deer as in other ungulates, however, likely vary as functions of sex, habitat structure, social groupings, nutrition, reproductive status, specific individual experiences, and other factors (200).

In many places throughout the western United States, people are building houses in areas occupied by deer seasonally or year-round. Mule deer and black-tailed deer do become habituated to the presence of humans and may even become nuisances when they start to feed on ornamental landscaping. However, the general effect of the development of home sites is to reduce deer use in areas immediately adjacent to the structures. Smith and others found reductions in mule deer use close to rural houses in northern California (201). Within a 90-meter radius of the homes, deer use averaged only 61% of overall mean deer use (201). Similar data developed locally could be integrated with GIS analyses to buffer areas around home sites and used to predict reductions in carrying capacity. Vogel also noted that the number of mule deer seen per study plot in Montana dropped as a curvilinear function of the number of houses within 900 meters (202). The greatest adverse effect on mule deer occurred when houses were evenly distributed. Therefore, concentrating new home sites in already developed areas might lessen the impact on deer (202).

SUMMARY

Mule and black-tailed deer are prevalent throughout western Canada, Mexico, and the United States. They are important as big game species and play significant roles in ecosystem structure and function. Small-bodied ruminants like mule and black-tailed deer must feed on forages in which the nutrients are more concentrated and more digestible. Selection of specific dietary items by mule and black-tailed deer is varied and complex. Rocky Mountain mule deer eat at least 673 species of plants. Mule and black-tailed deer traditionally have been thought of as browsers, particularly during winter, relying primarily on twigs and other vegetative parts of woody plants. However, given access to seasonally abundant, nutritious, herbaceous plants of high digestibility, deer will select these species in preference to browse species of lower digestibility.

Most mule deer females breed for the first time at 1.5 years of age and bear their first fawns when they turn 2. Breeding among mule deer occurs in mid- to late autumn and early winter, with exact dates varying by subspecies and location. The mean length of gestation in mule deer is 200 to 208 days. The average number of fetuses per yearling female ranges from less than 0.66 to 1.71, and the average for adult females from 0.92 to 1.96. Adult females outnumber adult males in most populations by more than 2:1. Home-range size in mule and black-tailed deer has ranged from 39 hectares for a yearling female black-tailed deer in western Oregon to 3,379 hectares for an adult female mule deer in the open plains of South Dakota, reflecting general trends between closed and open habitats.

Populations of mule and black-tailed deer increased from the 1920s to the 1960s, when there were about 4.7 million mule and black-tailed deer in the United States. A period of decline in deer numbers began as early as 1950 in some areas and lasted until about 1980. The long-term prospects for mule and black-tailed deer remain a function of land development in the West.

Mule and black-tailed deer are adapted to a wide variety of habitats throughout western North America. They occur in all life zones except the arctic, tropics, and most xeric deserts. Fire and logging are generally favorable for producing high-quality habitat for mule and black-tailed deer. Old-growth, mature forest is important for black-tailed deer, however, in British Columbia and southeast Alaska in providing snow-free areas for foraging during winter. Although direct habitat improvement projects may provide an illusion of actively benefiting deer, efforts directed at coordinating deer habitat management with other land uses such as livestock grazing, timber management, roads and recreation, and home site development have far greater potential to provide benefits to mule and black-tailed deer.

LITERATURE CITED

1. LOOMIS, J., M. CREEL, and J. COOPER. 1989. Economic benefits of deer in California: Hunting and viewing values. Institute of Ecology Report 32. University of California, Davis, CA.
2. SWANSON, C. S., M. THOMAS, and D. M. DONNELLY. 1989. Economic value of big game hunting in southeast Alaska. Resource Bulletin RM-16. U.S. Forest Service, Rocky Mountain Forest and Range Experiment Station, Fort Collins, CO.

3. CONOVER, M. R. 1997. Wildlife management by metropolitan residents in the United States: Practices, perceptions, costs, and values. *Wildlife Society Bulletin* 25:306311.

4. AUSTIN, D. D., and P. J. URNESS. 1993. Evaluating production losses due to depredating big game. *Wildlife Society Bulletin* 21:397–401.

5. AUSTIN, D. D., P. J. URNESS, and D. DUERSCH. 1998. Alfalfa hay crop loss due to mule deer depredation. *Journal of Range Management* 51:29–31.

6. BANDY, P. J., and R. D. TABER. 1974. Forest and wildlife management: Conflict and coordination. Pages 21–26 *in* H. C. Black, ed., *Wildlife and Forest Management in the Pacific Northwest.* Oregon State University, Corvallis.

7. ROMIN, L. A., and J. A. BISSONETTE. 1996. Deer–vehicle collisions: Status of state monitoring activities and mitigation efforts. *Wildlife Society Bulletin* 24:127–132.

8. HANLEY, T. A. 1996. Potential role of deer (Cervidae) as ecological indicators of forest management. *Forest Ecology and Management* 88:199–204.

9. WALLIS DE VRIES, M. 1995. Large herbivores and the design of large-scale nature reserves in western Europe. *Conservation Biology* 9:25–33.

10. HOBBS, N. T. 1996. Modification of ecosystems by ungulates. *Journal of Wildlife Management* 60:695–713.

11. COWAN, I. McT. 1956. What and where are the mule and black-tailed deer? Pages 335–359 *in* W. P. Taylor, ed., *The Deer of North America.* Stackpole Books, Harrisburg, PA.

12. WALLMO, O. C. 1981. Mule and black-tailed deer distribution and habitats. Pages 1–15 in O. C. Wallmo, ed., *Mule and Black-Tailed Deer of North America.* University of Nebraska Press, Lincoln.

13. MACKIE, R. J., K. L. HAMLIN, and D. F. PAC. 1982. Mule deer. Pages 862–877 *in* J. A. Chapman and G. A. Feldhamer, eds., *Wild Mammals of North America—Biology, Management, Economics.* Johns Hopkins University Press, Baltimore, MD.

14. ANDERSON, A. E., and O. C. WALLMO. 1984. *Odocoileus hemionus. Mammalian Species* 219:1–9.

15. CRONIN, M. A. 1992. Intraspecific variation in mitochondrial DNA of North American cervids. *Journal of Mammalogy* 73:70–82.

16. TOMICH, P. Q. 1986. *Mammals in Hawaii,* 2nd ed. Bishop Museum Press, Honolulu, HA.

17. TELFER, T. C. 1988. Status of black-tailed deer on Kauai. *Transactions Western Section of The Wildlife Society* 24:53–60.

18. WHITEHEAD, G. K. 1972. *Deer of the World.* Viking Press, New York.

19. DEMMENT, M. W., and P. J. VAN SOEST. 1983. *Body Size, Digestive Capacity, and Feeding Strategies of Herbivores.* Winrock International, Morrilton, AK.

20. HOFFMAN, R. R. 1985. Digestive physiology of the deer: Their morphophysiological specialisation and adaptation. Pages 393–407 *in* P. F. Fennessy and K. R. Drew, eds., *Biology of Deer Production.* Royal Society of New Zealand.

21. PUTMAN, R. 1988. *The Natural History of Deer.* Comstock Publishing Associates, Cornell University Press, Ithaca, NY.

22. KLEIBER, M. 1961. *The Fire of Life.* Wiley, New York.

23. CROUCH, G. L. 1981. Coniferous forest habitats: Part 1, food habits and nutrition. Pages 423–433 *in* O. C. Wallmo, ed., *Mule and Black-Tailed Deer of North America.* University of Nebraska Press, Lincoln.

24. URNESS, P. J. 1981. Desert and chaparral habitats: Part 1, food habits and nutrition. Pages 347–365 in O. C. Wallmo, ed., *Mule and Black-Tailed Deer of North America.* University of Nebraska Press, Lincoln.

25. WALLMO, O. C., and W. L. REGELIN. 1981. Rocky Mountain and Intermountain habitats: Part 1, food habits and nutrition. Pages 387–398 *in* O. C. Wallmo, ed., *Mule and Black-Tailed Deer of North America.* University of Nebraska Press, Lincoln.

26. KUFELD, R. C., O. C. WALLMO, and C. FEDDEMA. 1973. Foods of the Rocky Mountain mule deer. USDA Forest Service Research Paper RM-111. U.S. Forest Service, Rocky Mountain Forest and Range Experiment Station, Fort Collins, CO.

27. WALLMO, O. C. 1978. Mule and black-tailed deer. Pages 31–41 *in* J. L. Schmidt and D. L. Gilbert, eds., *Big Game of North America: Ecology and Management.* Stackpole Books, Harrisburg, PA.

28. MAIN, M. B., and B. R. COBLENTZ. 1990. Sexual segregation in ungulates: A critique. *Wildlife Society Bulletin* 18:204–210.

29. GILL, R. B. 1976. Mule deer management myths and the mule deer population decline. Pages 99–106 *in* G. W. Workman and J. B. Low, eds., *Mule Deer Decline in the West.* College of Natural Resources, Utah State University, Logan.

30. HANLEY, T. A., C. T. ROBBINS, and D. E. SPALINGER. 1989. Forest habitats and the nutritional ecology of Sitka black-tailed deer: A research synthesis with implications for forest management. Forest Service General Technical Report PNW-GTR-230. U.S. Forest Service, Pacific Northwest Research Station, Portland, OR.

31. TABER, R. D., and R. F. DASMANN. 1958. The black-tailed deer of the chaparral chaparral: Its life history and management in the North Coast Range of California. Bulletin No. 8. California Department of Fish and Game.

32. BEALE, D. M., and N. W. DARBY. 1991. Diet composition of mule deer in mountain brush habitat of southwestern Utah. Publication 91–14:1–70. Utah Division of Wildlife Resources, Salt Lake City.

33. URNESS, P. J. 1969. Nutritional analyses and *in vitro* digestibility of mistletoes browsed by deer in Arizona. *Journal of Wildlife Management* 33:499–505.

34. KRAUSMAN, P. R., A. J. KUENZI, R. C. ETCHBERGER, K. R. RAUTENSTRAUCH, L. L. ORDWAY, and J. J. HERVERT. 1997. Diets of desert mule deer. *Journal of Range Management* 50:513–522.

35. ALLDREDGE A. W., J. F. LIPSCOMB, and F. W. WICKER. 1974. Forage intake rates of mule deer estimated with fallout cesium-137. *Journal of Wildlife Management* 38:508–516.

36. NORDAN, H. C., I. McT. COWAN, and A. J. WOOD. 1968. Nutritional requirements and growth of black-tailed deer, *Odocoileus hemionus columbianus,* in captivity. Pages 89–96 *in* M. A. Crawford, ed., *Comparative Nutrition of Wild Animals.* Symposia of the Zoological Society of London, No. 21. Academic Press, New York.

37. BANDY, P. J., I. McT. COWAN, and A. J. WOOD. 1970. Comparative growth in four races of black-tailed deer (*Odocoileus hemionus*). Part I. Growth in body weight. *Canadian Journal of Zoology* 48:1401–1410.

38. NORDAN, H. C., I. McT. COWAN, and A. J. WOOD. 1970. The feed intake and heat production of the young black-tailed deer (*Odocoileus hemionus columbianus*). *Canadian Journal of Zoology* 48:275–282.

39. SADLEIR, R. M. F. S. 1982. Energy consumption and subsequent partitioning in lactating black tailed deer. *Canadian Journal of Zoology* 60:382–386.

40. GEIST, V. 1981. Behavior: Adaptive strategies in mule deer. Pages 157–223 *in* O. C. Wallmo, ed., *Mule and Black-Tailed Deer of North America.* University of Nebraska Press, Lincoln.

41. ESTES, R. D. 1972. The role of the vomeronasal organ in mammalian reproduction. *Mammalia* 36:315–341.

42. THOMAS, D. C., and I. D. SMITH. 1973. Reproduction in a wild black-tailed deer fawn. *Journal of Mammalogy* 54:302–303.

43. NELLIS, C. H., J. L. THIESSEN, and C. A. PRENTICE. 1976. Pregnant fawn and quintuplet mule deer. *Journal of Wildlife Management* 40:795–796.

44. DIXON, J. S. 1934. A study of the life history and food habits of mule deer in California. *California Fish and Game* 20:1–146.

45. ROBINETTE, W. L., N. V. HANCOCK, and D. A. JONES. 1977. The Oak Creek mule deer herd in Utah. Publication 77–15:1–148. Utah State Division of Wildlife Resources, Salt Lake City.

46. BOWYER, R. T. 1991. Timing of parturition and lactation in southern mule deer. *Journal of Mammalogy* 72:138–145.

47. ANDERSON, A. E. 1981. Morphological and physiological characteristics. Pages 27–97 *in* O. C. Wallmo, ed., *Mule and Black-Tailed Deer of North America.* University of Nebraska Press, Lincoln.

48. RACHLOW, J. L., and R. T. BOWYER. 1991. Interannual variation in timing and synchrony of parturition in Dall's sheep. *Journal of Mammalogy* 72:487–492.

49. BERGER, J. 1992. Facilitation of reproductive synchrony by gestation adjustment in gregarious mammals: A new hypothesis. *Ecology* 73:323–329.

50. McCULLOUGH, D. R., D. S. PINE, D. L. WHITMORE, T. M. MANSFIELD, and R. H. DECKER. 1990. Linked sex harvest strategy for big game management with a test case on black-tailed deer. *Wildlife Monographs* 112:1–41.

51. ROBINETTE, W. L., J. S. GASHWILER, J. B. LOW, and D. A. JONES. 1957. Differential mortality by sex and age among mule deer. *Journal of Wildlife Management* 21:1–16.

52. VERME, L. J. 1969. Reproductive patterns of white-tailed deer related to nutritional plane. *Journal of Wildlife Management* 33:881–887.

53. VERME, L. J., and J. J. OZOGA. 1981. Sex ratio of white-tailed deer and the estrus cycle. *Journal of Wildlife Management* 45:710–715.

54. MCCULLOUGH, D. R. 1979. *The George Reserve Deer Herd: Population Ecology of a K-Selected Species.* University of Michigan Press, Ann Arbor.

55. KIE, J. G., and M. WHITE. 1985. Population dynamics of white-tailed deer *(Odocoileus virginianus)* on the Welder Wildlife Refuge, Texas. *Southwestern Naturalist* 30:105–118.

56. TRIVERS, R. L., and D. E. WILLARD. 1973. Natural selection of parental ability to vary the sex ratio. *Science* 179:90–92.

57. CLUTTON-BROCK, T. H., S. D. ALBON, and F. E. GUINESS. 1982. *Red Deer: Behavior and Ecology of Two Sexes.* University of Chicago Press, Chicago.

58. CLUTTON-BROCK, T. H., F. E. GUINESS, and S. D. ALBON. 1986. Great expectations: Dominance, breeding success, and offspring sex ratios. *Animal Behavior* 34:460–471.

59. HOEFS, M., and U. NOWLAN. 1994. Distorted sex ratios in young ungulates: The role of nutrition. *Journal of Mammalogy* 75:631–636.

60. SMITH, B. L., R. L. ROBBINS, and S. A. ANDERSON. 1996. Adaptive sex ratios: Another example? *Journal of Mammalogy* 77:818–825.

61. BIRGERSSEN, B. 1998. Adaptive adjustment of the sex ratio: More data and considerations from a fallow deer population. *Behavioral Ecology* 9:404–408.

62. DEGAYNER, E. J., and P. A. JORDAN. 1987. Skewed fetal sex ratios in white-tailed deer: Evidence and evolutionary speculations. Pages 178–188 *in* Christen M. Wemmer, ed., *Biology and Management of the Cervidae.* Research Symposia of the National Zoological Park, Smithsonian Museum. Smithsonian Institution Press, Washington, DC.

63. COWAN, I. McT. 1950. Some vital statistics of big game on over-stocked mountain range. *Transactions of the North American Wildlife Conference* 15:581–588.

64. MARTINKA, C. J. 1978. Ungulates populations in relation to wilderness in Glacier National Park, Montana. *Transactions North American Wildlife and Natural Resources Conference* 43:351–357.

65. FLOOK, D. R. 1970. Causes and implications of an observed sex differential in the survival of wapiti. Report Series Number 11, Canadian Wildlife Service, Ottawa.

66. WHITE, M. 1973. The whitetail deer of the Arkansas National Wildlife Refuge. *Texas Journal of Science* 24:457–489.

67. ROSS, R. C. 1934. Age and fecundity of mule deer *(Odocoileus hemionus hemionus). Journal of Mammalogy* 15:72.

68. COWAN, I. McT. 1956. Life and times of the coast black-tailed deer. Pages 523–617 *in* W. P. Taylor, ed., *The Deer of North America.* Stackpole Books, Harrisburg, PA.

69. MACKIE, R. J., D. F. PAC, K. L. HAMLIN, and G. L. DUSEK. 1998. Ecology and management of mule deer and white-tailed deer in Montana. Federal Aid Project W-120–R. Montana Fish, Wildlife, and Parks, Helena.

70. COWAN, I. McT., and A. J. WOOD. 1955. The growth rate of the black-tailed deer *(Odocoileus hemionus columbianus). Journal of Wildlife Management* 19:331–336.

71. ROBINETTE, W. L., C. H. BAER, R. E. PILLMORE, and C. E. KNITTLE. 1973. Effects of nutritional change on captive mule deer. *Journal of Wildlife Management* 37:312–326.

72. ANDERSON, A. E., D. E. MEDIN, and D. C. BOWDEN. 1974. Growth and morphometry of the carcass, selected bones, organs and glands of mule deer. *Wildlife Monographs* 39:1–122.

73. POND, C. M. 1978. Morphological aspects and the ecological and mechanical consequences of fat deposition in wild vertebrates. *Annual Review of Ecology and Systematics* 9:519–570.

74. LERESCHE, R. E., U. S. SEAL, P. D. KARNS, and A. W. FRANZMANN. 1974. A review of blood chemistry of moose and other Cervidae with emphasis on nutritional assessment. *La Naturaliste Canadien* 101:263–290.

75. ANDERSON, A. E., D. E. MEDIN, and D. C. BOWDEN. 1972. Indices of carcass fat in a Colorado mule deer herd. *Journal of Wildlife Management* 36:579–594.

76. HERVERT, J. J., and P. R. KRAUSMAN. 1986. Desert mule deer use of water developments in Arizona. *Journal of Wildlife Management* 50:670–676.

77. HAZAM, J. E., and P. R. KRAUSMAN. 1988. Measuring water consumption of desert mule deer. *Journal of Wildlife Management* 52:528–534.

78. TAYLOR, C. R. 1969. The eland and the oryx. *Scientific American* 220:88–95.

79. SEVERSON, K. E., and A. L. MEDINA. 1983. Deer and elk habitat management in the southwest. *Journal of Range Management,* Monograph No. 2.

80. LAUTIER, J. K., T. V. DAILEY, and R. D. BROWN. 1988. Effect of water restriction on feed intake of white-tailed deer. *Journal of Wildlife Management* 52:602–606.

81. HANSON, W. R., and C. Y. McCULLOCH. 1955. Factors influencing the distribution of mule deer on Arizona rangelands. *Transactions of the North American Wildlife Conference* 20:568–588.

82. SWANK, W. G. 1958. The mule deer in Arizona chaparral. Game Bulletin Number 3. Arizona Game and Fish Department, Phoenix.

83. WOOD, J. E., T. S. BICKLE, W. EVANS, J. C. GERMANY, and V. W. HOWARD, JR. 1970. The Fort Stanton mule deer herd. Agricultural Experiment Station Bulletin Number 567. New Mexico State University, Albuquerque.

84. BOROSKI, B. B., and A. S. MOSSMAN. 1996. Distribution of mule deer in relation to water sources in northern California. *Journal of Wildlife Management* 60:770–776.

85. BOWYER, R. T. 1984. Sexual segregation in southern mule deer. *Journal of Mammalogy* 65:410–417.

86. MAIN, M. B., and B. R. COBLENTZ. 1996. Sexual segregation in Rocky Mountain mule deer. *Journal of Wildlife Management* 60:497–507.

87. FOX, K. B., and P. R. KRAUSMAN. 1994. Fawning habitat of desert mule deer. *Southwestern Naturalist* 39:269–275. General Technical Report INT-206. United States Forest Service, Intermountain Research Station, Ogden, UT.

88. BROYLES, B. 1995. Desert wildlife water developments: Questioning the use in the Southwest. *Wildlife Society Bulletin* 23:663–675.

89. KRAUSMAN, P. R., and B. CZECH. 1998. Water developments and desert ungulates. Pages 138–154 *in* R. Pearlman, ed., *Proceedings of a Symposium on Environmental, Economic, and Legal Issues Related to Rangeland Water Development.* Center for the Study of Law, Arizona State University, Tempe.

90. THORNE, E. T. 1975. Normal body temperature of pronghorn antelope and mule deer. *Journal of Mammalogy* 56:691–698.

91. SARGENT, G. A., L. E. EBERHARDT, and J. M. PEEK. 1994. Thermoregulation by mule deer (*Odocoileus hemionus*) in arid rangelands of southcentral Washington. *Journal of Mammalogy* 75:536–544.

92. LOVELESS, C. M. 1967. Ecological characteristics of a mule deer winter range. Technical Bulletin Number 20. Colorado Game, Fish, and Parks Department, Denver.

93. PARKER, K. L. 1988. Effects of heat, cold, and rain on coastal black-tailed deer. *Canadian Journal of Zoology* 66:2475–2483.

94. LENT, P. C. 1974. Mother–infant relationships in ungulates. Pages 1–55 *in* V. Geist and F. Walther, eds., *The Behaviour of Ungulates and Its Relation to Management.* Publications New Series No. 24. IUCN, Morges, Switzerland.

95. BOWYER, R. T., J. G. KIE, and V. VAN BALLENBERGHE. 1998. Habitat selection by neonatal black-tailed deer: Climate, forage, or risk of predation? *Journal of Mammalogy* 79:415–425.

96. BOWYER, R. T., J. G. KIE, and V. VAN BALLENBERGHE. 1996. Sexual segregation in black-tailed deer: Effects of scale. *Journal of Wildlife Management* 60:10–17.

97. MAIN, M. B., F. W. WECKERLY, and V. C. BLEICH. 1996. Sexual segregation in ungulates: New directions for research. *Journal of Mammalogy* 77:449–461.

98. KIE, J. G., and R. T. BOWYER. 1999. Sexual segregation in white-tailed deer: Density dependent changes in use of space, habitat selection, and dietary niche. *Journal of Mammalogy* 80:in press.

99. BUNNELL, F. L., and A. S. HARESTAD. 1983. Dispersal and dispersion of black-tailed deer: Models and observations. *Journal of Mammalogy* 64:201–209.

100. HARESTAD, A. S., and F. L. BUNNELL. 1983. Dispersal of a yearling male black-tailed deer. *Northwest Science* 57:45–48.

101. SCARBROUGH, D. L., and P. R. KRAUSMAN. 1988. Sexual segregation by desert mule deer. *Southwestern Naturalist* 33:157–165.

102. NICHOLSON, M. C., R. T. BOWYER, and J. G. KIE. 1997. Habitat selection and survival of mule deer: Tradeoffs associated with migration. *Journal of Mammalogy* 78:483–504.

103. EINARSEN, A. S. 1956. Life of the mule deer deer. Pages 363–429 *in* W. P. Taylor, ed., *The Deer of North America.* Stackpole Books, Harrisburg, PA.

104. BOROSKI, B. B. 1998. Development and testing of a wildlife-habitat relationships model for Columbian black-tailed deer, Trinity County, California. Ph.D. dissertation. University of California, Berkeley.

105. RAPPAPORT, A. G., J. M. MITCHELL, and J. G. NAGY. 1977. Mitigating the impacts to wildlife from socioeconomic developments. *Transactions North American and Natural Resources Conference* 42:169–178.

106. MILLER, F. L. 1970. Distribution patterns of black-tailed deer (*Odocoileus hemionus columbianus*) in relation to environment. *Journal of Mammalogy* 51:248–260.

107. SEVERSON, K. E., and A. V. CARTER. 1978. Movements and habitat use by mule deer in the Northern Great Plains, South Dakota. *Proceedings International Rangeland Congress* 1:466–468.

108. RILEY, S. J., and A. R. DOOD. 1984. Summer movements, home range, habitat use, and behavior of mule deer fawns. *Journal of Wildlife Management.* 48:1302–1310.

109. LOFT, E. R., J. G. KIE, and J. W. MENKE. 1993. Grazing in the Sierra Nevada: Home range and space use patterns of mule deer as influenced by cattle. *California Fish and Game* 79:145–166.

110. KIE, J. G., and B. B. BOROSKI. 1995. The effects of cattle grazing on black-tailed deer during winter on the Tehama Wildlife Management Area. Final Report PSW-89-CL-030, U.S. Department of Agriculture, Forest Service, Pacific Southwest Research Station, Fresno, CA.

111. MÜLLER-SCHWARTZ, D. 1971. Pheromones in black-tailed deer (*Odocoileus hemionus columbianus*). *Animal Behavior* 19:141–152.

112. MÜLLER-SCHWARTZ, D., and C. MÜLLER-SCHWARTZ. 1975. Subspecies specificity of response to a mammalian social odor. *Journal of Chemical Ecology* 1:125–131.

113. WECKERLY, F. W. 1992. Territoriality in North American deer: A call for a common definition. *Wildlife Society Bulletin* 20:228–231.

114. OZOGA, J. J., L. J. VERME, and C. S. BIENZ. 1982. Parturition behavior and territoriality in white-tailed deer: Impact on neonatal mortality. *Journal of Wildlife Management* 46:1–11.

115. GAVIN, T. A., L. H. SURING, P. A. VOHS, JR., and E. C. MESLOW. 1984. Population characteristics, spatial organization, and natural mortality in the Columbian white-tailed deer. *Wildlife Monographs* 91:1–41.

116. MILLER, F. L. 1974. Four types of territoriality observed in a herd of black-tailed deer. Pages 644–660 *in* V. Geist and F. Walther, eds., *The Behaviour of Ungulates and Its Relation to Management.* Publications New Series No. 24. IUCN, Morges, Switzerland.

117. DUSEK, G. L., R. J. MACKIE, J. D. HERRIGES, JR., and B. B. COMPTON. 1989. Population ecology of white-tailed deer along the lower Yellowstone River. *Wildlife Monographs* 108.

118. BEIER, P., and D. R. MCCULLOUGH. 1990. Factors influencing white-tailed deer activity patterns and habitat use. *Wildlife Monographs* 109:1–51.

119. RUE, L. L. III. 1978. *The Deer of North America.* Outdoor Life Books. Crown, New York.

120. JULANDER, O., and J. B. LOW. 1976. A historical account and present status of the mule deer in the West. Pages 3–19 *in* G. W. Workman and J. B. Low, eds., *Mule Deer Decline in the West: A Symposium.* Utah State University, Logan.

121. TREFETHEN, J. B. 1975. *An American Crusade for Wildlife.* Winchester Press, Boone and Crockett Club, New York.

122. DISILVESTRO, R. L. 1989. *The Endangered Kingdom: The Struggle to Save America's Wildlife.* John Wiley and Sons, New York.

123. CZECH, B. 1997. The Endangered Species Act, American democracy, and an omnibus role for public policy. Ph.D. dissertation. University of Arizona, Tucson.

124. CONNOLLY, G. E. 1981. Trends in populations and harvests. Pages 225–243 *in* O. C. Wallmo, ed., *Mule and Black-Tailed Deer of North America.* University of Nebraska Press, Lincoln.

125. GRUELL, G. E. 1986. Post-1990 mule deer irruptions in the Intermountain West: Principal cause and influences. *USDA Forest Service General Technical Report* INT-206. Ogden, UT.

126. REED, D. F. 1981. Conflicts with civilization. Pages 509–535 *in* O. C. Wallmo, ed., *Mule and Black-Tailed Deer of North America.* University of Nebraska Press, Lincoln.

127. PYNE, S. J. 1982. *Fire in America—A Cultural History of Wildland and Rural Fire.* Princeton University Press, Princeton, NJ.

128. CARPENTER, L. 1998. *Deer in the West. Proceedings of the Western States and Provinces Deer and Elk Workshop.* Arizona Game and Fish Department, Phoenix.

129. RASMUSSEN, D. I. 1941. Biotic communities of Kaibab Plateau, Arizona. *Ecological Monographs* 3:229–275.

130. CAUGHLEY, G., and A. R. E. SINCLAIR. 1994. *Wildlife Ecology and Management.* Blackwell Scientific Publications, Boston, MA.

131. CAUGHLEY, G. 1970. Eruption of ungulate populations, with special emphasis on Himalayan thar in New Zealand. *Ecology* 51:54–72.

132. CAUGHLEY, G. 1977. *Analysis of Vertebrate Populations.* John Wiley and Sons, New York.

133. DOMAN, E. V., and D. I. RASMUSSEN. 1944. Supplemental winter feeding of mule deer in northern Utah. *Journal of Wildlife Management* 8:317–338.

134. MACKIE, R. J., K. L. HAMLIN, D. F. PAC, G. L. DUSEK, and A. K. WOOD. 1990. Compensation in free-ranging deer populations. *Transactions North American Wildlife and Natural Resources Conference* 55:518–533.

135. BARTMANN, R. M., G. C. WHITE, and L. H. CARPENTER. 1992. Compensatory mortality in a Colorado mule deer population. *Wildlife Monographs* 121:39.

136. CONNOLLY, G. E. 1981. Limiting factors and population regulation. Pages 245–285 *in* O. C. Wallmo, ed., *Mule and Black-Tailed Deer of North America.* University of Nebraska Press, Lincoln.

137. MURIE, A. 1940. Ecology of the coyote in the Yellowstone. Fauna of National Parks Bulletin No. 4. Government Printing Office, Washington, DC.

138. HORN, E. E. 1941. Some coyote–wildlife relationships. *Transactions North American Wildlife and Natural Resources Conference* 6:283–287.

139. TRAINER, C. E. 1975. Direct causes of mortality in mule deer fawns during summer and winter periods on Steens Mountain, Oregon. *Proceedings Western Association of Game and Fish Commissioners* 55:163–170.

140. AUSTIN, D. D., P. J. URNESS, and M. L. WOLFE. 1977. The influence of predator control on two adjacent wintering deer herds. *Great Basin Naturalist* 37:101–102.

141. RICHENS, V. B. 1967. Characteristics of mule deer herds and their range in northeastern Utah. *Journal of Wildlife Management* 31:651–666.

142. HORNOCKER, M. G. 1970. An analysis of mountain lion predation upon mule deer and elk in the Idaho primitive area. *Wildlife Monographs* 21.

143. SHAW, H. G. 1977. Impact of mountain lion (*Felis concolor azteca*) on mule deer and cattle in northwestern Arizona. Pages 17–32 *in* R. L. Phillips and C. Jonkel, eds., *Proceedings of the 1975 Predator Symposium.* University of Montana, Missoula.

144. LEOPOLD, A. S., T. RINEY, R. MCCAIN, and L. TEVIS JR. 1951. The Jawbone deer herd. Fish and Game Bulletin No. 4. California Department of Fish and Game.

145. CRAIGHEAD, J. J. 1995. *The Grizzly Bears of Yellowstone: Their Ecology in the Yellowstone Ecosystem, 1959–1992.* Island Press, Washington, DC.

146. KLEIN, D. R., and S. T. OLSON. 1960. Natural mortality patterns of deer in southeast Alaska. *Journal of Wildlife Management* 24:80–88.

147. KIE, J. G., C. J. EVANS, E. R. LOFT, and J. W. MENKE. 1991. Foraging behavior by mule deer: The influence of cattle grazing. *Journal of Wildlife Management* 55:665–674.

147a. KIE, J.G. 1999. Optimal foraging and risk of predation: Effects on behavior and social structure in ungulates. *Journal of Mammalogy* 80:in press.

148. MCNAMARA, J. M., and A. I. HOUSTON. 1987. Starvation and predation as factors limiting population size. *Ecology* 68:1515–1519.

149. SINCLAIR, A. R. E., and P. ARCESE. 1995. Population consequences of predation-sensitive foraging: The Serengeti wildebeest. *Ecology* 76:882–891.

150. LOFT, E. R., J. W. MENKE, and J. G. KIE, and R. C. BERTRAM. 1987. Influence of cattle stocking rate on the structural profile of deer hiding cover. *Journal of Wildlife Management* 51:655–664.

151. LONGHURST, W. M., A. S. LEOPOLD, and R. F. DASMANN. 1952. A survey of California deer herds: Their ranges and management problems. Fish and Game Bulletin No. 6. California Department of Fish and Game.

152. HIBLER, C. P. 1981. Diseases. Pages 129–155 *in* O. C. Wallmo, ed., *Mule and Black-Tailed Deer of North America.* University of Nebraska Press, Lincoln.

153. McLean, D. D. 1940. The deer of California with particular reference to the Rocky Mountain mule deer. *California Fish and Game* 26:139–166.

154. Rosen, M. N., O. A. Brunetti, and A. I. Bischoff. 1951. An epizootic of foot rot in California deer. *Transactions North American Wildlife and Natural Resources Conference* 16:164–177.

155. Honess, R. F., and K. Winter. 1956. Diseases of wildlife in Wyoming. Bulletin 9. Wyoming Game and Fish Department, Cheyenne.

156. Anderson, R. C. 1972. The ecological relationships of meningeal worm and native cervids in North America. *Journal of Wildlife Diseases* 8:304–310.

157. Brown, E. R. 1961. The black-tailed deer of western Washington. Biological Bulletin No. 13. Washington Department of Game, Olympia.

158. McTaggart-Cowan, I. 1946. Parasites, diseases, injuries and anomalies of the Columbian black-tailed deer, *Odocoileus hemionus columbianus* (Richardson) in British Columbia. *Canadian Journal of Research* 24:71–103.

159. Mattfeld, G. F. 1973. The effect of snow on the energy expenditure of waling white-tailed deer. *Proceedings Northeastern Fish and Wildlife Conference* 30:327–343.

160. Gilbert, P. F., O. C. Wallmo, and R. B. Gill. 1970. Effect of snow depth on mule deer in Middle Park, Colorado. *Journal of Wildlife Management* 34:15–23.

161. Mackie, R. J. 1981. Interspecific relationships. Pages 487–507 *in* O. C. Wallmo, ed., *Mule and Black-Tailed Deer of North America.* University of Nebraska Press, Lincoln.

162. Rautenstrauch, K. R., and P. R. Krausman. 1989. Preventing mule deer drowning in the Mohawk Canal, Arizona. *Wildlife Society Bulletin* 17:280–286.

163. Krausman, P. R., and R. C. Etchberger. 1995. Response of desert ungulates to a water project in Arizona. *Journal of Wildlife Management* 59:292–300.

164. Carpenter, L. H., and R. B. Gill. 1987. Antler point regulations: The good, the bad, and the ugly. *Proceedings of the Western Association of Fish and Wildlife Agencies* 67:94–107.

165. McCullough, D. R. 1984. Lessons from the George Reserve. Pages 211–242 *in* L. K. Hall, ed., *White-Tailed Deer: Ecology and Management.* Stackpole Books, Harrisburg, PA.

166. Teer, J. G. 1984. Lessons from the Llano Basin, Texas. Pages 261–260 in L. K. Hall, ed., *White-Tailed Deer: Ecology and Management.* Stackpole Books, Harrisburg, PA.

167. Giles, R. H. 1978. *Wildlife Management.* W. H. Freeman, San Francisco, CA.

168. Hanley, T. A. 1984. Relationships between black-tailed deer and their habitat. Forest Service General Technical Report PNW-GTR-168. U.S. Forest Service, Pacific Northwest Research Station, Portland, OR.

169. Czech, B. 1996. Challenges to establishing and implementing sound natural fire policy. *Renewable Resources Journal* 14:14–19.

170. Mackie, R. J. 1978. Impacts of livestock grazing on wild ungulates. *Transactions North American Wildlife and Natural Resources Conference* 43:462–476.

171. Wagner, F. H. 1978 Livestock grazing and the wildlife industry. Pages 121–145 *in* H. P. Brokaw, ed., *Wildlife and America.* Council on Environmental Quality, U.S. Government Printing Office, Washington, DC.

172. Hood, R. E., and J. M. Inglis. 1974. Behavioral responses of white-tailed deer to intensive ranching operation. *Journal of Wildlife Management* 38:488–498.

173. Wilson, L. O. 1977. Guidelines and recommendations for design and modification of livestock watering developments to facilitate safe use by wildlife. T/N 305. U.S. Bureau of Land Management, Washington, DC.

174. Rodgers, K. J., P. F. Ffolliott, and D. R. Patton. 1978. Home range and movement of five mule deer in a semidesert grass-shrub community. Research Note RM-355. U.S. Forest Service, Fort Collins, CO.

175. Holechek, J. 1981. Brush control impacts on rangeland wildlife. *Journal of Soil and Water Conservation* 36:265–269.

176. Holechek, J. 1982. Managing rangelands for mule deer. *Rangelands* 4:25–28.

177. Longhurst, W. M., E. O. Garton, H. F. Heady, and G. E. Connolly. 1976. The California deer decline and possibilities for restoration. *Cal-Neva Wildlife Transactions* 23:74–103.

178. Holechek, J. 1980. Livestock grazing impacts on rangeland ecosystems. *Journal of Soil and Water Conservation* 35:162–164.

179. URNESS, P. J. 1982. Livestock as tools for managing big game range in the Intermountain West. Pages 20–31 in J. M. Peek and P. D. Dalke, eds., *Wildlife–Livestock Relationships Symposium,* Proceedings No. 10. University of Idaho, Moscow.

180. URNESS, P. J. 1990. Livestock as manipulators of mule deer winter habitats in northern Utah. Pages 25–40 in K. E. Severson, Technical Coordinator. Can livestock be used as a tool to enhance wildlife habitat? USDA Forest Service General Technical Report RM-194. U.S. Forest Service, Rocky Mountain Forest and Range Experiment Station, Fort Collins, CO.

181. LONGHURST, W. M., R. E. HAFENFELD, and G. E. CONNOLLY. 1982. Deer–livestock interrelationships in the western states. Pages 409–420 in J. M. Peek, and P. D. Dalke, eds., *Wildlife–Livestock Relationships Symposium.* Proceedings Number 10. University of Idaho, Moscow.

182. SEVERSON, K. E., Technical Coordinator. 1990. Can livestock be used as a tool to enhance wildlife habitat? General Technical Report RM-194. U.S. Forest Service, Rocky Mountain Forest and Range Experiment Station, Fort Collins, CO.

183. WILLMS, W. A., A. MCLEAN, and R. RITCEY. 1979. Interactions between mule deer and cattle on big sagebrush range in British Columbia. *Journal of Range Management* 32:299–304.

184. REINER, R. J., and P. J. URNESS. 1982. Effect of grazing horses managed as manipulators of big game winter range. *Journal of Range Management* 35:567–571.

185. LOFT, E. R., J. W. MENKE, and J. G. KIE. 1991. Habitat shifts by mule deer: The influence of cattle grazing. *Journal of Wildlife Management* 55:16–26.

186. KIE, J. G., V. C. BLEICH, A. L. MEDINA, J. D. YOAKUM, and J. W. THOMAS. 1994. Managing rangelands for wildlife. Pages 663–688 in T. A. Bookhout, ed., *Wildlife Management Techniques,* 5th ed. The Wildlife Society, Bethesda, MD.

187. FFOLLIOTT, P. F., and CLARY, W. P. 1972. A selected and annotated bibliography of understory–overstory relationships. Technical Bulletin 198. University of Arizona Agricultural Experiment Station, Tucson.

188. LEOPOLD, A. S. 1978. Wildlife and forest practice. Pages 108–120 in H. P. Brokaw, ed., *Wildlife and America.* Council on Environmental Quality, Washington, DC.

189. REYNOLDS, H. G. 1966. Use of openings in spruce-fir forests of Arizona by elk, deer, and cattle. Research Note RM-66. U.S. Forest Service, Rocky Mountain Forest and Range Experiment Station, Fort Collins, CO.

190. THOMAS, J. W., H. BLACK, JR., R. J. SCHERZINGER, and R. J. PEDERSEN. 1979. Deer and elk. Pages 104–127 in *Wildlife Habitats in Managed Forests—The Blue Mountains of Oregon and Washington.* Agricultural Handbook No. 553. U.S. Department of Agriculture, Forest Service, Portland, OR.

191. ARMLEDER, H. M., R. J. DAWSON, and R. N. THOMPSON. 1986. Handbook for timber and mule deer management coordination on winter ranges in the Cariboo Forest Region. Land Management Handbook Number 13. British Columbia Ministry of Forests, Williams Lake, British Columbia, Canada.

192. PERRY, C., and R. OVERLY. 1977. Impact of roads on big game distributions in portions of the Blue Mountains of Washington, 1972–1973. Applied Research Bulletin Number 11. Washington Department of Game, Olympia.

193. ROST, G. R., and J. A. BAILEY. 1979. Distribution of mule deer and elk in relation to roads. *Journal of Wildlife Management* 43:634–641.

194. LIVEZEY, K. B. 1991. Home range, habitat use, disturbance, and mortality of Columbian black-tailed deer in Mendocino National Forest. *California Fish and Game* 77:201–209.

195. LEHNERT, M. E. and J. A. BISSONETTE. 1998. Effectiveness of highway crossing structures at reducing deer–vehicle collisions. *Wildlife Society Bulletin* 25:809–819.

196. POMERANTZ, G. A., D. J. DECKER, G. R. GOFF, and K. G. PURDY. 1988. Assessing impact of recreation on wildlife: A classification scheme. *Wildlife Society Bulletin* 16:58–62.

197. YARMOLOY, C., M. BAYER, and V. GEIST. 1988. Behavior responses and reproduction of mule deer, *Odocoileus hemionus,* does following experimental harassment with an all-terrain vehicle. *Canadian Field-Naturalist* 102:425–429.

198. CORNETT, D. C., W. M. LONGHURST, R. E. HAFENFELD, T. P. HEMKER, and W. A. WILLIAMS. 1979. Evaluation of the potential impact of proposed recreation development on the Mineral King deer herd. Pages 474–480 *in* G. A. Swanson, Technical Coordinator, *The Mitigation Symposium: A National Workshop on Mitigating Losses of Fish and Wildlife Habitats.* General Technical Report RM-65. U.S. Forest Service, Fort Collins, CO.

199. FREDDY, D. J., W. M. BRONAUGH, and M. C. FOWLER. 1986. Response of mule deer to disturbances by persons afoot and snowmobiles. *Wildlife Society Bulletin* 14:63–68.

200. ALTMAN, M. 1958. The flight distance in free-ranging big game. *Journal of Wildlife Management* 22:207–209.

201. SMITH, D. O., M. CONNOR, and E. R. LOFT. 1989. The distribution of winter mule deer use around homesites. *Transactions Western Section of The Wildlife Society* 25:77–80.

202. VOGEL, W. O. 1989. Response of deer to density and distribution of housing in Montana. *Wildlife Society Bulletin* 17:406–413.

Caribou

Arthur Thompson Bergerud

INTRODUCTION

In 1497 Sebastian Cabot discovered "Prime Vista" (Newfoundland) and found "white bearers and stagges farre greater than ours" (1). Thus began the modern journey with caribou. However, indigenous people of Beringia had been hunting caribou for at least 23,000 years as they do today and the Clovis people in New Mexico knew of caribou 12,000 years ago (2, 3, 4, 5). Early explorers of northern Canada, such as Hearne, Franklin, Back, and Tyrell, were staggered at the numbers of caribou in the summer massing and the animals they saw migrating in the fall and spring (6, 7, 8, 9). These massed columns of "La Foule" (the throng) remain today and are unequaled with the exception of wildebeest.

In the first scientific census of the Forty-mile herd in 1921, Murie said that the herd covered a strip 96 kilometers wide, the main body in 64 kilometers and scattered bands for 32 kilometers (10). The herd took 20 days to pass. During 8 days, 1,500 passed per day and in the remaining 12 days 100 came to the crossing per day. These observations yield 568,000 animals, but Murie claimed based on subsequent experience, the herd could have numbered near a million.

Historical Perspective

Caribou have roamed the Arctic for more than 2 million years. The earliest evidence is a tooth from western Alaska 1.2 to 1.8 million years ago (11). Banfield believed that the ancestral home for caribou could have been Alaska and Harrington concurs (2, 5). Røed and Whitten identified 18 alleles at the transferrin polymorphic locus in Alaskan caribou and only 11 from the Eurasian subspecies, which is consistent with a longer evolutionary time frame for the species in North America than Asia (12).

Tundra caribou still occupy their pristine range. Early estimates of the pristine numbers of barren-ground caribou on the mainland in the Northwest Territories varied from 1.8 million to 2.5 million in 1900 (13, 14). A recent estimate of these populations was 1.2 million (15). The seven largest herds in the North have had maximum estimates of 2.7 million or 0.8 per square kilometer (16). The George River herd, the largest herd in the world in the 1980s, at 650,000, declined when it reached a density of 1.3 per square kilometer (17). This herd reached a similar abundance from 1873 to 1877 and the herd occupied a similar size range in the 1880s as in the 1980s (18). The estimates of Anderson and Banfield agree with current knowledge of maximum herd sizes and carrying capacity estimates of 1.0 to 1.5 per square kilometer. In Alaska, Skoog felt that numbers his-

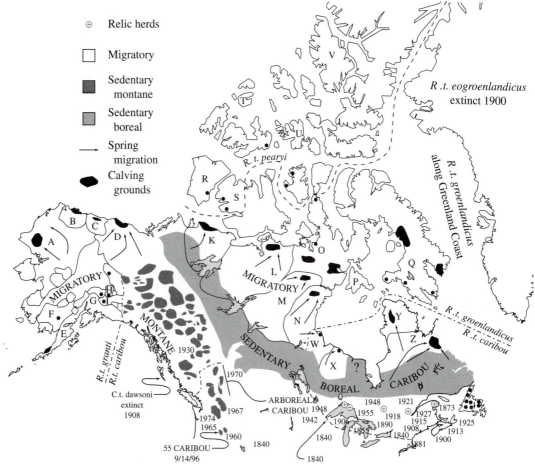

Figure 31–1 The present distribution of caribou subspecies, ecotypes, and the approximate disappearance dates of some caribou since European contact. Known calving grounds for the migratory ecotype are shown as are the primary directions of pregnant females that generally migrate at right angles to the treeline when returning to calving grounds north of trees and most denning wolves. Letters represent herds mentioned in Table 31–1. (Major Refs.: 235, 2, 74, 236, 237, 238, 146, 230, 21, 33).

torically had never exceeded 2 million and more likely had remained far below 2 million (19). The total Alaskan population in 1993 based on well-designed censuses of 31 herds was 880,000 and in 1995 the estimate was 959,000 (20, 21). Thus the migratory barren-ground caribou in Alaska and Canada exist in numbers today as they did in the 1890s.

The woodland caribou that live south of the Arctic treeline (excluding New-foundland) are extinct in the Lake States, New England, the Maritimes, and much of Ontario (Figure 31–1). Recent census results of 22 herds of woodland caribou residing in undisturbed habitats indicate a density one-twentieth of the tundra herds (22). The range for woodland caribou in 1850 was 5,000,000 square kilometers and there may have been only 300,000 of these more sedentary animals at the end of the Little Ice Age. There are now about 100,000 woodland caribou and an estimated 3.5 million caribou in North America (Table 31–1).

Value

The native peoples of the North have hunted caribou since the ice retreated and they remain hunters today. The caribou's presence was solely responsible for human settlement where otherwise there would have been no inhabitants (23). Native peoples still used spears at water

TABLE 31–1

Recent Population Estimates of Caribou in North America

Subspecies/Ecotype Herd Name (Location)	Population Estimate (1000s) and (year)	Trend in Population	References
R. t. granti (Alaska)			
Migratory[a] and Small Bodied (Arctic Slope)			
Western Arctic (A)	463 (1996)	I (Increasing)	30
Teshekpuk (B)	28 (1993)	I	21
Central Arctic (C)	18 (1995)	D (Decreasing)	21
Porcupine (D)	152 (1994)	D	21
Subtotal	661		
Migratory and Large Bodied (Alaska Range)			
Northern Peninsula (E)	12 (1995)	D	21
Mulchatna (F)	200 (1994)	I	21
Denali (G)	2.3 (1995)	S (Stable)	21
Delta (H)	4.7 (1995)	S	21
Nelchina (I)	50 (1995)	I	21
Fortymile (J)	22.6 (1995)	S	21
Other herds	18.7 (1990-1996)	I-D	21
Subtotal	310.3		
R. t. groenlandicus[b]			
Northwest Territories (migratory)			
Bluenose (K)	120 (1987)	? (Unknown)	15
Bathurst (L)	352 (1990)	?	15
Beverly (M)	277 (1994)	?	231
Kaminuriak (N)	496 (1994)	I	231
NE Mainland (O)	110 (1983)	?	15
Southampton/Coats Islands (P)	14 (1991)	?	43
Baffin Island (Q)	230 (1991)	?	15
Greenland	20 (1980s)	?	144, 34
Subtotal	1649		
R. t. peary (Arctic Islands)			
Bank Island (R)	1.0 (1991)	D	33
Victoria Island (north end) (S)	0.1 (1993)	D	33
Patrick/Melville Islands (T)	1.1 (1997)	D	33
Bathurst Island (U)	0.1 (1991)	D	33
Ellesmere/Axil Heiberg and Devon Islands (V)	Near extinction ?	D	33
Subtotal	<3		
R. t caribou			
Migratory			
Cape Churchill (W)	2.2 (1988)	?	
Pen Islands (X)	11 (1994)	I	230
Leaf River (Y)	260 (1991)	I	230
George River (Z)	400 (1993)	D	232
Newfoundland	100 (1998)	I	Mercer (personal communication)
Subtotal	773		
Sedentary/Montane			
Yukon	32 (1978–1997)	I	230
British Columbia[c]	17 (1996)	I	230
Northwest Territories	15 (1991)	?	146
Alberta	0.8 (1998)	D	230
	78–97		
Subtotal	55		
Sedentary/Boreal			
Alberta	4.5 (1980-1998)	D	230
Saskatchewan	2.4 (1985)	D	166
Manitoba	2 (1970-1990)	S	230
Ontario	10 (1996)	D	230

Subspecies/Ecotype Herd Name (Location)	Population Estimate (1000s) and (year)		Trend in Population	References
Northwest Territories	5	(1991)	S	146
Quebec	10	(1996)	D	233
Labrador	<2	(1998)	D	Schaeffer (personal communication)
Subtotal	51			
Grand total	3,502			

ᵃMigratory herds migrate and calve at common calving grounds; sedentary herds disperse at calving (131). Some herds classified migratory may have have components that calve dispersed when in mountains without level plateaus.

ᵇPopulation estimates for animals in the Northwest Territories are subject to large sample errors and generally have low precision and accuracy (231).

ᶜ2,300 caribou live in southeastern British Columbia in deep snow and depend on arboreal lichens (230).

crossings in the 1950s and there was starvation of humans due to a lack of caribou as late as the 1950s (23, 13, 24). In March 1958 the Naskapi of Labrador were still on the land as they were 4,000 years ago (25, 26). The native peoples believed in a caribou god, who lived on a high mountain to the northeast (the Torngats) called Cah-tee-which-ou-op (Caribou House). When the god did not send the caribou south the Naskapi starved (27).

The retreat from the land accelerated when dog teams were replaced by skidoos in the 1960s and bush planes brought supplies and located the caribou. Banfield estimated a harvest of 100,000 from the barren-ground herds in central Canada in the 1940s from his population estimate of 670,000 animals (13). Calef estimated a harvest of 20,000 from these same herds including Baffin Island for the 1970s (24). When the George River herd included 15,000 caribou in 1958 the harvest averaged 885 animals (1955 to 1965) and animals were only available to settlements on the Labrador and Ungava Bay coasts (28). But when the herds reached 600,000 and had spread to the east coast of the Hudson Bay the harvest reached 30,000 with the residents of both coasts taking 20,000 individuals and sport hunt taking 9,000 animals. The total harvest (native, resident, and sport) in Alaska was estimated in 1989 at 23,000 from a population of 729,000 within 914,000 square kilometers (29). The total harvest in 1994–1995 was estimated at 33,000 (21). More recently the estimated subsistence harvest of the largest Alaskan herd, the Western Arctic (463,000 animals in 1996), was estimated at 20,000 animals (30). Caribou today is the meat of choice of possibly 100,000 northern residents and for many the hunt and the love of the land are still essential elements of their identity.

The worth of caribou is not measured in carcasses or dollars but in aesthetics and biological diversity of ancient food chains. Calef said: "They quicken the country not merely by adding animation and excitement themselves but by carrying along a host of other creatures, the wolf, the fox, the raven. The empty tundra may appear a drab and barren place, but let one caribou trot onto the skyline of an esker and the land comes alive" (24, p. 15). There are productive lichen pastures on 1.8 million square kilometers of Canada between the closed-canopy forests and the arctic prairies, the "land-of-little-sticks." Only caribou have evolved to graze these lichen pastures; it would be a barren land indeed if caribou were gone.

TAXONOMY AND DISTRIBUTION

Caribou are a medium-size cervid with palmated antlers, larger than the deer but smaller than moose or elk (Figure 31–2). Males commonly weigh 120 to 200 kilograms and females 80 to 140 kilograms, but size varies greatly between populations in North

Figure 31–2 Caribou are a medium-size cervid with palmated antlers. The forward-projecting brow tine is called the "shovel." (Photograph by A.T. Bergerud.)

America with the largest races in the south and western mountains (31, 32). The smallest races are the polar populations on the northern arctic islands where green forage is not available until mid-July; Gunn reported that Peary males weighed only 66 to 92 kilograms (33).

Taxonomy

All members of the genus *Rangifer* are considered the species *tarandus* and probably are physiologically capable of interbreeding. Banfield recognized nine extant subspecies, six in North America (2). During the Wisconsin Ice Age, that reached a maximum 18,000 years ago, caribou persisted in Beringia (a refugium in Alaska, the ancestors to *R. t. granti*), the periglacial refugium south of the Laurentide Ice Sheet (the ancestors to *R. t. caribou*), and the Queen Elizabeth Islands (the ancestors to *R. t. pearyi*). Although Pearyland in northern Greenland remained free of ice, Meldgaard and Harington stated that *pearyi* may have survived the Ice Age on Banks Island (34, 5). As the ice retreated 18,000 to 10,000 years ago, the caribou dispersed into the newly exposed land surface as herbaceous plants developed, probably within 1,000 years post-ice.

Barren-ground caribou may have ancestors from three refugia (Beringia, Banks Island, and south of the ice) (35). Black spruce from south of the ice spread north and reached the MacKenzie Delta by 9,500 to 9,000 years ago and woodland caribou ancestors could have reached this latitude much sooner (36). There were no tree species in Beringia except possible scattered aspen; this refugium was arid and dominated by herbaceous species, woolly mammoths, horses, bison, saiga antelopes, and other grazing herbivores (4). The subspecies that arose from the caribou progenitors of this refugium have a longer evolutionary history of living on the tundra than the ancestors from south of the ice.

Distribution

Wildlife biologists manage and census caribou in North America on a herd basis. Discrete herds are generally recognized if the females in a general area have a distinct calving ground. Females migrate and aggregate on these grounds annually, and these calving grounds are the focus of the annual movement cycle. Herds are commonly been named on the basis of geographical features near or on the calving ground, such as the Bluenose herd calving near Bluenose Lake in the Northwest Territories (37). These common calving grounds occur for herds in Alaska, the Northwest Territories mainland, northern Ungava, the Island of Newfoundland, and on Baffin Island (Figure 31–1). The nine largest herds in North America are of this type (Figure 31–1, herds A, D, F, K, L, M, N, Y, Z) and numbered 2.7 million animals from 1987 to 1996 or 78% of all caribou in North America. The animals that calve in the mountains of the Yukon and British Columbia and in the boreal forest south of treeline do not migrate to calving grounds but generally calve solitary or in small, dispersed groups. For these woodland caribou, herds are recognized for management purposes on the basis of winter aggregations. However, these animals are so scattered during the nonsnow season that distinct distributions are not evident. Females in these distributions migrate but the migrations are short, the animals commonly scatter in many directions from the winter ranges, and the total area occupied at calving is greater than at any other season (38, 39).

When the migratory herds are low in number they restrict their range to optimum areas above treeline that include the calving ground and the key summer range of vascular plants; this key area is the center of habitation (19). As the herd increases so do the fall movements away from the center of habitation to the lichen winter pastures. The distance between these two distributions varies with population size (40). Hence fall and winter densities are ameliorated by range expansion, whereas summer distributions cannot increase as dramatically with herd growth. This winter expansion reduces overgrazing for herds that have room to expand south of the treeline. Severe overgrazing of lichens results only when winter movements are restricted in insular situations or for herds with limited winter pastures such as in Norway (41, 42, 43).

These range expansions are density dependent based on the abundance of forage supplies on summer and fall ranges. As the George River herd increased from 160,000 in 1973 to 600,000 in 1984 the distances between calving and rutting locations also increased. In contrast there was no increase in the distances traveled between the annual rutting ranges and the mid-January lichen ranges as the herd increased (A. T. Bergerud, unpublished data). The animals used different winter ranges, but these were stochastic responses in seeking reduced snow depths, the norm for winter movements (44, 45, 46, 47, 48).

LIFE HISTORY

Diet Selection

Caribou have catholic tastes (49). However, they prefer green vascular plants and mushrooms. When these plants cannot be secured they are opportunistic and relative availability is the prime factor in utilization. Their diet is similar throughout their range, even at the plants species level, no doubt because of the similarity of the forage resources between regions. Extremely important foods such as reindeer lichens, dwarf birch, sedges, cottongrass and horsetail are heavily used in Eurasia and North America.

As the spring thaw advances, caribou quickly change their diet to include new green growth. Preferences follow the primary production progression. In mountainous regions, caribou keep abreast of new plant phenology by altitudinal shifts and visiting receding snowfields. Sedges and grasses are the most important plant groups in the spring diet, frequently compromising 50% of the diet. Many growing plants are used when new and most nutritious, such as catkins of willows, larch needles, the leaves of alder, and Solomon's seal.

The most important plant materials in the summer diet are the green leaves of deciduous shrubs, especially willow, birch, and blueberry. Caribou are fastidious feeders that strip the leaves from woody stems and pick only new sprouts and finer stem tips of grasses, sedges, herbs, and horsetail. In late summer the animals actively search for mushrooms.

After the first hard frost the deciduous leaves quickly become unavailable and animals switch to terrestrial lichens and the leaves of evergreen shrubs. The fruticose reindeer lichens and Iceland moss are eaten along with *Sterocaulon* species. Many shrubs of the *Ericacea* family are consumed after snowfall. Crowberry, sedges, and horsetails are also heavily used if they can be found beneath snow cover by digging feeding craters. If a sun crust builds that can support the animal's mass, they will seek arboreal lichens on conifers or as litter on the snowcover. In southeastern British Columbia, where snows are too deep to detect the ground flora, animals do not dig food craters but depend on arboreal lichens. Lichens are the primary food in most areas because of their widespread distribution and relative abundance.

Energy Cycle

Diet selection can be combined with measurements of dry matter intake and species digestibility to estimate metabolizable energy intake. These estimates can be compared to energy expenditures derived from activity schedules and body requirements to estimate the annual cycle of fat, body condition, and energy balance (Figure 31–3).

The annual cycle of energy intake and expenditure differs among years. Snow depths and the energy spent cratering varies between winters. Another key variable is the spring phenology. In the last weeks of gestation fetal growth accelerates and the energy budget of females is critical if calves are going to accumulate sufficient birth mass to cope with the extrinsic environment and females are going to have adequate milk production. Annual changes in the level of harassment and parasitism by biting insects during July and August affect the activity budgets of feeding, traveling, resting, and running (Figure 31–3).

Metabolizable energy intake reaches an annual low for females in spring as calving approaches before the flush of new growth high in soluable nitrogen and phosphorus. However, for males who do not migrate to calving grounds but rather follow the green phenology in spring migration, the quality of the diet improves rapidly in May (50, 16). Over several spring seasons the diet was more nutritious for a woodland caribou herd than for the migratory tundra herds (16).

Reproductive Strategy

Breeding. The breeding season of caribou is brief. In the Kaminuriak herd, 80% of conceptions occurred in 11 days (51). More than 50% of the mating I observed in Newfoundland occurred during October 11–17 (52). Nikolaevskii observed that reindeer breeding was short and regular where the females are well fed but prolonged where the females were at different levels of nutrition (53). Females in the best condition bred first followed by younger and older females (54, 55). The brief breeding period for caribou is likely due to normalizing selection in which early and late born calves the next spring have higher mortality rates than those born synchronously (56, 57).

Gestation. Gestation for caribou ranges from 225 to 235 days (19). Two Alaskan herds had a mean breeding date of October 6 and a peak calving date of May 24 or a 231-day gestation (19). In Newfoundland in 1958, the peak breeding was October 12–13, and more than 50% of the calves were born May 28–29 or a 228-day gestation (52, 58). In late snow years (also late spring phenology) calves can be born a few days later (58, 59). This supports the idea of a critical birth mass and that natural selection has resulted in some flexibility (possibly 3–6 days) in the development time relative to neonate survival (60). Small neonates have higher prenatal mortality rates then larger calves and may be more susceptible to predation, other things being equal (61, 62, 63, 57, 64).

Calving. Caribou calve coincident with the flushing of green vegetation. Woodland herds residing in the boreal forest calve in the third and fourth weeks of May (65, 66), whereas the sedentary mountain herds in British Columbia calve in the first

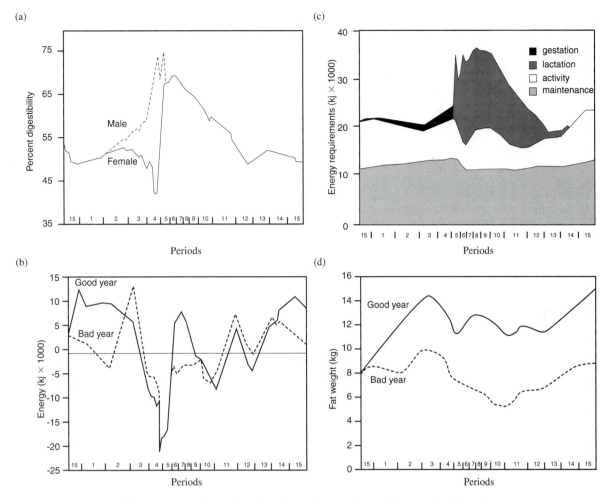

Figure 31–3 Annual cycles for the Porcupine caribou herd. Life cycle periods: 1 = midwinter, 1/11–2/20; 2 = late winter, 2/21–3/31; 3 = spring, 4/1–30; 4 = spring migration, 5/1–19; 5 = pre-calving, 5/20–31; 6 = calving, 6/1–10; 7 = post-calving, 6/11–20; 8 = movement, 2/21–30; 9, 10, and 11 early, mid, and late summer, respectively, 7/1–9/7; 12 = fall, 9/8–10/7; 13 = rut, 10/8–31; 14 = late fall, 11/1–30; 15 = early winter, 12/1–1/10. (a) Percent digestibility of forage, showing advantage of males lagging behind females in spring migration. (b) Energy requirement of adult females in good and bad years (bad year = deeper snow, shorter growing season, more insect harassment). (c) Energy balance of adult females (d) Total fat mass of adult females in good and bad years. (From Ref. 48.)

week of June (67, 68, 69, 70). In Alaska the large-bodied interior herds of the Alaska Range (Mulchatna, Denali, Delta, Nelchina herds) generally calve in the second and third week of May (19, 64, 55). The smaller bodied animals that calve on the north slope of the Brooks Range or the Arctic coastal plain have peak parturition about June 4–5 (71, 72, 59). The migratory barren-ground herds in the Northwest Territories (Bluenose, Bathurst, Beverly, Kaminuriak) calve June 9–17 and Peary caribou on the Arctic Islands calve even later in the third week in June (46, 73, 74). The dates of peak calving are thus correlated with the start of the growing season. Southern herds that calve earlier do so about 10 days after the start of plant growth. However, the most northern herds do not have the luxury of waiting till after plant growth begins due to the short growing season, so their peak calving dates commonly precede the start of new growth by 10 days.

Calving dates within herds generally do not vary greatly between years (71, 19, 58). But herds in poor nutrition generally calve later than well-nourished herds in the same region (75, 76). The George River herd in Ungava in the 1970s calved about June 4, but in the 1980s the summer range became overgrazed and calving became progressively later (16, 77, 78). When George River stock were captured in 1984 and held on a high plane diet for 3 years, the calving dates again reverted to the June 4 peak (79). The later calving of the free-ranging animals was due to a reduction in summer mass gain so that many females reached their "target" conception mass about 8 days later (80). In general females in good condition reach their target estrous mass first in the fall and thus calve earlier than animals in poorer condition (81).

Calves throughout the Holarctic generally are born in a 2-week interval. The calving interval for 14 herds in North America was about 17 days. More than 90% of the calves in the Western Arctic herd and the herds in Newfoundland were born within 10 days (71, 58). The shortest interval is in Svalbard where 90% of caribou were born June 2–9 (82). Calving in caribou occurs in a shorter time span than for other ungulates in North America with the exception of antelope, and also is more concise than that of tropical ungulates (83).

Calving could be synchronized to minimize energy problems for females and offspring (84, 85, 86, 42). Another theory is that it reduces predation of vulnerable progeny in open habitat (87, 52, 56, 88). Caribou females with calves of similar age form nursery herds shortly after calving that increase vigilance and escape time and may swamp predators, and provide the selfish herd advantage (89, 90). However, the population with the greatest calving synchrony inhabits Svalbard where there are no predators (91). Additionally woodland caribou also calve in a short interval even though they are generally alone or in small groups of two to three animals (92). Caribou calves are cryptic and have a partially developed hiding response that reduces conspicuousness before joining nursing groups (91).

Calving synchrony could be due to a combination of energy considerations and predator avoidance. Several herds in Norway and Svalbard that show calving synchrony without predators have severe nutritional problems. Large numbers of calves die during summer and winter in Norway and during winter in Svalbard (42, 82). For these European herds, being born at the very best moment nutritionally would be a selective dictate. In North America, calves born early and of small body mass would be laggards in escape flights and calves born late would be conspicuous to predators by their small body size in summer and winter, and in winter they would be at a disadvantage in deep snows.

Breeding Age. Domestic reindeer protected from predators and selected for early body mass commonly conceive as fawns in their first fall, but free-ranging female caribou in North America reach puberty at either 16 or 28 months of age depending on their body mass at the end of their second summer (19, 93, 58, 82). Males are physiologically capable of breeding at 16 months but are socially castrated through intrasexual selection and probably do not generally succeed until their fourth fall (94, 52). Reimers pioneered the concept that the fall mass of females rather than age was the primary prerequisite in reaching estrus (95). Yearlings that conceived in the George River herd at 16 to 17 months weighed about 79 kilograms in April 1980 whereas nonpregnant yearlings averaged 12 kilograms or less (96). Similar results have been shown for adult females in the Central Arctic herd and in the Northwest Territories (59, 97). When fecundity is plotted against mass the curves are sigmoid rising rapidly to upper asymptotes of about 95% pregnant. These upper asymptotes occurred at 85 kilograms in the Northwest Territories, 90 kilograms for the George River herd in Ungava, and at 58 kilograms for smaller polar caribou on the Arctic islands (97, 98; A. T. Bergerud, unpublished data). There is no postreproductive life for females and they continue to conceive and carry calves to term up to 17 years of age.

Dauphine showed the potential for skipping pregnancies (99). Lactating females reach their lowest mass in late June whereas nonlactating females are already gaining mass. If the summer range is overgrazed or summer growing conditions reduced, a lactating female may not make a target conception mass in fall and skip a pregnancy. If females do not calve or are released from their lactation expenditures by neonatal death they can devote all of their energy balance to their own condition. Thus, in suboptimum forage conditions more females should reach estrus in years of high calf mortality than in summers when calf survival is high and most females continue to lactate to the fall—an apparent conflict of this year's investment versus the potential for next year.

Sex Ratios. The proportion of males in caribou populations changes as males grow older and experience differential mortality. The fetal sex ratio was 51% males in Alaska, 45% males for Peary caribou, 49% males for barren-ground caribou in the Northwest Territories, 51% for the migratory George River herd (S. Luttich, Newfoundland Government, and A. T. Bergerud, unpublished data), and 50.1% males for domestic reindeer; these studies are close to the theoretical 50 : 50 prediction of Fisher's principle (19, 100, 101, 102). Because fetal resorption is uncommon in caribou and probably not sex specific, the sex ratio at birth should be balanced (103, 99, 104). However, the sex ratio of the living at birth favors males: 51% Alaska, 53% in the Northwest Territories, 52% Newfoundland, 51% for the Denali herd, 53% for the Porcupine herd, and 56% for the George River herd in Ungava (S. Luttich, Newfoundland Government, and A. T. Bergerud, unpublished data) (19, 46, 105, 57, 64, 62). These samples are not significantly different from a balanced ratio, but the consistent results favoring males indicates that slightly more males are born viable than females. More females than males die in the first 48 hours of life because females weigh less than males and thus are susceptible to a variety of neonatal decimating factors (61).

Following birth, males suffer higher mortality in their first summer and winter than females. The proportion of males at 5 to 6 months of age was 38% in Newfoundland (105). Lynx took more male calves than females because they were more active and wandered farther from their females. In British Columbia, wolves and bears killed more male calves than females during their first summer of life (106). In the Northwest Territories wolves killed more males in the first 2 weeks of life, whereas other mortality factors took more females (61). In contrast, where summer predation was low, the sex ratio at 5 to 6 months favored males, 53% for the George River herd (S. Luttich, Newfoundland Government and A. T. Bergerud, unpublished data) and 55% for the Nelchina herd (19).

Calves in some herds continue to die at a greater rate than adults in their first winter (73, 107, 40, 108). When this occurs it appears that males are most susceptible. The proportion of males in the George River herd at 16 months of age in rutting aggregations was 46%, significantly less than the 53% when these cohorts were only 5 to 6 months. The sex ratio was 35 males : 65 females for 212 yearlings that drowned at Limestone Falls in September 1984 (109). Because, in this herd, there was little or no winter starvation, more males probably died from wolf predation than females over winter (16). In Norway, where there is little predation but where calves starve during winter, the ratio at 1.5 years was 41 males : 59 females (42).

Male caribou continue to die at much higher rates than females as they age (110, 111, 82, 112). The sex ratio for animals older than 16 months of age for 18 lightly hunted herds in North America averaged 35.7% males (40). The percentage of males of the George River herd was 35.8% for animals older than 16 months and 37.0% for animals greater than 28 months of age (S. Luttich, Newfoundland Government, and A. T. Bergerud, unpublished data) and showed no significant trend with time as the herd increased to high numbers (109). The quinary sex ratio (animals greater than 10 years) was 6% males in the Kaminuriak herd, 20% in the hunted Nelchina herd, and 16% males for the drowned sample at Limestone Falls (110, 19).

The major cause of death for caribou in North America is predation (40). But in areas were there are few or no predators, the males still die at greater rates than females in winter from food shortages. Ratios of the living in areas with very little predation from Norway, Svalbard, and South Georgia ranged from 30% to 36% males (42, 82, 112). Males enter the winter in a reduced condition following mass loss during the rut, and if starvation occurs it is the largest males who were most active in breeding that pay the price for their increased reproductive fitness.

Longevity

The maximum longevity for females is 18 to 20 years in captivity but ecological longevity would be shorter (113). Males 14 to 15 years old have been reported by Tyler from Svalbard where there are no predators (82). But in large age arrays from North America the maximum life span for males was 11 to 13 years and 15 to 16 years for females (110).

Males simply age faster as demonstrated by more rapid wear of the dentine heights on the molarform teeth (110). In a polygynous breeding system where there is intrasexual selection for rapid growth and early breeding, the contribution to fitness is reduced in older males by accumulation of deleterious attributes on which selection can no longer cull (31, 114). However females continue to contribute genes to new generations up and until death.

Physiology and Growth

Caribou have evolved in an environment of cold, long winters and a short growing season, are well adapted to the rigors of the arctic north and rarely freeze or starve to death. Maintenance of a constant temperature can be a problem when the internal body core and the extrinsic environment differ by as much as 100°C. Two major defenses against the cold are a well-insulated fur coat that traps and holds a layer of air next to the skin and a restriction of respiratory heat loss. Caribou have a countercurrent heat exchange in their nasal cavity so that 70% of the heat and 80% of the water added in inspired air in the nose and lungs is regained upon expiration (115).

Caribou reduce energy expenditure in the winter by a voluntary reduction in appetite that reduces the resting metabolic rate. Basal metabolism accounts for between 50% and 70% of the normal energy expenditure of adults (116, 117). Reduced food intake is partially compensated for by use of endogenous fat reserves leading to an annual mass and fat cycle and also reduced activity in winter to reduce energy expenditures (99, 82, 112).

Another important physiologic adaptation to life in the north is the ability of caribou to subsist on low protein intake. Caribou can recycle urea, a waste product of the blood that contains nitrogen. They can return 60% of the urea to their rumens where the nitrogen can be used again. This ability to recycle waste nitrogen allows them to live on the low intake of protein available in the winter on a lichen diet.

Birth mass varies among populations and among years within populations. Males are generally 1 to 2 kilograms heavier than females. Neonates range from 3 kilograms in Svalbard to 9 kilograms in the mountains of British Columbia (82, 91, 68). Neonate mass is correlated with the mass of females the previous fall and preparturition (118, 58, 119, 60, 31, 59).

Caribou birth mass is less than that predicted based on other cervids, and less than that predicted based on the hypothesis that females in large bodied populations reallocate more resources from maintenance to reproduction (120, 103, 31). Robbins and Robbins suggest that female caribou/reindeer invest less in their calves than other ungulates because of the inadequate winter food supplies and the need to optimize maternal survival relative to reproduction (120). Postpartum females must balance the need to replenish fat and protein reserves to ensure their survival and future progeny with producing high-quality milk to ensure survival of a current calf (48).

Annual variation in birth mass appears related to late winter forage conditions and probably occurs in the last 6 weeks of gestation (105, 75). A very late spring may not allow calves to reach birth mass sufficient for survival (63). Recently discussions of birth viability noted reduced survival in winters with heavy snows, but what is likely the key is the availability of green forage under the snow in late April and May for females. Also late snow cover in the mountains can restrict the areas available for dispersed calving and facilitate searching for predators (121, 57).

Sexual Dimorphism. Caribou are more sexually dimorphic in body size than other cervids and within the species large bodied populations are more dimorphic than small bodied populations (31). With longer growing seasons, males have steeper growth curves than females and continue to grow until 5 to 6 years, whereas females generally halt most growth at first reproduction. The sexual dimorphic index at 6 years (males/females mass) varies from 2.1 for the large bodied Nelchina caribou herd to 1.6 for the small bodied Kaminuriak caribou herd (19, 99).

Sexual dimorphism may be the result of sexual selection or natural selection (122, 123, 124). Males in a polygynous society enhance their intrasexual advantage in reproductive fitness by continued body growth and the development of large antlers. Females maximize their fitness by early reproduction and by using their available energy on fetal growth and lactation. There is much less restraint on the size of males than females and when introduced to favorable island habitats some males have reached mass nearly double the norm for their genotype.

Antler Morphology and Growth. Female caribou, unlike other cervids, commonly have antlers and for both sexes there is considerable more palmation in the main tines than for deer and elk who have digitate main beams and tines (Figure 31–4). Initiation dates for growth, velvet shedding, and antler casting are hormonally controlled and vary between sex and age classes (Table 31–2).

Males have an anterior tine low and in front of the face (the brow). Above this tine is the bez tine generally comprising several points, above the bez and where the beam sweeps more vertically is commonly the rear tine, and on the "tops" there can be several tines commonly palmated. The antlers reach their maximum development at 4 to 6 years in Newfoundland and 7 to 9 years of age in Alaska (52, 19). Males over 10 years have regressing antlers but animals of this age are only common in herds with little predation or starvation (52).

Antlers of females are commonly 25 to 45 centimeters long and are correlated with body size and the size of male antlers within populations (2, 31). For each centimeter in length of male antlers, females can expect 0.4 centimeter in the length of the

Figure 31–4 Caribou are unique in the deer family, in that both males (left) and females (right) grow antlers, although the size is larger in males. (Photograph by A.T. Bergerud.)

TABLE 31–2
Annual Antler Cycle of Caribou

	Approximate Dates		
	Start of Growth	*Shed Velvet*	*Cast Antlers*
Mature males	Mostly April	September	November and later
Young males	May/June	September	Late winter
Yearling females	May/June	October	May/June
Pregnant females	2 weeks postpartum	11–2 weeks prior estrus	Parturition ± 2 weeks
Nonpregnant females	May	September	May/April
Male calves	June/July	October/November	May/June

Source: Based on Refs. 19, 46, and 234.

main beam. The length of antlers of females increased rapidly from 1 to 3 years of age but showed no further increase in length from 3 to 13 years (45, 125). Pedicel size and antler mass of females increase with age (126, 125).

There are major differences in antler shapes and sizes among populations. Western woodland caribou (i.e., montane) and barren-ground migratory herds have longer antlers with many palmated points on the tops and relatively fewer points on the bezes. Woodland and Peary males commonly have several centimeters between the emergence of the brow and bez (high bez) while the migratory tundra herds have these points arising more proximally (low bez). Antlers of barren-ground males had the most damage to the tips of the tines from fighting, whereas woodland males had the most damage on the bezes (31). Brief antler clashes were more frequent in large herds where there were many males than in the small groups of woodland caribou where one male was clearly dominant (31).

Butler developed the display weapon hypothesis to explain the different antler morphologies (31). Factors involved were female availability, male encounter rates, and male familiarity. Barren-ground caribou in large herds fought less with other males and the most antler contact was on the tops of the antlers. Barren-ground males are less familiar with each other than woodland males because of the large herds and high mobility, but encounter rates are frequent because individual males cannot deny access to several females as do woodland males. Given the short synchronous estrous interval in caribou, barren-ground males rely more on visual assessment of rivals based on terminal antler mass rather than prolonged fighting (31). Woodland caribou breed in smaller groups where dominant males have more control but when challenged fight harder with more contact on the proximal extremities with the brows and bezes acting as defensive shields. Woodland caribou males are more commonly killed in fighting and from locked antlers when the bezes slip through the wide spacing of the brow and bez tines of the opponent (105). Locked antlers are less common in migratory barren-ground males. Interestingly, the small-bodied Peary caribou, living in the polar desert on the arctic islands and rutting in small groups, have widely spaced brows and bezes and commonly lock antlers (F. Miller, Canadian Wildlife Service, personal communication).

When the summer range of the George River herd became overgrazed the antler mass of mature males declined from 3.8 kilograms in 1978 to 2.4 kilograms in 1986 to 1988. However, the antlers retained their heights. The maintenance of lengths despite reduced forage is consistent with the display weapons hypothesis.

BEHAVIOR

Behaviors in caribou other than those involved in propagating the species are a product of the extrinsic environment. Important extrinsic variables for caribou are the visibility of the habitat, which is a continuum from the tundra to closed-canopy forest, the presence of predators and biting/parasitic insects, and the slow growing phytomass subject to trampling and availability by snow cover (56, 91). The basic sequence is that animals inhabiting open habitats in the presence of predators adopt a gregarious lifestyle (127). These aggregations must remain mobile, to be able to obtain necessary nutrients. The harsh, treeless archipelago, Svalbard, has been predator free since caribou arrived 8,000 to 5,000 years ago. These caribou have short legs and do not aggregate or migrate (82, 128). In the absence of predation on these islands, at least in recent millennia, natural selection through starvation in a harsh polar desert has produced animals and behaviors unlike the other subspecies that currently are in an adaptive race with natural predators.

Social Structure

Caribou have developed an open, tolerant social structure in the interaction between the two sexes and across age arrays. An individual may join a group for a few hours of feeding and resting and then wander off. Animals tagged while together are later found widely separated (129, 130). Individual distances between caribou are low and animals commonly feed in parallel, shoulder-to-shoulder. Animals use each other to achieve efficient use of the environment such as location of feeding sites, breaking trails, locating conspecifics, sharing insect pests, viligance in detecting predators, and escape flights.

Dominance interactions occur when there is individual competition for a preferred and restricted commodity like early greens, digging feeding craters in snow, calves attempting thief-sucking, and female defense of their progeny. Overt threats include head-low, head-high, antler presentation, hoof strikes, and rearing and flailing when contestants are evenly matched and unknown to each other. In large herds dominance is based on visual recognition of sex and age classes, commonly without overt threats. But when motivation is high, dominance can be reversed, such as when an antlerless female defends her calf against an antlered female. In small populations it is obvious individuals recognize each other and abide by the outcomes of past encounters (31). Mature females generally are at the front of columns, because they are more perceptive of the stimuli of the extrinsic environment.

Males commonly fight in the rutting season and mortality can be as high as 2% (105). Normally, subordinate males turn aside as dominant males approach, but if one male is too slow and their antlers touch a fight will result (52, 31).

Spacing Ecotypes

Antipredator behavior has resulted in two suites of behavior relative to migration and aggregating behavior: the sedentary and migratory ecotypes. With the sedentary ecotype, females that live year-round south of the Arctic treeline migrate from winter ranges of low snow depth and disperse to spring calving sites, where they generally calve alone or in small groups. In contrast, females with the migratory ecotype make long migrations to calving grounds north of the treeline where they aggregate at calving locations by the thousands (131, 16).

Females of the sedentary ecotype are more dispersed at calving than other times of the year (38, 39). These females seek island refuges or shorelines or calve in large open peatlands or above the alpine treeline. These habitats may have a late phenology and also a low phytomass (67, 132, 92). Locations are selected to reduce risk by "spacing out" from the travel routes of predators and away from conspecifics and alternative prey such as moose that could attract predators (133, 131, 22, 16, 67, 92, 121, 132). These female–calf pairs are in effect hiding and maintain a low cruising radius if undetected (38, 39). These females show fidelity to former calving areas (134).

The females of the migratory ecotype use available space to leave many of their predators behind, especially wolves. They migrate to calving grounds that are at maximum distance from treeline consistent with remaining on a substrate with some bare ground for crypsis of their brown calves (less than 90% snow cover). The majority of wolves den relatively close to treeline where alternative prey species are more common (135, 136). Thus, this ecotype "spaces away" from many of its predators and alternative prey. The great advantage that caribou have over wolves is that their precocial calves can follow their dams within hours after birth. Wolves remain tied to their den sites for 2 months while the pups develop.

Wolf density on distant calving grounds may be only 20% of that on winter ranges (50). Additionally, 35 radiotagged wolves on the late winter range of the Bathurst herd followed the herd north in the spring migration, but 26 left the migrating caribou and denned an average of 375 kilometers from the calving ground and only 2 of 35 wolves reached the vicinity of the calving area (137).

On the calving ground of the Porcupine herd and adjacent foothills, researchers located the dens of barren-ground grizzly bears, the nest sites of golden eagles, and the territories of six wolf packs. All predators were concentrated in the foothills, leaving the coastal plains and the calving ground with reduced predator numbers (138). If wolves did den on calving grounds where there are generally few other prey species, they would be left without sufficient prey to rear their pups when the caribou left shortly after calving (46). Thus, the vast space ("the miles beyond measure"), the reduced occupied space of the caribou, resulting from their clumping, and the caribous' constant mobility are the skein of *Rangifer's* antipredator strategy (139).

The boundary line between migratory and sedentary ecotypes west of the Rocky Mountains is not at the Arctic treeline but further south at the hydrocline that results in open water at calving (16). The common characteristics of the area between treeline and the northern edge of the dispersed ecotype at calving that I believe makes this region unsuitable for calving are the lack of open water, the lack of mountains, and the lack of extensive tundra or peatlands for open vistas.

Water escape is the great equalizer in the adaptive race between the caribou and wolf. Cartwright noted "when pursued in the summer they [caribou] always make for the nearest water, in which no land animals have the least chance with them" (140). The migratory ecotype faces frozen lakes at the treeline and beyond at calving; their equalizer is the space to migrate away from the denning wolves. But even migratory caribou use water escape in summer; Seton noted on his trip to the arctic prairies in 1907 that he saw several crippled caribou on islands that were recovering and were relatively safe in their "hospital" (1).

The open water line proposed between the two strategies west of the Rocky Mountains follows the line Anderson drew between woodland caribou and barren-ground caribou (14). The distinction of these subspecies is then partially explained by these two different suites of antipredator behavior, and differences in antler and pelage morphology that have come about by differences in mobility and aggregation sizes and evolutionary inertia lingering from being isolated in different refugia (31).

The spacing ecotype approach provides an explanation for some of the confusion in the literature that notes that the caribou in northern Ungava are classified as woodland caribou but act like barren-ground caribou migrating to calving grounds north of trees (2). These caribou dispersed from the woodland stock that persisted south of the ice, as their skull measurements suggest, but they adopted the migratory suite when colonizing the vast treeless space in northern Ungava. One can visualize a natural selection sequence as females colonized the landscapes vacated by the ice. The females that calved the furthest south would encounter more predators than animals that were farther north of treeline foraging on a reduced phytomass. Calving females could compensate for the lack of the protection of open water for escape by moving even further north from the more diversified southern fauna of alternative prey and predators. But there was a nutritional trade-off. Ultimately these females congregated on the habitats with the fewest predators (the calving grounds). Such aggregated females would have an added survival advantage for their young of increased vigilance and sharing their neonate vulnerability with their neighbors.

Figure 31–5 The migrating columns of caribou are unique in North America, equaled only by the wildebeest of Africa. (Photograph by A.T. Bergerud.)

Movements

The caribou is the classic migratory terrestrial mammal, moving north in the spring, and south in their "Endless March" (Figure 31–5), (Figure 31–6), (141). However, the only marked directional goal-oriented movement is that of females of both ecotypes moving to calving areas as parturition approaches. Autumn migration is far less clear and when herds are reduced in numbers they may not move south at all (46, 142). When they do go south there is little fidelity to areas previously used.

There is an annual cycle of movement but it does not fit the classical fall/spring pattern; movement for the migratory ecotype is greatest in the summer and least in the winter and the reverse occurs for the sedentary ecotype. The George River herd traveled further in fall than the Porcupine herd, but at that time the George River herd had overgrazed its summer range, whereas the Porcupine had not (78, 143, 80, 77, 16, 48). The George River herd may have traveled further because of the need for nutrients after they left the overgrazed summer pastures and also there was a reduced phytomass on the fall pastures (16). The basic annual movement pattern of these two large migratory herds appears to be related to seeking nutrients instead of trying to get somewhere. The greatest mobility is in July and August when plants are most nutritious and there is a greater gain for high grading. The least movement in the biological year is in midwinter when the animals depend on the less nutritious lichen phytomass and need to reduce energy expenditures for metabolic conservation. This midwinter stall is not directly related to snow depths because snows continue to accumulate in March and April when mobility increases.

Communication

Caribou, as a gregarious species inhabiting open vistas, constantly use visual and vocal signals to communicate. The greatest commotion is the bawling and grunting of calves and females as they try to maintain contact while traveling and when they are disturbed. Females stimulate newborns to follow by rapidly lowering and raising their head, grunting, turning, and trotting. Calves have an innate following response and imprint on the first moving stimulus they perceive following birth.

Caribou have poor visual acuity and commonly fail to recognize humans who remain motionless. They rely on scent as the final verification of identity. Caribou that have been alerted on the basis of sound or sight commonly give an excitation leap with a wheeze snort when they receive a strong scent stimulus.

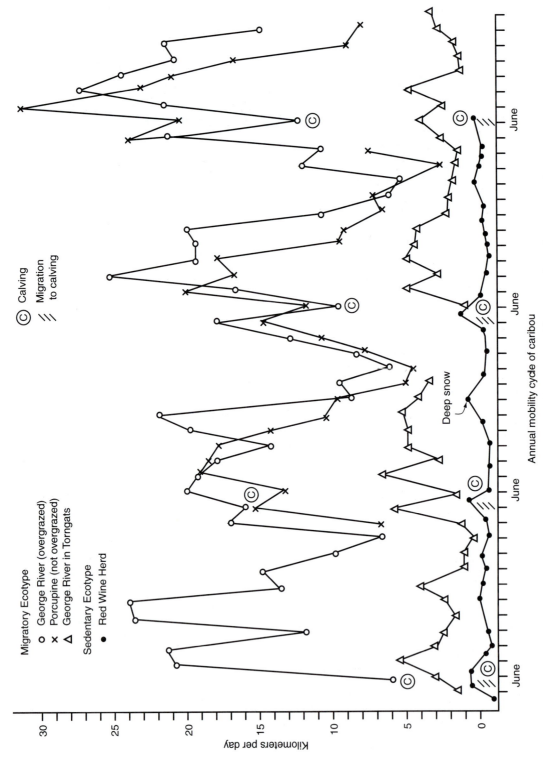

Figure 31–6 The annual mobility cycle compared between the migratory ecotype on overgrazed summer range (George River herd), and ranges that have not been overgrazed (Porcupine herd), between the George River herd when traveling in smaller groups in the Torngat Mountains, and the sedentary ecotype (Red Wine herd). (Data from Refs. 39, and 108, and personal files for the George River Herd.)

Home Range

Caribou show the most fidelity to their calving range; females of the sedentary ecotype remain near calving sites for several weeks if not disturbed by predators (19, 46, 74). The migratory ecotypes commonly return to the same general calving ground, but show little philopatry to the specific sites where they calved previously. A general "safe area," such as a large treeless area, is recognized but not a specific location within the area.

Woodland caribou are frequently found on the same wintering range, whereas migrating tundra animals commonly shift winter ranges as snow cover varies. This behavior of sedentary animals allows them to return to upland areas, which generally will have the least snow, regardless of annual variations in total snowfall. If the snow cover increases as the winter progresses then winter range may become more restricted (45, 39).

POPULATION DYNAMICS

Population Trends and Current Status

There were more than 3.3 million caribou in North America in the 1980s (Table 31–1). In that decade the migratory herds had generally increased in North America and Eurasia, whereas the sedentary herds had declined on both landmasses (144, 134). More recently the montane sedentary caribou have increased, although part of the gain is likely greater census effort, while the boreal sedentary ecotype has continued to decline (145, 146). Examples of changes based on solid census data include herds in Alaska increasing from 295,000 in 1977 to 880,000 in 1993 ($\lambda = 1.07$) and the George River herd increasing from 4,700 in 1955 (likely a low estimate) to the largest herd in the world at 643,000 by 1984 ($\lambda = 1.18$) (147, 17). The most spectacular decline has been the Peary caribou on the Queen Elizabeth Islands, declining from 26,000 in 1961 down to less than 3,000 in 27 years ($\lambda = 0.92$) (148, 33). The Peary decline resulted from a series of crashes in years when warm temperatures caused extensive ground ice and starvation losses and major reductions in fecundity (149, 150).

On the northern edge of the species range (the Queen Elizabeth Islands, and Greenland), the limiting factor for the species is the extreme weather that can result in severe ground-fast ice when temperatures rise in the winter. In Greenland the isolated populations on the West Coast face overgrazing problems and starvation because wolves have been absent for 4,000 years (151). The Greenland populations have a history in the past 9,000 years of extinctions from weather and reduced forage and then recolonization from Ellesmere Island and Baffin Island (152, 34).

On the southern edge of the species range biological diversity of the mammalian fauna limits abundance and distribution of caribou and not food resources or habitat. Caribou introduced to Michipicoten Island, Ontario, have prospered in a habitat dominated by deciduous forests where there are no white-tailed deer or wolves. Caribou cannot coexist with high numbers of deer because they are highly susceptible to meningeal worm of white-tailed deer (153). The reintroduction of caribou to ranges with deer has failed largely because of this disease transmission (154, 155). Also the prey biomass of white-tailed deer and moose generally result in wolf populations too high (greater than 10 per 1,000 square kilometers) for caribou to persist (Figure 31–7) (156, 157). Wolves are scarce west of the Great Plains but are replaced by mountain lions as a major predator (158). Reintroduced wolves in Idaho could create problems for the endangered Selkirk caribou herd; ironically, one endangered species may contribute to the demise of another.

Biologists commonly want to reintroduce caribou to Maine, Minnesota, and Idaho, the southern edge of their range; the rationale is that they were there at the time of settlement in the 1800s and are needed for ecological integrity and to restore a pristine fauna. But caribou were farther south at the end of the Little Ice Age, about 1860, than they had been in previous warmer cycles.

(a)

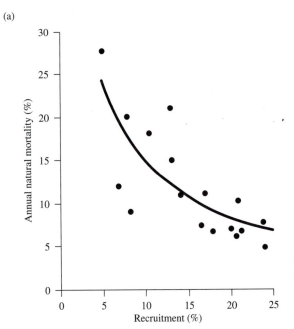

Figure 31–7 The annual natural mortality rates of adults is correlated with annual recruitment (percent calves in herd at 5 to 6 months or short yearlings at 10 to 11 months) for 17 herds that totalled 1,380,000 animals (a). Annual natural mortality and recruitment are correlated with wolf densities in North America where wolves had been counted as of 1986 (b)(From Ref. 131, used with permission of Elsevier Science.)

(b)

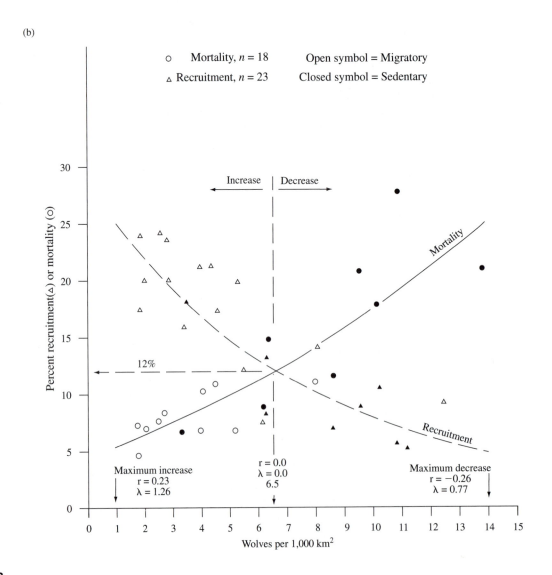

Productivity and Recruitment

The productivity of caribou is low compared to the other cervids in North America because caribou have one calf per year and fawns and most yearlings commonly are not pregnant (103). The mean pregnancy of females ranged from 82.5% for animals greater than 1.5 years to 85% for animals greater than 3 years (40, 81). Variation between herds was low because variable nutritional aspects primarily affect the age of puberty, whereas sexually mature females generally conceive in most years (105, 159, 40, 81). A maximum range for herds that have not experienced massive starvation is demonstrated by the George River herd: 89% were pregnant in April 1980 prior to range degradation but only 67 to 69% were pregnant from 1986 to 1992 with many 1.5- and 2.5-year-old females not conceiving after the summer range deteriorated (96, 78, 63). The lowest pregnancy percentages in the literature are for Peary caribou in years they starved; 7.1% were pregnant in April 1974 and 6.7% were pregnant in April 1975 (149, 150).

It has been difficult to separate the effects of adverse winter weather (snow depths) and the availability of nutritious spring forage on calf survival and conception rates because environmental factors are interrelated. However a natural experiment occurred in 1991 when Mount Pinatuba erupted in the Philippines, reducing solar radiation. Spring 1992 was the coldest on record in the eastern and western arctic, but winter snow depths in 1991 to 1992 were not extreme (16, 160, 161, 162). The birth mass of calves and survival were reduced in 1992 (63, 16, 160). Natality in Ungava and Alaska was reduced in 1993 (163, 161). Conversely in winter 1990-1991, the snow cover in Alaska exceeded levels in 18 other seasons but natality in 1992 for the Delta herd was the highest recorded (96%) (161, 164). Thus, the dominant factor in productivity is not winter snow cover per se, but the availability of a nutritious diet in April and May.

Population growth in North American caribou herds is highly correlated with the annual recruitment to the herd. For 22 sedentary herds the recruitment needed to balance adult mortality was 15.6% (22). Two woodland herds on the southern boundary of their range in Alberta and Saskatchewan were thought to be declining when recruitment in March was 18 to 28 calves per 100 females (165, 166). For migratory herds in Alaska 25 calves per 100 females were needed to balance adult losses (20).

Mortality Factors

The mortality rate of young calves from predation can be high in situations where wolves and bears are common (167, 67, 68, 70, 57, 64, 161). Mortality of 2- to 3-week-old calves in British Columbia herds varied from 56% to 80% and most deaths were from wolves and bears (168, 40). The loss of calves in five migratory herds at about 30 days was 42% (169). The mean mortality of 147 calves of radiocollared females in Denali National Park to 120 days of age in 5 years (1987–1991) was 66% (extremes 43% to 93%); 38% were killed by wolves, 26% by bears, 17% by unknown predators, 9% died from prenatal factors, 3% died of accidents, and 7% of the losses were undetermined (162). For the stable Forty-Mile herd in Alaska during 1994 to 1995 the loss of calves up to 12 months was attributed to wolves (27%), grizzly bears (23%), other predators (10%), and 10% nonpredator mortality (10%)(29% survived). Adult deaths during the same period were caused by wolves (9%), bears (0.5%), hunting (7.7%), and other factors (1%) (163). In Newfoundland 27% of calves born from 1957 to 1967 were killed within 2 weeks of life by lynx (105). In this case lynx had become abundant after the introduction of snowshoe hares. The lynx switched to caribou when hares had cyclic lows (170).

At one time it was believed that large numbers of neonates died from hypothermia in the Northwest Territories (46, 171). But this hypothesis was laid to rest by Miller and others, who studied the cause of death of calves (107a). They found, however, that none of 287 calf deaths on the Beverly calving grounds in the Northwest Territories were weather related and probably were the result of surplus killing by wolves (172, 61, 107a, 173). There may be an interaction because wolves are more successful in snowstorms and conditions of poor visibility, but the calves killed in these surplus killings were not doomed in the absence of wolves (46).

There is disagreement as to whether calves die at greater rates than adults during winter. Calves have lower fat reserves than adults, allocating nutrients to growth rather than condition, but there is little evidence of winter starvation for free-ranging herds. If snows are deep there can be surplus killing by wolves of calves and adults (174, 162). For the herds with good summer nutrition that have high neonatal predation rates, the remaining calves have gone through a culling process and possibly may have winter mortality rates similar to adults. But for those large migratory herds such as in the Northwest Territories where the majority of the calves survive to autumn, calves remain more susceptible than adults to wolf predation until they have completed their second summer of growth (175, 107, 40, 32).

The mortality rate of adults can be estimated on the basis of the recruitment needed to stabilize population growth by balancing adult mortality. If it takes 25 calves per 100 females to balance adult losses, and if the ratio of adult males to females is 1 : 2 (35.7% males for 18 herds), then an expected annual adult mortality rate is 16% [25/(100 females + 56 males)] (40). The stabilizing recruitment (R_s) value of 25 calves/100 females included only herds that calved south of the Arctic treeline in alpine tundra. These herds experience greater calf mortality than the migratory herds that shift to calving grounds a mean distance of 200 kilometers north of the Arctic treeline (131). By combining herds calving south and north of treeline, R_s is closer to 12%, a more inclusive estimate of natural adult mortality.

Adult males should have mortality rates generally twice that of females given a sex ratio of 1 : 2 for adults and equality between the sexes in the yearlings recruited (19, 110). Based on a loss of 12%, one could expect a mortality rate of 16% for males and 8% for females. Mech and others found that wolves killed 51 males and 43 females in 6 years; given the 1 : 2 sex ratio available, males would have died at twice the female rate (174). In the Western Arctic herd, Davis and Valkenburg found a natural mortality of males to be 13% and for females the natural rate was 7% (176). Males should be more vulnerable in March and April when females in migratory herds start north and leave the males alone closer to treeline and wolves.

Adult female mortality rates have been estimated at 16% for the Beverly Herd from 1980 to 1997, 11% for females in the George River herd, 10.4% for the Delta herd in Alaska from 1979 to 1992, 15.6% for the Porcupine herd from 1982 to 1992, 12.2% for the George River herd from 1984 to 1992, and 15% for nine herds in British Columbia (97, 109, 20, 108, 177, 168).

Total adult mortality was estimated at 12% for the Forty-Mile herd when numbers were stable (163). The mortality rate for two woodland herds that were not legally hunted was 12% to 16% when numbers were stable or declining (165, 166).

When the mortality of females exceeds 15%, it does not follow that male mortality should be 30%. Recently the Red Wine herd declined from predation from more than 600 animals to about 150 animals. This approaches the maximum predicted decline rate of caribou that have died from factors other than massive winter starvation ($\lambda = 0.77$) (22). Females in the Red Wine herd had a mortality rate of 29%, but males in the population increased from 37% to 45%. Mortality rates for females in the George River herd increased from 5% in 1984 to 17% by 1992, but the male mortality rate did not increase proportionally (177, 16). In this case females had special problems in energy balance in spring and summer during lactation (178, 177, 77).

Predation can be density dependent and regulating for sedentary caribou herds (179, 180, 22, 168). But this regulating effect is less likely for the migratory herds (181). The mortality rates of adults from wolves for these migratory herds has been estimated at 5 to 7% generally less than recruitment (13, 182, 183). However, for sedentary herds, declines from predation have occurred when caribou densities exceeded 0.06 per square kilometer (22) At this caribou density of 0.06 per square kilometer, R was 15% to 16%; that approximates the recruitment needed to balance the mortality of adults for sedentary herds. Hence, 0.06 animals per square kilometer could be considered the stabilizing density (D_s) that resulted in R_s. This D_s results because females space out at calving and display site fidelity; fewer females are farther apart, and at higher numbers closer to each other. This sequence moderates the searching success of predators with rareness acting as a damping mechanism (184).

When caribou are introduced into systems without predators, such as islands, they generally increase at rates approaching r_{max} of 1.28 (40, 185). At higher densities these insular populations later crash from malnutrition (186, 41). But it cannot be assumed in the absence of data that the starvation resulted from only lack of winter foods. They could have reduced summer foods and entered winter predisposed to death (16). Major declines of the insular herd on the Slate Islands resulted primarily from summer over-grazing even though the actual deaths occurred in late winter. The animals simply did not enter winter with sufficient reserves to maintain mass with a winter reduction of lichens. Starvation weights would depend on the initial fall masses plus the slope of decline during winter (16). The George River herd reached densities higher than 10 per square kilometer on the summer range and degraded vascular food supplies that increased mortality rates and lowered fecundity. However, winter lichen supplies remained abundant and winter starvation did not occur, but females did die of malnutrition in June and July (146, 187). Winter lichen supplies do not set the carrying capacity for caribou (19, 188). But when summer densities exceed two per square kilometer of summer range one can expect declines in body size and demography (189, 190). The Western Arctic herd increased from 75,000 in 1976 to 343,000 in 1989 ($\lambda = 1.12$) but after 1990 when densities exceeded 2 per square kilometer of *summer range* the herd only increased at 2% per year to 1996 (30). The Porcupine herd declined from 178,000 in 1989 to 160,000 in 1992 when density on the summer range exceeded two per kilometer square (108, 21).

Emigration and Immigration

There is no consensus at this time on the role of animal exchange between herds or between calving areas that would alter demography (191, 192). Before the advent of radiotelemetry, Skoog and Kelsall believed there were exchanges between herds (19, 46). It is recognized that animals that calve at one calving ground may shift to an area used by wintering animals from a separate calving population and then return to their own calving grounds.

However, I believe there can be major exchange between herds that inhabit large areas of adjacent tundra, such as the Keewatin Region of Northwest Territories and northern Ungava. Some major changes in population estimates cannot be explained by faulty census techniques or by internal herd growth (193). Kelsall documented a large-scale exodus in July 1956 (46). A significant number of radiocollared animals have switched calving grounds between the Leaf and George River herds (63, 16).

Because females select calving areas to avoid predation, one could expect them to change areas if the distribution of wolves changed (191, 16). Such shifts would be a major stabilizing mechanism, leaving a high-density predator habitat and swamping a low-density one, unless wolves could also alter their distributions quickly. However, at this time most population estimates from Alaska, Ungava, and the Northwest Territories assume discrete calving and postcalving aggregations without factoring in emigration–immigration possibilities.

POPULATION MANAGEMENT

The basic equations for population growth (the finite rate of increase λ) are $\lambda = N_2/N_1$ and/or $\lambda = (1 - R)/(1 - M)$, where R equals recruitment and M equals total mortality. Both R and M are expressed as a percentage of the herd and R should be measured when the mortality rate of the new generation equals that of adults (194). The most successful census technique for N_2 and N_1 for the migratory ecotype has been the aerial photocensus technique used for the huge postcalving aggregations that form in July in response to mosquito harassment (195, 196). The technique has recently been used for the George River herd (197). Care must be taken to evaluate herd exchanges, as happened recently when many of the animals of the Central Arctic herd were in postcalving aggregations of the Porcupine herd (108).

The recruitment/mortality (*R/M*) approach to evaluating population trends is possibly the best technique to gauge changes for the sedentary herds, although a reliable winter aerial quadrat census has been developed for some herds in the Yukon (198). It is also possible to spray paint open-ranging caribou and use a Petersen mark–resight technique (199). If the counts are made from fixed-wing aircraft only, calves can be separated from adults. If the hunting mortality of the population is known it can be incorporated into the *R/M* equation (161). For example, if natural mortality is 12% the equation becomes $\lambda = 0.88 (1 - M_h)/1 - R)$, where M_h = hunting mortality. This equation can also be altered to calculate natural mortality when hunting mortality has been ascertained with the formula $M_n = 1 - [(1 - R)/(1 - M_h)]$.

The basic tools available for managing caribou are hunting and regulation of natural mortality rates. These mortalities are additive, a lesson that was learned in the 1970s in Alaska when the harvest remained high at the same time wolf numbers were recovering, a deadly combination taking from both sides of the $R = M$ equation. These combined losses resulted in major declines of several herds in Alaska (159, 167).

Since the 1970s decline in Alaska, harvests generally have been less than 5% and fluctuating wolf numbers have allowed a recovery of all the major herds. However, for the two large migratory herds in Alaska and the seven on the mainland of Northwest Territories, nearly all the harvest is by native peoples and difficult to legislate. A major success story was the curtailment of the harvest of the Western Arctic herd in the 1970s, from 30,000 animals annually to less than 3,500 animals, most of which were males (200). This resulted in the herd only declining to a low of 75,000 animals by 1976 (a rate of decline from 1970 to 1976 of = 0.82). By 1993 the herd was counted at 450,000 animals, ($\lambda = 1.11$) (20). There are now two cooperative management boards, one for the Porcupine herd and the second for the Beverly/Kaminuriak herds, comprised of native representatives and government biologists. Traditional knowledge is incorporated into management programs. Because indigenous people are the primary users of caribou and wolves, it will be necessary for them to make the decisions relative to their own harvests of caribou and the allowable size of wolf populations.

HABITAT REQUIREMENTS

The primary habitat requirement of caribou is sufficient space to effectively carry out their antipredator tactics. Especially important is having sufficient space to reduce risk when calves are young by "spacing away" and "spacing out." The requirements are analogous to space needed by ground nesting ducks and grouse (201). When habitat becomes less continuous and nests are in "islands" of habitat, then the predator knows where to search. If logging results in islands of available lichens, then the caribou will of necessity be there, and so will the wolves. Disturbances, such as logging roads, seismic lines, and hydro rights-of-way, may increase effective searching for the predators. Caribou must have their "miles without measure" to make their "endless march" successfully with wolves.

Caribou do not have specific food and cover requirements and winter successfully as far north as land extends that has some vegetation. Caribou generally winter on lichen pastures with reduced snow depths (45, 48). The assumption was that they wintered in these habitats because of the ease of finding food but now biologists know that these low snow profiles reduce predation risk and that wolves can switch from moose to caribou as snow levels build (174, 162, 202). Annual mortality rates of radiocollared females were correlated with winter severity in the Forty-Mile herd in 7 years ($r = 0.798$) and for the Delta herd in 16 years ($r = 0.682$). But most of these deaths were from wolf predation, because bone marrow samples did not indicate starvation (32, 163, 161, 20).

Habitat Requirements

The greater success of wolves in deep snow may depend on relative foot-load-ings and sinking depths when the chase is on. The foot-loadings, not to be confused with hoof-loadings, of caribou and wolves are similar when all four legs are on the ground, 100 grams per square centimeter, and for caribou increase with body size (203, 204). Despite the favorable ratio of sinking depths between caribou and wolves, when caribou flee they are striking the snow with only one or two feet at a time increasing the loadings two to four times (204). For wolves with their shorter legs, the load is more evenly distributed in pursuit. Biologists can no longer state unequivocally that caribou select winter ranges of reduced snow primarily to reduce the energy for cratering and improve forage intake. Shallow snow also reduces risk. Further, the caribou practices of traveling on frozen lakes with shallow snow and bedding offshore on lake ice have energy-saving aspects and a predator detection and escape component.

Caribou in rut select areas more for the security they provide from predators than the optimal forage the areas may provide (205). Now biologists can add to the low-risk habitats of spring and autumn the selection of low-risk habitat in winter. The conclusion that is emerging, contrary to optimal foraging theory, is that habitat selection is first relative to survival and predation and only secondary to the ease of finding nutritious forage.

The habitat niche can be visualized as a series of decreasing space areas as necessary requirements increase. The major ecological factors are food resources, snow depths, insects, and predation risk (16). For example, in the winter distribution of caribou the largest space is based on the distribution of lichens. Caribou will not be found where there is not any food, but a lot of lichen ranges will not be occupied. Within this lichen niche they will select areas of low snow cover where there is a relative greater abundance of food. Within this even smaller space they may further restrict their distribution to areas with the least predation risk. At calving, females will optimally forage within those selections available that do not compromise safety (67). On warm windless days in July the first priority will be habitats that provide maximum insect relief, such as windswept slopes, low vegetation, areas of late insect hatch, and snow fields. The second order is to find forage in the vicinity of these relief habitats when harassment abates. The phytomass in the vicinity of these low-risk and relief habitats will generally be reduced because of their disproportionate occupancy, and this will occur over a wide range of densities because small herds will remain in preferred habitats longer than larger groupings.

HABITAT AND PREDATION MANAGEMENT

Forest fires were blamed for caribou declines due to reduced lichen supplies, but caribou declines did not occur simultaneously with fire chronologies (206, 207, 208, 209). An alternative explanation is that forest fires may set forest succession back to a more deciduous state that favors increases in deer and moose (207). Increased moose densities may lead to increased wolf densities, which could cause a caribou decline.

Biologists for the past 40 years have thought that the abundance of lichens in winter in conjunction with caribou densities determined winter physical condition. However, the hypothesis is not now supported by two extensive studies of winter condition (99, 125). Range studies now emphasize the energy budget (intake versus expenditure) and not simply forage intake (116, 179, 48, 77).

Extensive fires are necessary to renew the nutrients and energy flow and to maintain the integrity of the boreal forest and taiga (210, 211, 212, 213, 214, 215). Forest fires provide the heterogeneity and visibility that caribou require to cope with changing snow cover and predation risk. Forest fires reduce organic litter, favoring the reestablishment of lichens (215). Caribou have evolved in this fire-driven ecosystem. Forest fire protection is a short-sighted management policy and is generally not justified in habitats where the successional stage returns rapidly to a lichen woodland without a secondary stage of paper birch and balsam fir that benefit moose. Thomas and others concluded that there was no justification for fire suppression in relation to caribou conservation (216) because fire is an ecosystem process and not a catastrophic event (221).

Biologists are concerned about the continuing decline of the sedentary populations. Research is attempting to reduce predation by habitat alterations to lessen or segregate the prey base in favor of caribou rather than moose. Habitat mosaics and natural heterogeneity and the integration of silviculture systems are being incorporated into management plans for caribou and forests (218, 219, 220, 221, 222).

During the 1970s biologists became concerned about the impacts of economic development on caribou in their pristine habitat. However, it was the biologists that were disturbed seeing the pristine wilderness bridged by roads, dams, and mines (142). The caribou were just as adaptable as other cervids to developments. Reindeer are the only cervid that has been domesticated. In fact, some of the most spectacular increases in caribou have been the "disturbed" herds. In 1956 I found the last 71 animals of the Avalon herd in Newfoundland in their most remote range but by the 1980s the herd had increased to several thousand and grazed within sight of well-traveled highway (223, 224). Also in 1957 there were no published maps of the interior of Newfoundland when I made the first census of caribou and found only 4,500 animals (105). In the years that followed, roads and dams were built and the forests were logged, yet the herd increased to more than 53,000 from 1989 to 1991 (142, 15). As development increased the skidoo provided access for trappers which led to overharvesting the lynx and removing the most important limiting factor for caribou in that area (105, 170).

The most intense development in Alaska occurred when an oil field was built on the calving ground of the Central Arctic herd. The herd increased from 5,500 in 1978 to 18,000 in 1988 ($\lambda = 1.11$) and then to 23,400 in 1992 ($\lambda = 1.09$). A control herd in the pristine wilderness to the west of the Central Arctic herd, the Teshekpuk herd, had growth rates of $\lambda = 1.15$ and $\lambda = 1.14$, respectively, during equivalent periods. To the east of the Central Arctic herd another control herd, the Porcupine herd, that also calves on the Coastal Plains undisturbed by development, increased by 1.05 and 0.97 during these same periods (225). The Central Arctic herd prospered because it was initially at low densities. Additionally, development reduced predation pressure and the animals benefited from improved insect relief (berms, pipelines) (226). The major detrimental change was that females with neonates shifted a few kilometers from the traffic on heavily used roads. Cronin and others concluded that resource extraction (i.e., oil) and caribou populations can be compatible when managed properly (225).

Another herd that encountered disturbances was the Delta herd, Alaska, where military exercises and overflights occurred on calving grounds (227). This herd increased even though disturbed, but is sensitive to wolf predation and overhunting (167, 161). Caribou may alter their activity budgets slightly in response to human-induced stimuli, but these energetic costs are minuscule compared to the stochastic responses elicited when caribou must interact with the natural varying abundance of predators and insect tormentors (228).

The myth that there is a balance of nature is deep in our psyche. Charles Elton, the father of ecology, said " it is assumed that an undisturbed animal community lives in a certain harmony . . . the balance of nature . . . the picture has the advantage of being an intelligible and apparently logical result of natural selection in producing the best possible world for each species. It has the disadvantage of being untrue" (229, quoting Elton). Predation by humans and natural predators are the primary limiting factor for North American caribou. When wolves have been effectively reduced caribou have increased. Coyotes and bears now threaten the endangered relic population on the Gaspe Peninsula as do wolves for the less than 20 animals remaining in Pukaskwa National Park, Ontario (235, 16). Caribou can live with humans into the next millennium if the earth's climate will "cooperate." In the past 45 years, the mean June temperature has increased 2°C in the western Arctic (163). Unfortunately, the story of evolution is the tale of extinctions. There is no balance of nature to guarantee the continued persistence of caribou populations if climatic trends operate too swiftly for natural selection to keep abreast.

Habitat and Predation Management

Caribou have roamed the North American Arctic for at least 2 million years and the ancestral home could have been Alaska. During the Wisconsin Ice Age caribou persisted in Beringia in Alaska/Yukon, south of the ice, and probably on Banks Island, Northwest Territories. At the time of European contact the migratory herds that migrated to the tundra probably numbered 3-4 million animals, whereas the sedentary caribou that lived south of trees would have been far less abundant. In the past 100 years the numbers of migratory caribou have fluctuated but have maintained their pristine range and numbers. However, the sedentary woodland animals have declined and contracted their range as a result of natural predation and hunting, and now number 100,000 animals. The Peary caribou of the Arctic Island are nearly extinct. The major limiting factor for caribou has been predation, particularly from wolves. To reduce predation risk migratory females migrate to calving grounds on the tundra to space away from alternative prey and wolves denning further south. Females of the sedentary ecotype disperse at calving to reduce contact with predators. This spring movement of females is the only true migration that caribou make because it is a purposeful trek involving philopatry and traditions. The behavior of caribou is a product of the extrinsic environment: of the visibility of the habitat, the ratio of tundra to closed canopy, the presence of predators and biting/parasite insects, and the slow growing phytomass subject to trampling. The basic sequence is that caribou inhabiting this open habitat in the presence of predators adopted a gregarious life style. These aggregations must remain mobile, high grading nutrients as a result of density dependent cropping of the sparse Arctic vegetation. There is an annual cycle of mobility with reduced travel rates at calving, fall breeding and in midwinter when snows are deep. Rates accelerate in the summer and fall and for migratory herds can exceed 15 kilometers/day whereas sedentary herds may travel only 1-2 kilometers/day. The size of the range occupied by the migratory herds varies with population size. When the herds are reduced they constrict their range to include the calving grounds, tundra foraging areas, and areas along the treeline. When numbers increase the herds expand their range further into the taiga. These contractions and expansions are a density dependent response to the abundance of summer and fall forage supplies and commonly occurs at one animal/square kilometer. In the winter animals select open lichen woodlands with low snow depths probably to reduce predation risk. Caribou can and have successfully coped with economic development in the Arctic when these disturbances did not increase predation or exploitation. The native peoples of the north have been hunting caribou since the ice retreated and remain hunters today with a love of the vast spaces that caribou need. An ancient food chain still exists. Increasingly, resource management decisions are theirs. But global warming is in our hands to the south and is threatening the entire system. Especially vulnerable are the Peary caribou to the north, facing increased freezing and thawing, and woodland caribou to the south, facing range expansions of deer and moose to the north, bringing with them increased predation by wolves.

LITERATURE CITED

1. SETON, E. T. 1927. *Lives of Game Animals,* Vol. 3, Pt. 1. Doubleday Page, New York.

2. BANFIELD, A. W. F. 1961. A revision of the reindeer and caribou genus *Rangifer.* Naturalist Museum of Canadian Bulletin 277.

3. IRVING, L., 1975. Opening remarks. *1st International Reindeer and Caribou Symposium,* Fairbanks, AK, pp. 1–2.

4. PIELOU, F. C. 1992. *After the Ice Age.* University of Chicago Press, Chicago.

5. HARINGTON, C. R. 1998. Ancient caribou, its evolution and place as one of the "big 4" of the Beringian Mammoth steppe fauna. *North America Caribou Workshop* 8:in press.

6. HEARNE S. 1795. *A Journey from Prince of Wales Fort in Hudson Bay to the North Ocean in the Years 1769, 1770, 1771, 1772.* London.

7. FRANKLIN, J. 1823. *Narrative of a Journey to the Shores of the Polar Sea in the Years 1819, 20, 21, and 22.* Zoology appendix: Mammals and birds by Joseph Sabine. J. Murray, London.

8. BACK, SIR G. 1836. *Narrative of the Arctic land expedition to the mouth of the Great Fish River, and along the shores of the Arctic Ocean, in the years 1833, 1834, and 1835.* Appendix No. 1 by J. Richardson. J. Murray, London.

9. TYRELL, J. W. 1902. Exploratory survey between Great Slave Lake and Hudson Bay districts of MacKenzie and Keewatin. Annual Report Canadian Department Interior, Appendix 26 to Report Survey General, pp. 98–155 and 207–239.

10. MURIE, O. J. 1935. Alaska-Yukon caribou. U.S. Bur. Biology Survey. North American Fauna 54.

11. GUTHRIE, R. D., and J. V. MATTHEWS. 1971. The Cape Deceit fauna early pleistocene mammalian assemblage from the Alaskan Arctic. *Quarternary Research* 1:474–510.

12. RØED, K. H., and K. R. WHITTEN. 1986. Transferrin variation and evolution of Alaskan reindeer and caribou. *Rangifer Special Issue* 1:247–251.

13. BANFIELD, A. W. F. 1954. Preliminary investigation of the barren-ground caribou. Canadian Wildlife Service Wildlife Management Bulletin Service 1 No. 19A. Canadian Wildlife Service, Ottawa.

14. ANDERSON, R. M. 1938. The present status and distribution of big game mammals of Canada. *Transactions North American Wildlife Conference,* 3:390–406.

15. FERGUSON, M. A. D., and L. GAUTHIER. 1992. Status and trends of *Rangifer tarandus* and *Ovibos moschatus* populations in Canada. *Rangifer* 12:127–141.

16. BERGERUD, A. T. 1996. Evolving perspectives on caribou population dynamics. Have we got it right yet? *Rangifer Special Issue* 9:95–115.

17. CRÊTE, M., L.-P. RIVEST, D. LE HÉNAFF, and S. N. LUTTICH 1991. Adapting sampling plans to caribou distribution on calving grounds. *Rangifer Special Issue* 7:137–150.

18. MORNEAU, C., and S. PAYETTE. 1998. A dendroecological method to evaluate past caribou (*Rangifer tarandus* L.) activity. *Ecoscience* 5:64–76.

19. SKOOG, R. O. 1968. Ecology of the caribou (*Rangifer tarandus granti*) in Alaska. Thesis. University of Calif., Berkeley.

20. VALKENBURG, P., J. L. DAVIS, J. M. VERHOEF, R. D. BOERTJE, M. E. MCNAY, R. M. EAGAN, D. J. REED, C. L. GARDNER, and R. W. TOBEY. 1996. Population decline in the Delta herd with references to other Alaskan herds. *Rangifer Special Issue* 9:53–62.

21. VALKENBURG, P. 1998. Herd size, distribution, harvest, management issues, and research priorities relevant to caribou herds in Alaska. *Rangifer Special Issue* 10:125–129.

22. BERGERUD, A. T. 1992. Rareness as an antipredator strategy to reduce predation risk for moose and caribou. Pages 1008–1021 *in* D. R. McCullough and R. H. Barrett, eds., *Wildlife 2001: Populations.* Elsevier Applied Science, New York.

23. CLARKE, C. H. D. 1940. A biological investigation of the Thelon Game Sanctuary. Naturalist Museum of Canada Bulletin 96.

24. CALEF, G. W. 1981. *Caribou and the Barren-Lands.* Firefly Books Limited, Toronto.

25. JORDAN, R. 1978. Archaeological investigations of the Hamilton Inlet Labrador eskimo: Social and economic responded to European contact. *Arctic Anthropology* 15:1–8, 175–185.

26. SAMSON, G. 1978. Preliminary cultural sequence and palaeo-environmental reconstruction of the Indian House Lake Region, Noveau-Quebec. *Arctic Anthropology.* 15:186–205.

27. ELTON, C. 1942. *Voles, Mice and Lemmings, Problems in Population Dynamics.* University of Oxford, London.

28. BERGERUD, A. T. 1967. Management of Labrador caribou. *Journal of Wildlife Management* 31:621–642.

29. DAVIS, J. L., and P. VALKENBURG. 1991. A review of caribou population dynamics in Alaska emphasizing limiting factors, theory and management implications. *North America Caribou Workshop* 4:184–207.

30. DAU, J., J. COADY, S. MACHIDA, and L. A. AYRES. 1998. The western arctic caribou herd: Current status and management issues. North America Caribou Workshop 8 : in press.

31. BUTLER, H. E. 1986. Mating strategies of woodland caribou (*Rangifer tarandus caribou*). Thesis. University of Alberta, Calgary.

Literature Cited

32. VALKENBURG, P. 1997. Investigation of regulating and limiting factors in the Delta Caribou Herd. Alaska Department Fish and Game Fed. Aid in Wildlife Restoration, Progress Report W-24–2, W-24–3, W-24–4 Study 3.34:1–45.

33. GUNN, A., F. L. MILLER, and J. NISH. 1998. Status of endangered and threatened caribou on Canada's Arctic islands. *North America Caribou Workshop,* 8:in press.

34. MELDGAARD, M. 1986. The Greenland caribou—Zoogeography, taxonomy, and population dynamics. *Bioscience* 20:2–88.

35. RØED, K. H., and D. C. THOMAS. 1990. Transferrin variation and evolution of Canadian barren-ground caribou. *Rangifer Special Issue* 3:385–389.

36. RITCHIE, J. C. 1987. *Postglacial Vegetation of Canada.* Cambridge University Press, New York.

37. THOMAS, D. C. 1969. Population estimates and distribution of barren-ground caribou in Mackenzie District, NWT, Saskatchewan and Alberta March to May 1967. Canadian Wildlife Service Series No. 9, Canadian Wildlife Service, Ottawa.

38. HATLER, D. F. 1986. Studies of radio-collared caribou in the Spatsizi Wilderness Park, British Columbia 1980–84. Wildlife Branch Report 12. Spatsizi Association for Biological Research, Smithers, B. C.

39. BROWN, W. K. 1986. The ecology of a woodland caribou herd in central Labrador. Thesis. University of Waterloo, Ontario.

40. BERGERUD, A. T., 1980. A review of the population dynamics of caribou and wild reindeer in North America. *International Reindeer/Caribou Symposium* 2:556–581.

41. KLEIN, D. R. 1968, The introduction, increase and crash of reindeer on St. Matthew Island. *Journal of Wildlife Management.* 32:350–367.

42. SKOGLAND, T. 1985. The effects of density dependent resource limitation on the demography of wild reindeer. *Journal of Animal Ecology* 54:359–374.

43. OUELLET, J-P, D. C. HEARD, and R. MULDERS. 1996. Population ecology of caribou populations without predators: Southampton and Coat's Island herds. *Rangifer Special Issue* 9:17–26.

44. HARRINGTON, F., and S. LUTTICH. 1991. Migration patterns of George River caribou. *North America Caribou Workshop,* 4:237–248.

45. BERGERUD, A. T., 1974. Relative abundance of food in winter for Newfoundland caribou. *Oikos* 25:379–387.

46. KELSALL, J. P. 1968. The migratory barren-ground caribou of Canada. Queen's Printer, Ottawa.

47. THOMAS, D. C. 1991. Adaptations of barren-ground caribou to snow and burns. *North American Caribou Workshop,* 4 : 482–500.

48. RUSSELL, D. E., A. M. MARTELL, and W. A. C. NIXON. 1993. Range ecology of the Porcupine herd in Canada. *Rangifer Special Issue* 8:1–167.

49. BERGERUD, A. T., 1977. Diets of caribou. Pages 243–266 *in Diets of Mammals,* Vol. 1. CRC Press, Cleveland, OH.

50. HEARD, D. C., T. M. WILLIAMS, and D. A. MELTON. 1996. The relationship between food intake and predation risk in migratory caribou and implications to caribou and wolf population dynamics. *Rangifer Special Issue* 9:37–44.

51. DAUPHINE, JR., T. C., and R. L. MCCLURE. 1974. Synchronous mating in Canadian barren-ground caribou. *Journal of Wildlife Management* 38:54–66.

52. BERGERUD, A. T., 1974. Rutting behaviour of Newfoundland caribou. Pages 395–435 *in* V. Geist and F. Walther, eds., *The Behaviour of Ungulates and Its Relationship to Management.* International Union for Conservation of Nature and Natural Resources.

53. NIKOLAEVSKII, L. D. 1961. General outline of the anatomy and physiology of reindeer. Pages 5–56 *in* PS. Zhigunov, ed., *Reindeer Husbandry.* Transactions for U.S. Deptartment of International and National Science Foundation by Israel program for scientific translations, 1968, Jerusalem.

54. BASKIN, L. M. 1970. *Reindeer, Ecology and Behaviour.* Nauka, AS, USSR, Moscow.

55. ADAMS, L. G., and B. W. DALE. 1998. Timing and synchrony of parturition in Alaska caribou. *Journal of Mammalogy* 79:287–294.

56. BERGERUD, A. T., 1974. The role of the environment in the aggregation, movement and disturbance behaviour of caribou. Pages 552–584 *in* V. Geist and F. Walther, eds., *The Behaviour of Ungulates and Its Relationship to Management.* Union for Conservation of Nature and Natural Resources.

57. ADAMS, L. G., F. J. SINGER, and B. W. DALE. 1995. Caribou calf mortality Denali National Park, Alaska. *Journal of Wildlife Management* 59:584–594.

58. BERGERUD, A. T., 1975. The reproductive season in Newfoundland caribou. *Canadian Journal of Zoology* 53:1213–1221.

59. CAMERON, R. D., W. T. SMITH, S. G. FANCY, K. L. GEHART, and R. G. WHITE. 1993. Calving success of female caribou in relation to body weight. *Canadian Journal of Zoology* 71:480–486.

60. SKOGLAND, T. 1984. The effects of food and maternal conditions on fetal growth and size in wild reindeer. *Rangifer* 4:39–46.

61. MILLER, F. L., E. BROUGHTON, and A. GUNN. 1985. Mortality of migratory barren-ground caribou on the calving ground of the Beverly Herd, District of Keewatin, Northwest Territories 1981–83. Canadian Wildlife Service Project Completion Report CWSC 3678. Canadian Wildlife Service, Alberta.

62. WHITTEN, K. R., G. W. GARNER, F. J. MAUER, and R. B. HARRIS. 1992. Productivity and early calf survival in the Porcupine Herd. *Journal of Wildlife Management* 56:201–212.

63. COUTURIER, S., C. REHAUME, H. CREPEAU, L.-P. RIVEST, and S. LUTTICH. 1996. Calving photocensus of the Riviere George Caribou Herd and comparison with an independent census. *Rangifer Special Issue* 9:283–296.

64. ADAMS, L. G., and L. D. MECH. 1995. Wolf predation on caribou calves in Denali National Park, Alaska. Pages 245–260 *in* L. N. Carbyn, S. H. Fritts, and D. R. Seip, eds., *Ecology and Conservation of Wolves in a Changing World.* Canadian Circumpolar Institute, University of Alberta, Edmonton.

65. SHOESMITH, M. V. 1978. Social organization of wapiti and woodland caribou. Thesis. University of Manitoba, Winnipeg.

66. FULLER, T. K., and L. B. KEITH. 1981. Woodland caribou population dynamics in northeastern Alberta. *Journal of Wildlife Management* 45:197–213.

67. BERGERUD, A. T., H. E. BUTLER, and D. R. MILLER. 1984. Antipredator tactics of calving caribou: Dispersion in mountains. *Canadian Journal of Zoology* 62:1566–1575.

68. PAGE, R. E. 1985. Early caribou calf mortality in northwest British Columbia. Thesis. Department of Biology, University of Victoria, British Columbia.

69. EDMONDS, E. J. 1987. Population status, distribution and movements of woodland caribou in west central Alberta. *Canadian Journal of Zoology* 66:817–826.

70. SEIP, D. R. 1992. Factors limiting woodland caribou populations and their relationships with wolves and moose in southeastern British Columbia. *Canadian Journal of Zoology* 70:1494–1503.

71. LENT, P. 1964. Calving and related behaviour in the barren-ground caribou. Thesis. University of Alberta, Edmonton.

72. FANCY, S. G., and K. R. WHITTEN. 1991. Selection of calving sites by Porcupine herd caribou. *Canadian Journal of Zoology* 69:1736–1743.

73. PARKER, G. R. 1972. Biology of the Kaminuriak population of barren-ground caribou, Part 1, total numbers, mortality, recruitment, and seasonal distribution. Canadian Wildlife Service Report Series No. 20, Canadian Wildlife Service, Ottawa.

74. FLECK, E. S., and A. GUNN. 1982. Characteristics of three barren ground caribou calving grounds in the Northwest Territories. Progress Report No. 7, Northwest Territories Wildlife Service, Yellowknife.

75. SKOGLAND, T. 1983. The effect of density-dependent resource limitation on size in wild reindeer. *Oecologia* 60:156–168.

76. REIMERS, E., D. R. KLEIN, and R. SORUMJARD. 1983. Calving time, growth rate, and body size of Norwegian reindeer on different ranges. *Arctic and Alpine Research* 15:107–118.

77. MANSEAU, M. 1996. Relation reciproque entre les caribous et al vegetation des aires d'estivage: Le cas du troupeau de caribous de la Rivere George. Ph.D. thesis. University of Laval, Quebec City, Quebec, Canada.

78. COUTURIER, S., J. BRUNELLE, J. VANDAL, and G. ST-MARTIN. 1990. Changes in population dynamics of the George River Herd, 1976–1987. *Arctic* 43:9–20.

79. CRÊTE, M., J. HUOT, R. NAULT, and R. PATENAUDE. 1993. Reproduction, growth and body composition of Riviere George caribou in captivity. *Arctic* 46:189–196.

80. CRÊTE, M., and J. HUOT 1993. Regulation of a large herd of migratory caribou: Summer nutrition affects calf growth and body reserves of dams. *Canadian Journal of Zoology* 71:2291–2296.

81. ADAMS, L. G. and B. W. DALE. 1998. Reproduction performance of female caribou. *Journal of Wildlife Management* 62:1184–1195.

82. TYLER, N. J. C. 1987. Natural limitation of the abundance of the high Arctic Svalbard reindeer. Ph.D. thesis. University of Cambridge, Cambridge, UK.

83. RUTBERG, A. T. 1987. Adaptive hypotheses of birth synchrony in ruminants: An interspecific test. *American Naturalist* 130:692–710.

84. BUNNELL, F. L. 1980. Factors controlling lambing period of Dall's sheep. *Canadian Journal of Zoology* 58:1027–1031.

85. LEUTHOLD, W., and B. LEUTHOLD. 1975. Patterns of social grouping in ungulates of Tsavo National Park, Kenya. *Journal of Zoology London.* 175:405–420.

86. SADLEIR, R. M. F. S. 1973. *The Reproduction of Vertebrates.* Academic Press, New York.

87. KRUUK, H. 1972. *The Spotted Hyena: A Study of Predation and Social Behavior.* University of Chicago Press, Chicago.

88. ESTES, R. D. 1976. The significance of breeding synchrony in the wildebeest. *East African Wildlife Journal* 14:135–152.

89. HAMILTON, W. D. 1971. Geometry for the selfish herd. *Journal of Theoretical Biology.* 31: 295–311.

90. CUMMING, H. G. 1975. Clumping behaviour and predation with special reference to caribou. Biology Paper of University of Alaska, Special. Report No. 1, 474–497.

91. SKOGLAND T. 1989. Comparative social organization of wild reindeer in relation to food, mates, and predator avoidance. *Advances in Ethology* 29: 1–74.

92. BERGERUD, A. T., R. FERGUSON, and H. E. BUTLER 1990. Spring migration and dispersion of woodland caribou at calving. *Animal Behaviour* 39:360–368.

93. McEWAN, E. H., and R. E. WHITEHEAD. 1972. Reproduction in female reindeer and caribou. *Canadian Journal of Zoology* 50:43–46.

94. BERGERUD, A. T. 1973. Movements and rutting behaviour of caribou (*Rangifer tarandus*) at Mount Albert, Quebec. *Canadian Field Naturalist* 87:357–367.

95. REIMERS, E. 1983. Reproduction in wild reindeer in Norway. *Canadian Journal of Zoology* 61:211–217.

96. PARKER, G. R. 1981. Physical and reproduction parameters of an expanding woodland caribou population (*Rangifer tarandus caribou*) in northern Labrador. *Canadian Journal of Zoology* 59:1929–1940.

97. THOMAS, D. C., and S. J. BARRY. 1990. Age specific fecundity of the Beverly Herd of barren-ground caribou. *Rangifer Special Issue* 3:257–263.

98. THOMAS, D. C. 1982. Relationship between fertility and fat reserves of Peary caribou. *Canadian Journal of Zoology* 60:597–602.

99. DAUPHINE, JR., T. C. 1976. Biology of the Kaminuriak population of barren-ground caribou, Part 4 Growth, reproduction, and energy reserves. Report Service No. 38. Canadian Wildlife Service, Ottawa.

100. THOMAS, D. C., C. S. BARRY, and H. P. KILIAAN. 1989. Fetal sex ratios in caribou: Maternal age and condition effects. *Journal Wildlife Management* 53:885–890.

101. REIMERS, E., and D. LENVIK. 1997. Fetal sex ratio in relation to maternal mass and age in reindeer. *Canadian Journal of Zoology* 75:648–150.

102. FISHER, R. A. 1930. *The Genetical Theory of Natural Selection.* Clarendon Press, Oxford.

103. BUNNELL, 1987. Reproductive tactics of cervidae and their relationship to habitat. Pages 145–167. M. Wemmer, ed., *Biology and Management of the Cervidae.* Smithsonian Institution Press, Washington, DC.

104. THOMAS, D. C., and H. P. L. KILIAAN. 1998. Fire-caribou relationships, (1) Physical characteristics of the Beverly herd 1980–87. Technical Report Series No. 309. Canadian Wildlife Service Prairie and Northern Region, Edmonton, Alberta.

105. BERGERUD, A. T., 1971. The population dynamics of Newfoundland caribou. Wildlife Monograph No. 25.

106. BERGERUD, A. T., and J. P. ELLIOT. 1986. Dynamics of caribou and wolves in northern British Columbia. *Canadian Journal of Zoology* 64:1515–1529.

107. MILLER, D. R. 1975. Observations of wolf predation on barren-ground caribou in winter. *International Reindeer and Caribou Symposium 2.*

107a. MILLER, F. L., A. GUNN, and E. BROUGHTON. 1985. Surplus killing as exemplified by wolf predation on newborn calves. *Canadian Journal of Zoology* 63:295–300.

108. FANCY, S. G., K. R. WHITTEN, and D. E. RUSSELL. 1994. Demography of the Porcupine caribou herd 1983–1992. *Canadian Journal of Zoology* 72:840–846.

109. MESSIER, F., J. HUOT, D. LE HENAFF, and S. N. LUTTICH. 1988. Demography of the George River caribou herd: Evidence of population regulation for forage exploitation and range expansion. *Arctic* 41:279–287.

110. MILLER, F. L. 1974. Biology of the Kaminuriak population of barren-ground caribou, Part 2, Dentition as an indicator of age and sex; composition and socialization of the population. Canadian Wildlife Service Report Ser. No. 31, Canadian Wildlife Service, Ottawa.

111. DE BIE, S., and S. E. VAN WIEREN. 1980. Mortality patterns in wild reindeer on Edgeoya (Svalbard). *International Reindeer/Caribou Symposium,* 2:605–610.

112. LEADER-WILLIAMS, N. 1988. *Reindeer on South Georgia.* Cambridge University Press, Cambridge, UK.

113. MCEWAN, E. H. 1963. Seasonal annuali in the cementum of teeth of barren-ground caribou. *Canadian Journal of Zoology* 41:111–113.

114. MEDAWAR, P. B. 1957. *The Uniqueness of the Individual.* Methuen, London.

115. BLIX, A. S., and H. K. JOHNSON. 1983. Aspects of nasal heat exchange in resting reindeer. *Journal of Physiology (London)* 340:445–454.

116. BOERTJE, R. D. 1985. An energy model for adult female caribou of the Denali Herd, Alaska. *Journal of Range Management* 38: 468–473.

117. FANCY, S. G. 1986. Daily energy budgets of caribou: A simulation approach. Thesis. University of Alaska, Fairbanks.

118. ELORANTA, E., and M. NIEMINEN. 1986. Calving of the experimental reindeer herd in Kaamanen during 1970–1985. *Rangifer Special Issue* 1:115–121.

119. ROGNMO, A., K. A. MARKUSSEN, E. JACOBSEN, H. J. GRAV, and A. S. BLIX. 1983. Effects of improved nutrition in pregnant reindeer on milk quality, calf birth weight, growth and mortality. *Rangifer* 3:10–18.

120. ROBBINS, C. T., and B. L. ROBBINS. 1979. Fetal and neonatal growth patterns and maternal reproductive effort in ungulates and subungulates. *American Naturalist* 114:101–116.

121. BERGERUD, A. T., and R. E. PAGE. 1987. Displacement and dispersion of parturient caribou at calving as antipredator tactics. *Canadian Journal of Zoology* 65:1597–1606.

122. DARWIN, C. 1871. Descent of man and selection in relation to sex. John Murra, UK.

123. SELANDER, R. K. 1966. Sexual dimorphism and differential niche utilization in birds. *Condor* 68:113–151.

124. RALLS, K. 1976. Sexual dimorphism in mammals: Avian dodels and unanswered questions. *American Naturalist* 111:917–938.

125. THOMAS, D. C., and H. P. L. KILIAAN. 1998. Fire-caribou relationships: (II) Fecundity and condition of the Beverly Herd. Technical Report Series No. 310. Canadian Wildlife Service Prairie and Northern Region, Edmonton, Alberta.

126. BANFIELD, A. W. F. 1960. The use of caribou antler pedicels for age determination. *Journal of Wildlife Management* 24:99–102.

127. BOVING, P. S., and E. POST 1997. Vigilance and foraging behaviour of female caribou in relation to predation risk. *Rangifer* 17:55–63.

128. TYLER, N. J., and C. ORISLAND. 1989. Why don't Svalbard reindeer migrate? *Holarctic Ecology* 12:369–376.

129. VALKENBURG, P., J. L. DAVIS, and R. D. BOERTJE. 1983. Social organization and seasonal range fidelity of Alaska's Western Arctic caribou—Preliminary findings. *Acta Zoology Fennica* 175:125–126.

130. FANCY, S. G., R. F. FARNELL, and L. F. PANK. 1990. Social bonds among adult barren-ground caribou. *Journal of Mammalogy* 71:458–460.

131. BERGERUD, A. T., 1988. Caribou, wolves and man. *Trends in Ecology and Evolution* 3:68–71.

132. FERGUSON, S. H., A. T. BERGERUD, and J. R. FERGUSON. 1988. Predation risk and habitat selection in the persistence of remnant caribou population. *Oecologia* 76:236–245.

133. BERGERUD, A. T., 1985. Antipredator strategies of caribou: Dispersion along shorelines. *Canadian Journal of Zoology* 63:1324–1329.

134. BROWN, W. K., J. HUOT, P. LAMOTHE, S. LUTTICH, M. PARÉ, G. ST. MARTIN, and J. B. THEBERGE. 1986. The distribution and movement patterns of four woodland caribou herds in Quebec and Labrador. *Rangifer Special Issue* 1:43–49.

135. HEARD, D. C., and T. M. WILLIAMS. 1992. Distribution of wolf dens on migratory caribou range in the Northwest Territories, Canada. *Canadian Journal of Zoology* 70:1504–1510.

136. BIBIKOV, N. G., N. G. OVSYANNIKOV, and A. N. FILIMONOV. 1983. The status and management of the wolf population in the USSR. *Acta Zoology Fennica* 174: 269–271.

137. HEARD, D. C., E. S. FLECK, and G. W. CALEF. 1988. The wolf on the barren-grounds in central Northwest Territories. NWT Renewable Resources Unpublished Report, Yellowknife.

138. GARNER, G. W., and P. E. REYNOLDS, eds. 1986. Final report baseline study of the fish, wildlife, and habitats. Section 1002C. U.S. Department of the Interior, U.S. Fish and Wildlife Service, Region 7, Anchorage, AK.

139. CALEF, G. W. 1976. Numbers beyond counting, miles beyond measure. *Audubon* 78:42–61.

140. TOWNSEND, C. C., ed. 1911. *Captain Cartwright and His Labrador Journal*. Williams and Norgate, London.

141. RYAN, B. 1996. The endless march. *Equinox* 90:27–37.

142. BERGERUD, A. T., R. D. JAKIMCHUK, and D. R. CARRUTHERS. 1984. The buffalo of the north: Caribou (*Rangifer tarandus*) and human developments. *Arctic* 37:7–22.

143. CRÊTE, M., C. MORNEAU, and R. NAULT 1990. Biomase et especes de lichens terrestres disponibles pour le caribou dans le nord du Quebec. *Canadian Journal of Botany* 68:2047–2053.

144. WILLIAMS, T. M., and D. C. HEARD. 1986. World status of wild *Rangifer tarandus* population. *Rangifer Special Issue.* 1:19–28.

145. DARBY, W. R. and L. S. DUQUETTE. 1986. Woodland caribou and forestry in northern Ontario, Canada. *Rangifer Special Issue.* 1:87–93.

146. EDMONDS, E. J. 1991. Status of woodland caribou in western North America. *Rangifer Special Issue* 7:91–197.

147. BANFIELD, A. W. F. and J. S. TENER. 1958. A preliminary study of Ungava caribou. *Journal of Mammalogy* 39:560–573.

148. TENER, J. S. 1963. Queen Elizabeth Islands game survey. Canadian Wildlife Service Occasional Paper 4, Ottawa.

149. PARKER, G. R., D. C. THOMAS, E. BROUGHTON, and D. R. GRAY. 1975. Crashes of muskox and Peary caribou populations in 1973–74 on the Parry Islands, Arctic Canada. Canadian Wildlife Service Progress Note 56, Canadian Wildlife Service, Ottawa.

150. THOMAS, D. C., R. H. RUSSELL, E. BROUGHTON, and P. L. MADORE. 1976. Investigations of Peary caribou populations on Canadian Arctic Islands, March–April 1975. Canadian Wildlife Service Progress Note 64. Canadian Wildlife Service, Ottawa.

151. KLEIN, D. R., and M. MELDGAARD, S. G. FANCY. 1987. Factors determining leg length in *Rangifer tarandus. Journal of Mammalogy* 68:642–655.

152. VIBE, C. 1967. Arctic animals in relation to climatic fluctuations. *Meddr Gronland* 170:1–227.

153. ANDERSON, R. C. 1972. The ecological relationships of meningeal worm and the native cervids in North America. *Journal of Wildlife Diseases* 8:304–310.

154. BERGERUD, A. T., and MERCER 1989. Caribou introductions in eastern North America. Wildlife Soc. Bulletin 17:111–120.

155. LANKESTER, M. W., and D. FONG. 1989. Distribution of Elaphostrongyline nematodes (Metastronyloidea,Protostrongylidae) in cervidea and possible effects of moving *Rangifer* spp. into and within North America. *Alces* 25:133–145.

156. KEITH, L. B. 1983. Population dynamics of wolves. Pages 66–77 *in* L. N. Carbyn, ed., *Wolves in Canada and Alaska: Their Status, Biology and Management*. Canadian Wildlife Service Report Series No. 45, Ottawa.

157. FULLER, T. K. 1989. Population dynamics of wolves in north-central Minnesota. *Wildlife Monograph* 105.

158. ZAGER, R., L. S. MILLS, W. WAKKINEN, and D. TALLMON. 1996. Woodland caribou: A conservation dilemma. http://www.umich.edu/~esupdate/backissues/zager.html.

159. BERGERUD, A. T., 1978. Caribou. Pages 83–101 in J. L. Schmidt and D. L. Gibert, eds., *Big Game of North America: Ecology and Management.* Stackpole Books, Harrisburg, PA.

160. GRIFFITH, B. D. C. DOUGLAS, D. E. RUSSELL, R. G. WHITE, T. R. MCCABE, and K. R. WHITTEN. 1998. Effects of recent climate warming on caribou habitat and calf survival: Implications for management. *North America Caribou Workshop,* Whitehorse 8:in press.

161. BOERTJE, R. D., P. VALKENBURG, and M. E. MCNAY. 1996. Increases in moose, caribou, and wolves following wolf control in Alaska. *Journal of Wildlife Management* 60:474–489.

162. MECH, L. D., L. G. ADAMS, T. J. MEIER, J. W. BURCH, and B. W. DALE. 1998. The wolves of Denali. University of Minnesota Press, Minneapolis.

163. BOERTJE, R. D., G. L. GARDNER, and P. VALKENBURG. 1995. Factors limiting the Fortymile caribou herd. Fed. Aid in Wildlife Restoration Progress. Report W-24-3. Alaska Department of Fish and Game, Juneau.

164. FINSTAD, G. L., and A. K. PRICHARD. 1998. Climatic influence on forage quality, growth and reproduction of reindeer on the Seward Peninsula II: Reindeer growth and reproduction. *North America Caribou Workshop,* Whitehorse 8:in press.

165. STUART-SMITH, A. K., C. J. A. BRADSHAW S. BOUTIN, D. M. HEBERT, and A. B. RIPPIN. 1997. Woodland caribou relative to landscape patterns in northeastern Alberta. *Journal of Wildlife Management* 61:622–633.

166. RETTIE, W. J., and F. MESSIER. 1998. Dynamics of woodland caribou populations at the southern limit of their range in Saskatchewan. *Canadian Journal of Zoology* 76: 251–259.

167. GASAWAY, W. C., R. O. STEPHENSON, J. L. DAVIS, P. E. K. SHEPHERD, and O. E. BURRIS. 1983. Interrelationships of wolves, prey and man in Interior Alaska. *Wildlife Monograph* 84.

168. SEIP, D. R., and D. B. CICHOWSKI. 1996. Population ecology of caribou in British Columbia. *Rangifer Special Issue* 9:73–80.

169. BERGERUD, A. T., and W. B. BALLARD. 1989. Wolf predation on the Nelchina caribou herd: A reply. *Journal of Wildlife Management* 53:251–259.

170. BERGERUD, A. T. 1983. Prey switching in a simple ecosystem. *Scientific American* 249:130–141.

171. MILLER, F., and A. GUNN. 1986. Effects of adverse weather on neonatal calf survival—a review. *Rangifer Special Issue* 1:211–217.

172. MILLER, F. L., E. BROUGHTON, and A. GUNN. 1983. Mortality of migratory barren-ground caribou calves, Northwest Territories, Canada. *Acta Zoology Fennica* 175:155–156.

173. MILLER, F. L., E. BROUGHTON, and A. GUNN. 1988. Mortality of migratory barren-ground caribou on the calving grounds of the Beverley Herd, Northwest Territories. Canadian Wildlife Service Occasional Paper No 66, Ottawa.

174. MECH, L. D., T. J. MEIER, J. W. BURCH, and L. G. ADAMS. 1995. Patterns of prey selection by wolves in Denali National Park, Alaska. Pages 231–244 in L. N. Carbyn, S. H. Fritts, and D. R. Seip, eds., *Ecology and Conservation of Wolves in a Changing World.* Canadian Circumpolar Institute, University of Alberta, Edmonton.

175. MILLER, F. L., and E. BROUGHTON. 1974. Calf mortality on the calving grounds of Kaminuriak caribou. Canadian Wildlife Service Report Service No. 26, Canadian Wildlife Service, Ottawa.

176. DAVIS, J. L., and P. VALKENBURG. 1985. Qualitative and quantitative aspects of natural mortality in the Western Arctic caribou herd. Federal Aid in Wildlife Restoration Final Report Project W-17–11, W-21-2, W-22-1, W-22-2, and W-22–3. Alaska Department of Fish and Game, Juneau, AK.

177. CRÊTE, M., S. COUTURIER, B. J. HEARN, and T. E. CHUBBS. 1996. Relative contribution of decreased productivity and survival to recent changes in demographic trends of Riviere George Caribou Herd. *Rangifer Special Issue* 9:27–36.

178. CAMPS, L., and A. LINDERS. 1989. Summer activity budgets, nutrition and energy balance of George River female caribou. Thesis. Katholieke Universiteit Nijmegen, Netherlands.

179. MESSIER, F. 1991. Detection of density-dependent effects on caribou numbers from a series of census data. *Rangifer Special Issue* 7:36–49.

180. SCHAEFER, J. A., A. M. VEITCH, F. H. HARRINGTON, W. K. BROWN, J. B. THEBENGE, and S. N. LUTTICH. 1999. Demography of decline of the Red Wine Mountains Caribou herd. *Journal of Wildlife Management* 63:580–587.

181. DALE, B. W., L. G. ADAMS, and R. T. BOWYER. 1994. Functional response of wolves preying on barren ground caribou in a multiple-prey ecosystem. *Journal of Animal Ecology* 63:644–652.

182. BALLARD, W. B., L. A. AYRES, P. R. KRAUSMAN, D. J. READ, and S. G. FANCY. 1997. Ecology of wolves in relation to a migratory caribou herd in Northwest Alaska. *Wildlife Monograph* 135.

183. HAYES, R. D., and D. E. RUSSELL. 1998. Why wolves cannot regulate the Porcupine herd: A predation rate model. *North America Caribou Workshop* 8:in press.

184. TAYLOR, J. S. 1976. The advantage of spacing out. *Journal of Theoretical Biology* 59:485–490.

185. HEARD, D. C. 1990. The intrinsic rate of reindeer and caribou in Arctic environments. *Rangifer Special Issue* 3:169–173.

186. SCHEFFER, V. B. 1951. The rise and fall of a reindeer herd. *Scientific Monthly* 73:356–362.

187. HEARN, B. J., S. N. LUTTICH, M. CRÊTE, and M. B. BERGER. 1990. Survival of radio-collared caribou (*Rangifer tarandus caribou*) from the George River Herd, Nouveau-Quebec-Labrador. *Canadian Journal of Zoology* 68:642–655.

188. BERGERUD, A. T., 1974. The decline of caribou in North America following settlement. *Journal of Wildlife Management* 38:757–770.

189. COUTURIER, S., D. VANDAL, G. ST. MARTIN, and D. FISET. 1989. Suivide la condition physique des caribous de la Riviére George. Ministere du Loisir, de la Chasseet de la Peche, Direction Regionale du Nouveau. Quebec, Ontario, Canada.

190. VALKENBURG, P. D., W. PITCHER, D. J. REED, E. F. BECKER, J. R. DAU, D. N. LARISEN, and J. L. DAVIS. 1990. Density dependent responses in mandible length, calving dates, and recruitment in three Alaskan caribou herds. North American Caribou Workshop 4.

191. GUNN, A., and F. L. MILLER. 1986. Traditional behaviour and fidelity to calving grounds by barren-ground caribou. *Rangifer Special Issue* 1:151–158.

192. VALKENBURG, P., and J. L. DAVIS. 1986. Caribou distribution of Alaska's Steese-Fortymile caribou herd: A case of fidelity. *Rangifer Special Issue* 1:315–323.

193. HEARD, D. C., and G. W. CALEF. 1986. Population dynamics of the Kaminuriak herd 1968–1985. *Rangifer Special Issue* 1:159–166.

194. HICKEY, J. J. 1955. Some American population research on gallinaceous birds. Pages 326–396 *in* A. Wolfson, ed., *Recent Studies in Avian Biology.* University of Illinois Press, Urbana.

195. DAVIS, J. L., P. VALKENBURG, and S. J. HARBO. 1979. Refinement of the photo-direct count-extrapolation technique. Federal Aid in Wildlife Restoration Final Report Project W-17–11. Alaska Department Fish and Game, Juneau.

196. VALKENBURG, P. D., A. ANDERSON, J. L. DAVIS, and D. J. REED. 1985. Evaluation of an aerial photocensus techniques for caribou based on radio telemetry. *North America Caribou Workshop,* 2:17–20.

197. RUSSELL, J., S. COUTRUIER, L. G. SOPUCK, and K. OVASKA. 1996. Post-calving photo-census of the Riviere George caribou herd in July 1993. *Rangifer Special Issue* 9:319–330.

198. FARNELL, R., and D. A. GAUTHIER. 1988. Utility of the stratified random quadrant sampling census technique for woodland caribou in the Yukon. *Caribou Workshop,* 3:119–132.

199. MAHONEY, S. P., J. A. VIRGL, D. W. FONG, A. M. MACCHARLES, and M. MCGRATH. 1998. Evaluation of a mark–resighting technique for woodland caribou in Newfoundland. *Journal of Wildlife Management* 62:1227–1235.

200. DAVIS, J. L., P. VALKENBURG, and H. V. REYNOLDS. 1980. Population dynamics of Alaska's western Arctic caribou herd. *International Reindeer/Caribou Symposium,* 2:595–604.

201. BERGERUD, A. T., 1990. Rareness as an antipredator strategy to reduce predation risk. *International Union of Game Biologists Congress* 19:15–25.

202. DALE, B. W., L. G. ADAMS, and R. T. BOWYER. 1995. Winter wolf predation in a multiple ungulate prey system, Gates of the Arctic National Park, Alaska. Pages 223–230 *in* L. N. Carbyn, S. H. Fritts, and D. R. Seip, eds., *Ecology and Conservation of Wolves in a Changing World.* Canadian Circumpolar Institute, University of Alberta, Edmonton.

203. NOVIKOV, G. A. 1956. Carnivorous mammals of the fauna of the USSR. Academy of Sciences of the USSR. Translated by Israel program for scientific translations, 1962, Jerusalem.

204. NIEMINEN, M. 1990. Hoof and foot loads for reindeer (*Rangifer tarandus*). *Rangifer Special Issue* 3:249–254.

205. BUTLER, H. E. 1988. Habitat selection of rutting caribou. *North America Caribou Workshop,* 3.

206. LEOPOLD, A. S., and F. F. DARLING. 1953. *Wildlife in Alaska.* The Ronald Press Co., New York.

207. EDWARDS, R. Y. 1954. Fire and the decline of a mountain caribou herd. *Journal of Wildlife Management* 18:521–526.

208. SCOTTER, G. W. 1964. Effects of forest fires on the winter range of barren-ground caribou in northern Saskatchewan. Canadian Wildlife Service Wildlife Management Bulletin Service 1.

209. BERGERUD, A. T., 1983. The natural population control of caribou. Pages 14–61 *in Symposium on the Natural Regulation of Wildlife Populations.* Forest, Wildlife and Range Experiment Station, University of Idaho, Moscow.

210. HEINSELMEN, M. L. 1973. Fire in the virgin forests of the Boundary Waters Canoe Area. Minnesota Quaternary Research. 3:329–382.

211. JOHNSON, E. A. 1979. Fire recurrence in the subarctic and its implications on vegetation composition. *Canadian Journal of Botany* 57:1374–1379.

212. JOHNSON, E. A., and J. S. ROWE. 1975. Fire in the subarctic wintering ground of the Beverly herd. *American Midland Naturalist* 94:1–14.

213. KERSHAW, K. A., and W. R. ROUSE. 1976. The impact of fire on forest and tundra exosystems. ALUR 75-76-63, Arctic Land Use Res. Program. Department of Indian and Northern Affairs Information, Canada.

214. MILLER, D. R. 1976. Wildfire and caribou on the taiga exosystem of north-central Canada. Ph.D. thesis. University of Idaho, Moscow.

215. MILLER, D. R. 1998. Lichens, wildfire, and caribou on the taiga ecosystem of Northcentral Canada. *North America Caribou Workshop,* 8:in press.

216. THOMAS, D. C., and BEVERLY AND QAMANIRJUAG CARIBOU MANAGEMENT BOARD. 1996. A fire suppression model for forested range of the Beverly and Qamanirjuag herds of caribou. *Rangifer Special Issue* 9:343–350.

217. DRYNESS, C. T., L. A. VIERECK, and K. VAN CLEVE. 1986. Fire in taiga communities in interior Alaska. Pages 74–86 *in* Van Cleve, K., Chapin, F. S., Flanagan, P. W. Viereck, L. and Dryness, C. T., eds., *Forest Ecosystems in the Alaska Taiga.* Springer-Verlag, New York.

218. CUMMING, H. G. 1996. Managing for caribou survival in a partitioned habitat. *Rangifer Special Issue* 9:171–179.

219. CICHOWSKI, D. B. 1996. Managing woodland caribou in west-central British Columbia. *Rangifer Special Issue* 9:119–126.

220. RACEY, G. D., and E. R. ARMSTRONG. 1996. Towards a caribou habitat management strategy for northwestern Ontario: Running the gaunlet. *Rangifer Special Issue* 9:159–169.

221. ARMLEDER, H. M., and S. K. STEVENSON. 1996. Using alternative silvicultural systems to integrate mountain caribou and timber management in British Columbia. *Rangifer Special Issue* 9:141–148.

222. SIMPSON, K., J. P. KELSALL, and M. LEUNG. 1996. Integrated management of mountain caribou and forestry in southern British Columbia. *Rangifer Special Issue* 9:153–158.

223. BERGERUD, A. T., M. NOLAN, K. CURNEW, and W. E. MERCER. 1983. Growth of the Avalon Caribou Herd. *Journal of Wildlife Management* 47:989–998.

224. MERCER, E, S. MAHONEY, K. CURNEW, and C. FINLAY. 1985. Distribution and abundance of insular Newfoundland caribou and the effects of human activities. *North America Caribou Workshop.* 2:15–32.

225. CRONIN, M. A., W. B. BALLARD, J. D. BRYAN, B. J. PIERSON, and J. D. MCKENDICK. 1998. Northern Alaska oil fields and caribou: A commentary. *Biology Conservation* 83:195–208.

226. ROBY, D. D. 1978. Behavioral patterns of barren-ground caribou of central Arctic herd adjacent to the Trans-Alaska oil pipeline. MSc. thesis. University of Alaska, Fairbanks.

227. DAVIS, J. L., P. VALKENBURG, and R. D. BOERTJE. 1985. Disturbance and the Delta Caribou Herd. *North America Caribou Workshop,* 1:2–6.

228. MAIER, J. A., S. M. MURPHY, R. G. WHITE, and M. D. SMITH. 1998. Responses of caribou to overflights by low-altitude jet aircraft. *Journal of Wildlife Management* 62:752–766.

229. CONNELL, J. A., and W. P. SOUSA. 1983. On the evidence needed to judge ecological stability or persistence. *American Naturalist* 121:789–824.

230. RACEY, G. D., and B. DALTON, eds. 1998. North American Caribou Conference 7.

231. THOMAS, D. C. 1998. Needed: Less counting of caribou and more ecology. *Rangifer Special Issue* 10:15–23.

232. COUTURIER, S., R. COURTOIS, H. CREPEAU, L.-P. RIVERST, and S. LUTTICH., 1996. Calving photo census of the Riviere George caribou herd and comparison with an independent census. *Rangifer Special Issue* 9: 283–296.

233. CRÊTE, M., and M. DESROSIERS. 1995. Range expansion of coyotes, *Canis latrans,* threatens a remnant herd of caribou, *Rangifer tarandus,* in southeastern Quebec. *Canadian Field Naturalist* 109:227–235.

234. CRINGON, A. T. 1956. Some aspects of the biology of caribou and a study of the woodland caribou range on the Slate Islands, Lake Superior, Ontario. Thesis, University of Toronto, Canada.

235. FERGUSON, M. 1989. Baffin Island. Pages 141–149 in E. Hall, ed. *People and Caribou in the Northwest Territories.* Government of the Northwest Territories, Yellowknife, Northwest Territories.

236. GUNN, A. 1989. Beverly Herd. Peary caribou. Pages 117–121 *in* E. Hall, ed., *People and Caribou in the Northwest Territories.* Department of Renewable Resources, Government of Northwest Territories, Yellowknife.

237. WILLIAMS, T. M. 1989. Northeastern mainland. Pages 131–133 *in* E. Hall, ed. *People and Caribou in the Northwest Territories.* Government of the Northwestern Territories. Yellowknife.

238. WILLIAMS, T. M., and D. C. HEARD. 1986. World status of wild *Rangifer tarandus* population. *Rangifer Special Issue* 1:19–28.

North American Elk

Michael J. Wisdom and John G. Cook

INTRODUCTION

Few species of North American ungulates rival the ecological, social, aesthetic, and economic stature of elk. The species' large size, its delectable meat, and the impressive antlers of mature males all make for compelling notoriety. Moreover, the tendency of elk to form large herds, to pioneer and exploit a variety of habitats, and to move expediently across diverse and hostile landscapes in response to changes in weather, forage, and human disturbances further contribute to the species' notable stature. Finally, the potential for elk to compete with livestock, to modify vegetative succession, and to damage agricultural crops make the species an obvious source of controversy for farmers, ranchers, and public land managers. Combined with the public's keen interest in hunting and viewing elk, and the subsistence and spiritual functions that elk provide to tribal nations, this species demands an intensity of management commanded by few other wild ungulates. Accordingly, our chapter describes the fundamental concepts of elk ecology and management in relation to population and habitat considerations that often are challenging and sometimes impossible to accommodate, given the contrasting desires of the many vested interests who are affected by this unique resource.

Historical Perspective and Value

Historically, elk were one of the most common native ungulates in North America (1). Their meat, teeth, hide, and antlers were prized by native Americans and early European settlers (2). Nearly all tribal nations in the continental United States traded for elk meat and body parts, and most tribes hunted elk, attesting to the species' historical abundance and extensive distribution (2, 1).

Today, elk are one of the most highly valued species economically, socially, and aesthetically of all North American wildlife (3, 4). Sales of elk hunting licenses and permits generate millions of dollars in annual revenues to state wildlife agencies in the western United States, and elk viewing attracts thousands of visitors to areas such as the National Elk Refuge in Wyoming (3, 5, 6, 7). Moreover, nearly all land use plans for national forests in the western United States contain explicit standards regarding the management of elk habitat, and many partnerships have been established between public and private landowners to ensure that elk management objectives are met on public lands without causing problems on adjacent private lands (8, 9). Finally, elk are beneficiaries of one of the largest and most active conservation organizations, The Rocky Mountain Elk Foundation, attesting to the species' widespread popularity among hunters and conservationists.

Elk, a member of the Cervidae family, are part of the red deer species complex having circumpolar distribution that spans large areas of North America, Europe, and Asia (1). Six subspecies occupied North American biomes from the Atlantic to Pacific Coasts before European settlement: Eastern elk, Manitoban elk, Rocky Mountain elk, Merriam elk, Tule elk, and Roosevelt elk (Figure 32–1) (1).

With the exception of the Rocky Mountain and Roosevelt subspecies in the West, elk are now distributed in small, disjunct populations across much of their former range (Figure 32–2) (1, 10). Declines in elk numbers, subsequent reductions in their distribution, and ultimately the extirpation of populations and entire subspecies occurred across large expanses of North America as European settlers moved west in the 17th, 18th, and 19th centuries. By the early 1800s, the Eastern subspecies was extinct (1). Tule elk, once abundant in valleys and bottomlands of California, were reduced to ≤100 animals by 1875 (11). The Merriam subspecies was extirpated from the Southwest by 1906 (1). By the early 1900s, Manitoban elk of the northern prairies were close to extinction (1). During the same period, populations of Rocky Mountain and Roosevelt elk were reduced to scattered, isolated pockets within their former range (12). Excessive hunting, overgrazing by livestock, and conversion of habitat to agriculture and cities led to these broad-scale declines and extirpations (1, 13).

Description

Elk are among the largest of the North American Cervidae, second only to moose. Largest of the subspecies is Roosevelt elk, with mature males often weighing 320 to 500 kilograms, and mature females averaging 265 to 285 kilograms (14, 1). Smallest

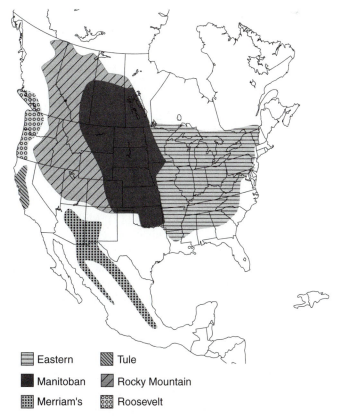

Legend:
- Eastern
- Manitoban
- Merriam's
- Tule
- Rocky Mountain
- Roosevelt

Figure 32–1 Historical distribution of North American elk. (From Ref. 1, p. 24.)

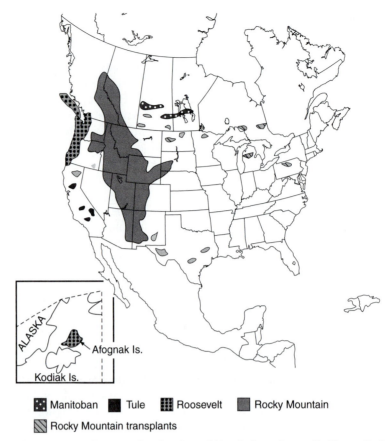

Figure 32–2 Current distribution of North American elk. (From Ref. 1, p. 25, and Ref. 57, p. 8.)

of the subspecies is Tule elk, with mature bulls and cows often weighing <190 and <160 kilograms, respectively (12). Weights of the other subspecies usually range from 300 to 350 kilograms for males and 225 to 275 kilograms for females (12, 1). Birth weights range from 7 to 27 kilograms; yearlings can exceed 200 kilograms entering their first breeding season (1.5 years old).

Antler growth of male elk is highly variable and largely determined by nutritive status (15). Antlers of mature (≥5-year-old) males often achieve notable beam width and length; for example, antlers of the phenotypical specimen of the Rocky Mountain subspecies that was described by Murie had a beam width 111 centimeters and beam length of 126 centimeters (12). Antlers of mature males also have impressive branching, typically having six or more points on each side (15). By contrast, the limited antler growth of yearling (1.5-year-old) males usually consists of a pair of single spikes. Males of subsequent age (≥2 but <5 years old) usually achieve an intermediate antler size, depending on nutritive status.

Other notable characteristics of elk phenotype include pelage and dentition. Color, density, and other characteristics of the elk coat can change markedly from summer to winter in response to seasonal changes in weather, as described by Bubenik (15). The dental pattern of both sexes is uniquely characterized by a pair of incisiform canine teeth present in the forward part of the upper jaw; these teeth, known as "elk teeth," "buglers," or "whistlers," are highly prized by hunters and native Americans. Additional details about elk phenotype and morphology were described by Murie, Bryant and Maser, and Bubenik (12, 1, 15).

Diet Selection

Diet selection by large herbivores is influenced by many plant and animal characteristics. Plant characteristics include abundance, nutritive content, and levels of chemical toxins. Animal characteristics reflect anatomic and digestive differences that affect each species' ability to acquire and digest different types of foods. For example, small-bodied ruminants such as deer tend to require and are adapted to select for highly nutritious forage (i.e., concentrate feeders). Large-bodied ruminants such as bison, by contrast, are better adapted for digesting highly fibrous, less nutritious forage (i.e., roughage feeders) (16, 17).

Elk are intermediate feeders. Compared to concentrate feeders, elk digest fibrous forage more effectively because they retain food particles in their rumen longer, which allows for more complete digestion (18, 19, 20). The larger muzzle of elk provides for comparatively large bite sizes and forage intake rates, but makes selecting for highly nutritious forage difficult. Although not well studied, elk probably have a lower capacity to retain and digest fibrous food particles, and undoubtedly are more selective of nutritious plants than roughage feeders such as cattle or bison. Classically, these relations would suggest that elk diets would contain greater amounts of grasses and lesser amounts of broad-leaved plants than deer and vice versa compared to cattle. Grasses are generally high in fiber that is unprotected by various plant defensive compounds; thus, this forage type is relatively digestible by ruminant species with relatively large ruminoreticular capacities.

Differences in the capability of ruminant species to tolerate plant defense compounds also may account for dietary differences. Wild ruminants have evolved physiologic adaptations to counter some defense compounds. For example, North American deer produce certain salivary proteins (proline proteins) that bind chemically with tannins and therefore reduce the toxic and digestion-inhibiting effects of tannins. Moreover, deer can increase production of these compounds, in response to increasing tannin consumption, by increasing size and output of their salivary glands. Elk undoubtedly have this capability but cattle and sheep do not (21). Consequently, elk would be expected to consume more plant species with tannins, such as forbs and shrubs, than would cattle or sheep.

Being intermediate feeders, elk tend to be dietary opportunists. Exhaustive reviews of elk diets such as those by Kufeld and Cook demonstrate the profound variety of plant species in elk diets (22, 23). Nevertheless, elk often exhibit strong selection and avoidance of certain plant species, consuming plants of high nutritive value and avoiding plants having toxic or digestion-inhibiting effects (24). During winter, particularly when snow is present, elk may be forced to eat whatever is available. Such diversity in elk diets among seasons and ecological settings precludes overly simplistic perceptions that elk and cattle diets are equivalent or that elk are primarily grass grazers.

In general, seasonal changes in plant nutritive quality and availability result in like changes in forage selection by elk. Diverse grasses, forbs, and shrubs are consumed during spring and early summer because most forages are abundant and nutritious (25, 23). Later in summer and early fall, diets often shift to forbs and shrubs. Both of these vegetation classes typically remain more succulent and nutritious late into the growing season compared to grasses. Exceptions occur on elk ranges where fall "green-up" occurs. Fall rains often initiate grass growth, and nutritive content of this vegetation is high (26). During winter, snow conditions strictly govern the availability of forage in many ecological settings, and thus the dietary choices that elk can make. Those species of vegetation that persist throughout the winter (e.g., certain grasses, sedges, shrubs, or trees) typically dominate elk diets (22, 23).

The capability of elk to select and efficiently digest a wide variety of forages allows them to successfully pioneer and occupy a diversity of environments; consequently, elk may out-compete other species of ungulates in a variety of environments (25). Also, the effects of long-term herbivory by elk may shift the composition and structure of vegetation away from a dominance of high-quality forages to a dominance of variable-quality forages (27, 28). Populations of intermediate and roughage feeders would be less negatively influenced by these herbivory effects than concentrate feeders such as deer and pronghorn. Long-term herbivory by elk may change long-term trajectories of vegetative succession following episodic disturbance (e.g., fire, timber harvest), altering ecosystem structure, productivity, and habitat quality for a variety of organisms (29). Finally, the dietary choices of elk can overlap closely with dietary selection by cattle, making the two species potential competitors for available forage on some sympatric ranges (25).

Reproduction

Elk are polygamous breeders in a "harem" style mating system. Dominant males seek groups of females and attempt to hold these groups until breeding has terminated. Harems apparently are retained primarily due to acceptance of the dominant male by the females, dominant males repulsing other males, and to some extent, by herding of females by males. Such a system places great demands on the males (30). Successful reproduction after the rut depends on the female's capability to avoid winter starvation and predation, and to obtain adequate nutrients for rapid development of offspring. The dual challenge of predator avoidance and obtaining adequate nutrition has fundamentally influenced behavior, anatomy, and physiology through evolution over eons.

Elk are spontaneous ovulators, meaning that females ovulate in the absence of males. Ovulation can begin in early September and occurs at 18- to 25-day intervals (average 21 days) until conception occurs (15). Initiation of ovulation apparently is cued to changes in day length. However, delays in ovulation often result from inadequate nutrition and/or condition (e.g., fat levels), the effects of nursing, and perhaps the absence of males (31).

Elk are first capable of breeding as yearlings. Pregnancy rates of yearling females may range from 0% to as high as 80% (15, 32). This wide range is a function of nutrition, body size, and condition. Hudson and others indicated that female elk must achieve 65 to 70% of their adult mass before they will breed (32). This requires a relatively high nutritional plane during their first 1.5 years of life. By the time females are 2.5 years old, most will be of adequate body size for breeding. Adult female elk are capable of breeding each year without pauses, but the probability of pregnancy begins to decline after about 12 years of age to about 20% at 16 to 18 years (15).

Female elk generally carry a single fetus during gestation; twinning occurs in less than 1% of births, and triplets are rare (31). Fetal growth is quite slow during late fall and winter, such that nutritional requirements of the mother are held to a minimum during this period of nutritional deficiency. Winter nutritional deficiencies apparently have little effect on calf viability at parturition. Use of maternal protein and energy helps to maintain the fetus during periods of winter food shortages, and reduced fetal growth in winter can be compensated to some extent by lengthened gestation (33, 34). Fetal losses can occur due to severe nutritional deficiencies in winter; however, the nature and extent of these losses are poorly understood.

Fetal growth begins to accelerate by early March, and the majority of fetal tissues accrue during the third trimester (March through May). Calf body mass and viability at birth apparently are influenced by nutritional conditions in spring (35, 36). Thorne and others showed that inadequate nutrition during winter and spring induces fetal death, stillbirths, and nonviable live births (37).

Gestation averages 255 days and ranges from 247 to 262 days (31, 15). Breeding generally occurs at a time that, given the length of gestation, will result in calves being born at the most optimum time of year (e.g., mid-May through early June) that maximizes their probability of survival. Calves born appreciably earlier face a greater probability of death via hypothermia due to cold, wet weather; calves born appreciably later face a greater chance of death their first winter due to inadequate growth before winter (36).

Birth weights can vary markedly. Robbins and others reported average birth weights of 21 kilograms from well-fed captive females of the Rocky Mountain subspecies; Cook and others and Thorne and others reported calves as small as 7 kilograms (38, 39, 37). Average birth weights of wild Rocky Mountain calves typically range from 14 to 18 kilograms (40, 41, 42). Stussy reported weights ranging from 8.6 to 26 kilograms for calves less than 2 days old in Roosevelt elk herds (43). Calves less than 15 kilograms are less robust than larger calves, and calves less than 12 kilograms at birth have a significantly lower probability of survival (37).

For calves, the summer and autumn period is key for developing adequate body size and energy reserves for survival during winter. Being born early, born large, and having access to ample milk early in summer and forage of good quality in late summer and early autumn provide advantages for survival in winter. Such conditions also provide important benefits to the mothers. Early-born, large calves should reach nutritional independence earlier, thereby freeing the mother to build her reserves for the coming winter. In ecological settings in which forage conditions are marginal or inadequate, females may have to periodically forego breeding to recover from the year-to-year demands of calf rearing (44, 45, 23).

Longevity

Life span of elk is highly variable and determined by a variety of factors that include predation, hunting, severe weather, and mechanisms of density dependence (46). Under unhunted conditions, female elk have been documented to live more than 15 years (47, 15). Male elk, however, typically live less than 10 years even without hunting (48). Under hunted conditions, most male elk live less than 5 years, and sometimes less than 2 years. In the absence of hunting, the relatively short life span of males is due largely to their high expenditure of energy during the fall rut; with energy reserves drained during the rut, male elk are vulnerable to starvation and exposure during severe winter weather when forage is nonexistent or quantity and quality are low.

BEHAVIOR

Key aspects of elk behavior include their affinity to form large and dynamic herds, to move long distances quickly in response to changing environmental conditions, and to communicate in highly vocal and observable ways, especially during rut. Herding behavior varies strongly by season and gender. During winter, herds of migratory elk often concentrate on limited winter range, can number in the thousands, and are composed of both sexes.

By contrast, herd size of nonmigratory elk, or of migratory elk on summer range, can vary from a few animals to 50 to 100 animals; such herds are typically composed of females and calves, with a smaller percentage of yearling and 2-year-old males. Older males, especially mature males, may lead a solitary existence on summer range before the rut or may form small groups. Geist contended that herd size of elk and other ungulates generally increases with openness of terrain (30). In open areas such as grasslands, herd size presumably is larger as a behavioral defense against predators.

Herd dynamics and behavior of elk change dramatically during rut, when dominant males gather harems of females under a polygamous mating system. Cohesion and size of such harems is determined largely by the extent to which a male can defend a group of females from other males. A male's defense of his harem is time consuming, exhaustive, and not always successful; if enough younger males simultaneously challenge the dominant male, one or more of the younger males may be successful in servicing females.

Sparring, bugling, urine spraying, wallowing, and antler rubbing on vegetation are prominent behaviors of males during rut; these behaviors are used to establish and maintain a dominance hierarchy among males that dictates breeding opportunities (30). Because dominant males have more opportunities for breeding, the intensity and frequency of sparring, bugling, and other dominance displays are more pronounced than those of younger, subdominant males. Interestingly, while males vie for establishment and defense of harems, the movements and activities of each harem are dictated by a lead female.

Following completion of the rut, migratory herds form as animals begin their journey to winter range (49). Migration can involve hundreds of animals, with movements sometimes encompassing hundreds of kilometers. During migration and on summer range, herds composed of females, calves, and young males are dominated by a lead female, whose decisions determine the herd's movements. Contact sounds such as "barking" and "knuckle-cracking" serve to maintain cohesiveness of the herd and communicate appropriate responses to changing environmental conditions or perceived threats (30). Body posturing such as the "warning gait" also serves to alert the herd to potential threats (30).

The spring calving period represents another time in which herd dynamics and elk behavior change markedly. Pregnant females leave the herd a few hours or days before giving birth; this is followed by a hiding period, in which a calf is in seclusion with its mother, which can last from a few days to 3 weeks (50, 51, 52). Females with calves eventually regroup after seclusion, forming larger herds on summer range as described earlier. Detailed accounts of female–calf behavior during the birth and lactation periods were described by Altmann, Geist, and Harper (53, 50, 30, 54).

POPULATION DYNAMICS

Population Trends and Current Status

Seton estimated numbers of North American elk at 10 million before the arrival of European settlers (55). McCabe also believed that population size was significant, numbering in the range of at least several million (2). By the early 20th century, these numbers were reduced to less than 60,000 (56). Since then, elk numbers have increased to more than 700,000, with an estimated population growth rate of 2% per year from 1977 to 1987 (25). This estimated growth rate may be conservative. A recent survey of states and provinces suggested that the number of elk in North America has nearly doubled from 1975 to 1995 (57). Currently, elk are present in 23 states and 7 provinces, with nearly all jurisdictions reporting positive rates of population growth and expanding distributions (57).

Although the current status of elk appears secure, the species faces a variety of threats. For example, elk often made expansive seasonal movements in response to annual changes in weather and the availability of forage before European settlement (12, 49). Many seasonal ranges used historically by elk are now privately owned and managed exclusively for agriculture, livestock, and timber production (58, 59). Elk use of these areas is not tolerated or tolerated minimally. In other cases, habitats used historically by elk have been converted to cities, industrial developments, and recreation areas (60). The result is that elk now reside in "ecologically incomplete" or "ecologically compressed" habitats across much of their historical range (Figure 32–3) (61). This shortage of year-round, ecologically complete habi-

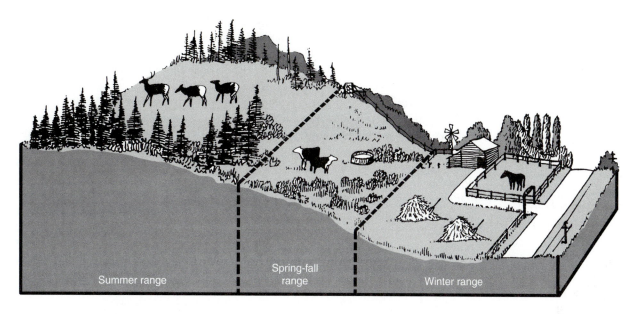

Figure 32–3 Elk produced on publicly owned summer range often face inadequate winter and spring–fall ranges due to conflicts with urban development, livestock management, and crop production on private lands. (From Ref. 25.)

tats for elk poses significant management problems throughout elk range in North America (61, 59, 25).

Another threat is the emergence of elk ranching. Domestic elk and red deer have transmitted diseases to wild elk and livestock (62, 63). For example, bovine tuberculosis was discovered in domestic elk on game ranches in Alberta, Colorado, and Montana; infected elk in Alberta subsequently transmitted the disease to cattle (62).

Red deer and domestic elk also have escaped from game ranches, raising concerns about hybridization with and subsequent genetic effects on wild elk (64, 65, 66). Some states and provinces have enacted strict legislation to control, monitor, and reduce such problems; others have banned game ranching of elk and red deer altogether. Holistic programs of disease control and prevention are needed if elk ranching is to be compatible with management of wild elk and livestock (65).

Productivity and Recruitment

Productivity and recruitment of elk often are defined in terms of three basic variables: (1) pregnancy rates, expressed as the percentage of cows that are pregnant following completion of the breeding season; (2) calf production, expressed as the number of calves born:100 females at or soon after parturition; and (3) calf recruitment, expressed as the number of calves recruited:100 females at a specified recruitment age. The first birthday of a calf's life or other specified date following the calf's first winter is typically identified as recruitment age.

Pregnancy rates of female elk can vary considerably; mean or median rates can range from 0 to 90%. However, most of this variation is associated with yearling females; pregnancy rates for adult females typically range from 65 to 90% (67, p. 149).

Many factors affect pregnancy rates. Poor nutritional plane of females entering the rut, which is influenced directly by poor summer or poor winter forage conditions, can reduce pregnancy rates, especially for yearling females (23). Moreover, females still in lactation during the rut typically have lower pregnancy rates (32, 68).

Calf production and recruitment also can vary widely, often ranging from 10 to 80 calves born or recruited : 100 females for a given population. Late summer, fall, or winter nutrition of elk calves and severe weather can have strong effects on winter survival (69, 70). Predation on newborn calves can also be high and presumably contributes to reduced calf : female ratios (42, 71, 67).

Empirical evidence suggests that productivity and recruitment of elk and red deer decline with increasing population density (72, 73, 27, 74, 75). This relation is consistent with negative feedback mechanisms between ungulate densities and forage conditions that have been postulated in theory and demonstrated empirically for a variety of wild ungulates (76, 77, 78, 79, 80, 81, 82, 83, 84, 85, 36, 86, 87). However, the theory and evidence for density dependence have largely been postulated or demonstrated for ungulate populations in the absence of high predator densities. When predators are abundant, ungulate densities may be regulated by predation below levels that prevent negative feedback mechanisms from operating (88, 67). Instead, productivity and recruitment may vary in less predictable ways in relation to stochastic weather events and fluctuating densities of predators, especially if additive mortality from hunting occurs (89, 90, 91).

Mortality Factors

The main source of elk mortality is from hunting by humans (67). Predation by cougar, wolf, and bear, especially on recently born calves, is a controversial but lesser source of mortality (42, 71, 67). Severe winter weather can also result in high mortality of elk, especially calves (70, 92). Other potential sources of mortality are diseases, the cumulative effects of poor nutrition in summer and winter, weather and predation, late parturition, low birth mass, and other neonatal problems as they interact with other mortality factors (65, 23). These sources of elk mortality tend to cause higher rates of death in the youngest and oldest age classes, with annual mortality rates exhibiting a U-shaped pattern that is typical of most ungulates (Figure 32–4) (93).

Of all the sources of elk mortality, the effect of predation on population growth is a particularly controversial subject, given the paucity of empirical research. Simulation of predation effects from translocated wolves in the Greater Yellowstone ecosystem suggested that elk numbers would decline an average of 10 to 25% over the next 100 years (94). However, Boyce concluded from these simulations that "there is no combination of [management] choices where wolves have devastating consequences to elk populations in the greater Yellowstone area" (94, p. 130). Aspects of these predictions were later validated or demonstrated to be reasonable and plausible (95). While results of such simulations are compelling, the dearth of empirical knowledge regarding dynamics of predator and elk populations begs for long-term research.

Each source of elk mortality affects each life stage differently. For example, hunter harvest typically increases mortality of adults, whereas severe winter weather and predation often increase mortality of juveniles (96). Consequently, an important question for elk managers is whether factors that affect juvenile mortality have a stronger influence on changes in population growth than factors affecting adult mortality (75). This question is important for at least three reasons: (1) Densities of predators such as puma and wolf are increasing in many areas of the western United States; (2) effects of predation on mortality of juvenile ungulates in temperate regions, and these effects on population dynamics, are largely unknown but hotly debated, with little empirical support from which to infer any effects with certainty; and (3) the relative benefit to population growth from reducing hunter harvest of adult females to compensate for increased predation on juveniles has not been quantified for elk (71).

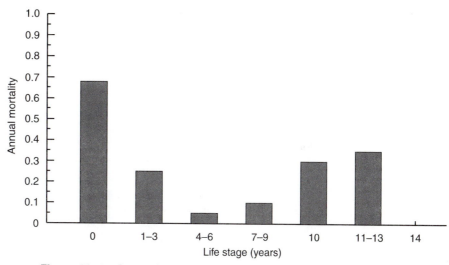

Figure 32–4 Generalized age-specific annual mortality of female elk. (From Ref. 46.)

Consequently, to further elucidate the relative effects of juvenile versus adult mortality on growth rate of elk populations, elasticities of population growth (finite rate of increase or λ) were calculated in relation to juvenile recruitment (mean annual fecundity combined with estimates of calf survival to age 1), and survival of yearling (≥1 but <2 years of age), early adult (≥2 but <8 years old), and late adult (≥8 years of age) life stages of the female portion of an example elk population (Figure 32–5a) (97). Elasticities represent the proportionate change in the finite rate of increase to one-at-a-time, proportionate changes in vital rates for each life stage (98). Results for elk and other ungulates in temperate regions suggest that population growth is more sensitive to changes in adult survival (and thus to changes in adult mortality) than to changes in fecundity or juvenile survival (Figure 32–5a and b); hence, predation or weather that primarily affects recruitment of juveniles has less proportionate effect on population growth than a like change in survival of adult females that could be achieved by hunting (97).

Recruitment of juvenile ungulates in temperate regions, however, is highly variable, implying that empirical effects of juvenile recruitment on population growth often are stronger than empirical effects of adult survival (75). Nonetheless, the elasticities suggest that even small changes in adult survival, such as that achieved through changes in hunting regulations, will cause strong changes in overall population growth. Consequently, population managers will be challenged to provide opportunities to hunt adult females under conditions of low or variable survival of juveniles if the overall goal is to maintain stationary or positive rates of population growth.

Emigration and Immigration

Elk have excellent dispersal capabilities, sometimes moving hundreds of kilometers in a matter of days (49). Young males are especially prone to long-distance movements; one male radiocollared in Montana was ultimately relocated near Kansas City, having moved more than 1,500 kilometers. Such movements demonstrate the high mobility of elk in response to changes in food, weather, and breeding condition (49).

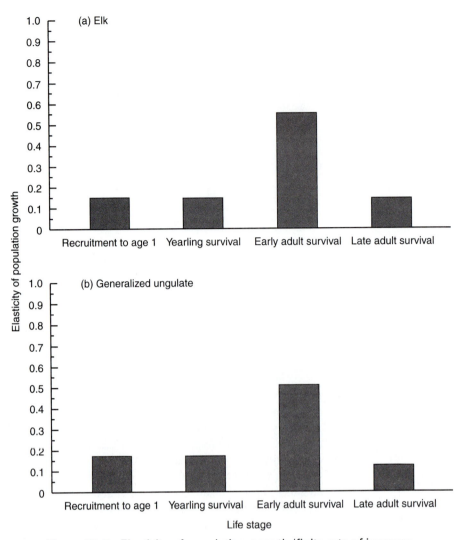

Figure 32–5 Elasticity of population growth (finite rate of increase or λ) to one-at-a-time, proportionate changes in vital rates using a generalized population matrix model for elk (calculated from vital rates summarized by Ref. 219) and (b) using a population matrix model of vital rates generalized by Gaillard and others for 16 species of ungulates (75). Elasticities were calculated specifically in relation to vital rates for juvenile recruitment (estimates of fecundity combined with estimates of calf survival to age 1), and for survival of yearling (≥1 but <2 years of age), early adult (≥2 but <8 years old), and late adult (≥8 years of age) life stages of the female portion of the population. Methods of calculating elasticities by life stage of ungulates followed that of Gaillard and others and Wisdom (78, 97).

POPULATION MANAGEMENT

Population size, trend, distribution, sex ratio, and age structure are key variables in the management of elk populations. Nearly all state wildlife agencies have explicit objectives or standards regarding each of these variables. Objectives and standards are developed in relation to strategies designed to satisfy diverse and often disparate interests of hunters, farmers, ranchers, elk viewers, tribal nations, and federal and private landowners (Table 32–1).

TABLE 32–1

Population Variables of Interest to Elk Managers, Example Management Objectives Related to Each Variable, the Associated Management Strategies that Could Be Used to Achieve the Objective, and the Rationale for Each Strategy

Population Variable	Management Objective	Management Strategy[a]	Rationale for Strategy[b]
1. Population size and trend	1. Control or change population size or trend.	1a. Focus management on hunter harvest of adult females.	1a. Changes in elk population growth are most sensitive to changes in survival of adult females (Figure 32–5a).
		1b. Actively manage population densities of nonhuman predators or account for predation effects.	1b. Changes in elk population growth are moderately sensitive to changes in survival of calves (Figure 32–5a), which can be affected by densities of predators (46).
		1c. Manage carrying capacity through forage manipulations (see 2a) and by accounting for elk security needs in relation to human presence and activities (see 2b).	1c. Elk population size presumably will change with large-scale changes in year-round quantity and quality of forages, particularly when combined with changes in security or escapement from human presence and activities (23, 25).
2. Population distribution on public lands.	2. Maintain population distribution on public lands.	2a. Enhance forage through burning, seeding, fertilization, and other silvicultural or rangeland techniques on public lands.	2a. Elk presumably select areas containing a diverse spatial and temporal mix of high-quantity and high-quality forages due to associated foraging efficiencies and nutritional benefits (25).
		2b. Increase security from humans through road obliteration, road closures, and other restrictions on human access and activities on public lands; and through management for escape cover on public lands.	2b. Elk select areas that provide escapement and security from human presence and activities (201).
3. Population distribution on private lands.	3. Shift population distribution away from private lands.	3. Combine strategies 2a and 2b with a series of special hunts on private lands that span long periods and result in continuous sources of human disturbance of elk on the targeted private lands.	3. Elk avoid areas subjected to high levels of human presence and activities as indexed by open road density and rates of traffic, with high avoidance during hunting seasons (172, 187, 183, 99).

continued

TABLE 32–1

Population Variables of Interest to Elk Managers, Example Management Objectives Related to Each Variable, the Associated Management Strategies that Could Be Used to Achieve the Objective, and the Rationale for Each Strategy—continued

Population Variable	Management Objective	Management Strategy [a]	Rationale for Strategy [b]
4. Sex ratio and age structure	4a. Increase the ratio of mature males:100 females	4a. Allow hunter harvest of spike-antlered males only and/or place strong constraints on harvest of mature males with use of permit-only hunting of branched-antlered males.	4a. Harvest of only young males facilitates survival of mature males but maintains ample hunting opportunities for males. Increasing the ratio of mature males : females better approximates the historical age structure of males and facilitates effective breeding (103).
	4b. Increase the ratio of calves: 100 females.	4b. Reduce the density of adult females under the presumption that calf recruitment is density dependent and will increase with an overall reduction in population density.	4b. Survival of juveniles is density dependent for most ungulate species in temperate environments, including red deer and elk (75).
		4c. Reduce densities of predators under the presumption that predators regulate calf recruitment.	4c. Survival of juvenile ungulates in North America can be regulated by predation.

[a]Management strategies are not inclusive but intended to identify the management actions that would likely need to be considered in any holistic approach for meeting the associated management objective.
[b]Rationales include example citations that support the strategy and rationale.

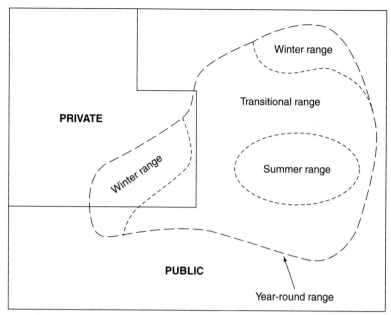

Figure 32–6 Seasonal ranges of elk often encompass areas as large as or larger than a watershed (areas ≥20,000 ha), whereas year-round ranges can encompass areas larger than a sub-basin (areas ≥300,000 ha); these large spatial scales require holistic planning, management, and monitoring of a dauntingly large array of habitat conditions, ecological processes, and human activities.

Management strategies for population variables will be most effective if addressed holistically at large geographic scales that encompass one or more herds of elk and their collective seasonal or year-round ranges (Figure 32–6). On a seasonal basis, such geographic scales often encompass areas more than 20,000 hectares (25). On a year-round basis, such geographic scales can encompass areas more than 300,000 hectares.

Population Size and Trend

Effective management of population size and trend must address or account for at least three major factors: (1) ecological carrying capacity (K) as influenced by forage conditions, security conditions, and severe weather; (2) hunter harvest, especially harvest of adult females; and (3) predation rates and corresponding predator densities. A variety of techniques have been identified for effective management of forage and security to maintain or increase carrying capacity, and for effective mediation of severe weather, as outlined in the following sections, in Table 32–1, and in other literature (e.g., 25, 23).

Within a given level of K, focusing management on hunter harvest of adult females is important in achieving objectives for population size and trend, based on elasticity results (Figure 32–5a). For example, a 5% reduction in adult survival (based on a combined elasticity of 0.71 for the early and late stages of adult survival, Figure 32–5a) would be expected to yield a 3.6% decline in population growth, whereas this same reduction in juvenile recruitment (based on an elasticity of 0.15, Figure 32–5a) would yield only a 0.8% decline. Interestingly, a similar pattern of elasticities and potential effects on population growth was calculated by Gaillard and others, based on generalized vital rates synthesized from demographic studies of 16 species of ungulates (Figure 32–5b) (75).

Elasticity results suggest that managers who want to modify population size or trend could do so most efficiently through manipulation of factors affecting survival of adult females, such as hunting. By contrast, manipulation of factors chiefly affecting juveniles, such as reducing predation rates or mediating effects of severe weather, would have less proportionate effect on population growth and therefore require greater per-unit change in juvenile survival to achieve the desired change in overall population growth.

Population Distribution

Management to achieve the desired spatial distribution of elk at large geographic scales, such as scales encompassing multiple herds of elk and their collective, year-round ranges, is a key consideration in meeting disparate elk management goals on public versus private lands. Often, the goal of state wildlife agencies is to maintain a large population size on public lands to maximize hunting and viewing opportunities, yet simultaneously minimize the time that elk spend on private lands to prevent damage to livestock forage, agricultural crops, fences, and other commodities (8, 9).

A number of management efforts are considered effective in achieving such disparate spatial goals (Table 32–1). Maximizing the time that elk spend on public lands can likely be achieved by increasing security from human presence and activities, by increasing elk nutrition through a variety of forage enhancements, and through proper regulation of hunting activities and harvest (99, 23, 25, 100). Such strategies are plausibly more effective when combined with hunting regulations specifically designed for private lands that increase the likelihood of elk avoidance of these lands.

Sex Ratio and Age Structure

Two common goals of elk managers are to increase the ratio of mature males : 100 females and the ratio of calves : 100 females. These two goals are qualitatively analogous to goals of most elk managers that call for increasing the mean age of males and decreasing the mean age of females.

Historically, male : female ratios probably exceeded 25 mature (greater than 5-year-old) males : 100 females, based on sex ratios documented for elk populations that are largely unhunted (48). Such high ratios of mature males : females typically do not occur in hunted herds of elk. Instead, ratios are typically less than 10, and often less than 5, mature males : 100 females (Figure 32–7).

Figure 32–7 Maintaining large, mature males in elk herds requires good nutrition and careful harvest management such that appreciable numbers of males can attain older age. (Photograph by A. Tiedemann.)

Increasing the ratio of mature males : females is desirable because hunters are willing to pay substantially higher prices for the opportunity to hunt mature males versus younger males or females (101). In addition, mature males are highly desired and sought by the public for viewing and photographic opportunities (3). Finally, high ratios of mature males : females may benefit population performance of elk through earlier, compressed conception and birth dates of calves (102, 103). Earlier, compressed conception and birth dates have been shown to increase survival of red deer calves; specifically, juvenile mortality was estimated to increase by 1% for each day a calf was born past the median birth date (36).

Two mechanisms may account for increased survival of calves whose birth dates are earlier and more synchronous. First, earlier birth dates coincide more closely to the period of highest forage quality in temperate regions, which facilitates optimal lactation of females and highest growth of calves before forage quality declines during fall and winter (36). And second, more synchronous births may cause "predator swamping," whereby a large pulse of prey (calves) becomes available over a short time period, overwhelming the capacity of predators and allowing a greater percentage of calves to survival beyond the period immediately following birth, when juveniles are most vulnerable to predation.

In testimony to the importance of mature males, all states and provinces in western North America now have one or more hunting seasons designed to conserve mature males. The most common strategy is to allow harvest of spike-antlered (yearling) males only, or to combine spike-antlered hunts with limited hunting of branch-antlered males.

A high ratio of calves : females, such as more than 50 calves : 100 females at calf recruitment age, typically signifies a productive, growing population of elk that is composed of a relatively young age structure of females. Alternatively, a low ratio of calves : females, such as less than 30 calves : 100 females, indicates an unproductive population whose growth rate may be declining. Obviously, most elk managers desire high calf : female ratios, but trade-offs with population density may exist that are not always recognized and considered in overall management (72, 73, 27, 74, 75). For example, if calf production is density dependent in a negative manner with overall population density, then management can feature a high density of animals or high production of young, but not both (Table 32–1). The exception would be habitat enhancements that substantially increase K, thereby allowing an abundance of animals to increase simultaneously with high production of calves. Even then, however, calf production and survival should be expected to decline after population density approaches or reaches K.

Other factors, such as predation or severe weather, can also regulate calf : female ratios and overall density. Under these conditions, management to reduce predation rates or to mediate effects of weather may increase calf : female ratios and population density until K is approached or reached (Table 32–1).

The potential for elk productivity to be influenced by a variety of factors and mechanisms suggests that population managers investigate and understand such processes as part of management planning and implementation. Otherwise the achievement of goals for calf production and survival in relation to other population goals may be unclear, infeasible, or unsuccessful, with little empirical evidence on which to make improvements or justify the productivity goals.

Methods to Control Populations

On public forests and rangelands, human tolerance of elk is relatively high, with population densities of elk that often are correspondingly high. On private lands, especially agricultural pastures and rangelands, tolerance for elk often is low. In such cases, a variety of population control measures are used to maintain low elk densities. Efforts to control elk damage on privately owned ranges include live trapping and transport of elk away from problem areas, fencing to exclude animals from private ranges, feeding to attract and hold elk on public winter ranges, government payments to landowners who experience significant economic damage, special hunts to reduce overall numbers, and leasing or acquisition of additional winter range to maintain winter carrying capacity (104). Methods to induce temporary infertility also have been examined (105). In addition, a variety of habitat strategies can be combined with such population control efforts to maintain desired elk numbers on public lands and minimize elk use of private lands (Table 32–1, 25).

HABITAT REQUIREMENTS AND MANAGEMENT

Requirements of elk can be described in two basic ways: need for adequate nutrition and need for adequate security. Adequate nutrition is provided by the proper landscape components of forage quantity, forage quality, water, and their spatial arrangement at geographic scales as large or larger than the seasonal ranges of elk. Adequate security is provided by the proper landscape components that provide refuge or escapement from human presence and activities at geographic scales that affect population distribution of a herd or multiple herds on a seasonal basis.

Nutritional Requirements

Although ranges occupied by elk usually contain abundant vegetation most of the year, the potential for undernutrition or malnutrition is high. Due to inadequate nutrient content or inadequate nutrient availability during digestion, elk may face considerable challenges to obtain sufficient nutrients to survive during periods of harsh weather and successfully reproduce during relatively short periods of nutritional abundance. Understanding nutritional requirements is necessary for understanding how vegetative conditions may influence dynamics of elk populations, for evaluating K and habitat quality, and for assessing the need for and results of various habitat improvements (Figure 32–8).

Estimates of energy and protein requirements in various seasons are summarized next from requirements calculated using a factorial approach (23). Calculation was

Figure 32–8 Tractable elk raised in captivity provide opportunities for research, particularly for testing hypotheses that are difficult to assess using more conventional methods with wild animals. (Photograph by J. Cook.)

Habitat Requirements and Management

necessary because specific estimates determined by research are unavailable for many seasons and productive stages. Estimates of mineral requirements are from the National Research Council's guidelines for cattle, again because such information generally is unavailable for elk (106). We present estimates of crude protein and energy requirements on a metabolic weight (MW) basis, which is calculated simply as body mass raised to the 0.75 power ($BW^{0.75}$). This standardizes estimates of requirements based on the amount of metabolically active tissues in the animal's body and enhances comparisons among animals of markedly different size among or within species (see Ref. 31 for more discussion of this concept).

Energy requirements vary markedly among seasons and age classes primarily as a function of growth in young animals and production stage (e.g., gestation and lactation) in adults. For adult animals in maintenance mode, such as during winter, energy needs reflect requirements of physiologic homeostasis. These requirements primarily involve energy for cellular respiration and maintenance, activity, and thermoregulation.

Gestation increases energy requirements due to the accumulation of energy contained in the tissues of the growing fetus and the energy required to grow the fetus. Most accretion of fetal tissues and the mother's energy requirements occurs after the first 150 days of gestation. In late winter, the energy costs of gestation are about 10 kilocalories of metabolizable energy (ME) per kilogram of metabolic body mass daily; this peaks at about 25 during the third trimester in April and May (23).

Lactation induces considerably greater energy requirements than gestation: About four times more energy is required for lactation than for pregnancy (107). Metabolizable energy required for lactation is a function of the energy contained in milk, the efficiency of energy conversion from food the mother consumes to the milk that she produces, and the amount of milk produced. Milk production peaks about 1 month after parturition and wanes gradually thereafter. During much of lactation, elk also must replace fat and muscle catabolized during the previous winter. Elk may lose 25% or more of their weight during severe winters and can be expected to lose 5 to 15% during mild to normal winters.

Metabolizable energy requirements markedly vary among seasons (Figure 32–9), particularly when tissue replacement requirements are considered. Requirements in summer are nearly twice that in winter or even during the third trimester of gestation. Estimates of ME requirements from other literature sources (Table 32–2) generally corroborate predictions in Figure 32–9. Estimates of ME requirements during lactation for elk from Robbins and others and for black-tailed deer from Sadlier are lower than that predicted for elk in Figure 32–9 (38, 108). Both studies, however, were conducted under penned conditions and do not reflect the greater energy requirements of free-ranging animals.

Elk calves have the genetic potential to quadruple their body mass during the first 5 months of life (Figure 32–10). Such rapid growth requires considerable ME (about 300 kilocalories of ME per kilogram MW daily) to provide the energy contained in the growing tissues, the "fuel" to grow the tissues, and the energy required to maintain body temperature and maintain other homeostatic functions. Metabolizable energy needs decline after midautumn due to a decline in growth rates as juveniles enter an energy conservation mode (109, 69).

Estimating protein needs is more difficult than estimating ME needs due to variations in amino acid composition of dietary protein, complex relations between protein usability and energy intake, and variations in total food intake (110).

For adults in maintenance mode, protein is required to replace endogenous protein lost in feces (e.g., originating as digestive enzymes and sloughed cells of the lining of the intestinal tract), protein lost in urine, and protein lost through dermal (e.g., hair) sloughing. Requirements for gestation primarily include protein deposited in the fetal body and peaks at about 1 gram of crude protein per kilogram of metabolic weight during the last month of gestation. Similarly, protein requirements for lactation primarily include protein included in milk. Daily crude protein requirements are about 7 and 3 grams of crude protein per kilogram of metabolic mass for lactation at 30 and 100

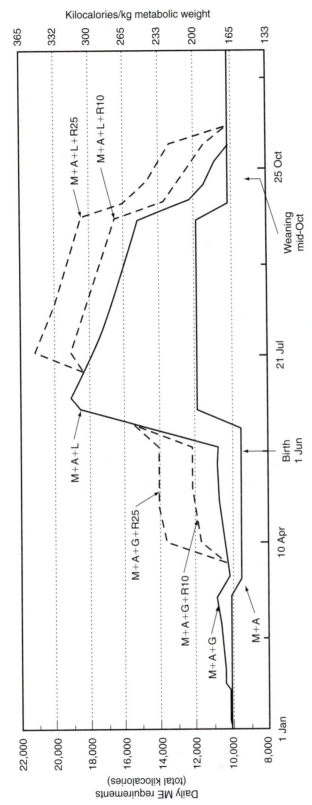

Figure 32-9 Estimated daily metabolizable energy (ME) requirements for an adult female elk to meet maintenance (M), activity (A), gestation (G), lactation (L), and winter-catabolism replacement (R10 and R25) needs during each season. Data are presented for a 236 kilogram female and on a metabolic-weight basis. R10 and R25 are based on 10 and 25% winter mass loss, assuming a normal winter and an unusually harsh winter, respectively. The replacement increments were based on total ME requirements divided by 165 days; replacement tissue accretion was assumed to occur primarily in April, May, and mid-July through early November. The M+A+G and M+A+L lines indicate estimates of spring, summer, and autumn ME requirements after a mild winter assuming that winter ME intake met requirements for winter. (From Ref. 23.)

TABLE 32–2

Estimates of Daily Metabolizable Energy Requirements of Mature Gravid or Lactating Female Ruminants[a]

Species	Maintenance[b]	Midgestation[c]	Late Gestation[d] Midlactation[e]	Early to Lactation[f]	Mid- to Late	Reference
Elk	165	165	170	300	265	Figure 32–9
Elk	145	150	195	285	275	31[g]
Elk	—	132	149	250	250	208[h]
Elk	132	—	—	—	—	209[i]
Elk	168	—	—	—	—	Cook (unpublished data)[j]
Elk	173	—	—	—	—	151[k]
Red deer	136	—	—	—	—	210[l]
Caribou	205	205	225	—	—	211[m]
Caribou	160	162	—	235	220	212[n]
Caribou	200	—	—	—	—	213[o]
Roe deer	166	—	—	—	—	214[p]
White-tailed deer	—	131	—	—	—	215[q]
Black-tailed deer	—	—	—	273	—	108[r]
Moose	131	—	—	—	—	216[s]
Cattle	138	147	173	[———227———]		106[t]
Sheep	100	110	157	243	157	115[u]

Source: From Ref. 23.

[a]Units are kilocalories per kilogram metabolic weight. Metabolizable energy requirements to replace winter-catabolized tissues are not reflected in estimates.

[b]Early winter period before gestation requirements become important.

[c]February and March assuming a June 1 birth date.

[d]April and May assuming a June 1 birth date.

[e]June and July assuming a June 1 birth date.

[f]August and September assuming a June 1 birth date.

[g]From Figure 11–2 on page 159 of Ref. 31, converted total daily ME requirements for a 250 kilogram cow to requirements on a MW basis.

[h]Calculated from reported intake and energy content of food during each period. In this study, penned elk gained an average of 15 kilograms during lactation. We subtracted ME associated with this gain from total energy intake to estimate requirements for live weight maintenance during lactation. This calculation assumed that about half of gain was due to increasing intake and ME required for remaining gain (9.3 kilocalories per kilogram of gain, 209) was subtracted from observed ME intake to estimate ME requirements for lactation.

[i]Study used penned subadult nongravid females averaging 234 kilograms of body mass.

[j]Study used penned 3-year-old nongravid females averaging 200 kilograms of body mass.

[k]Estimated requirements included ME for activity of free-ranging animals.

[l]Requirement for nongravid adult females held indoors during winter.

[m]Calculated from net energy requirements presented for caribou assuming ME use efficiency of 0.65 (106). Study included energy requirement increments for activity of free-ranging animals.

[n]Calculated to incorporate weight homeostasis. Data from this study were based on a model that included energy requirements for activity.

[o]Estimate for penned caribou and reindeer.

[p]Study used mix of nongravid females and males held in pens.

[q]Study used penned mature, gravid white-tailed deer does.

[r]Study used penned lactating deer and data value is metabolizable energy consumption at the peak of lactation.

[s]Study used penned male and female yearling ($n=2$) and adult ($n=8$) moose.

[t]Calculated from data for 550-kilogram cattle cow (106).

[u]Calculated from data for 70-kilogram ewe with a single lamb (115).

days postpartum, respectively (23). Summer/autumn protein requirements to replace muscle mass catabolized during winter depend on the amount of weight lost during this period and protein content of lost mass. Requirements presented here assume a protein:fat catabolism ratio of 40:60 for 10 and 25% winter weight loss levels (111).

As with ME, estimates of crude protein requirements vary considerably among seasons in response to changes in production stage (Figure 32–11). For example, females in winter need about 5 grams per kilogram of MW per day to meet the needs of maintenance and gestation. Lactation during summer markedly increases crude protein requirements to about 15 grams per kilogram of MW per day. These estimates are consistent with those from other sources (Table 32–3).

Figure 32–10 Rapid growth of calves generally requires the greatest weight-specific nutritional levels of any stage of life. Slow growth during the calf's first summer and autumn can increase probability of death by predation or winter starvation and can permanently "stunt" body size. (Photograph by M. J. Wisdom.)

TABLE 32–3
Estimates of Grams of Crude Protein Required Daily per Kilogram of Metabolic Weight of Mature Gravid or Lactating Female Ruminants.[a]

Species	Maintenance[b]	Midgestation	Late Gestation[d]	Early to Midlactation[e]	Mid- to Late Lactation[f]	Reference
Elk	4.9	5.5	6.0	13.5	12.0	23 (see Figure 32–11)
Elk	4.2	—	—	—	—	151
Caribou	5.1	—	—	—	—	213
Roe deer	4.1	—	—	—	—	214
White-tailed deer	4.8	—	—	—	—	217[g]
Moose	3.9	—	—	—	—	218
Cattle	5.6	5.8	6.9	[————12.0————]		106[h]
Sheep	4.7	5.4	8.0	13.8	8.0	115[i]

Source: From Ref. 23.

[a]Crude protein requirements to replace winter-catabolized tissues are excluded from estimates.

[b]Early winter period before gestation requirements become important.

[c]February and March assuming a June 1 birth date.

[d]April and May assuming a June 1 birth date.

[e]June and July assuming a June 1 birth date.

[f]August and September assuming a June 1 birth date.

[g]Estimates determined using yearling does during summer.

[h]Calculated from data for a 550-kilogram cow.

[i]Calculated from data for a 70-kilogram female with a single lamb.

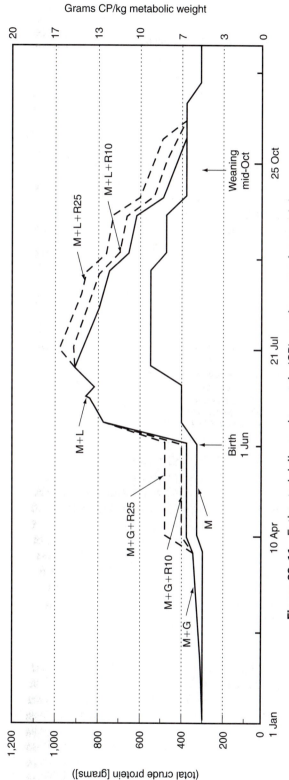

Figure 32–11 Estimated daily crude protein (CP) requirements for an adult female elk weighing 236 kilograms to meet maintenance (M), gestation (G), lactation (L), and winter-catabolism replacement (R10 and R25) needs during each season. R10 and R25 are based on 10 and 25% winter weight loss. The replacement increments were based on total CP requirements divided by 165 days; replacement tissue accretion was assumed to occur primarily in April, May, and mid-July through early November. The M+G and M+L lines indicate estimates of spring, summer, and autumn CP requirements after a mild winter. (From Ref. 23.)

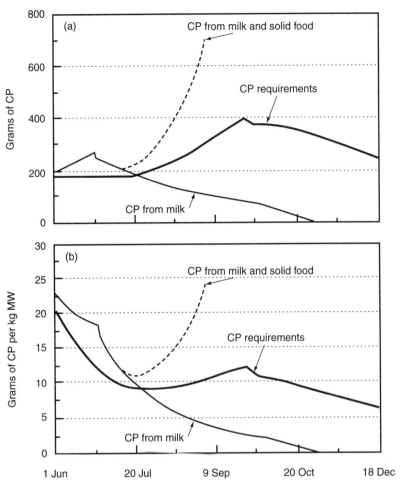

Figure 32–12 Predicted total daily crude protein (CP) requirements and intake of elk calves over summer and autumn (a), and predicted CP requirements and intake expressed on a metabolic weight (MW) basis (b). Estimated intake of CP is based on data of Robbins and others (38). The levels of crude protein requirements indicated for midsummer may be underestimated to some extent. (From Ref. 23.)

Protein requirements in very young calves are high because considerable protein is required in new tissues generated from rapid growth. Crude protein needs begin to decline after the neonatal period due to switching to a fat accretion mode in early autumn and due to reduced growth in late autumn (Figure 32–12). Published estimates of crude protein requirements for juveniles are variable and most are restricted to autumn. Studies by Smith and others and Ullrey and others indicated crude protein requirements of juveniles in autumn were 50 to 100% higher than those given in Figure 32–12 (112, 113). However, Verme and Ozoga, National Research Council's nutritional guidelines, and data of Cook and others support estimates of requirements presented here (i.e., Figure 32–12) (114, 115, 106, 69).

The estimates of crude protein and ME requirements presented were converted to nutrient concentrations in forage by Cook to provide a more practical estimate of requirements (23). This required dividing total daily requirements by daily forage dry-matter intake rates; energy content was expressed as digestible energy (DE), rather than ME, to enhance comparisons of dietary energy levels reported in a variety of field studies. Accordingly, ME was converted to DE by dividing ME by 0.82 (106).

Habitat Requirements and Management

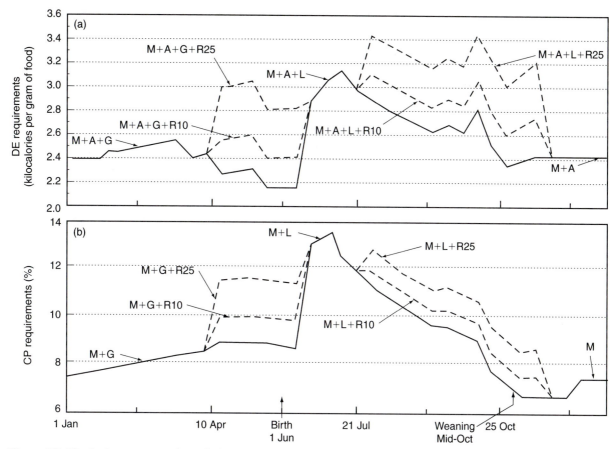

Figure 32–13 In A, concentration of digestible energy (DE) and, in B, concentration of crude protein in forage, expressed as kilocalories per gram of food and percent of food, required to satisfy daily metabolizable energy requirements of adult female elk for maintenance (M), activity (A), gestation (G), lactation (L), and replacement of winter-catabolized tissues assuming either 10% (R10) or 25% (R25) winter weight loss.

Estimated DE concentration in forage ranged from 2.3 to 3.0 kilocalories per gram of food. Replacement of winter-catabolized tissue increased forage concentration requirements appreciably during summer and early autumn (Figure 32–13). In calves from August through autumn, DE content of forage required above that provided by milk ranged from 2.6 to 3.0 kilocalories per gram of forage before late autumn and 2.3 kilocalories by the beginning of winter (Figure 32–14). Crude protein requirements ranged from 7% in winter to 12 to 13% during summer. Calf requirements for protein through August were low because milk apparently is able to supply much of their needs. Forage protein levels required by elk calves increased to 9% after milk intake declined, and then declined to about 7% at the beginning of winter (Figure 32–14). These forage-concentration estimates are based on a number of important assumptions discussed in detail by Cook (23).

Minerals play a key role in many aspects of animal production, yet they have been largely overlooked in nutritional studies of most wild ungulates. Specific roles of minerals in ruminant physiology are highly variable and often interactive, and excessive mineral intake can be toxic. Estimates of mineral requirements (Table 32–4) are based on beef cattle requirements and thus serve as approximations for female elk and their calves (106).

Requirements of certain minerals for antler growth for Cervidae are relatively high, especially compared to cattle (Table 32–4). Calcium comprises about 22% and phosphorus 11% of antlers, with the remainer largely compised of protein (110). Estimates of calcium and phosphorus forage concentration requirements for antler growth range from 0.40 to 0.64% and 0.26 to 0.54% (dry-matter basis), respectively (110, 116).

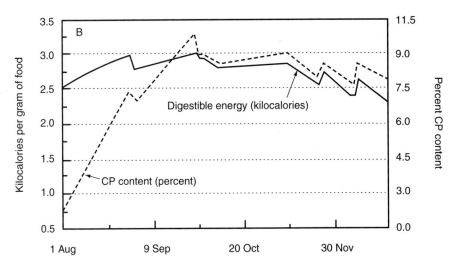

Figure 32–14 Concentration of digestible energy (DE) in forage (kilocalories per gram of food) and crude protein (CP) in forage required to satisfy daily metabolizable energy and protein requirements of calves from early August through mid-December. The estimates of requirements are based on data collected from captive calves and therefore do not completely account for activity levels of free-ranging calves. These requirements are in addition to energy and protein acquired from milk. (From Ref. 23.)

TABLE 32–4

Mineral Requirements and Maximum Tolerable Levels of Essential Minerals in Diets of Beef Cattle

	Requirement		
Mineral[a]	Suggested Value	Range[b]	Maximum Tolerable Level
Calcium, %	—	0.16–0.58	2
Cobalt, ppm	0.10	0.07–0.11	5
Copper, ppm	8.00	4–10	115
Iodine, ppm	0.5	0.20–2.0	50
Iron, ppm	50.00	50–100	1000
Magnesium, %	0.10	0.05–0.25	0.40
Manganese, ppm	40.00	20–50	1000
Molybdenum, ppm	—	—	6
Phosphorus, %	—	0.17–0.39	1
Potassium, %	0.65	0.5–0.7	3
Selenium, ppm	0.20	0.05–0.30	2
Sodium, %	0.08	0.06–0.10	10[c]
Chlorine, %	—	—	—
Sulfur, %	0.10	0.08–0.15	0.40
Zinc, ppm	30.00	20–40	500

[a]Units are percent or parts per million (ppm) content in diets.

[b]The listing of ranges recognizes that requirements for most minerals are affected by a variety of dietary and animal factors. For example, the high calcium levels are most applicable for young lactating cows; the low calcium levels are most applicable for mature animals in some stage of maintenance.

[c]Ten percent sodium chloride (salt).

Source: From Ref. 106.

Habitat Requirements and Management

Assessing the nutritional status of elk, and applying the appropriate mix of land treatments to overcome nutritional deficiencies, is a complex challenge. The following points should be considered:

1. Assessing the nutritional status of elk is most accurate when diet quality is evaluated in tandem with the quality of available forage. Inferring the nutritional status of elk based only on the quality of available forage can be misleading because elk and other ungulates select a diet higher in quality than that generally available (117, 118, 119, 120).

2. Many forage improvements will increase forage quality but reduce quantity in the short term, and increase forage quantity but reduce quality thereafter. For example, burning and fertilization may improve the nutritive value of some forages immediately after treatment, but quality usually declines within 2 years. Continued growth of post-treatment vegetation then results in increased biomass. Maintenance of a patchy mix of habitats, some high in forage biomass and some high in quality, therefore requires diverse temporal applications of forage improvements that span multiple seasons and years. Applying this mix of treatments in a "patchy" manner throughout a seasonal herd range is likely to maintain the desired spatial distribution of animals across public–private ownerships, commensurate with the distribution of forage improvements.

3. Forage treatments must be applied over extremely large areas, commensurate with the seasonal or year-round ranges of an elk herd, to ensure a high probability of achieving distribution, productivity, or abundance goals for the targeted population. Treatments must also be sustained over long time periods, perhaps 5 to 10 years or longer, to improve the probability of achieving desired population goals. Because of these scale and cost : benefit limitations, effective forage treatments and related habitat improvements generally are restricted to large-scale activities that occur for other purposes, primarily timber harvest and livestock grazing. Other treatments such as prescribed burning or fertilization will not have a high probability of success unless they are integrated with timber harvest, livestock grazing, and other commodity activities that typically have strong, large-scale effects on forage conditions.

4. It is difficult to specify the quantitative, interactive effects of various combinations of forage enhancements and other habitat treatments on elk nutrition, due to the lack of research on multivariable effects. General, single-variable effects of the most common treatments to improve nutrition are summarized briefly as follows. More detailed effects were described by Witmer and others, Wisdom and Thomas, and Cook (121, 25, 23).

Clear cutting, partial cuts, and thinning are examples of silvicultural treatments that allow increased sunlight to penetrate the forest floor, facilitating the growth of forage for elk. As a result, biomass of forage increases significantly following such treatments, and often continues to increase or remain high until the canopy closes (122, 123, 121). Timber harvest, however, may or may not enhance forage quality, but can change the timing of plant phenology (124, 125, 126). For example, forest cover can delay plant phenology 2 to 3 weeks compared to that in clear-cuts (127). Presumably, forage in openings may be high in biomass but relatively low in nutritive value during drought conditions on late summer and fall ranges (23). During the same time period, however, such forages often may still be growing and of high nutritive value under closed canopy forests (128).

Consequently, timber harvest designed to maintain a variety of seral stages, particularly the early and late stages, with patchy arrangement of these stages across the seasonal or year-round range of an elk herd, provides a diverse amount and distribution of forage species in varying phenological stages. Such landscape conditions presumably function as high-quality habitat for elk based on habitat modeling concepts and predictions (129, 130, 131).

On productive rangelands that were grazed historically by native herbivores, systems of cattle grazing can be designed to enhance forage conditions for elk. For example, grazing systems are being used by state wildlife agencies to enhance forage for elk and control elk distribution. Notable examples are those in Oregon and Montana (132, 133, 134, 135, 136, 137). Deferred, rest-rotation grazing by cattle is used during late spring and summer to "condition" grasses for later use by elk.

These case examples employ similar grazing strategies that are largely responsible for their success: (1) Cattle are removed from the early grazed pastures before or just after midgrowing season to allow for sufficient regrowth of grasses; this regrowth is unobstructed by litter and thus provides elk with efficient foraging access to higher quality plant parts during late summer, fall and/or winter; (2) cattle use of forage is light to moderate, further allowing for regrowth of forage for elk and maintenance of healthy range condition; (3) sufficient grazing rest is provided to facilitate survival and recovery of preferred grasses; (4) cattle are used primarily to "condition" grasses for elk; beef production is more of a benefit than a goal in itself.

Deferred, rest-rotation grazing systems designed specifically for elk work best on relatively productive rangelands such as those of the Pacific Northwest and northern Rocky Mountains. These systems will likely not be effective on rangelands in arid ecosystems of the Great Basin and desert Southwest, due to inherent limitations in site productivity. In these areas, both the retention of vegetation residue and the regrowth of desirable forage often are inadequate to provide the desired benefits to elk and maintain the desired composition, quantity, and vigor of forages.

Fire dramatically changes the composition, quantity, and quality of forage available to ungulates (138, 139, 140). Short-term effects of large-scale fires appear beneficial to forage and foraging conditions for many ungulates, including elk (141, 120, 104). Burning removes accumulated plant litter, thereby making current, nutritious growth more accessible (142, 143). Consequently, the regrowth of vegetation up to 2 years after burning is usually desired by elk, presumably due to greater foraging efficiency and higher forage quality (144, 142, 39). Biomass of herbs and shrubs may increase thereafter (145, 146, 39, 140).

On the negative side, forage quality may not increase after burning, although diet quality may be enhanced (120). Also, ungulate use usually declines quickly after burning (138). Moreover, long-term effects of burning on the composition and productivity of forages are highly variable and often unpredictable and/or undesirable (139). For example, burning can significantly reduce nitrogen levels through volatilization, although grazing animals may partially mitigate the effect (147). Much depends on the specific objectives and outcomes that are desired, the frequency and intensity of such fires, the scale of application, and the long-term effect on site productivity (148, 140).

Key forage species often are seeded or planted immediately after a wildfire or prescribed burn to establish nutritious forage for elk. Like burning, such treatments are generally considered beneficial, but few studies have documented specific benefits to elk nutrition. Moreover, seedings and plantings designed to control erosion may be largely unpalatable to elk.

In general, the effects are likely beneficial if the seedings or plantings provide a more balanced mix of forage classes and species than would be present otherwise. In western Oregon, for example, seeding of grasses and legumes is considered beneficial to elk nutrition because these forage classes complement a naturally high composition of shrubs (149, 130). In eastern Oregon, by contrast, grasses and forbs often dominate the forage base (27). Hence, shrubs of high nutritive value like curl-leaf mountain-mahogany and antelope bitterbrush often are planted to diversify the forage base.

Nitrogen is a critical element to production of protein in the physiology of ruminants. Most rangeland soils are moderately to severely deficient in nitrogen (150, 141). Not surprisingly, nitrogen is a critical chemical element in meeting the nutritional needs of elk (151, 144). Phosphorus and sulfur also are deficient on many rangelands, and both play important roles in physiologic functions of ruminants (152, 141, 153, 110).

Increasing the content of nitrogen, phosphorus, and/or sulfur in deficient soils can increase the protein content and digestible energy of herbage available to ungulates (141). Dramatic changes in elk distribution presumably can be achieved with the application of such fertilizers in deficient areas (8).

Beneficial effects of nitrogen fertilization, however, can be highly variable between vegetation communities (154). Use by elk, and presumably the nutritional benefits to them, is usually limited to 2 years or less unless fertilization is repeated (155). Repeated applications of nitrogen also can change the species composition of grasses, reduce or eliminate legumes, and make some soils more acidic (141, 156, 157).

Trace elements, essential to protein and energy metabolism in ruminants, are sometimes deficient in rangeland soils (110, 158). For example, selenium is deficient throughout much of the Pacific Northwest and California (110). Masupu found lower blood levels of selenium in elk, pronghorn, and bighorn sheep populations than that required by livestock (159).

Mineral supplementation can improve performance of elk and other ungulates if (1) deficient minerals are identified accurately; (2) deficient minerals are administered at the proper time, rate, and mix; (3) other minerals not known to be deficient are not included in supplementation efforts; and (4) supplementation is distributed at a scale large enough to ensure that animals locate and use the treated areas.

Winter damage by elk on privately owned farms and rangelands is often predictable and significant unless preventive steps are taken (104, 160). One approach is supplemental or replacement feeding with alfalfa hay or pellets. Although not as desirable as the acquisition or leasing of additional winter range, feeding is effective in keeping elk away from private land (161, 162). This is especially true when feeding is combined with the removal of problem animals on private land, either through live trapping and transplanting, hazing, or special hunts.

Feeding should meet the nutritional requirements of wintering elk; forage in the form of alfalfa hay should contain more than 10% crude protein and be more than 58 to 60% digestible (23). The rate of supplementation depends on the availability and quality of natural forage. If natural forage is limited, alfalfa hay fed at rates of 5.4 kilograms/elk/day appear sufficient (163). This assumes that alfalfa of high quality is used as the supplement (e.g., 15% crude protein and 66% digestibility of hay used in Ref. 163).

Feeding is not a panacea for resolving winter range problems (164). Once initiated, it is difficult to reduce elk dependence on such rations. Numbers of elk residing in the area during the nonwinter period may increase as well. Acquisition of additional winter ranges, improvements in the quality of natural forage on such ranges, and reductions in elk numbers to winter carrying capacity all are ecologically superior solutions.

Water is considered limiting to elk in many arid areas of western North America (2). Elk may concentrate near water sources in extremely dry areas (165). Moreover, some transplants of elk may have been ineffective in part from a shortage of free-standing water (166, 167). Increasing the distribution and availability of water on many of the driest rangelands of the western United States will likely enhance elk use of such areas, especially during dry seasons or years.

Further development of water sources on most other elk ranges, however, can be a "double-edged sword" if livestock have access to the water. Improving the distribution of water, that is, making water more evenly and readily available throughout a pasture, will also result in a more even distribution of livestock (150). This may increase the potential for competition with elk, or reduce elk use in favor of cattle use, for two reasons. First, most water developments are associated with roads, and elk avoid areas near roads open to motorized traffic (Figure 32–15). Second, livestock use is usually highest within 1.6 kilometers of water; this is the zone of most direct competition between elk and livestock, which can result in reduced quantity or quality of forage or elk avoidance of such areas when cattle are present (150, 25).

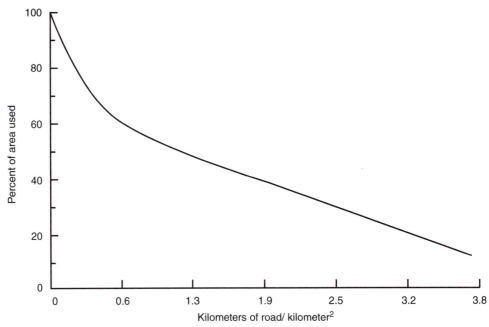

Figure 32–15 Elk distribution in relation to the density of roads open to motorized vehicles. (From Ref. 172.)

Security Requirements

Elk consistently select habitats in a manner that suggests selection for greater security from predators, especially from humans. Examples include the tendency of elk to select dense stands of cover, to remain close to cover when using openings, to avoid open or gentle terrain, and to avoid areas near roads open to motorized vehicles (open roads) (168, 169, 130, 131, 170, 171, 172). In addition, elk use combinations of these selection patterns to further enhance security, especially during the hunting season, when selection for greater security can improve an elk's chance of survival (169, 99, 173, 174).

Of the selection patterns listed above, the tendency of elk to avoid areas near open roads has been the most studied and well documented, as summarized in the following points. First, elk consistently and dramatically avoid areas near open roads across a variety of seasons, landscape conditions, and geographic regions (175, 176, 177, 178, 179, 180, 181, 182, 99, 169). Second, avoidance can occur at distances of up to 2.9 kilometers away from roads (171, 184). Third, the magnitude of open road avoidance increases with increasing rate of traffic, as documented by Wisdom and inferred from earlier elk research (183, 178, 180, 182).

In addition to these avoidance effects, higher densities of open roads can result in lower survival of elk during the hunting season and nonhunting seasons (174, 185). Moreover, Wisdom and Thomas demonstrated a significant reduction in elk K with increasing density of open roads as part of forage allocation modeling (25).

Road effects have been addressed in a generalized, composite model of elk selection in relation to density of open roads and imbedded in elk habitat models (129, 172, 186, 187, 130, 131). Application of elk–road density relations for management (Figure 32–14) should be done judiciously with the following points in mind: (1) Effects of open roads on elk K can be significant, as illustrated by an example application of Wisdom and Thomas and by empirical findings of Cole and others; (2) effects of open road density may be over- or underestimated by the Lyon model (Figure 32–15) because a given density of roads can be spatially distributed within watersheds in a variety of patterns, each of which results in a different availability of area in relation to distance bands from open roads; (3) avoidance effects in relation to open roads may be stronger during hunting seasons; and (4) survival of hunted animals may be lower in areas of higher road density (25, 185, 172, 184, 188, 169, 99, 174).

Habitat Requirements and Management

Efforts to provide escapement and increased security for elk involve managing a combination of vegetation characteristics, human access, and human activities. The following points are specifically important. First, manage for close juxtaposition of openings with cover; elk stay close to cover-forage edges when in openings, presumably in part because cover for escapement is nearby (130, 131). Second, manage a portion of cover areas to produce characteristics of hiding or escape cover (per definitions of Refs. 129 and 168); such cover may be particularly important during hunting seasons to reduce elk vulnerability to harvest (99). Third, manage for patches of cover greater than 100 hectares; such patches are thought to facilitate escapement of elk from hunters (189). Fourth, reduce the density of open roads through road obliteration, road closures, and other restrictions on human access and activities; reduced density of open roads presumably increases elk K and elk survival (187, 25, 174, 155).

Do Elk Need Thermal Cover?

A number of habitat models or summaries of elk ecology have emphasized the thermoregulatory benefits of dense forest cover having specific structural characteristics, often referred to as "thermal cover" (129, 121, 130, 131). This emphasis is intuitively reasonable, because dense forest cover can moderate harsh weather conditions, and elk often select for these structural characteristics (190, 191, 181, 192, 193, 194).

Four studies have tested the thermal cover hypothesis for Cervidae in North America. Three studies were conducted using deer, two in Maine and the third in Colorado (195, 196, 197, 198, 199). The fourth study was conducted using elk calves and yearlings in northeast Oregon (200). In all cases, no evidence was found that thermal cover provided thermoenergetic benefits of sufficient magnitude to enhance the condition and/or survival probability of deer or elk during winter. The elk study also evaluated the value of thermal cover during summer; again, no significant positive effects were found.

These studies collectively indicate that the value of thermal cover probably is of minor consequence in the context of animal performance and by extension, population dynamics in temperate zones of North America (200). This is because the weather-moderating effects of thermal cover are too small, occur too infrequently, or are too variable to provide meaningful benefits in relation to the thermoregulatory capabilities of deer and elk. Cook and others concluded that findings of the thermal cover studies indicate that elk biologists should refocus their attention to the influences of forest management on (1) forage resources and related forage production potential of forest successional stages and (2) cover characteristics related to security requirements of elk, particularly cover effects on elk vulnerability to harvest (200).

INTEGRATED HABITAT AND POPULATION MANAGEMENT

Traditionally the subjects of habitat and population management of elk have been treated separately. Until now, this separation has been convenient and clear: Landowners manage elk habitat, and state wildlife agencies manage elk populations. Recently, however, two paradigms have emerged that integrate management of habitat and populations: that of managing according to sustained yield theory and that of managing to minimize elk vulnerability to harvest (79, 99).

Elk Vulnerability to Harvest

Lyon and Christensen defined elk vulnerability as a "measure of elk susceptibility to being killed during the hunting season" (201, p. 3). Managing to minimize elk vulnerability to harvest requires explicit consideration of a variety of hunting, human access, and landscape factors in an integrated manner (Table 32–5) (100). Those factors considered to most significantly increase the vulnerability of elk to harvest are high density of hunters, high density of roads, loss or absence of large cover blocks, and gentle terrain (202, 203, 173, 174, 189, 170).

TABLE 32–5

Management Problems That Increase Elk Vulnerability to Harvest and the Corresponding Management Remedies

Management Problems	Management Remedies
1. Increasing density of roads	1. Design roads to minimize impacts. Close roads permanently or temporarily. Enforce road closures.
2. Increasing density of hunters	2. Restrict hunter numbers.
3. Decreasing amounts of cover	3. Control stand configuration, juxtaposition, and size through modifications in timber management program.
4. Fragmentation of cover into smaller patches	4. Retain adequate "escape cover" in the form of stands several hundred or more acres.
5. No restriction on antler class in male harvest	5. Impose regulations on what can be taken—such as allowing the kill of spike-antlered males only.
6. Setting of open seasons that include the rutting period	6. Ensure that open seasons do not include the rutting period.
7. Improving technology	7. Preclude "modern weapons."
8. Long open seasons	8. Shorten the open season.
9. Relatively gentle terrain	9. Decrease road density, maintain more cover, increase size of cover patches, decrease hunter numbers.
10. Increasing number of hunter days	10. Related to items 2 and 8. Reduce numbers and/or reduce length of hunting season.

Source: From Ref. 100, p. 319.

Often, management to reduce or minimize elk susceptibility to harvest focuses on goals to facilitate survival of older males; a confounding plethora of hunting, access, topographic, and vegetative factors can affect such goals, and population managers often are challenged to identify optimal combinations of these factors (100). As a result, the relations of elk vulnerability with a comprehensive set of environmental conditions have been synthesized in a meta-analysis and a resulting software application by Vales (204). Use of this software allows population and habitat managers to manipulate population and environmental conditions simultaneously or separately as part of "gaming" scenarios.

For example, a desired post-hunt ratio of mature males:100 females can be identified as the goal, and various combinations of road density, hunter density, season length, type of season, and other landscape and hunting factors can be modeled as meeting the goal. Alternatively, a specific road density, hunter density, season length, and physiography can be identified as part of current conditions, and resulting changes in sex and age characteristics of a population are predicted. Use of such gaming tools to meet goals related to elk harvest provides a common dialogue and link between population and land managers, and ultimately helps facilitate successful implementation of diverse population and land management goals.

Sustained Yield Theory

Managing according to sustained yield theory embraces the concepts of integrated management of population size, animal density, hunter harvest, habitat conditions, and overall K to meet goals for harvest, nutrition, animal performance, population performance, and other habitat or population factors of interest (Figure 32–16) (79). The basic premise of sustained yield theory is that the larger the population is in relation to ecological K, the lower the nutritional plane of individual animals and the lower their reproductive output. Slow growth rates and relatively high mortality of juveniles, delayed puberty, reduced pregnancy rates, predominance of older age classes, and slow or zero population growth should predominate as populations approach ecological K. Maintaining an elk population well below K should result in a more vigorous herd with a greater level of reproductive success. For most populations, there is a point below K at which the total number of animals recruited each year into the population is maximized (Figure 32–16); this is the point of maximum sustained yield (MSY).

Hunting is a primary tool to control populations with respect to MSY and K. However, this does not mean that hunted populations are held at MSY. To do so pre-

Integrated Habitat and Population Management

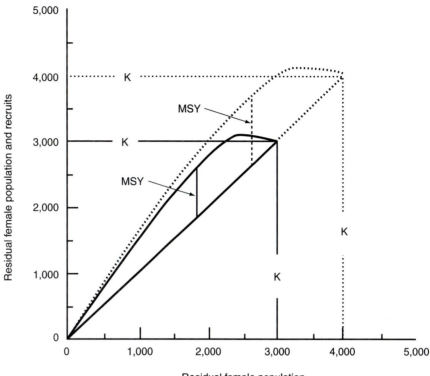

Figure 32–16 Carrying capacity (*K*) in relation to herd productivity (number of juveniles recruited into the population at age 1) of a hypothetical elk population. Maximum sustained yield (MSY) is the point at which the largest number of juveniles are recruited. Solid lines denote relations when *K* = 3,000 females; dotted lines represent relations when *K* = 4,000 females. These hypothesized relations illustrate two major points: (1) that reductions in the residual herd below *K* to the point at which MSY occurs produce the highest number of juveniles recruited annually due to improved nutritional conditions; and (2) that increased *K* results in both a higher population size and a higher MSY; in this case, increasing *K* from 3,000 to 4,000 animals results in a rise in MSY from 600 to 900 animals. Maintaining population size near MSY also may reduce elk damage to forage from overgrazing and elk use of private lands. These hypothesized relations were based on data from white-tailed deer, with adjustments made to account for elk reproductive characteristics (79, 23).

supposes that MSY and *K* are known and that an explicit decision has been made by society regarding MSY trade-offs between high density, unproductive populations versus reduced-density, productive populations. In practice, many political and biological factors are weighed to set hunting levels, none of which may explicitly reflect considerations emanating from the MSY concept.

Although often not recognized, habitat improvement projects provide an alternative approach for managing in the context of MSY and *K*. If *K* is changed appreciably through habitat manipulation, the point of MSY can be expected to change as well (Figure 32–16). Viewed from this perspective, the problem of spatial scale becomes more apparent (Figure 32–6); planting, burning, or fertilizing several hundred hectares at a time may have very little impact on *K* for a herd spread across hundreds of square kilometers. Although such projects perhaps are useful on winter range where animals are concentrated, meaningful improvements during seasons in which elk are not concentrated must be large scale. Costs of large-scale efforts will likely be prohibitive, unless commodity-based habitat manipulations such as timber harvest or livestock grazing that pay for themselves can also be designed to improve habitat for elk.

Aspects of the MSY theory also have implications regarding elk as disturbance agents with important effects on ecosystem dynamics. Elk populations throughout much of the western United States have increased substantially during the last 20 years (57). Such large populations undoubtedly have important effects on vegetative structure, composition, and productivity, and, in turn, on a variety of organisms and processes sensitive to changes in vegetation (e.g., *Wildlife Society Bulletin,* 1998, Vol. 26, No. 3; 205). Combined with this are important declines in elk herd productivity across substantial portions of the western United States (27, 74). Changes in vegetative composition and structure and declining productivity are predictable results of high-density populations at or near carrying capacity (206, 79). Both the increasing effects of herbivores on vegetation and declining productivity of elk herds that are becoming apparent in the western United States are likely to pose difficult challenges for elk managers early in the next century (207, 27, 205). A better understanding of the interrelations among habitat needs, population densities, productivity, distribution, herbivory effects, hunting levels, and predation will be required to make sound, science-based decisions.

SUMMARY

Elk were one of the most common and widely distributed of the wild ungulates in North America before European settlement. Declines in elk numbers, subsequent reductions in their distribution, and ultimately the extirpation of populations and entire subspecies occurred across large expanses of North America as European settlers moved west in the 17th, 18th, and 19th centuries. Today, populations of elk have recovered in large areas of the United States; distribution remains fragmented, however, especially in eastern North America.

Although the future of elk appears secure, managers of this important resource are faced with a number of controversial issues. One is the issue of density dependence. Population size and density of elk have increased significantly in the western United States during the past 20 years, and increased densities are suspected of contributing to significant changes in the composition, structure, and successional trajectories of vegetation (27, 74). If population dynamics of elk adhere to classical theory of density dependence then population size and density of elk ultimately may decline in many areas of western North America in response to degraded forage conditions associated with past high animal densities (206, 79).

A complicating and interesting twist in the density dependence scenario is the future role that predators may play in controlling or regulating elk populations. Puma and wolf populations are increasing across large areas of the western United States, and higher densities of these predators may change population dynamics of elk in ways not predicted by either density dependence or predator–prey theory.

While most admirers of elk prefer high abundance and density of animals, management to reduce density, particularly with acknowledgment of increasing population growth of predators, may enhance elk reproductive success and production of trophy-sized animals, reduce landowner conflicts, and prevent or mitigate detrimental impacts to wildland vegetation (79, 27). Increasing forage quality and quantity through habitat improvements provides some potential to concurrently maintain density and productivity at reasonable levels, but such improvements will have to be extensive and integrated with compatible strategies of predator management. The importance of these elk density–habitat–predator issues currently is unclear, but may present elk managers with difficult decisions in the 21st century.

LITERATURE CITED

1. Bryant, L. D., and C. Maser. 1982. Classification and distribution. Pages 1–59 *in* J. W. Thomas and D. E. Toweill, eds., *Elk of North America, Ecology and Management.* Stackpole Books, Harrisburg, PA.

2. MᴄCᴀʙᴇ, R. E. 1982. Elk and Indians: Historical values and perspectives. Pages 61–123 *in* W. Thomas and D. E. Toweill, eds., *Elk of North America, Ecology and Management*. Stackpole Books, Harrisburg, PA.

3. Pᴏᴛᴛᴇʀ, D. R. 1982. Recreational use of elk. Pages 509–559 *in* W. Thomas and D. E. Toweill, eds., *Elk of North America, Ecology and Management*. Stackpole Books, Harrisburg, PA.

4. Bᴏʟᴏɴ, N. A. 1994. Estimates of the values of elk in the Blue Mountains of Oregon & Washington: Evidence from existing literature. Pacific Northwest Research Station General Technical Report PNW-GTR-316, U.S. Department of Agriculture Forest Service, Portland, OR.

5. Dᴜꜰꜰɪᴇʟᴅ, J., and J. Hᴏʟʟɪᴍᴀɴ. 1988. The net economic value of elk hunting in Montana. Montana Department of Fish, Wildlife, and Parks, Helena.

6. Lᴏᴏᴍɪs, J., J. Cᴏᴏᴘᴇʀ, and S. Aʟʟᴇɴ. 1988. The Montana elk hunting experience: A contingent valuation assessment of economic benefits to hunters. Montana Department of Fish, Wildlife, and Parks, Helena.

7. Bʀᴏᴏᴋs, R., C. S. Sᴡᴀɴsᴏɴ, and J. Dᴜꜰꜰɪᴇʟᴅ. 1991. Total economic value of elk in Montana: An emphasis on hunting values. Pages 186–195 *in* A. G. Christensen, L. J. Lyon, T. N. Lonner, compilers, *Proceedings of Elk Vulnerability Symposium*. Montana State University, Bozeman.

8. Gᴇʀʀᴀɴs, J. 1992. The habitat partnership program in Colorado. *Rangelands* 14:84–87.

9. Bʀᴏᴄᴄɪ, J. 1993. The vulnerability blues, solving conflicts in the Blue Mountains. *Bugle* 10:57–65.

10. Wɪᴛᴍᴇʀ, G. W., 1990. Re-introduction of elk in the United States. *Journal of the Pennsylvania Academy of Science* 64:131–135.

11. MᴄCᴜʟʟᴏᴜɢʜ, D. R. 1969. The Tule elk: Its history, behavior, and ecology. University of California Publication Zoology, Vol. 88. University of California Press, Berkeley.

12. Mᴜʀɪᴇ, O. J. 1951. *The Elk of North America*. Stackpole Books, Harrisburg, PA.

13. Sᴋᴏᴠʟɪɴ, J. M., 1982. Habitat requirements and evaluations. Pages 369–413 *in* J. W. Thomas and D. E. Toweill, eds., *Elk of North America, Ecology and Management*. Stackpole Books, Harrisburg, PA.

14. Bᴏʏᴅ, R. J. 1978. American elk. Pages 11–29 *in* J. L. Schmidt and D. L. Gilbert, eds., *Big Game of North America, Ecology and Management*. Stackpole Books, Harrisburg, PA.

15. Bᴜʙᴇɴɪᴋ, A. B. 1982. Physiology. Pages 125–179 *in* J. W. Thomas and D. E. Toweill, eds., *Elk of North America, Ecology and Management*. Stackpole Books, Harrisburg, PA.

16. Hᴏꜰᴍᴀɴɴ, R. R. 1988. Anatomy of the gastro-intestinal tract. Pages 14–43 *in* D. C. Church, ed., *The Ruminant Animal: Digestive Physiology and Nutrition*. Prentice Hall, Englewood Cliffs, NJ.

17. Rᴏʙʙɪɴs, C. T., D. E. Sᴘᴀʟɪɴɢᴇʀ, and W. ᴠᴀɴ Hᴏᴠᴇɴ. 1995. Adaptation of ruminants to browse and grass diets: Are anatomical-based browser-grazer interpretations valid? *Oecologia* 103:208–213.

18. Mᴏᴜʟᴅ, E. D., and C. T. Rᴏʙʙɪɴs. 1982. Digestive capabilities in elk compared to white-tailed deer. *Journal of Wildlife Management* 46:22–29.

19. Bᴀᴋᴇʀ, D. L., and N. T. Hᴏʙʙs. 1987. Strategies of digestion: Digestive efficiency and retention time of forage diets in montane ungulates. *Canadian Journal of Zoology* 65:1978–1984.

20. Bᴀᴋᴇʀ, D. L., and D. R. Hᴀɴsᴇɴ. 1985. Comparative digestion of grass in mule deer and elk. *Journal of Wildlife Management* 49:77–79.

21. Rᴏʙʙɪɴs, C. T., T. Hᴀɴʟᴇʏ, A. E. Hᴀɢᴇʀᴍᴀɴ, O. Hᴊᴇʟᴊᴏʀᴅ, D. L. Bᴀᴋᴇʀ, C. C. Sᴄʜᴡᴀʀᴛᴢ, and W. W. Mᴀᴜᴛᴢ. 1987. Role of tannins in defending plants against ruminants: Reduction in protein availability. *Ecology* 68:98–107.

22. Kᴜꜰᴇʟᴅ, T. P. 1973. Foods eaten by the Rocky Mountain elk. *Journal of Range Management* 26:106–112.

23. Cᴏᴏᴋ, J. G. In press. Nutrition and food habits. *in* D. E. Toweill and J. W. Thomas, eds., *Ecology and Management of North American Elk*. Wildlife Management Institute, Washington, DC.

24. Mᴇʀʀɪʟʟ, E. H. 1994. Summer foraging ecology of wapiti (*Cervus elaphus nelsoni*) in the Mount Saint Helens blast zone. *Canadian Journal of Zoology* 72:303–311.

25. Wɪsᴅᴏᴍ, M. J., and J. W. Tʜᴏᴍᴀs. 1996. Elk. Pages 157–181 *in* P. R. Krausman, ed., *Rangeland Wildlife*. The Society for Range Management, Denver, CO.

26. Bʀʏᴀɴᴛ, L. D. 1993. Quality of bluebunch wheatgrass (*Agropyron spicatum*) as a winter range forage for Rocky Mountain elk (*Cervus elaphus nelsoni*) in the Blue Mountains of Oregon. Ph.D. thesis. Oregon State University, Corvallis.

27. Iʀᴡɪɴ, L. L., J. G. Cᴏᴏᴋ, R. A. Rɪɢɢs, and J. M. Sᴋᴏᴠʟɪɴ. 1994. Effects of long-term grazing by big game and livestock in the Blue Mountains forest ecosystems. Forest Service General Technical Report PNW-GTR-325. U.S. Department of Agriculture, Portland, OR.

28. RIGGS, R. A., A. R. TIEDEMANN, J. G. COOK, T. M. TIBBS, P. J. EDGERTON, M. VAVRA, W. C. KRUEGER, F. C. HALL, L. D. BRYANT, L. L. IRWIN, and T. DELCURTO. 1998. Modification of mixed-conifer forests by ruminant herbivores in the Blue Mountains Ecological Province. Unpublished report. USDA Forest Service Pacific Northwest Research Station, 1401 Gekeler Lane, La Grande, OR.

29. deCALESTA, D. S., and S. L. STOUT. 1997. Relative deer density and sustainability: A conceptual framework for integrating deer management with ecosystem management. *Wildlife Society Bulletin* 25:252–258.

30. GEIST, V. 1982. Adaptive behavioral strategies. Pages 219–277 *in* J. W. Thomas and D. E. Toweill, eds., *Elk of North America, Ecology and Management.* Stackpole Books, Harrisburg, PA.

31. HAIGH, J. C., and R. J. HUDSON. 1993. *Farming Wapiti and Red Deer.* Mosby-Year Book, St. Louis, MO.

32. HUDSON, R. J., H. M. KOZAK, J. Z. ADAMCZEWSKI, and C. D. OLSEN. 1991. Reproductive performance of wapiti (*Cervus elaphus nelsoni*). *Small Ruminant Research* 4:19–28.

33. HOLLAND, M. D., and K. G. ODDE. 1992. Factors affecting calf birth weight: A review. *Theriogenology* 38:769–798.

34. VERME, L. J. 1965. Reproduction studies on penned white-tailed deer. *Journal of Wildlife Management* 29:74–79.

35. SMITH, B. L., R. L. ROBBINS, and S. H. ANDERSON. 1997. Early development of supplementally-fed, free-ranging elk. *Journal of Wildlife Management* 61:26–38.

36. CLUTTON-BROCK, M. MAJOR, S. D. ALBON, and F. E. GUINNESS. 1987. Early development and population dynamics in red deer. I. Density-dependent effects on juvenile survival. *Journal of Animal Ecology* 56:53–57.

37. THORNE, E. T., R. E. DEAN, and W. G. HEPWORTH. 1976. Nutrition during gestation in relation to successful reproduction in elk. *Journal of Wildlife Management* 40:330–335.

38. ROBBINS, C. T., R. S. PODBIELANCIK-NORMAN, D. L. WILSON, and E. D. MOULD. 1981. Growth and nutrient consumption of elk calves compared to other ungulate species. *Journal of Wildlife Management* 45:172–186.

39. COOK, J. G., L. L. IRWIN, L. D. BRYANT, R. A. RIGGS, D. A. HENGEL, and J. W. THOMAS. 1994. Studies of elk biology in northeast Oregon. NCASI, Corvallis, OR.

40. JOHNSON, D. E. 1951. Biology of the elk calf, *Cervus canadensis nelsoni. Journal of Wildlife Management* 15:396–410.

41. SMITH, B. L. 1994. Population regulation of the Jackson elk herd. Ph.D. thesis. University of Wyoming, Laramie.

42. SCHLEGEL, M. 1976. Factors affecting calf elk survival in north central Idaho, a progress report. *Proceedings of the Western Association State Game Fish Commission* 56:342–355.

43. STUSSY, R. J. 1993. The effects of forage improvement practices on Roosevelt elk in the Oregon Coast Range. M.S. thesis. Oregon State University, Corvallis.

44. LEE, P. C., P. MAJLUF, and I. J. GORDON. 1991. Growth, weaning, and maternal investment from a comparative perspective. *Journal of Zoology (London)* 225:99–114.

45. CAMERON, R. D. 1994. Reproductive pauses by female caribou. *Journal of Mammalogy* 75:10–13.

46. TABER, R. D., K. RAEDEKE, and D. A. McCAUGHRAN. 1982. Population characteristics. Pages 279–298 *in* J. W. Thomas and D. E. Toweill, eds., *Elk of North America, Ecology and Management.* Stackpole Books, Harrisburg, PA.

47. HINES, W. W., and J. C. LEMOS. 1979. Reproductive performance by two age classes of male Roosevelt elk in southwestern Oregon. Wildlife Research Report Number 8. Oregon Department of Fish and Wildlife, Portland.

48. COLE, G. F. 1969. The elk of Grand Teton and southern Yellowstone National Parks. Office of Natural Science Studies Research Report GTRR-N-1. U.S. Department of Interior National Park Service, Jackson Hole, WY.

49. ADAMS, A. W. 1982. Migration. Pages 301–321 *in* J. W. Thomas and D. E. Toweill, eds., *Elk of North America, Ecology and Management.* Stackpole Books, Harrisburg, PA.

50. ALTMANN, M. 1956. Patterns of herd behavior in free-ranging elk of Wyoming, *Cervus canadensis nelsoni. Zoologica* 41:65–71.

51. ALTMANN, M., 1963. Naturalistic studies of maternal care in moose and elk. Pages 233–253 *in* H. L. Rheingold, ed., *Maternal Behavior in Mammals.* John Wiley and Sons, New York.

52. KNIGHT, R. R. 1970. The Sun River elk herd. *Wildlife Monographs* 23:11–66.

53. ALTMANN, M. 1952. Social behavior of elk, *Cervus canadensis nelsoni,* in the Jackson Hole area of Wyoming. *Behavior* 4:116–143.

54. HARPER, J. A. 1985. Ecology and management of Roosevelt elk in Oregon. Research report. Oregon Department of Fish and Wildlife, Portland.

55. SETON, E. T. 1927. *Lives of Game Animals,* Vol. 3, Part 1. Doubleday, Page, and Co., Garden City, NJ.

56. JACKSON, H. H. T. 1944. Big game resources of the United States. Bureau of Sport Fishing and Wildlife Research Report 8. U.S. Department of Interior, Washington, DC.

57. BUNNELL, S. D. 1997. *Status of Elk in North America, 1975–1995.* The Rocky Mountain Elk Foundation, Missoula, MT.

58. VAVRA, M., M. McINNIS, and D. SHEEHY. 1989. Implications of dietary overlap to management of free-ranging large herbivores. *Proceedings of the Western Section of American Society of Animal Science* 40:489–495.

59. VAVRA, M. 1992. Livestock and big game forage relationships. *Rangelands* 14:57–59.

60. HENDERSON, R. E., and A. O'HERREN. 1992. Winter ranges for elk and deer: Victims of uncontrolled subdivision? *Western Wildlands* 18:20–25.

61. COLE, G. F. 1971. An ecological rationale for the natural or artificial regulation of native ungulates in parks. *Transactions of North American Wildlife and Natural Resource Conference* 36:417–425.

62. LANKA, R. P., and R. J. GUENZEL. 1991. Game farms: what are the implications for North American elk? Pages 285–291 *in* A. G. Christensen, L. J. Lyon, and T. N. Lonner, compilers, *Proceedings of Elk Vulnerability Symposium.* Montana State University, Bozeman.

63. LANKA, R. P., E. T. THORNE, and R. J. GUENZEL. 1992. Game farms, wild ungulates, and disease in Western North America. *Western Wildlands* 18:2–7.

64. GEIST, V. 1985. Game ranching: Threat to wildlife conservation in North America. *Wildlife Society Bulletin* 13:594–598.

65. ROFFE, T., and B. SMITH. 1992. Tuberculosis: Will it infect wild elk? *Bugle* 9:86–92.

66. WILLIAMS, T. 1992. The elk-ranch boom. *Audubon* 94:14–20.

67. BUNNELL, F. L. 1987. Reproductive tactics of Cervidae and their relationships to habitat. Pages 145–167 *in* C. M. Wemmer, ed., *Biology and Management of the Cervidae. Proceedings of Research Symposium of the National Zoological Park.* Smithsonian Institute Press, Washington, DC.

68. TRAINER, C. E. 1971. The relationship of physical conditions and fertility of female Roosevelt elk (*Cervus canadensis roosevelti*) in Oregon. Thesis. Oregon State University, Corvallis.

69. COOK, J. G., L. J. QUINLAN, L. L. IRWIN, L. D. BRYANT, R. A. RIGGS, and J. W. THOMAS. 1996. Nutrition-growth relations of elk calves during late summer and fall. *Journal of Wildlife Management* 60:528–541.

70. SAUER, J. R., and M. S. BOYCE. 1979. Time series analysis of the National Elk Refuge census. Pages 9–12 *in* M. S. Boyce and L. D. Hayden-Wing, eds., *North American Elk: Ecology, Behavior, and Management.* University of Wyoming, Laramie.

71. LINNELL, J. D. C., R. AANES, and R. ANDERSEN. 1995. Who killed Bambi? The role of predation in the neonatal mortality of temperate ungulates. *Wildlife Biology* 1:209–223.

72. ALBON, S. D., B. MITCHELL, and B. W. STAINES. 1983. Fertility and body weight in female red deer: A density dependent relationship. *Journal of Animal Ecology* 52:969–980.

73. SAUER, J. R., and M. S. BOYCE. 1983. Density dependence and survival of elk in northwestern Montana. *Journal of Wildlife Management* 47:31–37.

74. COOK, J. G., R. A. RIGGS, A. R. TIEDEMANN, L. L. IRWIN, and L. D. BRYANT. 1995. Large herbivore–vegetative feedback relationships in the Blue Mountains ecoregion. Pages 155–159 *in* W. D. Edge and S. L. Olson-Edge, eds., *Proceedings on Sustaining Rangeland Ecosystems Symposium.* SR 953. Oregon State University, Corvallis.

75. GAILLARD, J., M. FESTA-BIANCHET, and N. G. YOCCOZ. 1998. Population dynamics of large herbivores: Variable recruitment with constant adult survival. *Trends in Ecology and Evolution* 13:58–63.

76. CAUGHLEY, G., 1981. Overpopulation. Pages 7–19 *in* Jewell, P. A., S. Holt, and D. Hart, eds., *Problems in Management of Locally Abundant Wild Mammals.* Academic Press, New York.

77. CAUGHLEY, G., and C. J. KREBS. 1983. Are big mammals simply little mammals writ large? *Oecologia (Berline)* 59:7–17.

78. SINCLAIR, A. R., E. DUBLIN, and M. MORNER. 1985. Population regulation of Serengeti wildebeest: A test of the food hypothesis. *Oecologia* 65:266–268.

79. McCullough, D. R. 1984. Lessons from the George Reserve, Michigan. Pages 211–242 *in* L. K. Halls, ed., *White-Tailed Deer: Ecology and Management.* Stackpole Books, Harrisburg, PA.

80. Fowler, C. W. 1987. A review of density dependence in populations of large mammals. Pages 401–441 *in* H. C. Genoways, ed. *Current Mammalogy.* Plenum Press, New York.

81. Klein, D. R. 1968. The introduction, increase, and crash of reindeer on St. Matthew Island. *Journal of Wildlife Management* 32:350–367.

82. Caughley, G. 1970. Eruption of ungulate populations, with emphasis on Himalayan thar. *Ecology* 51:53–72.

83. Geist, V. 1971. *Mountain sheep: A Study of Behavior and Evolution.* University of Chicago Press, Chicago, IL.

84. Guinness, F. E., T. H. Clutton-Brock, and S. D. Albon. 1978. Factors affecting calf mortality in red deer (*Cervus elaphus*). *Journal of Animal Ecology* 47:817–832.

85. Skogland, T. 1985. The effects of density-dependent resource limitations on the demography of wild reindeer. *Journal of Animal Ecology* 54:359–374.

86. Houston, D. B., and V. Stevens. 1988. Resource limitation in mountain goats: A test by experimental cropping. *Canadian Journal of Zoology* 66:228–238.

87. Messier, F. 1991. The significance of limiting and regulating factors on the demography of moose and white-tailed deer. *Journal of Animal Ecology* 60:377–393.

88. Gasaway, W. C., R. D. Boertje, D. V. Gvangaard, D. G. Kellyhouse, R. O. Stephenson, and D. G. Larsen. 1992. The role of predation in limiting moose at low densities in Alaska and Yukon and implications for conservation. *Wildlife Monographs* 120:1–59.

89. Gasaway, W. C., R. O. Stephensen, J. L. Davis, P. E. K. Shepard, and O. E. Burris. 1983. Interrelationships of wolves, prey, and man in interior Alaska. *Wildlife Monographs* 84:1–50.

90. Messier, F., and M. Crete. 1985. Moose–wolf dynamics and the natural regulation of moose populations. *Oecologia* 65:503–512.

91. Skogland, T. 1991. What are the effects of predators on large ungulate populations? *Oikos* 61:401–411.

92. Houston, D. B. 1982. *The Northern Yellowstone Elk: Ecology and Management.* Macmillan, New York.

93. Caughley, G. 1977. *Analysis of Vertebrate Populations.* John Wiley, London, UK.

94. Boyce, M. S. 1991. Wolf recovery for Yellowstone National Park: A simulation model. Pages 123–1138 *in* D. R. McCullough and R. H. Barrett, eds., *Wildlife 2001: Populations.* Elsevier Science, London, UK.

95. Boyce, M. S. 1995. Anticipating consequences of wolves in Yellowstone: Model validation. Pages 199–209 *in* L. N. Carbyn, S. H. Fritts, and D. R. Seip, eds., *Ecology and Conservation of Wolves in a Changing World.* Occasional Publication No. 35. Canadian Circumpolar Institute, Edmonton, Alberta, Canada.

96. Smith, B. L., and S. H. Anderson. 1998. Juvenile survival and population regulation of the Jackson elk herd. *Journal of Wildlife Management* 62:1036–1045.

97. Wisdom, M. J. 1999. Life-stage importance and associated management implications for an example elk population. Unpublished report. U.S. Department of Agriculture Forest Service, Pacific Northwest Research Station, LaGrande, OR.

98. Caswell, H. 1989. *Matrix Population Models.* Sinauer Associates, Sunderland, MA.

99. Christensen, A. G., L. J. Lyon, and T. N. Lonner, eds. 1991. *Proceedings of Elk Vulnerability Symposium.* Montana State University, Bozeman.

100. Thomas, J. W. 1991. Elk vulnerability—A conference perspective. Pages 318–319 *in* A. G. Christensen, L. J. Lyon, and T. N. Lonner, compilers, *Proceedings of the Elk Vulnerability Symposium.* Montana State University, Bozeman.

101. Fried, B. M., R. M. Adams, R. P. Berrens, and O. Bergland. 1995. Willingness to pay for a change in elk hunting quality. *Wildlife Society Bulletin* 23:680–686.

102. Hines, W. W., J. C. Lemos, and N. A. Hartmann. 1985. Male breeding efficiency in Roosevelt elk of southwestern Oregon. Wildlife Research Report Number 15. Oregon Department of Fish and Wildlife, Portland.

103. Noyes, J. H., B. K. Johnson, L. D. Bryant, S. L. Findholt, and J. W. Thomas. 1996. Effects of bull age on conception dates and pregnancy rates of cow elk. *Journal of Wildlife Management* 60:508–517.

104. LYON, J. L., and A. L. WARD. 1982. Elk and land management. Pages 443–477 *in* J. W. Thomas and D. E. Toweill, eds., *Elk of North America, Ecology and Management.* Stackpole Books, Harrisburg, PA.

105. GARROTT, R. A., J. G. COOK, M. M. BERNOCO, J. F. KIRKPATRICK, L. L. CADWELL, S. CHERRY, and B. TILLER. 1998. Antibody response of elk immunized with porcine zona pellucida. *Journal of Wildlife Diseases* 34:539–546.

106. NATIONAL RESEARCH COUNCIL. 1984. *Nutrient Requirements of Cattle,* 6th ed. National Academy Press, Washington, DC.

107. PRICE, M. A., and R. G. WHITE. 1985. Growth and development. Pages 183–214 *in* R. J. Hudson and R. G. White, eds., *Bioenergetics of Wild Herbivores.* CRC Press, Boca Raton, FL.

108. SADLIER, R. M. R. S. 1982. Energy consumption and subsequent partitioning in lactating black-tailed deer. *Canadian Journal of Zoology* 60:382–386.

109. RENECKER, L. A., and W. M. SAMUEL. 1991. Growth and seasonal weight changes as they relate to spring and autumn set points in mule deer. *Canadian Journal of Zoology* 69:744–747.

110. ROBBINS, C. T. 1983. *Wildlife Feeding and Nutrition.* Academic Press, Orlando, FL.

111. TORBIT, S. C., L. H. CARPENTER, D. M. SWIFT, and A. W. ALLDREDGE. 1985. Differential loss of fat and protein by mule deer during winter. *Journal of Wildlife Management* 49:80–85.

112. SMITH, S. H., J. B. HOLTER, H. H. HAYES, and H. SILVER. 1975. Protein requirement of white-tailed deer fawns. *Journal of Wildlife Management* 39:582–589.

113. ULLREY, D. E., W. G. YOUATT, H. E. JOHNSON, L. D. FAY, and B. L. BRADLEY. 1967. Protein requirement of white-tailed deer fawns. *Journal of Wildlife Management* 31:679–685.

114. VERME, L., and J. J. OZOGA. 1980. Influence of protein-energy intake on deer fawns in autumn. *Journal of Wildlife Management* 44:305–314.

115. NATIONAL RESEARCH COUNCIL. 1985. *Nutrient requirements of sheep,* 6th ed. National Academy Press, Washington, DC.

116. ULLREY, D. E. 1983. Nutrition and antler development in white-tailed deer. Pages 49–59 *in* R. D. Brown, ed., *Antler Development in Cervidae.* Caesar Kleberg Wildlife Research Institute, Kingsville, TX.

117. HOBBS, N. T., D. L. BAKER, J. E. ELLIS, and D. M. SWIFT. 1981. Composition and quality of elk diets in Colorado. *Journal of Wildlife Management* 45:156–171.

118. BAKER, D. L., and N. T. HOBBS. 1982. Composition and quality of elk summer diets in Colorado. *Journal of Wildlife Management* 46:694–703.

119. ROWLAND, M. M., ALDREDGE, A. W., ELLIS, J. E., WEBER, B. J., and G. C. WHITE. 1983. Comparative winter diets of elk in New Mexico. *Journal of Wildlife Management* 47:924–932.

120. HOBBS, N. T., and R. A. SPOWART. 1984. Effects of prescribed fire on nutrition of mountain sheep and mule deer during winter and spring. *Journal of Wildlife Management* 48:551–560.

121. WITMER, G. W., M. J. WISDOM, E. P. HARSHMAN, R. J. ANDERSON, C. CAREY, M. P. KUTTEL, I. D. LUMAN, J. A. ROCEHELLE, R. W. SCHARPF, and D. SMITHEY. 1985. Deer and elk. Pages 231–258 *in* E. R. Browne, ed., *Management of Wildlife and Fish Habitats in Forests of Western Oregon and Washington,* Vol. 1. Forest Service, Pacific Northwest Region Publication Number R6-F&WL-192-1985, U.S. Department of Agriculture, Portland, OR.

122. PACE, C. P. 1958. Herbage production and composition under an immature ponderosa pine stand in the Black Hills. *Journal of Range Management* 11:238–243.

123. MCCONNELL, B. R., and J. G. SMITH. 1970. Response of understory vegetation to ponderosa pine thinning in eastern Washington. *Journal of Range Management* 23:208–212.

124. REGELIN, W. L., O. C. WALLMO, J. NAGY, and D. R. DIETZ. 1974. Effect of logging on forage values from deer in Colorado. *Journal of Forestry* 72:282–285.

125. HANLEY, T. A., R. G. CATES, B. VAN HORNE, and J. D.. MCKENDRICK. 1987. Forest stand age-related differences in apparent nutritional quality of forage for deer in southeastern Alaska. Forest Service General Technical Report INT-222. U.S. Department of Agriculture, Ogden, UT.

126. VAN HORNE, B., T. A. HANLEY, R. G. CATES, J. D. MCKENDRICK, and J. D. HORNER. 1988. Influence of seral stage and season on leaf chemistry of southeastern Alaska deer forage. *Canadian Journal of Forest Research* 18:90–99.

127. KRUEGER, J. K., and D. J. BEDUNAH. 1988. Influence of forest site on total nonstructural carbohydrate reserves of pinegrass, elk sedge, and snowberry. *Journal of Range Management* 41:144–149.

128. Happe, P. J., K. J. Jenkins, E. E. Starkey, and S. H. Sharrow. 1990. Nutritional quality and tannin astringency of browse in clearcuts and old-growth forests. *Journal of Wildlife Management* 54:557–566.

129. Thomas, J. W., H. Black, Jr., R. J. Scherzinger, and R. J. Pedersen. 1979. Elk. Pages 104–127 *in* J. W. Thomas, ed., *Wildlife Habitats in Managed Forests, the Blue Mountains of Oregon and Washington*. Handbook 553. U.S. Department of Agriculture, Washington, DC.

130. Wisdom, M. J., L. R. Bright, C. G. Carey, W. W. Hines, R. J. Pedersen, D. A. Smithey, J. W. Thomas, and G. W. Witmer. 1986. A model to evaluate elk habitat in western Oregon. Forest Service Report R-6-F&WL-216-1986. U.S. Department of Agriculture, Portland, OR.

131. Thomas, J. W., D. A. Leckenby, M. Henjum, R. J. Pedersen, and L. D. Bryant. 1988. Habitat effectiveness index for elk on Blue Mountain winter ranges. Forest Service, General Technical Report PNW-GTR-218. U.S. Department of Agriculture, Pacific Northwest Research Station, Portland, OR.

132. Anderson, E. W., and R. J. Scherzinger. 1975. Improving quality of winter forage for elk by cattle grazing. *Journal of Range Management* 28:120–125.

133. Anderson, E. W., D. L. Franzen, and J. E. Melland. 1990. Prescribed grazing to benefit watershed–wildlife–livestock. *Rangelands* 12:105–111.

134. Anderson, E. W., D. L. Franzen, and J. E. Melland. 1990. Forage quality as influenced by prescribed grazing. Pages 56–70 *in* K. E. Severson, ed., *Can Livestock Be Used as a Tool to Enhance Wildlife Habitat?* Forest Service General Technical Report RM. U.S. Department of Agriculture, Washington, DC.

135. Frisina, M. R., and F. G. Morin. 1991. Grazing private and public land to improve the Fleecer Elk Winter Range. *Rangelands* 13:291–294.

136. Alt, K. L., M. R. Frisina, and F. J. King. 1992. Coordinated management of elk and cattle, a perspective—Wall Creek Wildlife Management Area. *Rangelands* 14:12–15.

137. Frisina, M. R. 1992. Elk habitat use within a rest-rotation grazing system. *Rangelands* 14:93–96.

138. Wright, H. A., and A. W. Bailey. 1982. *Fire Ecology of the United States and Southern Canada*. John Wiley and Sons, New York.

139. Bailey, A. W. 1988. Understanding fire ecology for range management. Pages 527–556 *in* P. T. Tueller, ed., *Vegetation Science Applications for Rangeland Analysis and Management*. Kluwer Academic, Boston, Massachusetts, USA.

140. Riggs, R. A., S. C. Bunting, and S. E. Daniels. 1996. Prescribed fire. Pages 295–320 *in* P. R. Krausman, ed., *Rangeland Wildlife*. The Society for Range Management, Denver, CO.

141. Vallentine, J. F. 1989. *Range Development and Improvements*. Academic Press, San Diego, CA.

142. Canon, S. K., P. J. Urness, and N. V. DeByle. 1987. Habitat selection, foraging behavior, and dietary nutrition of elk in burned aspen forest. *Journal of Range Management* 40:433–437.

143. Jourdonnais, C. S., and D. J. Bedunah. 1990. Prescribed fire and cattle grazing on an elk winter range in Montana. *Wildlife Society Bulletin* 18:232–240.

144. Nelson, J. R., and T. A. Leege. 1982. Nutritional requirements and food habits. Pages 323–367 *in* J. W. Thomas and D. E. Toweill, eds., *Elk of North America, Ecology and Management*. Stackpole Books, Harrisburg, PA.

145. Leege, T. A. 1979. Effects of repeated burns on northern Idaho browse. *Northwest Science* 53:107–113.

146. Peek, J. M., R. A. Riggs, and J. L. Lauer. 1979. Evaluation of fall burning on bighorn sheep winter range. *Journal of Range Management* 32:430–432.

147. Hobbs, N. T., D. S. Schimel, C. E. Owensby, and D. Ojima. 1991. Fire and grazing in the tallgrass prairie: Contingent effects on nitrogen budgets. *Ecology* 72:1374–1382.

148. Sanders, K., and J. Durham, eds., 1985. *Proceedings of Rangeland Fire Effects Symposium*. U.S. Department of Interior, Bureau of Land Management, Boise, ID.

149. Ramsey, K. J., and W. C. Krueger. 1986. Grass-legume seeding to improve winter forage for Roosevelt elk: A literature review. Report 763. Oregon State University Agriculture Experiment Station, Corvallis.

150. Holechek, J. L., R. D. Pieper, and C. H. Herbel. 1989. *Range Management Principles and Practices*. Prentice Hall, Englewood Cliffs, NJ.

151. Hobbs, N. T., D. L. Baker, J. E. Ellis, D. M. Swift, and R. A. Green. 1982. Energy- and nitrogen-based estimates of elk winter-range carrying capacity. *Journal of Wildlife Management* 46:12–21.

152. DIETZ, D. R. 1970. Animal production and forage quality—Definitions and components of forage quality. Pages 1–9 *in* H. A. Paulsen, Jr., and E. H. Reif, eds., *Proceedings of Range and Wildlife Habitat Evaluation—A Symposium.* Forest Service Miscellaneous Publication No. 1147. U.S. Department of Agriculture, Washington, DC.

153. MAUTZ, W. W. 1978. Nutrition and carrying capacity. Pages 321–348 *in* J. L. Schmidt and D. L. Gilbert, eds., *Big Game of North America.* Stackpole Books, Harrisburg, PA.

154. MERESZCAK, I. M., W. C. KRUEGER, and M. VAVRA. 1981. Effects of range improvements on Roosevelt winter nutrition. *Journal of Range Management* 34:184–187.

155. SKOVLIN, J. M., P. J. EDGERTON, and B. R. MCCONNELL. 1983. Elk use of winter range as affected by cattle grazing, fertilizing, and burning in Southeastern Washington. *Journal of Range Management* 36:184–189.

156. DUNCAN, D. A., N. K. MCDOUGALD, and J. R. LARSON. 1985. Effects of two nitrogen and a sulfur fertilizer on yield on native clover. California Agriculture Technology Institute, California State University, Fresno.

157. BERG, W. A. 1986. Effect of 20 years of low N rate pasture fertilization on soil acidity. *Journal of Range Management* 39:122–124.

158. HUSTON, J. E., and W. E. PINCHAK. 1991. Range animal nutrition. Pages 27–63 *in* R. D. Heitschmidt and J. W. Stuth, eds., *Grazing Management, An Ecological Perspective.* Timber Press, Portland, OR.

159. MASUPU, K. V. 1990. Selenium blood levels in free-ranging antelope (*Antilocapra americana*), elk (*Cervus elaphus* spp.), bighorn sheep (*Ovis canadensis* spp.) from California, seven western states, Mexico, and British Columbia, Canada. Thesis. University of California, Davis.

160. deCALESTA, D. S., and G. W. WITMER. 1994. Elk. Pages D41–D50 *in* R. M. Timm, ed., *Prevention and Control of Wildlife Damage,* 2nd ed. University of Nebraska Press, Lincoln.

161. EMERSON, K. 1988. Elkhorn Wildlife Area: Artificial feeding of elk. Pages 50–54 *in* M. Zahn, J. Pierce, and R. Johnson, eds., *Proceedings of Western States and Provinces 1988 Elk Workshop.* Washington Department of Wildlife, Olympia.

162. MCKEEL, R. 1988. Status of elk winter feeding in Yakima and Kattites Counties, Washington. Pages 26–29 *in* M. Zahn, J. Pierce, and R. Johnson, eds., *Proceedings of Western States and Provinces 1988 Elk Workshop.* Washington Department of Wildlife, Olympia.

163. THORNE, T., and G. BUTLER. 1976. Comparison of pelleted, cubed, and alfalfa hay as winter feed for elk. Wyoming Game and Fish Department, Laramie.

164. KIMBALL, J. F., JR., and M. L. WOLFE. 1985. Elk management opportunities in northern Utah: To feed or not to feed. Pages 191–197 *in* G. R. Workman, ed. *Proceeding of Western Elk Management Symposium.* Utah State University, Logan.

165. MCCORQUODALE, S. M., K. F. RAEDEKE, and R. D. TABER. 1986. Elk habitat use patterns in the shrub-steppe of Washington. *Journal of Wildlife Management* 50:664–669.

166. CARPENTER, J., and N. SILVY. 1991. Movements and habitat relations of a small elk population in West Texas. Pages 64–73 *in* A. G. Christensen, L. J. Lyon, and T. N. Lonner, compilers, *Proceedings of Elk Vulnerability Symposium.* Montana State University, Bozeman.

167. CARPENTER, J., and W. GRANT. 1991. The use of computer modeling to examine the population dynamics of a small elk herd in West Texas. Pages 123–125 *in* A. G. Christensen, L. J. Lyon, and T. N. Lonner, compilers, *Proceedings of Elk Vulnerability Symposium.* Montana State University, Bozeman.

168. LYON, J. L. 1985. Elk and cattle on the national forests: A simple question of allocation or a complex management problem? *Western Wildlands* 11:16–19.

169. UNSWORTH, J. W., L. KUCK, E. O. GARTON, and B. R. BUTTERFIELD. 1998. Elk habitat selection on the Clearwater National Forest, Idaho. *Journal of Wildlife Management* 62:1255–1263.

170. EDGE, W. D., and C. L. MARCUM. 1991. Topography ameliorates the effects of roads and human disturbance on elk. Pages 132–137 *in* A. G. Christensen, L. J. Lyon, and T. N. Lonner, compilers, *Proceedings of the Elk Vulnerability Symposium.* Montana State University, Bozeman.

171. LYON, J. L. 1979. Habitat effectiveness for elk as influenced by roads and cover. *Journal of Forestry* 79:658–660.

172. LYON, J. L. 1983. Road density models describing habitat effectiveness for elk. *Journal of Forestry* 81:592–595.

173. LEPTICH, D. J., and P. ZAGER. 1991. Road access management effects on elk mortality and population dynamics. Pages 126–137 *in* A. G. Christensen, L. J. Lyon, and T. N. Lonner, compilers, *Proceedings of Elk Vulnerability Symposium.* Montana State University, Bozeman.

174. UNSWORTH, J. W., L. KUCK, M. D. SCOTT, and E. O. GARTON. 1993. Elk mortality in the Clearwater Drainage of northeastern Idaho. *Journal of Wildlife Management* 57:495–502.

175. HIEB, S. R., ed. 1976. *Proceedings of the Elk–Logging–Roads Symposium.* Forest, Wildlife, and Range Experiment Station, University of Idaho, Moscow.

176. HERSHEY, T. J., and T. A. LEEGE. 1976. Influences of logging on elk on summer range in north-central Idaho. Pages 73–80 *in* S. R. Hieb, ed. *Proceedings of the Elk-Logging-Roads Symposium.* Forest, Wildlife, and Range Experiment Station, University of Idaho, Moscow.

177. MARCUM, C. L. 1975. Summer-fall habitat selection and use by a western Montana elk herd. Ph.D. thesis. University of Montana, Missoula.

178. PERRY, C., and R. OVERLY. 1977. Impact of roads on big game distributions in portions of the Blue Mountains of Washington, 1972–1973. Applied Research Bulletin. Washington Department of Game, Olympia.

179. MORGANTINI, L. E., and R. J. HUDSON. 1979. Human disturbance and habitat selection. Pages 132–139 *in* M. S. Boyce and L. D. Hayden-Wing, eds., *North American Elk: Ecology, Behavior, and Management.* University of Wyoming Press, Laramie.

180. ROST, G. R., and J. A. BAILEY. 1979. Distribution of mule deer and elk in relation to roads. *Journal of Wildlife Management* 43:634–641.

181. IRWIN, L. L., and J. M. PEEK. 1983. Elk habitat use relative to forest succession in Idaho. *Journal of Wildlife Management* 47:664–672.

182. WITMER, G. W., and D. S. deCALESTA. 1985. Effect of forest roads on habitat use by Roosevelt elk. *Northwest Science* 59:122–125.

183. WISDOM, M. J. 1998. Assessing life-stage importance and resource selection for conservation of selected vertebrates. Ph.D. thesis. University of Idaho, Moscow.

184. ROWLAND, M. M., M. J. WISDOM, B. K. JOHNSON, and J. G. KIE. 1999. Elk distribution and modeling in relation to roads. Unpublished report. U.S. Department of Agriculture Forest Service, Pacific Northwest Research Station, 1401 Gekeler Lane, LaGrande, OR.

185. COLE, E. K., M. D. POPE, and R. G. ANTHONY. 1998. Effects of road management on movement and survival of Roosevelt elk. *Journal of Wildlife Management* 61:1115–1126.

186. LEEGE, T. A., 1984. Guidelines for evaluating and managing summer elk habitat in northern Idaho. Wildlife Bulletin No. 11. Idaho Department of Fish and Game, Boise.

187. LYON, J. L., T. N. LONNER, J. P. WEIGAND, C. L. MARCUM, W. D. EDGE, J. D. JONES, D. W. McCLEERY, and L. L. HICKS. 1985. *Coordinating Elk and Timber Management, Final Report of the Montana Cooperative Elk-Logging Study,* 1970–1985. Montana Department of Fish, Wildlife, and Parks, Helena.

188. WITMER, G. W. 1981. Roosevelt elk habitat use in the Oregon Coast Range. Ph.D. thesis. Oregon State University, Corvallis.

189. HILLIS, J. M., M. J. THOMPSON, J. E. CANFIELD, L. J. LYON, C. L. MARCUM, P. M. DOLAN, and D. W. McCLEEREY. 1991. Defining elk security: The Hillis paradigm. Pages 38–43 *in* A. G. Christensen, L. J. Lyon, and T. N. Lonner, compilers, *Proceedings of Elk Vulnerability Symposium.* Montana State University, Bozeman.

190. REIFSNYDER, W. E., and H. W. LULL. 1965. Radiant energy in relation to forests. Forest Service General Technical Bulletin 1344. U.S. Department of Agriculture, Washington, DC.

191. HERSHEY, T. J., and T. A. LEEGE. 1982. Elk movements and habitat use on a managed forest in north-central Idaho. Wildlife Bulletin Number 10. Idaho Department of Fish and Game, Boise.

192. LECKENBY, D. A. 1984. Elk use and availability of cover and forage habitat components in the Blue Mountains, Northeast Oregon, 1976–1982. Wildlife Research Report 14. Oregon Department of Fish and Wildlife.

193. SMITHEY, D. A., M. J. WISDOM, and W. W. HINES. 1985. Roosevelt elk and black-tailed deer response to habitat change related to old-growth conversion in southwestern Oregon. Pages 41–55 *in* R. W. Nelson, ed., *Proceedings of the Western States and Provinces 1988 Elk Workshop.* Washington Department of Wildlife, Olympia.

194. ZAHN, M. 1985. Use of thermal cover by elk on a western Washington summer range. Dissertation. University of Washington, Seattle.

195. ROBINSON, W. L. 1960. Test of shelter requirements of penned white-tailed deer. *Journal of Wildlife Management* 24:364–371.

196. GILBERT, F. F., and M. C. BATEMAN. 1983. Some effects of winter shelter conditions on white-tailed deer, *Odocoileus virginianus. Canadian Field Naturalist* 97:391–400.

197. FREDDY, D. J., 1984. Quantifying capacity of winter ranges to support deer—Evaluation of thermal cover used by deer. Pages 21–25 *in* Wildlife Research Report. Colorado Division of Wildlife, Denver.

198. FREDDY, D. J., 1985. Quantifying capacity of winter ranges to support deer—Evaluation of thermal cover used by deer. Pages 13–36 *in* Wildlife Research Report. Colorado Division of Wildlife, Denver.

199. FREDDY, D. J., 1986. Quantifying capacity of winter ranges to support deer—Evaluation of thermal cover used by deer. Pages 9–18 *in* Wildlife Research Report. Colorado Division of Wildlife, Denver.

200. COOK, J. G., L. L. IRWIN, L. D. BRYANT, R. A. RIGGS, and J. W. THOMAS. 1998. Relations of forest cover and condition of elk: A test of the thermal cover hypothesis in summer and winter. *Wildlife Monographs* 141:1–61.

201. LYON, J. L., and A. G. CHRISTENSEN. 1992. A partial glossary of elk management terms. Forest Service INT-288. U.S. Department of Agriculture, Washington, DC.

202. VALES, D. J., V. L. COGGINS, P. MATTHEWS, and R. A. RIGGS. 1991. Analyzing options for improving bull : cow ratios of Rocky Mountain elk populations in northeast Oregon. Pages 174–181 *in* A. G. Christensen, L. J. Lyon, and T. N. Lonner, compilers, *Proceedings of Elk Vulnerability Symposium.* Montana State University, Bozeman.

203. HURLEY, M. A., and G. A. SARGEANT. 1991. Effects of hunting and land management on elk habitat use, movement patterns, and mortality in western Montana. *In Proceedings of the Elk Vulnerability Symposium.* Montana State University, Bozeman.

204. VALES, D. J., 1996. User's manual for ELKVULN, an elk vulnerability, hunter, and population projection program, version 1.00. Department of Fish and Wildlife Resources, University of Idaho, Moscow.

205. MCSHEA, W. J., H. B. UNDERWOOD, and J. H. RAPPOLE, eds., 1997. *The Science of Overabundance: Deer Ecology and Population Management.* Smithsonian Institution Press, Washington, DC.

206. CAUGHLEY, G., 1980. What is this thing called carrying capacity? Pages 2–8 *in* M. S. Boyce and L. D. Hayden-Wing, eds., *North American Elk: Ecology, Behavior and Management.* University of Wyoming, Laramie.

207. KAY, C. E. 1990. Yellowstone's northern elk herd: A critical evaluation of the "natural-regulation" paradigm. Ph.D. thesis. Utah State University, Logan.

208. ROBBINS, C. T., R. S. PODBIELANCIK-NORMAN, D. L. WILDON, and E. D. MOULD. 1981. Growth and nutrient consumption of elk calves compared to other ungulate species. *Journal of Wildlife Management* 45:172–186.

209. JIANG, A., and R. J. HUDSON. 1992. Estimating forage intake and energy requirements of free-ranging wapiti (*Cervus elaphus*). *Canadian Journal of Zoology* 70:675–679.

210. KAY, R. N. B., and B. W. STAINES. 1981. The nutrition of the red deer (*Cervus elaphus*). *Nutrition Abstract Review Series B* 51:601–622.

211. ADAMCZEWSKI, J. Z., R. J. HUDSON, and C. C. GATES. 1993. Winter energy balance and activity of female caribou on Coats Island, Northwest Territories: The relative importance of foraging and body reserves. *Canadian Journal of Zoology* 71:1221–1229.

212. BOERTJE, R. D. 1985. An energy model for adult female caribou of the Denali Herd, Alaska. *Journal of Range Management* 38:468–473.

213. MCEWAN, E. H., and P. E. WHITEHEAD. 1970. Seasonal changes in the energy and nitrogen intake in reindeer and caribou. *Canadian Journal of Zoology* 48:905–913.

214. PAPAGEORGIOU, M., C. NEOPHYTOU, A. SPAIS, and C. VAVALEKAS. 1981. Food preferences, and protein and energy requirements for maintenance of roe deer. *Journal of Wildlife Management* 45:728–732.

215. ULLREY, D. E., W. G. YOUATT, H. E. JOHNSON, L. D. FAY, B. L. SCHOEPKE, and W. T. MAGEE. 1970. Digestible and metabolizable energy requirements for winter maintenance of Michigan white-tailed does. *Journal of Wildlife Management* 34:863–869.

216. SCHWARTZ, C. C., M. E. HUBBERT, and A. W. FRANZMANN. 1988. Energy requirements of adult moose for winter maintenance. *Journal of Wildlife Management* 52:26–33.

217. HOLTER, J. B., H. H. HAYES, and S. H. SMITH. 1979. Protein requirement of yearling white-tailed deer. *Journal of Wildlife Management* 43:972–979.

218. SCHWARTZ, C. C., W. L. REGELIN, and A. W. FRANZMANN. 1987. Protein digestion in moose. *Journal of Wildlife Management* 352–357.

CHAPTER

33

Exotics

Elizabeth Cary Mungall

INTRODUCTION

An exotic animal or plant means different things to different people. The Indian blackbuck antelope in a Texas pasture, the ring-necked pheasant in a South Dakota cornfield, the brown trout in a cold stream in British Columbia, and the water hyacinth caught on a boat in Florida are all exotics. All are species living outside their normal area of distribution. Whether intended or unintended, human intervention is implied in their relocation. Technically, this makes domestic livestock (and most human beings) exotics in North America, but the word is rarely applied to domestics or people. Animals under the closely controlled conditions of zoos are also excluded.

The exotics to be dealt with here are large mammalian wildlife species in North America. They constitute a wide variety of ungulates, mainly in the families Cervidae and Bovidae. Numerous kinds have been tried in North America (Table 33–1). Even though North America's principal big game animal, the white-tailed deer, is approximated at more than 15 million animals, the more than 250,000 exotic ungulates on more than 1 million hectares are still a concern to managers charged with safeguarding habitat for native wildlife (1, 2). Exotics are often added to ranges that already support native wildlife rather than to areas that would be clear of natives even if the exotics did not exist. Therefore, competition between exotics and native wildlife is always a consideration. Some exotics and natives coexist adequately as long as forage is sufficient, while others never make good combinations. Managers have a great responsibility to assess the options before introducing exotics and to keep animals within carrying capacity afterward. This responsibility is among the aspects stressed in the introduction guidelines issued by the world conservation coordinating organization International Union for Conservation of Nature and Natural Resources (3).

This chapter provides an overview of the situation regarding exotic ungulates in North America. General considerations, such as the variety of reasons motivating people to keep exotics, are discussed as well as environmental consequences that should be evaluated when formulating guidelines for involvement with introductions or introduced species. Ten of the most numerous species and groups of species are reviewed separately, while an extensive table demonstrates the scope of exotics activity in different parts of North America and the multitude of species used.

There are four principal reasons why people keep exotics. In the frequency in which they operate, these perceived "pros" are enjoyment or curiosity, revenue or fee hunting, preservation or reintroduction, and a willingness to provide homes for surplus animals.

The first, enjoyment or curiosity, is the same motivating force that drives humans to set up zoos. Sometimes the enjoyment comes from the animals directly by viewing or personal hunting experiences. Sometimes it comes indirectly from the reactions of other people when they find out what interesting animals an owner has. From the time of the ancient Egyptians to that of one of the first major zoos in the Americas (the Aztec king Moctezuma's zoo found by Cortés in 1521) to the late 1800s when societies of civic-minded citizens founded the first of the big North American zoos, cultures with surplus resources have set up zoos (4). Where people control property suitable for herbivores, these individuals tend to enjoy seeing animals on the land as long as it is not reserved for cultivation. If there is no conspicuous wildlife, they add some. If animals are there, owners add more. Additions give a more varied assortment for viewing.

With white-tailed deer, mule deer, and American elk prominent among North America's native game, proponents of releases argue that many areas should be appropriate for other hoofed species, too, especially exotic deer. Such arguments can quickly become emotionally based. If there is no large native "game" animal because conditions have changed, then there are always people looking for an animal to introduce. As Decker points out, this argument is likely to be raised when people consider cases where a species (e.g., bighorn sheep) may have been extirpated or where hoofed species may never have occupied the area (e.g., some southwestern U.S. regions of arid brush or desert) or where land use practices may have changed conditions such that native wildlife can no longer maintain itself (e.g., forestry, farming, recreation) (5). Whichever applies, the multiplicity of adaptations offered by all the species around the world make an alluring array.

As with domestic stock that is an owner's property to do with as he or she pleases, exotics are often easier to obtain and easier to work with than native game with its public ownership and government oversight. Consequently, added animals are likely to be exotics.

The second element likely to prompt a person to keep exotics is the desire to make money. Attracted by high prices commanded for trophy animals of some species (e.g., greater than $1,000 for quality axis deer, blackbuck antelope, or various "super exotics") or a steady income from selling breeding stock, many people consider raising exotics (6). However, raising exotics is not a "get rich quick" business (7). Many animals must be supported to raise a few trophies, ranch hands are needed to feed or catch stock, fences have to be built and maintained, and taxes have to be paid.

The people who make money from exotics ranching are likely to be those who already have a ranch with fencing, barns, and other facilities, and especially those people who add exotics to a working livestock operation. Then, exotics can produce supplemental income when prices for domestic livestock are low but feed prices are high. Exotics became popular in Texas during the "seven years drought" of the 1950s (8). With ranchers able to set their own seasons and bag limits, fee hunting for exotics helped some ranchers avoid losing their land.

The late 1970s brought new ways to make money. First came game ranching reminiscent of the culling operations publicized from Africa throughout the 1970s but with exotics instead of native wildlife. In the 1980s, promotion for health-conscious consumers was added to gourmet fare in upscale restaurants as tests came out showing exotics meats to be a low-fat, low-cholesterol product (9). Once harvesting and marketing were worked out, the commercial meat outlets offered ranchers a realistic chance to make their operations efficient. Finally, excess animals could be removed, for cash, in practically any number and combination of species, sexes, and ages, and at any season. Reducing populations to carrying capacity became practical. Increasingly into the 1990s, similarly innovative ranchers looked at deer farming successes in New Zealand and showed how people could practice more intensive husbandry for high-density exotics farming. When choosing species to be used for exotics farming, ability to use grass is an important consideration because it is difficult to provide large amounts of browse on a sustained basis. Using large cervids such as red deer or red deer/elk hybrids, velvet from "harvested" antlers can even eclipse meat sales for some owners.

Preserving vanishing species or increasing numbers to support reintroduction programs for native ranges can also be motivations for exotics owners. There are ranchers who have spent many years and thousands of dollars of their personal money to create and maintain herds of endangered species with no other return than the satisfaction of being personally involved in wildlife conservation. At differing levels of involvement, private breeding efforts on Texas ranches with blackbuck antelope, Arabian oryx, Père David's deer, and barasingha are examples of owners working with foreign wildlife which they have made available or would like to make available for reintroduction into native habitat (6).

Other ranchers have entered into formal agreements with zoos as part of the American Zoo Association's "Species Survival Program." Realizing that they lack the space to raise the numbers needed to perpetuate all species indefinitely, zoo professionals overcame their wary attitude toward game ranches enough to propose a mutual pact. It put the space that ranchers have into the service of the zoos' conservation vision. Animals targeted for ranches, rather than for Species Survival Program space in the network of cooperating zoos, are from vanishing species that are hardy enough to survive and increase on the range and that are outgrowing the space that zoos can offer. This usually means endangered or threatened species, but not those down to so few individuals in captivity that their continued survival is precarious. Scimitar-horned oryx and Grévy's zebra were the first species released to ranchers under this program.

Animal brokers, watching the fortunes of ungulates wax and wane, look wistfully at what easy availability of breeding stock has done for some of the species that have appealed to exotics owners and wish that the same forces could be harnessed for other species that are declining toward extinction (e.g., Sömmering's gazelle). Species can decline to such low numbers that government regulations hamper involvement from the private sector.

In a few cases, ranchers have opened up their land to surplus stock mainly because they knew that the animals would be euthanized if no home could be found. Because of their interest in animals and their knowledge of animal husbandry and business, ranchers are often invited to sit on zoo boards. Thus, they are often the first to hear when their zoo has animals it cannot place but cannot keep. Such situations probably have been more significant than the actual numbers of animals involved would imply in fueling the beginnings of exotics activity in North America (6).

NEGATIVE CONSIDERATIONS

There are 10 common reasons for opposing introduction of exotic big game species in North America (10, 5, 11):

1. Exotics can cause problems when added to areas that are already fully stocked.
2. Exotics may displace native wildlife.
3. Exotics can change a site away from its pristine or representative condition by disturbing the ecological balance.
4. Managers should concentrate on preserving the native fauna and flora rather than promoting the worldwide trend toward biotic uniformity.

Negative Considerations

5. Money spent on exotics might be better spent on programs for native species.
6. Exotics enterprises promote commercialization of wildlife that may jeopardize conservation.
7. Populations of exotics can be difficult to control.
8. Exotics may introduce new diseases or parasites.
9. Exotics might interbreed with native wildlife, thus altering the gene pool.
10. Exotics can cause serious damage to crops, forests, and habitat of native species.

The first three points concentrate on ecological relationships. If a site is already at carrying capacity, then there is nowhere for additional animals with similar requirements to live without creating conflicts. The range may become overused. A site preserved as an example of a certain type of habitat may be changed beyond recall. Exotics may outcompete natives (12, 13). The danger of exotics displacing natives can be very real. First released in the state of Maryland in 1916, exotic sika deer have become increasingly more numerous compared to native white-tailed deer as indicated by hunter kills along the eastern shore (12). Whitetails on the Edwards Plateau of Texas showed relatively lower body condition compared to axis, fallow, and sika deer harvested from the same range during the summer and winter stress periods (14).

Points four through six deal with our obligation toward our native heritage. Should we allow the uniqueness of ecosystems to be diluted by adding foreign elements? This can only support a trend toward world homogeneity. Also, why allocate limited resources to exotics when much can usually be done to maintain or improve conditions for native species? There is also the fear that creating markets for hunting rights or animal parts will be counter to wildlife conservation efforts. We may lose our view of wildlife as a common trust if exotics used like livestock are perceptually grouped with native wildlife.

The last four points are specific complaints that often carry the most weight in arguments against exotics. Exotics can be difficult to control. Sooner or later, exotics escape. Floods breach fencing. Animals make holes. Poachers cut wire. With the natural tendency of some species to colonize rapidly (e.g., aoudad), exotics populations can flourish even if not wanted. When some of the early stocking efforts were being made, owners looked at the devastation caused in New Zealand by exotic deer and vowed never to release animals so small or secretive that they would be difficult to shoot out if necessary (8). However, today's exotics ranchers seem to have forgotten this self-imposed restriction. Small creatures like dik-dik, Chinese water deer, and muntjac are showing up on sale lists in spite of the way that some have demonstrated their ability to escape and increase. In England, for example, the particularly secretive Reeves's muntjac has spread throughout most of the south and the Midlands (15).

Another disadvantage of exotics relates to diseases and parasites. Although a few parasites previously unrecorded in North America have been found from exotics (e.g., two nematode species from blackbuck), strict quarantine procedures on even-hoofed imports have done much to keep out infections like foot-and-mouth disease and rinderpest (16). After quarantine, imports must go to a post-quarantine facility for the rest of their lives, and only offspring can be released on rangeland. This filtering has presumably reduced release of infected animals from zoos. However, shipments among private owners were not regulated until recently, and the potential for disease exchange between captive and wild cervids is a special concern (17).

Regardless of prophylactic measures, there have been problems. A particularly vexing problem has been with tuberculosis. Canada closed its border to cervid shipments from the United States after an elk shipment contaminated a Canadian herd. This touched off a wave of investigations in the United States between 1991 and 1995 that identified tuberculosis in 31 captive deer and elk herds involving 16 states (18). The result has been new testing requirements when shipping cervids.

Another problem posed by exotics is hybridization. Other than wild boar that promptly bred with feral domestics, the early exotics like axis deer, fallow deer, blackbuck antelope, and nilgai antelope were too distantly related to cross with anything in North America, including each other. This changed with release of red deer and mouflon. Red deer genes mixed with most of the captive elk populations on exotics ranches in Texas. Similarly, mouflon bred with any sheep, so diluting the mouflon stock that Texas census takers have had difficulty treating the "mouflon" category consistently (19).

Hybridization was a major point of contention when Wyoming blocked permitting for exotics ranching and Alberta instituted gene screening as part of the protocol for importation of elk (20). There could be hybridization if ranch animals were to get out and breed with the state's native wildlife. A native population is adapted by local selective pressures to function efficiently under local conditions. Ecologists warn against adding genes from another population, subspecies, or species. When animals under the more intensive management of a deer farm are artificially selected for production requirements, the effect of unintended hybridization with natives could be even more far reaching.

Differing Governmental Views

With each region having different natural and political environments, there is no agreement in regard to exotics (21). In most Canadian provinces, laws do not allow game ranching involving sport hunting as practiced in the United States, so exotics interest has centered on game farming (1). The provinces of British Columbia and Alberta have changed laws to favor exotics, and fallow farming has become a serious industry (B. Morgan, wildlife consultant, personal communication). In the United States, some states have a thriving deer farming industry (e.g., Washington State, New York, Texas, and Florida) or have acquired species to introduce for hunting (e.g., New Mexico). Other states discourage any exotics activity (e.g., Arizona, Kentucky, and Wyoming). In spite of the game department's mandate in Texas that encourages only native wildlife, Texas has growing numbers of exotics and exotics ranches. In addition, the weather is generally favorable for tropical species, and almost all Texas land is under private control so owners can do what they like with non-native animals.

EXAMPLE EXOTICS

Whether because of environmental conditions, land use practices, or human attitudes, exotics are important in some places but are not an issue in others. Provinces or states with few exotics can look at other regions with high densities as examples of what happens when exotics are released. The issue in areas with popular exotics programs is how to respond to exotics on rangeland or to their potential spread rather than whether they should be introduced (Table 33–1). In many areas, exotics are already entrenched. The foregoing positive and negative aspects still apply, but the emphasis shifts to gathering information necessary for effective management. The philosophical stance taken in regard to the pros and cons can then guide management's response. In each case, do we just live with the results? Do we encourage them? Do we limit or reverse them? Whatever course is taken should be backed by biological background on the animals involved.

The following material reviews what is known about some of the most numerous exotics and their interactions with natives (e.g., aoudad or Barbary sheep, axis deer, fallow deer, ibex, Indian blackbuck antelope, mouflon, nilgai antelope, red deer, sika deer, and the wild boar-feral hog complex). For more information on these plus other exotics, see Mungall and Sheffield (6).

Aoudad

The aoudad or Barbary sheep from North Africa is a link between sheep and goats (22, 23). Aoudad behave more like sheep but are built more like goats. Like sheep, aoudad males are not malodorous and they lack the typical goat chin beard even though they

TABLE 33–1

Origins, Locations, and Population Estimates for Major Exotics Concentrations in North America[a]

Animal	Origin	Major North American Locations	Estimated Number	Sources
Deer				
Axis deer	Indian and Sri Lanka	California	350	12
		Florida	6,500–9,500	M. Brady and W. Fewox, personal communication, 1998
		Hawaii	4,535–5,235	28
		Hawaii	>500	67
		Louisiana	700–800	R. Pesson, personal communication, 1998
		Mississippi	300	68
		Texas	55,424	2
Barasingha	India	Florida	50	W. Fewox, personal communication, 1998
		Louisiana	50	R. Pesson, personal communication, 1998
		Mississippi	16	68
		Texas	881	2
Brocket deer	S. America	Texas	c	2
Chevrotain	Africa?	Texas	c	2
Chinese water deer	Asia	Florida	15–20	W. Fewox, personal communication, 1998
		Mississippi	9	68
		Texas	14	2
Elk's deer	Asia	Texas	130	2
Elk, Rocky Mountain (wapiti)	N. America	Hawaii	85	69
Fallow deer (European and possibly a few Persian hybrids)	Europe and Asia Minor	Saskatchewan (Canada total)	600 (20,000)	68 1
		Alabama	200–300	68
		California	350	12
		Florida	2,000	W. Fewox, personal communication, 1998
		Georgia	500	12
		Kentucky	200–300	12
		Louisiana	700–800	R. Pesson personal communication, 1998
		Mississippi	26	12
		New York	>2,000	70
		New York	>500	J. Kerckerinck, personal communication, 1998
		Oregon	200–400	68
		Texas	27,177	2
		Utah	75	68
		Virginia	500	68
Hog deer	Asia	Texas	69	2
Muntjac	Asia	Florida	40–50	W. Fewox, personal communication, 1998
		Mississippi	12	68
		Texas	120	2
Père David's deer	China	Florida	70	W. Fewox, personal communication, 1998
		Texas	291	2
Red deer	Europe	Canada	5,000	1
		Florida	4,000	W. Fewox, personal communication, 1998
		Louisiana	100	R. Pesson, personal communication, 1998
		Mississippi	15	68

continued

TABLE 33–1

Origins, Locations, and Population Estimates for Major Exotics Concentrations in North America[a]—*continued*

Animal	Origin	Major North American Locations	Estimated Number	Sources
Red deer (other)	Europe?	Texas	4,802	2
Red deer-elk hybrids	Hybrid	Florida	53	71
		Texas	3,975	2
Rusa deer	Asia	Texas	23	34
Sambar	Asia	Florida	75–100	D. Kosin, personal communication, 1997
		Texas	82	34
Sika deer	Asia	Florida	1,500	W. Fewox, personal communication, 1998
		Louisiana	75	R. Pesson, personal communication, 1998
		Maryland	4,300	T. Mathews, personal communication, 1997
		Texas	11,966	2
		Utah	3	68
		Virginia	700	68
Other deer		Texas	949	2
Antelopes				
Addax	North Africa	Florida	5	71
		Texas	1,824	2
Indian blackbuck antelope	India	Florida	200	W. Fewox, personal communication, 1998
		Louisiana	300	R. Pesson, personal communication, 1998
		Mississippi	60	68
		Texas	35,328	2
Blesbok	Africa	Texas	57	2
Bongo	Africa	Florida	21	71
		Texas	1	34
Bontebok	Africa	Florida	16	71
		Texas	36	2
Bushbuck	Africa	Texas	c	2
Dik-dik	Africa	Florida	15–20	W. Fewox, personal communication, 1998
		Texas	37	2
Duiker	Africa	Texas	c	2
Eland, common	Africa	Florida	50	W. Fewox, personal communication, 1998
		Mississippi	5	68
		Texas	919	2
Gazelle, dama	Africa	Florida	26	71
		Texas	91	2
Gazelle, dorcas	Africa	Texas	c	2
Gazelle, Grant's	Africa	Texas	104	2
Gazelle, Persian goitered	Asia	Texas	c	2
Gazelle, slender-horned	Africa	Florida	8	71
		Texas	c	2
Gazelle, Thomson's	Africa	Texas	143	2
Gazelles, other		Texas	99	2
Gerenuk	Africa	Florida	13	71
Hartebeest	Africa	Texas	c	2
Impala	Africa	Florida	13	71
		Texas	341	2
Kudu	Africa	Florida	13	71
		Texas	341	2
Lechwe	Africa	Florida	13	71
		Texas	537	2

TABLE 33–1

Origins, Locations, and Population Estimates for Major Exotics Concentrations in North America[a]—*continued*

Animal	Origin	Major North American Locations	Estimated Number	Sources
Nilgai	India	Florida	40	W. Fewox, personal communication, 1998
		Texas	8,153	2
Nyala	Africa	Florida	24	71
		Texas	49	2
Oryx, Arabian	Arabia	Texas	288	2
Oryx, East African (both beisa and fringe-eared)	East Africa	Texas	26	2
Oryx, scimitar-horned	Africa	Florida	40–50	W. Fewox, personal communication, 1998
		Mississippi	6	68
		Texas	2,145	2
Oryx, South African (gemsbok)	Africa	New Mexico	2,000	R. Valdez, personal communication, 1998
		Texas	456	2
Roan antelope	Africa	Florida	12	71
		Texas	c	2
Sable antelope	Africa	Florida	4	71
		Texas	215	2
Sitatunga	Africa	Texas	93	2
Springbok	South Africa	Florida	3	71
		Texas	87	2
Suni	Africa	Texas	c	2
Waterbuck	Africa	Florida	14	71
		Texas	432	2
		Utah	2	68
Wildebeest, black	Africa	Texas	c	2
Wildebeest, blue	Africa	Texas	c	2
Wildebeest, white bearded	Africa	Texas	32	2
Wildebeest, other or unspecified	Africa	Louisiana	1	R. Pesson, personal communication, 1998
		Texas	166	2
Other antelopes		Texas	134	2
Cattle				
African Cape buffalo	Africa	Texas	6	34
Banteng and gaur	Asia	Florida	43	71
		Texas	22	2
Forest buffalo (dwarf buffalo)	Africa	Florida	21	71
Water buffalo	Asia	Florida	750	W. Fewox, personal communication, 1998
		Texas	30	34
Watusi	Africa	Florida	9	71
		Texas	370	2
Yak	Asia	Florida	2	71
		Texas	11	34
Zebu	India	Texas	38	2
Other buffalo		Texas	86	2
Goats and relatives				
Catalina goats	Domestic	Florida	170	71
		Texas	1,714	2
Chamois	Europe	Texas	c	2
Ibex, Iranian (Persian ibex, wild goat)	Asia	New Mexico	600	68

continued

TABLE 33–1
Origins, Locations, and Population Estimates for Major Exotics Concentrations in North America[a]—*continued*

Animal	Origin	Major North American Locations	Estimated Number	Sources
Ibex, Siberian	Asia	New Mexico	75	68
Ibex, (unspecified)	Asia, some Africa?	Florida	10	71
		Texas	1,042	2
Ibex hybrids	Texas	Texas	1,697	2
Markhor	Asia	Texas	221	2
Tahr, Himalayan	Asia	California	>200	68
		Texas	165	34
		Utah	3	68
Tur, Caucasian	Asia	Texas	c	2
Other goats		Texas	51	2
Sheep and relatives				
Aoudad (Barbary sheep)	Africa	California	1,200	68
		California	Eliminated	12
		Colorado	120	68
		Florida	26	71
		New Mexico	4,000	68
		Texas	12,292	2
Argali	Asia	Texas	c	2
Barbados	Africa	Texas	3,917	2
Corsican sheep (mouflon hybrids)	Texas	Florida	500	W. Fewox, personal communication, 1998
		Texas	4,953	2
Texas Dall	Texas	Florida	45	W. Fewox, personal communication, 1998
		Texas	c	2
Four-horned sheep	Asia and Europe	Florida	3	71
		Texas	491	2
Mouflon	Europe	Colorado	30	68
		Florida	40	W. Fewox, personal communication, 1998
		Texas	7,574	2
		Utah	25	68
Red sheep	Asia	Texas	924	2
Urial	Asia	Texas	c	2
Other sheep		Texas	277	2
Pigs				
Wild boar, Eurasian[d]	Europe and Asia	Canada	2,000	1
		Colorado	25	68
		Kentucky	300	5
		New Hampshire	>1,000	T. Walski, personal communication, 1998
		North Carolina	700–900	5
		Tennessee	2,300	5
		Texas	1,417	6
Giraffes and relative				
Giraffe	Africa	Florida	14	71
		Texas	96	2
Okapi	Africa	Florida	3	71
Camels				
Bactrian	Asia	Texas	16	2
Dromedary	Middle East	Texas	43	2
Camel (unspecified)		Florida	2	71

TABLE 33–1

Origins, Locations, and Population Estimates for Major Exotics Concentrations in North America[a]—*continued*

Animal	Origin	Major North American Locations	Estimated Number	Sources
Llamas and relatives				
Alpaca	S. America	Texas	48	2
Guanaco	S. America	Texas	49	34
Llama	S. America	Florida	5	71
		Mississippi	2	68
		Oregon	18,000	68
		Texas	12	2
Vicuna	S. America	Texas	[c]	2
Other	S. America	Texas	1,327	2
Zebra				
Zebra, Chapman's	Africa	Texas	75	2
Zebra, Damara	Africa	Texas	[c]	2
Zebra, Grant's	Africa	Texas	445	2
Zebra, Grévy's	Africa	Florida	11	71
		Texas	98	2
Zebra, mountain	Africa	Texas	[c]	2
Other zebras	Africa	Florida	9	71
		Texas	28	2
		Utah	1	68
Zebra-donkey hybrids	Hybrid	Florida	2	71
Tapirs				
Tapir, Malayan	Asia	Texas	[c]	2
Tapir, other	Central or S. America	Texas	[c]	2
Rhinoceroses				
Rhinoceros, black	Africa	Texas	13	2
Rhinoceros, white (square-lipped rhinoceros)	Africa	Florida	5	71
		Texas	21	2
Other rhinoceroses		Texas	27	2
Hippopotamuses				
Hippopotamus	Africa	Texas	[c]	2
Hippopotamus, pygmy	Africa	Texas	2	34
Elephants				
Elephant	Asia or Africa	Texas	18	2
Other exotics (Chamois, suni, Malayan tapir and other tapirs combined for reporting privacy)		Texas	32	2
Totals	Canada	Canada	27,000	1
	U.S.A	Alabama	200–300	68
		California	>900	68, 12
		Colorado	175	68
		Florida	16,487–19,542	71
				D. Kosin, personal communication, 1997
				M. Brady, personal communication, 1998
				W. Fewox, personal communication, 1998
		Georgia	500	12

continued

TABLE 33–1

Origins, Locations, and Population Estimates for Major Exotics Concentrations in North America[a]

Animal	Origin	Major North American Locations	Estimated Number	Sources
		Hawaii	>500– >5,235	28 67
		Kentucky	500–600	5, 12
		Louisiana	1,926–2,126	R. Pesson, personal communication, 1998
		Maryland	4,300	5, 12
		Mississippi	457	68
		New Hampshire	>1,000	T. Walski, personal communication, 1998
		New Mexico	6,675	68 R. Valdez, personal communication, 1998
		New York	>500	J. Kerckerinck, personal communication, 1998
		North Carolina	700–900	5
		Oregon	18,200– 18,400	
		Tennessee	2,300	5
		Texas	>199,449	6, 34, 2
		Utah	109	68
		Virginia	1,200	68

[a]Some novel domestics are included.

[b]One species = sp., more than one species = spp., one or more species = (spp).

[c]Source suppresses some Texas figures to safeguard owner privacy.

[d] In most, if not all areas, wild boar have interbred with feral swine. Feral swine are free ranging in at least 18 states, forming a population estimated at 500,000 to more than 1 million spread through the southeastern United States from Virginia south across Florida (36,000), west through Texas (nearly 121,700) and into Arizona (200–400), as well as large concentrations in California (30,500) and Hawaii (>100,000) (5, 54, 61, 6, 64).

have long fringes of hair from throat to chest and around the forelegs (Figure 33–1). The main part of the coat is short, sandy-colored hair. Horns curve back and out in a simple arc. Except for larger horn bases and body size, male and female aoudad look alike. Female horns grow 30 to 51 centimeters with records of nearly 69 centimeters. Male horns are 35 to 81 centimeters with trophies of 84 to more than 91 centimeters. Males and females reach 145 kilograms and 63 kilograms, respectively.

Aoudad are cropped for meat production, but their meat is not consistently as good as deer venison. Thus, the main economic use is hunting and live sale of breeding stock.

Aoudad are notorious colonizers. The animals establish themselves in dry country wherever females find a rocky outcrop for their chief bedding site, shelter in bad weather, and escape cover. Females, their young, and subadult males travel together in tight groups (23). Adult males stay as bachelors in groups of four to six until the rut brings them in with the females. About half the conceptions are achieved during the rut in mid-September to mid-November, but there is some breeding all year (24).

Life history information is summarized in Mungall and Sheffield (6). Gestation lasts about 5 months, and most births occur in spring. Neonates (i.e., singleton, twins, or sometimes triplets) stay right beside the mother. Females mature sexually at least by 18 months and sometimes younger. Males take 11 to 15 months or more but rarely breed until adulthood because subadult males are subordinate to adult (i.e., horned) females. Males take more than 5 years to reach prime breeding weight. Old for wild aoudad is 10 to 13 years, while the zoo record is 24.

Aoudad are highly resistant to parasites in comparison to sheep, but, like domestic sheep, are tolerant of crowding (24, 25). High evaporation rates during critical spring and summer transmission periods are credited with the low parasite loads in Palo Duro Canyon, Texas. However, a population decline there coincided with an outbreak

Example Exotics

Figure 33–1 The aoudad sheep, endemic to northern Africa, now resides in California, Colorado, New Mexico, Texas, and Florida. (Photograph from New Mexico, courtesy of New Mexico Department of Game and Fish.)

of the nematode, arterial worm (23). In wetter areas the large stomach worm, likely picked up from domestic goats or sheep, can be devastating (6).

Aoudad will take at least a bite or two of almost anything. The consistency with which high-fiber plants show up on food lists demonstrates how suited aoudad are to desert terrain. Forbs are highly preferred and eaten in quantity when available. One New Mexico study found a seasonal high of 55% forbs (23). Either grass or browse can make up the bulk of the diet, with browse being more highly preferred. At most, grass consumption registers only half the intake, but aoudad are definitely good at scrounging for grass if forage is scarce.

Aoudad are potential competitors of mule deer, desert bighorn sheep, white-tailed deer, and other exotics because of the aoudad's habitat flexibility, large group size, and aggressive nature. A dietary overlap index of 74% for aoudad and mule deer in Palo Duro Canyon, Texas, had browse showing the greatest similarity in use (23). Habitat partitioning was judged to be lowering the chances for interspecific competition between aoudad and mule deer in Largo Canyon, New Mexico (23). Studies on white-tailed deer and aoudad in the Edwards Plateau of Texas indicated that forbs would be the real problem for these species when inhabiting the same areas. Both used much browse, but aoudad also consumed grass (26).

Axis Deer

The axis deer is the most popular of the exotics. This Indian native, also called chital or spotted deer in its homeland, is one of the few deer that retains its crisply spotted coat through adulthood (Figure 33–2). A white throat bib distinguishes axis from spotted fallow and sika deer. All axis are colored alike, but only bucks carry antlers. Lengths for the simple but graceful three-point antlers run 56 to 68 centimeters with large examples reaching 76 to 91.5 centimeters. Bucks weigh 65 to 113 kilograms. Does weigh 43 to 66 kilograms.

Large carcass size and excellent meat qualities make axis a favorite for farmers and hunters (6). Nevertheless, their use as a commercial meat animal is severely limited by their intolerance to cold. Large-scale axis farming has been most successful in Texas and Florida.

Figure 33–2 Distribution of axis deer in North America is limited to warmer climates because this species is endemic to South India. (Photograph from Texas by E. Cary Mungall, courtesy of Dave and Liz Broomfield, Lazy B Ranch.)

Other endearing aspects of axis are their diurnal habits, their sociability, their relatively placid nature, and the asynchrony of their antler cycle. Some bucks are likely to be in hard antler at any time of year. The highest proportion of Texas bucks come into hard antler between May and December. During the Texas rut, which runs from mid-May to August and peaks in June and July, bucks drift restlessly from group to group. Violent fights develop when one rutting buck interferes with another (27, 28).

Reproductive data are summarized in Mungall and Sheffield (6). Females reach sexual maturity by about 1 year old and continue to cycle all year until bred. A newborn lies hidden between nursings for 10 to 20 days before starting to follow the dam with the herd. Unlike most hoofed mothers, axis are remarkably tolerant of strange fawns, sometimes even letting one nurse. Twins are extremely rare. Female axis frequently rebreed 4 to 5 months after parturition. With gestation lasting 7.5 months, one fawn per year is the norm. Axis deer generally live 9 to 13 years, with maximum longevity in zoo-type close confinement being 18 to 22.

Axis deer seem to have a high natural resistance to parasites (27). Axis parasite infection is usually infrequent and light. The main diseases of concern have been white muscle disease and malignant catarrhal fever. The latter makes it hard to keep axis on pastures previously devoted to domestic sheep, a noted carrier species for one strain of this disease.

Axis diets remain relatively constant all year unless their staple type of forage, grass, declines in quantity or quality (29, 30, 26, 31). Then, they switch dependence easily to browse, and fruits can be an important extra. Fresh grass sprouts are favored and sedge sprouts may also be used when young. Their selection of forbs when available and their increased use of browse during seasonal stress periods makes axis deer potentially competitive for forage with white-tailed deer (26, 32).

Blackbuck Antelope

Indian blackbuck antelope are so popular in Texas that their numbers rival those in the animal's native Indian subcontinent: 29,000 to 38,000 in India and 30,390 to 35,328 in Texas (33, 34, 2). As its name indicates, bucks turn dark as they mature. Active males stay dark even during summer. This makes the white eye rings and other white patches that they share with the honey-colored females and young stand out conspicuously (Figure 33–3).

Figure 33–3 Blackbuck antelope numbers in Texas rival those of the animal's native Indian subcontinent. (Photograph by C. Mungall, courtesy of Fossil Rim Wildlife Ranch.)

Corkscrew horns rise in a tall "V" over the buck's head. Adult horns measure a minimum of 33 centimeters in a straight line from base to tip. Trophies can go 56 to 66 centimeters or more measured in a straight line from base to tip. Adult bucks average 37 kilograms and seldom exceed 56, while females average 27 and range between 19 and 33.

Blackbuck are cropped for meat in Texas where populations are large. This helps ranchers keep herds within carrying capacity. Meat is rated second only to axis deer venison, but the bigger axis yields more for the same harvest effort (6). Cold sensitivity prevents commercial use farther north. When nutritious forage or supplements are always available, blackbuck with simple shelters can be kept even in areas where temperatures go below −18°C, but north of the Texas Rolling Plains winter horn freezing can become a problem (25, 35).

Mungall studied blackbuck behavior in Texas (35). Blackbuck occur as females and young in female groups, young and adult bucks in bachelor associations, and territorial males. Where large numbers inhabit open country, males and females also form mixed groups. When a female group enters a territory, the animals often loiter forming a "pseudo-harem" with the resident buck. A female stays alone only if she has a new fawn lying nearby. A fawn becomes increasingly more active during its initial 3 to 4 weeks until it becomes part of its mother's group. The attraction between two like-aged fawns can be so close that they seem like twins, but blackbuck do not twin. Bucks mature physically at 2 to 2.5 years and can become territorial thereafter. Weeks or months as a territorial male alternate with recovery periods as a bachelor.

Mungall also documented reproductive behavior (35). Mating and fawning go on at all seasons with peaks likely twice a year. The 5-month gestation period plus the 1-month interval typical before rebreeding put fawning on a predictable 6-month cycle for two fawns a year. Female sexual maturity comes as early as 8 months, but first fawning usually comes as the doe approaches 2. A male can try to breed at 18 months but ordinarily has too much bachelor competition to succeed until becoming territorial. Life span can normally reach 10 to 12 years, the limit being about 15.

The main health problem for blackbuck is the occasional winter with repeated ice and snow. These tropical animals start to die if their natural forage is cut off for 2 or 3 days. Supplemental feeding as a hedge against winterkills needs to be started in summer and continued until spring green-up. Supplemental feeding has to be planned carefully, however, because blackbuck are poor competitors when faced with a multitude of other species at feeders. Parasites that particularly affect blackbuck are the North American deer fluke, which is a particular difficulty in wetter areas. Coccidiosis kills many immature blackbuck and is worst where blackbuck density is high (35).

The blackbuck's preference for short to midlength grasses complements its use of open plains habitat (36). Browsing augments grazing, and forbs are eaten when available. Nevertheless, blackbuck are so flexible in their diet that they can switch to browse or even forbs as a staple when the other forage classes decline in quantity and quality (6). A few seasonal items like rainlilies and mesquite beans are relished when they appear (35).

Fallow Deer

European fallow deer may have been the first of the hoofed wildlife species introduced in North America. George Washington imported the first fallow deer for his Mount Vernon estate in Virginia (37). Assumed to have originated in the Mediterranean region of southern Europe and Asia Minor, fallow deer have been traded since the time of the ancient Phoenicians and Romans (38). They occur in all inhabited continents and are the most widely distributed deer in the world (25, 39, 38). Through generations as a semi-tame park deer, several color forms have developed (38). Most common in North America are the white (i.e., born tan, dark eyes), black (i.e., two-tone chocolate color), and menil (i.e., light tan coat with white spots). The common-colored fallow (i.e., white spots on a rusty coat, Figure 33–4) so characteristic in Europe are less common in North America and are the most easily confused with other species of deer. To distinguish fallow deer from axis or sika deer, look for the prominent Adam's apple.

A special attraction of fallow is the large, palmate antlers. Hungary, a consistent producer of large antlers, has had trophies averaging more than 71 centimeters long with palms more than 18 centimeters across (38). In North America, antlers of 58 centimeters or better with palm widths of at least 9 centimeters are desirable (37). Bucks weigh about 79 to 102 kilograms in good condition but lose about 9 to 22 kilograms during the rut (6). Females weigh roughly 36 to 41 kilograms.

Breeding lines incorporating genes of the larger but rare Persian fallow deer are being developed on North American deer farms. Using Persian fallow semen on European fallow does is said to add 45 to 68 kilograms to carcass size. Both races are cold hardy over a wide range of temperate zone conditions, so fallow are a species of choice in northern regions of the United States and into Canada. These farms join a long tradition of fallow farms in Europe where venison is traditional autumn fare.

Figure 33–4 Fallow deer are the most widely distributed deer in the world, having found utility for deer farming and ranching. (Photograph from New York by J. von Kerckerinck zur Borg, courtesy of Lucky Star Ranch.)

Example Exotics

Chapman and Chapman describe the many aspects of reproduction in fallow deer (38). Fallow are in breeding condition from September to February. Bachelor herds break up in preparation for the breeding season. Bucks migrate back to their traditional rutting grounds, occupied all year by the females and young. Having made a scrape in or near the woods, a resident buck urinates and scoops mud onto himself, strengthening his pungent odor. By rut's peak in October, about half of mature males have a "stand." Most fawns are born between late May and the end of June after the 7.5 month gestation period. As pregnant does go off into cover to fawn, the big female groups split into small groups of half a dozen or fewer. Single fawns are the norm. For the first few weeks, the fawn lies hidden between the mother's nursing visits. By two months, fawns are with their mothers in the female herds. Females mature sexually at 16 months. Theoretically, bucks are capable of breeding at 14 months, but are inhibited for the next few years by older, higher ranking males. Captive fallow may live 11 to 15 years, and the record is 25 (25).

Fallow are great wanderers, probably a reflection of their forest-edge habitat. Mungall and Sheffield discuss the sometimes-conflicting food habits data for these adaptable deer (6). They eat large amounts of grass and browse. Forbs are also heavily used, sometimes even more than grass. During drought in Texas, fallow are prone to eating, and to becoming addicted to, fruits of the prickly pear cactus. Unfortunately, the spines in quantity mean a slow, painful death for fallow (40).

Competition can limit fallow. All fallow, except hard-antler bucks, are timid at feeders. Only white-tailed deer get less food from feeders when present in a crowd of mixed species composition. Fallow rely more on browse than axis and sika and experienced a die-off when browse amounts were insufficient for the white-tailed deer and fallow together in a 39-hectare enclosure (12). Forb use can surpass that of white-tailed deer, and the browse selections made by fallow are important white-tailed deer foods (32).

Ibex

"Ibex" as a North American exotic usually refers to the Iranian "ibex" (Figure 33–5), which really is not an ibex at all (41). Also called Persian ibex, pasang, bezoar goat, or wild goat, this is the progenitor of the domestic goat. Thus, it is hardly surprising that Iranian ibex hybridize readily with domestic goats. Breeding to a purebred sire, ranchers have capitalized on this facility to develop huntable herds, circumventing the difficulty in obtaining purebred females (6). True ibex, in the form of Siberian ibex, also have been tried as range exotics and as a male put with domestic nannies. Reproduction has been low in both cases, so Texas crossbreeding trials were terminated in favor of the quicker Iranian results (6). In New Mexico, Siberian ibex released in the Canadian River Canyon languished compared to the Iranian ibex expanding from the Florida Mountains (42, 43).

Both Iranian ibex and true ibex are muscular golden to brown to gray goat forms, 45 to 90 kilograms, with a typical goat beard. While the females, which are smaller, have short, crescent horns about 15 to 38 centimeters long, males have huge cartwheels of up to about 150 centimeters. Iranian ibex have a ragged keel on the front horn surface whereas true ibex have knobs. Hybrids tend to have larger horn bases so their profile tends to look more hook-shaped.

The hybrids have often been marketed as "YO ibex" after the YO Ranch in Texas that first developed these crossbreeding programs. Promotion for hunting has been so successful that "ibex" in Texas are of very mixed lineage with no conservation value. Meat of the malodorous males is unappealing, so there is no game farming.

The following focuses on Iranian ibex and its hybrids as given in Mungall and Sheffield (6). Ibex are highly gregarious, herds of 20 to 30 being likely and more where good feeding brings groups together. Males banding together in large groups stay apart from the somewhat smaller groups of females, subadults, and young until the rut. Scent and nuzzling behavior from younger males during the weeks preceding the rut help

Figure 33–5 The Iranian ibex is adapted to mountainous terrain and readily interbreeds with domestic goats. (Photograph from New Mexico, courtesy of New Mexico Department of Game and Fish.)

bring females into breeding condition. Older males then take over and separate estrous females from the flock. In the mixed herds of Texas, ibex are fall breeders, kids usually arriving during February through March. In native habitat, rainfall times rut and birth seasons.

The expectant mother goes off alone soon before parturition, and her kid or kids lie apart for a few days before becoming their mother's inseparable companions. Iranian mothers band together with other mothers, the older kids forming a subunit as they play with each other. Both sexes mature sexually at about 18 months, but males need the physical and social ability of a 5-year-old to compete successfully at mating time (44, 41). With an admixture of domestic goat, females can breed as early as 9 months and can raise twins, triplets, and occasionally even quadruplets. Gestation lasts 5 to 5.5 months. Ibex and their hybrids in exotics programs suffer from numerous parasites, ear mites being particularly bothersome. Estimated ranch and native life spans fall between 10 and 13 years.

Ibex are mountain animals able to withstand cold weather and extreme variations between heat and cold. Iranian ibex eat mainly grasses until these die back in the dry season. Then they eat almost any vegetation, especially anything green. To get green twigs and fleshy leaves, they rear up against tree trunks or climb out onto branches. Both fresh leaves and fallen ones are consumed, and forbs are a welcome supplement. Exotic ibex and their hybrids use steep bluffs and brush-studded hillsides.

Mouflon

The European mouflon from the islands of Corsica and Sardinia is the smallest of the wild sheep and one of the most colorful. Black and white markings show brightly against the brown background of coarse hair, rather than wool, which forms the outer coat. Besides the characteristic white socks and belly, older males develop a white saddle patch in winter (Figure 33–6). Rams and ewes weigh as much as 40 to 55 kilograms and about 35 kilograms, respectively. Some native ewes, mainly on Corsica, grow short, thin horns up to 18 centimeters in length (45). Exotic ewes in Texas are hornless. Mouflon rams grow 50 to 74 centimeter horns with broomed ends and rarely go above 94 centimeters. Typical horns make a three-quarter curl and point in near the tips. Horns that point out bring suspicions of hybridization with domestic goats.

Example Exotics

Figure 33–6 The mouflon, from the islands of Corsica and Sardinia, is one of the smallest, most colorful of wild sheep. (Figure 33–6 reprinted from *Applied Animal Behaviour Science*, vol. 29, B.E. McClelland, Courtship and agonistic behavior in mouflon sheep, p. 68, 1991, with permission from Elsevier Science.)

Breeders have often added domestic blood to increase horn size for the hunting market (6). Mouflon cross readily with any domestic sheep. As a result, all different gradations of horns, hair type, and coloration are found. Considering the facility of this interbreeding, some taxonomists contend that native mouflon may be the remnant of an ancient domestication rather than a truly wild race (46). However, mouflon are usually considered to be ancestral to today's domestic breeds. Many of the Texas "mouflon" with black bellies and black horns are results of breeding with Barbados sheep. The YO Ranch in Texas popularized these as "Corsican" sheep. Other marketing innovations include the white Corsican strain with the confusing nickname "Texas Dall" and a black variety advertised as "Hawaiian black sheep."

As Pfeffer documented for native mouflon on Corsica, large flocks of mouflon form during winter rather than during the rut (45). The males leave in spring and form all-male groups while the females and young go off in small parties. Then the sheep scatter for the fall rut. Single rams, or occasionally two or three together, wander from one cluster of females to another checking the ewes.

Gestation takes 5 months. Single births are the usual situation, so frequent twinning is assumed to indicate hybridization with other sheep. Females mature sexually as early as 7 months, but most wild mouflon are 3 years before having their first lamb. After staying near her hidden newborn for about 3 days, the mother rejoins the flock with her lamb at heel.

Mungall and Sheffield summarize information on maturation, breeding, and longevity (6). Males achieve sexual maturity at 1.5, are 2.5 before they can dominate a ewe sufficiently to mate, and are 5.5 before they gain the experience and restraint for consistent success. Both sexes are active producers at least until 9 or 10 years, about the limit in the wild. The captive longevity record is 19 years.

Mouflon are more parasite resistant than is usual for sheep (25). Native mouflon carry common sheep flukes but appear less affected than their domestic relatives.

Mouflon eat most of the plant species they come across, especially if green and tender, but grass is the staple (29). Mouflon seem able to include bits of poisonous plants in their diet as the tight flocks feed across a pasture (47). Mouflon can live mainly on browse in regions where grass is largely unavailable or where grass declines seasonally in quantity and quality (45, 29). Where there is a high browse line, mouflon accept fallen leaves (45). Both mouflon and white-tailed deer have a fondness for forbs (29).

Nilgai Antelope

Although nilgai are Asia's largest antelope, they carry some of the smallest horns. Sharp cones of 15 to 20 centimeters are typical. Occasionally, a female grows horns. Another contrast peculiar to this native of the Indian subcontinent is that, like the Indian blackbuck antelope, all nilgai are born brown, but males turn progressively darker as they mature (Figure 33–7). Steel gray is characteristic, with black suggesting older males in prime breeding condition (48). High shoulders give nilgai a sloping backline. Bulls weigh 110 to 290 kilograms up to a maximum of 306. Cows weigh 110 to 213.

Nilgai have excellent meat (49, 9). With their large size and frequent rate of twins and even triplets, nilgai would seem an obvious choice for meat production. However, their aggressive nature (e.g., bulls in small pastures will attack animals of other species, including humans) limits use to large ranches with animals that can be harvested from the open range. "South Texas antelope" is a diversification of the product line for exotics meats, but game farming is unlikely.

Sheffield and others included social dynamics in their study of nilgai in South Texas (48). The nilgai's big dung piles are found in dry, open forest and scrub. Adept at sheltering even in small patches of brush until time to slip out for grazing, nilgai groups remain small. Bulls and cows stay separate most of the year. Mature males are solitary or wander two to five together. Young males form bachelor groups of their own. Lone bulls enter cow–calf groups briefly for breeding. As births peak, lone females become common. For nearly a month, the young lie hidden between nursings, but then join cow–calf groups of up to 10. Calving seems to peak just before or during seasonal rains, but breeding groups and newborns are seen at any time.

Figure 33–7 Nilgai are Asia's largest antelope, but they carry some of the smallest horns. (Photograph of nilgai male in Texas by E. Cary Mungall, courtesy of International Wildlife Park.)

Example Exotics

Mungall and Sheffield condense reproductive and survivorship data (6). Gestation lasts about 8.25 months. Half the births for fully adult cows can be twins, and some are triplets. Well-nourished females can breed at about a year and a half, but few give birth before they are 3. Males mature sexually by 2.5 years at most but take 4 years to become fully adult and competitive. Maximum life span is about 12 to 13 years for native nilgai or ranch exotics and about 22 in zoos.

Nilgai from the large population in South Texas are impressively disease free (48). No sign of vector-borne disease has been found. Neither external nor internal parasites have occurred in heavy loads, although young animals tended to carry a greater burden of the nematodes that are the nilgai's principal internal parasites.

Winterkill can be a problem for nilgai. These tropical antelope start to die within days of prolonged snow or ice cutting off the food supply. However, their high reproductive potential coupled with immigration from adjoining areas rebuild populations quickly (48).

Food habits are summarized in Sheffield and others (48). Nilgai prefer grass but also need substantial amounts of browse or other supplements to gain a more nutritious diet. During spring when grasses are fresh and forbs are available, browse can almost drop out of the diet. Flowers, fruits, acorns, and mesquite beans may reach a larger proportion of the diet than leaves.

A few key aspects of diet overlap can create competition with other ungulates during drought and seasonal lows in grass quantity and quality (48). Ordinarily, food habits differences allow nilgai and white-tailed deer to do well on the same range. Nilgai prefer grass while whitetails prefer forbs. Both nilgai and whitetails eat leaves, but nilgai use of mast-like mesquite beans and acorns is usually more of a problem for deer than availability of leaves. In limited space, the taller nilgai can outcompete deer for their staple browse at the same time that the nilgai are making heavy use of the forbs on which whitetails also depend. With abundant grasses, nilgai and cattle cohabit compatibly. Cattle add forbs and browse to grazing during South Texas summers but generally eat less selectively of both plant kinds and parts.

Red Deer

The European red deer, reddish brown in summer and grayer in winter, is the principal large deer of Europe and a major deer farm species (Figure 33–8). Typical antlers curve inward with five or six points trending toward the front, including a bez tine just above

Figure 33–8 Red deer are farmed extensively in Europe and New Zealand, and to a lesser extent in North America. (Photograph of red deer male in eastern Canada by R. DesJardins, courtesy of Ferme de cerfs de l'Estrie Inc.)

the brow point. In prime stags, the forked top becomes a crown. In eastern Europe, stags can weigh up to about 300 kilograms with antlers carrying 10 or more points on a side and growing as long as 124.5 centimeters. In contrast, good Scottish stags can weigh 126 kilograms with simpler antlers of 89 centimeters (39, 50). Exotic stags in Texas grow to about 272 kilograms with five to seven point antlers of 66 to more than 100 centimeters (6). Hinds weigh 74 to 100 kilograms. Forage quality governs size.

Their large body and antler size combined with tasty venison have made red deer a traditional fall favorite in Europe. New Zealand capitalized on rampant populations of free-ranging exotic red deer and other antlered species to develop a deer farming industry. North America responded with similar experiments. Quality red deer breeding stock has been difficult to get because purebreds are not so numerous as for the common exotics and because regulations on importation are strict to prevent introduction of diseases. Red deer numbers have remained low in Texas because they do poorly in the heat of southwestern ranges and many of the early red deer released bred with the closely related North American elk also being stocked (6). Although unintended, this hybridization was so widespread that it became hard to find a pure red deer population in Texas. The elk-red deer hybridization situation became so confused that in 1981 the Texas legislature took all elk east of the Pecos River, east of their traditional native habitat, out of state regulation and designated them privately owned. This simplified management of the mixed herds because they are concentrated east of this division.

In their wild condition, red deer are true harem holders. Clutton-Brock and others detail reproductive dynamics from a long-term study in Scotland (50). Most of the year, stags and hinds live in separate groups. Toward fall, the male groups fragment, and stags head for their traditional rutting grounds. A stag holds 10 or more hinds within his territory. Males first become fertile as young as 16 months but need to be at least 5 years old for the weight and experience to be successful harem masters. Stags compete effectively for about 5 years. Females need a body mass of 57 to 77 kilograms to become sexually active, which is achieved between 16 and 40 months, depending on forage conditions.

Nearly three-quarters of a population may conceive during 2 weeks in the middle of the rut (50). This concentrates births, nearly always single, in late spring after a 7.5- to 8.5-month gestation. Survival is poor for early and late calves (50). The calf lies out for its first 1 or 2 weeks before following its mother back to her group. Culling strategies need to account for the reluctance of hinds to relocate out of their maternal home range.

Few wild red deer live as long as 12 years although up to 26 has been reported (51, 50). Keeping well-nourished farm deer on green pasture, which is less abrasive on the teeth, can extend the expected life span to 14 or 15 years (51, 6). As for elk, watery wallows should allow red deer to maintain acceptable body temperature in hot North American climates and so to thrive in a wider variety of North American habitats (6). Wallowing also offers some relief from the flies that plague red deer.

Open forests with ample understory are the favored habitat of red deer. Where available, the red deer diet staple is browse (51). Where browse is scarce, grasses can rise to more than 90% of the diet during part of the year, an important facility for deer farming (51).

Sika Deer

Sika deer are to the Far East as white-tailed deer are to North America (Figure 33–9). The common deer of the countryside, sika inhabited forests all the way from the plains into the mountains. The forms most familiar in North America are the Formosan sika and the series of other island forms known collectively as Japanese sika. Formosan and Japanese sika have interbred, resulting in variety within herds ranging from rusty-coated sika with off-white flecks to black flecks on brown to no spots at all. Mainland forms succeed each other through China and Korea into Manchuria, with the northernmost largest: Manchurian sika and Dybowski's deer (39). Males grow a shaggy winter neck ruff. Distinguishing feature for all these sika forms is the powder-puff rump patch of long, white hairs that flashes out of its inconspicuous covering when sika get excited.

Figure 33–9 Sika deer are aggressive among themselves and toward other species and survive well even on overgrazed range. (Photograph of male sika deer in Virginia by Brian Eyler, courtesy of Maryland Cooperative Fish and Wildlife Research Unit and Chincoteague National Wildlife Refuge.)

The biggest sika, Dybowski's males and females weigh 70 to 110 kilograms and 45 to 50 kilograms, respectively. The male's antlers can grow as long as 87 centimeters. The common sika exotics weigh 45 to 80 kilograms for males and grow small 28 to 48 centimeter antlers. Antler asymmetry and breakage in fights are both frequent. Native Formosan and Japanese antler records slightly exceed 48 and 71 centimeters, respectively. Brow tine, intermediate tine, and top fork is the basic antler pattern, with a fifth and occasionally a sixth point added in large bucks.

Large-scale sika farming has not developed in North America because sika meat can be stronger tasting and less tender than axis venison. However, fat distributed throughout the muscle tissue produces juicier meat than for red deer that deposit fat on the surface of the muscle. The discovery that sika will breed with elk has led to development of the "American silk" for the deer farmer who wants to get going fast by stretching the investment dollar with less expensive breeding stock than either elk or red deer, wants fast weight gains, and does not mind juggling parentage percentages (6).

Kiddie observed sika deer extensively as exotics in New Zealand (52). He concludes that sika often travel in family parties of doe, fawn, and yearling. For 6 to 8 weeks starting in September, bucks set up territories and assemble harems. A prime buck can collect as many as 12 females. Two or three smaller bucks mixed in as "helpers" roar and investigate challenges and leave the master freer to attend to breeding. The territorial buck digs a series of scrapes surrounding his territory, urinates in them, and whines as he wallows in his pits.

Mungall and Sheffield give reproductive and longevity statistics (6). The main birth season comes from April to May, births continuing into August, after a gestation of roughly eight months. Twins are rare. For the first 2 or 3 weeks, the fawn lies quietly in the forest between nursings. Life span under zoo conditions can go 15 to 18 years, with more than 25 possible.

Sika are exceptionally hardy deer. Fat deposited under the skin lets them withstand winter weather better than tropical deer. But more than this, they are aggressive with other species, and they can survive on overgrazed ranges (52). They can live mainly on grass, mainly on browse, or on first one and then the other with massive shifts from one season to the next (29, 26). These shifts to the most available vegetation that provides adequate nutrition rest plants heavily used during a limited period, a successful survival strategy (32, 52). When six adults each of sika and white-tailed deer were put into a 39-hectare Texas enclosure and left, the sika increased to 62 while the whitetails died out (12).

Wild Boar and Feral Hog

The wild boar of Eurasia and North Africa is different from all other hoofed exotics (6). It is an omnivore instead of a herbivore. It is destructive and predatory. It can become secretive and dangerous when hunted. Also, it has interbred so readily with its derivative, the domestic hog, that it is arguable whether the wild form imported into North America even exists any longer as a free-ranging animal. Some regions have a higher percentage of animals showing wild-type characteristics, including the horizontal streaks and spots in the juvenile coat, but, generally, domestic hog morphology is most prevalent (Figure 33–10).

Domestic swine revert to a wild existence particularly easily, and well-established feral hog populations with no exposure to any of the wild boar released in North America are still noted for deviation from the domestic condition. Black is the predominant coat color, and the feral hogs in these established populations tend to have coarser hair, lighter, more streamlined bodies, and larger tusks (53, 54). This makes speculation about prevalence of true wild boar influence in today's hybrid populations difficult to verify.

Wild boar, probably most from European breeding lines, were introduced starting as early as the 1890s and continuing at least into the 1930s (55, 53, 6). States as scattered as North Carolina, Texas, and California have all had releases. These animals readily interbred with feral hogs. Their descendants are part of a vast feral hog population common to moist habitats throughout the southern United States. The state of New Hampshire in the northern United States also has a growing population of wild

Figure 33–10 This sow and her piglets are from a hybrid wild boar-feral hog population in Tennessee. (Photograph courtesy of Tennessee Wildlife Resources Agency.)

Example Exotics

boar. These animals are largely concentrated within a hunting preserve and, for many years, were more isolated from domestic influences (55, 56).

In the following discussion, "wild boar" will be used where the information is from native wild boar populations and "wild hog" will be used for either feral hogs or possible hybrids free ranging in North America. Except for a few traits, like the particular aggressiveness of purebred wild boar and the greater litter size of feral hogs compared to purebred wild boar, wild boar, feral hog, and their hybrids all share great similarity in the way they exploit their environment (53, 54, 57). Thus, the behavior and food habits information can be taken as representative for all three forms. The wild type is stressed as the foundation from which variations have arisen.

The purebred wild boar is heavy shouldered and short legged with dark, bristly hair and sharp tusks. Its tapered snout ends in a flexible disk for rooting. Big boars range from 113 to 136 kilograms. North of Vladivostok, Russian boars can grow to be monsters of 350 kilograms. Winter-supplemented park specimens in New Hampshire, probably purebred, average 36 to 40 kilograms with a maximum male weight of 123 (55). Weights in the patently hybrid population in North Carolina and Tennessee average about 103 kilograms for free-ranging animals (53).

A wild hog carcass can make a memorable barbecue, but other uses get mixed reactions. Some ranches and parks shoot wild hogs on sight or mount costly control operations. Other properties derive revenue by selling hunting rights or by live-trapping for sale to hunting areas. Sometimes, sport hunting is an integral part of control efforts. Feral hogs are game animals in at least parts of six states: California, Florida, Hawaii, North Carolina, Tennessee, and West Virginia. The first three register the largest harvests (54). Estimates range from 10,000 feral hogs taken annually in Hawaii to a record high for Florida of 84,100 in 1974 (54). It takes a special sort of hunter to enjoy chasing a pack of hounds to face a quarry that may charge out with slashing tusks (53). There is serious risk. The way disturbance such as hunting induces wild hogs to become nocturnal in their habits adds to the challenge. Nevertheless, parks and private areas within regions where hog hunting is a familiar activity often receive an enthusiastic response when offering hog hunting opportunities. Hunting areas lacking abundant escape cover may require restocking to support demand, especially where hunting with dogs is permitted (54, 58).

In their undisturbed daily routine, big boars remain apart even for bedding while as many as 15 sows and youngsters may pile on top of each other to rest (47). Wild boar "sounders" are made up of one or more sows, their present young, and their previous year's young of at least a year old. Sows with piglets band together, but they run off their older offspring until the piglets gain in coordination and independence. Male yearlings finally separate, usually of their own accord, and go off to join other growing males. In November to January, boars collect harems of three to as many as eight sows.

With adequate food availability, wild boar males and females mature sexually as early as 8 months of age. Gestation lasts 4 to 4.5 months, first-time mothers carrying for the shorter time and producing fewer piglets than experienced mothers (47). In North Carolina, wild hog litters usually range from three to eight piglets, but four or five is most common (53). Mean fetal counts for feral hogs in the United States ranged from 5.1 to 8.4 before adjustment for stillbirths and similar losses; 12 was the indicated maximum litter size for feral hogs, whereas domestic swine can give birth to twice as many young. When feeding is plentiful, older sows often have two litters a year. Most sows, both as introduced animals and in native habitat, farrow between March and May. As studied in the United States, feral hogs can and do reproduce at any time of year. However, there tend to be two birth peaks, indications being that different sows contribute to each of the two peaks (54).

For birth, the sow relocates to a quiet spot in dense cover and makes a nest. Her piglets stay piled in this for 1 to 3 weeks before following her on foraging expeditions (54). A sow remains in prime condition from 5 until 10 years old. Male prime starts at 8. Maximum life span expected for the wild boar, either free ranging or in captivity, is 15, although rare individuals live to about 20 (25, 6). In contrast, wild hog populations develop a very young age structure in areas with heavy hunting pressure. Few individuals in these populations live past 2 or 3 (59, 54).

Like feral hogs in general, wild hogs in a hybrid population on the Texas Gulf Coast had large numbers of parasites such as swine kidney worm, lungworms, roundworms, hookworms, and stomach worms (54, 59). Whipworms are also frequently noted (60). Directly or indirectly, internal parasites contribute to mortality although opinions conflict as to extent. Infestation with ticks and lice varies with location, year, and season and seems to have little impact on the health of the host (60, 59, 54). Tests for trichinosis in Texas were negative (59). Pseudorabies and swine brucellosis are potential problems for feral hogs as well as for domestic pigs in all swine areas, with transfer of infection to or from domestics being a major concern (61).

Wild boar are hardy generalists. As long as they have a little cover, plus places for their mud baths, they will live in almost any kind of habitat. Lacking brush, a gully having a moist corner with summer shade and a dry corner with winter sun is sufficient. When their needs for food and cover are met locally, wild boar are sedentary. Otherwise, they travel widely in search of food.

Springer along the Texas coast and Henry and Conley in the southern Appalachian mountains of Tennessee are among the investigators who have studied food habits for hybrid wild boar-feral hog populations (59, 62). Grasses and forbs are important forage, but the hog's simple stomach is not as efficient at assimilating green plants as is a ruminant's multichambered digestive system. Acorns and hickory nuts are the top two foods in North Carolina and Tennessee. Fruits like grapes also can comprise more than half the diet. Crops like sorghum are such favorites that a hog will travel 6 miles for feasting. Predator or scavenger, wild hogs take small animals like mice, snakes, grasshoppers, fish, and bird nests with contents. Digging with tusks, wild hogs devour quantities of earthworms and plant roots.

Feeding competition with native species like white-tailed deer, wild turkey, black bears, squirrels, chipmunks, and the pig-like collared peccary is minimal except for acorns and assorted nuts, grapes, and mushrooms (55, 62, 59). High dietary overlap with collared peccary is balanced out by different habitat preferences as long as hog densities remain low (63).

Faced with wild hog populations, managers have to assess advantages and disadvantages before developing management guidelines. First is rooting activity. It can loosen the soil, accelerate erosion, set back plant succession, favor forb production, expose food for birds, bury tree seeds, and reduce populations of tree parasites destructive to timber production. Rooting also reduces populations of earthworms that aerate soil less violently, favors invasion by exotic plants, destroys crops, roads, and fences, and can worry outdoor recreation users. General feeding and trampling can further damage crops and local flora and fauna. The impact on endangered species is apt to be greatest on islands and has been an acute concern in Hawaii where so many of the native species are endemics (57, 64). Wild hogs eat snakes, which favors ground-nesting birds, but wild hogs will also prey on a nest if they happen across one. Studying with dummy nests led Henry to conclude that wild hogs replace nest destruction that would occur from other predators if no wild hogs were present (65). Jones considered the wild boar more beneficial than harmful to ground-bird populations because grouse were plentiful in the vicinity of the original North Carolina introduction and because turkeys were most numerous in the areas used by these wild boar and their descendants (53). Potential competition with other wildlife for mast has already been mentioned. Wild hogs remain a reservoir for diseases that can infect livestock or even humans. Sport hunting creates economic opportunities as well as helping in control programs. Control is often a matter of keeping numbers within carrying capacity rather than eliminating the wild hog population. Thus, hog control is a continuing program in many areas in spite of the drain on labor and support budgets.

CONCLUSION

To sum up using the sentiments of noted ungulate specialist Fritz R. Walther, do not overstock, do not hybridize, and do not introduce just anything (66). Controlling predators, using management practices to reduce parasites, and furnishing supplemental feed can result very easily in populations exceeding carrying capacity. Managers do well to keep in mind that maintaining animals year-round from a trough is not the goal of game ranching. Some hybridization has been intentional, but much has not. This is damaging to everyone dealing with exotics because it alienates segments of the scientific and zoo communities. Members of the latter groups tend to distrust anyone in the exotics industry because scientists and zoo professionals see the frequency of hybrids as indicative of a failure to appreciate the integrity of species. Without committed stewardship, whole populations can lose their conservation value. As for introductions, species need to be evaluated carefully before being released in a new environment. Some species are not fit to thrive in the habitat available. In other cases, the well-being of endemic species is jeopardized by additions.

To justify considering an introduction, at least one of the following should be true: The candidate species is rare, or at least seriously declining, in its native habitat, there is no equivalent endemic so that a truly "exotic" species lays claim on humankind's aesthetic sense, or the species provides unmatched qualities for human use (e.g., spectacular trophies for gun or camera, unique attractiveness for tourist schemes, or superior market qualities for farming).

SUMMARY

Exotics are animals or plants living outside the normal distributional area of their species. Whether intended or unintended, human intervention is implied. Presently in North America, there are more than 250,000 hoofed exotics, the types of exotics discussed here. Whether within fences, free ranging, or, more recently, being farmed, numbers of exotics are continuing to grow. This creates a serious responsibility for managers to safeguard the productivity of land and the native species that depend on it.

Motivations for keeping exotics fall into four general categories: enjoyment or curiosity, fee hunting or other revenue, preservation or reintroduction, and providing living space for surplus animals. Negative considerations include concerns dealing with ecological relationships, obligation toward local heritage, difficulty of controlling exotics, disease risks, hybridization with native wildlife, and damage potential to resources such as crops. Reflecting this multitude of aspects, policies regarding exotics in the different states and provinces vary considerably.

Ten of the most numerous exotics are discussed: aoudad or Barbary sheep, axis deer, fallow deer, ibex, Indian blackbuck antelope, mouflon, nilgai antelope, red deer, sika deer, and wild boar. These illustrate the variety of species, factors for success, economic options, patterns of forage use, and potential for competition. As the histories of these species demonstrate, the results of releases have been too consequential and too long-lasting, for further species to be added without valid justification.

LITERATURE CITED

1. Haigh, J. C., and R. J. Hudson. 1993. *Farming Wapiti and Red Deer.* Mosby, Chicago.
2. Anonymous. 1996. Exotic hoofstock survey. Texas Agricultural Statistics Service and the Exotic Wildlife Association, Ingram, TX.
3. Species Survival Commission, the Commission on Ecology, and the Commission on Environmental Policy, Law and Administration. 1987. The IUCN position statement on translocation of living organisms: Introductions, re-introductions and re-stocking. International Union for Conservation of Nature and Natural Resources, Gland, Switzerland.

4. MANN, W. M. 1934. *Wild Animals In and Out of the Zoo,* VI, Smithsonian Scientific Series, Smithsonian Institution Series, New York.

5. DECKER, E. 1978. Exotics. Pages 249–256 *in* J. L. Schmidt and D. L. Gilbert, eds., *Big Game of North America: Ecology and Management.* Stackpole Books, Harrisburg, PA.

6. MUNGALL, E. C., and W. J. SHEFFIELD. 1994. *Exotics on the range: The Texas example.* Texas A&M University Press, College Station.

7. RAMSEY, C. W. 1972. Exotic game ranching . . . expensive, experimental. *Texas Agricultural Progress* 18:9–12.

8. SCHREINER, C., III. 1968. Uses of exotic animals in a commercial hunting program. Pages 13–26 *in Symposium: Introduction of Exotic Animals: Ecological and Socioeconomic Considerations.* Caesar Kleberg Research Program in Wildlife Ecology, Texas A&M University, College Station.

9. SHEFFIELD, W. J. 1985. Exotics' meat makes health news. *Exotic Wildlife Association Newsletter* 17(10):2–4.

10. CRAIGHEAD, F. C., and R. F. DASMANN. 1966. Exotic big game on public lands. Bureau of Land Management, U.S. Department of the Interior, Washington, DC.

11. GEIST, V. 1988. How markets in wildlife meats and parts, and the sale of hunting privileges, jeopardize wildlife conservation. *Conservation Biology* 2:15–26.

12. FELDHAMER, G. A., and W. E. ARMSTRONG. 1993. Interspecific competition between four exotic species and native artiodactyls in the United States. *Transactions North American Wildlife and Natural Resources Conference* 58:468–478.

13. DEMARAIS, S., J. T. BACCUS, and M. S. TRAWEEK, JR. 1998. Nonindigenous ungulates in Texas: Long-term population trends and possible competitive mechanisms. *Transactions North American Wildlife and Natural Resources Conference* 63:49–55.

14. RICHARDSON, M. L., and S. DEMARAIS. 1992. Parasites and condition of coexisting populations of white-tailed and exotic deer in south-central Texas. *Journal of Wildlife Disease* 28:485–489.

15. HARRIS, R. A., and K. R. DUFF. 1970. *Wild Deer in Britain.* Taplinger Publishing Company, New York.

16. THORNTON, J. E., T. J. GALVIN, R. R. BELL, and C. W. RAMSEY. 1973. Parasites of the blackbuck antelope (*Antilope cervicapra*) in Texas. *Journal of Wildlife Disease* 9:160–162.

17. MILLER, M. W., and E. T. THORNE. 1993. Captive cervids as potential sources of disease for North America's wild cervid populations: Avenues, implications, and preventive management. *Transactions North American Wildlife and Natural Resources Conference* 58:460–467.

18. ANONYMOUS. 1995. Tuberculosis testing program started for exotic Cervidae. *Animal Health Matters* Fall:7.

19. HARMEL, D. E. 1980. Statewide census of exotic big game animals. Job No. 21, Federal Aid Project No. W-109-R-3. Job Performance Report. Texas Parks and Wildlife Department, Austin.

20. STEVENSON, R. E. 1991. Big game farming in Alberta. Pages 516–517 *in* L. A. Renecker and R. J. Hudson, eds., *Wildlife Production: Conservation and Sustainable Development.* Agricultural and Forestry Experiment Station, University of Alaska, Fairbanks.

21. WHITE, R. J. 1987. *In* R. Valdez, ed., *Big Game Ranching in the United States.* Wild Sheep and Goat International, Mesilla, NM.

22. GEIST, V. 1971. *Mountain Sheep: A Study in Behavior and Evolution.* University of Chicago Press, Chicago.

23. SIMPSON, C. D., ed. 1980. *Proceedings of the Symposium on Ecology and Management of Barbary Sheep.* Department of Range and Wildlife Management, Texas Tech University, Lubbock.

24. OGREN, H. A. 1965. Barbary sheep. Bulletin No. 13. New Mexico Department of Game and Fish, Santa Fe.

25. CRANDALL, L. S. 1968. *The Management of Wild Mammals in Captivity.* University of Chicago Press, Chicago.

26. BUTTS, G. L., M. J. ANDEREGG, W. E. ARMSTRONG, D. E. HARMEL, C. W. RAMSEY, and S. H. SOROLA. 1982. Food habits of five exotic ungulates on Kerr Wildlife Management Area, Texas. Technical Series No. 30. Texas Parks and Wildlife Department, Austin.

27. ABLES, E. D., ed. 1977. *The Axis Deer in Texas.* Kleberg Studies in Natural Resources. College Station, Texas.

28. GRAF, W., and L. NICHOLS, JR. 1966. The axis deer in Hawaii. *Journal of the Bombay Natural History Society* 63:629–734.

29. KELLEY, J. A. 1970. Food habits of four exotic big-game animals on a Texas "Hill Country" ranch. Thesis. Texas A&I University, Kingsville.

30. BERWICK, S. W. 1974. The community of wild ruminants in the Gir Forest ecosystem, India. Dissertation. Yale University, New Haven, CT.

31. HENKE, S. E., S. DEMARAIS, and J. A. PFISTER. 1988. Digestive capacity and diets of white-tailed deer and exotic ruminants. *Journal of Wildlife Management* 52:595–598.

32. JACKLEY, J. J. 1991. Dietary overlap among axis, fallow, sika, and white-tailed deer in the Edwards Plateau Region of Texas. Thesis. Texas Tech University, Lubbock.

33. RAHMANI, A. R. 1991. Present distribution of the blackbuck Antilope cervicapra Linn. in India, with special emphasis on the lesser known populations. *Journal of the Bombay Natural History Society* 88:35–46.

34. TRAWEEK, M. S. 1995. Statewide census of exotic big game animals. Project No. 21, Federal Aid Project No. W-127-R-3, Performance Report. Texas Parks and Wildlife Department, Austin.

35. MUNGALL, E. C. 1978. *The Indian blackbuck antelope: A Texas view.* Kleberg Studies in Natural Resources. Texas A&M University, College Station.

36. RANJITSINH, M. K. 1989. *The Indian Blackbuck.* Natraj Publishers, Dehra Dun, India.

37. TEMPLE, T. B. 1976. *Records of Exotics,* I. Thompson B. Temple, Mountain Home, TX.

38. CHAPMAN, D., and N. CHAPMAN. 1975. *Fallow Deer: Their History, Distribution and Biology.* Terence Dalton, Lavenham, Suffolk, UK.

39. WHITEHEAD, G. K. 1972. *Deer of the World.* Constable & Company Ltd., London.

40. PRIOUR, D., and E. C. MUNGALL. 1985. Fallow deer addiction to prickly pear. *Exotic Wildlife Association Newsletter* 17:5–6.

41. VALDEZ, R. 1985. *Lords of the Pinnacles—Wild Goats of the World.* Wild Sheep and Goat International, Mesilla, NM.

42. WOOD, J. E., R. J. WHITE, and J. L. DURHAM. 1970. Investigations preliminary to the release of exotic ungulates in New Mexico. Bulletin No. 13. New Mexico Department of Game & Fish, Albuquerque.

43. MORRISON, B. 1987. New Mexico's exotic wildlife program: Its past, present and future. Pages 36–39 *in* T. B. Temple, ed., *Records of Exotics,* IV. Temple, Mountain Home, TX.

44. JAMSHEED, R. 1976. *Big Game Animals of Iran (Persia).*

45. PFEFFER, P. 1967. Le mouflon de Corse (*Ovis ammon musimon* Schreber, 1782); position systématique, écologie et éthologie comparées. *Mammalia,* Suppl. Vol. 31.

46. VALDEZ, R. 1982. *The Wild Sheep of the World.* Wild Sheep and Goat International, Mesilla, NM.

47. GRZIMEK, B. 1972. *Grzimek's Animal Life Encyclopedia, Mammals IV,* Vol. 13. Van Nostrand Reinhold Company, New York.

48. SHEFFIELD, W. J., B. A. FALL, and BENNETT A. BROWN. 1983. *The nilgai antelope in Texas.* Kleberg Studies in Natural Resources, Texas A&M University, College Station.

49. ABLES, E. D., Z. L. CARPENTER, L. QUARRIER, and W. J. SHEFFIELD. 1973. Carcass and meat characteristics of nilgai antelope. Leaflet B-1,130. Texas Agricultural Experiment Station, Texas A&M University, College Station.

50. CLUTTON-BROCK, T. H., F. E. GUINNESS, and S. D. ALBON. 1982. *Red Deer: Behavior and Ecology of Two Sexes.* University of Chicago Press, Chicago.

51. MITCHELL, B., B. W. STAINES, and D. WELCH. 1977. *Ecology of red deer: A research review relevant to their management in Scotland.* Institute of Terrestrial Ecology, Natural Environment Research Council, Cambridge, UK.

52. KIDDIE, D. G. 1962. The sika deer (*Cervus nippon*) in New Zealand. Information Series No. 44. New Zealand Forest Service, Wellington.

53. JONES, P. 1959. *The European Wild Boar in North Carolina.* North Carolina Wildlife Resources Commission, Raleigh.

54. SWEENEY, J. M., and J. R. SWEENEY. 1982. Feral hog: Sus scrofa. Pages 1099–1113 *in* J. A. Chapman and G. A. Feldhamer, eds., *Wild Mammals of North America: Biology, Management, and Economics.* Johns Hopkins University Press, Baltimore, MD.

55. SILVER, H. 1957. A history of New Hampshire game and furbearers, Survey Report No. 6. New Hampshire Fish and Game Department, Concord, New Hampshire.

56. HUGHES, R. 1991. Wild boar proves formidable foe. *Eagle Times,* August 18, pp. 1 & 6.

57. GRAVES, H. B. 1984. Behavior and ecology of wild and feral swine (*Sus scrofa*). *Journal of Animal Science* 58:482–492.

58. ANDERSON, S. J., and C. P. STONE. 1993. Snaring to control feral pigs *Sus scrofa* in a remote Hawaiian rain forest. *Biological Conservation* 63:195–201.

59. SPRINGER, M. D. 1975. Food habits of wild hogs on the Texas Gulf Coast. Thesis. Texas A&M University, College Station.

59. SPRINGER, M. D. 1977. Ecologic and economic aspects of wild hogs in Texas. Pages 37–46 *in* G. W. Wood, ed., *Research & Management of Wild Hog Populations: Proceedings of a Symposium.* Belle W. Baruch Forest Service Institute, Clemson University, Georgetown, SC.

60. HENRY, V. G. and R. H. CONLEY. 1970. Some parasites of European wild hogs in the southern Appalachians. *Journal of Wildlife Management* 34:913–917.

61. ANONYMOUS. 1992. Wild pigs: hidden danger for farmers and hunters. Agriculture Information Bulletin No. 620. U.S. Department of Agriculture Animal and Plant Health Inspection Service, Hyattsville, MD.

62. HENRY, V. G., and R. H. CONLEY. 1972. Fall foods of European wild hogs in the southern Appalachians. *Journal of Wildlife Management* 36:854–860.

63. ILSE, L. M. 1993. Resource partitioning in sympatric populations of feral hogs and collared peccaries in South Texas. Thesis. Texas A&M University–Kingsville.

64. ROYTE, E. 1995. On the brink: Hawaii's vanishing species. *National Geographic* 188:2–37.

65. HENRY, V. G. 1969. Predation on dummy nests of ground-nesting birds in the southern Appalachians. *Journal of Wildlife Management* 33:169–172.

66. WALTHER, F. R. 1997. Department of Wildlife and Fisheries, Texas A&M University, Retired, Personal communication.

67. WARING, G. H. 1997. Preliminary study of the behavior and ecology of axis deer on Maui, Hawaii. Hawaii Ecosystems at Risk (HEAR) project. http://www.hear.org/AlienSpeciesInHawaii/waringreports/axisdeer.htm.

68. TEER, J. G. 1991. Non-native large ungulates in North America. Pages 55–66 *in* L. A. Renecker and R. J. Hudson, eds. *Wildlife Production: Conservation and Sustainable Development.* Agricultural and Forestry Experiment Station, University of Alaska, Fairbanks.

69. SMITH, K. C., and A. PEISCHEL. 1991. Intensive wapiti production in the tropics: Hawaii. Pages 503–515 *in* L. A. Renecker and R. J. Hudson, eds., *Wildlife Production: Conservation and Sustainable Development.* Agricultural and Forestry Experiment Station, University of Alaska, Fairbanks.

70. KERCHERINCK, J. VON, ZUR BORG. 1987. *Deer Farming in North America: The Conquest of a New Frontier.* Phanter Press, Rhinebeck, NY.

71. BELDEN, R. C. 1992. Exotic ungulates in Florida. Study No. 7559, Federal No. W-41,XXXI, Final Project Report. Florida Game & Fresh Water Fish Commission, Gainesville. (Published 1994 without tables in *Proceedings Southeastern Fish and Wildlife Association.*)

Common and Scientific Names of Animals Mentioned in the Text

COMMON NAMES	SCIENTIFIC NAMES
Addax	*Addax nasomaculatus*
Agouti	*Dasyprocta aguti*
Alligator	*Alligator mississippiensis*
Alpaca	*Lama pacos*
Anteater	*Myrmecophagidae*
Anteater, giant	*Myrmecophaga jubata*
Antelope	*Antilocapra americana*
Antelope, blackbuck	*Antilope cervicapra*
Antelope, Indian blackbuck	*Antilope cervicapra*
Antelope, pronghorn	*Antilocapra americana*
Antelope, roan	*Hippotragus equinus*
Antelope, sable	*Hippotragus niger*
Antelope, saiga	*Saiga tatarica*
Aoudad	*Ammotragus lervia*
Argali	*Ovis Ammon*
Armadillo	*Dasypus novemcinctus*
Armadillo, giant	*Priodontes gigas*
Armadillo, nine banded	*Dasypus novemcinctos*
Bacteria	*Pasteurella* spp.
Bacteria, yersiniosis	*Yersinia pseudotuberculosis*
Bactrian	*Camelus bactrianus*
Badger	*Taxidea taxus*
Banteng	*Bos banteng*
Barasingha	*Cervus duvauceli*
Barbados	*Ovis aries*
Bear, American black	*Ursus americanus*
Bear, Asian black	*Ursus thibetanus*
Bear, black	*Ursus americanus*
Bear, brown	*Ursus arctos*
Bear, European brown	*Ursus arctos*
Bear, grizzly	*Ursus arctos horribilis*
Bear, polar	*Ursus maritimus*
Bear, sloth	*Melursus ursinus*
Bear, spectacled	*Tremarctos ornatus*
Bear, sun	*Helarctos malayanus*
Beaver	*Castor canadensis*
Bees	*Apis mellifera*
Beetles	*Coleoptera*
Beluga	*Delphinapterus leucas*

Bighorn	*Ovis canadensis*
Bighorn, California	*Ovis canadensis californiana*
Bighorn, desert	*Ovis canadensis cremnobates*
	Ovis canadensis mexicana
	Ovis canadensis nelsoni
	Ovis canadensis weemsi
Bighorn, Rocky mountain	*Ovis canadensis canadensis*
Bison	*Bison bison*
	Bison antiquus
	Bison occidentalis
Bison, European	*Bos [=Bison] bonasus*
Bison, North American	*Bos [=Bison] bison*
Bison, plains	*Bison b. bison*
Bison, wood	*Bison b. athabascae*
Blesbok	*Damaliscus dorcas philippsi*
Boar, European wild	*Sus scrofa ferus*
Bobcat	*Lynx rufus*
Bontebok	*Damaliscus dorcas dorcus*
Bovids	*Bovidae*
Buffalo	*Bison Bison*
Brocket, gray	*Mazama gouazoubira*
Buffalo, African	*Syncerus caffer*
Buffalo, African cape	*Syncerus caffer caffer*
Buffalo, water	*Bubalus bubalis*
Bushbuck	*Tragelaphus scriptus*
Camel	*Camelops hesternus*
Caprids	*Caprini*
Capybara	*Hydrochoerus hydrochaeris*
Caribou	*Rangifer tarandus*
Caribou, barren ground	*Rangifer arcticus*
Caribou, woodland	*Rangifer caribou*
Cattle	*Bos taurus*
Cervids	*Cervidae*
Chamois	*Rupicapra rupicapras*
Cheetah	*Acinonyx jubatus*
Chuckwalla	*Sauromalus obesus*
Coatimundi	*Nasua norica*
Copybora	*Hydrochoeris hydrochaeris*
Cougar	*Felis concolor*
Cow, domestic	*Bos taurus*
Coyote	*Canis latrans*
Dall, Texas (white Corsican)	*Ovis hybrids*
Deer, axis	*Axis axis*
Deer, black-tailed	*Odocoileus hemionus columbianus*
Deer, brocket	*Mazama americana*
Deer, burro	*Odocoileus hemionus eremicus*
Deer, California mule	*Odocoileus hemionus californicus*
Deer, Cedros Island mule	Odocoileus hemionus cerrosensis
Deer, Chinese water	*Hydropotes inermis*
Deer, Columbian black-tailed	*Odocoileus hemionus columbianus*
Deer, Columbian whitetail	*Odocoileus hemionus leucurus*
Deer, desert mule	*Odocoileus hemionus crooki*
Deer, Eld's	*Cervus eldi*
Deer, European fallow	*Dama dama dama*
Deer, fallow	*Dama dama*
Deer, Hog	*Axis porcinus*
Deer, Inyo mule	*Odocoileus hemionus inyoensis*
Deer, mule	*Odocoileus hemionus*
Deer, Peninsula mule	*Odocoileus hemionus peninsulae*
Deer, Père David's	*Elaphurus davidianus*
Deer, Persian Fallow	*Dama dama mesopotamica*

Deer, red	*Cervus elaphus*
Deer, red brocket	*Mazama americona*
Deer, Rocky Mountain mule	*Odocoileus hemionus hemionus*
Deer, Rusa	*Cervus timorensis*
Deer, Sambar	*Cervus unicolor*
Deer, Sika	*Cervus nippon*
Deer, Sitka black-tailed	*Odocoileus hemionus sitkensis*
Deer, small Florida key	*Odocoileus virginianus clavium*
Deer, Southern mule	*Odocoileus hemionus fuliginatus*
Deer, Tiburon Island mule	*Odocoileus hemionus sheldoni*
Deer, white-tailed	*Odocoileus virginianus*
Dhole	*Cuon alpinus*
Dik-dik	*Madoqua* spp.
Dog	*Canis familiaris or Canidae*
Dog, African hunting	*Lycaon pictus*
Dog, bush	*Speothos venaticus*
Dog, domestic	*Canis familiaris*
Donkey	*Equius asious*
Dromedary	*Camelus dromedarius*
Eagle	*Accipitrinae*
Eagle, golden	*Aquila chrysaetos*
Eland, common	*Taurotragus oryx*
Elk	*Cervus elaphus*
Elk, American	*Cervus canadensis*
Elk, Eastern	*Cervus elaphus canadensis*
Elk, Manitoban	*Cervus elaphus manitobensis*
Elk, Merriam	*Cervus elaphus merriami*
Elk, North American	*Cervus elaphus*
Elk, Rocky Mountain	*Cervus elaphus nelsoni*
Elk, Roosevelt	*Cervus elaphus roosevelti*
Elk, tule	*Cervus elaphus nannodes*
Fluke	*Fasiola hepatica*
Fluke, liver	*Metorchis conjunctus*
Fluke, North American deer	*Fascioloides magna*
Fly, louse	*Neolipoptena ferrisi*
Fox, gray	*Urocyon cinereoargenteus*
Fox, red	*Vulpes vulpes*
Gaur	*Bos gaurus*
Gazelle, dorcas	*Gazella dorcas*
Gazelle, Grant's	*Gazella granti*
Gazelle, Persian goitered	*Gazella subgutturosa subgutturosa*
Gazelle, slender-horned	*Gazella leptoceros*
Gazelle, Sommering's	*Gazella soemmeringi*
Gazelle, Thomson's	*Gazella thomsoni*
Giraffes	*Giraffa camelopardalis*
Goats, African pygmy	*Capra hircus*
Goats, catalina	*Capra hircus*
Goat, Harrington's mountain	*Oreamnos harringtoni*
Goat, mountain	*Oreamnos americanus*
Goral	*Nemorhaedus goral*
Guanaco	*Lama guanicöe*
Hare	*Lepus* spp.
Hare, snowshoe	*Lepus americanus*
Hartebeest	*Alcelaphus* sp.
Heartworm, dog	*Dirofilaria immitis*
Arterial worm	*Elaeophora schneideri*
Hippopotamus	*Hippopotamus amphibius*
Hippopotamus, pygmy	*Choeropsis liberiensis*
Hog, feral	*Sus scrofa*
Hookworm, dog	*Ancylostoma caninum*
Horse	*Equus* sp.

Ibex, alpine	*Capra ibex ibex*
Ibex, Persian	*Capra aegagrus*
Ibex, Siberian	*Capra ibex sibirica*
Iguanas	*Iguanidae*
Impala	*Aepyceros melampus*
Jaguar	*Panthera onca*
Jaguarundi	*Herpailurus yaguarondi*
Kangaroos, grey	*Macropus giganteus*
Kangaroos, red	*Macropus rufus*
Kudu, greater	*Strepsiceros strepsiceros*
Leopard	*Panthera pardus*
Lice, dog	*Trichodectes canis*
Lion	*Panthera leo*
Lion, African	*Panthera leo*
Lion, mountain	*Puma concolor*
Lion, sea	*Eumetopias jubatus*
Llama	*Lama glama*
Lungworm	*Protostrongylus* sp.
Lynx	*Lynx canadensis*
Lynx, Iberian	*Lynx pardinus*
Mammoth	*Mammuthus columbi*
	Mammuthus jeffersonii
	Mammuthus primigenius
Mange	*Psoroptes* spp.
Markhor	*Capra falconeri*
Marmot	*Marmota flaviventris*
Martens	*Martes americana*
Mastodon	*Mammut americanum*
Mites	*Sarcoptes scabei*
Monster, gila	*Heloderma suspectum*
Moose	*Alces alces*
Moose, Alaska/Yukon	*Alces alces gigas*
Moose, eastern	*Alces alces americanus*
Moose, northwestern	*Alces alces andersoni*
Moose, Yellowstone or Shira's	*Alces alces shirasi*
Moth, cutworm	*Euxoa auxiliaris*
Mouflon	*Ovis musimon*
Mouflon, Mediterrranean	*Ovis gmelini*
Mouse, multimammate	*Mastomys natalensis*
Muntjac	*Muntiacus* spp.
Muskox	*Ovibos moschatus*
Muskox, barren ground	*Ovibos moschatus moschatus*
Muskox, Hudson Bay	*Ovibos moschatus niphoecus*
Muskox, white-faced	*Ovibos moschatus wardi*
Narwhale	*Monodon monoceros*
Nilgai	*Boselaphus tragocamelus*
Nyala	*Tragelaphus angasi*
Ocelot	*Leopardus pardalis*
Opossum	*Didelphis virginiana*
Opossum, white eared	*Didelphis albivenh*
Oryx, Arabian	*Oryx leucoryx*
Oryx, East African	*Oryx beisa*
Oryx, scimitar-horned	*Oryx dammah*
Oryx, South African (gemsbok)	*Oryx gazella*
Ovids	*Ovini*
Owl	*Strigiformes*
Owl, spotted	*Strix occdientalis*
Paca	*Agouti paca*
Paloverde	*Cervus elaphus*
Panda, giant	*Ailuropoda melanoleuca*
Panther, Florida	*Felis concolor coryi*

Peccary	*Tayassu tajacu*
Peccary, chacoan	*Catagonus wagneri*
Peccary, collared	*Dicotyles tajacu*
Peccary, white-lipped	*Tayassu pecari*
Pheasant, ring-necked	*Phasianus colchicus*
Pig, domestic	*Sus scrofa domesticus*
Pig, feral	*Sus scrofa*
Pig, wild	*Sus scrofa*
Porcupine	*Erethizon dorsatum*
Pronghorn	*Antilocapra americana*
Pronghorn, American	*Antilocapra americana americana*
Pronghorn, Mexican	*Antilocapra americana mexicana*
Pronghorn, Oregon	*Antilocapra americana oregona*
Pronghorn, Peninsular	*Antilocapra americana peninsularis*
Pronghorn, Sonoran	*Antilocapra americana sonoriensis*
Puma	*Puma concolor*
Puma, yuma	*Puma concolor browni*
Rabbit, Brazilian	*Sylvilagus brasilensis*
Rabbit, marsh	*Sylvilagus palustris*
Raccoon	*Procyon lotor*
Rattlesnake	*Crotalus* spp.
Reindeer	*Rangifer tarandus*
Rhinoceros, black	*Diceros bicornis*
Rhinoceros, white	*Ceratotherium simum*
Ringtail	*Bassariscus astutus*
Roebucks	*Capreolus capreolus*
Sambar	*Cervus unicolor*
Seals	*Callorhinus ursinus*
	Pagophilus groenlandicus
	Phoca vitulina
	Pusa hispida
Seal, bearded	*Erignathus barbatus*
Seal, harp	*Pagophilus groeenlandicus*
Seal, hooded	*Cystophora cristata*
Seal, ringed	*Phoca hispida*
Serow	*Capricornis sumatrensis*
Sheep	*Caprinae*
Sheep, Barbary	*Ammotraqus lervia*
Sheep, bighorn	*Ovis canadensis*
Sheep, corsican	*Ovis hybrids*
Sheep, Dall	*Ovis dalli*
Sheep, desert bighorn	*Ovis canadensis mexicana*
Sheep, domestic	*Ovis aries*
Sheep, four-horned	*Ovis aries*
Sheep, Mexican bighorn	*Ovis canadensis mexicana*
Sheep, mouflon	*Ovis musimon*
Sheep, mountain	*Ovis canadensis*
Sheep, red	*Ovis orientalis*
Sheep, Rocky Mountain bighorn	*Ovis canadensis*
Sheep, snow	*Ovis nivicola*
Sheep, Stone's	*Ovis dalli stonei*
Sika	*Cervus nippon*
Sitatunga	*Tragelaphus spekei*
Skunk	*Mephitis mephitis or Spilogale* spp.
Sloths	*Bradypodidae*
Sloth, ground	*Glossotherium* spp.
	Glyptotherium spp.
	Megalonyx spp.
	Nothrotheriops spp.
Snake	*Ophidia–Serpentes*
Springbok	*Antidorcas marsupialis*

Suni	*Nesotragus moschatus*
Tahr	*Hemitragus jemlahicus*
Tahr, Himalayan	*Hemitragus jemlahicus*
Takin	*Budorcas takin*
Tamandua	*Tamandua tetradactyla*
Tamondua	*Tamondua mexicana*
Tapeworm	*Eucestoda*
Tapeworm, hydatid	*Echinococcus gramulosus*
Tapeworm, hydatid cyst	*Echinoccocus granulosis*
Tapir	*Tapirus terrestris*
Tapir, Malayan	*Tapirus indicus*
Tick	*Ixodidae* and *Argasidae*
Tick, winter	*Dermacentor albipictus*
Tiger	*Panthera tigris*
Trout, brown	*Salmo trutta*
Trout, cutthroat	*Oncorhynchus clarki*
Tuberculosis, bovine	*Cobacterium bovis*
Tur, Caucasian	*Capra* spp.
Turkey	*Meleagris gallopavo*
Turkey, wild	*Meleagris gallopavo*
Turtle, river	*Dodocremis* spp.
Urial	*Ovis orientalis*
Vicuna	*Lama vicugna*
Voles, meadow	*Microtus pennsylvanicus*
Vulture, turkey	*Cathartes aura*
Walrus	*Odobenus rosmarus*
Wapiti	*Cervus elaphus*
Warbler, golden-cheeked	*Dendroica chrysoparia*
Waterbuck	*Kobus ellipsiprymnus*
Watusi	*Bos taurus*
Whales	*Balaena mysticetus*
	Eschrichtius robustus
	Megaptera novaeangliae
	Monodon monoceros
Whale, beluga	*Delphinapteruss leucas*
Wildebeest, black	*Connochaetes gnou*
Wildebeest, blue	*Connochaetes taurinus taurinus*
Wildebeest, white bearded	*Connochaetes taurinus albojubatus*
Wildboar, Eurasian	*Sus scrofa*
Wolf	*Canis lupus*
Wolf, Alaskan	*Canis lupus occidentalis*
Wolf, Algonquin park	*Canis lupus lycaom*
Wolf, Arctic Island	*Canis lupus arctos*
Wolf, Ethiopian	*Canis simensis*
Wolf, gray	*Canis lupus*
Wolf, Mexican	*Canis lupus baileyi*
Wolf, Minnesota	*Canis lupus nubilus*
Wolf, red	*Canis rufus*
Wolverine	*Gulo gulo*
Worm, arterial	*Elaeophora schneideri*
Worm, meningeal	*Parelaphostrongylus tenuis*
Worm, round	*Nematoda*
Worm, large stomach	*Haemonchus contortus*
Yaks	*Bos grunniens*
Zebu	*Bos indicus*
Zebra, Chapman's	*Equus quagga antiquorum*
Zebra, Damara	*Equus quagga antiquorum*
Zebra, Grant's	*Equus quagga böhmi*
Zebra, Grévy's	*Equus grevyi*
Zebra, mountain	*Equus zebra*

Index

A

abundance. *See* population size
activity patterns
    of black bears, 396–97
    of pumas, 359
    of wolves, 327
adaptation, 3
adaptive radiation, 5
aesthetic values, 40–41, 262
age ratios, 65
    estimates of, 75–76
    and harvest management, 205–6
    of pumas, 362
    See also longevity
age structures, of North American elk, 708–9
agency review, of harvest management, 207
age-structured population models, 91–96
alkaloids, 123
allometry, 7
altruistism, 262
American bison. *See* bison
American black bears. *See* black bears
American elk. *See* North American elk
American Indians. *See* tribal big game management; tribes
anthrax, 455–56
anti-hunters, 53–54
Antilocapridae Gray. *See* pronghorn
antler morphology and growth, caribou, 669–70
aoudad, 740, 746–47
apparent digestibility, 116
artificial enclosures, 250
attitude-based approaches, to human dimensions research, 216–18
attrition, of bison, 455
automobile accidents, and white-tailed deer, 615
axis deer, 747–48

B

Barbary sheep, 740, 746–47
basal metabolic rate (BMR), 124–25
bears. *See* black bears; brown bears; polar bears
beef industry, and bison, 56
behavior
    bighorn sheep, 523–26
    bison, 452–54
    black bears, 396–98
    black-tailed deer, 637–39
    brown bears, 415–17
    caribou, 670–75
    collared peccary, 436–37
    Dall's sheep, 499–501
    jaguars, 383–84
    moose, 584–87
    mountain goats, 472–79
    mule deer, 637–39
    muskoxen, 549–51
    North American elk, 699–700
    polar bears, 415–17
    pronghorn, 564–65
    pumas, 356–60
    white-tailed deer, 608–13
    wolves, 327–29
    See also behavioral ecology
behavioral carrying capacity, 146–47
behavioral ecology, 175, 185–86
    breeding strategies, 182–85
    foraging strategies, 175–79
    predator avoidance, 179–81
    social systems, 181–82
    See also behavior
big game management. *See* conflict resolution; habitat management; harvest management; human dimensions research; population management; predation management; tribal big game management; wildlife management
big game ranching, 260, 273

for conservation and sport hunting, 265–72
    in North America, 262–65
    overview, 260–62
big game species. *See* specific species
bighorn sheep, 517, 533–34
    behavior, 523–26
    distribution, 517–18
    habitat requirements, 529–32
    hybridization in, 30
    life history, 518–23
    management, 532–33
    population dynamics, 526–29
    taxonomy, 15–16, 517–18
biological carrying capacity, and harvest management, 196–98
biological diversity, 58
biological species, 4–5, 27
birth rates, 66
    estimates of, 76
    of bighorn sheep, 527–28
bison, 447, 462
    behavior, 452–54
    distribution, 449–51
    habitat management, 461–62
    habitat requirements, 460–61
    historical accounts, 447–49
    hybridization in, 30
    life history, 451–52
    recovery and management, 56
    population dynamics, 455–58
    population management, 458–60
    taxonomy of, 14–15, 449–51
black bears, 389–90, 402
    behavior, 396–98
    distribution, 390–91
    habitat management, 402
    habitat requirements, 398–99
    hybridization in, 29
    life history and population dynamics, 391–96
    population management, 399–401
    taxonomy of, 17–18, 390–91
blackbuck antelope, 748–50

771

population density. *See* density
density-dependent population growth,
    88–90
density-independent population growth,
    85–87
density vagueness, 162
descent, versus adaptation, 3
desert, tribal big game cultures in, 282
Desert tradition, 281–82
diet
    bighorn sheep, 518–19
    bison, 452
    black bears, 391–92
    black-tailed deer, 631–32
    brown bears, 412–13
    caribou, 663–64
    Dall's sheep, 496–97
    jaguars, 382–83
    moose, 582–83
    mountain goats, 475–79
    mule deer, 631–32
    North American elk, 697–98
    polar bears, 412–13
    pumas, 353
    white-tailed deer, 604–5
    wolves, 323–25
    See also foraging strategies;
        nutritional ecology
differential vulnerability, in harvest
        management data analysis, 203
digestion kinetics, of herbivores,
        116–18
dimorphism, of caribou, 669
direct counts, 76
diseases
    affecting black bears, 394
    affecting bison, 455–57
    affecting white-tailed deer, 614–15
dispersal, 66–67
    estimates of, 79
    pumas, 359–60
    See also distribution; emigration;
        immigration; movements
dispersal phenotype individuals, 6
distribution, 206
    bighorn sheep, 517–18
    bison, 450–51
    black bears, 390–91
    brown bears, 409–12
    caribou, 662–63
    Dall's sheep, 492–94
    jaguars, 379–80
    moose, 582
    mountain goats, 467–68
    muskoxen, 547–48
    North American elk, 695–96, 708
    polar bears, 409–12
    pronghorn, 561–62
    white-tailed deer, 602–4
    wolves, 322–23
diversity, biological and sociological, 58

DNA, genetic variation in, 241–43
DNA fingerprinting, 245
DNA markers
    as ecological tools, 250–51
    mitochondrial, 247–49
    nuclear, 243–47
"Doctrine of Wise Use," 192
dominionistic values, 42–43
dynamic systems, patterns of regulation
    in, 169–70

E

ecologically based carrying capacities,
    141–47
ejidos, 284, 286
emigration, 66–67
    brown bears, 420
    caribou, 679
    estimates of, 79
    moose, 590
    North American elk, 703
    polar bears, 420
    pumas, 364
    See also dispersal; distribution;
        immigration; movements
endogenous urinary nitrogen (EUN),
    125
energy cycle, of caribou, 664
environmental variation, 98
European settlement values, 47
evolution, of collared peccary, 429
evolutionary species, 5
exotics, 736, 761
    examples of, 740, 741–46
        aoudad, 740, 746–47
        axis deer, 747–48
        blackbuck antelope, 748–50
        fallow deer, 750–51
        ibex, 751–52
        mouflon, 752–54
        nilgai antelope, 754–55
        red deer, 755–56
        sika deer, 756–58
        wild boar and feral hog, 758–60
    motivations for keeping, 737–38
    negative considerations, 738–40
experimental paradigm, 160–61
exploitation
    of bison, 448–49
    during colonization, 298
extermination, of bison, 448–49

F

fallow deer, 750–51
farming, 262
fat deposition and mobilization, of
        black-tailed and mule deer, 634

fecundity. *See* birth rates
feeding rate, of herbivores, constraints
        on, 109–11
fee hunting, 262
Felidae Gray. *See* jaguars; pumas
female-defense polygyny, 184
fences
    and bison, 462
    and pronghorn, 573
feral hog, 758–60
fire, and bison, 454, 461
fluoride toxicosis, 458
forage processing, constraints on,
        111–15
foraging strategies, 175–79
    of collared peccary, 431–33
    See also diet; nutritional ecology
foregut fermenters, 112
founder effects, 238, 239

G

game farm, 260
gene flow, reduced, 239–40
genealogical concordance, 28
genetic drift, 236–38
genetic markers. *See* molecular
        markers
genetic stocks, 249–50
genetic variation, 99, 233, 251–52
    in DNA, 241–43
    mitochondrial DNA markers,
        247–49
    nuclear DNA markers, 243–47
    population subdivision and reduced
        gene flow, 239–40
    processes contributing to, 234–38
    processes eroding, 238–39
    in proteins, 241
    of white-tailed deer, 602–4
    and wildlife management, 249–51
gestation
    black-tailed deer, 632–33
    caribou, 664
    mule deer, 632–33
gray wolves. *See* wolves
grazing lawns, 121–22
grizzly bears. *See* brown bears
group stability, of bison, 453
growth
    bighorn sheep, 522–23
    black-tailed deer, 634
    caribou, 668
    collared peccary, 435
    herbivores, 125–27
    mule deer, 634
    wolves, 326–27
    See also physiology; population
        growth rates
gut capacity, of herbivores, 111–15

L

large mammals. *See specific species*
leaf stripping, 121
leisure activities, 50–52
life history, 582–84
    bighorn sheep, 518–23
    bison, 451–52
    black bears, 391–96
    black-tailed deer, 631–36
    brown bears, 412–15
    caribou, 663–70
    collared peccary, 431–35
    Dall's sheep, 495–96
    jaguars, 382–83
    mountain goats, 468–71
    mule deer, 631–36
    muskoxen, 548–49
    North American elk, 697–99
    polar bears, 412–15
    pronghorn, 562–64
    pumas, 353–56
    white-tailed deer, 604–8
    wolves, 323–27
live animal sales, 262–63
livestock
    and mule and black-tailed deer, 644–45
    and pronghorn, 574
longevity
    bison, 452
    black bears, 394
    black-tailed deer, 634
    brown bears, 414–15
    caribou, 668
    moose, 583
    mule deer, 634
    North American elk, 699
    polar bears, 414–15
    pronghorn, 562
    white-tailed deer, 606
    wolves, 326
    See also age ratios

M

MacKenzie Bison Sanctuary, 460
mail surveys, in harvest management, 202
maintenance phenotype individuals, 6
major histocompatibility complex (MHC), 244–45
management. *See* conflict resolution; habitat management; harvest management; human dimensions research; population management; predation management; tribal big game management; wildlife management

mandatory reporting, and harvest management, 200
mark trees, 397–98
meaning-based approaches, to human dimensions research, 218–21
mestizos, 284
metabolic fecal nitrogen (MFN), 125
metabolism, of herbivores, 119
Mexico
    big game industry in, 264–65
    political status of tribes in, 283–84
    tribal big game cultures in, 282–83
    tribal big game management in, 286
migration
    moose, 586–87
    white-tailed deer, 610–11
    See also emigration; immigration
minimum-impact carrying capacity, 148
mitochondrial DNA markers, 247–49
models. *See* population models
molecular markers
    as ecological tools, 250–51
    mitochondrial DNA, 247–49
    nuclear DNA, 243–47
monitoring, of pronghorn, 571–72
Montana Governor's Council, 224–26
moose, 578, 593–94
    behavior, 584–87
    habitat management, 592–93
    habitat requirements, 591–92
    historical perspective, 578–79
    life history, 582–84
    population dynamics, 587–90
    population management, 591
    taxonomy and distribution, 580–82
    taxonomy of, 13–14
    value, 579
moralistic values, 43–44
mortality factors
    bison, 455
    black bears, 394
    black-tailed deer, 640–41
    brown bears, 419–20
    caribou, 677–79
    moose, 589–90
    mule deer, 640–41
    North American elk, 702–3
    polar bears, 419–20
    pronghorn, 566
    pumas, 363–64
    white-tailed deer, 614–16
    wolves, 331–33
mortality rates, 66
    estimates of, 77–79
    bighorn sheep, 528
    pumas, 362–63
mouflon, 752–54
mountain goats, 467, 485
    behavior, 472–79
    distribution, 467–68

    habitat relationships and management, 483–85
    life history, 468–71
    population dynamics, 479–80
    population management, 481–83
    taxonomy, 15, 467–68
mountain lions. *See* pumas
mountain sheep. *See* bighorn sheep; Dall's sheep
movements
    bighorn sheep, 524–25
    bison, 453
    black bears, 396
    black-tailed deer, 637–38
    brown bears, 416–17
    caribou, 673
    jaguars, 384
    moose, 586–87
    mountain goats, 474–75
    mule deer, 637–38
    polar bears, 416–17
    pronghorn, 565
    pumas, 359
    white-tailed deer, 609–10
    wolves, 327–28
    See also dispersal; distribution; emigration; immigration
mule deer, 629, 648
    behavior, 637–39
    habitat management, 643–48
    habitat requirements, 642
    hybridization in, 30–34
    life history, 631–36
    population dynamics, 639–41
    population management, 641–42
    taxonomy and description, 12, 629–30
muskoxen, 545, 556
    behavior, 549–51
    distribution, 547–48
    habitat management, 555–56
    habitat requirements, 554–55
    historical perspective, 545–46
    life history, 548–49
    population dynamics, 551–53
    population management, 553–54
    taxonomy, 15, 547–48
    value, 547

N

Native Americans. *See* tribal big game management; tribes
natural mortality, 77
naturalistic values, 39–40
negativistic values, 44
New World deer. *See* black-tailed deer; caribou; moose; mule deer; white-tailed deer
nilgai antelope, 754–55

North American bison. *See* bison
North American elk, 694, 726
    behavior, 699–700
    distribution, 695–96
    habitat requirements and
        management, 710–23
    historical perspective and value, 694
    hybridization in, 30
    integrated habitat and population
        management, 723–26
    life history, 697–99
    population dynamics, 700–703
    population management, 704–10
    taxonomy of, 10–11, 695–96
North American mountain goats. *See*
    mountain goats
North American mountain sheep. *See*
    bighorn sheep; Dall's sheep
North Americans, contemporary, and
    value of large mammals, 52–57
nuclear DNA markers, 243–47
nuisance bear management, 400–401
nutrient metabolism, of herbivores, 119
nutrient requirements, of pronghorn,
    563–64
nutritional ecology, herbivory, 108–9,
    131–32
    body condition and condition
        indices, 127–30
    digestion kinctics, 116–18
    feeding rate constraints, 109–11
    forage processing constraints,
        111–15
    habitat assessment, 130–31
    nutrient metabolism efficiency, 119
    nutritional requirements, 123–27
    plant defenses, 120–23
    See also diet; foraging strategies
nutritional requirements, of North
    American elk, 710–21

O

observability, 69–70
Old World deer. *See* North American
    elk
opening day impact, 200
open mark-recapture models, 76
open populations, 69
optimum carrying capacity, 148–49
overcompensation, and population
    regulation, 161–62

P

panthers. *See* pumas
    affecting bison, 457–58
    affecting black bears, 394
    affecting white-tailed deer, 614–15

passage rate, of herbivores, 111–15
pasturella, 457
patch dynamics, 454
phenolics, 123
phenotypic variation, 99
philopatry, of pumas, 359–60
phylogenetic species, 28
physiology
    bighorn sheep, 522–23
    bison, 449–50
    black-tailed deer, 634–36
    brown bears, 414–15
    caribou, 668
    collared peccary, 435
    moose, 583–84
    mule deer, 634–36
    polar bears, 414–15
    pronghorn, 562
    white-tailed deer, 606–8
    wolves, 326–27
    See also growth
plant defenses
    chemical, 122–23
    mechanical, 120–22
Plesiometecarpalinae Brooke. *See*
    North American elk
polar bears, 409, 424
    behavior, 415–17
    distribution, 409–12
    habitat requirements and
        management, 422–23
    hybridization in, 29
    life history, 412–15
    population dynamics, 417–20
    population management, 420–22
    taxonomy of, 17–18, 409–12
population bottlenecks, 238–39
population census. *See* census
population density. *See* density
population distribution. *See* distribution
population dynamics
    bighorn sheep, 526–29
    bison, 455–58
    black bears, 391–96
    black-tailed deer, 639–41
    brown bears, 417–20
    caribou, 675–79
    collared peccary, 437–39
    Dall's sheep, 501–5
    moose, 587–90
    mountain goats, 479–80
    mule deer, 639–41
    muskoxen, 551–53
    North American elk, 700–703
    polar bears, 417–20
    pronghorn, 565–67
    pumas, 360–64
    white-tailed deer, 613–16
    wolves, 329–37
population estimation, of collared
    peccary, 438–39

population growth
    density-dependent, 88–90
    density-independent, 85–87
population growth rates, 67–68
    estimates of, 80
    pumas, 364
    wolves, 330–31
population indices, 74–75
    for black bears, 394–95
population management
    bighorn sheep, 532
    bison, 56, 458–60
    black bears, 399–401
    black-tailed deer, 641–42
    brown bears, 420–22
    caribou, 679–80
    collared peccary, 439–40
    Dall's sheep, 505–6
    jaguars, 384–86
    moose, 591
    mountain goats, 481–83
    mule deer, 641–42
    muskoxen, 553–54
    North American elk, 704–10,
        723–26
    polar bears, 420–22
    pronghorn, 568–69
    pumas, 366–70
    white-tailed deer, 616–17
    wolves, 56–57, 335–37
    See also harvest management
population manipulation, using
    harvests, 204–6
population models, 84–85, 105
    age-structured, 91–96
    and data, 100–101
    for census, 195–96
    density-dependent population
        growth, 88–90
    density-independent population
        growth, 85–87
    philosophy of, 104–5
    stochastic, 96–100
population regulation, 156–59,
    170–71
    and consumer-resource interactions,
        164–69
    and overcompensation, 161–62
    and time lags, 163–64
    in dynamic systems, 169–70
    of bighorn sheep, 528–29
    testing for, 159–61
populations, irruptive, 149
population size, 64–65
    during colonization, 298
    estimates of, 68–75
    of North American elk, 707–8
    See also population dynamics
population subdivisions, 239–40
population trends, 613
    bighorn sheep, 526–27